Biology 2e

Part 2: Chapters 28–47

SENIOR CONTRIBUTING AUTHORS
MARY ANN CLARK, TEXAS WESLEYAN UNIVERSITY
JUNG CHOI, GEORGIA INSTITUTE OF TECHNOLOGY
MATTHEW DOUGLAS, GRAND RAPIDS COMMUNITY COLLEGE

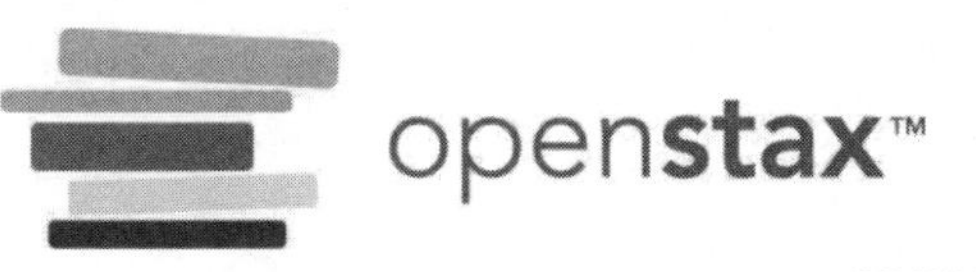

ISBN: 978-1-50669-985-1

OpenStax
Rice University
6100 Main Street MS-375
Houston, Texas 77005

To learn more about OpenStax, visit https://openstax.org.
Individual print copies and bulk orders can be purchased through our website.

HARDCOVER BOOK ISBN-13	**978-1-947172-51-7**
PAPERBACK BOOK ISBN-13	**978-1-947172-95-1**
B&W PAPERBACK BOOK ISBN-13	**978-1-50669-985-1**
DIGITAL VERSION ISBN-13	**978-1-947172-52-4**
ENHANCED TEXTBOOK ISBN-13	**978-1-947172-53-1**
ORIGINAL PUBLICATION YEAR	**2018**

9 8 7 6 5 4 3 2

Printed by
XanEdu
4750 Venture Drive, Suite 400
Ann Arbor, MI 48108
800-562-2147
www.xanedu.com

OpenStax

OpenStax provides free, peer-reviewed, openly licensed textbooks for introductory college and Advanced Placement® courses and low-cost, personalized courseware that helps students learn. A nonprofit ed tech initiative based at Rice University, we're committed to helping students access the tools they need to complete their courses and meet their educational goals.

Rice University

OpenStax, OpenStax CNX, and OpenStax Tutor are initiatives of Rice University. As a leading research university with a distinctive commitment to undergraduate education, Rice University aspires to path-breaking research, unsurpassed teaching, and contributions to the betterment of our world. It seeks to fulfill this mission by cultivating a diverse community of learning and discovery that produces leaders across the spectrum of human endeavor.

Philanthropic Support

OpenStax is grateful for our generous philanthropic partners, who support our vision to improve educational opportunities for all learners.

Laura and John Arnold Foundation

Arthur and Carlyse Ciocca Charitable Foundation

Ann and John Doerr

Bill & Melinda Gates Foundation

Girard Foundation

Google Inc.

The William and Flora Hewlett Foundation

Rusty and John Jaggers

The Calvin K. Kazanjian Economics Foundation

Charles Koch Foundation

Leon Lowenstein Foundation, Inc.

The Maxfield Foundation

Burt and Deedee McMurtry

Michelson 20MM Foundation

National Science Foundation

The Open Society Foundations

Jumee Yhu and David E. Park III

Brian D. Patterson USA-International Foundation

The Bill and Stephanie Sick Fund

Robin and Sandy Stuart Foundation

The Stuart Family Foundation

Tammy and Guillermo Treviño

Contents

UNIT 2 THE CELL

UNIT 3 GENETICS

CHAPTER 11 Meiosis and Sexual Reproduction

CHAPTER 12 Mendel's Experiments and Heredity

CHAPTER 13 Modern Understandings of Inheritance

CHAPTER 14 DNA Structure and Function

UNIT 4 EVOLUTIONARY PROCESSES

UNIT 5 BIOLOGICAL DIVERSITY

CHAPTER 43

UNIT 8 ECOLOGY

CHAPTER 44

CHAPTER 45

CHAPTER 28
Invertebrates

Figure 28.1 Nearly 97 percent of animal species are invertebrates, including this sea star (*Novodinia antillensis*) (credit: NOAA's National Ocean Service)

Chapter Outline

INTRODUCTION A brief look at any magazine pertaining to our natural world, such as *National Geographic*, would show a rich variety of vertebrates, especially mammals and birds. To most people, these are the animals that attract our attention. Concentrating on vertebrates, however, gives us a rather biased and limited view of animal diversity, because it ignores nearly 97 percent of the animal kingdom—the invertebrates—animals that lack a cranium and a defined vertebral column or spine.

The invertebrate animal phyla exhibit an enormous variety of cells and tissues adapted for specific purposes, and frequently these tissues are unique to their phyla. These specializations show the range of cellular differentiation possible within the clade Opisthokonta, which has both unicellular and multicellular members. Cellular and structural specializations include cuticles for protection, spines and tiny harpoons for defense, toothy structures for feeding, and wings for flight. An exoskeleton may be adapted for movement or for the attachment of muscles as in the clams and insects. Secretory cells can produce venom, mucus, or digestive enzymes. The body plans of some phyla, such as those of the molluscs, annelids, arthropods, and echinoderms, have been modified and adapted throughout evolution to produce thousands of different forms. Perhaps you will find it amazing that an enormous number of both aquatic and terrestrial invertebrates—perhaps millions of species—have not yet been scientifically classified. As a result, the phylogenetic relationships among the invertebrates are constantly being updated as new information is collected about the organisms of each phylum.

28.1 Phylum Porifera

By the end of this section, you will be able to do the following:

- Describe the organizational features of the simplest multicellular organisms
- Explain the various body forms and bodily functions of sponges

As we have seen, the vast majority of invertebrate animals do *not* possess a defined bony vertebral endoskeleton, or a bony cranium. However, one of the most ancestral groups of deuterostome invertebrates, the Echinodermata, do produce tiny skeletal "bones" called *ossicles* that make up a true **endoskeleton**, or internal skeleton, covered by an epidermis.

We will start our investigation with the simplest of all the invertebrates—animals sometimes classified within the clade Parazoa ("beside the animals"). This clade currently includes only the phylum Placozoa (containing a single species, *Trichoplax adhaerens*), and the phylum **Porifera**, containing the more familiar sponges (Figure 28.2). The split between the Parazoa and the Eumetazoa (all animal clades above Parazoa) likely took place over a billion years ago.

We should reiterate here that the Porifera do not possess "true" tissues that are embryologically homologous to those of all other derived animal groups such as the insects and mammals. This is because they do not create a true gastrula during embryogenesis, and as a result do not produce a true endoderm or ectoderm. But even though they are not considered to have true tissues, they do have specialized cells that perform specific functions *like tissues* (for example, the external "pinacoderm" of a sponge acts like our epidermis). Thus, *functionally*, the poriferans can be said to have tissues; however, these tissues are likely not *embryologically* homologous to our own.

Sponge larvae (e.g, parenchymula and amphiblastula) are flagellated and able to swim; however, adults are non-motile and spend their life attached to a substratum. Since water is vital to sponges for feeding, excretion, and gas exchange, their body structure facilitates the movement of water through the sponge. Various canals, chambers, and cavities enable water to move through the sponge to allow the exchange of food and waste as well as the exchange of gases to nearly all body cells.

Figure 28.2 Sponges. Sponges are members of the phylum Porifera, which contains the simplest invertebrates. (credit: Andrew Turner)

Morphology of Sponges

There are at least 5,000 named species of sponges, likely with thousands more yet to be classified. The morphology of the simplest sponges takes the shape of an irregular cylinder with a large central cavity, the **spongocoel**, occupying the inside of the cylinder (Figure 28.3). Water enters into the spongocoel through numerous pores, or *ostia*, that create openings in the body wall. Water entering the spongocoel is expelled via a large common opening called the **osculum**. However, we should note that sponges exhibit a range of diversity in body forms, including variations in the size and shape of the spongocoel, as well as the number and arrangement of feeding chambers within the body wall. In some sponges, multiple feeding chambers open off of a central spongocoel and in others, several feeding chambers connecting to one another may lie between the entry pores and the spongocoel.

While sponges do not exhibit true tissue-layer organization, they do have a number of functional "tissues" composed of different cell types specialized for distinct functions. For example, epithelial-like cells called *pinacocytes* form the outermost body, called a *pinacoderm*, that serves a protective function similar that of our epidermis. Scattered among the pinacoderm are the ostia that allow entry of water into the body of the sponge. These pores have given the sponges their phylum name Porifera—pore-bearers. In some sponges, ostia are formed by *porocytes*, single tube-shaped cells that act as valves to regulate the flow of water into the spongocoel. In other sponges, ostia are formed by folds in the body wall of the sponge. Between the outer layer and the feeding chambers of the sponge is a jelly-like substance called the **mesohyl**, which contains collagenous fibers. Various cell types reside within the mesohyl, including **amoebocytes**, the "stem cells" of sponges, and *sclerocytes*, which produce skeletal materials. The gel-like consistency of mesohyl acts like an endoskeleton and maintains the tubular morphology of sponges.

The feeding chambers inside the sponge are lined by choanocytes ("collar cells"). The structure of a choanocyte is critical to its function, which is to generate a *directed* water current through the sponge and to trap and ingest microscopic food particles by phagocytosis. These feeding cells are similar in appearance to unicellular *choanoflagellates* (Protista). This similarity suggests that sponges and choanoflagellates are closely related and likely share common ancestry. The body of the choanocyte is embedded in mesohyl and contains all the organelles required for normal cell function. Protruding into the "open space" inside the feeding chamber is a mesh-like collar composed of microvilli with a single flagellum in the center of the column. The beating of the flagella from all choanocytes draws water into the sponge through the numerous ostia, into the spaces lined by choanocytes, and eventually out through the osculum (or osculi, if the sponge consists of a colony of attached sponges). Food particles, including waterborne bacteria and unicellular organisms such as algae and various animal-like protists, are trapped by the sieve-like collar of the choanocytes, slide down toward the body of the cell, and are ingested by phagocytosis. Choanocytes also serve another surprising function: They can differentiate into sperm for sexual reproduction, at which time they become dislodged from the mesohyl and leave the sponge with expelled water through the osculum.

LINK TO LEARNING

Watch this video to see the movement of water through the sponge body.

Click to view content (https://www.openstax.org/l/filter_sponges)

The **amoebocytes** (derived from stem-cell-like archaeocytes), are so named because they move throughout the mesohyl in an amoeba-like fashion. They have a variety of functions: In addition to delivering nutrients from choanocytes to other cells within the sponge, they also give rise to eggs for sexual reproduction. (The eggs remain in the mesohyl, whereas the sperm cells are released into the water.) The amoebocytes can differentiate into other cell types of the sponge, such as collenocytes and lophocytes, which produce the collagen-like protein that support the mesohyl. Amoebocytes can also give rise to sclerocytes, which produce *spicules* (skeletal spikes of silica or calcium carbonate) in some sponges, and spongocytes, which produce the protein spongin in the majority of sponges. These different cell types in sponges are shown in Figure 28.3.

VISUAL CONNECTION

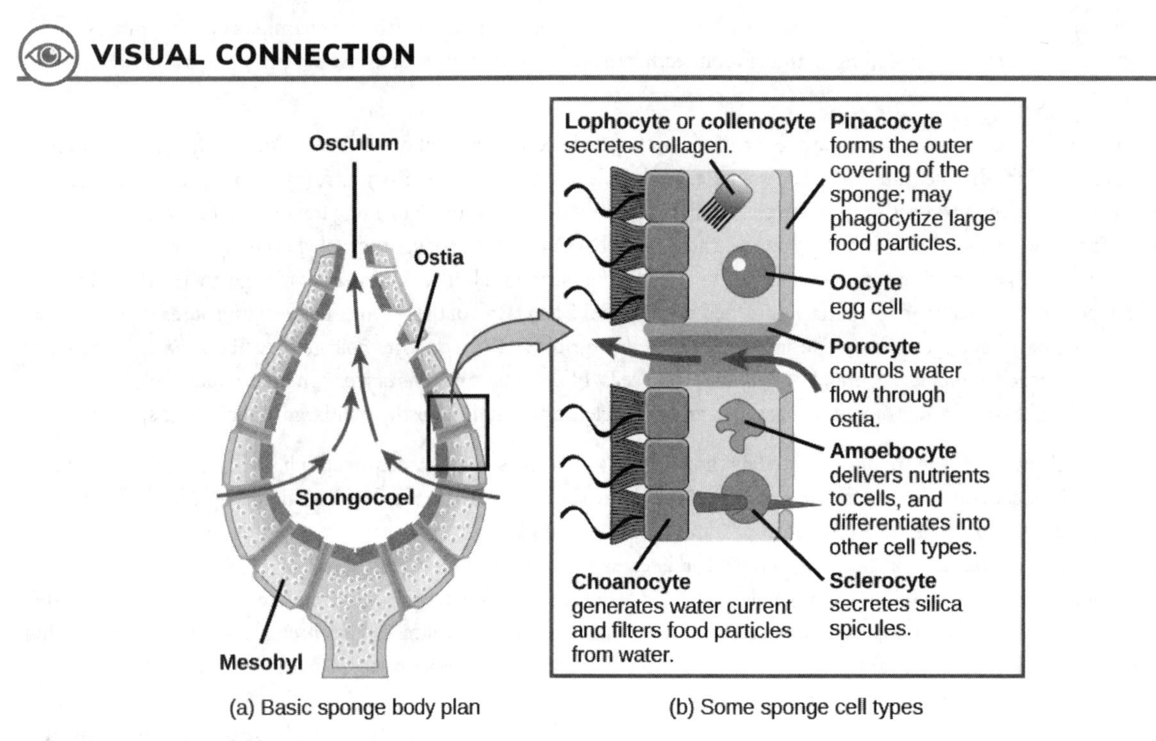

Figure 28.3 Simple sponge body plan and cell types. The sponge's (a) basic body plan and (b) some of the specialized cell types found in sponges are shown.

Which of the following statements is false?

a. Choanocytes have flagella that propel water through the body.
b. Pinacocytes can transform into any cell type.
c. Lophocytes secrete collagen.
d. Porocytes control the flow of water through pores in the sponge body.

LINK TO LEARNING

Take an up-close tour (http://openstax.org/l/sponge_ride) through the sponge and its cells.

As we've seen, most sponges are supported by small bone-like *spicules* (usually tiny pointed structures made of calcium carbonate or silica) in the mesohyl. Spicules provide support for the body of the sponge, and may also deter predation. The presence and composition of spicules form the basis for differentiating three of the four classes of sponges (Figure 28.4). Sponges in class Calcarea produce calcium carbonate spicules and no spongin; those in class Hexactinellida produce six-rayed siliceous (glassy) spicules and no spongin; and those in class Demospongia contain spongin and may or may not have spicules; if present, those spicules are siliceous. Sponges in this last class have been used as bath sponges. Spicules are most conspicuously present in the glass sponges, class Hexactinellida. Some of the spicules may attain gigantic proportions. For example, relative to typical glass sponge spicules, whose size generally ranges from 3 to 10 mm, some of the *basal spicules* of the hexactinellid *Monorhaphis chuni* are enormous and grow up to 3 meters long! The glass sponges are also unusual in that most of their body cells are fused together to form a *multinucleate syncytium*. Because their cells are interconnected in this way, the hexactinellid sponges have no mesohyl. A fourth class of sponges, the Sclerospongiae, was described from species discovered in underwater tunnels. These are also called coralline sponges after their multilayered calcium carbonate skeletons. Dating based on the rate of deposition of the skeletal layers suggests that some of these sponges are hundreds of years old.

Figure 28.4 Several classes of sponges. (a) *Clathrina clathrus* belongs to class Calcarea, (b) *Staurocalyptus* spp. (common name: yellow Picasso sponge) belongs to class Hexactinellida, and (c) *Acarnus erithacus* belongs to class Demospongia. (credit a: modification of work by Parent Géry; credit b: modification of work by Monterey Bay Aquarium Research Institute, NOAA; credit c: modification of work by Sanctuary Integrated Monitoring Network, Monterey Bay National Marine Sanctuary, NOAA)

LINK TO LEARNING

Use the Interactive Sponge Guide (http://openstax.org/l/id_sponges) to identify species of sponges based on their external form, mineral skeleton, fiber, and skeletal architecture.

Physiological Processes in Sponges

Sponges, despite being simple organisms, regulate their different physiological processes through a variety of mechanisms. These processes regulate their metabolism, reproduction, and locomotion.

Digestion

Sponges lack complex digestive, respiratory, circulatory, and nervous systems. Their food is trapped as water passes through the ostia and out through the osculum. Bacteria smaller than 0.5 microns in size are trapped by choanocytes, which are the principal cells engaged in feeding, and are ingested by phagocytosis. However, particles that are larger than the ostia may be phagocytized at the sponge's surface by pinacocytes. In some sponges, amoebocytes transport food from cells that have ingested food particles to those that do not. In sponges, in spite of what looks like a large digestive cavity, all digestion is *intracellular*. The limit of this type of digestion is that food particles must be smaller than individual sponge cells.

All other major body functions in the sponge (gas exchange, circulation, excretion) are performed by diffusion between the cells that line the openings within the sponge and the water that is passing through those openings. All cell types within the sponge obtain oxygen from water through diffusion. Likewise, carbon dioxide is released into seawater by diffusion. In addition, nitrogenous waste produced as a byproduct of protein metabolism is excreted via diffusion by individual cells into the water as it passes through the sponge.

Some sponges host green algae or cyanobacteria as endosymbionts within archeocytes and other cells. It may be a surprise to learn that there are nearly 150 species of carnivorous sponges, which feed primarily on tiny crustaceans, snaring them through sticky threads or hooked spicules!

Although there is no specialized nervous system in sponges, there is intercellular communication that can regulate events like contraction of the sponge's body or the activity of the choanocytes.

Reproduction

Sponges reproduce by sexual as well as asexual methods. The typical means of asexual reproduction is either fragmentation (during this process, a piece of the sponge breaks off, settles on a new substrate, and develops into a new individual), or budding (a genetically identical outgrowth grows from the parent and eventually detaches or remains attached to form a colony). An atypical type of asexual reproduction is found only in freshwater sponges and occurs through the formation of *gemmules*. **Gemmules** are environmentally resistant structures produced by adult sponges (e.g., in the freshwater sponge *Spongilla*). In gemmules, an inner layer of archeocytes (amoebocytes) is surrounded by a pneumatic cellular layer that may be reinforced with spicules. In freshwater sponges, gemmules may survive hostile environmental conditions like changes in temperature, and then serve to recolonize the habitat once environmental conditions improve and stabilize. Gemmules are capable of attaching to a substratum and generating a new sponge. Since gemmules can withstand harsh environments, are resistant to desiccation, and remain dormant for long periods, they are an excellent means of colonization for a sessile organism.

Sexual reproduction in sponges occurs when gametes are generated. Oocytes arise by the differentiation of amoebocytes and are retained within the spongocoel, whereas spermatozoa result from the differentiation of choanocytes and are ejected via the osculum. Sponges are **monoecious** (hermaphroditic), which means that one individual can produce both gametes (eggs and sperm) simultaneously. In some sponges, production of gametes may occur throughout the year, whereas other sponges may show sexual cycles depending upon water temperature. Sponges may also become *sequentially hermaphroditic*, producing oocytes first and spermatozoa later. This temporal separation of gametes produced by the same sponge helps to encourage cross-fertilization and genetic diversity. Spermatozoa carried along by water currents can fertilize the oocytes borne in the mesohyl of other sponges. Early larval development occurs within the sponge, and free-swimming larvae (such as flagellated *parenchymula*) are then released via the osculum.

Locomotion

Sponges are generally sessile as adults and spend their lives attached to a fixed substratum. They do not show movement over large distances like other free-swimming marine invertebrates. However, sponge cells are capable of creeping along substrata via *organizational plasticity*, i.e., rearranging their cells. Under experimental conditions, researchers have shown that sponge cells spread on a physical support demonstrate a leading edge for directed movement. It has been speculated that this localized creeping movement may help sponges adjust to microenvironments near the point of attachment. It must be noted, however, that this pattern of movement has been documented in laboratories, it remains to be observed in natural sponge habitats.

LINK TO LEARNING

Watch this BBC video (http://openstax.org/l/sea_sponges) showing the array of sponges seen along the Cayman Wall during a submersible dive.

28.2 Phylum Cnidaria

By the end of this section, you will be able to do the following:

- Compare structural and organization characteristics of Porifera and Cnidaria
- Describe the progressive development of tissues and their relevance to animal complexity
- Identify the two general body forms found in the Cnidaria
- Describe the identifying features of the major cnidarian classes

Phylum **Cnidaria** includes animals that exhibit radial or biradial symmetry and are diploblastic, meaning that they develop from two embryonic layers, ectoderm and endoderm. Nearly all (about 99 percent) cnidarians are marine species.

Whereas the defining cell type for the sponges is the choanocyte, the defining cell type for the cnidarians is the **cnidocyte**, or stinging cell. These cells are located around the mouth and on the tentacles, and serve to capture prey or repel predators. Cnidocytes have large stinging organelles called **nematocysts**, which usually contain barbs at the base of a long coiled thread. The outer wall of the cell has a hairlike projection called a *cnidocil*, which is sensitive to tactile stimulation. If the cnidocils are touched, the hollow threads evert with enormous acceleration, approaching 40,000 times that of gravity. The microscopic threads then either entangle the prey or instantly penetrate the flesh of the prey or predator, releasing toxins (including neurotoxins and pore-forming toxins that can lead to cell lysis) into the target, thereby immobilizing it or paralyzing it (see Figure 28.5).

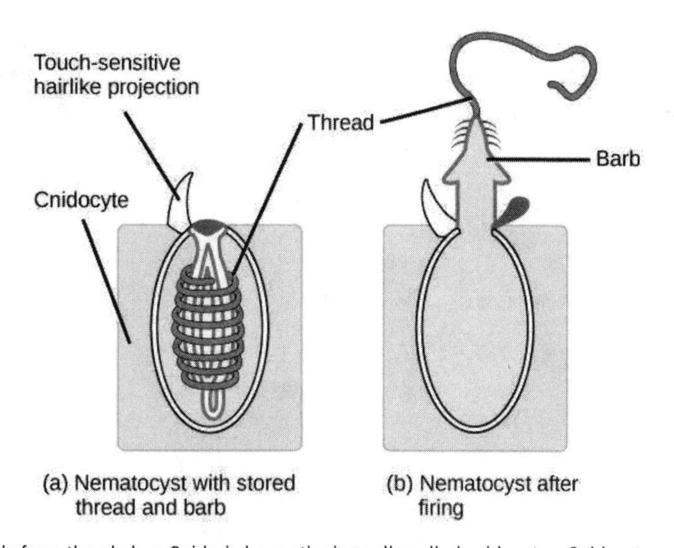

Figure 28.5 Cnidocytes. Animals from the phylum Cnidaria have stinging cells called cnidocytes. Cnidocytes contain large organelles called (a) nematocysts that store a coiled thread and barb, the nematocyst. When the hairlike cnidocil on the cell surface is touched, even lightly, (b) the thread, barb, and a toxin are fired from the organelle.

LINK TO LEARNING

View this video (https://www.openstax.org/l/nematocyst) animation showing two anemones engaged in a battle.

Two distinct body plans are found in Cnidarians: the polyp or tuliplike "stalk" form and the medusa or "bell" form. (Figure 28.6). An example of the polyp form is found in the genus *Hydra*, whereas the most typical form of medusa is found in the group called the "sea jellies" (jellyfish). Polyp forms are sessile as adults, with a single opening (the mouth/anus) to the digestive cavity facing up with tentacles surrounding it. Medusa forms are motile, with the mouth and tentacles hanging down from an umbrella-shaped bell.

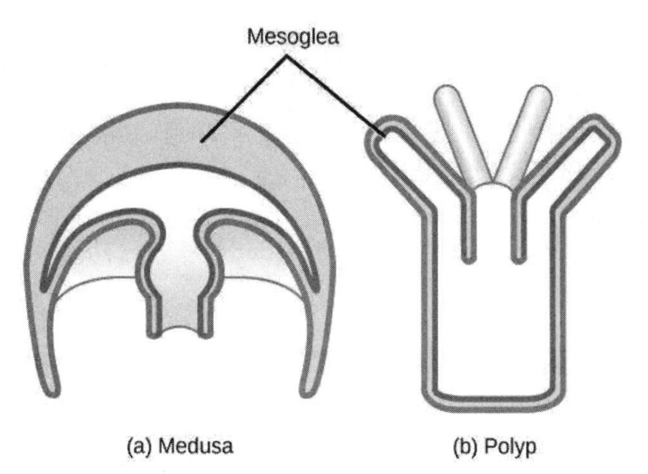

Figure 28.6 Cnidarian body forms. Cnidarians have two distinct body plans, the medusa (a) and the polyp (b). All cnidarians have two membrane layers, with a jelly-like mesoglea between them.

Some cnidarians are **dimorphic**, that is, they exhibit both body plans during their life cycle. In these species, the polyp serves as the asexual phase, while the medusa serves as the sexual stage and produces gametes. However, both body forms are diploid.

An example of cnidarian dimorphism can be seen in the colonial hydroid *Obelia*. The sessile asexual colony has two types of polyps, shown in Figure 28.7. The first is the *gastrozooid*, which is adapted for capturing prey and feeding. In *Obelia*, all polyps are connected through a common digestive cavity called a *coenosarc*. The other type of polyp is the *gonozooid*, adapted for the asexual budding and the production of sexual medusae. The reproductive buds from the gonozooid break off and mature into free-swimming medusae, which are either male or female (dioecious). Each medusa has either several testes or several ovaries in which meiosis occurs to produce sperm or egg cells. Interestingly, the gamete-producing cells do not arise within the gonad itself, but migrate into it from the tissues in the gonozooid. This separate origin of gonad and gametes is common throughout the eumetazoa. The gametes are released into the surrounding water, and after fertilization, the zygote develops into a blastula, which soon develops into a ciliated, bilaterally symmetrical planula larva. The planula swims freely for a while, but eventually attaches to a substrate and becomes a single polyp, from which a new colony of polyps is formed by budding.

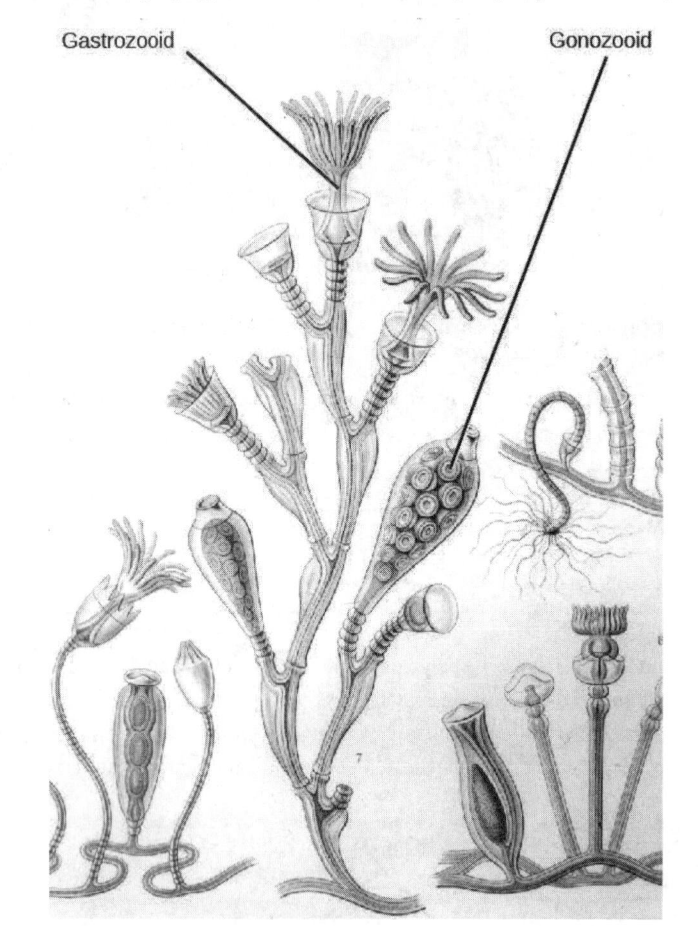

Figure 28.7 *Obelia.* The colonial sessile form of *Obelia geniculata* has two types of polyps: gastrozooids, which are adapted for capturing prey, and gonozooids, which asexually bud to produce medusae.

LINK TO LEARNING

Click here to follow an *Obelia* life cycle (http://openstax.org/l/obelia) animation.

All cnidarians are diploblastic and thus have two "epithelial" layers in the body that are derived from the endoderm and ectoderm of the embryo. The outer layer (from ectoderm) is called the *epidermis* and lines the outside of the animal, whereas the inner layer (from endoderm) is called the *gastrodermis* and lines the digestive cavity. In the planula larva, a layer of ectoderm surrounds a solid mass of endoderm, but as the polyp develops, the digestive or gastrovascular cavity opens within the endoderm. A non-living, jelly-like mesoglea lies between these two epithelial layers. In terms of cellular complexity, cnidarians show the presence of differentiated cell types in each tissue layer, such as nerve cells, contractile epithelial cells, enzyme-secreting cells, and nutrient-absorbing cells, as well as the presence of intercellular connections. However, with a few notable exceptions such as *statocysts* and *rhopalia* (see below), the development of organs or organ systems is not advanced in this phylum.

The nervous system is rudimentary, with nerve cells organized in a network scattered across the body. This **nerve net** may show the presence of groups of cells that form nerve plexi (singular: plexus) or nerve cords. Organization of the nervous system in the motile medusa is more complex than that of the sessile polyp, with a nerve ring around the edge of the medusa bell that controls the action of the tentacles. Cnidarian nerve cells show mixed characteristics of motor and sensory neurons. The predominant signaling molecules in these primitive nervous systems are peptides, which perform both excitatory and inhibitory functions.

Despite the simplicity of the nervous system, it is remarkable that it coordinates the complicated movement of the tentacles, the drawing of captured prey to the mouth, the digestion of food, and the expulsion of waste.

The gastrovascular cavity has only one opening that serves as both a mouth and an anus; this arrangement is called an incomplete digestive system. In the gastrovascular cavity, extracellular digestion occurs as food is taken into the gastrovascular cavity, enzymes are secreted into the cavity, and the cells lining the cavity absorb nutrients. However, some intracellular digestion also occurs. The gastrovascular cavity distributes nutrients throughout the body of the animal, with nutrients passing from the digestive cavity across the mesoglea to the epidermal cells. Thus, this cavity serves both digestive and circulatory functions.

Cnidarian cells exchange oxygen and carbon dioxide by diffusion between cells in the epidermis and water in the environment, and between cells in the gastrodermis and water in the gastrovascular cavity. The lack of a circulatory system to move dissolved gases limits the thickness of the body wall and necessitates a non-living mesoglea between the layers. In the cnidarians with a thicker mesoglea, a number of canals help to distribute both nutrients and gases. There is neither an excretory system nor organs, and nitrogenous wastes simply diffuse from the cells into the water outside the animal or into the gastrovascular cavity.

The phylum Cnidaria contains about 10,000 described species divided into two monophyletic clades: the Anthozoa and the Medusozoa. The Anthozoa include the corals, sea fans, sea whips, and the sea anemones. The Medusozoa include several classes of Cnidaria in two clades: The Hydrozoa include sessile forms, some medusoid forms, and swimming colonial forms like the Portuguese man-of-war. The other clade contains various types of jellies including both Scyphozoa and Cubozoa. The Anthozoa contain only sessile polyp forms, while the Medusozoa include species with both polyp and medusa forms in their life cycle.

Class Anthozoa

The class Anthozoa ("flower animals") includes sea anemones (Figure 28.8), sea pens, and corals, with an estimated number of 6,100 described species. Sea anemones are usually brightly colored and can attain a size of 1.8 to 10 cm in diameter. Individual animals are cylindrical in shape and are attached directly to a substrate.

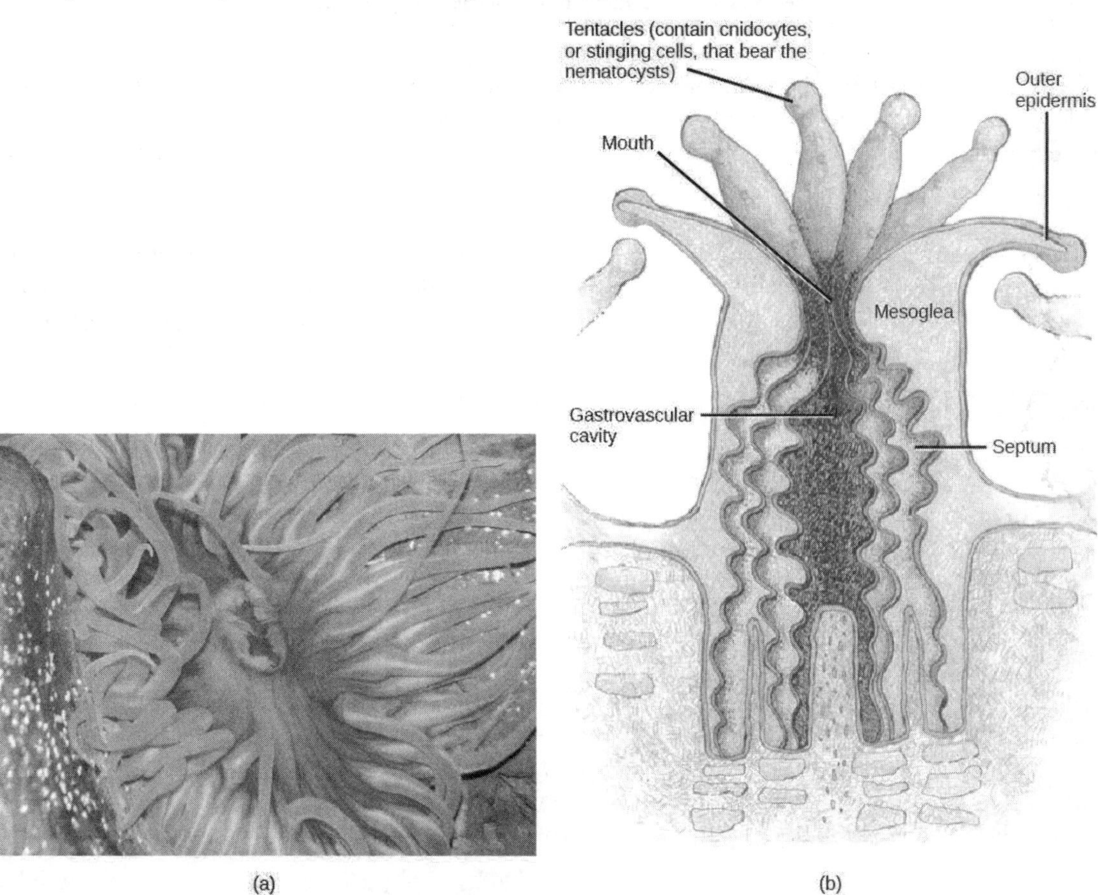

Figure 28.8 Sea anemone. The sea anemone is shown (a) photographed and (b) in a diagram illustrating its morphology. (credit a: modification of work by "Dancing With Ghosts"/Flickr; credit b: modification of work by NOAA)

The mouth of a sea anemone is surrounded by tentacles that bear cnidocytes. The slit-like mouth opening and flattened pharynx are lined with ectoderm. This structure of the pharynx makes anemones bilaterally symmetrical. A ciliated groove called a siphonoglyph is found on two opposite sides of the pharynx and directs water into it. The pharynx is the muscular part of the digestive system that serves to ingest as well as egest food, and may extend for up to two-thirds the length of the body before opening into the gastrovascular cavity. This cavity is divided into several chambers by longitudinal **septa** called *mesenteries*. Each mesentery consists of a fold of gastrodermal tissue with a layer of mesoglea between the sheets of gastrodermis. Mesenteries do not divide the gastrovascular cavity completely, and the smaller cavities coalesce at the pharyngeal opening. The adaptive benefit of the mesenteries appears to be an increase in surface area for absorption of nutrients and gas exchange, as well as additional mechanical support for the body of the anemone.

Sea anemones feed on small fish and shrimp, usually by immobilizing their prey with nematocysts. Some sea anemones establish a mutualistic relationship with hermit crabs when the crab seizes and attaches them to their shell. In this relationship, the anemone gets food particles from prey caught by the crab, and the crab is protected from the predators by the stinging cells of the anemone. Some species of anemone fish, or clownfish, are also able to live with sea anemones because they build up an acquired immunity to the toxins contained within the nematocysts and also secrete a protective mucus that prevents them from being stung.

The structure of coral polyps is similar to that of anemones, although the individual polyps are usually smaller and part of a colony, some of which are massive and the size of small buildings. Coral polyps feed on smaller planktonic organisms, including algae, bacteria, and invertebrate larvae. Some anthozoans have symbiotic associations with dinoflagellate algae called zooxanthellae. The mutually beneficial relationship between zooxanthellae and modern corals—which provides the algae with shelter—gives coral reefs their colors and supplies both organisms with nutrients. This complex mutualistic association began more than 210 million years ago, according to a new study by an international team of scientists. That this symbiotic relationship arose during a time of massive worldwide coral-reef expansion suggests that the interconnection of algae and coral is crucial for the health of coral reefs, which provide habitat for roughly one-fourth of all marine life. Reefs are threatened by a trend in ocean warming that has caused corals to expel their zooxanthellae algae and turn white, a process called coral bleaching.

Anthozoans remain *polypoid* (note that this term is easily confused with "polyploid") throughout their lives and can reproduce asexually by budding or fragmentation, or sexually by producing gametes. Male or female gametes produced by a polyp fuse to give rise to a free-swimming planula larva. The larva settles on a suitable substratum and develops into a sessile polyp.

Class Scyphozoa

Class Scyphozoa ("cup animals") includes all (and only) the marine jellies, with about 200 known species. The medusa is the prominent stage in the life cycle, although there is a polyp stage in the life cycle of most species. Most jellies range from 2 to 40 cm in length but the largest scyphozoan species, *Cyanea capillata*, can reach a size of two meters in diameter. Scyphozoans display a characteristic bell-like morphology (Figure 28.9).

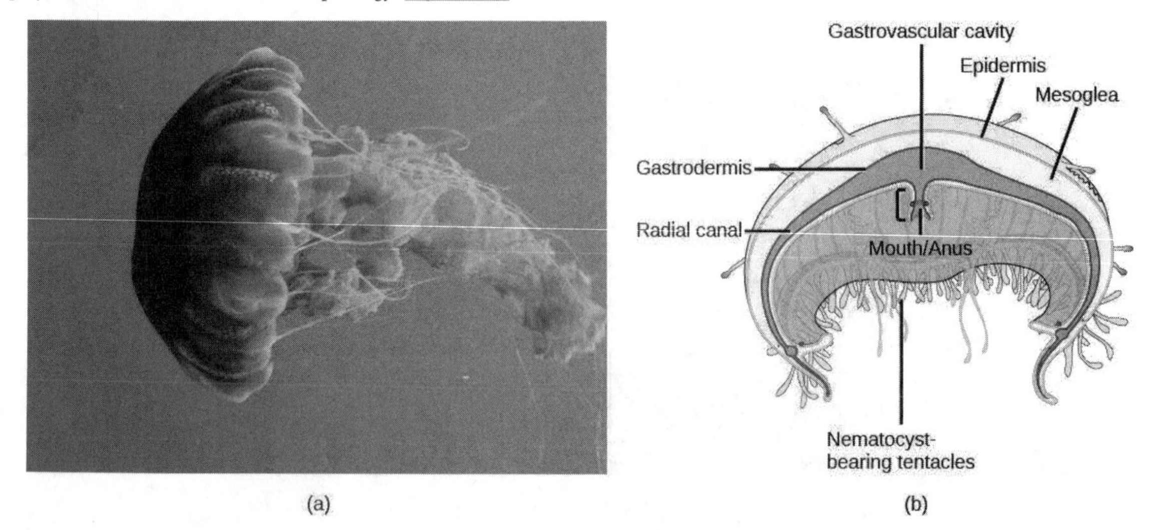

Figure 28.9 A sea jelly. A jelly is shown (a) photographed and (b) in a diagram illustrating its morphology. (credit a: modification of work by "Jimg944"/Flickr; credit b: modification of work by Mariana Ruiz Villareal)

In the sea jelly, a mouth opening is present on the underside of the animal, surrounded by hollow tentacles bearing nematocysts. Scyphozoans live most of their life cycle as free-swimming, solitary carnivores. The mouth leads to the gastrovascular cavity, which may be sectioned into four interconnected sacs, called *diverticuli*. In some species, the digestive system may branch further into *radial canals*. Like the septa in anthozoans, the branched gastrovascular cells serve two functions: to increase the surface area for nutrient absorption and diffusion, and to support the body of the animal.

In scyphozoans, nerve cells are organized in a nerve net that extends over the entire body, with a nerve ring around the edge of the bell. Clusters of sensory organs called rhopalia may be present in pockets in the edge of the bell. Jellies have a ring of muscles lining the dome of the body, which provides the contractile force required to swim through water, as well as to draw in food from the water as they swim. Scyphozoans have separate sexes. The gonads are formed from the gastrodermis and gametes are expelled through the mouth. Planula larvae are formed by external fertilization; they settle on a substratum in a polypoid form. These polyps may bud to form additional polyps or begin immediately to produce medusa buds. In a few species, the planula larva may develop directly into the medusa. The life cycle (Figure 28.10) of most scyphozoans includes both sexual medusoid and asexual polypoid body forms.

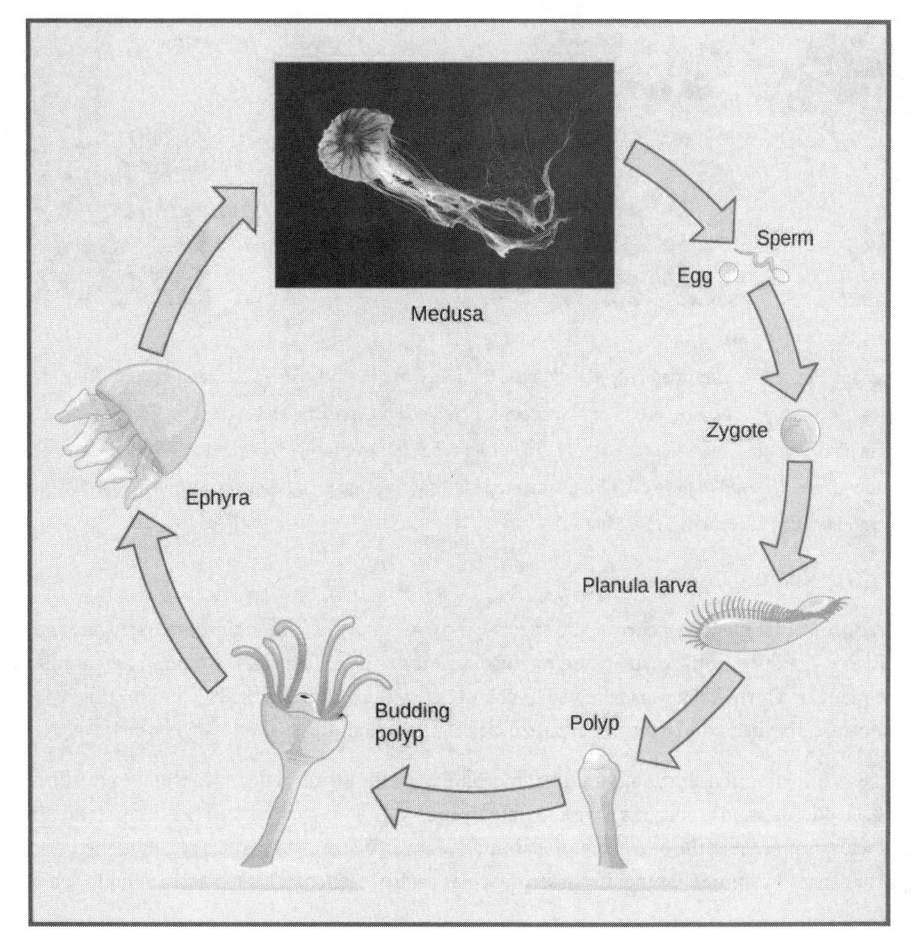

Figure 28.10 Scyphozoan life cycle. The lifecycle of most jellyfish includes two stages: the medusa stage and the polyp stage. The polyp reproduces asexually by budding, and the medusa reproduces sexually. (credit "medusa": modification of work by Francesco Crippa)

Class Cubozoa

This class includes jellies that have a box-shaped medusa, or a bell that is square in cross-section, and are colloquially known as "box jellyfish." These species may achieve sizes of 15 to 25 cm, but typically members of the Cubozoa are not as large as those of the Scyphozoa. However, cubozoans display overall morphological and anatomical characteristics that are similar to those of the scyphozoans. A prominent difference between the two classes is the arrangement of tentacles. The cubozoans contain muscular pads called **pedalia** at the corners of the square bell canopy, with one or more tentacles attached to each pedalium. In some cases, the digestive system may extend into the pedalia. Nematocysts may be arranged in a spiral configuration along the tentacles; this arrangement helps to effectively subdue and capture prey. Cubozoans include the most venomous of all the cnidarians (Figure 28.11).

These animals are unusual in having image-forming eyes, including a cornea, lens, and retina. Because these structures are made from a number of interactive tissues, they can be called *true organs*. Eyes are located in four clusters between each pair of pedalia. Each cluster consists of four simple eye spots plus two image-forming eyes oriented in different directions. How images formed by these very complex eyes are processed remains a mystery, since cubozoans have extensive nerve nets but no distinct brain. Nontheless, the presence of eyes helps the cubozoans to be active and effective hunters of small marine animals like worms, arthropods, and fish.

Cubozoans have separate sexes and fertilization occurs inside the female. Planula larvae may develop inside the female or be released, depending on species. Each planula develops into a polyp. These polyps may bud to form more polyps to create a colony; each polyp then transforms into a single medusa.

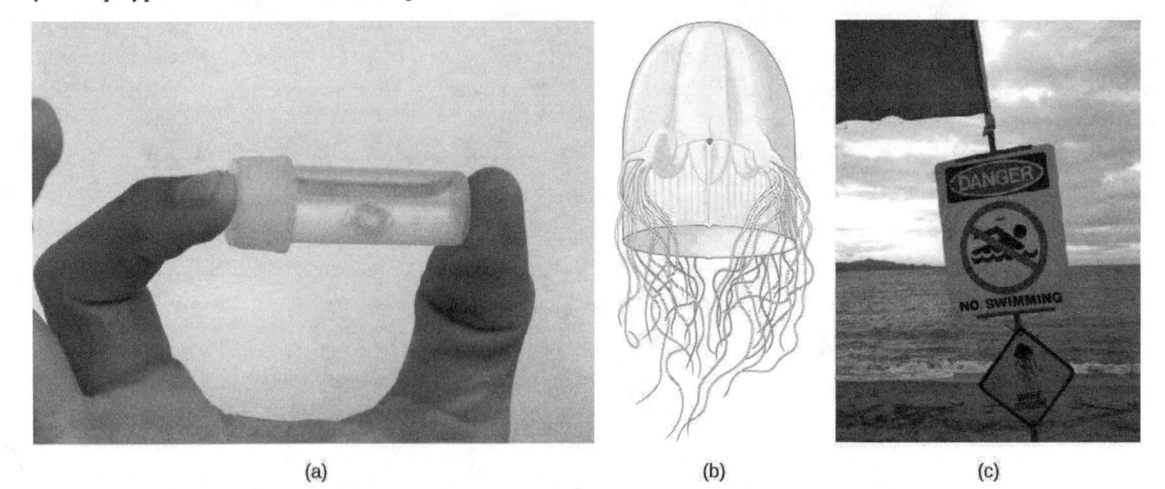

Figure 28.11 A cubozoan. The (a) tiny cubozoan jelly *Malo kingi* is thimble-shaped and, like all cubozoan jellies, (b) has four muscular pedalia to which the tentacles attach. *M. kingi* is one of two species of jellies known to cause Irukandji syndrome, a condition characterized by excruciating muscle pain, vomiting, increased heart rate, and psychological symptoms. Two people in Australia, where Irukandji jellies are most commonly found, are believed to have died from Irukandji stings. (c) A sign on a beach in northern Australia warns swimmers of the danger. (credit c: modification of work by Peter Shanks)

Class Hydrozoa

Hydrozoa is a diverse group that includes nearly 3,200 species; most are marine, although some freshwater species are known (Figure 28.12). Most species exhibit both polypoid and medusoid forms in their lifecycles, although the familiar *Hydra* has only the polyp form. The medusoid form has a muscular veil or **velum** below the margin of the bell and for this reason is called a *hydromedusa*. In contrast, the medusoid form of Scyphozoa lacks a velum and is termed a *scyphomedusa*.

The polyp form in these animals often shows a cylindrical morphology with a central gastrovascular cavity lined by the gastrodermis. The gastrodermis and epidermis have a simple layer of mesoglea sandwiched between them. A mouth opening, surrounded by tentacles, is present at the oral end of the animal. Many hydrozoans form sessile, branched colonies of specialized polyps that share a common, branching gastrovascular cavity (coenosarc), such as is found in the colonial hydroid *Obelia*.

Free-floating colonial species called **siphonophores** contain both medusoid and polypoid individuals that are specialized for feeding, defense, or reproduction. The distinctive rainbow-hued float of the Portuguese man o' war (*Physalia physalis*) creates a pneumatophore with which it regulates buoyancy by filling and expelling carbon monoxide gas. At first glance, these complex superorganisms appear to be a single organism; but the reality is that even the tentacles are actually composed of zooids laden with nematocysts. Thus, although it superficially resembles a typical medusozoan jellyfish, *P. physalis* is a free-floating hydrozoan *colony*; each specimen is made up of many hundreds of organisms, each specialized for a certain function, including motility and buoyancy, feeding, reproduction and defense. Although they are carnivorous and feed on many soft bodied marine animals, *P. physalis* lack stomachs and instead have specialized polyps called gastrozooids that they use to digest their prey in the open water.

Physalia has male and female colonies, which release their gametes into the water. The zygote develops into a single individual, which then buds asexually to form a new colony. Siphonophores include the largest known floating cnidarian colonies such as

Praya dubia, whose chain of zoids can get up to 50 meters (165 feet) long. Other hydrozoan species are solitary polyps (*Hydra*) or solitary hydromedusae (*Gonionemus*). One defining characteristic shared by the hydrozoans is that their gonads are derived from epidermal tissue, whereas in all other cnidarians they are derived from gastrodermal tissue.

(a) *Obelia*

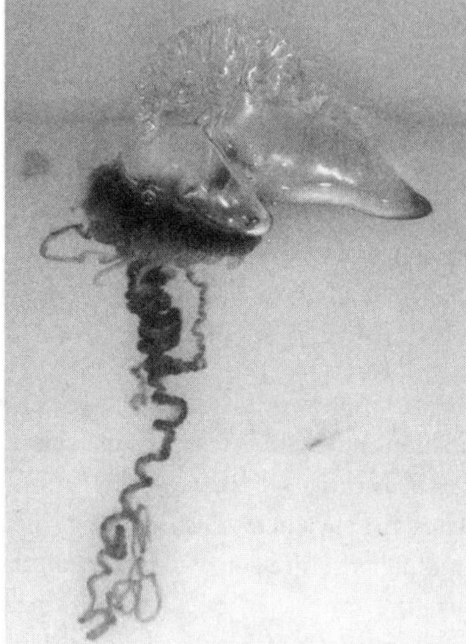

(b) *Physalia physalis* (Portuguese Man O' War)

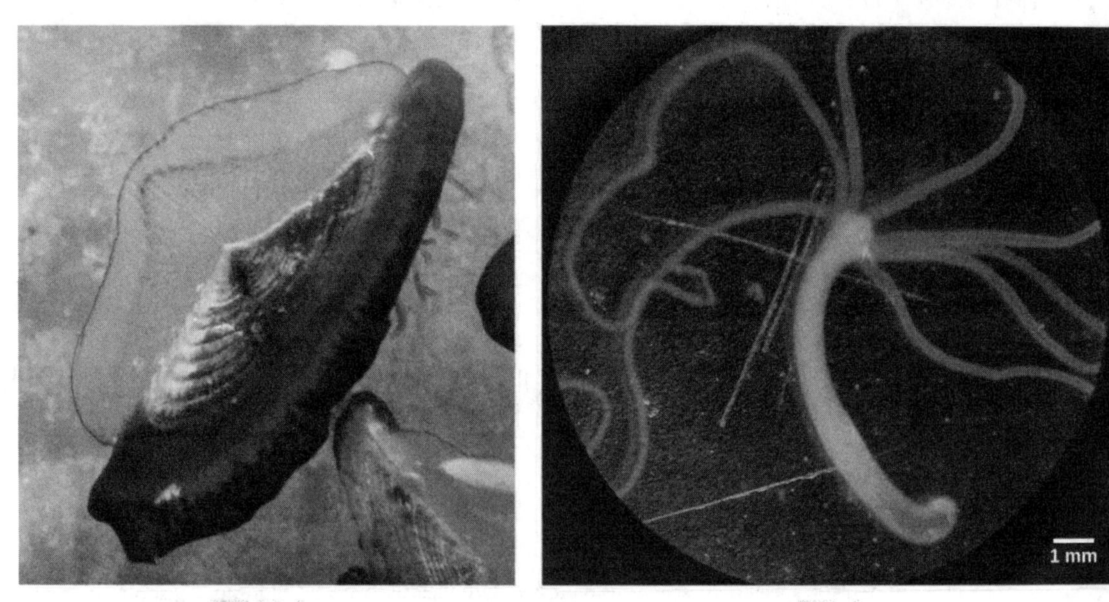

(c) *Velella bae* (d) *Hydra*

Figure 28.12 Hydrozoans. The polyp colony *Obelia* (a), siphonophore colonies *Physalia* (b) *physalis*, known as the Portuguese man o' war and *Velella bae* (c), and the solitary polyp *Hydra* (d) have different body shapes but all belong to the family Hydrozoa. (credit b: modification of work by NOAA; scale-bar data from Matt Russell)

28.3 Superphylum Lophotrochozoa: Flatworms, Rotifers, and Nemerteans

By the end of this section, you will be able to do the following:

- Describe the unique anatomical and morphological features of flatworms, rotifers, and Nemertea
- Identify an important extracoelomic cavity found in Nemertea
- Explain the key features of Platyhelminthes and their importance as parasites

Animals belonging to superphylum Lophotrochozoa are triploblastic (have three germ layers) and unlike the cnidarians, they possess an embryonic mesoderm sandwiched between the ectoderm and endoderm. These phyla are also bilaterally symmetrical, meaning that a longitudinal section will divide them into right and left sides that are superficially symmetrical. In these phyla, we also see the beginning of cephalization, the evolution of a concentration of nervous tissues and sensory organs in the head of the organism—exactly where a mobile bilaterally symmetrical organism first encounters its environment.

Lophotrochozoa are also protostomes, in which the blastopore, or the point of invagination of the ectoderm (outer germ layer), becomes the mouth opening into the alimentary canal. This developmental pattern is called protostomy or "first mouth." Protostomes include acoelomate, pseudocoelomate, and eucoelomate phyla. The **coelom** is a cavity that separates the ectoderm from the endoderm. In acoelomates, a *solid mass* of mesoderm is sandwiched between the ectoderm and endoderm and does not form a cavity or "coelom," leaving little room for organ development; in pseudocoelomates, there is a cavity or *pseudocoelom* that replaces the blastocoel (the cavity within the blastula), but it is only lined by mesoderm on the outside of the cavity, leaving the gut tube and organs unlined; in eucoelomates, the cavity that obliterates the blastocoel as the coelom develops is lined both on the outside of the cavity (parietal layer) *and* also on the inside of the cavity, around the gut tube and the internal organs (visceral layer).

Eucoelmate protostomes are schizocoels, in which mesoderm-producing cells typically migrate into the blastocoel during gastrulation and multiply to form a solid mass of cells. Cavities then develop within the cell mass to form the coelom. Since the forming body cavity splits the mesoderm, this protostomic coelom is termed a **schizocoelom**. As we will see later in this chapter, chordates, the phylum to which we belong, generally develop a coelom by enterocoely: pouches of mesoderm pinch off the invaginating primitive gut, or *archenteron*, and then fuse to form a complete coelom. We should note here that a eucoelomate can form its "true coelom" by either schizocoely or enterocoely. The process that produces the coelom is different and of taxonomic importance, but the result is the same: a complete, mesodermally lined coelom.

Most organisms placed in the superphylum Lophotrochozoa possess either a *lophophore* feeding apparatus or a *trochophore* larvae (thus the contracted name, "lopho-trocho-zoa"). The lophophore is a feeding structure composed of a set of ciliated tentacles surrounding the mouth. A trochophore is a free-swimming larva characterized by two bands of cilia surrounding a top-like body. Some of the phyla classified as Lophotrochozoa may be missing one or both of these defining structures. Nevertheless their placement with the Lophotrochozoa is upheld when ribosomal RNA and other gene sequences are compared. The systematics of this complex group is still unclear and much more work remains to resolve the cladistic relationships among them.

Phylum Platyhelminthes

The flatworms are acoelomate organisms that include many free-living and parasitic forms. The flatworms possess neither a lophophore nor trochophore larvae, although the larvae of one group of flatworms, the Polycladida (named after its many-branched digestive tract), are considered to be homologous to trochophore larvae. Spiral cleavage is also seen in the polycladids and other basal flatworm groups. The developmental pattern of some of the free-living forms is obscured by a phenomenon called "*blastomere anarchy*," in which a sort of temporary feeding larva forms, followed by a regrouping of cells within the embryo that gives rise to a second-stage embryo. However, both the monophyly of the flatworms and their placement in the Lophotrochozoa has been supported by molecular analyses.

The Platyhelminthes consist of two monophyletic lineages: the Catenulida and the Rhabditophora. The Catenulida, or "chain worms," is a small clade of just over 100 species. These worms typically reproduce asexually by budding. However, the offspring do not fully detach from the parents and the formation resembles a chain in appearance. All of the flatworms discussed here are part of the Rhabditophora ("rhabdite bearers"). **Rhabdites** are rodlike structures discharged in the mucus produced by some free-living flatworms; Eucoelmate protostomes are schizocoels, in which mesoderm-producing cells typically migrate into the blastocoel during gastrulation likely serve in both defense and to provide traction for ciliary gliding along the substrate. Unlike free-living flatworms, many species of trematodes and cestodes are parasitic, including important parasites of humans.

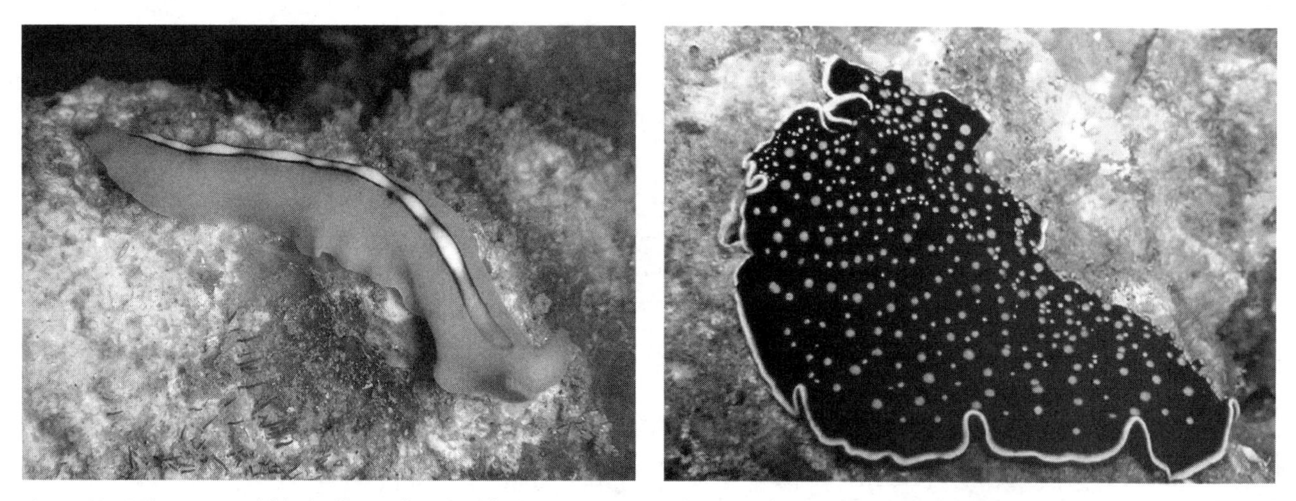

Figure 28.13 Flatworms exhibit significant diversity. (a) A blue Pseudoceros flatworm *(Pseudoceros bifurcus)*; (b) gold speckled flatworm *(Thysanozoon nigropapillosum)*. (credit a: modification of work by Stephen Childs; b: modification of work by Pril Fish.)

Flatworms have three embryonic tissue layers that give rise to epidermal tissues (from ectoderm), the lining of the digestive system (from endoderm), and other internal tissues (from mesoderm). The epidermal tissue is a single layer of cells or a layer of fused cells (**syncytium**) that covers two layers of muscle, one circular and the other longitudinal. The mesodermal tissues include mesenchymal cells that contain collagen and support secretory cells that produce mucus and other materials at the surface. Because flatworms are acoelomates, the mesodermal layer forms a solid mass between the outer epidermal surface and the cavity of the digestive system.

Physiological Processes of Flatworms

The free-living species of flatworms are predators or scavengers. Parasitic forms feed by absorbing nutrients provided by their hosts. Most flatworms, such as the planarian shown in Figure 28.14, have a branching gastrovascular cavity rather than a complete digestive system. In such animals, the "mouth" is also used to expel waste materials from the digestive system, and thus also serves as an anus. (A few species may have a second anal pore or opening.) The gut may be a simple sac or highly branched. Digestion is primarily extracellular, with digested materials taken into the cells of the gut lining by phagocytosis. One parasitic group, the tapeworms (cestodes), lacks a digestive system altogether, and absorb digested food from the host.

Flatworms have an excretory system with a network of tubules attached to **flame cells**, whose cilia beat to direct waste fluids concentrated in the tubules out of the body through excretory pores. The system is responsible for the regulation of dissolved salts and the excretion of nitrogenous wastes. The nervous system consists of a pair of lateral nerve cords running the length of the body with transverse connections between them. Two large cerebral ganglia—concentrations of nerve cell bodies at the anterior end of the worm—are associated with photosensory and chemosensory cells. There is neither a circulatory nor a respiratory system, with gas and nutrient exchange dependent on diffusion and cell-to-cell junctions. This necessarily limits the thickness of the body in these organisms, constraining them to be "flat" worms. Most flatworm species are *monoecious* (both male and female reproductive organs are found in the same individual), and fertilization is typically internal. Asexual reproduction by fission is common in some groups.

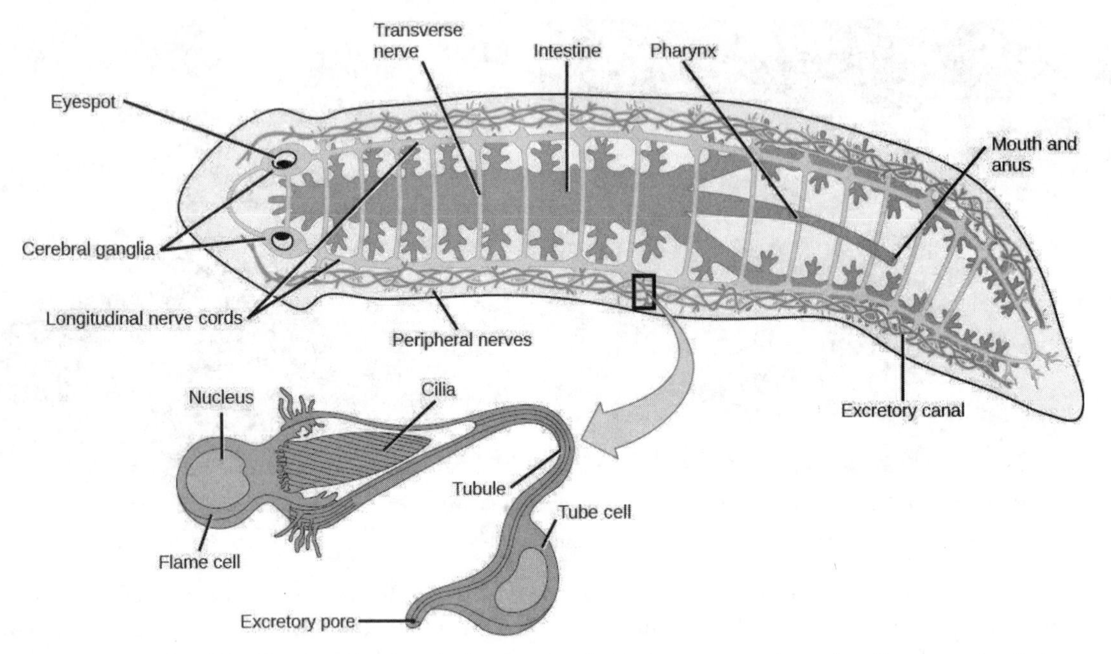

Figure 28.14 *Planaria*, a free-living flatworm. The planarian is a flatworm that has a gastrovascular cavity with one opening that serves as both mouth and anus. The excretory system is made up of flame cells and tubules connected to excretory pores on both sides of the body. The nervous system is composed of two interconnected nerve cords running the length of the body, with cerebral ganglia and eyespots at the anterior end.

Diversity of Flatworms

The flatworms have been traditionally divided into four classes: Turbellaria, Monogenea, Trematoda, and Cestoda (Figure 28.15). However, the relationships among members of these classes has recently been reassessed, with the turbellarians in particular now viewed as paraphyletic, since its descendants may also include members of the other three classes. Members of the clade or class Rhabditophora are now dispersed among multiple orders of Platyhelminthes, the most familiar of these being the Polycladida, which contains the large marine flatworms; the Tricladida (which includes *Dugesia* ["planaria"] and *Planaria* and its relatives); and the major parasitic orders: Monogenea (fish ectoparasites), Trematoda (flukes), and Cestoda (tapeworms), which together form a monophyletic clade.

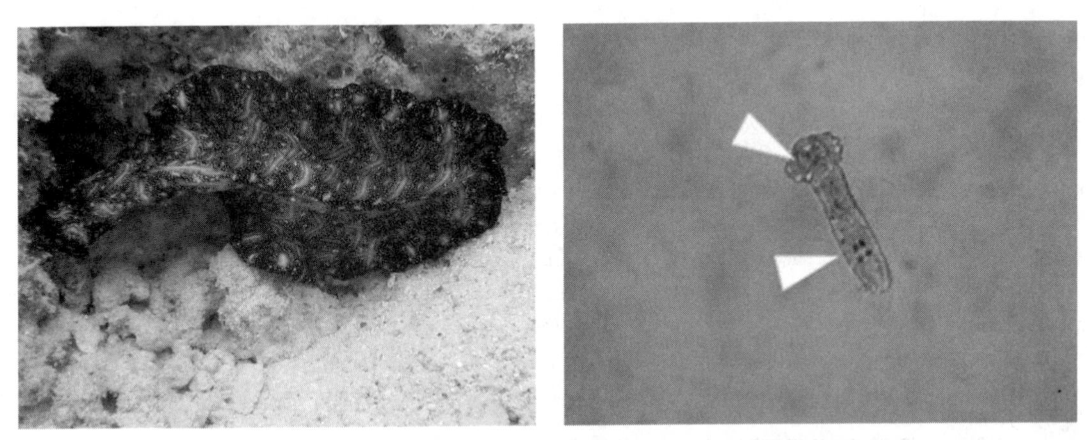

(a) Class Turbellaria

(b) Class Monogenea

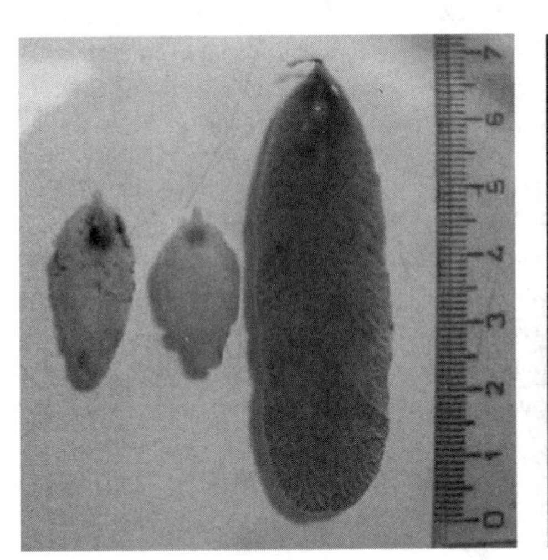

(c) Class Trematoda

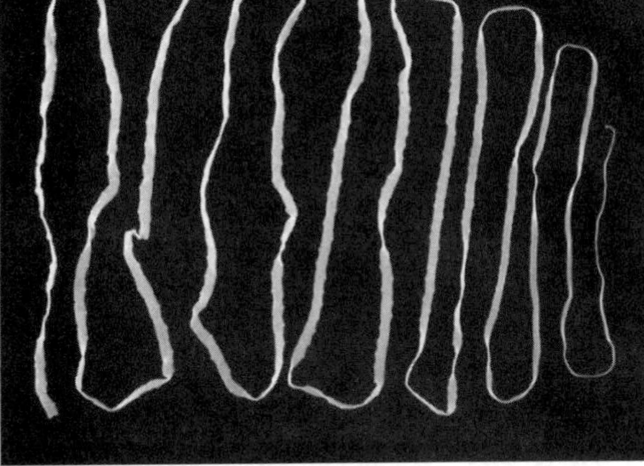

(d) Class Cestoda

Figure 28.15 Traditional flatworm classes. Phylum Platyhelminthes was previously divided into four classes. (a) Class Turbellaria includes the free-living polycladid Bedford's flatworm (*Pseudobiceros bedfordi*), which is about 8 to 10 cm in length. (b) The parasitic class Monogenea includes *Dactylogyrus* spp, commonly called gill flukes, which are about 0.2 mm in length and have two anchors, indicated by arrows used to attach the parasite on to the gills of host fish. (c) The class Trematoda includes *Fascioloides magna* (right) and *Fasciola hepatica* (two specimens on left, also known as the common liver fluke). (d) Class Cestoda includes tapeworms such as this *Taenia saginata*, infects both cattle and humans, and can reach 4 to 10 meters in length; the specimen shown here is about four meters long. (credit a: modification of work by Jan Derk; credit d: modification of work by CDC)

Most free-living flatworms are marine polycladids, although tricladid species live in freshwater or moist terrestrial environments, and there are a number of members from other orders in both environments. The ventral epidermis of free-living flatworms is ciliated, which facilitates their locomotion. Some free-living flatworms are capable of remarkable feats of regeneration in which an individual may regrow its head or tail after being severed, or even several heads if the planaria is cut lengthwise.

The monogeneans are *ectoparasites*, mostly of fish, with simple life cycles that consist of a free-swimming larva that attaches to a fish, prior to its transformation to the ectoparasitic adult form. The parasite has only one host and that host is usually very specific. The worms may produce enzymes that digest the host tissues, or they may simply graze on surface mucus and skin particles. Most monogeneans are hermaphroditic, but the male gametes develop first and so cross-fertilization is quite common.

The trematodes, or flukes, are internal parasites of mollusks and many other groups, including humans. Trematodes have complex life cycles that involve a primary host in which sexual reproduction occurs, and one or more secondary hosts in which asexual reproduction occurs. The primary host is usually a vertebrate and the secondary host is almost always a mollusk, in which multiple larvae are produced asexually. Trematodes, which attached internally to the host via an oral and medial sucker, are responsible for serious human diseases including schistosomiasis, caused by several species of the blood fluke, *Schistosoma* spp. Various forms of schistosomiasis infect an estimated 200 million people in the tropics, leading to organ damage, secondary infection by bacteria, and chronic symptoms like fatigue. Infection occurs when the human enters the water and metacercaria larvae, released from the snail host, locate and penetrate the skin. The parasite infects various organs in the body and feeds on red blood cells before reproducing.

Many of the eggs are released in feces and find their way into a waterway, where they are able to reinfect the snail host. The eggs, which have a barb on them, can damage the vascular system of the human host, causing ulceration, abscesses, and bloody diarrhea, wherever they reside, thereby allowing other pathogens to cause secondary infections. In fact, it is the parasite's eggs that produce most of the main ill effects of schistosomiasis. Many eggs do not make the transit through the veins of the host for elimination, and are swept by blood flow back to the liver and other locations, where they can cause severe inflammation. In the liver, the errant eggs may impede circulation and cause cirrhosis. Control is difficult in impoverished areas in unsanitary, crowded conditions, and prognosis is poor in people with heavy infections of *Schistosoma japonicum*, without early treatment.

The cestodes, or tapeworms, are also internal parasites, mainly of vertebrates (Figure 28.16). Tapeworms, such as those of *Taenia* spp, live in the intestinal tract of the primary host and remain fixed using a sucker or hooks on the anterior end, or scolex, of the tapeworm body, which is essentially a colony of similar subunits called proglottids. Each proglottid may contain an excretory system with flame cells, along with reproductive structures, both male and female. Because they are so long and flat, tapeworms do not need a digestive system; instead, they absorb nutrients from the food matter surrounding them in the host's intestine by diffusion.

Proglottids are produced at the scolex and gradually migrate to the end of the tapeworm; at this point, they are "mature" and all structures except fertilized eggs have degenerated. Most reproduction occurs by cross-fertilization between different worms in the same host, but may also occur between proglottids. The mature proglottids detach from the body of the worm and are released into the feces of the organism. The eggs are eaten by an intermediate host, typically another vertebrate. The juvenile worm infects the intermediate host and takes up residence, usually in muscle tissue. When the muscle tissue is consumed by the primary host, the cycle is completed. There are several tapeworm parasites of humans that are transmitted by eating uncooked or poorly cooked pork, beef, or fish.

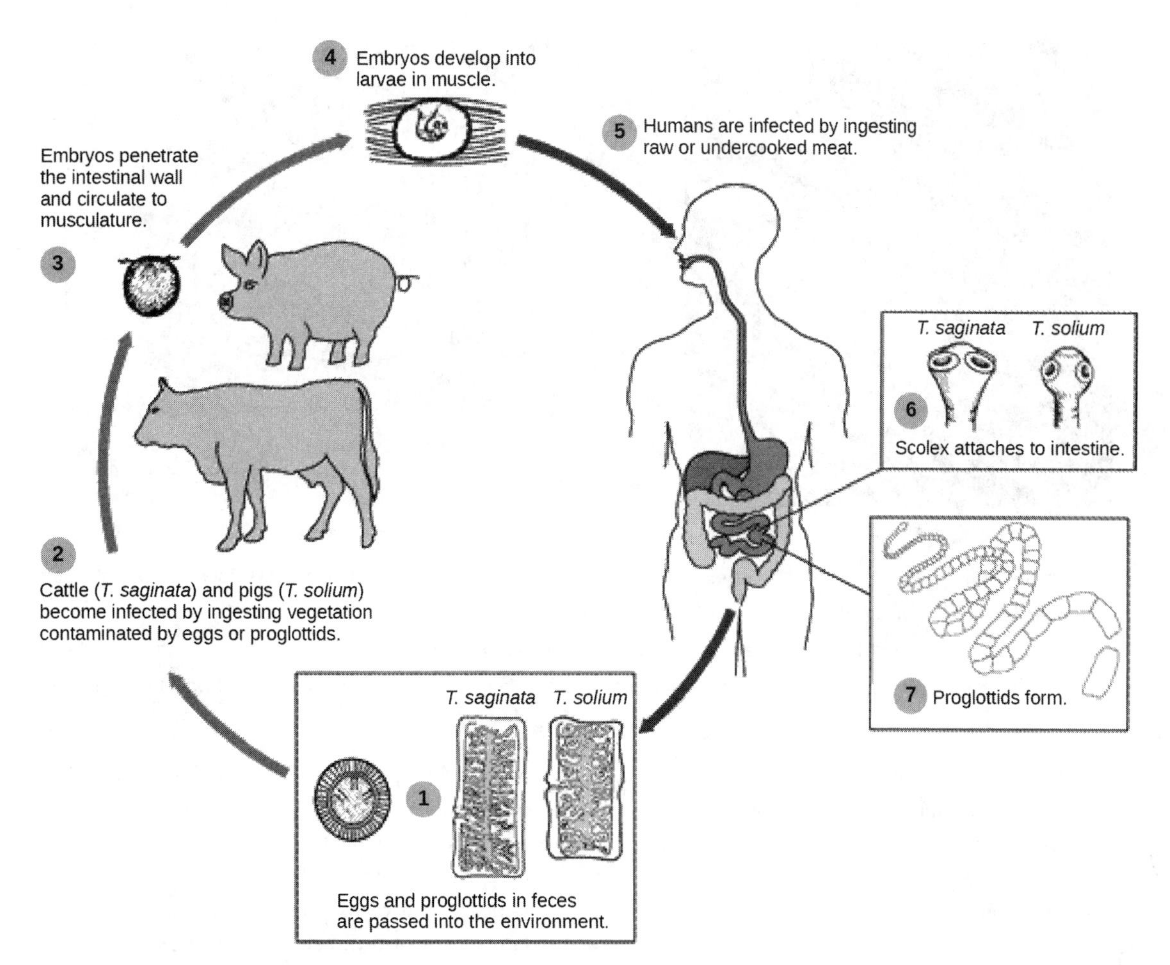

Figure 28.16 Tapeworm life cycle. Tapeworm (*Taenia* spp.) infections occur when humans consume raw or undercooked infected meat. (credit: modification of work by CDC)

Phylum Rotifera

The rotifers ("wheel-bearer") belong to a group of microscopic (about 100 μm to 2 mm) mostly aquatic animals that get their name from the **corona**—a pair of ciliated feeding structures that appear to rotate when viewed under the light microscope (Figure 28.17). Although their taxonomic status is currently in flux, one treatment places the rotifers in three classes: Bdelloidea, Monogononta, and Seisonidea. In addition, the parasitic "spiny headed worms" currently in phylum Acanthocephala, appear to be modified rotifers and will probably be placed into the group in the near future. Undoubtedly the rotifers will continue to be revised as more phylogenetic evidence becomes available.

The pseudocoelomate body of a rotifer is remarkably complex for such a small animal (roughly the size of a *Paramecium*) and is divided into three sections: a *head* (which contains the corona), a *trunk* (which contains most of the internal organs), and the *foot*. A cuticle, rigid in some species and flexible in others, covers the body surface. They have both skeletal muscle associated with locomotion and visceral muscles associated with the gut, both composed of single cells. Rotifers are typically free-swimming or planktonic (drifting) organisms, but the toes or extensions of the foot can secrete a sticky material to help them adhere to surfaces. The head contains a number of eyespots and a bilobed "brain," with nerves extending into the body.

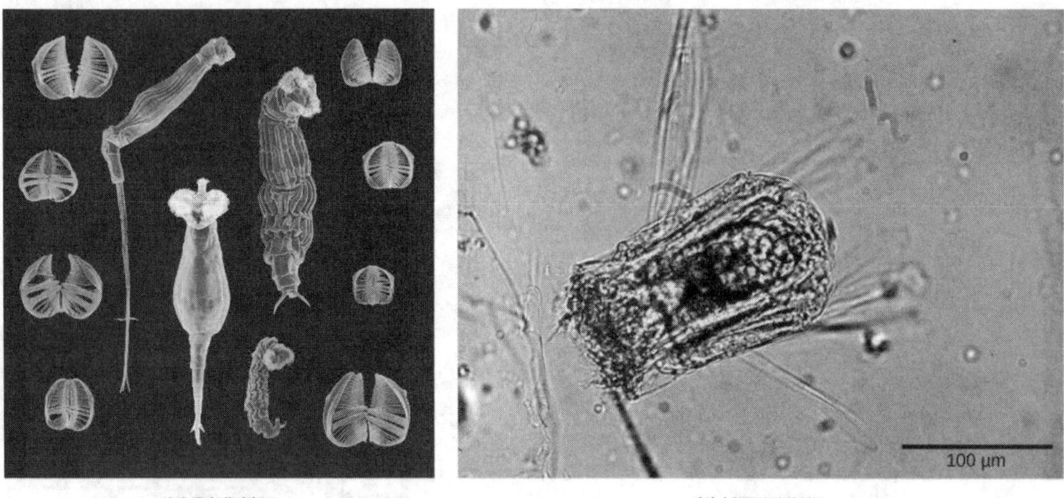

Figure 28.17 Rotifers. Shown are examples from two of the three classes of rotifer. (a) Species from the class Bdelloidea are characterized by a large corona. The whole animals in the center of this scanning electron micrograph are shown surrounded by several sets of jaws from the mastax of rotifers. (b) *Polyarthra*, from the largest rotifer class Monogononta, has a smaller corona than bdelloid rotifers, and a single gonad, which give the class its name. (credit a: modification of work by Diego Fontaneto; credit b: modification of work by U.S. EPA; scale-bar data from Cory Zanker)

Rotifers are commonly found in freshwater and some saltwater environments throughout the world. As filter feeders, they will eat dead material, algae, and other microscopic living organisms, and are therefore very important components of aquatic food webs. A rotifer's food is directed toward the mouth by the current created from the movement of the coronal cilia. The food particles enter the mouth and travel first to the **mastax**—a muscular pharynx with toothy jaw-like structures. Examples of the jaws of various rotifers are seen in Figure 28.17a. Masticated food passes near digestive and salivary glands, into the stomach, and then to the intestines. Digestive and excretory wastes are collected in a cloacal bladder before being released out the anus.

LINK TO LEARNING

Watch this video (http://openstax.org/l/rotifers) to see rotifers feeding.

About 2,200 species of rotifers have been identified. Figure 28.18 shows the anatomy of a rotifer belonging to class Bdelloidea. Some rotifers are dioecious organisms and exhibit sexual dimorphism (males and females have different forms). In many dioecious species, males are short-lived and smaller with no digestive system and a single testis. Many rotifer species exhibit haplodiploidy, a method of sex determination in which a fertilized egg develops into a female and an unfertilized egg develops into a male. However, reproduction in the bdelloid rotifers is exclusively parthenogenetic and appears to have been so for millions of years: Thus, all bdelloid rotifers and their progeny are female! The bdelloids may compensate for this genetic insularity by borrowing genes from the DNA of other species. Up to 10% of a bdelloid genome comprises genes imported from related species. Some rotifer eggs are capable of extended dormancy for protection during harsh environmental conditions.

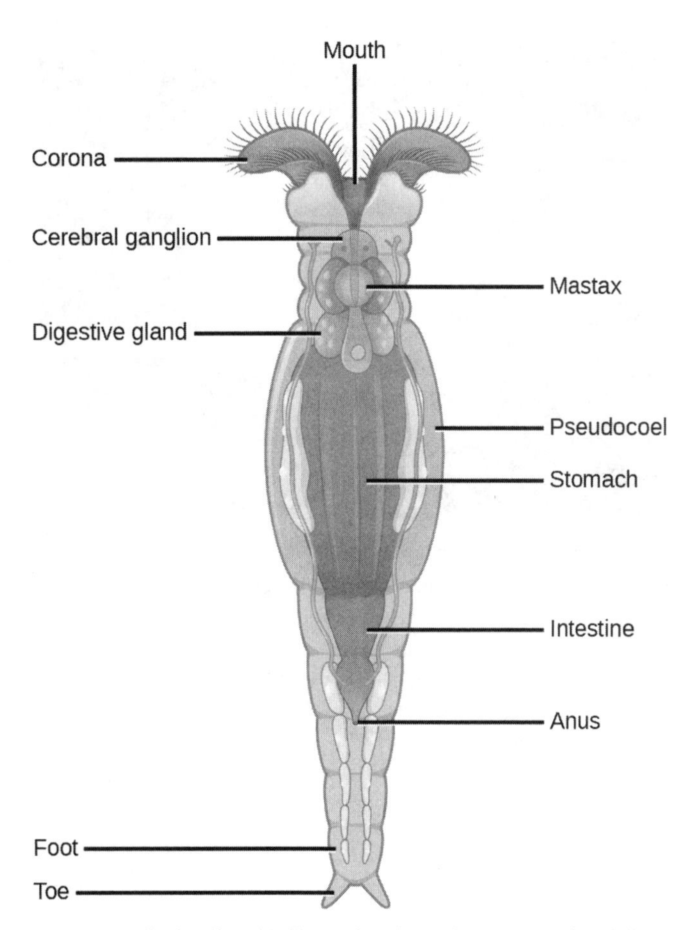

Figure 28.18 A bdelloid rotifer. This illustration shows the anatomy of a bdelloid rotifer.

Phylum Nemertea

The Nemertea are colloquially known as *ribbon worms* or proboscis worms. Most species of phylum Nemertea are marine and predominantly benthic (bottom dwellers), with an estimated 900 known species. However, nemerteans have been recorded in freshwater and very damp terrestrial habitats as well. Most nemerteans are carnivores, feeding on worms, clams, and crustaceans. Some species are scavengers, and some, like *Malacobdella grossa*, have also evolved commensal relationships with mollusks. Economically important species have at times devastated commercial fishing of clams and crabs. Nemerteans have almost no predators and two species are sold as fish bait.

Morphology

Nemerteans vary in size from 1 cm to several meters. They show bilateral symmetry and remarkable contractile properties. Because of their contractility, they can change their morphological presentation in response to environmental cues. Animals in phylum Nemertea are soft and unsegmented animals, with a morphology like a flattened tube. (Figure 28.19).

Figure 28.19 A proboscis worm. The proboscis worm (*Parborlasia corrugatus*) is a scavenger that combs the sea floor for food. The species is a member of the phylum Nemertea. The specimen shown here was photographed in the Ross Sea, Antarctica. (credit: Henry Kaiser, National Science Foundation)

A unique characteristic of this phylum is the presence of an eversible proboscis enclosed in a pocket called a **rhynchocoel** (not part of the animal's actual coelom). The proboscis is located dorsal to the gut and serves as a harpoon or tentacle for food capture. In some species it is ornamented with barbs. The rhynchocoel is a fluid-filled cavity that extends from the head to nearly two-thirds of the length of the gut in these animals (Figure 28.20). The proboscis may be extended by hydrostatic pressure generated by contraction of muscle of the rhynchocoel and retracted by a retractor muscle attached to the rear wall of the rhynchocoel.

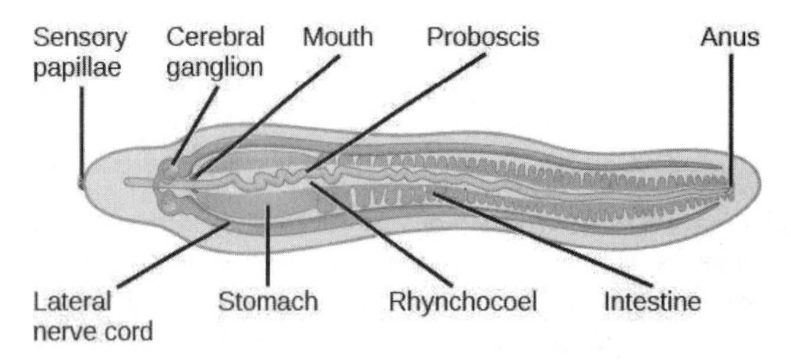

Figure 28.20 The anatomy of a Nemertean is shown.

LINK TO LEARNING

Watch this video (https://www.openstax.org/l/nemertean) to see a nemertean attack a polychaete with its proboscis.

Digestive System

The nemerteans, which are primarily predators of annelids and crustaceans, have a well-developed digestive system. A mouth opening that is ventral to the rhynchocoel leads into the foregut, followed by the intestine. The intestine is present in the form of diverticular pouches and ends in a rectum that opens via an anus. Gonads are interspersed with the intestinal diverticular pouches and open outward via genital pores.

Nemerteans are sometimes classified as acoels, but because their closed circulatory system is derived from the coelomic cavity of the embryo, they may be regarded as coelomic. Their circulatory system consists of a closed loop formed by a connected pair of lateral blood vessels. Some species may also have a dorsal vessel or cross-connecting vessels in addition to lateral ones. Although the circulatory fluid contains cells, it is often colorless. However, the blood cells of some species bear hemoglobin as well as other yellow or green pigments. The blood vessels are contractile, although there is usually no regular circulatory pathway, and movement of blood is also facilitated by the contraction of muscles in the body wall. The circulation of fluids in the rhychocoel is more or less independent of the blood circulation, although blind branches from the blood vessels into the rhyncocoel wall can mediate exchange of materials between them. A pair of **protonephridia**, or excretory tubules, is present in these animals to facilitate osmoregulation. Gaseous exchange occurs through the skin.

Nervous System

Nemerteans have a "brain" composed of four ganglia situated at the anterior end, around the rhynchocoel. Paired longitudinal nerve cords emerge from the brain ganglia and extend to the posterior end. Additional nerve cords are found in some species. Interestingly, the brain can contain hemoglobin, which acts as an oxygen reserve. *Ocelli* or eyespots are present in pairs, in multiples of two in the anterior portion of the body. It is speculated that the eyespots originate from neural tissue and not from the epidermis.

Reproduction

Nemerteans, like flatworms, have excellent powers of regeneration, and asexual reproduction by fragmentation is seen in some species. Most animals in phylum Nemertea are dioecious, although freshwater species may be hermaphroditic. Stem cells that become gametes aggregate within gonads placed along the digestive tract. Eggs and sperm are released into the water, and fertilization occurs externally. Like most lophotrochozoan protostomes, cleavage is spiral, and development is usually direct, although some species have a trochophore-like larva, in which a young worm is constructed from a series of imaginal discs that begin as invaginations from the body surface of the larva.

28.4 Superphylum Lophotrochozoa: Molluscs and Annelids

By the end of this section, you will be able to do the following:

- Describe the unique anatomical and morphological features of molluscs and annelids
- Describe the formation of the coelom
- Identify an important extracoelomic cavity in molluscs
- Describe the major body regions of Mollusca and how they vary in different molluscan classes
- Discuss the advantages of true body segmentation
- Describe the features of animals classified in phylum Annelida

The annelids and the mollusks are the most familiar of the lophotrochozoan protostomes. They are also more "typical" lophotrochozoans, since both groups include aquatic species with trochophore larvae, which unite both taxa in common ancestry. These phyla show how a flexible body plan can lead to biological success in terms of abundance and species diversity. The phylum Mollusca has the second greatest number of species of all animal phyla with nearly 100,000 described extant species, and about 80,000 described extinct species. In fact, it is estimated that about 25 percent of all known marine species are mollusks! The annelids and mollusca are both bilaterally symmetrical, cephalized, triploblastic, schizocoelous eucoeolomates They include animals you are likely to see in your backyard or on your dinner plate!

Phylum Mollusca

The name "Mollusca" means "soft" body, since the earliest descriptions of molluscs came from observations of "squishy," unshelled cuttlefish. Molluscs are predominantly a marine group of animals; however, they are also known to inhabit freshwater as well as terrestrial habitats. This enormous phylum includes chitons, tusk shells, snails, slugs, nudibranchs, sea butterflies, clams, mussels, oysters, squids, octopuses, and nautiluses. Molluscs display a wide range of morphologies in each class and subclass, but share a few key characteristics (Figure 28.21). The chief locomotor structure is usually a muscular **foot**. Most internal organs are contained in a region called the **visceral mass**. Overlying the visceral mass is a fold of tissue called the **mantle**; within the cavity formed by the mantle are respiratory structures called **gills**, that typically fold over the visceral mass. The mouths of most mollusks, except bivalves (e.g., clams) contain a specialized feeding organ called a **radula**, an abrasive tonguelike structure. Finally, the mantle secretes a calcium-carbonate-hardened shell in most molluscs, although this is greatly reduced in the class Cephalopoda, which contains the octopuses and squids.

VISUAL CONNECTION

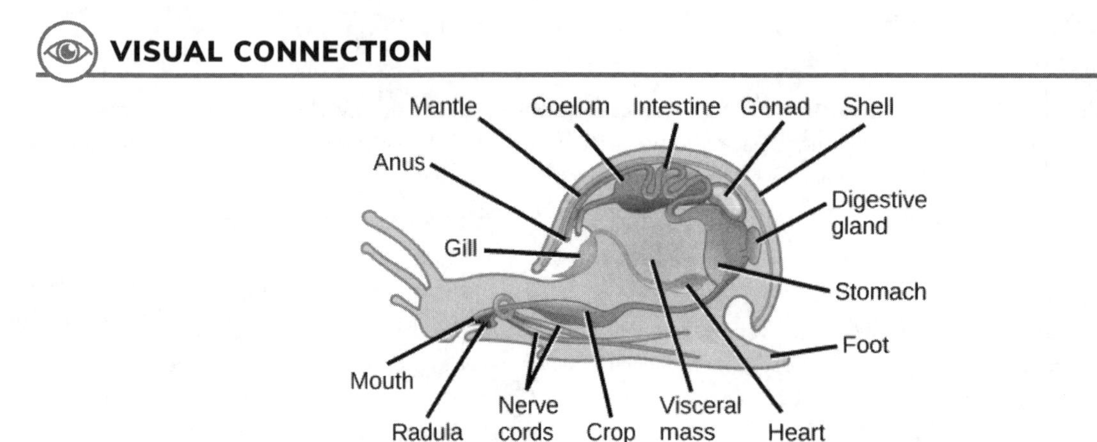

Figure 28.21 Molluscan body regions. There are many species and variations of molluscs; this illustration shows the anatomy of an aquatic gastropod. In a terrestrial gastropod, the mantle cavity itself would serve as a respiratory organ.

Which of the following statements about the anatomy of a mollusc is false?

1. Most molluscs have a radula for grinding food.
2. A digestive gland is connected to the stomach.
3. The tissue beneath the shell is called the mantle.
4. The digestive system includes a gizzard, a stomach, a digestive gland, and the intestine.

The muscular **foot** is the ventral-most organ, whereas the **mantle** is the limiting dorsal organ that folds over the **visceral mass**. The foot, which is used for locomotion and anchorage, varies in shape and function, depending on the type of mollusk under study. In shelled mollusks, the foot is usually the same size as the opening of the shell. The foot is both retractable and extendable. In the class Cephalopoda ("head-foot"), the foot takes the form of a funnel for expelling water at high velocity from the mantle cavity; and the anterior margin of the foot has been modified into a circle of *arms* and *tentacles*.

The visceral mass is present above the foot, in the visceral hump. This mass contains digestive, nervous, excretory, reproductive, and respiratory systems. Molluscan species that are exclusively aquatic have gills that extend into the mantle cavity, whereas some terrestrial species have "lungs" formed from the lining of the mantle cavity. Mollusks are schizocoelous eucoelomates, but the coelomic cavity in adult animals has been largely reduced to a cavity around the heart. However, a reduced coelom sometimes surrounds the gonads, part of the kidneys, and intestine as well. This overall coelomic reduction makes the mantle cavity the major internal body chamber.

Most mollusks have a special rasp-like organ, the **radula**, which bears chitinous filelike teeth. The radula is present in all groups except the bivalves, and serves to shred or scrape food before it enters the digestive tract. The **mantle** (also known as the *pallium*) is the dorsal epidermis in mollusks; all mollusks except some cephalopods are specialized to secrete a calcareous shell that protects the animal's soft body.

Most mollusks are dioecious animals and fertilization occurs externally, although this is not the case in terrestrial mollusks, such as snails and slugs, or in cephalopods. In most aquatic mollusks, the zygote hatches and produces a **trochophore larva**, with several bands of cilia around a toplike body, and an additional apical tuft of cilia. In some species, the trochophore may be followed by additional larval stages, such as a **veliger** larvae, before the final metamorphosis to the adult form. Most cephalopods develop directly into small versions of their adult form.

Classification of Phylum Mollusca

Phylum Mollusca comprises a very diverse group of organisms that exhibits a dramatic variety of forms, ranging from chitons to snails to squids, the latter of which typically show a high degree of intelligence. This variability is a consequence of modification of the basic body regions, especially the foot and mantle. The phylum is organized into eight classes: Caudofoveata, Solenogastres, Monoplacophora, Polyplacophora, Gastropoda, Cephalopoda, Bivalvia, and Scaphopoda. Although each molluscan class appears to be monophyletic, their relationship to one another is unclear and still being reviewed.

Both the Caudofoveata and the Solenogastres include shell-less, worm-like animals primarily found in benthic marine habitats.

Although these animals lack a calcareous shell, they get some protection from calcareous spicules embedded in a cuticle that covers their epidermis. The mantle cavity is reduced, and both groups lack eyes, tentacles, and nephridia (excretory organs). The Caudofoveata possess a radula, but the Solenogastres do not have a radula or gills. The foot is also reduced in the Solenogastres and absent from the Caudofoveata.

Long thought to be extinct, the first living specimens of Monoplacophora, *Neopilina galatheae,* were discovered in 1952 on the ocean bottom near the west coast of Costa Rica. Today there are over 25 described species. Members of class Monoplacophora ("bearing one plate") possess a single, cap-like shell that covers the dorsal body. The morphology of the shell and the underlying animal can vary from circular to ovate. They have a simple radula, a looped digestive system, multiple pairs of excretory organs, and a pair of gonads. Multiple gills are located between the foot and the edge of the mantle.

Animals in class Polyplacophora ("bearing many plates") are commonly known as "chitons" and bear eight limy plates that make up the dorsal shell (Figure 28.22). These animals have a broad, ventral foot that is adapted for suction onto rocks and other substrates, and a mantle that extends beyond the edge of the shell. Calcareous spines on the exposed mantle edge provide protection from predators. Respiration is facilitated by multiple pairs of gills in the mantle cavity. Blood from the gills is collected in a posterior heart, and then sent to the rest of the body in a **hemocoel**—an *open circulation system* in which the blood is contained in connected chambers surrounding various organs rather than within individual blood vessels. The radula, which has teeth composed of an ultra-hard magnetite, is used to scrape food organisms off rocky surfaces. Chiton teeth have been shown to exhibit the greatest hardness and stiffness of any biomineral material reported to date, being as much as three-times harder than human enamel and the calcium carbonate-based shells of mollusks.

The nervous system is rudimentary with only buccal or "cheek" ganglia present at the anterior end. Multiple tiny sensory structures, including photosensors, extend from the mantle into channels in the upper layer of the shell. These structures are called **esthetes** and are unique to the chitons. Another sensory structure under the radula is used to sample the feeding environment. A single pair of nephridia is used for the excretion of nitrogenous wastes.

Figure 28.22 A chiton. This chiton from the class Polyplacaphora has the eight-plated shell for which its class is named. (credit: Jerry Kirkhart)

Class Bivalvia ("two-valves") includes clams, oysters, mussels, scallops, geoducks, and shipworms. Some bivalves are almost microscopic, while others, in the genus *Tridacna,* may be one meter in length and weigh 225 kilograms. Members of this class are found in marine as well as freshwater habitats. As the name suggests, bivalves are enclosed in two-part valves or shells (Figure 28.23a) fused on the dorsal side by hinge ligaments as well as shell teeth on the ventral side that keep the two halves aligned. The two shells, which consist of an outer organic layer, a middle prismatic layer, and a very smooth nacreous layer, are joined at the oldest part of the shell called the umbo. Anterior and posterior adductor and abductor muscles close and open the shell respectively.

The overall body of the bivalve is laterally flattened; the foot is wedge-shaped; and the head region is poorly developed (with no obvious mouth). Bivalves are filter-feeders, and a radula is absent in this class of mollusks. The mantle cavity is fused along the edges except for openings for the foot and for the intake and expulsion of water, which is circulated through the mantle cavity by the actions of the incurrent and excurrent siphons. During water intake by the incurrent siphon, food particles are captured by the paired posterior gills (ctenidia) and then carried by the movement of cilia forward to the mouth. Excretion and osmoregulation are performed by a pair of nephridia. Eyespots and other sensory structures are located along the edge of the mantle in some species. The "eyes" are especially conspicuous in scallops (Figure 28.23b). Three pairs of connected ganglia regulate activity of different body structures.

Figure 28.23 Bivalves. These mussels (a), found in the intertidal zone in Cornwall, England, show the bivalve shell. The scallop *Argopecten irradians* (b) has a fluted shell and conspicuous eyespots. (credit (a): Mark A. Wilson. credit (b) Rachael Norris and Marina Freudzon. https://commons.wikimedia.org/w/index.php?curid=17251065 (http://openstax.org/l/scallop_eyes))

One of the functions of the mantle is to secrete the shell. Some bivalves, like oysters and mussels, possess the unique ability to secrete and deposit a calcareous nacre or "mother of pearl" around foreign particles that may enter the mantle cavity. This property has been commercially exploited to produce pearls.

LINK TO LEARNING

Watch the animations of bivalves feeding: View the process in clams (http://openstax.org/l/clams) and mussels (http://openstax.org/l/mussels) at these sites.

More than half of molluscan species are in the class Gastropoda ("stomach foot"), which includes well-known mollusks like snails, slugs, conchs, cowries, limpets, and whelks. Aquatic gastropods include both marine and freshwater species, and all terrestrial mollusks are gastropods. Gastropoda includes shell-bearing species as well as species without shells. Gastropod bodies are asymmetrical and usually present a coiled shell (Figure 28.24a). Shells may be *planospiral* (like a garden hose wound up), commonly seen in garden snails, or *conispiral*, (like a spiral staircase), commonly seen in marine conches. Cowrie shells have a polished surface because the mantle extends up over the top of the shell as it is secreted.

Figure 28.24 Gastropods. Snails(a) and slugs(b) are both gastropods, but slugs lack a shell. (credit a: modification of work by Murray Stevenson; credit b: modification of work by Rosendahl)

A key characteristic of some gastropods is the embryonic development of **torsion**. During this process, the mantle and visceral mass are rotated around the perpendicular axis over the center of the foot to bring the anal opening forward just behind the head (Figure 28.25), creating a very peculiar situation. The left gill, kidney, and heart atrium are now on the right side, whereas the original right gill, kidney, and heart atrium are on the left side. Even stranger, the nerve cords have been twisted and contorted into a figure-eight pattern. Because of the space made available by torsion in the mantle cavity, the animal's sensitive head end can now be withdrawn into the protection of the shell, and the tougher foot (and sometimes the protective covering or **operculum**) forms a barrier to the outside. The strange arrangement that results from torsion poses a serious sanitation problem by creating the possibility of wastes being washed back over the gills, causing **fouling.** There is actually no really perfect explanation for the embryonic development of torsion, and some groups that formerly exhibited torsion in their ancestral groups are now known to have reversed the process.

Gastropods also have a foot that is modified for crawling. Most gastropods have a well-defined head with tentacles and eyes. A complex radula is used to scrape up food particles. In aquatic gastropods, the mantle cavity encloses the gills (ctenidia), but in land gastropods, the mantle itself is the major respiratory structure, acting as a kind of lung. **Nephridia** ("kidneys") are also found in the mantle cavity.

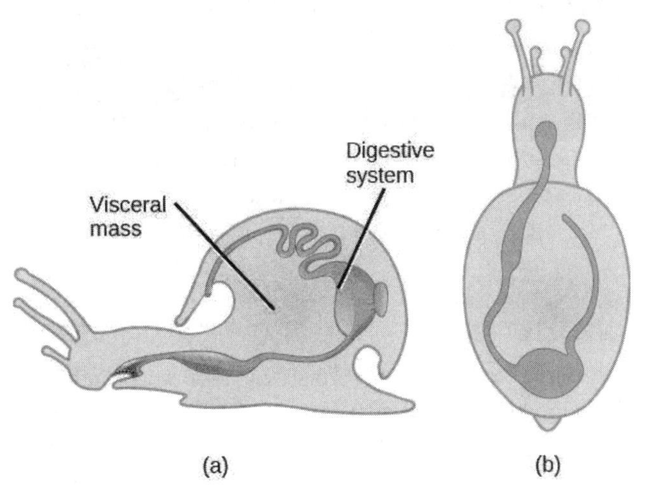

Figure 28.25 Torsion in gastropods. During embryonic development of some gastropods, the visceral mass undergoes torsion, or counterclockwise rotation of the visceral anatomical features. As a result, the anus of the adult animal is located over the head. Although torsion is always counterclockwise, the shell may coil in either direction; thus coiling of a shell is *not* the same as torsion of the visceral mass.

Everyday Connection

Can Snail Venom Be Used as a Pharmacological Painkiller?

Marine snails of the genus *Conus* (Figure 28.26) attack prey with a venomous stinger, modified from the radula. The toxin released, known as conotoxin, is a peptide with internal disulfide linkages. Conotoxins can bring about paralysis in humans, indicating that this toxin attacks neurological targets. Some conotoxins have been shown to block neuronal ion channels. These findings have led researchers to study conotoxins for possible medical applications.

Conotoxins are an exciting area of potential pharmacological development, since these peptides may be possibly modified and used in specific medical conditions to inhibit the activity of specific neurons. For example, conotoxins or modifications of them may be used to induce paralysis in muscles in specific health applications, similar to the use of botulinum toxin. Since the entire spectrum of conotoxins, as well as their mechanisms of action, is not completely known, the study of their potential applications is still in its infancy. Most research to date has focused on their use to treat neurological diseases. They have also shown some efficacy in relieving chronic pain, and the pain associated with conditions like sciatica and shingles. The study and use of biotoxins—toxins derived from living organisms—are an excellent example of the application of biological science to modern medicine.

Figure 28.26 *Conus.* Members of the genus *Conus* produce neurotoxins that may one day have medical uses. The tube above the eyes is a siphon used both to circulate water over the gills and to sample the water for chemical evidence of prey nearby. Note the eyes below the siphon. The proboscis, through which the venomous harpoon is projected, is located between the eyes. (credit: David Burdick, NOAA)

Class Cephalopoda ("head foot" animals), includes octopuses, squids, cuttlefish, and nautiluses. Cephalopods include both animals with shells as well as animals in which the shell is reduced or absent. In the shell-bearing *Nautilus,* the spiral shell is multi-chambered. These chambers are filled with gas or water to regulate buoyancy. A siphuncle runs through the chambers, and it is this tube that regulates the amount of water and gases (nitrogen, carbon dioxide, and oxygen mixture) present in the chambers. Ammonites and other nautiloid shells are commonly seen in the fossil record. The shell structure in squids and cuttlefish is reduced and is present internally in the form of a squid pen and cuttlefish bone, respectively. Cuttle bone is sold in pet stores to help smooth the beaks of birds and also to provide birds such as egg-laying chickens and quail with an inexpensive natural source of calcium carbonate. Examples of cephalopods are shown in Figure 28.27.

Cephalopods can display vivid and rapidly changing coloration, almost like flashing neon signs. Typically these flashing displays are seen in squids and octopuses, where they may be used for camouflage and possibly as signals for mating displays. We should note, however, that researchers are not entirely sure if squid can actually see color, or see color in the same way as we do. We know that pigments in the skin are contained in special pigment cells (**chromatophores**), which can expand or contract to produce different color patterns. But chromatophores can only make yellow, red, brown, and black pigmentation; however, underneath them is a whole different set of elements called **iridophores** and **leucophores** that reflect light and can make blue, green, and white. It is possible that squid skin might actually be able to detect some light on its own, without even needing its eyes!

All animals in this class are carnivorous predators and have beak-like jaws in addition to the radula. Cephalopods include the most intelligent of the mollusks, and have a well-developed nervous system along with image-forming eyes. Unlike other mollusks, they have a closed circulatory system, in which the blood is entirely contained in vessels rather than in a hemocoel.

The foot is lobed and subdivided into arms and tentacles. Suckers with chitinized rings are present on the arms and tentacles of octopuses and squid. Siphons are well developed and the expulsion of water is used as their primary mode of locomotion, which resembles jet propulsion. Gills (ctenidia) are attached to the wall of the mantle cavity and are serviced by large blood vessels, each with its own heart. A pair of nephridia is present within the mantle cavity for water balance and excretion of nitrogenous wastes. Cephalopods such as squids and octopuses also produce sepia or a dark ink, which contains melanin. The ink gland is located between the gills and can be released into the excurrent water stream. Ink clouds can be used either as a "smoke screen" to hide the animal from predators during a quick attempt at escape, or to create a fake image to distract predators.

Cephalopods are dioecious. Members of a species mate, and the female then lays the eggs in a secluded and protected niche. Females of some species care for the eggs for an extended period of time and may end up dying during that time period. While most other aquatic mollusks produce trochophore larvae, cephalopod eggs develop directly into a juvenile without an intervening larval stage.

Figure 28.27 Cephalopods. The (a) nautilus, (b) giant cuttlefish, (c) reef squid, and (d) blue-ring octopus are all members of the class Cephalopoda. (credit a: modification of work by J. Baecker; credit b: modification of work by Adrian Mohedano; credit c: modification of work by Silke Baron; credit d: modification of work by Angell Williams)

Members of class Scaphopoda ("boat feet") are known colloquially as "tusk shells" or "tooth shells," as evident when examining *Dentalium*, one of the few remaining scaphopod genera (Figure 28.28). Scaphopods are usually buried in sand with the anterior opening exposed to water. These animals have a single conical shell, which is open on both ends. The head is not well developed, but the mouth, containing a radula, opens among a group of tentacles that terminate in ciliated bulbs used to catch and manipulate prey. Scaphopods also have a foot similar to that seen in bivalves. Ctenidia are absent in these animals; the mantle cavity forms a tube open at both ends and serves as the respiratory structure in these animals.

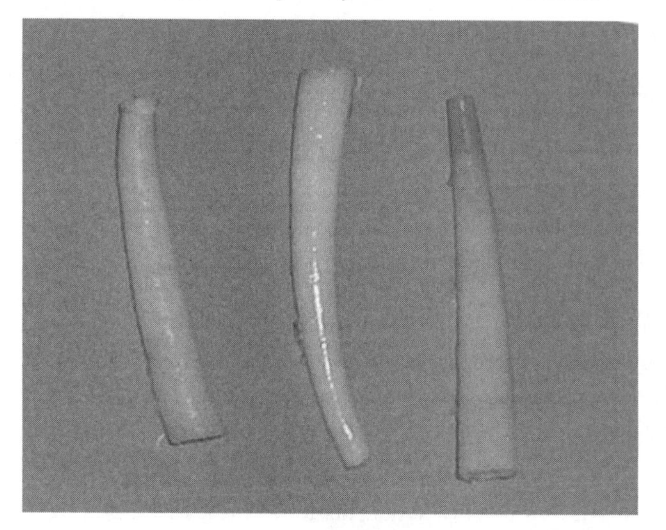

Figure 28.28 Tooth shells. *Antalis vulgaris* shows the classic Dentaliidae shape that gives these animals their common name of "tusk shell." (credit: Georges Jansoone)

Phylum Annelida

Phylum Annelida comprises the true, segmented worms. These animals are found in marine, terrestrial, and freshwater habitats, but the presence of water or humidity is a critical factor for their survival in terrestrial habitats. The annelids are often called "segmented worms" due to their key characteristic of **metamerism**, or true segmentation. Approximately 22,000 species have been described in phylum Annelida, which includes polychaete worms (marine annelids with multiple appendages), and oligochaetes (earthworms and leeches). Some animals in this phylum show parasitic and commensal symbioses with other species in their habitat.

Morphology

Annelids display bilateral symmetry and are worm-like in overall morphology. The name of the phylum is derived from the Latin word *annullus*, which means a small ring, an apt description of the ring-like segmentation of the body. Annelids have a body plan with metameric segmentation, in which several internal and external morphological features are repeated in each body segment. Metamerism allows animals to become bigger by adding "compartments," while making their movement more efficient. The overall body can be divided into head, body, and pygidium (or tail). During development, the segments behind the head arise sequentially from a growth region anterior to the pygidium, a pattern called **teloblastic growth**. In the Oligochaetes, the **clitellum** is a reproductive structure that generates mucus to aid sperm transfer and also produces a "cocoon," within which fertilization occurs; it appears as a permanent, fused band located on the anterior third of the animal (Figure 28.29).

Figure 28.29 The clitellum of an earthworm. The clitellum, seen here as a protruding segment with different coloration than the rest of the body, is a structure that aids in oligochaete reproduction. (credit: Rob Hille)

Anatomy

The epidermis is protected by a collagenous, external cuticle, which is much thinner than the cuticle found in the ecdysozoans and does not require periodic shedding for growth. Circular as well as longitudinal muscles are located interior to the epidermis. Chitinous bristles called **setae** (or chaetae) are anchored in the epidermis, each with its own muscle. In the polychaetes, the setae are borne on paired appendages called **parapodia**.

Most annelids have a well-developed and complete digestive system. Feeding mechanisms vary widely across the phylum. Some polychaetes are filter-feeders that use feather-like appendages to collect small organisms. Others have tentacles, jaws, or an eversible pharynx to capture prey. Earthworms collect small organisms from soil as they burrow through it, and most leeches are blood-feeders armed with teeth or a muscular proboscis. In earthworms, the digestive tract includes a mouth, muscular pharynx, esophagus, crop, and muscular gizzard. The gizzard leads to the intestine, which ends in an anal opening in the terminal segment. A cross-sectional view of a body segment of an earthworm is shown in Figure 28.30; each segment is limited by a membranous septum that divides the coelomic cavity into a series of compartments.

Most annelids possess a closed circulatory system of dorsal and ventral blood vessels that run parallel to the alimentary canal as well as capillaries that service individual tissues. In addition, the dorsal and ventral vessels are connected by transverse loops in every segment. Some polychaetes and leeches have an open system in which the major blood vessels open into a hemocoel. In many species, the blood contains hemoglobin, but not contained in cells. Annelids lack a well-developed respiratory system, and gas exchange occurs across the moist body surface. In the polychaetes, the parapodia are highly vascular and serve as respiratory structures. Excretion is facilitated by a pair of metanephridia (a type of primitive "kidney" that consists of a convoluted tubule and an open, ciliated funnel) that is present in every segment toward the ventral side. Annelids show well-developed nervous systems with a ring of fused ganglia present around the pharynx. The nerve cord is ventral in position and bears enlarged nodes or ganglia in each segment.

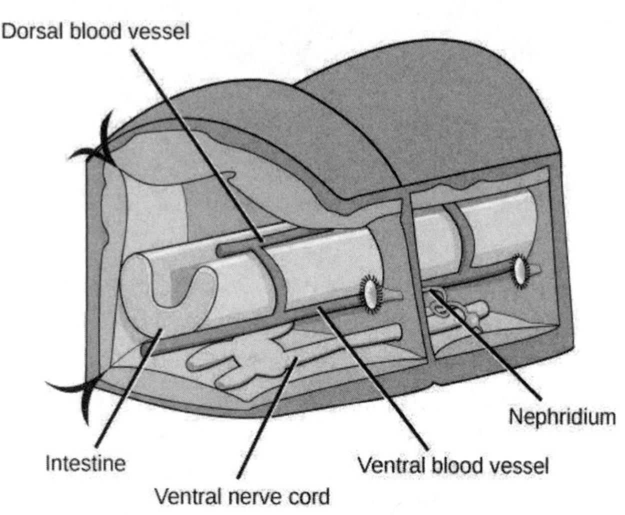

Figure 28.30 Segmental anatomy of an earthworm. This schematic drawing shows the basic anatomy of annelids in a cross-sectional view.

Annelids may be either monoecious with permanent gonads (as in earthworms and leeches) or dioecious with temporary or seasonal gonads (as in polychaetes). However, cross-fertilization is preferred even in hermaphroditic animals. Earthworms may show simultaneous mutual fertilization when they are aligned for copulation. Some leeches change their sex over their reproductive lifetimes. In most polychaetes, fertilization is external and development includes a trochophore larva, which then metamorphoizes to the adult form. In oligochaetes, fertilization is typically internal and the fertilized eggs develop in a cocoon produced by the clitellum; development is direct. Polychaetes are excellent regenerators and some even reproduce asexually by budding or fragmentation.

LINK TO LEARNING

This combination video and animation (http://shapeoflife.org/video/animation/annelid-animation-body-plan) provides a close-up look at annelid anatomy.

Classification of Phylum Annelida

Phylum Annelida contains the class Polychaeta (the polychaetes) and the class Oligochaeta (the earthworms, leeches, and their relatives). The earthworms and the leeches form a monophyletic clade within the polychaetes, which are therefore paraphyletic as a group.

There are more than 22,000 different species of annelids, and more than half of these are marine polychaetes ("many bristles"). In the polychaetes, bristles are arranged in clusters on their parapodia—fleshy, flat, paired appendages that protrude from each segment. Many polychaetes use their parapodia to crawl along the sea floor, but others are adapted for swimming or floating. Some are sessile, living in tubes. Some polychaetes live near hydrothermal vents. These deepwater tubeworms have no digestive tract, but have a symbiotic relationship with bacteria living in their bodies.

Earthworms are the most abundant members of the class Oligochaeta ("few bristles"), distinguished by the presence of a permanent clitellum as well as the small number of reduced chaetae on each segment. (Recall that oligochaetes do not have parapodia.) The oligochaete subclass Hirudinea, includes leeches such as the medicinal leech, *Hirudo medicinalis*, which is effective at increasing blood circulation and breaking up blood clots, and thus can be used to treat some circulatory disorders and cardiovascular diseases. Their use goes back thousands of years. These animals produce a seasonal clitellum, unlike the permanent clitellum of other oligochaetes. A significant difference between leeches and other annelids is the lack of setae and the development of suckers at the anterior and posterior ends, which are used to attach to the host animal. Additionally, in leeches, the segmentation of the body wall may not correspond to the internal segmentation of the coelomic cavity. This adaptation possibly helps the leeches to elongate when they ingest copious quantities of blood from host vertebrates, a condition in which they are said to be “engorged.” The subclass Brachiobdella includes tiny leechlike worms that attach themselves to the gills or body surface of crayfish.

Figure 28.31 Annelid groups. The (a) earthworm, (b) leech, and (c) featherduster are all annelids. The earthworm and leech are oligochaetes, while the featherduster worm is a tube-dwelling filter-feeding polychaete. (credit a: modification of work by S. Shepherd; credit b: modification of work by "Sarah G..."/Flickr; credit c: modification of work by Chris Gotschalk, NOAA)

28.5 Superphylum Ecdysozoa: Nematodes and Tardigrades

By the end of this section, you will be able to do the following:

- Describe the structural organization of nematodes
- Describe the importance of Caenorhabditis elegans in research
- Describe the features of Tardigrades

Superphylum Ecdysozoa

The superphylum Ecdysozoa contains an incredibly large number of species. This is because it contains two of the most diverse animal groups: phylum Nematoda (the roundworms) and phylum Arthropoda (the arthropods). The most prominent distinguishing feature of ecdysozoans is the cuticle—a tough, but flexible exoskeleton that protects these animals from water loss, predators, and other dangers of the external environment. One small phylum within the Ecdysozoa, with exceptional resistance to desiccation and other environmental hazards, is the Tardigrada. The nematodes, tardigrades, and arthropods all belong to the superphylum Ecdysozoa, which is believed to be **monophyletic**—a clade consisting of all evolutionary descendants from one common ancestor. All members of this superphylum periodically go through a molting process that culminates in ecdysis—the actual shedding of the old exoskeleton. (The term "ecdysis" translates roughly as "take off" or "strip.") During the molting process, old cuticle is replaced by a new cuticle, which is secreted beneath it, and which will last until the next growth period.

Phylum Nematoda

The Nematoda, like other members of the superphylum Ecdysozoa, are triploblastic and possess an embryonic mesoderm that is sandwiched between the ectoderm and endoderm. They are also bilaterally symmetrical, meaning that a longitudinal section will divide them into right and left sides that are superficially symmetrical. In contrast with flatworms, nematodes are pseudocoelomates and show a tubular morphology and circular cross-section. Nematodes include both free-living and parasitic forms.

In 1914, N.A. Cobb said, "In short, if all the matter in the universe except the nematodes were swept away, our world would still be dimly recognizable, and if, as disembodied spirits, we could then investigate it, we should find its mountains, hills, vales, rivers, lakes and oceans represented by a thin film of nematodes…" To paraphrase Cobb, nematodes are so abundant that if all the non-nematode matter of the biosphere were removed, there would still remain a shadow of the former world outlined by nematodes![1] The phylum Nematoda includes more than 28,000 species with an estimated 16,000 being parasitic in nature. However, nematologists believe there may be over one million unclassified species.

The name Nematoda is derived from the Greek word "Nemos," which means "thread," and includes all true roundworms. Nematodes are present in all habitats, typically with each species occurring in great abundance. The free-living nematode, *Caenorhabditis elegans*, has been extensively used as a model system for many different avenues of biological inquiry in laboratories all over the world.

1Stoll, N. R., "This wormy world. 1947," *Journal of Parasitology* 85(3) (1999): 392-396.

Morphology

The cylindrical body form of the nematodes is seen in Figure 28.32. These animals have a complete digestive system with a distinct mouth and anus, whereas only one opening is present in the digestive tract of flatworms. The mouth opens into a muscular pharynx and intestine, which leads to a rectum and anal opening at the posterior end. The epidermis can be either a single layer of cells or a **syncytium**—a multinucleated tissue that in this case is formed by the fusion of many single cells. The cuticle of nematodes is rich in collagen and a polymer called **chitin**, which forms a protective armor outside the epidermis. The cuticle extends into both ends of the digestive tract, the pharynx, and rectum. In the head, an anterior mouth opening is composed of three (or six) "lips" as well as teeth derived from the cuticle (in some species). Some nematodes may present other modifications of the cuticle such as rings, head shields, or warts. These external rings, however, do not reflect *true* internal body segmentation, which as we have seen is a hallmark of phylum Annelida. The attachment of the muscles of nematodes differs from that of most animals: they have a longitudinal layer only, and their direct attachment to the dorsal and ventral nerve cords creates a strong muscular contraction that results in a whiplike, almost spastic, body movement.

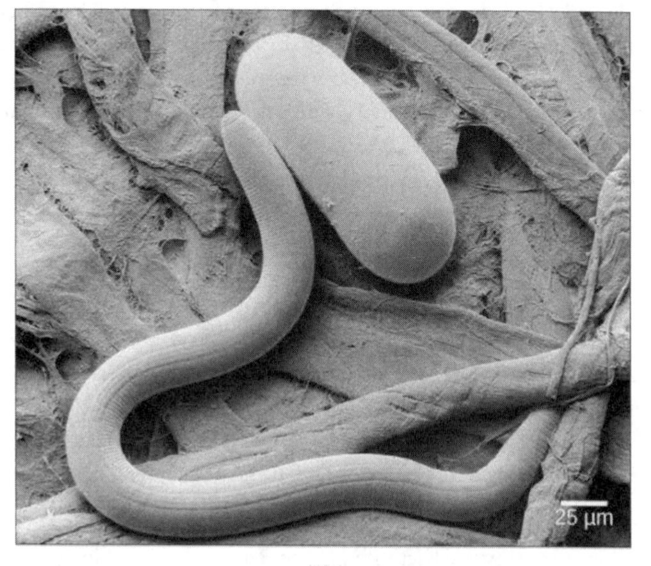

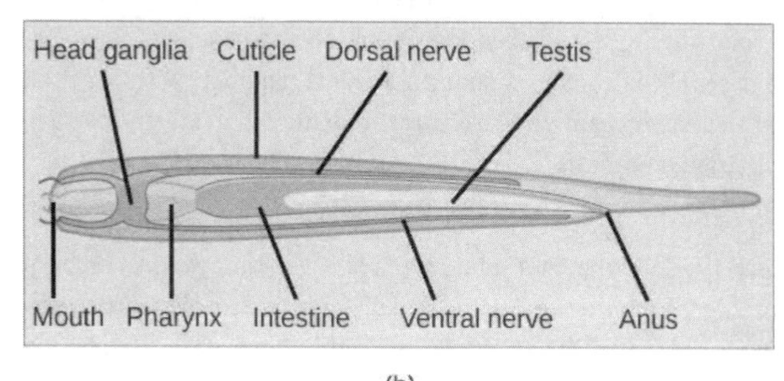

(b)

Figure 28.32 Nematode morphology. Scanning electron micrograph shows (a) the soybean cyst nematode (*Heterodera glycines*) and a nematode egg. (b) A schematic representation shows the anatomy of a typical nematode. (credit a: modification of work by USDA ARS; scale-bar data from Matt Russell)

Excretory System

In nematodes, specialized excretory systems are not well developed. Nitrogenous wastes, largely in the form of *ammonia*, are released directly across the body wall. In some nematodes, osmoregulation and salt balance are performed by simple excretory cells or glands that may be connected to paired canals that release wastes through an anterior pore. In marine nematodes, the excretory cells are called **renette cells**, which are unique to nematodes.

Nervous system

Most nematodes have four longitudinal nerve cords that run along the length of the body in dorsal, ventral, and lateral positions. The ventral nerve cord is better developed than the dorsal and lateral cords. Nonetheless, all nerve cords fuse at the anterior end, to form a **pharyngeal nerve ring** around the pharynx, which acts as the head ganglion or the "brain" of the roundworm. A similar fusion forms a posterior ganglion at the tail. In *C. elegans*, the nervous system accounts for nearly one-third of the total number of cells in the animal!

Reproduction

Nematodes employ a variety of reproductive strategies ranging from monoecious to dioecious to parthenogenetic, depending upon the species. *C. elegans* is a mostly monoecious species with both self-fertilizing hermaphrodites and some males. In the hermaphrodites, ova and sperm develop at different times in the same gonad. Ova are contained in a uterus and amoeboid sperm are contained in a spermatheca ("sperm receptacle"). The uterus has an external opening known as the vulva. The female genital pore is near the middle of the body, whereas the male genital pore is nearer to the tip. In anatomical males, specialized structures called *copulatory spicules* at the tail of the male keep him in place and open the vulva of the female into which the amoeboid sperm travel into the spermatheca.

Fertilization is internal, and embryonic development starts very soon after fertilization. The embryo is released from the vulva during the gastrulation stage. The embryonic development stage lasts for 14 hours; development then continues through four successive larval stages with molting and ecdysis taking place between each stage—L1, L2, L3, and L4—ultimately leading to the development of a young adult worm. Adverse environmental conditions such as overcrowding or lack of food can result in the formation of an intermediate larval stage known as the *dauer larva*. An unusual feature of some nematodes is eutely: the body of a given species contains a specific number of cells as the consequence of a rigid developmental pathway.

Everyday Connection

C. elegans: The Model System for Linking Developmental Studies with Genetics

If biologists wanted to research how nicotine dependence develops in the body, how lipids are regulated, or observe the attractant or repellant properties of certain odors, they would clearly need to design three very different experiments. However, they might only need one subject of study: *Caenorhabditis elegans*. The nematode *C. elegans* was brought into the focus of mainstream biological research by Dr. Sydney Brenner. Since 1963, Dr. Brenner and scientists worldwide have used this animal as a model system to study many different physiological and developmental mechanisms.

C. elegans is a free-living nematode found in soil. Only about a millimeter long, it can be cultured on agar plates (10,000 worms/plate!), feeding on the common intestinal bacterium *Escherichia coli* (another long-term resident of biological laboratories worldwide), and therefore can be readily grown and maintained in a laboratory. The biggest asset of this nematode is its transparency, which helps researchers to observe and monitor changes within the animal with ease. It is also a simple organism with about 1,000 cells and a genome of only 20,000 genes. Its chromosomes are organized into five pairs of autosomes plus a pair of sex chromosomes, making it an ideal candidate with which to study genetics. Since every cell can be visualized and identified, this organism is useful for studying cellular phenomena like cell-to-cell interactions, cell-fate determinations, cell division, apoptosis (cell death), and intracellular transport.

Another tremendous asset is the short life cycle of this worm (Figure 28.33). It takes only three days to achieve the "egg to adult to daughter egg"; therefore, the developmental consequences of genetic changes can be quickly identified. The total life span of *C. elegans* is two to three weeks; hence, age-related phenomena are also easy to observe. There are two sexes in this species: hermaphrodites (XX) and males (XO). However, anatomical males are relatively infrequently obtained from matings between hermaphrodites, since their XO chromosome composition requires meiotic nondisjunction when both parents are XX. Another feature that makes *C. elegans* an excellent experimental model is that the position and number of the 959 cells present in adult hermaphrodites of this organism is *constant*. This feature is extremely significant when studying cell differentiation, cell-to-cell communication, and apoptosis. Lastly, *C. elegans* is also amenable to genetic manipulations using molecular methods, rounding off its usefulness as a model system.

Biologists worldwide have created information banks and groups dedicated to research using *C. elegans*. Their findings have led, for example, to better understandings of cell communication during development, neuronal signaling, and insight into lipid regulation (which is important in addressing health issues like the development of obesity and diabetes). In recent years, studies have enlightened the medical community with a better understanding of polycystic kidney disease. This simple organism has led biologists to complex and significant findings, growing the field of science in ways that touch the everyday world.

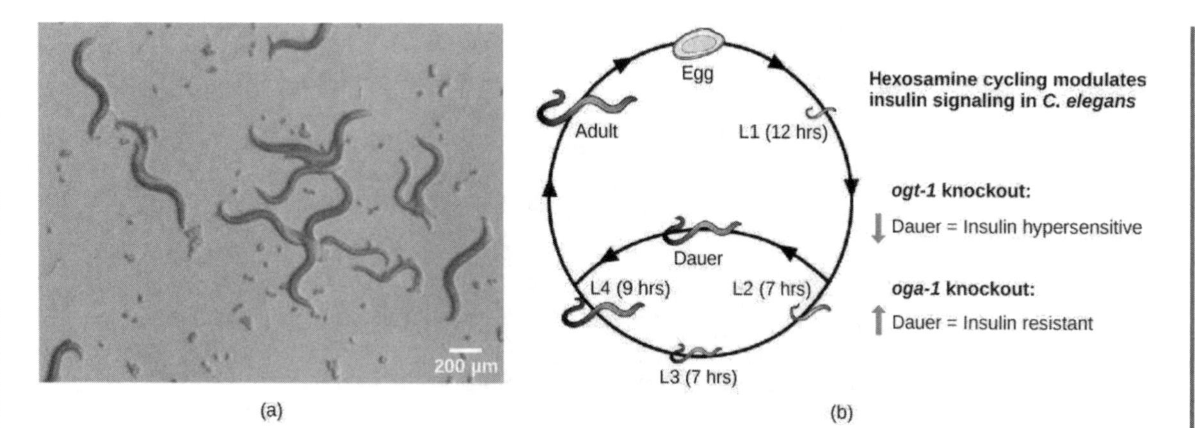

Figure 28.33 *Caenorhabditis elegans.* (a) This light micrograph shows the bodies of a group of roundworms. These hermaphrodites consist of exactly 959 cells. (b) The life cycle of *C. elegans* has four juvenile stages (L1 through L4) and an adult stage. Under ideal conditions, the nematode spends a set amount of time at each juvenile stage, but under stressful conditions, it may enter a *dauer state* that does not age significantly and is somewhat analogous to the diapausing state of some insects. (credit a: modification of work by "snickclunk"/Flickr: credit b: modification of work by NIDDK, NIH; scale-bar data from Matt Russell)

Parasitic Nematodes

A number of common parasitic nematodes serve as prime examples of parasitism (endoparasitism). These economically and medically important animals exhibit complex life cycles that often involve multiple hosts, and they can have significant medical and veterinary impacts. Here is a partial list of nasty nematodes: Humans may become infected by *Dracunculus medinensis,* known as guinea worms, when they drink unfiltered water containing copepods (Figure 28.34), an intermediate crustacean host. Hookworms, such as *Ancylostoma* and *Necator,* infest the intestines and feed on the blood of mammals, especially of dogs, cats, and humans. Trichina worms (*Trichinella*) are the causal organism of trichinosis in humans, often resulting from the consumption of undercooked pork; *Trichinella* can infect other mammalian hosts as well. *Ascaris,* a large intestinal roundworm, steals nutrition from its human host and may create physical blockage of the intestines. The filarial worms, such as *Dirofilaria* and *Wuchereria,* are commonly vectored by mosquitoes, which pass the infective agents among mammals through their blood-sucking activity. One species, *Wuchereria bancrofti,* infects the lymph nodes of over 120 million people worldwide, usually producing a non-lethal but deforming condition called *elephantiasis.* In this disease, parts of the body often swell to gigantic proportions due to obstruction of lymphatic drainage, inflammation of lymphatic tissues, and resulting edema. *Dirofilaria immitis,* a blood-infective parasite, is the notorious dog heartworm species.

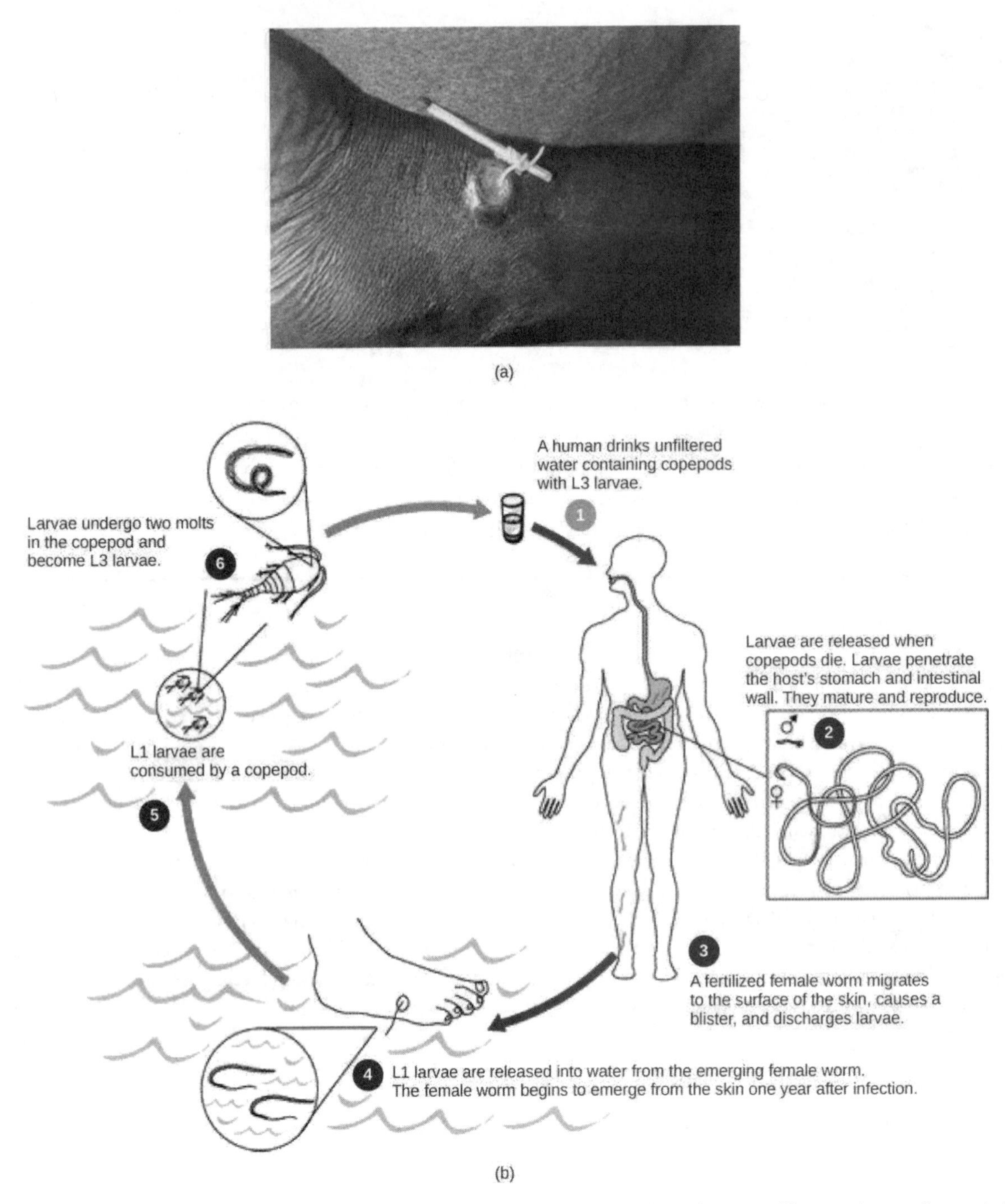

Figure 28.34 Life cycle of the guinea worm. The guinea worm *Dracunculus medinensis* infects about 3.5 million people annually, mostly in Africa. (a) Here, the worm is wrapped around a stick so it can be slowly extracted. (b) Infection occurs when people consume water contaminated by infected copepods, but this can easily be prevented by simple filtration systems. (credit: modification of work by CDC)

Phylum Tardigrada

The tardigrades ("slow-steppers") comprise a phylum of inconspicuous little animals living in marine, freshwater, or damp terrestrial environments throughout the world. They are commonly called "water bears" because of their plump bodies and the large claws on their stubby legs. There are over 1,000 species, most of which are less than 1 mm in length. A chitinous cuticle covers the body surface and may be divided into plates (Figure 28.35). Tardigrades are known for their ability to enter a state called **cryptobiosis**, which provides them with are resistance to multiple environmental challenges, including desiccation, very low temperatures, vacuum, high pressure, and radiation. They can suspend their metabolic activity for years, and survive the loss of up to 99% of their water content. Their remarkable resistance has recently been attributed to unique proteins that replace water in their cells and protect their internal cell structure and their DNA from damage.

Figure 28.35 Scanning electronmicrograph of *Milnesium tardigradum.* (credit: Schokraie E, Warnken U, Hotz-Wagenblatt A, Grohme MA, Hengherr S, et al. (2012) -https://commons.wikimedia.org/w/index.php?curid=22716809 (http://openstax.org/l/waterbear))

Morphology and Physiology

Tardigrades have cylindrical bodies, with four pairs of legs terminating in a number of claws. The cuticle is periodically shed, including the cuticular covering of the claws. The first three pairs of legs are used for walking, and the posterior pair for clinging to the substrate. A circular mouth leads to a muscular pharynx and salivary glands. Tardigrades feed on plants, algae, or small animals. Plant cells are pierced with a chitinous stylet and the cellular contents are then sucked into the gut by the muscular pharynx. Bands of single muscle cells are attached to the various points of the epidermis and extend into the legs to provide ambulatory movement. The major body cavity is a *hemocoel,* but there are no specialized circulatory structures for moving the blood, nor are there specialized respiratory structures. Malpighian tubules in the hemocoel remove metabolic wastes and transport them to the gut. A dorsal brain is connected to a ventral nerve cord with segmental ganglia associated with the appendages. Sensory structures are greatly reduced, but there is a pair of simple eyespots on the head, and sensory cilia or bristles concentrated toward the head end of the animal.

Reproduction

Most tardigrades are dioecious, and males and females each have a single gonad. Mating usually occurs at the time of a molt and fertilization is external. Eggs may be deposited in a molted cuticle or attached to other objects. Development is direct, and the animal may molt a dozen times during its lifetime. In tardigrades, like nematodes, development produces a fixed number of cells, with the actual number of cells being dependent on the species. Further growth occurs by enlarging the cells, not by multiplying them.

28.6 Superphylum Ecdysozoa: Arthropods

By the end of this section, you will be able to do the following:

- Compare the internal systems and appendage specializations of phylum Arthropoda
- Discuss the environmental importance of arthropods
- Discuss the reasons for arthropod success and abundance

The superphylum Ecdysozoa also includes the phylum Arthropoda, one of the most successful clades of animals on the planet. Arthropods are coelomate organisms characterized by a sturdy chitinous exoskeleton and jointed appendages. There are well over a million arthropod species described, and systematists believe that there are millions of species awaiting proper classification. Like other Ecdysozoa, all arthropods periodically go through the physiological process of **molting**, followed by **ecdysis** (the actual shedding of the exoskeleton), as they grow. Arthropods are eucoelomate, protostomic organisms, often with incredibly complicated life cycles.

Phylum Arthropoda

The name “arthropoda” means “jointed feet.” The name aptly describes the invertebrates included in this phylum. Arthropods have probably always dominated the animal kingdom in terms of number of species and likely will continue to do so: An estimated 85 percent of all known species are included in this phylum! In effect, life on Earth could conceivably be called the Age of Arthropods beginning nearly 500 million years ago.

The principal characteristics of all the animals in this phylum are the structural and functional segmentation of the body and the presence of jointed appendages. Arthropods have an exoskeleton made principally of chitin—a waterproof, tough polysaccharide composed of N-acetylglucosamine. Phylum Arthropoda is the most speciose clade in the animal world (Table 28.1), and insects form the single largest class within this phylum. For comparison, refer to the approximate numbers of species in the phyla listed below.

Phylum	# species
Ctenophora	100
Porifera	5,000
Cnidaria	11,000
Platyhelminthes	25,000
Rotifera	2,000
Nemertea	1,200
Annelida	22,000
Mollusca	112,000
Nematoda	28,000+
Tardigrades	>1,000
Arthropoda	1,134,000
Echinodermata	7,000
Chordata	100,000

Table 28.1

Phylum Arthropoda includes animals that have been successful in colonizing terrestrial, aquatic, and aerial habitats. This phylum is further classified into five subphyla: Trilobita (trilobites, all extinct), Chelicerata (horseshoe crabs, spiders, scorpions, ticks, mites, and daddy longlegs or harvestmen), Myriapoda (millipedes, centipedes, and their relatives), Crustacea (crabs, lobsters, crayfish, isopods, barnacles, and some zooplankton), and Hexapoda (insects and their six-legged relatives). Trilobites, an extinct group of arthropods found chiefly in the pre-Cambrian Era (about 500 million years ago), are probably most closely related to the Chelicerata. These are identified based on their fossils; they were quite diverse and radiated significantly into thousands of species before their complete extinction at the end of the Permian about 240 million years ago (Figure 28.36).

Figure 28.36 A trilobite. Trilobites, like the one in this fossil, are an extinct group of arthropods. Their name "trilobite" refers to the *three longitudinal lobes* making up the body: right pleural lobe, axial lobe, and left pleural lobe (credit: Kevin Walsh).

Morphology

Characteristic features of the arthropods include the presence of jointed appendages, body segmentation, and chitinized exoskeleton. Fusion of adjacent groups of segments gave rise to functional body regions called tagmata (singular = tagma). Tagmata may be in the form of a head, thorax, and abdomen, or a cephalothorax and abdomen, or a head and trunk, depending on the taxon. Commonly described tagmata may be composed of different numbers of segments; for example, the head of most insects results from the fusion of six ancestral segments, whereas the "head" of another arthropod may be made of fewer ancestral segments, due to independent evolutionary events. Jointed arthropod appendages, often in segmental pairs, have been specialized for various functions: sensing their environment (antennae), capturing and manipulating food (mandibles and maxillae), as well as for walking, jumping, digging, and swimming.

In the arthropod body, a central cavity, called the **hemocoel** (or blood cavity), is present, and the hemocoel fluids are moved by contraction of regions of the tubular dorsal blood vessel called "hearts." Groups of arthropods also differ in the organs used for nitrogenous waste excretion, with crustaceans possessing *green glands* and insects using *Malpighian tubules*, which work in conjunction with the hindgut to reabsorb water while ridding the body of nitrogenous waste. The nervous system tends to be distributed among the segments, with larger ganglia in segments with sensory structures or appendages. The ganglia are connected by a ventral nerve cord.

Respiratory systems vary depending on the group of arthropod. Insects and myriapods use a series of tubes (tracheae) that branch through the body, ending in minute tracheoles. These fine respiratory tubes perform gas exchange *directly* between the air and cells within the organism. The major tracheae open to the surface of the cuticle via apertures called **spiracles.** We should note that these tracheal systems of ventilation have evolved independently in hexapods, myriapods, and arachnids. Although the tracheal system works extremely well in terrestrial environments, it also works well in freshwater aquatic environments: In fact, numerous species of aquatic insects in both immature and adult stages possess tracheal systems. However, although there are insects that live on the surface of marine environments, none is strictly marine—meaning that they complete their entire metamorphosis *in* salt water.

In contrast, aquatic crustaceans utilize gills, terrestrial chelicerates employ book lungs, and aquatic chelicerates use book gills (Figure 28.37). The book lungs of arachnids (scorpions, spiders, ticks, and mites) contain a vertical stack of hemocoel wall tissue that somewhat resembles the pages of a book. Between each of the "pages" of tissue is an air space. This allows both sides of the tissue to be in contact with the air at all times, greatly increasing the efficiency of gas exchange. The gills of crustaceans are filamentous structures that exchange gases with the surrounding water.

The **cuticle** is the hard "covering" of an arthropod. It is made up of two layers: the *epicuticle*, which is a thin, waxy, water-resistant outer layer containing no chitin, and the layer beneath it, the *chitinous procuticle*, which itself is composed of an *exocuticle* and a lower *endocuticle*, all supported ultimately by a *basement membrane*. The exoskeleton is very protective (it is sometimes difficult to squish a big beetle!), but does not sacrifice flexibility or mobility. Both the exocuticle (which is secreted *before* a molt), and an endocuticle, (which is secreted *after* a molt), are composed of chitin bound with a protein; chitin is insoluble in water, alkalis, and weak acids. The procuticle is not only flexible and lightweight, but also provides protection against dehydration and other biological and physical stresses. Some hexapods, such as the crustaceans, add calcium salts to their exoskeleton, which increases the strength of the cuticle, but does reduce its flexibility. In some cases, such as lobsters, the amount of calcium salt deposited within the chitin is extreme. In order to grow, the arthropod must "shed" the exoskeleton during the physiological process called **molting**, following by the actual stripping of the outer cuticle, called **ecdysis** ("to strip off"). At first, this seems to be a dangerous method of growth, because while the new exoskeleton is hardening, the animal is vulnerable to predation; however, molting and ecdysis also allow for growth and change in morphology, as well as for great diversification in size, simply because the numbers of molts can be modified through evolution.

The characteristic morphology of representative animals from each subphylum is described below.

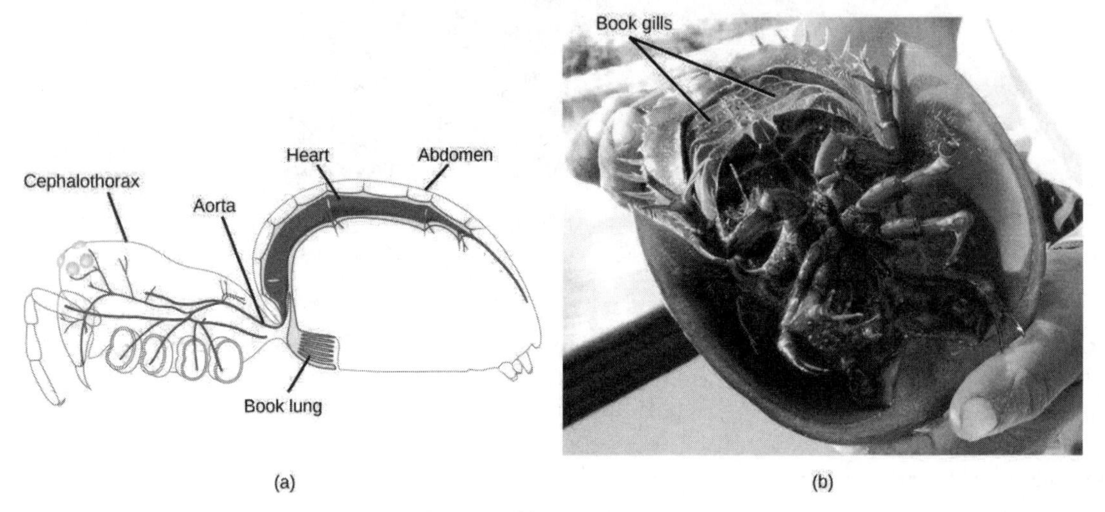

Figure 28.37 Arthropod respiratory structures. The book lungs of (a) arachnids are made up of alternating air pockets and hemocoel tissue shaped like a stack of books (hence the name, "book lung"). The book gills of (b) horseshoe crabs are similar to book lungs but are external so that gas exchange can occur with the surrounding water. (credit a: modification of work by Ryan Wilson based on original work by John Henry Comstock; credit b: modification of work by Angel Schatz)

Subphylum Chelicerata

This subphylum includes animals such as horseshoe crabs, sea spiders, spiders, mites, ticks, scorpions, whip scorpions, and harvestmen. Chelicerates are predominantly terrestrial, although some freshwater and marine species also exist. An estimated 77,000 species of chelicerates can be found in almost all terrestrial habitats.

The body of chelicerates is divided into two tagmata: prosoma and opisthosoma, which are basically the equivalents of a *cephalothorax* (usually smaller) and an *abdomen* (usually larger). A distinct "head" tagma is not usually discernible. The phylum derives its name from the first pair of appendages: the **chelicerae** (Figure 28.38), which serve as specialized clawlike or fanglike mouthparts. We should note here that chelicerae are actually *modified legs*, but they are not the exact serial equivalent of mandibles, which are the modified leglike chewing mouthparts of insects and crustaceans: The chelicerae are borne on the *first segment* making up the prosoma, whereas the mandibles are embryonically on the *fourth segment* of the mandibulate head. The chelicerates have secondarily lost their antennae and hence do not have them. Some of the functions of the antennae (such as touch) are now performed by the second pair of appendages— the **pedipalps**, which may also be used for general sensing the environment as well as the manipulation of food. In some species, such as sea spiders, an additional pair of derived leg appendages, called **ovigers,** is present between the chelicerae and pedipalps. Ovigers are used for grooming and by males to carry eggs. In spiders, the chelicerae are often modified and terminate in fangs that inject venom into their prey before feeding (Figure 28.39).

Figure 28.38 Chelicerae. The chelicerae (first set of appendages) are well developed in the scorpion. (credit: Kevin Walsh)

Most chelicerates ingest food using a preoral cavity formed by the chelicerae and pedipalps. Some chelicerates may secrete digestive enzymes to pre-digest food before ingesting it. Parasitic chelicerates like ticks and mites have evolved blood-sucking apparatuses. Members of this subphylum have an open circulatory system with a heart that pumps blood into the hemocoel. Aquatic species, like horseshoe crabs, have gills, whereas terrestrial species have either tracheae or book lungs for gaseous exchange. Chelicerate hemolymph contains hemocyanin a copper-containing oxygen transport protein.

Figure 28.39 Spider. The trapdoor spider, like all spiders, is a member of the subphylum Chelicerata. (credit: Marshal Hedin)

The nervous system in chelicerates consists of a brain and two ventral nerve cords. Chelicerates are dioecious, meaning that the sexes are separate. These animals use external fertilization as well as internal fertilization strategies for reproduction, depending upon the species and its habitat. Parental care for the young ranges from absolutely none to relatively prolonged care.

LINK TO LEARNING

Visit this site (http://openstax.org/l/arthropodstory) to click through a lesson on arthropods, including interactive habitat maps, and more.

Subphylum Myriapoda

Subphylum Myriapoda comprises arthropods with numerous legs. Although the name is misleading, suggesting that thousands of legs are present in these invertebrates, the number of legs typically varies from 10 to 750. This subphylum includes 16,000 species; the most commonly found examples are millipedes and centipedes. Virtually all myriapods are terrestrial animals and prefer a humid environment. Ancient myriapods (or myriapod-like arthropods) from the Silurian to the Devonian grew up to 10 feet in length (three meters). Unfortunately, they are all extinct!

Myriapods are typically found in moist soils, decaying biological material, and leaf litter. Subphylum Myriapoda is divided into four classes: Chilopoda, Symphyla, Diplopoda, and Pauropoda. Centipedes like *Scutigera coleoptrata* (Figure 28.40) are classified as chilopods. These animals bear one pair of legs per segment, mandibles as mouthparts, and are somewhat dorsoventrally flattened. The legs in the first segment are modified to form forcipules (poison claws) that deliver poison to prey like spiders and cockroaches, as these animals are all predatory. Symphyla are similar to centipedes, but lack the poison claws and are vegetarian. Millipedes bear two pairs of legs per diplosegment—a feature that results from the embryonic fusion of adjacent pairs of body segments. These arthropods are usually rounder in cross-section than centipedes, and are herbivores or detritivores. Millipedes have visibly more numbers of legs as compared to centipedes, although they do *not* have a thousand legs (Figure 28.40b). The Pauropods are similar to millipedes, but have fewer segments.

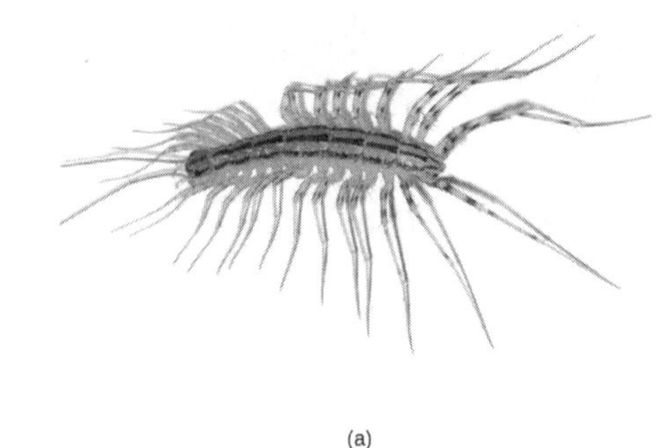

(a) (b)

Figure 28.40 Myriapods. The centipede *Scutigera coleoptrata* (a) has up to 15 pairs of legs. The North American millipede *Narceus americanus* (b) bears many legs, although not a thousand, as its name might suggest. (credit a: modification of work by Bruce Marlin; credit b: modification of work by Cory Zanker)

Subphylum Crustacea

Crustaceans are the most dominant aquatic (both freshwater and marine) arthropods, with the total number of marine crustaceans standing at about 70,000 species. Krill, shrimp, lobsters, crabs, and crayfish are examples of crustaceans (Figure 28.41). However, there are also a number of terrestrial crustacean species as well: Terrestrial species like the wood lice (*Armadillidium* spp), also called pill bugs, roly-polies, potato bugs, or isopods, are also crustaceans. Nonetheless, the number of terrestrial species in this subphylum is relatively low.

(a) (b) (c)

Figure 28.41 Crustaceans. The (a) crab and (b) shrimp krill are both aquatic crustaceans. The pill bug *Armadillidium* is a terrestrial crustacean. (credit a: modification of work by William Warby; credit b: modification of work by Jon Sullivan credit c: modification of work by Franco Folini. https://commons.wikimedia.org/w/index.php?curid=789616 (http://openstax.org/l/pillbug))

Crustaceans typically possess two pairs of antennae, mandibles as mouthparts, and *biramous* ("two branched") appendages, which means that their legs are formed in two parts called endopods and exopods, which appear superficially distinct from the *uniramous* ("one branched") legs of myriapods and hexapods (Figure 28.42). Since biramous appendages are also seen in the trilobites, biramous appendages represent the ancestral condition in the arthropods. Currently, we describe various arthropods as having uniramous or biramous appendages, but these are descriptive only, and do not necessarily reflect evolutionary relationships other than that all jointed legs of arthropods share common ancestry.

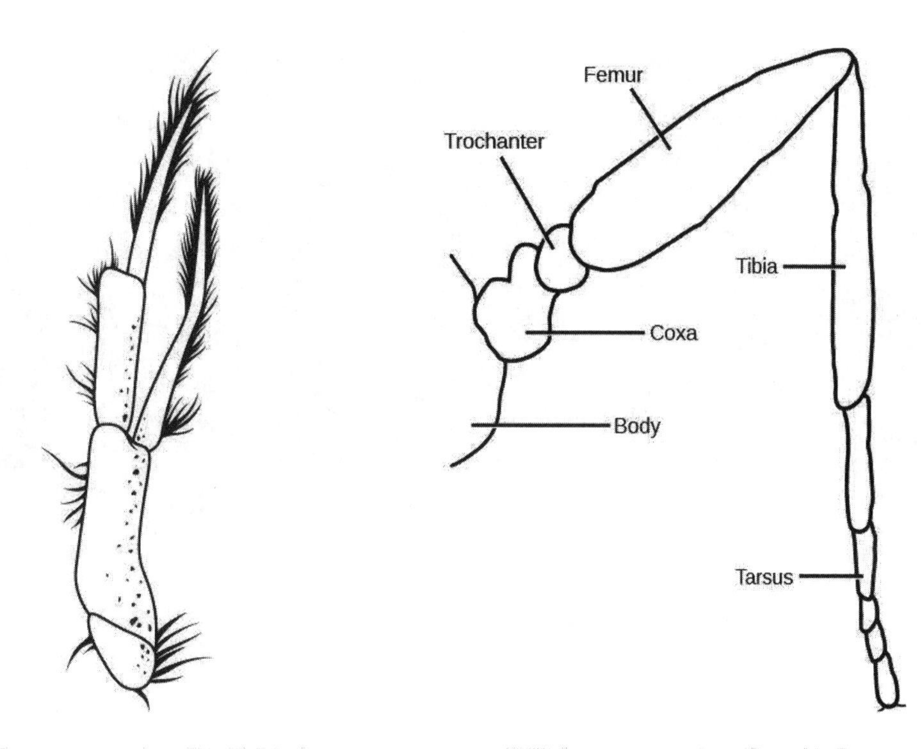

Figure 28.42 Arthropod appendages. Arthropods may have (a) biramous (two-branched) appendages or (b) uniramous (one-branched) appendages. (credit b: modification of work by Nicholas W. Beeson)

In most crustaceans, the head and thorax is fused to form a **cephalothorax** (Figure 28.43), which is covered by a plate called the **carapace**, thus producing a body plan comprising two tagmata: **cephalophorax** and **abdomen**. Crustaceans have a chitinous exoskeleton that is shed by molting and ecdysis whenever the animal requires an increase in size or the next stage of development. The exoskeletons of many aquatic species are also infused with calcium carbonate, which makes them even stronger than those of other arthropods. Crustaceans have an open circulatory system where blood is pumped into the hemocoel by the dorsally located heart. Hemocyanin is the major respiratory pigment present in crustaceans, but hemoglobin is found in a few species and both are dissolved in the hemolymph rather than carried in cells.

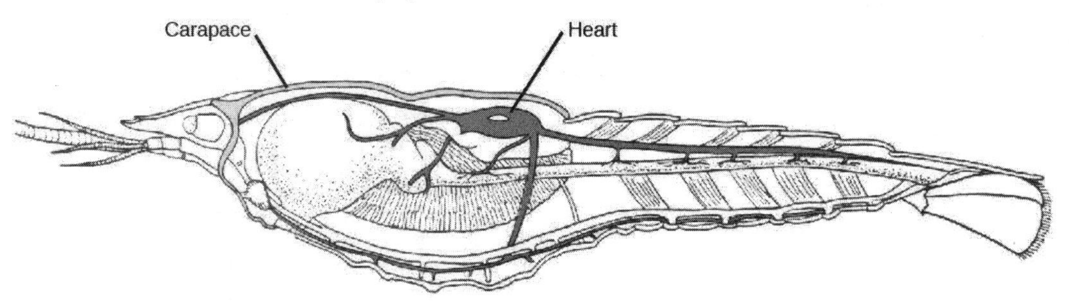

Figure 28.43 Crustacean anatomy. The crayfish is an example of a crustacean. It has a carapace around the cephalothorax and the heart in the dorsal thorax area. (credit: Jane Whitney)

As in the chelicerates, most crustaceans are dioecious. However, some species like barnacles may be hermaphrodites. **Serial hermaphroditism**, where the gonad can switch from producing sperm to ova, is also exhibited in some species. Fertilized eggs may be held within the female of the species or may be released in the water. Terrestrial crustaceans seek out damp spaces in their habitats to lay eggs.

Larval stages—nauplius or zoea—are seen in the early development of aquatic crustaceans. A cypris larva is also seen in the early development of barnacles (Figure 28.44).

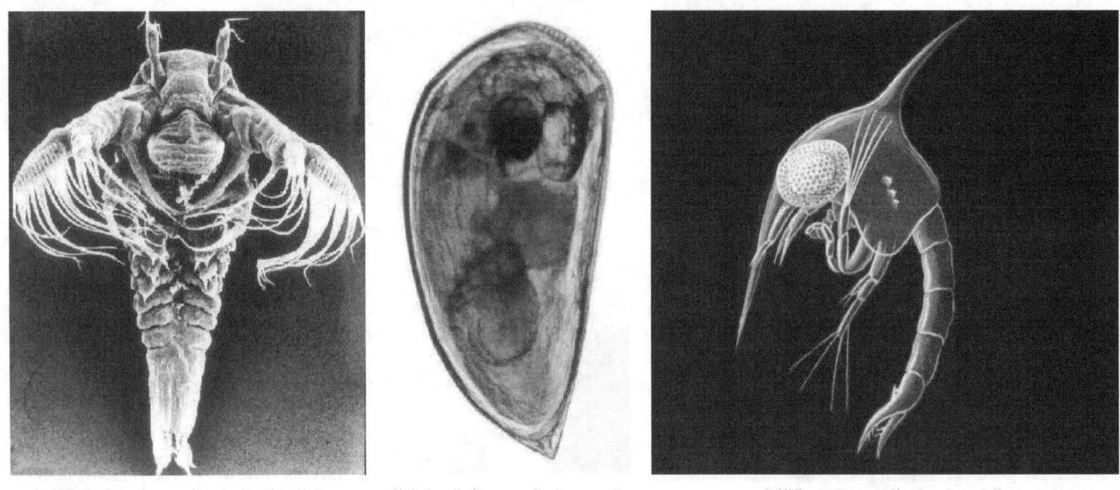

Figure 28.44 Crustacean larvae. All crustaceans go through different larval stages. Shown are (a) the nauplius larval stage of a tadpole shrimp, (b) the cypris larval stage of a barnacle, and (c) the zoea larval stage of a green crab. (credit a: modification of work by USGS; credit b: modification of work by Mª. C. Mingorance Rodríguez; credit c: modification of work by B. Kimmel based on original work by Ernst Haeckel)

Crustaceans possess a brain formed by the fusion of the first three segmental ganglia, as well as two compound eyes. A ventral nerve cord connects additional segmental ganglia. Most crustaceans are carnivorous, but herbivorous and detritivorous species, and even endoparasitic species are known. A highly evolved endoparasitic species, such as *Sacculina* spp, parasitizes its crab host and ultimately destroys it after it forces the host to incubate the parasite's eggs! Crustaceans may also be cannibalistic when extremely high populations of these organisms are present.

Subphylum Hexapoda

The insects comprise the largest class of arthropods in terms of species diversity as well as in terms of biomass—at least in terrestrial habitats.

The name Hexapoda describes the presence of six legs (three pairs) in these animals, which differentiates them from other groups of arthropods that have different numbers of legs. In some cases, however, the number of legs has been evolutionarily reduced, or the legs have been highly modified to accommodate specific conditions, such as endoparasitism. Hexapod bodies are organized into three tagmata: head, thorax, and abdomen. Individual segments of the head have mouthparts derived from jointed legs, and the thorax has three pairs of jointed appendages, and also wings, in most derived groups. For example, in the **pterygotes** (winged insects), in addition to a pair of jointed legs on all three segments comprising the thorax: prothorax, mesothorax, and metathorax.

Appendages found on other body segments are also evolutionarily derived from modified legs. Typically, the head bears an upper "lip" or labrum and mandibles (or derivation of mandibles) that serve as mouthparts; maxillae, and a lower "lip" called a labium: both of which manipulate food. The head also has one pair of sensory antennae, as well as sensory organs such as a pair of compound eyes, ocelli (simple eyes), and numerous sensory hairs. The abdomen usually has 11 segments and bears external reproductive apertures. The subphylum Hexapoda includes some insects that are winged (such as fruit flies) and others that are secondarily wingless (such as fleas). The only order of "primitively wingless" insects is the Thysanura, the bristletails. All other orders are winged or are descendants of formally winged insects.

The evolution of wings is a major, unsolved mystery. Unlike vertebrates, whose "wings" are simply preadaptations of "arms" that served as the structural foundations for the evolution of functional wings (this has occurred independently in pterosaurs, dinosaurs [birds], and bats), the evolution of wings in insects is a what we call a *de novo* (new) development that has given the pteryogotes domination over the Earth. Winged insects existed over 425 million years ago, and by the Carboniferous, several orders of winged insects (**Paleoptera**), most of which are now extinct, had evolved. There is good physical evidence that Paleozoic nymphs with thoracic winglets (perhaps hinged, former gill covers of semi-aquatic species) used these devices on land to elevate the thoracic temperature (the thorax is where the legs are located) to levels that would enable them to escape predators faster, find more food resources and mates, and disperse more easily. The thoracic winglets (which can be found on fossilized insects preceding the advent of truly winged insects) could have easily been selected for thermoregulatory purposes prior to reaching a size that would have allowed them the capacity for gliding or actual flapping flight. Even modern insects with broadly attached wings, such as butterflies, use the basal one-third of their wings (the area next to the thorax) for thermoregulation, and the outer two-thirds for flight, camouflage, and mate selection.

Many of the common insects we encounter on a daily basis—including ants, beetles, cockroaches, butterflies, crickets and flies—are examples of Hexapoda. Among these, adult ants, beetles, flies, and butterflies develop by complete metamorphosis from grub-like or caterpillar-like larvae, whereas adult cockroaches and crickets develop through a gradual or incomplete metamorphosis from wingless immatures. All growth occurs during the juvenile stages. Adults do not grow further (but may become larger) after their final molt. Variations in wing, leg, and mouthpart morphology all contribute to the enormous variety seen in the insects. Insect variability was also encouraged by their activity as pollinators and their coevolution with flowering plants. Some insects, especially termites, ants, bees, and wasps, are eusocial, meaning that they live in large groups with individuals assigned to specific roles or castes, like queen, drone, and worker. Social insects use **pheromones**—external chemical signals—to communicate and maintain group structure as well as a cohesive colony.

VISUAL CONNECTION

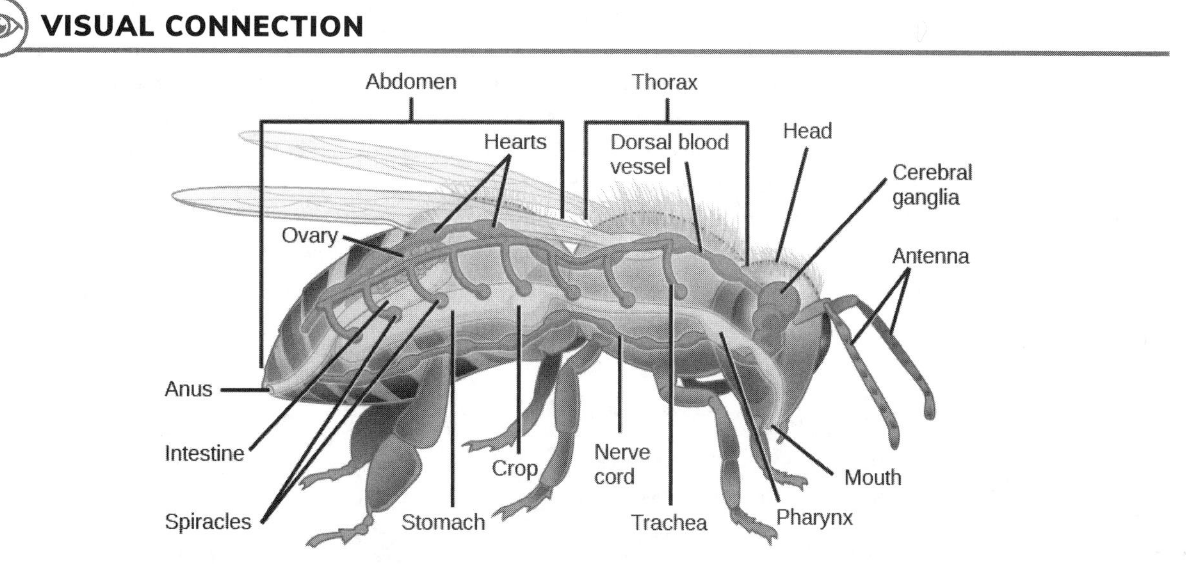

Figure 28.45 Insect anatomy. In this basic anatomy of a hexapod insect, note that insects have a well-developed digestive system (yellow), a respiratory system (blue), a circulatory system (red), and a nervous system (purple). Note the multiple "hearts" and the segmental ganglia.

Which of the following statements about insects is false?

a. Insects have both dorsal and ventral blood vessels.
b. Insects have spiracles, openings that allow air to enter into the tracheal system.
c. The trachea is part of the digestive system.
d. Most insects have a well-developed digestive system with a mouth, crop, and intestine.

28.7 Superphylum Deuterostomia

By the end of this section, you will be able to do the following:

- Describe the distinguishing characteristics of echinoderms
- Describe the distinguishing characteristics of chordates

The phyla Echinodermata and Chordata (the phylum that includes humans) both belong to the superphylum Deuterostomia. Recall that protostomes and deuterostomes differ in certain aspects of their embryonic development, and they are named based on which opening of the **archenteron** (primitive gut tube) develops first. The word deuterostome comes from the Greek word meaning "mouth second," indicating that the mouth develops as a secondary structure opposite the location of the blastopore, which becomes the anus. In protostomes ("mouth first"), the first embryonic opening becomes the mouth, and the second opening becomes the anus.

There are a series of other developmental characteristics that differ between protostomes and deuterostomes, including the type of early cleavage (embryonic cell division) and the mode of formation of the coelom of the embryo: Protosomes typically exhibit spiral mosaic cleavage whereas deuterostomes exhibit radial regulative cleavage. In deuterostomes, the endodermal lining of the archenteron usually forms buds called **coelomic pouches** that expand and ultimately obliterate the embryonic blastocoel (the cavity within the blastula and early gastrula) to become the **embryonic mesoderm**, the third germ layer. This happens when the mesodermal pouches become separated from the invaginating endodermal layer forming the archenteron, then expand and fuse to form the coelomic cavity. The resulting coelom is termed an **enterocoelom**. The archenteron develops into the alimentary canal, and a mouth opening is formed by invagination of ectoderm at the pole opposite the blastopore of the gastrula. The blastopore forms the anus of the alimentary system in the juvenile and adult forms. Cleavage in most deuterostomes is also *indeterminant*, meaning that the developmental fates of early embryonic cells are not decided at that point of embryonic development (this is why we could potentially clone most deuterostomes, including ourselves).

The deuterostomes consist of two major clades—the Chordata and the Ambulacraria. The Chordata include the vertebrates and two invertebrate subphyla, the urochordates and the cephalochordates. The Ambulacraria include the echinoderms and the hemichordates, which were once considered to be a chordate subphylum (Figure 28.46). The two clades, in addition to being deuterostomes, have some other interesting features in common. As we have seen, the vast majority of invertebrate animals do *not* possess a defined bony vertebral endoskeleton, or a bony cranium. However, one of the most ancestral groups of deuterostome invertebrates, the Echinodermata, do produce tiny skeletal "bones" called *ossicles* that make up a true **endoskeleton**, or internal skeleton, covered by an epidermis. The Hemichordata (acorn worms and pterobranchs) will not be covered here, but share with the echinoderms a three-part (tripartite) coelom, similar larval forms, and a derived metanephridium that rids the animals of nitrogenous wastes. They also share pharyngeal slits with the chordates (Figure 28.46). In addition, hemichordates have a dorsal nerve cord in the midline of the epidermis, but lack a neural tube, a true notochord and the endostyle and post-anal tail characteristic of chordates.

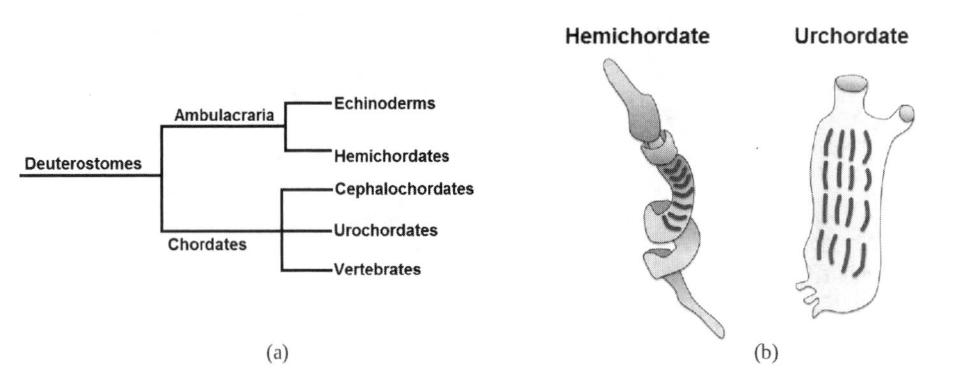

Figure 28.46 Ambulacraria and Chordata. (a) The major deuterostome taxa. (b) pharyngeal slits in hemichordates and urochordates. (credit a MAC; credit b modification of Gill Slits By Own work by Zebra.element [Public domain], via Wikimedia Commons)

Phylum Echinodermata

Echinodermata are named after their "prickly skin" (from the Greek "echinos" meaning "prickly" and "dermos" meaning "skin"). This phylum is a collection of about 7,000 described living species of exclusively marine, bottom-dwelling organisms. Sea stars (Figure 28.47), sea cucumbers, sea urchins, sand dollars, and brittle stars are all examples of echinoderms.

Morphology and Anatomy

Despite the adaptive value of bilaterality for most free-living cephalized animals, adult echinoderms exhibit pentaradial symmetry (with "arms" typically arrayed in multiples of five around a central axis). Echinoderms have an endoskeleton made of calcareous ossicles (small bony plates), covered by the epidermis. For this reason, it is an endoskeleton like our own, not an exoskeleton like that of arthropods. The ossicles may be fused together, embedded separately in the connective tissue of the dermis, or be reduced to minute spicules of bone as in sea cucumbers. The spines for which the echinoderms are named are connected to some of the plates. The spines may be moved by small muscles, but they can also be locked into place for defense. In some species, the spines are surrounded by tiny stalked claws called **pedicellaria**, which help keep the animal's surface clean of debris, protect *papulae* used in respiration, and sometimes aid in food capture.

The endoskeleton is produced by dermal cells, which also produce several kinds of pigments, imparting vivid colors to these animals. In sea stars, fingerlike projections (papillae) of dermal tissue extend through the endoskeleton and function as gills.

Some cells are glandular, and may produce toxins. Each arm or section of the animal contains several different structures: for example, digestive glands, gonads, and the tube feet that are unique to the echinoderms. In echinoderms like sea stars, every arm bears two rows of *tube feet* on the oral side, running along an external ambulacral groove. These tube feet assist in locomotion, feeding, and chemical sensations, as well as serve to attach some species to the substratum.

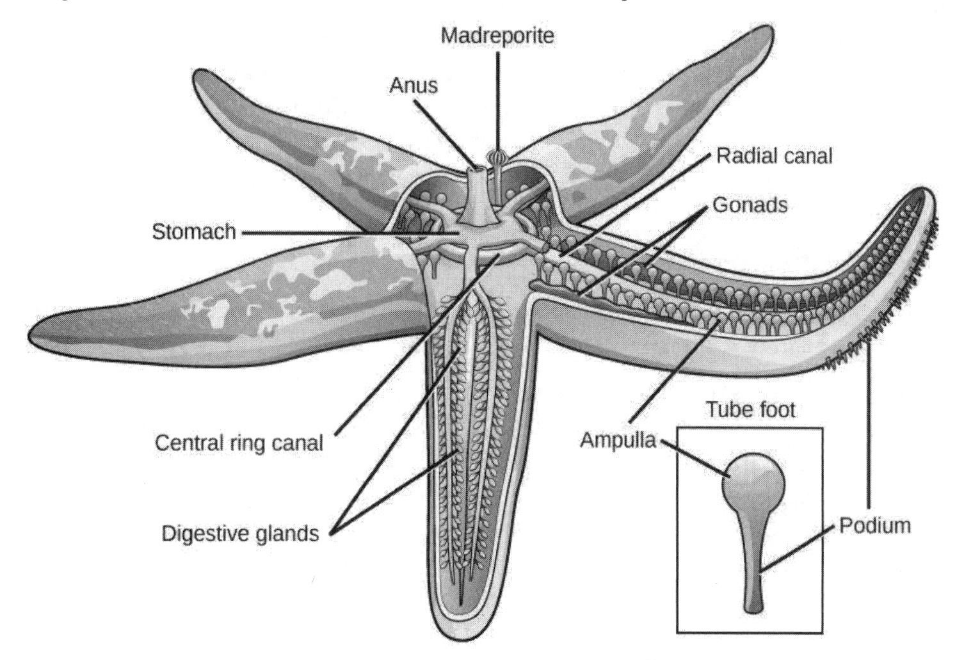

Figure 28.47 Anatomy of a sea star. This diagram of a sea star shows the pentaradial pattern typical of adult echinoderms, and the water vascular system that is their defining characteristic.

Water Vascular and Hemal Systems

Echinoderms have a unique **ambulacral (water vascular) system**, derived from part of the **coelom**, or "body cavity." The water vascular system consists of a central ring canal and radial canals that extend along each arm. Each radial canal is connected to a double row of **tube feet**, which project through holes in the endoskeleton, and function as tactile and ambulatory structures. These tube feet can extend or retract based on the volume of water present in the system of that arm, allowing the animal to move and also allowing it to capture prey with their suckerlike action. Individual tube feet are controlled by bulblike **ampullae**. Seawater enters the system through an **aboral madreporite** (opposite the oral area where the mouth is located) and passes to the ring canal through a short **stone canal**. Water circulating through these structures facilitates gaseous exchange and provides a hydrostatic source for locomotion and prey manipulation. A **hemal system**, consisting of oral, gastric, and aboral rings, as well as other vessels running roughly parallel to the water vascular system, circulates nutrients. Transport of nutrients and gases is shared by the water vascular and hemal systems in addition to the visceral body cavity that surrounds the major organs.

Nervous System

The nervous system in these animals is a relatively simple, comprising a circumoral nerve ring at the center and five radial nerves extending outward along the arms. In addition, several networks of nerves are located in different parts of the body. However, structures analogous to a brain or large ganglia are not present in these animals. Depending on the group, echinoderms may have well-developed sensory organs for touch and chemoreception (e.g., within the tube feet and on tentacles at the tips of the arms), as well as photoreceptors and statocysts.

Digestive and Excretory Systems

A mouth, located on the oral (ventral) side, opens through a short esophagus to a large, baglike stomach. The so-called "cardiac" stomach can be everted through the mouth during feeding (for example, when a starfish everts its stomach into a bivalve prey item to digest the animal—*alive*—within its own shell!) There are masses of digestive glands (**pyloric caeca**) in each arm, running dorsally along the arms and overlying the reproductive glands below them. After passing through the pyloric caeca in each arm, the digested food is channeled to a small anus, if one exists.

Podocytes—cells specialized for ultrafiltration of bodily fluids—are present near the center of the echinoderm disc, at the junction of the water vascular and hemal systems. These podocytes are connected by an internal system of canals to the

madreporite, where water enters the stone canal. The adult echinoderm typically has a spacious and fluid-filled coelom. Cilia aid in circulating the fluid within the body cavity, and lead to the fluid-filled **papulae**, where the exchange of oxygen and carbon dioxide takes place, as well as the secretion of nitrogenous waste such as ammonia, by diffusion.

Reproduction

Echinoderms are dioecious, but males and females are indistinguishable apart from their gametes. Males and females release their gametes into water at the same time and fertilization is external. The early larval stages of all echinoderms (e.g., the *bipinnaria* of asteroid echinoderms such as sea stars) have bilateral symmetry, although each class of echinoderms has its own larval form. The radially symmetrical adult forms from a cluster of cells in the larva. Sea stars, brittle stars, and sea cucumbers may also reproduce asexually by **fragmentation**, as well as regenerate body parts lost in trauma, even when over 75 percent of their body mass is lost!

Classes of Echinoderms

This phylum is divided into five extant classes: Asteroidea (sea stars), Ophiuroidea (brittle stars), Echinoidea (sea urchins and sand dollars), Crinoidea (sea lilies or feather stars), and Holothuroidea (sea cucumbers) (Figure 28.48).

The most well-known echinoderms are members of class Asteroidea, or sea stars. They come in a large variety of shapes, colors, and sizes, with more than 1,800 species known so far. The key characteristic of sea stars that distinguishes them from other echinoderm classes includes thick arms that extend from a central disk from which various body organs branch into the arms. At the end of each arm are simple eye spots and tentacles that serve as touch receptors. Sea stars use their rows of tube feet not only for gripping surfaces but also for grasping prey. Most sea stars are carnivores and their major prey are in the phylum Mollusca. By manipulating its tube feet, a sea star can open molluscan shells. Sea stars have two stomachs, one of which can protrude through their mouths and secrete digestive juices into or onto prey, even before ingestion. A sea star eating a clam can partially open the shell, and then evert its stomach into the shell, introducing digestive enzymes into the interior of the mollusk. This process can both weaken the strong adductor (closing) muscles of a bivalve and begin the process of digestion.

LINK TO LEARNING

Explore the sea star's body plan (http://openstax.org/l/sea_star) up close, watch one move across the sea floor, and see it devour a mussel.

Brittle stars belong to the class Ophiuroidea ("snake-tails"). Unlike sea stars, which have plump arms, brittle stars have long, thin, flexible arms that are sharply demarcated from the central disk. Brittle stars move by lashing out their arms or wrapping them around objects and pulling themselves forward. Their arms are also used for grasping prey. The water vascular system in ophiuroids is not used for locomotion.

Sea urchins and sand dollars are examples of Echinoidea ("prickly"). These echinoderms do not have arms, but are hemispherical or flattened with five rows of tube feet that extend through five rows of pores in a continuous internal shell called a *test.* Their tube feet are used to keep the body surface clean. Skeletal plates around the mouth are organized into a complex multipart feeding structure called *"Aristotle's lantern."* Most echinoids graze on algae, but some are suspension feeders, and others may feed on small animals or organic *detritus*—the fragmentary remains of plants or animals.

Sea lilies and feather stars are examples of Crinoidea. Sea lilies are *sessile*, with the body attached to a stalk, but the feather stars can actively move about using leglike *cirri* that emerge from the aboral surface. Both types of crinoid are suspension feeders, collecting small food organisms along the ambulacral grooves of their feather-like arms. The "feathers" consisted of branched arms lined with tube feet. The tube feet are used to move captured food toward the mouth. There are only about 600 extant species of crinoids, but they were far more numerous and abundant in ancient oceans. Many crinoids are deep-water species, but feather stars typically inhabit shallow areas, especially in substropical and tropical waters.

Sea cucumbers of class Holothuroidea exhibit an extended oral-aboral axis. These are the only echinoderms that demonstrate "functional" bilateral symmetry as adults, because the extended oral-aboral axis compels the animal to lie horizontally rather than stand vertically. The tube feet are reduced or absent, except on the side on which the animal lies. They have a single gonad and the digestive tract is more typical of a bilaterally symmetrical animal. A pair of gill-like structures called **respiratory trees** branch from the posterior gut; muscles around the cloaca pump water in and out of these trees. There are clusters of tentacles around the mouth. Some sea cucumbers feed on detritus, while others are suspension feeders, sifting out small organisms with their oral tentacles. Some species of sea cucumbers are unique among the echinoderms in that cells containing *hemoglobin* circulate in the coelomic fluid, the water vascular system and/or the hemal system.

Figure 28.48 Classes of echinoderms. Different members of Echinodermata include the (a) sea star of class Asteroidea, (b) the brittle star of class Ophiuroidea, (c) the sea urchins of class Echinoidea, (d) the sea lilies belonging to class Crinoidea, and (e) sea cucumbers, representing class Holothuroidea. (credit a: modification of work by Adrian Pingstone; credit b: modification of work by Joshua Ganderson; credit c: modification of work by Samuel Chow; credit d: modification of work by Sarah Depper; credit e: modification of work by Ed Bierman)

Phylum Chordata

Animals in the phylum Chordata share five key features that appear at some stage of their development: a notochord, a dorsal hollow nerve cord, pharyngeal slits, a post-anal tail, and an endostyle/thyroid gland that secretes iodinated hormones. In some groups, some of these traits are present only during embryonic development. In addition to containing vertebrate classes, the phylum Chordata contains two clades of "invertebrates": Urochordata (tunicates, salps, and larvaceans) and Cephalochordata (lancelets). Most tunicates live on the ocean floor and are suspension feeders. Lancelets are suspension feeders that feed on phytoplankton and other microorganisms. The invertebrate chordates will be discussed more extensively in the following chapter.

KEY TERMS

amoebocyte sponge cell with multiple functions, including nutrient delivery, egg formation, sperm delivery, and cell differentiation

Annelida phylum of vermiform animals with metamerism

archenteron primitive gut cavity within the gastrula that opens outward via the blastopore

Arthropoda phylum of animals with jointed appendages

biramous referring to two branches per appendage

captacula tentacle-like projection that is present in tusks shells to catch prey

cephalothorax fused head and thorax in some species

chelicera modified first pair of appendages in subphylum Chelicerata

choanocyte (also, collar cell) sponge cell that functions to generate a water current and to trap and ingest food particles via phagocytosis

Chordata phylum of animals distinguished by their possession of a notochord, a dorsal, hollow nerve cord, an endostyle, pharyngeal slits, and a post-anal tail at some point in their development

clitellum specialized band of fused segments, which aids in reproduction

Cnidaria phylum of animals that are diploblastic and have radial symmetry

cnidocyte specialized stinging cell found in Cnidaria

conispiral shell shape coiled around a horizontal axis

corona wheel-like structure on the anterior portion of the rotifer that contains cilia and moves food and water toward the mouth

ctenidium specialized gill structure in mollusks

cuticle (animal) the tough, external layer possessed by members of the invertebrate class Ecdysozoa that is periodically molted and replaced

cypris larval stage in the early development of crustaceans

Echinodermata phylum of deuterostomes with spiny skin; exclusively marine organisms

enterocoelom coelom formed by fusion of coelomic pouches budded from the endodermal lining of the archenteron

epidermis outer layer (from ectoderm) that lines the outside of the animal

extracellular digestion food is taken into the gastrovascular cavity, enzymes are secreted into the cavity, and the cells lining the cavity absorb nutrients

gastrodermis inner layer (from endoderm) that lines the digestive cavity

gastrovascular cavity opening that serves as both a mouth and an anus, which is termed an incomplete digestive system

gemmule structure produced by asexual reproduction in freshwater sponges where the morphology is inverted

hemocoel internal body cavity seen in arthropods

hermaphrodite referring to an animal where both male and female gonads are present in the same individual

invertebrata (also, invertebrates) category of animals that do not possess a cranium or vertebral column

madreporite pore for regulating entry and exit of water into the water vascular system

mantle (also, pallium) specialized epidermis that encloses all visceral organs and secretes shells

mastax jawed pharynx unique to the rotifers

medusa free-floating cnidarian body plan with mouth on underside and tentacles hanging down from a bell

mesoglea non-living, gel-like matrix present between ectoderm and endoderm in cnidarians

mesohyl collagen-like gel containing suspended cells that perform various functions in the sponge

metamerism series of body structures that are similar internally and externally, such as segments

Mollusca phylum of protostomes with soft bodies and no segmentation

nacre calcareous secretion produced by bivalves to line the inner side of shells as well as to coat intruding particulate matter

nauplius larval stage in the early development of crustaceans

nematocyst harpoon-like organelle within cnidocyte with pointed projectile and poison to stun and entangle prey

Nematoda phylum of worm-like animals that are triploblastic, pseudocoelomates that can be free-living or parasitic

Nemertea phylum of dorsoventrally flattened protostomes known as ribbon worms

osculum large opening in the sponge's body through which water leaves

ostium pore present on the sponge's body through which water enters

oviger additional pair of appendages present on some arthropods between the chelicerae and pedipalps

parapodium fleshy, flat, appendage that protrudes in pairs from each segment of polychaetes

pedipalp second pair of appendages in Chelicerata

pilidium larval form found in some nemertine species

pinacocyte epithelial-like cell that forms the outermost layer of sponges and encloses a jelly-like substance called mesohyl

planospiral shell shape coiled around a vertical axis

planuliform larval form found in phylum Nemertea

polymorphic possessing multiple body plans within the lifecycle of a group of organisms

polyp stalk-like sessile life form of a cnidarians with mouth and tentacles facing upward, usually sessile but may be able to glide along surface

Porifera phylum of animals with no true tissues, but a

porous body with rudimentary endoskeleton
radula tongue-like organ with chitinous ornamentation
rhynchocoel cavity present above the mouth that houses the proboscis
schizocoelom coelom formed by groups of cells that split from the endodermal layer
sclerocyte cell that secretes silica spicules into the mesohyl
seta/chaeta chitinous projection from the cuticle
siphon tubular structure that serves as an inlet for water into the mantle cavity
spicule structure made of silica or calcium carbonate that provides structural support for sponges
spongocoel central cavity within the body of some sponges
trochophore first of the two larval stages in mollusks
uniramous referring to one branch per appendage
veliger second of the two larval stages in mollusks
water vascular system system in echinoderms where water is the circulatory fluid
zoea larval stage in the early development of crustaceans

CHAPTER SUMMARY

28.1 Phylum Porifera

Animals included in phylum Porifera are parazoans because they do not show the formation of true embryonically derived tissues, although they have a number of specific cell types and "functional" tissues such as pinacoderm. These organisms show very simple organization, with a rudimentary endoskeleton of spicules and spongin fibers. Glass sponge cells are connected together in a multinucleated syncytium. Although sponges are very simple in organization, they perform most of the physiological functions typical of more complex animals.

28.2 Phylum Cnidaria

Cnidarians represent a more complex level of organization than Porifera. They possess outer and inner tissue layers that sandwich a noncellular mesoglea between them. Cnidarians possess a well-formed digestive system and carry out extracellular digestion in a digestive cavity that extends through much of the animal. The mouth is surrounded by tentacles that contain large numbers of cnidocytes—specialized cells bearing nematocysts used for stinging and capturing prey as well as discouraging predators. Cnidarians have separate sexes and many have a lifecycle that involves two distinct morphological forms—medusoid and polypoid—at various stages in their life cycles. In species with both forms, the medusa is the sexual, gamete-producing stage and the polyp is the asexual stage. Cnidarian species include individual or colonial polypoid forms, floating colonies, or large individual medusa forms (sea jellies).

28.3 Superphylum Lophotrochozoa: Flatworms, Rotifers, and Nemerteans

This section describes three phyla of relatively simple invertebrates: one acoelomate, one pseudocoelomate, and one eucoelomate. Flatworms are acoelomate, triploblastic animals. They lack circulatory and respiratory systems, and have a rudimentary excretory system. This digestive system is incomplete in most species, and absent in tapeworms. There are four traditional groups of flatworms, the largely free-living turbellarians, which include polycladid marine worms and tricladid freshwater species, the ectoparasitic monogeneans, and the endoparasitic trematodes and cestodes. Trematodes have complex life cycles involving a molluscan secondary host and a primary host in which sexual reproduction takes place. Cestodes, or tapeworms, infect the digestive systems of their primary vertebrate hosts.

Rotifers are microscopic, multicellular, mostly aquatic organisms that are currently under taxonomic revision. The group is characterized by the ciliated, wheel-like corona, located on their head. Food collected by the corona is passed to another structure unique to this group of organisms—the mastax or jawed pharynx.

The nemerteans are probably simple eucoelomates. These ribbon-shaped animals also bear a specialized proboscis enclosed within a rhynchocoel. The development of a closed circulatory system derived from the coelom is a significant difference seen in this species compared to other phyla described here. Alimentary, nervous, and excretory systems are more developed in the nemerteans than in the flatworms or rotifers. Embryonic development of nemertean worms proceeds via a planuliform or trochophore-like larval stage.

28.4 Superphylum Lophotrochozoa: Molluscs and Annelids

Phylum Mollusca is a large, group of protostome schizocoelous invertebrates that occupy marine, freshwater, and terrestrial habitats. Mollusks can be divided into seven classes, each of which exhibits variations on the basic molluscan body plan. Two defining features are the mantle, which secretes a protective calcareous shell in many species, and the radula, a rasping feeding organ found in most classes. Some mollusks have evolved a reduced shell, and others have no radula. The mantle also covers the body and forms a mantle cavity, which is quite distinct from the coelomic cavity—typically reduced to the area surrounding the heart, kidneys, and intestine. In aquatic mollusks, respiration is facilitated by gills (ctenidia) in the mantle cavity. In terrestrial mollusks, the mantle cavity itself serves as an organ of gas exchange. Mollusks also have a muscular foot, which is modified in various ways for locomotion or food capture. Most mollusks have separate sexes. Early development in aquatic species occurs via one or more larval stages, including a trochophore larva, that precedes a veliger larva in some groups.

Phylum Annelida includes vermiform, segmented animals. Segmentation is metameric (i.e., each segment is partitioned internally as well as externally, with various structures repeated in each segment). These animals have well-developed neuronal, circulatory, and digestive systems. The two major groups of annelids are the polychaetes, which have parapodia with multiple bristles, and oligochaetes, which have no parapodia and fewer bristles or no bristles. Oligochaetes, which include earthworms and leeches, have a specialized band of segments known as a clitellum, which secretes a cocoon and protects gametes during reproduction. The leeches do not have full internal segmentation. Reproductive strategies include separate sexes, hermaphroditism, and serial hermaphroditism. Polychaetes typically have trochophore larvae, while the oligochaetes develop more directly.

28.5 Superphylum Ecdysozoa: Nematodes and Tardigrades

The defining feature of the Ecdysozoa is a collagenous/chitinous cuticle that covers the body, and the necessity to molt the cuticle periodically during growth. Nematodes are roundworms, with a pseudocoel body cavity. They have a complete digestive system, a differentiated nervous system, and a rudimentory excretory system. The phylum includes free-living species like *Caenorhabditis elegans* as well as many species of endoparasitic organisms such as *Ascaris* spp. They include dioeceous as well as hermaphroditic species. Embryonic development proceeds via several larval stages, and most adults have a fixed number of cells.

The tardigrades, sometimes called "water bears," are a widespread group of tiny animals with a segmented cuticle covering the epidermis and four pairs of clawed legs. Like the nematodes, they are pseudocoelomates and have a fixed number of cells as adults. Specialized proteins enable them to enter cryptobiosis, a kind of suspended animation during which they can resist a number of adverse environmental conditions.

28.6 Superphylum Ecdysozoa: Arthropods

Arthropods represent the most successful animal phylum on Earth, both in terms of the number of species and the number of individuals. As members of the Ecdysozoa, all arthropods have a protective chitinous cuticle that must be periodically molted and shed during development or growth. Arthropods are characterized by a segmented body as well as the presence of jointed appendages. In the basic body plan, a pair of appendages is present per body segment. Within the phylum, traditional classification is based on mouthparts, body subdivisions, number of appendages, and modifications of appendages present. In aquatic arthropods, the chitinous exoskeleton may be calcified. Gills, tracheae, and book lungs facilitate respiration. Unique larval stages are commonly seen in both aquatic and terrestrial groups of arthropods.

28.7 Superphylum Deuterostomia

Echinoderms are deuterostome marine organisms, whose adults show five-fold symmetry. This phylum of animals has a calcareous endoskeleton composed of ossicles, or body plates. Epidermal spines are attached to some ossicles and serve in a protective capacity. Echinoderms possess a water-vascular system that serves both for respiration and for locomotion, although other respiratory structures such as papulae and respiratory trees are found in some species. A large aboral madreporite is the point of entry and exit for sea water pumped into the water vascular system. Echinoderms have a variety of feeding techniques ranging from predation to suspension feeding. Osmoregulation is carried out by specialized cells known as podocytes associated with the hemal system.

The characteristic features of the Chordata are a notochord, a dorsal hollow nerve cord, pharyngeal slits, a post-anal tail, and an endostyle/thyroid that secretes iodinated hormones. The phylum Chordata contains two clades of invertebrates: Urochordata (tunicates, salps, and larvaceans) and Cephalochordata (lancelets), together with the vertebrates in the Vertebrata. Most tunicates live on the ocean floor and are suspension feeders. Lancelets are suspension feeders that feed on phytoplankton and other microorganisms. The sister taxon of the Chordates is the Ambulacraria, which includes both the Echinoderms and the hemichordates, which share pharyngeal slits with the chordates.

VISUAL CONNECTION QUESTIONS

1. Figure 28.3 Which of the following statements is false?
    a. Choanocytes have flagella that propel water through the body.
    b. Pinacocytes can transform into any cell type.
    c. Lophocytes secrete collagen.
    d. Porocytes control the flow of water through pores in the sponge body.

2. Figure 28.21 Which of the following statements about the anatomy of a mollusk is false?
    a. Mollusks have a radula for grinding food.
    b. A digestive gland is connected to the stomach.
    c. The tissue beneath the shell is called the mantle.
    d. The digestive system includes a gizzard, a stomach, a digestive gland, and the intestine.

3. Figure 28.45 Which of the following statements about insects is false?
    a. Insects have both dorsal and ventral blood vessels.
    b. Insects have spiracles, openings that allow air to enter into the tracheal system.
    c. The trachea is part of the digestive system.
    d. Most insects have a well-developed digestive system with a mouth, crop, and intestine.

REVIEW QUESTIONS

4. Mesohyl contains:
    a. a polysaccharide gel and dead cells.
    b. a collagen-like gel and suspended cells for various functions.
    c. spicules composed of silica or calcium carbonate.
    d. multiple pores.

5. The large central opening in the parazoan body is called the:
    a. gemmule.
    b. spicule.
    c. ostia.
    d. osculum.

6. Most sponge body plans are slight variations on a simple tube-within-a-tube design. Which of the following is a key limitation of sponge body plans?
    a. Sponges lack the specialized cell types needed to produce more complex body plans.
    b. The reliance on osmosis/diffusion requires a design that maximizes the surface area to volume ratio of the sponge.
    c. Choanocytes must be protected from the hostile exterior environment.
    d. Spongin cannot support heavy bodies.

7. Cnidocytes are found in _____.
    a. phylum Porifera
    b. phylum Nemertea
    c. phylum Nematoda
    d. phylum Cnidaria

8. Cubozoans are ________.
    a. polyps
    b. medusoids
    c. polymorphs
    d. sponges

9. While collecting specimens, a marine biologist finds a sessile Cnidarian. The medusas that bud from it swim by contracting a ring of muscle in their bells. To which class does this specimen belong?
    a. Class Hydrozoa
    b. Class Cubozoa
    c. Class Scyphozoa
    d. Class Anthozoa

10. Which group of flatworms are primarily ectoparasites of fish?
    a. monogeneans
    b. trematodes
    c. cestodes
    d. turbellarians

11. The rhynchocoel is a ________.
    a. circulatory system
    b. fluid-filled cavity
    c. primitive excretory system
    d. proboscis

12. Annelids have (a):
    a. pseudocoelom.
    b. true coelom.
    c. no coelom.
    d. none of the above

13. A mantle and mantle cavity are present in:
    a. phylum Echinodermata.
    b. phylum Adversoidea.
    c. phylum Mollusca.
    d. phylum Nemertea.

14. How does segmentation enhance annelid locomotion?
    a. Segmentation creates repeating body structures so the entire organism functions in synchrony.
    b. Segmentation allows specialization of different body regions.
    c. Neural segmentation allows annelids to localize sensations.
    d. Muscle contractions can be localized to specific regions of the body to coordinate movement.

15. The embryonic development in nematodes can have up to _________ larval stages.
    a. one
    b. two
    c. three
    d. four

16. The nematode cuticle contains _____.
    a. glucose
    b. skin cells
    c. chitin
    d. nerve cells

17. Crustaceans are _____.
    a. ecdysozoans
    b. nematodes
    c. arachnids
    d. parazoans

18. Flies are_______.
    a. chelicerates
    b. hexapods
    c. arachnids
    d. crustaceans

19. Which of the following is **not** a key advantage provided by the exoskeleton of terrestrial arthropods?
    a. Prevents dessication
    b. Protects internal tissue
    c. Provides mechanical support
    d. Grows with the arthropod throughout its life

20. Echinoderms have _____.
    a. triangular symmetry
    b. radial symmetry
    c. hexagonal symmetry
    d. pentaradial symmetry

21. The circulatory fluid in echinoderms is _____.
    a. blood
    b. mesohyl
    c. water
    d. saline

22. Which of the following features does not distinguish humans as a member of phylum Chordata?
    a. Human embryos undergo indeterminate cleavage.
    b. A spinal cord runs along an adult human's dorsal side.
    c. Human embryos exhibit pharyngeal arches and gill slits.
    d. The human coccyx forms from an embryonic tail.

23. The sister taxon of the Chordata is the _____.
    a. Mollusca
    b. Arthropoda
    c. Ambulacraria
    d. Rotifera

CRITICAL THINKING QUESTIONS

24. Describe the different cell types and their functions in sponges.
25. Describe the feeding mechanism of sponges and identify how it is different from other animals.
26. Explain the function of nematocysts in cnidarians.
27. Compare the structural differences between Porifera and Cnidaria.
28. Compare the differences in sexual reproduction between Porifera and Cubozoans. How does the difference in fertilization provide an evolutionary advantage to the Cubozoans?
29. How does the tapeworm body plan support widespread dissemination of the parasite?
30. Describe the morphology and anatomy of mollusks.
31. What are the anatomical differences between nemertines and mollusks?
32. How does a change in the circulatory system organization support the body designs in cephalopods compared to other mollusks?
33. Enumerate features of *Caenorhabditis elegans* that make it a valuable model system for biologists.
34. What are the different ways in which nematodes can reproduce?

35. Why are tardigrades essential to recolonizing habits following destruction or mass extinction?
36. Describe the various superclasses that phylum Arthropoda can be divided into.
37. Compare and contrast the segmentation seen in phylum Annelida with that seen in phylum Arthropoda.
38. How do terrestrial arthropods of the subphylum Hexapoda impact the world's food supply? Provide at least two positive and two negative effects.
39. Describe the different classes of echinoderms using examples.

CHAPTER 29
Vertebrates

Figure 29.1 Examples of critically endangered vertebrate species include (a) the Siberian tiger (*Panthera tigris*), (b) the mountain gorilla (*Gorilla beringei*), and (c) the harpy eagle (*Harpia harpyja*). (The harpy eagle is considered "near threatened" globally, but critically endangered in much of its former range in Mexico and Central America.) (credit a: modification of work by Dave Pape; credit b: modification of work by Dave Proffer; credit c: modification of work by Haui Ared)

INTRODUCTION Vertebrates are among the most recognizable organisms of the Animal Kingdom, and more than 62,000 vertebrate species have been identified. The vertebrate species now living represent only a small portion of the vertebrates that have existed in the past. The best-known extinct vertebrates are the dinosaurs, a unique group of reptiles, some of which reached sizes not seen before or after in *terrestrial* animals. In fact, they were the dominant terrestrial animals for 150 million years, until most of them died out in a mass extinction near the end of the Cretaceous period (except for the feathered theropod ancestors of modern birds, whose direct descendents now number nearly 10,000 species). Although it is not known with certainty what caused this mass extinction (not only of dinosaurs, but of many other groups of organisms), a great deal is known about the anatomy of the dinosaurs and early birds, given the preservation of numerous skeletal elements, nests, eggs, and embryos in the fossil record.

Chapter Outline

29.1 Chordates

By the end of this section, you will be able to do the following:

- Describe the distinguishing characteristics of chordates
- Identify the derived characters of craniates that sets them apart from other chordates
- Describe the developmental fate of the notochord in vertebrates

The vertebrates exhibit two major innovations in their evolution from the invertebrate chordates. These innovations may be associated with the whole genome duplications that resulted in a quadruplication of the basic chordate genome, including the *Hox* gene loci that regulate the placement of structures along the three axes of the body. One of the first major steps was the emergence of the quadrupeds in the form of the amphibians. A second step was the evolution of the amniotic egg, which, similar to the evolution of pollen and seeds in plants, freed terrestrial animals from their dependence on water for fertilization and embryonic development. Within the amniotes, modifications of keratinous epidermal structures have given rise to scales, claws, hair, and feathers. The scales of reptiles sealed their skins against water loss, while hair and feathers provided insulation to support the evolution of endothermy, as well as served other functions such as camouflage and mate attraction in the vertebrate lineages that led to birds and mammals.

Currently, a number of vertebrate species face extinction primarily due to habitat loss and pollution. According to the International Union for the Conservation of Nature, more than 6,000 vertebrate species are classified as threatened. Amphibians and mammals are the classes with the greatest percentage of threatened species, with 29 percent of all amphibians and 21 percent of all mammals classified as threatened. Attempts are being made around the world to prevent the extinction of threatened species. For example, the Biodiversity Action Plan is an international program, ratified by 188 countries, which is designed to protect species and habitats.

Vertebrates are members of the kingdom Animalia and the phylum Chordata (Figure 29.2). Recall that animals that possess bilateral symmetry can be divided into two groups—protostomes and deuterostomes—based on their patterns of embryonic development. The deuterostomes, whose name translates as "second mouth," consist of two major phyla: Echinodermata and Chordata. Echinoderms are invertebrate marine animals that have pentaradial symmetry and a spiny body covering, a group that includes sea stars, sea urchins, and sea cucumbers. The most conspicuous and familiar members of Chordata are vertebrates, but this phylum also includes two groups of invertebrate chordates.

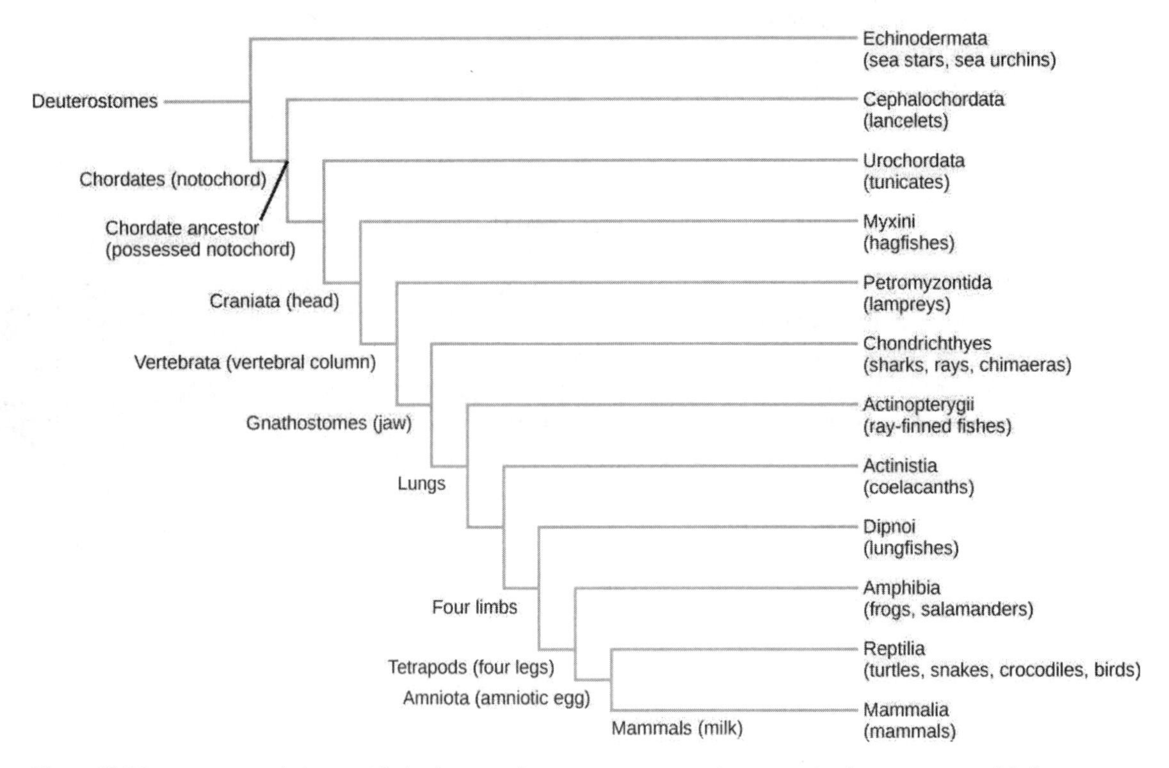

Figure 29.2 Deuterostome phylogeny. All chordates are deuterostomes possessing a notochord at some stage of their life cycle.

Characteristics of Chordata

Animals in the phylum **Chordata** share five key chacteristics that appear at some stage during their development: a notochord, a dorsal hollow (tubular) nerve cord, pharyngeal gill arches or slits, a post-anal tail, and an endostyle/thyroid gland (Figure 29.3). In some groups, some of these key chacteristics are present only during embryonic development.

The chordates are named for the **notochord**, which is a flexible, rod-shaped mesodermal structure that is found in the embryonic stage of all chordates and in the adult stage of some chordate species. It strengthened with glycoproteins similar to cartilage and covered with a collagenous sheath. The notocord is located between the digestive tube and the nerve cord, and provides rigid skeletal support as well as a flexible location for attachment of axial muscles. In some chordates, the notochord acts as the primary axial support of the body throughout the animal's lifetime. However, in vertebrates (craniates), the notochord is present only during embryonic development, at which time it induces the development of the neural tube and serves as a support for the developing embryonic body. The notochord, however, is not found in the postembryonic stages of vertebrates; at this point, it has been replaced by the vertebral column (that is, the spine).

VISUAL CONNECTION

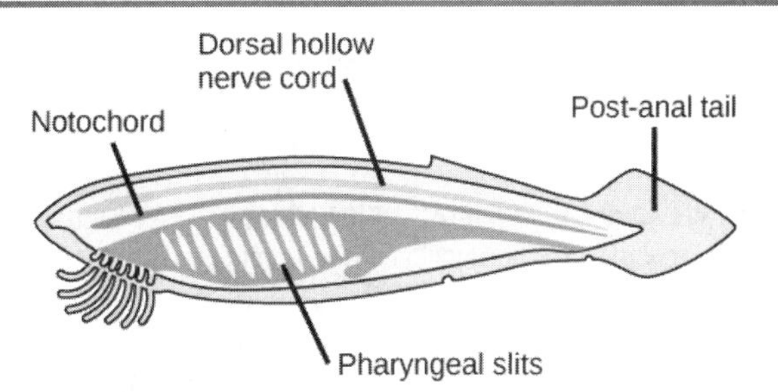

Figure 29.3 Chordate features. In chordates, four common features appear at some point during development: a notochord, a dorsal hollow nerve cord, pharyngeal slits, and a post-anal tail. The endostyle is embedded in the floor of the pharynx.

Which of the following statements about common features of chordates is true?

1. The dorsal hollow nerve cord is part of the chordate central nervous system.
2. In vertebrate fishes, the pharyngeal slits become the gills.
3. Humans are not chordates because humans do not have a tail.
4. Vertebrates do not have a notochord at any point in their development; instead, they have a vertebral column.
5. The endostyle secretes steroid hormones.

The **dorsal hollow nerve cord** is derived from ectoderm that rolls into a hollow tube during development. In chordates, it is located dorsally to the notochord. In contrast, the nervous system in protostome animal phyla is characterized by solid nerve cords that are located either ventrally and/or laterally to the gut. In vertebrates, the neural tube develops into the brain and spinal cord, which together comprise the central nervous system (CNS). The peripheral nervous system (PNS) refers to the peripheral nerves (including the cranial nerves) lying outside of the brain and spinal cord.

Pharyngeal slits are openings in the pharynx (the region just posterior to the mouth) that extend to the outside environment. In organisms that live in aquatic environments, pharyngeal slits allow for the exit of water that enters the mouth during feeding. Some invertebrate chordates use the pharyngeal slits to filter food out of the water that enters the mouth. The endostyle is a strip of ciliated mucus-producing tissue in the floor of the pharynx. Food particles trapped in the mucus are moved along the endostyle toward the gut. The endostyle also produces substances similar to thyroid hormones and is homologous with the thyroid gland in vertebrates. In vertebrate fishes, the pharyngeal slits are modified into gill supports, and in jawed fishes, into jaw supports. In tetrapods (land vertebrates), the slits are highly modified into components of the ear, and tonsils and thymus glands. In other vertebrates, pharyngeal arches, derived from all three germ layers, give rise to the oral jaw from the first pharyngeal arch, with the second arch becoming the hyoid and jaw support.

The **post-anal tail** is a posterior elongation of the body, extending beyond the anus. The tail contains skeletal elements and muscles, which provide a source of locomotion in aquatic species, such as fishes. In some terrestrial vertebrates, the tail also helps with balance, courting, and signaling when danger is near. In humans and other great apes, the post-anal tail is reduced to a vestigial coccyx ("tail bone") that aids in balance during sitting.

LINK TO LEARNING

Click for a video discussing the evolution of chordates and five characteristics that they share.

Click to view content (https://www.openstax.org/l/chordate_evol)

Chordates and the Evolution of Vertebrates

Two clades of chordates are invertebrates: Cephalochordata and Urochordata. Members of these groups also possess the five distinctive features of chordates at some point during their development.

Cephalochordata

Members of **Cephalochordata** possess a notochord, dorsal hollow tubular nerve cord, pharyngeal slits, endostyle/thyroid gland, and a post-anal tail in the adult stage (Figure 29.4). The notochord extends into the head, which gives the subphylum its name. Although the neural tube also extends into the head region, there is no well-defined brain, and the nervous system is centered around a hollow nerve cord lying above the notochord. Extinct members of this subphylum include *Pikaia*, which is the oldest known cephalochordate. Excellently preserved *Pikaia* fossils were recovered from the Burgess shales of Canada and date to the middle of the Cambrian age, making them more than 500 million years old. Its anatomy of *Pikaia* closely resembles that of the extant lancelet in the genus *Branchiostoma*.

The lancelets are named for their bladelike shape. Lancelets are only a few centimeters long and are usually found buried in sand at the bottom of warm temperate and tropical seas. Cephalochordates are suspension feeders. A water current is created by cilia in the mouth, and is filtered through oral tentacles. Water from the mouth then enters the pharyngeal slits, which filter out food particles. The filtered water collects in a gill chamber called the **atrium** and exits through the **atriopore**. Trapped food particles are caught in a stream of mucus produced by the endostyle in a ventral ciliated fold (or groove) of the pharynx and carried to the gut. Most gas exchange occurs across the body surface. Sexes are separate and gametes are released into the water through the atriopore for external fertilization.

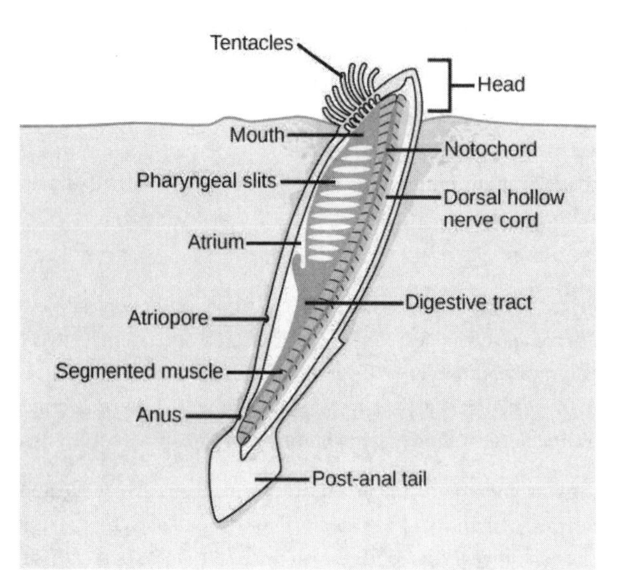

Figure 29.4 Cephalochordate anatomy. In the lancelet and other cephalochordates, the notochord extends into the head region. Adult lancelets retain all five key characteristics of chordates: a notochord, a dorsal hollow nerve cord, pharyngeal slits, an endostyle, and a post-anal tail.

Urochordata

The 1,600 species of **Urochordata** are also known as **tunicates** (Figure 29.5). The name tunicate derives from the cellulose-like carbohydrate material, called the **tunic**, which covers the outer body of tunicates. Although tunicates are classified as chordates, the adults do *not* have a notochord, a dorsal hollow nerve cord, or a post-anal tail, although they do have pharyngeal slits and an endostyle. The "tadpole" larval form, however, possesses all five structures. Most tunicates are hermaphrodites; their larvae hatch from eggs inside the adult tunicate's body. After hatching, a tunicate larva (possessing all five chordate features) swims for a few days until it finds a suitable surface on which it can attach, usually in a dark or shaded location. It then attaches via the head to the surface and undergoes metamorphosis into the adult form, at which point the notochord, nerve cord, and tail disappear, leaving the pharyngeal gill slits and the endostyle as the two remaining features of its chordate morphology.

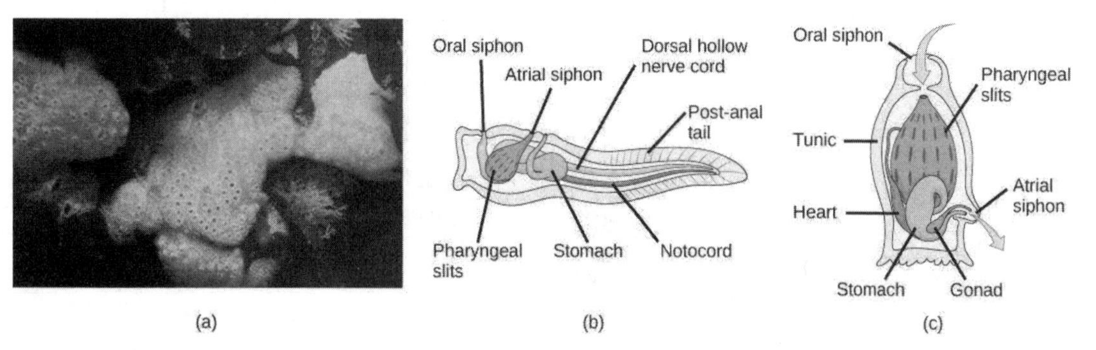

Figure 29.5 Urochordate anatomy. (a) This photograph shows a colony of the tunicate *Botrylloides violaceus*. (b) The larval stage of the tunicate possesses all of the features characteristic of chordates: a notochord, a dorsal hollow nerve cord, pharyngeal slits, an endostyle, and a post-anal tail. (c) In the adult stage, the notochord, nerve cord, and tail disappear, leaving just the pharyngeal slits and endostyle. (credit: modification of work by Dann Blackwood, USGS)

Adult tunicates may be either solitary or colonial forms, and some species may reproduce by budding. Most tunicates live a sessile existence on the ocean floor and are *suspension feeders*. However, chains of thaliacean tunicates called *salps* (Figure 29.6) can swim actively while feeding, propelling themselves as they move water through the pharyngeal slits. The primary foods of tunicates are plankton and detritus. Seawater enters the tunicate's body through its incurrent siphon. Suspended material is filtered out of this water by a mucous net produced by the endostyle and is passed into the intestine via the action of cilia. The anus empties into the excurrent siphon, which expels wastes and water. Tunicates are found in shallow ocean waters around the world.

Figure 29.6 Salps. These colonial tunicates feed on phytoplankton. Salps are sequential hermaphrodites, with younger female colonies fertilized by older male colonies. (credit: Oregon Department of Fish & Wildlife via Wikimedia Commons)

Subphylum Vertebrata (Craniata)

A **cranium** is a bony, cartilaginous, or fibrous structure surrounding the brain, jaw, and facial bones (Figure 29.7). Most bilaterally symmetrical animals have a head; of these, those that have a cranium comprise the clade Craniata/Vertebrata, which includes the primitively jawless Myxini (hagfishes), Petromyzontida (lampreys), and all of the organisms called "vertebrates." (We should note that the Myxini have a cranium but lack a backbone.)

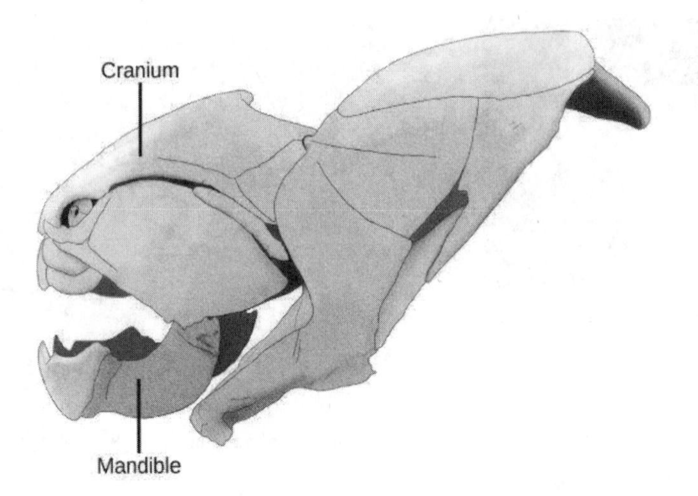

Figure 29.7 A craniate skull. The subphylum **Craniata (or Vertebrata)**, including this placoderm fish (*Dunkleosteus* sp.), are characterized by the presence of a cranium, mandible, and other facial bones. (credit: "Steveoc 86"/Wikimedia Commons)

Members of the phylum Craniata/Vertebrata display the five characteristic features of the chordates; however, members of this group also share derived characteristics that distinguish them from invertebrate chordates. Vertebrates are named for the vertebral column, composed of **vertebrae**—a series of separate, irregularly shaped bones joined together to form a backbone (Figure 29.8). Initially, the vertebrae form in segments around the embryonic notochord, but eventually replace it in adults. In most derived vertebrates, the notochord becomes the *nucleus pulposus* of the intervertebral discs that cushion and support adjacent vertebrae.

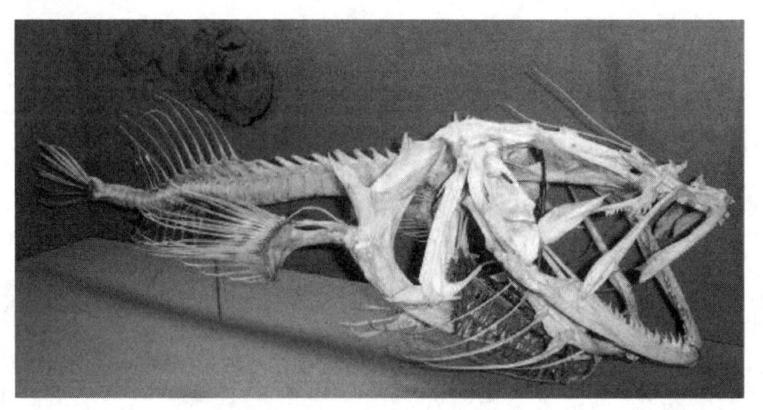

Figure 29.8 A vertebrate skeleton. Vertebrata are characterized by the presence of a backbone, such as the one that runs through the middle of this fish. All vertebrates are in the Craniata clade and have a cranium. (credit: Ernest V. More; taken at Smithsonian Museum of Natural History, Washington, D.C.)

The relationship of the vertebrates to the invertebrate chordates has been a matter of contention, but although these cladistic relationships are still being examined, it appears that the Craniata/Vertebrata are a monophyletic group that shares the five basic chordate characteristics with the other two subphyla, Urochordata and Cephalochordata. Traditional phylogenies place the cephalochordates as a sister clade to the chordates, a view that has been supported by most current molecular analyses. This hypothesis is further supported by the discovery of a fossil in China from the genus *Haikouella*. This organism seems to be an intermediate form between cephalochordates and vertebrates. The *Haikouella* fossils are about 530 million years old and appear similar to modern lancelets. These organisms had a brain and eyes, as do vertebrates, but lack the skull found in craniates.[1] This evidence suggests that vertebrates arose during the Cambrian explosion.

Vertebrates are the largest group of chordates, with more than 62,000 living species, which are grouped based on anatomical and physiological traits. More than one classification and naming scheme is used for these animals. Here we will consider the *traditional groups* Agnatha, Chondrichthyes, Osteichthyes, Amphibia, Reptilia, Aves, and Mammalia, which constitute classes in the subphylum Vertebrata/Craniata. Virtually all modern cladists classify birds within Reptilia, which correctly reflects their

1Chen, J. Y., Huang, D. Y., and Li, C. W., "An early Cambrian craniate-like chordate," *Nature* 402 (1999): 518–522, doi:10.1038/990080.

evolutionary heritage. Thus, we now have the nonavian reptiles and the avian reptiles in our reptilian classification. We consider them separately only for convenience. Further, we will consider hagfishes and lampreys together as jawless fishes, the **Agnatha**, although emerging classification schemes separate them into chordate jawless fishes (the hagfishes) and vertebrate jawless fishes (the lampreys).

Animals that possess jaws are known as **gnathostomes**, which means "jawed mouth." Gnathostomes include fishes and tetrapods. **Tetrapod** literally means "four-footed," which refers to the phylogenetic history of various land vertebrates, even though in some of the tetrapods, the limbs may have been modified for purposes other than walking. Tetrapods include amphibians, reptiles, birds, and mammals, and technically could also refer to the extinct fishlike groups that gave rise to the tetrapods. Tetrapods can be further divided into two groups: amphibians and amniotes. Amniotes are animals whose eggs contain four extraembryonic membranes (yolk sac, amnion, chorion, and allantois) that provide nutrition and a water-retaining environment for their embryos. Amniotes are adapted for terrestrial living, and include mammals, reptiles, and birds.

29.2 Fishes

By the end of this section, you will be able to do the following:

- Describe the difference between jawless and jawed fishes
- Discuss the distinguishing features of sharks and rays compared to other modern fishes

Modern fishes include an estimated 31,000 species, by far the most of all clades within the Vertebrata. Fishes were the earliest vertebrates, with jawless species being the earliest forms and jawed species evolving later. They are active feeders, rather than sessile, suspension feeders. The Agnatha (jawless fishes)—the hagfishes and lampreys—have a distinct cranium and complex sense organs including eyes, that distinguish them from the invertebrate chordates, the urochordates and cephalochordates.

Jawless Fishes: Superclass Agnatha

Jawless fishes (Agnatha) are craniates representing an ancient vertebrate lineage that arose over 550 million years ago. In the past, hagfishes and lampreys were sometimes recognized as separate clades within the Agnatha, primarily because lampreys were regarded as true vertebrates, whereas hagfishes were not. However, recent molecular data, both from rRNA and mtDNA, as well as embryological data, provide strong support for the hypothesis that living agnathans—previously called *cyclostomes*—are monophyletic, and thus share recent common ancestry. The discussion below, for convenience, separates the modern "cyclostomes" into the class Myxini and class Petromyzontida. The defining features of the living jawless fishes are the lack of jaws and lack of paired lateral appendages (fins). They also lack internal ossification and scales, although these are not defining features of the clade.

Some of the earliest jawless fishes were the armored ostracoderms (which translates to "shell-skin"): vertebrate fishes encased in bony armor—unlike present-day jawless fishes, which lack bone in their scales. Some ostracoderms, also unlike living jawless fishes, may have had paired fins. We should note, however, that the "ostracoderms" represent an assemblage of heavily armored extinct jawless fishes that may not form a natural evolutionary group. Fossils of the genus *Haikouichthys* from China, with an age of about 530 million years, show many typical vertebrate characteristics including paired eyes, auditory capsules, and rudimentary vertebrae.

Class Myxini: Hagfishes

The class **Myxini** includes at least 70 species of hagfishes—eel-like scavengers that live on the ocean floor and feed on living or dead invertebrates, fishes, and marine mammals (Figure 29.9). Although they are almost completely blind, sensory barbels around the mouth help them locate food by smell and touch. They feed using keratinized teeth on a movable cartilaginous plate in the mouth, which rasp pieces of flesh from their prey. These feeding structures allow the gills to be used exclusively for respiration, *not* for filter feeding as in the urochordates and cephalochordates. Hagfishes are entirely marine and are found in oceans around the world, except for the polar regions. Unique slime glands beneath the skin release a milky mucus (through surface pores) that upon contact with water becomes incredibly slippery, making the animal almost impossible to hold. This slippery mucus thus allows the hagfish to escape from the grip of predators. Hagfish can also twist their bodies into a knot, which provides additional leverage to feed. Sometimes hagfish enter the bodies of dead animals and eat carcasses from the inside out! Interestingly, they do not have a stomach!

Figure 29.9 Hagfish. Pacific hagfish are scavengers that live on the ocean floor. (credit: Linda Snook, NOAA/CBNMS)

Hagfishes have a cartilaginous skull, as well as a fibrous and cartilaginous skeleton, but the major supportive structure is the notochord that runs the length of the body. In hagfishes, the notochord is *not* replaced by the vertebral column, as it is in true vertebrates, and thus they may (morphologically) represent a sister group to the true vertebrates, making them the most basal clade among the skull-bearing chordates.

Class Petromyzontida: Lampreys

The class **Petromyzontida** includes approximately 40 species of lampreys, which are superficially similar to hagfishes in size and shape. However, lampreys possess extrinsic eye muscles, at least two semicircular canals, and a true cerebellum, as well as simple vertebral elements, called *arcualia*—cartilaginous structures arranged above the notochord. These features are also shared with the *gnathostomes*—vertebrates with jawed mouths and paired appendages (see below). Lampreys also have a dorsal tubular nerve cord with a well-differentiated brain, a small cerebellum, and 10 pairs of nerves. The classification of lampreys is still debated, but they clearly represent one of the oldest divergences of the vertebrate lineage. Lampreys lack paired appendages, as do the hagfishes, although they have one or two fleshy dorsal fins. As adults, lampreys are characterized by a rasping tongue within a toothed, funnel-like sucking mouth. Many species have a parasitic stage of their life cycle during which they are fish ectoparasites (some call them predators because they attack and eventually fall off) (Figure 29.10).

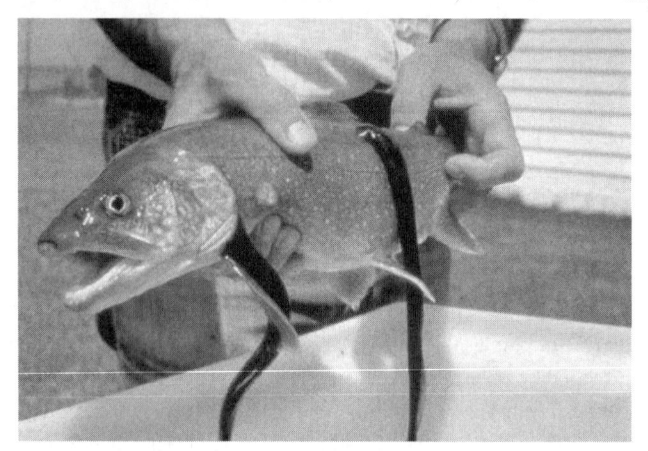

Figure 29.10 Lamprey. These parasitic sea lampreys, *Petromyzon marinus*, attach by suction to their lake trout host, and use their rough tongues to rasp away flesh in order to feed on the trout's blood. (credit: USGS)

Lampreys live primarily in coastal and freshwater environments, and have a worldwide distribution, except for the tropics and polar regions. Some species are marine, but all species spawn in fresh water. Interestingly, northern lampreys in the family Petromyzontidae, have the highest number of chromosomes (164 to 174) among the vertebrates. Eggs are fertilized externally, and the larvae (called *ammocoetes*) differ greatly from the adult form, closely resembling the adult cephalocordate *amphioxus*. After spending three to 15 years as suspension feeders in rivers and streams, they attain sexual maturity. Shortly afterward, the adults swim upstream, reproduce, and die within days.

Gnathostomes: Jawed Fishes

Gnathostomes, or "jaw-mouths," are vertebrates that possess true jaws—a milestone in the evolution of the vertebrates. In fact, one of the most significant developments in early vertebrate evolution was the development of the jaw: a hinged structure attached to the cranium that allows an animal to grasp and tear its food. Jaws were probably derived from the first pair of gill arches supporting the gills of jawless fishes.

Early gnathostomes also possessed two sets of paired fins, allowing the fishes to maneuver accurately. Pectoral fins are typically located on the anterior body, and pelvic fins on the posterior. Evolution of the jaw and paired fins permitted gnathostomes to expand their food options from the scavenging and suspension feeding of jawless fishes to active predation. The ability of gnathostomes to exploit new nutrient sources probably contributed to their replacing most jawless fishes during the Devonian period. Two early groups of gnathostomes were the *acanthodians* and *placoderms* (Figure 29.11), which arose in the late Silurian period and are now extinct. Most modern fishes are gnathostomes that belong to the clades Chondrichthyes and Osteichthyes (which include the class Actinoptertygii and class Sarcopterygii).

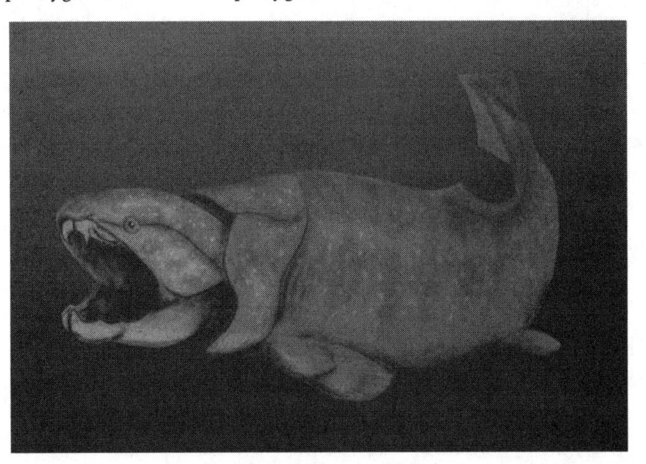

Figure 29.11 A placoderm. *Dunkleosteus* was an enormous placoderm from the Devonian period, 380 to 360 million years ago. It measured up to 10 meters in length and weighed up to 3.6 tons. Its head and neck were armored with heavy bony plates. Although *Dunkleosteus* had no true teeth, the edge of the jaw was armed with sharp bony blades. (credit: Nobu Tamura)

Class Chondrichthyes: Cartilaginous Fishes

The class Chondrichthyes (about 1,000 species) is a morphologically diverse clade, consisting of subclass Elasmobranchii (sharks [Figure 29.12], rays, and skates, together with the obscure and critically endangered sawfishes), and a few dozen species of fishes called *chimaeras*, or "ghost sharks" in the subclass Holocephali. Chondrichthyes are jawed fishes that possess paired fins and a skeleton made of cartilage. This clade arose approximately 370 million years ago in the early or middle Devonian. They are thought to be descended from the placoderms, which had endoskeletons made of bone; thus, the lighter cartilaginous skeleton of Chondrichthyes is a secondarily derived evolutionary development. Parts of shark skeleton are strengthened by granules of calcium carbonate, but this is not the same as bone.

Most cartilaginous fishes live in marine habitats, with a few species living in fresh water for a part or all of their lives. Most sharks are carnivores that feed on live prey, either swallowing it whole or using their jaws and teeth to tear it into smaller pieces. Sharks have abrasive skin covered with tooth-like scales called placoid scales. Shark teeth probably evolved from rows of these scales lining the mouth. A few species of sharks and rays, like the enormous whale shark (Figure 29.13), are suspension feeders that feed on plankton. The sawfishes have an extended rostrum that looks like a double-edged saw. The rostrum is covered with electrosensitive pores that allow the sawfish to detect slight movements of prey hiding in the muddy sea floor. The teeth in the rostrum are actually modified tooth-like structures called denticles, similar to scales.

Figure 29.12 Shark. Hammerhead sharks tend to school during the day and hunt prey at night. (credit: Masashi Sugawara)

Sharks have well-developed sense organs that aid them in locating prey, including a keen sense of smell and the ability to detect electromagnetic fields. Electroreceptors called **ampullae of Lorenzini** allow sharks to detect the electromagnetic fields that are produced by all living things, including their prey. (Electroreception has only been observed in aquatic or amphibious animals and sharks have perhaps the most sensitive electroreceptors of any animal.) Sharks, together with most fishes and aquatic and larval amphibians, also have a row of sensory structures called the **lateral line**, which is used to detect movement and vibration in the surrounding water, and is often considered to be functionally similar to the sense of "hearing" in terrestrial vertebrates. The lateral line is visible as a darker stripe that runs along the length of a fish's body. Sharks have no mechanism for maintaining neutral buoyancy and must swim continuously to stay suspended in the water. Some must also swim in order to ventilate their gills but others have muscular pumps in their mouths to keep water flowing over the gills.

Figure 29.13 Whale shark in the Georgia Aquarium. Whale sharks are filter-feeders and can grow to be over 10 meters long. Whale sharks, like most other sharks, are ovoviviparous. (credit: modified from Zac Wolf [Own work] [CC BY-SA 2.5 (http://creativecommons.org/licenses/by-sa/2.5 (http://openstax.org/l/CCSA))], via Wikimedia Commons)

Sharks reproduce sexually, and eggs are fertilized internally. Most species are *ovoviviparous*: The fertilized egg is retained in the oviduct of the mother's body and the embryo is nourished by the egg yolk. The eggs hatch in the uterus, and young are born alive and fully functional. Some species of sharks are oviparous: They lay eggs that hatch outside of the mother's body. Embryos are protected by a shark egg case or "mermaid's purse" (Figure 29.14) that has the consistency of leather. The shark egg case has tentacles that snag in seaweed and give the newborn shark cover. A few species of sharks, e.g., tiger sharks and hammerheads, are viviparous: the yolk sac that initially contains the egg yolk and transfers its nutrients to the growing embryo becomes attached to the oviduct of the female, and nutrients are transferred directly from the mother to the growing embryo. In both viviparous and ovoviviparous sharks, gas exchange uses this yolk sac transport.

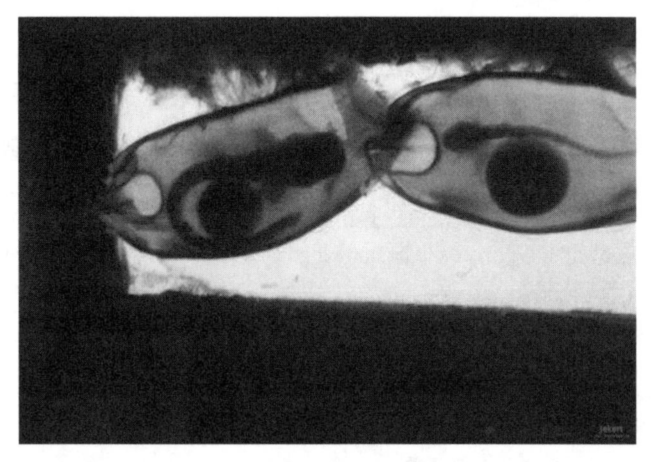

Figure 29.14 Shark egg cases. Shark embryos are clearly visible through these transparent egg cases. The round structure is the yolk that nourishes the growing embryo. (credit: Jek Bacarisas)

In general, the Chondrichthyes have a fusiform or dorsoventrally flattened body, a *heterocercal* caudal fin or tail (unequally sized fin lobes, with the tail vertebrae extending into the larger upper lobe) paired pectoral and pelvic fins (in males these may be modified as claspers), exposed gill slits (*elasmobranch*), and an intestine with a spiral valve that condenses the length of the intestine. They also have three pairs of semicircular canals, and excellent senses of smell, vibration, vision, and electroreception. A very large lobed liver produces *squalene oil* (a lightweight biochemical precursor to steroids) that serves to aid in buoyancy (because with a specific gravity of 0.855, it is lighter than that of water).

Rays and skates comprise more than 500 species. They are closely related to sharks but can be distinguished from sharks by their flattened bodies, pectoral fins that are enlarged and fused to the head, and gill slits on their ventral surface (Figure 29.15). Like sharks, rays and skates have a cartilaginous skeleton. Most species are marine and live on the sea floor, with nearly a worldwide distribution.

Unlike the stereotypical sharks and rays, the Holocephali (chimaeras or ratfish) have a *diphycercal* tail (equally sized fin lobes, with the tail vertebrae located between them), lack scales (lost secondarily in evolution), and have teeth modified as grinding plates that are used to feed on mollusks and other invertebrates (Figure 29.15b). Unlike sharks with elasmobranch or naked gills, chimaeras have four pairs of gills covered by an operculum. Many species have a pearly iridescence and are extremely pretty.

Figure 29.15 Cartilaginous fish. (a) Stingray. This stingray blends into the sandy bottom of the ocean floor. A spotted ratfish (b) Hydrolagus colliei credit a "Sailn1"/Flickr; (credit: a "Sailn1"/Flickr b: Linda Snook / MBNMS [Public domain], via Wikimedia Commons.)

Osteichthyes: Bony Fishes

Members of the clade **Osteichthyes**, also called bony fishes, are characterized by a bony skeleton. The vast majority of present-day fishes belong to this group, which consists of approximately 30,000 species, making it the largest class of vertebrates in existence today.

Nearly all bony fishes have an *ossified skeleton* with specialized bone cells (osteocytes) that produce and maintain a calcium phosphate matrix. This characteristic has been reversed only in a few groups of Osteichthyes, such as sturgeons and paddlefish, which have primarily cartilaginous skeletons. The skin of bony fishes is often covered by overlapping scales, and glands in the skin secrete mucus that reduces drag when swimming and aids the fish in osmoregulation. Like sharks, bony fishes have a lateral line system that detects vibrations in water.

All bony fishes use gills to breathe. Water is drawn over gills that are located in chambers covered and ventilated by a protective, muscular flap called the operculum. Many bony fishes also have a **swim bladder**, a gas-filled organ derived as a pouch from the gut. The swim bladder helps to control the buoyancy of the fish. In most bony fish, the gases of the swim bladder are exchanged directly with the blood. The swim bladder is believed to be homologous to the lungs of lungfish and the lungs of land vertebrates.

Bony fishes are further divided into two extant clades: Class **Actinopterygii** (ray-finned fishes) and Class **Sarcopterygii** (lobe-finned fishes).

Actinopterygii (Figure 29.16a), the ray-finned fishes, include many familiar fishes—tuna, bass, trout, and salmon among others—and represent about half of all vertebrate species. Ray-finned fishes are named for the fan of slender bones that supports their fins.

In contrast, the fins of Sarcopterygii (Figure 29.16b) are fleshy and lobed, supported by bones that are similar in type and arrangement to the bones in the limbs of early tetrapods. The few extant members of this clade include several species of lungfishes and the less familiar coelacanths, which were thought to be extinct until living specimens were discovered between Africa and Madagascar. Currently, two species of coelocanths have been described.

Figure 29.16 Osteichthyes. The (a) sockeye salmon and (b) coelacanth are both bony fishes of the Osteichthyes clade. The coelacanth, sometimes called a lobe-finned fish, was thought to have gone extinct in the Late Cretaceous period, 100 million years ago, until one was discovered in 1938 near the Comoros Islands between Africa and Madagascar. (credit a: modification of work by Timothy Knepp, USFWS; credit b: modification of work by Robbie Cada)

29.3 Amphibians

By the end of this section, you will be able to do the following:

- Describe the important difference between the life cycle of amphibians and the life cycles of other vertebrates
- Distinguish between the characteristics of Urodela, Anura, and Apoda
- Describe the evolutionary history of amphibians

Amphibians are vertebrate tetrapods ("four limbs"), and include frogs, salamanders, and caecilians. The term "amphibian" loosely translates from the Greek as "dual life," which is a reference to the metamorphosis that many frogs and salamanders undergo and the unique mix of aquatic and terrestrial phases that are required in their life cycle. In fact, they cannot stray far from water because their reproduction is intimately tied to aqueous environments. Amphibians evolved during the Devonian period and were the earliest terrestrial tetrapods. They represent an evolutionary transition from water to land that occurred over many millions of years. Thus, the Amphibia are the only living true vertebrates that have made a transition from water to land in both their ontogeny (life development) and phylogeny (evolution). They have not changed much in morphology over the past 350 million years!

LINK TO LEARNING

Watch this series of five Animal Planet videos on tetrapod evolution:

- 1: The evolution from fish to earliest tetrapod
  Click to view content (https://www.openstax.org/l/tetrapod_evol1)
- 2: Fish to Earliest Tetrapod
  Click to view content (https://www.openstax.org/l/tetrapod_evol2)
- 3: The discovery of coelacanth and *Acanthostega* fossils

Click to view content (https://www.openstax.org/l/tetrapod_evol3)

- 4: The number of fingers on "legs"
  Click to view content (https://www.openstax.org/l/tetrapod_evol4)
- 5: Reconstructing the environment of early tetrapods
  Click to view content (https://www.openstax.org/l/tetrapod_evol5)

Characteristics of Amphibians

As tetrapods, most amphibians are characterized by four well-developed limbs. In some species of salamanders, hindlimbs are reduced or absent, but all caecilians are (secondarily) limbless. An important characteristic of extant amphibians is a moist, permeable skin that is achieved via mucus glands. Most water is taken in across the skin rather than by drinking. The skin is also one of three respiratory surfaces used by amphibians. The other two are the lungs and the buccal (mouth) cavity. Air is taken first into the mouth through the nostrils, and then pushed by positive pressure into the lungs by elevating the throat and closing the nostrils.

All extant adult amphibians are carnivorous, and some terrestrial amphibians have a sticky tongue used to capture prey. Amphibians also have multiple small teeth at the edge of the jaws. In salamanders and caecilians, teeth are present in both jaws, sometimes in multiple rows. In frogs and toads, teeth are seen only in the upper jaw. Additional teeth, called **vomerine teeth**, may be found in the roof of the mouth. Amphibian teeth are *pedicellate*, which means that the root and crown are calcified, separated by a zone of noncalcified tissue.

Amphibians have image-forming eyes and color vision. Ears are best developed in frogs and toads, which vocalize to communicate. Frogs use separate regions of the inner ear for detecting higher and lower sounds: the *papilla amphibiorum*, which is sensitive to frequencies below 10,000 hertz and unique to amphibians, and the *papilla basilaris*, which is sensitive to higher frequencies, including mating calls, transmitted from the eardrum through the stapes bone. Amphibians also have an extra bone in the ear, the operculum, which transmits low-frequency vibrations from the forelimbs and shoulders to the inner ear, and may be used for the detection of seismic signals.

Evolution of Amphibians

The fossil record provides evidence of the first tetrapods: now-extinct amphibian species dating to nearly 400 million years ago. Evolution of tetrapods from lobe-finned freshwater fishes (similar to coelacanths and lungfish) represented a significant change in body plan from one suited to organisms that respired and swam in water, to organisms that breathed air and moved onto land; these changes occurred over a span of 50 million years during the Devonian period.

Aquatic tetrapods of the Devonian period include *Ichthyostega* and *Acanthostega*. Both were aquatic, and may have had both gills and lungs. They also had four limbs, with the skeletal structure of limbs found in present-day tetrapods, including amphibians. However, the limbs could not be pulled in under the body and would not have supported their bodies well out of water. They probably lived in shallow freshwater environments, and may have taken brief terrestrial excursions, much like "walking" catfish do today in Florida. In *Ichthyostega*, the forelimbs were more developed than the hind limbs, so it might have dragged itself along when it ventured onto land. What preceded *Acanthostega* and *Ichthyostega*?

In 2006, researchers published news of their discovery of a fossil of a "tetrapod-like fish," *Tiktaalik roseae*, which seems to be a morphologically "intermediate form" between sarcopterygian fishes having feet-like fins and early tetrapods having true limbs (Figure 29.17). *Tiktaalik* likely lived in a shallow water environment about 375 million years ago.[2] Tiktaalik also had gills and lungs, but the loss of some gill elements gave it a neck, which would have allowed its head to move sideways for feeding. The eyes were on top of the head. It had fins, but the attachment of the fin bones to the shoulder suggested they might be weight-bearing. Tiktaalik preceded *Acanthostega* and *Ichthyostega*, with their four limbs, by about 10 million years and is considered to be a true intermediate clade between fish and amphibians.

2Daeschler, E. B., Shubin, N. H., and Jenkins, F. J. "A Devonian tetrapod-like fish and the evolution of the tetrapod body plan," *Nature* 440 (2006): 757–763, doi:10.1038/nature04639, http://www.nature.com/nature/journal/v440/n7085/abs/nature04639.html (http://openstax.org/l/tetrapod) .

Figure 29.17 Tiktaalik. The recent fossil discovery of *Tiktaalik roseae* suggests evidence for an animal intermediate to finned fish and legged tetrapods, sometimes called a "fishapod." (credit: Zina Deretsky, National Science Foundation)

The early tetrapods that moved onto land had access to new nutrient sources and relatively few predators. This led to the widespread distribution of tetrapods during the early Carboniferous period, a period sometimes called the "age of the amphibians."

Modern Amphibians

Amphibia comprises an estimated 6,770 extant species that inhabit tropical and temperate regions around the world. All living species are classified in the subclass Lissamphibia ("smooth-amphibian"), which is divided into three clades: **Urodela** ("tailed"), the salamanders; **Anura** ("tail-less"), the frogs; and **Apoda** ("legless ones"), the caecilians.

Urodela: Salamanders

Salamanders are amphibians that belong to the order Urodela (or Caudata). These animals are probably the most similar to ancestral amphibians. Living salamanders (Figure 29.18) include approximately 620 species, some of which are aquatic, others terrestrial, and some that live on land only as adults. Most adult salamanders have a generalized tetrapod body plan with four limbs and a tail. The placement of their legs makes it difficult to lift their bodies off the ground and they move by bending their bodies from side to side, called *lateral undulation*, in a fish-like manner while "walking" their arms and legs fore-and-aft. It is thought that their gait is similar to that used by early tetrapods. The majority of salamanders are lungless, and respiration occurs through the skin or through external gills in aquatic species. Some terrestrial salamanders have primitive lungs; a few species have both gills and lungs. The giant Japanese salamander, the largest living amphibian, has additional folds in its skin that enlarge its respiratory surface.

Most salamanders reproduce using an unusual process of internal fertilization of the eggs. Mating between salamanders typically involves an elaborate and often prolonged courtship. Such a courtship ends in the deposition of sperm by the males in a packet called a **spermatophore**, which is subsequently picked up by the female, thus ultimately fertilization is internal. All salamanders except one, the fire salamander, are oviparous. Aquatic salamanders lay their eggs in water, where they develop into legless larvae called *efts*. Terrestrial salamanders lay their eggs in damp nests, where the eggs are guarded by their mothers. These embryos go through the larval stage and complete metamorphosis before hatching into tiny adult forms. One aquatic salamander, the Mexican *axolotl*, never leaves the larval stage, becoming sexually mature without metamorphosis.

Figure 29.18 Salamander. Most salamanders have legs and a tail, but respiration varies among species. (credit: Valentina Storti)

LINK TO LEARNING

View River Monsters: Fish With Arms and Hands? (http://openstax.org/l/river_monster) to see a video about an unusually large salamander species.

Anura: Frogs

Frogs (Figure 29.19) are amphibians that belong to the order Anura or Salientia ("jumpers"). Anurans are among the most diverse groups of vertebrates, with approximately 5,965 species that occur on all of the continents except Antarctica. Anurans, ranging from the minute New Guinea frog at 7 mm to the huge goliath frog at 32 cm from tropical Africa, have a body plan that is more specialized for movement. Adult frogs use their hind limbs and their arrow-like endoskeleton to jump accurately to capture prey on land. Tree frogs have hands adapted for grasping branches as they climb. In tropical areas, "flying frogs" can glide from perch to perch on the extended webs of their feet. Frogs have a number of modifications that allow them to avoid predators, including skin that acts as camouflage. Many species of frogs and salamanders also release defensive chemicals that are poisonous to predators from glands in the skin. Frogs with more toxic skins have bright warning (*aposematic*) coloration.

Figure 29.19 Tree frog. The Australian green tree frog is a nocturnal predator that lives in the canopies of trees near a water source.

Frog eggs are fertilized externally, and like other amphibians, frogs generally lay their eggs in moist environments. Although amphibian eggs are protected by a thick jelly layer, they would still dehydrate quickly in a dry environment. Frogs demonstrate a great diversity of parental behaviors, with some species laying many eggs and exhibiting little parental care, to species that carry eggs and tadpoles on their hind legs or embedded in their backs. The males of Darwin's frog carry tadpoles in their vocal sac. Many tree frogs lay their eggs off the ground in a folded leaf located over water so that the tadpoles can drop into the water as they hatch.

The life cycle of most frogs, as other amphibians, consists of two distinct stages: the larval stage followed by metamorphosis to an adult stage. However, the eggs of frogs in the genus *Eleutherodactylus* develop directly into little froglets, guarded by a parent. The larval stage of a frog, the tadpole, is often a filter-feeding herbivore. Tadpoles usually have gills, a lateral line system, longfinned tails, and lack limbs. At the end of the tadpole stage, frogs undergo metamorphosis into the adult form (Figure 29.20). During this stage, the gills, tail, and lateral line system disappear, and four limbs develop. The jaws become larger and are suited for carnivorous feeding, and the digestive system transforms into the typical short gut of a predator. An eardrum and air-breathing lungs also develop. These changes during metamorphosis allow the larvae to move onto land in the adult stage.

Figure 29.20 Amphibian metapmorphosis. A juvenile frog metamorphoses into a frog. Here, the frog has started to develop limbs, but its tadpole tail is still evident.

Apoda: Caecilians

An estimated 185 species comprise the **caecilians**, a group of amphibians that belong to the order Apoda. They have no limbs, although they evolved from a legged vertebrate ancestor. The complete lack of limbs makes them resemble earthworms. This resemblance is enhanced by folds of skin that look like the segments of an earthworm. However, unlike earthworms, they have teeth in both jaws, and feed on a variety of small organisms found in soil, including earthworms! Caecilians are adapted for a burrowing or aquatic lifestyle, and they are nearly blind, with their tiny eyes sometimes covered by skin. Although they have a single lung, they also depend on cutaneous respiration. These animals are found in the tropics of South America, Africa, and Southern Asia. In the caecelians, the only amphibians in which the males have copulatory structures, fertilization is internal. Some caecilians are oviparous, but most bear live young. In these cases, the females help nourish their young with tissue from their oviduct before birth and from their skin after birth.

EVOLUTION CONNECTION

The Paleozoic Era and the Evolution of Vertebrates

When the vertebrates arose during the Paleozoic Era (542 to 251 MYA), the climate and geography of Earth was vastly different. The distribution of landmasses on Earth were also very different from that of today. Near the equator were two large supercontinents, **Laurentia** and **Gondwana**, which included most of today's continents, but in a radically different configuration (Figure 29.21). At this time, sea levels were very high, probably at a level that hasn't been reached since. As the Paleozoic progressed, glaciations created a cool global climate, but conditions warmed near the end of the first half of the Paleozoic. During the latter half of the Paleozoic, the landmasses began moving together, with the initial formation of a large northern block called **Laurasia**, which contained parts of what is now North America, along with Greenland, parts of Europe, and Siberia. Eventually, a single supercontinent, called **Pangaea**, was formed, starting in the latter third of the Paleozoic. Glaciations then began to affect Pangaea's climate and the distribution of vertebrate life.

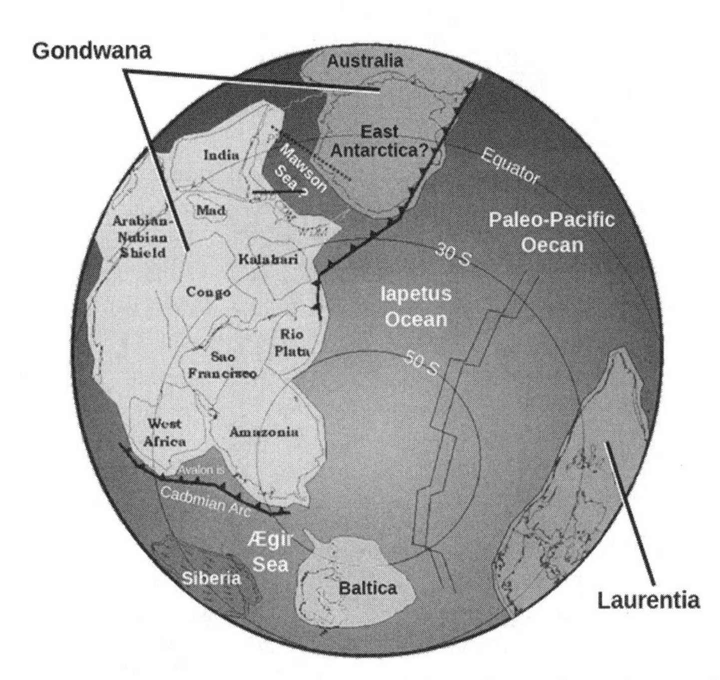

Figure 29.21 Paleozoic continents. During the Paleozoic Era, around 550 million years ago, the continent Gondwana formed. Both Gondwana and the continent Laurentia were located near the equator.

During the early Paleozoic, the amount of carbon dioxide in the atmosphere was much greater than it is today. This may have begun to change later, as land plants became more common. As the roots of land plants began to infiltrate rock and soil began to form, carbon dioxide was drawn out of the atmosphere and became trapped in the rock. This reduced the levels of carbon dioxide and increased the levels of oxygen in the atmosphere, so that by the end of the Paleozoic, atmospheric conditions were similar to those of today.

As plants became more common through the latter half of the Paleozoic, microclimates began to emerge and ecosystems began to change. As plants and ecosystems continued to grow and become more complex, vertebrates moved from the water to land. The presence of shoreline vegetation may have contributed to the movement of vertebrates onto land. One hypothesis suggests that the fins of aquatic vertebrates were used to maneuver through this vegetation, providing a precursor to the movement of fins on land and the further development of limbs. The late Paleozoic was a time of diversification of vertebrates, as amniotes emerged and became two different lines that gave rise, on one hand, to synapsids and mammals, and, on the other hand, to the codonts, reptiles, dinosaurs, and birds. Many marine vertebrates became extinct near the end of the Devonian period, which ended about 360 million years ago, and both marine and terrestrial vertebrates were decimated by a mass extinction in the early Permian period about 250 million years ago.

LINK TO LEARNING

View Earth's Paleogeography: Continental Movements Through Time (http://openstax.org/l/paleogeography) to see changes in Earth as life evolved.

29.4 Reptiles

By the end of this section, you will be able to do the following:

- Describe the main characteristics of amniotes
- Explain the difference between anapsids, synapsids, and diapsids, and give an example of each
- Identify the characteristics of reptiles
- Discuss the evolution of reptiles

The reptiles (including dinosaurs and birds) are distinguished from amphibians by their terrestrially adapted egg, which is supported by four *extraembryonic membranes*: the yolk sac, the amnion, the chorion, and the allantois (Figure 29.22). The chorion and amnion develop from folds in the body wall, and the yolk sac and allantois are extensions of the midgut and hindgut respectively. The amnion forms a fluid-filled cavity that provides the embryo with its own internal aquatic environment. The evolution of the extraembryonic membranes led to less dependence on water for development and thus allowed the amniotes to branch out into drier environments.

In addition to these membranes, the eggs of birds, reptiles, and a few mammals have shells. An **amniote embryo** was then enclosed in the amnion, which was in turn encased in an extra-embryonic coelom contained within the chorion. Between the shell and the chorion was the albumin of the egg, which provided additional fluid and cushioning. This was a significant development that further distinguishes the amniotes from amphibians, which were and continue to be restricted to moist environments due their shell-less eggs. Although the shells of various reptilian amniotic species vary significantly, they all permit the retention of water and nutrients for the developing embryo. The egg shells of bird (avian reptiles) are hardened with calcium carbonate, making them rigid, but fragile. The shells of most nonavian reptile eggs, such as turtles, are leathery and require a moist environment. Most mammals do not lay eggs (except for monotremes such as the echindnas and platypuses). Instead, the embryo grows within the mother's body, with the placenta derived from two of the extraembryonic membranes.

Characteristics of Amniotes

The **amniotic egg** is the key characteristic of amniotes. In amniotes that lay eggs, the shell of the egg provides protection for the developing embryo while being permeable enough to allow for the exchange of carbon dioxide and oxygen. The *albumin*, or egg white, outside of the chorion provides the embryo with water and protein, whereas the fattier egg yolk contained in the yolk sac provides nutrients for the embryo, as is the case with the eggs of many other animals, such as amphibians. Here are the functions of the extraembryonic membranes:

1. Blood vessels in the **yolk sac** transport yolk nutrients to the circulatory system of the embryo.
2. The **chorion** facilitates exchange of oxygen and carbon dioxide between the embryo and the egg's external environment.
3. The **allantois** stores nitrogenous wastes produced by the embryo and also facilitates respiration.
4. The **amnion** protects the embryo from mechanical shock and supports hydration.

In mammals, the yolk sac is very reduced, but the embryo is still cushioned and enclosed within the amnion. The *placenta,* which transports nutrients and functions in gas exchange and waste management, is derived from the chorion and allantois.

VISUAL CONNECTION

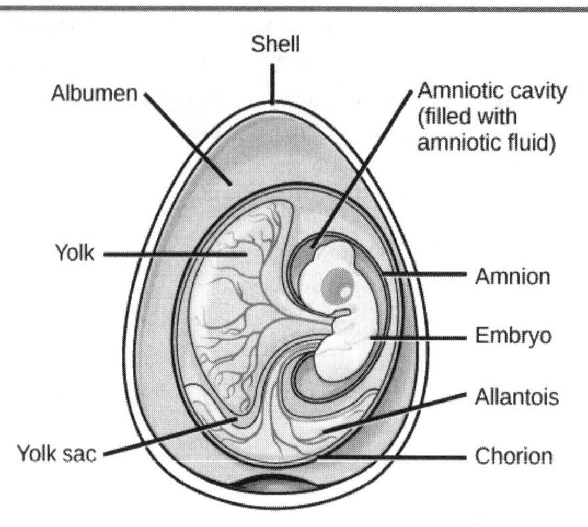

Figure 29.22 An amniotic egg. The key features of an amniotic egg are shown.

Which of the following statements about the parts of an egg are false?

1. The allantois stores nitrogenous waste and facilitates respiration.
2. The chorion facilitates gas exchange.
3. The yolk provides food for the growing embryo.
4. The amniotic cavity is filled with albumen.

Additional derived characteristics of amniotes include a waterproof skin, accessory keratinized structures, and costal (rib) ventilation of the lungs.

Evolution of Amniotes

The first amniotes evolved from tetrapod ancestors approximately 340 million years ago during the Carboniferous period. The early amniotes quickly diverged into two main lines: *synapsids* and *sauropsids*. **Synapsids** included the *therapsids*, a clade from which mammals evolved. **Sauropsids** were further divided into *anapsids* and *diapsids*. Diapsids gave rise to the reptiles, including the dinosaurs and birds. The key differences between the synapsids, anapsids, and diapsids are the structures of the skull and the number of *temporal fenestrae* ("windows") behind each eye (Figure 29.23). **Temporal fenestrae** are post-orbital openings in the skull that allow muscles to expand and lengthen. Anapsids have no temporal fenestrae, synapsids have one (fused ancestrally from two fenestrae), and diapsids have two (although many diapsids such as birds have highly modified diapsid skulls). Anapsids include extinct organisms and traditionally included turtles. However, more recent molecular and fossil evidence clearly shows that turtles arose within the diapsid line and secondarily lost the temporal fenestrae; thus they appear to be anapsids because modern turtles do not have fenestrae in the temporal bones of the skull. The canonical diapsids include dinosaurs, birds, and all other extinct and living reptiles.

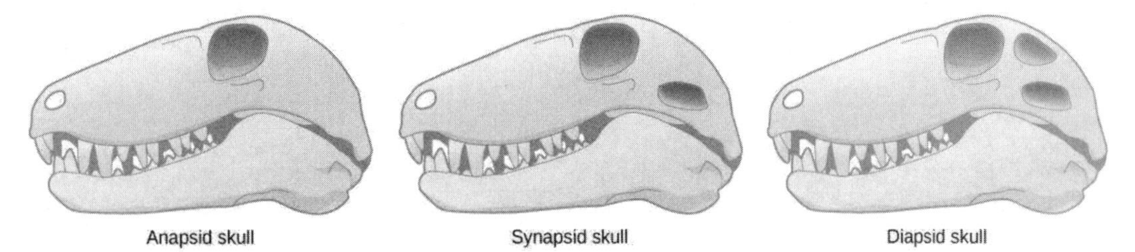

Figure 29.23 Amniote skulls. Compare the skulls and temporal fenestrae of anapsids, synapsids, and diapsids. Anapsids have no openings, synapsids have one opening, and diapsids have two openings.

The diapsids in turn diverged into two groups, the *Archosauromorpha* ("ancient lizard form") and the *Lepidosauromorpha* ("scaly lizard form") during the Mesozoic period (Figure 29.24). The **lepidosaurs** include modern lizards, snakes, and tuataras. The **archosaurs** include modern crocodiles and alligators, and the extinct ichthyosaurs ("fish lizards" superficially resembling dolphins), pterosaurs ("winged lizard"), dinosaurs ("terrible lizard"), and birds. (We should note that clade Dinosauria includes birds, which evolved from a branch of maniraptoran theropod dinosaurs in the Mesozoic.)

The evolutionarily derived characteristics of amniotes include the amniotic egg and its four extraembryonic membranes, a thicker and more waterproof skin, and rib ventilation of the lungs (ventilation is performed by drawing air into and out of the lungs by muscles such as the costal rib muscles and the diaphragm).

VISUAL CONNECTION

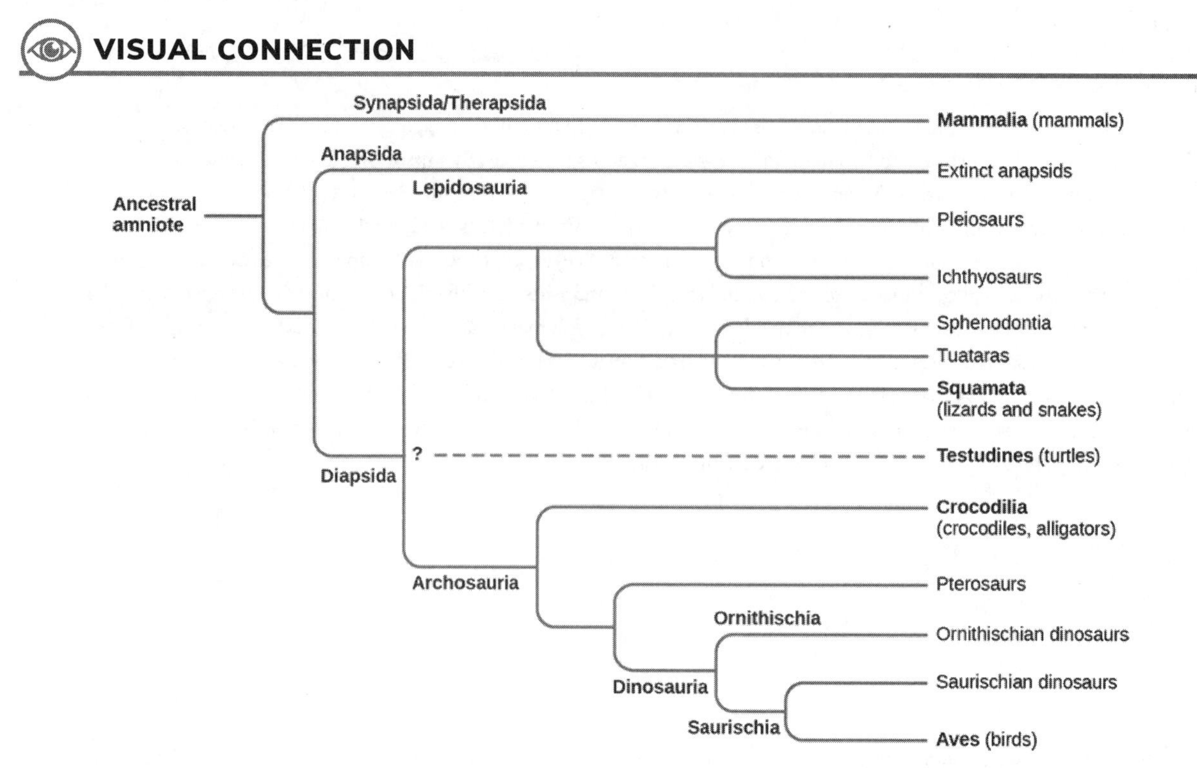

Figure 29.24 Amniote phylogeny. This chart shows the evolution of amniotes. The placement of Testudines (turtles) is currently still debated.

Question: Members of the order Testudines have an anapsid-like skull without obvious temporal fenestrae. However, molecular studies clearly indicate that turtles descended from a diapsid ancestor. Why might this be the case?

In the past, the most common division of amniotes has been into the classes Mammalia, Reptilia, and Aves. However, both birds and mammals are descended from different amniote branches: the synapsids giving rise to the therapsids and mammals, and the diapsids giving rise to the lepidosaurs and archosaurs. We will consider both the birds and the mammals as groups distinct from reptiles for the purpose of this discussion with the understanding that this does not accurately reflect phylogenetic history and relationships.

Characteristics of Reptiles

Reptiles are tetrapods. Limbless reptiles—snakes and legless lizards—are classified as tetrapods because they are descended from four-limbed ancestors. Reptiles lay calcareous or leathery eggs enclosed in shells on land. Even aquatic reptiles return to the land to lay eggs. They usually reproduce sexually with internal fertilization. Some species display ovoviviparity, with the eggs remaining in the mother's body until they are ready to hatch. In ovoviviparous reptiles, most nutrients are supplied by the egg yolk, while the chorioallantois assists with respiration. Other species are viviparous, with the offspring born alive, with their development supported by a yolk sac-placenta, a chorioallantoic-placenta, or both.

One of the key adaptations that permitted reptiles to live on land was the development of their *scaly skin*, containing the protein keratin and waxy lipids, which reduced water loss from the skin. A number of keratinous epidermal structures have emerged in the descendants of various reptilian lineages and some have become defining characters for these lineages: scales, claws, nails, horns, feathers, and hair. Their occlusive skin means that reptiles cannot use their skin for respiration, like amphibians, and thus all amniotes breathe with lungs. All reptiles grow throughout their lives and regularly shed their skin, both to accommodate their growth and to rid themselves of ectoparasites. Snakes tend to shed the entire skin at one time, but other reptiles shed their skins in patches.

Reptiles ventilate their lungs using various muscular mechanisms to produce *negative pressure* (low pressure) within the lungs that allows them to expand and draw in air. In snakes and lizards, the muscles of the spine and ribs are used to expand or contract the rib cage. Since walking or running interferes with this activity, the squamates cannot breathe effectively while running. Some squamates can supplement rib movement with buccal pumping through the nose, with the mouth closed. In crocodilians, the lung chamber is expanded and contracted by moving the liver, which is attached to the pelvis. Turtles have a special problem with breathing, because their rib cage cannot expand. However, they can change the pressure around the lungs by pulling their limbs in and out of the shell, and by moving their internal organs. Some turtles also have a posterior respiratory sac that opens off the hindgut that aids in the diffusion of gases.

Most reptiles are **ectotherms**, animals whose main source of body heat comes from the environment; however, some crocodilians maintain elevated *thoracic* temperatures and thus appear to be at least regional **endotherms**. This is in contrast to true endotherms, which use heat produced by metabolism and muscle contraction to regulate body temperature over a very narrow temperature range, and thus are properly referred to as **homeotherms**. Reptiles have behavioral adaptations to help regulate body temperature, such as basking in sunny places to warm up through the absorption of solar radiation, or finding shady spots or going underground to minimize the absorption of solar radiation, which allows them to cool down and prevent overheating. The advantage of ectothermy is that metabolic energy from food is not required to heat the body; therefore, reptiles can survive on about 10 percent of the calories required by a similarly sized endotherm. In cold weather, some reptiles such as the garter snake *brumate*. **Brumation** is similar to hibernation in that the animal becomes less active and can go for long periods without eating, but differs from hibernation in that brumating reptiles are not asleep or living off fat reserves. Rather, their metabolism is slowed in response to cold temperatures, and the animal is very sluggish.

Evolution of Reptiles

Reptiles originated approximately 300 million years ago during the Carboniferous period. One of the oldest known amniotes is *Casineria*, which had both amphibian and reptilian characteristics. One of the earliest undisputed reptile fossils was *Hylonomus*, a lizardlike animal about 20 cm long. Soon after the first amniotes appeared, they diverged into three groups—synapsids, anapsids, and diapsids—during the Permian period. The Permian period also saw a second major divergence of diapsid reptiles into stem archosaurs (predecessors of thecodonts, crocodilians, dinosaurs, and birds) and lepidosaurs (predecessors of snakes and lizards). These groups remained inconspicuous until the Triassic period, when the archosaurs became the dominant terrestrial group possibly due to the extinction of large-bodied anapsids and synapsids during the Permian-Triassic extinction. About 250 million years ago, archosaurs radiated into the pterosaurs and both saurischian "lizard hip" and ornithischian "bird-hip" dinosaurs (see below).

Although they are sometimes mistakenly called dinosaurs, the **pterosaurs** were distinct from true dinosaurs (Figure 29.25). Pterosaurs had a number of adaptations that allowed for flight, including hollow bones (birds also exhibit hollow bones, a case of convergent evolution). Their wings were formed by membranes of skin that attached to the long, fourth finger of each arm and extended along the body to the legs.

Figure 29.25 Pterosaurs. Pterosaurs, such as this *Quetzalcoatlus*, which existed from the late Triassic to the Cretaceous period (230 to 65.5 million years ago), possessed wings but are not believed to have been capable of powered flight. Instead, they may have been able to soar after launching from cliffs. (credit: Mark Witton, Darren Naish)

Archosaurs: Dinosaurs

Dinosaurs ("fearfully-great lizard") include the Saurischia ("lizard-hipped") with a simple, three-pronged pelvis, and

Ornithischia ("bird-hipped") dinosaurs with a more complex pelvis, superficially similar to that of birds. However, it is a fact that birds evolved from the saurischian "lizard hipped" lineage, *not* the ornithischian "bird hip" lineage. Dinosaurs and their theropod descendants, the birds, are remnants of what was formerly a hugely diverse group of reptiles, some of which like *Argentinosaurus* were nearly 40 meters (130 feet) in length and weighed at least 80,000 kg (88 tons). They were the largest land animals to have lived, challenging and perhaps exceeding the great blue whale in size, but probably not weight—which could be greater than 200 tons.

Herrerasaurus, a bipedal dinosaur from Argentina, was one of the earliest dinosaurs that walked upright with the legs positioned directly below the pelvis, rather than splayed outward to the sides as in the crocodilians. The Ornithischia were all herbivores, and sometimes evolved into crazy shapes, such as ankylosaur "armored tanks" and horned dinosaurs such as *Triceratops*. Some, such as *Parasaurolophus*, lived in great herds and may have amplified their species-specific calls through elaborate crests on their heads.

Both the ornithischian and saurischian dinosaurs provided parental care for their broods, just as crocodilians and birds do today. The end of the age of dinosaurs came about 65 million years ago, during the Mesozoic, coinciding with the impact of a large asteroid (that produced the Chicxulub crater) in what is now the Yucatan Peninsula of Mexico. Besides the immediate environmental disasters associated with this asteroid impacting the Earth at about 45,000 miles per hour, the impact may also have helped generate an enormous series of volcanic eruptions that changed the distribution and abundance of plant life worldwide, as well as its climate. At the end of the Triassic, massive volcanic activity across North America, South America, Africa, and southwest Europe ultimately would lead to the break-up of Pangea and the opening of the Atlantic Ocean. The formerly incredibly diverse dinosaurs (save for the evolving birds) met their extinction during this time period.

Archosaurs: Pterosaurs

More than 200 species of pterosaurs have been described, and in their day, beginning about 230 million years ago, they were the undisputed rulers of the Mesozoic skies for over 170 million years. Recent fossils suggest that hundreds of pterosaur species may have lived during any given period, dividing up the environment much like birds do today. Pterosaurs came in amazing sizes and shapes, ranging in size from that of a small song bird to that of the enormous *Quetzalcoatlus northropi*, which stood nearly 6 meters (19 feet) high and had a wingspan of nearly 14 meters (40 feet). This monstrous pterosaur, named after the Aztec god Quetzalcoatl, the feathered flying serpent that contributed largely to the creation of humankind, may have been the largest flying animal that ever evolved!

Some male pterosaurs apparently had brightly colored crests that may have served in sexual displays; some of these crests were much higher than the actual head! Pterosaurs had ultralight skeletons, with a *pteroid bone*, unique to pterosaurs, that strengthened the forewing membrane. Much of their wing span was exaggerated by a greatly elongated fourth finger that supported perhaps half of the wing. It is tempting to relate to them in terms of bird characteristics, but in reality, their proportions were decidedly not like birds at all. For example, it is common to find specimens, such as *Quetzalcoatlus*, with a head and neck region that together was three to four *times* as large as the torso. In addition, unlike the feathered bird wing, the reptilian wing had a layer of muscles, connective tissue, and blood vessels, all reinforced with a webbing of fibrous cords.

In contrast to the aerial pterosaurs, the dinosaurs were a diverse group of terrestrial reptiles with more than 1,000 species classified to date. Paleontologists continue to discover new species of dinosaurs. Some dinosaurs were quadrupeds (Figure 29.26); others were bipeds. Some were carnivorous, whereas others were herbivorous. Dinosaurs laid eggs, and a number of nests containing fossilized eggs, with intact embryos, have been found. It is not known with certainty whether dinosaurs were homeotherms or facultative endotherms. However, given that modern birds are endothermic, the dinosaurs that were the immediate ancestors to birds likely were endothermic as well. Some fossil evidence exists for dinosaurian parental care, and comparative biology supports this hypothesis since the archosaur birds and crocodilians both display extensive parental care.

(a) (b)

Figure 29.26 Ornithischian and saurischian Dinosaurs. *Edmontonia* was an armored dinosaur that lived in the Late Cretaceous period, 145.5 to 65.6 million years ago. Herrerrasaurus and Eoraptor (b) were late Triassic saurischian dinosaurs dating to about 230 million years ago. (credit: a Mariana Ruiz Villareal b Zach Tirrell from Plymouth, USA, Dino Origins)

Dinosaurs dominated the Mesozoic era, which was known as the "Age of Reptiles." The dominance of dinosaurs lasted until the end of the Cretaceous, the last period of the Mesozoic era. The Cretaceous-Tertiary extinction resulted in the loss of most of the large-bodied animals of the Mesozoic era. Birds are the only living descendants of one of the major clades of theropod dinosaurs.

LINK TO LEARNING

Visit this site to see a video (http://openstax.org/l/K-T_extinction) discussing the hypothesis that an asteroid caused the Cretaceous-Triassic (KT) extinction.

Modern Reptiles

Class Reptilia includes many diverse species that are classified into four living clades. There are the 25 species of Crocodilia, two species of Sphenodontia, approximately 9,200 Squamata species, and about 325 species of Testudines.

Crocodilia

Crocodilia ("small lizard") arose as a distinct lineage by the middle Triassic; extant species include alligators, crocodiles, gharials, and caimans. Crocodilians (Figure 29.27) live throughout the tropics and subtropics of Africa, South America, Southern Florida, Asia, and Australia. They are found in freshwater, saltwater, and brackish habitats, such as rivers and lakes, and spend most of their time in water. Crocodiles are descended from terrestrial reptiles and can still walk and run well on land. They often move on their bellies in a swimming motion, propelled by alternate movements of their legs. However, some species can lift their bodies off the ground, pulling their legs in under the body with their feet rotated to face forward. This mode of locomotion takes a lot of energy, and seems to be used primarily to clear ground obstacles. Amazingly, some crocodiles can also gallop, pushing off with their hind legs and moving their hind and forelegs alternately in pairs. Galloping crocodiles have been clocked at speeds over 17 kph and, over short distances, in an ambush situation, they can easily chase down most humans if they are taken by surprise. However, they are short distance runners, not interested in a long chase, and most fit humans can probably outrun them in a sprint (assuming they respond quickly to the ambush!).

Figure 29.27 A crocodilian. Crocodilians, such as this Siamese crocodile (*Crocodylus siamensis*), provide parental care for their offspring. (credit: Keshav Mukund Kandhadai)

Sphenodontia

Sphenodontia ("wedge tooth") arose in the early Mesozoic era, when they had a moderate radiation, but now are represented by only two living species: *Sphenodon punctatus* and *Sphenodon guntheri,* both found on offshore islands in New Zealand (Figure 29.28). The common name "tuatara" comes from a Maori word describing the crest along its back. Tuataras have a primitive diapsid skull with biconcave vertebrae. They measure up to 80 centimeters and weigh about 1 kilogram. Although superficially similar to an iguanid lizard, several unique features of the skull and jaws clearly define them and distinguish this group from the Squamata. They have no external ears. Tuataras briefly have a third (parietal) eye—with a lens, retina, and cornea—in the middle of the forehead. The eye is visible only in very young animals; it is soon covered with skin. Parietal eyes can sense light, but have limited color discrimination. Similar light-sensing structures are also seen in some other lizards. In their jaws, tuataras have two rows of teeth in the upper jaw that bracket a single row of teeth in the lower jaw. These teeth are actually projections from the jawbones, and are not replaced as they wear down.

Figure 29.28 A tuatara. This tuatara from New Zealand may resemble a lizard but belongs to a distinct lineage, the Sphenodontidae family. (credit: Sid Mosdell)

Squamata

The **Squamata** ("scaly or having scales") arose in the late Permian, and extant species include lizards and snakes. Both are found on all continents except Antarctica. Lizards and snakes are most closely related to tuataras, both groups having evolved from a lepidosaurian ancestor. Squamata is the largest extant clade of reptiles (Figure 29.29).

Figure 29.29 A chameleon. This Jackson's chameleon (*Trioceros jacksonii*) blends in with its surroundings.

Most lizards differ from snakes by having four limbs, although these have often been lost or significantly reduced in at least 60 lineages. Snakes lack eyelids and external ears, which are both present in lizards. There are about 6,000 species of lizards, ranging in size from tiny chameleons and geckos, some of which are only a few centimeters in length, to the Komodo dragon, which is about 3 meters in length.

Some lizards are extravagantly decorated with spines, crests, and frills, and many are brightly colored. Some lizards, like chameleons (Figure 29.29), can change their skin color by redistributing pigment within *chromatophores* in their skins. Chameleons change color both for camouflage and for social signaling. Lizards have multiple-colored oil droplets in their retinal cells that give them a good range of color vision. Lizards, unlike snakes, can focus their eyes by changing the shape of the lens. The eyes of chameleons can move independently. Several species of lizards have a "hidden" *parietal eye*, similar to that in the tuatara. Both lizards and snakes use their tongues to sample the environment and a pit in the roof of the mouth, Jacobson's organ, is used to evaluate the collected sample.

Most lizards are carnivorous, but some large species, such as iguanas, are herbivores. Some predatory lizards are ambush predators, waiting quietly until their prey is close enough for a quick grab. Others are patient foragers, moving slowly through their environment to detect possible prey. Lizard tongues are long and sticky and can be extended at high speed for capturing insects or other small prey. Traditionally, the only venomous lizards are the Gila monster and the beaded lizard. However, venom glands have also been identified in several species of monitors and iguanids, but the venom is not injected directly and should probably be regarded as a toxin delivered with the bite.

Specialized features of the jaw are related to adaptations for feeding that have evolved to feed on relatively large prey (even though some current species have reversed this trend). Snakes are thought to have descended from either burrowing or aquatic lizards over 100 million years ago (Figure 29.30). They include about 3,600 species, ranging in size from 10 centimeter-long thread snakes to 10 meter-long pythons and anacondas. All snakes are legless, except for boids (e.g., boa constrictors), which have vestigial hindlimbs in the form of *pelvic spurs*. Like caecilian amphibians, the narrow bodies of most snakes have only a single functional lung. All snakes are carnivorous and eat small animals, birds, eggs, fish, and insects.

Figure 29.30 A nonvenomous snake. The garter snake belongs to the genus *Thamnophis*, the most widely distributed reptile genus in North America. (credit: Steve Jurvetson)

Most snakes have a skull that is very flexible, involving eight rotational joints. They also differ from other squamates by having mandibles (lower jaws) without either bony or ligamentous attachment anteriorly. Having this connection via skin and muscle allows for great dynamic expansion of the gape and independent motion of the two sides—both advantages in swallowing big prey. Most snakes are nonvenomous and simply swallow their prey alive, or subdue it by constriction before swallowing it. Venomous snakes use their venom both to kill or immobilize their prey, and to help digest it.

Although snakes have no eyelids, their eyes are protected with a transparent scale. Their retinas have both rods and cones, and like many animals, they do not have receptor pigments for red light. Some species, however, can see in the ultraviolet, which allows them to track ultraviolet signals in rodent trails. Snakes adjust focus by moving their heads. They have lost both external and middle ears, although their inner ears are sensitive to ground vibrations. Snakes have a number of sensory structures that assist in tracking prey. In pit vipers, like rattlesnakes, a sensory pit between the eye and nostrils is sensitive to infrared ("heat") emissions from warm-blooded prey. A row of similar pits is located on the upper lip of boids. As noted above, snakes also use **Jacobson's organ** for detecting olfactory signals.

Testudines

The turtles, terrapins, and tortoises are members of the clade **Testudines** ("having a shell") (Figure 29.31), and are characterized by a bony or cartilaginous shell. The shell in turtles is not just an epidermal covering, but is incorporated *into* the skeletal system. The dorsal shell is called the **carapace** and includes the backbone and ribs; the ventral shell is called the **plastron**. Both shells are covered with keratinous plates or **scutes**, and the two shells are held together by a bridge. In some turtles, the plastron is hinged to allow the head and legs to be withdrawn under the shell.

The two living groups of turtles, Pleurodira and Cryptodira, have significant anatomical differences and are most easily recognized by how they retract their necks. The more common Cryptodira retract their neck in a vertical S-curve; they appear to simply pull their head backward when retracting. Less common Pleurodira ("side-neck") retract their neck with a horizontal curve, basically folding their neck to the side.

The Testudines arose approximately 200 million years ago, predating crocodiles, lizards, and snakes. There are about 325 living species of turtles and tortoises. Like other reptiles, turtles are ectotherms. All turtles are oviparous, laying their eggs on land, although many species live in or near water. None exhibit parental care. Turtles range in size from the speckled padloper tortoise at 8 centimeters (3.1 inches) to the leatherback sea turtle at 200 centimeters (over 6 feet). The term "turtle" is sometimes used to describe only those species of Testudines that live in the sea, with the terms "tortoise" and "terrapin" used to refer to species that live on land and in fresh water, respectively.

Figure 29.31 A tortoise. The African spurred tortoise (*Geochelone sulcata*) lives at the southern edge of the Sahara Desert. It is the third largest tortoise in the world. (credit: Jim Bowen)

29.5 Birds

By the end of this section, you will be able to do the following:

- Describe the evolutionary history of birds
- Describe the derived characteristics in birds that facilitate flight

With over 10,000 identified species, the birds are the most speciose of the land vertebrate classes. Abundant research has shown that birds are really an extant clade that evolved from maniraptoran theropod dinosaurs about 150 million years ago. Thus, even though the most obvious characteristic that *seems* to set birds apart from other extant vertebrates is the presence of feathers, we now know that feathers probably appeared in the common ancestor of both ornithischian and saurischian lineages of dinosaurs. Feathers in these clades are also homologous to reptilian scales and mammalian hair, according to the most recent research. While the wings of vertebrates like bats function without feathers, birds rely on feathers, and wings, along with other modifications of body structure and physiology, for flight, as we shall see.

Characteristics of Birds

Birds are endothermic, and more specifically, **homeothermic**—meaning that they usually maintain an elevated and constant body temperature, which is significantly above the average body temperature of most mammals. This is, in part, due to the fact that active flight—especially the hovering skills of birds such as hummingbirds—requires enormous amounts of energy, which in turn necessitates a high metabolic rate. Like mammals (which are also endothermic and homeothermic and covered with an insulating pelage), birds have several different types of feathers that together keep "heat" (infrared energy) within the core of the body, away from the surface where it can be lost by radiation and convection to the environment.

Modern birds produce two main types of feathers: *contour feathers* and *down feathers*. **Contour feathers** have a number of parallel *barbs* that branch from a *central shaft*. The barbs in turn have microscopic branches called *barbules* that are linked together by minute hooks, making the vane of a feather a strong, flexible, and uninterrupted surface. In contrast, the barbules of **down feathers** do not interlock, making these feathers especially good for insulation, trapping air in spaces between the loose, interlocking barbules of adjacent feathers to decrease the rate of heat loss by convection and radiation. Certain parts of a bird's body are covered in down feathers, and the base of other feathers has a downy portion, whereas newly hatched birds are covered almost entirely in down, which serves as an excellent coat of insulation, increasing the *thermal boundary layer* between the skin and the outside environment.

Feathers not only provide insulation, but also allow for flight, producing the lift and thrust necessary for flying birds to become and stay airborne. The feathers on a wing are *flexible,* so the feathers at the end of the wing separate as air moves over them, reducing the drag on the wing. Flight feathers are also asymmetrical and curved, so that air flowing over them generates lift. Two types of flight feathers are found on the wings, *primary feathers* and *secondary feathers* (Figure 29.32). **Primary feathers** are located at the tip of the wing and provide thrust as the bird moves its wings downward, using the pectoralis major muscles.

Secondary feathers are located closer to the body, in the forearm portion of the wing, and provide lift. In contrast to primary and secondary feathers, contour feathers are found on the body, where they help reduce form drag produced by wind resistance against the body during flight. They create a smooth, aerodynamic surface so that air moves swiftly over the bird's body, preventing turbulence and creating ideal aerodynamic conditions for efficient flight.

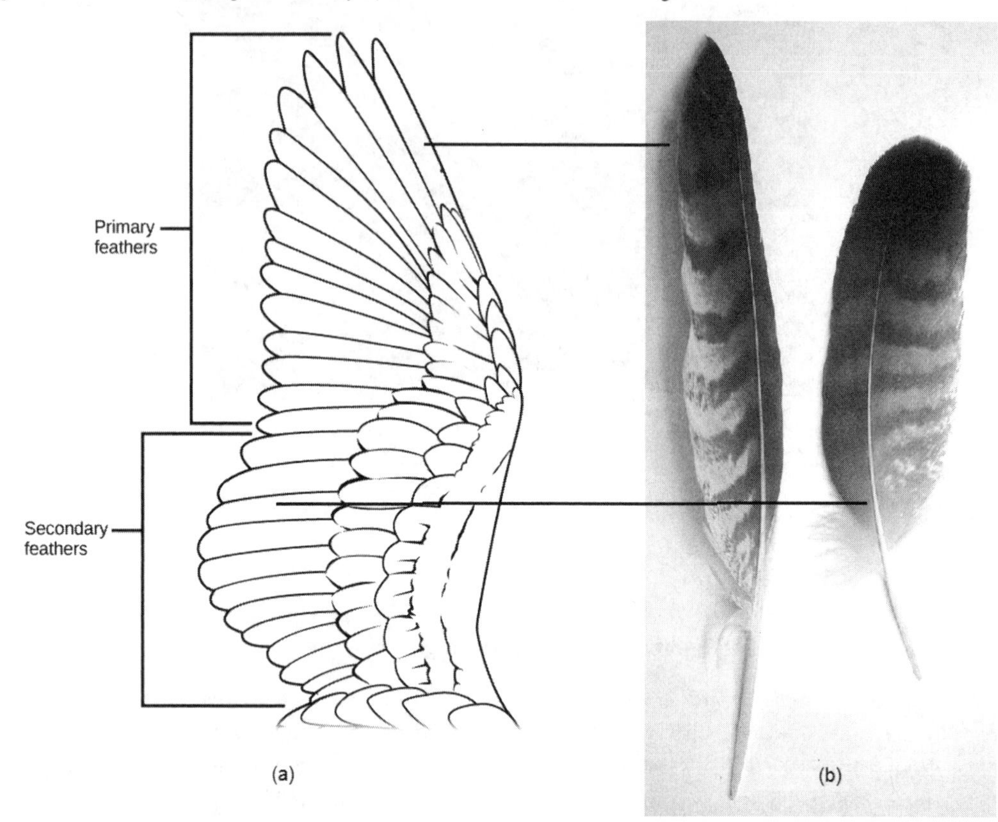

Figure 29.32 Flight feathers. (a) Primary feathers are located at the wing tip and provide thrust; secondary feathers are located close to the body and provide lift. (b) Primary and secondary feathers from a common buzzard (*Buteo buteo*). (Credit b: Mod. from S. Seyfert https://commons.wikimedia.org/w/index.php?curid=613813 (http://openstax.org/l/buzzard_feathers))

Flapping of the entire wing occurs primarily through the actions of the chest muscles: Specifically, the contraction of the pectoralis major muscles moves the wings downward (downstroke), whereas contraction of the supracoracoideus muscles moves the wings upward (upstroke) via a tough tendon that passes over the coracoid bone and the top of the humerus. Both muscles are attached to the keel of the sternum, and these are the muscles that humans eat on holidays (this is why the back of the bird offers little meat!). These muscles are highly developed in birds and account for a higher percentage of body mass than in most mammals. The flight muscles attach to a blade-shaped keel projecting ventrally from the sternum, like the keel of a boat. The sternum of birds is deeper than that of other vertebrates, which accommodates the large flight muscles. The flight muscles of birds who are active flyers are rich with oxygen-storing *myoglobin*. Another skeletal modification found in most birds is the fusion of the two clavicles (collarbones), forming the **furcula** or wishbone. The furcula is flexible enough to bend and provide support to the shoulder girdle during flapping.

An important requirement for flight is a low body weight. As body weight increases, the muscle output required for flying increases. The largest living bird is the ostrich, and while it is much smaller than the largest mammals, it is secondarily flightless. For birds that do fly, reduction in body weight makes flight easier. Several modifications are found in birds to reduce body weight, including *pneumatization of bones*. **Pneumatic bones** (Figure 29.33) are bones that are hollow, rather than filled with tissue; *cross struts* of bone called *trabeculae* provide structural reinforcement. Pneumatic bones are not found in all birds, and they are more extensive in large birds than in small birds. Not all bones of the skeleton are pneumatic, although the skulls of almost all birds are. The jaw is also lightened by the replacement of heavy jawbones and teeth with a beak made of keratin (just as hair, scales, and feathers are).

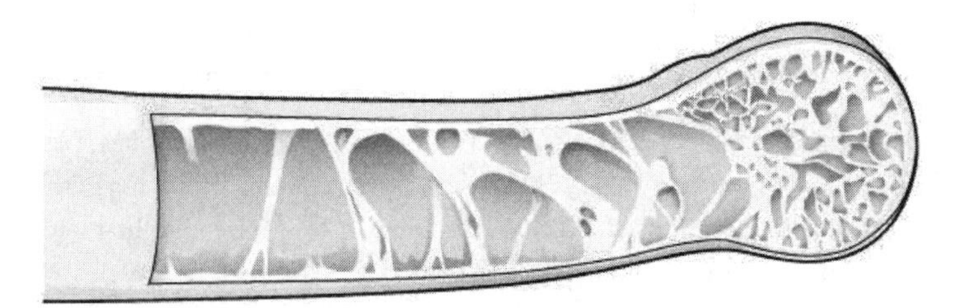

Figure 29.33 Pneumatic bone. Many birds have hollow, pneumatic bones, which make flight easier.

Other modifications that reduce weight include the lack of a urinary bladder. Birds possess a **cloaca**, an external body cavity into which the intestinal, urinary, and genital orifices empty in reptiles, birds, and the monotreme mammals. The cloaca allows water to be reabsorbed from waste back into the bloodstream. Thus, uric acid is *not* eliminated as a liquid but is concentrated into **urate salts**, which are expelled along with fecal matter. In this way, water is not held in a urinary bladder, which would increase body weight. In addition, the females of most bird species only possess one functional (left) ovary rather than two, further reducing body mass.

The respiratory system of birds is dramatically different from that of reptiles and mammals, and is well adapted for the high metabolic rate required for flight. To begin, the air spaces of pneumatic bone are sometimes connected to **air sacs** in the body cavity, which replace coelomic fluid and also lighten the body. These air sacs are also connected to the path of airflow through the bird's body, and function in respiration. Unlike mammalian lungs in which air flows in two directions, as it is breathed in and out, diluting the concentration of oxygen, airflow through bird lungs is unidirectional (Figure 29.34). Gas exchange occurs in "air capillaries" or microscopic air passages within the lungs. The arrangement of air capillaries in the lungs creates a counter-current exchange system with the pulmonary blood. In a counter-current system, the air flows in one direction and the blood flows in the opposite direction, producing a favorable diffusion gradient and creating an efficient means of gas exchange. This very effective oxygen-delivery system of birds supports their higher metabolic activity. In effect, ventilation is provided by the **parabronchi** (minimally expandible lungs) with thin air sacs located among the visceral organs and the skeleton. A **syrinx** (voice box) resides near the junction of the trachea and bronchi. The syrinx, however, is *not* homologous to the mammalian larynx, which resides within the upper part of the trachea.

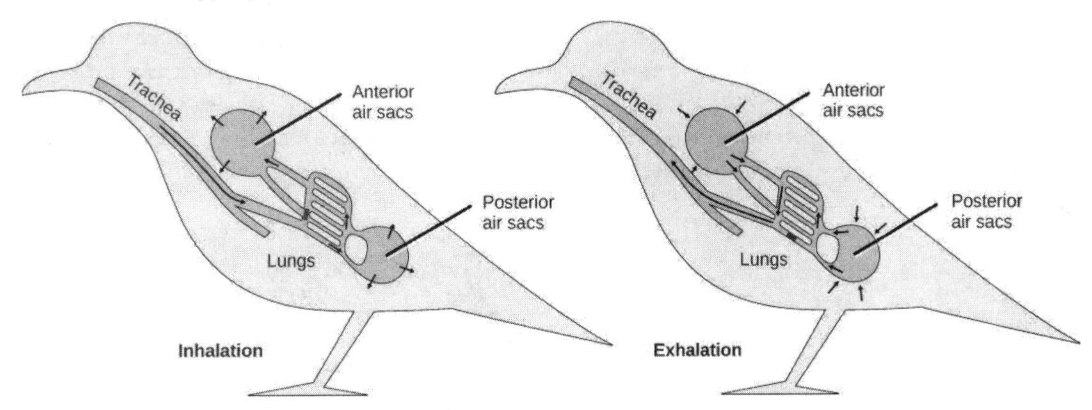

Figure 29.34 Air flow in bird lungs. Avian respiration is an efficient system of gas exchange with air flowing *unidirectionally*. A full ventilation cycle takes two breathing cycles. During the first inhalation, air passes from the trachea into posterior air sacs, then during the first exhalation into the lungs. The second inhalation moves the air in the lungs to the anterior air sacs, and the second exhalation moves the air in the anterior air sacs out of the body. Overall, each inhalation moves air into the air sacs, while each exhalation moves fresh air through the lungs and "used" air out of the body. The air sacs are connected to the hollow interior of bones. (credit: modification of work by L. Shyamal)

Beyond the unique characteristics discussed above, birds are also unusual vertebrates because of a number of other features. First, they typically have an elongate (very "dinosaurian") S-shaped neck, but a short tail or *pygostyle*, produced from the fusion of the caudal vertebrae. Unlike mammals, birds have only one occipital condyle, allowing them extensive movement of the head and neck. They also have a very thin epidermis without sweat glands, and a specialized **uropygial gland** or sebaceous "preening gland" found at the dorsal base of the tail. This gland is an essential to **preening** (a virtually continuous activity) in most birds because it produces an oily substance that birds use to help waterproof their feathers as well as keep them flexible for flight. A number of birds, such as pigeons, parrots, hawks, and owls, lack a uropygial gland but have specialized feathers that "disintegrate" into a powdery down, which serves the same purpose as the oils of the uropygial gland.

Like mammals, birds have 12 pairs of cranial nerves, and a very large cerebellum and optic lobes, but only a single bone in the middle ear called the **columella** (the stapes in mammals). They have a closed circulatory system with two atria and two ventricles, but rather than a "left-bending" aortic arch like that of mammals, they have a "right-bending" aortic arch, and nucleated red blood cells (unlike the enucleated red blood cells of mammals).

All these unique and highly derived characteristics make birds one of the most conspicuous and successful groups of vertebrate animals, filling a range of ecological niches, and ranging in size from the tiny bee hummingbird of Cuba (about 2 grams) to the ostrich (about 140,000 grams). Their large brains, keen senses, and the abilities of many species to imitate vocalization and use tools make them some of the most intelligent vertebrates on Earth.

Evolution of Birds

Thanks to amazing new fossil discoveries in China, the evolutionary history of birds has become clearer, even though bird bones do not fossilize as well as those of other vertebrates. As we've seen earlier, birds are highly modified diapsids, but rather than having two fenestrations or openings in their skulls behind the eye, the skulls of modern birds are so specialized that it is difficult to see any trace of the original diapsid condition.

Birds belong to a group of diapsids called the **archosaurs**, which includes three other groups: living crocodilians, pterosaurs, and dinosaurs. Overwhelming evidence shows that birds evolved within the clade Dinosauria, which is further subdivided into two groups, the Saurischia ("lizard hips") and the Ornithischia ("bird hips"). Despite the names of these groups, it was not the bird-hipped dinosaurs that gave rise to modern birds. Rather, Saurischia diverged into two groups: One included the long-necked herbivorous dinosaurs, such as *Apatosaurus*. The second group, bipedal predators called theropods, gave rise to birds. This course of evolution is highlighted by numerous similarities between late (maniraptoran) theropod fossils and birds, specifically in the structure of the hip and wrist bones, as well as the presence of the wishbone, formed by the fusion of the clavicles.

The clade **Neornithes** includes the avian crown group, which comprises all living birds and the descendants from their most recent common maniraptoran ancestor. One well-known and important fossil of an animal that appears "intermediate" between dinosaurs and birds is *Archaeopteryx* (Figure 29.35), which is from the Jurassic period (200 to 145 MYA). *Archaeopteryx* has characteristics of both maniraptoran dinosaurs and modern birds. Some scientists propose classifying it as a bird, but others prefer to classify it as a dinosaur. Traits in skeletons of *Archaeopteryx* like those of a dinosaur included a jaw with teeth and a long bony tail. Like birds, it had feathers modified for flight, both on the forelimbs and on the tail, a trait associated only with birds among modern animals. Fossils of older feathered dinosaurs exist, but the feathers may not have had the characteristics of modern flight feathers.

Figure 29.35 *Archaeopteryx*. (a) *Archaeopteryx* lived in the late Jurassic period around 150 million years ago. It had cuplike thecodont teeth like a dinosaur, but had (b) flight feathers like modern birds, which can be seen in this fossil. Note the claws on the wings, which are still found in a number of birds, such as the newborn chicks of the South American Hoatzin.

The Evolution of Flight in Birds

There are two basic hypotheses that explain how flight may have evolved in birds: the *arboreal ("tree") hypothesis* and the *terrestrial ("land") hypothesis*. The **arboreal hypothesis** posits that tree-dwelling precursors to modern birds jumped from branch to branch using their feathers for gliding before becoming fully capable of flapping flight. In contrast to this, the **terrestrial hypothesis** holds that running (perhaps pursuing active prey such as small cursorial animals) was the stimulus for flight. In this scenario, wings could be used to capture prey and were preadapted for balance and flapping flight. Ostriches, which are large flightless birds, hold their wings out when they run, possibly for balance. However, this condition may represent a behavioral relict of the clade of flying birds that were their ancestors. It seems more likely that small feathered arboreal dinosaurs, were capable of gliding (and flapping) from tree to tree and branch to branch, improving the chances of escaping enemies, finding mates, and obtaining prey such as flying insects. This early flight behavior would have also greatly increased the opportunity for species dispersal.

Although we have a good understanding of how feathers and flight may have evolved, the question of how endothermy evolved in birds (and other lineages) remains unanswered. Feathers provide insulation, but this is only beneficial for *thermoregulatory purposes* if body heat is being produced internally. Similarly, internal heat production is only viable for the evolution of endothermy if *insulation is present* to retain that infrared energy. It has been suggested that one or the other—feathers or endothermy—evolved first in response to some other selective pressure (e.g., the ability to be active at night, provide camouflage, repel water, or serve as signals for mate selection). It seems probable that feathers and endothermy coevolved together, the improvement and evolutionary advancement of feathers reinforcing the evolutionary advancement of endothermy, and so on.

During the Cretaceous period (145 to 66 MYA), a group known as the Enantiornithes was the dominant bird type (Figure 29.36). Enantiornithes means "opposite birds," which refers to the fact that certain bones of the shoulder are joined differently than the way the bones are joined in modern birds. Like *Archaeopteryx*, these birds retained teeth in their jaws, but did have a shortened tail, and at least some fossils have preserved "fans" of tail feathers. These birds formed an evolutionary lineage separate from that of modern birds, and they did not survive past the Cretaceous. Along with the Enantiornithes, however, another group of birds—the Ornithurae ("bird tails"), with a short, fused tail or **pygostyle**—emerged from the evolutionary line that includes modern birds. This clade was also present in the Cretaceous.

After the extinction of Enantiornithes, the **Ornithurae** became the dominant birds, with a large and rapid radiation occurring after the extinction of the dinosaurs during the Cenozoic era (66 MYA to the present). Molecular analysis based on very large data sets has produced our current understanding of the relationships among living birds. There are three major clades: the Paleognathae, the Galloanserae, and the Neoaves. The Paleognathae ("old jaw") or ratites (polyphyletic) are a group of flightless birds including ostriches, emus, rheas, and kiwis. The Galloanserae include pheasants, ducks, geese and swans. The Neoaves ("new birds") includes all other birds. The Neoaves themselves have been distributed among five clades:[3] Strisores (nightjars, swifts, and hummingbirds), Columbaves (turacos, bustards, cuckoos, pigeons, and doves), Gruiformes (cranes), Aequorlitornithes (diving birds, wading birds, and shorebirds), and Inopinaves (a very large clade of land birds including hawks, owls, woodpeckers, parrots, falcons, crows, and songbirds). Despite the current classification scheme, it is important to understand that phylogenetic revisions, even for the extant birds, are still taking place.

3Prum, RO et al. 2015. A comprehensive phylogeny of birds (Aves) using targeted next-generation DNA sequencing. Nature 526: 569 - 573. http://dx.doi.org/10.1038/nature15697 (http://openstax.org/l/bird_phylogeny)

Figure 29.36 Enantiornithean bird. *Shanweiniao cooperorum* was a species of Enantiornithes that did not survive past the Cretaceous period. (credit: Nobu Tamura)

CAREER CONNECTION

Veterinarian

Veterinarians are concerned with diseases, disorders, and injuries in animals, primarily vertebrates. They treat pets, livestock, and animals in zoos and laboratories. Veterinarians often treat dogs and cats, but also take care of birds, reptiles, rabbits, and other animals that are kept as pets. Veterinarians that work with farms and ranches care for pigs, goats, cows, sheep, and horses.

Veterinarians are required to complete a degree in veterinary medicine, which includes taking courses in comparative zoology, animal anatomy and physiology, microbiology, and pathology, among many other courses in chemistry, physics, and mathematics.

Veterinarians are also trained to perform surgery on many different vertebrate species, which requires an understanding of the vastly different anatomies of various species. For example, the stomach of ruminants like cows has four "compartments" versus one compartment for non-ruminants. As we have seen, birds also have unique anatomical adaptations that allow for flight, which requires additional training and care.

Some veterinarians conduct research in academic settings, broadening our knowledge of animals and medical science. One area of research involves understanding the transmission of animal diseases to humans, called **zoonotic diseases**. For example, one area of great concern is the transmission of the avian flu virus to humans. One type of avian flu virus, H5N1, is a highly pathogenic strain that has been spreading in birds in Asia, Europe, Africa, and the Middle East. Although the virus does not cross over easily to humans, there have been cases of bird-to-human transmission. More research is needed to understand how this virus can cross the species barrier and how its spread can be prevented.

29.6 Mammals

By the end of this section, you will be able to do the following:

- Name and describe the distinguishing features of the three main groups of mammals
- Describe the likely line of evolutionary descent that produced mammals
- List some derived features that may have arisen in response to mammals' need for constant, high-level metabolism
- Identify the major clades of eutherian mammals

Mammals, comprising about 5,200 species, are vertebrates that possess hair and mammary glands. Several other characteristics are distinctive to mammals, including certain features of the jaw, skeleton, integument, and internal anatomy. Modern mammals belong to three clades: monotremes, marsupials, and eutherians (or placental mammals).

Characteristics of Mammals

The presence of **hair**, composed of the protein **keratin**, is one of the most obvious characteristics of mammals. Although it is not very extensive or obvious on some species (such as whales), hair has many important functions for most mammals. Mammals are endothermic, and hair traps a boundary layer of air close to the body, retaining heat generated by metabolic activity. Along with insulation, hair can serve as a sensory mechanism via specialized hairs called *vibrissae*, better known as whiskers. Vibrissae attach to nerves that transmit information about tactile vibration produced by sound sensation, which is particularly useful to nocturnal or burrowing mammals. Hair can also provide protective coloration or be part of social signaling, such as when an animal's hair stands "on end" to warn enemies, or possibly to make the mammal "look bigger" to predators.

Unlike the skin of birds, the integument (skin) of mammals, includes a number of different types of secretory glands. **Sebaceous glands** produce a lipid mixture called *sebum* that is secreted onto the hair and skin, providing water resistance and lubrication for hair. Sebaceous glands are located over most of the body. **Eccrine glands** produce sweat, or perspiration, which is mainly composed of water, but also contains metabolic waste products, and sometimes compounds with antibiotic activity. In most mammals, eccrine glands are limited to certain areas of the body, and some mammals do not possess them at all. However, in primates, especially humans, sweat glands are located over most of the body surface and figure prominently in regulating the body temperature through evaporative cooling. **Apocrine glands**, or *scent glands*, secrete substances that are used for chemical communication, such as in skunks. **Mammary glands** produce milk that is used to feed newborns. In both monotremes and eutherians, both males and females possess mammary glands, while in marsupials, mammary glands have been found only in some opossums. Mammary glands likely are modified sebaceous or eccrine glands, but their evolutionary origin is not entirely clear.

The skeletal system of mammals possesses many unique features. The lower jaw of mammals consists of only one bone, the **dentary**, and the jaw hinge connects the dentary to the squamosal (flat) part of the temporal bone in the skull. The jaws of other vertebrates are composed of several bones, including the quadrate bone at the back of the skull and the articular bone at the back of the jaw, with the jaw connected between the quadrate and articular bones. In the ear of other vertebrates, vibrations are transmitted to the inner ear by a single bone, the *stapes*. In mammals, the quadrate and articular bones have moved into the middle ear (Figure 29.37). The malleus is derived from the articular bone, whereas the incus originated from the quadrate bone. This arrangement of jaw and ear bones aids in distinguishing fossil mammals from fossils of other synapsids.

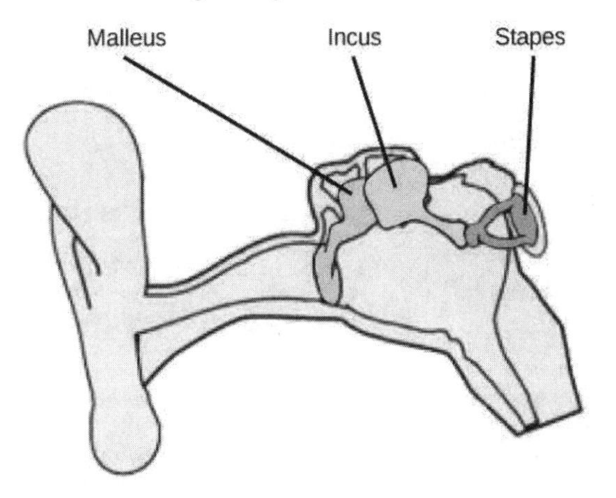

Figure 29.37 Mammalian ear bones. Bones of the mammalian middle ear are modified from bones of the jaw and skull in reptiles. The stapes is found in other vertebrates (e.g., the columella of birds) whereas in mammals, the malleus and incus are derived from the articular and quadrate bones, respectively. (credit: NCI)

The adductor muscles that close the jaw comprise two major muscles in mammals: the *temporalis* and the *masseter*. Working together, these muscles permit up-and-down and side-to-side movements of the jaw, making chewing possible—which is unique to mammals. Most mammals have *heterodont teeth*, meaning that they have different types and shapes of teeth (incisors, canines, premolars, and molars) rather than just one type and shape of tooth. Most mammals are also **diphyodonts**, meaning that they have two sets of teeth in their lifetime: deciduous or "baby" teeth, and permanent teeth. Most other vertebrates with teeth are *polyphyodonts*, that is, their teeth are replaced throughout their entire life.

Mammals, like birds, possess a four-chambered heart; however, the hearts of birds and mammals are an example of convergent evolution, since mammals clearly arose independently from different groups of tetrapod ancestors. Mammals also have a specialized group of cardiac cells (fibers) located in the walls of their right atrium called the sinoatrial node, or pacemaker, which determines the rate at which the heart beats. Mammalian erythrocytes (red blood cells) do *not* have nuclei, whereas the erythrocytes of other vertebrates are nucleated.

The kidneys of mammals have a portion of the nephron called the loop of Henle or nephritic loop, which allows mammals to produce urine with a high concentration of solutes—higher than that of the blood. Mammals lack a renal portal system, which is a system of veins that moves blood from the hind or lower limbs and region of the tail to the kidneys. Renal portal systems are present in all other vertebrates except jawless fishes. A urinary bladder is present in all mammals.

Unlike birds, the skulls of mammals have two occipital condyles, bones at the base of the skull that articulate with the first vertebra, as well as a secondary palate at the rear of the pharynx that helps to separate the pathway of swallowing from that of breathing. Turbinate bones (chonchae in humans) are located along the sides of the nasal cavity, and help warm and moisten air as it is inhaled. The pelvic bones are fused in mammals, and there are typically seven cervical vertebrae (except for some edentates and manatees). Mammals have movable eyelids and fleshy external ears (pinnae), quite unlike the naked external auditory openings of birds. Mammals also have a muscular diaphragm that is lacking in birds.

Mammalian brains also have certain characteristics that differ from the brains of other vertebrates. In some, but not all mammals, the *cerebral cortex*, the outermost part of the cerebrum, is highly convoluted and folded, allowing for a greater surface area than is possible with a smooth cortex. The optic lobes, located in the midbrain, are divided into two parts in mammals, while other vertebrates possess a single, undivided lobe. Eutherian mammals also possess a specialized structure, the corpus callosum, which links the two cerebral hemispheres together. The corpus callosum functions to integrate motor, sensory, and cognitive functions between the left and right cerebral cortexes.

Evolution of Mammals

Mammals are synapsids, meaning they have a single, ancestrally fused, postorbital opening in the skull. They are the only living synapsids, as earlier forms became extinct by the Jurassic period. The early non-mammalian synapsids can be divided into two groups, the pelycosaurs and the therapsids. Within the therapsids, a group called the cynodonts are thought to have been the ancestors of mammals (Figure 29.38).

Figure 29.38 Cynodont. Cynodonts ("dog teeth"), which first appeared in the Late Permian period 260 million years ago, are thought to be the ancestors of modern mammals. Holes in the upper jaws of cynodonts suggest that they had whiskers, which might also indicate the presence of hair. (credit: Nobu Tamura)

As with birds, a key characteristic of synapsids is endothermy, rather than the ectothermy seen in many other vertebrates (such as fish, amphibians, and most reptiles). The increased metabolic rate required to internally modify body temperature likely went hand-in-hand with changes to certain skeletal structures that improved food processing and ambulation. The later synapsids, which had more evolved characteristics unique to mammals, possess cheeks for holding food and heterodont teeth, which are specialized for chewing, mechanically breaking down food to speed digestion, and releasing the energy needed to produce heat. Chewing also requires the ability to breathe at the same time, which is facilitated by the presence of a *secondary palate* (comprising the bony palate and the posterior continuation of the soft palate). The secondary palate separates the area of the mouth where chewing occurs from the area above where respiration occurs, allowing breathing to proceed uninterrupted while the animal is chewing. A secondary palate is not found in pelycosaurs but *is* present in cynodonts and mammals. The jawbone also shows changes from early synapsids to later ones. The zygomatic arch, or cheekbone, is present in mammals and advanced therapsids such as cynodonts, but is not present in pelycosaurs. The presence of the zygomatic arch suggests the presence of masseter muscles, which close the jaw and function in chewing.

In the appendicular skeleton, the shoulder girdle of therian mammals is modified from that of other vertebrates in that it does not possess a procoracoid bone or an interclavicle, and the scapula is the dominant bone.

Mammals evolved from therapsids in the late Triassic period, as the earliest known mammal fossils are from the early Jurassic period, some 205 million years ago. One group of transitional mammals was the **morganucodonts**, small nocturnal insectivores. The jaws of morganucodonts were "transitional," with features of both reptilian and mammalian jaws (Figure 29.39). Like modern mammals, the morganucodonts had differentiated teeth and were diphyodonts. Mammals first began to diversify in the Mesozoic era, from the Jurassic to the Cretaceous periods. Even some small gliding mammals appear in the fossil record during this time period. However, most of the Jurassic mammals were extinct by the end of the Mesozoic. During the Cretaceous period, another radiation of mammals began and continued through the Cenozoic era, about 65 million years ago.

Figure 29.39 A morganucodont. This morganucodont *Megazotrodon*, an extinct basal mammal, may have been nocturnal and insectivorous. Inset: Jaw of a morganucodont, showing a double hinge, one between the dentary and squamosal and one between the articular (yellow) and quadrate (blue) bones. In living mammals, the articular and quadrate bones have been incorporated into the middle ear. (Credit: By Nordelch [Megazostrodon Natural History Museum] Wikimedia Commons. Credit inset: Mod from Philcha. https://commons.wikimedia.org/w/index.php? curid=3631949 (http://openstax.org/l/jaw_joint))

Living Mammals

There are three major groups of living mammals: *monotremes* (*prototheria*), *marsupials* (*metatheria*), and *placental* (*eutheria*) mammals. The eutherians and the marsupials together comprise a clade of therian mammals, with the monotremes forming a sister clade to both metatherians and eutherians.

There are very few living species of **monotremes**: the platypus and four species of echidnas, or spiny anteaters. The leathery-beaked platypus belongs to the family **Ornithorhynchidae** ("bird beak"), whereas echidnas belong to the family **Tachyglossidae** ("sticky tongue") (Figure 29.40). The platypus and one species of echidna are found in Australia, and the other species of echidna are found in New Guinea. Monotremes are unique among mammals because they lay eggs, rather than giving birth to live young. The shells of their eggs are not like the hard shells of birds, but have a leathery shell, similar to the shells of reptile eggs. Monotremes retain their eggs through about two-thirds of the developmental period, and then lay them in nests. A yolk-sac placenta helps support development. The babies hatch in a fetal state and complete their development in the nest, nourished by milk secreted by mammary glands opening directly to the skin. Monotremes, except for young platypuses, do not have teeth. Body temperature in the three monotreme species is maintained at about 30°C, considerably lower than the average body temperature of marsupial and placental mammals, which are typically between 35 and 38°C.

Figure 29.40 Egg-laying mammals. (a) The platypus, a monotreme, possesses a leathery beak and lays eggs rather than giving birth to live young. (b) The echidna is another monotreme, with long hairs modified into spines. (credit b: modification of work by Barry Thomas)

Over 2/3 of the approximately 330 living species of marsupials are found in Australia, with the rest, nearly all various types of opossum, found in the Americas, especially South America. Australian marsupials include the kangaroo, koala, bandicoot, Tasmanian devil (Figure 29.41), and several other species. Like monotremes, the embryos of marsupials are nourished during a short gestational period (about a month in kangaroos) by a yolk-sac placenta, but with no intervening egg shell. Some marsupial embryos can enter an embryonic diapause, and delay implantation, suspending development until implantation is completed. Marsupial young are also effectively fetal at birth. Most, but not all, species of marsupials possess a pouch in which the very premature young reside, receiving milk and continuing their development. In kangaroos, the young joeys continue to nurse for about a year and a half.

Figure 29.41 A marsupial mammal. The Tasmanian devil is one of several marsupials native to Australia. (credit: Wayne McLean)

Eutherians (placentals) are the most widespread and numerous of the mammals, occurring throughout the world. Eutherian mammals are sometimes called "placental mammals" because all species possess a complex **chorioallantoic placenta** that connects a fetus to the mother, allowing for gas, fluid, and nutrient exchange. There are about 4,000 species of placental mammals in 18 to 20 orders with various adaptations for burrowing, flying, swimming, hunting, running, and climbing. In the evolutionary sense, they have been incredibly successful in form, diversity, and abundance. The eutherian mammals are classified in two major clades, the Atlantogenata and the Boreoeutheria. The Atlantogeneta include the Afrotheria (e.g., elephants, hyraxes, and manatees) and the Xenarthra (anteaters, armadillos, and sloths). The Boreoeutheria contain two large groups, the Euarchontoglires and the Laurasiatheria. Familiar orders in the Euarchontoglires are the Scandentia (tree shrews), Rodentia (rats, mice, squirrels, porcupines), Lagomorpha (rabbits and hares), and the Primates (including humans). Major Laurasiatherian orders include the Perissodactyla (e.g., horses and rhinos), the Cetartiodactyla (e.g., cows, giraffes, pigs, hippos, and whales), the Carnivora (e.g., cats, dogs, and bears), and the Chiroptera (bats and flying foxes). The two largest orders are the rodents (2,000 species) and bats (about 1,000 species), which together constitute approximately 60 percent of all eutherian species.

29.7 The Evolution of Primates

By the end of this section, you will be able to do the following:

- Describe the derived features that distinguish primates from other animals
- Describe the defining features of the major groups of primates
- Identify the major hominin precursors to modern humans
- Explain why scientists are having difficulty determining the true lines of descent in hominids

Order Primates of class Mammalia includes lemurs, tarsiers, monkeys, apes, and humans. Non-human primates live primarily in the tropical or subtropical regions of South America, Africa, and Asia. They range in size from the mouse lemur at 30 grams (1 ounce) to the mountain gorilla at 200 kilograms (441 pounds). The characteristics and evolution of primates are of particular interest to us as they allow us to understand the evolution of our own species.

Characteristics of Primates

All primate species possess adaptations for climbing trees, as they all descended from tree-dwellers. This arboreal heritage of primates has resulted in hands and feet that are adapted for climbing, or brachiation (swinging through trees using the arms). These adaptations include, but are not limited to: 1) a rotating shoulder joint, 2) a big toe that is widely separated from the other toes (except humans) and thumbs sufficiently separated from fingers to allow for gripping branches, and 3) **stereoscopic vision**, two overlapping fields of vision from the eyes, which allows for the perception of depth and gauging distance. Other characteristics of primates are brains that are larger than those of most other mammals, claws that have been modified into flattened nails, typically only one offspring per pregnancy, and a trend toward holding the body upright.

Order Primates is divided into two groups: Strepsirrhini ("turned-nosed") and Haplorhini ("simple-nosed") primates. Strepsirrhines, also called the wet-nosed primates, include prosimians like the bush babies and pottos of Africa, the lemurs of Madagascar, and the lorises of Southeast Asia. Haplorhines, or dry-nosed primates, include tarsiers (Figure 29.42) and simians (New World monkeys, Old World monkeys, apes, and humans). In general, strepsirrhines tend to be nocturnal, have larger olfactory centers in the brain, and exhibit a smaller size and smaller brain than anthropoids. Haplorhines, with a few exceptions, are diurnal, and depend more on their vision. Another interesting difference between the strepsirrhines and haplorhines is that strepsirrhines have the enzymes for making vitamin C, while haplorhines have to get it from their food.

Figure 29.42 A Philippine tarsier. This tarsier, *Carlito syrichta*, is one of the smallest primates—about 5 inches long, from nose to the base of the tail. The tail is not shown, but is about twice the length of the body. Note the large eyes, each of which is about the same size as the animal's brain, and the long hind legs. (credit: mtoz (http://creativecommons.org/licenses/by-sa/2.0 (http://openstax.org/l/CCSA_2)), via Wikimedia Commons)

Evolution of Primates

The first primate-like mammals are referred to as proto-primates. They were roughly similar to squirrels and tree shrews in size and appearance. The existing fossil evidence (mostly from North Africa) is very fragmented. These proto-primates remain largely mysterious creatures until more fossil evidence becomes available. Although genetic evidence suggests that primates diverged from other mammals about 85 MYA, the oldest known primate-like mammals with a relatively robust fossil record date to about 65 MYA. Fossils like the proto-primate *Plesiadapis* (although some researchers do not agree that *Plesiadapis* was a proto-primate) had some features of the teeth and skeleton in common with true primates. They were found in North America and Europe in the Cenozoic and went extinct by the end of the Eocene.

The first true primates date to about 55 MYA in the Eocene epoch. They were found in North America, Europe, Asia, and Africa. These early primates resembled present-day prosimians such as lemurs. Evolutionary changes continued in these early primates, with larger brains and eyes, and smaller muzzles being the trend. By the end of the Eocene epoch, many of the early prosimian species went extinct due either to cooler temperatures or competition from the first monkeys.

Anthropoid monkeys evolved from prosimians during the Oligocene epoch. By 40 million years ago, evidence indicates that monkeys were present in the New World (South America) and the Old World (Africa and Asia). New World monkeys are also called Platyrrhini—a reference to their broad noses (Figure 29.43). Old World monkeys are called Catarrhini—a reference to their narrow, downward-pointed noses. There is still quite a bit of uncertainty about the origins of the New World monkeys. At the time the platyrrhines arose, the continents of South American and Africa had drifted apart. Therefore, it is thought that monkeys arose in the Old World and reached the New World either by drifting on log rafts or by crossing land bridges. Due to this reproductive isolation, New World monkeys and Old World monkeys underwent separate adaptive radiations over millions of years. The New World monkeys are all arboreal, whereas Old World monkeys include both arboreal and ground-dwelling species. The arboreal habits of the New World monkeys are reflected in the possession of prehensile or grasping tails by most species. The tails of Old World monkeys are never prehensile and are often reduced, and some species have ischial callosities—thickened patches of skin on their seats.

Figure 29.43 A New World monkey. The howler monkey is native to Central and South America. It makes a call that sounds like a lion roaring. (credit: Xavi Talleda)

Apes evolved from the catarrhines in Africa midway through the Cenozoic, approximately 25 million years ago. Apes are generally larger than monkeys and they do not possess a tail. All apes are capable of moving through trees, although many species spend most their time on the ground. When walking quadrupedally, monkeys walk on their palms, while apes support the upper body on their knuckles. Apes are more intelligent than monkeys, and they have larger brains relative to body size. The apes are divided into two groups. The lesser apes comprise the family **Hylobatidae**, including gibbons and siamangs. The great apes include the genera ***Pan*** (chimpanzees and bonobos) ***Gorilla*** (gorillas), ***Pongo*** (orangutans), and ***Homo*** (humans) (Figure 29.44).

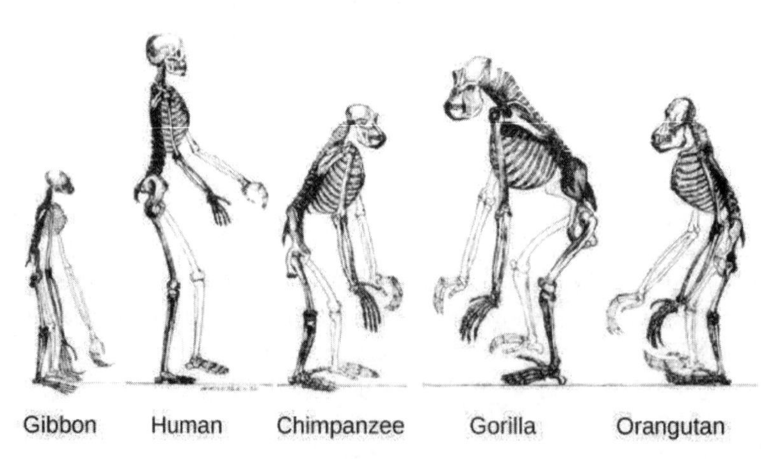

Figure 29.44 Primate skeletons. All great apes have a similar skeletal structure. (credit: modification of work by Tim Vickers)

The very arboreal gibbons are smaller than the great apes; they have low sexual dimorphism (that is, the sexes are not markedly different in size), although in some species, the sexes differ in color; and they have relatively longer arms used for swinging through trees (Figure 29.45a). Two species of orangutan are native to different islands in Indonesia: Borneo (*P. pygmaeus*) and Sumatra (*P. abelii*). A third orangutan species, *Pongo tapanuliensis*, was reported in 2017 from the Batang Toru forest in Sumatra. Orangutans are arboreal and solitary. Males are much larger than females and have cheek and throat pouches when mature. Gorillas all live in Central Africa. The eastern and western populations are recognized as separate species, *G. berengei* and *G. gorilla*. Gorillas are strongly sexually dimorphic, with males about twice the size of females. In older males, called silverbacks, the hair on the back turns white or gray. Chimpanzees (Figure 29.45b) are the species considered to be most closely related to humans. However, the species most closely related to the chimpanzee is the bonobo. Genetic evidence suggests that chimpanzee and human lineages separated 5 to 7 MYA, while chimpanzee (*Pan troglodytes*) and bonobo (*Pan paniscus*) lineages separated about 2 MYA. Chimpanzees and bonobos both live in Central Africa, but the two species are separated by the Congo River, a significant geographic barrier. Bonobos are slighter than chimpanzees, but have longer legs and more hair on their heads. In chimpanzees, white tail tufts identify juveniles, while bonobos keep their white tail tufts for life. Bonobos also have higher-pitched voices than chimpanzees. Chimpanzees are more aggressive and sometimes kill animals from other groups, while bonobos are not known to do so. Both chimpanzees and bonobos are omnivorous. Orangutan and gorilla diets also include foods from multiple sources, although the predominant food items are fruits for orangutans and foliage for gorillas.

Figure 29.45 Lesser and great apes. This white-cheeked gibbon (a) is a lesser ape. In gibbons of this species, females and infants are buff and males are black. This young chimpanzee (b) is one of the great apes. It possesses a relatively large brain and has no tail. (credit a: MAC. credit b: modification of work by Aaron Logan)

Human Evolution

The family Hominidae of order Primates includes the hominoids: the great apes and humans (Figure 29.46). Evidence from the fossil record and from a comparison of human and chimpanzee DNA suggests that humans and chimpanzees diverged from a common hominoid ancestor approximately six million years ago. Several species evolved from the evolutionary branch that includes humans, although our species is the only surviving member. The term hominin is used to refer to those species that evolved after this split of the primate line, thereby designating species that are more closely related to humans than to chimpanzees. A number of marker features differentiate humans from the other hominoids, including bipedalism or upright posture, increase in the size of the brain, and a fully opposable thumb that can touch the little finger. Bipedal hominins include several groups that were probably part of the modern human lineage—*Australopithecus*, *Homo habilis*, and *Homo erectus*—and several non-ancestral groups that can be considered "cousins" of modern humans, such as Neanderthals and Denisovans.

Determining the true lines of descent in hominins is difficult. In years past, when relatively few hominin fossils had been recovered, some scientists believed that considering them in order, from oldest to youngest, would demonstrate the course of evolution from early hominins to modern humans. In the past several years, however, many new fossils have been found, and it is clear that there was often more than one species alive at any one time and that many of the fossils found (and species named) represent hominin species that died out and are not ancestral to modern humans.

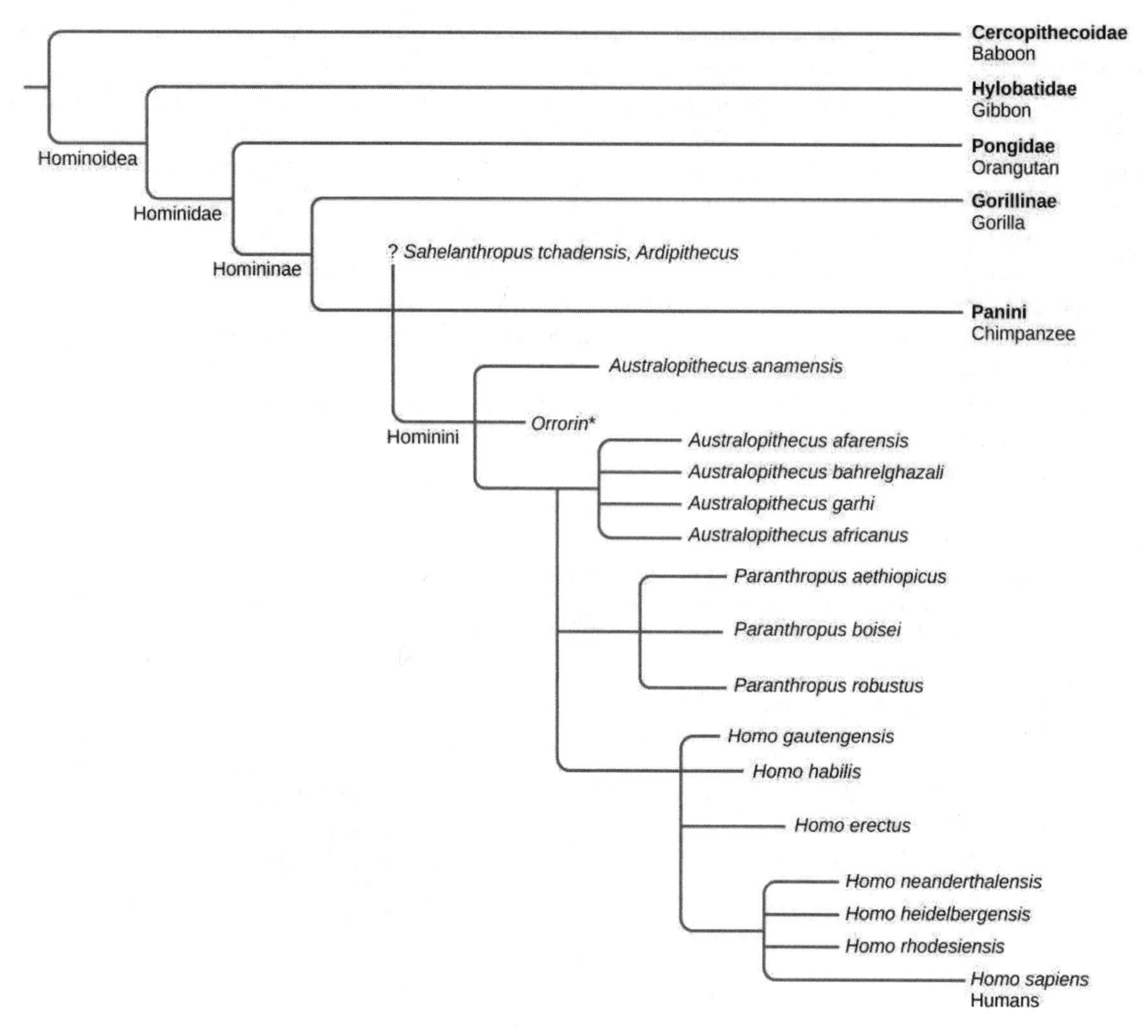

Figure 29.46 Hominin phylogeny. This chart shows evolutionary relationship among Hominins and hypothesized relation to modern humans. (*still debated phylogeny position).

Very Early Hominins

Three species of very early hominids have made news in the late 20th and early 21st centuries: *Ardipithecus*, *Sahelanthropus*, and *Orrorin*. The youngest of the three species, *Ardipithecus*, was discovered in the 1990s, and dates to about 4.4 MYA. Although the bipedality of the early specimens was uncertain, several more specimens of *Ardipithecus* were discovered in the intervening years and demonstrated that the organism was bipedal. Two different species of *Ardipithecus* have been identified, *A. ramidus* and *A. kadabba*, whose specimens are older, dating to 5.6 MYA. However, the status of this genus as a human ancestor is uncertain.

The oldest of the three, *Sahelanthropus tchadensis*, was discovered in 2001-2002 and has been dated to nearly seven million years ago. There is a single specimen of this genus, a skull that was a surface find in Chad. The fossil, informally called "Toumai," is a mosaic of primitive and evolved characteristics, and it is unclear how this fossil fits with the picture given by molecular data, namely that the line leading to modern humans and modern chimpanzees apparently bifurcated about six million years ago. It is not thought at this time that this species was an ancestor of modern humans.

A younger (c. 6 MYA) species, *Orrorin tugenensis*, is also a relatively recent discovery, found in 2000. There are several specimens of *Orrorin*. Some features of *Orrorin* are more similar to those of modern humans than are the australopithicenes, although *Orrorin* is much older. If *Orrorin* is a human ancestor, then the australopithicenes may not be in the direct human lineage. Additional specimens of these species may help to clarify their role.

Early Hominins: Genus *Australopithecus*

Australopithecus ("southern ape") is a genus of hominin that evolved in eastern Africa approximately four million years ago and went extinct about two million years ago. This genus is of particular interest to us as it is thought that our genus, genus *Homo*, evolved from a common ancestor shared with *Australopithecus* about two million years ago (after likely passing through some transitional states). *Australopithecus* had a number of characteristics that were more similar to the great apes than to modern humans. For example, sexual dimorphism was more exaggerated than in modern humans. Males were up to 50 percent larger than females, a ratio that is similar to that seen in modern gorillas and orangutans. In contrast, modern human males are approximately 15 to 20 percent larger than females. The brain size of *Australopithecus* relative to its body mass was also smaller than in modern humans and more similar to that seen in the great apes. A key feature that *Australopithecus* had in common with modern humans was bipedalism, although it is likely that *Australopithecus* also spent time in trees. Hominin footprints, similar to those of modern humans, were found in Laetoli, Tanzania and dated to 3.6 million years ago. They showed that hominins at the time of *Australopithecus* were walking upright.

There were a number of *Australopithecus* species, which are often referred to as *australopiths*. *Australopithecus anamensis* lived about 4.2 million years ago. More is known about another early species, *Australopithecus afarensis*, which lived between 3.9 and 2.9 million years ago. This species demonstrates a trend in human evolution: the reduction of the dentition and jaw in size. *A. afarensis* (Figure 29.47a) had smaller canines and molars compared to apes, but these were larger than those of modern humans. Its brain size was 380 to 450 cubic centimeters, approximately the size of a modern chimpanzee brain. It also had prognathic jaws, which is a relatively longer jaw than that of modern humans. In the mid-1970s, the fossil of an adult female *A. afarensis* was found in the Afar region of Ethiopia and dated to 3.24 million years ago (Figure 29.48). The fossil, which is informally called "Lucy," is significant because it was the most complete australopith fossil found, with 40 percent of the skeleton recovered.

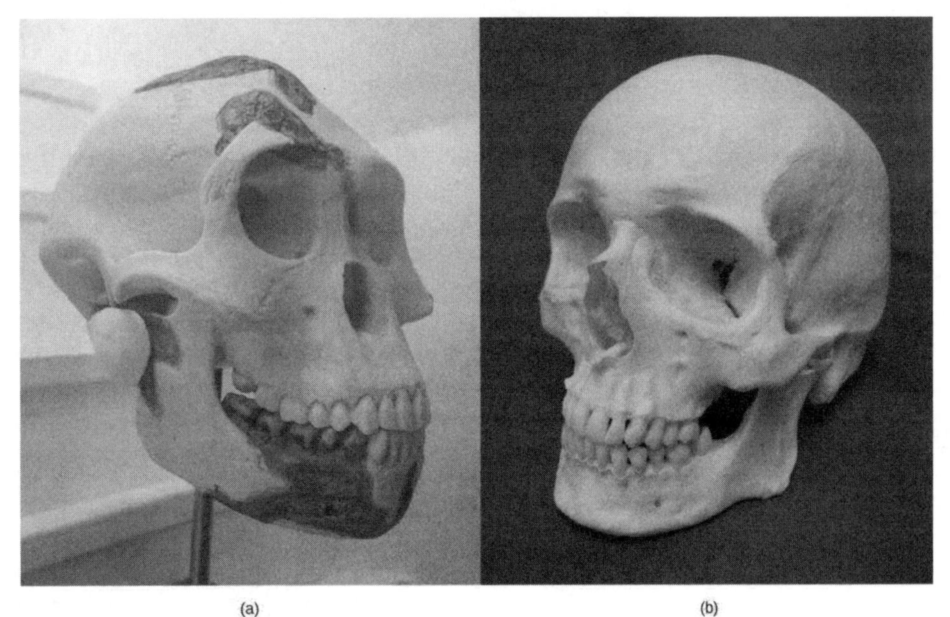

Figure 29.47 Australopithicene and modern human skulls. The skull of (a) *Australopithecus afarensis*, an early hominid that lived between two and three million years ago, resembled that of (b) modern humans but was smaller with a sloped forehead, larger teeth, and a prominent jaw.

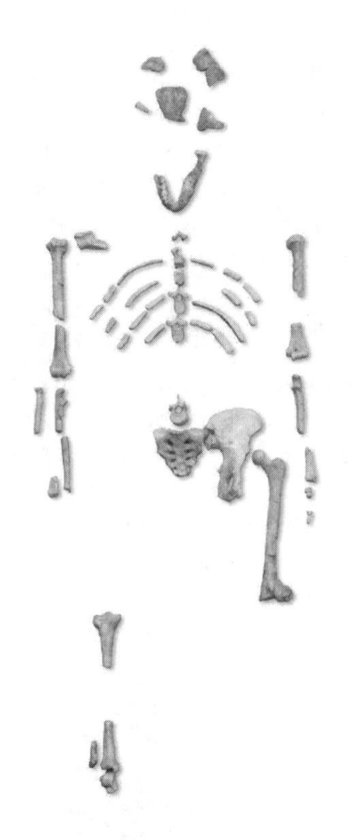

Figure 29.48 Lucy. This adult female *Australopithecus afarensis* skeleton, nicknamed Lucy, was discovered in the mid-1970s. (credit: "120"/Wikimedia Commons)

Australopithecus africanus lived between two and three million years ago. It had a slender build and was bipedal, but had robust arm bones and, like other early hominids, may have spent significant time in trees. Its brain was larger than that of *A. afarensis* at 500 cubic centimeters, which is slightly less than one-third the size of modern human brains. Two other species, *Australopithecus bahrelghazali* and *Australopithecus garhi*, have been added to the roster of australopiths in recent years. A. *bahrelghazali* is unusual in being the only australopith found in Central Africa.

A Dead End: Genus *Paranthropus*

The australopiths had a relatively slender build and teeth that were suited for soft food. In the past several years, fossils of hominids of a different body type have been found and dated to approximately 2.5 million years ago. These hominids, of the genus *Paranthropus*, were muscular, stood 1.3 to 1.4 meters tall, and had large grinding teeth. Their molars showed heavy wear, suggesting that they had a coarse and fibrous vegetarian diet as opposed to the partially carnivorous diet of the australopiths. *Paranthropus* includes *Paranthropus robustus* of South Africa, and *Paranthropus aethiopicus* and *Paranthropus boisei* of East Africa. The hominids in this genus went extinct more than one million years ago and are not thought to be ancestral to modern humans, but rather members of an evolutionary branch on the hominin tree that left no descendants.

Early Hominins: Genus *Homo*

The human genus, *Homo*, first appeared between 2.5 and three million years ago. For many years, fossils of a species called *H. habilis* were the oldest examples in the genus *Homo*, but in 2010, a new species called *Homo gautengensis* was discovered and may be older. Compared to *A. africanus*, *H. habilis* had a number of features more similar to modern humans. *H. habilis* had a jaw that was less prognathic than the australopiths and a larger brain, at 600 to 750 cubic centimeters. However, *H. habilis* retained some features of older hominin species, such as long arms. The name *H. habilis* means "handy man," which is a reference to the stone tools that have been found with its remains.

LINK TO LEARNING

Watch this video about Smithsonian paleontologist Briana Pobiner explaining the link between hominin eating of meat and evolutionary trends.

Click to view content (https://www.openstax.org/l/diet_detective)

H. erectus appeared approximately 1.8 million years ago (Figure 29.49). It is believed to have originated in East Africa and was the first hominin species to migrate out of Africa. Fossils of *H. erectus* have been found in India, China, Java, and Europe, and were known in the past as "Java Man" or "Peking Man." *H. erectus* had a number of features that were more similar to modern humans than those of *H. habilis*. *H. erectus* was larger in size than earlier hominins, reaching heights up to 1.85 meters and weighing up to 65 kilograms, which are sizes similar to those of modern humans. Its degree of sexual dimorphism was less than in earlier species, with males being 20 to 30 percent larger than females, which is close to the size difference seen in our own species. *H. erectus* had a larger brain than earlier species at 775 to 1,100 cubic centimeters, which compares to the 1,130 to 1,260 cubic centimeters seen in modern human brains. *H. erectus* also had a nose with downward-facing nostrils similar to modern humans, rather than the forward-facing nostrils found in other primates. Longer, downward-facing nostrils allow for the warming of cold air before it enters the lungs and may have been an adaptation to colder climates. Artifacts found with fossils of *H. erectus* suggest that it was the first hominin to use fire, hunt, and have a home base. *H. erectus* is generally thought to have lived until about 50,000 years ago.

Figure 29.49 *Homo erectus. Homo erectus* had a prominent brow and a nose that pointed downward rather than forward.

Humans: *Homo sapiens*

A number of species, sometimes called archaic *Homo sapiens*, apparently evolved from *H. erectus* starting about 500,000 years ago. These species include *Homo heidelbergensis*, *Homo rhodesiensis*, and *Homo neanderthalensis*. These archaic *H. sapiens* had a brain size similar to that of modern humans, averaging 1,200 to 1,400 cubic centimeters. They differed from modern humans by having a thick skull, a prominent brow ridge, and a receding chin. Some of these species survived until 30,000 to 10,000 years ago, overlapping with modern humans (Figure 29.50).

Figure 29.50 Neanderthal. The *Homo neanderthalensis* used tools and may have worn clothing.

There is considerable debate about the origins of anatomically modern humans or ***Homo sapiens sapiens***. As discussed earlier, *H. erectus* migrated out of Africa and into Asia and Europe in the first major wave of migration about 1.5 million years ago. It is thought that modern humans arose in Africa from *H. erectus* and migrated out of Africa about 100,000 years ago in a second major migration wave. Then, modern humans replaced *H. erectus* species that had migrated into Asia and Europe in the first wave.

This evolutionary timeline is supported by molecular evidence. One approach to studying the origins of modern humans is to examine mitochondrial DNA (mtDNA) from populations around the world. Because a fetus develops from an egg containing its mother's mitochondria (which have their own, non-nuclear DNA), mtDNA is passed entirely through the maternal line. Mutations in mtDNA can now be used to estimate the timeline of genetic divergence. The resulting evidence suggests that all modern humans have mtDNA inherited from a common ancestor that lived in Africa about 160,000 years ago. Another approach to the molecular understanding of human evolution is to examine the Y chromosome, which is passed from father to son. This evidence suggests that all men today inherited a Y chromosome from a male that lived in Africa about 140,000 years ago.

The study of mitochondrial DNA led to the identification of another human species or subspecies, the Denisovans. DNA from teeth and finger bones suggested two things. First, the mitochondrial DNA was different from that of both modern humans and Neanderthals. Second, the genomic DNA suggested that the Denisovans shared a common ancestor with the Neanderthals. Genes from both Neanderthals and Denisovans have been identified in modern human populations, indicating that interbreeding among the three groups occurred over part of their range.

KEY TERMS

Acanthostega one of the earliest known tetrapods
Actinopterygii ray-finned fishes
allantois membrane of the egg that stores nitrogenous wastes produced by the embryo; also facilitates respiration
amnion membrane of the egg that protects the embryo from mechanical shock and prevents dehydration
amniote animal that produces a terrestrially adapted egg protected by amniotic membranes
Amphibia frogs, salamanders, and caecilians
ampulla of Lorenzini sensory organ that allows sharks to detect electromagnetic fields produced by living things
anapsid animal having no temporal fenestrae in the cranium
anthropoid monkeys, apes, and humans
Anura frogs
apocrine gland scent gland that secretes substances that are used for chemical communication
Apoda caecilians
Archaeopteryx transition species from dinosaur to bird from the Jurassic period
archosaur modern crocodilian or bird, or an extinct pterosaur or dinosaur
Australopithecus genus of hominins that evolved in eastern Africa approximately four million years ago
brachiation movement through trees branches via suspension from the arms
brumation period of much reduced metabolism and torpor that occurs in any ectotherm in cold weather
caecilian legless amphibian that belongs to the clade Apoda
Casineria one of the oldest known amniotes; had both amphibian and reptilian characteristics
Catarrhini clade of Old World monkeys
Cephalochordata chordate clade whose members possess a notochord, dorsal hollow nerve cord, pharyngeal slits, and a post-anal tail in the adult stage
Chondrichthyes jawed fish with paired fins and a skeleton made of cartilage
Chordata phylum of animals distinguished by their possession of a notochord, a dorsal hollow nerve cord, pharyngeal slits, and a post-anal tail at some point during their development
chorion membrane of the egg that surrounds the embryo and yolk sac
contour feather feather that creates an aerodynamic surface for efficient flight
Craniata clade composed of chordates that possess a cranium; includes Vertebrata together with hagfishes
cranium bony, cartilaginous, or fibrous structure surrounding the brain, jaw, and facial bones
Crocodilia crocodiles and alligators
cutaneous respiration gas exchange through the skin
dentary single bone that comprises the lower jaw of mammals
diapsid animal having two temporal fenestrae in the cranium
diphyodont refers to the possession of two sets of teeth in a lifetime
dorsal hollow nerve cord hollow, tubular structure derived from ectoderm, which is located dorsal to the notochord in chordates
down feather feather specialized for insulation
eccrine gland sweat gland
Enantiornithes dominant bird group during the Cretaceous period
eutherian mammal mammal that possesses a complex placenta, which connects a fetus to the mother; sometimes called placental mammals
flight feather feather specialized for flight
frog tail-less amphibian that belongs to the clade Anura
furcula wishbone formed by the fusing of the clavicles
gnathostome jawed fish
Gorilla genus of gorillas
hagfish eel-like jawless fish that live on the ocean floor and are scavengers
heterodont tooth different types of teeth that are modified for different purposes
hominin species that are more closely related to humans than chimpanzees
hominoid pertaining to great apes and humans
Homo genus of humans
Homo sapiens sapiens anatomically modern humans
Hylobatidae family of gibbons
Hylonomus one of the earliest reptiles
lamprey jawless fish characterized by a toothed, funnel-like, sucking mouth
lancelet member of Cephalochordata; named for its blade-like shape
lateral line sense organ that runs the length of a fish's body; used to detect vibration in the water
lepidosaur modern lizards, snakes, and tuataras
mammal one of the groups of endothermic vertebrates that possesses hair and mammary glands
mammary gland in female mammals, a gland that produces milk for newborns
marsupial one of the groups of mammals that includes the kangaroo, koala, bandicoot, Tasmanian devil, and several other species; young develop within a pouch
monotreme egg-laying mammal
Myxini hagfishes
Neognathae birds other than the Paleognathae
Neornithes modern birds
notochord flexible, rod-shaped support structure that is found in the embryonic stage of all chordates and in the

adult stage of some chordates

Ornithorhynchidae clade that includes the duck-billed platypus

Osteichthyes bony fish

ostracoderm one of the earliest jawless fish covered in bone

Paleognathae ratites; flightless birds, including ostriches and emus

Pan genus of chimpanzees and bonobos

Petromyzontidae clade of lampreys

pharyngeal slit opening in the pharynx

Platyrrhini clade of New World monkeys

Plesiadapis oldest known primate-like mammal

pneumatic bone air-filled bone

Pongo genus of orangutans

post-anal tail muscular, posterior elongation of the body extending beyond the anus in chordates

primary feather feather located at the tip of the wing that provides thrust

Primates order of lemurs, tarsiers, monkeys, apes, and humans

prognathic jaw long jaw

prosimian division of primates that includes bush babies and pottos of Africa, lemurs of Madagascar, and lorises of Southeast Asia

salamander tailed amphibian that belongs to the clade Urodela

Sarcopterygii lobe-finned fish

sauropsid reptile or bird

sebaceous gland in mammals, a skin gland that produce a lipid mixture called *sebum*

secondary feather feather located at the base of the wing that provides lift

Sphenodontia clade of tuataras

Squamata clade of lizards and snakes

stereoscopic vision two overlapping fields of vision from the eyes that produces depth perception

swim bladder in fishes, a gas filled organ that helps to control the buoyancy of the fish

synapsid mammal having one temporal fenestra

Tachyglossidae clade that includes the echidna or spiny anteater

tadpole larval stage of a frog

temporal fenestra non-orbital opening in the skull that may allow muscles to expand and lengthen

Testudines order of turtles

tetrapod phylogenetic reference to an organism with a four-footed evolutionary history; includes amphibians, reptiles, birds, and mammals

theropod dinosaur group ancestral to birds

tunicate sessile chordate that is a member of Urochordata

Urochordata clade composed of tunicates

Urodela salamanders

vertebral column series of separate bones joined together as a backbone

Vertebrata members of the phylum Chordata that possess a backbone

CHAPTER SUMMARY

29.1 Chordates

The five characteristic features of chordates present during some time of their life cycles are a notochord, a dorsal hollow tubular nerve cord, pharyngeal slits, endostyle/thyroid gland, and a post-anal tail. Chordata contains two clades of invertebrates: Urochordata (tunicates) and Cephalochordata (lancelets), together with the vertebrates in the Vertebrata/Craniata. Lancelets are suspension feeders that feed on phytoplankton and other microorganisms. Most tunicates live on the ocean floor and are suspension feeders. Which of the two invertebrate chordate clades is more closely related to the vertebrates continues to be debated. Vertebrata is named for the vertebral column, which is a feature of almost all members of this clade. The name Craniata (organisms with a cranium) is considered to be synonymous with Vertebrata.

29.2 Fishes

The earliest vertebrates that diverged from the invertebrate chordates were the agnathan jawless fishes, whose extant members include the hagfishes and lampreys. Hagfishes are eel-like scavengers that feed on dead invertebrates and other fishes. Lampreys are characterized by a toothed, funnel-like sucking mouth, and most species are parasitic or predaceous on other fishes. Fishes with jaws (gnathostomes) evolved later. Jaws allowed early gnathostomes to exploit new food sources.

Gnathostomes include the cartilaginous fishes and the bony fishes, as well as all other tetrapods (amphibians, reptiles, mammals). Cartilaginous fishes include sharks, rays, skates, and ghost sharks. Most cartilaginous fishes live in marine habitats, with a few species living in fresh water for part or all of their lives. The vast majority of present-day fishes belong to the clade Osteichthyes, which consists of approximately 30,000 species. Bony fishes (Osteichthyes) can be divided into two clades: Actinopterygii (ray-finned fishes, virtually all extant species) and Sarcopterygii (lobe-finned fishes, comprising fewer than 10 extant species, but form the sister group of the tetrapods).

29.3 Amphibians

As tetrapods, most amphibians are characterized by four well-developed limbs, although some species of salamanders and all caecilians are limbless. The most important characteristic of extant amphibians is a moist, permeable skin used for cutaneous respiration, although lungs are found in the adults of many species.

All amphibians are carnivores and possess many small teeth. The fossil record provides evidence of amphibian species, now extinct, that arose over 400 million years ago as the first tetrapods. Living Amphibia can be divided into three classes: salamanders (Urodela), frogs (Anura), and caecilians (Apoda). In the majority of amphibians, development occurs in two distinct stages: a gilled aquatic larval stage that metamorphoses into an adult stage, acquiring lungs and legs, and losing the tail in Anurans. A few species in all three clades bypass a free-living larval stage. Various levels of parental care are seen in the amphibians.

29.4 Reptiles

The amniotes are distinguished from amphibians by the presence of a terrestrially adapted egg protected by four extra-embryonic membranes. The amniotes include reptiles, birds, and mammals. The early amniotes diverged into two main lines soon after the first amniotes arose. The initial split was into synapsids (mammals) and sauropsids. Sauropsids can be further divided into anapsids and diapsids (crocodiles, dinosaurs, birds, and modern reptiles).

Reptiles are tetrapods that ancestrally had four limbs; however, a number of extant species have secondarily lost them or greatly reduced them over evolutionary time. For example, limbless reptiles (e.g., snakes) are classified as tetrapods, because they descended from ancestors with four limbs. One of the key adaptations that permitted reptiles to live on land was the development of scaly skin containing the protein keratin, which prevented water loss from the skin. Reptilia includes four living clades of nonavian organisms: Crocodilia (crocodiles and alligators), Sphenodontia (tuataras), Squamata (lizards and snakes), and Testudines (turtles). Currently, this classification is paraphyletic, leaving out the birds, which are now classified as avian reptiles in the class Reptilia.

29.5 Birds

Birds are the most speciose group of land vertebrates and display a number of adaptations related to their ability to fly, which were first present in their theropod (maniraptoran) ancestors. Birds are endothermic (and homeothermic), meaning they have a very high metabolism that produces a considerable amount of heat, as well as structures such as feathers that allow them to retain their own body heat. These adaptations are used to regulate their internal temperature, making it largely independent of ambient thermal conditions.

Birds have feathers, which allow for insulation and flight, as well as for mating and warning signals. Flight feathers have a broad and continuously curved vane that produces lift. Some birds have pneumatic bones containing air spaces that are sometimes connected to air sacs in the body cavity. Airflow through bird lungs travels in one direction, creating a counter-current gas exchange with the blood.

Birds are highly modified diapsids and belong to a group called the archosaurs. Within the archosaurs, birds are most likely evolved from theropod (maniraptoran) dinosaurs. One of the oldest known fossils (and best known) of a "dinosaur-bird" is that of *Archaeopteryx*, which is dated from the Jurassic period. Modern birds are now classified into three groups: Paleognathae, Galloanserae, and Neoaves.

29.6 Mammals

Mammals are vertebrates that possess hair and mammary glands. The mammalian integument includes various secretory glands, including sebaceous glands, eccrine glands, apocrine glands, and mammary glands.

Mammals are synapsids, meaning that they have a single opening in the skull behind the eye. Mammals probably evolved from therapsids in the late Triassic period, as the earliest known mammal fossils are from the early Jurassic period. A key characteristic of synapsids is endothermy, and most mammals are homeothermic.

There are three groups of mammals living today: monotremes, marsupials, and eutherians. Monotremes are unique among mammals as they lay eggs, rather than giving birth to young. Marsupials give birth to very immature young, which typically complete their development in a pouch. Eutherian mammals are sometimes called placental mammals, because all species possess a complex placenta that connects a fetus to the mother, allowing for gas, fluid, and nutrient exchange. All mammals nourish their young with milk, which is derived from modified sweat or sebaceous glands.

29.7 The Evolution of Primates

All primate species possess adaptations for climbing trees and probably descended from arboreal ancestors, although not all living species are arboreal. Other characteristics of primates are brains that are larger, relative to body size, than those of other mammals, claws that have been modified into flattened nails, typically only one young per pregnancy, stereoscopic vision, and a trend toward holding the body upright. Primates are divided into two groups: strepsirrhines, which include most prosimians, and haplorhines, which include simians. Monkeys evolved from prosimians during the Oligocene epoch. The simian line includes both platyrrhine and catarrhine branches. Apes evolved from catarrhines in Africa during the Miocene epoch. Apes are divided into the lesser apes and the greater apes. Hominins include those groups that gave rise to our own species, such as *Australopithecus* and *H. erectus*, and those groups that can be considered "cousins" of humans, such as Neanderthals and Denisovans. Fossil evidence shows that hominins at the time of *Australopithecus* were walking upright, the first evidence of bipedal hominins. A number of species, sometimes called archaic *H. sapiens*, evolved from *H. erectus* approximately 500,000 years ago. There is considerable debate about the origins of anatomically modern humans or *H. sapiens sapiens*, and the discussion will continue, as new evidence from fossil finds and genetic analysis emerges.

VISUAL CONNECTION QUESTIONS

1. Figure 29.3 Which of the following statements about common features of chordates is true?
   a. The dorsal hollow nerve cord is part of the chordate central nervous system.
   b. In vertebrate fishes, the pharyngeal slits become the gills.
   c. Humans are not chordates because humans do not have a tail.
   d. Vertebrates do not have a notochord at any point in their development; instead, they have a vertebral column.

2. Figure 29.22 Which of the following statements about the parts of an amniotic egg are false?
   a. The allantois stores nitrogenous waste and facilitates respiration.
   b. The chorion facilitates gas exchange.
   c. The yolk provides food for the growing embryo.
   d. The amniotic cavity is filled with albumen.

3. Figure 29.24 Members of the order Testudines have an anapsid-like skull without obvious temporal fenestrae. However, molecular studies indicate that turtles descended from a diapsid ancestor. Why might this be the case?

REVIEW QUESTIONS

4. Which of the following is *not* contained in phylum Chordata?
   a. Cephalochordata
   b. Echinodermata
   c. Urochordata
   d. Vertebrata

5. Which group of invertebrates is most closely related to vertebrates?
   a. cephalochordates
   b. echinoderms
   c. arthropods
   d. urochordates

6. Hagfish, lampreys, sharks, and tuna are all chordates that can also be classified into which group?
   a. Craniates
   b. Vertebrates
   c. Cartilaginous fish
   d. Cephalocordata

7. Members of Chondrichthyes differ from members of Osteichthyes by having (a) _______.
   a. jaw
   b. bony skeleton
   c. cartilaginous skeleton
   d. two sets of paired fins

8. Members of Chondrichthyes are thought to be descended from fishes that had _______.
   a. a cartilaginous skeleton
   b. a bony skeleton
   c. mucus glands
   d. slime glands

9. A marine biologist catches a species of fish she has never seen before. Upon examination, she determines that the species has a predominantly cartilaginous skeleton and a swim bladder. If its pectoral fins are not fused with its head, to which category of fish does the specimen belong?
   a. Rays
   b. Osteichthyes
   c. Sharks
   d. Hagfish

10. Which of the following is *not* true of *Acanthostega*?
    a. It was aquatic.
    b. It had gills.
    c. It had four limbs.
    d. It laid shelled eggs.

11. Frogs belong to which order?
    a. Anura
    b. Urodela
    c. Caudata
    d. Apoda

12. During the Mesozoic period, diapsids diverged into______.
    a. pterosaurs and dinosaurs
    b. mammals and reptiles
    c. lepidosaurs and archosaurs
    d. Testudines and Sphenodontia

13. Squamata includes______.
    a. crocodiles and alligators
    b. turtles
    c. tuataras
    d. lizards and snakes

14. Which of the following reptile groups gave rise to modern birds?
    a. Lepidosaurs
    b. Pterosaurs
    c. Anapsids
    d. Archosaurs

15. A bird or feathered dinosaur is _______.
    a. Neornithes
    b. *Archaeopteryx*
    c. Enantiornithes
    d. Paleognathae

16. Which of the following feather types helps to reduce drag produced by wind resistance during flight?
    a. Flight feathers
    b. Primary feathers
    c. Secondary feathers
    d. Contour feathers

17. Eccrine glands produce _______.
    a. sweat
    b. lipids
    c. scents
    d. milk

18. Monotremes include:
    a. kangaroos.
    b. koalas.
    c. bandicoots.
    d. platypuses.

19. The evolution of which of the following features of mammals is hardest to trace through the fossil record?
    a. Jaw structure
    b. Mammary glands
    c. Middle ear structure
    d. Development of hair

20. Which of the following is *not* an anthropoid?
    a. Lemurs
    b. Monkeys
    c. Apes
    d. Humans

21. Which of the following is part of a clade believed to have died out, leaving no descendants?
    a. *Paranthropus robustus*
    b. *Australopithecus africanus*
    c. *Homo erectus*
    d. *Homo sapiens sapiens*

22. Which of the following human traits is not a shared characteristic of primates?
    a. Hip structure supporting bipedalism
    b. Detection and processing of three-color vision
    c. Nails at the end of each digit
    d. Enlarged brain area associated with vision, and reduced area associated with smell

CRITICAL THINKING QUESTIONS

23. What are the characteristic features of the chordates?
24. What is the structural advantage of the notochord in the human embryo? Be sure to compare the notochord with the corresponding structure in adults.
25. What can be inferred about the evolution of the cranium and vertebral column from examining hagfishes and lampreys?
26. Why did gnathostomes replace most agnathans?
27. Explain why frogs are restricted to a moist environment.
28. Describe the differences between the larval and adult stages of frogs.
29. Describe how metamorphosis changes the structures involved in gas exchange over the life cycle of animals in the clade Anura, and what evolutionary advantage this change provides.
30. Describe the functions of the three extra-embryonic membranes present in amniotic eggs.
31. What characteristics differentiate lizards and snakes?
32. Based on how reptiles thermoregulate, which climates would you predict to have the highest reptile population density, and why?
33. Explain why birds are thought to have evolved from theropod dinosaurs.

34. Describe three skeletal adaptations that allow for flight in birds.
35. How would the chest structure differ between ostriches, penguins, and terns?
36. Describe three unique features of the mammalian skeletal system.
37. Describe three characteristics of the mammalian brain that differ from other vertebrates.
38. How did the evolution of jaw musculature allow mammals to spread?
39. How did archaic *Homo sapiens* differ from anatomically modern humans?
40. Why is it so difficult to determine the sequence of hominin ancestors that have led to modern *Homo sapiens*?

CHAPTER 30
Plant Form and Physiology

Figure 30.1 A locust leaf consists of leaflets arrayed along a central midrib. Each leaflet is a complex photosynthetic machine, exquisitely adapted to capture sunlight and carbon dioxide. An intricate vascular system supplies the leaf with water and minerals, and exports the products of photosynthesis. (credit: modification of work by Todd Petit)

Chapter Outline

INTRODUCTION Plants are as essential to human existence as land, water, and air. Without plants, our day-to-day lives would be impossible because without oxygen from photosynthesis, aerobic life cannot be sustained. From providing food and shelter to serving as a source of medicines, oils, perfumes, and industrial products, plants provide humans with numerous valuable resources.

When you think of plants, most of the organisms that come to mind are vascular plants. These plants have tissues that conduct food and water, and most of them have seeds. Seed plants are divided into gymnosperms and angiosperms. Gymnosperms include the needle-leaved conifers—spruce, fir, and pine—as well as less familiar plants, such as ginkgos and cycads. Their seeds are not enclosed by a fleshy fruit. Angiosperms, also called flowering plants, constitute the majority of seed plants. They include broadleaved trees (such as maple, oak, and elm), vegetables (such as potatoes, lettuce, and carrots), grasses, and plants known for the beauty of their flowers (roses, irises, and daffodils, for example).

While individual plant species are unique, all share a common structure: a plant body consisting of stems, roots, and leaves. They all transport water, minerals, and sugars produced through photosynthesis through the plant body in a similar manner. All plant species also respond to environmental factors, such as light, gravity, competition, temperature, and predation.

30.1 The Plant Body

By the end of this section, you will be able to do the following:

- Describe the shoot organ system and the root organ system
- Distinguish between meristematic tissue and permanent tissue
- Identify and describe the three regions where plant growth occurs
- Summarize the roles of dermal tissue, vascular tissue, and ground tissue
- Compare simple plant tissue with complex plant tissue

Like animals, plants contain cells with organelles in which specific metabolic activities take place. Unlike animals, however, plants use energy from sunlight to form sugars during photosynthesis. In addition, plant cells have cell walls, plastids, and a large central vacuole: structures that are not found in animal cells. Each of these cellular structures plays a specific role in plant structure and function.

LINK TO LEARNING

Watch *Botany Without Borders* (http://openstax.org/l/botany_wo_bord) , a video produced by the Botanical Society of America about the importance of plants.

Plant Organ Systems

In plants, just as in animals, similar cells working together form a tissue. When different types of tissues work together to perform a unique function, they form an organ; organs working together form organ systems. Vascular plants have two distinct organ systems: a shoot system, and a root system. The **shoot system** consists of two portions: the vegetative (non-reproductive) parts of the plant, such as the leaves and the stems, and the reproductive parts of the plant, which include flowers and fruits. The shoot system generally grows above ground, where it absorbs the light needed for photosynthesis. The **root system**, which supports the plants and absorbs water and minerals, is usually underground. Figure 30.2 shows the organ systems of a typical plant.

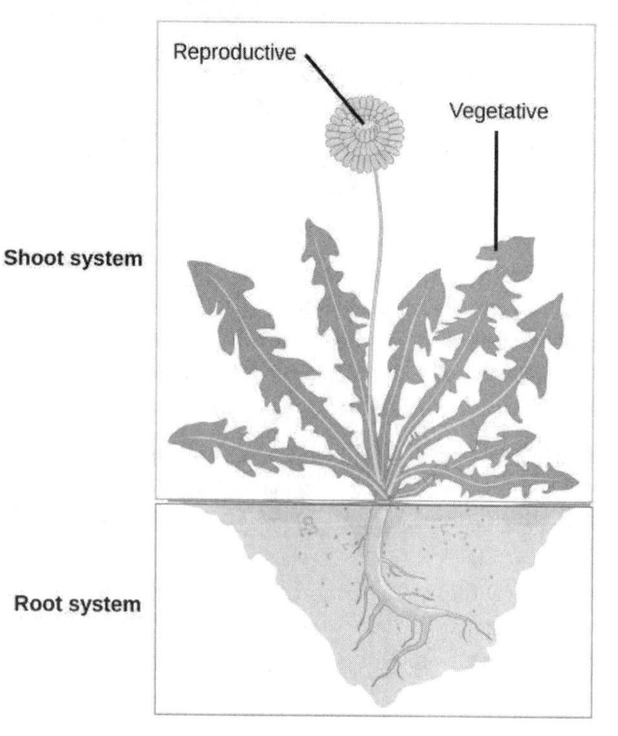

Figure 30.2 The shoot system of a plant consists of leaves, stems, flowers, and fruits. The root system anchors the plant while absorbing water and minerals from the soil.

Plant Tissues

Plants are multicellular eukaryotes with tissue systems made of various cell types that carry out specific functions. Plant tissue systems fall into one of two general types: meristematic tissue, and permanent (or non-meristematic) tissue. Cells of the meristematic tissue are found in **meristems**, which are plant regions of continuous cell division and growth. **Meristematic tissue** cells are either undifferentiated or incompletely differentiated, and they continue to divide and contribute to the growth of the plant. In contrast, **permanent tissue** consists of plant cells that are no longer actively dividing.

Meristematic tissues consist of three types, based on their location in the plant. **Apical meristems** contain meristematic tissue located at the tips of stems and roots, which enable a plant to extend in length. **Lateral**

meristems facilitate growth in thickness or girth in a maturing plant. **Intercalary meristems** occur only in monocots, at the bases of leaf blades and at nodes (the areas where leaves attach to a stem). This tissue enables the monocot leaf blade to increase in length from the leaf base; for example, it allows lawn grass leaves to elongate even after repeated mowing.

Meristems produce cells that quickly differentiate, or specialize, and become permanent tissue. Such cells take on specific roles and lose their ability to divide further. They differentiate into three main types: dermal, vascular, and ground tissue. **Dermal tissue** covers and protects the plant, and **vascular tissue** transports water, minerals, and sugars to different parts of the plant. **Ground tissue** serves as a site for photosynthesis, provides a supporting matrix for the vascular tissue, and helps to store water and sugars.

Secondary tissues are either simple (composed of similar cell types) or complex (composed of different cell types). Dermal tissue, for example, is a simple tissue that covers the outer surface of the plant and controls gas exchange. Vascular tissue is an example of a complex tissue, and is made of two specialized conducting tissues: xylem and phloem. Xylem tissue transports water and nutrients from the roots to different parts of the plant, and includes three different cell types: vessel elements and tracheids (both of which conduct water), and xylem parenchyma. Phloem tissue, which transports organic compounds from the site of photosynthesis to other parts of the plant, consists of four different cell types: sieve cells (which conduct photosynthates), companion cells, phloem parenchyma, and phloem fibers. Unlike xylem conducting cells, phloem conducting cells are alive at maturity. The xylem and phloem always lie adjacent to each other (Figure 30.3). In stems, the xylem and the phloem form a structure called a **vascular bundle**; in roots, this is termed the **vascular stele** or **vascular cylinder**.

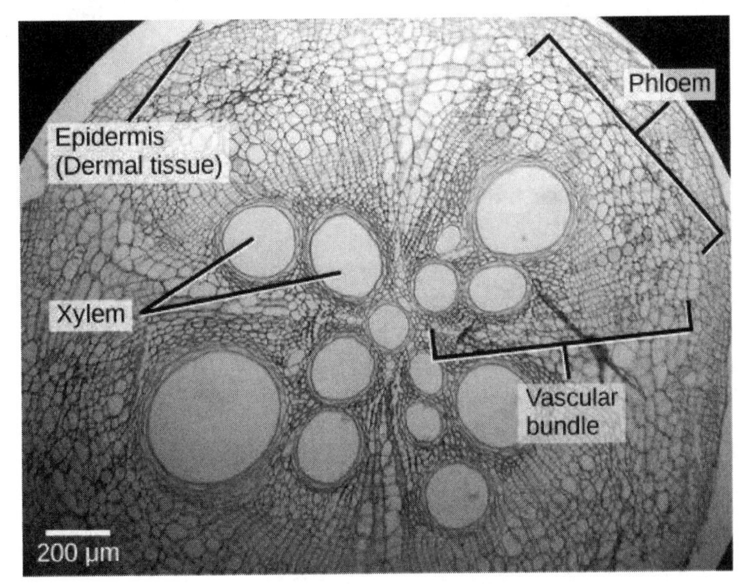

Figure 30.3 This light micrograph shows a cross section of a squash (*Curcurbita maxima*) stem. Each teardrop-shaped vascular bundle consists of large xylem vessels toward the inside and smaller phloem cells toward the outside. Xylem cells, which transport water and nutrients from the roots to the rest of the plant, are dead at functional maturity. Phloem cells, which transport sugars and other organic compounds from photosynthetic tissue to the rest of the plant, are living. The vascular bundles are encased in ground tissue and surrounded by dermal tissue. (credit: modification of work by "(biophotos)"/Flickr; scale-bar data from Matt Russell)

30.2 Stems

By the end of this section, you will be able to do the following:

- Describe the main function and basic structure of stems
- Compare and contrast the roles of dermal tissue, vascular tissue, and ground tissue
- Distinguish between primary growth and secondary growth in stems
- Summarize the origin of annual rings
- List and describe examples of modified stems

Stems are a part of the shoot system of a plant. They may range in length from a few millimeters to hundreds of meters, and also vary in diameter, depending on the plant type. Stems are usually above ground, although the stems of some plants, such as the potato, also grow underground. Stems may be herbaceous (soft) or woody in nature. Their main function is to provide support to the plant, holding leaves, flowers and buds; in some cases, stems also store food for the plant. A stem may be unbranched, like that of a palm tree, or it may be highly branched, like that of a magnolia tree. The stem of the plant connects the roots to the leaves, helping to transport absorbed water and minerals to different parts of the plant. It also helps to transport the products of photosynthesis, namely sugars, from the leaves to the rest of the plant.

Plant stems, whether above or below ground, are characterized by the presence of nodes and internodes (Figure 30.4). **Nodes** are points of attachment for leaves, aerial roots, and flowers. The stem region between two nodes is called an **internode**. The stalk that extends from the stem to the base of the leaf is the petiole. An **axillary bud** is usually found in the axil—the area between the base of a leaf and the stem—where it can give rise to a branch or a flower. The apex (tip) of the shoot contains the apical meristem within the **apical bud**.

Figure 30.4 Leaves are attached to the plant stem at areas called nodes. An internode is the stem region between two nodes. The petiole is the stalk connecting the leaf to the stem. The leaves just above the nodes arose from axillary buds.

Stem Anatomy

The stem and other plant organs arise from the ground tissue, and are primarily made up of simple tissues formed from three types of cells: parenchyma, collenchyma, and sclerenchyma cells.

Parenchyma cells are the most common plant cells (Figure 30.5). They are found in the stem, the root, the inside of the leaf, and the pulp of the fruit. Parenchyma cells are responsible for metabolic functions, such as photosynthesis, and they help repair and heal wounds. Some parenchyma cells also store starch.

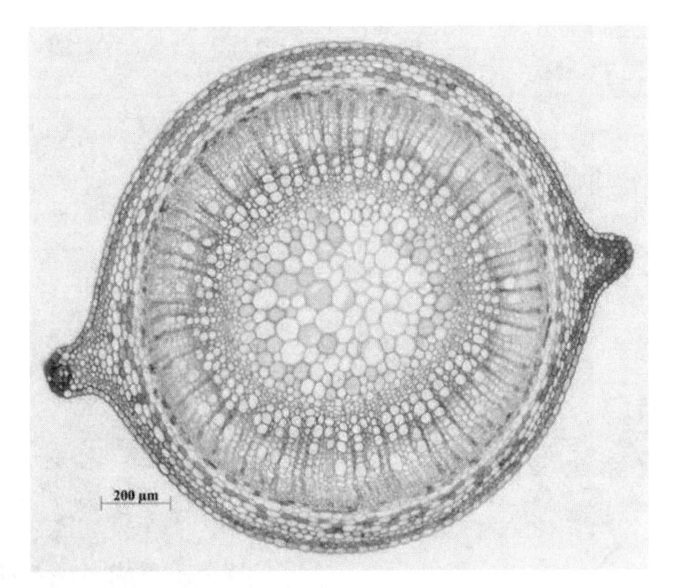

Figure 30.5 The stem of common St John's Wort (*Hypericum perforatum*) is shown in cross section in this light micrograph. The central pith (greenish-blue, in the center) and peripheral cortex (narrow zone 3–5 cells thick just inside the epidermis) are composed of parenchyma cells. Vascular tissue composed of xylem (red) and phloem tissue (green, between the xylem and cortex) surrounds the pith. (credit: Rolf-Dieter Mueller)

Collenchyma cells are elongated cells with unevenly thickened walls (Figure 30.6). They provide structural support, mainly to the stem and leaves. These cells are alive at maturity and are usually found below the epidermis. The "strings" of a celery stalk are an example of collenchyma cells.

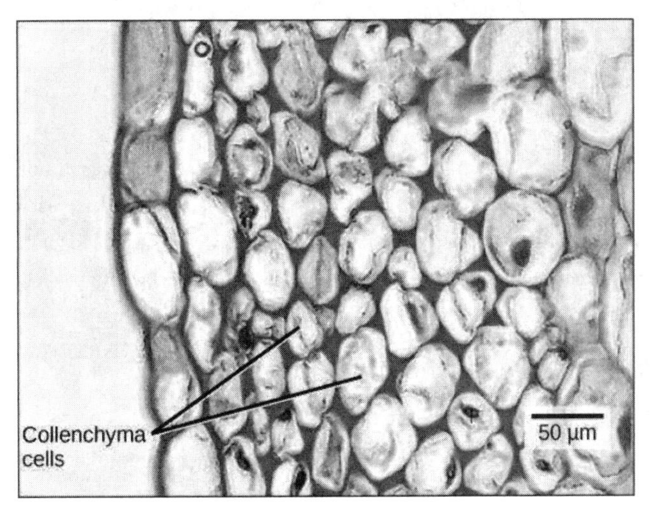

Figure 30.6 Collenchyma cell walls are uneven in thickness, as seen in this light micrograph. They provide support to plant structures. (credit: modification of work by Carl Szczerski; scale-bar data from Matt Russell)

Sclerenchyma cells also provide support to the plant, but unlike collenchyma cells, many of them are dead at maturity. There are two types of sclerenchyma cells: fibers and sclereids. Both types have secondary cell walls that are thickened with deposits of lignin, an organic compound that is a key component of wood. Fibers are long, slender cells; sclereids are smaller-sized. Sclereids give pears their gritty texture. Humans use sclerenchyma fibers to make linen and rope (Figure 30.7).

VISUAL CONNECTION

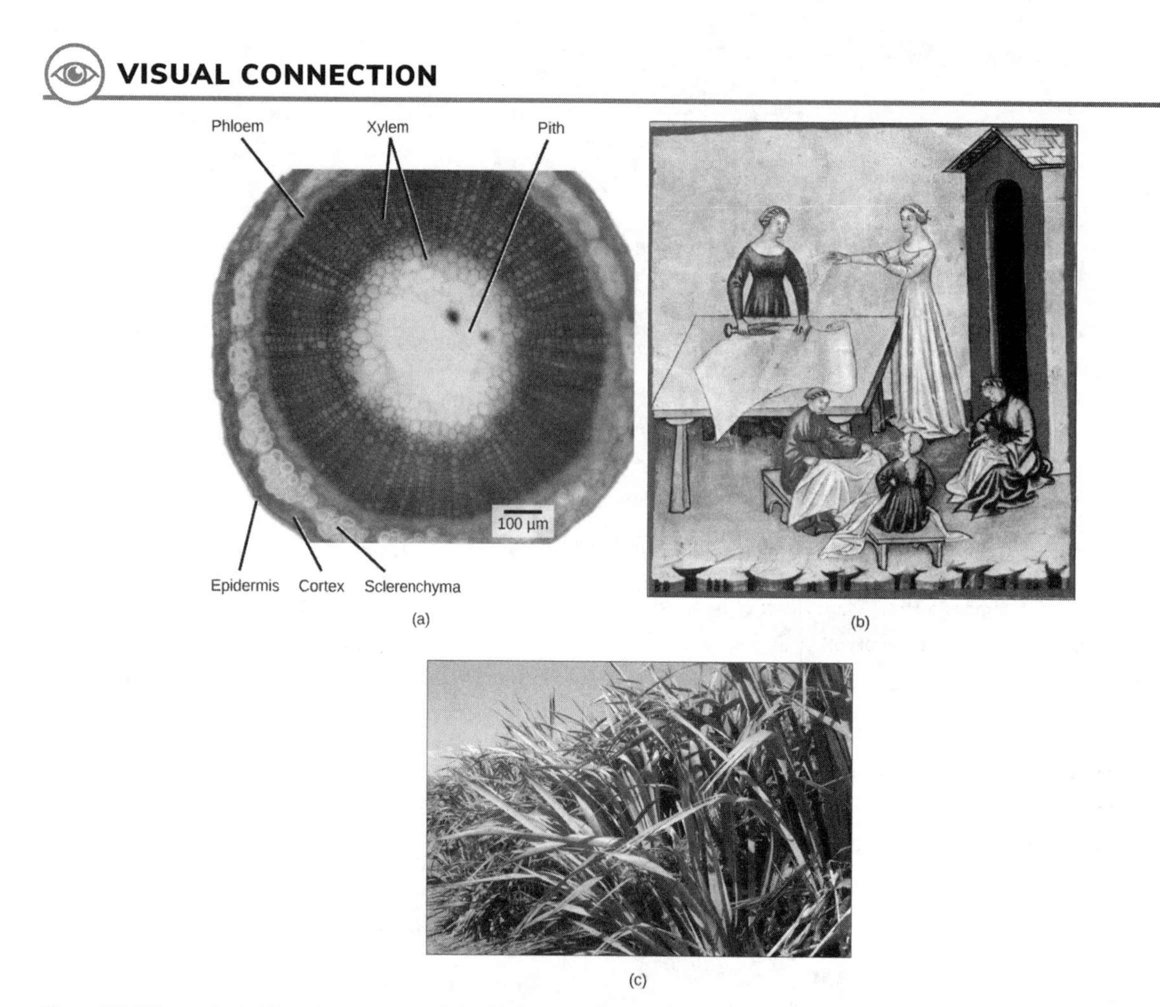

Figure 30.7 The central pith and outer cortex of the (a) flax stem are made up of parenchyma cells. Inside the cortex is a layer of sclerenchyma cells, which make up the fibers in flax rope and clothing. Humans have grown and harvested flax for thousands of years. In (b) this drawing, fourteenth-century women prepare linen. The (c) flax plant is grown and harvested for its fibers, which are used to weave linen, and for its seeds, which are the source of linseed oil. (credit a: modification of work by Emmanuel Boutet based on original work by Ryan R. MacKenzie; credit c: modification of work by Brian Dearth; scale-bar data from Matt Russell)

Which layers of the stem are made of parenchyma cells?

a. cortex and pith
b. phloem
c. sclerenchyma
d. xylem

Like the rest of the plant, the stem has three tissue systems: dermal, vascular, and ground tissue. Each is distinguished by characteristic cell types that perform specific tasks necessary for the plant's growth and survival.

Dermal Tissue

The dermal tissue of the stem consists primarily of **epidermis**, a single layer of cells covering and protecting the underlying tissue. Woody plants have a tough, waterproof outer layer of cork cells commonly known as **bark**, which further protects the plant from damage. Epidermal cells are the most numerous and least differentiated of the cells in the epidermis. The epidermis of a leaf also contains openings known as stomata, through which the exchange of gases takes place (Figure 30.8). Two cells, known as **guard cells**, surround each leaf stoma, controlling its opening and closing and thus regulating the uptake of carbon dioxide and the release of oxygen and water vapor. **Trichomes** are hair-like structures on the epidermal surface. They help to reduce **transpiration** (the loss of water by aboveground plant parts), increase solar reflectance, and store compounds that defend the leaves against predation by herbivores.

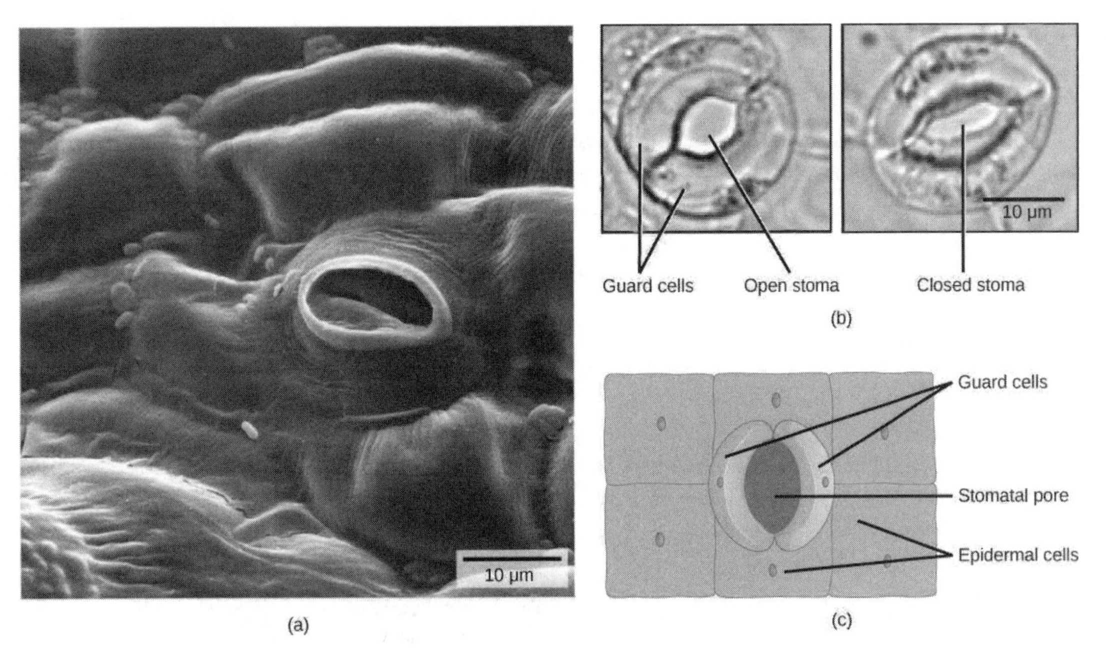

Figure 30.8 Openings called stomata (singular: stoma) allow a plant to take up carbon dioxide and release oxygen and water vapor. The (a) colorized scanning-electron micrograph shows a closed stoma of a dicot. Each stoma is flanked by two guard cells that regulate its (b) opening and closing. The (c) guard cells sit within the layer of epidermal cells. (credit a: modification of work by Louisa Howard, Rippel Electron Microscope Facility, Dartmouth College; credit b: modification of work by June Kwak, University of Maryland; scale-bar data from Matt Russell)

Vascular Tissue

The xylem and phloem that make up the vascular tissue of the stem are arranged in distinct strands called vascular bundles, which run up and down the length of the stem. When the stem is viewed in cross section, the vascular bundles of dicot stems are arranged in a ring. In plants with stems that live for more than one year, the individual bundles grow together and produce the characteristic growth rings. In monocot stems, the vascular bundles are randomly scattered throughout the ground tissue (Figure 30.9).

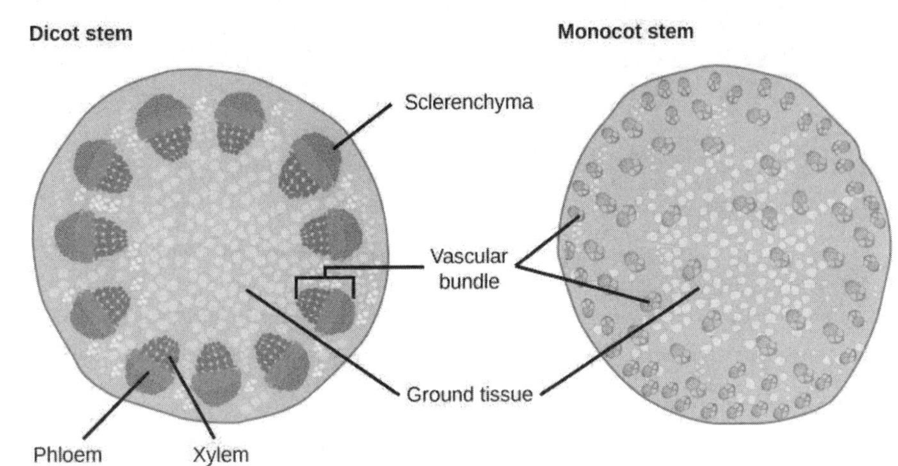

Figure 30.9 In (a) dicot stems, vascular bundles are arranged around the periphery of the ground tissue. The xylem tissue is located toward the interior of the vascular bundle, and phloem is located toward the exterior. Sclerenchyma fibers cap the vascular bundles. In (b) monocot stems, vascular bundles composed of xylem and phloem tissues are scattered throughout the ground tissue.

Xylem tissue has three types of cells: xylem parenchyma, tracheids, and vessel elements. The latter two types conduct water and are dead at maturity. **Tracheids** are xylem cells with thick secondary cell walls that are lignified. Water moves from one tracheid to another through regions on the side walls known as pits, where secondary walls are absent. **Vessel elements** are xylem cells with thinner walls; they are shorter than tracheids. Each vessel element is connected to the next by means of a perforation plate at the end walls of the element. Water moves through the perforation plates to travel up the plant.

Phloem tissue is composed of sieve-tube cells, companion cells, phloem parenchyma, and phloem fibers. A series of **sieve-tube cells** (also called sieve-tube elements) are arranged end to end to make up a long sieve tube, which transports organic substances such as sugars and amino acids. The sugars flow from one sieve-tube cell to the next through perforated sieve plates, which are found at the end junctions between two cells. Although still alive at maturity, the nucleus and other cell components of the sieve-tube cells have disintegrated. **Companion cells** are found alongside the sieve-tube cells, providing them with metabolic support. The companion cells contain more ribosomes and mitochondria than the sieve-tube cells, which lack some cellular organelles.

Ground Tissue

Ground tissue is mostly made up of parenchyma cells, but may also contain collenchyma and sclerenchyma cells that help support the stem. The ground tissue towards the interior of the vascular tissue in a stem or root is known as **pith**, while the layer of tissue between the vascular tissue and the epidermis is known as the **cortex**.

Growth in Stems

Growth in plants occurs as the stems and roots lengthen. Some plants, especially those that are woody, also increase in thickness during their life span. The increase in length of the shoot and the root is referred to as **primary growth**, and is the result of cell division in the shoot apical meristem. **Secondary growth** is characterized by an increase in thickness or girth of the plant, and is caused by cell division in the lateral meristem. Figure 30.10 shows the areas of primary and secondary growth in a plant. Herbaceous plants mostly undergo primary growth, with hardly any secondary growth or increase in thickness. Secondary growth or "wood" is noticeable in woody plants; it occurs in some dicots, but occurs very rarely in monocots.

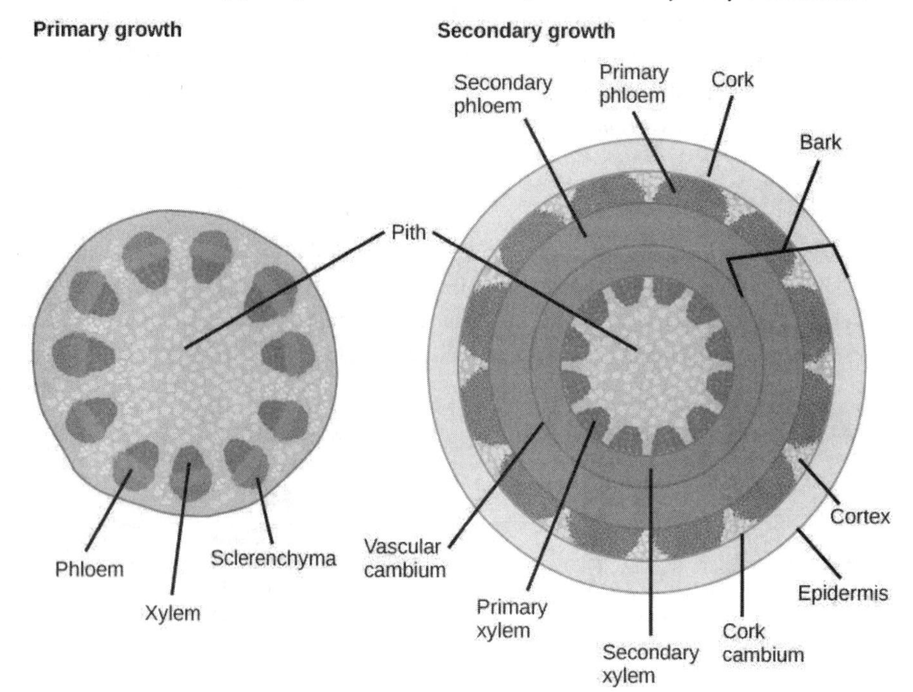

Figure 30.10 In woody plants, primary growth is followed by secondary growth, which allows the plant stem to increase in thickness or girth. Secondary vascular tissue is added as the plant grows, as well as a cork layer. The bark of a tree extends from the vascular cambium to the epidermis.

Some plant parts, such as stems and roots, continue to grow throughout a plant's life: a phenomenon called indeterminate growth. Other plant parts, such as leaves and flowers, exhibit determinate growth, which ceases when a plant part reaches a particular size.

Primary Growth

Most primary growth occurs at the apices, or tips, of stems and roots. Primary growth is a result of rapidly dividing cells in the apical meristems at the shoot tip and root tip. Subsequent cell elongation also contributes to primary growth. The growth of shoots and roots during primary growth enables plants to continuously seek water (roots) or sunlight (shoots).

The influence of the apical bud on overall plant growth is known as apical dominance, which diminishes the growth of axillary buds that form along the sides of branches and stems. Most coniferous trees exhibit strong apical dominance, thus producing the typical conical Christmas tree shape. If the apical bud is removed, then the axillary buds will start forming lateral branches. Gardeners make use of this fact when they prune plants by cutting off the tops of branches, thus encouraging the axillary buds to grow out, giving the plant a bushy shape.

LINK TO LEARNING

Watch this BBC Nature video (http://openstax.org/l/motion_plants) showing how time-lapse photography captures plant growth at high speed.

Secondary Growth

The increase in stem thickness that results from secondary growth is due to the activity of the lateral meristems, which are lacking in herbaceous plants. Lateral meristems include the vascular cambium and, in woody plants, the cork cambium (see Figure 30.10). The vascular cambium is located just outside the primary xylem and to the interior of the primary phloem. The cells of the vascular cambium divide and form secondary xylem (tracheids and vessel elements) to the inside, and secondary phloem (sieve elements and companion cells) to the outside. The thickening of the stem that occurs in secondary growth is due to the formation of secondary phloem and secondary xylem by the vascular cambium, plus the action of cork cambium, which forms the tough outermost layer of the stem. The cells of the secondary xylem contain lignin, which provides hardiness and strength.

In woody plants, cork cambium is the outermost lateral meristem. It produces cork cells (bark) containing a waxy substance known as suberin that can repel water. The bark protects the plant against physical damage and helps reduce water loss. The cork cambium also produces a layer of cells known as phelloderm, which grows inward from the cambium. The cork cambium, cork cells, and phelloderm are collectively termed the **periderm**. The periderm substitutes for the epidermis in mature plants. In some plants, the periderm has many openings, known as **lenticels**, which allow the interior cells to exchange gases with the outside atmosphere (Figure 30.11). This supplies oxygen to the living and metabolically active cells of the cortex, xylem, and phloem.

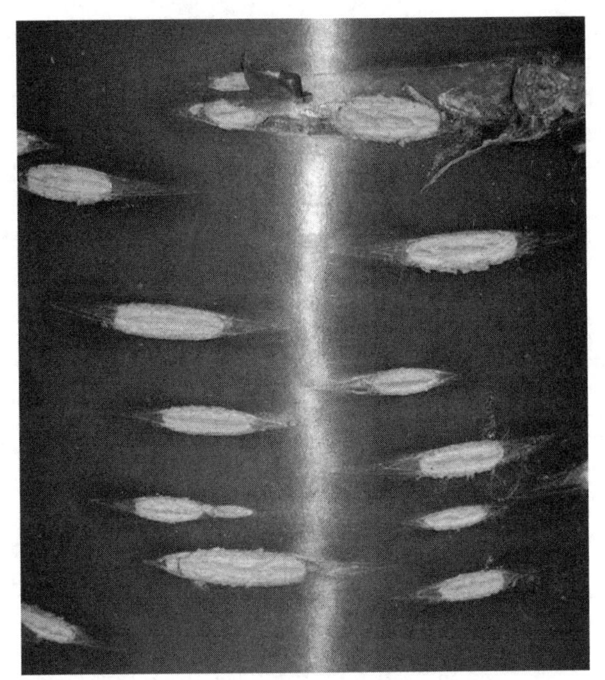

Figure 30.11 Lenticels on the bark of this cherry tree enable the woody stem to exchange gases with the surrounding atmosphere. (credit: Roger Griffith)

Annual Rings

The activity of the vascular cambium gives rise to annual growth rings. During the spring growing season, cells of the secondary xylem have a large internal diameter and their primary cell walls are not extensively thickened. This is known as early wood, or spring wood. During the fall season, the secondary xylem develops thickened cell walls, forming late wood, or autumn wood, which is denser than early wood. This alternation of early and late wood is due largely to a seasonal decrease in the number of vessel elements and a seasonal increase in the number of tracheids. It results in the formation of an annual ring, which can be seen as a circular ring in the cross section of the stem (Figure 30.12). An examination of the number of annual rings and their nature (such as their size and cell wall thickness) can reveal the age of the tree and the prevailing climatic conditions during each season.

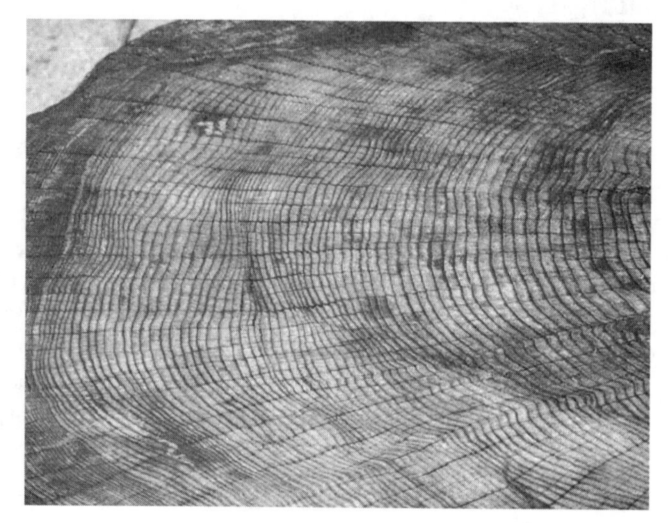

Figure 30.12 The rate of wood growth increases in summer and decreases in winter, producing a characteristic ring for each year of growth. Seasonal changes in weather patterns can also affect the growth rate—note how the rings vary in thickness. (credit: Adrian Pingstone)

Stem Modifications

Some plant species have modified stems that are especially suited to a particular habitat and environment (Figure 30.13). A **rhizome** is a modified stem that grows horizontally underground and has nodes and internodes. Vertical shoots may arise from the buds on the rhizome of some plants, such as ginger and ferns. **Corms** are similar to rhizomes, except they are more rounded and fleshy (such as in gladiolus). Corms contain stored food that enables some plants to survive the winter. **Stolons** are stems that run almost parallel to the ground, or just below the surface, and can give rise to new plants at the nodes. **Runners** are a type of stolon that runs above the ground and produces new clone plants at nodes at varying intervals: strawberries are an example. **Tubers** are modified stems that may store starch, as seen in the potato (*Solanum* sp.). Tubers arise as swollen ends of stolons, and contain many adventitious or unusual buds (familiar to us as the "eyes" on potatoes). A **bulb**, which functions as an underground storage unit, is a modification of a stem that has the appearance of enlarged fleshy leaves emerging from the stem or surrounding the base of the stem, as seen in the iris.

Figure 30.13 Stem modifications enable plants to thrive in a variety of environments. Shown are (a) ginger (*Zingiber officinale*) rhizomes, (b) a carrion flower (*Amorphophallus titanum*) corm, (c) Rhodes grass (*Chloris gayana*) stolons, (d) strawberry (*Fragaria ananassa*) runners, (e) potato (*Solanum tuberosum*) tubers, and (f) red onion (*Allium*) bulbs. (credit a: modification of work by Maja Dumat; credit c: modification of work by Harry Rose; credit d: modification of work by Rebecca Siegel; credit e: modification of work by Scott Bauer, USDA ARS; credit f:

modification of work by Stephen Ausmus, USDA ARS)

LINK TO LEARNING

Watch botanist Wendy Hodgson, of Desert Botanical Garden in Phoenix, Arizona, explain how agave plants were cultivated for food hundreds of years ago in the Arizona desert in this video: (http://openstax.org/l/ancient_crop) *Finding the Roots of an Ancient Crop.*

Some aerial modifications of stems are tendrils and thorns (Figure 30.14). **Tendrils** are slender, twining strands that enable a plant (like a vine or pumpkin) to seek support by climbing on other surfaces. **Thorns** are modified branches appearing as sharp outgrowths that protect the plant; common examples include roses, Osage orange, and devil's walking stick.

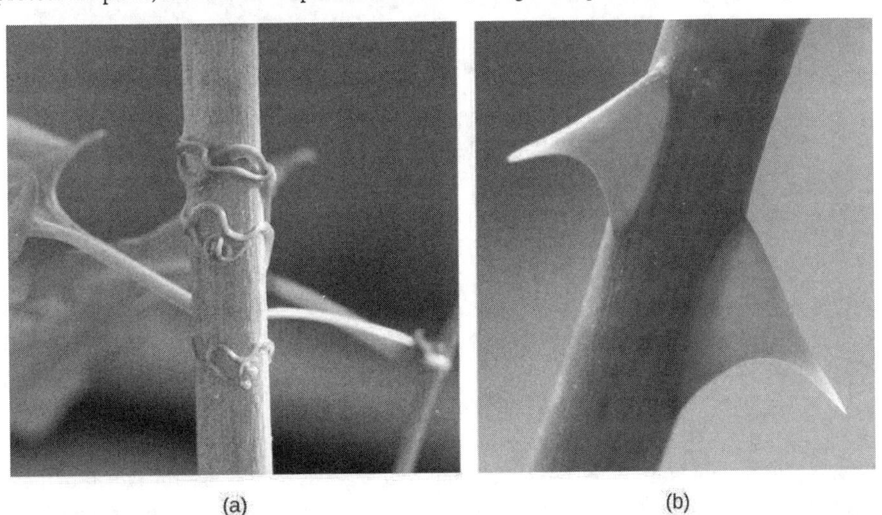

Figure 30.14 Found in southeastern United States, (a) buckwheat vine (*Brunnichia ovata*) is a weedy plant that climbs with the aid of tendrils. This one is shown climbing up a wooden stake. (b) Thorns are modified branches. (credit a: modification of work by Christopher Meloche, USDA ARS; credit b: modification of work by "macrophile"/Flickr)

30.3 Roots

By the end of this section, you will be able to do the following:

- Identify the two types of root systems
- Describe the three zones of the root tip and summarize the role of each zone in root growth
- Describe the structure of the root
- List and describe examples of modified roots

The roots of seed plants have three major functions: anchoring the plant to the soil, absorbing water and minerals and transporting them upwards, and storing the products of photosynthesis. Some roots are modified to absorb moisture and exchange gases. Most roots are underground. Some plants, however, also have **adventitious roots**, which emerge above the ground from the shoot.

Types of Root Systems

Root systems are mainly of two types (Figure 30.15). Dicots have a tap root system, while monocots have a fibrous root system. A **tap root system** has a main root that grows down vertically, and from which many smaller lateral roots arise. Dandelions are a good example; their tap roots usually break off when trying to pull these weeds, and they can regrow another shoot from the remaining root. A tap root system penetrates deep into the soil. In contrast, a **fibrous root system** is located closer to the soil surface, and forms a dense network of roots that also helps prevent soil erosion (lawn grasses are a good example, as are wheat, rice, and corn). Some plants have a combination of tap roots and fibrous roots. Plants that grow in dry areas often have deep root systems, whereas plants growing in areas with abundant water are likely to have shallower root systems.

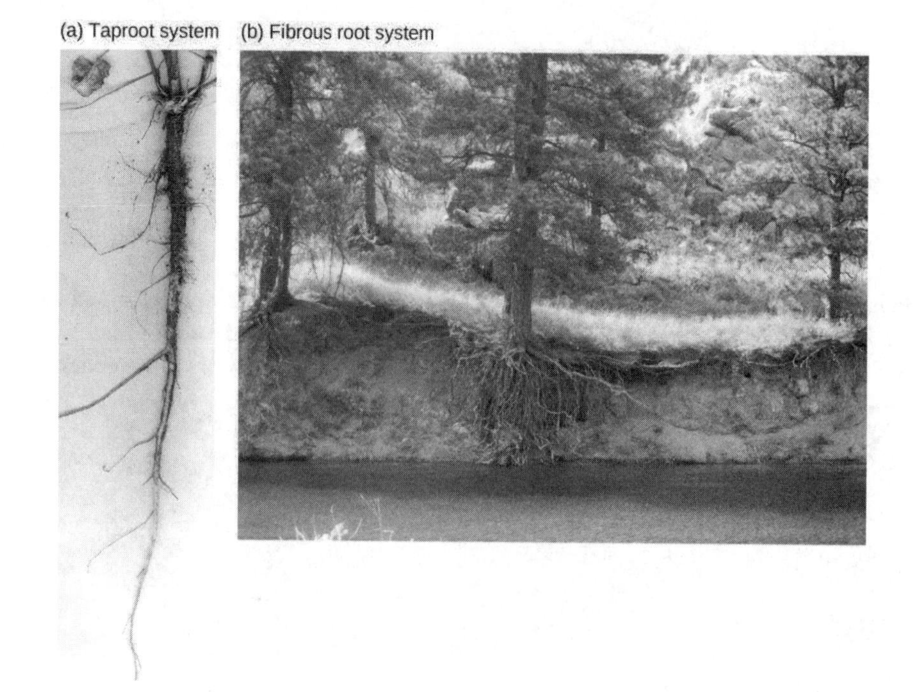

Figure 30.15 (a) Tap root systems have a main root that grows down, while (b) fibrous root systems consist of many small roots. (credit b: modification of work by "Austen Squarepants"/Flickr)

Root Growth and Anatomy

Root growth begins with seed germination. When the plant embryo emerges from the seed, the radicle of the embryo forms the root system. The tip of the root is protected by the **root cap**, a structure exclusive to roots and unlike any other plant structure. The root cap is continuously replaced because it gets damaged easily as the root pushes through soil. The root tip can be divided into three zones: a zone of cell division, a zone of elongation, and a zone of maturation and differentiation (Figure 30.16). The zone of cell division is closest to the root tip; it is made up of the actively dividing cells of the root meristem. The zone of elongation is where the newly formed cells increase in length, thereby lengthening the root. Beginning at the first root hair is the zone of cell maturation where the root cells begin to differentiate into special cell types. All three zones are in the first centimeter or so of the root tip.

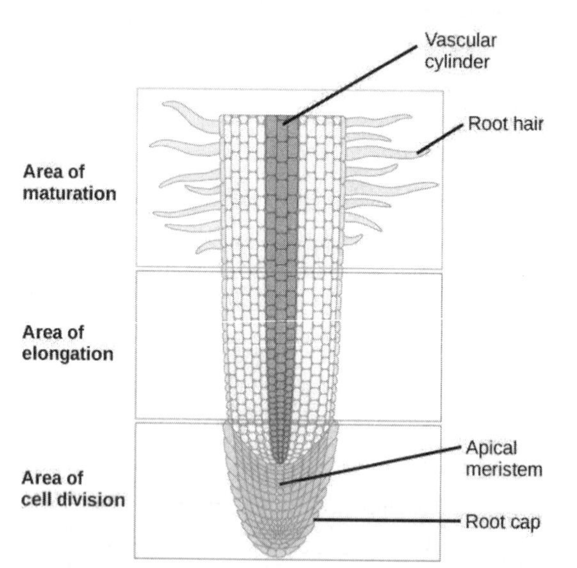

Figure 30.16 A longitudinal view of the root reveals the zones of cell division, elongation, and maturation. Cell division occurs in the apical meristem.

The root has an outer layer of cells called the epidermis, which surrounds areas of ground tissue and vascular tissue. The epidermis provides protection and helps in absorption. **Root hairs**, which are extensions of root epidermal cells, increase the surface area of the root, greatly contributing to the absorption of water and minerals.

Inside the root, the ground tissue forms two regions: the cortex and the pith (Figure 30.17). Compared to stems, roots have lots of cortex and little pith. Both regions include cells that store photosynthetic products. The cortex is between the epidermis and the vascular tissue, whereas the pith lies between the vascular tissue and the center of the root.

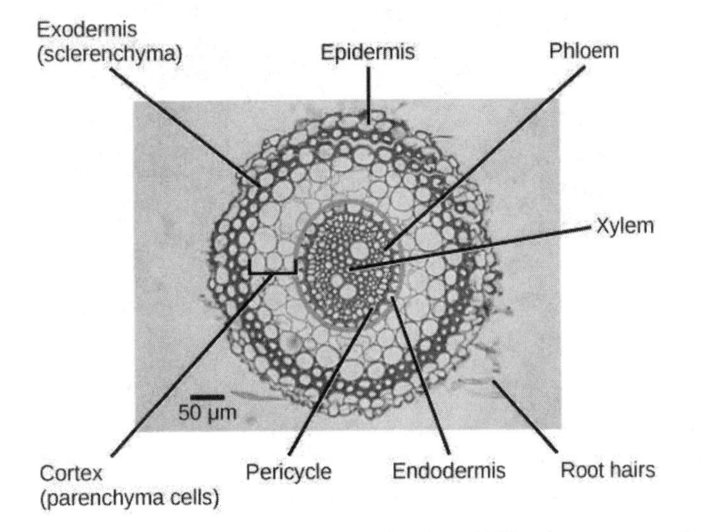

Figure 30.17 Staining reveals different cell types in this light micrograph of a wheat (*Triticum*) root cross section. Sclerenchyma cells of the exodermis and xylem cells stain red, and phloem cells stain blue. Other cell types stain black. The stele, or vascular tissue, is the area inside endodermis (indicated by a green ring). Root hairs are visible outside the epidermis. (credit: scale-bar data from Matt Russell)

The vascular tissue in the root is arranged in the inner portion of the root, which is called the **stele** (Figure 30.18). A layer of cells known as the **endodermis** separates the stele from the ground tissue in the outer portion of the root. The endodermis is exclusive to roots, and serves as a checkpoint for materials entering the root's vascular system. A waxy substance called suberin is present on the walls of the endodermal cells. This waxy region, known as the **Casparian strip**, forces water and solutes to cross the plasma membranes of endodermal cells instead of slipping between the cells. This ensures that only materials required by the root pass through the endodermis, while toxic substances and pathogens are generally excluded. The outermost cell layer of the root's vascular tissue is the **pericycle**, an area that can give rise to lateral roots. In dicot roots, the xylem and phloem of the stele are arranged alternately in an X shape, whereas in monocot roots, the vascular tissue is arranged in a ring around the pith.

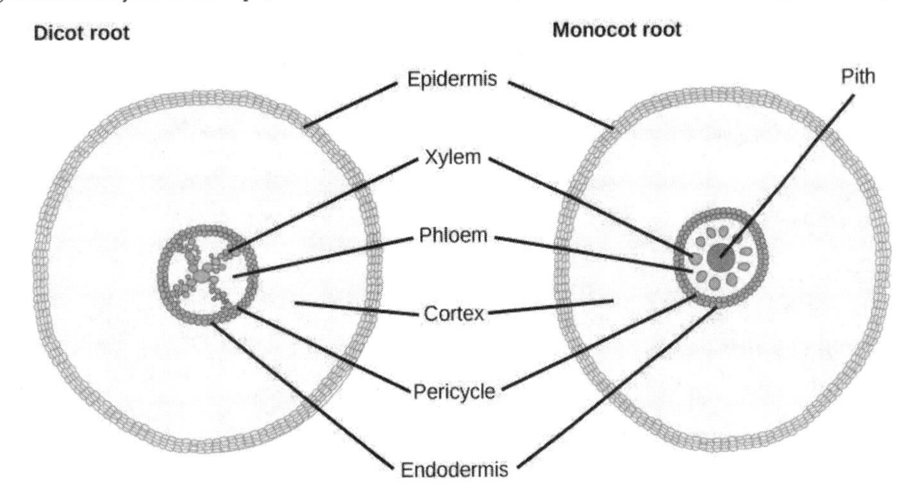

Figure 30.18 In (left) typical dicots, the vascular tissue forms an X shape in the center of the root. In (right) typical monocots, the phloem cells and the larger xylem cells form a characteristic ring around the central pith.

Root Modifications

Root structures may be modified for specific purposes. For example, some roots are bulbous and store starch. Aerial roots and prop roots are two forms of aboveground roots that provide additional support to anchor the plant. Tap roots, such as carrots, turnips, and beets, are examples of roots that are modified for food storage (Figure 30.19).

Figure 30.19 Many vegetables are modified roots.

Epiphytic roots enable a plant to grow on another plant. For example, the epiphytic roots of orchids develop a spongy tissue to absorb moisture. The banyan tree (*Ficus* sp.) begins as an epiphyte, germinating in the branches of a host tree; aerial roots develop from the branches and eventually reach the ground, providing additional support (Figure 30.20). In screwpine (*Pandanus* sp.), a palm-like tree that grows in sandy tropical soils, aboveground prop roots develop from the nodes to provide additional support.

Figure 30.20 The (a) banyan tree, also known as the strangler fig, begins life as an epiphyte in a host tree. Aerial roots extend to the ground and support the growing plant, which eventually strangles the host tree. The (b) screwpine develops aboveground roots that help support the plant in sandy soils. (credit a: modification of work by "psyberartist"/Flickr; credit b: modification of work by David Eikhoff)

30.4 Leaves

By the end of this section, you will be able to do the following:

- Identify the parts of a typical leaf
- Describe the internal structure and function of a leaf
- Compare and contrast simple leaves and compound leaves
- List and describe examples of modified leaves

Leaves are the main sites for photosynthesis: the process by which plants synthesize food. Most leaves are usually green, due to the presence of chlorophyll in the leaf cells. However, some leaves may have different colors, caused by other plant pigments that mask the green chlorophyll.

The thickness, shape, and size of leaves are adapted to the environment. Each variation helps a plant species maximize its chances of survival in a particular habitat. Usually, the leaves of plants growing in tropical rainforests have larger surface areas than those of plants growing in deserts or very cold conditions, which are likely to have a smaller surface area to minimize water loss.

Structure of a Typical Leaf

Each leaf typically has a leaf blade called the **lamina**, which is also the widest part of the leaf. Some leaves are attached to the plant stem by a **petiole**. Leaves that do not have a petiole and are directly attached to the plant stem are called **sessile** leaves. Small green appendages usually found at the base of the petiole are known as **stipules**. Most leaves have a midrib, which travels the length of the leaf and branches to each side to produce veins of vascular tissue. The edge of the leaf is called the margin. Figure 30.21 shows the structure of a typical eudicot leaf.

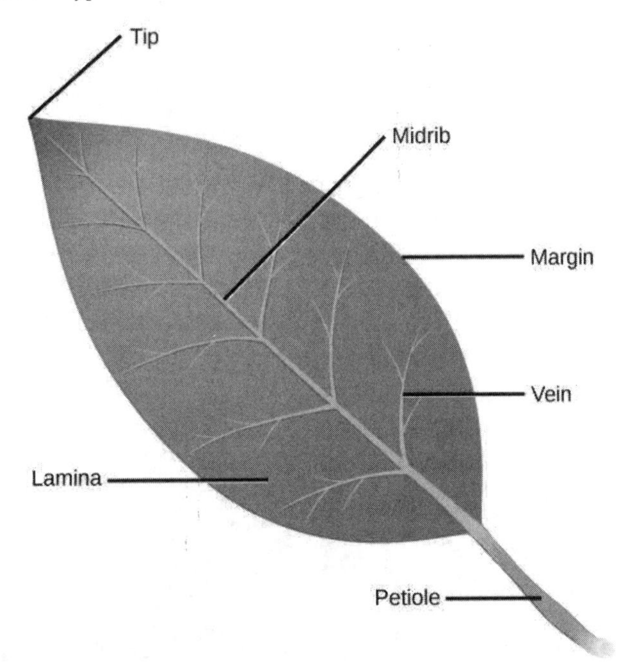

Figure 30.21 Deceptively simple in appearance, a leaf is a highly efficient structure.

Within each leaf, the vascular tissue forms veins. The arrangement of veins in a leaf is called the **venation** pattern. Monocots and dicots differ in their patterns of venation (Figure 30.22). Monocots have parallel venation; the veins run in straight lines across the length of the leaf without converging at a point. In dicots, however, the veins of the leaf have a net-like appearance, forming a pattern known as reticulate venation. One extant plant, the *Ginkgo biloba*, has dichotomous venation where the veins fork.

Figure 30.22 (a) Tulip (*Tulipa*), a monocot, has leaves with parallel venation. The netlike venation in this (b) linden (*Tilia cordata*) leaf distinguishes it as a dicot. The (c) *Ginkgo biloba* tree has dichotomous venation. (credit a photo: modification of work by

"Drewboy64"/Wikimedia Commons; credit b photo: modification of work by Roger Griffith; credit c photo: modification of work by "geishaboy500"/Flickr; credit abc illustrations: modification of work by Agnieszka Kwiecień)

Leaf Arrangement

The arrangement of leaves on a stem is known as **phyllotaxy**. The number and placement of a plant's leaves will vary depending on the species, with each species exhibiting a characteristic leaf arrangement. Leaves are classified as either alternate, spiral, or opposite. Plants that have only one leaf per node have leaves that are said to be either alternate—meaning the leaves alternate on each side of the stem in a flat plane—or spiral, meaning the leaves are arrayed in a spiral along the stem. In an opposite leaf arrangement, two leaves arise at the same point, with the leaves connecting opposite each other along the branch. If there are three or more leaves connected at a node, the leaf arrangement is classified as **whorled**.

Leaf Form

Leaves may be simple or compound (Figure 30.23). In a **simple leaf**, the blade is either completely undivided—as in the banana leaf—or it has lobes, but the separation does not reach the midrib, as in the maple leaf. In a **compound leaf**, the leaf blade is completely divided, forming leaflets, as in the locust tree. Each leaflet may have its own stalk, but is attached to the rachis. A **palmately compound leaf** resembles the palm of a hand, with leaflets radiating outwards from one point. Examples include the leaves of poison ivy, the buckeye tree, or the familiar houseplant *Schefflera* sp. (common name "umbrella plant"). **Pinnately compound leaves** take their name from their feather-like appearance; the leaflets are arranged along the midrib, as in rose leaves (*Rosa* sp.), or the leaves of hickory, pecan, ash, or walnut trees.

Figure 30.23 Leaves may be simple or compound. In simple leaves, the lamina is continuous. The (a) banana plant (*Musa* sp.) has simple leaves. In compound leaves, the lamina is separated into leaflets. Compound leaves may be palmate or pinnate. In (b) palmately compound leaves, such as those of the horse chestnut (*Aesculus hippocastanum*), the leaflets branch from the petiole. In (c) pinnately compound leaves, the leaflets branch from the midrib, as on a scrub hickory (*Carya floridana*). The (d) honey locust has double compound leaves, in which leaflets branch from the veins. (credit a: modification of work by "BazzaDaRambler"/Flickr; credit b: modification of work by Roberto Verzo; credit c: modification of work by Eric Dion; credit d: modification of work by Valerie Lykes)

Leaf Structure and Function

The outermost layer of the leaf is the epidermis; it is present on both sides of the leaf and is called the upper and lower epidermis, respectively. Botanists call the upper side the adaxial surface (or adaxis) and the lower side the abaxial surface (or abaxis). The epidermis helps in the regulation of gas exchange. It contains stomata (Figure 30.24): openings through which the exchange of gases takes place. Two guard cells surround each stoma, regulating its opening and closing.

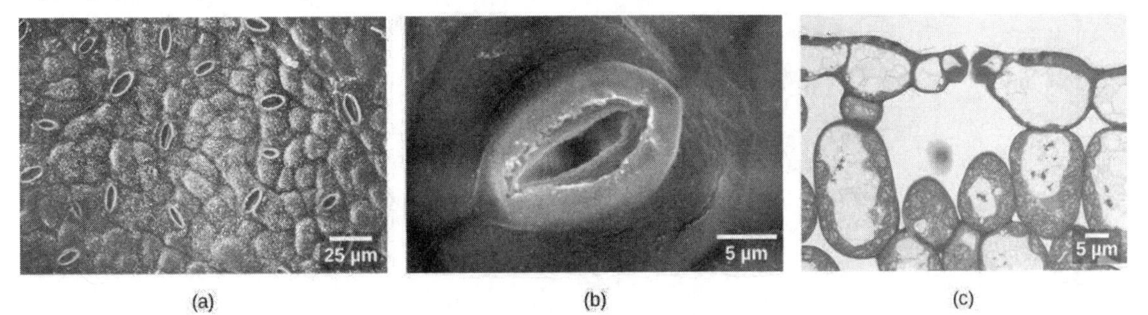

Figure 30.24 Visualized at 500x with a scanning electron microscope, several stomata are clearly visible on (a) the surface of this sumac (*Rhus glabra*) leaf. At 5,000x magnification, the guard cells of (b) a single stoma from lyre-leaved sand cress (*Arabidopsis lyrata)* have the appearance of lips that surround the opening. In this (c) light micrograph cross-section of an *A. lyrata* leaf, the guard cell pair is visible along with the large, sub-stomatal air space in the leaf. (credit: modification of work by Robert R. Wise; part c scale-bar data from Matt Russell)

The epidermis is usually one cell layer thick; however, in plants that grow in very hot or very cold conditions, the epidermis may be several layers thick to protect against excessive water loss from transpiration. A waxy layer known as the **cuticle** covers the leaves of all plant species. The cuticle reduces the rate of water loss from the leaf surface. Other leaves may have small hairs (trichomes) on the leaf surface. Trichomes help to deter herbivory by restricting insect movements, or by storing toxic or bad-tasting compounds; they can also reduce the rate of transpiration by blocking air flow across the leaf surface (Figure 30.25).

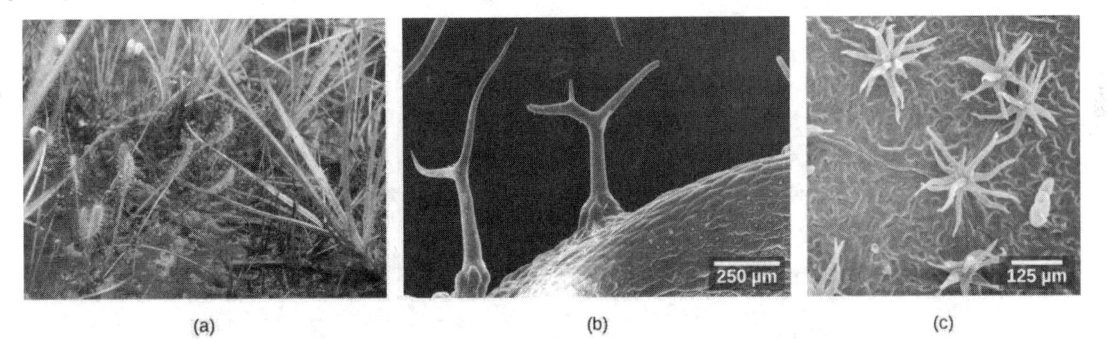

Figure 30.25 Trichomes give leaves a fuzzy appearance as in this (a) sundew (*Drosera* sp.). Leaf trichomes include (b) branched trichomes on the leaf of *Arabidopsis lyrata* and (c) multibranched trichomes on a mature *Quercus marilandica* leaf. (credit a: John Freeland; credit b, c: modification of work by Robert R. Wise; scale-bar data from Matt Russell)

Below the epidermis of dicot leaves are layers of cells known as the mesophyll, or "middle leaf." The mesophyll of most leaves typically contains two arrangements of parenchyma cells: the palisade parenchyma and spongy parenchyma (Figure 30.26). The palisade parenchyma (also called the palisade mesophyll) has column-shaped, tightly packed cells, and may be present in one, two, or three layers. Below the palisade parenchyma are loosely arranged cells of an irregular shape. These are the cells of the spongy parenchyma (or spongy mesophyll). The air space found between the spongy parenchyma cells allows gaseous exchange between the leaf and the outside atmosphere through the stomata. In aquatic plants, the intercellular spaces in the spongy parenchyma help the leaf float. Both layers of the mesophyll contain many chloroplasts. Guard cells are the only epidermal cells to contain chloroplasts.

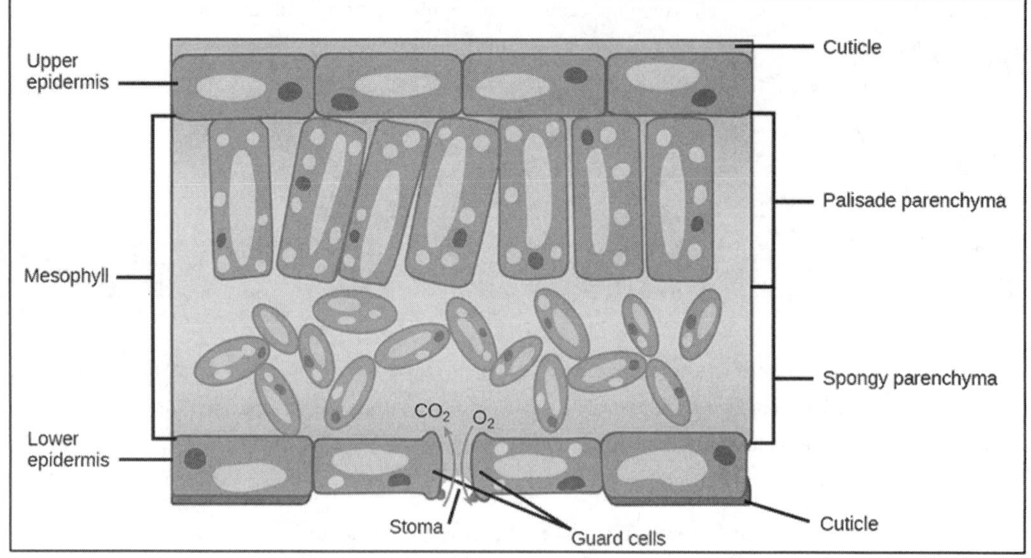

(a)

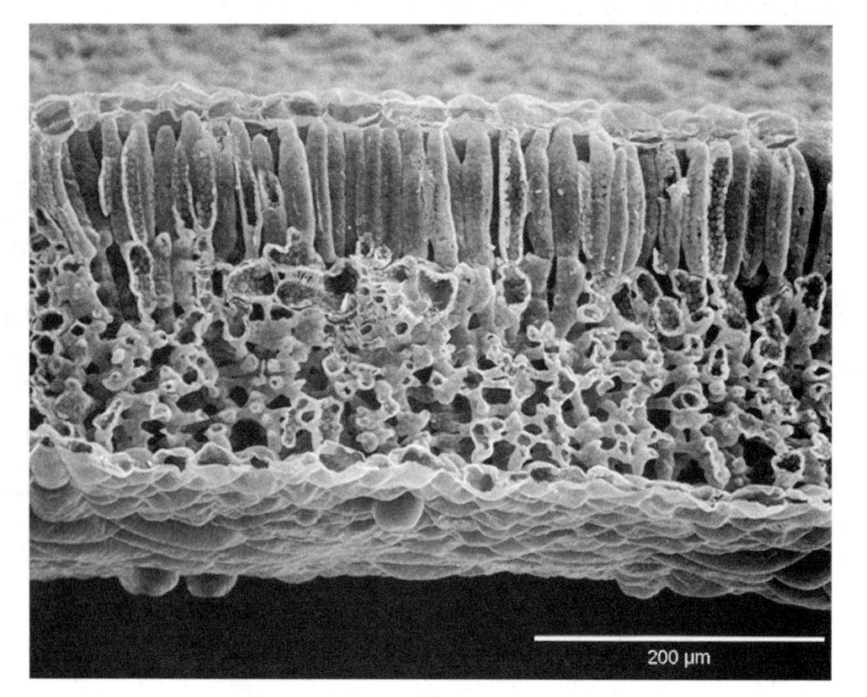

(b)

Figure 30.26 In the (a) leaf drawing, the central mesophyll is sandwiched between an upper and lower epidermis. The mesophyll has two layers: an upper palisade layer comprised of tightly packed, columnar cells, and a lower spongy layer, comprised of loosely packed, irregularly shaped cells. Stomata on the leaf underside allow gas exchange. A waxy cuticle covers all aerial surfaces of land plants to minimize water loss. These leaf layers are clearly visible in the (b) scanning electron micrograph. The numerous small bumps in the palisade parenchyma cells are chloroplasts. Chloroplasts are also present in the spongy parenchyma, but are not as obvious. The bumps protruding from the lower surface of the leave are glandular trichomes, which differ in structure from the stalked trichomes in Figure 30.25. (credit b: modification of work by Robert R. Wise)

Like the stem, the leaf contains vascular bundles composed of xylem and phloem (Figure 30.27). The xylem consists of tracheids and vessels, which transport water and minerals to the leaves. The phloem transports the photosynthetic products from the leaf to the other parts of the plant. A single vascular bundle, no matter how large or small, always contains both xylem and phloem tissues.

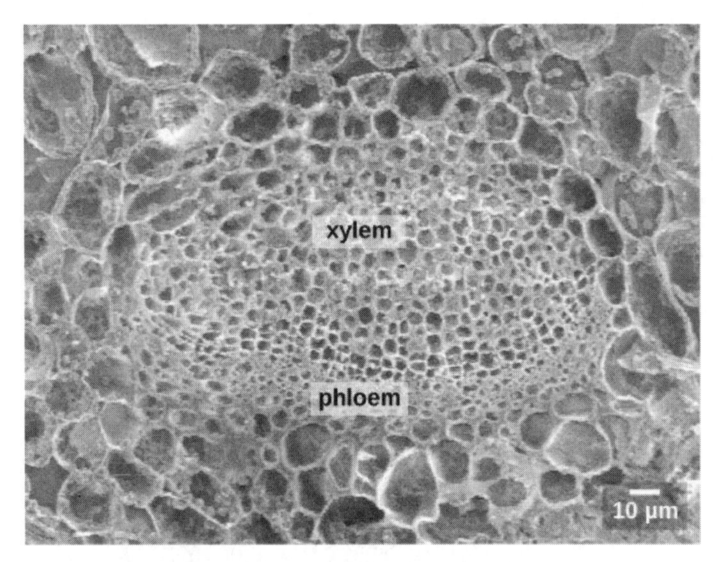

Figure 30.27 This scanning electron micrograph shows xylem and phloem in the leaf vascular bundle from the lyre-leaved sand cress (*Arabidopsis lyrata*). (credit: modification of work by Robert R. Wise; scale-bar data from Matt Russell)

Leaf Adaptations

Coniferous plant species that thrive in cold environments, like spruce, fir, and pine, have leaves that are reduced in size and needle-like in appearance. These needle-like leaves have sunken stomata and a smaller surface area: two attributes that aid in reducing water loss. In hot climates, plants such as cacti have leaves that are reduced to spines, which in combination with their succulent stems, help to conserve water. Many aquatic plants have leaves with wide lamina that can float on the surface of the water, and a thick waxy cuticle on the leaf surface that repels water.

LINK TO LEARNING

Watch "The Pale Pitcher Plant" episode of the video (http://openstax.org/l/plants_cool_too) series *Plants Are Cool, Too,* a Botanical Society of America video about a carnivorous plant species found in Louisiana.

EVOLUTION CONNECTION

Plant Adaptations in Resource-Deficient Environments

Roots, stems, and leaves are structured to ensure that a plant can obtain the required sunlight, water, soil nutrients, and oxygen resources. Some remarkable adaptations have evolved to enable plant species to thrive in less than ideal habitats, where one or more of these resources is in short supply.

In tropical rainforests, light is often scarce, since many trees and plants grow close together and block much of the sunlight from reaching the forest floor. Many tropical plant species have exceptionally broad leaves to maximize the capture of sunlight. Other species are epiphytes: plants that grow on other plants that serve as a physical support. Such plants are able to grow high up in the canopy atop the branches of other trees, where sunlight is more plentiful. Epiphytes live on rain and minerals collected in the branches and leaves of the supporting plant. Bromeliads (members of the pineapple family), ferns, and orchids are examples of tropical epiphytes (Figure 30.28). Many epiphytes have specialized tissues that enable them to efficiently capture and store water.

Figure 30.28 One of the most well known bromeliads is Spanish moss (*Tillandsia usneoides*), seen here in an oak tree. (credit: Kristine Paulus)

Some plants have special adaptations that help them to survive in nutrient-poor environments. Carnivorous plants, such as the Venus flytrap and the pitcher plant (Figure 30.29), grow in bogs where the soil is low in nitrogen. In these plants, leaves are modified to capture insects. The insect-capturing leaves may have evolved to provide these plants with a supplementary source of much-needed nitrogen.

Figure 30.29 The (a) Venus flytrap has modified leaves that can capture insects. When an unlucky insect touches the trigger hairs inside the leaf, the trap suddenly closes. The opening of the (b) pitcher plant is lined with a slippery wax. Insects crawling on the lip slip and fall into a pool of water in the bottom of the pitcher, where they are digested by bacteria. The plant then absorbs the smaller molecules. (credit a: modification of work by Peter Shanks; credit b: modification of work by Tim Mansfield)

Many swamp plants have adaptations that enable them to thrive in wet areas, where their roots grow submerged underwater. In these aquatic areas, the soil is unstable and little oxygen is available to reach the roots. Trees such as mangroves (*Rhizophora* sp.) growing in coastal waters produce aboveground roots that help support the tree (Figure 30.30). Some species of mangroves, as well as cypress trees, have pneumatophores: upward-growing roots containing pores and pockets of tissue specialized for gas exchange. Wild rice is an aquatic plant with large air spaces in the root cortex. The air-filled tissue—called aerenchyma—provides a path for oxygen to diffuse down to the root tips, which are embedded in oxygen-poor bottom sediments.

Figure 30.30 The branches of (a) mangrove trees develop aerial roots, which descend to the ground and help to anchor the trees. (b) Cypress trees and some mangrove species have upward-growing roots called pneumatophores that are involved in gas exchange. Aquatic plants such as (c) wild rice have large spaces in the root cortex called aerenchyma, visualized here using scanning electron microscopy. (credit a: modification of work by Roberto Verzo; credit b: modification of work by Duane Burdick; credit c: modification of work by Robert R. Wise)

LINK TO LEARNING

Watch *Venus Flytraps: Jaws of Death*, an extraordinary BBC close-up of the Venus flytrap in action.

Click to view content (https://www.openstax.org/l/venus_flytrap)

30.5 Transport of Water and Solutes in Plants

By the end of this section, you will be able to do the following:

- Define water potential and explain how it is influenced by solutes, pressure, gravity, and the matric potential
- Describe how water potential, evapotranspiration, and stomatal regulation influence how water is transported in plants
- Explain how photosynthates are transported in plants

The structure of plant roots, stems, and leaves facilitates the transport of water, nutrients, and photosynthates throughout the plant. The phloem and xylem are the main tissues responsible for this movement. Water potential, evapotranspiration, and stomatal regulation influence how water and nutrients are transported in plants. To understand how these processes work, we must first understand the energetics of water potential.

Water Potential

Plants are phenomenal hydraulic engineers. Using only the basic laws of physics and the simple manipulation of potential energy, plants can move water to the top of a 116-meter-tall tree (Figure 30.31a). Plants can also use hydraulics to generate enough force to split rocks and buckle sidewalks (Figure 30.31b). Plants achieve this because of water potential.

Figure 30.31 With heights nearing 116 meters, (a) coastal redwoods (*Sequoia sempervirens*) are the tallest trees in the world. Plant roots can easily generate enough force to (b) buckle and break concrete sidewalks, much to the dismay of homeowners and city maintenance departments. (credit a: modification of work by Bernt Rostad; credit b: modification of work by Pedestrians Educating Drivers on Safety, Inc.)

Water potential is a measure of the potential energy in water. Plant physiologists are not interested in the energy in any one particular aqueous system, but are very interested in water movement between two systems. In practical terms, therefore, water potential is the difference in potential energy between a given water sample and pure water (at atmospheric pressure and ambient temperature). Water potential is denoted by the Greek letter ψ (*psi*) and is expressed in units of pressure (pressure is a form of energy) called **megapascals** (MPa). The potential of pure water ($\Psi_w^{\text{pure H2O}}$) is, by convenience of definition, designated a value of zero (even though pure water contains plenty of potential energy, that energy is ignored). Water potential values for the water in a plant root, stem, or leaf are therefore expressed relative to $\Psi_w^{\text{pure H2O}}$.

The water potential in plant solutions is influenced by solute concentration, pressure, gravity, and factors called matrix effects. Water potential can be broken down into its individual components using the following equation:

$$\Psi_{\text{system}} = \Psi_{\text{total}} = \Psi_s + \Psi_p + \Psi_g + \Psi_m$$

where Ψ_s, Ψ_p, Ψ_g, and Ψ_m refer to the solute, pressure, gravity, and matric potentials, respectively. "System" can refer to the water potential of the soil water (Ψ^{soil}), root water (Ψ^{root}), stem water (Ψ^{stem}), leaf water (Ψ^{leaf}) or the water in the atmosphere ($\Psi^{\text{atmosphere}}$): whichever aqueous system is under consideration. As the individual components change, they raise or lower the total water potential of a system. When this happens, water moves to equilibrate, moving from the system or compartment with a higher water potential to the system or compartment with a lower water potential. This brings the difference in water potential between the two systems ($\Delta\Psi$) back to zero ($\Delta\Psi = 0$). Therefore, for water to move through the plant from the soil to the air (a process called transpiration), Ψ^{soil} must be > $\Psi^{\text{root}} > \Psi^{\text{stem}} > \Psi^{\text{leaf}} > \Psi^{\text{atmosphere}}$.

Water only moves in response to $\Delta\Psi$, not in response to the individual components. However, because the individual components influence the total Ψ_{system}, by manipulating the individual components (especially Ψ_s), a plant can control water movement.

Solute Potential

Solute potential (Ψ_s), also called osmotic potential, is related to the solute concentration (in molarity). That relationship is given by the van 't Hoff equation: $\Psi_s = -M i RT$; where M is the molar concentration of the solute, *i* is the van 't Hoff factor (the ratio of the amount of particles in the solution to amount of formula units dissolved), R is the ideal gas constant, and T is temperature in Kelvin degrees. The solute potential is negative in a plant cell and zero in distilled water. Typical values for cell cytoplasm are –0.5 to –1.0 MPa. Solutes reduce water potential (resulting in a negative Ψ_w) by consuming some of the potential energy available in the water. Solute molecules can dissolve in water because water molecules can bind to them via hydrogen bonds; a hydrophobic molecule like oil, which cannot bind to water, cannot go into solution. The energy in the hydrogen bonds between solute molecules and water is no longer available to do work in the system because it is tied up in the bond. In other words, the amount of available potential energy is reduced when solutes are added to an aqueous system. Thus, Ψ_s decreases with increasing solute concentration. Because Ψ_s is one of the four components of Ψ_{system} or Ψ_{total}, a decrease in Ψ_s will cause a decrease in Ψ_{total}. The internal water potential of a plant cell is more negative than pure water because of the cytoplasm's high solute content (Figure 30.32). Because of this difference in water potential water will move from the soil into a plant's root cells via the process of osmosis. This is why solute potential is sometimes called osmotic potential.

Plant cells can metabolically manipulate Ψ_s (and by extension, Ψ_{total}) by adding or removing solute molecules. Therefore, plants have control over Ψ_{total} via their ability to exert metabolic control over Ψ_s.

VISUAL CONNECTION

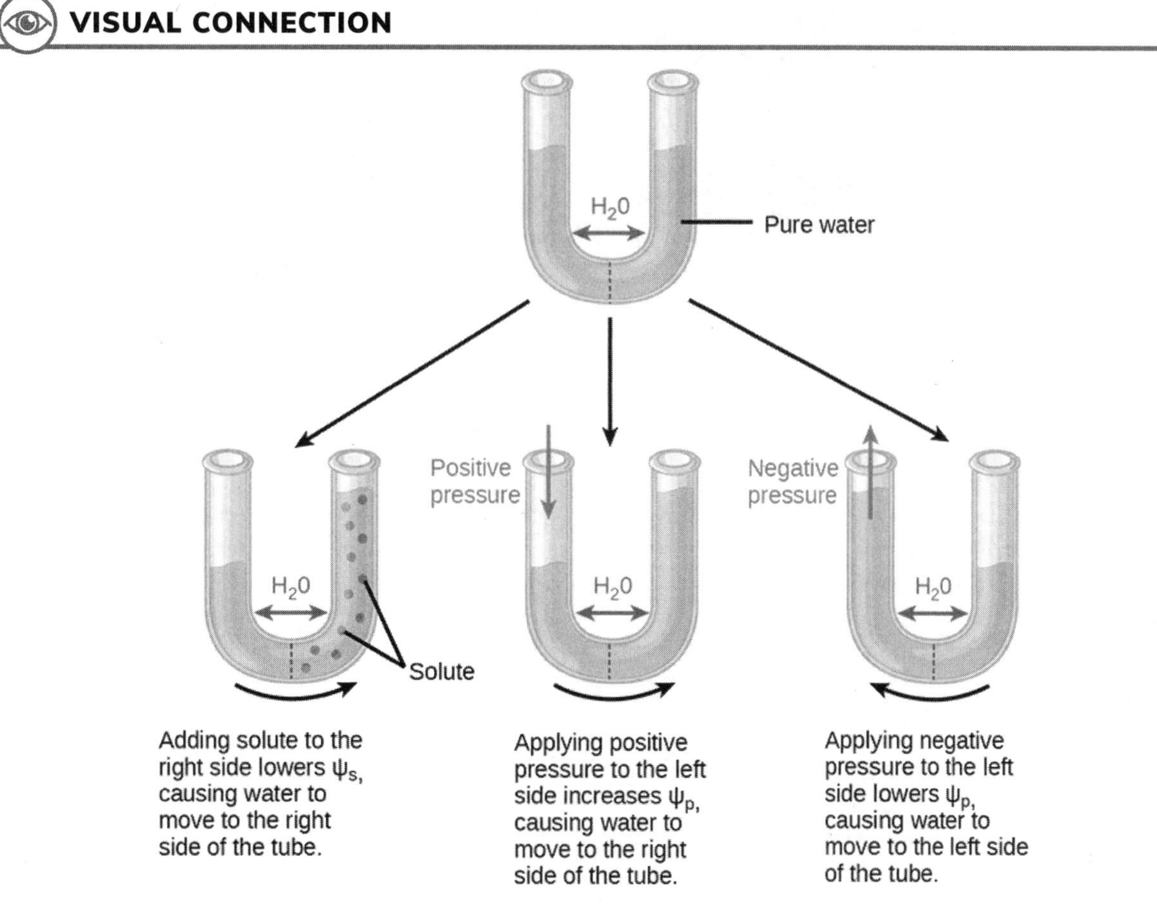

Figure 30.32 In this example with a semipermeable membrane between two aqueous systems, water will move from a region of higher to lower water potential until equilibrium is reached. Solutes (Ψ_s), pressure (Ψ_p), and gravity (Ψ_g) influence total water potential for each side of the tube (Ψ_{total} $^{right\ or\ left}$), and therefore, the difference between Ψ_{total} on each side ($\Delta\Psi$). (Ψ_m , the potential due to interaction of water with solid substrates, is ignored in this example because glass is not especially hydrophilic). Water moves in response to the difference in water potential between two systems (the left and right sides of the tube).

Positive water potential is placed on the left side of the tube by increasing Ψ_p such that the water level rises on the right side. Could you equalize the water level on each side of the tube by adding solute, and if so, how?

Pressure Potential

Pressure potential (Ψ_p), also called turgor potential, may be positive or negative (Figure 30.32). Because pressure is an expression of energy, the higher the pressure, the more potential energy in a system, and vice versa. Therefore, a positive Ψp (compression) increases Ψ_{total}, and a negative Ψ_p (tension) decreases Ψ_{total}. Positive pressure inside cells is contained by the cell wall, producing turgor pressure. Pressure potentials are typically around 0.6–0.8 MPa, but can reach as high as 1.5 MPa in a well-watered plant. A Ψ_p of 1.5 MPa equates to 210 pounds per square inch (1.5 MPa x 140 lb/in^{-2} MPa^{-1} = 210 lb/in^{-2}). As a comparison, most automobile tires are kept at a pressure of 30–34 psi. An example of the effect of turgor pressure is the wilting of leaves and their restoration after the plant has been watered (Figure 30.33). Water is lost from the leaves via transpiration (approaching Ψ_p = 0 MPa at the wilting point) and restored by uptake via the roots.

A plant can manipulate Ψ_p via its ability to manipulate Ψ_s and by the process of osmosis. If a plant cell increases the cytoplasmic solute concentration, Ψ_s will decline, Ψ_{total} will decline, the $\Delta\Psi$ between the cell and the surrounding tissue will decline, water will move into the cell by osmosis, and Ψ_p will increase. Ψ_p is also under indirect plant control via the opening and closing of stomata. Stomatal openings allow water to evaporate from the leaf, reducing Ψ_p and Ψ_{total} of the leaf and increasing Ψ between the water in the leaf and the petiole, thereby allowing water to flow from the petiole into the leaf.

Figure 30.33 When (a) total water potential (Ψ_{total}) is lower outside the cells than inside, water moves out of the cells and the plant wilts. When (b) the total water potential is higher outside the plant cells than inside, water moves into the cells, resulting in turgor pressure (Ψ_p) and keeping the plant erect. (credit: modification of work by Victor M. Vicente Selvas)

Gravity Potential

Gravity potential (Ψ_g) is always negative to zero in a plant with no height. It always removes or consumes potential energy from the system. The force of gravity pulls water downwards to the soil, reducing the total amount of potential energy in the water in the plant (Ψ_{total}). The taller the plant, the taller the water column, and the more influential Ψ_g becomes. On a cellular scale and in short plants, this effect is negligible and easily ignored. However, over the height of a tall tree like a giant coastal redwood, the gravitational pull of -0.1 MPa m^{-1} is equivalent to an extra 1 MPa of resistance that must be overcome for water to reach the leaves of the tallest trees. Plants are unable to manipulate Ψ_g.

Matric Potential

Matric potential (Ψ_m) is always negative to zero. In a dry system, it can be as low as -2 MPa in a dry seed, and it is zero in a water-saturated system. The binding of water to a matrix always removes or consumes potential energy from the system. Ψ_m is similar to solute potential because it involves tying up the energy in an aqueous system by forming hydrogen bonds between the water and some other component. However, in solute potential, the other components are soluble, hydrophilic solute molecules, whereas in Ψ_m, the other components are insoluble, hydrophilic molecules of the plant cell wall. Every plant cell has a cellulosic cell wall and the cellulose in the cell walls is hydrophilic, producing a matrix for adhesion of water: hence the name matric potential. Ψ_m is very large (negative) in dry tissues such as seeds or drought-affected soils. However, it quickly goes to zero as the seed takes up water or the soil hydrates. Ψ_m cannot be manipulated by the plant and is typically ignored in well-watered roots, stems, and leaves.

Movement of Water and Minerals in the Xylem

Solutes, pressure, gravity, and matric potential are all important for the transport of water in plants. Water moves from an area of higher total water potential (higher Gibbs free energy) to an area of lower total water potential. Gibbs free energy is the energy associated with a chemical reaction that can be used to do work. This is expressed as $\Delta\Psi$.

Transpiration is the loss of water from the plant through evaporation at the leaf surface. It is the main driver of water movement in the xylem. Transpiration is caused by the evaporation of water at the leaf–atmosphere interface; it creates negative pressure (tension) equivalent to -2 MPa at the leaf surface. This value varies greatly depending on the vapor pressure deficit, which can be negligible at high relative humidity (RH) and substantial at low RH. Water from the roots is pulled up by this tension. At night, when stomata shut and transpiration stops, the water is held in the stem and leaf by the adhesion of water to the cell walls of the xylem vessels and tracheids, and the cohesion of water molecules to each other. This is called the cohesion–tension theory of sap ascent.

Inside the leaf at the cellular level, water on the surface of mesophyll cells saturates the cellulose microfibrils of the primary cell wall. The leaf contains many large intercellular air spaces for the exchange of oxygen for carbon dioxide, which is required for photosynthesis. The wet cell wall is exposed to this leaf internal air space, and the water on the surface of the cells evaporates into the air spaces, decreasing the thin film on the surface of the mesophyll cells. This decrease creates a greater tension on the water in the mesophyll cells (Figure 30.34), thereby increasing the pull on the water in the xylem vessels. The xylem vessels and tracheids are structurally adapted to cope with large changes in pressure. Rings in the vessels maintain their tubular shape, much like the rings on a vacuum cleaner hose keep the hose open while it is under pressure. Small perforations between vessel elements reduce the number and size of gas bubbles that can form via a process called cavitation. The formation of gas bubbles in xylem interrupts the continuous stream of water from the base to the top of the plant, causing a break termed an embolism in the flow of xylem sap. The taller the tree, the greater the tension forces needed to pull water, and the more cavitation events. In larger trees, the resulting embolisms can plug xylem vessels, making them nonfunctional.

VISUAL CONNECTION

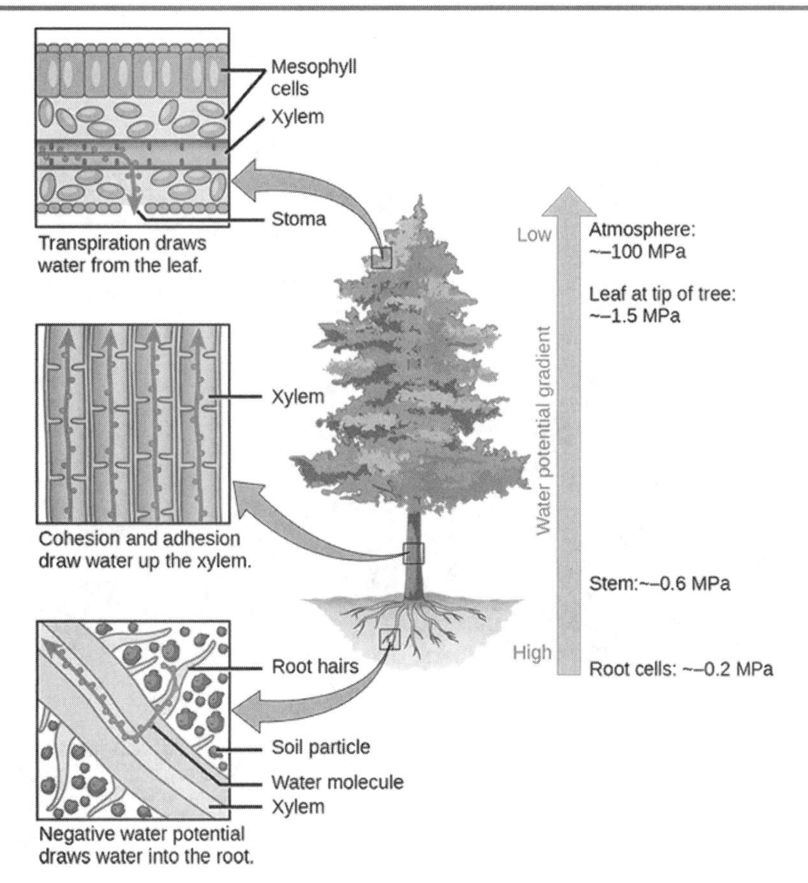

Figure 30.34 The cohesion–tension theory of sap ascent is shown. Evaporation from the mesophyll cells produces a negative water potential gradient that causes water to move upwards from the roots through the xylem.

Which of the following statements is false?

a. Negative water potential draws water into the root hairs. Cohesion and adhesion draw water up the xylem. Transpiration draws water from the leaf.
b. Negative water potential draws water into the root hairs. Cohesion and adhesion draw water up the phloem. Transpiration draws water from the leaf.
c. Water potential decreases from the roots to the top of the plant.
d. Water enters the plants through root hairs and exits through stoma.

Transpiration—the loss of water vapor to the atmosphere through stomata—is a passive process, meaning that metabolic energy in the form of ATP is not required for water movement. The energy driving transpiration is the difference in energy between the water in the soil and the water in the atmosphere. However, transpiration is tightly controlled.

Control of Transpiration

The atmosphere to which the leaf is exposed drives transpiration, but also causes massive water loss from the plant. Up to 90 percent of the water taken up by roots may be lost through transpiration.

Leaves are covered by a waxy **cuticle** on the outer surface that prevents the loss of water. Regulation of transpiration, therefore, is achieved primarily through the opening and closing of stomata on the leaf surface. Stomata are surrounded by two specialized cells called guard cells, which open and close in response to environmental cues such as light intensity and quality, leaf water status, and carbon dioxide concentrations. Stomata must open to allow air containing carbon dioxide and oxygen to diffuse into the leaf for photosynthesis and respiration. When stomata are open, however, water vapor is lost to the external environment, increasing the rate of transpiration. Therefore, plants must maintain a balance between efficient photosynthesis and water loss.

Plants have evolved over time to adapt to their local environment and reduce transpiration (Figure 30.35). Desert plant (xerophytes) and plants that grow on other plants (epiphytes) have limited access to water. Such plants usually have a much thicker waxy cuticle than those growing in more moderate, well-watered environments (mesophytes). Aquatic plants (hydrophytes) also have their own set of anatomical and morphological leaf adaptations.

Figure 30.35 Plants are suited to their local environment. (a) Xerophytes, like this prickly pear cactus (*Opuntia sp.*) and (b) epiphytes such as this tropical *Aeschynanthus perrottetii* have adapted to very limited water resources. The leaves of a prickly pear are modified into spines, which lowers the surface-to-volume ratio and reduces water loss. Photosynthesis takes place in the stem, which also stores water. (b) *A. perottetii* leaves have a waxy cuticle that prevents water loss. (c) Goldenrod (*Solidago sp.*) is a mesophyte, well suited for moderate environments. (d) Hydrophytes, like this fragrant water lily (*Nymphaea odorata*), are adapted to thrive in aquatic environments. (credit a: modification of work by Jon Sullivan; credit b: modification of work by L. Shyamal/Wikimedia Commons; credit c: modification of work by

Huw Williams; credit d: modification of work by Jason Hollinger)

Xerophytes and epiphytes often have a thick covering of trichomes or of stomata that are sunken below the leaf's surface. Trichomes are specialized hair-like epidermal cells that secrete oils and substances. These adaptations impede air flow across the stomatal pore and reduce transpiration. Multiple epidermal layers are also commonly found in these types of plants.

Transportation of Photosynthates in the Phloem

Plants need an energy source to grow. In seeds and bulbs, food is stored in polymers (such as starch) that are converted by metabolic processes into sucrose for newly developing plants. Once green shoots and leaves are growing, plants are able to produce their own food by photosynthesizing. The products of photosynthesis are called photosynthates, which are usually in the form of simple sugars such as sucrose.

Structures that produce photosynthates for the growing plant are referred to as **sources**. Sugars produced in sources, such as leaves, need to be delivered to growing parts of the plant via the phloem in a process called **translocation**. The points of sugar delivery, such as roots, young shoots, and developing seeds, are called **sinks**. Seeds, tubers, and bulbs can be either a source or a sink, depending on the plant's stage of development and the season.

The products from the source are usually translocated to the nearest sink through the phloem. For example, the highest leaves will send photosynthates upward to the growing shoot tip, whereas lower leaves will direct photosynthates downward to the roots. Intermediate leaves will send products in both directions, unlike the flow in the xylem, which is always unidirectional (soil to leaf to atmosphere). The pattern of photosynthate flow changes as the plant grows and develops. Photosynthates are directed primarily to the roots early on, to shoots and leaves during vegetative growth, and to seeds and fruits during reproductive development. They are also directed to tubers for storage.

Translocation: Transport from Source to Sink

Photosynthates, such as sucrose, are produced in the mesophyll cells of photosynthesizing leaves. From there they are translocated through the phloem to where they are used or stored. Mesophyll cells are connected by cytoplasmic channels called plasmodesmata. Photosynthates move through these channels to reach phloem sieve-tube elements (STEs) in the vascular bundles. From the mesophyll cells, the photosynthates are loaded into the phloem STEs. The sucrose is actively transported against its concentration gradient (a process requiring ATP) into the phloem cells using the electrochemical potential of the proton gradient. This is coupled to the uptake of sucrose with a carrier protein called the sucrose-H^+ symporter.

Phloem STEs have reduced cytoplasmic contents, and are connected by a sieve plate with pores that allow for pressure-driven bulk flow, or translocation, of phloem sap. Companion cells are associated with STEs. They assist with metabolic activities and produce energy for the STEs (Figure 30.36).

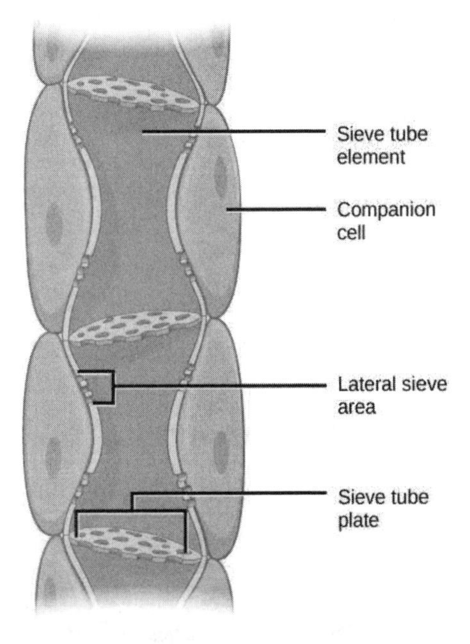

Figure 30.36 Phloem is comprised of cells called sieve-tube elements. Phloem sap travels through perforations called sieve tube plates.

Neighboring companion cells carry out metabolic functions for the sieve-tube elements and provide them with energy. Lateral sieve areas connect the sieve-tube elements to the companion cells.

Once in the phloem, the photosynthates are translocated to the closest sink. Phloem sap is an aqueous solution that contains up to 30 percent sugar, minerals, amino acids, and plant growth regulators. The high percentage of sugar decreases Ψ_s, which decreases the total water potential and causes water to move by osmosis from the adjacent xylem into the phloem tubes, thereby increasing pressure. This increase in total water potential causes the bulk flow of phloem from source to sink (Figure 30.37). Sucrose concentration in the sink cells is lower than in the phloem STEs because the sink sucrose has been metabolized for growth, or converted to starch for storage or other polymers, such as cellulose, for structural integrity. Unloading at the sink end of the phloem tube occurs by either diffusion or active transport of sucrose molecules from an area of high concentration to one of low concentration. Water diffuses from the phloem by osmosis and is then transpired or recycled via the xylem back into the phloem sap.

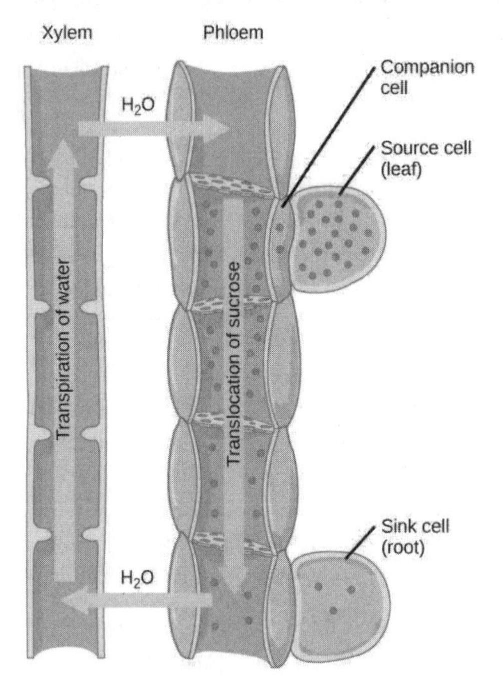

Figure 30.37 Sucrose is actively transported from source cells into companion cells and then into the sieve-tube elements. This reduces the water potential, which causes water to enter the phloem from the xylem. The resulting positive pressure forces the sucrose-water mixture down toward the roots, where sucrose is unloaded. Transpiration causes water to return to the leaves through the xylem vessels.

30.6 Plant Sensory Systems and Responses

By the end of this section, you will be able to do the following:

- Describe how red and blue light affect plant growth and metabolic activities
- Discuss gravitropism
- Understand how hormones affect plant growth and development
- Describe thigmotropism, thigmonastism, and thigmogenesis
- Explain how plants defend themselves from predators and respond to wounds

Animals can respond to environmental factors by moving to a new location. Plants, however, are rooted in place and must respond to the surrounding environmental factors. Plants have sophisticated systems to detect and respond to light, gravity, temperature, and physical touch. Receptors sense environmental factors and relay the information to effector systems—often through intermediate chemical messengers—to bring about plant responses.

Plant Responses to Light

Plants have a number of sophisticated uses for light that go far beyond their ability to photosynthesize low-molecular-weight sugars using only carbon dioxide, light, and water. **Photomorphogenesis** is the growth and development of plants in response to light. It allows plants to optimize their use of light and space. **Photoperiodism** is the ability to use light to track time. Plants can tell the time of day and time of year by sensing and using various wavelengths of sunlight. **Phototropism** is a directional response that allows plants to grow towards, or even away from, light.

The sensing of light in the environment is important to plants; it can be crucial for competition and survival. The response of plants to light is mediated by different photoreceptors, which are comprised of a protein covalently bonded to a light-absorbing pigment called a **chromophore**. Together, the two are called a chromoprotein.

The red/far-red and violet-blue regions of the visible light spectrum trigger structural development in plants. Sensory photoreceptors absorb light in these particular regions of the visible light spectrum because of the quality of light available in the daylight spectrum. In terrestrial habitats, light absorption by chlorophylls peaks in the blue and red regions of the spectrum. As light filters through the canopy and the blue and red wavelengths are absorbed, the spectrum shifts to the far-red end, shifting the plant community to those plants better adapted to respond to far-red light. Blue-light receptors allow plants to gauge the direction and abundance of sunlight, which is rich in blue–green emissions. Water absorbs red light, which makes the detection of blue light essential for algae and aquatic plants.

The Phytochrome System and the Red/Far-Red Response

The **phytochromes** are a family of chromoproteins with a linear tetrapyrrole chromophore, similar to the ringed tetrapyrrole light-absorbing head group of chlorophyll. Phytochromes have two photo-interconvertible forms: Pr and Pfr. Pr absorbs red light (~667 nm) and is immediately converted to Pfr. Pfr absorbs far-red light (~730 nm) and is quickly converted back to Pr. Absorption of red or far-red light causes a massive change to the shape of the chromophore, altering the conformation and activity of the phytochrome protein to which it is bound. Pfr is the physiologically active form of the protein; therefore, exposure to red light yields physiological activity. Exposure to far-red light inhibits phytochrome activity. Together, the two forms represent the phytochrome system (Figure 30.38).

The phytochrome system acts as a biological light switch. It monitors the level, intensity, duration, and color of environmental light. The effect of red light is reversible by immediately shining far-red light on the sample, which converts the chromoprotein to the inactive Pr form. Additionally, Pfr can slowly revert to Pr in the dark, or break down over time. In all instances, the physiological response induced by red light is reversed. The active form of phytochrome (Pfr) can directly activate other molecules in the cytoplasm, or it can be trafficked to the nucleus, where it directly activates or represses specific gene expression.

Once the phytochrome system evolved, plants adapted it to serve a variety of needs. Unfiltered, full sunlight contains much more red light than far-red light. Because chlorophyll absorbs strongly in the red region of the visible spectrum, but not in the far-red region, any plant in the shade of another plant on the forest floor will be exposed to red-depleted, far-red-enriched light. The preponderance of far-red light converts phytochrome in the shaded leaves to the Pr (inactive) form, slowing growth. The nearest non-shaded (or even less-shaded) areas on the forest floor have more red light; leaves exposed to these areas sense the red light, which activates the Pfr form and induces growth. In short, plant shoots use the phytochrome system to grow away from shade and towards light. Because competition for light is so fierce in a dense plant community, the evolutionary advantages of the phytochrome system are obvious.

In seeds, the phytochrome system is not used to determine direction and quality of light (shaded versus unshaded). Instead, is it used merely to determine if there is any light at all. This is especially important in species with very small seeds, such as lettuce. Because of their size, lettuce seeds have few food reserves. Their seedlings cannot grow for long before they run out of fuel. If they germinated even a centimeter under the soil surface, the seedling would never make it into the sunlight and would die. In the dark, phytochrome is in the Pr (inactive form) and the seed will not germinate; it will only germinate if exposed to light at the surface of the soil. Upon exposure to light, Pr is converted to Pfr and germination proceeds.

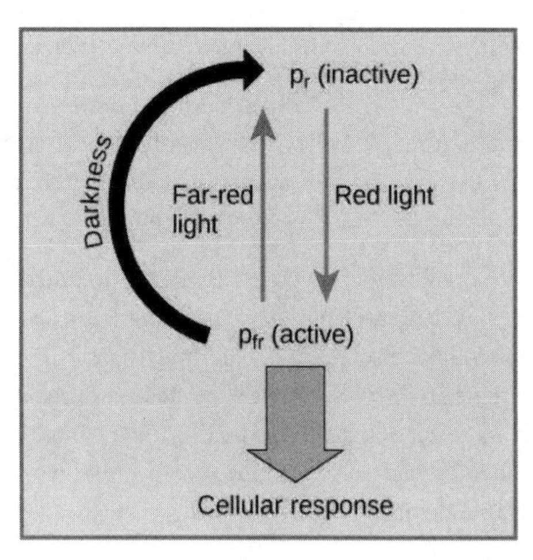

Figure 30.38 The biologically inactive form of phytochrome (Pr) is converted to the biologically active form Pfr under illumination with red light. Far-red light and darkness convert the molecule back to the inactive form.

Plants also use the phytochrome system to sense the change of season. Photoperiodism is a biological response to the timing and duration of day and night. It controls flowering, setting of winter buds, and vegetative growth. Detection of seasonal changes is crucial to plant survival. Although temperature and light intensity influence plant growth, they are not reliable indicators of season because they may vary from one year to the next. Day length is a better indicator of the time of year.

As stated above, unfiltered sunlight is rich in red light but deficient in far-red light. Therefore, at dawn, all the phytochrome molecules in a leaf quickly convert to the active Pfr form, and remain in that form until sunset. In the dark, the Pfr form takes hours to slowly revert back to the Pr form. If the night is long (as in winter), all of the Pfr form reverts. If the night is short (as in summer), a considerable amount of Pfr may remain at sunrise. By sensing the Pr/Pfr ratio at dawn, a plant can determine the length of the day/night cycle. In addition, leaves retain that information for several days, allowing a comparison between the length of the previous night and the preceding several nights. Shorter nights indicate springtime to the plant; when the nights become longer, autumn is approaching. This information, along with sensing temperature and water availability, allows plants to determine the time of the year and adjust their physiology accordingly. Short-day (long-night) plants use this information to flower in the late summer and early fall, when nights exceed a critical length (often eight or fewer hours). Long-day (short-night) plants flower during the spring, when darkness is less than a critical length (often eight to 15 hours). Not all plants use the phytochrome system in this way. Flowering in day-neutral plants is not regulated by daylength.

CAREER CONNECTION

Horticulturalist

The word "horticulturist" comes from the Latin words for garden (*hortus*) and culture (*cultura*). This career has been revolutionized by progress made in the understanding of plant responses to environmental stimuli. Growers of crops, fruit, vegetables, and flowers were previously constrained by having to time their sowing and harvesting according to the season. Now, horticulturists can manipulate plants to increase leaf, flower, or fruit production by understanding how environmental factors affect plant growth and development.

Greenhouse management is an essential component of a horticulturist's education. To lengthen the night, plants are covered with a blackout shade cloth. Long-day plants are irradiated with red light in winter to promote early flowering. For example, fluorescent (cool white) light high in blue wavelengths encourages leafy growth and is excellent for starting seedlings. Incandescent lamps (standard light bulbs) are rich in red light, and promote flowering in some plants. The timing of fruit ripening can be increased or delayed by applying plant hormones. Recently, considerable progress has been made in the development of plant breeds that are suited to different climates and resistant to pests and transportation damage. Both crop yield and quality have increased as a result of practical applications of the knowledge of plant responses to external stimuli and hormones.

Horticulturists find employment in private and governmental laboratories, greenhouses, botanical gardens, and in the production or research fields. They improve crops by applying their knowledge of genetics and plant physiology. To prepare for a horticulture career, students take classes in botany, plant physiology, plant pathology, landscape design, and plant breeding. To complement these traditional courses, horticulture majors add studies in economics, business, computer science, and communications.

The Blue Light Responses

Phototropism—the directional bending of a plant toward or away from a light source—is a response to blue wavelengths of light. Positive phototropism is growth towards a light source (Figure 30.39), while negative phototropism (also called skototropism) is growth away from light.

The aptly-named **phototropins** are protein-based receptors responsible for mediating the phototropic response. Like all plant photoreceptors, phototropins consist of a protein portion and a light-absorbing portion, called the chromophore. In phototropins, the chromophore is a covalently-bound molecule of flavin; hence, phototropins belong to a class of proteins called flavoproteins.

Other responses under the control of phototropins are leaf opening and closing, chloroplast movement, and the opening of stomata. However, of all responses controlled by phototropins, phototropism has been studied the longest and is the best understood.

In their 1880 treatise *The Power of Movements in Plants*, Charles Darwin and his son Francis first described phototropism as the bending of seedlings toward light. Darwin observed that light was perceived by the tip of the plant (the apical meristem), but that the response (bending) took place in a different part of the plant. They concluded that the signal had to travel from the apical meristem to the base of the plant.

Figure 30.39 Azure bluets (*Houstonia caerulea*) display a phototropic response by bending toward the light. (credit: Cory Zanker)

In 1913, Peter Boysen-Jensen demonstrated that a chemical signal produced in the plant tip was responsible for the bending at the base. He cut off the tip of a seedling, covered the cut section with a layer of gelatin, and then replaced the tip. The seedling bent toward the light when illuminated. However, when impermeable mica flakes were inserted between the tip and the cut base, the seedling did not bend. A refinement of the experiment showed that the signal traveled on the shaded side of the seedling. When the mica plate was inserted on the illuminated side, the plant did bend towards the light. Therefore, the chemical signal was a growth stimulant because the phototropic response involved faster cell elongation on the shaded side than on the illuminated side. We now know that as light passes through a plant stem, it is diffracted and generates phototropin activation across the stem. Most activation occurs on the lit side, causing the plant hormone indole acetic acid (IAA) to accumulate on the shaded side. Stem cells elongate under influence of IAA.

Cryptochromes are another class of blue-light absorbing photoreceptors that also contain a flavin-based chromophore. Cryptochromes set the plants' 24-hour activity cycle, also know as its circadian rhythem, using blue light cues. There is some evidence that cryptochromes work together with phototropins to mediate the phototropic response.

LINK TO LEARNING

Use the navigation menu in the left panel of this website (http://openstax.org/l/plnts_n_motion) to view images of plants in motion.

Plant Responses to Gravity

Whether or not they germinate in the light or in total darkness, shoots usually sprout up from the ground, and roots grow downward into the ground. A plant laid on its side in the dark will send shoots upward when given enough time. Gravitropism ensures that roots grow into the soil and that shoots grow toward sunlight. Growth of the shoot apical tip upward is called **negative gravitropism**, whereas growth of the roots downward is called **positive gravitropism**.

Amyloplasts (also known as **statoliths**) are specialized plastids that contain starch granules and settle downward in response to gravity. Amyloplasts are found in shoots and in specialized cells of the root cap. When a plant is tilted, the statoliths drop to the new bottom cell wall. A few hours later, the shoot or root will show growth in the new vertical direction.

The mechanism that mediates gravitropism is reasonably well understood. When amyloplasts settle to the bottom of the gravity-sensing cells in the root or shoot, they physically contact the endoplasmic reticulum (ER), causing the release of calcium ions from inside the ER. This calcium signaling in the cells causes polar transport of the plant hormone IAA to the bottom of the cell. In roots, a high concentration of IAA inhibits cell elongation. The effect slows growth on the lower side of the root, while cells develop normally on the upper side. IAA has the opposite effect in shoots, where a higher concentration at the lower side of the shoot stimulates cell expansion, causing the shoot to grow up. After the shoot or root begin to grow vertically, the amyloplasts return to their normal position. Other hypotheses—involving the entire cell in the gravitropism effect—have been proposed to explain why some mutants that lack amyloplasts may still exhibit a weak gravitropic response.

Growth Responses

A plant's sensory response to external stimuli relies on chemical messengers (hormones). Plant hormones affect all aspects of plant life, from flowering to fruit setting and maturation, and from phototropism to leaf fall. Potentially every cell in a plant can produce plant hormones. They can act in their cell of origin or be transported to other portions of the plant body, with many plant responses involving the synergistic or antagonistic interaction of two or more hormones. In contrast, animal hormones are produced in specific glands and transported to a distant site for action, and they act alone.

Plant hormones are a group of unrelated chemical substances that affect plant morphogenesis. Five major plant hormones are traditionally described: auxins (particularly IAA), cytokinins, gibberellins, ethylene, and abscisic acid. In addition, other nutrients and environmental conditions can be characterized as growth factors.

Auxins

The term auxin is derived from the Greek word *auxein*, which means "to grow." **Auxins** are the main hormones responsible for cell elongation in phototropism and gravitropism. They also control the differentiation of meristem into vascular tissue, and promote leaf development and arrangement. While many synthetic auxins are used as herbicides, IAA is the only naturally occurring auxin that shows physiological activity. Apical dominance—the inhibition of lateral bud formation—is triggered by auxins produced in the apical meristem. Flowering, fruit setting and ripening, and inhibition of **abscission** (leaf falling) are other plant responses under the direct or indirect control of auxins. Auxins also act as a relay for the effects of the blue light and red/far-red responses.

Commercial use of auxins is widespread in plant nurseries and for crop production. IAA is used as a rooting hormone to promote growth of adventitious roots on cuttings and detached leaves. Applying synthetic auxins to tomato plants in greenhouses promotes normal fruit development. Outdoor application of auxin promotes synchronization of fruit setting and dropping to coordinate the harvesting season. Fruits such as seedless cucumbers can be induced to set fruit by treating unfertilized plant flowers with auxins.

Cytokinins

The effect of cytokinins was first reported when it was found that adding the liquid endosperm of coconuts to developing plant embryos in culture stimulated their growth. The stimulating growth factor was found to be **cytokinin**, a hormone that promotes cytokinesis (cell division). Almost 200 naturally occurring or synthetic cytokinins are known to date. Cytokinins are most abundant in growing tissues, such as roots, embryos, and fruits, where cell division is occurring. Cytokinins are known to delay senescence in leaf tissues, promote mitosis, and stimulate differentiation of the meristem in shoots and roots. Many effects on plant development are under the influence of cytokinins, either in conjunction with auxin or another hormone. For example, apical dominance seems to result from a balance between auxins that inhibit lateral buds, and cytokinins that promote bushier growth.

Gibberellins

Gibberellins (GAs) are a group of about 125 closely related plant hormones that stimulate shoot elongation, seed germination, and fruit and flower maturation. GAs are synthesized in the root and stem apical meristems, young leaves, and seed embryos. In urban areas, GA antagonists are sometimes applied to trees under power lines to control growth and reduce the frequency of pruning.

GAs break dormancy (a state of inhibited growth and development) in the seeds of plants that require exposure to cold or light to germinate. Abscisic acid is a strong antagonist of GA action. Other effects of GAs include gender expression, seedless fruit development, and the delay of senescence in leaves and fruit. Seedless grapes are obtained through standard breeding methods and contain inconspicuous seeds that fail to develop. Because GAs are produced by the seeds, and because fruit development and stem elongation are under GA control, these varieties of grapes would normally produce small fruit in compact clusters. Maturing grapes are routinely treated with GA to promote larger fruit size, as well as looser bunches (longer stems), which reduces the instance of mildew infection (Figure 30.40).

Figure 30.40 In grapes, application of gibberellic acid increases the size of fruit and loosens clustering. (credit: Bob Nichols, USDA)

Abscisic Acid

The plant hormone **abscisic acid** (ABA) was first discovered as the agent that causes the abscission or dropping of cotton bolls. However, more recent studies indicate that ABA plays only a minor role in the abscission process. ABA accumulates as a response to stressful environmental conditions, such as dehydration, cold temperatures, or shortened day lengths. Its activity counters many of the growth-promoting effects of GAs and auxins. ABA inhibits stem elongation and induces dormancy in lateral buds.

ABA induces dormancy in seeds by blocking germination and promoting the synthesis of storage proteins. Plants adapted to temperate climates require a long period of cold temperature before seeds germinate. This mechanism protects young plants from sprouting too early during unseasonably warm weather in winter. As the hormone gradually breaks down over winter, the seed is released from dormancy and germinates when conditions are favorable in spring. Another effect of ABA is to promote the development of winter buds; it mediates the conversion of the apical meristem into a dormant bud. Low soil moisture causes an increase in ABA, which causes stomata to close, reducing water loss in winter buds.

Ethylene

Ethylene is associated with fruit ripening, flower wilting, and leaf fall. Ethylene is unusual because it is a volatile gas (C_2H_4). Hundreds of years ago, when gas street lamps were installed in city streets, trees that grew close to lamp posts developed twisted, thickened trunks and shed their leaves earlier than expected. These effects were caused by ethylene volatilizing from the lamps.

Aging tissues (especially senescing leaves) and nodes of stems produce ethylene. The best-known effect of the hormone, however, is the promotion of fruit ripening. Ethylene stimulates the conversion of starch and acids to sugars. Some people store unripe fruit, such as avocadoes, in a sealed paper bag to accelerate ripening; the gas released by the first fruit to mature will speed up the maturation of the remaining fruit. Ethylene also triggers leaf and fruit abscission, flower fading and dropping, and promotes germination in some cereals and sprouting of bulbs and potatoes.

Ethylene is widely used in agriculture. Commercial fruit growers control the timing of fruit ripening with application of the gas. Horticulturalists inhibit leaf dropping in ornamental plants by removing ethylene from greenhouses using fans and ventilation.

Nontraditional Hormones

Recent research has discovered a number of compounds that also influence plant development. Their roles are less understood than the effects of the major hormones described so far.

Jasmonates play a major role in defense responses to herbivory. Their levels increase when a plant is wounded by a predator, resulting in an increase in toxic secondary metabolites. They contribute to the production of volatile compounds that attract natural enemies of predators. For example, chewing of tomato plants by caterpillars leads to an increase in jasmonic acid levels, which in turn triggers the release of volatile compounds that attract predators of the pest.

Oligosaccharins also play a role in plant defense against bacterial and fungal infections. They act locally at the site of injury, and can also be transported to other tissues. **Strigolactones** promote seed germination in some species and inhibit lateral apical development in the absence of auxins. Strigolactones also play a role in the establishment of mycorrhizae, a mutualistic association of plant roots and fungi. Brassinosteroids are important to many developmental and physiological processes. Signals between these compounds and other hormones, notably auxin and GAs, amplifies their physiological effect. Apical dominance, seed germination, gravitropism, and resistance to freezing are all positively influenced by hormones. Root growth and fruit dropping are inhibited by steroids.

Plant Responses to Wind and Touch

The shoot of a pea plant winds around a trellis, while a tree grows on an angle in response to strong prevailing winds. These are examples of how plants respond to touch or wind.

The movement of a plant subjected to constant directional pressure is called **thigmotropism**, from the Greek words *thigma* meaning "touch," and *tropism* implying "direction." Tendrils are one example of this. The meristematic region of tendrils is very touch sensitive; light touch will evoke a quick coiling response. Cells in contact with a support surface contract, whereas cells on the opposite side of the support expand (Figure 30.14). Application of jasmonic acid is sufficient to trigger tendril coiling without a mechanical stimulus.

A **thigmonastic** response is a touch response independent of the direction of stimulus Figure 30.24. In the Venus flytrap, two modified leaves are joined at a hinge and lined with thin fork-like tines along the outer edges. Tiny hairs are located inside the trap. When an insect brushes against these trigger hairs, touching two or more of them in succession, the leaves close quickly, trapping the prey. Glands on the leaf surface secrete enzymes that slowly digest the insect. The released nutrients are absorbed by the leaves, which reopen for the next meal.

Thigmomorphogenesis is a slow developmental change in the shape of a plant subjected to continuous mechanical stress. When trees bend in the wind, for example, growth is usually stunted and the trunk thickens. Strengthening tissue, especially xylem, is produced to add stiffness to resist the wind's force. Researchers hypothesize that mechanical strain induces growth and differentiation to strengthen the tissues. Ethylene and jasmonate are likely involved in thigmomorphogenesis.

LINK TO LEARNING

Use the menu at the left to navigate to three short movies: (http://openstax.org/l/nastic_mvmt) a Venus fly trap capturing prey, the progressive closing of sensitive plant leaflets, and the twining of tendrils.

Defense Responses against Herbivores and Pathogens

Plants face two types of enemies: herbivores and pathogens. Herbivores both large and small use plants as food, and actively chew them. Pathogens are agents of disease. These infectious microorganisms, such as fungi, bacteria, and nematodes, live off of the plant and damage its tissues. Plants have developed a variety of strategies to discourage or kill attackers.

The first line of defense in plants is an intact and impenetrable barrier. Bark and the waxy cuticle can protect against predators. Other adaptations against herbivory include thorns, which are modified branches, and spines, which are modified leaves. They discourage animals by causing physical damage and inducing rashes and allergic reactions. A plant's exterior protection can be compromised by mechanical damage, which may provide an entry point for pathogens. If the first line of defense is breached, the plant must resort to a different set of defense mechanisms, such as toxins and enzymes.

Secondary metabolites are compounds that are not directly derived from photosynthesis and are not necessary for respiration or plant growth and development. Many metabolites are toxic, and can even be lethal to animals that ingest them. Some metabolites are alkaloids, which discourage predators with noxious odors (such as the volatile oils of mint and sage) or repellent tastes (like the bitterness of quinine). Other alkaloids affect herbivores by causing either excessive stimulation (caffeine is one example) or the lethargy associated with opioids. Some compounds become toxic after ingestion. For instance, glycol cyanide in the cassava root releases cyanide only upon ingestion; the nearly 500 million humans who rely on cassava for nutrition must be certain to process the root properly before eating.

Mechanical wounding and predator attacks activate defense and protection mechanisms both in the damaged tissue and at sites farther from the injury location. Some defense reactions occur within minutes: others over several hours. The infected and surrounding cells may die, thereby stopping the spread of infection.

Long-distance signaling elicits a systemic response aimed at deterring the predator. As tissue is damaged, jasmonates may promote the synthesis of compounds that are toxic to predators. Jasmonates also elicit the synthesis of volatile compounds that attract parasitoids, which are insects that spend their developing stages in or on another insect, and eventually kill their host. The plant may activate abscission of injured tissue if it is damaged beyond repair.

KEY TERMS

abscisic acid (ABA) plant hormone that induces dormancy in seeds and other organs

abscission physiological process that leads to the fall of a plant organ (such as leaf or petal drop)

adventitious root aboveground root that arises from a plant part other than the radicle of the plant embryo

apical bud bud formed at the tip of the shoot

apical meristem meristematic tissue located at the tips of stems and roots; enables a plant to extend in length

auxin plant hormone that influences cell elongation (in phototropism), gravitropism, apical dominance, and root growth

axillary bud bud located in the axil: the stem area where the petiole connects to the stem

bark tough, waterproof, outer epidermal layer of cork cells

bulb modified underground stem that consists of a large bud surrounded by numerous leaf scales

Casparian strip waxy coating that forces water to cross endodermal plasma membranes before entering the vascular cylinder, instead of moving between endodermal cells

chromophore molecule that absorbs light

collenchyma cell elongated plant cell with unevenly thickened walls; provides structural support to the stem and leaves

companion cell phloem cell that is connected to sieve-tube cells; has large amounts of ribosomes and mitochondria

compound leaf leaf in which the leaf blade is subdivided to form leaflets, all attached to the midrib

corm rounded, fleshy underground stem that contains stored food

cortex ground tissue found between the vascular tissue and the epidermis in a stem or root

cryptochrome protein that absorbs light in the blue and ultraviolet regions of the light spectrum

cuticle waxy protective layer on the leaf surface

cuticle waxy covering on the outside of the leaf and stem that prevents the loss of water

cytokinin plant hormone that promotes cell division

dermal tissue protective plant tissue covering the outermost part of the plant; controls gas exchange

endodermis layer of cells in the root that forms a selective barrier between the ground tissue and the vascular tissue, allowing water and minerals to enter the root while excluding toxins and pathogens

epidermis single layer of cells found in plant dermal tissue; covers and protects underlying tissue

ethylene volatile plant hormone that is associated with fruit ripening, flower wilting, and leaf fall

fibrous root system type of root system in which the roots arise from the base of the stem in a cluster, forming a dense network of roots; found in monocots

gibberellin (GA) plant hormone that stimulates shoot elongation, seed germination, and the maturation and dropping of fruit and flowers

ground tissue plant tissue involved in photosynthesis; provides support, and stores water and sugars

guard cells paired cells on either side of a stoma that control stomatal opening and thereby regulate the movement of gases and water vapor

intercalary meristem meristematic tissue located at nodes and the bases of leaf blades; found only in monocots

internode region between nodes on the stem

jasmonates small family of compounds derived from the fatty acid linoleic acid

lamina leaf blade

lateral meristem meristematic tissue that enables a plant to increase in thickness or girth

lenticel opening on the surface of mature woody stems that facilitates gas exchange

megapascal (MPa) pressure units that measure water potential

meristem plant region of continuous growth

meristematic tissue tissue containing cells that constantly divide; contributes to plant growth

negative gravitropism growth away from Earth's gravity

node point along the stem at which leaves, flowers, or aerial roots originate

oligosaccharin hormone important in plant defenses against bacterial and fungal infections

palmately compound leaf leaf type with leaflets that emerge from a point, resembling the palm of a hand

parenchyma cell most common type of plant cell; found in the stem, root, leaf, and in fruit pulp; site of photosynthesis and starch storage

pericycle outer boundary of the stele from which lateral roots can arise

periderm outermost covering of woody stems; consists of the cork cambium, cork cells, and the phelloderm

permanent tissue plant tissue composed of cells that are no longer actively dividing

petiole stalk of the leaf

photomorphogenesis growth and development of plants in response to light

photoperiodism occurrence of plant processes, such as germination and flowering, according to the time of year

phototropin blue-light receptor that promotes phototropism, stomatal opening and closing, and other responses that promote photosynthesis

phototropism directional bending of a plant toward a light source

phyllotaxy arrangement of leaves on a stem

phytochrome plant pigment protein that exists in two reversible forms (Pr and Pfr) and mediates morphologic changes in response to red light

pinnately compound leaf leaf type with a divided leaf blade consisting of leaflets arranged on both sides of the midrib

pith ground tissue found towards the interior of the vascular tissue in a stem or root

positive gravitropism growth toward Earth's gravitational center

primary growth growth resulting in an increase in length of the stem and the root; caused by cell division in the shoot or root apical meristem

rhizome modified underground stem that grows horizontally to the soil surface and has nodes and internodes

root cap protective cells covering the tip of the growing root

root hair hair-like structure that is an extension of epidermal cells; increases the root surface area and aids in absorption of water and minerals

root system belowground portion of the plant that supports the plant and absorbs water and minerals

runner stolon that runs above the ground and produces new clone plants at nodes

sclerenchyma cell plant cell that has thick secondary walls and provides structural support; usually dead at maturity

secondary growth growth resulting in an increase in thickness or girth; caused by the lateral meristem and cork cambium

sessile leaf without a petiole that is attached directly to the plant stem

shoot system aboveground portion of the plant; consists of nonreproductive plant parts, such as leaves and stems, and reproductive parts, such as flowers and fruits

sieve-tube cell phloem cell arranged end to end to form a sieve tube that transports organic substances such as sugars and amino acids

simple leaf leaf type in which the lamina is completely undivided or merely lobed

sink growing parts of a plant, such as roots and young leaves, which require photosynthate

source organ that produces photosynthate for a plant

statolith (also, **amyloplast**) plant organelle that contains heavy starch granules

stele inner portion of the root containing the vascular tissue; surrounded by the endodermis

stipule small green structure found on either side of the leaf stalk or petiole

stolon modified stem that runs parallel to the ground and can give rise to new plants at the nodes

strigolactone hormone that promotes seed germination in some species and inhibits lateral apical development in the absence of auxins

tap root system type of root system with a main root that grows vertically with few lateral roots; found in dicots

tendril modified stem consisting of slender, twining strands used for support or climbing

thigmomorphogenesis developmental response to touch

thigmonastic directional growth of a plant independent of the direction in which contact is applied

thigmotropism directional growth of a plant in response to constant contact

thorn modified stem branch appearing as a sharp outgrowth that protects the plant

tracheid xylem cell with thick secondary walls that helps transport water

translocation mass transport of photosynthates from source to sink in vascular plants

transpiration loss of water vapor to the atmosphere through stomata

trichome hair-like structure on the epidermal surface

tuber modified underground stem adapted for starch storage; has many adventitious buds

vascular bundle strands of stem tissue made up of xylem and phloem

vascular stele strands of root tissue made up of xylem and phloem

vascular tissue tissue made up of xylem and phloem that transports food and water throughout the plant

venation pattern of veins in a leaf; may be parallel (as in monocots), reticulate (as in dicots), or dichotomous (as in *Gingko biloba*)

vessel element xylem cell that is shorter than a tracheid and has thinner walls

water potential (Ψ_w) the potential energy of a water solution per unit volume in relation to pure water at atmospheric pressure and ambient temperature

whorled pattern of leaf arrangement in which three or more leaves are connected at a node

CHAPTER SUMMARY

30.1 The Plant Body

A vascular plant consists of two organ systems: the shoot system and the root system. The shoot system includes the aboveground vegetative portions (stems and leaves) and reproductive parts (flowers and fruits). The root system supports the plant and is usually underground. A plant is composed of two main types of tissue: meristematic tissue and permanent tissue. Meristematic tissue consists of actively dividing cells found in root and shoot tips. As growth occurs, meristematic tissue differentiates into permanent tissue, which is categorized as either simple or complex. Simple tissues are made up of similar cell types; examples include dermal tissue and ground tissue. Dermal tissue provides the outer covering of the plant. Ground tissue is responsible for photosynthesis; it also supports vascular tissue and may store water and sugars. Complex tissues are made up of different cell types. Vascular tissue, for example, is made up of xylem and phloem cells.

30.2 Stems

The stem of a plant bears the leaves, flowers, and fruits. Stems are characterized by the presence of nodes (the points of attachment for leaves or branches) and internodes (regions between nodes).

Plant organs are made up of simple and complex tissues. The stem has three tissue systems: dermal, vascular, and ground tissue. Dermal tissue is the outer covering of the plant. It contains epidermal cells, stomata, guard cells, and trichomes. Vascular tissue is made up of xylem and phloem tissues and conducts water, minerals, and photosynthetic products. Ground tissue is responsible for photosynthesis and support and is composed of parenchyma, collenchyma, and sclerenchyma cells.

Primary growth occurs at the tips of roots and shoots, causing an increase in length. Woody plants may also exhibit secondary growth, or increase in thickness. In woody plants, especially trees, annual rings may form as growth slows at the end of each season. Some plant species have modified stems that help to store food, propagate new plants, or discourage predators. Rhizomes, corms, stolons, runners, tubers, bulbs, tendrils, and thorns are examples of modified stems.

30.3 Roots

Roots help to anchor a plant, absorb water and minerals, and serve as storage sites for food. Taproots and fibrous roots are the two main types of root systems. In a taproot system, a main root grows vertically downward with a few lateral roots. Fibrous root systems arise at the base of the stem, where a cluster of roots forms a dense network that is shallower than a taproot. The growing root tip is protected by a root cap. The root tip has three main zones: a zone of cell division (cells are actively dividing), a zone of elongation (cells increase in length), and a zone of maturation (cells differentiate to form different kinds of cells). Root vascular tissue conducts water, minerals, and sugars. In some habitats, the roots of certain plants may be modified to form aerial roots or epiphytic roots.

30.4 Leaves

Leaves are the main site of photosynthesis. A typical leaf consists of a lamina (the broad part of the leaf, also called the blade) and a petiole (the stalk that attaches the leaf to a stem). The arrangement of leaves on a stem, known as phyllotaxy, enables maximum exposure to sunlight. Each plant species has a characteristic leaf arrangement and form. The pattern of leaf arrangement may be alternate, opposite, or spiral, while leaf form may be simple or compound. Leaf tissue consists of the epidermis, which forms the outermost cell layer, and mesophyll and vascular tissue, which make up the inner portion of the leaf. In some plant species, leaf form is modified to form structures such as tendrils, spines, bud scales, and needles.

30.5 Transport of Water and Solutes in Plants

Water potential (Ψ) is a measure of the difference in potential energy between a water sample and pure water. The water potential in plant solutions is influenced by solute concentration, pressure, gravity, and matric potential. Water potential and transpiration influence how water is transported through the xylem in plants. These processes are regulated by stomatal opening and closing. Photosynthates (mainly sucrose) move from sources to sinks through the plant's phloem. Sucrose is actively loaded into the sieve-tube elements of the phloem. The increased solute concentration causes water to move by osmosis from the xylem into the phloem. The positive pressure that is produced pushes water and solutes down the pressure gradient. The sucrose is unloaded into the sink, and the water returns to the xylem vessels.

30.6 Plant Sensory Systems and Responses

Plants respond to light by changes in morphology and activity. Irradiation by red light converts the photoreceptor phytochrome to its far-red light-absorbing form—Pfr. This form controls germination and flowering in response to length of day, as well as triggers photosynthesis in dormant plants or those that just emerged from the soil. Blue-light receptors, cryptochromes, and phototropins are responsible for phototropism. Amyloplasts, which contain heavy starch granules, sense gravity. Shoots exhibit negative gravitropism, whereas roots exhibit positive gravitropism. Plant hormones—naturally occurring compounds synthesized in small amounts—can act both in the cells that produce them and in distant tissues and organs. Auxins are responsible for apical dominance, root growth, directional growth toward light, and many other growth responses. Cytokinins stimulate cell division and counter apical dominance in shoots. Gibberellins inhibit dormancy of seeds and promote stem growth. Abscisic acid induces dormancy in seeds and buds, and protects plants from excessive water loss by promoting stomatal closure. Ethylene gas speeds up fruit ripening and dropping of leaves. Plants respond to touch by rapid movements (thigmotropy and thigmonasty) and slow differential growth (thigmomorphogenesis). Plants have evolved defense mechanisms against predators and pathogens. Physical barriers like bark and spines protect tender tissues. Plants also have chemical defenses, including toxic secondary metabolites and hormones, which elicit additional defense mechanisms.

VISUAL CONNECTION QUESTIONS

1. Figure 30.7 Which layers of the stem are made of parenchyma cells?
   A. cortex and pith
   B. phloem
   C. sclerenchyma
   D. xylem

2. Figure 30.32 Positive water potential is placed on the left side of the tube by increasing Ψ_p such that the water level rises on the right side. Could you equalize the water level on each side of the tube by adding solute, and if so, how?

3. Figure 30.34 Which of the following statements is false?
   a. Negative water potential draws water into the root hairs. Cohesion and adhesion draw water up the xylem. Transpiration draws water from the leaf.
   b. Negative water potential draws water into the root hairs. Cohesion and adhesion draw water up the phloem. Transpiration draws water from the leaf.
   c. Water potential decreases from the roots to the top of the plant.
   d. Water enters the plants through root hairs and exits through stoma.

REVIEW QUESTIONS

4. Plant regions of continuous growth are made up of _______.
   a. dermal tissue
   b. vascular tissue
   c. meristematic tissue
   d. permanent tissue

5. Which of the following is the major site of photosynthesis?
   a. apical meristem
   b. ground tissue
   c. xylem cells
   d. phloem cells

6. Stem regions at which leaves are attached are called _______.
   a. trichomes
   b. lenticels
   c. nodes
   d. internodes

7. Which of the following cell types forms most of the inside of a plant?
   a. meristem cells
   b. collenchyma cells
   c. sclerenchyma cells
   d. parenchyma cells

8. Tracheids, vessel elements, sieve-tube cells, and companion cells are components of _______.
   a. vascular tissue
   b. meristematic tissue
   c. ground tissue
   d. dermal tissue

9. The primary growth of a plant is due to the action of the _______.
   a. lateral meristem
   b. vascular cambium
   c. apical meristem
   d. cork cambium

10. Which of the following is an example of secondary growth?
    a. increase in length
    b. increase in thickness or girth
    c. increase in root hairs
    d. increase in leaf number

11. Secondary growth in stems is usually seen in _______.
    a. monocots
    b. dicots
    c. both monocots and dicots
    d. neither monocots nor dicots

12. Roots that enable a plant to grow on another plant are called _______.
    a. epiphytic roots
    b. prop roots
    c. adventitious roots
    d. aerial roots

13. The _______ forces selective uptake of minerals in the root.
    a. pericycle
    b. epidermis
    c. endodermis
    d. root cap

14. Newly-formed root cells begin to form different cell types in the _______.
    a. zone of elongation
    b. zone of maturation
    c. root meristem
    d. zone of cell division

15. The stalk of a leaf is known as the _______.
    a. petiole
    b. lamina
    c. stipule
    d. rachis

16. Leaflets are a characteristic of _______ leaves.
    a. alternate
    b. whorled
    c. compound
    d. opposite

17. Cells of the _______ contain chloroplasts.
    a. epidermis
    b. vascular tissue
    c. stomata
    d. mesophyll

18. Which of the following is most likely to be found in a desert environment?
    a. broad leaves to capture sunlight
    b. spines instead of leaves
    c. needle-like leaves
    d. wide, flat leaves that can float

19. When stomata open, what occurs?
    a. Water vapor is lost to the external environment, increasing the rate of transpiration.
    b. Water vapor is lost to the external environment, decreasing the rate of transpiration.
    c. Water vapor enters the spaces in the mesophyll, increasing the rate of transpiration.
    d. Water vapor enters the spaces in the mesophyll, decreasing the rate of transpiration.

20. Which cells are responsible for the movement of photosynthates through a plant?
    a. tracheids, vessel elements
    b. tracheids, companion cells
    c. vessel elements, companion cells
    d. sieve-tube elements, companion cells

21. The main photoreceptor that triggers phototropism is a _______.
    a. phytochrome
    b. cryptochrome
    c. phototropin
    d. carotenoid

22. Phytochrome is a plant pigment protein that:
    a. mediates plant infection
    b. promotes plant growth
    c. mediates morphological changes in response to red and far-red light
    d. inhibits plant growth

23. A mutant plant has roots that grow in all directions. Which of the following organelles would you expect to be missing in the cell?
    a. mitochondria
    b. amyloplast
    c. chloroplast
    d. nucleus

24. After buying green bananas or unripe avocadoes, they can be kept in a brown bag to ripen. The hormone released by the fruit and trapped in the bag is probably:
    a. abscisic acid
    b. cytokinin
    c. ethylene
    d. gibberellic acid

25. A decrease in the level of which hormone releases seeds from dormancy?
    a. abscisic acid
    b. cytokinin
    c. ethylene
    d. gibberellic acid

26. A seedling germinating under a stone grows at an angle away from the stone and upward. This response to touch is called _______.
    a. gravitropism
    b. thigmonasty
    c. thigmotropism
    d. skototropism

CRITICAL THINKING QUESTIONS

27. What type of meristem is found only in monocots, such as lawn grasses? Explain how this type of meristematic tissue is beneficial in lawn grasses that are mowed each week.

28. Which plant part is responsible for transporting water, minerals, and sugars to different parts of the plant? Name the two types of tissue that make up this overall tissue, and explain the role of each.

29. Describe the roles played by stomata and guard cells. What would happen to a plant if these cells did not function correctly?
30. Compare the structure and function of xylem to that of phloem.
31. Explain the role of the cork cambium in woody plants.
32. What is the function of lenticels?
33. Besides the age of a tree, what additional information can annual rings reveal?
34. Give two examples of modified stems and explain how each example benefits the plant.
35. Compare a tap root system with a fibrous root system. For each type, name a plant that provides a food in the human diet. Which type of root system is found in monocots? Which type of root system is found in dicots?
36. What might happen to a root if the pericycle disappeared?
37. How do dicots differ from monocots in terms of leaf structure?
38. Describe an example of a plant with leaves that are adapted to cold temperatures.
39. The process of bulk flow transports fluids in a plant. Describe the two main bulk flow processes.
40. Owners and managers of plant nurseries have to plan lighting schedules for a long-day plant that will flower in February. What lighting periods will be most effective? What color of light should be chosen?
41. What are the major benefits of gravitropism for a germinating seedling?
42. Fruit and vegetable storage facilities are usually refrigerated and well ventilated. Why are these conditions advantageous?
43. Stomata close in response to bacterial infection. Why is this response a mechanism of defense for the plant? Which hormone is most likely to mediate this response?

CHAPTER 31
Soil and Plant Nutrition

Figure 31.1 For this (a) squash seedling (*Cucurbita maxima*) to develop into a mature plant bearing its (b) fruit, numerous nutritional requirements must be met. (credit a: modification of work by Julian Colton; credit b: modification of work by "Wildfeuer"/Wikimedia Commons)

INTRODUCTION Cucurbitaceae is a family of plants first cultivated in Mesoamerica, although several species are native to North America. The family includes many edible species, such as squash and pumpkin, as well as inedible gourds. In order to grow and develop into mature, fruit-bearing plants, many requirements must be met and events must be coordinated. Seeds must germinate under the right conditions in the soil; therefore, temperature, moisture, and soil quality are important factors that play a role in germination and seedling development. Soil quality and climate are significant to plant distribution and growth. The young seedling will eventually grow into a mature plant, and the roots will absorb nutrients and water from the soil. At the same time, the aboveground parts of the plant will absorb carbon dioxide from the atmosphere and use energy from sunlight to produce organic compounds through photosynthesis. This chapter will explore the complex dynamics between plants and soils, and the adaptations that plants have evolved to make better use of nutritional resources.

Chapter Outline

31.1 Nutritional Requirements of Plants

By the end of this section, you will be able to do the following:

- Describe how plants obtain nutrients
- List the elements and compounds required for proper plant nutrition
- Describe an essential nutrient

Plants are unique organisms that can absorb nutrients and water through their root system, as well as carbon dioxide from the atmosphere. Soil quality and climate are the major determinants of plant distribution and growth. The combination of soil nutrients, water, and carbon dioxide, along with sunlight, allows plants to grow.

The Chemical Composition of Plants

Since plants require nutrients in the form of elements such as carbon and potassium, it is important to understand the chemical composition of plants. The majority of volume in a plant cell is water; it typically comprises 80 to 90 percent of the plant's total weight. Soil is the water source for land plants, and can be an abundant source of water, even if it appears dry. Plant roots absorb water from the soil through root hairs and transport it up to the leaves through the xylem. As water vapor is lost from the leaves, the process of transpiration and the polarity of water molecules (which enables them to form hydrogen bonds) draws more water from the roots up through the plant to the leaves (Figure 31.2). Plants need water to support cell structure, for metabolic functions, to carry nutrients, and for photosynthesis.

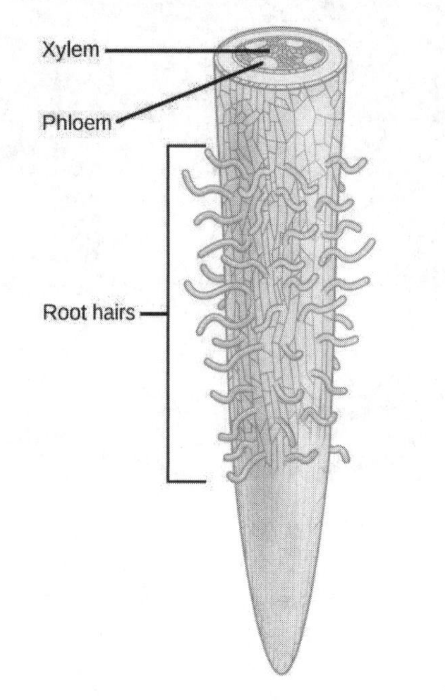

Figure 31.2 Water is absorbed through the root hairs and moves up the xylem to the leaves.

Plant cells need essential substances, collectively called nutrients, to sustain life. Plant nutrients may be composed of either organic or inorganic compounds. An **organic compound** is a chemical compound that contains carbon, such as carbon dioxide obtained from the atmosphere. Carbon that was obtained from atmospheric CO_2 composes the majority of the dry mass within most plants. An **inorganic compound** does not contain carbon and is not part of, or produced by, a living organism. Inorganic substances, which form the majority of the soil solution, are commonly called minerals: those required by plants include nitrogen (N) and potassium (K) for structure and regulation.

Essential Nutrients

Plants require only light, water, and about 20 elements to support all their biochemical needs: these 20 elements are called essential nutrients (Table 31.1). For an element to be regarded as **essential**, three criteria are required: 1) a plant cannot complete its life cycle without the element; 2) no other element can perform the function of the element; and 3) the element is directly involved in plant nutrition.

Essential Elements for Plant Growth

Macronutrients	Micronutrients
Carbon (C)	Iron (Fe)
Hydrogen (H)	Manganese (Mn)
Oxygen (O)	Boron (B)

Table 31.1

Macronutrients	Micronutrients
Nitrogen (N)	Molybdenum (Mo)
Phosphorus (P)	Copper (Cu)
Potassium (K)	Zinc (Zn)
Calcium (Ca)	Chlorine (Cl)
Magnesium (Mg)	Nickel (Ni)
Sulfur (S)	Cobalt (Co)
	Sodium (Na)
	Silicon (Si)

Table 31.1

Macronutrients and Micronutrients

The essential elements can be divided into two groups: macronutrients and micronutrients. Nutrients that plants require in larger amounts are called **macronutrients**. About half of the essential elements are considered macronutrients: carbon, hydrogen, oxygen, nitrogen, phosphorus, potassium, calcium, magnesium and sulfur. The first of these macronutrients, carbon (C), is required to form carbohydrates, proteins, nucleic acids, and many other compounds; it is therefore present in all macromolecules. On average, the dry weight (excluding water) of a cell is 50 percent carbon. As shown in Figure 31.3, carbon is a key part of plant biomolecules.

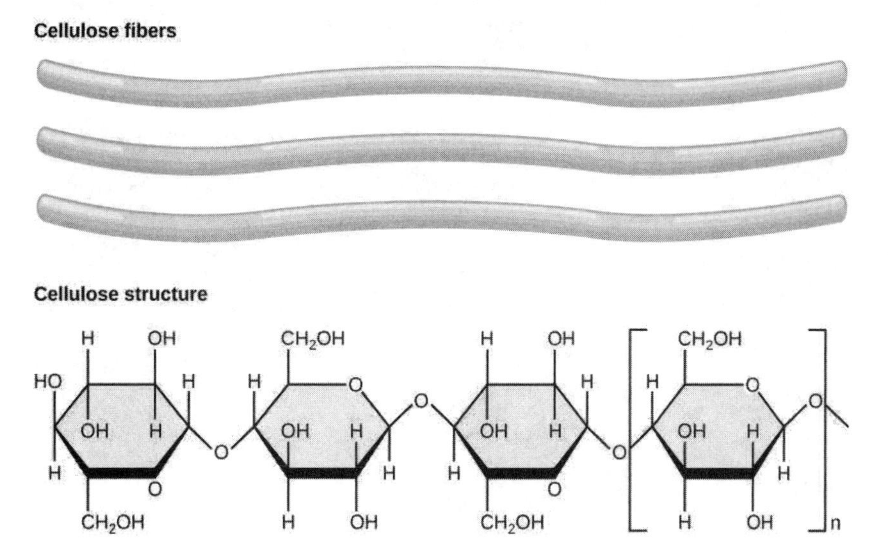

Figure 31.3 Cellulose, the main structural component of the plant cell wall, makes up over thirty percent of plant matter. It is the most abundant organic compound on earth.

The next most abundant element in plant cells is nitrogen (N); it is part of proteins and nucleic acids. Nitrogen is also used in the synthesis of some vitamins. Hydrogen and oxygen are macronutrients that are part of many organic compounds, and also form water. Oxygen is necessary for cellular respiration; plants use oxygen to store energy in the form of ATP. Phosphorus (P), another macromolecule, is necessary to synthesize nucleic acids and phospholipids. As part of ATP, phosphorus enables food energy to be converted into chemical energy through oxidative phosphorylation. Likewise, light energy is converted into chemical energy during photophosphorylation in photosynthesis, and into chemical energy to be extracted during respiration. Sulfur is part of certain amino acids, such as cysteine and methionine, and is present in several coenzymes. Sulfur also plays a role in photosynthesis as part of the electron transport chain, where hydrogen gradients play a key role in the conversion of light energy into ATP. Potassium (K) is important because of its role in regulating stomatal opening and closing. As the openings for gas exchange, stomata help maintain a healthy water balance; a potassium ion pump supports this process.

Magnesium (Mg) and calcium (Ca) are also important macronutrients. The role of calcium is twofold: to regulate nutrient transport, and to support many enzyme functions. Magnesium is important to the photosynthetic process. These minerals, along with the micronutrients, which are described below, also contribute to the plant's ionic balance.

In addition to macronutrients, organisms require various elements in small amounts. These **micronutrients**, or trace elements, are present in very small quantities. They include boron (B), chlorine (Cl), manganese (Mn), iron (Fe), zinc (Zn), copper (Cu), molybdenum (Mo), nickel (Ni), silicon (Si), and sodium (Na).

Deficiencies in any of these nutrients—particularly the macronutrients—can adversely affect plant growth (Figure 31.4). Depending on the specific nutrient, a lack can cause stunted growth, slow growth, or chlorosis (yellowing of the leaves). Extreme deficiencies may result in leaves showing signs of cell death.

LINK TO LEARNING

Visit this website (http://openstax.org/l/plant_mineral) to participate in an interactive experiment on plant nutrient deficiencies. You can adjust the amounts of N, P, K, Ca, Mg, and Fe that plants receive . . . and see what happens.

Figure 31.4 Nutrient deficiency is evident in the symptoms these plants show. This (a) grape tomato suffers from blossom end rot caused by calcium deficiency. The yellowing in this (b) *Frangula alnus* results from magnesium deficiency. Inadequate magnesium also leads to (c) intervenal chlorosis, seen here in a sweetgum leaf. This (d) palm is affected by potassium deficiency. (credit c: modification of work by Jim Conrad; credit d: modification of work by Malcolm Manners)

Everyday Connection

Hydroponics

Figure 31.5 Plant physiologist Ray Wheeler checks onions being grown using hydroponic techniques. The other plants are Bibb lettuce (left) and radishes (right). Credit: NASA

Hydroponics is a method of growing plants in a water-nutrient solution instead of soil. Since its advent, hydroponics has developed into a growing process that researchers often use. Scientists who are interested in studying plant nutrient deficiencies can use hydroponics to study the effects of different nutrient combinations under strictly controlled conditions. Hydroponics has also developed as a way to grow flowers, vegetables, and other crops in greenhouse environments. You might find hydroponically grown produce at your local grocery store. Today, many lettuces and tomatoes in your market have been hydroponically grown.

31.2 The Soil

By the end of this section, you will be able to do the following:

- Describe how soils are formed
- Explain soil composition
- Describe a soil profile

Plants obtain inorganic elements from the soil, which serves as a natural medium for land plants. **Soil** is the outer loose layer that covers the surface of Earth. Soil quality is a major determinant, along with climate, of plant distribution and growth. Soil quality depends not only on the chemical composition of the soil, but also the topography (regional surface features) and the presence of living organisms. In agriculture, the history of the soil, such as the cultivating practices and previous crops, modify the characteristics and fertility of that soil.

Soil develops very slowly over long periods of time, and its formation results from natural and environmental forces acting on mineral, rock, and organic compounds. Soils can be divided into two groups: **organic soils** are those that are formed from sedimentation and primarily composed of organic matter, while those that are formed from the weathering of rocks and are primarily composed of inorganic material are called **mineral soils**. Mineral soils are predominant in terrestrial ecosystems, where soils may be covered by water for part of the year or exposed to the atmosphere.

Soil Composition

Soil consists of these major components (Figure 31.6):

- inorganic mineral matter, about 40 to 45 percent of the soil volume
- organic matter, about 5 percent of the soil volume
- water and air, about 50 percent of the soil volume

The amount of each of the four major components of soil depends on the amount of vegetation, soil compaction, and water present in the soil. A good healthy soil has sufficient air, water, minerals, and organic material to promote and sustain plant life.

VISUAL CONNECTION

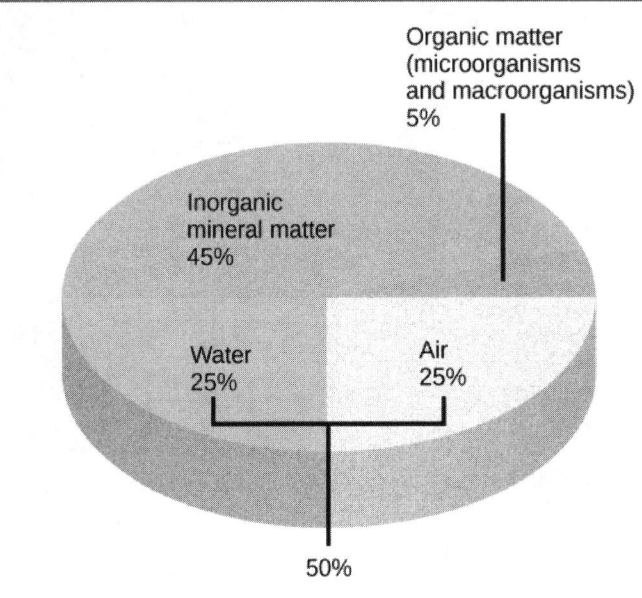

Figure 31.6 The four major components of soil are shown: inorganic minerals, organic matter, water, and air.

Soil compaction can result when soil is compressed by heavy machinery or even foot traffic. How might this compaction change the soil composition?

The organic material of soil, called **humus**, is made up of microorganisms (dead and alive), and dead animals and plants in varying stages of decay. Humus improves soil structure and provides plants with water and minerals. The inorganic material of soil consists of rock, slowly broken down into smaller particles that vary in size. Soil particles that are 0.1 to 2 mm in diameter are **sand**. Soil particles between 0.002 and 0.1 mm are called **silt**, and even smaller particles, less than 0.002 mm in diameter, are called **clay**. Some soils have no dominant particle size and contain a mixture of sand, silt, and humus; these soils are called **loams**.

LINK TO LEARNING

Explore this interactive map (http://openstax.org/l/soil_survey_map) from the USDA's National Cooperative Soil Survey to access soil data for almost any region in the United States.

Soil Formation

Soil formation is the consequence of a combination of biological, physical, and chemical processes. Soil should ideally contain 50 percent solid material and 50 percent pore space. About one-half of the pore space should contain water, and the other half should contain air. The organic component of soil serves as a cementing agent, returns nutrients to the plant, allows soil to store moisture, makes soil tillable for farming, and provides energy for soil microorganisms. Most soil microorganisms—bacteria, algae, or fungi—are dormant in dry soil, but become active once moisture is available.

Soil distribution is not homogenous because its formation results in the production of layers; together, the vertical section of a soil is called the **soil profile**. Within the soil profile, soil scientists define zones called horizons. A **horizon** is a soil layer with distinct physical and chemical properties that differ from those of other layers. Five factors account for soil formation: parent material, climate, topography, biological factors, and time.

Parent Material

The organic and inorganic material in which soils form is the **parent material**. Mineral soils form directly from the weathering of **bedrock**, the solid rock that lies beneath the soil, and therefore, they have a similar composition to the original rock. Other soils form in materials that came from elsewhere, such as sand and glacial drift. Materials located in the depth of the soil are relatively unchanged compared with the deposited material. Sediments in rivers may have different characteristics, depending on whether the stream moves quickly or slowly. A fast-moving river could have sediments of rocks and sand, whereas a slow-moving river could have fine-textured material, such as clay.

Climate

Temperature, moisture, and wind cause different patterns of weathering and therefore affect soil characteristics. The presence of moisture and nutrients from weathering will also promote biological activity: a key component of a quality soil.

Topography

Regional surface features (familiarly called "the lay of the land") can have a major influence on the characteristics and fertility of a soil. Topography affects water runoff, which strips away parent material and affects plant growth. Steeps soils are more prone to erosion and may be thinner than soils that are relatively flat or level.

Biological factors

The presence of living organisms greatly affects soil formation and structure. Animals and microorganisms can produce pores and crevices, and plant roots can penetrate into crevices to produce more fragmentation. Plant secretions promote the development of microorganisms around the root, in an area known as the **rhizosphere**. Additionally, leaves and other material that fall from plants decompose and contribute to soil composition.

Time

Time is an important factor in soil formation because soils develop over long periods. Soil formation is a dynamic process. Materials are deposited over time, decompose, and transform into other materials that can be used by living organisms or deposited onto the surface of the soil.

Physical Properties of the Soil

Soils are named and classified based on their horizons. The soil profile has four distinct layers: 1) O horizon; 2) A horizon; 3) B horizon, or subsoil; and 4) C horizon, or soil base (Figure 31.7). The **O horizon** has freshly decomposing organic matter—humus—at its surface, with decomposed vegetation at its base. Humus enriches the soil with nutrients and enhances soil moisture retention. Topsoil—the top layer of soil—is usually two to three inches deep, but this depth can vary considerably. For instance, river deltas like the Mississippi River delta have deep layers of topsoil. Topsoil is rich in organic material; microbial processes occur there, and it is the "workhorse" of plant production. The **A horizon** consists of a mixture of organic material with inorganic products of weathering, and it is therefore the beginning of true mineral soil. This horizon is typically darkly colored because of the presence of organic matter. In this area, rainwater percolates through the soil and carries materials from the surface. The **B horizon** is an accumulation of mostly fine material that has moved downward, resulting in a dense layer in the soil. In some soils, the B horizon contains nodules or a layer of calcium carbonate. The **C horizon**, or soil base, includes the parent material, plus the organic and inorganic material that is broken down to form soil. The parent material may be either created in its natural place, or transported from elsewhere to its present location. Beneath the C horizon lies bedrock.

VISUAL CONNECTION

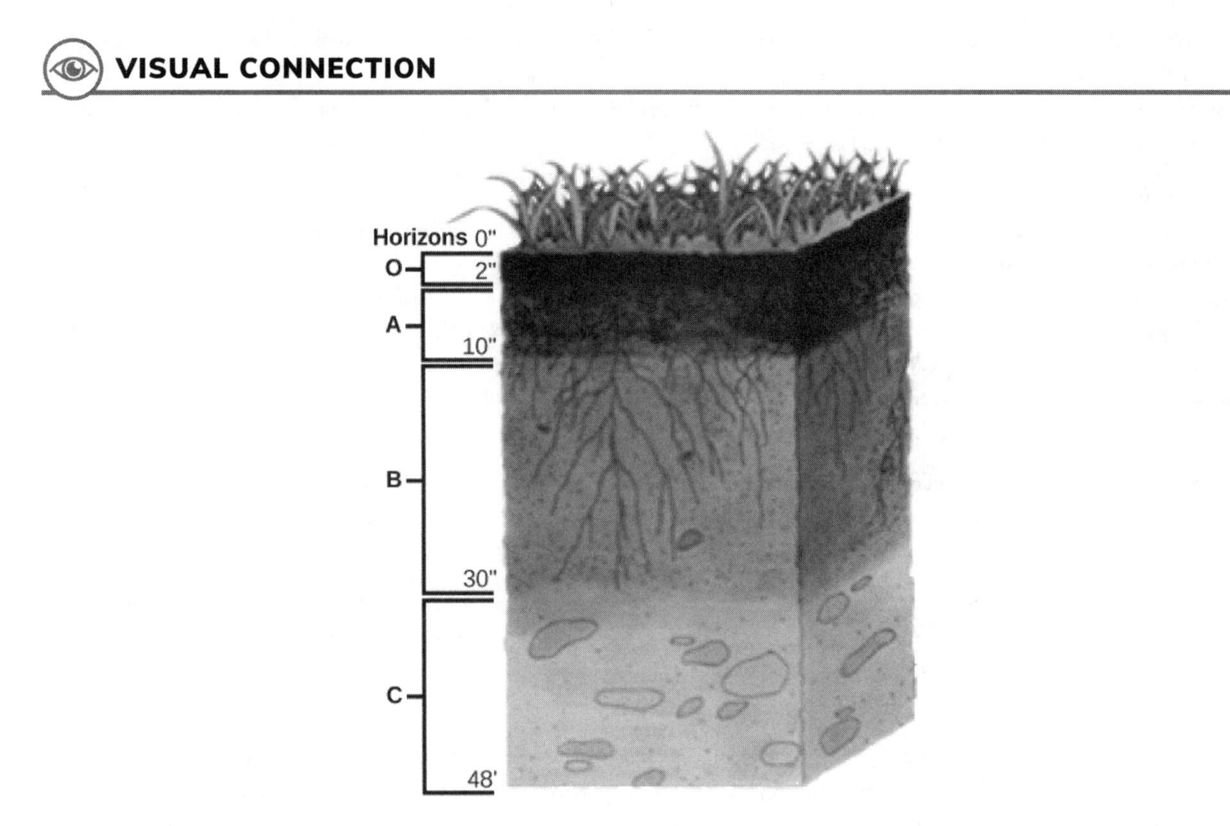

Figure 31.7 This soil profile shows the different soil layers (O horizon, A horizon, B horizon, and C horizon) found in typical soils. (credit: modification of work by USDA)

Which horizon is considered the topsoil, and which is considered the subsoil?

Some soils may have additional layers, or lack one of these layers. The thickness of the layers is also variable, and depends on the factors that influence soil formation. In general, immature soils may have O, A, and C horizons, whereas mature soils may display all of these, plus additional layers (Figure 31.8).

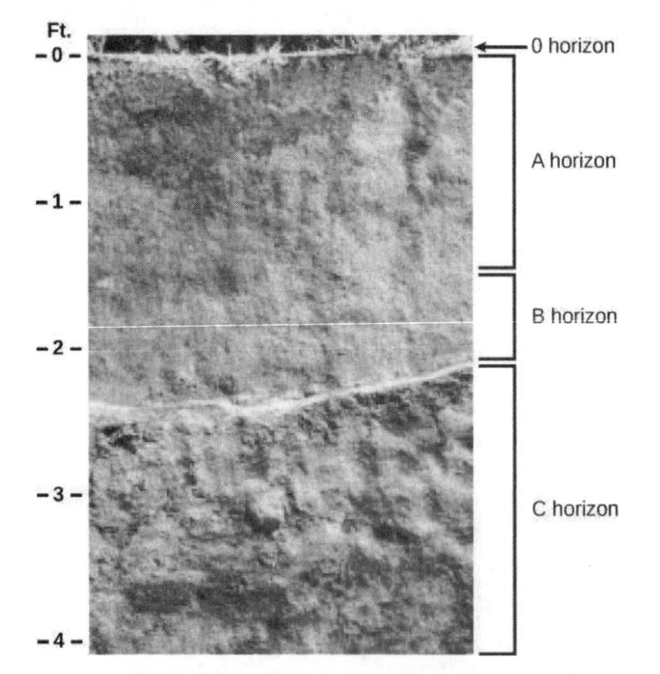

Figure 31.8 The San Joaquin soil profile has an O horizon, A horizon, B horizon, and C horizon. (credit: modification of work by USDA)

CAREER CONNECTION

Soil Scientist

A soil scientist studies the biological components, physical and chemical properties, distribution, formation, and morphology of soils. Soil scientists need to have a strong background in physical and life sciences, plus a foundation in mathematics. They may work for federal or state agencies, academia, or the private sector. Their work may involve collecting data, carrying out research, interpreting results, inspecting soils, conducting soil surveys, and recommending soil management programs.

Figure 31.9 This soil scientist is studying the horizons and composition of soil at a research site. (credit: USDA)

Many soil scientists work both in an office and in the field. According to the United States Department of Agriculture (USDA): "a soil scientist needs good observation skills to analyze and determine the characteristics of different types of soils. Soil types are complex and the geographical areas a soil scientist may survey are varied. Aerial photos or various satellite images are often used to research the areas. Computer skills and geographic information systems (GIS) help the scientist to analyze the multiple facets of geomorphology, topography, vegetation, and climate to discover the patterns left on the landscape."[1] Soil scientists play a key role in understanding the soil's past, analyzing present conditions, and making recommendations for future soil-related practices.

31.3 Nutritional Adaptations of Plants

By the end of this section, you will be able to do the following:

- Understand the nutritional adaptations of plants
- Describe mycorrhizae
- Explain nitrogen fixation

Plants obtain food in two different ways. Autotrophic plants can make their own food from inorganic raw materials, such as carbon dioxide and water, through photosynthesis in the presence of sunlight. Green plants are included in this group. Some plants, however, are heterotrophic: they are totally parasitic and lacking in chlorophyll. These plants, referred to as holo-parasitic plants, are unable to synthesize organic carbon and draw all of their nutrients from the host plant.

1National Resources Conservation Service / United States Department of Agriculture. "Careers in Soil Science." http://openstax.org/l/NRCS (http://openstax.org/l/NRCS)

Plants may also enlist the help of microbial partners in nutrient acquisition. Particular species of bacteria and fungi have evolved along with certain plants to create a mutualistic symbiotic relationship with roots. This improves the nutrition of both the plant and the microbe. The formation of nodules in legume plants and mycorrhization can be considered among the nutritional adaptations of plants. However, these are not the only type of adaptations that we may find; many plants have other adaptations that allow them to thrive under specific conditions.

LINK TO LEARNING

This video (http://openstax.org/l/basic_photosyn) reviews basic concepts about photosynthesis. In the left panel, click each tab to select a topic for review.

Nitrogen Fixation: Root and Bacteria Interactions

Nitrogen is an important macronutrient because it is part of nucleic acids and proteins. Atmospheric nitrogen, which is the diatomic molecule N_2, or dinitrogen, is the largest pool of nitrogen in terrestrial ecosystems. However, plants cannot take advantage of this nitrogen because they do not have the necessary enzymes to convert it into biologically useful forms. However, nitrogen can be "fixed," which means that it can be converted to ammonia (NH_3) through biological, physical, or chemical processes. As you have learned, biological nitrogen fixation (BNF) is the conversion of atmospheric nitrogen (N_2) into ammonia (NH_3), exclusively carried out by prokaryotes such as soil bacteria or cyanobacteria. Biological processes contribute 65 percent of the nitrogen used in agriculture. The following equation represents the process:

$$N_2 + 16\,ATP + 8\,e^- + 8\,H^+ \rightarrow 2NH_3 + 16\,ADP + 16\,Pi + H_2$$

The most important source of BNF is the symbiotic interaction between soil bacteria and legume plants, including many crops important to humans (Figure 31.10). The NH_3 resulting from fixation can be transported into plant tissue and incorporated into amino acids, which are then made into plant proteins. Some legume seeds, such as soybeans and peanuts, contain high levels of protein, and serve among the most important agricultural sources of protein in the world.

VISUAL CONNECTION

Figure 31.10 Some common edible legumes—like (a) peanuts, (b) beans, and (c) chickpeas—are able to interact symbiotically with soil bacteria that fix nitrogen. (credit a: modification of work by Jules Clancy; credit b: modification of work by USDA)

Farmers often rotate corn (a cereal crop) and soy beans (a legume), planting a field with each crop in alternate seasons. What advantage might this crop rotation confer?

Soil bacteria, collectively called **rhizobia**, symbiotically interact with legume roots to form specialized structures called **nodules**, in which nitrogen fixation takes place. This process entails the reduction of atmospheric nitrogen to ammonia, by means of the enzyme **nitrogenase**. Therefore, using rhizobia is a natural and environmentally friendly way to fertilize plants, as opposed to chemical fertilization that uses a nonrenewable resource, such as natural gas. Through symbiotic nitrogen fixation, the plant benefits from using an endless source of nitrogen from the atmosphere. The process simultaneously contributes to soil fertility because the plant root system leaves behind some of the biologically available nitrogen. As in any symbiosis, both organisms benefit from the interaction: the plant obtains ammonia, and bacteria obtain carbon compounds generated through photosynthesis, as well as a protected niche in which to grow (Figure 31.11).

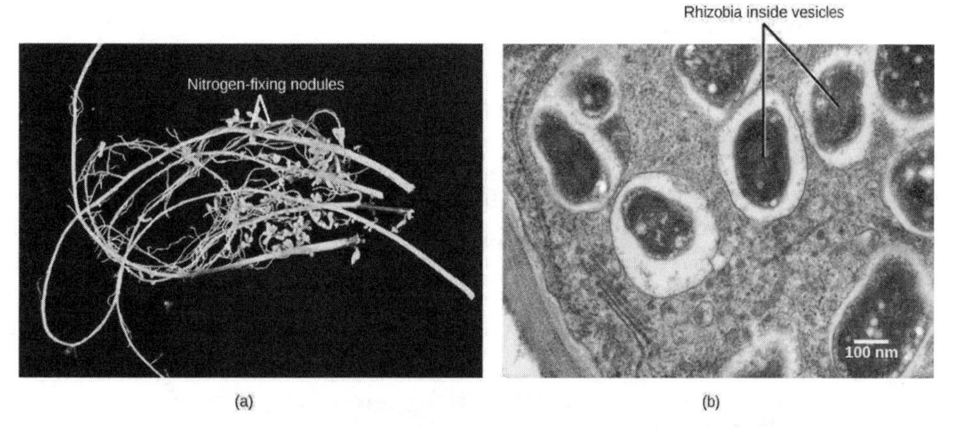

Figure 31.11 Soybean roots contain (a) nitrogen-fixing nodules. Cells within the nodules are infected with *Bradyrhyzobium japonicum,* a rhizobia or "root-loving" bacterium. The bacteria are encased in (b) vesicles inside the cell, as can be seen in this transmission electron micrograph. (credit a: modification of work by USDA; credit b: modification of work by Louisa Howard, Dartmouth Electron Microscope Facility; scale-bar data from Matt Russell)

Mycorrhizae: The Symbiotic Relationship between Fungi and Roots

A nutrient depletion zone can develop when there is rapid soil solution uptake, low nutrient concentration, low diffusion rate, or low soil moisture. These conditions are very common; therefore, most plants rely on fungi to facilitate the uptake of minerals from the soil. Fungi form symbiotic associations called mycorrhizae with plant roots, in which the fungi actually are integrated into the physical structure of the root. The fungi colonize the living root tissue during active plant growth.

Through mycorrhization, the plant obtains mainly phosphate and other minerals, such as zinc and copper, from the soil. The fungus obtains nutrients, such as sugars, from the plant root (Figure 31.12). Mycorrhizae help increase the surface area of the plant root system because hyphae, which are narrow, can spread beyond the nutrient depletion zone. Hyphae can grow into small soil pores that allow access to phosphorus that would otherwise be unavailable to the plant. The beneficial effect on the plant is best observed in poor soils. The benefit to fungi is that they can obtain up to 20 percent of the total carbon accessed by plants. Mycorrhizae functions as a physical barrier to pathogens. It also provides an induction of generalized host defense mechanisms, and sometimes involves production of antibiotic compounds by the fungi.

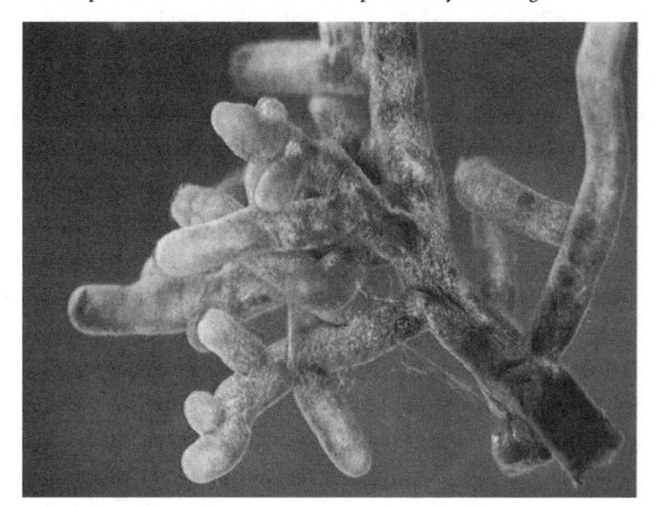

Figure 31.12 Root tips proliferate in the presence of mycorrhizal infection, which appears as off-white fuzz in this image. (credit: modification of work by Nilsson et al., BMC Bioinformatics 2005)

There are two types of mycorrhizae: ectomycorrhizae and endomycorrhizae. Ectomycorrhizae form an extensive dense sheath around the roots, called a mantle. Hyphae from the fungi extend from the mantle into the soil, which increases the surface area for water and mineral absorption. This type of mycorrhizae is found in forest trees, especially conifers, birches, and oaks. Endomycorrhizae, also called arbuscular mycorrhizae, do not form a dense sheath over the root. Instead, the fungal mycelium is embedded within the root tissue. Endomycorrhizae are found in the roots of more than 80 percent of terrestrial plants.

Nutrients from Other Sources

Some plants cannot produce their own food and must obtain their nutrition from outside sources. This may occur with plants that are parasitic or saprophytic. Some plants are mutualistic symbionts, epiphytes, or insectivorous.

Plant Parasites

A **parasitic plant** depends on its host for survival. Some parasitic plants have no leaves. An example of this is the dodder (Figure 31.13), which has a weak, cylindrical stem that coils around the host and forms suckers. From these suckers, cells invade the host stem and grow to connect with the vascular bundles of the host. The parasitic plant obtains water and nutrients through these connections. The plant is a total parasite (a holoparasite) because it is completely dependent on its host. Other parasitic plants (hemiparasites) are fully photosynthetic and only use the host for water and minerals. There are about 4,100 species of parasitic plants.

Figure 31.13 The dodder is a holoparasite that penetrates the host's vascular tissue and diverts nutrients for its own growth. Note that the vines of the dodder, which has white flowers, are beige. The dodder has no chlorophyll and cannot produce its own food. (credit: "Lalithamba"/Flickr)

Saprophytes

A **saprophyte** is a plant that does not have chlorophyll and gets its food from dead matter, similar to bacteria and fungi (note that fungi are often called saprophytes, which is incorrect, because fungi are not plants). Plants like these use enzymes to convert organic food materials into simpler forms from which they can absorb nutrients (Figure 31.14). Most saprophytes do not directly digest dead matter: instead, they parasitize fungi that digest dead matter, or are mycorrhizal, ultimately obtaining photosynthate from a fungus that derived photosynthate from its host. Saprophytic plants are uncommon; only a few species are described.

Figure 31.14 Saprophytes, like this Dutchmen's pipe (*Monotropa hypopitys)*, obtain their food from dead matter and do not have chlorophyll. (credit: modification of work by Iwona Erskine-Kellie)

Symbionts

A **symbiont** is a plant in a symbiotic relationship, with special adaptations such as mycorrhizae or nodule formation. Fungi also form symbiotic associations with cyanobacteria and green algae (called lichens). Lichens can sometimes be seen as colorful growths on the surface of rocks and trees (Figure 31.15). The algal partner (phycobiont) makes food autotrophically, some of which it shares with the fungus; the fungal partner (mycobiont) absorbs water and minerals from the environment, which are made available to the green alga. If one partner was separated from the other, they would both die.

Figure 31.15 Lichens, which often have symbiotic relationships with other plants, can sometimes be found growing on trees. (credit: "benketaro"/Flickr)

Epiphytes

An **epiphyte** is a plant that grows on other plants, but is not dependent upon the other plant for nutrition (Figure 31.16). Epiphytes have two types of roots: clinging aerial roots, which absorb nutrients from humus that accumulates in the crevices of trees; and aerial roots, which absorb moisture from the atmosphere.

Figure 31.16 These epiphyte plants grow in the main greenhouse of the *Jardin des Plantes* in Paris.

Insectivorous Plants

An **insectivorous** plant has specialized leaves to attract and digest insects. The Venus flytrap is popularly known for its insectivorous mode of nutrition, and has leaves that work as traps (Figure 31.17). The minerals it obtains from prey compensate for those lacking in the boggy (low pH) soil of its native North Carolina coastal plains. There are three sensitive hairs in the center of each half of each leaf. The edges of each leaf are covered with long spines. Nectar secreted by the plant attracts flies to the leaf. When a fly touches the sensory hairs, the leaf immediately closes. Next, fluids and enzymes break down the prey and minerals are absorbed by the leaf. Since this plant is popular in the horticultural trade, it is threatened in its original habitat.

Figure 31.17 A Venus flytrap has specialized leaves to trap insects. (credit: "Selena N. B. H."/Flickr)

KEY TERMS

A horizon consists of a mixture of organic material with inorganic products of weathering

B horizon soil layer that is an accumulation of mostly fine material that has moved downward

bedrock solid rock that lies beneath the soil

C horizon layer of soil that contains the parent material, and the organic and inorganic material that is broken down to form soil; also known as the soil base

clay soil particles that are less than 0.002 mm in diameter

epiphyte plant that grows on other plants but is not dependent upon other plants for nutrition

horizon soil layer with distinct physical and chemical properties, which differs from other layers depending on how and when it was formed

humus organic material of soil; made up of microorganisms, dead animals, and plants in varying stages of decay

inorganic compound chemical compound that does not contain carbon; it is not part of or produced by a living organism

insectivorous plant plant that has specialized leaves to attract and digest insects

loam soil that has no dominant particle size

macronutrient nutrient that is required in large amounts for plant growth; carbon, hydrogen, oxygen, nitrogen, phosphorus, potassium, calcium, magnesium, and sulfur

micronutrient nutrient required in small amounts; also called trace element

mineral soil type of soil that is formed from the weathering of rocks and inorganic material; composed primarily of sand, silt, and clay

nitrogenase enzyme that is responsible for the reduction of atmospheric nitrogen to ammonia

nodules specialized structures that contain *Rhizobia* bacteria where nitrogen fixation takes place

O horizon layer of soil with humus at the surface and decomposed vegetation at the base

organic compound chemical compound that contains carbon

organic soil type of soil that is formed from sedimentation; composed primarily of organic material

parasitic plant plant that is dependent on its host for survival

parent material organic and inorganic material in which soils form

rhizobia soil bacteria that symbiotically interact with legume roots to form nodules and fix nitrogen

rhizosphere area of soil affected by root secretions and microorganisms

sand soil particles between 0.1–2 mm in diameter

saprophyte plant that does not have chlorophyll and gets its food from dead matter

silt soil particles between 0.002 and 0.1 mm in diameter

soil outer loose layer that covers the surface of Earth

soil profile vertical section of a soil

symbiont plant in a symbiotic relationship with bacteria or fungi

CHAPTER SUMMARY

31.1 Nutritional Requirements of Plants

Plants can absorb inorganic nutrients and water through their root system, and carbon dioxide from the environment. The combination of organic compounds, along with water, carbon dioxide, and sunlight, produce the energy that allows plants to grow. Inorganic compounds form the majority of the soil solution. Plants access water though the soil. Water is absorbed by the plant root, transports nutrients throughout the plant, and maintains the structure of the plant. Essential elements are indispensable elements for plant growth. They are divided into macronutrients and micronutrients. The macronutrients plants require are carbon, nitrogen, hydrogen, oxygen, phosphorus, potassium, calcium, magnesium, and sulfur. Important micronutrients include iron, manganese, boron, molybdenum, copper, zinc, chlorine, nickel, cobalt, silicon, and sodium.

31.2 The Soil

Plants obtain mineral nutrients from the soil. Soil is the outer loose layer that covers the surface of Earth. Soil quality depends on the chemical composition of the soil, the topography, the presence of living organisms, the climate, and time. Agricultural practice and history may also modify the characteristics and fertility of soil. Soil consists of four major components: 1) inorganic mineral matter, 2) organic matter, 3) water and air, and 4) living matter. The organic material of soil is made of humus, which improves soil structure and provides water and minerals. Soil inorganic material consists of rock slowly broken down into smaller particles that vary in size, such as sand, silt, and loam.

Soil formation results from a combination of biological, physical, and chemical processes. Soil is not homogenous because its formation results in the production of layers called a soil profile. Factors that affect soil formation include: parent material, climate, topography, biological factors, and time. Soils are classified based on their horizons, soil particle size, and proportions. Most soils have four distinct horizons: O, A, B, and C.

31.3 Nutritional Adaptations of Plants

Atmospheric nitrogen is the largest pool of available nitrogen in terrestrial ecosystems. However, plants cannot use this nitrogen because they do not have the necessary enzymes. Biological nitrogen fixation (BNF) is the conversion of atmospheric nitrogen to ammonia. The most important source of BNF is the symbiotic interaction between soil bacteria and legumes. The bacteria form nodules on the legume's roots in which nitrogen fixation takes place. Fungi form symbiotic associations (mycorrhizae) with plants, becoming integrated into the physical structure of the root. Through mycorrhization, the plant obtains minerals from the soil and the fungus obtains photosynthate from the plant root. Ectomycorrhizae form an extensive dense sheath around the root, while endomycorrhizae are embedded within the root tissue. Some plants—parasites, saprophytes, symbionts, epiphytes, and insectivores—have evolved adaptations to obtain their organic or mineral nutrition from various sources.

VISUAL CONNECTION QUESTIONS

1. Figure 31.6 Soil compaction can result when soil is compressed by heavy machinery or even foot traffic. How might this compaction change the soil composition?
2. Figure 31.7 Which horizon is considered the topsoil, and which is considered the subsoil?
3. Figure 31.10 Farmers often rotate corn (a cereal crop) and soy beans (a legume) planting a field with each crop in alternate seasons. What advantage might this crop rotation confer?

REVIEW QUESTIONS

4. For an element to be regarded as essential, all of the following criteria must be met, except:
   a. No other element can perform the function.
   b. The element is directly involved in plant nutrition.
   c. The element is inorganic.
   d. The plant cannot complete its lifecycle without the element.

5. The nutrient that is part of carbohydrates, proteins, and nucleic acids, and that forms biomolecules, is _______.
   a. nitrogen
   b. carbon
   c. magnesium
   d. iron

6. Most _______ are necessary for enzyme function.
   a. micronutrients
   b. macronutrients
   c. biomolecules
   d. essential nutrients

7. What is the main water source for land plants?
   a. rain
   b. soil
   c. biomolecules
   d. essential nutrients

8. Which factors affect soil quality?
   a. chemical composition
   b. history of the soil
   c. presence of living organisms and topography
   d. all of the above

9. Soil particles that are 0.1 to 2 mm in diameter are called _______.
   a. sand
   b. silt
   c. clay
   d. loam

10. A soil consists of layers called _______ that taken together are called a _______.
    a. soil profiles : horizon
    b. horizons : soil profile
    c. horizons : humus
    d. humus : soil profile

11. What is the term used to describe the solid rock that lies beneath the soil?
    a. sand
    b. bedrock
    c. clay
    d. loam

12. Which process produces an inorganic compound that plants can easily use?
    a. photosynthesis
    b. nitrogen fixation
    c. mycorrhization
    d. Calvin cycle

13. Through mycorrhization, a plant obtains important nutrients such as _______.
    a. phosphorus, zinc, and copper
    b. phosphorus, zinc, and calcium
    c. nickel, calcium, and zinc
    d. all of the above

14. What term describes a plant that requires nutrition from a living host plant?
    a. parasite
    b. saprophyte
    c. epiphyte
    d. insectivorous

15. What is the term for the symbiotic association between fungi and cyanobacteria?
    a. lichen
    b. mycorrhizae
    c. epiphyte
    d. nitrogen-fixing nodule

CRITICAL THINKING QUESTIONS

16. What type of plant problems result from nitrogen and calcium deficiencies?
17. Research the life of Jan Babtista van Helmont. What did the van Helmont experiment show?
18. List two essential macronutrients and two essential micro nutrients.
19. Describe the main differences between a mineral soil and an organic soil.
20. Name and briefly explain the factors that affect soil formation.
21. Describe how topography influences the characteristics and fertility of a soil.
22. Why is biological nitrogen fixation an environmentally friendly way of fertilizing plants?
23. What is the main difference, from an energy point of view, between photosynthesis and biological nitrogen fixation?
24. Why is a root nodule a nutritional adaptation of a plant?

CHAPTER 32
Plant Reproduction

Figure 32.1 Plants that reproduce sexually often achieve fertilization with the help of pollinators such as (a) bees, (b) birds, and (c) butterflies. (credit a: modification of work by John Severns; credit b: modification of work by Charles J. Sharp; credit c: modification of work by "Galawebdesign"/Flickr)

INTRODUCTION Plants have evolved different reproductive strategies for the continuation of their species. Some plants reproduce sexually, and others asexually, in contrast to animal species, which rely almost exclusively on sexual reproduction. Plant sexual reproduction usually depends on pollinating agents, while asexual reproduction is independent of these agents. Flowers are often the showiest or most strongly scented part of plants. With their bright colors, fragrances, and interesting shapes and sizes, flowers attract insects, birds, and animals to serve their pollination needs. Other plants pollinate via wind or water; still others self-pollinate.

Chapter Outline

32.1 Reproductive Development and Structure

By the end of this section, you will be able to do the following:

- Describe the two stages of a plant's lifecycle
- Compare and contrast male and female gametophytes and explain how they form in angiosperms
- Describe the reproductive structures of a plant
- Describe the components of a complete flower
- Describe the development of microsporangium and megasporangium in gymnosperms

Sexual reproduction takes place with slight variations in different groups of plants. Plants have two distinct stages in their lifecycle: the gametophyte stage and the sporophyte stage. The haploid **gametophyte** produces the male and female gametes by mitosis in distinct multicellular structures. Fusion of the male and females gametes forms the diploid zygote, which develops into the **sporophyte**. After reaching maturity, the diploid sporophyte produces spores by meiosis, which in turn divide by mitosis to produce the haploid gametophyte. The new gametophyte produces gametes, and the cycle continues. This is the alternation of generations, and is typical of plant reproduction (Figure 32.2).

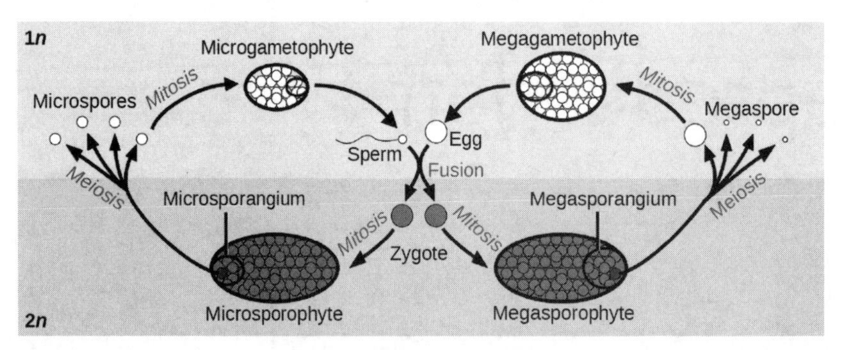

Figure 32.2 The alternation of generations in angiosperms is depicted in this diagram. (credit: modification of work by Peter Coxhead)

The life cycle of higher plants is dominated by the sporophyte stage, with the gametophyte borne on the sporophyte. In ferns, the gametophyte is free-living and very distinct in structure from the diploid sporophyte. In bryophytes, such as mosses, the haploid gametophyte is more developed than the sporophyte.

During the vegetative phase of growth, plants increase in size and produce a shoot system and a root system. As they enter the reproductive phase, some of the branches start to bear flowers. Many flowers are borne singly, whereas some are borne in clusters. The flower is borne on a stalk known as a receptacle. Flower shape, color, and size are unique to each species, and are often used by taxonomists to classify plants.

Sexual Reproduction in Angiosperms

The lifecycle of angiosperms follows the alternation of generations explained previously. The haploid gametophyte alternates with the diploid sporophyte during the sexual reproduction process of angiosperms. Flowers contain the plant's reproductive structures.

Flower Structure

A typical flower has four main parts—or whorls—known as the calyx, corolla, androecium, and gynoecium (Figure 32.3). The outermost whorl of the flower has green, leafy structures known as sepals. The sepals, collectively called the calyx, help to protect the unopened bud. The second whorl is comprised of petals—usually, brightly colored—collectively called the corolla. The number of sepals and petals varies depending on whether the plant is a monocot or dicot. In monocots, petals usually number three or multiples of three; in dicots, the number of petals is four or five, or multiples of four and five. Together, the calyx and corolla are known as the **perianth**. The third whorl contains the male reproductive structures and is known as the androecium. The **androecium** has stamens with anthers that contain the microsporangia. The innermost group of structures in the flower is the **gynoecium**, or the female reproductive component(s). The carpel is the individual unit of the gynoecium and has a stigma, style, and ovary. A flower may have one or multiple carpels.

VISUAL CONNECTION

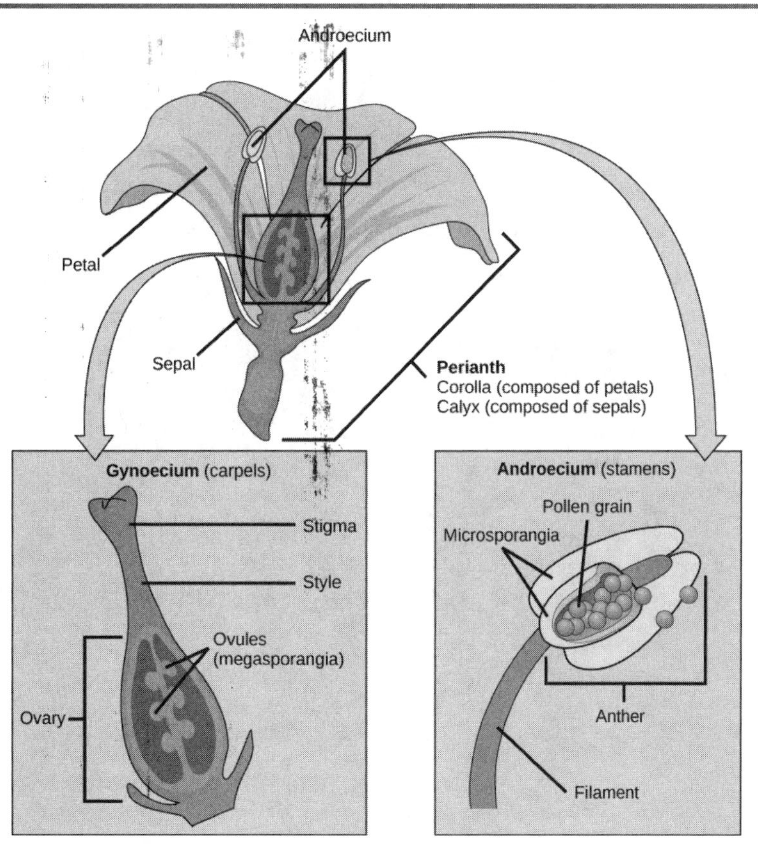

Figure 32.3 The four main parts of the flower are the calyx, corolla, androecium, and gynoecium. The androecium is the sum of all the male reproductive organs, and the gynoecium is the sum of the female reproductive organs. (credit: modification of work by Mariana Ruiz Villareal)

If the anther is missing, what type of reproductive structure will the flower be unable to produce? What term is used to describe an incomplete flower lacking the androecium? What term describes an incomplete flower lacking a gynoecium?

If all four whorls (the calyx, corolla, androecium, and gynoecium) are present, the flower is described as complete. If any of the four parts is missing, the flower is known as incomplete. Flowers that contain both an androecium and a gynoecium are called perfect, androgynous or hermaphrodites. There are two types of incomplete flowers: staminate flowers contain only an androecium, and carpellate flowers have only a gynoecium (Figure 32.4).

Figure 32.4 The corn plant has both staminate (male) and carpellate (female) flowers. Staminate flowers, which are clustered in the tassel at the tip of the stem, produce pollen grains. Carpellate flowers are clustered in the immature ears. Each strand of silk is a stigma. The corn kernels are seeds that develop on the ear after fertilization. Also shown is the lower stem and root.

If both male and female flowers are borne on the same plant, the species is called monoecious (meaning "one home"): examples are corn and pea. Species with male and female flowers borne on separate plants are termed dioecious, or "two homes," examples of which are *C. papaya* and *Cannabis*. The ovary, which may contain one or multiple ovules, may be placed above other flower parts, which is referred to as superior; or, it may be placed below the other flower parts, referred to as inferior (Figure 32.5).

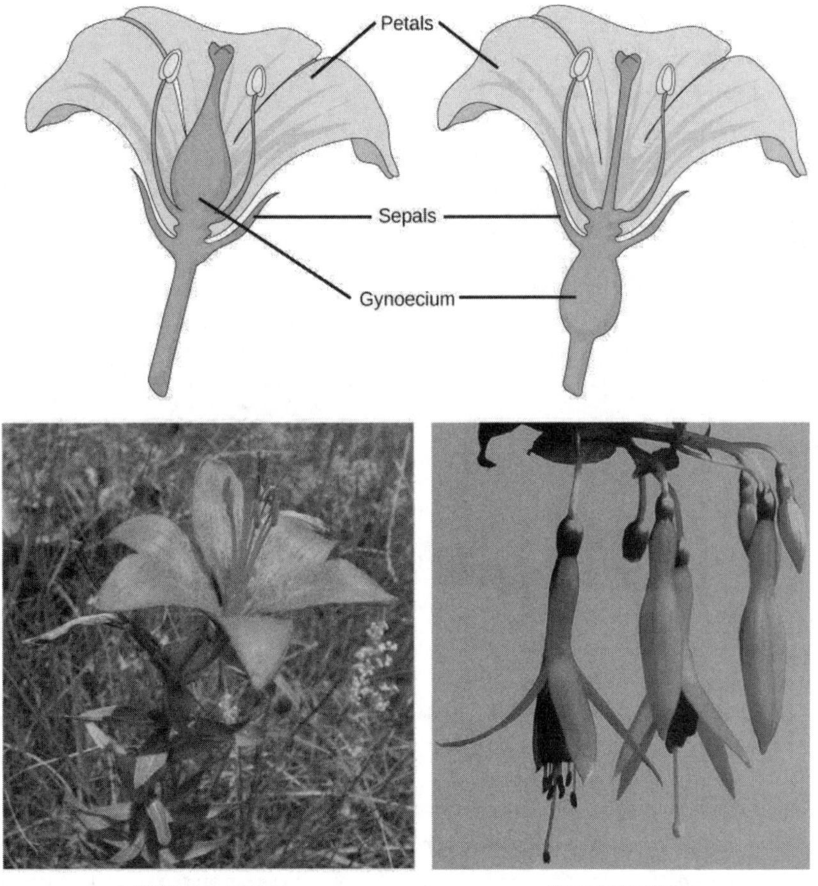

Figure 32.5 The (a) lily is a superior flower, which has the ovary above the other flower parts. (b) Fuchsia is an inferior flower, which has the ovary beneath other flower parts. (credit a photo: modification of work by Benjamin Zwittnig; credit b photo: modification of work by "Koshy Koshy"/Flickr)

Male Gametophyte (The Pollen Grain)

The male gametophyte develops and reaches maturity in an immature anther. In a plant's male reproductive organs, development of pollen takes place in a structure known as the **microsporangium** (Figure 32.6). The microsporangia, which are usually bilobed, are pollen sacs in which the microspores develop into pollen grains. These are found in the anther, which is at the end of the stamen—the long filament that supports the anther.

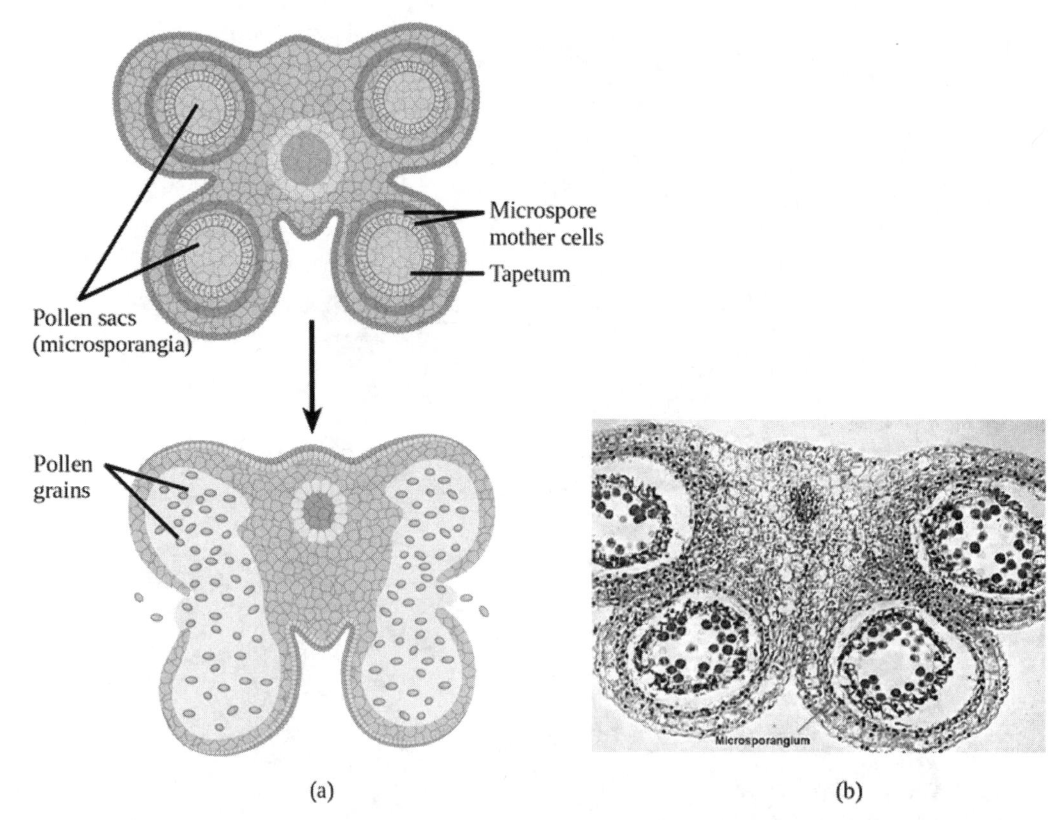

Figure 32.6 Shown is (a) a cross section of an anther at two developmental stages. The immature anther (top) contains four microsporangia, or pollen sacs. Each microsporangium contains hundreds of microspore mother cells that will each give rise to four pollen grains. The tapetum supports the development and maturation of the pollen grains. Upon maturation of the pollen (bottom), the pollen sac walls split open and the pollen grains (male gametophytes) are released, as shown in the (b) micrograph of an immature lily anther. In these scanning electron micrographs, pollen sacs are ready to burst, releasing their grains. (credit a: modification of work by LibreTexts; b: modification of work by Robert R. Wise; scale-bar data from Matt Russell)

Within the microsporangium, each of the microspore mother cells divides by meiosis to give rise to four microspores, each of which will ultimately form a pollen grain (Figure 32.7). An inner layer of cells, known as the tapetum, provides nutrition to the developing microspores and contributes key components to the pollen wall. Mature pollen grains contain two cells: a generative cell and a pollen tube cell. The generative cell is contained within the larger pollen tube cell. Upon germination, the tube cell forms the pollen tube through which the generative cell migrates to enter the ovary. During its transit inside the pollen tube, the generative cell divides to form two male gametes (sperm cells). Upon maturity, the microsporangia burst, releasing the pollen grains from the anther.

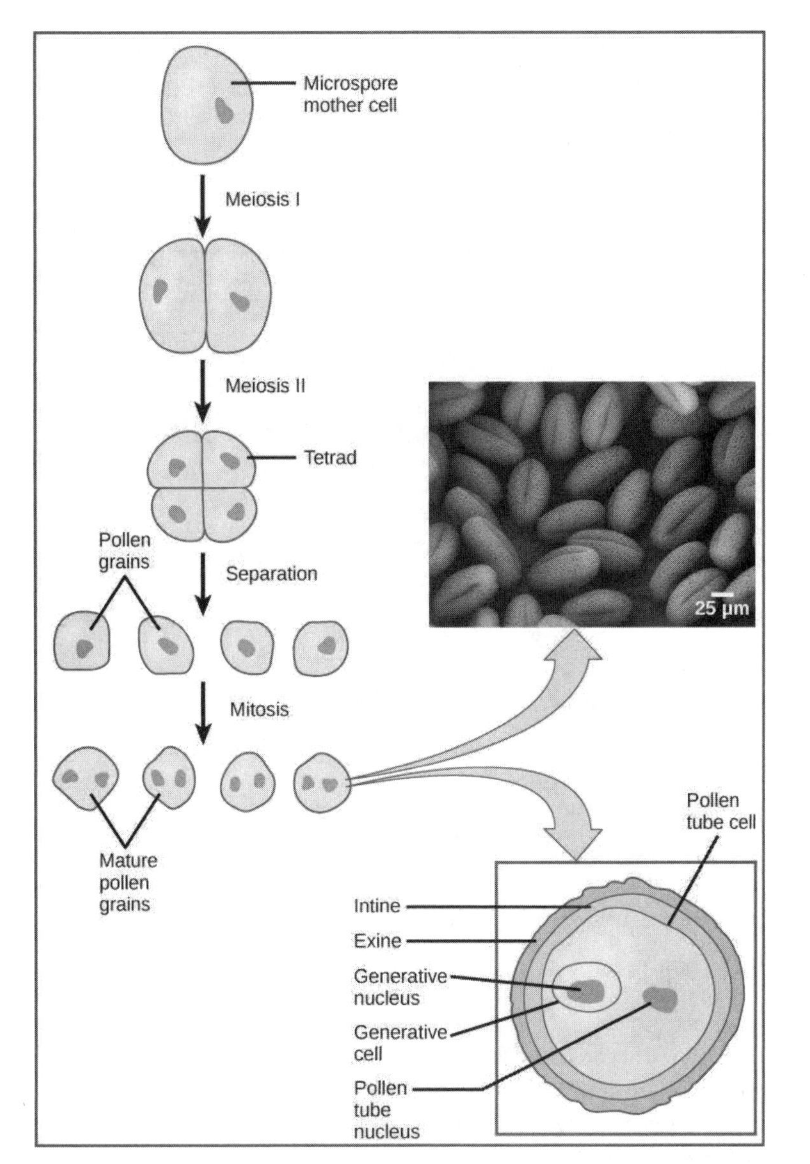

Figure 32.7 Pollen develops from the microspore mother cells. The mature pollen grain is composed of two cells: the pollen tube cell and the generative cell, which is inside the tube cell. The pollen grain has two coverings: an inner layer (intine) and an outer layer (exine). The inset scanning electron micrograph shows *Arabidopsis lyrata* pollen grains. (credit "pollen micrograph": modification of work by Robert R. Wise; scale-bar data from Matt Russell)

Each pollen grain has two coverings: the **exine** (thicker, outer layer) and the **intine** (Figure 32.7). The exine contains sporopollenin, a complex waterproofing substance supplied by the tapetal cells. Sporopollenin allows the pollen to survive under unfavorable conditions and to be carried by wind, water, or biological agents without undergoing damage.

Female Gametophyte (The Embryo Sac)

While the details may vary between species, the overall development of the female gametophyte has two distinct phases. First, in the process of **megasporogenesis**, a single cell in the diploid **megasporangium**—an area of tissue in the ovules—undergoes meiosis to produce four megaspores, only one of which survives. During the second phase, **megagametogenesis**, the surviving haploid megaspore undergoes mitosis to produce an eight-nucleate, seven-cell female gametophyte, also known as the megagametophyte or embryo sac. Two of the nuclei—the **polar nuclei**—move to the equator and fuse, forming a single, diploid central cell. This central cell later fuses with a sperm to form the triploid endosperm. Three nuclei position themselves on the end of the embryo sac opposite the micropyle and develop into the **antipodal** cells, which later degenerate. The nucleus closest to the micropyle becomes the female gamete, or egg cell, and the two adjacent nuclei develop into **synergid** cells (Figure 32.8). The synergids help guide the pollen tube for successful fertilization, after which they disintegrate. Once fertilization is complete, the resulting diploid zygote develops into the embryo, and the fertilized ovule forms the other tissues of the seed.

A double-layered integument protects the megasporangium and, later, the embryo sac. The integument will develop into the seed coat after fertilization and protect the entire seed. The ovule wall will become part of the fruit. The integuments, while protecting the megasporangium, do not enclose it completely, but leave an opening called the **micropyle**. The micropyle allows the pollen tube to enter the female gametophyte for fertilization.

VISUAL CONNECTION

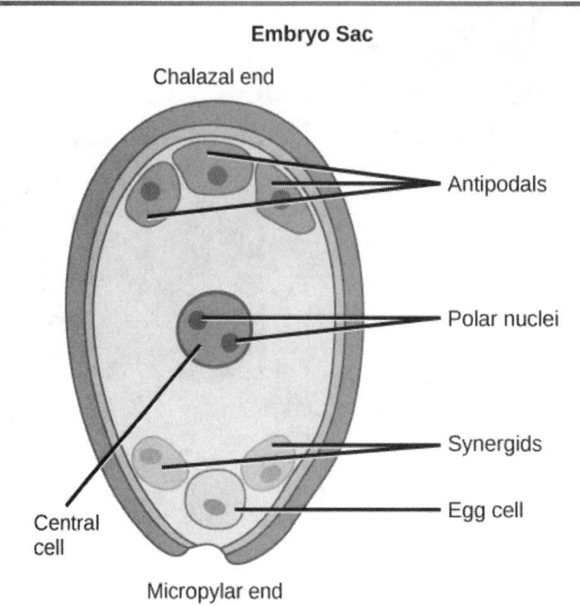

Figure 32.8 As shown in this diagram of the embryo sac in angiosperms, the ovule is covered by integuments and has an opening called a micropyle. Inside the embryo sac are three antipodal cells, two synergids, a central cell, and the egg cell.

An embryo sac is missing the synergids. What specific impact would you expect this to have on fertilization?

a. The pollen tube will be unable to form.
b. The pollen tube will form but will not be guided toward the egg.
c. Fertilization will not occur because the synergid is the egg.
d. Fertilization will occur but the embryo will not be able to grow.

Sexual Reproduction in Gymnosperms

As with angiosperms, the lifecycle of a gymnosperm is also characterized by alternation of generations. In conifers such as pines, the green leafy part of the plant is the sporophyte, and the cones contain the male and female gametophytes (Figure 32.9). The female cones are larger than the male cones and are positioned towards the top of the tree; the small, male cones are located in the lower region of the tree. Because the pollen is shed and blown by the wind, this arrangement makes it difficult for a gymnosperm to self-pollinate.

Figure 32.9 This image shows the lifecycle of a conifer. Pollen from male cones blows up into upper branches, where it fertilizes female cones. Examples are shown of female and male cones. (credit "female": modification of work by "Geographer"/Wikimedia Commons; credit "male": modification of work by Roger Griffith)

Male Gametophyte

A male cone has a central axis on which bracts, a type of modified leaf, are attached. The bracts are known as **microsporophylls** (Figure 32.10) and are the sites where microspores will develop. The microspores develop inside the microsporangium. Within the microsporangium, cells known as microsporocytes divide by meiosis to produce four haploid microspores. Further mitosis of the microspore produces two nuclei: the generative nucleus, and the tube nucleus. Upon maturity, the male gametophyte (pollen) is released from the male cones and is carried by the wind to land on the female cone.

LINK TO LEARNING

Watch this video to see a cedar releasing its pollen in the wind.

Click to view content (https://www.openstax.org/l/pollen_release)

Female Gametophyte

The female cone also has a central axis on which bracts known as **megasporophylls** (Figure 32.10) are present. In the female cone, megaspore mother cells are present in the megasporangium. The megaspore mother cell divides by meiosis to produce four haploid megaspores. One of the megaspores divides to form the multicellular female gametophyte, while the others divide to form the rest of the structure. The female gametophyte is contained within a structure called the archegonium.

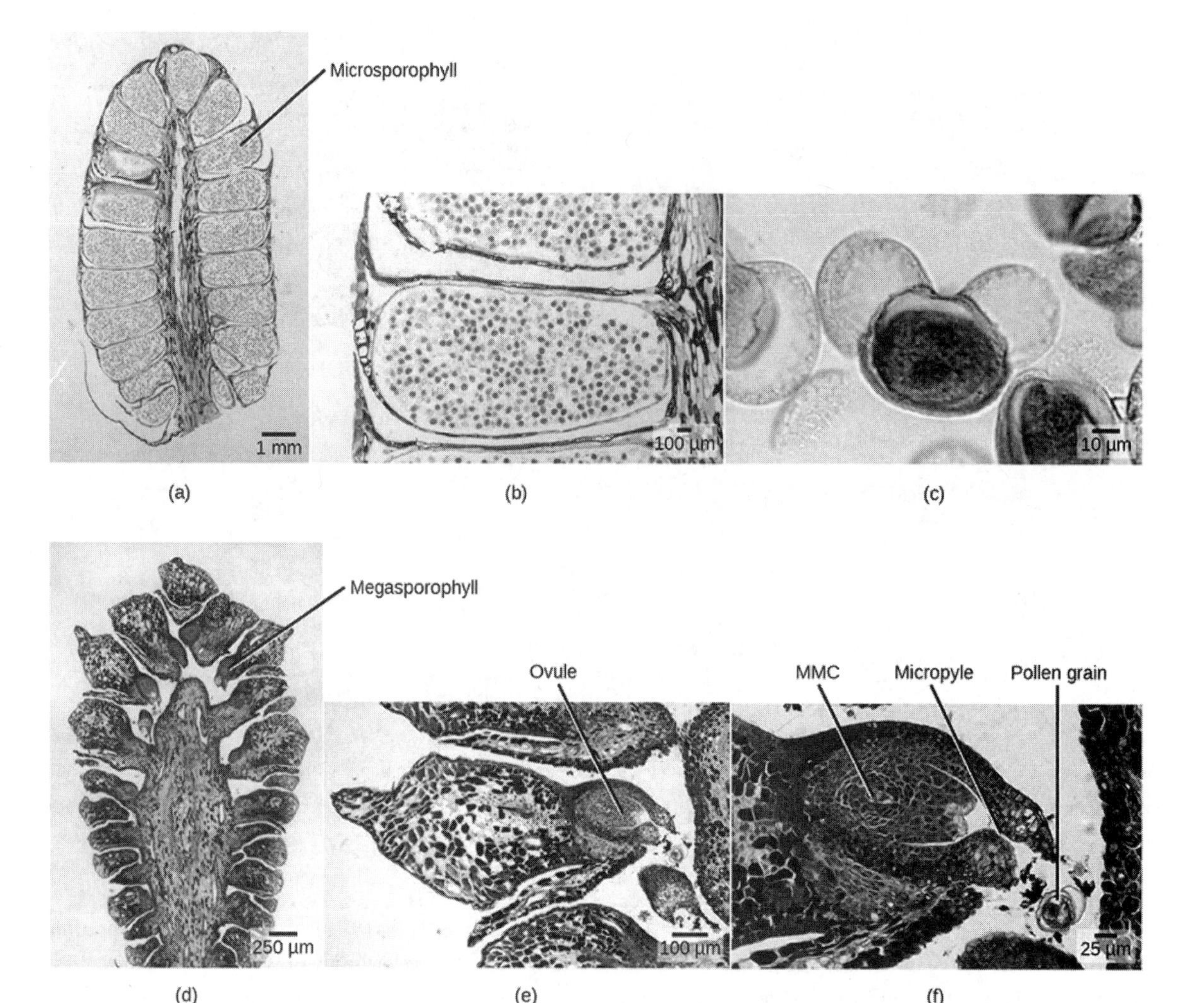

Figure 32.10 This series of micrographs shows male and female gymnosperm gametophytes. (a) This male cone, shown in cross section, has approximately 20 microsporophylls, each of which produces hundreds of male gametophytes (pollen grains). (b) Pollen grains are visible in this single microsporophyll. (c) This micrograph shows an individual pollen grain. (d) This cross section of a female cone shows portions of about 15 megasporophylls. (e) The ovule can be seen in this single megasporophyll. (f) Within this single ovule are the megaspore mother cell (MMC), micropyle, and a pollen grain. (credit: modification of work by Robert R. Wise; scale-bar data from Matt Russell)

Reproductive Process

Upon landing on the female cone, the tube cell of the pollen forms the pollen tube, through which the generative cell migrates towards the female gametophyte through the micropyle. It takes approximately one year for the pollen tube to grow and migrate towards the female gametophyte. The male gametophyte containing the generative cell splits into two sperm nuclei, one of which fuses with the egg, while the other degenerates. After fertilization of the egg, the diploid zygote is formed, which divides by mitosis to form the embryo. The scales of the cones are closed during development of the seed. The seed is covered by a seed coat, which is derived from the female sporophyte. Seed development takes another one to two years. Once the seed is ready to be dispersed, the bracts of the female cones open to allow the dispersal of seed; no fruit formation takes place because gymnosperm seeds have no covering.

Angiosperms versus Gymnosperms

Gymnosperm reproduction differs from that of angiosperms in several ways (Figure 32.11). In angiosperms, the female gametophyte exists in an enclosed structure—the ovule—which is within the ovary; in gymnosperms, the female gametophyte is present on exposed bracts of the female cone. Double fertilization is a key event in the lifecycle of angiosperms, but is completely absent in gymnosperms. The male and female gametophyte structures are present on separate male and female cones in gymnosperms, whereas in angiosperms, they are a part of the flower. Lastly, wind plays an important role in pollination in gymnosperms because pollen is blown by the wind to land on the female cones. Although many angiosperms are also wind-pollinated, animal pollination is more common.

(a) (b)

Figure 32.11 (a) Angiosperms are flowering plants, and include grasses, herbs, shrubs and most deciduous trees, while (b) gymnosperms are conifers. Both produce seeds but have different reproductive strategies. (credit a: modification of work by Wendy Cutler; credit b: modification of work by Lews Castle UHI)

LINK TO LEARNING

View an animation of the double fertilization process of angiosperms.

Click to view content (https://www.openstax.org/l/angiosperms)

32.2 Pollination and Fertilization

By the end of this section, you will be able to do the following:

- Describe what must occur for plant fertilization
- Explain cross-pollination and the ways in which it takes place
- Describe the process that leads to the development of a seed
- Define double fertilization

In angiosperms, **pollination** is defined as the placement or transfer of pollen from the anther to the stigma of the same flower or another flower. In gymnosperms, pollination involves pollen transfer from the male cone to the female cone. Upon transfer, the pollen germinates to form the pollen tube and the sperm for fertilizing the egg. Pollination has been well studied since the time of Gregor Mendel. Mendel successfully carried out self- as well as cross-pollination in garden peas while studying how characteristics were passed on from one generation to the next. Today's crops are a result of plant breeding, which employs artificial selection to produce the present-day cultivars. A case in point is today's corn, which is a result of years of breeding that started with its ancestor, teosinte. The teosinte that the ancient Mayans originally began cultivating had tiny seeds—vastly different from today's relatively giant ears of corn. Interestingly, though these two plants appear to be entirely different, the genetic difference between them is miniscule.

Pollination takes two forms: self-pollination and cross-pollination. **Self-pollination** occurs when the pollen from the anther is deposited on the stigma of the same flower, or another flower on the same plant. **Cross-pollination** is the transfer of pollen from the anther of one flower to the stigma of another flower on a different individual of the same species. Self-pollination occurs in flowers where the stamen and carpel mature at the same time, and are positioned so that the pollen can land on the flower's stigma. This method of pollination does not require an investment from the plant to provide nectar and pollen as food for pollinators.

LINK TO LEARNING

Explore this interactive website (http://openstax.org/l/pollination) to review self-pollination and cross-pollination.

Living species are designed to ensure survival of their progeny; those that fail become extinct. Genetic diversity is therefore required so that in changing environmental or stress conditions, some of the progeny can survive. Self-pollination leads to the production of plants with less genetic diversity, since genetic material from the same plant is used to form gametes, and eventually, the zygote. In contrast, cross-pollination—or out-crossing—leads to greater genetic diversity because the microgametophyte and megagametophyte are derived from different plants.

Because cross-pollination allows for more genetic diversity, plants have developed many ways to avoid self-pollination. In some species, the pollen and the ovary mature at different times. These flowers make self-pollination nearly impossible. By the time pollen matures and has been shed, the stigma of this flower is mature and can only be pollinated by pollen from another flower. Some flowers have developed physical features that prevent self-pollination. The primrose is one such flower. Primroses have evolved two flower types with differences in anther and stigma length: the pin-eyed flower has anthers positioned at the pollen tube's halfway point, and the thrum-eyed flower's stigma is likewise located at the halfway point. Insects easily cross-pollinate while seeking the nectar at the bottom of the pollen tube. This phenomenon is also known as heterostyly. Many plants, such as cucumber, have male and female flowers located on different parts of the plant, thus making self-pollination difficult. In yet other species, the male and female flowers are borne on different plants (dioecious). All of these are barriers to self-pollination; therefore, the plants depend on pollinators to transfer pollen. The majority of pollinators are biotic agents such as insects (like bees, flies, and butterflies), bats, birds, and other animals. Other plant species are pollinated by abiotic agents, such as wind and water.

Everyday Connection

Incompatibility Genes in Flowers

In recent decades, incompatibility genes—which prevent pollen from germinating or growing into the stigma of a flower—have been discovered in many angiosperm species. If plants do not have compatible genes, the pollen tube stops growing. Self-incompatibility is controlled by the S (sterility) locus. Pollen tubes have to grow through the tissue of the stigma and style before they can enter the ovule. The carpel is selective in the type of pollen it allows to grow inside. The interaction is primarily between the pollen and the stigma epidermal cells. In some plants, like cabbage, the pollen is rejected at the surface of the stigma, and the unwanted pollen does not germinate. In other plants, pollen tube germination is arrested after growing one-third the length of the style, leading to pollen tube death. Pollen tube death is due either to apoptosis (programmed cell death) or to degradation of pollen tube RNA. The degradation results from the activity of a ribonuclease encoded by the S locus. The ribonuclease is secreted from the cells of the style in the extracellular matrix, which lies alongside the growing pollen tube.

In summary, self-incompatibility is a mechanism that prevents self-fertilization in many flowering plant species. The working of this self-incompatibility mechanism has important consequences for plant breeders because it inhibits the production of inbred and hybrid plants.

Pollination by Insects

Bees are perhaps the most important pollinator of many garden plants and most commercial fruit trees (Figure 32.12). The most common species of bees are bumblebees and honeybees. Since bees cannot see the color red, bee-pollinated flowers usually have shades of blue, yellow, or other colors. Bees collect energy-rich pollen or nectar for their survival and energy needs. They visit flowers that are open during the day, are brightly colored, have a strong aroma or scent, and have a tubular shape, typically with the presence of a nectar guide. A **nectar guide** includes regions on the flower petals that are visible only to bees, and not to humans; it helps to guide bees to the center of the flower, thus making the pollination process more efficient. The pollen sticks to the bees' fuzzy hair, and when the bee visits another flower, some of the pollen is transferred to the second flower. Recently, there have been many reports about the declining population of honeybees. Many flowers will remain unpollinated and not bear seed if honeybees disappear. The impact on commercial fruit growers could be devastating.

Figure 32.12 Insects, such as bees, are important agents of pollination. (credit: modification of work by Jon Sullivan)

Many flies are attracted to flowers that have a decaying smell or an odor of rotting flesh. These flowers, which produce nectar, usually have dull colors, such as brown or purple. They are found on the corpse flower or voodoo lily (*Amorphophallus*), dragon arum (*Dracunculus*), and carrion flower (*Stapleia, Rafflesia*). The nectar provides energy, whereas the pollen provides protein. Wasps are also important insect pollinators, and pollinate many species of figs.

Butterflies, such as the monarch, pollinate many garden flowers and wildflowers, which usually occur in clusters. These flowers are brightly colored, have a strong fragrance, are open during the day, and have nectar guides to make access to nectar easier. The pollen is picked up and carried on the butterfly's limbs. Moths, on the other hand, pollinate flowers during the late afternoon and night. The flowers pollinated by moths are pale or white and are flat, enabling the moths to land. One well-studied example of a moth-pollinated plant is the yucca plant, which is pollinated by the yucca moth. The shape of the flower and moth have adapted in such a way as to allow successful pollination. The moth deposits pollen on the sticky stigma for fertilization to occur later. The female moth also deposits eggs into the ovary. As the eggs develop into larvae, they obtain food from the flower and developing seeds. Thus, both the insect and flower benefit from each other in this symbiotic relationship. The corn earworm moth and Gaura plant have a similar relationship (Figure 32.13).

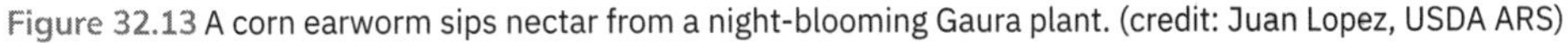

Figure 32.13 A corn earworm sips nectar from a night-blooming Gaura plant. (credit: Juan Lopez, USDA ARS)

Pollination by Bats

In the tropics and deserts, bats are often the pollinators of nocturnal flowers such as agave, guava, and morning glory. The flowers are usually large and white or pale-colored; thus, they can be distinguished from the dark surroundings at night. The flowers have a strong, fruity, or musky fragrance and produce large amounts of nectar. They are naturally large and wide-mouthed to accommodate the head of the bat. As the bats seek the nectar, their faces and heads become covered with pollen, which is then transferred to the next flower.

Pollination by Birds

Many species of small birds, such as the hummingbird (Figure 32.14) and sun birds, are pollinators for plants such as orchids and other wildflowers. Flowers visited by birds are usually sturdy and are oriented in such a way as to allow the birds to stay near the flower without getting their wings entangled in the nearby flowers. The flower typically has a curved, tubular shape, which allows access for the bird's beak. Brightly colored, odorless flowers that are open during the day are pollinated by birds. As a bird seeks energy-rich nectar, pollen is deposited on the bird's head and neck and is then transferred to the next flower it visits. Botanists have been known to determine the range of extinct plants by collecting and identifying pollen from 200-year-old bird specimens from the same site.

Figure 32.14 Hummingbirds have adaptations that allow them to reach the nectar of certain tubular flowers. (credit: Lori Branham)

Pollination by Wind

Most species of conifers, and many angiosperms, such as grasses, maples and oaks, are pollinated by wind. Pine cones are brown and unscented, while the flowers of wind-pollinated angiosperm species are usually green, small, may have small or no petals, and produce large amounts of pollen. Unlike the typical insect-pollinated flowers, flowers adapted to pollination by wind do not produce nectar or scent. In wind-pollinated species, the microsporangia hang out of the flower, and, as the wind blows, the lightweight pollen is carried with it (Figure 32.15). The flowers usually emerge early in the spring, before the leaves, so that the leaves do not block the movement of the wind. The pollen is deposited on the exposed feathery stigma of the flower (Figure 32.16).

Figure 32.15 A person knocks pollen from a pine tree.

Figure 32.16 These male (a) and female (b) catkins are from the goat willow tree (*Salix caprea*). Note how both structures are light and feathery to better disperse and catch the wind-blown pollen.

Pollination by Water

Some weeds, such as Australian sea grass and pond weeds, are pollinated by water. The pollen floats on water, and when it comes into contact with the flower, it is deposited inside the flower.

EVOLUTION CONNECTION

Pollination by Deception

Orchids are highly valued flowers, with many rare varieties (Figure 32.17). They grow in a range of specific habitats, mainly in the tropics of Asia, South America, and Central America. At least 25,000 species of orchids have been identified.

Figure 32.17 Certain orchids use food deception or sexual deception to attract pollinators. Shown here is a bee orchid (*Ophrys apifera*). (credit: David Evans)

Flowers often attract pollinators with food rewards, in the form of nectar. However, some species of orchid are an exception to this standard: they have evolved different ways to attract the desired pollinators. They use a method known as food deception, in which bright colors and perfumes are offered, but no food. *Anacamptis morio*, commonly known as the green-winged orchid, bears bright purple flowers and emits a strong scent. The bumblebee, its main pollinator, is attracted to the flower because of the strong scent—which usually indicates food for a bee—and in the process, picks up the pollen to be transported to another flower.

Other orchids use sexual deception. *Chiloglottis trapeziformis* emits a compound that smells the same as the pheromone emitted by a female wasp to attract male wasps. The male wasp is attracted to the scent, lands on the orchid flower, and in the process, transfers pollen. Some orchids, like the Australian hammer orchid, use scent as well as visual trickery in yet another sexual deception strategy to attract wasps. The flower of this orchid mimics the appearance of a female wasp and emits a pheromone. The male wasp tries to mate with what appears to be a female wasp, and in the process, picks up pollen, which it then transfers to the next counterfeit mate.

Double Fertilization

After pollen is deposited on the stigma, it must germinate and grow through the style to reach the ovule. The microspores, or the pollen, contain two cells: the pollen tube cell and the generative cell. The pollen tube cell grows into a pollen tube through which the generative cell travels. The germination of the pollen tube requires water, oxygen, and certain chemical signals. As it travels through the style to reach the embryo sac, the pollen tube's growth is supported by the tissues of the style. In the meantime, if the generative cell has not already split into two cells, it now divides to form two sperm cells. The pollen tube is guided by the chemicals secreted by the synergids present in the embryo sac, and it enters the ovule sac through the micropyle. Of the two sperm cells, one sperm fertilizes the egg cell, forming a diploid zygote; the other sperm fuses with the two polar nuclei, forming a triploid cell that develops into the **endosperm**. Together, these two fertilization events in angiosperms are known as **double fertilization** (Figure 32.18). After fertilization is complete, no other sperm can enter. The fertilized ovule forms the seed, whereas the tissues of the ovary become the fruit, usually enveloping the seed.

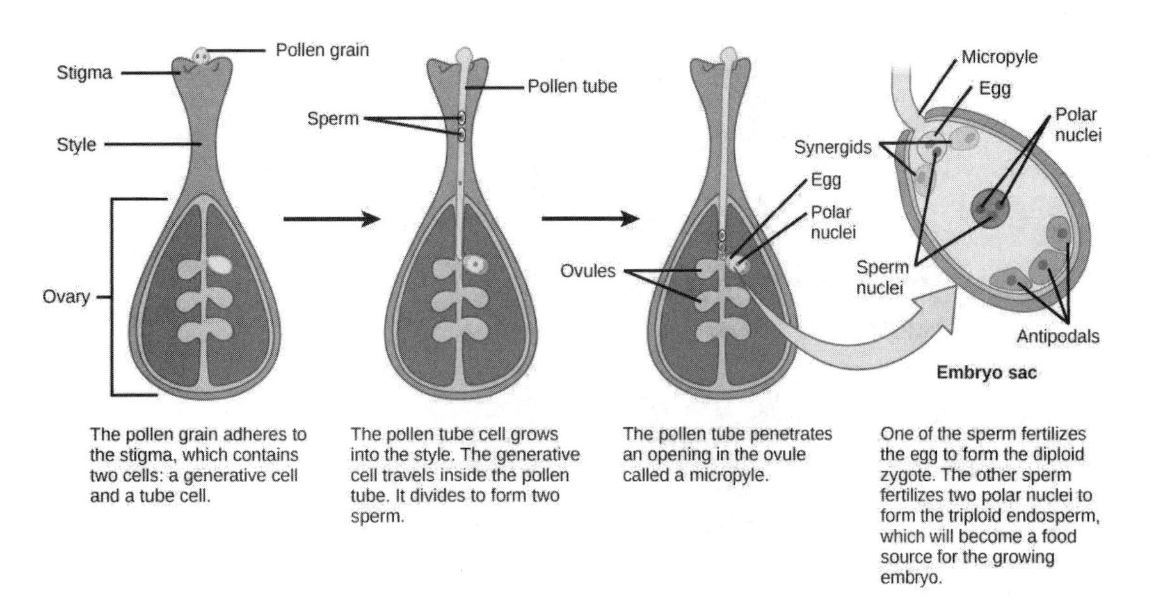

Figure 32.18 In angiosperms, one sperm fertilizes the egg to form the $2n$ zygote, and the other sperm fertilizes the central cell to form the $3n$ endosperm. This is called a double fertilization.

After fertilization, the zygote divides to form two cells: the upper cell, or terminal cell, and the lower, or basal, cell. The division of the basal cell gives rise to the **suspensor**, which eventually makes connection with the maternal tissue. The suspensor provides a route for nutrition to be transported from the mother plant to the growing embryo. The terminal cell also divides, giving rise to a globular-shaped proembryo (Figure 32.19a). In dicots (eudicots), the developing embryo has a heart shape, due to the presence of the two rudimentary **cotyledons** (Figure 32.19b). In non-endospermic dicots, such as *Capsella bursa*, the endosperm develops initially, but is then digested, and the food reserves are moved into the two cotyledons. As the embryo and cotyledons enlarge, they run out of room inside the developing seed, and are forced to bend (Figure 32.19c). Ultimately, the embryo and cotyledons fill the seed (Figure 32.19d), and the seed is ready for dispersal. Embryonic development is suspended after some time, and growth is resumed only when the seed germinates. The developing seedling will rely on the food reserves stored in the cotyledons until the first set of leaves begin photosynthesis.

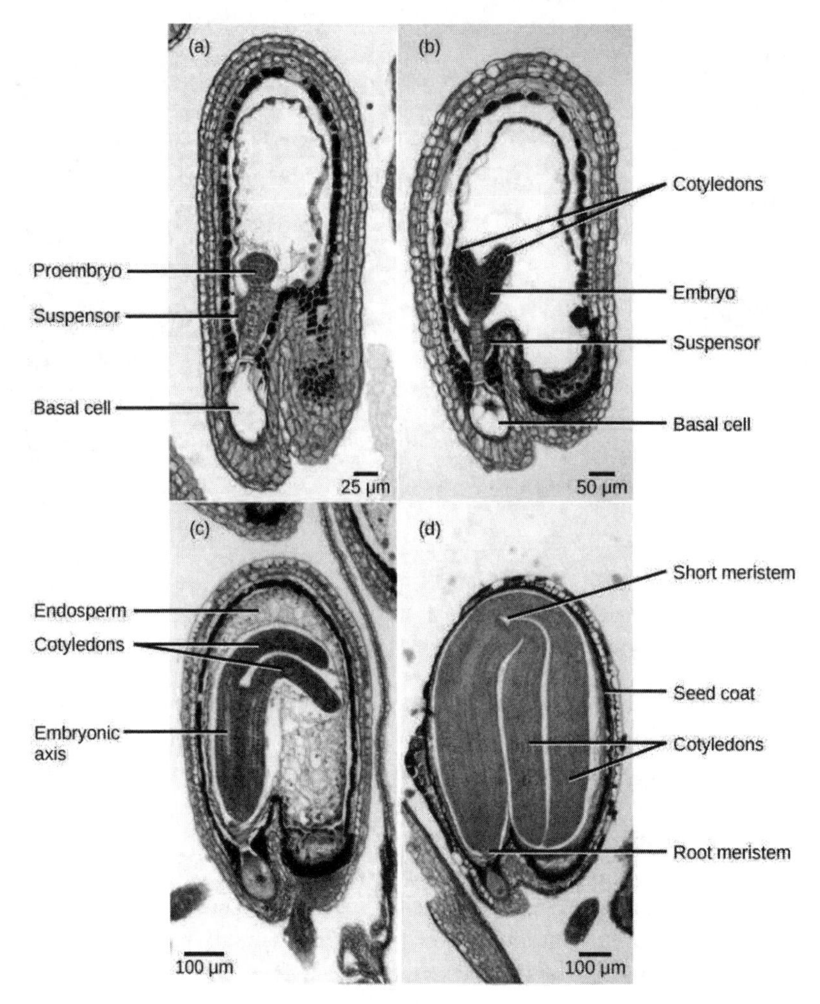

Figure 32.19 Shown are the stages of embryo development in the ovule of a shepherd's purse (*Capsella bursa*). After fertilization, the zygote divides to form an upper terminal cell and a lower basal cell. (a) In the first stage of development, the terminal cell divides, forming a globular pro-embryo. The basal cell also divides, giving rise to the suspensor. (b) In the second stage, the developing embryo has a heart shape due to the presence of cotyledons. (c) In the third stage, the growing embryo runs out of room and starts to bend. (d) Eventually, it completely fills the seed. (credit: modification of work by Robert R. Wise; scale-bar data from Matt Russell)

Development of the Seed

The mature ovule develops into the seed. A typical seed contains a seed coat, cotyledons, endosperm, and a single embryo (Figure 32.20).

VISUAL CONNECTION

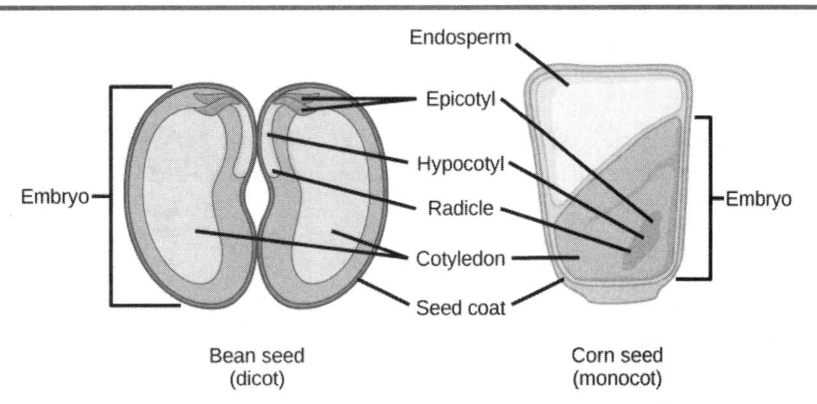

Figure 32.20 The structures of dicot and monocot seeds are shown. Dicots (left) have two cotyledons. Monocots, such as corn (right), have one cotyledon, called the scutellum; it channels nutrition to the growing embryo. Both monocot and dicot embryos have a plumule that forms the leaves, a hypocotyl that forms the stem, and a radicle that forms the root. The embryonic axis comprises everything between the plumule and the radicle, not including the cotyledon(s).

What of the following statements is true?

a. Both monocots and dicots have an endosperm.
b. The radicle develops into the root.
c. The plumule is part of the epicotyl.
d. The endosperm is part of the embryo.

The storage of food reserves in angiosperm seeds differs between monocots and dicots. In monocots, such as corn and wheat, the single cotyledon is called a **scutellum**; the scutellum is connected directly to the embryo via vascular tissue (xylem and phloem). Food reserves are stored in the large endosperm. Upon germination, enzymes are secreted by the **aleurone**, a single layer of cells just inside the seed coat that surrounds the endosperm and embryo. The enzymes degrade the stored carbohydrates, proteins and lipids, the products of which are absorbed by the scutellum and transported via a vasculature strand to the developing embryo. Therefore, the scutellum can be seen to be an absorptive organ, not a storage organ.

The two cotyledons in the dicot seed also have vascular connections to the embryo. In **endospermic dicots**, the food reserves are stored in the endosperm. During germination, the two cotyledons therefore act as absorptive organs to take up the enzymatically released food reserves, much like in monocots (monocots, by definition, also have endospermic seeds). Tobacco (*Nicotiana tabaccum*), tomato (*Solanum lycopersicum*), and pepper (*Capsicum annuum*) are examples of endospermic dicots. In **non-endospermic dicots**, the triploid endosperm develops normally following double fertilization, but the endosperm food reserves are quickly remobilized and moved into the developing cotyledon for storage. The two halves of a peanut seed (*Arachis hypogaea*) and the split peas (*Pisum sativum*) of split pea soup are individual cotyledons loaded with food reserves.

The seed, along with the ovule, is protected by a seed coat that is formed from the integuments of the ovule sac. In dicots, the seed coat is further divided into an outer coat known as the **testa** and inner coat known as the **tegmen**.

The embryonic axis consists of three parts: the plumule, the radicle, and the hypocotyl. The portion of the embryo between the cotyledon attachment point and the radicle is known as the **hypocotyl** (hypocotyl means "below the cotyledons"). The embryonic axis terminates in a **radicle** (the embryonic root), which is the region from which the root will develop. In dicots, the hypocotyls extend above ground, giving rise to the stem of the plant. In monocots, the hypocotyl does not show above ground because monocots do not exhibit stem elongation. The part of the embryonic axis that projects above the cotyledons is known as the **epicotyl**. The **plumule** is composed of the epicotyl, young leaves, and the shoot apical meristem.

Upon germination in dicot seeds, the epicotyl is shaped like a hook with the plumule pointing downwards. This shape is called the plumule hook, and it persists as long as germination proceeds in the dark. Therefore, as the epicotyl pushes through the tough and abrasive soil, the plumule is protected from damage. Upon exposure to light, the hypocotyl hook straightens out, the young foliage leaves face the sun and expand, and the epicotyl continues to elongate. During this time, the radicle is also growing and producing the primary root. As it grows downward to form the tap root, lateral roots branch off to all sides, producing the typical dicot tap root system.

In monocot seeds (Figure 32.21), the testa and tegmen of the seed coat are fused. As the seed germinates, the primary root emerges, protected by the root-tip covering: the **coleorhiza**. Next, the primary shoot emerges, protected by the **coleoptile**: the covering of the shoot tip. Upon exposure to light (i.e., when the plumule has exited the soil and the protective coleoptile is no longer needed), elongation of the coleoptile ceases and the leaves expand and unfold. At the other end of the embryonic axis, the primary root soon dies, while other, adventitious roots (roots that do not arise from the usual place – i.e., the root) emerge from the base of the stem. This gives the monocot a fibrous root system.

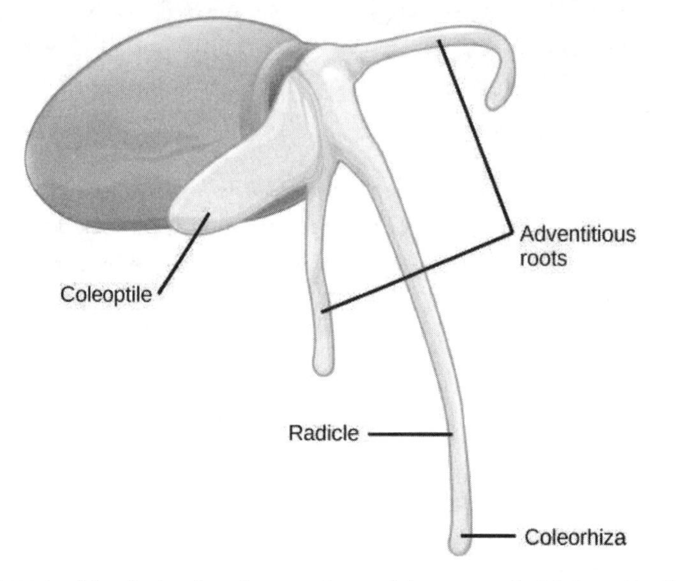

Figure 32.21 As this monocot grass seed germinates, the primary root, or radicle, emerges first, followed by the primary shoot, or coleoptile, and the adventitious roots.

Seed Germination

Many mature seeds enter a period of inactivity, or extremely low metabolic activity: a process known as **dormancy**, which may last for months, years, or even centuries. Dormancy helps keep seeds viable during unfavorable conditions. Upon a return to favorable conditions, seed germination takes place. Favorable conditions could be as diverse as moisture, light, cold, fire, or chemical treatments. After heavy rains, many new seedlings emerge. Forest fires also lead to the emergence of new seedlings. Some seeds require **vernalization** (cold treatment) before they can germinate. This guarantees that seeds produced by plants in temperate climates will not germinate until the spring. Plants growing in hot climates may have seeds that need a heat treatment in order to germinate, to avoid germination in the hot, dry summers. In many seeds, the presence of a thick seed coat retards the ability to germinate. **Scarification**, which includes mechanical or chemical processes to soften the seed coat, is often employed before germination. Presoaking in hot water, or passing through an acid environment, such as an animal's digestive tract, may also be employed.

Depending on seed size, the time taken for a seedling to emerge may vary. Species with large seeds have enough food reserves to germinate deep below ground, and still extend their epicotyl all the way to the soil surface. Seeds of small-seeded species usually require light as a germination cue. This ensures the seeds only germinate at or near the soil surface (where the light is greatest). If they were to germinate too far underneath the surface, the developing seedling would not have enough food reserves to reach the sunlight.

Development of Fruit and Fruit Types

After fertilization, the ovary of the flower usually develops into the fruit. Fruits are usually associated with having a sweet taste; however, not all fruits are sweet. Botanically, the term "fruit" is used for a ripened ovary. In most cases, flowers in which fertilization has taken place will develop into fruits, and flowers in which fertilization has not taken place will not. Some fruits develop from the ovary and are known as true fruits, whereas others develop from other parts of the female gametophyte and are known as accessory fruits. The fruit encloses the seeds and the developing embryo, thereby providing it with protection. Fruits are of many types, depending on their origin and texture. The sweet tissue of the blackberry, the red flesh of the tomato, the shell of the peanut, and the hull of corn (the tough, thin part that gets stuck in your teeth when you eat popcorn) are all fruits. As the fruit matures, the seeds also mature.

Fruits may be classified as simple, aggregate, multiple, or accessory, depending on their origin (Figure 32.22). If the fruit develops from a single carpel or fused carpels of a single ovary, it is known as a **simple fruit**, as seen in nuts and beans. An **aggregate fruit** is one that develops from more than one carpel, but all are in the same flower: the mature carpels fuse together to form the entire fruit, as seen in the raspberry. **Multiple fruit** develops from an inflorescence or a cluster of flowers. An example is the pineapple, where the flowers fuse together to form the fruit. **Accessory fruits** (sometimes called false fruits) are not derived from the ovary, but from another part of the flower, such as the receptacle (strawberry) or the hypanthium (apples and pears).

Figure 32.22 There are four main types of fruits. Simple fruits, such as these nuts, are derived from a single ovary. Aggregate fruits, like raspberries, form from many carpels that fuse together. Multiple fruits, such as pineapple, form from a cluster of flowers called an inflorescence. Accessory fruit, like the apple, are formed from a part of the plant other than the ovary. (credit "nuts": modification of work by Petr Kratochvil; credit "raspberries": modification of work by Cory Zanker; credit "pineapple": modification of work by Howie Le; credit "apple": modification of work by Paolo Neo)

Fruits generally have three parts: the **exocarp** (the outermost skin or covering), the **mesocarp** (middle part of the fruit), and the **endocarp** (the inner part of the fruit). Together, all three are known as the **pericarp**. The mesocarp is usually the fleshy, edible part of the fruit; however, in some fruits, such as the almond, the endocarp is the edible part. In many fruits, two or all three of the layers are fused, and are indistinguishable at maturity. Fruits can be dry or fleshy. Furthermore, fruits can be divided into dehiscent or indehiscent types. Dehiscent fruits, such as peas, readily release their seeds, while indehiscent fruits, like peaches, rely on decay to release their seeds.

Fruit and Seed Dispersal

The fruit has a single purpose: seed dispersal. Seeds contained within fruits need to be dispersed far from the mother plant, so they may find favorable and less competitive conditions in which to germinate and grow.

Some fruit have built-in mechanisms so they can disperse by themselves, whereas others require the help of agents like wind, water, and animals (Figure 32.23). Modifications in seed structure, composition, and size help in dispersal. Wind-dispersed fruit are lightweight and may have wing-like appendages that allow them to be carried by the wind. Some have a parachute-like structure to keep them afloat. Some fruits—for example, the dandelion—have hairy, weightless structures that are suited to dispersal by wind.

Seeds dispersed by water are contained in light and buoyant fruit, giving them the ability to float. Coconuts are well known for their ability to float on water to reach land where they can germinate. Similarly, willow and silver birches produce lightweight fruit that can float on water.

Animals and birds eat fruits, and the seeds that are not digested are excreted in their droppings some distance away. Some animals, like squirrels, bury seed-containing fruits for later use; if the squirrel does not find its stash of fruit, and if conditions are favorable, the seeds germinate. Some fruits, like the cocklebur, have hooks or sticky structures that stick to an animal's coat and are then transported to another place. Humans also play a big role in dispersing seeds when they carry fruits to new places and throw away the inedible part that contains the seeds.

All of the above mechanisms allow for seeds to be dispersed through space, much like an animal's offspring can move to a new location. Seed dormancy, which was described earlier, allows plants to disperse their progeny through time: something animals cannot do. Dormant seeds can wait months, years, or even decades for the proper conditions for germination and propagation of the species.

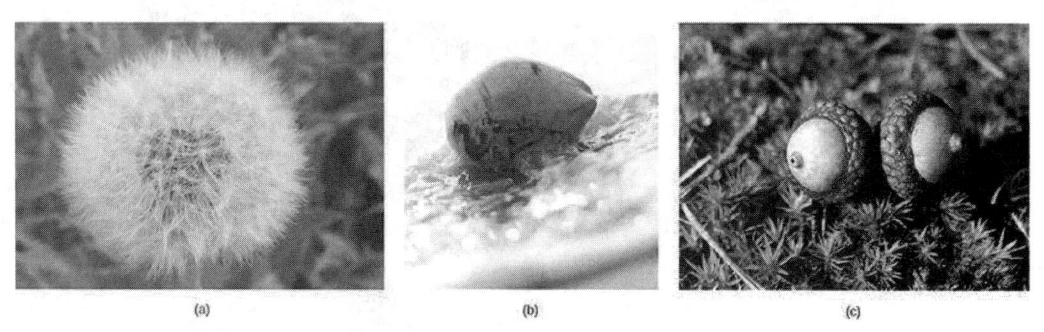

Figure 32.23 Fruits and seeds are dispersed by various means. (a) Dandelion seeds are dispersed by wind, the (b) coconut seed is dispersed by water, and the (c) acorn is dispersed by animals that cache and then forget it. (credit a: modification of work by "Rosendahl"/Flickr; credit b: modification of work by Shine Oa; credit c: modification of work by Paolo Neo)

32.3 Asexual Reproduction

By the end of this section, you will be able to do the following:

- Compare the mechanisms and methods of natural and artificial asexual reproduction
- Describe the advantages and disadvantages of natural and artificial asexual reproduction
- Discuss plant life spans

Many plants are able to propagate themselves using asexual reproduction. This method does not require the investment required to produce a flower, attract pollinators, or find a means of seed dispersal. Asexual reproduction produces plants that are genetically identical to the parent plant because no mixing of male and female gametes takes place. Traditionally, these plants survive well under stable environmental conditions when compared with plants produced from sexual reproduction because they carry genes identical to those of their parents.

Many different types of roots exhibit asexual reproduction (Figure 32.24). The corm is used by gladiolus and garlic. Bulbs, such as a scaly bulb in lilies and a tunicate bulb in daffodils, are other common examples. A potato is a stem tuber, while parsnip propagates from a taproot. Ginger and iris produce rhizomes, while ivy uses an adventitious root (a root arising from a plant part other than the main or primary root), and the strawberry plant has a stolon, which is also called a runner.

Figure 32.24 Different types of stems allow for asexual reproduction. (a) The corm of a garlic plant looks similar to (b) a tulip bulb, but the corm is solid tissue, while the bulb consists of layers of modified leaves that surround an underground stem. Both corms and bulbs can self-propagate, giving rise to new plants. (c) Ginger forms masses of stems called rhizomes that can give rise to multiple plants. (d) Potato plants form fleshy stem tubers. Each eye in the stem tuber can give rise to a new plant. (e) Strawberry plants form stolons: stems that grow at the soil surface or just below ground and can give rise to new plants. (credit a: modification of work by Dwight Sipler; credit c: modification of work by Albert Cahalan, USDA ARS; credit d: modification of work by Richard North; credit e: modification of work by Julie Magro)

Some plants can produce seeds without fertilization. Either the ovule or part of the ovary, which is diploid in nature, gives rise to a new seed. This method of reproduction is known as **apomixis**.

An advantage of asexual reproduction is that the resulting plant will reach maturity faster. Since the new plant is arising from an adult plant or plant parts, it will also be sturdier than a seedling. Asexual reproduction can take place by natural or artificial (assisted by humans) means.

Natural Methods of Asexual Reproduction

Natural methods of asexual reproduction include strategies that plants have developed to self-propagate. Many plants—like ginger, onion, gladioli, and dahlia—continue to grow from buds that are present on the surface of the stem. In some plants, such as the sweet potato, adventitious roots or runners can give rise to new plants (Figure 32.25). In *Bryophyllum* and kalanchoe, the leaves have small buds on their margins. When these are detached from the plant, they grow into independent plants; or, they may start growing into independent plants if the leaf touches the soil. Some plants can be propagated through cuttings alone.

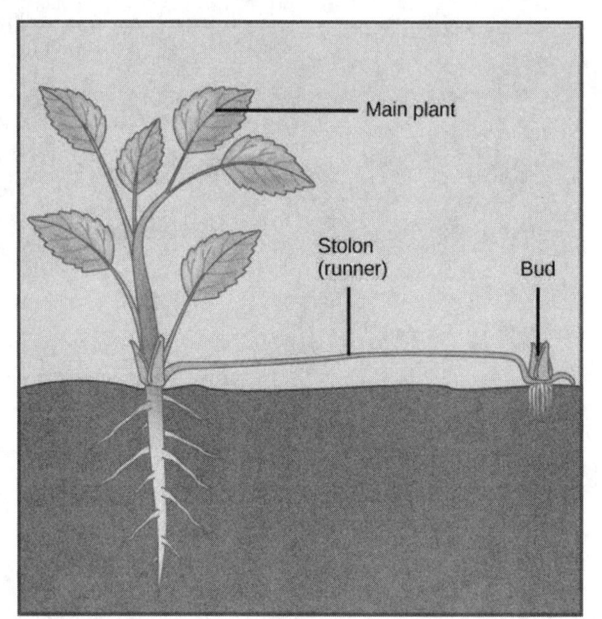

Figure 32.25 A stolon, or runner, is a stem that runs along the ground. At the nodes, it forms adventitious roots and buds that grow into a new plant.

Artificial Methods of Asexual Reproduction

These methods are frequently employed to give rise to new, and sometimes novel, plants. They include grafting, cutting, layering, and micropropagation.

Grafting

Grafting has long been used to produce novel varieties of roses, citrus species, and other plants. In **grafting**, two plant species are used; part of the stem of the desirable plant is grafted onto a rooted plant called the stock. The part that is grafted or attached is called the **scion**. Both are cut at an oblique angle (any angle other than a right angle), placed in close contact with each other, and are then held together (Figure 32.26). Matching up these two surfaces as closely as possible is extremely important because these will be holding the plant together. The vascular systems of the two plants grow and fuse, forming a graft. After a period of time, the scion starts producing shoots, and eventually starts bearing flowers and fruits. Grafting is widely used in viticulture (grape growing) and the citrus industry. Scions capable of producing a particular fruit variety are grafted onto root stock with specific resistance to disease.

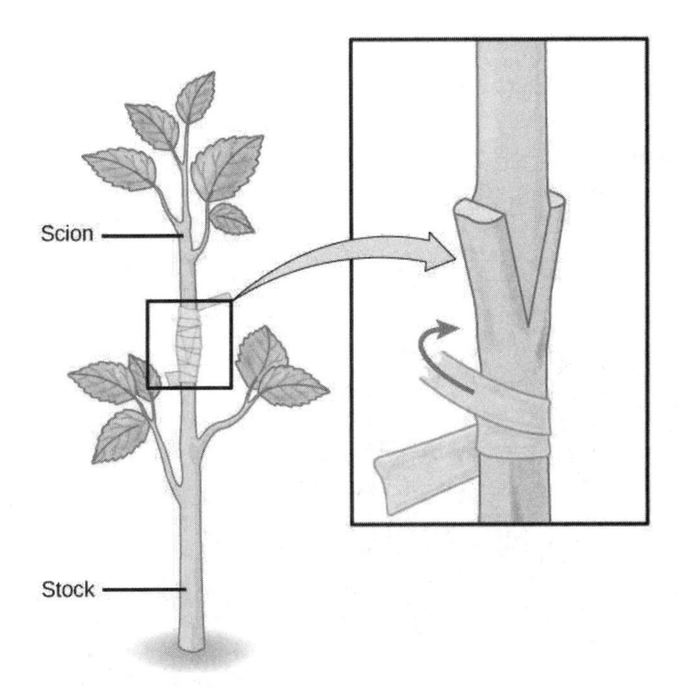

Figure 32.26 Grafting is an artificial method of asexual reproduction used to produce plants combining favorable stem characteristics with favorable root characteristics. The stem of the plant to be grafted is known as the scion, and the root is called the stock.

Cutting

Plants such as coleus and money plant are propagated through stem **cuttings**, where a portion of the stem containing nodes and internodes is placed in moist soil and allowed to root. In some species, stems can start producing a root even when placed only in water. For example, leaves of the African violet will root if kept in water undisturbed for several weeks.

Layering

Layering is a method in which a stem attached to the plant is bent and covered with soil. Young stems that can be bent easily without any injury are preferred. Jasmine and bougainvillea (paper flower) can be propagated this way (Figure 32.27). In some plants, a modified form of layering known as air layering is employed. A portion of the bark or outermost covering of the stem is removed and covered with moss, which is then taped. Some gardeners also apply rooting hormone. After some time, roots will appear, and this portion of the plant can be removed and transplanted into a separate pot.

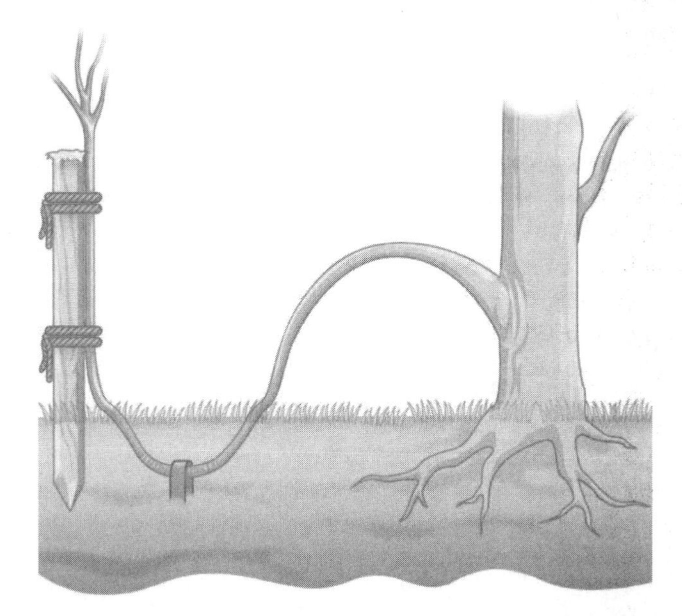

Figure 32.27 In layering, a part of the stem is buried so that it forms a new plant. (credit: modification of work by Pearson Scott Foresman, donated to the Wikimedia Foundation)

Micropropagation

Micropropagation (also called plant tissue culture) is a method of propagating a large number of plants from a single plant in a short time under laboratory conditions (Figure 32.28). This method allows propagation of rare, endangered species that may be difficult to grow under natural conditions, are economically important, or are in demand as disease-free plants.

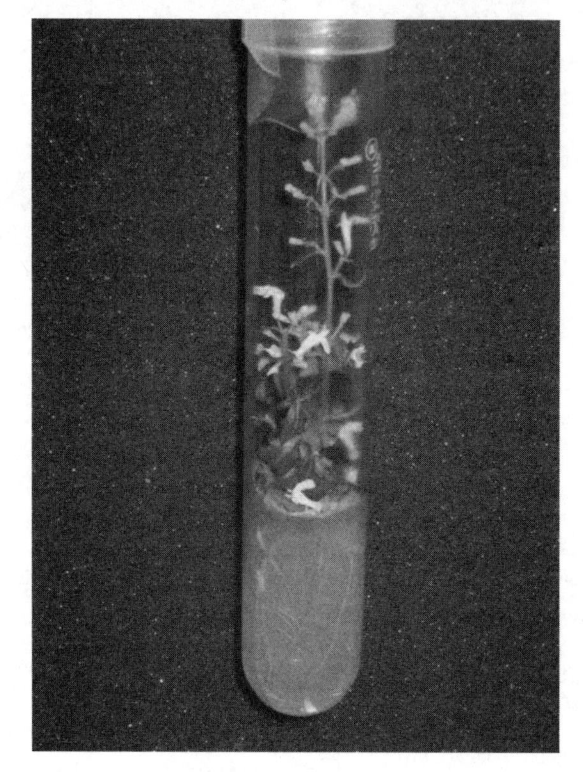

Figure 32.28 Micropropagation is used to propagate plants in sterile conditions. (credit: Nikhilesh Sanyal)

To start plant tissue culture, a part of the plant such as a stem, leaf, embryo, anther, or seed can be used. The plant material is thoroughly sterilized using a combination of chemical treatments standardized for that species. Under sterile conditions, the plant material is placed on a plant tissue culture medium that contains all the minerals, vitamins, and hormones required by the plant. The plant part often gives rise to an undifferentiated mass known as callus, from which individual plantlets begin to grow after a period of time. These can be separated and are first grown under greenhouse conditions before they are moved to field conditions.

Plant Life Spans

The length of time from the beginning of development to the death of a plant is called its life span. The life cycle, on the other hand, is the sequence of stages a plant goes through from seed germination to seed production of the mature plant. Some plants, such as annuals, only need a few weeks to grow, produce seeds and die. Other plants, such as the bristlecone pine, live for thousands of years. Some bristlecone pines have a documented age of 4,500 years (Figure 32.29). Even as some parts of a plant, such as regions containing meristematic tissue—the area of active plant growth consisting of undifferentiated cells capable of cell division—continue to grow, some parts undergo programmed cell death (apoptosis). The cork found on stems, and the water-conducting tissue of the xylem, for example, are composed of dead cells.

Figure 32.29 The bristlecone pine, shown here in the Ancient Bristlecone Pine Forest in the White Mountains of eastern California, has been known to live for 4,500 years. (credit: Rick Goldwaser)

Plant species that complete their lifecycle in one season are known as annuals, an example of which is *Arabidopsis*, or mouse-ear cress. Biennials such as carrots complete their lifecycle in two seasons. In a biennial's first season, the plant has a vegetative phase, whereas in the next season, it completes its reproductive phase. Commercial growers harvest the carrot roots after the first year of growth, and do not allow the plants to flower. Perennials, such as the magnolia, complete their lifecycle in two years or more.

In another classification based on flowering frequency, **monocarpic** plants flower only once in their lifetime; examples include bamboo and yucca. During the vegetative period of their life cycle (which may be as long as 120 years in some bamboo species), these plants may reproduce asexually and accumulate a great deal of food material that will be required during their once-in-a-lifetime flowering and setting of seed after fertilization. Soon after flowering, these plants die. **Polycarpic** plants form flowers many times during their lifetime. Fruit trees, such as apple and orange trees, are polycarpic; they flower every year. Other polycarpic species, such as perennials, flower several times during their life span, but not each year. By this means, the plant does not require all its nutrients to be channelled towards flowering each year.

As is the case with all living organisms, genetics and environmental conditions have a role to play in determining how long a plant will live. Susceptibility to disease, changing environmental conditions, drought, cold, and competition for nutrients are some of the factors that determine the survival of a plant. Plants continue to grow, despite the presence of dead tissue such as cork. Individual parts of plants, such as flowers and leaves, have different rates of survival. In many trees, the older leaves turn yellow and eventually fall from the tree. Leaf fall is triggered by factors such as a decrease in photosynthetic efficiency, due to shading by upper leaves, or oxidative damage incurred as a result of photosynthetic reactions. The components of the part to be shed are recycled by the plant for use in other processes, such as development of seed and storage. This process is known as nutrient recycling.

The aging of a plant and all the associated processes is known as **senescence**, which is marked by several complex biochemical changes. One of the characteristics of senescence is the breakdown of chloroplasts, which is characterized by the yellowing of leaves. The chloroplasts contain components of photosynthetic machinery such as membranes and proteins. Chloroplasts also contain DNA. The proteins, lipids, and nucleic acids are broken down by specific enzymes into smaller molecules and salvaged by the plant to support the growth of other plant tissues.

The complex pathways of nutrient recycling within a plant are not well understood. Hormones are known to play a role in senescence. Applications of cytokinins and ethylene delay or prevent senescence; in contrast, abscissic acid causes premature onset of senescence.

KEY TERMS

accessory fruit fruit derived from tissues other than the ovary

aggregate fruit fruit that develops from multiple carpels in the same flower

aleurone single layer of cells just inside the seed coat that secretes enzymes upon germination

androecium sum of all the stamens in a flower

antipodals the three cells away from the micropyle

apomixis process by which seeds are produced without fertilization of sperm and egg

coleoptile covering of the shoot tip, found in germinating monocot seeds

coleorhiza covering of the root tip, found in germinating monocot seeds

cotyledon fleshy part of seed that provides nutrition to the seed

cross-pollination transfer of pollen from the anther of one flower to the stigma of a different flower

cutting method of asexual reproduction where a portion of the stem contains nodes and internodes is placed in moist soil and allowed to root

dormancy period of no growth and very slow metabolic processes

double fertilization two fertilization events in angiosperms; one sperm fuses with the egg, forming the zygote, whereas the other sperm fuses with the polar nuclei, forming endosperm

endocarp innermost part of fruit

endosperm triploid structure resulting from fusion of a sperm with polar nuclei, which serves as a nutritive tissue for embryo

endospermic dicot dicot that stores food reserves in the endosperm

epicotyl embryonic shoot above the cotyledons

exine outermost covering of pollen

exocarp outermost covering of a fruit

gametophyte multicellular stage of the plant that gives rise to haploid gametes or spores

grafting method of asexual reproduction where the stem from one plant species is spliced to a different plant

gravitropism response of a plant growth in the same direction as gravity

gynoecium the sum of all the carpels in a flower

hypocotyl embryonic axis above the cotyledons

intine inner lining of the pollen

layering method of propagating plants by bending a stem under the soil

megagametogenesis second phase of female gametophyte development, during which the surviving haploid megaspore undergoes mitosis to produce an eight-nucleate, seven-cell female gametophyte, also known as the megagametophyte or embryo sac.

megasporangium tissue found in the ovary that gives rise to the female gamete or egg

megasporogenesis first phase of female gametophyte development, during which a single cell in the diploid megasporangium undergoes meiosis to produce four megaspores, only one of which survives

megasporophyll bract (a type of modified leaf) on the central axis of a female gametophyte

mesocarp middle part of a fruit

micropropagation propagation of desirable plants from a plant part; carried out in a laboratory

micropyle opening on the ovule sac through which the pollen tube can gain entry

microsporangium tissue that gives rise to the microspores or the pollen grain

microsporophyll central axis of a male cone on which bracts (a type of modified leaf) are attached

monocarpic plants that flower once in their lifetime

multiple fruit fruit that develops from multiple flowers on an inflorescence

nectar guide pigment pattern on a flower that guides an insect to the nectaries

non-endospermic dicot dicot that stores food reserves in the developing cotyledon

perianth (also, petal or sepal) part of the flower consisting of the calyx and/or corolla; forms the outer envelope of the flower

pericarp collective term describing the exocarp, mesocarp, and endocarp; the structure that encloses the seed and is a part of the fruit

plumule shoot that develops from the germinating seed

polar nuclei found in the ovule sac; fusion with one sperm cell forms the endosperm

pollination transfer of pollen to the stigma

polycarpic plants that flower several times in their lifetime

radicle original root that develops from the germinating seed

scarification mechanical or chemical processes to soften the seed coat

scion the part of a plant that is grafted onto the root stock of another plant

scutellum type of cotyledon found in monocots, as in grass seeds

self-pollination transfer of pollen from the anther to the stigma of the same flower

senescence process that describes aging in plant tissues

simple fruit fruit that develops from a single carpel or fused carpels

sporophyte multicellular diploid stage in plants that is formed after the fusion of male and female gametes

suspensor part of the growing embryo that makes connection with the maternal tissues

synergid type of cell found in the ovule sac that secretes chemicals to guide the pollen tube towards the egg
tegmen inner layer of the seed coat
testa outer layer of the seed coat
vernalization exposure to cold required by some seeds before they can germinate

CHAPTER SUMMARY

32.1 Reproductive Development and Structure

The flower contains the reproductive structures of a plant. All complete flowers contain four whorls: the calyx, corolla, androecium, and gynoecium. The stamens are made up of anthers, in which pollen grains are produced, and a supportive strand called the filament. The pollen contains two cells— a generative cell and a tube cell—and is covered by two layers called the intine and the exine. The carpels, which are the female reproductive structures, consist of the stigma, style, and ovary. The female gametophyte is formed from mitotic divisions of the megaspore, forming an eight-nuclei ovule sac. This is covered by a layer known as the integument. The integument contains an opening called the micropyle, through which the pollen tube enters the embryo sac.

The diploid sporophyte of angiosperms and gymnosperms is the conspicuous and long-lived stage of the life cycle. The sporophytes differentiate specialized reproductive structures called sporangia, which are dedicated to the production of spores. The microsporangium contains microspore mother cells, which divide by meiosis to produce haploid microspores. The microspores develop into male gametophytes that are released as pollen. The megasporangium contains megaspore mother cells, which divide by meiosis to produce haploid megaspores. A megaspore develops into a female gametophyte containing a haploid egg. A new diploid sporophyte is formed when a male gamete from a pollen grain enters the ovule sac and fertilizes this egg.

32.2 Pollination and Fertilization

For fertilization to occur in angiosperms, pollen has to be transferred to the stigma of a flower: a process known as pollination. Gymnosperm pollination involves the transfer of pollen from a male cone to a female cone. When the pollen of the flower is transferred to the stigma of the same or another flower on the same plant, it is called self-pollination. Cross-pollination occurs when pollen is transferred from one flower to another flower of another plant. Cross-pollination requires pollinating agents such as water, wind, or animals, and increases genetic diversity. After the pollen lands on the stigma, the tube cell gives rise to the pollen tube, through which the generative nucleus migrates. The pollen tube gains entry through the micropyle on the ovule sac. The generative cell divides to form two sperm cells: one fuses with the egg to form the diploid zygote, and the other fuses with the polar nuclei to form the endosperm, which is triploid in nature. This is known as double fertilization. After fertilization, the zygote divides to form the embryo and the fertilized ovule forms the seed. The walls of the ovary form the fruit in which the seeds develop. The seed, when mature, will germinate under favorable conditions and give rise to the diploid sporophyte.

32.3 Asexual Reproduction

Many plants reproduce asexually as well as sexually. In asexual reproduction, part of the parent plant is used to generate a new plant. Grafting, layering, and micropropagation are some methods used for artificial asexual reproduction. The new plant is genetically identical to the parent plant from which the stock has been taken. Asexually reproducing plants thrive well in stable environments.

Plants have different life spans, dependent on species, genotype, and environmental conditions. Parts of the plant, such as regions containing meristematic tissue, continue to grow, while other parts experience programmed cell death. Leaves that are no longer photosynthetically active are shed from the plant as part of senescence, and the nutrients from these leaves are recycled by the plant. Other factors, including the presence of hormones, are known to play a role in delaying senescence.

VISUAL CONNECTION QUESTIONS

1. Figure 32.3 If the anther is missing, what type of reproductive structure will the flower be unable to produce? What term is used to describe an incomplete flower lacking the androecium? What term describes an incomplete flower lacking a gynoecium?

2. Figure 32.8 An embryo sac is missing the synergids. What specific impact would you expect this to have on fertilization?
   a. The pollen tube will be unable to form.
   b. The pollen tube will form but will not be guided toward the egg.
   c. Fertilization will not occur because the synergid is the egg.
   d. Fertilization will occur but the embryo will not be able to grow.

3. Figure 32.20 What is the function of the cotyledon?
   a. It develops into the root.
   b. It provides nutrition for the embryo.
   c. It forms the embryo.
   d. It protects the embryo.

REVIEW QUESTIONS

4. In a plant's male reproductive organs, development of pollen takes place in a structure known as the _______.
   a. stamen
   b. microsporangium
   c. anther
   d. tapetum

5. The stamen consists of a long stalk called the filament that supports the _______.
   a. stigma
   b. sepal
   c. style
   d. anther

6. The _______ are collectively called the calyx.
   a. sepals
   b. petals
   c. tepals
   d. stamens

7. The pollen lands on which part of the flower?
   a. stigma
   b. style
   c. ovule
   d. integument

8. After double fertilization, a zygote and _______ form.
   a. an ovule
   b. endosperm
   c. a cotyledon
   d. a suspensor

9. The fertilized ovule gives rise to the _______.
   a. fruit
   b. seed
   c. endosperm
   d. embryo

10. What is the term for a fruit that develops from tissues other than the ovary?
    a. simple fruit
    b. aggregate fruit
    c. multiple fruit
    d. accessory fruit

11. The _______ is the outermost covering of a fruit.
    a. endocarp
    b. pericarp
    c. exocarp
    d. mesocarp

12. _______ is a useful method of asexual reproduction for propagating hard-to-root plants.
    a. grafting
    b. layering
    c. cuttings
    d. budding

13. Which of the following is an advantage of asexual reproduction?
    a. Cuttings taken from an adult plant show increased resistance to diseases.
    b. Grafted plants can more successfully endure drought.
    c. When cuttings or buds are taken from an adult plant or plant parts, the resulting plant will grow into an adult faster than a seedling.
    d. Asexual reproduction takes advantage of a more diverse gene pool.

14. Plants that flower once in their lifetime are known as _______.
    a. monoecious
    b. dioecious
    c. polycarpic
    d. monocarpic

15. Plant species that complete their lifecycle in one season are known as _______.
    a. biennials
    b. perennials
    c. annuals
    d. polycarpic

CRITICAL THINKING QUESTIONS

16. Describe the reproductive organs inside a flower.
17. Describe the two-stage lifecycle of plants: the gametophyte stage and the sporophyte stage.
18. Describe the four main parts, or whorls, of a flower.
19. Discuss the differences between a complete flower and an incomplete flower.
20. Why do some seeds undergo a period of dormancy, and how do they break dormancy?
21. Discuss some ways in which fruit seeds are dispersed.
22. What are some advantages of asexual reproduction in plants?
23. Describe natural and artificial methods of asexual reproduction in plants.
24. Discuss the life cycles of various plants.
25. How are plants classified on the basis of flowering frequency?

CHAPTER 33
The Animal Body: Basic Form and Function

Figure 33.1 An arctic fox is a complex animal, well adapted to its environment. It changes coat color with the seasons, and has longer fur in winter to trap heat. (credit: modification of work by Keith Morehouse, USFWS)

INTRODUCTION The arctic fox is an example of a complex animal that has adapted to its environment and illustrates the relationships between an animal's form and function. The structures of animals consist of primary tissues that make up more complex organs and organ systems. Homeostasis allows an animal to maintain a balance between its internal and external environments.

Chapter Outline

33.1 Animal Form and Function

By the end of this section, you will be able to do the following:

- Describe the various types of body plans that occur in animals
- Describe limits on animal size and shape
- Relate bioenergetics to body size, levels of activity, and the environment

Animals vary in form and function. From a sponge to a worm to a goat, an organism has a distinct body plan that limits its size and shape. Animals' bodies are also designed to interact with their environments, whether in the deep sea, a rainforest canopy, or the desert. Therefore, a large amount of information about the structure of an organism's body (anatomy) and the function of its cells, tissues and organs (physiology) can be learned by studying that organism's environment.

Body Plans

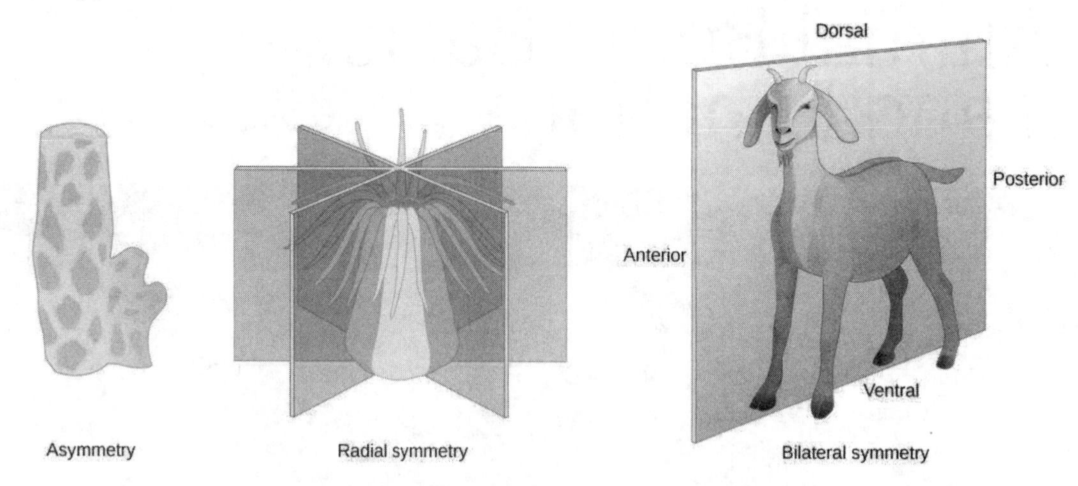

Figure 33.2 Animals exhibit different types of body symmetry. The sponge is asymmetrical, the sea anemone has radial symmetry, and the goat has bilateral symmetry.

Animal body plans follow set patterns related to symmetry. They are asymmetrical, radial, or bilateral in form as illustrated in Figure 33.2. **Asymmetrical** animals are animals with no pattern or symmetry; an example of an asymmetrical animal is a sponge. Radial symmetry, as illustrated in Figure 33.2, describes when an animal has an up-and-down orientation: any plane cut along its longitudinal axis through the organism produces equal halves, but not a definite right or left side. This plan is found mostly in aquatic animals, especially organisms that attach themselves to a base, like a rock or a boat, and extract their food from the surrounding water as it flows around the organism. Bilateral symmetry is illustrated in the same figure by a goat. The goat also has an upper and lower component to it, but a plane cut from front to back separates the animal into definite right and left sides. Additional terms used when describing positions in the body are anterior (front), posterior (rear), dorsal (toward the back), and ventral (toward the stomach). Bilateral symmetry is found in both land-based and aquatic animals; it enables a high level of mobility.

Limits on Animal Size and Shape

Animals with bilateral symmetry that live in water tend to have a **fusiform** shape: this is a tubular shaped body that is tapered at both ends. This shape decreases the drag on the body as it moves through water and allows the animal to swim at high speeds. Table 33.1 lists the maximum speed of various animals. Certain types of sharks can swim at fifty kilometers per hour and some dolphins at 32 to 40 kilometers per hour. Land animals frequently travel faster, although the tortoise and snail are significantly slower than cheetahs. Another difference in the adaptations of aquatic and land-dwelling organisms is that aquatic organisms are constrained in shape by the forces of drag in the water since water has higher viscosity than air. On the other hand, land-dwelling organisms are constrained mainly by gravity, and drag is relatively unimportant. For example, most adaptations in birds are for gravity not for drag.

Maximum Speed of Assorted Land & Marine Animals

Animal	Speed (kmh)	Speed (mph)
Cheetah	113	70
Quarter horse	77	48
Fox	68	42

Table 33.1

Animal	Speed (kmh)	Speed (mph)
Shortfin mako shark	50	31
Domestic house cat	48	30
Human	45	28
Dolphin	32–40	20–25
Mouse	13	8
Snail	0.05	0.03

Table 33.1

Most animals have an exoskeleton, including insects, spiders, scorpions, horseshoe crabs, centipedes, and crustaceans. Scientists estimate that, of insects alone, there are over 30 million species on our planet. The exoskeleton is a hard covering or shell that provides benefits to the animal, such as protection against damage from predators and from water loss (for land animals); it also provides for the attachments of muscles.

As the tough and resistant outer cover of an arthropod, the exoskeleton may be constructed of a tough polymer such as chitin and is often biomineralized with materials such as calcium carbonate. This is fused to the animal's epidermis. Ingrowths of the exoskeleton, called **apodemes**, function as attachment sites for muscles, similar to tendons in more advanced animals (Figure 33.3). In order to grow, the animal must first synthesize a new exoskeleton underneath the old one and then shed or molt the original covering. This limits the animal's ability to grow continually, and may limit the individual's ability to mature if molting does not occur at the proper time. The thickness of the exoskeleton must be increased significantly to accommodate any increase in weight. It is estimated that a doubling of body size increases body weight by a factor of eight. The increasing thickness of the chitin necessary to support this weight limits most animals with an exoskeleton to a relatively small size. The same principles apply to endoskeletons, but they are more efficient because muscles are attached on the outside, making it easier to compensate for increased mass.

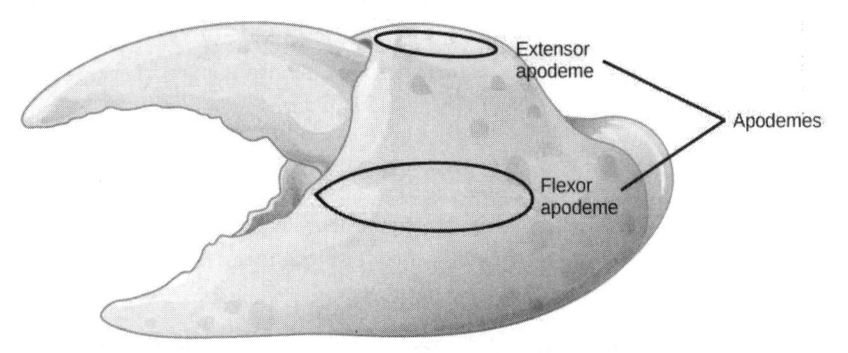

Figure 33.3 Apodemes are ingrowths on arthropod exoskeletons to which muscles attach. The apodemes on this crab leg are located above and below the fulcrum of the claw. Contraction of muscles attached to the apodemes pulls the claw closed.

An animal with an endoskeleton has its size determined by the amount of skeletal system it needs in order to support the other tissues and the amount of muscle it needs for movement. As the body size increases, both bone and muscle mass increase. The speed achievable by the animal is a balance between its overall size and the bone and muscle that provide support and movement.

Limiting Effects of Diffusion on Size and Development

The exchange of nutrients and wastes between a cell and its watery environment occurs through the process of diffusion. All living cells are bathed in liquid, whether they are in a single-celled organism or a multicellular one. Diffusion is effective over a specific distance and limits the size that an individual cell can attain. If a cell is a single-celled microorganism, such as an amoeba, it can satisfy all of its nutrient and waste needs through diffusion. If the cell is too large, then diffusion is ineffective and the center of the cell does not receive adequate nutrients nor is it able to effectively dispel its waste.

An important concept in understanding how efficient diffusion is as a means of transport is the surface to volume ratio. Recall that any three-dimensional object has a surface area and volume; the ratio of these two quantities is the surface-to-volume ratio. Consider a cell shaped like a perfect sphere: it has a surface area of $4\pi r^2$, and a volume of $(4/3)\pi r^3$. The surface-to-volume ratio of a sphere is $3/r$; as the cell gets bigger, its surface to volume ratio decreases, making diffusion less efficient. The larger the size of the sphere, or animal, the less surface area for diffusion it possesses.

The solution to producing larger organisms is for them to become multicellular. Specialization occurs in complex organisms, allowing cells to become more efficient at doing fewer tasks. For example, circulatory systems bring nutrients and remove waste, while respiratory systems provide oxygen for the cells and remove carbon dioxide from them. Other organ systems have developed further specialization of cells and tissues and efficiently control body functions. Moreover, surface-to-volume ratio applies to other areas of animal development, such as the relationship between muscle mass and cross-sectional surface area in supporting skeletons, and in the relationship between muscle mass and the generation and dissipation of heat.

LINK TO LEARNING

Visit this interactive site (http://openstax.org/l/nanoscopy) to see an entire animal (a zebrafish embryo) at the cellular and sub-cellular level. Use the zoom and navigation functions for a virtual nanoscopy exploration.

Animal Bioenergetics

All animals must obtain their energy from food they ingest or absorb. These nutrients are converted to adenosine triphosphate (ATP) for short-term storage and use by all cells. Some animals store energy for slightly longer times as glycogen, and others store energy for much longer times in the form of triglycerides housed in specialized adipose tissues. No energy system is one hundred percent efficient, and an animal's metabolism produces waste energy in the form of heat. If an animal can conserve that heat and maintain a relatively constant body temperature, it is classified as a warm-blooded animal and called an **endotherm**. The insulation used to conserve the body heat comes in the forms of fur, fat, or feathers. The absence of insulation in **ectothermic** animals increases their dependence on the environment for body heat.

The amount of energy expended by an animal over a specific time is called its metabolic rate. The rate is measured variously in joules, calories, or kilocalories (1000 calories). Carbohydrates and proteins contain about 4.5 to 5 kcal/g, and fat contains about 9 kcal/g. Metabolic rate is estimated as the **basal metabolic rate (BMR)** in endothermic animals at rest and as the **standard metabolic rate (SMR)** in ectotherms. Human males have a BMR of 1600 to 1800 kcal/day, and human females have a BMR of 1300 to 1500 kcal/day. Even with insulation, endothermal animals require extensive amounts of energy to maintain a constant body temperature. An ectotherm such as an alligator has an SMR of 60 kcal/day.

Energy Requirements Related to Body Size

Smaller endothermic animals have a greater surface area for their mass than larger ones (Figure 33.4). Therefore, smaller animals lose heat at a faster rate than larger animals and require more energy to maintain a constant internal temperature. This results in a smaller endothermic animal having a higher BMR, per body weight, than a larger endothermic animal.

Species		
Mass	35 g	4,500,000 g
Metabolic rate	890 mm^3 O_2/g body mass/hr	75 mm^3 O_2/g body mass/hr

Figure 33.4 The mouse has a much higher metabolic rate than the elephant. (credit "mouse": modification of work by Magnus Kjaergaard;

credit "elephant": modification of work by "TheLizardQueen"/Flickr)

Energy Requirements Related to Levels of Activity

The more active an animal is, the more energy is needed to maintain that activity, and the higher its BMR or SMR. The average daily rate of energy consumption is about two to four times an animal's BMR or SMR. Humans are more sedentary than most animals and have an average daily rate of only 1.5 times the BMR. The diet of an endothermic animal is determined by its BMR. For example: the type of grasses, leaves, or shrubs that an herbivore eats affects the number of calories that it takes in. The relative caloric content of herbivore foods, in descending order, is tall grasses > legumes > short grasses > forbs (any broad-leaved plant, not a grass) > subshrubs > annuals/biennials.

Energy Requirements Related to Environment

Animals adapt to extremes of temperature or food availability through torpor. **Torpor** is a process that leads to a decrease in activity and metabolism and allows animals to survive adverse conditions. Torpor can be used by animals for long periods, such as entering a state of **hibernation** during the winter months, in which case it enables them to maintain a reduced body temperature. During hibernation, ground squirrels can achieve an abdominal temperature of 0° C (32° F), while a bear's internal temperature is maintained higher at about 37° C (99° F).

If torpor occurs during the summer months with high temperatures and little water, it is called **estivation**. Some desert animals use this to survive the harshest months of the year. Torpor can occur on a daily basis; this is seen in bats and hummingbirds. While endothermy is limited in smaller animals by surface to volume ratio, some organisms can be smaller and still be endotherms because they employ daily torpor during the part of the day that is coldest. This allows them to conserve energy during the colder parts of the day, when they consume more energy to maintain their body temperature.

Animal Body Planes and Cavities

A standing vertebrate animal can be divided by several planes. A **sagittal plane** divides the body into right and left portions. A **midsagittal plane** divides the body exactly in the middle, making two equal right and left halves. A **frontal plane** (also called a coronal plane) separates the front from the back. A **transverse plane** (or, horizontal plane) divides the animal into upper and lower portions. This is sometimes called a cross section, and, if the transverse cut is at an angle, it is called an oblique plane. Figure 33.5 illustrates these planes on a goat (a four-legged animal) and a human being.

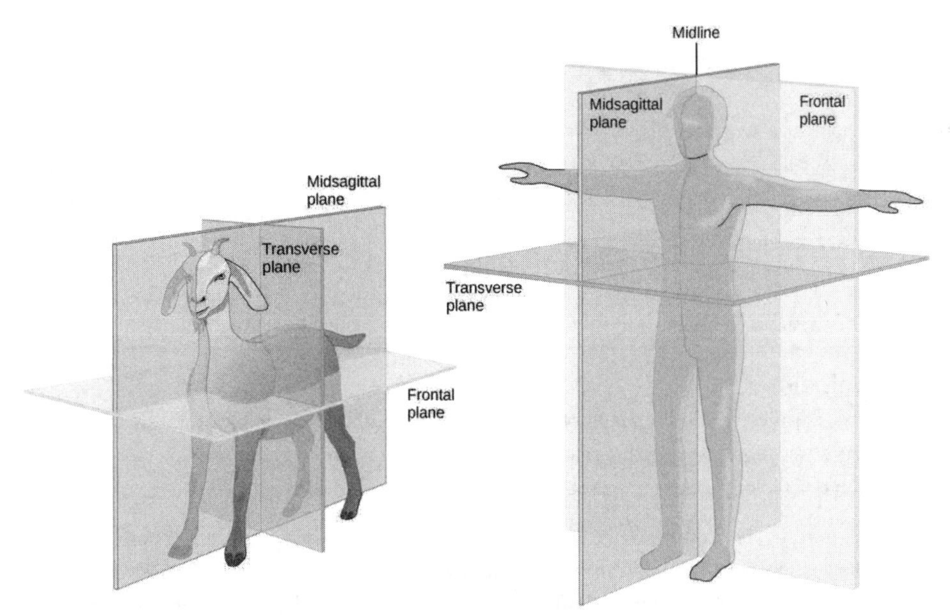

Figure 33.5 Shown are the planes of a quadrupedal goat and a bipedal human. The midsagittal plane divides the body exactly in half, into right and left portions. The frontal plane divides the front and back, and the transverse plane divides the body into upper and lower portions.

Vertebrate animals have a number of defined body cavities, as illustrated in Figure 33.6. Two of these are major cavities that contain smaller cavities within them. The **dorsal cavity** contains the cranial and the vertebral (or spinal) cavities. The **ventral cavity** contains the thoracic cavity, which in turn contains the pleural cavity around the lungs and the pericardial cavity, which surrounds the heart. The ventral cavity also contains the abdominopelvic cavity, which can be separated into the abdominal and the pelvic cavities.

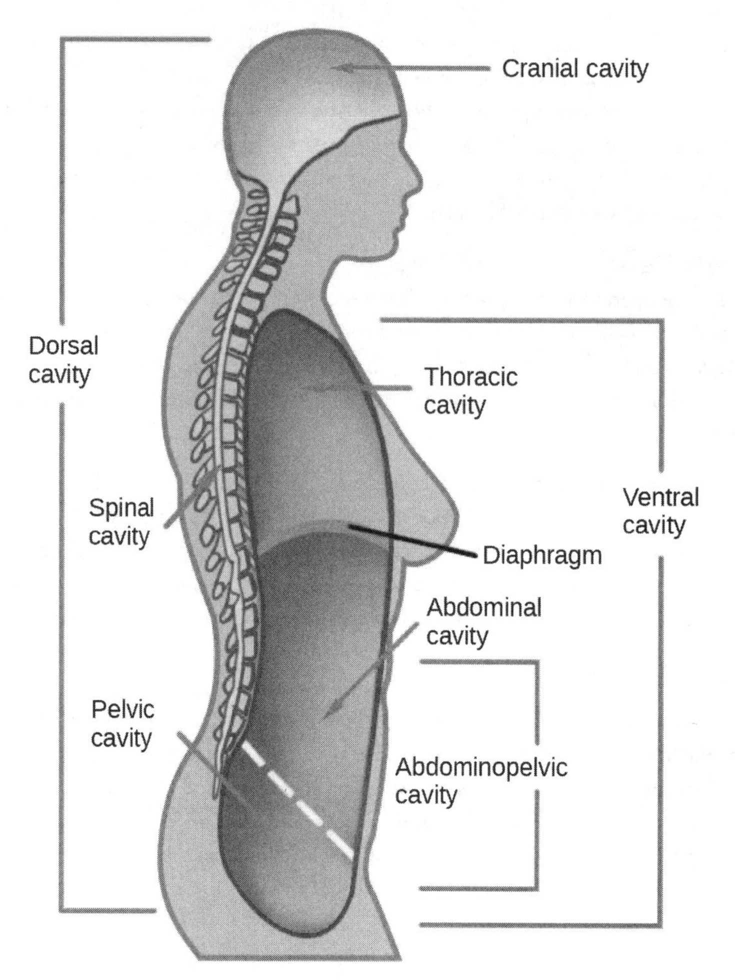

Figure 33.6 Vertebrate animals have two major body cavities. The dorsal cavity contains the cranial and the spinal cavity. The ventral cavity contains the thoracic cavity and the abdominopelvic cavity. The thoracic cavity is separated from the abdominopelvic cavity by the diaphragm. The abdominopelvic cavity is separated into the abdominal cavity and the pelvic cavity by an imaginary line parallel to the pelvis bones. (credit: modification of work by NCI)

CAREER CONNECTION

Physical Anthropologist

Physical anthropologists study the adaption, variability, and evolution of human beings, plus their living and fossil relatives. They can work in a variety of settings, although most will have an academic appointment at a university, usually in an anthropology department or a biology, genetics, or zoology department.

Nonacademic positions are available in the automotive and aerospace industries where the focus is on human size, shape, and anatomy. Research by these professionals might range from studies of how the human body reacts to car crashes to exploring how to make seats more comfortable. Other nonacademic positions can be obtained in museums of natural history, anthropology, archaeology, or science and technology. These positions involve educating students from grade school through graduate school. Physical anthropologists serve as education coordinators, collection managers, writers for museum publications, and as administrators. Zoos employ these professionals, especially if they have an expertise in primate biology; they work in collection management and captive breeding programs for endangered species. Forensic science utilizes physical anthropology expertise in identifying human and animal remains, assisting in determining the cause of death, and for expert testimony in trials.

33.2 Animal Primary Tissues

By the end of this section, you will be able to do the following:

- Describe epithelial tissues
- Discuss the different types of connective tissues in animals
- Describe three types of muscle tissues
- Describe nervous tissue

The tissues of multicellular, complex animals are four primary types: epithelial, connective, muscle, and nervous. Recall that tissues are groups of similar cells (cells carrying out related functions). These tissues combine to form organs—like the skin or kidney—that have specific, specialized functions within the body. Organs are organized into organ systems to perform functions; examples include the circulatory system, which consists of the heart and blood vessels, and the digestive system, consisting of several organs, including the stomach, intestines, liver, and pancreas. Organ systems come together to create an entire organism.

Epithelial Tissues

Epithelial tissues cover the outside of organs and structures in the body and line the lumens of organs in a single layer or multiple layers of cells. The types of epithelia are classified by the shapes of cells present and the number of layers of cells. Epithelia composed of a single layer of cells is called **simple epithelia**; epithelial tissue composed of multiple layers is called **stratified epithelia.** Table 33.2 summarizes the different types of epithelial tissues.

Different Types of Epithelial Tissues

Cell shape	Description	Location
squamous	flat, irregular round shape	simple: lung alveoli, capillaries; stratified: skin, mouth, vagina
cuboidal	cube shaped, central nucleus	glands, renal tubules
columnar	tall, narrow, nucleus toward base; tall, narrow, nucleus along cell	simple: digestive tract; pseudostratified: respiratory tract
transitional	round, simple but appear stratified	urinary bladder

Table 33.2

Squamous Epithelia

Squamous epithelial cells are generally round, flat, and have a small, centrally located nucleus. The cell outline is slightly irregular, and cells fit together to form a covering or lining. When the cells are arranged in a single layer (simple epithelia), they facilitate diffusion in tissues, such as the areas of gas exchange in the lungs and the exchange of nutrients and waste at blood capillaries.

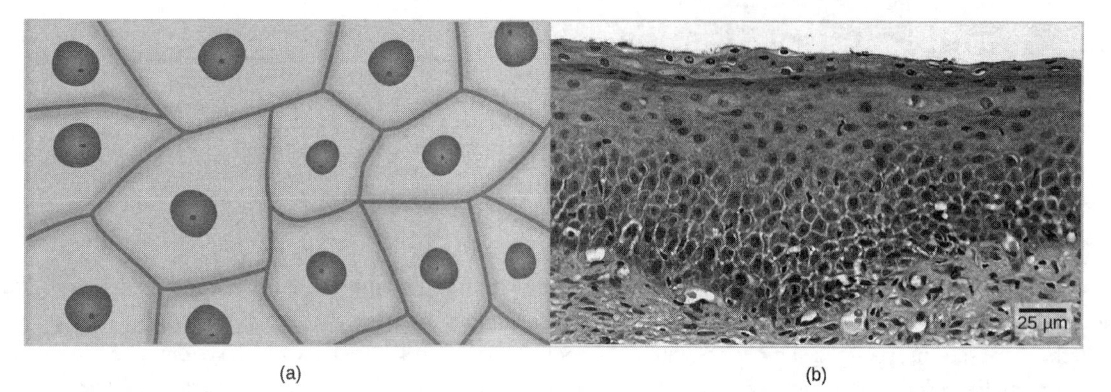

Figure 33.7 Squamous epithelia cells (a) have a slightly irregular shape, and a small, centrally located nucleus. These cells can be stratified into layers, as in (b) this human cervix specimen. (credit b: modification of work by Ed Uthman; scale-bar data from Matt Russell)

Figure 33.7a illustrates a layer of squamous cells with their membranes joined together to form an epithelium. Image Figure 33.7b illustrates squamous epithelial cells arranged in stratified layers, where protection is needed on the body from outside abrasion and damage. This is called a stratified squamous epithelium and occurs in the skin and in tissues lining the mouth and vagina.

Cuboidal Epithelia

Cuboidal epithelial cells, shown in Figure 33.8, are cube-shaped with a single, central nucleus. They are most commonly found in a single layer representing a simple epithelia in glandular tissues throughout the body where they prepare and secrete glandular material. They are also found in the walls of tubules and in the ducts of the kidney and liver.

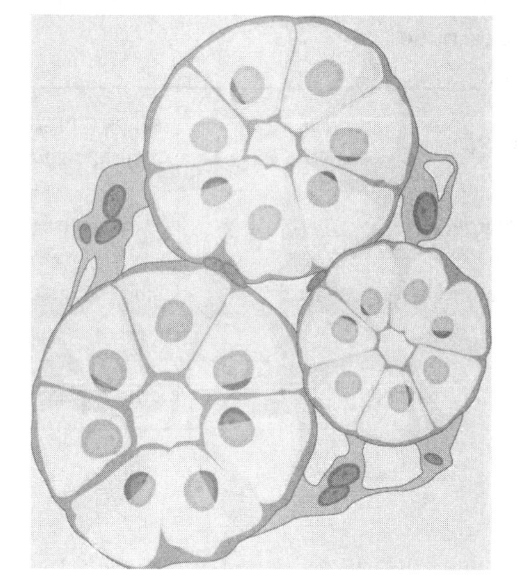

Figure 33.8 Simple cuboidal epithelial cells line tubules in the mammalian kidney, where they are involved in filtering the blood.

Columnar Epithelia

Columnar epithelial cells are taller than they are wide: they resemble a stack of columns in an epithelial layer, and are most commonly found in a single-layer arrangement. The nuclei of columnar epithelial cells in the digestive tract appear to be lined up at the base of the cells, as illustrated in Figure 33.9. These cells absorb material from the lumen of the digestive tract and prepare it for entry into the body through the circulatory and lymphatic systems.

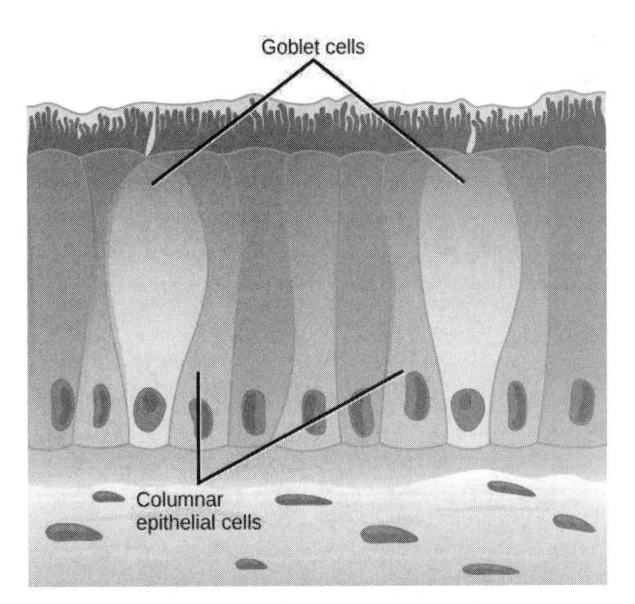

Figure 33.9 Simple columnar epithelial cells absorb material from the digestive tract. Goblet cells secrete mucus into the digestive tract lumen.

Columnar epithelial cells lining the respiratory tract appear to be stratified. However, each cell is attached to the base membrane of the tissue and, therefore, they are simple tissues. The nuclei are arranged at different levels in the layer of cells, making it appear as though there is more than one layer, as seen in Figure 33.10. This is called **pseudostratified**, columnar epithelia. This cellular covering has cilia at the apical, or free, surface of the cells. The cilia enhance the movement of mucus and trapped particles out of the respiratory tract, helping to protect the system from invasive microorganisms and harmful material that has been breathed into the body. Goblet cells are interspersed in some tissues (such as the lining of the trachea). The goblet cells contain mucus that traps irritants, which in the case of the trachea keep these irritants from getting into the lungs.

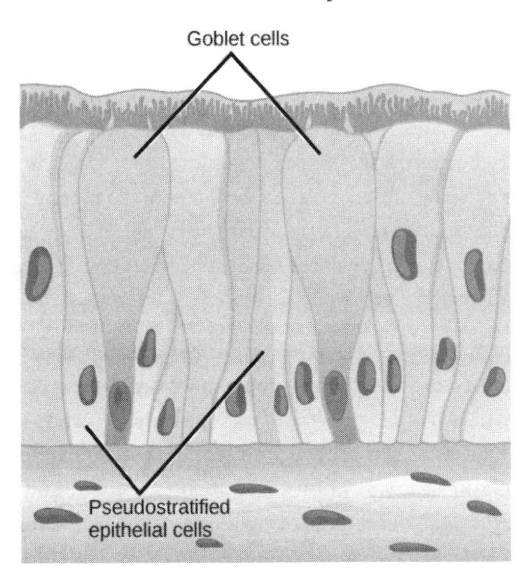

Figure 33.10 Pseudostratified columnar epithelia line the respiratory tract. They exist in one layer, but the arrangement of nuclei at different levels makes it appear that there is more than one layer. Goblet cells interspersed between the columnar epithelial cells secrete mucus into the respiratory tract.

Transitional Epithelia

Transitional or uroepithelial cells appear only in the urinary system, primarily in the bladder and ureter. These cells are arranged in a stratified layer, but they have the capability of appearing to pile up on top of each other in a relaxed, empty bladder, as illustrated in Figure 33.11. As the urinary bladder fills, the epithelial layer unfolds and expands to hold the volume of urine introduced into it. As the bladder fills, it expands and the lining becomes thinner. In other words, the tissue transitions from thick to thin.

VISUAL CONNECTION

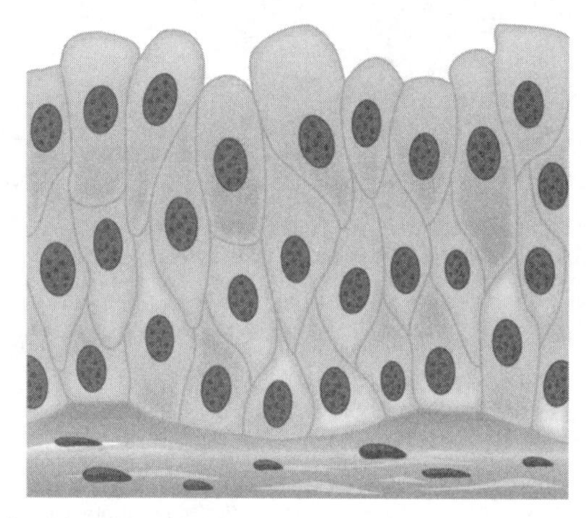

Figure 33.11 Transitional epithelia of the urinary bladder undergo changes in thickness depending on how full the bladder is.

Which of the following statements about types of epithelial cells is false?

a. Simple columnar epithelial cells line the tissue of the lung.
b. Simple cuboidal epithelial cells are involved in the filtering of blood in the kidney.
c. Pseudostratisfied columnar epithilia occur in a single layer, but the arrangement of nuclei makes it appear that more than one layer is present.
d. Transitional epithelia change in thickness depending on how full the bladder is.

Connective Tissues

Connective tissues are made up of a matrix consisting of living cells and a nonliving substance, called the ground substance. The ground substance is made of an organic substance (usually a protein) and an inorganic substance (usually a mineral or water). The principal cell of connective tissues is the fibroblast. This cell makes the fibers found in nearly all of the connective tissues. Fibroblasts are motile, able to carry out mitosis, and can synthesize whichever connective tissue is needed. Macrophages, lymphocytes, and, occasionally, leukocytes can be found in some of the tissues. Some tissues have specialized cells that are not found in the others. The **matrix** in connective tissues gives the tissue its density. When a connective tissue has a high concentration of cells or fibers, it has proportionally a less dense matrix.

The organic portion or protein fibers found in connective tissues are either collagen, elastic, or reticular fibers. Collagen fibers provide strength to the tissue, preventing it from being torn or separated from the surrounding tissues. Elastic fibers are made of the protein elastin; this fiber can stretch to one and one half of its length and return to its original size and shape. Elastic fibers provide flexibility to the tissues. Reticular fibers are the third type of protein fiber found in connective tissues. This fiber consists of thin strands of collagen that form a network of fibers to support the tissue and other organs to which it is connected. The various types of connective tissues, the types of cells and fibers they are made of, and sample locations of the tissues is summarized in Table 33.3.

Connective Tissues

Tissue	Cells	Fibers	Location
loose/areolar	fibroblasts, macrophages, some lymphocytes, some neutrophils	few: collagen, elastic, reticular	around blood vessels; anchors epithelia

Table 33.3

Tissue	Cells	Fibers	Location
dense, fibrous connective tissue	fibroblasts, macrophages	mostly collagen	irregular: skin; regular: tendons, ligaments
cartilage	chondrocytes, chondroblasts	hyaline: few: collagen fibrocartilage: large amount of collagen	shark skeleton, fetal bones, human ears, intervertebral discs
bone	osteoblasts, osteocytes, osteoclasts	some: collagen, elastic	vertebrate skeletons
adipose	adipocytes	few	adipose (fat)
blood	red blood cells, white blood cells	none	blood

Table 33.3

Loose/Areolar Connective Tissue

Loose connective tissue, also called areolar connective tissue, has a sampling of all of the components of a connective tissue. As illustrated in Figure 33.12, loose connective tissue has some fibroblasts; macrophages are present as well. Collagen fibers are relatively wide and stain a light pink, while elastic fibers are thin and stain dark blue to black. The space between the formed elements of the tissue is filled with the matrix. The material in the connective tissue gives it a loose consistency similar to a cotton ball that has been pulled apart. Loose connective tissue is found around every blood vessel and helps to keep the vessel in place. The tissue is also found around and between most body organs. In summary, areolar tissue is tough, yet flexible, and comprises membranes.

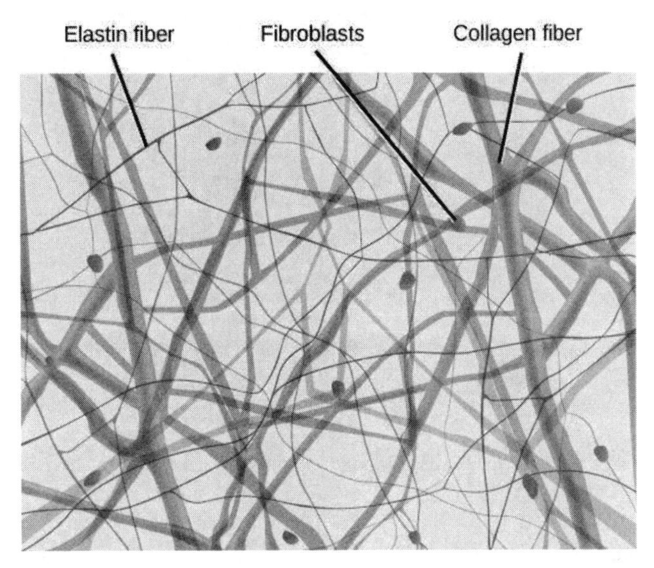

Figure 33.12 Loose connective tissue is composed of loosely woven collagen and elastic fibers. The fibers and other components of the connective tissue matrix are secreted by fibroblasts.

Fibrous Connective Tissue

Fibrous connective tissues contain large amounts of collagen fibers and few cells or matrix material. The fibers can be arranged irregularly or regularly with the strands lined up in parallel. Irregularly arranged fibrous connective tissues are found in areas of the body where stress occurs from all directions, such as the dermis of the skin. Regular fibrous connective tissue, shown in Figure 33.13, is found in tendons (which connect muscles to bones) and ligaments (which connect bones to bones).

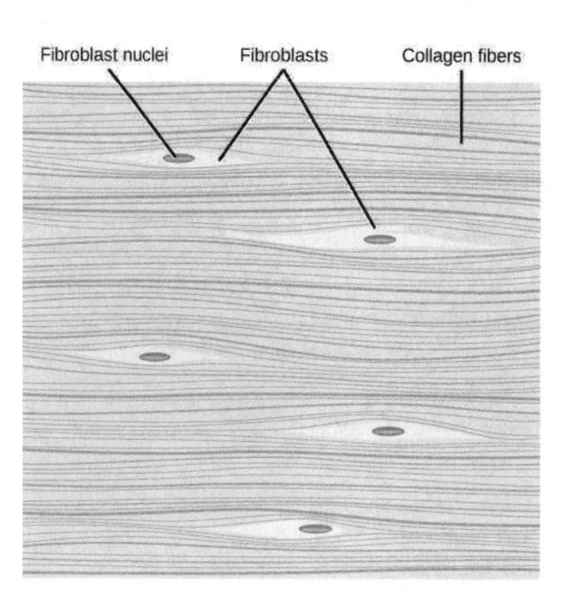

Figure 33.13 Fibrous connective tissue from the tendon has strands of collagen fibers lined up in parallel.

Cartilage

Cartilage is a connective tissue with a large amount of the matrix and variable amounts of fibers. The cells, called **chondrocytes**, make the matrix and fibers of the tissue. Chondrocytes are found in spaces within the tissue called **lacunae**.

A cartilage with few collagen and elastic fibers is hyaline cartilage, illustrated in Figure 33.14. The lacunae are randomly scattered throughout the tissue and the matrix takes on a milky or scrubbed appearance with routine histological stains. Sharks have cartilaginous skeletons, as does nearly the entire human skeleton during a specific pre-birth developmental stage. A remnant of this cartilage persists in the outer portion of the human nose. Hyaline cartilage is also found at the ends of long bones, reducing friction and cushioning the articulations of these bones.

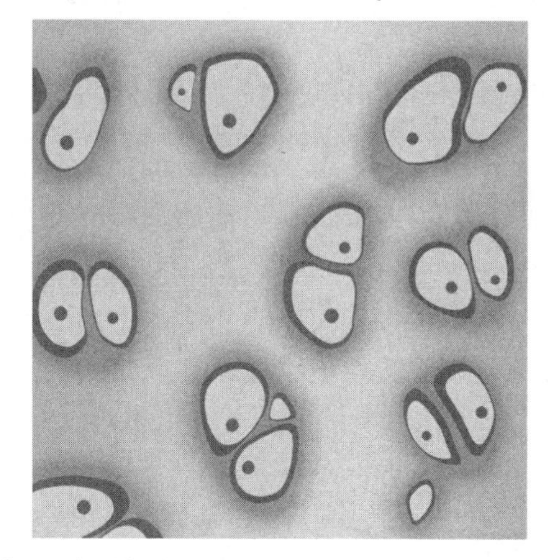

Figure 33.14 Hyaline cartilage consists of a matrix with cells called chondrocytes embedded in it. The chondrocytes exist in cavities in the matrix called lacunae.

Elastic cartilage has a large amount of elastic fibers, giving it tremendous flexibility. The ears of most vertebrate animals contain this cartilage as do portions of the larynx, or voice box. Fibrocartilage contains a large amount of collagen fibers, giving the tissue tremendous strength. Fibrocartilage comprises the intervertebral discs in vertebrate animals. Hyaline cartilage found in movable joints such as the knee and shoulder becomes damaged as a result of age or trauma. Damaged hyaline cartilage is replaced by fibrocartilage and results in the joints becoming "stiff."

Bone

Bone, or osseous tissue, is a connective tissue that has a large amount of two different types of matrix material. The organic matrix is similar to the matrix material found in other connective tissues, including some amount of collagen and elastic fibers. This gives strength and flexibility to the tissue. The inorganic matrix consists of mineral salts—mostly calcium salts—that give the tissue hardness. Without adequate organic material in the matrix, the tissue breaks; without adequate inorganic material in the matrix, the tissue bends.

There are three types of cells in bone: osteoblasts, osteocytes, and osteoclasts. Osteoblasts are active in making bone for growth and remodeling. Osteoblasts deposit bone material into the matrix and, after the matrix surrounds them, they continue to live, but in a reduced metabolic state as osteocytes. Osteocytes are found in lacunae of the bone. Osteoclasts are active in breaking down bone for bone remodeling, and they provide access to calcium stored in tissues. Osteoclasts are usually found on the surface of the tissue.

Bone can be divided into two types: compact and spongy. Compact bone is found in the shaft (or diaphysis) of a long bone and the surface of the flat bones, while spongy bone is found in the end (or epiphysis) of a long bone. Compact bone is organized into subunits called **osteons**, as illustrated in Figure 33.15. A blood vessel and a nerve are found in the center of the structure within the Haversian canal, with radiating circles of lacunae around it known as lamellae. The wavy lines seen between the lacunae are microchannels called **canaliculi**; they connect the lacunae to aid diffusion between the cells. Spongy bone is made of tiny plates called **trabeculae**; these plates serve as struts to give the spongy bone strength. Over time, these plates can break causing the bone to become less resilient. Bone tissue forms the internal skeleton of vertebrate animals, providing structure to the animal and points of attachment for tendons.

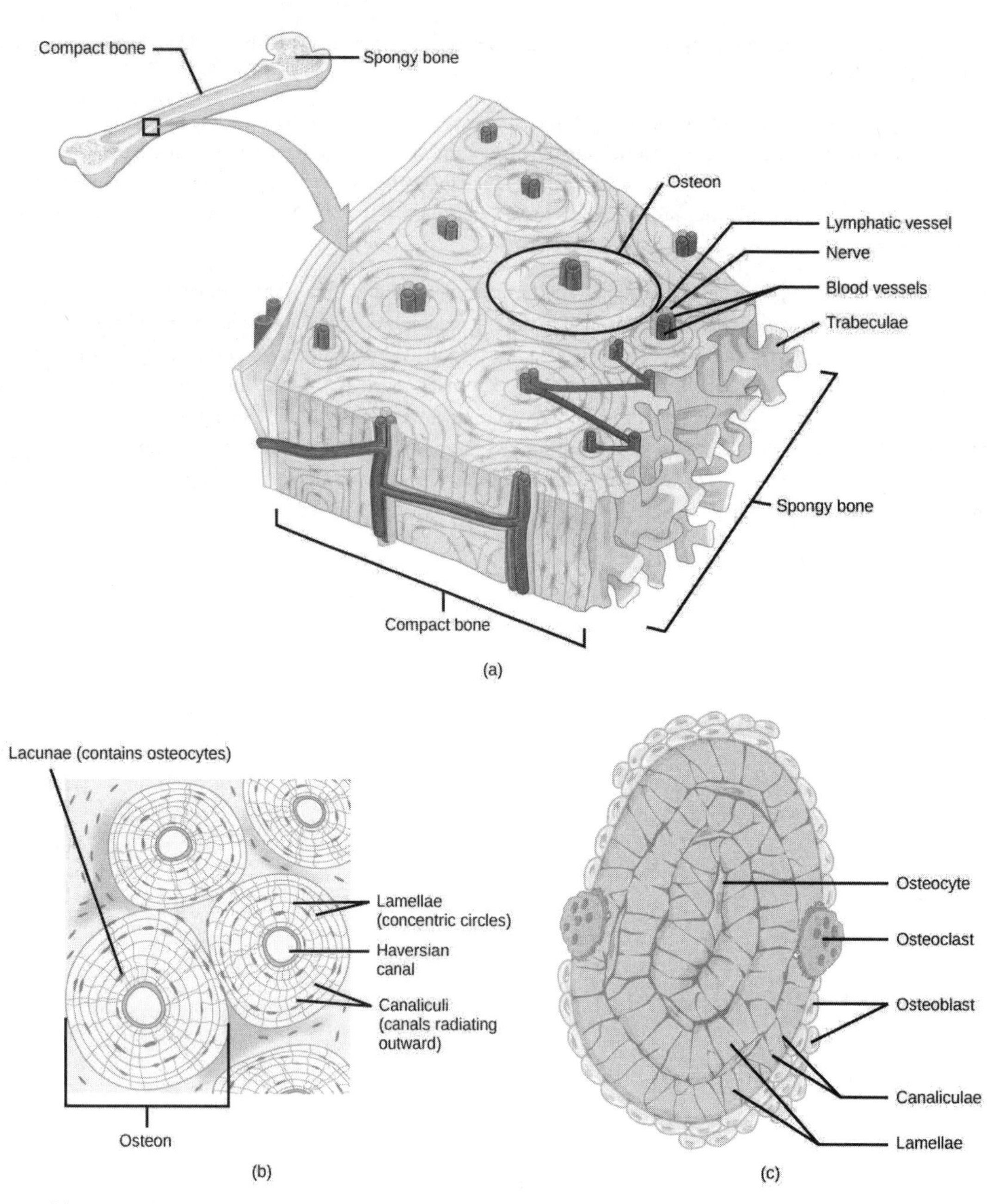

Figure 33.15 (a) Compact bone is a dense matrix on the outer surface of bone. Spongy bone, inside the compact bone, is porous with web-like trabeculae. (b) Compact bone is organized into rings called osteons. Blood vessels, nerves, and lymphatic vessels are found in the central Haversian canal. Rings of lamellae surround the Haversian canal. Between the lamellae are cavities called lacunae. Canaliculi are microchannels connecting the lacunae together. (c) Osteoblasts surround the exterior of the bone. Osteoclasts bore tunnels into the bone and osteocytes are found in the lacunae.

Adipose Tissue

Adipose tissue, or fat tissue, is considered a connective tissue even though it does not have fibroblasts or a real matrix and only has a few fibers. Adipose tissue is made up of cells called adipocytes that collect and store fat in the form of triglycerides, for energy metabolism. Adipose tissues additionally serve as insulation to help maintain body temperatures, allowing animals to be endothermic, and they function as cushioning against damage to body organs. Under a microscope, adipose tissue cells appear empty due to the extraction of fat during the processing of the material for viewing, as seen in Figure 33.16. The thin lines in the image are the cell membranes, and the nuclei are the small, black dots at the edges of the cells.

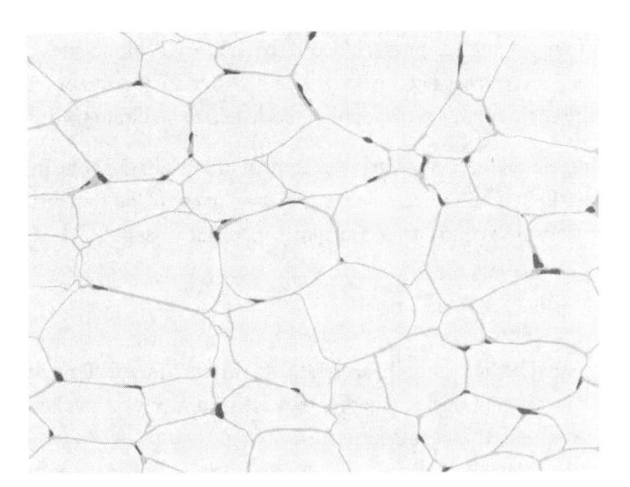

Figure 33.16 Adipose is a connective tissue is made up of cells called adipocytes. Adipocytes have small nuclei localized at the cell edge.

Blood

Blood is considered a connective tissue because it has a matrix, as shown in Figure 33.17. The living cell types are red blood cells (RBC), also called erythrocytes, and white blood cells (WBC), also called leukocytes. The fluid portion of whole blood, its matrix, is commonly called plasma.

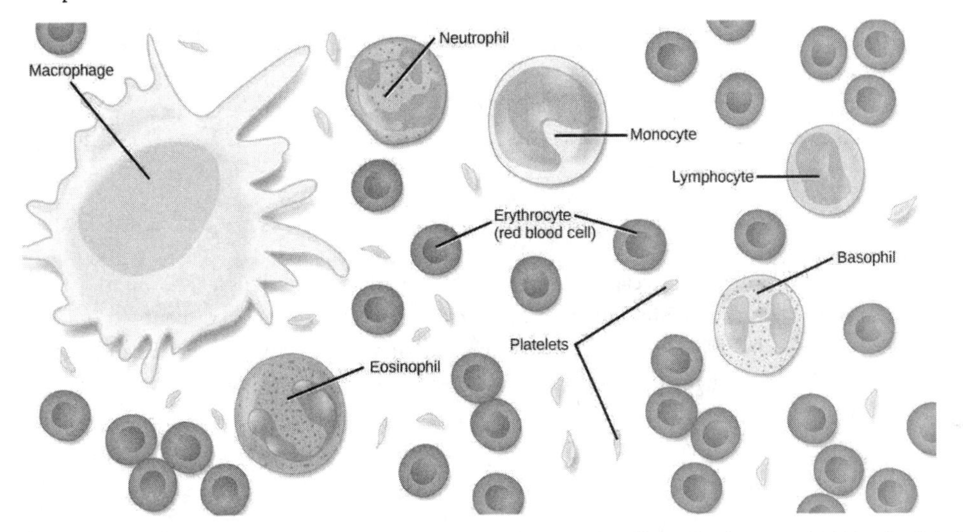

Figure 33.17 Blood is a connective tissue that has a fluid matrix, called plasma, and no fibers. Erythrocytes (red blood cells), the predominant cell type, are involved in the transport of oxygen and carbon dioxide. Also present are various leukocytes (white blood cells) involved in immune response.

The cell found in greatest abundance in blood is the erythrocyte. Erythrocytes are counted in millions in a blood sample: the average number of red blood cells in primates is 4.7 to 5.5 million cells per microliter. Erythrocytes are consistently the same size in a species, but vary in size between species. For example, the average diameter of a primate red blood cell is 7.5 µl, a dog is close at 7.0 µl, but a cat's RBC diameter is 5.9 µl. Sheep erythrocytes are even smaller at 4.6 µl. Mammalian erythrocytes lose their nuclei and mitochondria when they are released from the bone marrow where they are made. Fish, amphibian, and avian red blood cells maintain their nuclei and mitochondria throughout the cell's life. The principal job of an erythrocyte is to carry and deliver oxygen to the tissues.

Leukocytes are the predominant white blood cells found in the peripheral blood. Leukocytes are counted in the thousands in the blood with measurements expressed as ranges: primate counts range from 4,800 to 10,800 cells per µl, dogs from 5,600 to 19,200 cells per µl, cats from 8,000 to 25,000 cells per µl, cattle from 4,000 to 12,000 cells per µl, and pigs from 11,000 to 22,000 cells per µl.

Lymphocytes function primarily in the immune response to foreign antigens or material. Different types of lymphocytes make antibodies tailored to the foreign antigens and control the production of those antibodies. Neutrophils are phagocytic cells and they participate in one of the early lines of defense against microbial invaders, aiding in the removal of bacteria that has entered the body. Another leukocyte that is found in the peripheral blood is the monocyte. Monocytes give rise to phagocytic macrophages that clean up dead and damaged cells in the body, whether they are foreign or from the host animal. Two additional leukocytes in the blood are eosinophils and basophils—both help to facilitate the inflammatory response.

The slightly granular material among the cells is a cytoplasmic fragment of a cell in the bone marrow. This is called a platelet or thrombocyte. Platelets participate in the stages leading up to coagulation of the blood to stop bleeding through damaged blood vessels. Blood has a number of functions, but primarily it transports material through the body to bring nutrients to cells and remove waste material from them.

Muscle Tissues

There are three types of muscle in animal bodies: smooth, skeletal, and cardiac. They differ by the presence or absence of striations or bands, the number and location of nuclei, whether they are voluntarily or involuntarily controlled, and their location within the body. Table 33.4 summarizes these differences.

Types of Muscles

Type of Muscle	Striations	Nuclei	Control	Location
smooth	no	single, in center	involuntary	visceral organs
skeletal	yes	many, at periphery	voluntary	skeletal muscles
cardiac	yes	single, in center	involuntary	heart

Table 33.4

Smooth Muscle

Smooth muscle does not have striations in its cells. It has a single, centrally located nucleus, as shown in Figure 33.18. Constriction of smooth muscle occurs under involuntary, autonomic nervous control and in response to local conditions in the tissues. Smooth muscle tissue is also called non-striated as it lacks the banded appearance of skeletal and cardiac muscle. The walls of blood vessels, the tubes of the digestive system, and the tubes of the reproductive systems are composed of mostly smooth muscle.

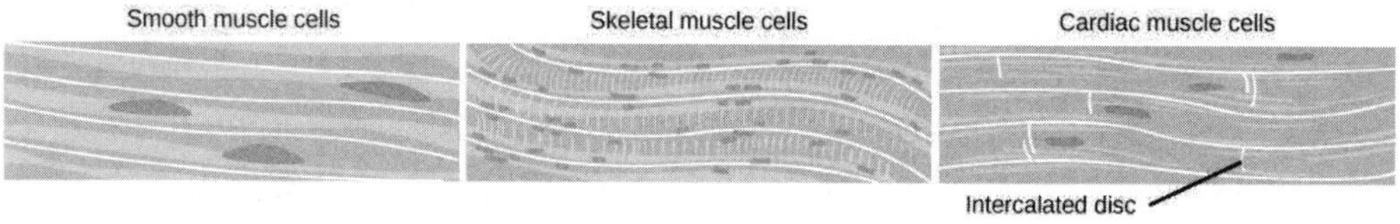

Figure 33.18 Smooth muscle cells do not have striations, while skeletal muscle cells do. Cardiac muscle cells have striations, but, unlike the multinucleate skeletal cells, they have only one nucleus. Cardiac muscle tissue also has intercalated discs, specialized regions running along the plasma membrane that join adjacent cardiac muscle cells and assist in passing an electrical impulse from cell to cell.

Skeletal Muscle

Skeletal muscle has striations across its cells caused by the arrangement of the contractile proteins actin and myosin. These muscle cells are relatively long and have multiple nuclei along the edge of the cell. Skeletal muscle is under voluntary, somatic nervous system control and is found in the muscles that move bones. Figure 33.18 illustrates the histology of skeletal muscle.

Cardiac Muscle

Cardiac muscle, shown in Figure 33.18, is found only in the heart. Like skeletal muscle, it has cross striations in its cells, but cardiac muscle has a single, centrally located nucleus. Cardiac muscle is not under voluntary control but can be influenced by the autonomic nervous system to speed up or slow down. An added feature to cardiac muscle cells is a line than extends along the end of the cell as it abuts the next cardiac cell in the row. This line is called an intercalated disc: it assists in passing electrical impulse efficiently from one cell to the next and maintains the strong connection between neighboring cardiac cells.

Nervous Tissues

Nervous tissues are made of cells specialized to receive and transmit electrical impulses from specific areas of the body and to send them to specific locations in the body. The main cell of the nervous system is the neuron, illustrated in Figure 33.19. The large structure with a central nucleus is the cell body of the neuron. Projections from the cell body are either dendrites specialized in receiving input or a single axon specialized in transmitting impulses. Some glial cells are also shown. Astrocytes regulate the chemical environment of the nerve cell, and oligodendrocytes insulate the axon so the electrical nerve impulse is transferred more efficiently. Other glial cells that are not shown support the nutritional and waste requirements of the neuron. Some of the glial cells are phagocytic and remove debris or damaged cells from the tissue. A nerve consists of neurons and glial cells.

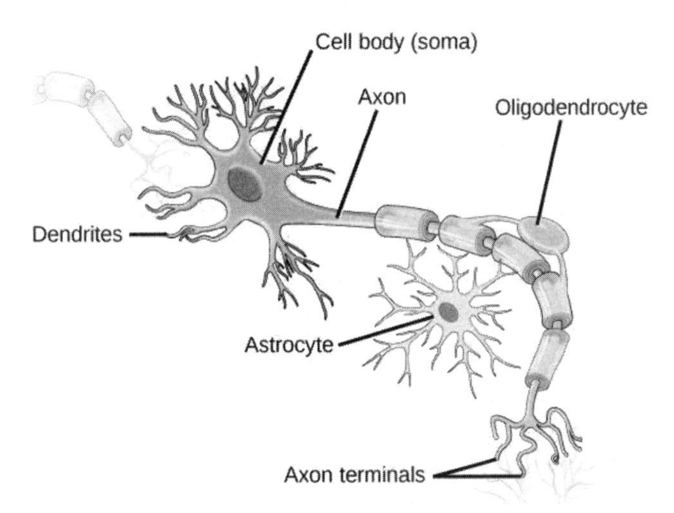

Figure 33.19 The neuron has projections called dendrites that receive signals and projections called axons that send signals. Also shown are two types of glial cells: astrocytes regulate the chemical environment of the nerve cell, and oligodendrocytes insulate the axon so the electrical nerve impulse is transferred more efficiently.

LINK TO LEARNING

Click through the interactive review (http://openstax.org/l/tissues) to learn more about epithelial tissues.

CAREER CONNECTION

Pathologist

A pathologist is a medical doctor or veterinarian who has specialized in the laboratory detection of disease in animals, including humans. These professionals complete medical school education and follow it with an extensive post-graduate residency at a medical center. A pathologist may oversee clinical laboratories for the evaluation of body tissue and blood samples for the detection of disease or infection. They examine tissue specimens through a microscope to identify cancers and other diseases. Some pathologists perform autopsies to determine the cause of death and the progression of disease.

33.3 Homeostasis

By the end of this section, you will be able to do the following:

- Define homeostasis
- Describe the factors affecting homeostasis
- Discuss positive and negative feedback mechanisms used in homeostasis
- Describe thermoregulation of endothermic and ectothermic animals

Animal organs and organ systems constantly adjust to internal and external changes through a process called homeostasis ("steady state"). These changes might be in the level of glucose or calcium in blood or in external temperatures. **Homeostasis** means to maintain dynamic equilibrium in the body. It is dynamic because it is constantly adjusting to the changes that the body's systems encounter. It is equilibrium because body functions are kept within specific ranges. Even an animal that is apparently inactive is maintaining this homeostatic equilibrium.

Homeostatic Process

The goal of homeostasis is the maintenance of equilibrium around a point or value called a **set point**. While there are normal fluctuations from the set point, the body's systems will usually attempt to go back to this point. A change in the internal or external environment is called a stimulus and is detected by a receptor; the response of the system is to adjust the deviation parameter toward the set point. For instance, if the body becomes too warm, adjustments are made to cool the animal. If the blood's glucose rises after a meal, adjustments are made to lower the blood glucose level by getting the nutrient into tissues that need it or to store it for later use.

Control of Homeostasis

When a change occurs in an animal's environment, an adjustment must be made. The receptor senses the change in the environment, then sends a signal to the control center (in most cases, the brain) which in turn generates a response that is signaled to an effector. The effector is a muscle (that contracts or relaxes) or a gland that secretes. Homeostatsis is maintained by negative feedback loops. Positive feedback loops actually push the organism further out of homeostasis, but may be necessary for life to occur. Homeostasis is controlled by the nervous and endocrine system of mammals.

Negative Feedback Mechanisms

Any homeostatic process that changes the direction of the stimulus is a **negative feedback loop**. It may either increase or decrease the stimulus, but the stimulus is not allowed to continue as it did before the receptor sensed it. In other words, if a level is too high, the body does something to bring it down, and conversely, if a level is too low, the body does something to make it go up. Hence the term negative feedback. An example is animal maintenance of blood glucose levels. When an animal has eaten, blood glucose levels rise. This is sensed by the nervous system. Specialized cells in the pancreas sense this, and the hormone insulin is released by the endocrine system. Insulin causes blood glucose levels to decrease, as would be expected in a negative feedback system, as illustrated in Figure 33.20. However, if an animal has not eaten and blood glucose levels decrease, this is sensed in another group of cells in the pancreas, and the hormone glucagon is released causing glucose levels to increase. This is still a negative feedback loop, but not in the direction expected by the use of the term "negative." Another example of an increase as a result of the feedback loop is the control of blood calcium. If calcium levels decrease, specialized cells in the parathyroid gland sense this and release parathyroid hormone (PTH), causing an increased absorption of calcium through the intestines and kidneys and, possibly, the breakdown of bone in order to liberate calcium. The effects of PTH are to raise blood levels of the element. Negative feedback loops are the predominant mechanism used in homeostasis.

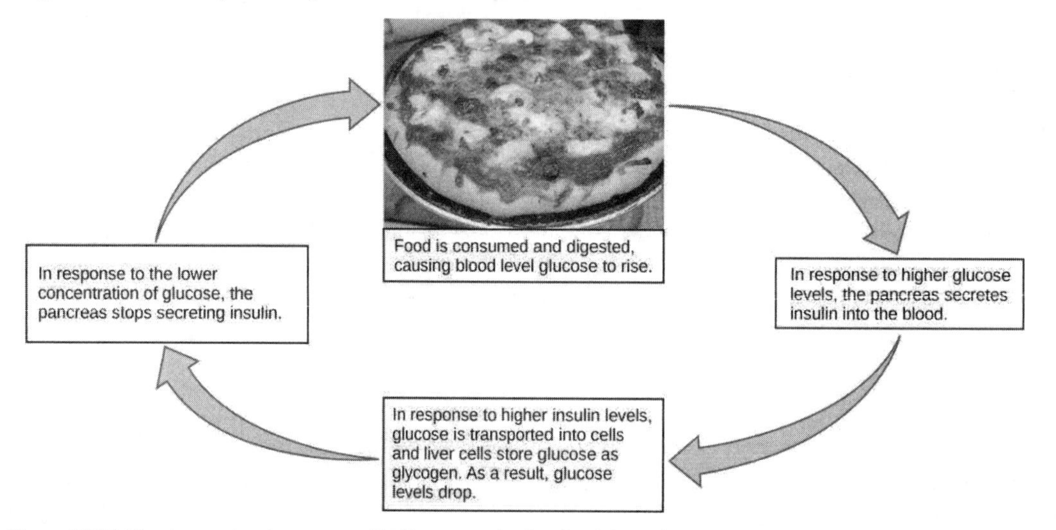

Figure 33.20 Blood sugar levels are controlled by a negative feedback loop. (credit: modification of work by Jon Sullivan)

Positive Feedback Loop

A **positive feedback loop** maintains the direction of the stimulus, possibly accelerating it. Few examples of positive feedback loops exist in animal bodies, but one is found in the cascade of chemical reactions that result in blood clotting, or coagulation.

As one clotting factor is activated, it activates the next factor in sequence until a fibrin clot is achieved. The direction is maintained, not changed, so this is positive feedback. Another example of positive feedback is uterine contractions during childbirth, as illustrated in Figure 33.21. The hormone oxytocin, made by the endocrine system, stimulates the contraction of the uterus. This produces pain sensed by the nervous system. Instead of lowering the oxytocin and causing the pain to subside, more oxytocin is produced until the contractions are powerful enough to produce childbirth.

VISUAL CONNECTION

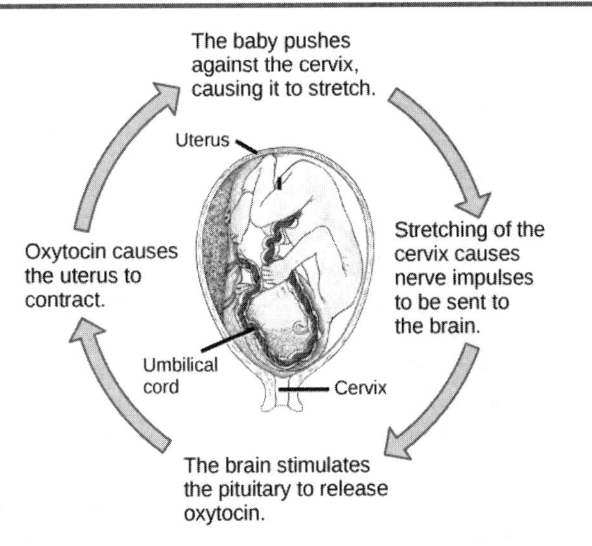

Figure 33.21 The birth of a human infant is the result of positive feedback.

State whether each of the following processes is regulated by a positive feedback loop or a negative feedback loop.

a. A person feels satiated after eating a large meal.
b. The blood has plenty of red blood cells. As a result, erythropoietin, a hormone that stimulates the production of new red blood cells, is no longer released from the kidney.

Set Point

It is possible to adjust a system's set point. When this happens, the feedback loop works to maintain the new setting. An example of this is blood pressure: over time, the normal or set point for blood pressure can increase as a result of continued increases in blood pressure. The body no longer recognizes the elevation as abnormal and no attempt is made to return to the lower set point. The result is the maintenance of an elevated blood pressure that can have harmful effects on the body. Medication can lower blood pressure and lower the set point in the system to a more healthy level. This is called a process of **alteration** of the set point in a feedback loop.

Changes can be made in a group of body organ systems in order to maintain a set point in another system. This is called **acclimatization**. This occurs, for instance, when an animal migrates to a higher altitude than that to which it is accustomed. In order to adjust to the lower oxygen levels at the new altitude, the body increases the number of red blood cells circulating in the blood to ensure adequate oxygen delivery to the tissues. Another example of acclimatization is animals that have seasonal changes in their coats: a heavier coat in the winter ensures adequate heat retention, and a light coat in summer assists in keeping body temperature from rising to harmful levels.

LINK TO LEARNING

Feedback mechanisms can be understood in terms of driving a race car along a track: watch a short video lesson on positive and negative feedback loops.

Click to view content (https://www.openstax.org/l/feedback_loops)

Homeostasis: Thermoregulation

Body temperature affects body activities. Generally, as body temperature rises, enzyme activity rises as well. For every ten degree centigrade rise in temperature, enzyme activity doubles, up to a point. Body proteins, including enzymes, begin to denature and lose their function with high heat (around 50°C for mammals). Enzyme activity will decrease by half for every ten degree centigrade drop in temperature, to the point of freezing, with a few exceptions. Some fish can withstand freezing solid and return to normal with thawing.

LINK TO LEARNING

Watch this Discovery Channel video on thermoregulation to see illustrations of this process in a variety of animals.

Click to view content (https://www.openstax.org/l/thermoregulate)

Endotherms and Ectotherms

Animals can be divided into two groups: some maintain a constant body temperature in the face of differing environmental temperatures, while others have a body temperature that is the same as their environment and thus varies with the environment. Animals that rely on external temperatures to set their body temperature are ectotherms. This group has been called cold-blooded, but the term may not apply to an animal in the desert with a very warm body temperature. In contrast to ectotherms, poikilotherms are animals with constantly varying internal temperatures. An animal that maintains a constant body temperature in the face of environmental changes is called a homeotherm. Endotherms are animals that rely on internal sources for maintenance of relatively constant body temperature in varying environmental temperatures. These animals are able to maintain a level of metabolic activity at cooler temperature, which an ectotherm cannot due to differing enzyme levels of activity. It is worth mentioning that some ectotherms and poikilotherms have relatively constant body temperatures due to the constant environmental temperatures in their habitats. These animals are so-called ectothermic homeotherms, like some deep sea fish species.

Heat can be exchanged between an animal and its environment through four mechanisms: radiation, evaporation, convection, and conduction (Figure 33.22). Radiation is the emission of electromagnetic "heat" waves. Heat comes from the sun in this manner and radiates from dry skin the same way. Heat can be removed with liquid from a surface during evaporation. This occurs when a mammal sweats. Convection currents of air remove heat from the surface of dry skin as the air passes over it. Heat will be conducted from one surface to another during direct contact with the surfaces, such as an animal resting on a warm rock.

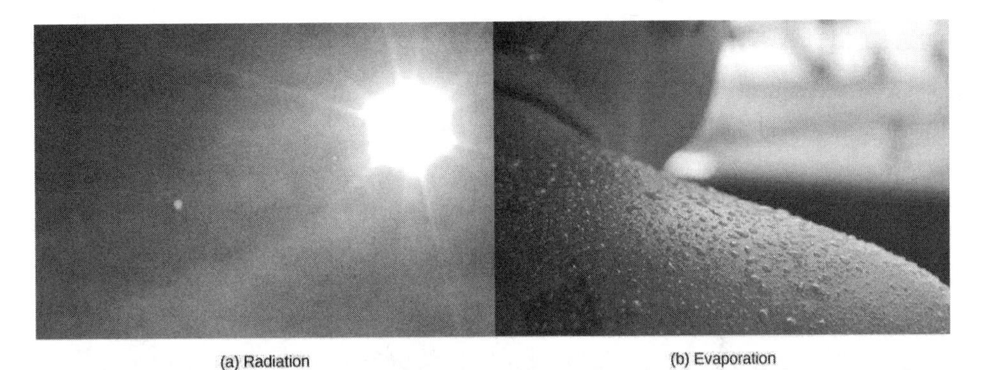

Figure 33.22 Heat can be exchanged by four mechanisms: (a) radiation, (b) evaporation, (c) convection, or (d) conduction. (credit b: modification of work by "Kullez"/Flickr; credit c: modification of work by Chad Rosenthal; credit d: modification of work by "stacey.d"/Flickr)

Heat Conservation and Dissipation

Animals conserve or dissipate heat in a variety of ways. In certain climates, endothermic animals have some form of insulation, such as fur, fat, feathers, or some combination thereof. Animals with thick fur or feathers create an insulating layer of air between their skin and internal organs. Polar bears and seals live and swim in a subfreezing environment and yet maintain a constant, warm, body temperature. The arctic fox, for example, uses its fluffy tail as extra insulation when it curls up to sleep in cold weather. Mammals have a residual effect from shivering and increased muscle activity: arrector pili muscles cause "goose bumps," causing small hairs to stand up when the individual is cold; this has the intended effect of increasing body temperature. Mammals use layers of fat to achieve the same end. Loss of significant amounts of body fat will compromise an individual's ability to conserve heat.

Endotherms use their circulatory systems to help maintain body temperature. Vasodilation brings more blood and heat to the body surface, facilitating radiation and evaporative heat loss, which helps to cool the body. Vasoconstriction reduces blood flow in peripheral blood vessels, forcing blood toward the core and the vital organs found there, and conserving heat. Some animals have adaptions to their circulatory system that enable them to transfer heat from arteries to veins, warming blood returning to the heart. This is called a countercurrent heat exchange; it prevents the cold venous blood from cooling the heart and other internal organs. This adaption can be shut down in some animals to prevent overheating the internal organs. The countercurrent adaption is found in many animals, including dolphins, sharks, bony fish, bees, and hummingbirds. In contrast, similar adaptations can help cool endotherms when needed, such as dolphin flukes and elephant ears.

Some ectothermic animals use changes in their behavior to help regulate body temperature. For example, a desert ectothermic animal may simply seek cooler areas during the hottest part of the day in the desert to keep from getting too warm. The same animals may climb onto rocks to capture heat during a cold desert night. Some animals seek water to aid evaporation in cooling them, as seen with reptiles. Other ectotherms use group activity such as the activity of bees to warm a hive to survive winter.

Many animals, especially mammals, use metabolic waste heat as a heat source. When muscles are contracted, most of the energy from the ATP used in muscle actions is wasted energy that translates into heat. Severe cold elicits a shivering reflex that generates heat for the body. Many species also have a type of adipose tissue called brown fat that specializes in generating heat.

Neural Control of Thermoregulation

The nervous system is important to **thermoregulation**, as illustrated in Figure 33.22. The processes of homeostasis and temperature control are centered in the hypothalamus of the advanced animal brain.

VISUAL CONNECTION

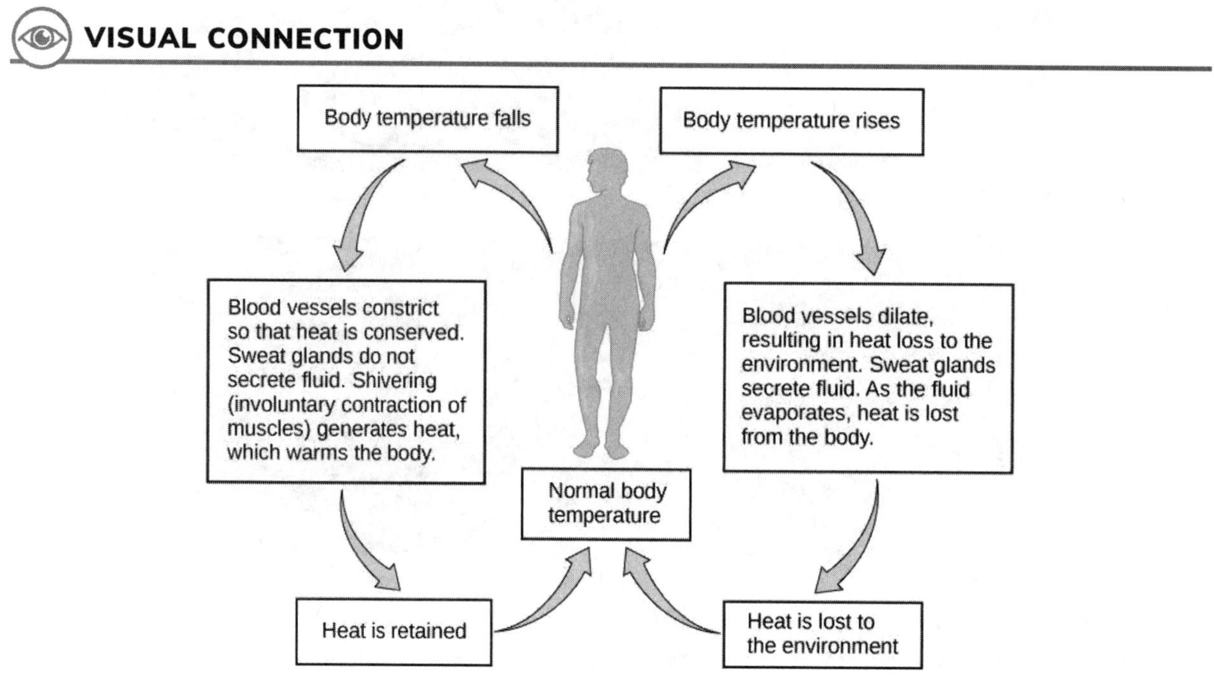

Figure 33.23 The body is able to regulate temperature in response to signals from the nervous system.

When bacteria are destroyed by leuckocytes, pyrogens are released into the blood. Pyrogens reset the body's thermostat to a higher temperature, resulting in fever. How might pyrogens cause the body temperature to rise?

The hypothalamus maintains the set point for body temperature through reflexes that cause vasodilation and sweating when the body is too warm, or vasoconstriction and shivering when the body is too cold. It responds to chemicals from the body. When a bacterium is destroyed by phagocytic leukocytes, chemicals called endogenous pyrogens are released into the blood. These pyrogens circulate to the hypothalamus and reset the thermostat. This allows the body's temperature to increase in what is commonly called a fever. An increase in body temperature causes iron to be conserved, which reduces a nutrient needed by bacteria. An increase in body heat also increases the activity of the animal's enzymes and protective cells while inhibiting the enzymes and activity of the invading microorganisms. Finally, heat itself may also kill the pathogen. A fever that was once thought to be a complication of an infection is now understood to be a normal defense mechanism.

KEY TERMS

acclimatization alteration in a body system in response to environmental change

alteration change of the set point in a homeostatic system

apodeme ingrowth of an animal's exoskeleton that functions as an attachment site for muscles

asymmetrical describes animals with no axis of symmetry in their body pattern

basal metabolic rate (BMR) metabolic rate at rest in endothermic animals

canaliculus microchannel that connects the lacunae and aids diffusion between cells

cartilage type of connective tissue with a large amount of ground substance matrix, cells called chondrocytes, and some amount of fibers

chondrocyte cell found in cartilage

columnar epithelia epithelia made of cells taller than they are wide, specialized in absorption

connective tissue type of tissue made of cells, ground substance matrix, and fibers

cuboidal epithelia epithelia made of cube-shaped cells, specialized in glandular functions

dorsal cavity body cavity on the posterior or back portion of an animal; includes the cranial and vertebral cavities

ectotherm animal incapable of maintaining a relatively constant internal body temperature

endotherm animal capable of maintaining a relatively constant internal body temperature

epithelial tissue tissue that either lines or covers organs or other tissues

estivation torpor in response to extremely high temperatures and low water availability

fibrous connective tissue type of connective tissue with a high concentration of fibers

frontal (coronal) plane plane cutting through an animal separating the individual into front and back portions

fusiform animal body shape that is tubular and tapered at both ends

hibernation torpor over a long period of time, such as a winter

homeostasis dynamic equilibrium maintaining appropriate body functions

lacuna space in cartilage and bone that contains living cells

loose (areolar) connective tissue type of connective tissue with small amounts of cells, matrix, and fibers; found around blood vessels

matrix component of connective tissue made of both living and nonliving (ground substances) cells

midsagittal plane plane cutting through an animal separating the individual into even right and left sides

negative feedback loop feedback to a control mechanism that increases or decreases a stimulus instead of maintaining it

osteon subunit of compact bone

positive feedback loop feedback to a control mechanism that continues the direction of a stimulus

pseudostratified layer of epithelia that appears multilayered, but is a simple covering

sagittal plane plane cutting through an animal separating the individual into right and left sides

set point midpoint or target point in homeostasis

simple epithelia single layer of epithelial cells

squamous epithelia type of epithelia made of flat cells, specialized in aiding diffusion or preventing abrasion

standard metabolic rate (SMR) metabolic rate at rest in ectothermic animals

stratified epithelia multiple layers of epithelial cells

thermoregulation regulation of body temperature

torpor decrease in activity and metabolism that allows an animal to survive adverse conditions

trabecula tiny plate that makes up spongy bone and gives it strength

transitional epithelia epithelia that can transition for appearing multilayered to simple; also called uroepithelial

transverse (horizontal) plane plane cutting through an animal separating the individual into upper and lower portions

ventral cavity body cavity on the anterior or front portion of an animal that includes the thoracic cavities and the abdominopelvic cavities

CHAPTER SUMMARY

33.1 Animal Form and Function

Animal bodies come in a variety of sizes and shapes. Limits on animal size and shape include impacts to their movement. Diffusion affects their size and development. Bioenergetics describes how animals use and obtain energy in relation to their body size, activity level, and environment.

33.2 Animal Primary Tissues

The basic building blocks of complex animals are four primary tissues. These are combined to form organs, which have a specific, specialized function within the body, such as the skin or kidney. Organs are organized together to perform common functions in the form of systems. The four primary tissues are epithelia, connective tissues, muscle tissues, and nervous tissues.

33.3 Homeostasis

Homeostasis is a dynamic equilibrium that is maintained in body tissues and organs. It is dynamic because it is constantly adjusting to the changes that the systems encounter. It is in equilibrium because body functions are kept within a normal range, with some fluctuations around a set point for the processes.

VISUAL CONNECTION QUESTIONS

1. Figure 33.11 Which of the following statements about types of epithelial cells is false?
   a. Simple columnar epithelial cells line the tissue of the lung.
   b. Simple cuboidal epithelial cells are involved in the filtering of blood in the kidney.
   c. Pseudostratisfied columnar epithilia occur in a single layer, but the arrangement of nuclei makes it appear that more than one layer is present.
   d. Transitional epithelia change in thickness depending on how full the bladder is.
2. Figure 33.21 State whether each of the following processes are regulated by a positive feedback loop or a negative feedback loop.
   a. A person feels satiated after eating a large meal.
   b. The blood has plenty of red blood cells. As a result, erythropoietin, a hormone that stimulates the production of new red blood cells, is no longer released from the kidney.
3. Figure 33.23 When bacteria are destroyed by leuckocytes, pyrogens are released into the blood. Pyrogens reset the body's thermostat to a higher temperature, resulting in fever. How might pyrogens cause the body temperature to rise?

REVIEW QUESTIONS

4. Which type of animal maintains a constant internal body temperature?
   a. endotherm
   b. ectotherm
   c. coelomate
   d. mesoderm
5. The symmetry found in animals that move swiftly is _______.
   a. radial
   b. bilateral
   c. sequential
   d. interrupted
6. What term describes the condition of a desert mouse that lowers its metabolic rate and "sleeps" during the hot day?
   a. turgid
   b. hibernation
   c. estivation
   d. normal sleep pattern
7. A plane that divides an animal into equal right and left portions is _______.
   a. diagonal
   b. midsagittal
   c. coronal
   d. transverse
8. A plane that divides an animal into dorsal and ventral portions is _______.
   a. sagittal
   b. midsagittal
   c. coronal
   d. transverse
9. The pleural cavity is a part of which cavity?
   a. dorsal cavity
   b. thoracic cavity
   c. abdominal cavity
   d. pericardial cavity
10. How could the increasing global temperature associated with climate change impact ectotherms?
    a. Ectotherm diversity will decrease in cool regions.
    b. Ectotherms will be able to be active all day in the tropics.
    c. Ectotherms will have to expend more energy to cool their body temperatures.
    d. Ectotherms will be able to expand into new habitats.

11. Although most animals are bilaterally symmetrical, a few exhibit radial symmetry. What is an advantage of radial symmetry?
    a. It confuses predators.
    b. It allows the animal to gather food from all sides.
    c. It allows the animal to undergo rapid, purposeful movement in any direction.
    d. It lets an animal use its dorsal surface to sense its environment.

12. Which type of epithelial cell is best adapted to aid diffusion?
    a. squamous
    b. cuboidal
    c. columnar
    d. transitional

13. Which type of epithelial cell is found in glands?
    a. squamous
    b. cuboidal
    c. columnar
    d. transitional

14. Which type of epithelial cell is found in the urinary bladder?
    a. squamous
    b. cuboidal
    c. columnar
    d. transitional

15. Which type of connective tissue has the most fibers?
    a. loose connective tissue
    b. fibrous connective tissue
    c. cartilage
    d. bone

16. Which type of connective tissue has a mineralized different matrix?
    a. loose connective tissue
    b. fibrous connective tissue
    c. cartilage
    d. bone

17. The cell found in bone that breaks it down is called an ________.
    a. osteoblast
    b. osteocyte
    c. osteoclast
    d. osteon

18. The cell found in bone that makes the bone is called an ________.
    a. osteoblast
    b. osteocyte
    c. osteoclast
    d. osteon

19. Plasma is the ________.
    a. fibers in blood
    b. matrix of blood
    c. cell that phagocytizes bacteria
    d. cell fragment found in the tissue

20. The type of muscle cell under voluntary control is the ________.
    a. smooth muscle
    b. skeletal muscle
    c. cardiac muscle
    d. visceral muscle

21. The part of a neuron that contains the nucleus is the
    a. cell body
    b. dendrite
    c. axon
    d. glial

22. Why are intercalated discs essential to the function of cardiac muscle?
    a. The discs maintain the barriers between the cells.
    b. The discs pass nutrients between cells.
    c. The discs ensure that all the cardiac muscle cells beat as a single unit.
    d. The discs control the heart rate.

23. When faced with a sudden drop in environmental temperature, an endothermic animal will:
    a. experience a drop in its body temperature
    b. wait to see if it goes lower
    c. increase muscle activity to generate heat
    d. add fur or fat to increase insulation

24. Which is an example of negative feedback?
    a. lowering of blood glucose after a meal
    b. blood clotting after an injury
    c. lactation during nursing
    d. uterine contractions during labor

25. Which method of heat exchange occurs during direct contact between the source and animal?
    a. radiation
    b. evaporation
    c. convection
    d. conduction

26. The body's thermostat is located in the ________.
    a. homeostatic receptor
    b. hypothalamus
    c. medulla
    d. vasodilation center

27. Which of the following is **not** true about acclimatization?
    a. Acclimatization allows animals to compensate for changes in their environment.
    b. Acclimatization improves function in a new environment.
    c. Acclimatization occurs when an animal tries to reestablish a homeostatic set point.
    d. Acclimatization is passed on to offspring of acclimated individuals.
28. Which of the following is **not** a way that ectotherms can change their body temperatures?
    a. Sweating for evaporative cooling.
    b. Adjusting the timing of their daily activities.
    c. Seek out or avoid direct sunlight.
    d. Huddle in a group.

CRITICAL THINKING QUESTIONS

29. How does diffusion limit the size of an organism? How is this counteracted?
30. What is the relationship between BMR and body size? Why?
31. Explain how using an open circulatory system constrains the size of animals.
32. Describe one key environmental constraint for ectotherms and one for endotherms. Why are they limited by different factors?
33. How can squamous epithelia both facilitate diffusion and prevent damage from abrasion?
34. What are the similarities between cartilage and bone?
35. Multiple sclerosis is a debilitating autoimmune disease that results in the loss of the insulation around neuron axons. What cell type is the immune system attacking, and how does this disrupt the transfer of messages by the nervous system?
36. When a person leads a sedentary life his skeletal muscles atrophy, but his smooth muscles do not. Why?
37. Why are negative feedback loops used to control body homeostasis?
38. Why is a fever a “good thing” during a bacterial infection?
39. How is a condition such as diabetes a good example of the failure of a set point in humans?
40. On a molecular level, how can endotherms produce their own heat by adjusting processes associated with cellular respiration? If needed, review Ch. 7 for details on respiration.

CHAPTER 34
Animal Nutrition and the Digestive System

Figure 34.1 For humans, fruits and vegetables are important in maintaining a balanced diet. (credit: modification of work by Julie Rybarczyk)

Chapter Outline

INTRODUCTION All living organisms need nutrients to survive. While plants can obtain the molecules required for cellular function through the process of photosynthesis, most animals obtain their nutrients by the consumption of other organisms. At the cellular level, the biological molecules necessary for animal function are amino acids, lipid molecules, nucleotides, and simple sugars. However, the food consumed consists of protein, fat, and complex carbohydrates. Animals must convert these macromolecules into the simple molecules required for maintaining cellular functions, such as assembling new molecules, cells, and tissues. The conversion of the food consumed to the nutrients required is a multistep process involving digestion and absorption. During digestion, food particles are broken down to smaller components, and later, they are absorbed by the body.

One of the challenges in human nutrition is maintaining a balance between food intake, storage, and energy expenditure. Imbalances can have serious health consequences. For example, eating too much food while not expending much energy leads to obesity, which in turn will increase the risk of developing illnesses such as type-2 diabetes and cardiovascular disease. The recent rise in obesity and related diseases makes understanding the role of diet and nutrition in maintaining good health all the more important.

34.1 Digestive Systems

By the end of this section, you will be able to do the following:

- Explain the processes of digestion and absorption
- Compare and contrast different types of digestive systems
- Explain the specialized functions of the organs involved in processing food in the body
- Describe the ways in which organs work together to digest food and absorb nutrients

Animals obtain their nutrition from the consumption of other organisms. Depending on their diet, animals can be classified into the following categories: plant eaters (herbivores), meat eaters (carnivores), and those that eat both plants and animals (omnivores). The nutrients and macromolecules present in food are not immediately accessible to the cells. There are a number of processes that modify food within the animal body in order to make the nutrients and organic molecules accessible for cellular function. As animals evolved in complexity of form and function, their digestive systems have also evolved to accommodate their various dietary needs.

Herbivores, Omnivores, and Carnivores

Herbivores are animals whose primary food source is plant-based. Examples of herbivores, as shown in Figure 34.2 include vertebrates like deer, koalas, and some bird species, as well as invertebrates such as crickets and caterpillars. These animals have evolved digestive systems capable of handling large amounts of plant material. Herbivores can be further classified into frugivores (fruit-eaters), granivores (seed eaters), nectivores (nectar feeders), and folivores (leaf eaters).

Figure 34.2 Herbivores, like this (a) mule deer and (b) monarch caterpillar, eat primarily plant material. (credit a: modification of work by Bill Ebbesen; credit b: modification of work by Doug Bowman)

Carnivores are animals that eat other animals. The word carnivore is derived from Latin and literally means "meat eater." Wild cats such as lions, shown in Figure 34.3a and tigers are examples of vertebrate carnivores, as are snakes and sharks, while invertebrate carnivores include sea stars, spiders, and ladybugs, shown in Figure 34.3b. Obligate carnivores are those that rely entirely on animal flesh to obtain their nutrients; examples of obligate carnivores are members of the cat family, such as lions and cheetahs. Facultative carnivores are those that also eat non-animal food in addition to animal food. Note that there is no clear line that differentiates facultative carnivores from omnivores; dogs would be considered facultative carnivores.

(a) (b)

Figure 34.3 Carnivores like the (a) lion eat primarily meat. The (b) ladybug is also a carnivore that consumes small insects called aphids. (credit a: modification of work by Kevin Pluck; credit b: modification of work by Jon Sullivan)

Omnivores are animals that eat both plant- and animal-derived food. In Latin, omnivore means to eat everything. Humans, bears (shown in Figure 34.4a), and chickens are example of vertebrate omnivores; invertebrate omnivores include cockroaches and crayfish (shown in Figure 34.4b).

(a) (b)

Figure 34.4 Omnivores like the (a) bear and (b) crayfish eat both plant and animal based food. (credit a: modification of work by Dave Menke; credit b: modification of work by Jon Sullivan)

Invertebrate Digestive Systems

Animals have evolved different types of digestive systems to aid in the digestion of the different foods they consume. The simplest example is that of a **gastrovascular cavity** and is found in organisms with only one opening for digestion. Platyhelminthes (flatworms), Ctenophora (comb jellies), and Cnidaria (coral, jelly fish, and sea anemones) use this type of digestion. Gastrovascular cavities, as shown in Figure 34.5a, are typically a blind tube or cavity with only one opening, the "mouth", which also serves as an "anus". Ingested material enters the mouth and passes through a hollow, tubular cavity. Cells within the cavity secrete digestive enzymes that breakdown the food. The food particles are engulfed by the cells lining the gastrovascular cavity.

The **alimentary canal**, shown in Figure 34.5b, is a more advanced system: it consists of one tube with a mouth at one end and an anus at the other. Earthworms are an example of an animal with an alimentary canal. Once the food is ingested through the mouth, it passes through the esophagus and is stored in an organ called the crop; then it passes into the gizzard where it is churned and digested. From the gizzard, the food passes through the intestine, the nutrients are absorbed, and the waste is eliminated as feces, called castings, through the anus.

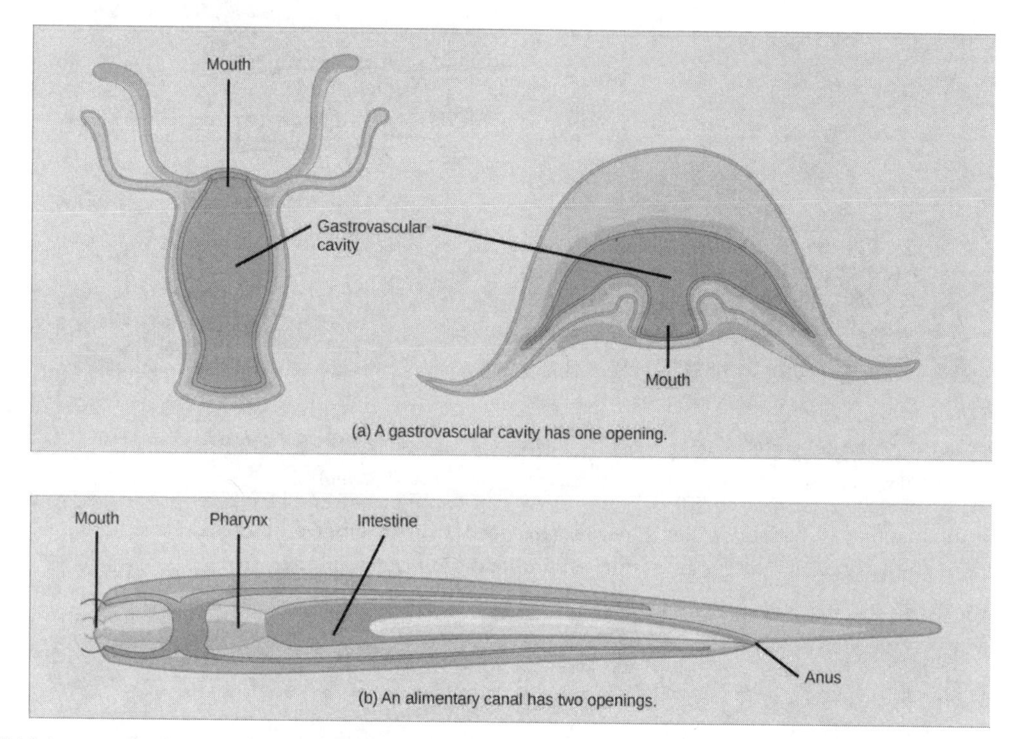

Figure 34.5 (a) A gastrovascular cavity has a single opening through which food is ingested and waste is excreted, as shown in this hydra and in this jellyfish medusa. (b) An alimentary canal has two openings: a mouth for ingesting food, and an anus for eliminating waste, as shown in this nematode.

Vertebrate Digestive Systems

Vertebrates have evolved more complex digestive systems to adapt to their dietary needs. Some animals have a single stomach, while others have multi-chambered stomachs. Birds have developed a digestive system adapted to eating unmasticated food.

Monogastric: Single-chambered Stomach

As the word **monogastric** suggests, this type of digestive system consists of one ("mono") stomach chamber ("gastric"). Humans and many animals have a monogastric digestive system as illustrated in Figure 34.6ab. The process of digestion begins with the mouth and the intake of food. The teeth play an important role in masticating (chewing) or physically breaking down food into smaller particles. The enzymes present in saliva also begin to chemically breakdown food. The esophagus is a long tube that connects the mouth to the stomach. Using peristalsis, or wave-like smooth muscle contractions, the muscles of the esophagus push the food towards the stomach. In order to speed up the actions of enzymes in the stomach, the stomach is an extremely acidic environment, with a pH between 1.5 and 2.5. The gastric juices, which include enzymes in the stomach, act on the food particles and continue the process of digestion. Further breakdown of food takes place in the small intestine where enzymes produced by the liver, the small intestine, and the pancreas continue the process of digestion. The nutrients are absorbed into the bloodstream across the epithelial cells lining the walls of the small intestines. The waste material travels on to the large intestine where water is absorbed and the drier waste material is compacted into feces; it is stored until it is excreted through the rectum.

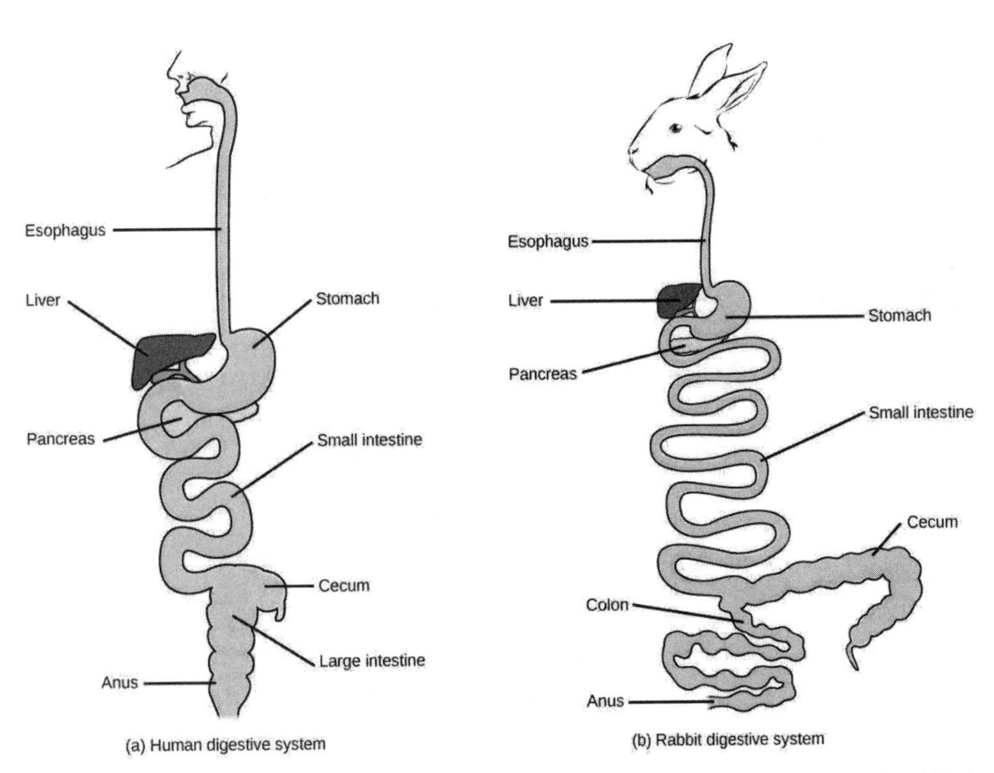

Figure 34.6 (a) Humans and herbivores, such as the (b) rabbit, have a monogastric digestive system. However, in the rabbit the small intestine and cecum are enlarged to allow more time to digest plant material. The enlarged organ provides more surface area for absorption of nutrients. Rabbits digest their food twice: the first time food passes through the digestive system, it collects in the cecum, and then it passes as soft feces called cecotrophes. The rabbit re-ingests these cecotrophes to further digest them.

Avian

Birds face special challenges when it comes to obtaining nutrition from food. They do not have teeth and so their digestive system, shown in Figure 34.7, must be able to process un-masticated food. Birds have evolved a variety of beak types that reflect the vast variety in their diet, ranging from seeds and insects to fruits and nuts. Because most birds fly, their metabolic rates are high in order to efficiently process food and keep their body weight low. The stomach of birds has two chambers: the **proventriculus**, where gastric juices are produced to digest the food before it enters the stomach, and the **gizzard**, where the food is stored, soaked, and mechanically ground. The undigested material forms food pellets that are sometimes regurgitated. Most of the chemical digestion and absorption happens in the intestine and the waste is excreted through the cloaca.

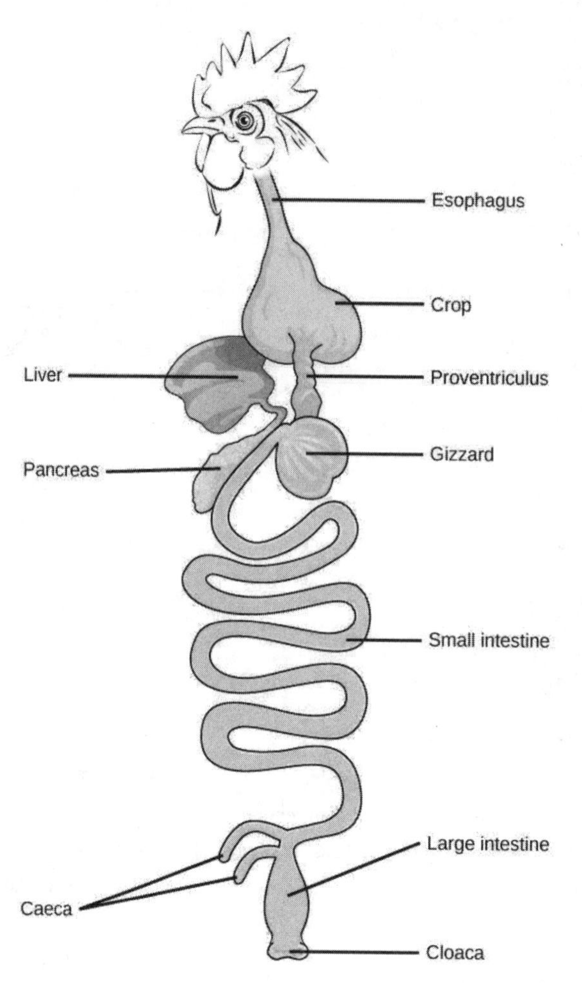

Figure 34.7 The avian esophagus has a pouch, called a crop, which stores food. Food passes from the crop to the first of two stomachs, called the proventriculus, which contains digestive juices that breakdown food. From the proventriculus, the food enters the second stomach, called the gizzard, which grinds food. Some birds swallow stones or grit, which are stored in the gizzard, to aid the grinding process. Birds do not have separate openings to excrete urine and feces. Instead, uric acid from the kidneys is secreted into the large intestine and combined with waste from the digestive process. This waste is excreted through an opening called the cloaca.

EVOLUTION CONNECTION

Avian Adaptations

Birds have a highly efficient, simplified digestive system. Recent fossil evidence has shown that the evolutionary divergence of birds from other land animals was characterized by streamlining and simplifying the digestive system. Unlike many other animals, birds do not have teeth to chew their food. In place of lips, they have sharp pointy beaks. The horny beak, lack of jaws, and the smaller tongue of the birds can be traced back to their dinosaur ancestors. The emergence of these changes seems to coincide with the inclusion of seeds in the bird diet. Seed-eating birds have beaks that are shaped for grabbing seeds and the two-compartment stomach allows for delegation of tasks. Since birds need to remain light in order to fly, their metabolic rates are very high, which means they digest their food very quickly and need to eat often. Contrast this with the ruminants, where the digestion of plant matter takes a very long time.

Ruminants

Ruminants are mainly herbivores like cows, sheep, and goats, whose entire diet consists of eating large amounts of **roughage** or fiber. They have evolved digestive systems that help them digest vast amounts of cellulose. An interesting feature of the ruminants' mouth is that they do not have upper incisor teeth. They use their lower teeth, tongue and lips to tear and chew their food. From the mouth, the food travels to the esophagus and on to the stomach.

To help digest the large amount of plant material, the stomach of the ruminants is a multi-chambered organ, as illustrated in Figure 34.8. The four compartments of the stomach are called the rumen, reticulum, omasum, and abomasum. These chambers contain many microbes that breakdown cellulose and ferment ingested food. The abomasum is the "true" stomach and is the equivalent of the monogastric stomach chamber where gastric juices are secreted. The four-compartment gastric chamber provides larger space and the microbial support necessary to digest plant material in ruminants. The fermentation process produces large amounts of gas in the stomach chamber, which must be eliminated. As in other animals, the small intestine plays an important role in nutrient absorption, and the large intestine helps in the elimination of waste.

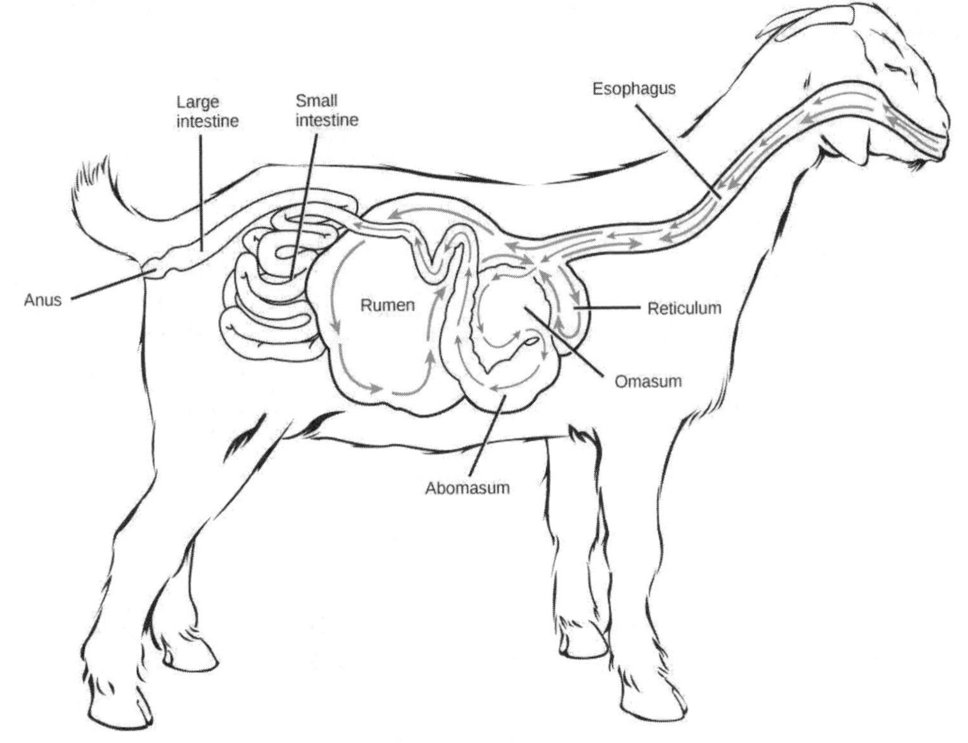

Figure 34.8 Ruminant animals, such as goats and cows, have four stomachs. The first two stomachs, the rumen and the reticulum, contain prokaryotes and protists that are able to digest cellulose fiber. The ruminant regurgitates cud from the reticulum, chews it, and swallows it into a third stomach, the omasum, which removes water. The cud then passes onto the fourth stomach, the abomasum, where it is digested by enzymes produced by the ruminant.

Pseudo-ruminants

Some animals, such as camels and alpacas, are pseudo-ruminants. They eat a lot of plant material and roughage. Digesting plant material is not easy because plant cell walls contain the polymeric sugar molecule cellulose. The digestive enzymes of these animals cannot breakdown cellulose, but microorganisms present in the digestive system can. Therefore, the digestive system must be able to handle large amounts of roughage and breakdown the cellulose. Pseudo-ruminants have a three-chamber stomach in the digestive system. However, their cecum—a pouched organ at the beginning of the large intestine containing many microorganisms that are necessary for the digestion of plant materials—is large and is the site where the roughage is fermented and digested. These animals do not have a rumen but have an omasum, abomasum, and reticulum.

Parts of the Digestive System

The vertebrate digestive system is designed to facilitate the transformation of food matter into the nutrient components that sustain organisms.

Oral Cavity

The oral cavity, or mouth, is the point of entry of food into the digestive system, illustrated in Figure 34.9. The food consumed is broken into smaller particles by mastication, the chewing action of the teeth. All mammals have teeth and can chew their food.

The extensive chemical process of digestion begins in the mouth. As food is being chewed, saliva, produced by the salivary glands, mixes with the food. Saliva is a watery substance produced in the mouths of many animals. There are three major glands that secrete saliva—the parotid, the submandibular, and the sublingual. Saliva contains mucus that moistens food and buffers the pH of the food. Saliva also contains immunoglobulins and lysozymes, which have antibacterial action to reduce tooth decay by inhibiting growth of some bacteria. Saliva also contains an enzyme called **salivary amylase** that begins the process of converting starches in the food into a disaccharide called maltose. Another enzyme called **lipase** is produced by the cells in the tongue. Lipases are a class of enzymes that can breakdown triglycerides. The lingual lipase begins the breakdown of fat components in the food. The chewing and wetting action provided by the teeth and saliva prepare the food into a mass called the **bolus** for swallowing. The tongue helps in swallowing—moving the bolus from the mouth into the pharynx. The pharynx opens to two passageways: the trachea, which leads to the lungs, and the esophagus, which leads to the stomach. The trachea has an opening called the glottis, which is covered by a cartilaginous flap called the epiglottis. When swallowing, the epiglottis closes the glottis and food passes into the esophagus and not the trachea. This arrangement allows food to be kept out of the trachea.

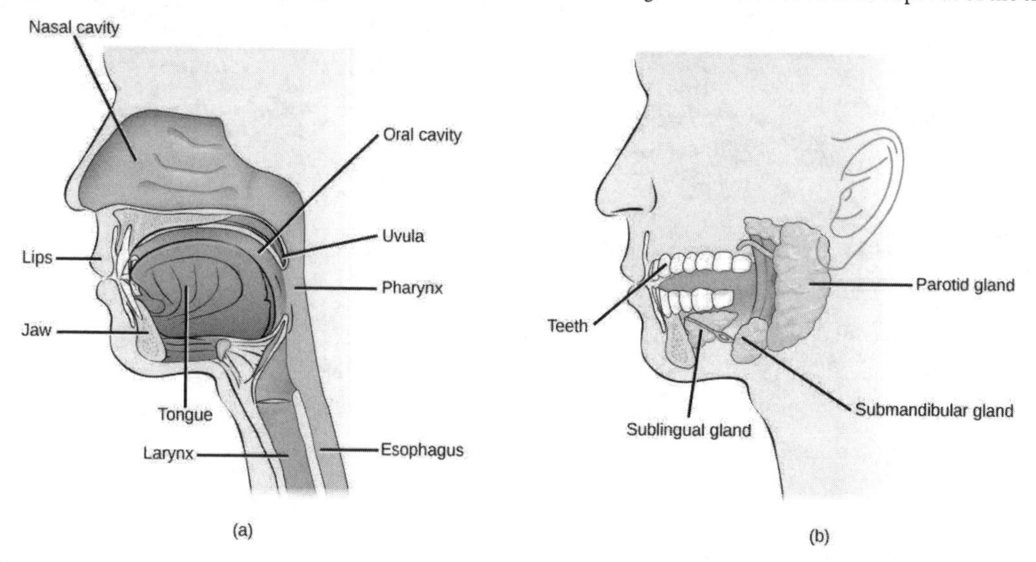

Figure 34.9 Digestion of food begins in the (a) oral cavity. Food is masticated by teeth and moistened by saliva secreted from the (b) salivary glands. Enzymes in the saliva begin to digest starches and fats. With the help of the tongue, the resulting bolus is moved into the esophagus by swallowing. (credit: modification of work by the National Cancer Institute)

Esophagus

The **esophagus** is a tubular organ that connects the mouth to the stomach. The chewed and softened food passes through the esophagus after being swallowed. The smooth muscles of the esophagus undergo a series of wave like movements called **peristalsis** that push the food toward the stomach, as illustrated in Figure 34.10. The peristalsis wave is unidirectional—it moves food from the mouth to the stomach, and reverse movement is not possible. The peristaltic movement of the esophagus is an involuntary reflex; it takes place in response to the act of swallowing.

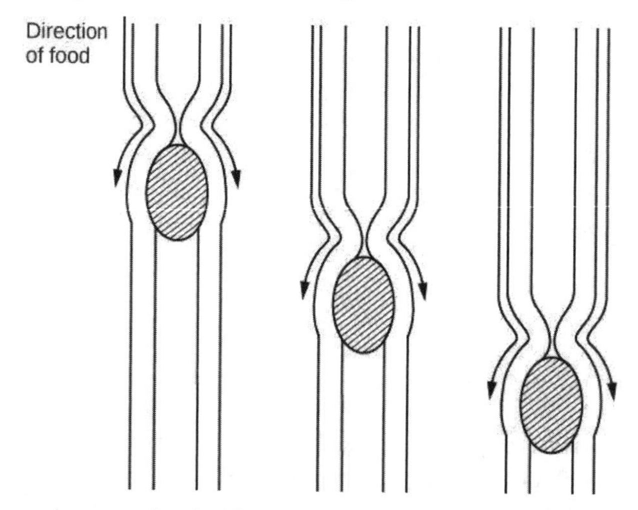

Figure 34.10 The esophagus transfers food from the mouth to the stomach through peristaltic movements.

A ring-like muscle called a **sphincter** forms valves in the digestive system. The gastro-esophageal sphincter is located at the stomach end of the esophagus. In response to swallowing and the pressure exerted by the bolus of food, this sphincter opens, and the bolus enters the stomach. When there is no swallowing action, this sphincter is shut and prevents the contents of the stomach from traveling up the esophagus. Many animals have a true sphincter; however, in humans, there is no true sphincter, but the esophagus remains closed when there is no swallowing action. Acid reflux or "heartburn" occurs when the acidic digestive juices escape into the esophagus.

Stomach

A large part of digestion occurs in the stomach, shown in Figure 34.11. The **stomach** is a saclike organ that secretes gastric digestive juices. The pH in the stomach is between 1.5 and 2.5. This highly acidic environment is required for the chemical breakdown of food and the extraction of nutrients. When empty, the stomach is a rather small organ; however, it can expand to up to 20 times its resting size when filled with food. This characteristic is particularly useful for animals that need to eat when food is available.

VISUAL CONNECTION

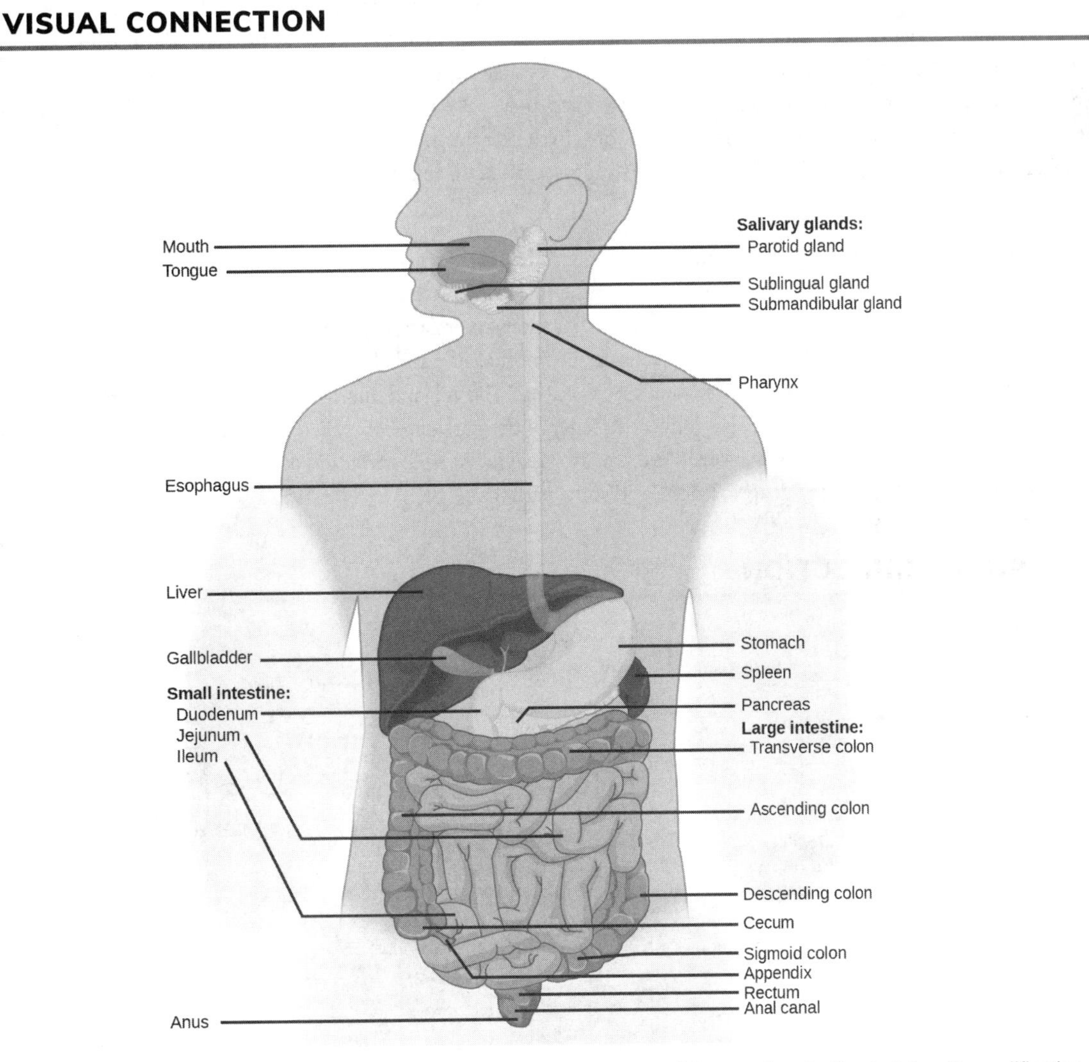

Figure 34.11 The human stomach has an extremely acidic environment where most of the protein gets digested. (credit: modification of work by Mariana Ruiz Villareal)

Which of the following statements about the digestive system is false?

a. Chyme is a mixture of food and digestive juices that is produced in the stomach.
b. Food enters the large intestine before the small intestine.
c. In the small intestine, chyme mixes with bile, which emulsifies fats.
d. The stomach is separated from the small intestine by the pyloric sphincter.

The stomach is also the major site for protein digestion in animals other than ruminants. Protein digestion is mediated by an enzyme called pepsin in the stomach chamber. **Pepsin** is secreted by the chief cells in the stomach in an inactive form called **pepsinogen**. Pepsin breaks peptide bonds and cleaves proteins into smaller polypeptides; it also helps activate more pepsinogen, starting a positive feedback mechanism that generates more pepsin. Another cell type—parietal cells—secrete hydrogen and chloride ions, which combine in the lumen to form hydrochloric acid, the primary acidic component of the stomach juices. Hydrochloric acid helps to convert the inactive pepsinogen to pepsin. The highly acidic environment also kills many microorganisms in the food and, combined with the action of the enzyme pepsin, results in the hydrolysis of protein in the food. Chemical digestion is facilitated by the churning action of the stomach. Contraction and relaxation of smooth muscles mixes the stomach contents about every 20 minutes. The partially digested food and gastric juice mixture is called **chyme**. Chyme passes from the stomach to the small intestine. Further protein digestion takes place in the small intestine. Gastric emptying occurs within two to six hours after a meal. Only a small amount of chyme is released into the small intestine at a time. The movement of chyme from the stomach into the small intestine is regulated by the pyloric sphincter.

When digesting protein and some fats, the stomach lining must be protected from getting digested by pepsin. There are two points to consider when describing how the stomach lining is protected. First, as previously mentioned, the enzyme pepsin is synthesized in the inactive form. This protects the chief cells, because pepsinogen does not have the same enzyme functionality of pepsin. Second, the stomach has a thick mucus lining that protects the underlying tissue from the action of the digestive juices. When this mucus lining is ruptured, ulcers can form in the stomach. Ulcers are open wounds in or on an organ caused by bacteria (*Helicobacter pylori*) when the mucus lining is ruptured and fails to reform.

Small Intestine

Chyme moves from the stomach to the small intestine. The **small intestine** is the organ where the digestion of protein, fats, and carbohydrates is completed. The small intestine is a long tube-like organ with a highly folded surface containing finger-like projections called the **villi**. The apical surface of each villus has many microscopic projections called microvilli. These structures, illustrated in Figure 34.12, are lined with epithelial cells on the luminal side and allow for the nutrients to be absorbed from the digested food and absorbed into the bloodstream on the other side. The villi and microvilli, with their many folds, increase the surface area of the intestine and increase absorption efficiency of the nutrients. Absorbed nutrients in the blood are carried into the hepatic portal vein, which leads to the liver. There, the liver regulates the distribution of nutrients to the rest of the body and removes toxic substances, including drugs, alcohol, and some pathogens.

VISUAL CONNECTION

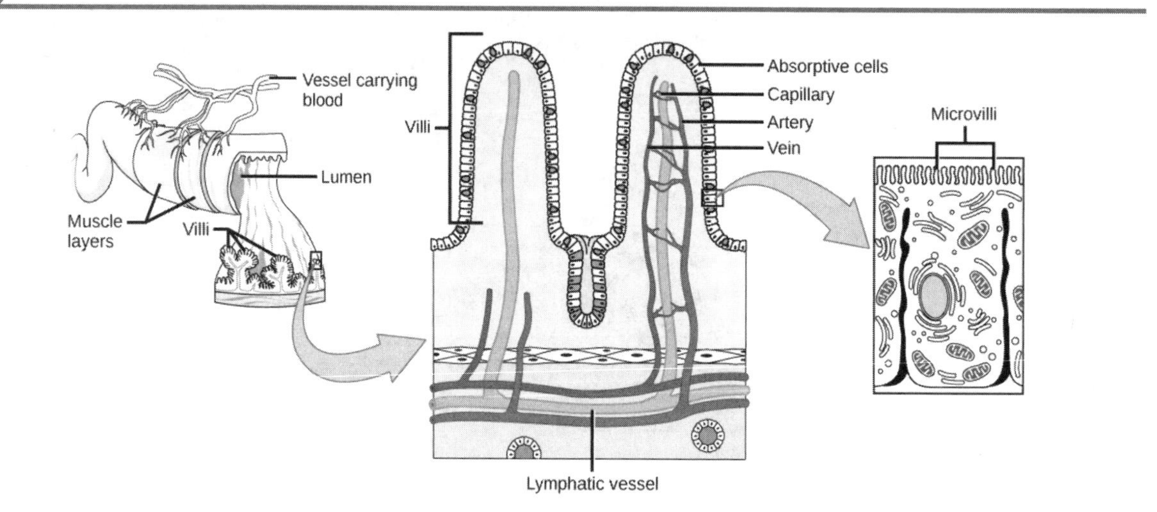

Figure 34.12 Villi are folds on the small intestine lining that increase the surface area to facilitate the absorption of nutrients.

Which of the following statements about the small intestine is false?

a. Absorptive cells that line the small intestine have microvilli, small projections that increase surface area and aid in the absorption of food.
b. The inside of the small intestine has many folds, called villi.
c. Microvilli are lined with blood vessels as well as lymphatic vessels.

d. The inside of the small intestine is called the lumen.

The human small intestine is over 6m long and is divided into three parts: the duodenum, the jejunum, and the ileum. The "C-shaped," fixed part of the small intestine is called the **duodenum** and is shown in Figure 34.11. The duodenum is separated from the stomach by the pyloric sphincter which opens to allow chyme to move from the stomach to the duodenum. In the duodenum, chyme is mixed with pancreatic juices in an alkaline solution rich in bicarbonate that neutralizes the acidity of chyme and acts as a buffer. Pancreatic juices also contain several digestive enzymes. Digestive juices from the pancreas, liver, and gallbladder, as well as from gland cells of the intestinal wall itself, enter the duodenum. **Bile** is produced in the liver and stored and concentrated in the gallbladder. Bile contains bile salts which emulsify lipids while the pancreas produces enzymes that catabolize starches, disaccharides, proteins, and fats. These digestive juices breakdown the food particles in the chyme into glucose, triglycerides, and amino acids. Some chemical digestion of food takes place in the duodenum. Absorption of fatty acids also takes place in the duodenum.

The second part of the small intestine is called the **jejunum**, shown in Figure 34.11. Here, hydrolysis of nutrients is continued while most of the carbohydrates and amino acids are absorbed through the intestinal lining. The bulk of chemical digestion and nutrient absorption occurs in the jejunum.

The **ileum**, also illustrated in Figure 34.11 is the last part of the small intestine and here the bile salts and vitamins are absorbed into the bloodstream. The undigested food is sent to the colon from the ileum via peristaltic movements of the muscle. The ileum ends and the large intestine begins at the ileocecal valve. The vermiform, "worm-like," appendix is located at the ileocecal valve. The appendix of humans secretes no enzymes and has an insignificant role in immunity.

Large Intestine

The **large intestine**, illustrated in Figure 34.13, reabsorbs the water from the undigested food material and processes the waste material. The human large intestine is much smaller in length compared to the small intestine but larger in diameter. It has three parts: the cecum, the colon, and the rectum. The cecum joins the ileum to the colon and is the receiving pouch for the waste matter. The colon is home to many bacteria or "intestinal flora" that aid in the digestive processes. The colon can be divided into four regions, the ascending colon, the transverse colon, the descending colon, and the sigmoid colon. The main functions of the colon are to extract the water and mineral salts from undigested food, and to store waste material. Carnivorous mammals have a shorter large intestine compared to herbivorous mammals due to their diet.

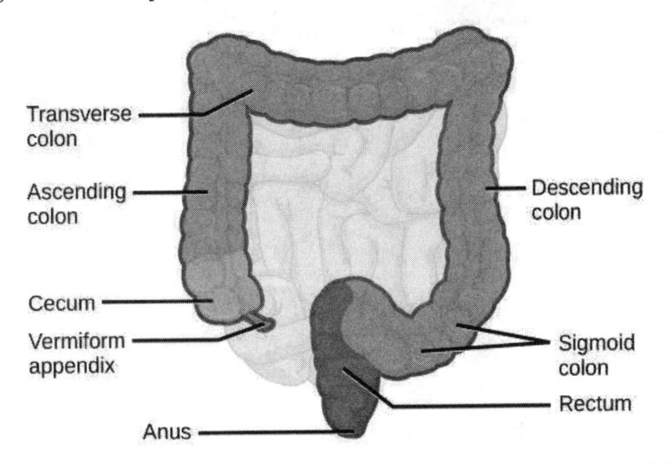

Figure 34.13 The large intestine reabsorbs water from undigested food and stores waste material until it is eliminated.

Rectum and Anus

The **rectum** is the terminal end of the large intestine, as shown in Figure 34.13. The primary role of the rectum is to store the feces until defecation. The feces are propelled using peristaltic movements during elimination. The **anus** is an opening at the far-end of the digestive tract and is the exit point for the waste material. Two sphincters between the rectum and anus control elimination: the inner sphincter is involuntary and the outer sphincter is voluntary.

Accessory Organs

The organs discussed above are the organs of the digestive tract through which food passes. Accessory organs are organs that add secretions (enzymes) that catabolize food into nutrients. Accessory organs include salivary glands, the liver, the pancreas, and the gallbladder. The liver, pancreas, and gallbladder are regulated by hormones in response to the food consumed.

The **liver** is the largest internal organ in humans and it plays a very important role in digestion of fats and detoxifying blood. The liver produces bile, a digestive juice that is required for the breakdown of fatty components of the food in the duodenum. The liver also processes the vitamins and fats and synthesizes many plasma proteins.

The **pancreas** is another important gland that secretes digestive juices. The chyme produced from the stomach is highly acidic in nature; the pancreatic juices contain high levels of bicarbonate, an alkali that neutralizes the acidic chyme. Additionally, the pancreatic juices contain a large variety of enzymes that are required for the digestion of protein and carbohydrates.

The **gallbladder** is a small organ that aids the liver by storing bile and concentrating bile salts. When chyme containing fatty acids enters the duodenum, the bile is secreted from the gallbladder into the duodenum.

34.2 Nutrition and Energy Production

By the end of this section, you will be able to do the following:

- Explain why an animal's diet should be balanced and meet the needs of the body
- Define the primary components of food
- Describe the essential nutrients required for cellular function that cannot be synthesized by the animal body
- Explain how energy is produced through diet and digestion
- Describe how excess carbohydrates and energy are stored in the body

Given the diversity of animal life on our planet, it is not surprising that the animal diet would also vary substantially. The animal diet is the source of materials needed for building DNA and other complex molecules needed for growth, maintenance, and reproduction; collectively these processes are called biosynthesis. The diet is also the source of materials for ATP production in the cells. The diet must be balanced to provide the minerals and vitamins that are required for cellular function.

Food Requirements

What are the fundamental requirements of the animal diet? The animal diet should be well balanced and provide nutrients required for bodily function and the minerals and vitamins required for maintaining structure and regulation necessary for good health and reproductive capability. These requirements for a human are illustrated graphically in Figure 34.14

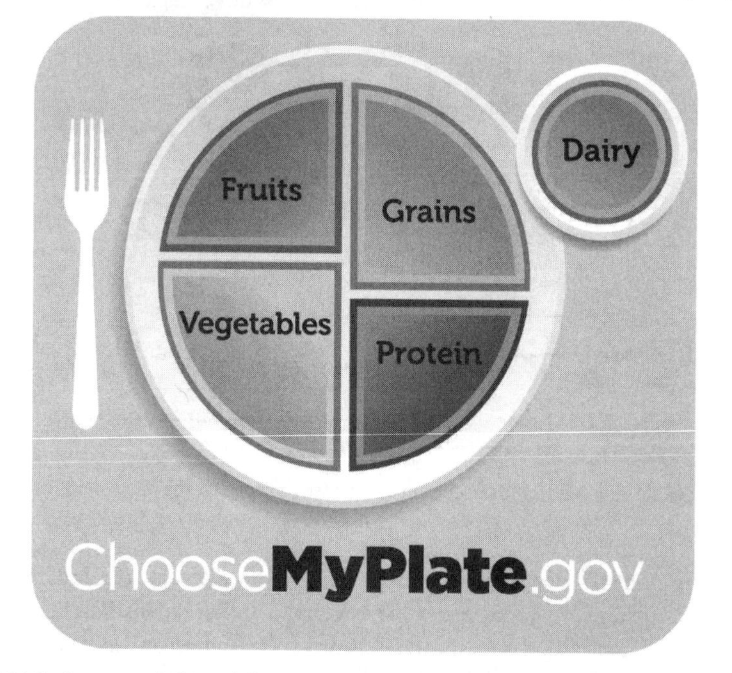

Figure 34.14 For humans, a balanced diet includes fruits, vegetables, grains, and protein. (credit: USDA)

LINK TO LEARNING

The first step in ensuring that you are meeting the food requirements of your body is an awareness of the food groups and the nutrients they provide. To learn more about each food group and the recommended daily amounts, explore this interactive site (http://openstax.org/l/food_groups) by the United States Department of Agriculture.

Everyday Connection

Let's Move! Campaign

Obesity is a growing epidemic and the rate of obesity among children is rapidly rising in the United States. To combat childhood obesity and ensure that children get a healthy start in life, first lady Michelle Obama has launched the Let's Move! campaign. The goal of this campaign is to educate parents and caregivers on providing healthy nutrition and encouraging active lifestyles to future generations. This program aims to involve the entire community, including parents, teachers, and healthcare providers to ensure that children have access to healthy foods—more fruits, vegetables, and whole grains—and consume fewer calories from processed foods. Another goal is to ensure that children get physical activity. With the increase in television viewing and stationary pursuits such as video games, sedentary lifestyles have become the norm. Learn more at https://letsmove.obamawhitehouse.archives.gov (http://openstax.org/l/Letsmove) .

Organic Precursors

The organic molecules required for building cellular material and tissues must come from food. Carbohydrates or sugars are the primary source of organic carbons in the animal body. During digestion, digestible carbohydrates are ultimately broken down into glucose and used to provide energy through metabolic pathways. Complex carbohydrates, including polysaccharides, can be broken down into glucose through biochemical modification; however, humans do not produce the enzyme cellulase and lack the ability to derive glucose from the polysaccharide cellulose. In humans, these molecules provide the fiber required for moving waste through the large intestine and a healthy colon. The intestinal flora in the human gut are able to extract some nutrition from these plant fibers. The excess sugars in the body are converted into glycogen and stored in the liver and muscles for later use. Glycogen stores are used to fuel prolonged exertions, such as long-distance running, and to provide energy during food shortage. Excess glycogen can be converted to fats, which are stored in the lower layer of the skin of mammals for insulation and energy storage. Excess digestible carbohydrates are stored by mammals in order to survive famine and aid in mobility.

Another important requirement is that of nitrogen. Protein catabolism provides a source of organic nitrogen. Amino acids are the building blocks of proteins and protein breakdown provides amino acids that are used for cellular function. The carbon and nitrogen derived from these become the building block for nucleotides, nucleic acids, proteins, cells, and tissues. Excess nitrogen must be excreted as it is toxic. Fats add flavor to food and promote a sense of satiety or fullness. Fatty foods are also significant sources of energy because one gram of fat contains nine calories. Fats are required in the diet to aid the absorption of fat-soluble vitamins and the production of fat-soluble hormones.

Essential Nutrients

While the animal body can synthesize many of the molecules required for function from the organic precursors, there are some nutrients that need to be consumed from food. These nutrients are termed **essential nutrients**, meaning they must be eaten, and the body cannot produce them.

The omega-3 alpha-linolenic acid and the omega-6 linoleic acid are essential fatty acids needed to make some membrane phospholipids. **Vitamins** are another class of essential organic molecules that are required in small quantities for many enzymes to function and, for this reason, are considered to be coenzymes. Absence or low levels of vitamins can have a dramatic effect on health, as outlined in Table 34.1 and Table 34.2. Both fat-soluble and water-soluble vitamins must be obtained from food. **Minerals**, listed in Table 34.3, are inorganic essential nutrients that must be obtained from food. Among their many functions, minerals help in structure and regulation and are considered cofactors. Certain amino acids also must be procured from food and cannot be synthesized by the body. These amino acids are the "essential" amino acids. The human body can synthesize only 11 of the 20 required amino acids; the rest must be obtained from food. The essential amino acids are listed in Table 34.4.

Water-soluble Essential Vitamins

Vitamin	Function	Deficiencies Can Lead To	Sources
Vitamin B_1 (Thiamine)	Needed by the body to process lipids, proteins, and carbohydrates; coenzyme removes CO_2 from organic compounds	Muscle weakness, Beriberi: reduced heart function, CNS problems	Milk, meat, dried beans, whole grains
Vitamin B_2 (Riboflavin)	Takes an active role in metabolism, aiding in the conversion of food to energy (FAD and FMN)	Cracks or sores on the outer surface of the lips (cheliosis); inflammation and redness of the tongue; moist, scaly skin inflammation (seborrheic dermatitis)	Meat, eggs, enriched grains, vegetables
Vitamin B_3 (Niacin)	Used by the body to release energy from carbohydrates and to process alcohol; required for the synthesis of sex hormones; component of coenzyme NAD^+ and $NADP^+$	Pellagra, which can result in dermatitis, diarrhea, dementia, and death	Meat, eggs, grains, nuts, potatoes
Vitamin B_5 (Pantothenic acid)	Assists in producing energy from foods (lipids, in particular); component of coenzyme A	Fatigue, poor coordination, retarded growth, numbness, tingling of hands and feet	Meat, whole grains, milk, fruits, vegetables
Vitamin B_6 (Pyridoxine)	The principal vitamin for processing amino acids and lipids; also helps convert nutrients into energy	Irritability, depression, confusion, mouth sores or ulcers, anemia, muscular twitching	Meat, dairy products, whole grains, orange juice
Vitamin B_7 (Biotin)	Used in energy and amino acid metabolism, fat synthesis, and fat breakdown; helps the body use blood sugar	Hair loss, dermatitis, depression, numbness and tingling in the extremities; neuromuscular disorders	Meat, eggs, legumes and other vegetables
Vitamin B_9 (Folic acid)	Assists the normal development of cells, especially during fetal development; helps metabolize nucleic and amino acids	Deficiency during pregnancy is associated with birth defects, such as neural tube defects and anemia	Leafy green vegetables, whole wheat, fruits, nuts, legumes
Vitamin B_{12} (Cobalamin)	Maintains healthy nervous system and assists with blood cell formation; coenzyme in nucleic acid metabolism	Anemia, neurological disorders, numbness, loss of balance	Meat, eggs, animal products
Vitamin C (Ascorbic acid)	Helps maintain connective tissue: bone, cartilage, and dentin; boosts the immune system	Scurvy, which results in bleeding, hair and tooth loss; joint pain and swelling; delayed wound healing	Citrus fruits, broccoli, tomatoes, red sweet bell peppers

Table 34.1

Fat-soluble Essential Vitamins

Vitamin	Function	Deficiencies Can Lead To	Sources
Vitamin A (Retinol)	Critical to the development of bones, teeth, and skin; helps maintain eyesight, enhances the immune system, fetal development, gene expression	Night-blindness, skin disorders, impaired immunity	Dark green leafy vegetables, yellow-orange vegetables, fruits, milk, butter
Vitamin D	Critical for calcium absorption for bone development and strength; maintains a stable nervous system; maintains a normal and strong heartbeat; helps in blood clotting	Rickets, osteomalacia, immunity	Cod liver oil, milk, egg yolk
Vitamin E (Tocopherol)	Lessens oxidative damage of cells and prevents lung damage from pollutants; vital to the immune system	Deficiency is rare; anemia, nervous system degeneration	Wheat germ oil, unrefined vegetable oils, nuts, seeds, grains
Vitamin K (Phylloquinone)	Essential to blood clotting	Bleeding and easy bruising	Leafy green vegetables, tea

Table 34.2

Figure 34.15 A healthy diet should include a variety of foods to ensure that needs for essential nutrients are met. (credit: Keith Weller, USDA ARS)

Minerals and Their Function in the Human Body

Mineral	Function	Deficiencies Can Lead To	Sources
*Calcium	Needed for muscle and neuron function; heart health; builds bone and supports synthesis and function of blood cells; nerve function	Osteoporosis, rickets, muscle spasms, impaired growth	Milk, yogurt, fish, green leafy vegetables, legumes
*Chlorine	Needed for production of hydrochloric acid (HCl) in the stomach and nerve function; osmotic balance	Muscle cramps, mood disturbances, reduced appetite	Table salt
Copper (trace amounts)	Required component of many redox enzymes, including cytochrome c oxidase; cofactor for hemoglobin synthesis	Copper deficiency is rare	Liver, oysters, cocoa, chocolate, sesame, nuts
Iodine	Required for the synthesis of thyroid hormones	Goiter	Seafood, iodized salt, dairy products
Iron	Required for many proteins and enzymes, notably hemoglobin, to prevent anemia	Anemia, which causes poor concentration, fatigue, and poor immune function	Red meat, leafy green vegetables, fish (tuna, salmon), eggs, dried fruits, beans, whole grains
*Magnesium	Required cofactor for ATP formation; bone formation; normal membrane functions; muscle function	Mood disturbances, muscle spasms	Whole grains, leafy green vegetables
Manganese (trace amounts)	A cofactor in enzyme functions; trace amounts are required	Manganese deficiency is rare	Common in most foods
Molybdenum (trace amounts)	Acts as a cofactor for three essential enzymes in humans: sulfite oxidase, xanthine oxidase, and aldehyde oxidase	Molybdenum deficiency is rare	
*Phosphorus	A component of bones and teeth; helps regulate acid-base balance; nucleotide synthesis	Weakness, bone abnormalities, calcium loss	Milk, hard cheese, whole grains, meats
*Potassium	Vital for muscles, heart, and nerve function	Cardiac rhythm disturbance, muscle weakness	Legumes, potato skin, tomatoes, bananas
Selenium (trace amounts)	A cofactor essential to activity of antioxidant enzymes like glutathione peroxidase; trace amounts are required	Selenium deficiency is rare	Common in most foods

Table 34.3

Mineral	Function	Deficiencies Can Lead To	Sources
*Sodium	Systemic electrolyte required for many functions; acid-base balance; water balance; nerve function	Muscle cramps, fatigue, reduced appetite	Table salt
Zinc (trace amounts)	Required for several enzymes such as carboxypeptidase, liver alcohol dehydrogenase, and carbonic anhydrase	Anemia, poor wound healing, can lead to short stature	Common in most foods

*Greater than 200mg/day required

Table 34.3

Essential Amino Acids

Amino acids that must be consumed	Amino acids anabolized by the body
isoleucine	alanine
leucine	selenocysteine
lysine	aspartate
methionine	cysteine
phenylalanine	glutamate
tryptophan	glycine
valine	proline
histidine*	serine
threonine	tyrosine
arginine*	asparagine

*The human body can synthesize histidine and arginine, but not in the quantities required, especially for growing children.

Table 34.4

Food Energy and ATP

Animals need food to obtain energy and maintain homeostasis. Homeostasis is the ability of a system to maintain a stable internal environment even in the face of external changes to the environment. For example, the normal body temperature of humans is 37°C (98.6°F). Humans maintain this temperature even when the external temperature is hot or cold. It takes energy to maintain this body temperature, and animals obtain this energy from food.

The primary source of energy for animals is carbohydrates, mainly glucose. Glucose is called the body's fuel. The digestible carbohydrates in an animal's diet are converted to glucose molecules through a series of catabolic chemical reactions.

Adenosine triphosphate, or ATP, is the primary energy currency in cells; ATP stores energy in phosphate ester bonds. ATP releases energy when the phosphodiester bonds are broken and ATP is converted to ADP and a phosphate group. ATP is produced by the oxidative reactions in the cytoplasm and mitochondrion of the cell, where carbohydrates, proteins, and fats undergo a series of metabolic reactions collectively called cellular respiration. For example, glycolysis is a series of reactions in which glucose is converted to pyruvic acid and some of its chemical potential energy is transferred to NADH and ATP.

ATP is required for all cellular functions. It is used to build the organic molecules that are required for cells and tissues; it provides energy for muscle contraction and for the transmission of electrical signals in the nervous system. When the amount of ATP is available in excess of the body's requirements, the liver uses the excess ATP and excess glucose to produce molecules called glycogen. Glycogen is a polymeric form of glucose and is stored in the liver and skeletal muscle cells. When blood sugar drops, the liver releases glucose from stores of glycogen. Skeletal muscle converts glycogen to glucose during intense exercise. The process of converting glucose and excess ATP to glycogen and the storage of excess energy is an evolutionarily important step in helping animals deal with mobility, food shortages, and famine.

Everyday Connection

Obesity

Obesity is a major health concern in the United States, and there is a growing focus on reducing obesity and the diseases it may lead to, such as type-2 diabetes, cancers of the colon and breast, and cardiovascular disease. How does the food consumed contribute to obesity?

Fatty foods are calorie-dense, meaning that they have more calories per unit mass than carbohydrates or proteins. One gram of carbohydrates has four calories, one gram of protein has four calories, and one gram of fat has nine calories. Animals tend to seek lipid-rich food for their higher energy content.

The signals of hunger ("time to eat") and satiety ("time to stop eating") are controlled in the hypothalamus region of the brain. Foods that are rich in fatty acids tend to promote satiety more than foods that are rich only in carbohydrates.

Excess carbohydrate and ATP are used by the liver to synthesize glycogen. The pyruvate produced during glycolysis is used to synthesize fatty acids. When there is more glucose in the body than required, the resulting excess pyruvate is converted into molecules that eventually result in the synthesis of fatty acids within the body. These fatty acids are stored in adipose cells—the fat cells in the mammalian body whose primary role is to store fat for later use.

It is important to note that some animals benefit from obesity. Polar bears and seals need body fat for insulation and to keep them from losing body heat during Arctic winters. When food is scarce, stored body fat provides energy for maintaining homeostasis. Fats prevent famine in mammals, allowing them to access energy when food is not available on a daily basis; fats are stored when a large kill is made or lots of food is available.

34.3 Digestive System Processes

By the end of this section, you will be able to do the following:

- Describe the process of digestion
- Detail the steps involved in digestion and absorption
- Define elimination
- Explain the role of both the small and large intestines in absorption

Obtaining nutrition and energy from food is a multistep process. For true animals, the first step is ingestion, the act of taking in food. This is followed by digestion, absorption, and elimination. In the following sections, each of these steps will be discussed in detail.

Ingestion

The large molecules found in intact food cannot pass through the cell membranes. Food needs to be broken into smaller particles so that animals can harness the nutrients and organic molecules. The first step in this process is **ingestion**. Ingestion is the process of taking in food through the mouth. In vertebrates, the teeth, saliva, and tongue play important roles in mastication (preparing the food into bolus). While the food is being mechanically broken down, the enzymes in saliva begin to chemically process the food as well. The combined action of these processes modifies the food from large particles to a soft mass that can be swallowed and can travel the length of the esophagus.

Digestion and Absorption

Digestion is the mechanical and chemical breakdown of food into small organic fragments. It is important to breakdown macromolecules into smaller fragments that are of suitable size for absorption across the digestive epithelium. Large, complex molecules of proteins, polysaccharides, and lipids must be reduced to simpler particles such as simple sugar before they can be absorbed by the digestive epithelial cells. Different organs play specific roles in the digestive process. The animal diet needs carbohydrates, protein, and fat, as well as vitamins and inorganic components for nutritional balance. How each of these components is digested is discussed in the following sections.

Carbohydrates

The digestion of carbohydrates begins in the mouth. The salivary enzyme amylase begins the breakdown of food starches into maltose, a disaccharide. As the bolus of food travels through the esophagus to the stomach, no significant digestion of carbohydrates takes place. The esophagus produces no digestive enzymes but does produce mucous for lubrication. The acidic environment in the stomach stops the action of the amylase enzyme.

The next step of carbohydrate digestion takes place in the duodenum. Recall that the chyme from the stomach enters the duodenum and mixes with the digestive secretion from the pancreas, liver, and gallbladder. Pancreatic juices also contain amylase, which continues the breakdown of starch and glycogen into maltose, a disaccharide. The disaccharides are broken down into monosaccharides by enzymes called **maltases**, **sucrases**, and **lactases**, which are also present in the brush border of the small intestinal wall. Maltase breaks down maltose into glucose. Other disaccharides, such as sucrose and lactose are broken down by sucrase and lactase, respectively. Sucrase breaks down sucrose (or "table sugar") into glucose and fructose, and lactase breaks down lactose (or "milk sugar") into glucose and galactose. The monosaccharides (glucose) thus produced are absorbed and then can be used in metabolic pathways to harness energy. The monosaccharides are transported across the intestinal epithelium into the bloodstream to be transported to the different cells in the body. The steps in carbohydrate digestion are summarized in Figure 34.16 and Table 34.5.

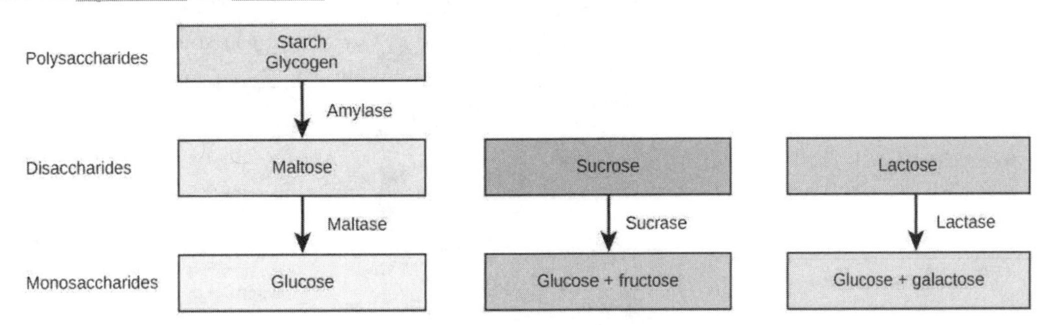

Figure 34.16 Digestion of carbohydrates is performed by several enzymes. Starch and glycogen are broken down into glucose by amylase and maltase. Sucrose (table sugar) and lactose (milk sugar) are broken down by sucrase and lactase, respectively.

Digestion of Carbohydrates

Enzyme	Produced By	Site of Action	Substrate Acting On	End Products
Salivary amylase	Salivary glands	Mouth	Polysaccharides (Starch)	Disaccharides (maltose), oligosaccharides
Pancreatic amylase	Pancreas	Small intestine	Polysaccharides (starch)	Disaccharides (maltose), monosaccharides
Oligosaccharidases	Lining of the intestine; brush border membrane	Small intestine	Disaccharides	Monosaccharides (e.g., glucose, fructose, galactose)

Table 34.5

Protein

A large part of protein digestion takes place in the stomach. The enzyme pepsin plays an important role in the digestion of proteins by breaking down the intact protein to peptides, which are short chains of four to nine amino acids. In the duodenum, other enzymes—**trypsin**, **elastase**, and **chymotrypsin**—act on the peptides reducing them to smaller peptides. Trypsin elastase, carboxypeptidase, and chymotrypsin are produced by the pancreas and released into the duodenum where they act on the chyme. Further breakdown of peptides to single amino acids is aided by enzymes called peptidases (those that breakdown peptides). Specifically, **carboxypeptidase**, **dipeptidase**, and **aminopeptidase** play important roles in reducing the peptides to free amino acids. The amino acids are absorbed into the bloodstream through the small intestines. The steps in protein digestion are summarized in Figure 34.17 and Table 34.6.

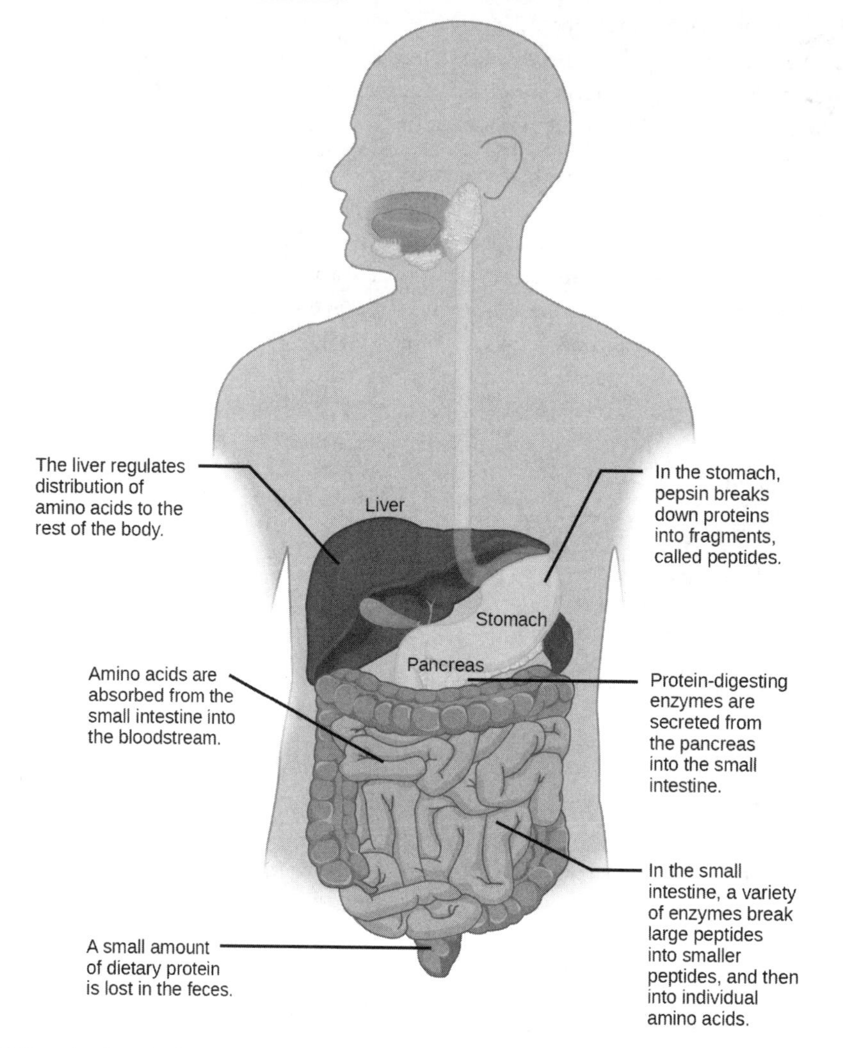

Figure 34.17 Protein digestion is a multistep process that begins in the stomach and continues through the intestines.

Digestion of Protein

Enzyme	Produced By	Site of Action	Substrate Acting On	End Products
Pepsin	Stomach chief cells	Stomach	Proteins	Peptides
Trypsin Elastase Chymotrypsin	Pancreas	Small intestine	Proteins	Peptides

Table 34.6

Enzyme	Produced By	Site of Action	Substrate Acting On	End Products
Carboxypeptidase	Pancreas	Small intestine	Peptides	Amino acids and peptides
Aminopeptidase Dipeptidase	Lining of intestine	Small intestine	Peptides	Amino acids

Table 34.6

Lipids

Lipid digestion begins in the stomach with the aid of lingual lipase and gastric lipase. However, the bulk of lipid digestion occurs in the small intestine due to pancreatic lipase. When chyme enters the duodenum, the hormonal responses trigger the release of bile, which is produced in the liver and stored in the gallbladder. Bile aids in the digestion of lipids, primarily triglycerides by emulsification. Emulsification is a process in which large lipid globules are broken down into several small lipid globules. These small globules are more widely distributed in the chyme rather than forming large aggregates. Lipids are hydrophobic substances: in the presence of water, they will aggregate to form globules to minimize exposure to water. Bile contains bile salts, which are amphipathic, meaning they contain hydrophobic and hydrophilic parts. Thus, the bile salts hydrophilic side can interface with water on one side and the hydrophobic side interfaces with lipids on the other. By doing so, bile salts emulsify large lipid globules into small lipid globules.

Why is emulsification important for digestion of lipids? Pancreatic juices contain enzymes called lipases (enzymes that breakdown lipids). If the lipid in the chyme aggregates into large globules, very little surface area of the lipids is available for the lipases to act on, leaving lipid digestion incomplete. By forming an emulsion, bile salts increase the available surface area of the lipids many fold. The pancreatic lipases can then act on the lipids more efficiently and digest them, as detailed in Figure 34.18. Lipases breakdown the lipids into fatty acids and glycerides. These molecules can pass through the plasma membrane of the cell and enter the epithelial cells of the intestinal lining. The bile salts surround long-chain fatty acids and monoglycerides forming tiny spheres called micelles. The micelles move into the brush border of the small intestine absorptive cells where the long-chain fatty acids and monoglycerides diffuse out of the micelles into the absorptive cells leaving the micelles behind in the chyme. The long-chain fatty acids and monoglycerides recombine in the absorptive cells to form triglycerides, which aggregate into globules and become coated with proteins. These large spheres are called **chylomicrons**. Chylomicrons contain triglycerides, cholesterol, and other lipids and have proteins on their surface. The surface is also composed of the hydrophilic phosphate "heads" of phospholipids. Together, they enable the chylomicron to move in an aqueous environment without exposing the lipids to water. Chylomicrons leave the absorptive cells via exocytosis. Chylomicrons enter the lymphatic vessels, and then enter the blood in the subclavian vein.

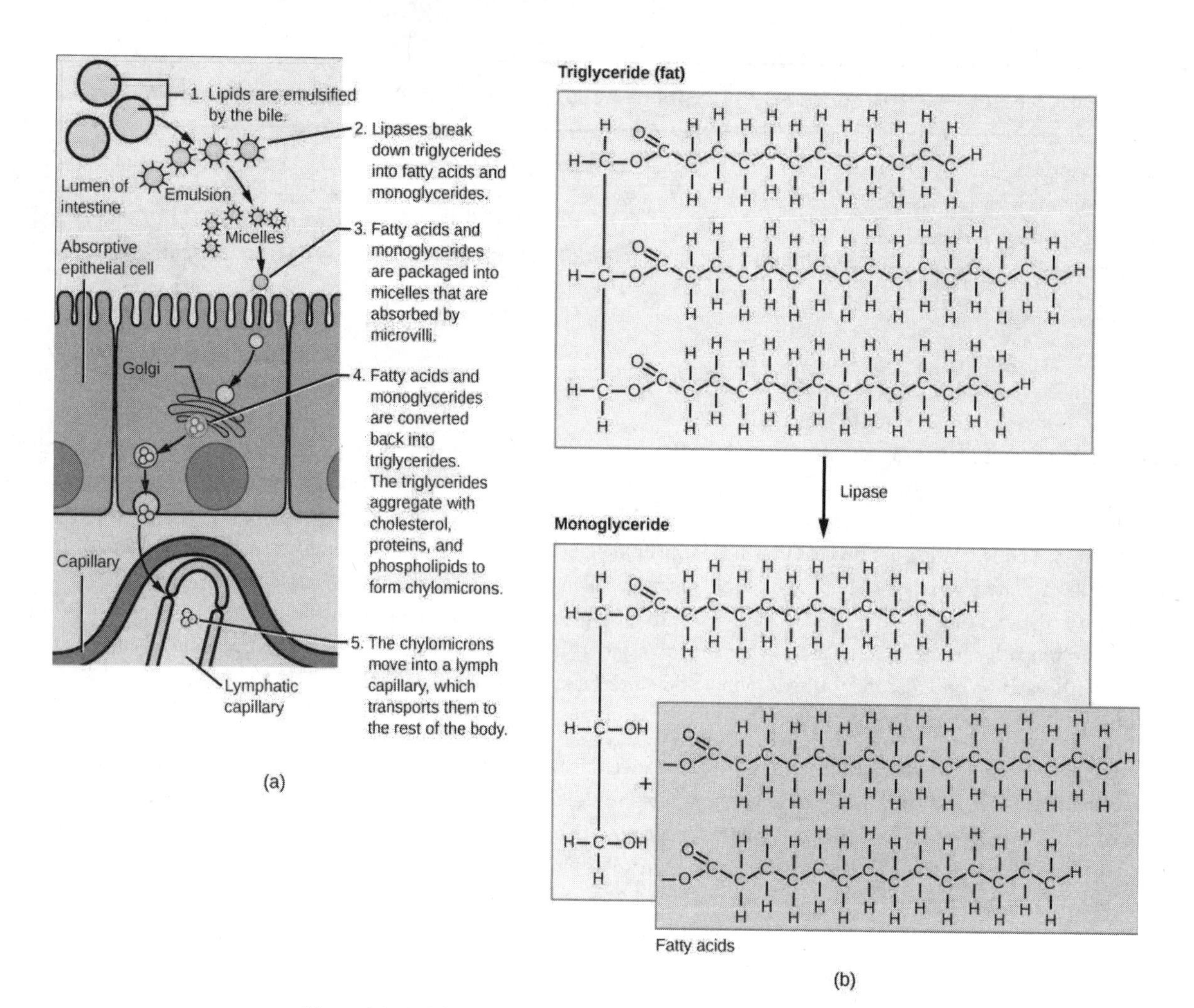

Figure 34.18 Lipids are digested and absorbed in the small intestine.

Vitamins

Vitamins can be either water-soluble or lipid-soluble. Fat soluble vitamins are absorbed in the same manner as lipids. It is important to consume some amount of dietary lipid to aid the absorption of lipid-soluble vitamins. Water-soluble vitamins can be directly absorbed into the bloodstream from the intestine.

LINK TO LEARNING

This website (http://openstax.org/l/digest_enzymes) has an overview of the digestion of protein, fat, and carbohydrates.

VISUAL CONNECTION

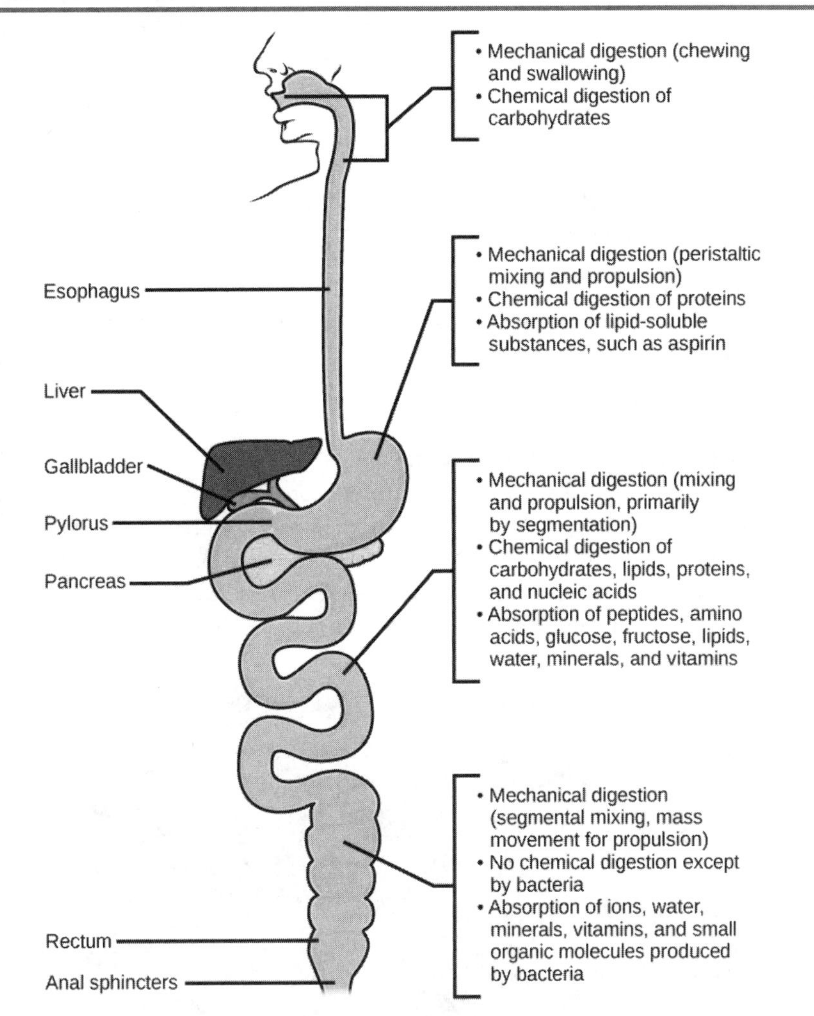

Figure 34.19 Mechanical and chemical digestion of food takes place in many steps, beginning in the mouth and ending in the rectum.

Which of the following statements about digestive processes is true?

a. Amylase, maltase, and lactase in the mouth digest carbohydrates.
b. Trypsin and lipase in the stomach digest protein.
c. Bile emulsifies lipids in the small intestine.
d. No food is absorbed until the small intestine.

Elimination

The final step in digestion is the elimination of undigested food content and waste products. The undigested food material enters the colon, where most of the water is reabsorbed. Recall that the colon is also home to the microflora called "intestinal flora" that aid in the digestion process. The semi-solid waste is moved through the colon by peristaltic movements of the muscle and is stored in the rectum. As the rectum expands in response to storage of fecal matter, it triggers the neural signals required to set up the urge to eliminate. The solid waste is eliminated through the anus using peristaltic movements of the rectum.

Common Problems with Elimination

Diarrhea and constipation are some of the most common health concerns that affect digestion. Constipation is a condition where the feces are hardened because of excess water removal in the colon. In contrast, if enough water is not removed from the feces, it results in diarrhea. Many bacteria, including the ones that cause cholera, affect the proteins involved in water reabsorption in the colon and result in excessive diarrhea.

Emesis

Emesis, or vomiting, is elimination of food by forceful expulsion through the mouth. It is often in response to an irritant that affects the digestive tract, including but not limited to viruses, bacteria, emotions, sights, and food poisoning. This forceful expulsion of the food is due to the strong contractions produced by the stomach muscles. The process of emesis is regulated by the medulla.

34.4 Digestive System Regulation

By the end of this section, you will be able to do the following:

- Discuss the role of neural regulation in digestive processes
- Explain how hormones regulate digestion

The brain is the control center for the sensation of hunger and satiety. The functions of the digestive system are regulated through neural and hormonal responses.

Neural Responses to Food

In reaction to the smell, sight, or thought of food, like that shown in Figure 34.20, the first response is that of salivation. The salivary glands secrete more saliva in response to stimulation by the autonomic nervous system triggered by food in preparation for digestion. Simultaneously, the stomach begins to produce hydrochloric acid to digest the food. Recall that the peristaltic movements of the esophagus and other organs of the digestive tract are under the control of the brain. The brain prepares these muscles for movement as well. When the stomach is full, the part of the brain that detects satiety signals fullness. There are three overlapping phases of gastric control—the cephalic phase, the gastric phase, and the intestinal phase—each requires many enzymes and is under neural control as well.

Figure 34.20 Seeing a plate of food triggers the secretion of saliva in the mouth and the production of HCL in the stomach. (credit: Kelly Bailey)

Digestive Phases

The response to food begins even before food enters the mouth. The first phase of ingestion, called the **cephalic phase**, is controlled by the neural response to the stimulus provided by food. All aspects—such as sight, sense, and smell—trigger the neural responses resulting in salivation and secretion of gastric juices. The gastric and salivary secretion in the cephalic phase can also take place due to the thought of food. Right now, if you think about a piece of chocolate or a crispy potato chip, the increase in salivation is a cephalic phase response to the thought. The central nervous system prepares the stomach to receive food.

The **gastric phase** begins once the food arrives in the stomach. It builds on the stimulation provided during the cephalic phase. Gastric acids and enzymes process the ingested materials. The gastric phase is stimulated by (1) distension of the stomach, (2) a decrease in the pH of the gastric contents, and (3) the presence of undigested material. This phase consists of local, hormonal, and neural responses. These responses stimulate secretions and powerful contractions.

The **intestinal phase** begins when chyme enters the small intestine triggering digestive secretions. This phase controls the rate of gastric emptying. In addition to gastrin emptying, when chyme enters the small intestine, it triggers other hormonal and neural events that coordinate the activities of the intestinal tract, pancreas, liver, and gallbladder.

Hormonal Responses to Food

The **endocrine system** controls the response of the various glands in the body and the release of hormones at the appropriate times.

One of the important factors under hormonal control is the stomach acid environment. During the gastric phase, the hormone **gastrin** is secreted by G cells in the stomach in response to the presence of proteins. Gastrin stimulates the release of stomach acid, or hydrochloric acid (HCl) which aids in the digestion of the proteins. However, when the stomach is emptied, the acidic environment need not be maintained and a hormone called **somatostatin** stops the release of hydrochloric acid. This is controlled by a negative feedback mechanism.

In the duodenum, digestive secretions from the liver, pancreas, and gallbladder play an important role in digesting chyme during the intestinal phase. In order to neutralize the acidic chyme, a hormone called **secretin** stimulates the pancreas to produce alkaline bicarbonate solution and deliver it to the duodenum. Secretin acts in tandem with another hormone called **cholecystokinin** (CCK). Not only does CCK stimulate the pancreas to produce the requisite pancreatic juices, it also stimulates the gallbladder to release bile into the duodenum.

LINK TO LEARNING

Visit this website (http://openstax.org/l/enteric_endo) to learn more about the endocrine system. Review the text and watch the animation of how control is implemented in the endocrine system.

Another level of hormonal control occurs in response to the composition of food. Foods high in lipids take a long time to digest. A hormone called **gastric inhibitory peptide** is secreted by the small intestine to slow down the peristaltic movements of the intestine to allow fatty foods more time to be digested and absorbed.

Understanding the hormonal control of the digestive system is an important area of ongoing research. Scientists are exploring the role of each hormone in the digestive process and developing ways to target these hormones. Advances could lead to knowledge that may help to battle the obesity epidemic.

KEY TERMS

alimentary canal tubular digestive system with a mouth and anus

aminopeptidase protease that breaks down peptides to single amino acids; secreted by the brush border of small intestine

anus exit point for waste material

bile digestive juice produced by the liver; important for digestion of lipids

bolus mass of food resulting from chewing action and wetting by saliva

carboxypeptidase protease that breaks down peptides to single amino acids; secreted by the brush border of the small intestine

carnivore animal that consumes animal flesh

cephalic phase first phase of digestion, controlled by the neural response to the stimulus provided by food

cholecystokinin hormone that stimulates the contraction of the gallbladder to release bile

chylomicron small lipid globule

chyme mixture of partially digested food and stomach juices

chymotrypsin pancreatic protease

digestion mechanical and chemical breakdown of food into small organic fragments

dipeptidase protease that breaks down peptides to single amino acids; secreted by the brush border of small intestine

duodenum first part of the small intestine where a large part of digestion of carbohydrates and fats occurs

elastase pancreatic protease

endocrine system system that controls the response of the various glands in the body and the release of hormones at the appropriate times

esophagus tubular organ that connects the mouth to the stomach

essential nutrient nutrient that cannot be synthesized by the body; it must be obtained from food

gallbladder organ that stores and concentrates bile

gastric inhibitory peptide hormone secreted by the small intestine in the presence of fatty acids and sugars; it also inhibits acid production and peristalsis in order to slow down the rate at which food enters the small intestine

gastric phase digestive phase beginning once food enters the stomach; gastric acids and enzymes process the ingested materials

gastrin hormone which stimulates hydrochloric acid secretion in the stomach

gastrovascular cavity digestive system consisting of a single opening

gizzard muscular organ that grinds food

herbivore animal that consumes a strictly plant diet

ileum last part of the small intestine; connects the small intestine to the large intestine; important for absorption of B-12

ingestion act of taking in food

intestinal phase third digestive phase; begins when chyme enters the small intestine triggering digestive secretions and controlling the rate of gastric emptying

jejunum second part of the small intestine

lactase enzyme that breaks down lactose into glucose and galactose

large intestine digestive system organ that reabsorbs water from undigested material and processes waste matter

lipase enzyme that chemically breaks down lipids

liver organ that produces bile for digestion and processes vitamins and lipids

maltase enzyme that breaks down maltose into glucose

mineral inorganic, elemental molecule that carries out important roles in the body

monogastric digestive system that consists of a single-chambered stomach

omnivore animal that consumes both plants and animals

pancreas gland that secretes digestive juices

pepsin enzyme found in the stomach whose main role is protein digestion

pepsinogen inactive form of pepsin

peristalsis wave-like movements of muscle tissue

proventriculus glandular part of a bird's stomach

rectum area of the body where feces is stored until elimination

roughage component of food that is low in energy and high in fiber

ruminant animal with a stomach divided into four compartments

salivary amylase enzyme found in saliva, which converts carbohydrates to maltose

secretin hormone which stimulates sodium bicarbonate secretion in the small intestine

small intestine organ where digestion of protein, fats, and carbohydrates is completed

somatostatin hormone released to stop acid secretion when the stomach is empty

sphincter band of muscle that controls movement of materials throughout the digestive tract

stomach saclike organ containing acidic digestive juices

sucrase enzyme that breaks down sucrose into glucose and fructose

trypsin pancreatic protease that breaks down protein

villi folds on the inner surface of the small intestine whose role is to increase absorption area

vitamin organic substance necessary in small amounts to sustain life

CHAPTER SUMMARY

34.1 Digestive Systems

Different animals have evolved different types of digestive systems specialized to meet their dietary needs. Humans and many other animals have monogastric digestive systems with a single-chambered stomach. Birds have evolved a digestive system that includes a gizzard where the food is crushed into smaller pieces. This compensates for their inability to masticate. Ruminants that consume large amounts of plant material have a multi-chambered stomach that digests roughage. Pseudo-ruminants have similar digestive processes as ruminants but do not have the four-compartment stomach. Processing food involves ingestion (eating), digestion (mechanical and enzymatic breakdown of large molecules), absorption (cellular uptake of nutrients), and elimination (removal of undigested waste as feces).

Many organs work together to digest food and absorb nutrients. The mouth is the point of ingestion and the location where both mechanical and chemical breakdown of food begins. Saliva contains an enzyme called amylase that breaks down carbohydrates. The food bolus travels through the esophagus by peristaltic movements to the stomach. The stomach has an extremely acidic environment. An enzyme called pepsin digests protein in the stomach. Further digestion and absorption take place in the small intestine. The large intestine reabsorbs water from the undigested food and stores waste until elimination.

34.2 Nutrition and Energy Production

Animal diet should be balanced and meet the needs of the body. Carbohydrates, proteins, and fats are the primary components of food. Some essential nutrients are required for cellular function but cannot be produced by the animal body. These include vitamins, minerals, some fatty acids, and some amino acids. Food intake in more than necessary amounts is stored as glycogen in the liver and muscle cells, and in fat cells. Excess adipose storage can lead to obesity and serious health problems. ATP is the energy currency of the cell and is obtained from the metabolic pathways. Excess carbohydrates and energy are stored as glycogen in the body.

34.3 Digestive System Processes

Digestion begins with ingestion, where the food is taken in the mouth. Digestion and absorption take place in a series of steps with special enzymes playing important roles in digesting carbohydrates, proteins, and lipids. Elimination describes removal of undigested food contents and waste products from the body. While most absorption occurs in the small intestines, the large intestine is responsible for the final removal of water that remains after the absorptive process of the small intestines. The cells that line the large intestine absorb some vitamins as well as any leftover salts and water. The large intestine (colon) is also where feces is formed.

34.4 Digestive System Regulation

The brain and the endocrine system control digestive processes. The brain controls the responses of hunger and satiety. The endocrine system controls the release of hormones and enzymes required for digestion of food in the digestive tract.

VISUAL CONNECTION QUESTIONS

1. Figure 34.11 Which of the following statements about the digestive system is false?
   a. Chyme is a mixture of food and digestive juices that is produced in the stomach.
   b. Food enters the large intestine before the small intestine.
   c. In the small intestine, chyme mixes with bile, which emulsifies fats.
   d. The stomach is separated from the small intestine by the pyloric sphincter.

2. Figure 34.12 Which of the following statements about the small intestine is false?
   a. Absorptive cells that line the small intestine have microvilli, small projections that increase surface area and aid in the absorption of food.
   b. The inside of the small intestine has many folds, called villi.
   c. Microvilli are lined with blood vessels as well as lymphatic vessels.
   d. The inside of the small intestine is called the lumen.

3. Figure 34.19 Which of the following statements about digestive processes is true?
    a. Amylase, maltase, and lactase in the mouth digest carbohydrates.
    b. Trypsin and lipase in the stomach digest protein.
    c. Bile emulsifies lipids in the small intestine.
    d. No food is absorbed until the small intestine.

REVIEW QUESTIONS

4. Which of the following is a pseudo-ruminant?
    a. cow
    b. pig
    c. crow
    d. horse

5. Which of the following statements is untrue?
    a. Roughage takes a long time to digest.
    b. Birds eat large quantities at one time so that they can fly long distances.
    c. Cows do not have upper teeth.
    d. In pseudo-ruminants, roughage is digested in the cecum.

6. The acidic nature of chyme is neutralized by _______.
    a. potassium hydroxide
    b. sodium hydroxide
    c. bicarbonates
    d. vinegar

7. The digestive juices from the liver are delivered to the _______.
    a. stomach
    b. liver
    c. duodenum
    d. colon

8. A scientist dissects a new species of animal. If the animal's digestive system has a single stomach with an extended small intestine, to which animal could the dissected specimen be closely related?
    a. lion
    b. snowshoe hare
    c. earthworm
    d. eagle

9. Which of the following statements is not true?
    a. Essential nutrients can be synthesized by the body.
    b. Vitamins are required in small quantities for bodily function.
    c. Some amino acids can be synthesized by the body, while others need to be obtained from diet.
    d. Vitamins come in two categories: fat-soluble and water-soluble.

10. Which of the following is a water-soluble vitamin?
    a. vitamin A
    b. vitamin E
    c. vitamin K
    d. vitamin C

11. What is the primary fuel for the body?
    a. carbohydrates
    b. lipids
    c. protein
    d. glycogen

12. Excess glucose is stored as _______.
    a. fat
    b. glucagon
    c. glycogen
    d. it is not stored in the body

13. Many distance runners "carb load" the day before a big race. How does this eating strategy provide an advantage to the runner?
    a. The carbohydrates cause the release of insulin.
    b. The excess carbohydrates are converted to fats, which have a higher calorie density.
    c. The glucose from the carbohydrates lets the muscles make excess ATP overnight.
    d. The excess carbohydrates can be stored in the muscles as glycogen.

14. Where does the majority of protein digestion take place?
    a. stomach
    b. duodenum
    c. mouth
    d. jejunum

15. Lipases are enzymes that breakdown _______.
    a. disaccharides
    b. lipids
    c. proteins
    d. cellulose

16. Which of the following conditions is most likely to cause constipation?
    a. bacterial infection
    b. dehydration
    c. ulcer
    d. excessive cellulose consumption

17. Which hormone controls the release of bile from the gallbladder
    a. pepsin
    b. amylase
    c. CCK
    d. gastrin

18. Which hormone stops acid secretion in the stomach?
    a. gastrin
    b. somatostatin
    c. gastric inhibitory peptide
    d. CCK

19. In the famous conditioning experiment, Pavlov demonstrated that his dogs started drooling in response to a bell sounding. What part of the digestive process did he stimulate?
    a. cephalic phase
    b. gastric phase
    c. intestinal phase
    d. elimination phase

CRITICAL THINKING QUESTIONS

20. How does the polygastric digestive system aid in digesting roughage?
21. How do birds digest their food in the absence of teeth?
22. What is the role of the accessory organs in digestion?
23. Explain how the villi and microvilli aid in absorption.
24. Name two components of the digestive system that perform mechanical digestion. Describe how mechanical digestion contributes to acquiring nutrients from food.
25. What are essential nutrients?
26. What is the role of minerals in maintaining good health?
27. Discuss why obesity is a growing epidemic.
28. There are several nations where malnourishment is a common occurrence. What may be some of the health challenges posed by malnutrition?
29. Generally describe how a piece of bread can power your legs as you walk up a flight of stairs.
30. In the 1990s fat-free foods became popular among people trying to lose weight. However, many dieticians now conclude that the fat-free trend made people less healthy and heavier. Describe how this could occur.
31. Explain why some dietary lipid is a necessary part of a balanced diet.
32. The gut microbiome (the bacterial colonies in the intestines) have become a popular area of study in biomedical research. How could varying gut microbiomes impact a person's nutrition?
33. Many mammals become ill if they drink milk as adults even though they could consume it as babies. What causes this digestive issue?
34. Describe how hormones regulate digestion.
35. Describe one or more scenarios where loss of hormonal regulation of digestion can lead to diseases.
36. A scientist is studying a model that has a mutation in the receptor for somatostatin that prevents hormone binding. How would this mutation affect the structure and function of the digestive system?

CHAPTER 35
The Nervous System

Figure 35.1 An athlete's nervous system is hard at work during the planning and execution of a movement as precise as a high jump. Parts of the nervous system are involved in determining how hard to push off and when to turn, as well as controlling the muscles throughout the body that make this complicated movement possible without knocking the bar down—all in just a few seconds. (credit: Scott Ray)

INTRODUCTION When you're reading this book, your nervous system is performing several functions simultaneously. The visual system is processing what is seen on the page; the motor system controls the turn of the pages (or click of the mouse); the prefrontal cortex maintains attention. Even fundamental functions, like breathing and regulation of body temperature, are controlled by the nervous system. A nervous system is an organism's control center: it processes sensory information from outside (and inside) the body and controls all behaviors—from eating to sleeping to finding a mate.

Chapter Outline

35.1 Neurons and Glial Cells

By the end of this section, you will be able to do the following:

- List and describe the functions of the structural components of a neuron
- List and describe the four main types of neurons
- Compare the functions of different types of glial cells

Nervous systems throughout the animal kingdom vary in structure and complexity, as illustrated by the variety of animals shown in Figure 35.2. Some organisms, like sea sponges, lack a true nervous system. Others, like jellyfish, lack a true brain and instead have a system of separate but connected nerve cells (neurons) called a "nerve net." Echinoderms such as sea stars have nerve cells that are bundled into fibers called nerves. Flatworms of the phylum Platyhelminthes have both a central nervous system (CNS), made up of a small "brain" and two nerve cords, and a peripheral nervous system (PNS) containing a system of nerves that extend throughout the body. The insect nervous system is more complex but also fairly decentralized. It contains a brain, ventral nerve cord, and ganglia (clusters of connected neurons). These ganglia can control movements and behaviors without input from the brain. Octopi may have the most complicated of invertebrate nervous systems—they have neurons that are organized in specialized lobes and eyes that are structurally similar to vertebrate species.

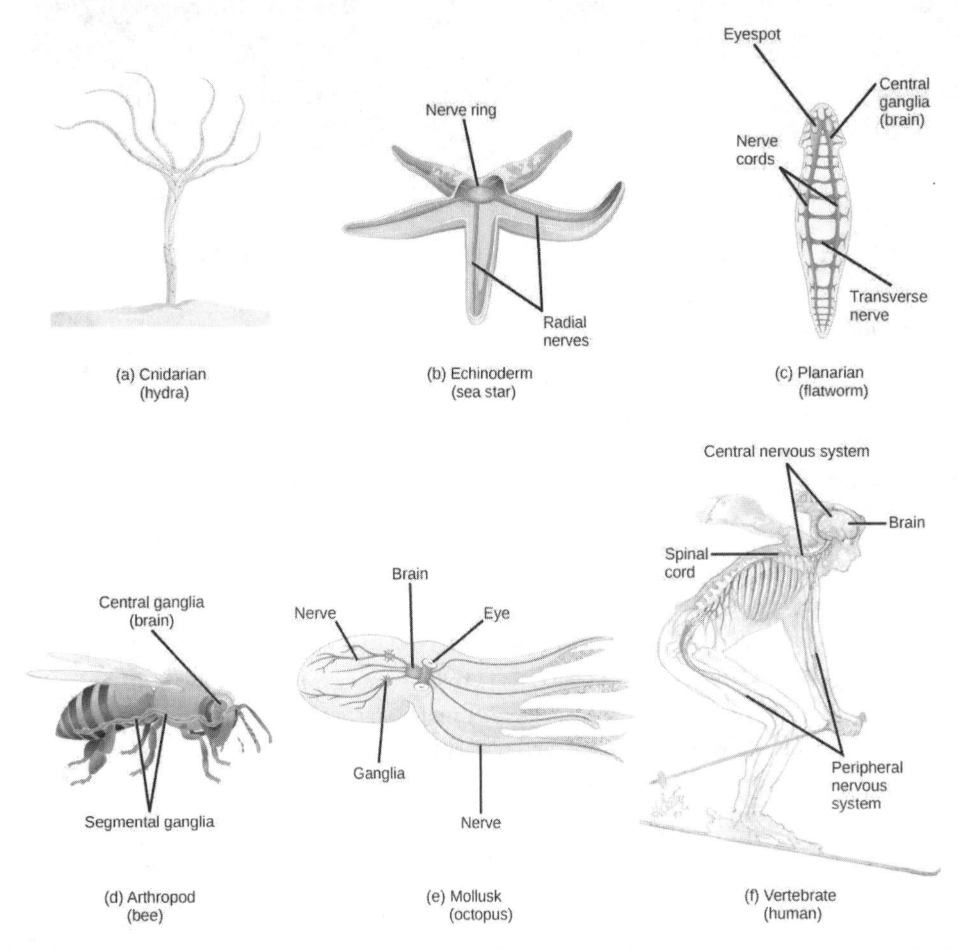

Figure 35.2 Nervous systems vary in structure and complexity. In (a) cnidarians, nerve cells form a decentralized nerve net. In (b) echinoderms, nerve cells are bundled into fibers called nerves. In animals exhibiting bilateral symmetry such as (c) planarians, neurons cluster into an anterior brain that processes information. In addition to a brain, (d) arthropods have clusters of nerve cell bodies, called peripheral ganglia, located along the ventral nerve cord. Mollusks such as squid and (e) octopi, which must hunt to survive, have complex brains containing millions of neurons. In (f) vertebrates, the brain and spinal cord comprise the central nervous system, while neurons extending into the rest of the body comprise the peripheral nervous system. (credit e: modification of work by Michael Vecchione, Clyde F.E. Roper, and Michael J. Sweeney, NOAA; credit f: modification of work by NIH)

Compared to invertebrates, vertebrate nervous systems are more complex, centralized, and specialized. While there is great diversity among different vertebrate nervous systems, they all share a basic structure: a CNS that contains a brain and spinal cord and a PNS made up of peripheral sensory and motor nerves. One interesting difference between the nervous systems of invertebrates and vertebrates is that the nerve cords of many invertebrates are located ventrally whereas the vertebrate spinal cords are located dorsally. There is debate among evolutionary biologists as to whether these different nervous system plans evolved separately or whether the invertebrate body plan arrangement somehow "flipped" during the evolution of vertebrates.

LINK TO LEARNING

Watch this video of biologist Mark Kirschner discussing the "flipping" phenomenon of vertebrate evolution.

Click to view content (https://www.openstax.org/l/vertebrate_evol)

The nervous system is made up of **neurons**, specialized cells that can receive and transmit chemical or electrical signals, and **glia**, cells that provide support functions for the neurons by playing an information processing role that is complementary to neurons. A neuron can be compared to an electrical wire—it transmits a signal from one place to another. Glia can be compared to the workers at the electric company who make sure wires go to the right places, maintain the wires, and take down wires that are broken. Although glia have been compared to workers, recent evidence suggests that they also usurp some of the signaling functions of neurons.

There is great diversity in the types of neurons and glia that are present in different parts of the nervous system. There are four major types of neurons, and they share several important cellular components.

Neurons

The nervous system of the common laboratory fly, *Drosophila melanogaster*, contains around 100,000 neurons, the same number as a lobster. This number compares to 75 million in the mouse and 300 million in the octopus. A human brain contains around 86 billion neurons. Despite these very different numbers, the nervous systems of these animals control many of the same behaviors—from basic reflexes to more complicated behaviors like finding food and courting mates. The ability of neurons to communicate with each other as well as with other types of cells underlies all of these behaviors.

Most neurons share the same cellular components. But neurons are also highly specialized—different types of neurons have different sizes and shapes that relate to their functional roles.

Parts of a Neuron

Like other cells, each neuron has a cell body (or soma) that contains a nucleus, smooth and rough endoplasmic reticulum, Golgi apparatus, mitochondria, and other cellular components. Neurons also contain unique structures, illustrated in Figure 35.3 for receiving and sending the electrical signals that make neuronal communication possible. **Dendrites** are tree-like structures that extend away from the cell body to receive messages from other neurons at specialized junctions called **synapses**. Although some neurons do not have any dendrites, some types of neurons have multiple dendrites. Dendrites can have small protrusions called dendritic spines, which further increase surface area for possible synaptic connections.

Once a signal is received by the dendrite, it then travels passively to the cell body. The cell body contains a specialized structure, the **axon hillock** that integrates signals from multiple synapses and serves as a junction between the cell body and an **axon**. An axon is a tube-like structure that propagates the integrated signal to specialized endings called **axon terminals**. These terminals in turn synapse on other neurons, muscle, or target organs. Chemicals released at axon terminals allow signals to be communicated to these other cells. Neurons usually have one or two axons, but some neurons, like amacrine cells in the retina, do not contain any axons. Some axons are covered with **myelin**, which acts as an insulator to minimize dissipation of the electrical signal as it travels down the axon, greatly increasing the speed of conduction. This insulation is important as the axon from a human motor neuron can be as long as a meter—from the base of the spine to the toes. The myelin sheath is not actually part of the neuron. Myelin is produced by glial cells. Along the axon there are periodic gaps in the myelin sheath. These gaps are called **nodes of Ranvier** and are sites where the signal is "recharged" as it travels along the axon.

It is important to note that a single neuron does not act alone—neuronal communication depends on the connections that neurons make with one another (as well as with other cells, like muscle cells). Dendrites from a single neuron may receive synaptic contact from many other neurons. For example, dendrites from a Purkinje cell in the cerebellum are thought to receive contact from as many as 200,000 other neurons.

VISUAL CONNECTION

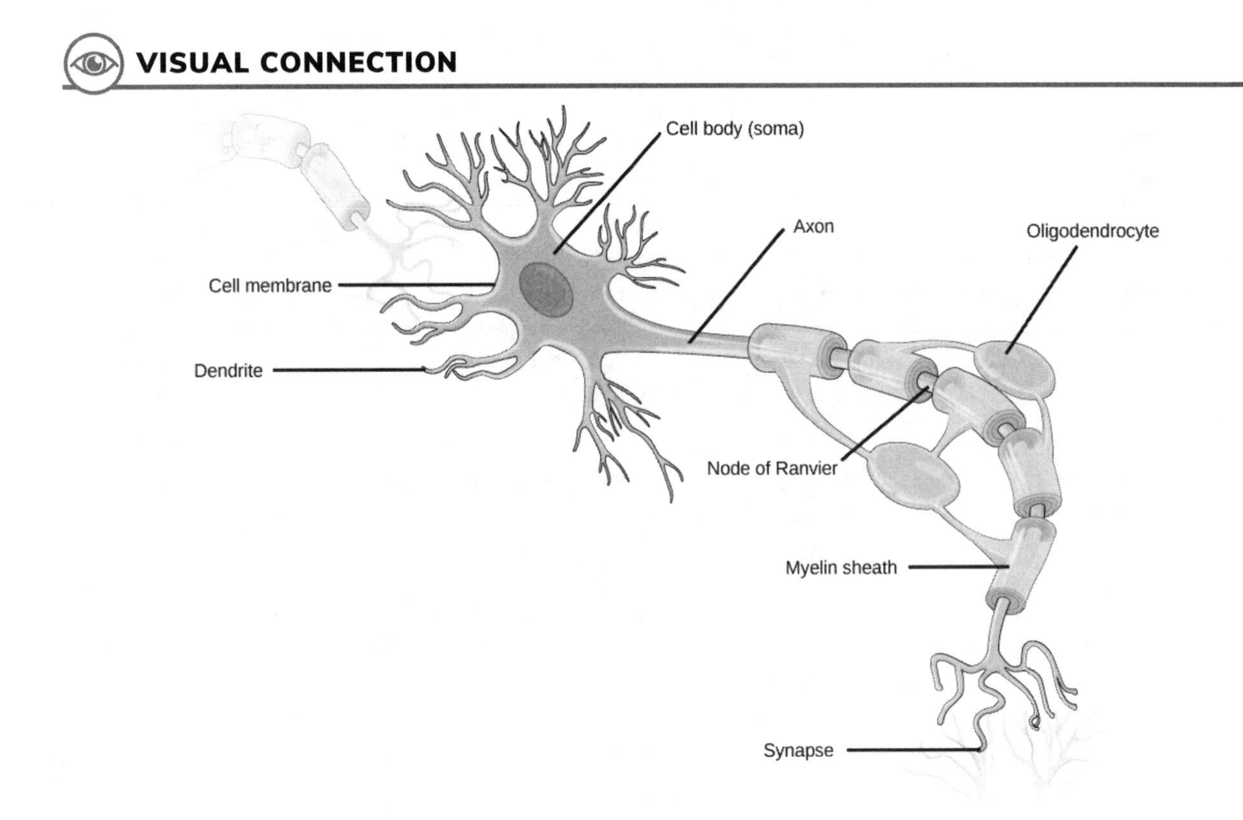

Figure 35.3 Neurons contain organelles common to many other cells, such as a nucleus and mitochondria. They also have more specialized structures, including dendrites and axons.

Which of the following statements is false?

a. The soma is the cell body of a nerve cell.
b. Myelin sheath provides an insulating layer to the dendrites.
c. Axons carry the signal from the soma to the target.
d. Dendrites carry the signal to the soma.

Types of Neurons

There are different types of neurons, and the functional role of a given neuron is intimately dependent on its structure. There is an amazing diversity of neuron shapes and sizes found in different parts of the nervous system (and across species), as illustrated by the neurons shown in Figure 35.4.

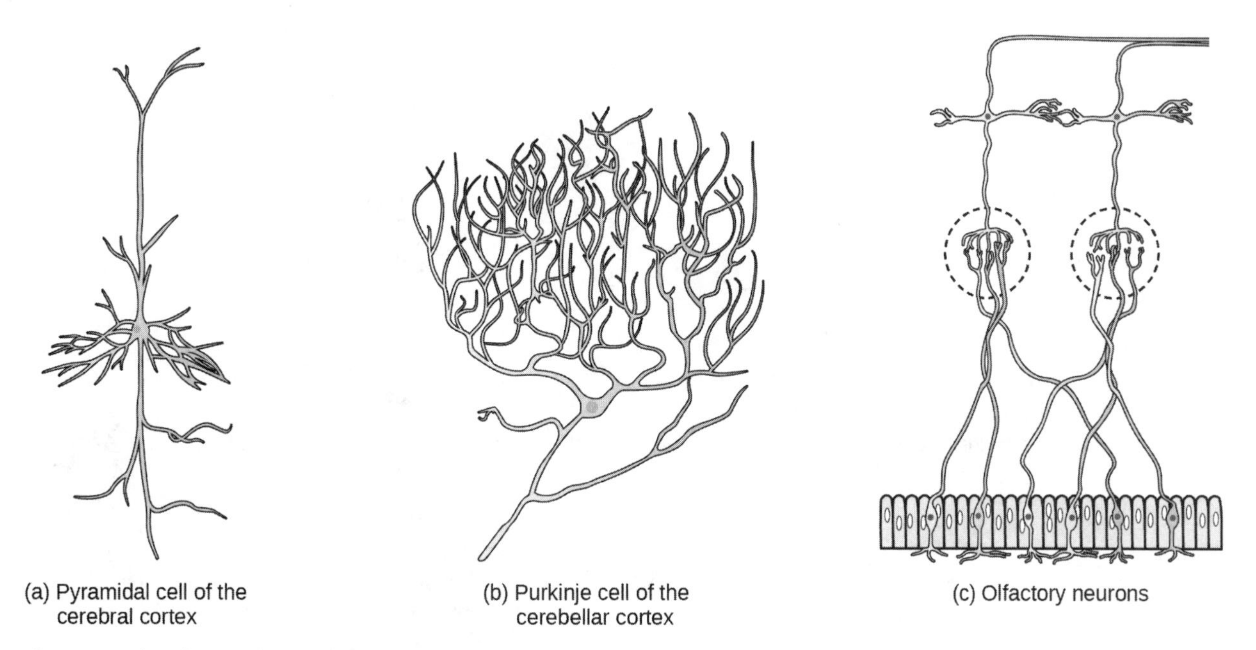

Figure 35.4 There is great diversity in the size and shape of neurons throughout the nervous system. Examples include (a) a pyramidal cell from the cerebral cortex, (b) a Purkinje cell from the cerebellar cortex, and (c) olfactory cells from the olfactory epithelium and olfactory bulb.

While there are many defined neuron cell subtypes, neurons are broadly divided into four basic types: unipolar, bipolar, multipolar, and pseudounipolar. Figure 35.5 illustrates these four basic neuron types. Unipolar neurons have only one structure that extends away from the soma. These neurons are not found in vertebrates but are found in insects where they stimulate muscles or glands. A bipolar neuron has one axon and one dendrite extending from the soma. An example of a bipolar neuron is a retinal bipolar cell, which receives signals from photoreceptor cells that are sensitive to light and transmits these signals to ganglion cells that carry the signal to the brain. Multipolar neurons are the most common type of neuron. Each multipolar neuron contains one axon and multiple dendrites. Multipolar neurons can be found in the central nervous system (brain and spinal cord). An example of a multipolar neuron is a Purkinje cell in the cerebellum, which has many branching dendrites but only one axon. Pseudounipolar cells share characteristics with both unipolar and bipolar cells. A pseudounipolar cell has a single process that extends from the soma, like a unipolar cell, but this process later branches into two distinct structures, like a bipolar cell. Most sensory neurons are pseudounipolar and have an axon that branches into two extensions: one connected to dendrites that receive sensory information and another that transmits this information to the spinal cord.

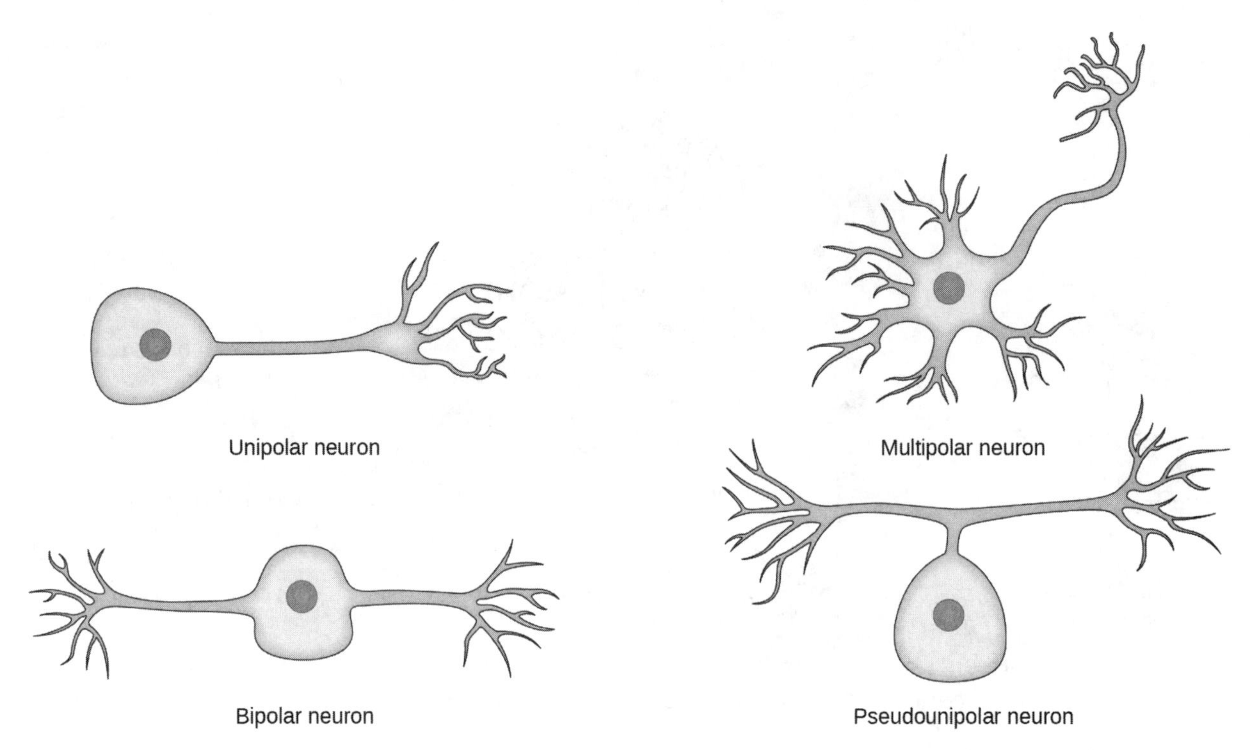

Figure 35.5 Neurons are broadly divided into four main types based on the number and placement of axons: (1) unipolar, (2) bipolar, (3) multipolar, and (4) pseudounipolar.

Everyday Connection

Neurogenesis

At one time, scientists believed that people were born with all the neurons they would ever have. Research performed during the last few decades indicates that neurogenesis, the birth of new neurons, continues into adulthood. Neurogenesis was first discovered in songbirds that produce new neurons while learning songs. For mammals, new neurons also play an important role in learning: about 1000 new neurons develop in the hippocampus (a brain structure involved in learning and memory) each day. While most of the new neurons will die, researchers found that an increase in the number of surviving new neurons in the hippocampus correlated with how well rats learned a new task. Interestingly, both exercise and some antidepressant medications also promote neurogenesis in the hippocampus. Stress has the opposite effect. While neurogenesis is quite limited compared to regeneration in other tissues, research in this area may lead to new treatments for disorders such as Alzheimer's, stroke, and epilepsy.

How do scientists identify new neurons? A researcher can inject a compound called bromodeoxyuridine (BrdU) into the brain of an animal. While all cells will be exposed to BrdU, BrdU will only be incorporated into the DNA of newly generated cells that are in S phase. A technique called immunohistochemistry can be used to attach a fluorescent label to the incorporated BrdU, and a researcher can use fluorescent microscopy to visualize the presence of BrdU, and thus new neurons, in brain tissue. Figure 35.6 is a micrograph which shows fluorescently labeled neurons in the hippocampus of a rat.

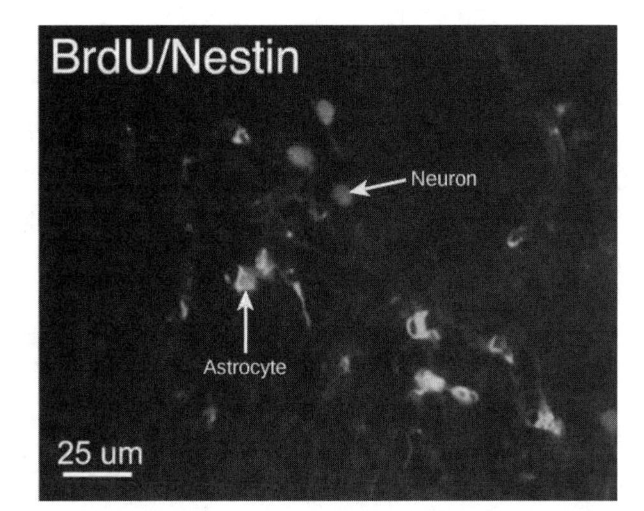

Figure 35.6 This micrograph shows fluorescently labeled new neurons in a rat hippocampus. Cells that are actively dividing have bromodoxyuridine (BrdU) incorporated into their DNA and are labeled in red. Cells that express glial fibrillary acidic protein (GFAP) are labeled in green. Astrocytes, but not neurons, express GFAP. Thus, cells that are labeled both red and green are actively dividing astrocytes, whereas cells labeled red only are actively dividing neurons. (credit: modification of work by Dr. Maryam Faiz, et. al., University of Barcelona; scale-bar data from Matt Russell)

LINK TO LEARNING

This site (http://openstax.org/l/neurogenesis) contains more information about neurogenesis, including an interactive laboratory simulation and a video that explains how BrdU labels new cells.

Glia

While glia are often thought of as the supporting cast of the nervous system, the number of glial cells in the brain actually outnumbers the number of neurons by a factor of ten. Neurons would be unable to function without the vital roles that are fulfilled by these glial cells. Glia guide developing neurons to their destinations, buffer ions and chemicals that would otherwise harm neurons, and provide myelin sheaths around axons. Scientists have recently discovered that they also play a role in responding to nerve activity and modulating communication between nerve cells. When glia do not function properly, the result can be disastrous—most brain tumors are caused by mutations in glia.

Types of Glia

There are several different types of glia with different functions, two of which are shown in Figure 35.7. **Astrocytes**, shown in Figure 35.8a make contact with both capillaries and neurons in the CNS. They provide nutrients and other substances to neurons, regulate the concentrations of ions and chemicals in the extracellular fluid, and provide structural support for synapses. Astrocytes also form the blood-brain barrier—a structure that blocks entrance of toxic substances into the brain. Astrocytes, in particular, have been shown through calcium imaging experiments to become active in response to nerve activity, transmit calcium waves between astrocytes, and modulate the activity of surrounding synapses. **Satellite glia** provide nutrients and structural support for neurons in the PNS. **Microglia** scavenge and degrade dead cells and protect the brain from invading microorganisms. **Oligodendrocytes**, shown in Figure 35.8b form myelin sheaths around axons in the CNS. One axon can be myelinated by several oligodendrocytes, and one oligodendrocyte can provide myelin for multiple neurons. This is distinctive from the PNS where a single **Schwann cell** provides myelin for only one axon as the entire Schwann cell surrounds the axon. **Radial glia** serve as scaffolds for developing neurons as they migrate to their end destinations. **Ependymal** cells line fluid-filled ventricles of the brain and the central canal of the spinal cord. They are involved in the production of cerebrospinal fluid, which serves as a cushion for the brain, moves the fluid between the spinal cord and the brain, and is a component for the choroid plexus.

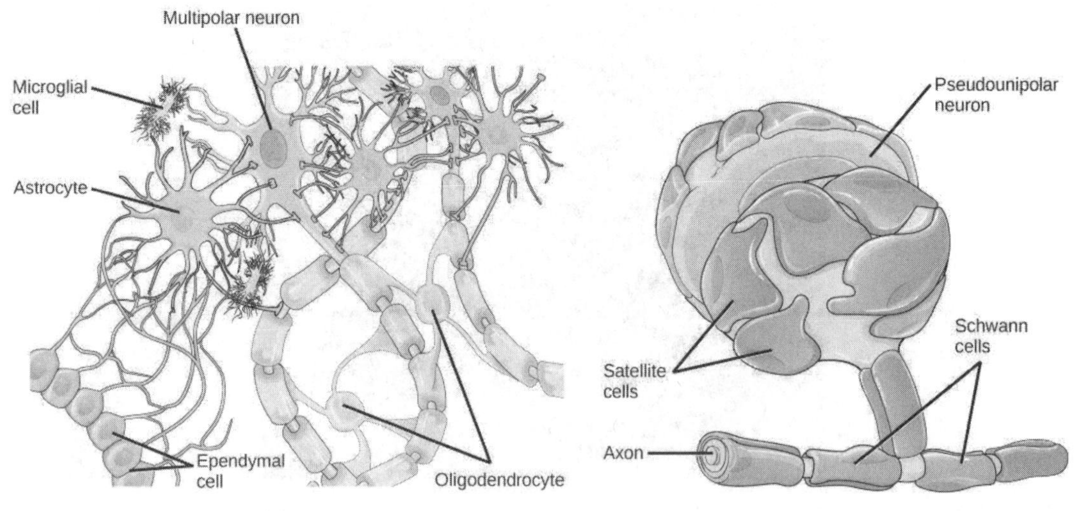

Figure 35.7 Glial cells support neurons and maintain their environment. Glial cells of the (a) central nervous system include oligodendrocytes, astrocytes, ependymal cells, and microglial cells. Oligodendrocytes form the myelin sheath around axons. Astrocytes provide nutrients to neurons, maintain their extracellular environment, and provide structural support. Microglia scavenge pathogens and dead cells. Ependymal cells produce cerebrospinal fluid that cushions the neurons. Glial cells of the (b) peripheral nervous system include Schwann cells, which form the myelin sheath, and satellite cells, which provide nutrients and structural support to neurons.

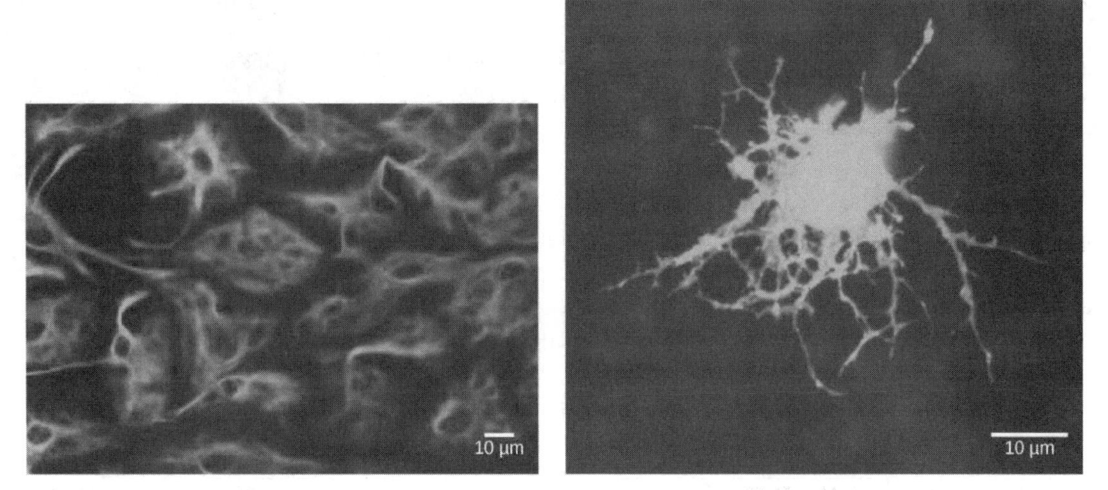

Figure 35.8 (a) Astrocytes and (b) oligodendrocytes are glial cells of the central nervous system. (credit a: modification of work by Uniformed Services University; credit b: modification of work by Jurjen Broeke; scale-bar data from Matt Russell)

35.2 How Neurons Communicate

By the end of this section, you will be able to do the following:

- Describe the basis of the resting membrane potential
- Explain the stages of an action potential and how action potentials are propagated
- Explain the similarities and differences between chemical and electrical synapses
- Describe long-term potentiation and long-term depression

All functions performed by the nervous system—from a simple motor reflex to more advanced functions like making a memory or a decision—require neurons to communicate with one another. While humans use words and body language to communicate, neurons use electrical and chemical signals. Just like a person in a committee, one neuron usually receives and synthesizes messages from multiple other neurons before "making the decision" to send the message on to other neurons.

Nerve Impulse Transmission within a Neuron

For the nervous system to function, neurons must be able to send and receive signals. These signals are possible because each neuron has a charged cellular membrane (a voltage difference between the inside and the outside), and the charge of this membrane can change in response to neurotransmitter molecules released from other neurons and environmental stimuli. To understand how neurons communicate, one must first understand the basis of the baseline or 'resting' membrane charge.

Neuronal Charged Membranes

The lipid bilayer membrane that surrounds a neuron is impermeable to charged molecules or ions. To enter or exit the neuron, ions must pass through special proteins called ion channels that span the membrane. Ion channels have different configurations: open, closed, and inactive, as illustrated in Figure 35.9. Some ion channels need to be activated in order to open and allow ions to pass into or out of the cell. These ion channels are sensitive to the environment and can change their shape accordingly. Ion channels that change their structure in response to voltage changes are called voltage-gated ion channels. Voltage-gated ion channels regulate the relative concentrations of different ions inside and outside the cell. The difference in total charge between the inside and outside of the cell is called the **membrane potential**.

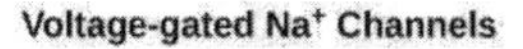

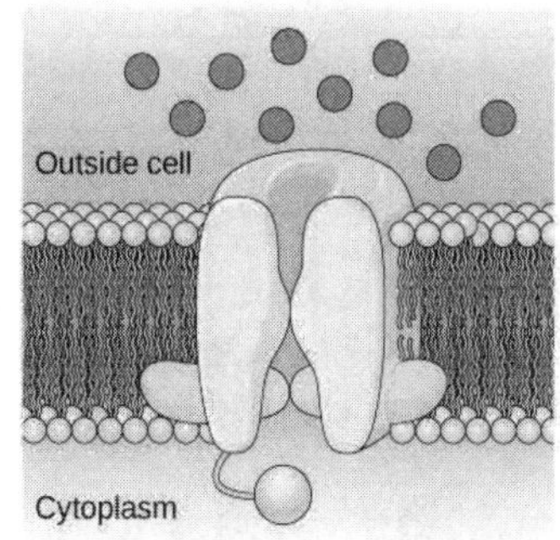

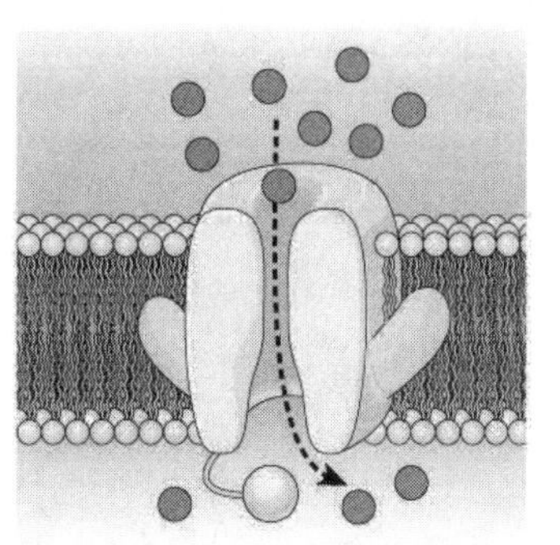

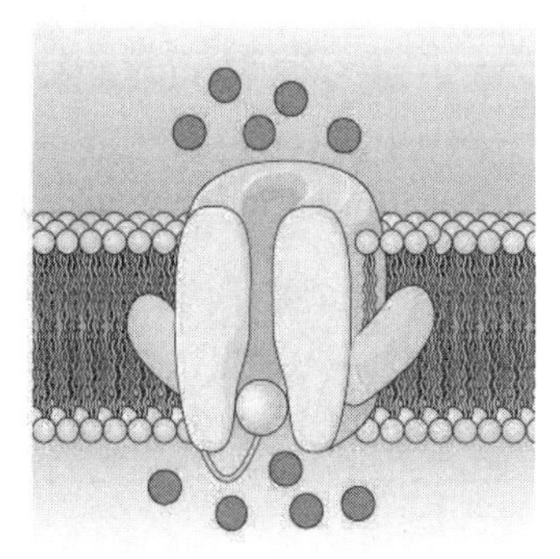

Figure 35.9 Voltage-gated ion channels open in response to changes in membrane voltage. After activation, they become inactivated for a brief period and will no longer open in response to a signal.

LINK TO LEARNING

This video discusses the basis of the resting membrane potential.

Click to view content (https://www.openstax.org/l/resting_neuron)

Resting Membrane Potential

A neuron at rest is negatively charged: the inside of a cell is approximately 70 millivolts more negative than the outside (–70 mV, note that this number varies by neuron type and by species). This voltage is called the resting membrane potential; it is caused by differences in the concentrations of ions inside and outside the cell. If the membrane were equally permeable to all ions, each type of ion would flow across the membrane and the system would reach equilibrium. Because ions cannot simply cross the membrane at will, there are different concentrations of several ions inside and outside the cell, as shown in Table 35.1. The difference in the number of positively charged potassium ions (K^+) inside and outside the cell dominates the resting membrane potential (Figure 35.10). When the membrane is at rest, K^+ ions accumulate inside the cell due to a net movement with the concentration gradient. The negative resting membrane potential is created and maintained by increasing the concentration of cations outside the cell (in the extracellular fluid) relative to inside the cell (in the cytoplasm). The negative charge within the cell is created by the cell membrane being more permeable to potassium ion movement than sodium ion movement. In neurons, potassium ions are maintained at high concentrations within the cell while sodium ions are maintained at high concentrations outside of the cell. The cell possesses potassium and sodium leakage channels that allow the two cations to diffuse down their concentration gradient. However, the neurons have far more potassium leakage channels than sodium leakage channels. Therefore, potassium diffuses out of the cell at a much faster rate than sodium leaks in. Because more cations are leaving the cell than are entering, this causes the interior of the cell to be negatively charged relative to the outside of the cell. The actions of the sodium potassium pump help to maintain the resting potential, once established. Recall that sodium potassium pumps brings two K^+ ions into the cell while removing three Na^+ ions per ATP consumed. As more cations are expelled from the cell than taken in, the inside of the cell remains negatively charged relative to the extracellular fluid. It should be noted that chloride ions (Cl^-) tend to accumulate outside of the cell because they are repelled by negatively-charged proteins within the cytoplasm.

Ion Concentration Inside and Outside Neurons

Ion	Extracellular concentration (mM)	Intracellular concentration (mM)	Ratio outside/ inside
Na^+	145	12	12
K+	4	155	0.026
Cl^-	120	4	30
Organic anions (A–)	—	100	

Table 35.1 The resting membrane potential is a result of different concentrations inside and outside the cell.

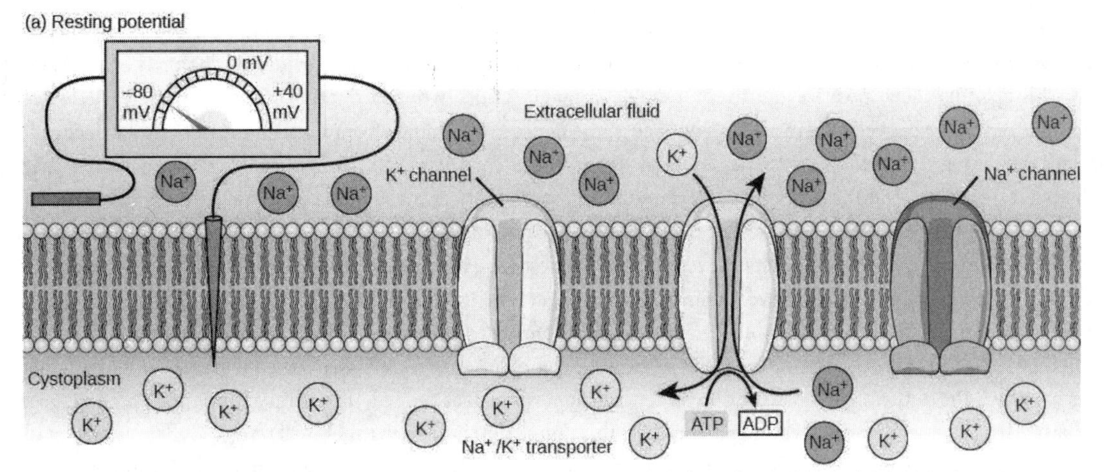

At the resting potential, all voltage-gated Na^+ channels and most voltage-gated K^+ channels are closed. The Na^+/K^+ transporter pumps K^+ ions into the cell and Na^+ ions out.

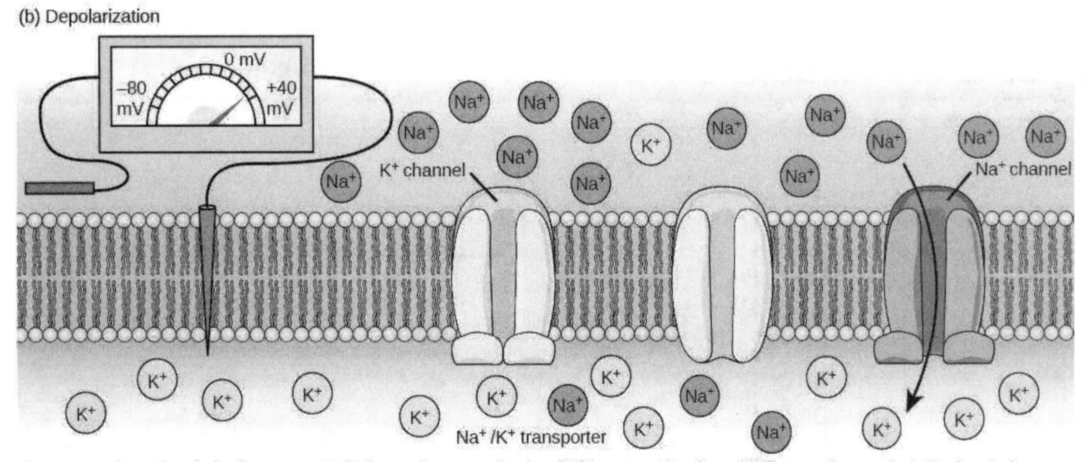

In response to a depolarization, some Na^+ channels open, allowing Na^+ ions to enter the cell. The membrane starts to depolarize (the charge across the membrane lessens). If the threshold of excitation is reached, all the Na^+ channels open.

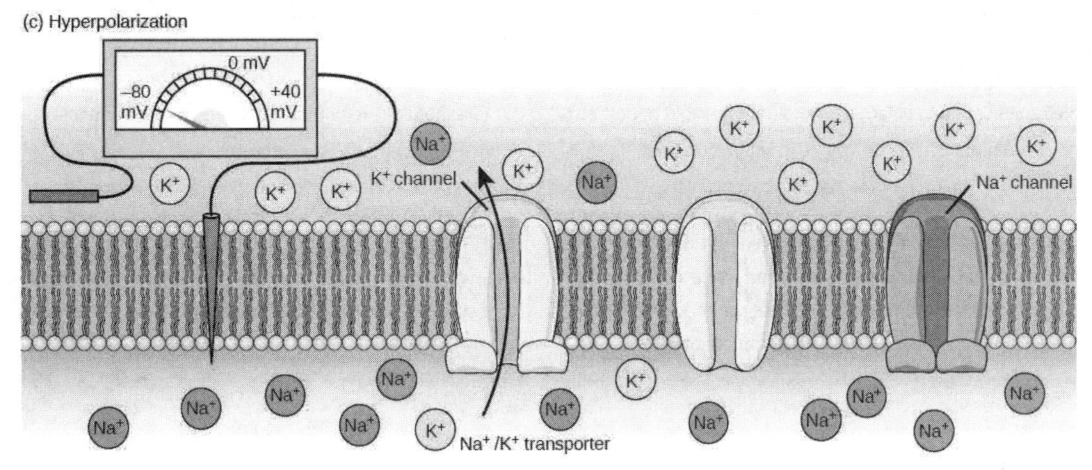

At the peak action potential, Na^+ channels close while K^+ channels open. K^+ leaves the cell, and the membrane eventually becomes hyperpolarized.

Figure 35.10 The (a) resting membrane potential is a result of different concentrations of Na^+ and K^+ ions inside and outside the cell. A nerve impulse causes Na^+ to enter the cell, resulting in (b) depolarization. At the peak action potential, K^+ channels open and the cell becomes (c) hyperpolarized.

Action Potential

A neuron can receive input from other neurons and, if this input is strong enough, send the signal to downstream neurons. Transmission of a signal between neurons is generally carried by a chemical called a neurotransmitter. Transmission of a signal within a neuron (from dendrite to axon terminal) is carried by a brief reversal of the resting membrane potential called an **action potential**. When neurotransmitter molecules bind to receptors located on a neuron's dendrites, ion channels open. At excitatory synapses, this opening allows positive ions to enter the neuron and results in **depolarization** of the membrane—a decrease in the difference in voltage between the inside and outside of the neuron. A stimulus from a sensory cell or another neuron depolarizes the target neuron to its threshold potential (-55 mV). Na^+ channels in the axon hillock open, allowing positive ions to enter the cell (Figure 35.10 and Figure 35.11). Once the sodium channels open, the neuron completely depolarizes to a membrane potential of about +40 mV. Action potentials are considered an "all-or nothing" event, in that, once the threshold potential is reached, the neuron always completely depolarizes. Once depolarization is complete, the cell must now "reset" its membrane voltage back to the resting potential. To accomplish this, the Na^+ channels close and cannot be opened. This begins the neuron's **refractory period**, in which it cannot produce another action potential because its sodium channels will not open. At the same time, voltage-gated K^+ channels open, allowing K^+ to leave the cell. As K^+ ions leave the cell, the membrane potential once again becomes negative. The diffusion of K^+ out of the cell actually **hyperpolarizes** the cell, in that the membrane potential becomes more negative than the cell's normal resting potential. At this point, the sodium channels will return to their resting state, meaning they are ready to open again if the membrane potential again exceeds the threshold potential. Eventually the extra K^+ ions diffuse out of the cell through the potassium leakage channels, bringing the cell from its hyperpolarized state, back to its resting membrane potential.

VISUAL CONNECTION

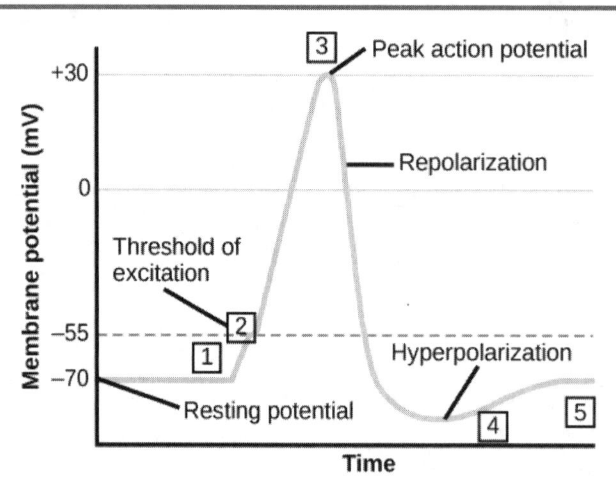

Figure 35.11 The formation of an action potential can be divided into five steps: (1) A stimulus from a sensory cell or another neuron causes the target cell to depolarize toward the threshold potential. (2) If the threshold of excitation is reached, all Na^+ channels open and the membrane depolarizes. (3) At the peak action potential, K^+ channels open and K^+ begins to leave the cell. At the same time, Na^+ channels close. (4) The membrane becomes hyperpolarized as K^+ ions continue to leave the cell. The hyperpolarized membrane is in a refractory period and cannot fire. (5) The K^+ channels close and the Na^+/K^+ transporter restores the resting potential.

Potassium channel blockers, such as amiodarone and procainamide, which are used to treat abnormal electrical activity in the heart, called cardiac dysrhythmia, impede the movement of K^+ through voltage-gated K^+ channels. Which part of the action potential would you expect potassium channels to affect?

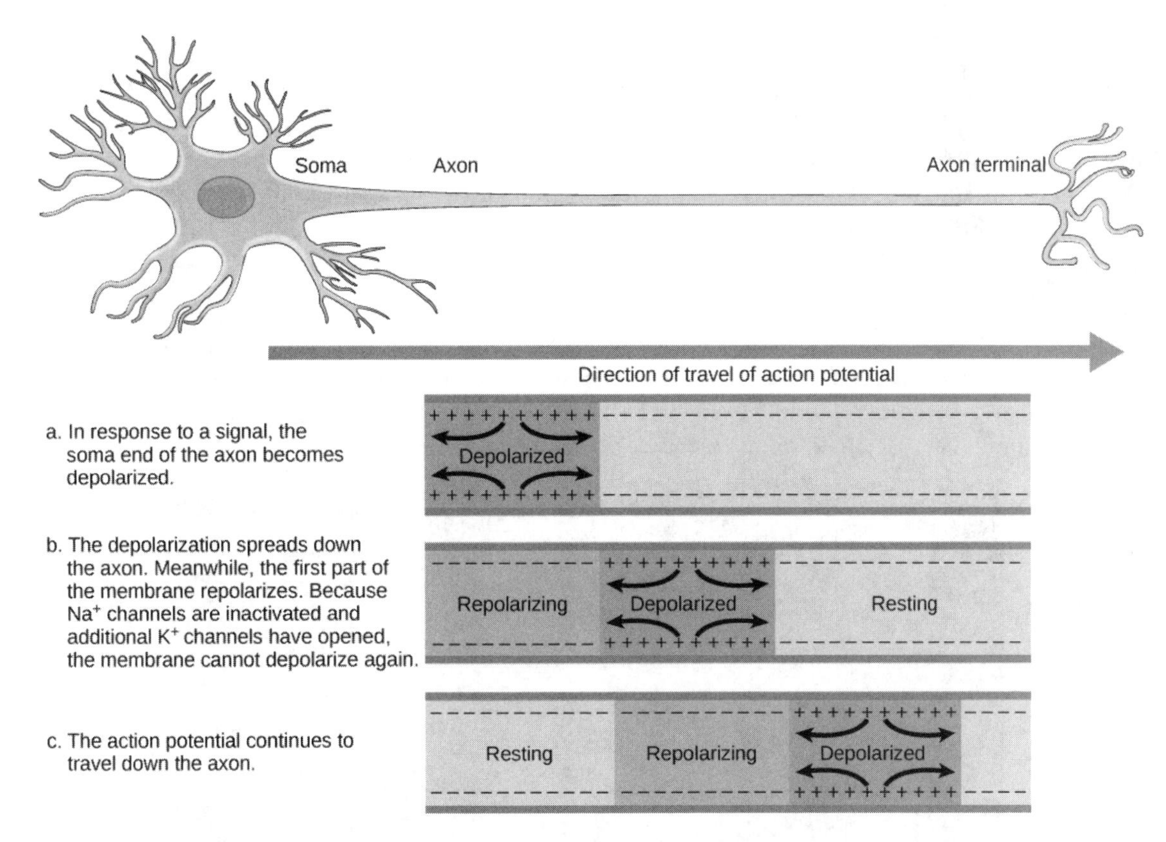

Figure 35.12 The action potential is conducted down the axon as the axon membrane depolarizes, then repolarizes.

LINK TO LEARNING

This video (http://openstax.org/l/actionpotential) presents an overview of action potential.

Myelin and the Propagation of the Action Potential

For an action potential to communicate information to another neuron, it must travel along the axon and reach the axon terminals where it can initiate neurotransmitter release. The speed of conduction of an action potential along an axon is influenced by both the diameter of the axon and the axon's resistance to current leak. Myelin acts as an insulator that prevents current from leaving the axon; this increases the speed of action potential conduction. In demyelinating diseases like multiple sclerosis, action potential conduction slows because current leaks from previously insulated axon areas. The nodes of Ranvier, illustrated in Figure 35.13 are gaps in the myelin sheath along the axon. These unmyelinated spaces are about one micrometer long and contain voltage-gated Na^+ and K^+ channels. Flow of ions through these channels, particularly the Na^+ channels, regenerates the action potential over and over again along the axon. This 'jumping' of the action potential from one node to the next is called **saltatory conduction**. If nodes of Ranvier were not present along an axon, the action potential would propagate very slowly since Na^+ and K^+ channels would have to continuously regenerate action potentials at every point along the axon instead of at specific points. Nodes of Ranvier also save energy for the neuron since the channels only need to be present at the nodes and not along the entire axon.

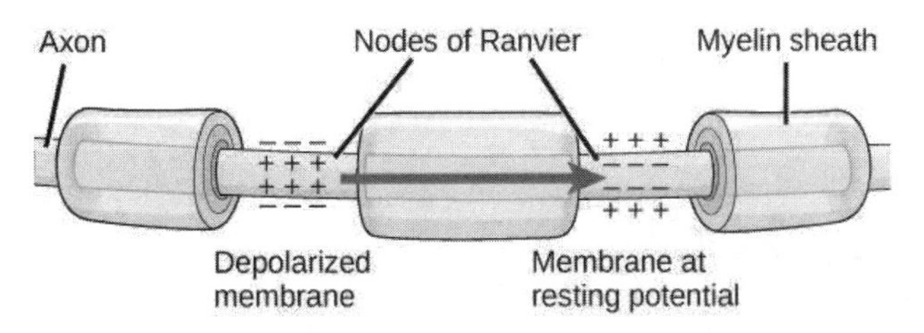

Figure 35.13 Nodes of Ranvier are gaps in myelin coverage along axons. Nodes contain voltage-gated K^+ and Na^+ channels. Action potentials travel down the axon by jumping from one node to the next.

Synaptic Transmission

The synapse or "gap" is the place where information is transmitted from one neuron to another. Synapses usually form between axon terminals and dendritic spines, but this is not universally true. There are also axon-to-axon, dendrite-to-dendrite, and axon-to-cell body synapses. The neuron transmitting the signal is called the presynaptic neuron, and the neuron receiving the signal is called the postsynaptic neuron. Note that these designations are relative to a particular synapse—most neurons are both presynaptic and postsynaptic. There are two types of synapses: chemical and electrical.

Chemical Synapse

When an action potential reaches the axon terminal it depolarizes the membrane and opens voltage-gated Na^+ channels. Na^+ ions enter the cell, further depolarizing the presynaptic membrane. This depolarization causes voltage-gated Ca^{2+} channels to open. Calcium ions entering the cell initiate a signaling cascade that causes small membrane-bound vesicles, called **synaptic vesicles**, containing neurotransmitter molecules to fuse with the presynaptic membrane. Synaptic vesicles are shown in Figure 35.14, which is an image from a scanning electron microscope.

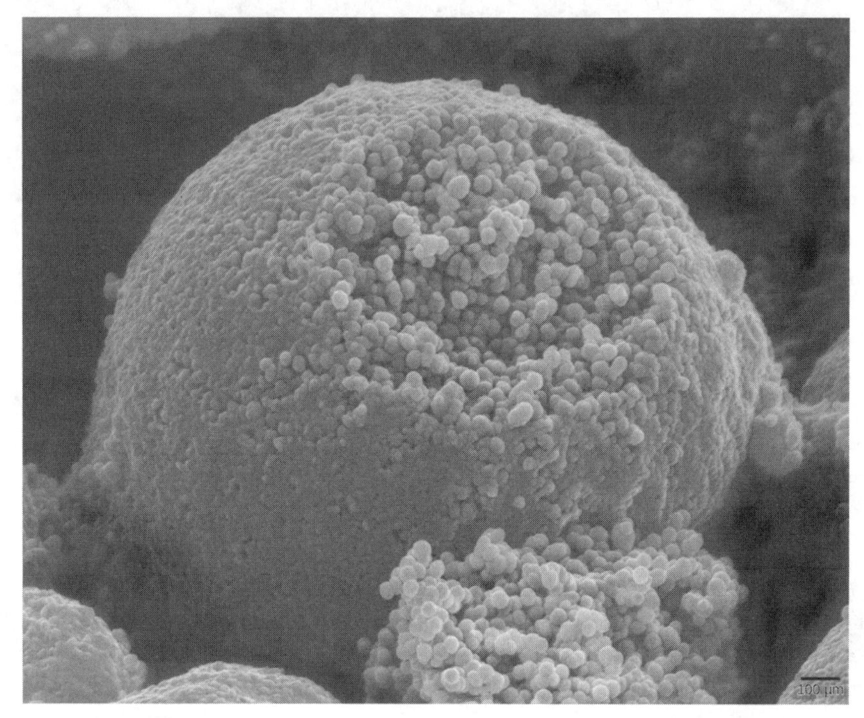

Figure 35.14 This pseudocolored image taken with a scanning electron microscope shows an axon terminal that was broken open to reveal synaptic vesicles (blue and orange) inside the neuron. (credit: modification of work by Tina Carvalho, NIH-NIGMS; scale-bar data from Matt Russell)

Fusion of a vesicle with the presynaptic membrane causes neurotransmitter to be released into the **synaptic cleft**, the extracellular space between the presynaptic and postsynaptic membranes, as illustrated in Figure 35.15. The neurotransmitter diffuses across the synaptic cleft and binds to receptor proteins on the postsynaptic membrane.

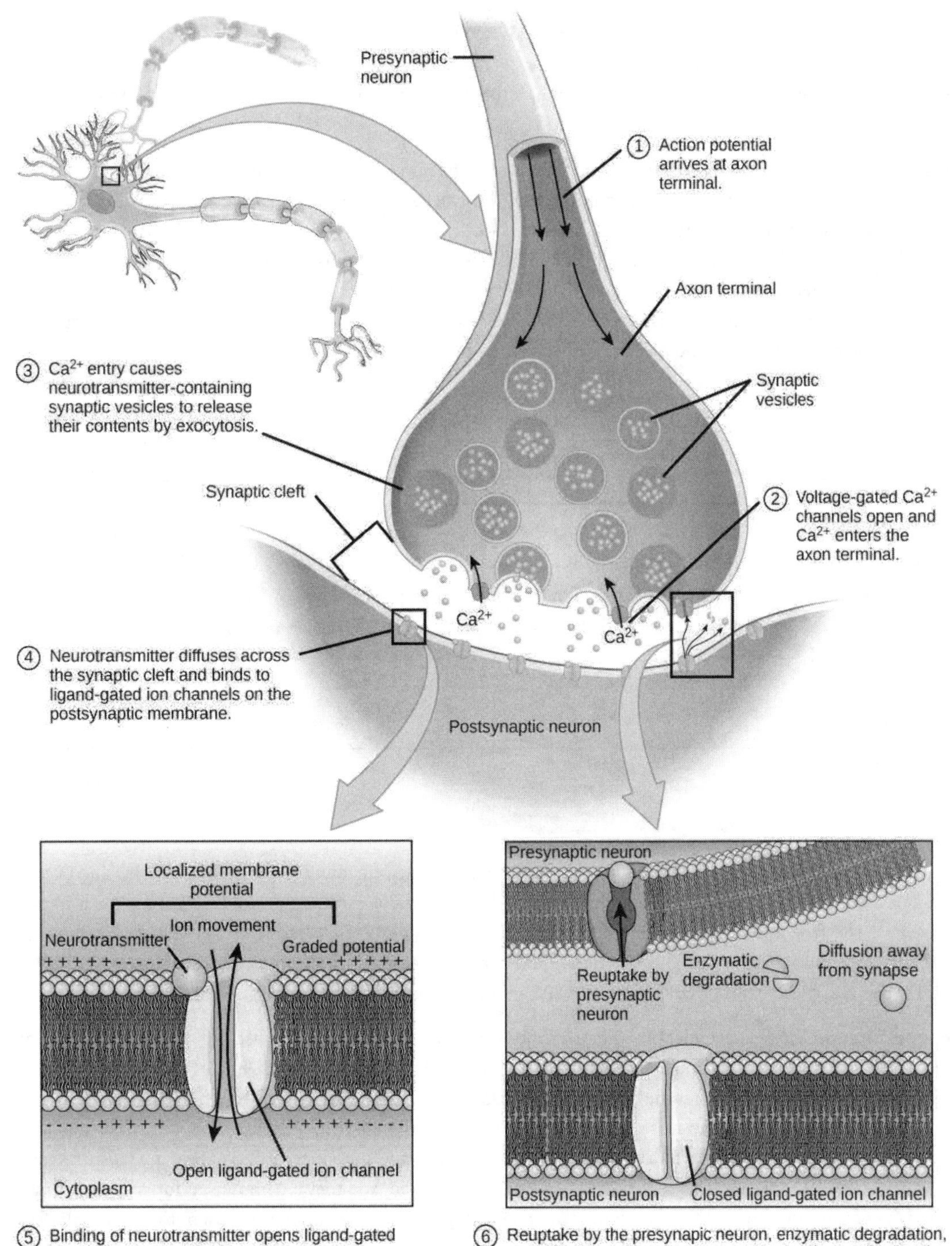

Figure 35.15 Communication at chemical synapses requires release of neurotransmitters. When the presynaptic membrane is depolarized, voltage-gated Ca^{2+} channels open and allow Ca^{2+} to enter the cell. The calcium entry causes synaptic vesicles to fuse with the membrane and release neurotransmitter molecules into the synaptic cleft. The neurotransmitter diffuses across the synaptic cleft and binds to ligand-gated ion channels in the postsynaptic membrane, resulting in a localized depolarization or hyperpolarization of the postsynaptic neuron.

The binding of a specific neurotransmitter causes particular ion channels, in this case ligand-gated channels, on the postsynaptic membrane to open. Neurotransmitters can either have excitatory or inhibitory effects on the postsynaptic membrane, as detailed in Table 35.1. For example, when acetylcholine is released at the synapse between a nerve and muscle (called the neuromuscular junction) by a presynaptic neuron, it causes postsynaptic Na^+ channels to open. Na^+ enters the postsynaptic cell and causes the postsynaptic membrane to depolarize. This depolarization is called an **excitatory postsynaptic**

potential (EPSP) and makes the postsynaptic neuron more likely to fire an action potential. Release of neurotransmitter at inhibitory synapses causes **inhibitory postsynaptic potentials (IPSPs)**, a hyperpolarization of the presynaptic membrane. For example, when the neurotransmitter GABA (gamma-aminobutyric acid) is released from a presynaptic neuron, it binds to and opens Cl^- channels. Cl^- ions enter the cell and hyperpolarizes the membrane, making the neuron less likely to fire an action potential.

Once neurotransmission has occurred, the neurotransmitter must be removed from the synaptic cleft so the postsynaptic membrane can "reset" and be ready to receive another signal. This can be accomplished in three ways: the neurotransmitter can diffuse away from the synaptic cleft, it can be degraded by enzymes in the synaptic cleft, or it can be recycled (sometimes called reuptake) by the presynaptic neuron. Several drugs act at this step of neurotransmission. For example, some drugs that are given to Alzheimer's patients work by inhibiting acetylcholinesterase, the enzyme that degrades acetylcholine. This inhibition of the enzyme essentially increases neurotransmission at synapses that release acetylcholine. Once released, the acetylcholine stays in the cleft and can continually bind and unbind to postsynaptic receptors.

Neurotransmitter Function and Location

Neurotransmitter	Example	Location
Acetylcholine	—	CNS and/or PNS
Biogenic amine	Dopamine, serotonin, norepinephrine	CNS and/or PNS
Amino acid	Glycine, glutamate, aspartate, gamma aminobutyric acid	CNS
Neuropeptide	Substance P, endorphins	CNS and/or PNS

Table 35.2

Electrical Synapse

While electrical synapses are fewer in number than chemical synapses, they are found in all nervous systems and play important and unique roles. The mode of neurotransmission in electrical synapses is quite different from that in chemical synapses. In an electrical synapse, the presynaptic and postsynaptic membranes are very close together and are actually physically connected by channel proteins forming gap junctions. Gap junctions allow current to pass directly from one cell to the next. In addition to the ions that carry this current, other molecules, such as ATP, can diffuse through the large gap junction pores.

There are key differences between chemical and electrical synapses. Because chemical synapses depend on the release of neurotransmitter molecules from synaptic vesicles to pass on their signal, there is an approximately one millisecond delay between when the axon potential reaches the presynaptic terminal and when the neurotransmitter leads to opening of postsynaptic ion channels. Additionally, this signaling is unidirectional. Signaling in electrical synapses, in contrast, is virtually instantaneous (which is important for synapses involved in key reflexes), and some electrical synapses are bidirectional. Electrical synapses are also more reliable as they are less likely to be blocked, and they are important for synchronizing the electrical activity of a group of neurons. For example, electrical synapses in the thalamus are thought to regulate slow-wave sleep, and disruption of these synapses can cause seizures.

Signal Summation

Sometimes a single EPSP is strong enough to induce an action potential in the postsynaptic neuron, but often multiple presynaptic inputs must create EPSPs around the same time for the postsynaptic neuron to be sufficiently depolarized to fire an action potential. This process is called **summation** and occurs at the axon hillock, as illustrated in Figure 35.16. Additionally, one neuron often has inputs from many presynaptic neurons—some excitatory and some inhibitory—so IPSPs can cancel out EPSPs and vice versa. It is the net change in postsynaptic membrane voltage that determines whether the postsynaptic cell has reached its threshold of excitation needed to fire an action potential. Together, synaptic summation and the threshold for excitation act as a filter so that random "noise" in the system is not transmitted as important information.

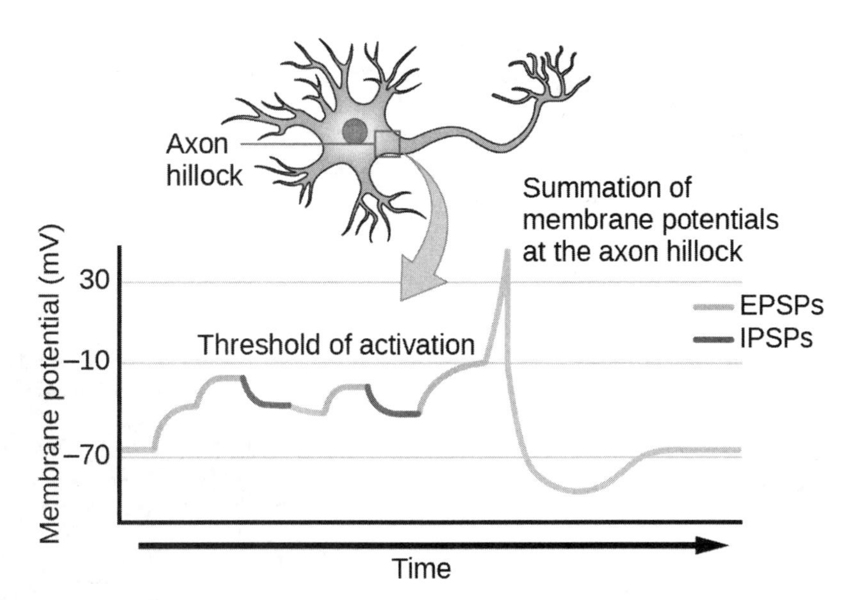

Figure 35.16 A single neuron can receive both excitatory and inhibitory inputs from multiple neurons, resulting in local membrane depolarization (EPSP input) and hyperpolarization (IPSP input). All these inputs are added together at the axon hillock. If the EPSPs are strong enough to overcome the IPSPs and reach the threshold of excitation, the neuron will fire.

Everyday Connection

Brain-computer interface

Amyotrophic lateral sclerosis (ALS, also called Lou Gehrig's Disease) is a neurological disease characterized by the degeneration of the motor neurons that control voluntary movements. The disease begins with muscle weakening and lack of coordination and eventually destroys the neurons that control speech, breathing, and swallowing; in the end, the disease can lead to paralysis. At that point, patients require assistance from machines to be able to breathe and to communicate. Several special technologies have been developed to allow "locked-in" patients to communicate with the rest of the world. One technology, for example, allows patients to type out sentences by twitching their cheek. These sentences can then be read aloud by a computer.

A relatively new line of research for helping paralyzed patients, including those with ALS, to communicate and retain a degree of self-sufficiency is called brain-computer interface (BCI) technology and is illustrated in Figure 35.17. This technology sounds like something out of science fiction: it allows paralyzed patients to control a computer using only their thoughts. There are several forms of BCI. Some forms use EEG recordings from electrodes taped onto the skull. These recordings contain information from large populations of neurons that can be decoded by a computer. Other forms of BCI require the implantation of an array of electrodes smaller than a postage stamp in the arm and hand area of the motor cortex. This form of BCI, while more invasive, is very powerful as each electrode can record actual action potentials from one or more neurons. These signals are then sent to a computer, which has been trained to decode the signal and feed it to a tool—such as a cursor on a computer screen. This means that a patient with ALS can use e-mail, read the Internet, and communicate with others by thinking of moving his or her hand or arm (even though the paralyzed patient cannot make that bodily movement). Recent advances have allowed a paralyzed locked-in patient who suffered a stroke 15 years ago to control a robotic arm and even to feed herself coffee using BCI technology.

Despite the amazing advancements in BCI technology, it also has limitations. The technology can require many hours of training and long periods of intense concentration for the patient; it can also require brain surgery to implant the devices.

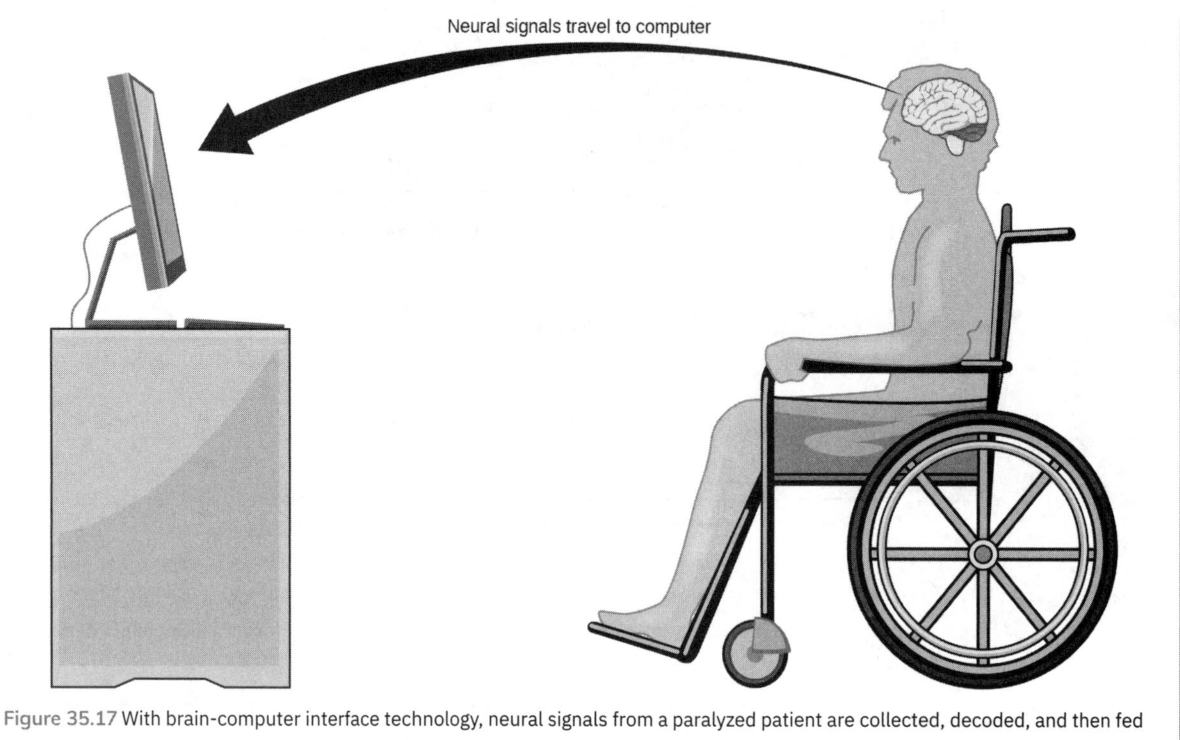

Figure 35.17 With brain-computer interface technology, neural signals from a paralyzed patient are collected, decoded, and then fed to a tool, such as a computer, a wheelchair, or a robotic arm.

LINK TO LEARNING

Watch this video (http://openstax.org/l/paralyzation) in which a paralyzed woman uses a brain-controlled robotic arm to bring a drink to her mouth, among other images of brain-computer interface technology in action.

Click to view content (https://www.openstax.org/l/paralyzation)

Synaptic Plasticity

Synapses are not static structures. They can be weakened or strengthened. They can be broken, and new synapses can be made. Synaptic plasticity allows for these changes, which are all needed for a functioning nervous system. In fact, synaptic plasticity is the basis of learning and memory. Two processes in particular, long-term potentiation (LTP) and long-term depression (LTD) are important forms of synaptic plasticity that occur in synapses in the hippocampus, a brain region that is involved in storing memories.

Long-term Potentiation (LTP)

Long-term potentiation (LTP) is a persistent strengthening of a synaptic connection. LTP is based on the Hebbian principle: cells that fire together wire together. There are various mechanisms, none fully understood, behind the synaptic strengthening seen with LTP. One known mechanism involves a type of postsynaptic glutamate receptor, called NMDA (N-Methyl-D-aspartate) receptors, shown in Figure 35.18. These receptors are normally blocked by magnesium ions; however, when the postsynaptic neuron is depolarized by multiple presynaptic inputs in quick succession (either from one neuron or multiple neurons), the magnesium ions are forced out allowing Ca ions to pass into the postsynaptic cell. Next, Ca^{2+} ions entering the cell initiate a signaling cascade that causes a different type of glutamate receptor, called AMPA (α-amino-3-hydroxy-5-methyl-4-isoxazolepropionic acid) receptors, to be inserted into the postsynaptic membrane, since activated AMPA receptors allow positive ions to enter the cell. So, the next time glutamate is released from the presynaptic membrane, it will have a larger excitatory effect (EPSP) on the postsynaptic cell because the binding of glutamate to these AMPA receptors will allow more positive ions into the cell. The insertion of additional AMPA receptors strengthens the synapse and means that the postsynaptic neuron is more likely to fire in response to presynaptic neurotransmitter release. Some drugs of abuse co-opt the LTP pathway, and this synaptic strengthening can lead to addiction.

Long-term Depression (LTD)

Long-term depression (LTD) is essentially the reverse of LTP: it is a long-term weakening of a synaptic connection. One mechanism known to cause LTD also involves AMPA receptors. In this situation, calcium that enters through NMDA receptors initiates a different signaling cascade, which results in the removal of AMPA receptors from the postsynaptic membrane, as illustrated in Figure 35.18. The decrease in AMPA receptors in the membrane makes the postsynaptic neuron less responsive to glutamate released from the presynaptic neuron. While it may seem counterintuitive, LTD may be just as important for learning and memory as LTP. The weakening and pruning of unused synapses allows for unimportant connections to be lost and makes the synapses that have undergone LTP that much stronger by comparison.

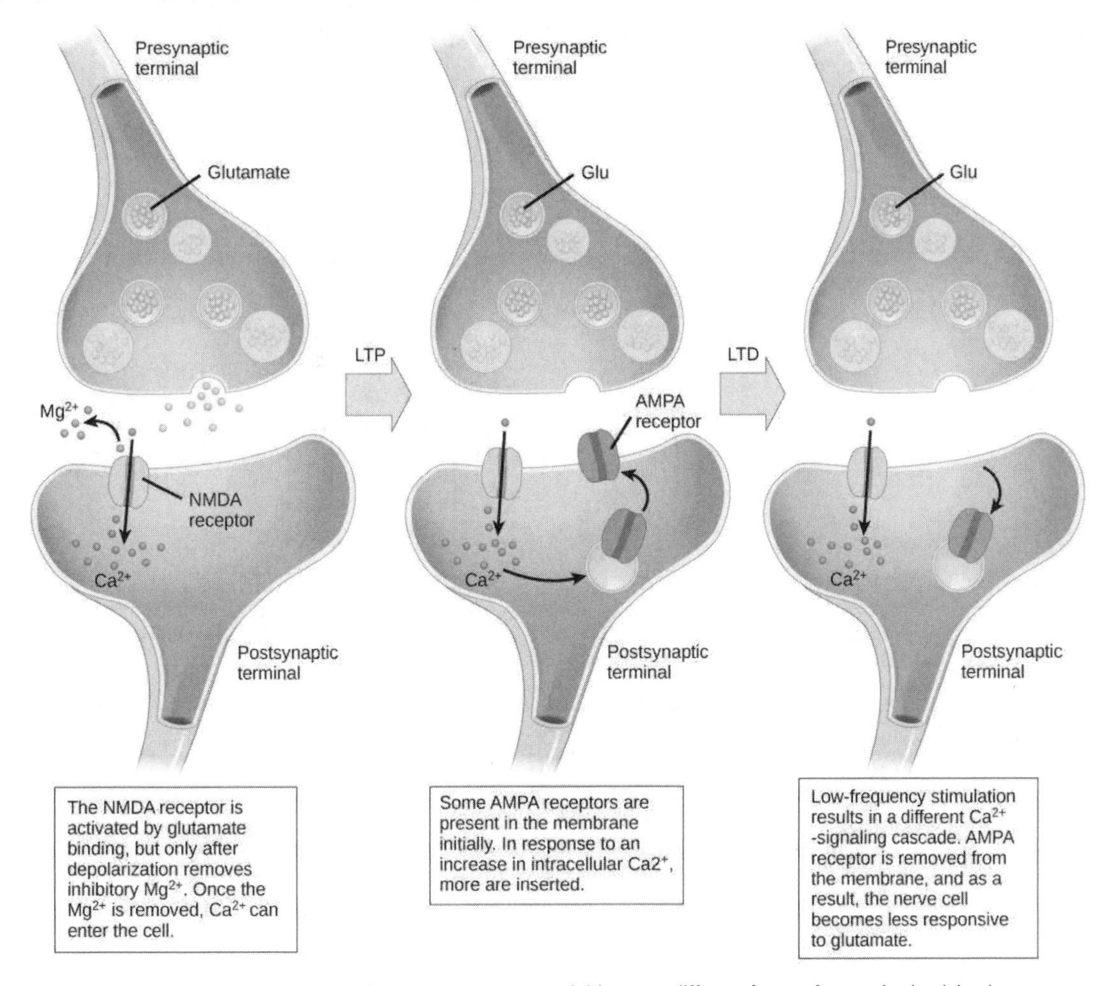

Figure 35.18 Calcium entry through postsynaptic NMDA receptors can initiate two different forms of synaptic plasticity: long-term potentiation (LTP) and long-term depression (LTD). LTP arises when a single synapse is repeatedly stimulated. This stimulation causes a calcium- and CaMKII-dependent cellular cascade, which results in the insertion of more AMPA receptors into the postsynaptic membrane. The next time glutamate is released from the presynaptic cell, it will bind to both NMDA and the newly inserted AMPA receptors, thus depolarizing the membrane more efficiently. LTD occurs when few glutamate molecules bind to NMDA receptors at a synapse (due to a low firing rate of the presynaptic neuron). The calcium that does flow through NMDA receptors initiates a different calcineurin and protein phosphatase 1-dependent cascade, which results in the endocytosis of AMPA receptors. This makes the postsynaptic neuron less responsive to glutamate released from the presynaptic neuron.

35.3 The Central Nervous System

By the end of this section, you will be able to do the following:

- Identify the spinal cord, cerebral lobes, and other brain areas on a diagram of the brain
- Describe the basic functions of the spinal cord, cerebral lobes, and other brain areas

The central nervous system (CNS) is made up of the brain, a part of which is shown in Figure 35.19 and spinal cord and is covered with three layers of protective coverings called **meninges** (from the Greek word for membrane). The outermost layer is the **dura**

mater (Latin for "hard mother"). As the Latin suggests, the primary function for this thick layer is to protect the brain and spinal cord. The dura mater also contains vein-like structures that carry blood from the brain back to the heart. The middle layer is the web-like **arachnoid mater**. The last layer is the **pia mater** (Latin for "soft mother"), which directly contacts and covers the brain and spinal cord like plastic wrap. The space between the arachnoid and pia maters is filled with **cerebrospinal fluid (CSF)**. CSF is produced by a tissue called **choroid plexus** in fluid-filled compartments in the CNS called **ventricles**. The brain floats in CSF, which acts as a cushion and shock absorber and makes the brain neutrally buoyant. CSF also functions to circulate chemical substances throughout the brain and into the spinal cord.

The entire brain contains only about 8.5 tablespoons of CSF, but CSF is constantly produced in the ventricles. This creates a problem when a ventricle is blocked—the CSF builds up and creates swelling and the brain is pushed against the skull. This swelling condition is called hydrocephalus ("water head") and can cause seizures, cognitive problems, and even death if a shunt is not inserted to remove the fluid and pressure.

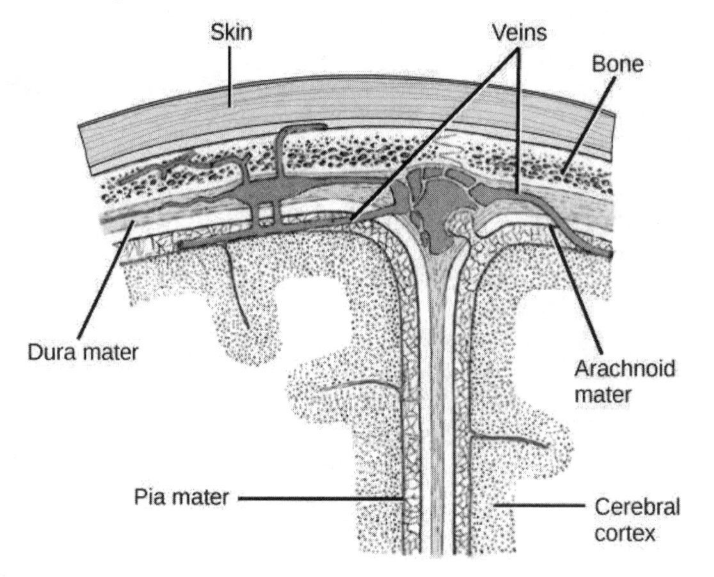

Figure 35.19 The cerebral cortex is covered by three layers of meninges: the dura, arachnoid, and pia maters. (credit: modification of work by Gray's Anatomy)

Brain

The brain is the part of the central nervous system that is contained in the cranial cavity of the skull. It includes the cerebral cortex, limbic system, basal ganglia, thalamus, hypothalamus, and cerebellum. There are three different ways that a brain can be sectioned in order to view internal structures: a sagittal section cuts the brain left to right, as shown in Figure 35.21b, a coronal section cuts the brain front to back, as shown in Figure 35.20a, and a horizontal section cuts the brain top to bottom.

Cerebral Cortex

The outermost part of the brain is a thick piece of nervous system tissue called the **cerebral cortex**, which is folded into hills called **gyri** (singular: gyrus) and valleys called **sulci** (singular: sulcus). The cortex is made up of two hemispheres—right and left—which are separated by a large sulcus. A thick fiber bundle called the **corpus callosum** (Latin: "tough body") connects the two hemispheres and allows information to be passed from one side to the other. Although there are some brain functions that are localized more to one hemisphere than the other, the functions of the two hemispheres are largely redundant. In fact, sometimes (very rarely) an entire hemisphere is removed to treat severe epilepsy. While patients do suffer some deficits following the surgery, they can have surprisingly few problems, especially when the surgery is performed on children who have very immature nervous systems.

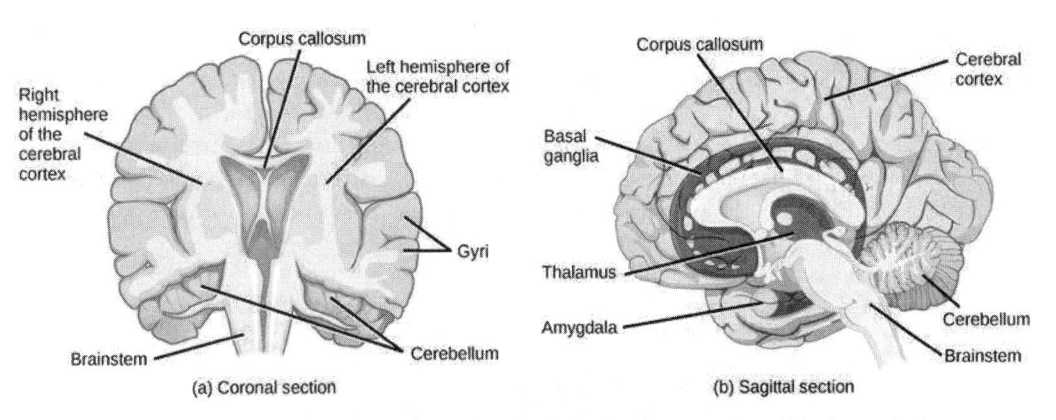

Figure 35.20 These illustrations show the (a) coronal and (b) sagittal sections of the human brain.

In other surgeries to treat severe epilepsy, the corpus callosum is cut instead of removing an entire hemisphere. This causes a condition called split-brain, which gives insights into unique functions of the two hemispheres. For example, when an object is presented to patients' left visual field, they may be unable to verbally name the object (and may claim to not have seen an object at all). This is because the visual input from the left visual field crosses and enters the right hemisphere and cannot then signal to the speech center, which generally is found in the left side of the brain. Remarkably, if a split-brain patient is asked to pick up a specific object out of a group of objects with the left hand, the patient will be able to do so but will still be unable to vocally identify it.

LINK TO LEARNING

See this website (http://openstax.org/l/split-brain) to learn more about split-brain patients and to play a game where you can model the split-brain experiments yourself.

Each cortical hemisphere contains regions called lobes that are involved in different functions. Scientists use various techniques to determine what brain areas are involved in different functions: they examine patients who have had injuries or diseases that affect specific areas and see how those areas are related to functional deficits. They also conduct animal studies where they stimulate brain areas and see if there are any behavioral changes. They use a technique called transcranial magnetic stimulation (TMS) to temporarily deactivate specific parts of the cortex using strong magnets placed outside the head; and they use functional magnetic resonance imaging (fMRI) to look at changes in oxygenated blood flow in particular brain regions that correlate with specific behavioral tasks. These techniques, and others, have given great insight into the functions of different brain regions but have also showed that any given brain area can be involved in more than one behavior or process, and any given behavior or process generally involves neurons in multiple brain areas. That being said, each hemisphere of the mammalian cerebral cortex can be broken down into four functionally and spatially defined lobes: frontal, parietal, temporal, and occipital. Figure 35.21 illustrates these four lobes of the human cerebral cortex.

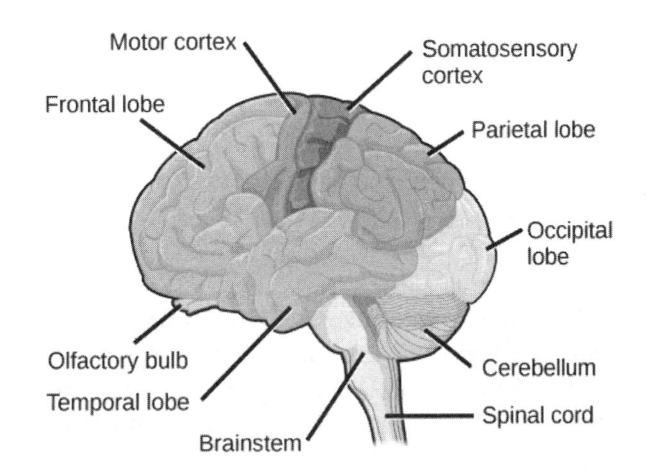

Figure 35.21 The human cerebral cortex includes the frontal, parietal, temporal, and occipital lobes.

The **frontal lobe** is located at the front of the brain, over the eyes. This lobe contains the olfactory bulb, which processes smells. The frontal lobe also contains the motor cortex, which is important for planning and implementing movement. Areas within the motor cortex map to different muscle groups, and there is some organization to this map, as shown in Figure 35.22. For example, the neurons that control movement of the fingers are next to the neurons that control movement of the hand. Neurons in the frontal lobe also control cognitive functions like maintaining attention, speech, and decision-making. Studies of humans who have damaged their frontal lobes show that parts of this area are involved in personality, socialization, and assessing risk.

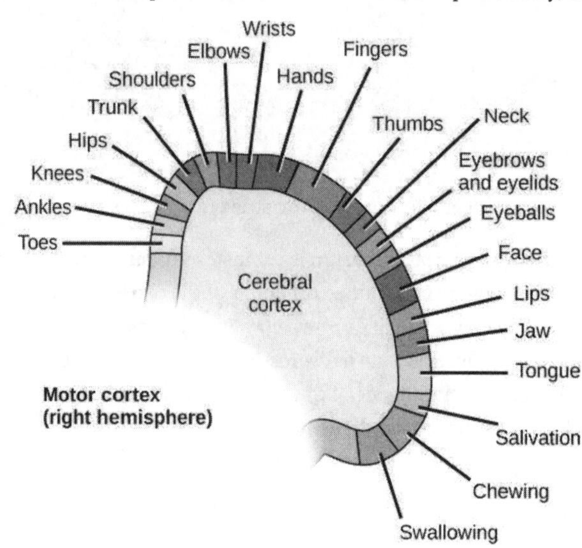

Figure 35.22 Different parts of the motor cortex control different muscle groups. Muscle groups that are neighbors in the body are generally controlled by neighboring regions of the motor cortex as well. For example, the neurons that control finger movement are near the neurons that control hand movement.

The **parietal lobe** is located at the top of the brain. Neurons in the parietal lobe are involved in speech and also reading. Two of the parietal lobe's main functions are processing **somatosensation**—touch sensations like pressure, pain, heat, cold—and processing **proprioception**—the sense of how parts of the body are oriented in space. The parietal lobe contains a somatosensory map of the body similar to the motor cortex.

The **occipital lobe** is located at the back of the brain. It is primarily involved in vision—seeing, recognizing, and identifying the visual world.

The **temporal lobe** is located at the base of the brain by your ears and is primarily involved in processing and interpreting sounds. It also contains the **hippocampus** (Greek for "seahorse")—a structure that processes memory formation. The hippocampus is illustrated in Figure 35.24. The role of the hippocampus in memory was partially determined by studying one famous epileptic patient, HM, who had both sides of his hippocampus removed in an attempt to cure his epilepsy. His seizures went away, but he could no longer form new memories (although he could remember some facts from before his surgery and could learn new motor tasks).

EVOLUTION CONNECTION

Cerebral Cortex

Compared to other vertebrates, mammals have exceptionally large brains for their body size. An entire alligator's brain, for example, would fill about one and a half teaspoons. This increase in brain to body size ratio is especially pronounced in apes, whales, and dolphins. While this increase in overall brain size doubtlessly played a role in the evolution of complex behaviors unique to mammals, it does not tell the whole story. Scientists have found a relationship between the relatively high surface area of the cortex and the intelligence and complex social behaviors exhibited by some mammals. This increased surface area is due, in part, to increased folding of the cortical sheet (more sulci and gyri). For example, a rat cortex is very smooth with very few sulci and gyri. Cat and sheep cortices have more sulci and gyri. Chimps, humans, and dolphins have even more.

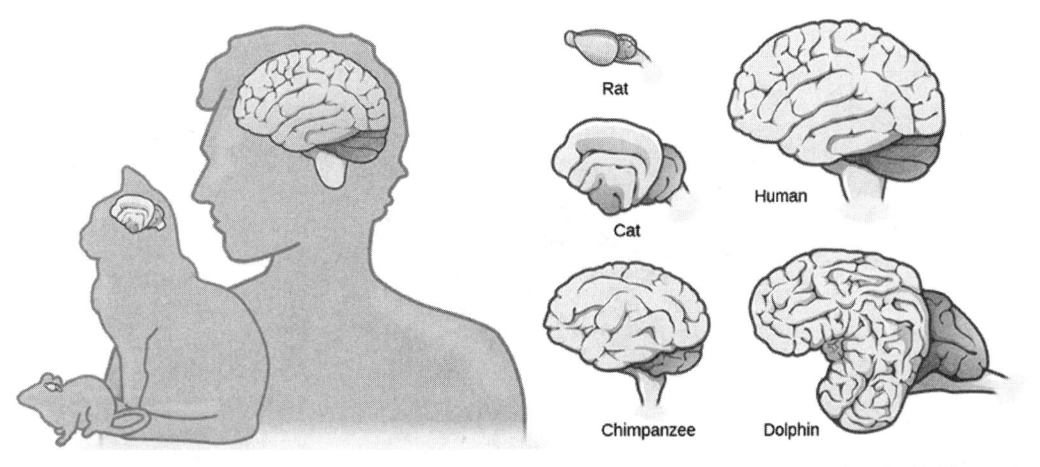

Figure 35.23 Mammals have larger brain-to-body ratios than other vertebrates. Within mammals, increased cortical folding and surface area is correlated with complex behavior.

Basal Ganglia

Interconnected brain areas called the **basal ganglia** (or **basal nuclei**), shown in Figure 35.20b, play important roles in movement control and posture. Damage to the basal ganglia, as in Parkinson's disease, leads to motor impairments like a shuffling gait when walking. The basal ganglia also regulate motivation. For example, when a wasp sting led to bilateral basal ganglia damage in a 25-year-old businessman, he began to spend all his days in bed and showed no interest in anything or anybody. But when he was externally stimulated—as when someone asked to play a card game with him—he was able to function normally. Interestingly, he and other similar patients do not report feeling bored or frustrated by their state.

Thalamus

The **thalamus** (Greek for "inner chamber"), illustrated in Figure 35.24, acts as a gateway to and from the cortex. It receives sensory and motor inputs from the body and also receives feedback from the cortex. This feedback mechanism can modulate conscious awareness of sensory and motor inputs depending on the attention and arousal state of the animal. The thalamus helps regulate consciousness, arousal, and sleep states. A rare genetic disorder called fatal familial insomnia causes the degeneration of thalamic neurons and glia. This disorder prevents affected patients from being able to sleep, among other symptoms, and is eventually fatal.

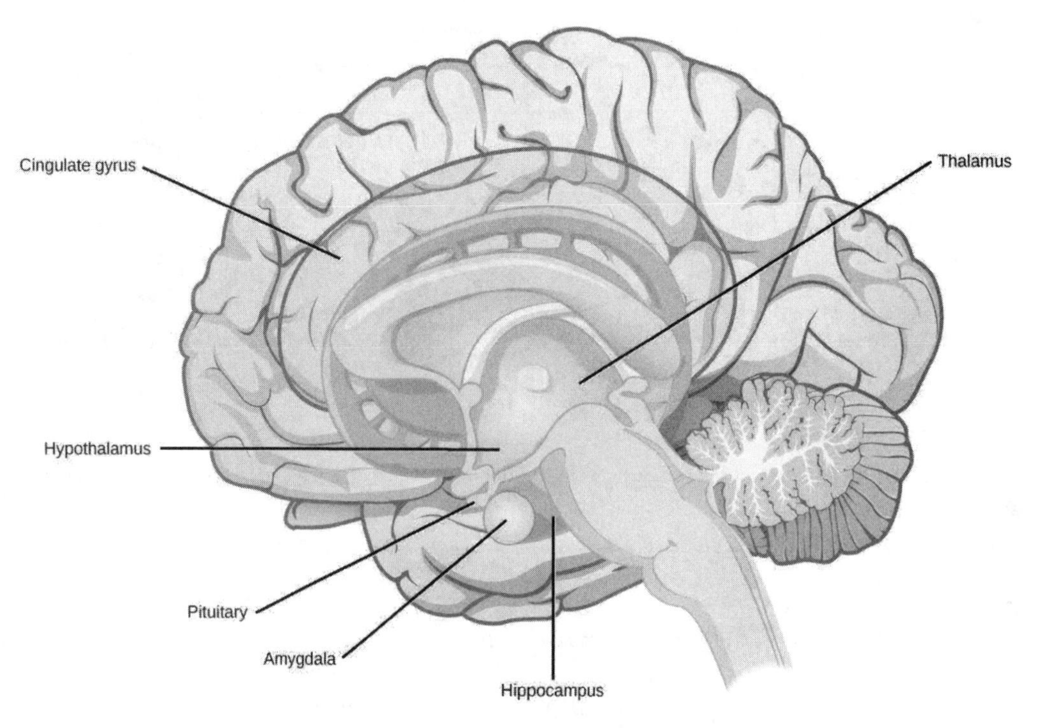

Figure 35.24 The limbic system regulates emotion and other behaviors. It includes parts of the cerebral cortex located near the center of the brain, including the cingulate gyrus and the hippocampus as well as the thalamus, hypothalamus, and amygdala.

Hypothalamus

Below the thalamus is the **hypothalamus**, shown in Figure 35.24. The hypothalamus controls the endocrine system by sending signals to the pituitary gland, a pea-sized endocrine gland that releases several different hormones that affect other glands as well as other cells. This relationship means that the hypothalamus regulates important behaviors that are controlled by these hormones. The hypothalamus is the body's thermostat—it makes sure key functions like food and water intake, energy expenditure, and body temperature are kept at appropriate levels. Neurons within the hypothalamus also regulate circadian rhythms, sometimes called sleep cycles.

Limbic System

The **limbic system** is a connected set of structures that regulates emotion, as well as behaviors related to fear and motivation. It plays a role in memory formation and includes parts of the thalamus and hypothalamus as well as the hippocampus. One important structure within the limbic system is a temporal lobe structure called the **amygdala** (Greek for "almond"), illustrated in Figure 35.24. The two amygdala are important both for the sensation of fear and for recognizing fearful faces. The **cingulate gyrus** helps regulate emotions and pain.

Cerebellum

The **cerebellum** (Latin for "little brain"), shown in Figure 35.21, sits at the base of the brain on top of the brainstem. The cerebellum controls balance and aids in coordinating movement and learning new motor tasks.

Brainstem

The **brainstem**, illustrated in Figure 35.21, connects the rest of the brain with the spinal cord. It consists of the midbrain, medulla oblongata, and the pons. Motor and sensory neurons extend through the brainstem allowing for the relay of signals between the brain and spinal cord. Ascending neural pathways cross in this section of the brain allowing the left hemisphere of the cerebrum to control the right side of the body and vice versa. The brainstem coordinates motor control signals sent from the brain to the body. The brainstem controls several important functions of the body including alertness, arousal, breathing, blood pressure, digestion, heart rate, swallowing, walking, and sensory and motor information integration.

Spinal Cord

Connecting to the brainstem and extending down the body through the spinal column is the **spinal cord**, shown in Figure 35.21. The spinal cord is a thick bundle of nerve tissue that carries information about the body to the brain and from the brain to the body. The spinal cord is contained within the bones of the vertebrate column but is able to communicate signals to and from the body through its connections with spinal nerves (part of the peripheral nervous system). A cross-section of the spinal cord looks like a white oval containing a gray butterfly-shape, as illustrated in Figure 35.25. Myelinated axons make up the "white matter" and neuron and glial cell bodies make up the "gray matter." Gray matter is also composed of interneurons, which connect two neurons each located in different parts of the body. Axons and cell bodies in the dorsal (facing the back of the animal) spinal cord convey mostly sensory information from the body to the brain. Axons and cell bodies in the ventral (facing the front of the animal) spinal cord primarily transmit signals controlling movement from the brain to the body.

The spinal cord also controls motor reflexes. These reflexes are quick, unconscious movements—like automatically removing a hand from a hot object. Reflexes are so fast because they involve local synaptic connections. For example, the knee reflex that a doctor tests during a routine physical is controlled by a single synapse between a sensory neuron and a motor neuron. While a reflex may only require the involvement of one or two synapses, synapses with interneurons in the spinal column transmit information to the brain to convey what happened (the knee jerked, or the hand was hot).

In the United States, there around 10,000 spinal cord injuries each year. Because the spinal cord is the information superhighway connecting the brain with the body, damage to the spinal cord can lead to paralysis. The extent of the paralysis depends on the location of the injury along the spinal cord and whether the spinal cord was completely severed. For example, if the spinal cord is damaged at the level of the neck, it can cause paralysis from the neck down, whereas damage to the spinal column further down may limit paralysis to the legs. Spinal cord injuries are notoriously difficult to treat because spinal nerves do not regenerate, although ongoing research suggests that stem cell transplants may be able to act as a bridge to reconnect severed nerves. Researchers are also looking at ways to prevent the inflammation that worsens nerve damage after injury. One such treatment is to pump the body with cold saline to induce hypothermia. This cooling can prevent swelling and other processes that are thought to worsen spinal cord injuries.

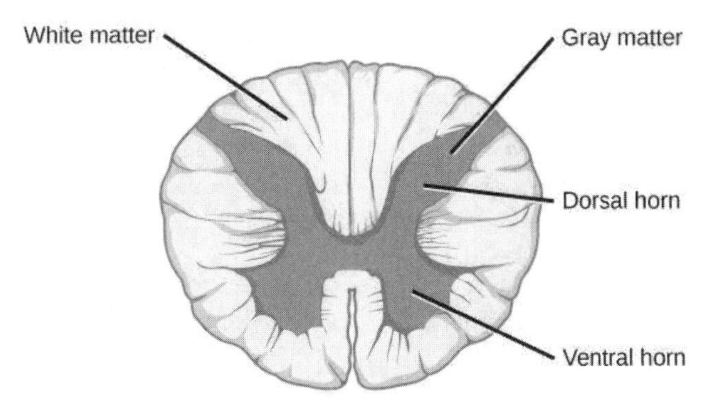

Figure 35.25 A cross-section of the spinal cord shows gray matter (containing cell bodies and interneurons) and white matter (containing axons).

35.4 The Peripheral Nervous System

By the end of this section, you will be able to do the following:

- Describe the organization and functions of the sympathetic and parasympathetic nervous systems
- Describe the organization and function of the sensory-somatic nervous system

The peripheral nervous system (PNS) is the connection between the central nervous system and the rest of the body. The CNS is like the power plant of the nervous system. It creates the signals that control the functions of the body. The PNS is like the wires that go to individual houses. Without those "wires," the signals produced by the CNS could not control the body (and the CNS would not be able to receive sensory information from the body either).

The PNS can be broken down into the **autonomic nervous system**, which controls bodily functions without conscious control, and the **sensory-somatic nervous system**, which transmits sensory information from the skin, muscles, and sensory organs to the CNS and sends motor commands from the CNS to the muscles.

Autonomic Nervous System

VISUAL CONNECTION

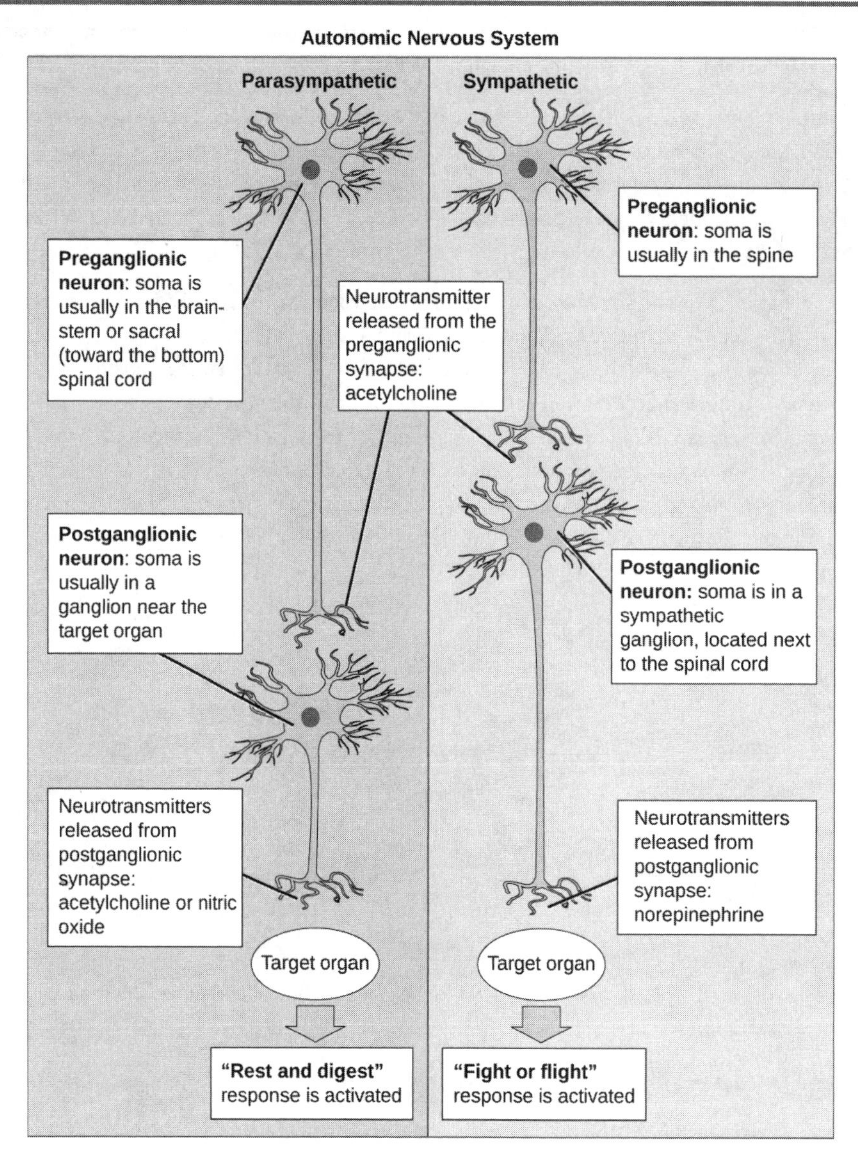

Figure 35.26 In the autonomic nervous system, a preganglionic neuron of the CNS synapses with a postganglionic neuron of the PNS. The postganglionic neuron, in turn, acts on a target organ. Autonomic responses are mediated by the sympathetic and the parasympathetic systems, which are antagonistic to one another. The sympathetic system activates the "fight or flight" response, while the parasympathetic system activates the "rest and digest" response.

Which of the following statements is false?

a. The parasympathetic pathway is responsible for resting the body, while the sympathetic pathway is responsible for preparing for an emergency.
b. Most preganglionic neurons in the sympathetic pathway originate in the spinal cord.
c. Slowing of the heartbeat is a parasympathetic response.
d. Parasympathetic neurons are responsible for releasing norepinephrine on the target organ, while sympathetic neurons are responsible for releasing acetylcholine.

The autonomic nervous system serves as the relay between the CNS and the internal organs. It controls the lungs, the heart, smooth muscle, and exocrine and endocrine glands. The autonomic nervous system controls these organs largely without conscious control; it can continuously monitor the conditions of these different systems and implement changes as needed. Signaling to the target tissue usually involves two synapses: a preganglionic neuron (originating in the CNS) synapses to a neuron in a ganglion that, in turn, synapses on the target organ, as illustrated in Figure 35.26. There are two divisions of the autonomic nervous system that often have opposing effects: the sympathetic nervous system and the parasympathetic nervous system.

Sympathetic Nervous System

The **sympathetic nervous system** is responsible for the "fight or flight" response that occurs when an animal encounters a dangerous situation. One way to remember this is to think of the surprise a person feels when encountering a snake ("snake" and "sympathetic" both begin with "s"). Examples of functions controlled by the sympathetic nervous system include an accelerated heart rate and inhibited digestion. These functions help prepare an organism's body for the physical strain required to escape a potentially dangerous situation or to fend off a predator.

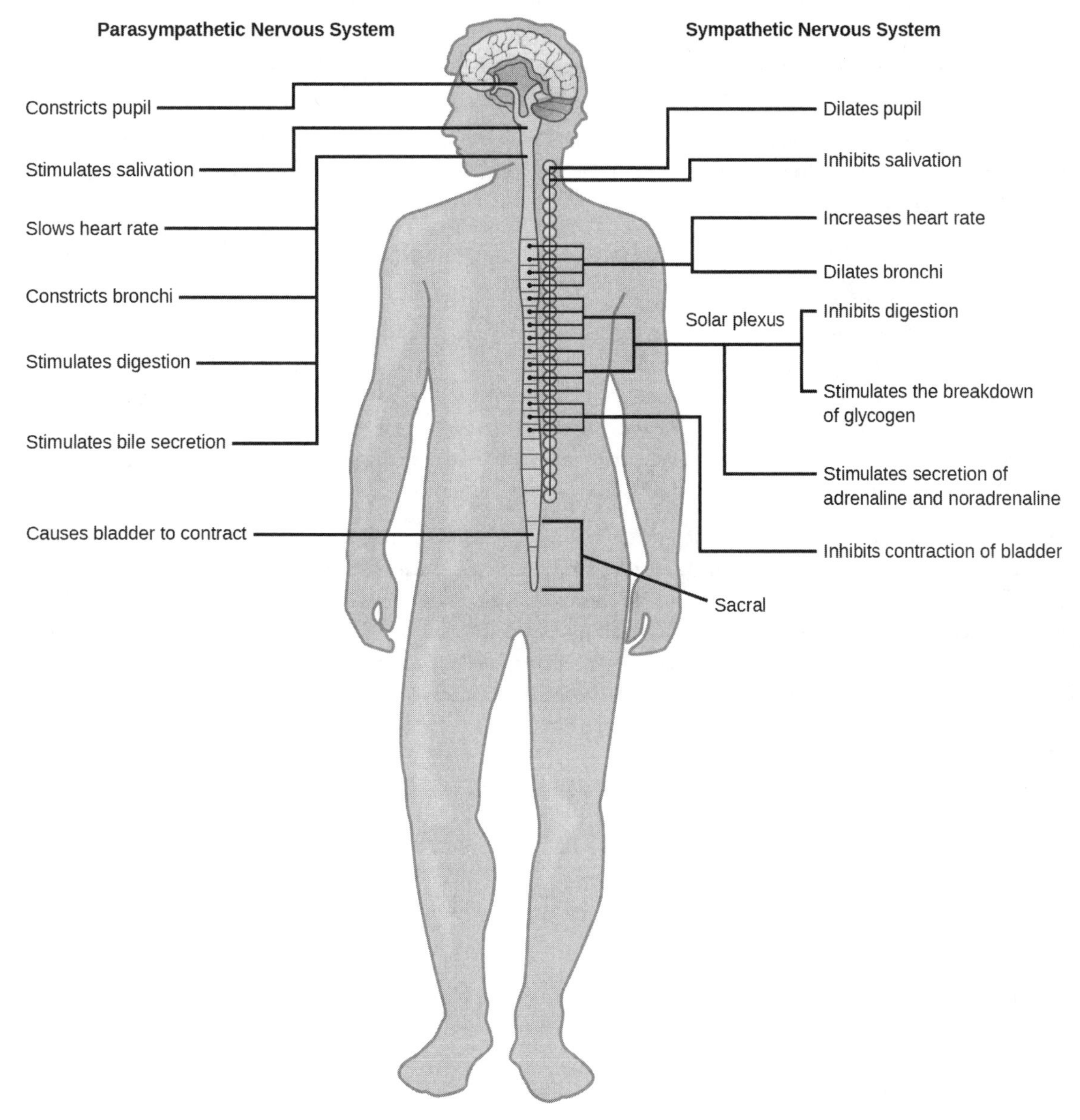

Figure 35.27 The sympathetic and parasympathetic nervous systems often have opposing effects on target organs.

Most preganglionic neurons in the sympathetic nervous system originate in the spinal cord, as illustrated in Figure 35.27. The axons of these neurons release **acetylcholine** on postganglionic neurons within sympathetic ganglia (the sympathetic ganglia form a chain that extends alongside the spinal cord). The acetylcholine activates the postganglionic neurons. Postganglionic neurons then release **norepinephrine** onto target organs. As anyone who has ever felt a rush before a big test, speech, or athletic event can attest, the effects of the sympathetic nervous system are quite pervasive. This is both because one preganglionic neuron synapses on multiple postganglionic neurons, amplifying the effect of the original synapse, and because the adrenal gland also releases norepinephrine (and the closely related hormone epinephrine) into the bloodstream. The physiological effects of this norepinephrine release include dilating the trachea and bronchi (making it easier for the animal to breathe), increasing heart rate, and moving blood from the skin to the heart, muscles, and brain (so the animal can think and run). The strength and speed of the sympathetic response helps an organism avoid danger, and scientists have found evidence that it may also increase LTP—allowing the animal to remember the dangerous situation and avoid it in the future.

Parasympathetic Nervous System

While the sympathetic nervous system is activated in stressful situations, the **parasympathetic nervous system** allows an animal to "rest and digest." One way to remember this is to think that during a restful situation like a picnic, the parasympathetic nervous system is in control ("picnic" and "parasympathetic" both start with "p"). Parasympathetic preganglionic neurons have cell bodies located in the brainstem and in the sacral (toward the bottom) spinal cord, as shown in Figure 35.27. The axons of the preganglionic neurons release acetylcholine on the postganglionic neurons, which are generally located very near the target organs. Most postganglionic neurons release acetylcholine onto target organs, although some release nitric oxide.

The parasympathetic nervous system resets organ function after the sympathetic nervous system is activated (the common adrenaline dump you feel after a 'fight-or-flight' event). Effects of acetylcholine release on target organs include slowing of heart rate, lowered blood pressure, and stimulation of digestion.

Sensory-Somatic Nervous System

The sensory-somatic nervous system is made up of cranial and spinal nerves and contains both sensory and motor neurons. Sensory neurons transmit sensory information from the skin, skeletal muscle, and sensory organs to the CNS. Motor neurons transmit messages about desired movement from the CNS to the muscles to make them contract. Without its sensory-somatic nervous system, an animal would be unable to process any information about its environment (what it sees, feels, hears, and so on) and could not control motor movements. Unlike the autonomic nervous system, which has two synapses between the CNS and the target organ, sensory and motor neurons have only one synapse—one ending of the neuron is at the organ and the other directly contacts a CNS neuron. Acetylcholine is the main neurotransmitter released at these synapses.

Humans have 12 **cranial nerves**, nerves that emerge from or enter the skull (cranium), as opposed to the spinal nerves, which emerge from the vertebral column. Each cranial nerve is accorded a name, which are detailed in Figure 35.28. Some cranial nerves transmit only sensory information. For example, the olfactory nerve transmits information about smells from the nose to the brainstem. Other cranial nerves transmit almost solely motor information. For example, the oculomotor nerve controls the opening and closing of the eyelid and some eye movements. Other cranial nerves contain a mix of sensory and motor fibers. For example, the glossopharyngeal nerve has a role in both taste (sensory) and swallowing (motor).

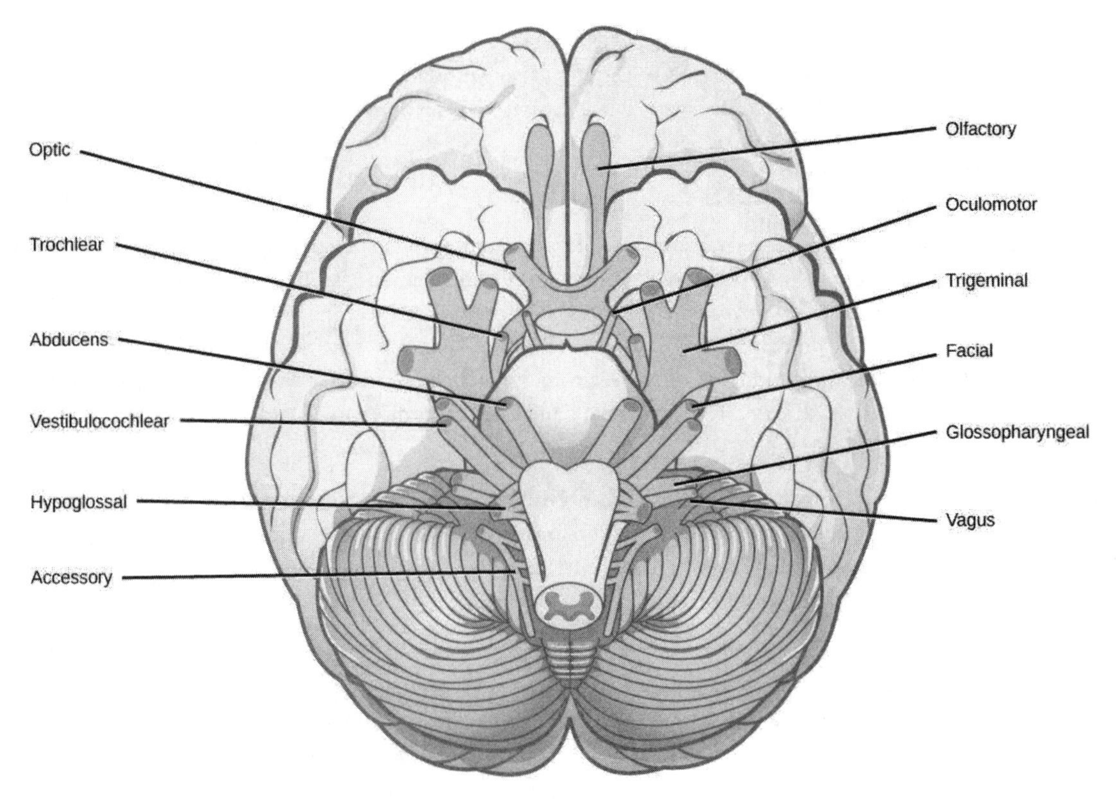

Figure 35.28 The human brain contains 12 cranial nerves that receive sensory input and control motor output for the head and neck.

Spinal nerves transmit sensory and motor information between the spinal cord and the rest of the body. Each of the 31 spinal nerves (in humans) contains both sensory and motor axons. The sensory neuron cell bodies are grouped in structures called dorsal root ganglia and are shown in Figure 35.29. Each sensory neuron has one projection—with a sensory receptor ending in skin, muscle, or sensory organs—and another that synapses with a neuron in the dorsal spinal cord. Motor neurons have cell bodies in the ventral gray matter of the spinal cord that project to muscle through the ventral root. These neurons are usually stimulated by interneurons within the spinal cord but are sometimes directly stimulated by sensory neurons.

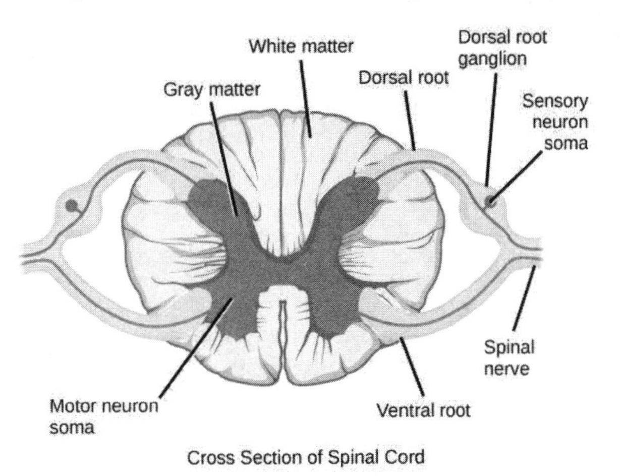

Figure 35.29 Spinal nerves contain both sensory and motor axons. The somas of sensory neurons are located in dorsal root ganglia. The somas of motor neurons are found in the ventral portion of the gray matter of the spinal cord.

35.5 Nervous System Disorders

By the end of this section, you will be able to do the following:

- Describe the symptoms, potential causes, and treatment of several examples of nervous system disorders

A nervous system that functions correctly is a fantastically complex, well-oiled machine—synapses fire appropriately, muscles move when needed, memories are formed and stored, and emotions are well regulated. Unfortunately, each year millions of people in the United States deal with some sort of nervous system disorder. While scientists have discovered potential causes of many of these diseases, and viable treatments for some, ongoing research seeks to find ways to better prevent and treat all of these disorders.

Neurodegenerative Disorders

Neurodegenerative disorders are illnesses characterized by a loss of nervous system functioning that are usually caused by neuronal death. These diseases generally worsen over time as more and more neurons die. The symptoms of a particular neurodegenerative disease are related to where in the nervous system the death of neurons occurs. Spinocerebellar ataxia, for example, leads to neuronal death in the cerebellum. The death of these neurons causes problems in balance and walking. Neurodegenerative disorders include Huntington's disease, amyotrophic lateral sclerosis, Alzheimer's disease and other types of dementia disorders, and Parkinson's disease. Here, Alzheimer's and Parkinson's disease will be discussed in more depth.

Alzheimer's Disease

Alzheimer's disease is the most common cause of dementia in the elderly. In 2012, an estimated 5.4 million Americans suffered from Alzheimer's disease, and payments for their care are estimated at $200 billion. Roughly one in every eight people age 65 or older has the disease. Due to the aging of the baby-boomer generation, there are projected to be as many as 13 million Alzheimer's patients in the United States in the year 2050.

Symptoms of Alzheimer's disease include disruptive memory loss, confusion about time or place, difficulty planning or executing tasks, poor judgment, and personality changes. Problems smelling certain scents can also be indicative of Alzheimer's disease and may serve as an early warning sign. Many of these symptoms are also common in people who are aging normally, so it is the severity and longevity of the symptoms that determine whether a person is suffering from Alzheimer's.

Alzheimer's disease was named for Alois Alzheimer, a German psychiatrist who published a report in 1911 about a woman who showed severe dementia symptoms. Along with his colleagues, he examined the woman's brain following her death and reported the presence of abnormal clumps, which are now called amyloid plaques, along with tangled brain fibers called neurofibrillary tangles. Amyloid plaques, neurofibrillary tangles, and an overall shrinking of brain volume are commonly seen in the brains of Alzheimer's patients. Loss of neurons in the hippocampus is especially severe in advanced Alzheimer's patients. Figure 35.30 compares a normal brain to the brain of an Alzheimer's patient. Many research groups are examining the causes of these hallmarks of the disease.

One form of the disease is usually caused by mutations in one of three known genes. This rare form of early onset Alzheimer's disease affects fewer than five percent of patients with the disease and causes dementia beginning between the ages of 30 and 60. The more prevalent, late-onset form of the disease likely also has a genetic component. One particular gene, apolipoprotein E (APOE) has a variant (E4) that increases a carrier's likelihood of getting the disease. Many other genes have been identified that might be involved in the pathology.

LINK TO LEARNING

Visit this website (http://openstax.org/l/alzheimers) for video links discussing genetics and Alzheimer's disease.

Unfortunately, there is no cure for Alzheimer's disease. Current treatments focus on managing the symptoms of the disease. Because decrease in the activity of cholinergic neurons (neurons that use the neurotransmitter acetylcholine) is common in Alzheimer's disease, several drugs used to treat the disease work by increasing acetylcholine neurotransmission, often by inhibiting the enzyme that breaks down acetylcholine in the synaptic cleft. Other clinical interventions focus on behavioral therapies like psychotherapy, sensory therapy, and cognitive exercises. Since Alzheimer's disease appears to hijack the normal aging process, research into prevention is prevalent. Smoking, obesity, and cardiovascular problems may be risk factors for the disease, so treatments for those may also help to prevent Alzheimer's disease. Some studies have shown that people who remain intellectually active by playing games, reading, playing musical instruments, and being socially active in later life have a reduced risk of developing the disease.

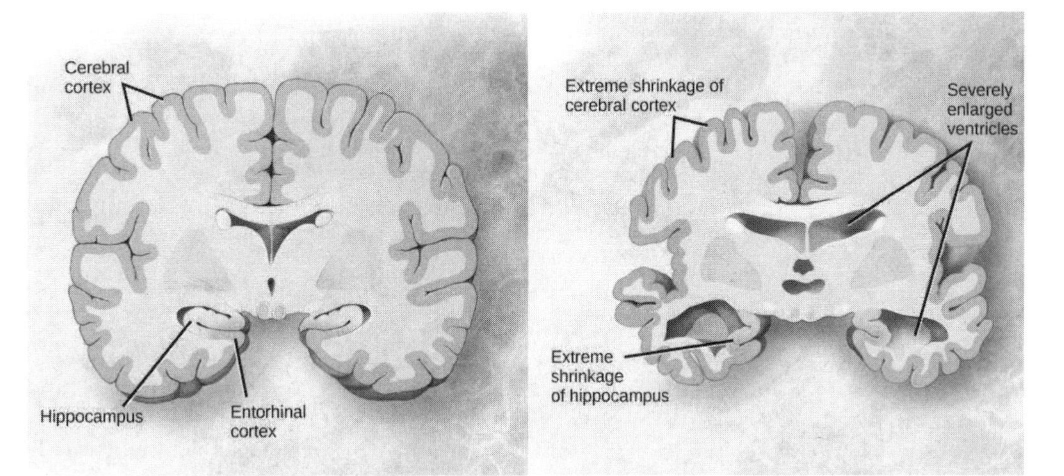

Figure 35.30 Compared to a normal brain (left), the brain from a patient with Alzheimer's disease (right) shows a dramatic neurodegeneration, particularly within the ventricles and hippocampus. (credit: modification of work by "Garrando"/Wikimedia Commons based on original images by ADEAR: "Alzheimer's Disease Education and Referral Center, a service of the National Institute on Aging")

Parkinson's Disease

Like Alzheimer's disease, **Parkinson's disease** is a neurodegenerative disease. It was first characterized by James Parkinson in 1817. Each year, 50,000-60,000 people in the United States are diagnosed with the disease. Parkinson's disease causes the loss of dopamine neurons in the substantia nigra, a midbrain structure that regulates movement. Loss of these neurons causes many symptoms including tremor (shaking of fingers or a limb), slowed movement, speech changes, balance and posture problems, and rigid muscles. The combination of these symptoms often causes a characteristic slow hunched shuffling walk, illustrated in Figure 35.31. Patients with Parkinson's disease can also exhibit psychological symptoms, such as dementia or emotional problems.

Although some patients have a form of the disease known to be caused by a single mutation, for most patients the exact causes of Parkinson's disease remain unknown: the disease likely results from a combination of genetic and environmental factors (similar to Alzheimer's disease). Post-mortem analysis of brains from Parkinson's patients shows the presence of Lewy bodies—abnormal protein clumps—in dopaminergic neurons. The prevalence of these Lewy bodies often correlates with the severity of the disease.

There is no cure for Parkinson's disease, and treatment is focused on easing symptoms. One of the most commonly prescribed drugs for Parkinson's is L-DOPA, which is a chemical that is converted into dopamine by neurons in the brain. This conversion increases the overall level of dopamine neurotransmission and can help compensate for the loss of dopaminergic neurons in the substantia nigra. Other drugs work by inhibiting the enzyme that breaks down dopamine.

Figure 35.31 Parkinson's patients often have a characteristic hunched walk.

Neurodevelopmental Disorders

Neurodevelopmental disorders occur when the development of the nervous system is disturbed. There are several different classes of neurodevelopmental disorders. Some, like Down Syndrome, cause intellectual deficits. Others specifically affect communication, learning, or the motor system. Some disorders like autism spectrum disorder and attention deficit/hyperactivity disorder have complex symptoms.

Autism

Autism spectrum disorder (ASD) is a neurodevelopmental disorder. Its severity differs from person to person. Estimates for the prevalence of the disorder have changed rapidly in the past few decades. Current estimates suggest that one in 88 children will develop the disorder. ASD is four times more prevalent in males than females.

LINK TO LEARNING

This video (http://openstax.org/l/autism) discusses possible reasons why there has been a recent increase in the number of people diagnosed with autism.

A characteristic symptom of ASD is impaired social skills. Children with autism may have difficulty making and maintaining eye contact and reading social cues. They also may have problems feeling empathy for others. Other symptoms of ASD include repetitive motor behaviors (such as rocking back and forth), preoccupation with specific subjects, strict adherence to certain rituals, and unusual language use. Up to 30 percent of patients with ASD develop epilepsy, and patients with some forms of the disorder (like Fragile X) also have intellectual disability. Because it is a spectrum disorder, other ASD patients are very functional and have good-to-excellent language skills. Many of these patients do not feel that they suffer from a disorder and instead think that their brains just process information differently.

Except for some well-characterized, clearly genetic forms of autism (like Fragile X and Rett's Syndrome), the causes of ASD are largely unknown. Variants of several genes correlate with the presence of ASD, but for any given patient, many different mutations in different genes may be required for the disease to develop. At a general level, ASD is thought to be a disease of "incorrect" wiring. Accordingly, brains of some ASD patients lack the same level of synaptic pruning that occurs in non-affected people. In the 1990s, a research paper linked autism to a common vaccine given to children. This paper was retracted when it was discovered that the author falsified data, and follow-up studies showed no connection between vaccines and autism.

Treatment for autism usually combines behavioral therapies and interventions, along with medications to treat other disorders common to people with autism (depression, anxiety, obsessive compulsive disorder). Although early interventions can help mitigate the effects of the disease, there is currently no cure for ASD.

Attention Deficit Hyperactivity Disorder (ADHD)

Approximately three to five percent of children and adults are affected by **attention deficit/hyperactivity disorder (ADHD)**. Like ASD, ADHD is more prevalent in males than females. Symptoms of the disorder include inattention (lack of focus), executive functioning difficulties, impulsivity, and hyperactivity beyond what is characteristic of the normal developmental stage. Some patients do not have the hyperactive component of symptoms and are diagnosed with a subtype of ADHD: attention deficit disorder (ADD). Many people with ADHD also show comorbitity, in that they develop secondary disorders in addition to ADHD. Examples include depression or obsessive compulsive disorder (OCD). Figure 35.32 provides some statistics concerning comorbidity with ADHD.

The cause of ADHD is unknown, although research points to a delay and dysfunction in the development of the prefrontal cortex and disturbances in neurotransmission. According to studies of twins, the disorder has a strong genetic component. There are several candidate genes that may contribute to the disorder, but no definitive links have been discovered. Environmental factors, including exposure to certain pesticides, may also contribute to the development of ADHD in some patients. Treatment for ADHD often involves behavioral therapies and the prescription of stimulant medications, which paradoxically cause a calming effect in these patients.

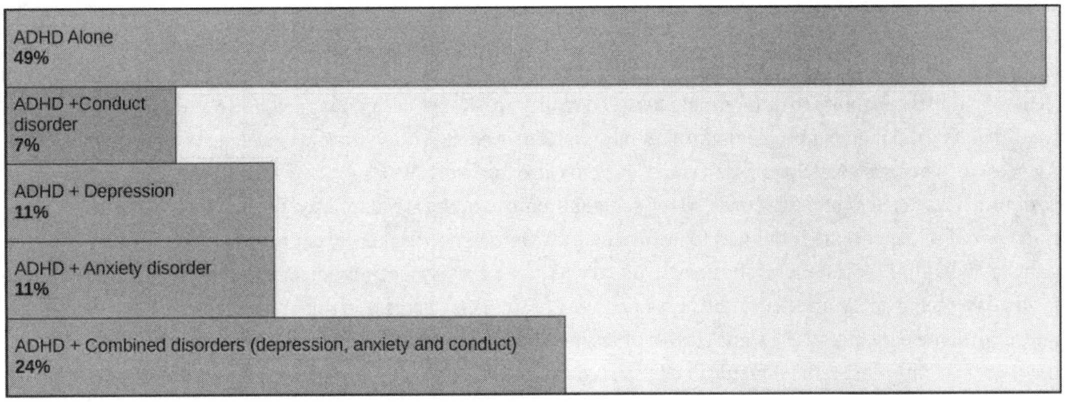

ADHD Alone
49%
ADHD +Conduct disorder
7%
ADHD + Depression
11%
ADHD + Anxiety disorder
11%
ADHD + Combined disorders (depression, anxiety and conduct)
24%

Figure 35.32 Many people with ADHD have one or more other neurological disorders. (credit "chart design and illustration": modification of work by Leigh Coriale; credit "data": Drs. Biederman and Faraone, Massachusetts General Hospital).

CAREER CONNECTION

Neurologist

Neurologists are physicians who specialize in disorders of the nervous system. They diagnose and treat disorders such as epilepsy, stroke, dementia, nervous system injuries, Parkinson's disease, sleep disorders, and multiple sclerosis. Neurologists are medical doctors who have attended college, medical school, and completed three to four years of neurology residency.

When examining a new patient, a neurologist takes a full medical history and performs a complete physical exam. The physical exam contains specific tasks that are used to determine what areas of the brain, spinal cord, or peripheral nervous system may be damaged. For example, to check whether the hypoglossal nerve is functioning correctly, the neurologist will ask the patient to move his or her tongue in different ways. If the patient does not have full control over tongue movements, then the hypoglossal nerve may be damaged or there may be a lesion in the brainstem where the cell bodies of these neurons reside (or there could be damage to the tongue muscle itself).

Neurologists have other tools besides a physical exam they can use to diagnose particular problems in the nervous system. If the patient has had a seizure, for example, the neurologist can use electroencephalography (EEG), which involves taping electrodes to the scalp to record brain activity, to try to determine which brain regions are involved in the seizure. In suspected stroke patients, a neurologist can use a computerized tomography (CT) scan, which is a type of X-ray, to look for bleeding in the brain or a possible brain tumor. To treat patients with neurological problems, neurologists can prescribe medications or refer the patient to a neurosurgeon for surgery.

LINK TO LEARNING

This website (http://openstax.org/l/neurologic_exam) allows you to see the different tests a neurologist might use to see what regions of the nervous system may be damaged in a patient.

Mental Illnesses

Mental illnesses are nervous system disorders that result in problems with thinking, mood, or relating with other people. These disorders are severe enough to affect a person's quality of life and often make it difficult for people to perform the routine tasks of daily living. Debilitating mental disorders plague approximately 12.5 million Americans (about 1 in 17 people) at an annual cost of more than $300 billion. There are several types of mental disorders including schizophrenia, major depression, bipolar disorder, anxiety disorders and phobias, post-traumatic stress disorders, and obsessive-compulsive disorder (OCD), among others. The American Psychiatric Association publishes the Diagnostic and Statistical Manual of Mental Disorders (or DSM), which describes the symptoms required for a patient to be diagnosed with a particular mental disorder. Each newly released version of the DSM contains different symptoms and classifications as scientists learn more about these disorders, their causes, and how they relate to each other. A more detailed discussion of two mental illnesses—schizophrenia and major depression—is given below.

Schizophrenia

Schizophrenia is a serious and often debilitating mental illness affecting one percent of people in the United States. Symptoms of the disease include the inability to differentiate between reality and imagination, inappropriate and unregulated emotional responses, difficulty thinking, and problems with social situations. People with schizophrenia can suffer from hallucinations and hear voices; they may also suffer from delusions. Patients also have so-called "negative" symptoms like a flattened emotional state, loss of pleasure, and loss of basic drives. Many schizophrenic patients are diagnosed in their late adolescence or early 20s. The development of schizophrenia is thought to involve malfunctioning dopaminergic neurons and may also involve problems with glutamate signaling. Treatment for the disease usually requires antipsychotic medications that work by blocking dopamine receptors and decreasing dopamine neurotransmission in the brain. This decrease in dopamine can cause Parkinson's disease-like symptoms in some patients. While some classes of antipsychotics can be quite effective at treating the disease, they are not a cure, and most patients must remain medicated for the rest of their lives.

Depression

Major depression affects approximately 6.7 percent of the adults in the United States each year and is one of the most common mental disorders. To be diagnosed with major depressive disorder, a person must have experienced a severely depressed mood lasting longer than two weeks along with other symptoms including a loss of enjoyment in activities that were previously enjoyed, changes in appetite and sleep schedules, difficulty concentrating, feelings of worthlessness, and suicidal thoughts. The exact causes of major depression are unknown and likely include both genetic and environmental risk factors. Some research supports the "classic monoamine hypothesis," which suggests that depression is caused by a decrease in norepinephrine and serotonin neurotransmission. One argument against this hypothesis is the fact that some antidepressant medications cause an increase in norepinephrine and serotonin release within a few hours of beginning treatment—but clinical results of these medications are not seen until weeks later. This has led to alternative hypotheses: for example, dopamine may also be decreased in depressed patients, or it may actually be an increase in norepinephrine and serotonin that causes the disease, and antidepressants force a feedback loop that decreases this release. Treatments for depression include psychotherapy, electroconvulsive therapy, deep-brain stimulation, and prescription medications. There are several classes of antidepressant medications that work through different mechanisms. For example, monoamine oxidase inhibitors (MAO inhibitors) block the enzyme that degrades many neurotransmitters (including dopamine, serotonin, norepinephrine), resulting in increased neurotransmitter in the synaptic cleft. Selective serotonin reuptake inhibitors (SSRIs) block the reuptake of serotonin into the presynaptic neuron. This blockage results in an increase in serotonin in the synaptic cleft. Other types of drugs such as norepinephrine-dopamine reuptake inhibitors and norepinephrine-serotonin reuptake inhibitors are also used to treat depression.

Other Neurological Disorders

There are several other neurological disorders that cannot be easily placed in the above categories. These include chronic pain conditions, cancers of the nervous system, epilepsy disorders, and stroke. Epilepsy and stroke are discussed below.

Epilepsy

Estimates suggest that up to three percent of people in the United States will be diagnosed with **epilepsy** in their lifetime. While there are several different types of epilepsy, all are characterized by recurrent seizures. Epilepsy itself can be a symptom of a brain injury, disease, or other illness. For example, people who have intellectual disability or ASD can experience seizures, presumably because the developmental wiring malfunctions that caused their disorders also put them at risk for epilepsy. For many patients, however, the cause of their epilepsy is never identified and is likely to be a combination of genetic and environmental factors. Often, seizures can be controlled with anticonvulsant medications. However, for very severe cases, patients may undergo brain surgery to remove the brain area where seizures originate.

Stroke

A stroke results when blood fails to reach a portion of the brain for a long enough time to cause damage. Without the oxygen supplied by blood flow, neurons in this brain region die. This neuronal death can cause many different symptoms—depending on the brain area affected— including headache, muscle weakness or paralysis, speech disturbances, sensory problems, memory loss, and confusion. Stroke is often caused by blood clots and can also be caused by the bursting of a weak blood vessel. Strokes are extremely common and are the third most common cause of death in the United States. On average one person experiences a stroke every 40 seconds in the United States. Approximately 75 percent of strokes occur in people older than 65. Risk factors for stroke include high blood pressure, diabetes, high cholesterol, and a family history of stroke. Smoking doubles the risk of stroke. Because a stroke is a medical emergency, patients with symptoms of a stroke should immediately go to the emergency room, where they can receive drugs that will dissolve any clot that may have formed. These drugs will not work if the stroke was caused by a burst blood vessel or if the stroke occurred more than three hours before arriving at the hospital. Treatment following a stroke can include blood pressure medication (to prevent future strokes) and (sometimes intense) physical therapy.

KEY TERMS

acetylcholine neurotransmitter released by neurons in the central nervous system and peripheral nervous system

action potential self-propagating momentary change in the electrical potential of a neuron (or muscle) membrane

Alzheimer's disease neurodegenerative disorder characterized by problems with memory and thinking

amygdala structure within the limbic system that processes fear

arachnoid mater spiderweb-like middle layer of the meninges that cover the central nervous system

astrocyte glial cell in the central nervous system that provide nutrients, extracellular buffering, and structural support for neurons; also makes up the blood-brain barrier

attention deficit hyperactivity disorder (ADHD) neurodevelopmental disorder characterized by difficulty maintaining attention and controlling impulses

autism spectrum disorder (ASD) neurodevelopmental disorder characterized by impaired social interaction and communication abilities

autonomic nervous system part of the peripheral nervous system that controls bodily functions

axon tube-like structure that propagates a signal from a neuron's cell body to axon terminals

axon hillock electrically sensitive structure on the cell body of a neuron that integrates signals from multiple neuronal connections

axon terminal structure on the end of an axon that can form a synapse with another neuron

basal ganglia interconnected collections of cells in the brain that are involved in movement and motivation; also known as basal nuclei

basal nuclei see basal ganglia

brainstem portion of the brain that connects with the spinal cord; controls basic nervous system functions like breathing, heart rate, and swallowing

cerebellum brain structure involved in posture, motor coordination, and learning new motor actions

cerebral cortex outermost sheet of brain tissue; involved in many higher-order functions

cerebrospinal fluid (CSF) clear liquid that surrounds the brain and spinal cord and fills the ventricles and central canal; acts as a shock absorber and circulates material throughout the brain and spinal cord

choroid plexus spongy tissue within ventricles that produces cerebrospinal fluid

cingulate gyrus helps regulate emotions and pain; thought to directly drive the body's conscious response to unpleasant experiences

corpus callosum thick fiber bundle that connects the cerebral hemispheres

cranial nerve sensory and/or motor nerve that emanates from the brain

dendrite structure that extends away from the cell body to receive messages from other neurons

depolarization change in the membrane potential to a less negative value

dura mater tough outermost layer that covers the central nervous system

ependymal cell that lines fluid-filled ventricles of the brain and the central canal of the spinal cord; involved in production of cerebrospinal fluid

epilepsy neurological disorder characterized by recurrent seizures

excitatory postsynaptic potential (EPSP) depolarization of a postsynaptic membrane caused by neurotransmitter molecules released from a presynaptic cell

frontal lobe part of the cerebral cortex that contains the motor cortex and areas involved in planning, attention, and language

glia (also, glial cells) cells that provide support functions for neurons

gyrus (plural: gyri) ridged protrusions in the cortex

hippocampus brain structure in the temporal lobe involved in processing memories

hyperpolarization change in the membrane potential to a more negative value

hypothalamus brain structure that controls hormone release and body homeostasis

inhibitory postsynaptic potential (IPSP) hyperpolarization of a postsynaptic membrane caused by neurotransmitter molecules released from a presynaptic cell

limbic system connected brain areas that process emotion and motivation

long-term depression (LTD) prolonged decrease in synaptic coupling between a pre- and postsynaptic cell

long-term potentiation (LTP) prolonged increase in synaptic coupling between a pre-and postsynaptic cell

major depression mental illness characterized by prolonged periods of sadness

membrane potential difference in electrical potential between the inside and outside of a cell

meninge membrane that covers and protects the central nervous system

microglia glia that scavenge and degrade dead cells and protect the brain from invading microorganisms

myelin fatty substance produced by glia that insulates axons

neurodegenerative disorder nervous system disorder characterized by the progressive loss of neurological functioning, usually caused by neuron death

neuron specialized cell that can receive and transmit electrical and chemical signals

nodes of Ranvier gaps in the myelin sheath where the

signal is recharged

norepinephrine neurotransmitter and hormone released by activation of the sympathetic nervous system

occipital lobe part of the cerebral cortex that contains visual cortex and processes visual stimuli

oligodendrocyte glial cell that myelinates central nervous system neuron axons

parasympathetic nervous system division of autonomic nervous system that regulates visceral functions during rest and digestion

parietal lobe part of the cerebral cortex involved in processing touch and the sense of the body in space

Parkinson's disease neurodegenerative disorder that affects the control of movement

pia mater thin membrane layer directly covering the brain and spinal cord

proprioception sense about how parts of the body are oriented in space

radial glia glia that serve as scaffolds for developing neurons as they migrate to their final destinations

refractory period period after an action potential when it is more difficult or impossible for an action potential to be fired; caused by inactivation of sodium channels and activation of additional potassium channels of the membrane

saltatory conduction "jumping" of an action potential along an axon from one node of Ranvier to the next

satellite glia glial cell that provides nutrients and structural support for neurons in the peripheral nervous system

schizophrenia mental disorder characterized by the inability to accurately perceive reality; patients often have difficulty thinking clearly and can suffer from delusions

Schwann cell glial cell that creates myelin sheath around a peripheral nervous system neuron axon

sensory-somatic nervous system system of sensory and motor nerves

somatosensation sense of touch

spinal cord thick fiber bundle that connects the brain with peripheral nerves; transmits sensory and motor information; contains neurons that control motor reflexes

spinal nerve nerve projecting between skin or muscle and spinal cord

sulcus (plural: sulci) indents or "valleys" in the cortex

summation process of multiple presynaptic inputs creating EPSPs around the same time for the postsynaptic neuron to be sufficiently depolarized to fire an action potential

sympathetic nervous system division of autonomic nervous system activated during stressful "fight or flight" situations

synapse junction between two neurons where neuronal signals are communicated

synaptic cleft space between the presynaptic and postsynaptic membranes

synaptic vesicle spherical structure that contains a neurotransmitter

temporal lobe part of the cerebral cortex that processes auditory input; parts of the temporal lobe are involved in speech, memory, and emotion processing

thalamus brain area that relays sensory information to the cortex

threshold of excitation level of depolarization needed for an action potential to fire

ventricle cavity within brain that contains cerebrospinal fluid

CHAPTER SUMMARY

35.1 Neurons and Glial Cells

The nervous system is made up of neurons and glia. Neurons are specialized cells that are capable of sending electrical as well as chemical signals. Most neurons contain dendrites, which receive these signals, and axons that send signals to other neurons or tissues. There are four main types of neurons: unipolar, bipolar, multipolar, and pseudounipolar neurons. Glia are non-neuronal cells in the nervous system that support neuronal development and signaling. There are several types of glia that serve different functions.

35.2 How Neurons Communicate

Neurons have charged membranes because there are different concentrations of ions inside and outside of the cell. Voltage-gated ion channels control the movement of ions into and out of a neuron. When a neuronal membrane is depolarized to at least the threshold of excitation, an action potential is fired. The action potential is then propagated along a myelinated axon to the axon terminals. In a chemical synapse, the action potential causes release of neurotransmitter molecules into the synaptic cleft. Through binding to postsynaptic receptors, the neurotransmitter can cause excitatory or inhibitory postsynaptic potentials by depolarizing or hyperpolarizing, respectively, the postsynaptic membrane. In electrical synapses, the action potential is directly communicated to the postsynaptic cell through gap junctions—large channel proteins that connect the pre-and postsynaptic membranes. Synapses are not static structures and can be strengthened and weakened. Two mechanisms of synaptic plasticity are long-term potentiation and long-term depression.

35.3 The Central Nervous System

The vertebrate central nervous system contains the brain and the spinal cord, which are covered and protected by three meninges. The brain contains structurally and functionally defined regions. In mammals, these include the cortex (which can be broken down into four primary functional lobes: frontal, temporal, occipital, and parietal), basal ganglia, thalamus, hypothalamus, limbic system, cerebellum, and brainstem—although structures in some of these designations overlap. While functions may be primarily localized to one structure in the brain, most complex functions, like language and sleep, involve neurons in multiple brain regions. The spinal cord is the information superhighway that connects the brain with the rest of the body through its connections with peripheral nerves. It transmits sensory and motor input and also controls motor reflexes.

35.4 The Peripheral Nervous System

The peripheral nervous system contains both the autonomic and sensory-somatic nervous systems. The autonomic nervous system provides unconscious control over visceral functions and has two divisions: the sympathetic and parasympathetic nervous systems. The sympathetic nervous system is activated in stressful situations to prepare the animal for a "fight or flight" response. The parasympathetic nervous system is active during restful periods. The sensory-somatic nervous system is made of cranial and spinal nerves that transmit sensory information from skin and muscle to the CNS and motor commands from the CNS to the muscles.

35.5 Nervous System Disorders

Some general themes emerge from the sampling of nervous system disorders presented above. The causes for most disorders are not fully understood—at least not for all patients—and likely involve a combination of nature (genetic mutations that become risk factors) and nurture (emotional trauma, stress, hazardous chemical exposure). Because the causes have yet to be fully determined, treatment options are often lacking and only address symptoms.

VISUAL CONNECTION QUESTIONS

1. Figure 35.3 Which of the following statements is false?
    a. The soma is the cell body of a nerve cell.
    b. Myelin sheath provides an insulating layer to the dendrites.
    c. Axons carry the signal from the soma to the target.
    d. Dendrites carry the signal to the soma.

2. Figure 35.11 Potassium channel blockers, such as amiodarone and procainamide, which are used to treat abnormal electrical activity in the heart, called cardiac dysrhythmia, impede the movement of K+ through voltage-gated K+ channels. Which part of the action potential would you expect potassium channels to affect?

3. Figure 35.26 Which of the following statements is false?
    a. The parasympathetic pathway is responsible for relaxing the body, while the sympathetic pathway is responsible for preparing for an emergency.
    b. Most preganglionic neurons in the sympathetic pathway originate in the spinal cord.
    c. Slowing of the heartbeat is a parasympathetic response.
    d. Parasympathetic neurons are responsible for releasing norepinephrine on the target organ, while sympathetic neurons are responsible for releasing acetylcholine.

REVIEW QUESTIONS

4. Neurons contain _______, which can receive signals from other neurons.
    a. axons
    b. mitochondria
    c. dendrites
    d. Golgi bodies

5. A(n) _______ neuron has one axon and one dendrite extending directly from the cell body.
    a. unipolar
    b. bipolar
    c. multipolar
    d. pseudounipolar

6. Glia that provide myelin for neurons in the brain are called _______.
    a. Schwann cells
    b. oligodendrocytes
    c. microglia
    d. astrocytes

7. Meningitis is a viral or bacterial infection of the brain. Which cell type is the first to have its function disrupted during meningitis?
    a. astrocytes
    b. microglia
    c. neurons
    d. satellite glia

8. For a neuron to fire an action potential, its membrane must reach _______.
   a. hyperpolarization
   b. the threshold of excitation
   c. the refractory period
   d. inhibitory postsynaptic potential

9. After an action potential, the opening of additional voltage-gated _______ channels and the inactivation of sodium channels, cause the membrane to return to its resting membrane potential.
   a. sodium
   b. potassium
   c. calcium
   d. chloride

10. What is the term for protein channels that connect two neurons at an electrical synapse?
    a. synaptic vesicles
    b. voltage-gated ion channels
    c. gap junction protein
    d. sodium-potassium exchange pumps

11. Which of the following molecules is **not** involved in the maintenance of the resting membrane potential?
    a. potassium cations
    b. ATP
    c. voltage-gated ion channels
    d. calcium cations

12. The _______ lobe contains the visual cortex.
    a. frontal
    b. parietal
    c. temporal
    d. occipital

13. The _______ connects the two cerebral hemispheres.
    a. limbic system
    b. corpus callosum
    c. cerebellum
    d. pituitary

14. Neurons in the _______ control motor reflexes.
    a. thalamus
    b. spinal cord
    c. parietal lobe
    d. hippocampus

15. Phineas Gage was a 19th century railroad worker who survived an accident that drove a large iron rod through his head. If the injury resulted in him becoming temperamental and capricious what part of his brain was damaged?
    a. frontal lobe
    b. hippocampus
    c. parietal lobe
    d. temporal lobe

16. Activation of the sympathetic nervous system causes:
    a. increased blood flow into the skin
    b. a decreased heart rate
    c. an increased heart rate
    d. increased digestion

17. Where are parasympathetic preganglionic cell bodies located?
    a. cerebellum
    b. brainstem
    c. dorsal root ganglia
    d. skin

18. _______ is released by motor nerve endings onto muscle.
    a. Acetylcholine
    b. Norepinephrine
    c. Dopamine
    d. Serotonin

19. Parkinson's disease is a caused by the degeneration of neurons that release _______.
    a. serotonin
    b. dopamine
    c. glutamate
    d. norepinephrine

20. _______ medications are often used to treat patients with ADHD.
    a. Tranquilizer
    b. Antibiotic
    c. Stimulant
    d. Anti-seizure

21. Strokes are often caused by _______.
    a. neurodegeneration
    b. blood clots or burst blood vessels
    c. seizures
    d. viruses

22. Why is it difficult to identify the cause of many nervous system disorders?
    a. The genes associated with the diseases are not known.
    b. There are no obvious defects in brain structure.
    c. The onset and display of symptoms varies between patients.
    d. all of the above

23. Why do many patients with neurodevelopmental disorders develop secondary disorders?
    a. Their genes predispose them to schizophrenia.
    b. Stimulant medications cause new behavioral disorders.
    c. Behavioral therapies only improve neurodevelopmental disorders.
    d. Dysfunction in the brain can affect many aspects of the body.

CRITICAL THINKING QUESTIONS

24. How are neurons similar to other cells? How are they unique?
25. Multiple sclerosis causes demyelination of axons in the brain and spinal cord. Why is this problematic?
26. Many neurons have only a single axon, but many terminals at the end of the axon. How does this end structure of the axon support its function?
27. How does myelin aid propagation of an action potential along an axon? How do the nodes of Ranvier help this process?
28. What are the main steps in chemical neurotransmission?
29. Describe how long-term potentiation can lead to a nicotine addiction.
30. What methods can be used to determine the function of a particular brain region?
31. What are the main functions of the spinal cord?
32. Alzheimer's disease involves three of the four lobes of the brain. Identify one of the involved lobes and describe the lobe's symptoms associated with the disease.
33. What are the main differences between the sympathetic and parasympathetic branches of the autonomic nervous system?
34. What are the main functions of the sensory-somatic nervous system?
35. Describe how the sensory-somatic nervous system reacts by reflex to a person touching something hot. How does this allow for rapid responses in potentially dangerous situations?
36. Scientists have suggested that the autonomic nervous system is not well-adapted to modern human life. How is the sympathetic nervous system an ineffective response to the everyday challenges faced by modern humans?
37. What are the main symptoms of Alzheimer's disease?
38. What are possible treatments for patients with major depression?

CHAPTER 36
Sensory Systems

Figure 36.1 This shark uses its senses of sight, vibration (lateral-line system), and smell to hunt, but it also relies on its ability to sense the electric fields of prey, a sense not present in most land animals. (credit: modification of work by Hermanus Backpackers Hostel, South Africa)

INTRODUCTION In more advanced animals, the senses are constantly at work, making the animal aware of stimuli—such as light, or sound, or the presence of a chemical substance in the external environment—and monitoring information about the organism's internal environment. All bilaterally symmetric animals have a sensory system, and the development of any species' sensory system has been driven by natural selection; thus, sensory systems differ among species according to the demands of their environments. The shark, unlike most fish predators, is electrosensitive—that is, sensitive to electrical fields produced by other animals in its environment. While it is helpful to this underwater predator, electrosensitivity is a sense not found in most land animals.

Chapter Outline

36.1 Sensory Processes

By the end of this section, you will be able to do the following:

- Identify the general and special senses in humans
- Describe three important steps in sensory perception
- Explain the concept of just-noticeable difference in sensory perception

Senses provide information about the body and its environment. Humans have five special senses: olfaction (smell), gustation (taste), equilibrium (balance and body position), vision, and hearing. Additionally, we possess general senses, also called somatosensation, which respond to stimuli like temperature, pain, pressure, and vibration. **Vestibular sensation**, which is an organism's sense of spatial orientation and balance, **proprioception** (position of bones, joints, and muscles), and the sense of limb position that is used to track **kinesthesia** (limb movement) are part of somatosensation. Although the sensory systems associated with these senses are very different, all share a common function: to convert a stimulus (such as light, or sound, or the position of the body) into an electrical signal in the nervous system. This process is called **sensory transduction**.

There are two broad types of cellular systems that perform sensory transduction. In one, a neuron works with a **sensory receptor**, a cell, or cell process that is specialized to engage with and detect a specific stimulus. Stimulation of the sensory receptor activates the associated afferent neuron, which carries information about the stimulus to the central nervous system. In the second type of sensory transduction, a sensory nerve ending responds to a stimulus in the internal or external environment: this neuron constitutes the sensory receptor. Free nerve endings can be stimulated by several different stimuli, thus showing little receptor specificity. For example, pain receptors in your gums and teeth may be stimulated by temperature changes, chemical stimulation, or pressure.

Reception

The first step in sensation is **reception**, which is the activation of sensory receptors by stimuli such as mechanical stimuli (being bent or squished, for example), chemicals, or temperature. The receptor can then respond to the stimuli. The region in space in which a given sensory receptor can respond to a stimulus, be it far away or in contact with the body, is that receptor's **receptive field**. Think for a moment about the differences in receptive fields for the different senses. For the sense of touch, a stimulus must come into contact with the body. For the sense of hearing, a stimulus can be a moderate distance away (some baleen whale sounds can propagate for many kilometers). For vision, a stimulus can be very far away; for example, the visual system perceives light from stars at enormous distances.

Transduction

The most fundamental function of a sensory system is the translation of a sensory signal to an electrical signal in the nervous system. This takes place at the sensory receptor, and the change in electrical potential that is produced is called the **receptor potential**. How is sensory input, such as pressure on the skin, changed to a receptor potential? In this example, a type of receptor called a **mechanoreceptor** (as shown in Figure 36.2) possesses specialized membranes that respond to pressure. Disturbance of these dendrites by compressing them or bending them opens gated ion channels in the plasma membrane of the sensory neuron, changing its electrical potential. Recall that in the nervous system, a positive change of a neuron's electrical potential (also called the membrane potential), depolarizes the neuron. Receptor potentials are graded potentials: the magnitude of these graded (receptor) potentials varies with the strength of the stimulus. If the magnitude of depolarization is sufficient (that is, if membrane potential reaches a threshold), the neuron will fire an action potential. In most cases, the correct stimulus impinging on a sensory receptor will drive membrane potential in a positive direction, although for some receptors, such as those in the visual system, this is not always the case.

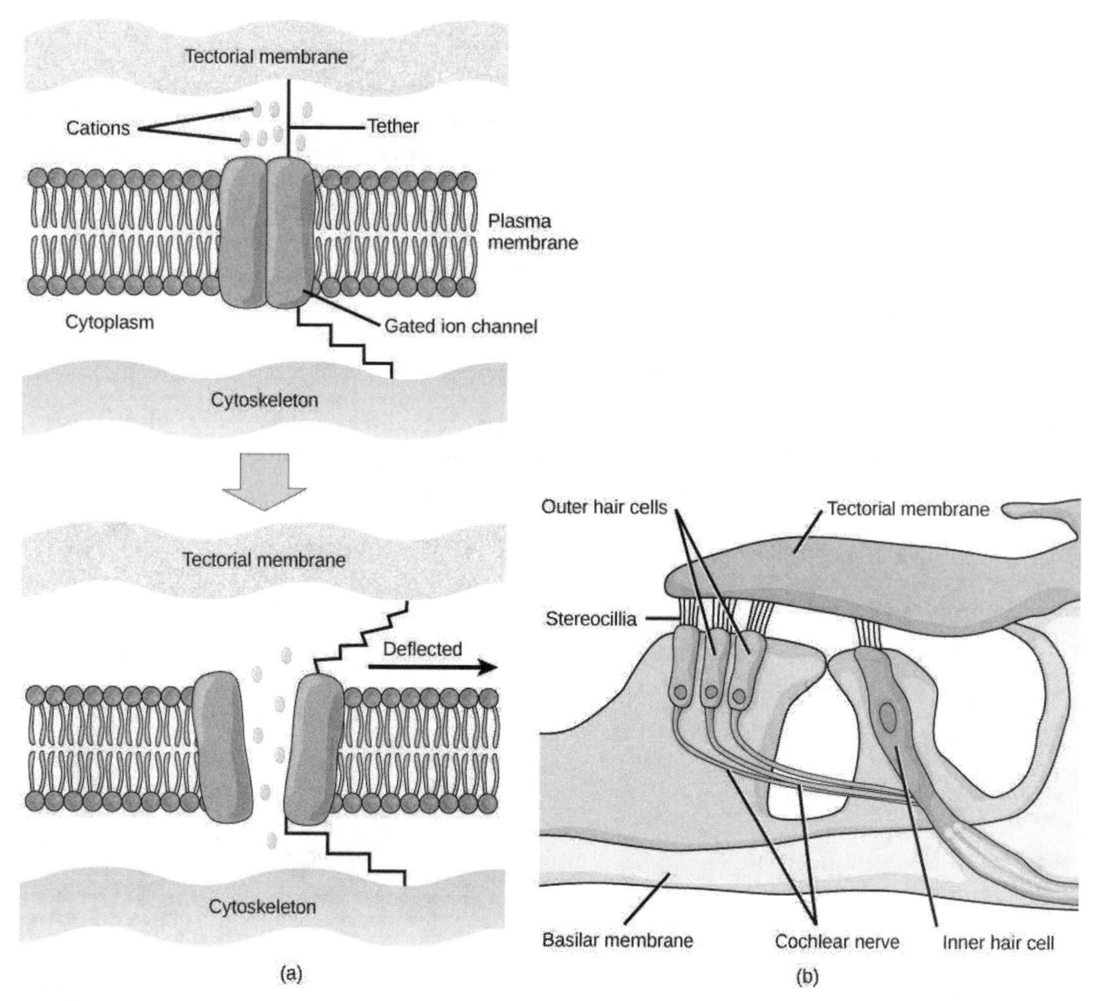

Figure 36.2 (a) Mechanosensitive ion channels are gated ion channels that respond to mechanical deformation of the plasma membrane. A mechanosensitive channel is connected to the plasma membrane and the cytoskeleton by hair-like tethers. When pressure causes the extracellular matrix to move, the channel opens, allowing ions to enter or exit the cell. (b) Stereocilia in the human ear are connected to mechanosensitive ion channels. When a sound causes the stereocilia to move, mechanosensitive ion channels transduce the signal to the cochlear nerve.

Sensory receptors for different senses are very different from each other, and they are specialized according to the type of stimulus they sense: they have receptor specificity. For example, touch receptors, light receptors, and sound receptors are each activated by different stimuli. Touch receptors are not sensitive to light or sound; they are sensitive only to touch or pressure. However, stimuli may be combined at higher levels in the brain, as happens with olfaction, contributing to our sense of taste.

Encoding and Transmission of Sensory Information

Four aspects of sensory information are encoded by sensory systems: the type of stimulus, the location of the stimulus in the receptive field, the duration of the stimulus, and the relative intensity of the stimulus. Thus, action potentials transmitted over a sensory receptor's afferent axons encode one type of stimulus, and this segregation of the senses is preserved in other sensory circuits. For example, auditory receptors transmit signals over their own dedicated system, and electrical activity in the axons of the auditory receptors will be interpreted by the brain as an auditory stimulus—a sound.

The intensity of a stimulus is often encoded in the rate of action potentials produced by the sensory receptor. Thus, an intense stimulus will produce a more rapid train of action potentials, and reducing the stimulus will likewise slow the rate of production of action potentials. A second way in which intensity is encoded is by the number of receptors activated. An intense stimulus might initiate action potentials in a large number of adjacent receptors, while a less intense stimulus might stimulate fewer receptors. Integration of sensory information begins as soon as the information is received in the CNS, and the brain will further process incoming signals.

Perception

Perception is an individual's interpretation of a sensation. Although perception relies on the activation of sensory receptors, perception happens not at the level of the sensory receptor, but at higher levels in the nervous system, in the brain. The brain distinguishes sensory stimuli through a sensory pathway: action potentials from sensory receptors travel along neurons that are dedicated to a particular stimulus. These neurons are dedicated to that particular stimulus and synapse with particular neurons in the brain or spinal cord.

All sensory signals, except those from the olfactory system, are transmitted though the central nervous system and are routed to the thalamus and to the appropriate region of the cortex. Recall that the thalamus is a structure in the forebrain that serves as a clearinghouse and relay station for sensory (as well as motor) signals. When the sensory signal exits the thalamus, it is conducted to the specific area of the cortex (Figure 36.3) dedicated to processing that particular sense.

How are neural signals interpreted? Interpretation of sensory signals between individuals of the same species is largely similar, owing to the inherited similarity of their nervous systems; however, there are some individual differences. A good example of this is individual tolerances to a painful stimulus, such as dental pain, which certainly differ.

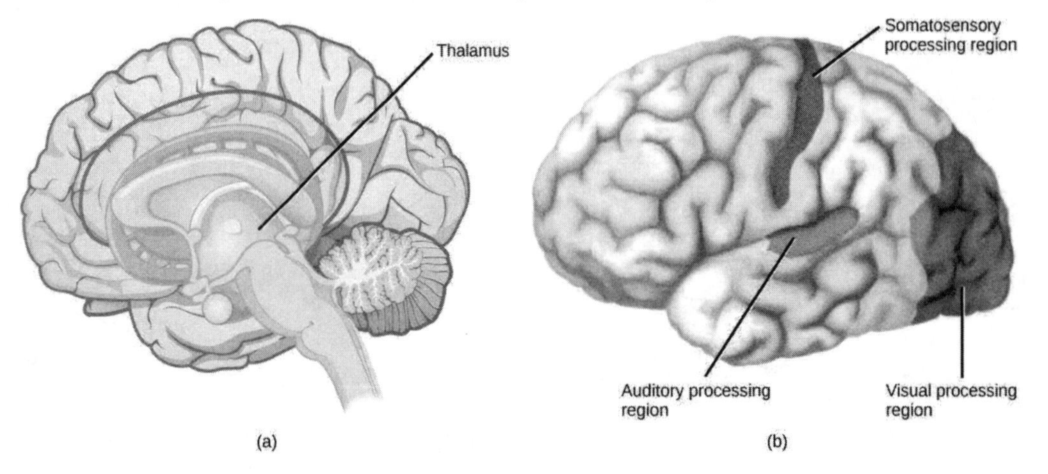

Figure 36.3 In humans, with the exception of olfaction, all sensory signals are routed from the (a) thalamus to (b) final processing regions in the cortex of the brain. (credit b: modification of work by Polina Tishina)

SCIENTIFIC METHOD CONNECTION

Just-Noticeable Difference

It is easy to differentiate between a one-pound bag of rice and a two-pound bag of rice. There is a one-pound difference, and one bag is twice as heavy as the other. However, would it be as easy to differentiate between a 20- and a 21-pound bag?

Question: What is the smallest detectible weight difference between a one-pound bag of rice and a larger bag? What is the smallest detectible difference between a 20-pound bag and a larger bag? In both cases, at what weights are the differences detected? This smallest detectible difference in stimuli is known as the just-noticeable difference (JND).

Background: Research background literature on JND and on Weber's Law, a description of a proposed mathematical relationship between the overall magnitude of the stimulus and the JND. You will be testing JND of different weights of rice in bags. Choose a convenient increment that is to be stepped through while testing. For example, you could choose 10 percent increments between one and two pounds (1.1, 1.2, 1.3, 1.4, and so on) or 20 percent increments (1.2, 1.4, 1.6, and 1.8).

Hypothesis: Develop a hypothesis about JND in terms of percentage of the whole weight being tested (such as "the JND between the two small bags and between the two large bags is proportionally the same," or ". . . is not proportionally the same.") So, for the first hypothesis, if the JND between the one-pound bag and a larger bag is 0.2 pounds (that is, 20 percent; 1.0 pound feels the same as 1.1 pounds, but 1.0 pound feels less than 1.2 pounds), then the JND between the 20-pound bag and a larger bag will also be 20 percent. (So, 20 pounds feels the same as 22 pounds or 23 pounds, but 20 pounds feels less than 24 pounds.)

Test the hypothesis: Enlist 24 participants, and split them into two groups of 12. To set up the demonstration, assuming a 10 percent increment was selected, have the first group be the one-pound group. As a counter-balancing measure against a systematic error, however, six of the first group will compare one pound to two pounds, and step down in weight (1.0 to 2.0, 1.0 to 1.9, and so on), while the other six will step up (1.0 to 1.1, 1.0 to 1.2, and so on). Apply the same principle to the 20-pound group (20 to 40, 20 to 38, and so on, and 20 to 22, 20 to 24, and so on). Given the large difference between 20 and 40 pounds, you may wish to use 30 pounds as your larger weight. In any case, use two weights that are easily detectable as different.

Record the observations: Record the data in a table similar to the table below. For the one-pound and 20-pound groups (base weights) record a plus sign (+) for each participant that detects a difference between the base weight and the step weight. Record a minus sign (-) for each participant that finds no difference. If one-tenth steps were not used, then replace the steps in the "Step Weight" columns with the step you are using.

Results of JND Testing (+ = difference; – = no difference)

Step Weight	One pound	20 pounds	Step Weight
1.1			22
1.2			24
1.3			26
1.4			28
1.5			30
1.6			32
1.7			34
1.8			36
1.9			38
2.0			40

Table 36.1

Analyze the data/report the results: What step weight did all participants find to be equal with one-pound base weight? What about the 20-pound group?

Draw a conclusion: Did the data support the hypothesis? Are the final weights proportionally the same? If not, why not? Do the findings adhere to Weber's Law? Weber's Law states that the concept that a just-noticeable difference in a stimulus is proportional to the magnitude of the original stimulus.

36.2 Somatosensation

By the end of this section, you will be able to do the following:

- Describe four important mechanoreceptors in human skin
- Describe the topographical distribution of somatosensory receptors between glabrous and hairy skin
- Explain why the perception of pain is subjective

Somatosensation is a mixed sensory category and includes all sensation received from the skin and mucous membranes, as well from as the limbs and joints. Somatosensation is also known as tactile sense, or more familiarly, as the sense of touch. Somatosensation occurs all over the exterior of the body and at some interior locations as well. A variety of receptor types—embedded in the skin, mucous membranes, muscles, joints, internal organs, and cardiovascular system—play a role.

Recall that the epidermis is the outermost layer of skin in mammals. It is relatively thin, is composed of keratin-filled cells, and has no blood supply. The epidermis serves as a barrier to water and to invasion by pathogens. Below this, the much thicker dermis contains blood vessels, sweat glands, hair follicles, lymph vessels, and lipid-secreting sebaceous glands (Figure 36.4). Below the epidermis and dermis is the subcutaneous tissue, or hypodermis, the fatty layer that contains blood vessels, connective tissue, and the axons of sensory neurons. The hypodermis, which holds about 50 percent of the body's fat, attaches the dermis to the bone and muscle, and supplies nerves and blood vessels to the dermis.

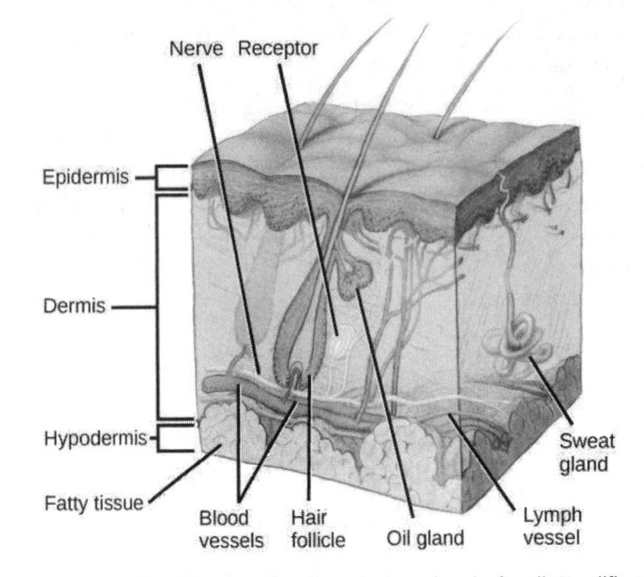

Figure 36.4 Mammalian skin has three layers: an epidermis, a dermis, and a hypodermis. (credit: modification of work by Don Bliss, National Cancer Institute)

Somatosensory Receptors

Sensory receptors are classified into five categories: mechanoreceptors, thermoreceptors, proprioceptors, pain receptors, and chemoreceptors. These categories are based on the nature of stimuli each receptor class transduces. What is commonly referred to as "touch" involves more than one kind of stimulus and more than one kind of receptor. Mechanoreceptors in the skin are described as encapsulated (that is, surrounded by a capsule) or unencapsulated (a group that includes free nerve endings). A **free nerve ending**, as its name implies, is an unencapsulated dendrite of a sensory neuron. Free nerve endings are the most common nerve endings in skin, and they extend into the middle of the epidermis. Free nerve endings are sensitive to painful stimuli, to hot and cold, and to light touch. They are slow to adjust to a stimulus and so are less sensitive to abrupt changes in stimulation.

There are three classes of mechanoreceptors: tactile, proprioceptors, and baroreceptors. Mechanoreceptors sense stimuli due to physical deformation of their plasma membranes. They contain mechanically gated ion channels whose gates open or close in response to pressure, touch, stretching, and sound." There are four primary tactile mechanoreceptors in human skin: Merkel's disks, Meissner's corpuscles, Ruffini endings, and Pacinian corpuscles; two are located toward the surface of the skin and two are located deeper. A fifth type of mechanoreceptor, Krause end bulbs, are found only in specialized regions. **Merkel's disks** (shown in Figure 36.5) are found in the upper layers of skin near the base of the epidermis, both in skin that has hair and on **glabrous** skin, that is, the hairless skin found on the palms and fingers, the soles of the feet, and the lips of humans and other primates. Merkel's disks are densely distributed in the fingertips and lips. They are slow-adapting, encapsulated nerve endings, and they respond to light touch. Light touch, also known as discriminative touch, is a light pressure that allows the location of a stimulus to be pinpointed. The receptive fields of Merkel's disks are small with well-defined borders. That makes them finely sensitive to edges and they come into use in tasks such as typing on a keyboard.

VISUAL CONNECTION

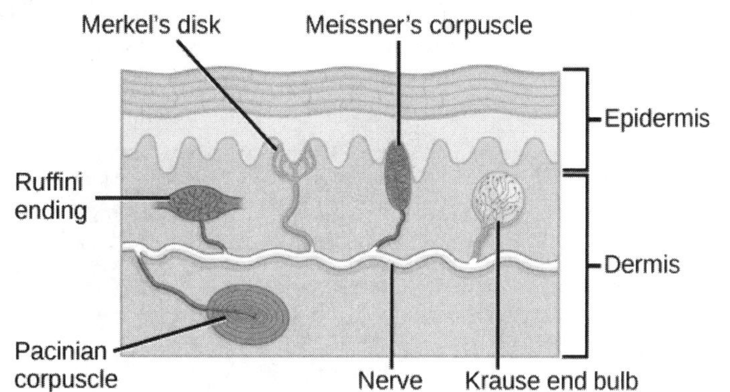

Figure 36.5 Four of the primary mechanoreceptors in human skin are shown. Merkel's disks, which are unencapsulated, respond to light touch. Meissner's corpuscles, Ruffini endings, Pacinian corpuscles, and Krause end bulbs are all encapsulated. Meissner's corpuscles respond to touch and low-frequency vibration. Ruffini endings detect stretch, deformation within joints, and warmth. Pacinian corpuscles detect transient pressure and high-frequency vibration. Krause end bulbs detect cold.

Which of the following statements about mechanoreceptors is false?

a. Pacinian corpuscles are found in both glabrous and hairy skin.
b. Merkel's disks are abundant on the fingertips and lips.
c. Ruffini endings are encapsulated mechanoreceptors.
d. Meissner's corpuscles extend into the lower dermis.

Meissner's corpuscles, (shown in Figure 36.6) also known as tactile corpuscles, are found in the upper dermis, but they project into the epidermis. They, too, are found primarily in the glabrous skin on the fingertips and eyelids. They respond to fine touch and pressure, but they also respond to low-frequency vibration or flutter. They are rapidly adapting, fluid-filled, encapsulated neurons with small, well-defined borders and are responsive to fine details. Like Merkel's disks, Meissner's corpuscles are not as plentiful in the palms as they are in the fingertips.

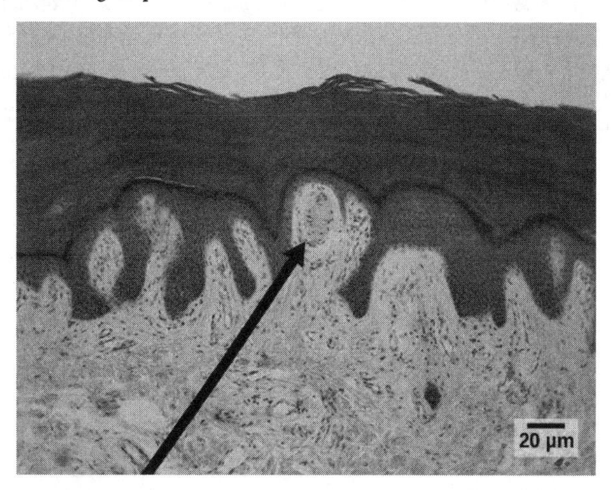

Figure 36.6 Meissner corpuscles in the fingertips, such as the one viewed here using bright field light microscopy, allow for touch discrimination of fine detail. (credit: modification of work by "Wbensmith"/Wikimedia Commons; scale-bar data from Matt Russell)

Deeper in the epidermis, near the base, are **Ruffini endings**, which are also known as bulbous corpuscles. They are found in both glabrous and hairy skin. These are slow-adapting, encapsulated mechanoreceptors that detect skin stretch and deformations within joints, so they provide valuable feedback for gripping objects and controlling finger position and movement. Thus, they also contribute to proprioception and kinesthesia. Ruffini endings also detect warmth. Note that these warmth detectors are situated deeper in the skin than are the cold detectors. It is not surprising, then, that humans detect cold stimuli before they detect warm stimuli.

Pacinian corpuscles (seen in Figure 36.7) are located deep in the dermis of both glabrous and hairy skin and are structurally similar to Meissner's corpuscles; they are found in the bone periosteum, joint capsules, pancreas and other viscera, breast, and genitals. They are rapidly adapting mechanoreceptors that sense deep transient (but not prolonged) pressure and high-frequency vibration. Pacinian receptors detect pressure and vibration by being compressed, stimulating their internal dendrites. There are fewer Pacinian corpuscles and Ruffini endings in skin than there are Merkel's disks and Meissner's corpuscles.

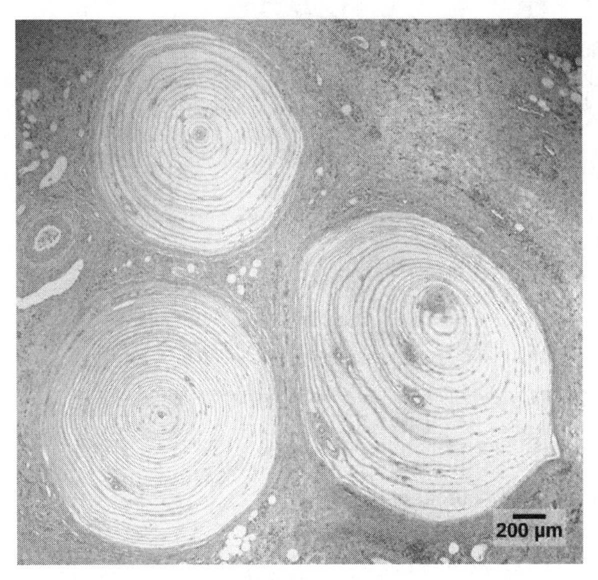

Figure 36.7 Pacinian corpuscles, such as these visualized using bright field light microscopy, detect pressure (touch) and high-frequency vibration. (credit: modification of work by Ed Uthman; scale-bar data from Matt Russell)

In proprioception, proprioceptive and kinesthetic signals travel through myelinated afferent neurons running from the spinal cord to the medulla. Neurons are not physically connected, but communicate via neurotransmitters secreted into synapses or "gaps" between communicating neurons. Once in the medulla, the neurons continue carrying the signals to the thalamus.

Muscle spindles are stretch receptors that detect the amount of stretch, or lengthening of muscles. Related to these are **Golgi tendon organs**, which are tension receptors that detect the force of muscle contraction. Proprioceptive and kinesthetic signals come from limbs. Unconscious proprioceptive signals run from the spinal cord to the cerebellum, the brain region that coordinates muscle contraction, rather than to the thalamus, like most other sensory information.

Baroreceptors detect pressure changes in an organ. They are found in the walls of the carotid artery and the aorta where they monitor blood pressure, and in the lungs where they detect the degree of lung expansion. Stretch receptors are found at various sites in the digestive and urinary systems.

In addition to these two types of deeper receptors, there are also rapidly adapting hair receptors, which are found on nerve endings that wrap around the base of hair follicles. There are a few types of hair receptors that detect slow and rapid hair movement, and they differ in their sensitivity to movement. Some hair receptors also detect skin deflection, and certain rapidly adapting hair receptors allow detection of stimuli that have not yet touched the skin.

Integration of Signals from Mechanoreceptors

The configuration of the different types of receptors working in concert in human skin results in a very refined sense of touch. The nociceptive receptors—those that detect pain—are located near the surface. Small, finely calibrated mechanoreceptors—Merkel's disks and Meissner's corpuscles—are located in the upper layers and can precisely localize even gentle touch. The large mechanoreceptors—Pacinian corpuscles and Ruffini endings—are located in the lower layers and respond to deeper touch. (Consider that the deep pressure that reaches those deeper receptors would not need to be finely localized.) Both the upper and lower layers of the skin hold rapidly and slowly adapting receptors. Both primary somatosensory cortex and secondary cortical areas are responsible for processing the complex picture of stimuli transmitted from the interplay of mechanoreceptors.

Density of Mechanoreceptors

The distribution of touch receptors in human skin is not consistent over the body. In humans, touch receptors are less dense in skin covered with any type of hair, such as the arms, legs, torso, and face. Touch receptors are denser in glabrous skin (the type found on human fingertips and lips, for example), which is typically more sensitive and is thicker than hairy skin (4 to 5 mm versus 2 to 3 mm).

How is receptor density estimated in a human subject? The relative density of pressure receptors in different locations on the body can be demonstrated experimentally using a two-point discrimination test. In this demonstration, two sharp points, such as two thumbtacks, are brought into contact with the subject's skin (though not hard enough to cause pain or break the skin). The subject reports if he or she feels one point or two points. If the two points are felt as one point, it can be inferred that the two points are both in the receptive field of a single sensory receptor. If two points are felt as two separate points, each is in the receptive field of two separate sensory receptors. The points could then be moved closer and retested until the subject reports feeling only one point, and the size of the receptive field of a single receptor could be estimated from that distance.

Thermoreception

In addition to Krause end bulbs that detect cold and Ruffini endings that detect warmth, there are different types of cold receptors on some free nerve endings: thermoreceptors, located in the dermis, skeletal muscles, liver, and hypothalamus, that are activated by different temperatures. Their pathways into the brain run from the spinal cord through the thalamus to the primary somatosensory cortex. Warmth and cold information from the face travels through one of the cranial nerves to the brain. You know from experience that a tolerably cold or hot stimulus can quickly progress to a much more intense stimulus that is no longer tolerable. Any stimulus that is too intense can be perceived as pain because temperature sensations are conducted along the same pathways that carry pain sensations.

Pain

Pain is the name given to **nociception**, which is the neural processing of injurious stimuli in response to tissue damage. Pain is caused by true sources of injury, such as contact with a heat source that causes a thermal burn or contact with a corrosive chemical. But pain also can be caused by harmless stimuli that mimic the action of damaging stimuli, such as contact with capsaicins, the compounds that cause peppers to taste hot and which are used in self-defense pepper sprays and certain topical medications. Peppers taste "hot" because the protein receptors that bind capsaicin open the same calcium channels that are activated by warm receptors.

Nociception starts at the sensory receptors, but pain, inasmuch as it is the perception of nociception, does not start until it is communicated to the brain. There are several nociceptive pathways to and through the brain. Most axons carrying nociceptive information into the brain from the spinal cord project to the thalamus (as do other sensory neurons) and the neural signal undergoes final processing in the primary somatosensory cortex. Interestingly, one nociceptive pathway projects not to the thalamus but directly to the hypothalamus in the forebrain, which modulates the cardiovascular and neuroendocrine functions of the autonomic nervous system. Recall that threatening—or painful—stimuli stimulate the sympathetic branch of the visceral sensory system, readying a fight-or-flight response.

LINK TO LEARNING

View this video (http://openstax.org/l/nociceptive) that animates the five phases of nociceptive pain.

Click to view content (https://www.openstax.org/l/nociceptive)

36.3 Taste and Smell

By the end of this section, you will be able to do the following:

- Explain in what way smell and taste stimuli differ from other sensory stimuli
- Identify the five primary tastes that can be distinguished by humans
- Explain in anatomical terms why a dog's sense of smell is more acute than a human's

Taste, also called **gustation**, and smell, also called **olfaction**, are the most interconnected senses in that both involve molecules of the stimulus entering the body and bonding to receptors. Smell lets an animal sense the presence of food or other animals—whether potential mates, predators, or prey—or other chemicals in the environment that can impact their survival. Similarly, the sense of taste allows animals to discriminate between types of foods. While the value of a sense of smell is obvious, what is the value of a sense of taste? Different tasting foods have different attributes, both helpful and harmful. For example, sweet-tasting substances tend to be highly caloric, which could be necessary for survival in lean times. Bitterness is associated with toxicity, and sourness is associated with spoiled food. Salty foods are valuable in maintaining homeostasis by helping the body retain water and by providing ions necessary for cells to function.

Tastes and Odors

Both taste and odor stimuli are molecules taken in from the environment. The primary tastes detected by humans are sweet, sour, bitter, salty, and umami. The first four tastes need little explanation. The identification of **umami** as a fundamental taste occurred fairly recently—it was identified in 1908 by Japanese scientist Kikunae Ikeda while he worked with seaweed broth, but it was not widely accepted as a taste that could be physiologically distinguished until many years later. The taste of umami, also known as savoriness, is attributable to the taste of the amino acid L-glutamate. In fact, monosodium glutamate, or MSG, is often used in cooking to enhance the savory taste of certain foods. What is the adaptive value of being able to distinguish umami? Savory substances tend to be high in protein.

All odors that we perceive are molecules in the air we breathe. If a substance does not release molecules into the air from its surface, it has no smell. And if a human or other animal does not have a receptor that recognizes a specific molecule, then that molecule has no smell. Humans have about 350 olfactory receptor subtypes that work in various combinations to allow us to sense about 10,000 different odors. Compare that to mice, for example, which have about 1,300 olfactory receptor types, and therefore probably sense more odors. Both odors and tastes involve molecules that stimulate specific chemoreceptors. Although humans commonly distinguish taste as one sense and smell as another, they work together to create the perception of flavor. A person's perception of flavor is reduced if he or she has congested nasal passages.

Reception and Transduction

Odorants (odor molecules) enter the nose and dissolve in the olfactory epithelium, the mucosa at the back of the nasal cavity (as illustrated in Figure 36.8). The **olfactory epithelium** is a collection of specialized olfactory receptors in the back of the nasal cavity that spans an area about 5 cm^2 in humans. Recall that sensory cells are neurons. An **olfactory receptor**, which is a dendrite of a specialized neuron, responds when it binds certain molecules inhaled from the environment by sending impulses directly to the olfactory bulb of the brain. Humans have about 12 million olfactory receptors, distributed among hundreds of different receptor types that respond to different odors. Twelve million seems like a large number of receptors, but compare that to other animals: rabbits have about 100 million, most dogs have about 1 billion, and bloodhounds—dogs selectively bred for their sense of smell—have about 4 billion. The overall size of the olfactory epithelium also differs between species, with that of bloodhounds, for example, being many times larger than that of humans.

Olfactory neurons are **bipolar neurons** (neurons with two processes from the cell body). Each neuron has a single dendrite buried in the olfactory epithelium, and extending from this dendrite are 5 to 20 receptor-laden, hair-like cilia that trap odorant molecules. The sensory receptors on the cilia are proteins, and it is the variations in their amino acid chains that make the receptors sensitive to different odorants. Each olfactory sensory neuron has only one type of receptor on its cilia, and the receptors are specialized to detect specific odorants, so the bipolar neurons themselves are specialized. When an odorant binds with a receptor that recognizes it, the sensory neuron associated with the receptor is stimulated. Olfactory stimulation is the only sensory information that directly reaches the cerebral cortex, whereas other sensations are relayed through the thalamus.

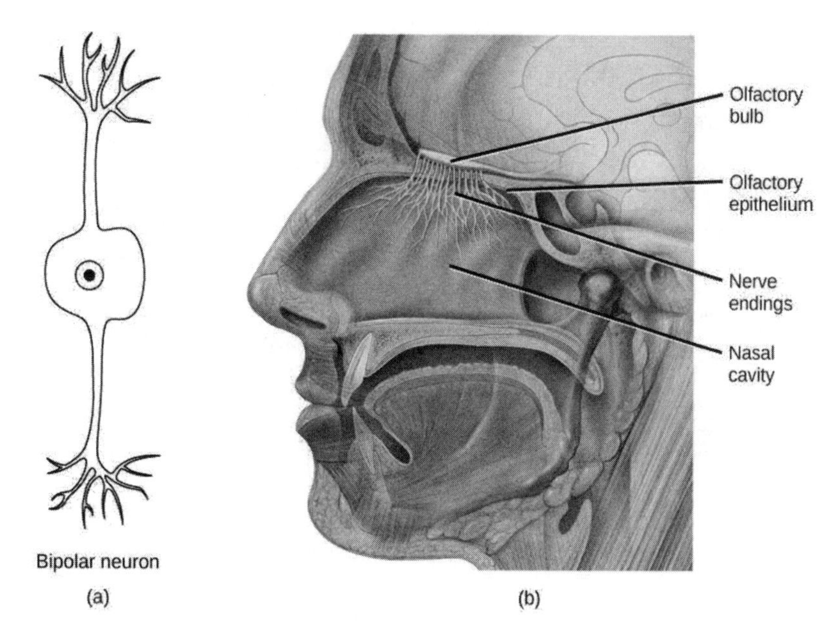

Figure 36.8 In the human olfactory system, (a) bipolar olfactory neurons extend from (b) the olfactory epithelium, where olfactory receptors are located, to the olfactory bulb. (credit: modification of work by Patrick J. Lynch, medical illustrator; C. Carl Jaffe, MD, cardiologist)

EVOLUTION CONNECTION

Pheromones

A **pheromone** is a chemical released by an animal that affects the behavior or physiology of animals of the same species. Pheromonal signals can have profound effects on animals that inhale them, but pheromones apparently are not consciously perceived in the same way as other odors. There are several different types of pheromones, which are released in urine or as glandular secretions. Certain pheromones are attractants to potential mates, others are repellants to potential competitors of the same sex, and still others play roles in mother-infant attachment. Some pheromones can also influence the timing of puberty, modify reproductive cycles, and even prevent embryonic implantation. While the roles of pheromones in many nonhuman species are important, pheromones have become less important in human behavior over evolutionary time compared to their importance to organisms with more limited behavioral repertoires.

The vomeronasal organ (VNO, or Jacobson's organ) is a tubular, fluid-filled, olfactory organ present in many vertebrate animals that sits adjacent to the nasal cavity. It is very sensitive to pheromones and is connected to the nasal cavity by a duct. When molecules dissolve in the mucosa of the nasal cavity, they then enter the VNO where the pheromone molecules among them bind with specialized pheromone receptors. Upon exposure to pheromones from their own species or others, many animals, including cats, may display the flehmen response (shown in Figure 36.9), a curling of the upper lip that helps pheromone molecules enter the VNO.

Pheromonal signals are sent, not to the main olfactory bulb, but to a different neural structure that projects directly to the amygdala (recall that the amygdala is a brain center important in emotional reactions, such as fear). The pheromonal signal then continues to areas of the hypothalamus that are key to reproductive physiology and behavior. While some scientists assert that the VNO is apparently functionally vestigial in humans, even though there is a similar structure located near human nasal cavities, others are researching it as a possible functional system that may, for example, contribute to synchronization of menstrual cycles in women living in close proximity.

Figure 36.9 The flehmen response in this tiger results in the curling of the upper lip and helps airborne pheromone molecules enter the vomeronasal organ. (credit: modification of work by "chadh"/Flickr)

Taste

Detecting a taste (gustation) is fairly similar to detecting an odor (olfaction), given that both taste and smell rely on chemical receptors being stimulated by certain molecules. The primary organ of taste is the taste bud. A **taste bud** is a cluster of gustatory receptors (taste cells) that are located within the bumps on the tongue called **papillae** (singular: papilla) (illustrated in Figure 36.11). There are several structurally distinct papillae. Filiform papillae, which are located across the tongue, are tactile, providing friction that helps the tongue move substances, and contain no taste cells. In contrast, fungiform papillae, which are located mainly on the anterior two-thirds of the tongue, each contain one to eight taste buds and also have receptors for pressure and temperature. The large circumvallate papillae contain up to 250 taste buds and form a V near the posterior margin of the tongue.

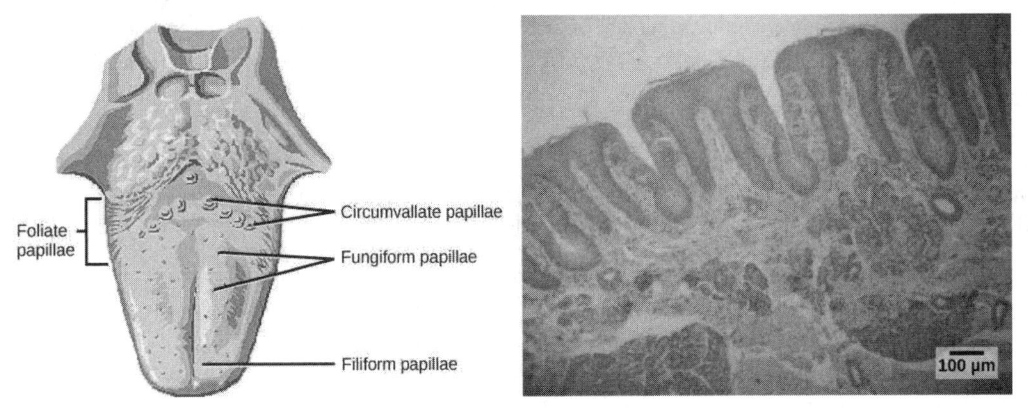

Figure 36.10 (a) Foliate, circumvallate, and fungiform papillae are located on different regions of the tongue. (b) Foliate papillae are prominent protrusions on this light micrograph. (credit a: modification of work by NCI; scale-bar data from Matt Russell)

In addition to those two types of chemically and mechanically sensitive papillae are foliate papillae—leaf-like papillae located in parallel folds along the edges and toward the back of the tongue, as seen in the Figure 36.10 micrograph. Foliate papillae contain about 1,300 taste buds within their folds. Finally, there are circumvallate papillae, which are wall-like papillae in the shape of an inverted "V" at the back of the tongue. Each of these papillae is surrounded by a groove and contains about 250 taste buds.

Each taste bud's taste cells are replaced every 10 to 14 days. These are elongated cells with hair-like processes called microvilli at the tips that extend into the taste bud pore (illustrated in Figure 36.11). Food molecules (**tastants**) are dissolved in saliva, and they bind with and stimulate the receptors on the microvilli. The receptors for tastants are located across the outer portion and front of the tongue, outside of the middle area where the filiform papillae are most prominent.

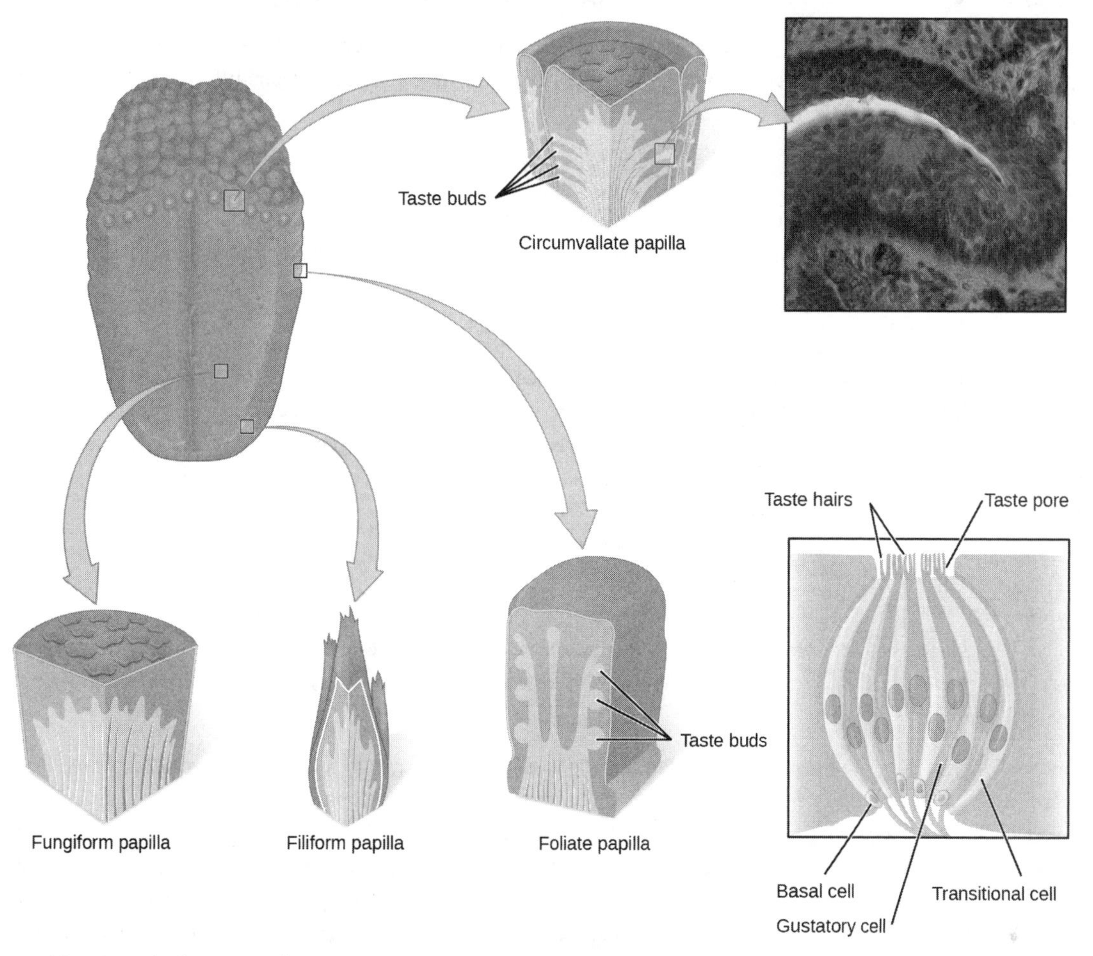

Figure 36.11 Pores in the tongue allow tastants to enter taste pores in the tongue. (credit: modification of work by Vincenzo Rizzo)

In humans, there are five primary tastes, and each taste has only one corresponding type of receptor. Thus, like olfaction, each receptor is specific to its stimulus (tastant). Transduction of the five tastes happens through different mechanisms that reflect the molecular composition of the tastant. A salty tastant (containing NaCl) provides the sodium ions (Na^+) that enter the taste neurons and excite them directly. Sour tastants are acids and belong to the thermoreceptor protein family. Binding of an acid or other sour-tasting molecule triggers a change in the ion channel and these increase hydrogen ion (H^+) concentrations in the taste neurons, thus depolarizing them. Sweet, bitter, and umami tastants require a G-protein coupled receptor. These tastants bind to their respective receptors, thereby exciting the specialized neurons associated with them.

Both tasting abilities and sense of smell change with age. In humans, the senses decline dramatically by age 50 and continue to decline. A child may find a food to be too spicy, whereas an elderly person may find the same food to be bland and unappetizing.

LINK TO LEARNING

View this animation (http://openstax.org/l/taste) that shows how the sense of taste works.

Smell and Taste in the Brain

Olfactory neurons project from the olfactory epithelium to the olfactory bulb as thin, unmyelinated axons. The **olfactory bulb** is composed of neural clusters called **glomeruli**, and each glomerulus receives signals from one type of olfactory receptor, so each glomerulus is specific to one odorant. From glomeruli, olfactory signals travel directly to the olfactory cortex and then to the frontal cortex and the thalamus. Recall that this is a different path from most other sensory information, which is sent directly to the thalamus before ending up in the cortex. Olfactory signals also travel directly to the amygdala, thereafter reaching the hypothalamus, thalamus, and frontal cortex. The last structure that olfactory signals directly travel to is a cortical center in the temporal lobe structure important in spatial, autobiographical, declarative, and episodic memories. Olfaction is finally processed by areas of the brain that deal with memory, emotions, reproduction, and thought.

Taste neurons project from taste cells in the tongue, esophagus, and palate to the medulla, in the brainstem. From the medulla, taste signals travel to the thalamus and then to the primary gustatory cortex. Information from different regions of the tongue is segregated in the medulla, thalamus, and cortex.

36.4 Hearing and Vestibular Sensation

By the end of this section, you will be able to do the following:

- Describe the relationship of amplitude and frequency of a sound wave to attributes of sound
- Trace the path of sound through the auditory system to the site of transduction of sound
- Identify the structures of the vestibular system that respond to gravity

Audition, or hearing, is important to humans and to other animals for many different interactions. It enables an organism to detect and receive information about danger, such as an approaching predator, and to participate in communal exchanges like those concerning territories or mating. On the other hand, although it is physically linked to the auditory system, the vestibular system is not involved in hearing. Instead, an animal's vestibular system detects its own movement, both linear and angular acceleration and deceleration, and balance.

Sound

Auditory stimuli are sound waves, which are mechanical, pressure waves that move through a medium, such as air or water. There are no sound waves in a vacuum since there are no air molecules to move in waves. The speed of sound waves differs, based on altitude, temperature, and medium, but at sea level and a temperature of 20° C (68° F), sound waves travel in the air at about 343 meters per second.

As is true for all waves, there are four main characteristics of a sound wave: frequency, wavelength, period, and amplitude. Frequency is the number of waves per unit of time, and in sound is heard as pitch. High-frequency (≥15.000Hz) sounds are higher-pitched (short wavelength) than low-frequency (long wavelengths; ≤100Hz) sounds. Frequency is measured in cycles per second, and for sound, the most commonly used unit is hertz (Hz), or cycles per second. Most humans can perceive sounds with frequencies between 30 and 20,000 Hz. Women are typically better at hearing high frequencies, but everyone's ability to hear high frequencies decreases with age. Dogs detect up to about 40,000 Hz; cats, 60,000 Hz; bats, 100,000 Hz; and dolphins 150,000 Hz, and American shad (*Alosa sapidissima*), a fish, can hear 180,000 Hz. Those frequencies above the human range are called **ultrasound**.

Amplitude, or the dimension of a wave from peak to trough, in sound is heard as volume and is illustrated in Figure 36.12. The sound waves of louder sounds have greater amplitude than those of softer sounds. For sound, volume is measured in decibels (dB). The softest sound that a human can hear is the zero point. Humans speak normally at 60 decibels.

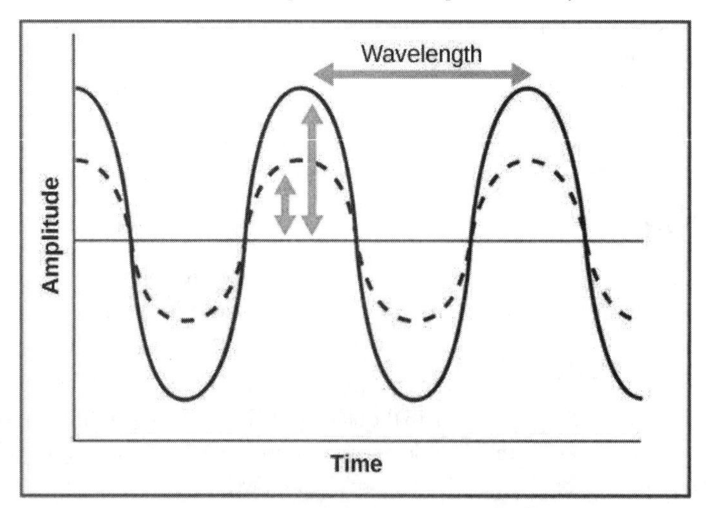

Figure 36.12 For sound waves, wavelength corresponds to pitch. Amplitude of the wave corresponds to volume. The sound wave shown

with a dashed line is softer in volume than the sound wave shown with a solid line. (credit: NIH)

Reception of Sound

In mammals, sound waves are collected by the external, cartilaginous part of the ear called the **pinna**, then travel through the auditory canal and cause vibration of the thin diaphragm called the **tympanum** or ear drum, the innermost part of the **outer ear** (illustrated in Figure 36.13). Interior to the tympanum is the **middle ear**. The middle ear holds three small bones called the **ossicles**, which transfer energy from the moving tympanum to the inner ear. The three ossicles are the **malleus** (also known as the hammer), the **incus** (the anvil), and **stapes** (the stirrup). The aptly named stapes looks very much like a stirrup. The three ossicles are unique to mammals, and each plays a role in hearing. The malleus attaches at three points to the interior surface of the tympanic membrane. The incus attaches the malleus to the stapes. In humans, the stapes is not long enough to reach the tympanum. If we did not have the malleus and the incus, then the vibrations of the tympanum would never reach the inner ear. These bones also function to collect force and amplify sounds. The ear ossicles are homologous to bones in a fish mouth: the bones that support gills in fish are thought to be adapted for use in the vertebrate ear over evolutionary time. Many animals (frogs, reptiles, and birds, for example) use the stapes of the middle ear to transmit vibrations to the middle ear.

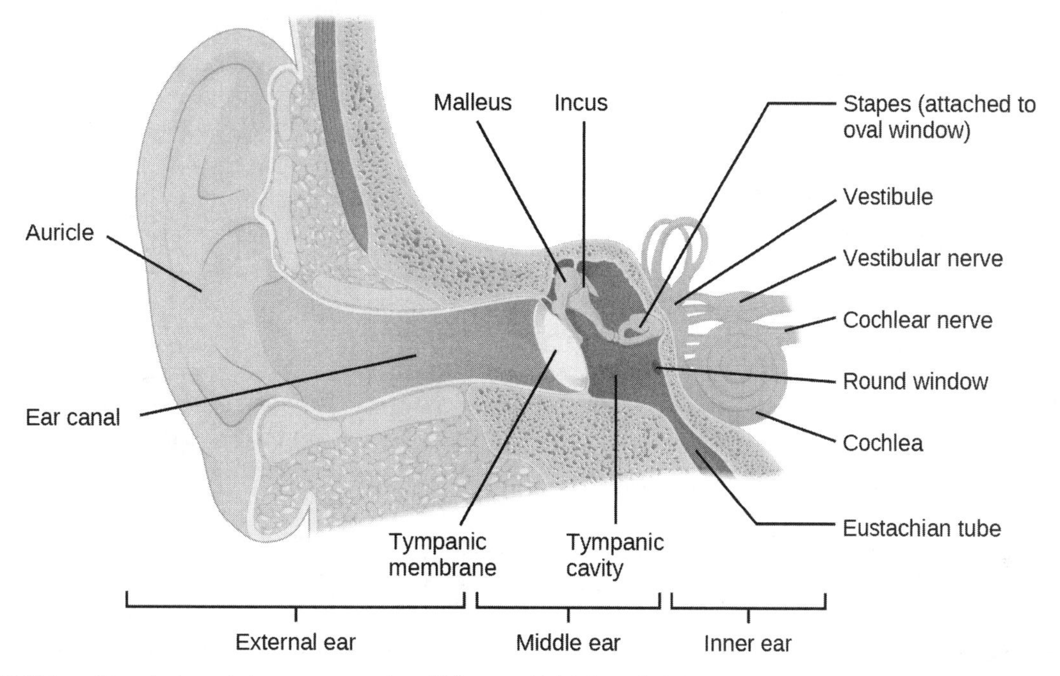

Figure 36.13 Sound travels through the outer ear to the middle ear, which is bounded on its exterior by the tympanic membrane. The middle ear contains three bones called ossicles that transfer the sound wave to the oval window, the exterior boundary of the inner ear. The organ of Corti, which is the organ of sound transduction, lies inside the cochlea.

Transduction of Sound

Vibrating objects, such as vocal cords, create sound waves or pressure waves in the air. When these pressure waves reach the ear, the ear transduces this mechanical stimulus (pressure wave) into a nerve impulse (electrical signal) that the brain perceives as sound. The pressure waves strike the tympanum, causing it to vibrate. The mechanical energy from the moving tympanum transmits the vibrations to the three bones of the middle ear. The stapes transmits the vibrations to a thin diaphragm called the **oval window**, which is the outermost structure of the **inner ear**. The structures of the inner ear are found in the **labyrinth**, a bony, hollow structure that is the most interior portion of the ear. Here, the energy from the sound wave is transferred from the stapes through the flexible oval window and to the fluid of the cochlea. The vibrations of the oval window create pressure waves in the fluid (perilymph) inside the cochlea. The **cochlea** is a whorled structure, like the shell of a snail, and it contains receptors for transduction of the mechanical wave into an electrical signal (as illustrated in Figure 36.14). Inside the cochlea, the **basilar membrane** is a mechanical analyzer that runs the length of the cochlea, curling toward the cochlea's center.

The mechanical properties of the basilar membrane change along its length, such that it is thicker, tauter, and narrower at the outside of the whorl (where the cochlea is largest), and thinner, floppier, and broader toward the apex, or center, of the whorl (where the cochlea is smallest). Different regions of the basilar membrane vibrate according to the frequency of the sound wave conducted through the fluid in the cochlea. For these reasons, the fluid-filled cochlea detects different wave frequencies (pitches) at different regions of the membrane. When the sound waves in the cochlear fluid contact the basilar membrane, it flexes back and forth in a wave-like fashion. Above the basilar membrane is the **tectorial membrane**.

VISUAL CONNECTION

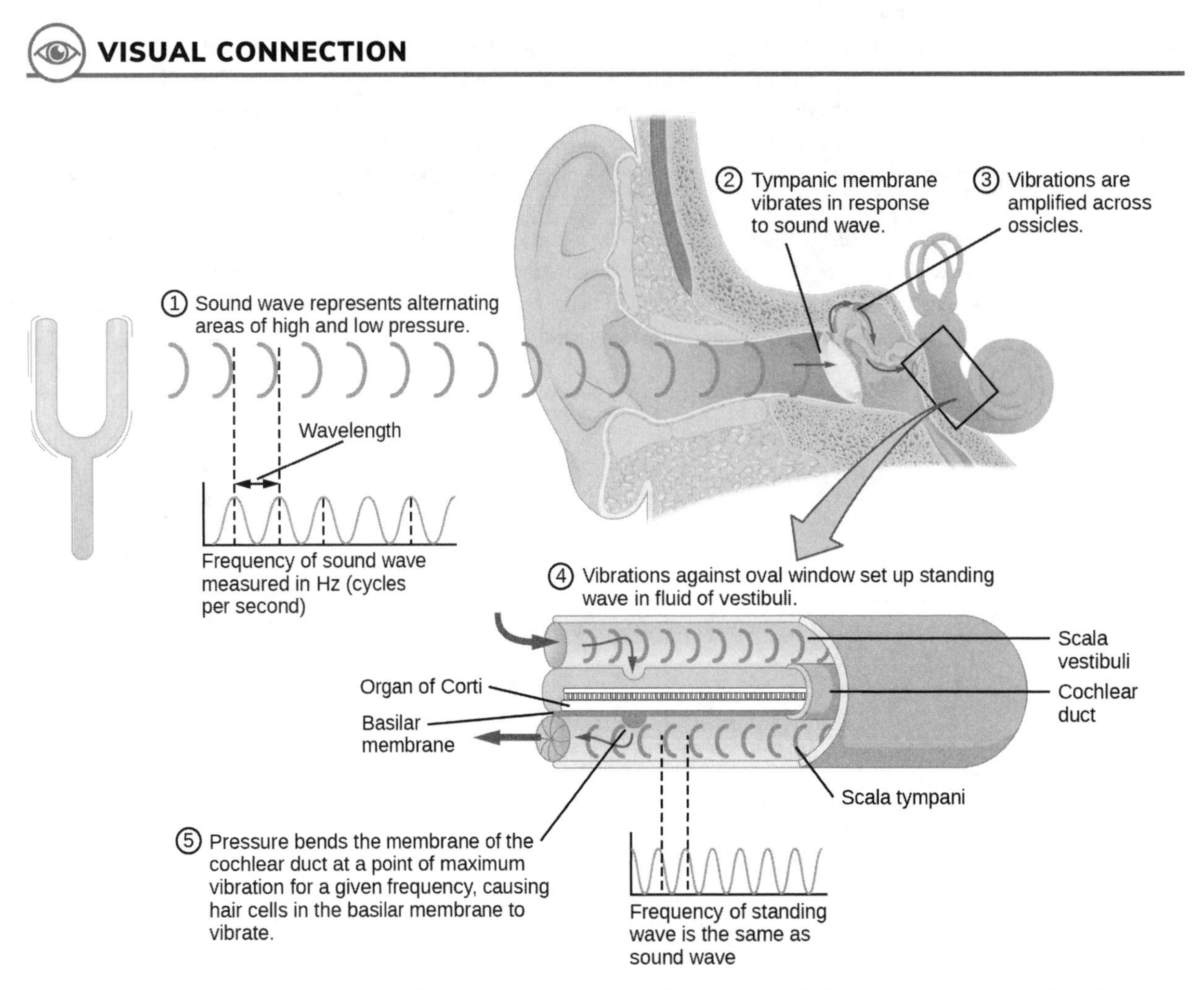

Figure 36.14 A sound wave causes the tympanic membrane to vibrate. This vibration is amplified as it moves across the malleus, incus, and stapes. The amplified vibration is picked up by the oval window causing pressure waves in the fluid of the scala vestibuli and scala tympani. The complexity of the pressure waves is determined by the changes in amplitude and frequency of the sound waves entering the ear.

Cochlear implants can restore hearing in people who have a nonfunctional cochlea. The implant consists of a microphone that picks up sound. A speech processor selects sounds in the range of human speech, and a transmitter converts these sounds to electrical impulses, which are then sent to the auditory nerve. Which of the following types of hearing loss would not be restored by a cochlear implant?

a. Hearing loss resulting from absence or loss of hair cells in the organ of Corti.
b. Hearing loss resulting from an abnormal auditory nerve.
c. Hearing loss resulting from fracture of the cochlea.
d. Hearing loss resulting from damage to bones of the middle ear.

The site of transduction is in the **organ of Corti** (spiral organ). It is composed of hair cells held in place above the basilar membrane like flowers projecting up from soil, with their exposed short, hair-like **stereocilia** contacting or embedded in the tectorial membrane above them. The inner hair cells are the primary auditory receptors and exist in a single row, numbering approximately 3,500. The stereocilia from inner hair cells extend into small dimples on the tectorial membrane's lower surface. The outer hair cells are arranged in three or four rows. They number approximately 12,000, and they function to fine tune incoming sound waves. The longer stereocilia that project from the outer hair cells actually attach to the tectorial membrane. All of the stereocilia are mechanoreceptors, and when bent by vibrations they respond by opening a gated ion channel (refer to Figure 36.15). As a result, the hair cell membrane is depolarized, and a signal is transmitted to the chochlear nerve. Intensity (volume) of sound is determined by how many hair cells at a particular location are stimulated.

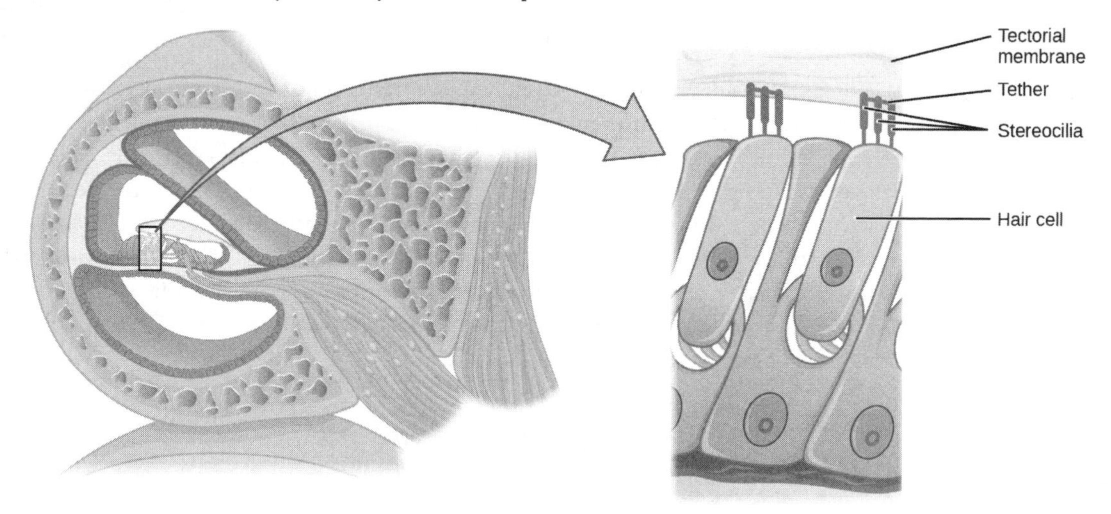

Figure 36.15 The hair cell is a mechanoreceptor with an array of stereocilia emerging from its apical surface. The stereocilia are tethered together by proteins that open ion channels when the array is bent toward the tallest member of their array, and closed when the array is bent toward the shortest member of their array.

The hair cells are arranged on the basilar membrane in an orderly way. The basilar membrane vibrates in different regions, according to the frequency of the sound waves impinging on it. Likewise, the hair cells that lay above it are most sensitive to a specific frequency of sound waves. Hair cells can respond to a small range of similar frequencies, but they require stimulation of greater intensity to fire at frequencies outside of their optimal range. The difference in response frequency between adjacent inner hair cells is about 0.2 percent. Compare that to adjacent piano strings, which are about six percent different. Place theory, which is the model for how biologists think pitch detection works in the human ear, states that high frequency sounds selectively vibrate the basilar membrane of the inner ear near the entrance port (the oval window). Lower frequencies travel farther along the membrane before causing appreciable excitation of the membrane. The basic pitch-determining mechanism is based on the location along the membrane where the hair cells are stimulated. The place theory is the first step toward an understanding of pitch perception. Considering the extreme pitch sensitivity of the human ear, it is thought that there must be some auditory "sharpening" mechanism to enhance the pitch resolution.

When sound waves produce fluid waves inside the cochlea, the basilar membrane flexes, bending the stereocilia that attach to the tectorial membrane. Their bending results in action potentials in the hair cells, and auditory information travels along the neural endings of the bipolar neurons of the hair cells (collectively, the auditory nerve) to the brain. When the hairs bend, they release an excitatory neurotransmitter at a synapse with a sensory neuron, which then conducts action potentials to the central nervous system. The cochlear branch of the vestibulocochlear cranial nerve sends information on hearing. The auditory system is very refined, and there is some modulation or "sharpening" built in. The brain can send signals back to the cochlea, resulting in a change of length in the outer hair cells, sharpening or dampening the hair cells' response to certain frequencies.

LINK TO LEARNING

Watch an animation (http://openstax.org/l/hearing) of sound entering the outer ear, moving through the ear structure, stimulating cochlear nerve impulses, and eventually sending signals to the temporal lobe.

Higher Processing

The inner hair cells are most important for conveying auditory information to the brain. About 90 percent of the afferent neurons carry information from inner hair cells, with each hair cell synapsing with 10 or so neurons. Outer hair cells connect to only 10 percent of the afferent neurons, and each afferent neuron innervates many hair cells. The afferent, bipolar neurons that convey auditory information travel from the cochlea to the medulla, through the pons and midbrain in the brainstem, finally reaching the primary auditory cortex in the temporal lobe.

Vestibular Information

The stimuli associated with the vestibular system are linear acceleration (gravity) and angular acceleration and deceleration. Gravity, acceleration, and deceleration are detected by evaluating the inertia on receptive cells in the vestibular system. Gravity is detected through head position. Angular acceleration and deceleration are expressed through turning or tilting of the head.

The vestibular system has some similarities with the auditory system. It utilizes hair cells just like the auditory system, but it excites them in different ways. There are five vestibular receptor organs in the inner ear: the utricle, the saccule, and three semicircular canals. Together, they make up what's known as the vestibular labyrinth that is shown in Figure 36.16. The utricle and saccule respond to acceleration in a straight line, such as gravity. The roughly 30,000 hair cells in the utricle and 16,000 hair cells in the saccule lie below a gelatinous layer, with their stereocilia projecting into the gelatin. Embedded in this gelatin are calcium carbonate crystals—like tiny rocks. When the head is tilted, the crystals continue to be pulled straight down by gravity, but the new angle of the head causes the gelatin to shift, thereby bending the stereocilia. The bending of the stereocilia stimulates the neurons, and they signal to the brain that the head is tilted, allowing the maintenance of balance. It is the vestibular branch of the vestibulocochlear cranial nerve that deals with balance.

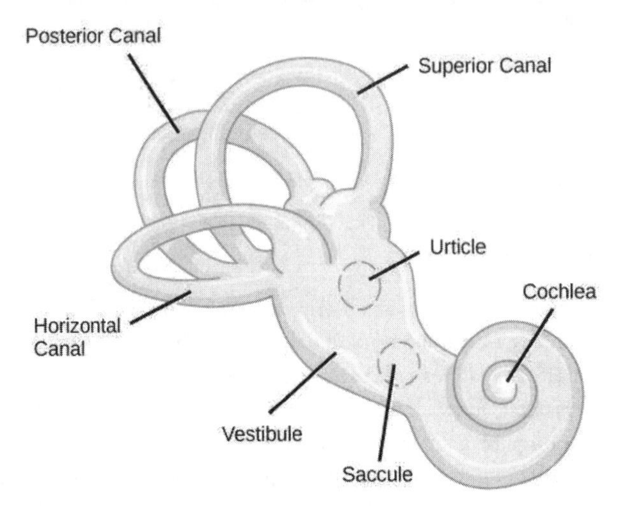

Figure 36.16 The structure of the vestibular labyrinth is shown. (credit: modification of work by NIH)

The fluid-filled **semicircular canals** are tubular loops set at oblique angles. They are arranged in three spatial planes. The base of each canal has a swelling that contains a cluster of hair cells. The hairs project into a gelatinous cap called the cupula and monitor angular acceleration and deceleration from rotation. They would be stimulated by driving your car around a corner, turning your head, or falling forward. One canal lies horizontally, while the other two lie at about 45 degree angles to the horizontal axis, as illustrated in Figure 36.16. When the brain processes input from all three canals together, it can detect angular acceleration or deceleration in three dimensions. When the head turns, the fluid in the canals shifts, thereby bending stereocilia and sending signals to the brain. Upon cessation accelerating or decelerating—or just moving—the movement of the fluid within the canals slows or stops. For example, imagine holding a glass of water. When moving forward, water may splash backwards onto the hand, and when motion has stopped, water may splash forward onto the fingers. While in motion, the water settles in the glass and does not splash. Note that the canals are not sensitive to velocity itself, but to changes in velocity, so moving forward at 60mph with your eyes closed would not give the sensation of movement, but suddenly accelerating or braking would stimulate the receptors.

Higher Processing

Hair cells from the utricle, saccule, and semicircular canals also communicate through bipolar neurons to the cochlear nucleus in the medulla. Cochlear neurons send descending projections to the spinal cord and ascending projections to the pons, thalamus, and cerebellum. Connections to the cerebellum are important for coordinated movements. There are also projections to the temporal cortex, which account for feelings of dizziness; projections to autonomic nervous system areas in the brainstem, which account for motion sickness; and projections to the primary somatosensory cortex, which monitors subjective measurements of the external world and self-movement. People with lesions in the vestibular area of the somatosensory cortex see vertical objects in the world as being tilted. Finally, the vestibular signals project to certain optic muscles to coordinate eye and head movements.

LINK TO LEARNING

Click through this interactive tutorial (http://openstax.org/l/ear_anatomy) to review the parts of the ear and how they function to process sound.

36.5 Vision

By the end of this section, you will be able to do the following:

- Explain how electromagnetic waves differ from sound waves
- Trace the path of light through the eye to the point of the optic nerve
- Explain tonic activity as it is manifested in photoreceptors in the retina

Vision is the ability to detect light patterns from the outside environment and interpret them into images. Animals are bombarded with sensory information, and the sheer volume of visual information can be problematic. Fortunately, the visual systems of species have evolved to attend to the most-important stimuli. The importance of vision to humans is further substantiated by the fact that about one-third of the human cerebral cortex is dedicated to analyzing and perceiving visual information.

Light

As with auditory stimuli, light travels in waves. The compression waves that compose sound must travel in a medium—a gas, a liquid, or a solid. In contrast, light is composed of electromagnetic waves and needs no medium; light can travel in a vacuum (Figure 36.17). The behavior of light can be discussed in terms of the behavior of waves and also in terms of the behavior of the fundamental unit of light—a packet of electromagnetic radiation called a photon. A glance at the electromagnetic spectrum shows that visible light for humans is just a small slice of the entire spectrum, which includes radiation that we cannot see as light because it is below the frequency of visible red light and above the frequency of visible violet light.

Certain variables are important when discussing perception of light. Wavelength (which varies inversely with frequency) manifests itself as hue. Light at the red end of the visible spectrum has longer wavelengths (and is lower frequency), while light at the violet end has shorter wavelengths (and is higher frequency). The wavelength of light is expressed in nanometers (nm); one nanometer is one billionth of a meter. Humans perceive light that ranges between approximately 380 nm and 740 nm. Some other animals, though, can detect wavelengths outside of the human range. For example, bees see near-ultraviolet light in order to locate nectar guides on flowers, and some non-avian reptiles sense infrared light (heat that prey gives off).

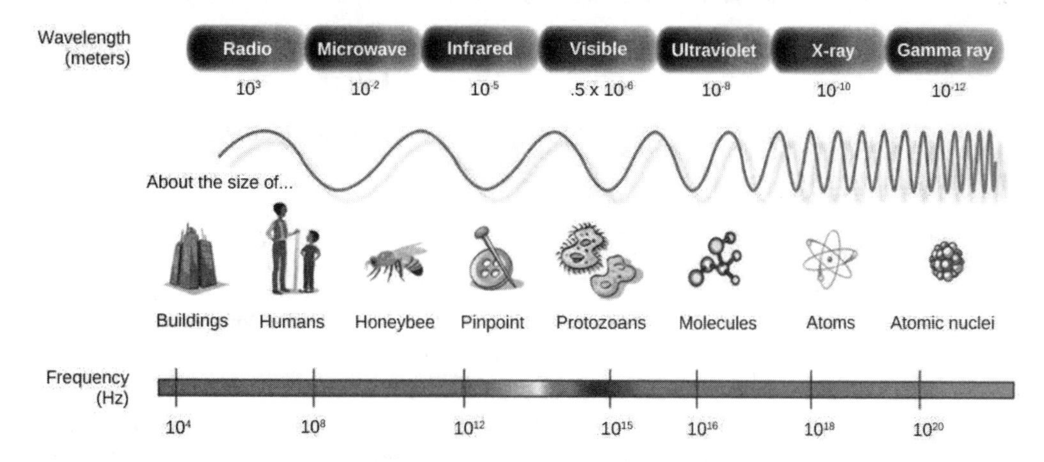

Figure 36.17 In the electromagnetic spectrum, visible light lies between 380 nm and 740 nm. (credit: modification of work by NASA)

Wave amplitude is perceived as luminous intensity, or brightness. The standard unit of intensity of light is the **candela**, which is approximately the luminous intensity of one common candle.

Light waves travel 299,792 km per second in a vacuum, (and somewhat slower in various media such as air and water), and those waves arrive at the eye as long (red), medium (green), and short (blue) waves. What is termed "white light" is light that is perceived as white by the human eye. This effect is produced by light that stimulates equally the color receptors in the human eye. The apparent color of an object is the color (or colors) that the object reflects. Thus a red object reflects the red wavelengths in mixed (white) light and absorbs all other wavelengths of light.

Anatomy of the Eye

The photoreceptive cells of the eye, where transduction of light to nervous impulses occurs, are located in the **retina** (shown in Figure 36.18) on the inner surface of the back of the eye. But light does not impinge on the retina unaltered. It passes through other layers that process it so that it can be interpreted by the retina (Figure 36.18b). The **cornea**, the front transparent layer of the eye, and the crystalline **lens**, a transparent convex structure behind the cornea, both refract (bend) light to focus the image on the retina. The **iris**, which is conspicuous as the colored part of the eye, is a circular muscular ring lying between the lens and cornea that regulates the amount of light entering the eye. In conditions of high ambient light, the iris contracts, reducing the size of the pupil at its center. In conditions of low light, the iris relaxes and the pupil enlarges.

VISUAL CONNECTION

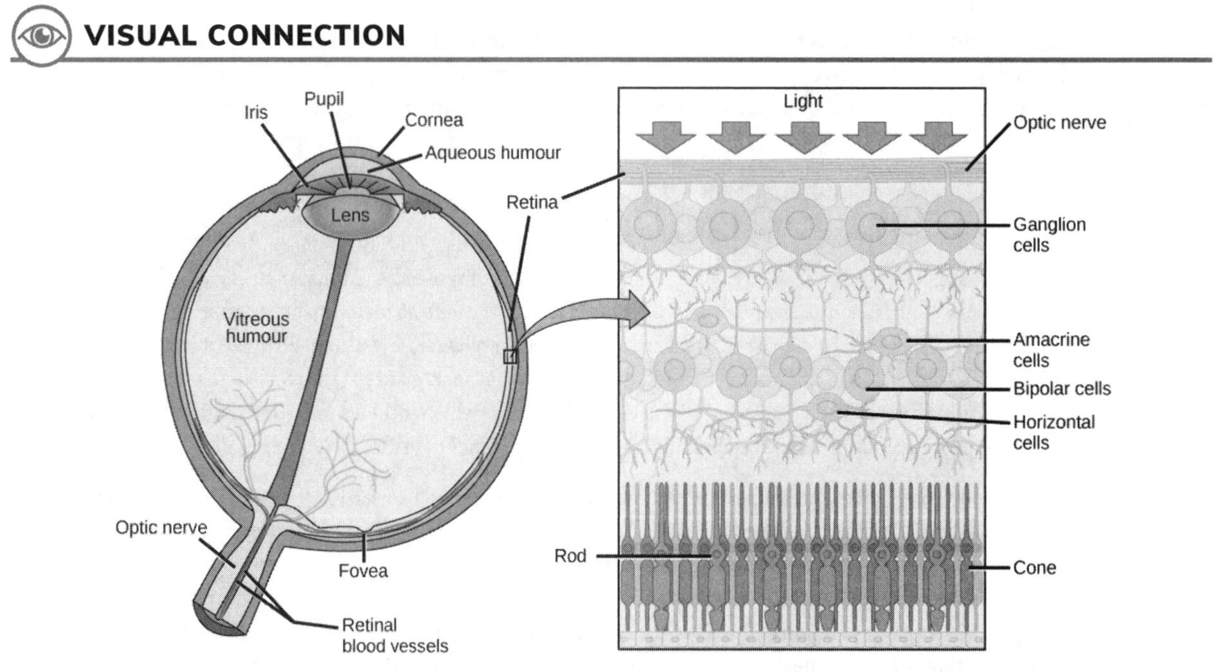

Figure 36.18 (a) The human eye is shown in cross section. (b) A blowup shows the layers of the retina.

Which of the following statements about the human eye is false?

a. Rods detect color, while cones detect only shades of gray.
b. When light enters the retina, it passes the ganglion cells and bipolar cells before reaching photoreceptors at the rear of the eye.
c. The iris adjusts the amount of light coming into the eye.
d. The cornea is a protective layer on the front of the eye.

The main function of the lens is to focus light on the retina and fovea centralis. The lens is dynamic, focusing and re-focusing light as the eye rests on near and far objects in the visual field. The lens is operated by muscles that stretch it flat or allow it to thicken, changing the focal length of light coming through it to focus it sharply on the retina. With age comes the loss of the flexibility of the lens, and a form of farsightedness called **presbyopia** results. Presbyopia occurs because the image focuses behind the retina. Presbyopia is a deficit similar to a different type of farsightedness called **hyperopia** caused by an eyeball that is too short. For both defects, images in the distance are clear but images nearby are blurry. **Myopia** (nearsightedness) occurs when an eyeball is elongated and the image focus falls in front of the retina. In this case, images in the distance are blurry but images nearby are clear.

There are two types of photoreceptors in the retina: **rods** and **cones**, named for their general appearance as illustrated in Figure 36.19. Rods are strongly photosensitive and are located in the outer edges of the retina. They detect dim light and are used primarily for peripheral and nighttime vision. Cones are weakly photosensitive and are located near the center of the retina. They respond to bright light, and their primary role is in daytime, color vision.

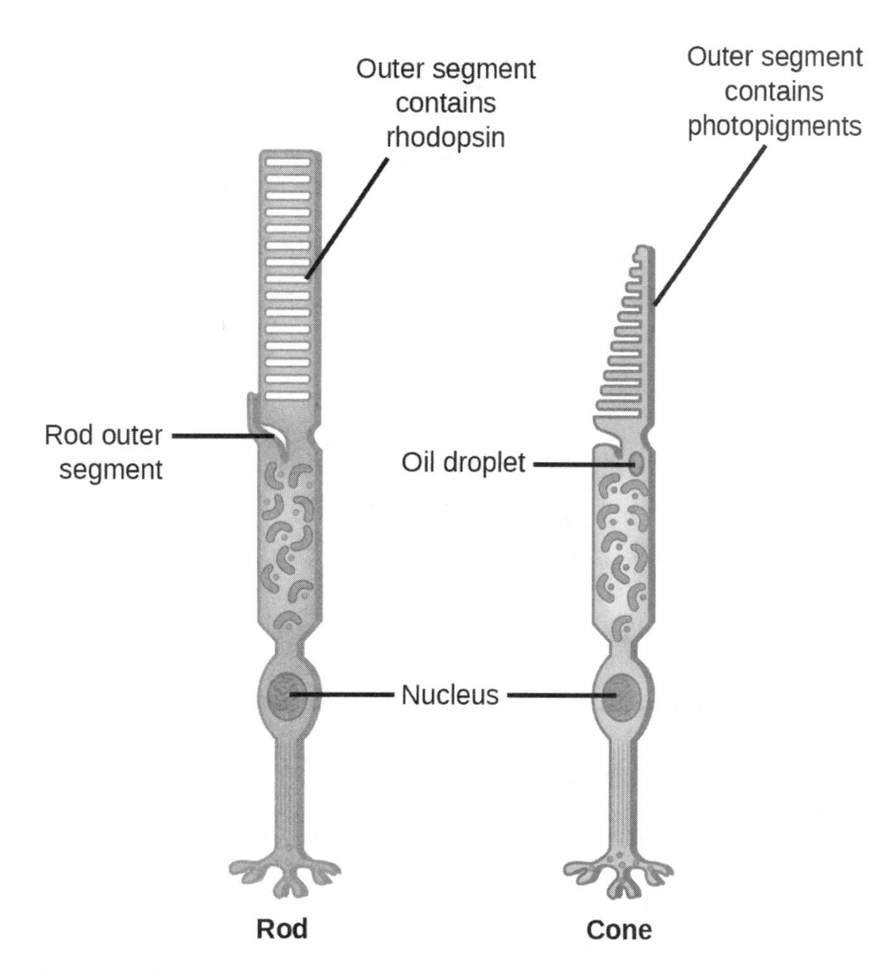

Figure 36.19 Rods and cones are photoreceptors in the retina. Rods respond in low light and can detect only shades of gray. Cones respond in intense light and are responsible for color vision. (credit: modification of work by Piotr Sliwa)

The **fovea** is the region in the center back of the eye that is responsible for acute vision. The fovea has a high density of cones. When you bring your gaze to an object to examine it intently in bright light, the eyes orient so that the object's image falls on the fovea. However, when looking at a star in the night sky or other object in dim light, the object can be better viewed by the peripheral vision because it is the rods at the edges of the retina, rather than the cones at the center, that operate better in low light. In humans, cones far outnumber rods in the fovea.

LINK TO LEARNING

Review the anatomical structure (http://openstax.org/l/eye_diagram) of the eye, clicking on each part to practice identification.

Transduction of Light

The rods and cones are the site of transduction of light to a neural signal. Both rods and cones contain photopigments. In vertebrates, the main photopigment, **rhodopsin**, has two main parts (Figure 36.20): an opsin, which is a membrane protein (in the form of a cluster of α-helices that span the membrane), and retinal—a molecule that absorbs light. When light hits a photoreceptor, it causes a shape change in the retinal, altering its structure from a bent (*cis*) form of the molecule to its linear (*trans*) isomer. This isomerization of retinal activates the rhodopsin, starting a cascade of events that ends with the closing of Na^+ channels in the membrane of the photoreceptor. Thus, unlike most other sensory neurons (which become depolarized by exposure to a stimulus) visual receptors become hyperpolarized and thus driven away from threshold (Figure 36.21).

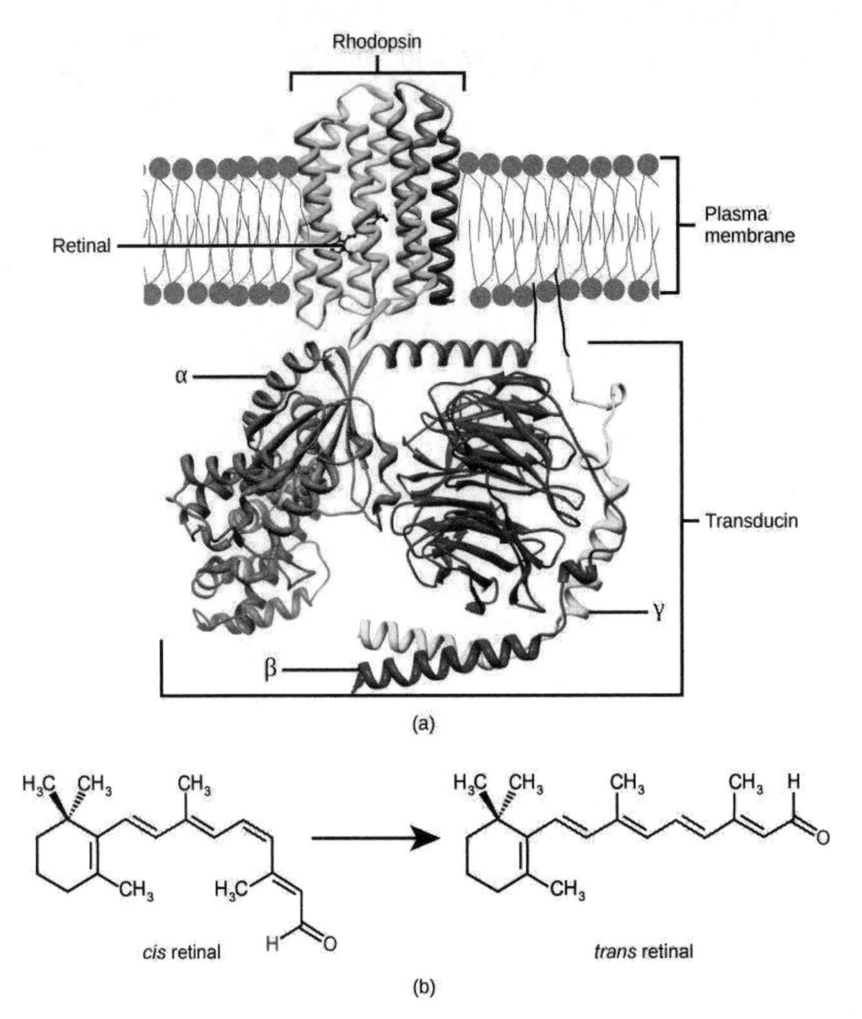

Figure 36.20 (a) Rhodopsin, the photoreceptor in vertebrates, has two parts: the trans-membrane protein opsin, and retinal. When light strikes retinal, it changes shape from (b) a *cis* to a *trans* form. The signal is passed to a G-protein called transducin, triggering a series of downstream events.

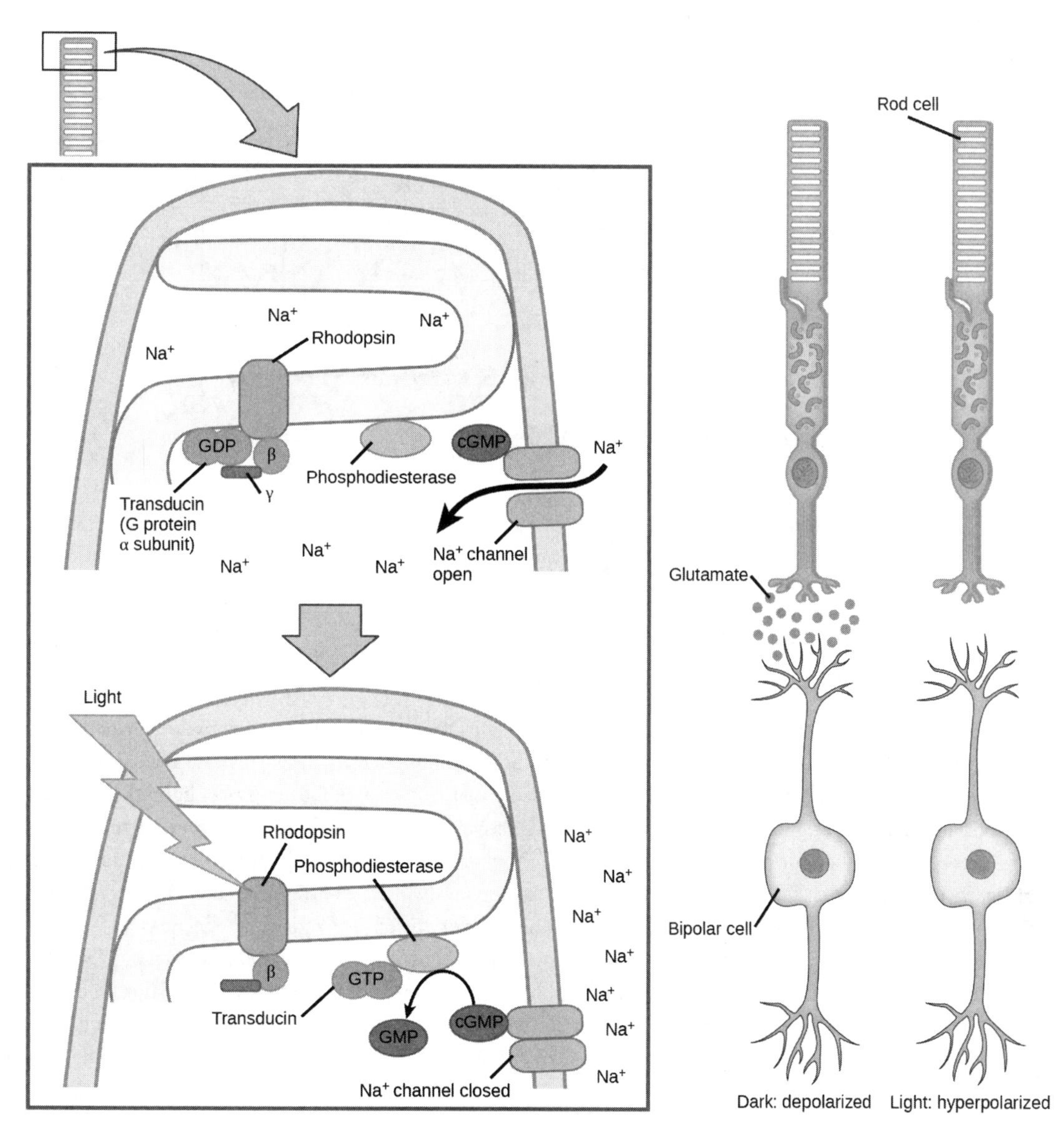

Figure 36.21 When light strikes rhodopsin, the G-protein transducin is activated, which in turn activates phosphodiesterase. Phosphodiesterase converts cGMP to GMP, thereby closing sodium channels. As a result, the membrane becomes hyperpolarized. The hyperpolarized membrane does not release glutamate to the bipolar cell.

Trichromatic Coding

There are three types of cones (with different photopsins), and they differ in the wavelength to which they are most responsive, as shown in Figure 36.22. Some cones are maximally responsive to short light waves of 420 nm, so they are called S cones ("S" for "short"); others respond maximally to waves of 530 nm (M cones, for "medium"); a third group responds maximally to light of longer wavelengths, at 560 nm (L, or "long" cones). With only one type of cone, color vision would not be possible, and a two-cone (dichromatic) system has limitations. Primates use a three-cone (trichromatic) system, resulting in full color vision.

The color we perceive is a result of the ratio of activity of our three types of cones. The colors of the visual spectrum, running from long-wavelength light to short, are red (700 nm), orange (600 nm), yellow (565 nm), green (497 nm), blue (470 nm), indigo (450 nm), and violet (425 nm). Humans have very sensitive perception of color and can distinguish about 500 levels of brightness, 200 different hues, and 20 steps of saturation, or about 2 million distinct colors.

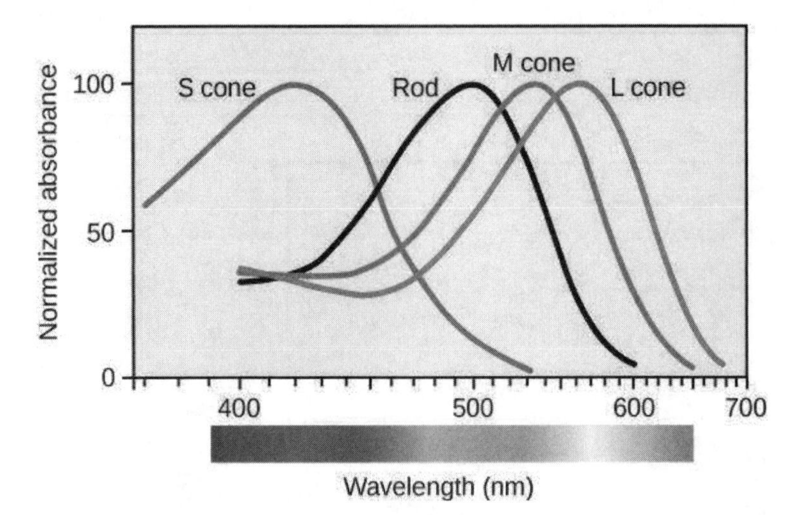

Figure 36.22 Human rod cells and the different types of cone cells each have an optimal wavelength. However, there is considerable overlap in the wavelengths of light detected.

Retinal Processing

Visual signals leave the cones and rods, travel to the bipolar cells, and then to ganglion cells. A large degree of processing of visual information occurs in the retina itself, before visual information is sent to the brain.

Photoreceptors in the retina continuously undergo **tonic activity**. That is, they are always slightly active even when not stimulated by light. In neurons that exhibit tonic activity, the absence of stimuli maintains a firing rate at a baseline; while some stimuli increase firing rate from the baseline, and other stimuli decrease firing rate. In the absence of light, the bipolar neurons that connect rods and cones to ganglion cells are continuously and actively inhibited by the rods and cones. Exposure of the retina to light hyperpolarizes the rods and cones and removes their inhibition of bipolar cells. The now active bipolar cells in turn stimulate the ganglion cells, which send action potentials along their axons (which leave the eye as the optic nerve). Thus, the visual system relies on change in retinal activity, rather than the absence or presence of activity, to encode visual signals for the brain. Sometimes horizontal cells carry signals from one rod or cone to other photoreceptors and to several bipolar cells. When a rod or cone stimulates a horizontal cell, the horizontal cell inhibits more distant photoreceptors and bipolar cells, creating lateral inhibition. This inhibition sharpens edges and enhances contrast in the images by making regions receiving light appear lighter and dark surroundings appear darker. Amacrine cells can distribute information from one bipolar cell to many ganglion cells.

You can demonstrate this using an easy demonstration to "trick" your retina and brain about the colors you are observing in your visual field. Look fixedly at Figure 36.23 for about 45 seconds. Then quickly shift your gaze to a sheet of blank white paper or a white wall. You should see an afterimage of the Norwegian flag in its correct colors. At this point, close your eyes for a moment, then reopen them, looking again at the white paper or wall; the afterimage of the flag should continue to appear as red, white, and blue. What causes this? According to an explanation called opponent process theory, as you gazed fixedly at the green, black, and yellow flag, your retinal ganglion cells that respond positively to green, black, and yellow increased their firing dramatically. When you shifted your gaze to the neutral white ground, these ganglion cells abruptly decreased their activity and the brain interpreted this abrupt downshift as if the ganglion cells were responding now to their "opponent" colors: red, white, and blue, respectively, in the visual field. Once the ganglion cells return to their baseline activity state, the false perception of color will disappear.

Figure 36.23 View this flag to understand how retinal processing works. Stare at the center of the flag (indicated by the white dot) for 45 seconds, and then quickly look at a white background, noticing how colors appear.

Higher Processing

The myelinated axons of ganglion cells make up the optic nerves. Within the nerves, different axons carry different qualities of the visual signal. Some axons constitute the magnocellular (big cell) pathway, which carries information about form, movement, depth, and differences in brightness. Other axons constitute the parvocellular (small cell) pathway, which carries information on color and fine detail. Some visual information projects directly back into the brain, while other information crosses to the opposite side of the brain. This crossing of optical pathways produces the distinctive optic chiasma (Greek, for "crossing") found at the base of the brain and allows us to coordinate information from both eyes.

Once in the brain, visual information is processed in several places, and its routes reflect the complexity and importance of visual information to humans and other animals. One route takes the signals to the thalamus, which serves as the routing station for all incoming sensory impulses except olfaction. In the thalamus, the magnocellular and parvocellular distinctions remain intact, and there are different layers of the thalamus dedicated to each. When visual signals leave the thalamus, they travel to the primary visual cortex at the rear of the brain. From the visual cortex, the visual signals travel in two directions. One stream that projects to the parietal lobe, in the side of the brain, carries magnocellular ("where") information. A second stream projects to the temporal lobe and carries both magnocellular ("where") and parvocellular ("what") information.

Another important visual route is a pathway from the retina to the **superior colliculus** in the midbrain, where eye movements are coordinated and integrated with auditory information. Finally, there is the pathway from the retina to the **suprachiasmatic nucleus** (SCN) of the hypothalamus. The SCN is a cluster of cells that is considered to be the body's internal clock, which controls our **circadian** (day-long) cycle. The SCN sends information to the pineal gland, which is important in sleep/wake patterns and annual cycles.

LINK TO LEARNING

View this interactive presentation (http://openstax.org/l/sense_of_sight) to review what you have learned about how vision functions.

KEY TERMS

audition sense of hearing

basilar membrane stiff structure in the cochlea that indirectly anchors auditory receptors

bipolar neuron neuron with two processes from the cell body, typically in opposite directions

candela (cd) unit of measurement of luminous intensity (brightness)

circadian describes a time cycle about one day in length

cochlea whorled structure that contains receptors for transduction of the mechanical wave into an electrical signal

cone weakly photosensitive, chromatic, cone-shaped neuron in the fovea of the retina that detects bright light and is used in daytime color vision

cornea transparent layer over the front of the eye that helps focus light waves

fovea region in the center of the retina with a high density of photoreceptors and which is responsible for acute vision

free nerve ending ending of an afferent neuron that lacks a specialized structure for detection of sensory stimuli; some respond to touch, pain, or temperature

glabrous describes the non-hairy skin found on palms and fingers, soles of feet, and lips of humans and other primates

glomerulus in the olfactory bulb, one of the two neural clusters that receives signals from one type of olfactory receptor

Golgi tendon organ muscular proprioceptive tension receptor that provides the sensory component of the Golgi tendon reflex

gustation sense of taste

hyperopia (also, farsightedness) visual defect in which the image focus falls behind the retina, thereby making images in the distance clear, but close-up images blurry

incus (also, anvil) second of the three bones of the middle ear

inner ear innermost part of the ear; consists of the cochlea and the vestibular system

iris pigmented, circular muscle at the front of the eye that regulates the amount of light entering the eye

kinesthesia sense of body movement

labyrinth bony, hollow structure that is the most internal part of the ear; contains the sites of transduction of auditory and vestibular information

lens transparent, convex structure behind the cornea that helps focus light waves on the retina

malleus (also, hammer) first of the three bones of the middle ear

mechanoreceptor sensory receptor modified to respond to mechanical disturbance such as being bent, touch, pressure, motion, and sound

Meissner's corpuscle (also, tactile corpuscle) encapsulated, rapidly-adapting mechanoreceptor in the skin that responds to light touch

Merkel's disk unencapsulated, slowly-adapting mechanoreceptor in the skin that responds to touch

middle ear part of the hearing apparatus that functions to transfer energy from the tympanum to the oval window of the inner ear

muscle spindle proprioceptive stretch receptor that lies within a muscle and that shortens the muscle to an optimal length for efficient contraction

myopia (also, nearsightedness) visual defect in which the image focus falls in front of the retina, thereby making images in the distance blurry, but close-up images clear

nociception neural processing of noxious (such as damaging) stimuli

odorant airborne molecule that stimulates an olfactory receptor

olfaction sense of smell

olfactory bulb neural structure in the vertebrate brain that receives signals from olfactory receptors

olfactory epithelium specialized tissue in the nasal cavity where olfactory receptors are located

olfactory receptor dendrite of a specialized neuron

organ of Corti in the basilar membrane, the site of the transduction of sound, a mechanical wave, to a neural signal

ossicle one of the three bones of the middle ear

outer ear part of the ear that consists of the pinna, ear canal, and tympanum and which conducts sound waves into the middle ear

oval window thin diaphragm between the middle and inner ears that receives sound waves from contact with the stapes bone of the middle ear

Pacinian corpuscle encapsulated mechanoreceptor in the skin that responds to deep pressure and vibration

papilla one of the small bump-like projections from the tongue

perception individual interpretation of a sensation; a brain function

pheromone substance released by an animal that can affect the physiology or behavior of other animals

pinna cartilaginous outer ear

presbyopia visual defect in which the image focus falls behind the retina, thereby making images in the distance clear, but close-up images blurry; caused by age-based changes in the lens

proprioception sense of limb position; used to track kinesthesia

pupil small opening though which light enters

reception receipt of a signal (such as light or sound) by sensory receptors

receptive field region in space in which a stimulus can activate a given sensory receptor

receptor potential membrane potential in a sensory receptor in response to detection of a stimulus

retina layer of photoreceptive and supporting cells on the inner surface of the back of the eye

rhodopsin main photopigment in vertebrates

rod strongly photosensitive, achromatic, cylindrical neuron in the outer edges of the retina that detects dim light and is used in peripheral and nighttime vision

Ruffini ending (also, bulbous corpuscle) slowly-adapting mechanoreceptor in the skin that responds to skin stretch and joint position

semicircular canal one of three half-circular, fluid-filled tubes in the vestibular labyrinth that monitors angular acceleration and deceleration

sensory receptor specialized neuron or other cells associated with a neuron that is modified to receive specific sensory input

sensory transduction conversion of a sensory stimulus into electrical energy in the nervous system by a change in the membrane potential

stapes (also, stirrup) third of the three bones of the middle ear

stereocilia in the auditory system, hair-like projections from hair cells that help detect sound waves

superior colliculus paired structure in the top of the midbrain, which manages eye movements and auditory integration

suprachiasmatic nucleus cluster of cells in the hypothalamus that plays a role in the circadian cycle

tastant food molecule that stimulates gustatory receptors

taste bud clusters of taste cells

tectorial membrane cochlear structure that lies above the hair cells and participates in the transduction of sound at the hair cells

tonic activity in a neuron, slight continuous activity while at rest

tympanum (also, tympanic membrane or ear drum) thin diaphragm between the outer and middle ears

ultrasound sound frequencies above the human detectable ceiling of approximately 20,000 Hz

umami one of the five basic tastes, which is described as "savory" and which may be largely the taste of L-glutamate

vestibular sense sense of spatial orientation and balance

vision sense of sight

CHAPTER SUMMARY

36.1 Sensory Processes

A sensory activation occurs when a physical or chemical stimulus is processed into a neural signal (sensory transduction) by a sensory receptor. Perception is an individual interpretation of a sensation and is a brain function. Humans have special senses: olfaction, gustation, equilibrium, and hearing, plus the general senses of somatosensation.

Sensory receptors are either specialized cells associated with sensory neurons or the specialized ends of sensory neurons that are a part of the peripheral nervous system, and they are used to receive information about the environment (internal or external). Each sensory receptor is modified for the type of stimulus it detects. For example, neither gustatory receptors nor auditory receptors are sensitive to light. Each sensory receptor is responsive to stimuli within a specific region in space, which is known as that receptor's receptive field. The most fundamental function of a sensory system is the translation of a sensory signal to an electrical signal in the nervous system.

All sensory signals, except those from the olfactory system, enter the central nervous system and are routed to the thalamus. When the sensory signal exits the thalamus, it is conducted to the specific area of the cortex dedicated to processing that particular sense.

36.2 Somatosensation

Somatosensation includes all sensation received from the skin and mucous membranes, as well as from the limbs and joints. Somatosensation occurs all over the exterior of the body and at some interior locations as well, and a variety of receptor types, embedded in the skin and mucous membranes, play a role.

There are several types of specialized sensory receptors. Rapidly adapting free nerve endings detect nociception, hot and cold, and light touch. Slowly adapting, encapsulated Merkel's disks are found in fingertips and lips, and respond to light touch. Meissner's corpuscles, found in glabrous skin, are rapidly adapting, encapsulated receptors that detect touch, low-frequency vibration, and flutter. Ruffini endings are slowly adapting, encapsulated receptors that detect skin stretch, joint activity, and warmth. Hair receptors are rapidly adapting nerve endings wrapped around the base of hair follicles that detect hair movement and skin deflection. Finally, Pacinian corpuscles are encapsulated, rapidly adapting receptors that detect transient pressure and high-frequency vibration.

36.3 Taste and Smell

There are five primary tastes in humans: sweet, sour, bitter, salty, and umami. Each taste has its own receptor type that responds only to that taste. Tastants enter the body and are dissolved in saliva. Taste cells are located within taste buds, which are found on three of the four types of papillae in the mouth.

Regarding olfaction, there are many thousands of odorants, but humans detect only about 10,000. Like taste receptors, olfactory receptors are each responsive to only one odorant. Odorants dissolve in nasal mucosa, where they excite their corresponding olfactory sensory cells. When these cells detect an odorant, they send their signals to the main olfactory bulb and then to other locations in the brain, including the olfactory cortex.

36.4 Hearing and Vestibular Sensation

Audition is important for territory defense, predation, predator defense, and communal exchanges. The vestibular system, which is not auditory, detects linear acceleration and angular acceleration and deceleration. Both the auditory system and vestibular system use hair cells as their receptors.

Auditory stimuli are sound waves. The sound wave energy reaches the outer ear (pinna, canal, tympanum), and vibrations of the tympanum send the energy to the middle ear. The middle ear bones shift and the stapes transfers mechanical energy to the oval window of the fluid-filled inner ear cochlea. Once in the cochlea, the energy causes the basilar membrane to flex, thereby bending the stereocilia on receptor hair cells. This activates the receptors, which send their auditory neural signals to the brain.

The vestibular system has five parts that work together to provide the sense of direction, thus helping to maintain balance. The utricle and saccule measure head orientation: their calcium carbonate crystals shift when the head is tilted, thereby activating hair cells. The semicircular canals work similarly, such that when the head is turned, the fluid in the canals bends stereocilia on hair cells. The vestibular hair cells also send signals to the thalamus and to the somatosensory cortex, but also to the cerebellum, the structure above the brainstem that plays a large role in timing and coordination of movement.

36.5 Vision

Vision is the only photo responsive sense. Visible light travels in waves and is a very small slice of the electromagnetic radiation spectrum. Light waves differ based on their frequency (wavelength = hue) and amplitude (intensity = brightness).

In the vertebrate retina, there are two types of light receptors (photoreceptors): cones and rods. Cones, which are the source of color vision, exist in three forms—L, M, and S—and they are differentially sensitive to different wavelengths. Cones are located in the retina, along with the dim-light, achromatic receptors (rods). Cones are found in the fovea, the central region of the retina, whereas rods are found in the peripheral regions of the retina.

Visual signals travel from the eye over the axons of retinal ganglion cells, which make up the optic nerves. Ganglion cells come in several versions. Some ganglion cell axons carry information on form, movement, depth, and brightness, while other axons carry information on color and fine detail. Visual information is sent to the superior colliculi in the midbrain, where coordination of eye movements and integration of auditory information takes place. Visual information is also sent to the suprachiasmatic nucleus (SCN) of the hypothalamus, which plays a role in the circadian cycle.

VISUAL CONNECTION QUESTIONS

1. Figure 36.5 Which of the following statements about mechanoreceptors is false?
   a. Pacini corpuscles are found in both glabrous and hairy skin.
   b. Merkel's disks are abundant on the fingertips and lips.
   c. Ruffini endings are encapsulated mechanoreceptors.
   d. Meissner's corpuscles extend into the lower dermis.
2. Figure 36.14 Cochlear implants can restore hearing in people who have a nonfunctional cochlea. The implant consists of a microphone that picks up sound. A speech processor selects sounds in the range of human speech, and a transmitter converts these sounds to electrical impulses, which are then sent to the auditory nerve. Which of the following types of hearing loss would not be restored by a cochlear implant?
   a. Hearing loss resulting from absence or loss of hair cells in the organ of Corti.
   b. Hearing loss resulting from an abnormal auditory nerve.
   c. Hearing loss resulting from fracture of the cochlea.
   d. Hearing loss resulting from damage to bones of the middle ear.

3. Figure 36.18 Which of the following statements about the human eye is false?
    a. Rods detect color, while cones detect only shades of gray.
    b. When light enters the retina, it passes the ganglion cells and bipolar cells before reaching photoreceptors at the rear of the eye.
    c. The iris adjusts the amount of light coming into the eye.
    d. The cornea is a protective layer on the front of the eye.

REVIEW QUESTIONS

4. Where does perception occur?
    a. spinal cord
    b. cerebral cortex
    c. receptors
    d. thalamus

5. If a person's cold receptors no longer convert cold stimuli into sensory signals, that person has a problem with the process of ________.
    a. reception
    b. transmission
    c. perception
    d. transduction

6. After somatosensory transduction, the sensory signal travels through the brain as a(n) _____ signal.
    a. electrical
    b. pressure
    c. optical
    d. thermal

7. Many people experience motion sickness while traveling in a car. This sensation results from contradictory inputs arising from which senses?
    a. Proprioception and Kinesthesia
    b. Somatosensation and Equilibrium
    c. Gustation and Vibration
    d. Vision and Vestibular System

8. _____ are found only in _____ skin, and detect skin deflection.
    a. Meissner's corpuscles; hairy
    b. Merkel's disks; glabrous
    c. hair receptors; hairy
    d. Krause end bulbs; hairy

9. If you were to burn your epidermis, what receptor type would you most likely burn?
    a. free nerve endings
    b. Ruffini endings
    c. Pacinian corpuscle
    d. hair receptors

10. Many diabetic patients are warned by their doctors to test their glucose levels by pricking the sides of their fingers rather than the pads. Pricking the sides avoids stimulating which receptor?
    a. Krause end bulbs
    b. Meissner's corpuscles
    c. Ruffini ending
    d. Nociceptors

11. Which of the following has the fewest taste receptors?
    a. fungiform papillae
    b. circumvallate papillae
    c. foliate papillae
    d. filiform papillae

12. How many different taste molecules do taste cells each detect?
    a. one
    b. five
    c. ten
    d. It depends on the spot on the tongue.

13. Salty foods activate the taste cells by _____.
    a. exciting the taste cell directly
    b. causing hydrogen ions to enter the cell
    c. causing sodium channels to close
    d. binding directly to the receptors

14. All sensory signals except _____ travel to the _____ in the brain before the cerebral cortex.
    a. vision; thalamus
    b. olfaction; thalamus
    c. vision; cranial nerves
    d. olfaction; cranial nerves

15. How is the ability to recognize the umami taste an evolutionary advantage?
    a. Umami identifies healthy foods that are low in salt and sugar.
    b. Umami enhances the flavor of bland foods.
    c. Umami identifies foods that might contain essential amino acids.
    d. Umami identifies foods that help maintain electrolyte balance.

16. In sound, pitch is measured in _____, and volume is measured in _____.
    a. nanometers (nm); decibels (dB)
    b. decibels (dB); nanometers (nm)
    c. decibels (dB); hertz (Hz)
    d. hertz (Hz); decibels (dB)

17. Auditory hair cells are indirectly anchored to the _____.
    a. basilar membrane
    b. oval window
    c. tectorial membrane
    d. ossicles

18. Which of the following are found both in the auditory system and the vestibular system?
    a. basilar membrane
    b. hair cells
    c. semicircular canals
    d. ossicles

19. Benign Paroxysmal Positional Vertigo is a disorder where some of the calcium carbonate crystals in the utricle migrate into the semicircular canals. Why does this condition cause periods of dizziness?
    a. The hair cells in the semicircular canals will be constantly activated.
    b. The hair cells in the semicircular canals will now be stimulated by gravity.
    c. The utricle will no longer recognize acceleration.
    d. There will be too much volume in the semicircular canals for them to detect motion.

20. Why do people over 55 often need reading glasses?
    a. Their cornea no longer focuses correctly.
    b. Their lens no longer focuses correctly.
    c. Their eyeball has elongated with age, causing images to focus in front of their retina.
    d. Their retina has thinned with age, making vision more difficult.

21. Why is it easier to see images at night using peripheral, rather than the central, vision?
    a. Cones are denser in the periphery of the retina.
    b. Bipolar cells are denser in the periphery of the retina.
    c. Rods are denser in the periphery of the retina.
    d. The optic nerve exits at the periphery of the retina.

22. A person catching a ball must coordinate her head and eyes. What part of the brain is helping to do this?
    a. hypothalamus
    b. pineal gland
    c. thalamus
    d. superior colliculus

23. A satellite is launched into space, but explodes after exiting the Earth's atmosphere. Which statement accurately reflects the observations made by an astronaut on a space walk outside the International Space Station during the explosion?
    a. The astronaut would see the explosion, but would not hear a boom.
    b. The astronaut will not sense the explosion.
    c. The astronaut will see the explosion, and then hear the boom.
    d. The astronaut will feel the concussive force of the explosion, but will not see it.

CRITICAL THINKING QUESTIONS

24. If a person sustains damage to axons leading from sensory receptors to the central nervous system, which step or steps of sensory perception will be affected?
25. In what way does the overall magnitude of a stimulus affect the just-noticeable difference in the perception of that stimulus?
26. Describe the difference in the localization of the sensory receptors for general and special senses in humans.
27. What can be inferred about the relative sizes of the areas of cortex that process signals from skin not densely innervated with sensory receptors and skin that is densely innervated with sensory receptors?
28. Many studies have demonstrated that women are able to tolerate the same painful stimuli for longer than men. Why don't all people experience pain the same way?
29. From the perspective of the recipient of the signal, in what ways do pheromones differ from other odorants?

30. What might be the effect on an animal of not being able to perceive taste?
31. A few recent cancer detection studies have used trained dogs to detect lung cancer in urine samples. What is the hypothesis behind this study? Why are dogs a better choice of detectors in this study than humans?
32. How would a rise in altitude likely affect the speed of a sound transmitted through air? Why?
33. How might being in a place with less gravity than Earth has (such as Earth's moon) affect vestibular sensation, and why?
34. How does the structure of the ear allow a person to determine where a sound originates?
35. How could the pineal gland, the brain structure that plays a role in annual cycles, use visual information from the suprachiasmatic nucleus of the hypothalamus?
36. How is the relationship between photoreceptors and bipolar cells different from other sensory receptors and adjacent cells?
37. Cataracts, the medical condition where the lens of the eye becomes cloudy, are a leading cause of blindness. Describe how developing a cataract would change the path of light through the eye.

CHAPTER 37
The Endocrine System

Figure 37.1 The process of amphibian metamorphosis, as seen in the tadpole-to-frog stages shown here, is driven by hormones. (credit "tadpole": modification of work by Brian Gratwicke)

Chapter Outline

INTRODUCTION An animal's endocrine system controls body processes through the production, secretion, and regulation of hormones, which serve as chemical "messengers" functioning in cellular and organ activity and, ultimately, maintaining the body's homeostasis. The endocrine system plays a role in growth, metabolism, and sexual development. In humans, common endocrine system diseases include thyroid disease and diabetes mellitus. In organisms that undergo metamorphosis, the process is controlled by the endocrine system. The transformation from tadpole to frog, for example, is complex and nuanced to adapt to specific environments and ecological circumstances.

37.1 Types of Hormones

By the end of this section, you will be able to do the following:

- List the different types of hormones
- Explain their role in maintaining homeostasis

Maintaining homeostasis within the body requires the coordination of many different systems and organs. Communication between neighboring cells, and between cells and tissues in distant parts of the body, occurs through the release of chemicals called hormones. Hormones are released into body fluids (usually blood) that carry these chemicals to their target cells. At the target cells, which are cells that have a receptor for a signal or ligand from a signal cell, the hormones elicit a response. The cells, tissues, and organs that secrete hormones make up the endocrine system. Examples of glands of the endocrine system include the adrenal glands, which produce hormones such as epinephrine and norepinephrine that regulate responses to stress, and the thyroid gland, which produces thyroid hormones that regulate metabolic rates.

Although there are many different hormones in the human body, they can be divided into three classes based on their chemical structure: lipid-derived, amino acid-derived, and peptide (peptide and proteins) hormones. One of the key distinguishing features of lipid-derived hormones is that they can diffuse across plasma membranes whereas the amino acid-derived and peptide hormones cannot.

Lipid-Derived Hormones (or Lipid-soluble Hormones)

Most **lipid hormones** are derived from cholesterol and thus are structurally similar to it, as illustrated in Figure 37.2. The primary class of lipid hormones in humans is the steroid hormones. Chemically, these hormones are usually ketones or alcohols; their chemical names will end in "-ol" for alcohols or "-one" for ketones. Examples of steroid hormones include estradiol, which is an **estrogen**, or female sex hormone, and testosterone, which is an androgen, or male sex hormone. These two hormones are released by the female and male reproductive organs, respectively. Other steroid hormones include aldosterone and cortisol, which are released by the adrenal glands along with some other types of androgens. Steroid hormones are insoluble in water, and they are transported by transport proteins in blood. As a result, they remain in circulation longer than peptide hormones. For example, cortisol has a half-life of 60 to 90 minutes, while epinephrine, an amino acid derived-hormone, has a half-life of approximately one minute.

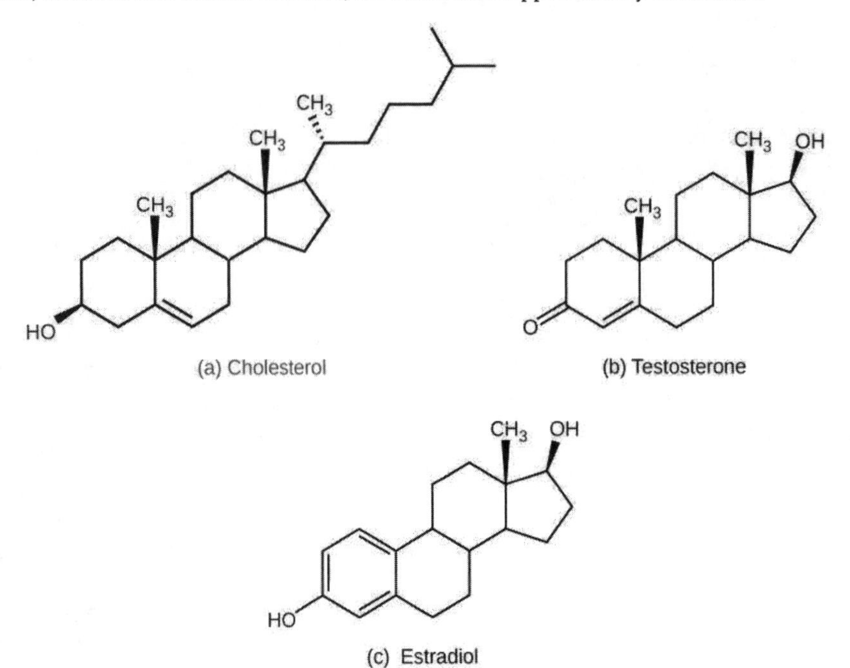

Figure 37.2 The structures shown here represent (a) cholesterol, plus the steroid hormones (b) testosterone and (c) estradiol.

Amino Acid-Derived Hormones

The **amino acid-derived hormones** are relatively small molecules that are derived from the amino acids tyrosine and tryptophan, shown in Figure 37.3. If a hormone is amino acid-derived, its chemical name will end in "-ine". Examples of amino acid-derived hormones include epinephrine and norepinephrine, which are synthesized in the medulla of the adrenal glands, and thyroxine, which is produced by the thyroid gland. The pineal gland in the brain makes and secretes melatonin which regulates sleep cycles.

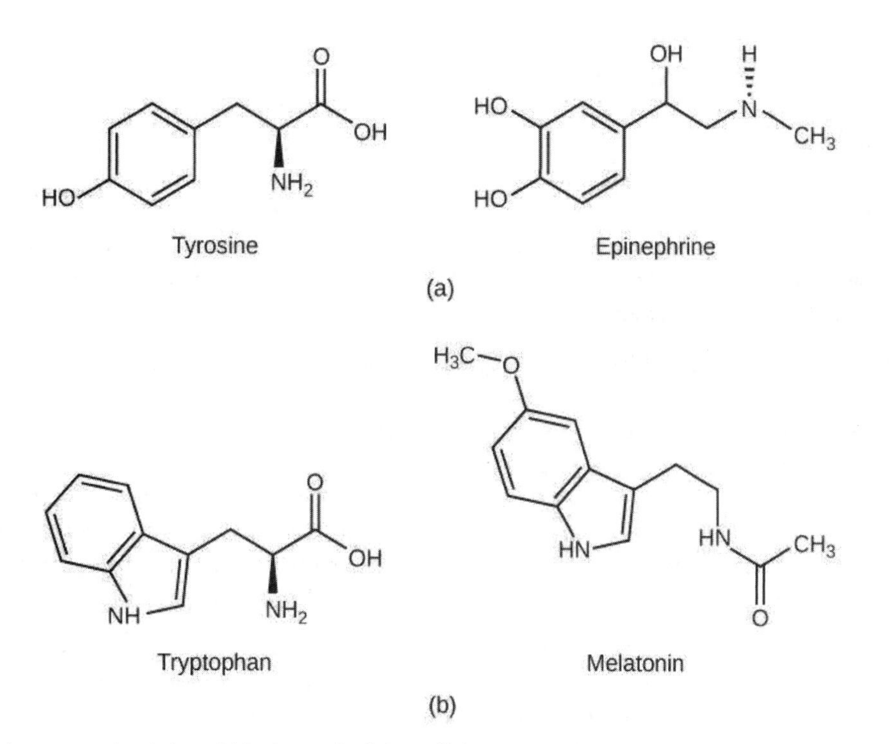

Figure 37.3 (a) The hormone epinephrine, which triggers the fight-or-flight response, is derived from the amino acid tyrosine. (b) The hormone melatonin, which regulates circadian rhythms, is derived from the amino acid tryptophan.

Peptide Hormones

The structure of **peptide hormones** is that of a polypeptide chain (chain of amino acids). The peptide hormones include molecules that are short polypeptide chains, such as antidiuretic hormone and oxytocin produced in the brain and released into the blood in the posterior pituitary gland. This class also includes small proteins, like growth hormones produced by the pituitary, and large glycoproteins such as follicle-stimulating hormone produced by the pituitary. Figure 37.4 illustrates these peptide hormones.

Secreted peptides like insulin are stored within vesicles in the cells that synthesize them. They are then released in response to stimuli such as high blood glucose levels in the case of insulin. Amino acid-derived and polypeptide hormones are water-soluble and insoluble in lipids. These hormones cannot pass through plasma membranes of cells; therefore, their receptors are found on the surface of the target cells.

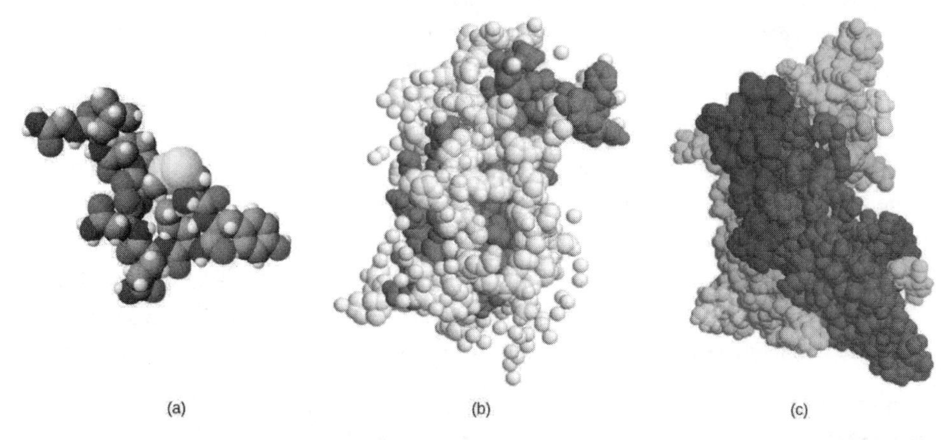

Figure 37.4 The structures of peptide hormones (a) oxytocin, (b) growth hormone, and (c) follicle-stimulating hormone are shown. These peptide hormones are much larger than those derived from cholesterol or amino acids.

CAREER CONNECTION

Endocrinologist

An endocrinologist is a medical doctor who specializes in treating disorders of the endocrine glands, hormone systems, and glucose and lipid metabolic pathways. An endocrine surgeon specializes in the surgical treatment of endocrine diseases and glands. Some of the diseases that are managed by endocrinologists: disorders of the pancreas (diabetes mellitus), disorders of the pituitary (gigantism, acromegaly, and pituitary dwarfism), disorders of the thyroid gland (goiter and Graves' disease), and disorders of the adrenal glands (Cushing's disease and Addison's disease).

Endocrinologists are required to assess patients and diagnose endocrine disorders through extensive use of laboratory tests. Many endocrine diseases are diagnosed using tests that stimulate or suppress endocrine organ functioning. Blood samples are then drawn to determine the effect of stimulating or suppressing an endocrine organ on the production of hormones. For example, to diagnose diabetes mellitus, patients are required to fast for 12 to 24 hours. They are then given a sugary drink, which stimulates the pancreas to produce insulin to decrease blood glucose levels. A blood sample is taken one to two hours after the sugar drink is consumed. If the pancreas is functioning properly, the blood glucose level will be within a normal range. Another example is the A1C test, which can be performed during blood screening. The A1C test measures average blood glucose levels over the past two to three months by examining how well the blood glucose is being managed over a long time.

Once a disease has been diagnosed, endocrinologists can prescribe lifestyle changes and/or medications to treat the disease. Some cases of diabetes mellitus can be managed by exercise, weight loss, and a healthy diet; in other cases, medications may be required to enhance insulin release. If the disease cannot be controlled by these means, the endocrinologist may prescribe insulin injections.

In addition to clinical practice, endocrinologists may also be involved in primary research and development activities. For example, ongoing islet transplant research is investigating how healthy pancreas islet cells may be transplanted into diabetic patients. Successful islet transplants may allow patients to stop taking insulin injections.

37.2 How Hormones Work

By the end of this section, you will be able to do the following:

- Explain how hormones work
- Discuss the role of different types of hormone receptors

Hormones mediate changes in target cells by binding to specific **hormone receptors**. In this way, even though hormones circulate throughout the body and come into contact with many different cell types, they only affect cells that possess the necessary receptors. Receptors for a specific hormone may be found on many different cells or may be limited to a small number of specialized cells. For example, thyroid hormones act on many different tissue types, stimulating metabolic activity throughout the body. Cells can have many receptors for the same hormone but often also possess receptors for different types of hormones. The number of receptors that respond to a hormone determines the cell's sensitivity to that hormone, and the resulting cellular response. Additionally, the number of receptors that respond to a hormone can change over time, resulting in increased or decreased cell sensitivity. In **up-regulation**, the number of receptors increases in response to rising hormone levels, making the cell more sensitive to the hormone and allowing for more cellular activity. When the number of receptors decreases in response to rising hormone levels, called **down-regulation**, cellular activity is reduced.

Receptor binding alters cellular activity and results in an increase or decrease in normal body processes. Depending on the location of the protein receptor on the target cell and the chemical structure of the hormone, hormones can mediate changes directly by binding to **intracellular hormone receptors** and modulating gene transcription, or indirectly by binding to cell surface receptors and stimulating signaling pathways.

Intracellular Hormone Receptors

Lipid-derived (soluble) hormones such as steroid hormones diffuse across the membranes of the endocrine cell. Once outside the cell, they bind to transport proteins that keep them soluble in the bloodstream. At the target cell, the hormones are released from the carrier protein and diffuse across the lipid bilayer of the plasma membrane of cells. The steroid hormones pass through the plasma membrane of a target cell and adhere to intracellular receptors residing in the cytoplasm or in the nucleus. The cell signaling pathways induced by the steroid hormones regulate specific genes on the cell's DNA. The hormones and receptor complex act as transcription regulators by increasing or decreasing the synthesis of mRNA molecules of specific genes. This, in turn, determines the amount of corresponding protein that is synthesized by altering gene expression. This protein can be used either to change the structure of the cell or to produce enzymes that catalyze chemical reactions. In this way, the steroid hormone regulates specific cell processes as illustrated in Figure 37.5.

VISUAL CONNECTION

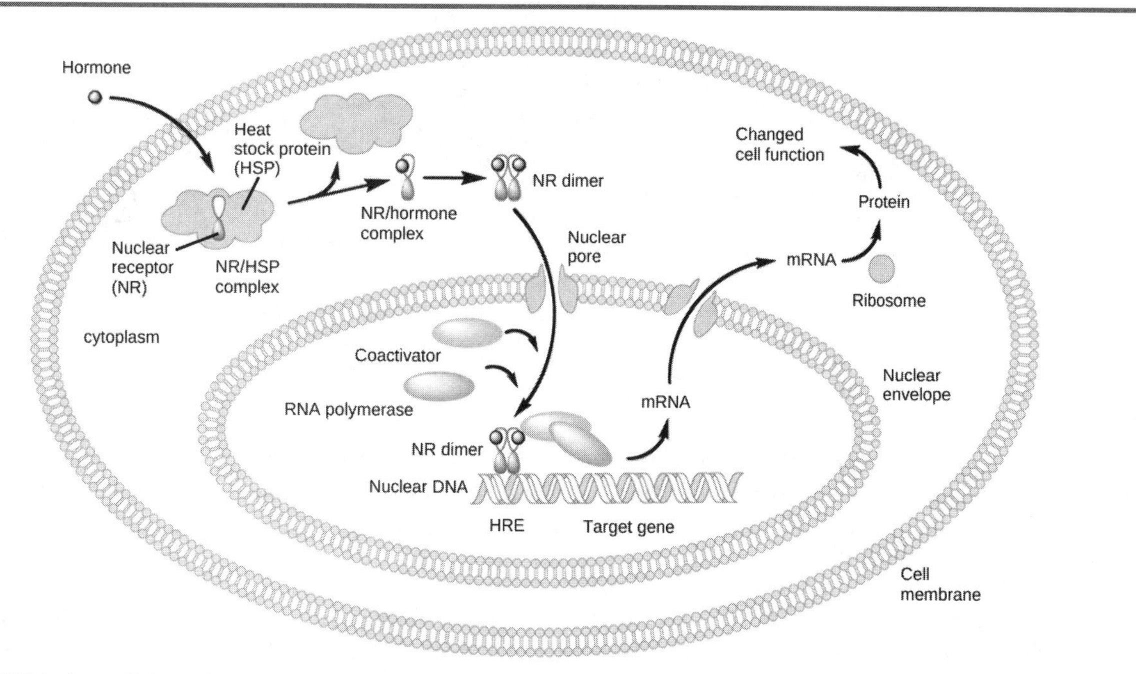

Figure 37.5 An intracellular nuclear receptor (NR) is located in the cytoplasm bound to a heat shock protein (HSP). Upon hormone binding, the receptor dissociates from the heat shock protein and translocates to the nucleus. In the nucleus, the hormone-receptor complex binds to a DNA sequence called a hormone response element (HRE), which triggers gene transcription and translation. The corresponding protein product can then mediate changes in cell function.

Heat shock proteins (HSP) are so named because they help refold misfolded proteins. In response to increased temperature (a "heat shock"), heat shock proteins are activated by release from the NR/HSP complex. At the same time, transcription of HSP genes is activated. Why do you think the cell responds to a heat shock by increasing the activity of proteins that help refold misfolded proteins?

Other lipid-soluble hormones that are not steroid hormones, such as vitamin D and thyroxine, have receptors located in the nucleus. While thyroxine is mostly hydrophobic, its passage across the membrane is dependent on transporter protein. Vitamin D diffuses across both the plasma membrane and the nuclear envelope. Once in the cell, both hormones bind to receptors in the nucleus. The hormone-receptor complex stimulates transcription of specific genes.

Plasma Membrane Hormone Receptors

Amino acid-derived hormones (with the exception of thyroxine) and polypeptide hormones are not lipid-derived (lipid-soluble) and therefore cannot diffuse through the plasma membrane of cells. Lipid insoluble hormones bind to receptors on the outer surface of the plasma membrane, via **plasma membrane hormone receptors**. Unlike steroid hormones, lipid insoluble hormones do not directly affect the target cell because they cannot enter the cell and act directly on DNA. Binding of these hormones to a cell surface receptor results in activation of a signaling pathway; this triggers intracellular activity and carries out the specific effects associated with the hormone. In this way, nothing passes through the cell membrane; the hormone that binds at the surface remains at the surface of the cell while the intracellular product remains inside the cell. The hormone that initiates the signaling pathway is called a **first messenger**, which activates a second messenger in the cytoplasm, as illustrated in Figure 37.6.

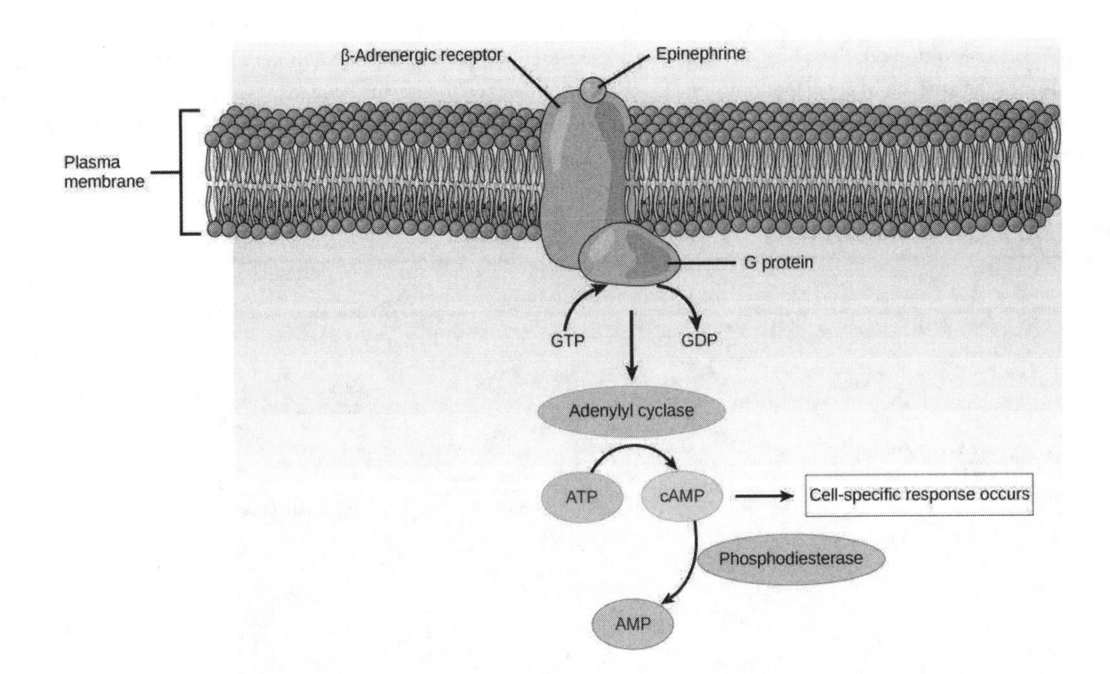

Figure 37.6 The amino acid-derived hormones epinephrine and norepinephrine bind to beta-adrenergic receptors on the plasma membrane of cells. Hormone binding to receptor activates a G-protein, which in turn activates adenylyl cyclase, converting ATP to cAMP. cAMP is a second messenger that mediates a cell-specific response. An enzyme called phosphodiesterase breaks down cAMP, terminating the signal.

One very important second messenger is cyclic AMP (cAMP). When a hormone binds to its membrane receptor, a **G-protein** that is associated with the receptor is activated; G-proteins are proteins separate from receptors that are found in the cell membrane. When a hormone is not bound to the receptor, the G-protein is inactive and is bound to guanosine diphosphate, or GDP. When a hormone binds to the receptor, the G-protein is activated by binding guanosine triphosphate, or GTP, in place of GDP. After binding, GTP is hydrolysed by the G-protein into GDP and becomes inactive.

The activated G-protein in turn activates a membrane-bound enzyme called **adenylyl cyclase**. Adenylyl cyclase catalyzes the conversion of ATP to cAMP. cAMP, in turn, activates a group of proteins called protein kinases, which transfer a phosphate group from ATP to a substrate molecule in a process called phosphorylation. The phosphorylation of a substrate molecule changes its structural orientation, thereby activating it. These activated molecules can then mediate changes in cellular processes.

The effect of a hormone is amplified as the signaling pathway progresses. The binding of a hormone at a single receptor causes the activation of many G-proteins, which activates adenylyl cyclase. Each molecule of adenylyl cyclase then triggers the formation of many molecules of cAMP. Further amplification occurs as protein kinases, once activated by cAMP, can catalyze many reactions. In this way, a small amount of hormone can trigger the formation of a large amount of cellular product. To stop hormone activity, cAMP is deactivated by the cytoplasmic enzyme **phosphodiesterase**, or PDE. PDE is always present in the cell and breaks down cAMP to control hormone activity, preventing overproduction of cellular products.

The specific response of a cell to a lipid insoluble hormone depends on the type of receptors that are present on the cell membrane and the substrate molecules present in the cell cytoplasm. Cellular responses to hormone binding of a receptor include altering membrane permeability and metabolic pathways, stimulating synthesis of proteins and enzymes, and activating hormone release.

37.3 Regulation of Body Processes

By the end of this section, you will be able to do the following:

- Explain how hormones regulate the excretory system
- Discuss the role of hormones in the reproductive system
- Describe how hormones regulate metabolism
- Explain the role of hormones in different diseases

Hormones have a wide range of effects and modulate many different body processes. The key regulatory processes that will be examined here are those affecting the excretory system, the reproductive system, metabolism, blood calcium concentrations, growth, and the stress response.

Hormonal Regulation of the Excretory System

Maintaining a proper water balance in the body is important to avoid dehydration or over-hydration (hyponatremia). The water concentration of the body is monitored by **osmoreceptors** in the hypothalamus, which detect the concentration of electrolytes in the extracellular fluid. The concentration of electrolytes in the blood rises when there is water loss caused by excessive perspiration, inadequate water intake, or low blood volume due to blood loss. An increase in blood electrolyte levels results in a neuronal signal being sent from the osmoreceptors in hypothalamic nuclei. The pituitary gland has two components: anterior and posterior. The anterior pituitary is composed of glandular cells that secrete protein hormones. The posterior pituitary is an extension of the hypothalamus. It is composed largely of neurons that are continuous with the hypothalamus.

The hypothalamus produces a polypeptide hormone known as **antidiuretic hormone (ADH)**, which is transported to and released from the posterior pituitary gland. The principal action of ADH is to regulate the amount of water excreted by the kidneys. As ADH (which is also known as vasopressin) causes direct water reabsorption from the kidney tubules, salts and wastes are concentrated in what will eventually be excreted as urine. The hypothalamus controls the mechanisms of ADH secretion, either by regulating blood volume or the concentration of water in the blood. Dehydration or physiological stress can cause an increase of osmolarity above 300 mOsm/L, which in turn, raises ADH secretion and water will be retained, causing an increase in blood pressure. ADH travels in the bloodstream to the kidneys. Once at the kidneys, ADH changes the kidneys to become more permeable to water by temporarily inserting water channels, aquaporins, into the kidney tubules. Water moves out of the kidney tubules through the aquaporins, reducing urine volume. The water is reabsorbed into the capillaries lowering blood osmolarity back toward normal. As blood osmolarity decreases, a negative feedback mechanism reduces osmoreceptor activity in the hypothalamus, and ADH secretion is reduced. ADH release can be reduced by certain substances, including alcohol, which can cause increased urine production and dehydration.

Chronic underproduction of ADH or a mutation in the ADH receptor results in **diabetes insipidus**. If the posterior pituitary does not release enough ADH, water cannot be retained by the kidneys and is lost as urine. This causes increased thirst, but water taken in is lost again and must be continually consumed. If the condition is not severe, dehydration may not occur, but severe cases can lead to electrolyte imbalances due to dehydration.

Another hormone responsible for maintaining electrolyte concentrations in extracellular fluids is **aldosterone**, a steroid hormone that is produced by the adrenal cortex. In contrast to ADH, which promotes the reabsorption of water to maintain proper water balance, aldosterone maintains proper water balance by enhancing Na^+ reabsorption and K^+ secretion from extracellular fluid of the cells in kidney tubules. Because it is produced in the cortex of the adrenal gland and affects the concentrations of minerals Na^+ and K^+, aldosterone is referred to as a **mineralocorticoid**, a corticosteroid that affects ion and water balance. Aldosterone release is stimulated by a decrease in blood sodium levels, blood volume, or blood pressure, or an increase in blood potassium levels. It also prevents the loss of Na^+ from sweat, saliva, and gastric juice. The reabsorption of Na^+ also results in the osmotic reabsorption of water, which alters blood volume and blood pressure.

Aldosterone production can be stimulated by low blood pressure, which triggers a sequence of chemical release, as illustrated in Figure 37.7. When blood pressure drops, the renin-angiotensin-aldosterone system (RAAS) is activated. Cells in the juxtaglomerular apparatus, which regulates the functions of the nephrons of the kidney, detect this and release **renin**. Renin, an enzyme, circulates in the blood and reacts with a plasma protein produced by the liver called angiotensinogen. When angiotensinogen is cleaved by renin, it produces angiotensin I, which is then converted into angiotensin II in the lungs. Angiotensin II functions as a hormone and then causes the release of the hormone aldosterone by the adrenal cortex, resulting in increased Na^+ reabsorption, water retention, and an increase in blood pressure. Angiotensin II in addition to being a potent vasoconstrictor also causes an increase in ADH and increased thirst, both of which help to raise blood pressure.

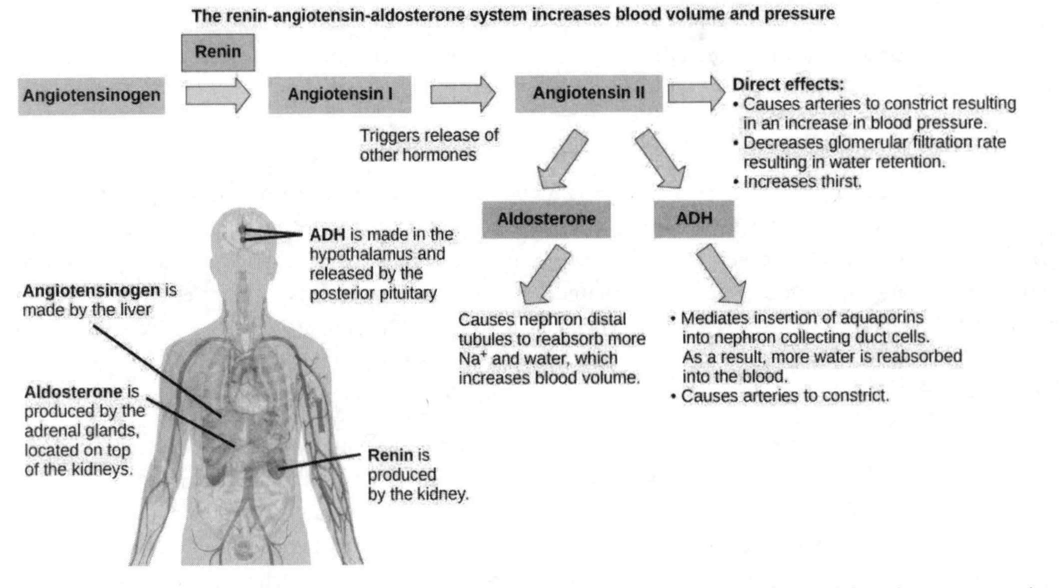

Figure 37.7 ADH and aldosterone increase blood pressure and volume. Angiotensin II stimulates release of these hormones. Angiotensin II, in turn, is formed when renin cleaves angiotensinogen. (credit: modification of work by Mikael Häggström)

Hormonal Regulation of the Reproductive System

Regulation of the reproductive system is a process that requires the action of hormones from the pituitary gland, the adrenal cortex, and the gonads. During puberty in both males and females, the hypothalamus produces gonadotropin-releasing hormone (GnRH), which stimulates the production and release of **follicle-stimulating hormone (FSH)** and luteinizing hormone (LH) from the anterior pituitary gland. These hormones regulate the gonads (testes in males and ovaries in females) and therefore are called **gonadotropins**. In both males and females, FSH stimulates gamete production and LH stimulates production of hormones by the gonads. An increase in gonad hormone levels inhibits GnRH production through a negative feedback loop.

Regulation of the Male Reproductive System

In males, FSH stimulates the maturation of sperm cells. FSH production is inhibited by the hormone inhibin, which is released by the testes. LH stimulates production of the sex hormones (**androgens**) by the interstitial cells of the testes and therefore is also called interstitial cell-stimulating hormone.

The most widely known androgen in males is testosterone. Testosterone promotes the production of sperm and masculine characteristics. The adrenal cortex also produces small amounts of testosterone precursor, although the role of this additional hormone production is not fully understood.

Everyday Connection

The Dangers of Synthetic Hormones

Figure 37.8 Professional baseball player Jason Giambi publically admitted to, and apologized for, his use of anabolic steroids supplied by a trainer. (credit: Bryce Edwards)

Some athletes attempt to boost their performance by using artificial hormones that enhance muscle performance. Anabolic steroids, a form of the male sex hormone testosterone, are one of the most widely known performance-enhancing drugs. Steroids are used to help build muscle mass. Other hormones that are used to enhance athletic performance include erythropoietin, which triggers the production of red blood cells, and human growth hormone, which can help in building muscle mass. Most performance enhancing drugs are illegal for nonmedical purposes. They are also banned by national and international governing bodies including the International Olympic Committee, the U.S. Olympic Committee, the National Collegiate Athletic Association, the Major League Baseball, and the National Football League.

The side effects of synthetic hormones are often significant and nonreversible, and in some cases, fatal. Androgens produce several complications such as liver dysfunctions and liver tumors, prostate gland enlargement, difficulty urinating, premature closure of epiphyseal cartilages, testicular atrophy, infertility, and immune system depression. The physiological strain caused by these substances is often greater than what the body can handle, leading to unpredictable and dangerous effects and linking their use to heart attacks, strokes, and impaired cardiac function.

Regulation of the Female Reproductive System

In females, FSH stimulates development of egg cells, called ova, which develop in structures called follicles. Follicle cells produce the hormone inhibin, which inhibits FSH production. LH also plays a role in the development of ova, induction of ovulation, and stimulation of estradiol and progesterone production by the ovaries, as illustrated in Figure 37.9. Estradiol and progesterone are steroid hormones that prepare the body for pregnancy. Estradiol produces secondary sex characteristics in females, while both estradiol and progesterone regulate the menstrual cycle.

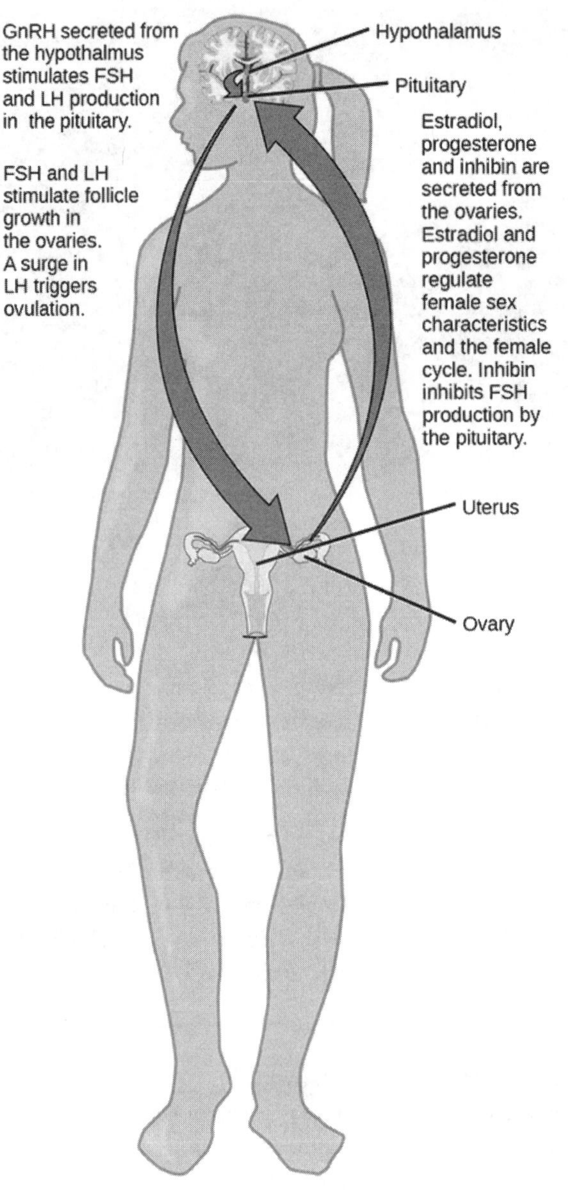

Figure 37.9 Hormonal regulation of the female reproductive system involves hormones from the hypothalamus, pituitary, and ovaries.

In addition to producing FSH and LH, the anterior portion of the pituitary gland also produces the hormone **prolactin (PRL)** in females. Prolactin stimulates the production of milk by the mammary glands following childbirth. Prolactin levels are regulated by the hypothalamic hormones **prolactin-releasing hormone (PRH)** and **prolactin-inhibiting hormone (PIH)**, which is now known to be dopamine. PRH stimulates the release of prolactin and PIH inhibits it.

The posterior pituitary releases the hormone **oxytocin**, which stimulates uterine contractions during childbirth. The uterine smooth muscles are not very sensitive to oxytocin until late in pregnancy when the number of oxytocin receptors in the uterus peaks. Stretching of tissues in the uterus and cervix stimulates oxytocin release during childbirth. Contractions increase in intensity as blood levels of oxytocin rise via a positive feedback mechanism until the birth is complete. Oxytocin also stimulates the contraction of myoepithelial cells around the milk-producing mammary glands. As these cells contract, milk is forced from the secretory alveoli into milk ducts and is ejected from the breasts in milk ejection ("let-down") reflex. Oxytocin release is stimulated by the suckling of an infant, which triggers the synthesis of oxytocin in the hypothalamus and its release into circulation at the posterior pituitary.

Hormonal Regulation of Metabolism

Blood glucose levels vary widely over the course of a day as periods of food consumption alternate with periods of fasting. Insulin and glucagon are the two hormones primarily responsible for maintaining homeostasis of blood glucose levels.

Additional regulation is mediated by the thyroid hormones.

Regulation of Blood Glucose Levels by Insulin and Glucagon

Cells of the body require nutrients in order to function, and these nutrients are obtained through feeding. In order to manage nutrient intake, storing excess intake and utilizing reserves when necessary, the body uses hormones to moderate energy stores. **Insulin** is produced by the beta cells of the pancreas, which are stimulated to release insulin as blood glucose levels rise (for example, after a meal is consumed). Insulin lowers blood glucose levels by enhancing the rate of glucose uptake and utilization by target cells, which use glucose for ATP production. It also stimulates the liver to convert glucose to glycogen, which is then stored by cells for later use. Insulin also increases glucose transport into certain cells, such as muscle cells and the liver. This results from an insulin-mediated increase in the number of glucose transporter proteins in cell membranes, which remove glucose from circulation by facilitated diffusion. As insulin binds to its target cell via insulin receptors and signal transduction, it triggers the cell to incorporate glucose transport proteins into its membrane. This allows glucose to enter the cell, where it can be used as an energy source. However, this does not occur in all cells: some cells, including those in the kidneys and brain, can access glucose without the use of insulin. Insulin also stimulates the conversion of glucose to fat in adipocytes and the synthesis of proteins. These actions mediated by insulin cause blood glucose concentrations to fall, called a hypoglycemic "low sugar" effect, which inhibits further insulin release from beta cells through a negative feedback loop.

LINK TO LEARNING

This animation describes the role of insulin and the pancreas in diabetes.

Click to view content (https://www.openstax.org/l/insulin)

Impaired insulin function can lead to a condition called **diabetes mellitus**, the main symptoms of which are illustrated in Figure 37.10. This can be caused by low levels of insulin production by the beta cells of the pancreas, or by reduced sensitivity of tissue cells to insulin. This prevents glucose from being absorbed by cells, causing high levels of blood glucose, or **hyperglycemia** (high sugar). High blood glucose levels make it difficult for the kidneys to recover all the glucose from nascent urine, resulting in glucose being lost in urine. High glucose levels also result in less water being reabsorbed by the kidneys, causing high amounts of urine to be produced; this may result in dehydration. Over time, high blood glucose levels can cause nerve damage to the eyes and peripheral body tissues, as well as damage to the kidneys and cardiovascular system. Oversecretion of insulin can cause **hypoglycemia**, low blood glucose levels. This causes insufficient glucose availability to cells, often leading to muscle weakness, and can sometimes cause unconsciousness or death if left untreated.

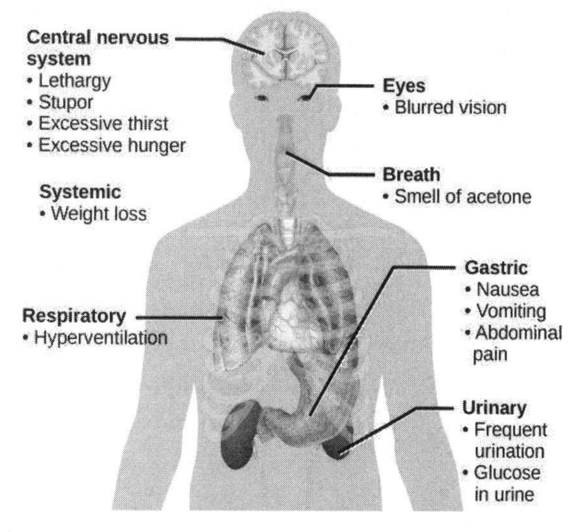

Figure 37.10 The main symptoms of diabetes are shown. (credit: modification of work by Mikael Häggström)

When blood glucose levels decline below normal levels, for example between meals or when glucose is utilized rapidly during exercise, the hormone **glucagon** is released from the alpha cells of the pancreas. Glucagon raises blood glucose levels, eliciting what is called a hyperglycemic effect, by stimulating the breakdown of glycogen to glucose in skeletal muscle cells and liver cells in a process called **glycogenolysis**. Glucose can then be utilized as energy by muscle cells and released into circulation by the liver cells. Glucagon also stimulates absorption of amino acids from the blood by the liver, which then converts them to glucose. This process of glucose synthesis is called **gluconeogenesis**. Glucagon also stimulates adipose cells to release fatty acids into the blood. These actions mediated by glucagon result in an increase in blood glucose levels to normal homeostatic levels. Rising blood glucose levels inhibit further glucagon release by the pancreas via a negative feedback mechanism. In this way, insulin and glucagon work together to maintain homeostatic glucose levels, as shown in Figure 37.11.

VISUAL CONNECTION

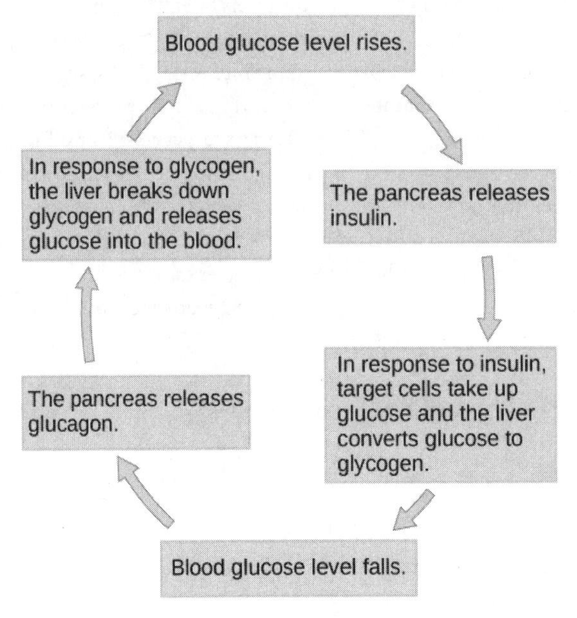

Figure 37.11 Insulin and glucagon regulate blood glucose levels.

Pancreatic tumors may cause excess secretion of glucagon. Type I diabetes results from the failure of the pancreas to produce insulin. Which of the following statement about these two conditions is true?

a. A pancreatic tumor and type I diabetes will have the opposite effects on blood sugar levels.
b. A pancreatic tumor and type I diabetes will both cause hyperglycemia.
c. A pancreatic tumor and type I diabetes will both cause hypoglycemia.
d. Both pancreatic tumors and type I diabetes result in the inability of cells to take up glucose.

Regulation of Blood Glucose Levels by Thyroid Hormones

The basal metabolic rate, which is the amount of calories required by the body at rest, is determined by two hormones produced by the thyroid gland: **thyroxine**, also known as tetraiodothyronine or T_4, and **triiodothyronine**, also known as T_3. These hormones affect nearly every cell in the body except for the adult brain, uterus, testes, blood cells, and spleen. They are transported across the plasma membrane of target cells and bind to receptors on the mitochondria resulting in increased ATP production. In the nucleus, T_3 and T_4 activate genes involved in energy production and glucose oxidation. This results in increased rates of metabolism and body heat production, which is known as the hormone's calorigenic effect.

T_3 and T_4 release from the thyroid gland is stimulated by **thyroid-stimulating hormone (TSH)**, which is produced by the anterior pituitary. TSH binding at the receptors of the follicle of the thyroid triggers the production of T_3 and T_4 from a glycoprotein called **thyroglobulin**. Thyroglobulin is present in the follicles of the thyroid, and is converted into thyroid hormones with the addition of iodine. Iodine is formed from iodide ions that are actively transported into the thyroid follicle from the bloodstream. A peroxidase enzyme then attaches the iodine to the tyrosine amino acid found in thyroglobulin. T_3 has three iodine ions attached, while T_4 has four iodine ions attached. T_3 and T_4 are then released into the bloodstream, with T_4 being released in much greater amounts than T_3. As T_3 is more active than T_4 and is responsible for most of the effects of thyroid hormones, tissues of the body convert T_4 to T_3 by the removal of an iodine ion. Most of the released T_3 and T_4 becomes attached to transport proteins in the bloodstream and is unable to cross the plasma membrane of cells. These protein-bound molecules are only released when blood levels of the unattached hormone begin to decline. In this way, a week's worth of reserve hormone is maintained in the blood. Increased T_3 and T_4 levels in the blood inhibit the release of TSH, which results in lower T_3 and T_4 release from the thyroid.

The follicular cells of the thyroid require iodides (anions of iodine) in order to synthesize T_3 and T_4. Iodides obtained from the diet are actively transported into follicle cells resulting in a concentration that is approximately 30 times higher than in blood. The typical diet in North America provides more iodine than required due to the addition of iodide to table salt. Inadequate iodine intake, which occurs in many developing countries, results in an inability to synthesize T_3 and T_4 hormones. The thyroid gland enlarges in a condition called **goiter**, which is caused by overproduction of TSH without the formation of thyroid hormone. Thyroglobulin is contained in a fluid called colloid, and TSH stimulation results in higher levels of colloid accumulation in the thyroid. In the absence of iodine, this is not converted to thyroid hormone, and colloid begins to accumulate more and more in the thyroid gland, leading to goiter.

Disorders can arise from both the underproduction and overproduction of thyroid hormones. **Hypothyroidism**, underproduction of the thyroid hormones, can cause a low metabolic rate leading to weight gain, sensitivity to cold, and reduced mental activity, among other symptoms. In children, hypothyroidism can cause cretinism, which can lead to mental retardation and growth defects. **Hyperthyroidism**, the overproduction of thyroid hormones, can lead to an increased metabolic rate and its effects: weight loss, excess heat production, sweating, and an increased heart rate. Graves' disease is one example of a hyperthyroid condition.

Hormonal Control of Blood Calcium Levels

Regulation of blood calcium concentrations is important for generation of muscle contractions and nerve impulses, which are electrically stimulated. If calcium levels get too high, membrane permeability to sodium decreases and membranes become less responsive. If calcium levels get too low, membrane permeability to sodium increases and convulsions or muscle spasms can result.

Blood calcium levels are regulated by **parathyroid hormone (PTH)**, which is produced by the parathyroid glands, as illustrated in Figure 37.12. PTH is released in response to low blood Ca^{2+} levels. PTH increases Ca^{2+} levels by targeting the skeleton, the kidneys, and the intestine. In the skeleton, PTH stimulates osteoclasts, which causes bone to be reabsorbed, releasing Ca^{2+} from bone into the blood. PTH also inhibits osteoblasts, reducing Ca^{2+} deposition in bone. In the intestines, PTH increases dietary Ca^{2+} absorption, and in the kidneys, PTH stimulates reabsorption of the CA^{2+}. While PTH acts directly on the kidneys to increase Ca^{2+} reabsorption, its effects on the intestine are indirect. PTH triggers the formation of calcitriol, an active form of vitamin D, which acts on the intestines to increase absorption of dietary calcium. PTH release is inhibited by rising blood calcium levels.

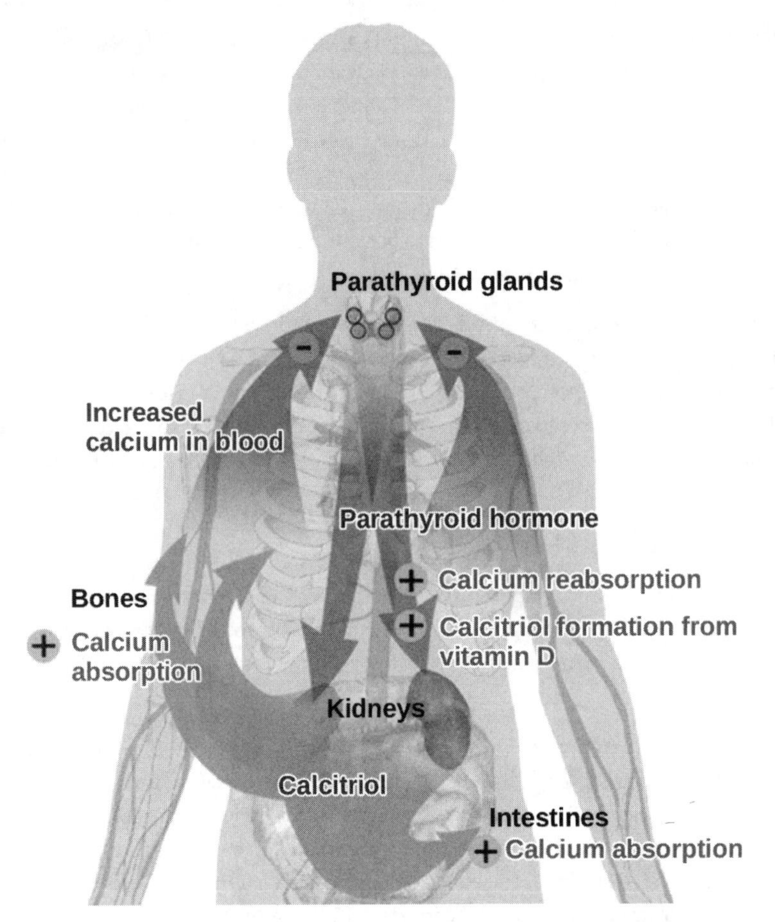

Figure 37.12 Parathyroid hormone (PTH) is released in response to low blood calcium levels. It increases blood calcium levels by targeting the skeleton, the kidneys, and the intestine. (credit: modification of work by Mikael Häggström)

Hyperparathyroidism results from an overproduction of parathyroid hormone. This results in excessive calcium being removed from bones and introduced into blood circulation, producing structural weakness of the bones, which can lead to deformation and fractures, plus nervous system impairment due to high blood calcium levels. Hypoparathyroidism, the underproduction of PTH, results in extremely low levels of blood calcium, which causes impaired muscle function and may result in tetany (severe sustained muscle contraction).

The hormone **calcitonin**, which is produced by the parafollicular or C cells of the thyroid, has the opposite effect on blood calcium levels as does PTH. Calcitonin decreases blood calcium levels by inhibiting osteoclasts, stimulating osteoblasts, and stimulating calcium excretion by the kidneys. This results in calcium being added to the bones to promote structural integrity. Calcitonin is most important in children (when it stimulates bone growth), during pregnancy (when it reduces maternal bone loss), and during prolonged starvation (because it reduces bone mass loss). In healthy nonpregnant, unstarved adults, the role of calcitonin is unclear.

Hormonal Regulation of Growth

Hormonal regulation is required for the growth and replication of most cells in the body. **Growth hormone (GH)**, produced by the anterior portion of the pituitary gland, accelerates the rate of protein synthesis, particularly in skeletal muscle and bones. Growth hormone has direct and indirect mechanisms of action. The first direct action of GH is stimulation of triglyceride breakdown (lipolysis) and release into the blood by adipocytes. This results in a switch by most tissues from utilizing glucose as an energy source to utilizing fatty acids. This process is called a **glucose-sparing effect**. In another direct mechanism, GH stimulates glycogen breakdown in the liver; the glycogen is then released into the blood as glucose. Blood glucose levels increase as most tissues are utilizing fatty acids instead of glucose for their energy needs. The GH mediated increase in blood glucose levels is called a **diabetogenic effect** because it is similar to the high blood glucose levels seen in diabetes mellitus.

The indirect mechanism of GH action is mediated by **insulin-like growth factors (IGFs)** or somatomedins, which are a family of growth-promoting proteins produced by the liver, which stimulates tissue growth. IGFs stimulate the uptake of amino acids from the blood, allowing the formation of new proteins, particularly in skeletal muscle cells, cartilage cells, and other target cells, as shown in Figure 37.13. This is especially important after a meal, when glucose and amino acid concentration levels are high in the blood. GH levels are regulated by two hormones produced by the hypothalamus. GH release is stimulated by **growth hormone-releasing hormone (GHRH)** and is inhibited by **growth hormone-inhibiting hormone (GHIH)**, also called somatostatin.

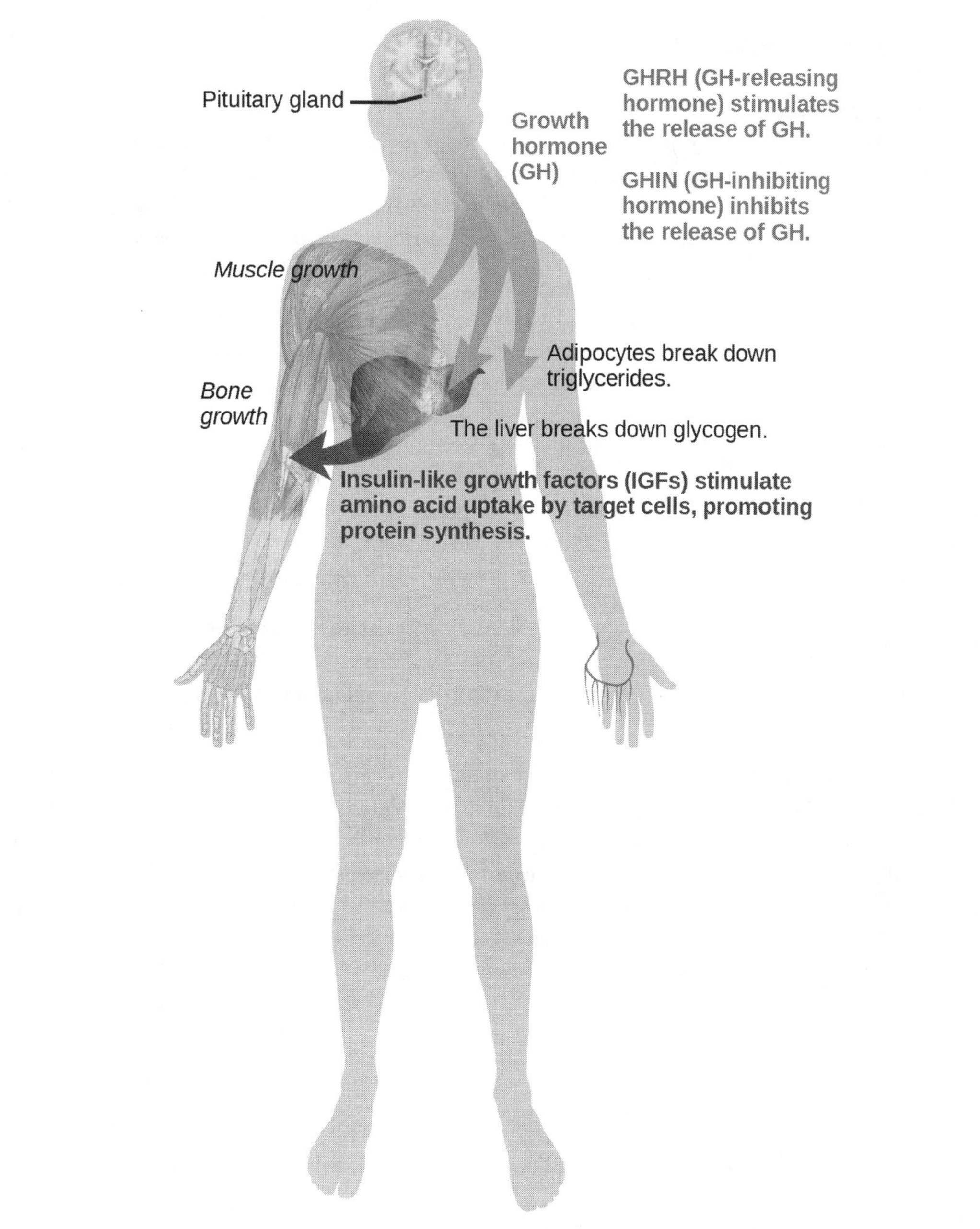

Figure 37.13 Growth hormone directly accelerates the rate of protein synthesis in skeletal muscle and bones. Insulin-like growth factor 1 (IGF-1) is activated by growth hormone and also allows formation of new proteins in muscle cells and bone. (credit: modification of work by Mikael Häggström)

A balanced production of growth hormone is critical for proper development. Underproduction of GH in adults does not appear to cause any abnormalities, but in children it can result in **pituitary dwarfism**, in which growth is reduced. Pituitary dwarfism is characterized by symmetric body formation. In some cases, individuals are under 30 inches in height. Oversecretion of growth hormone can lead to **gigantism** in children, causing excessive growth. In some documented cases, individuals can reach heights of over eight feet. In adults, excessive GH can lead to **acromegaly**, a condition in which there is enlargement of bones in the face, hands, and feet that are still capable of growth.

Hormonal Regulation of Stress

When a threat or danger is perceived, the body responds by releasing hormones that will ready it for the "fight-or-flight" response. The effects of this response are familiar to anyone who has been in a stressful situation: increased heart rate, dry mouth, and hair standing up.

EVOLUTION CONNECTION

Fight-or-Flight Response

Interactions of the endocrine hormones have evolved to ensure the body's internal environment remains stable. Stressors are stimuli that disrupt homeostasis. The sympathetic division of the vertebrate autonomic nervous system has evolved the fight-or-flight response to counter stress-induced disruptions of homeostasis. In the initial alarm phase, the sympathetic nervous system stimulates an increase in energy levels through increased blood glucose levels. This prepares the body for physical activity that may be required to respond to stress: to either fight for survival or to flee from danger.

However, some stresses, such as illness or injury, can last for a long time. Glycogen reserves, which provide energy in the short-term response to stress, are exhausted after several hours and cannot meet long-term energy needs. If glycogen reserves were the only energy source available, neural functioning could not be maintained once the reserves became depleted due to the nervous system's high requirement for glucose. In this situation, the body has evolved a response to counter long-term stress through the actions of the glucocorticoids, which ensure that long-term energy requirements can be met. The glucocorticoids mobilize lipid and protein reserves, stimulate gluconeogenesis, conserve glucose for use by neural tissue, and stimulate the conservation of salts and water. The mechanisms to maintain homeostasis that are described here are those observed in the human body. However, the fight-or-flight response exists in some form in all vertebrates.

The sympathetic nervous system regulates the stress response via the hypothalamus. Stressful stimuli cause the hypothalamus to signal the adrenal medulla (which mediates short-term stress responses) via nerve impulses, and the adrenal cortex, which mediates long-term stress responses, via the hormone **adrenocorticotropic hormone (ACTH)**, which is produced by the anterior pituitary.

Short-term Stress Response

When presented with a stressful situation, the body responds by calling for the release of hormones that provide a burst of energy. The hormones **epinephrine** (also known as adrenaline) and **norepinephrine** (also known as noradrenaline) are released by the adrenal medulla. How do these hormones provide a burst of energy? Epinephrine and norepinephrine increase blood glucose levels by stimulating the liver and skeletal muscles to break down glycogen and by stimulating glucose release by liver cells. Additionally, these hormones increase oxygen availability to cells by increasing the heart rate and dilating the bronchioles. The hormones also prioritize body function by increasing blood supply to essential organs such as the heart, brain, and skeletal muscles, while restricting blood flow to organs not in immediate need, such as the skin, digestive system, and kidneys. Epinephrine and norepinephrine are collectively called catecholamines.

LINK TO LEARNING

Watch this Discovery Channel animation (http://openstax.org/l/adrenaline) describing the flight-or-flight response.

Long-term Stress Response

Long-term stress response differs from short-term stress response. The body cannot sustain the bursts of energy mediated by epinephrine and norepinephrine for long times. Instead, other hormones come into play. In a long-term stress response, the hypothalamus triggers the release of ACTH from the anterior pituitary gland. The adrenal cortex is stimulated by ACTH to release steroid hormones called **corticosteroids**. Corticosteroids turn on transcription of certain genes in the nuclei of target cells. They change enzyme concentrations in the cytoplasm and affect cellular metabolism. There are two main corticosteroids: glucocorticoids such as **cortisol**, and mineralocorticoids such as aldosterone. These hormones target the breakdown of fat into fatty acids in the adipose tissue. The fatty acids are released into the bloodstream for other tissues to use for ATP production. The **glucocorticoids** primarily affect glucose metabolism by stimulating glucose synthesis. Glucocorticoids also have anti-inflammatory properties through inhibition of the immune system. For example, cortisone is used as an anti-inflammatory medication; however, it cannot be used long term as it increases susceptibility to disease due to its immune-suppressing effects.

Mineralocorticoids function to regulate ion and water balance of the body. The hormone aldosterone stimulates the reabsorption of water and sodium ions in the kidney, which results in increased blood pressure and volume.

Hypersecretion of glucocorticoids can cause a condition known as **Cushing's disease**, characterized by a shifting of fat storage areas of the body. This can cause the accumulation of adipose tissue in the face and neck, and excessive glucose in the blood. Hyposecretion of the corticosteroids can cause **Addison's disease**, which may result in bronzing of the skin, hypoglycemia, and low electrolyte levels in the blood.

37.4 Regulation of Hormone Production

By the end of this section, you will be able to do the following:

- Explain how hormone production is regulated
- Discuss the different stimuli that control hormone levels in the body

Hormone production and release are primarily controlled by negative feedback. In negative feedback systems, a stimulus elicits the release of a substance; once the substance reaches a certain level, it sends a signal that stops further release of the substance. In this way, the concentration of hormones in blood is maintained within a narrow range. For example, the anterior pituitary signals the thyroid to release thyroid hormones. Increasing levels of these hormones in the blood then give feedback to the hypothalamus and anterior pituitary to inhibit further signaling to the thyroid gland, as illustrated in Figure 37.14. There are three mechanisms by which endocrine glands are stimulated to synthesize and release hormones: humoral stimuli, hormonal stimuli, and neural stimuli.

VISUAL CONNECTION

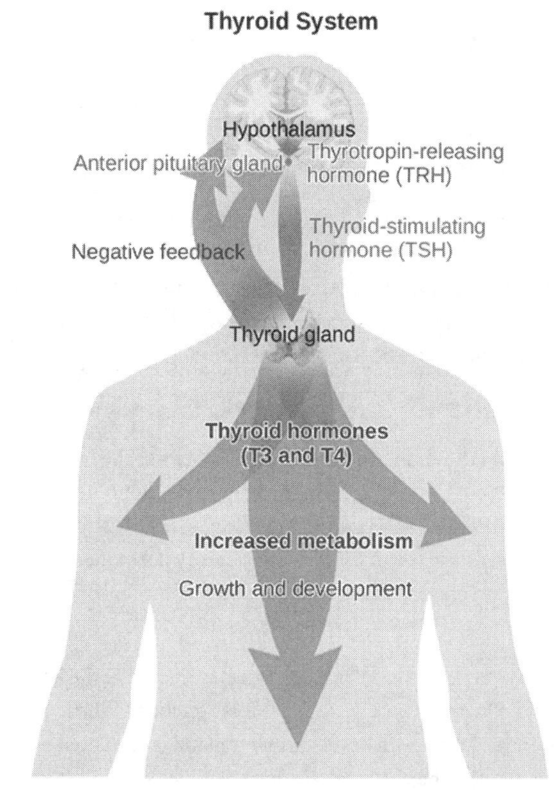

Figure 37.14 The anterior pituitary stimulates the thyroid gland to release thyroid hormones T_3 and T_4. Increasing levels of these hormones in the blood results in feedback to the hypothalamus and anterior pituitary to inhibit further signaling to the thyroid gland. (credit: modification of work by Mikael Häggström)

Hyperthyroidism is a condition in which the thyroid gland is overactive. Hypothyroidism is a condition in which the thyroid gland is underactive. Which of the conditions are the following two patients most likely to have?

Patient A has symptoms including weight gain, cold sensitivity, low heart rate, and fatigue.

Patient B has symptoms including weight loss, profuse sweating, increased heart rate, and difficulty sleeping.

Humoral Stimuli

The term "humoral" is derived from the term "humor," which refers to bodily fluids such as blood. A **humoral stimulus** refers to the control of hormone release in response to changes in extracellular fluids such as blood or the ion concentration in the blood. For example, a rise in blood glucose levels triggers the pancreatic release of insulin. Insulin causes blood glucose levels to drop, which signals the pancreas to stop producing insulin in a negative feedback loop.

Hormonal Stimuli

Hormonal stimuli refers to the release of a hormone in response to another hormone. A number of endocrine glands release hormones when stimulated by hormones released by other endocrine glands. For example, the hypothalamus produces hormones that stimulate the anterior portion of the pituitary gland. The anterior pituitary in turn releases hormones that regulate hormone production by other endocrine glands. The anterior pituitary releases the thyroid-stimulating hormone, which then stimulates the thyroid gland to produce the hormones T_3 and T_4. As blood concentrations of T_3 and T_4 rise, they inhibit both the pituitary and the hypothalamus in a negative feedback loop.

Neural Stimuli

In some cases, the nervous system directly stimulates endocrine glands to release hormones, which is referred to as **neural stimuli**. Recall that in a short-term stress response, the hormones epinephrine and norepinephrine are important for providing the bursts of energy required for the body to respond. Here, neuronal signaling from the sympathetic nervous system directly stimulates the adrenal medulla to release the hormones epinephrine and norepinephrine in response to stress.

37.5 Endocrine Glands

By the end of this section, you will be able to do the following:

- Describe the role of different glands in the endocrine system
- Explain how the different glands work together to maintain homeostasis

Both the endocrine and nervous systems use chemical signals to communicate and regulate the body's physiology. The endocrine system releases hormones that act on target cells to regulate development, growth, energy metabolism, reproduction, and many behaviors. The nervous system releases neurotransmitters or neurohormones that regulate neurons, muscle cells, and endocrine cells. Because the neurons can regulate the release of hormones, the nervous and endocrine systems work in a coordinated manner to regulate the body's physiology.

Hypothalamic-Pituitary Axis

The **hypothalamus** in vertebrates integrates the endocrine and nervous systems. The hypothalamus is an endocrine organ located in the diencephalon of the brain. It receives input from the body and other brain areas and initiates endocrine responses to environmental changes. The hypothalamus acts as an endocrine organ, synthesizing hormones and transporting them along axons to the posterior pituitary gland. It synthesizes and secretes regulatory hormones that control the endocrine cells in the anterior pituitary gland. The hypothalamus contains autonomic centers that control endocrine cells in the adrenal medulla via neuronal control.

The **pituitary gland**, sometimes called the hypophysis or "master gland" is located at the base of the brain in the sella turcica, a groove of the sphenoid bone of the skull, illustrated in Figure 37.15. It is attached to the hypothalamus via a stalk called the **pituitary stalk** (or infundibulum). The anterior portion of the pituitary gland is regulated by releasing or release-inhibiting hormones produced by the hypothalamus, and the posterior pituitary receives signals via neurosecretory cells to release hormones produced by the hypothalamus. The pituitary has two distinct regions—the anterior pituitary and the posterior pituitary—which between them secrete nine different peptide or protein hormones. The posterior lobe of the pituitary gland contains axons of the hypothalamic neurons.

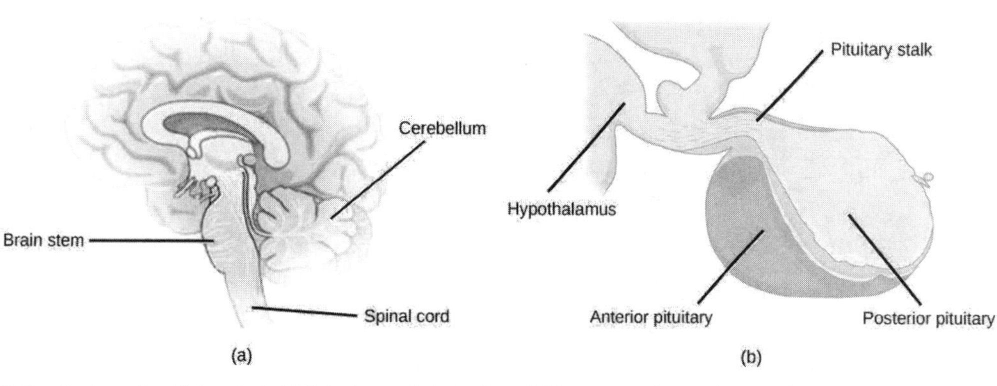

Figure 37.15 The pituitary gland is located at (a) the base of the brain and (b) connected to the hypothalamus by the pituitary stalk. (credit a: modification of work by NCI; credit b: modification of work by Gray's Anatomy)

Anterior Pituitary

The **anterior pituitary** gland, or adenohypophysis, is surrounded by a capillary network that extends from the hypothalamus, down along the infundibulum, and to the anterior pituitary. This capillary network is a part of the **hypophyseal portal system** that carries substances from the hypothalamus to the anterior pituitary and hormones from the anterior pituitary into the circulatory system. A portal system carries blood from one capillary network to another; therefore, the hypophyseal portal system allows hormones produced by the hypothalamus to be carried directly to the anterior pituitary without first entering the circulatory system.

The anterior pituitary produces seven hormones: growth hormone (GH), prolactin (PRL), thyroid-stimulating hormone (TSH), melanin-stimulating hormone (MSH), adrenocorticotropic hormone (ACTH), follicle-stimulating hormone (FSH), and luteinizing hormone (LH). Anterior pituitary hormones are sometimes referred to as tropic hormones, because they control the functioning of other organs. While these hormones are produced by the anterior pituitary, their production is controlled by regulatory hormones produced by the hypothalamus. These regulatory hormones can be releasing hormones or inhibiting hormones, causing more or less of the anterior pituitary hormones to be secreted. These travel from the hypothalamus through the hypophyseal portal system to the anterior pituitary where they exert their effect. Negative feedback then regulates how much of these regulatory hormones are released and how much anterior pituitary hormone is secreted.

Posterior Pituitary

The **posterior pituitary** is significantly different in structure from the anterior pituitary. It is a part of the brain, extending down from the hypothalamus, and contains mostly nerve fibers and neuroglial cells, which support axons that extend from the hypothalamus to the posterior pituitary. The posterior pituitary and the infundibulum together are referred to as the neurohypophysis.

The hormones antidiuretic hormone (ADH), also known as vasopressin, and oxytocin are produced by neurons in the hypothalamus and transported within these axons along the infundibulum to the posterior pituitary. They are released into the circulatory system via neural signaling from the hypothalamus. These hormones are considered to be posterior pituitary hormones, even though they are produced by the hypothalamus, because that is where they are released into the circulatory system. The posterior pituitary itself does not produce hormones, but instead stores hormones produced by the hypothalamus and releases them into the bloodstream.

Thyroid Gland

The **thyroid gland** is located in the neck, just below the larynx and in front of the trachea, as shown in Figure 37.16. It is a butterfly-shaped gland with two lobes that are connected by the **isthmus**. It has a dark red color due to its extensive vascular system. When the thyroid swells due to dysfunction, it can be felt under the skin of the neck.

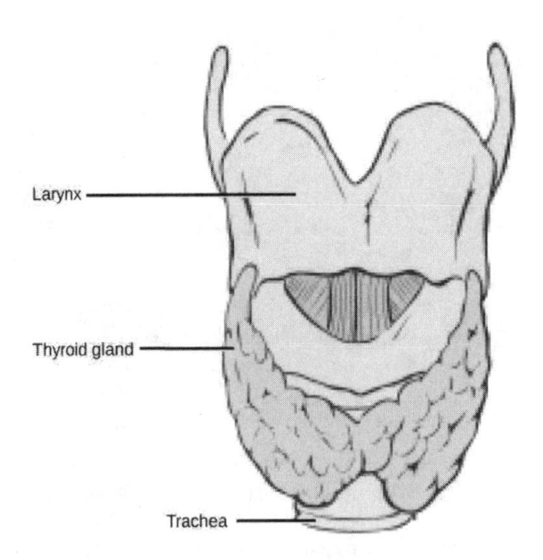

Figure 37.16 This illustration shows the location of the thyroid gland.

The thyroid gland is made up of many spherical thyroid follicles, which are lined with a simple cuboidal epithelium. These follicles contain a viscous fluid, called **colloid**, which stores the glycoprotein thyroglobulin, the precursor to the thyroid hormones. The follicles produce hormones that can be stored in the colloid or released into the surrounding capillary network for transport to the rest of the body via the circulatory system.

Thyroid follicle cells synthesize the hormone thyroxine, which is also known as T_4 because it contains four atoms of iodine, and triiodothyronine, also known as T_3 because it contains three atoms of iodine. Follicle cells are stimulated to release stored T_3 and T_4 by thyroid stimulating hormone (TSH), which is produced by the anterior pituitary. These thyroid hormones increase the rates of mitochondrial ATP production.

A third hormone, calcitonin, is produced by **parafollicular cells** of the thyroid either releasing hormones or inhibiting hormones. Calcitonin release is not controlled by TSH, but instead is released when calcium ion concentrations in the blood rise. Calcitonin functions to help regulate calcium concentrations in body fluids. It acts in the bones to inhibit osteoclast activity and in the kidneys to stimulate excretion of calcium. The combination of these two events lowers body fluid levels of calcium.

Parathyroid Glands

Most people have four **parathyroid glands**; however, the number can vary from two to six. These glands are located on the posterior surface of the thyroid gland, as shown in Figure 37.17. Normally, there is a superior gland and an inferior gland associated with each of the thyroid's two lobes. Each parathyroid gland is covered by connective tissue and contains many secretory cells that are associated with a capillary network.

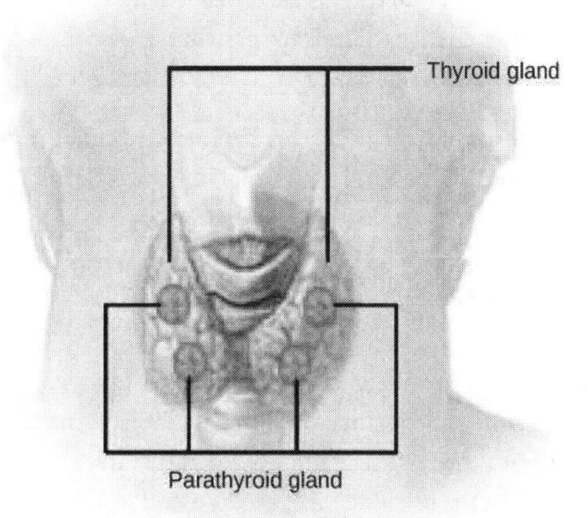

Figure 37.17 The parathyroid glands are located on the posterior of the thyroid gland. (credit: modification of work by NCI)

The parathyroid glands produce parathyroid hormone (PTH). PTH increases blood calcium concentrations when calcium ion levels fall below normal. PTH (1) enhances reabsorption of Ca^{2+} by the kidneys, (2) stimulates osteoclast activity and inhibits osteoblast activity, and (3) it stimulates synthesis and secretion of calcitriol by the kidneys, which enhances Ca^{2+} absorption by the digestive system. PTH is produced by chief cells of the parathyroid. PTH and calcitonin work in opposition to one another to maintain homeostatic Ca^{2+} levels in body fluids. Another type of cells, oxyphil cells, exist in the parathyroid but their function is not known. These hormones encourage bone growth, muscle mass, and blood cell formation in children and women.

Adrenal Glands

The **adrenal glands** are associated with the kidneys; one gland is located on top of each kidney as illustrated in Figure 37.18. The adrenal glands consist of an outer adrenal cortex and an inner adrenal medulla. These regions secrete different hormones.

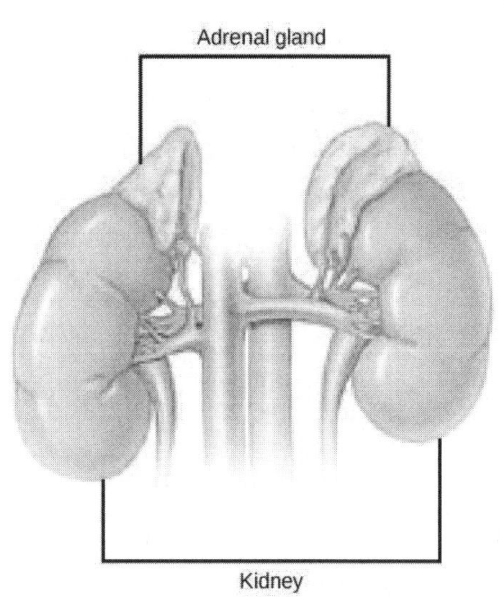

Figure 37.18 The location of the adrenal glands on top of the kidneys is shown. (credit: modification of work by NCI)

Adrenal Cortex

The **adrenal cortex** is made up of layers of epithelial cells and associated capillary networks. These layers form three distinct regions: an outer zona glomerulosa that produces mineralocorticoids, a middle zona fasciculata that produces glucocorticoids, and an inner zona reticularis that produces androgens.

The main mineralocorticoid is aldosterone, which regulates the concentration of Na^+ ions in urine, sweat, pancreas, and saliva. Aldosterone release from the adrenal cortex is stimulated by a decrease in blood concentrations of sodium ions, blood volume, or blood pressure, or by an increase in blood potassium levels.

The three main glucocorticoids are cortisol, corticosterone, and cortisone. The glucocorticoids stimulate the synthesis of glucose and gluconeogenesis (converting a non-carbohydrate to glucose) by liver cells and they promote the release of fatty acids from adipose tissue. These hormones increase blood glucose levels to maintain levels within a normal range between meals. These hormones are secreted in response to ACTH and levels are regulated by negative feedback.

Androgens are sex hormones that promote masculinity. They are produced in small amounts by the adrenal cortex in both males and females. They do not affect sexual characteristics and may supplement sex hormones released from the gonads.

Adrenal Medulla

The **adrenal medulla** contains large, irregularly shaped cells that are closely associated with blood vessels. These cells are innervated by preganglionic autonomic nerve fibers from the central nervous system.

The adrenal medulla contains two types of secretory cells: one that produces epinephrine (adrenaline) and another that produces norepinephrine (noradrenaline). Epinephrine is the primary adrenal medulla hormone accounting for 75 to 80 percent of its secretions. Epinephrine and norepinephrine increase heart rate, breathing rate, cardiac muscle contractions, blood pressure, and blood glucose levels. They also accelerate the breakdown of glucose in skeletal muscles and stored fats in adipose tissue.

The release of epinephrine and norepinephrine is stimulated by neural impulses from the sympathetic nervous system. Secretion of these hormones is stimulated by acetylcholine release from preganglionic sympathetic fibers innervating the adrenal medulla. These neural impulses originate from the hypothalamus in response to stress to prepare the body for the fight-or-flight response.

Pancreas

The **pancreas**, illustrated in Figure 37.19, is an elongated organ that is located between the stomach and the proximal portion of the small intestine. It contains both exocrine cells that excrete digestive enzymes and endocrine cells that release hormones. It is sometimes referred to as a heterocrine gland because it has both endocrine and exocrine functions.

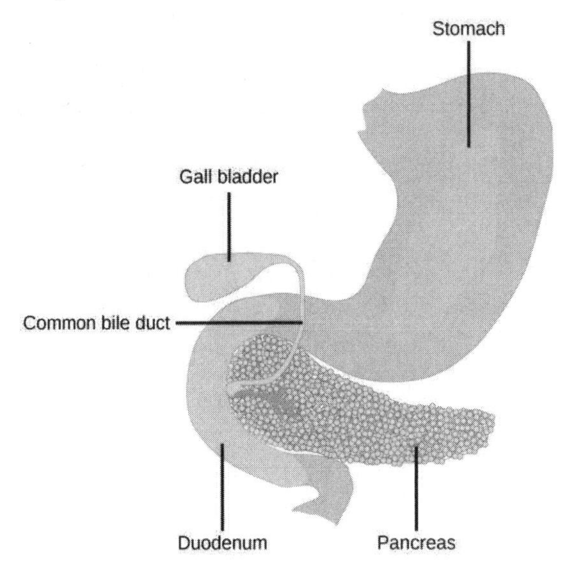

Figure 37.19 The pancreas is found underneath the stomach and points toward the spleen. (credit: modification of work by NCI)

The endocrine cells of the pancreas form clusters called pancreatic islets or the **islets of Langerhans**, as visible in the micrograph shown in Figure 37.20. The pancreatic islets contain two primary cell types: **alpha cells**, which produce the hormone glucagon, and **beta cells**, which produce the hormone insulin. These hormones regulate blood glucose levels. As blood glucose levels decline, alpha cells release glucagon to raise the blood glucose levels by increasing rates of glycogen breakdown and glucose release by the liver. When blood glucose levels rise, such as after a meal, beta cells release insulin to lower blood glucose levels by increasing the rate of glucose uptake in most body cells, and by increasing glycogen synthesis in skeletal muscles and the liver. Together, glucagon and insulin regulate blood glucose levels.

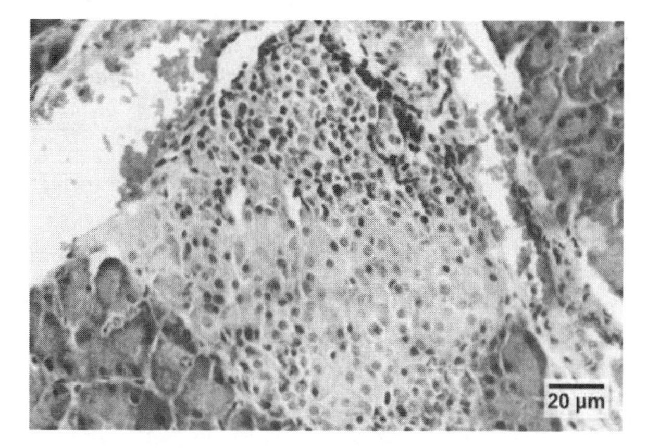

Figure 37.20 The islets of Langerhans are clusters of endocrine cells found in the pancreas; they stain lighter than surrounding cells. (credit: modification of work by Muhammad T. Tabiin, Christopher P. White, Grant Morahan, and Bernard E. Tuch; scale-bar data from Matt Russell)

Pineal Gland

The pineal gland produces melatonin. The rate of melatonin production is affected by the photoperiod. Collaterals from the visual pathways innervate the pineal gland. During the day photoperiod, little melatonin is produced; however, melatonin production increases during the dark photoperiod (night). In some mammals, melatonin has an inhibitory affect on reproductive functions by decreasing production and maturation of sperm, oocytes, and reproductive organs. Melatonin is an effective antioxidant, protecting the CNS from free radicals such as nitric oxide and hydrogen peroxide. Lastly, melatonin is involved in biological rhythms, particularly circadian rhythms such as the sleep-wake cycle and eating habits.

Gonads

The gonads—the male testes and female ovaries—produce steroid hormones. The testes produce androgens, testosterone being the most prominent, which allow for the development of secondary sex characteristics and the production of sperm cells. The ovaries produce estradiol and progesterone, which cause secondary sex characteristics and prepare the body for childbirth.

Endocrine Glands and their Associated Hormones

Endocrine Gland	Associated Hormones	Effect
Hypothalamus	releasing and inhibiting hormones	regulate hormone release from pituitary gland; produce oxytocin; produce uterine contractions and milk secretion in females
	antidiuretic hormone (ADH)	water reabsorption from kidneys; vasoconstriction to increase blood pressure
Pituitary (Anterior)	growth hormone (GH)	promotes growth of body tissues, protein synthesis; metabolic functions
	prolactin (PRL)	promotes milk production
	thyroid stimulating hormone (TSH)	stimulates thyroid hormone release
	adrenocorticotropic hormone (ACTH)	stimulates hormone release by adrenal cortex, glucocorticoids

Table 37.1

Endocrine Gland	Associated Hormones	Effect
	follicle-stimulating hormone (FSH)	stimulates gamete production (both ova and sperm); secretion of estradiol
	luteinizing hormone (LH)	stimulates androgen production by gonads; ovulation, secretion of progesterone
	melanocyte-stimulating hormone (MSH)	stimulates melanocytes of the skin increasing melanin pigment production.
Pituitary (Posterior)	antidiuretic hormone (ADH)	stimulates water reabsorption by kidneys
	oxytocin	stimulates uterine contractions during childbirth; milk ejection; stimulates ductus deferens and prostate gland contraction during emission
Thyroid	thyroxine, triiodothyronine	stimulate and maintain metabolism; growth and development
	calcitonin	reduces blood Ca^{2+} levels
Parathyroid	parathyroid hormone (PTH)	increases blood Ca^{2+} levels
Adrenal (Cortex)	aldosterone	increases blood Na^{+} levels; increases K^{+} secretion
	cortisol, corticosterone, cortisone	increase blood glucose levels; anti-inflammatory effects
Adrenal (Medulla)	epinephrine, norepinephrine	stimulate fight-or-flight response; increase blood gluclose levels; increase metabolic activities
Pancreas	insulin	reduces blood glucose levels
	glucagon	increases blood glucose levels
Pineal gland	melatonin	regulates some biological rhythms and protects CNS from free radicals
Testes	androgens	regulate, promote, increase or maintain sperm production; male secondary sexual characteristics
Ovaries	estrogen	promotes uterine lining growth; female secondary sexual characteristics
	progestins	promote and maintain uterine lining growth

Table 37.1

Organs with Secondary Endocrine Functions

There are several organs whose primary functions are non-endocrine but that also possess endocrine functions. These include the heart, kidneys, intestines, thymus, gonads, and adipose tissue.

The heart possesses endocrine cells in the walls of the atria that are specialized cardiac muscle cells. These cells release the hormone **atrial natriuretic peptide (ANP)** in response to increased blood volume. High blood volume causes the cells to be stretched, resulting in hormone release. ANP acts on the kidneys to reduce the reabsorption of Na^+, causing Na^+ and water to be excreted in the urine. ANP also reduces the amounts of renin released by the kidneys and aldosterone released by the adrenal cortex, further preventing the retention of water. In this way, ANP causes a reduction in blood volume and blood pressure, and reduces the concentration of Na^+ in the blood.

The gastrointestinal tract produces several hormones that aid in digestion. The endocrine cells are located in the mucosa of the GI tract throughout the stomach and small intestine. Some of the hormones produced include gastrin, secretin, and cholecystokinin, which are secreted in the presence of food, and some of which act on other organs such as the pancreas, gallbladder, and liver. They trigger the release of gastric juices, which help to break down and digest food in the GI tract.

While the adrenal glands associated with the kidneys are major **endocrine glands**, the kidneys themselves also possess endocrine function. Renin is released in response to decreased blood volume or pressure and is part of the renin-angiotensin-aldosterone system that leads to the release of aldosterone. Aldosterone then causes the retention of Na^+ and water, raising blood volume. The kidneys also release calcitriol, which aids in the absorption of Ca^{2+} and phosphate ions. **Erythropoietin (EPO)** is a protein hormone that triggers the formation of red blood cells in the bone marrow. EPO is released in response to low oxygen levels. Because red blood cells are oxygen carriers, increased production results in greater oxygen delivery throughout the body. EPO has been used by athletes to improve performance, as greater oxygen delivery to muscle cells allows for greater endurance. Because red blood cells increase the viscosity of blood, artificially high levels of EPO can cause severe health risks.

The **thymus** is found behind the sternum; it is most prominent in infants, becoming smaller in size through adulthood. The thymus produces hormones referred to as thymosins, which contribute to the development of the immune response.

Adipose tissue is a connective tissue found throughout the body. It produces the hormone **leptin** in response to food intake. Leptin increases the activity of anorexigenic neurons and decreases that of orexigenic neurons, producing a feeling of satiety after eating, thus affecting appetite and reducing the urge for further eating. Leptin is also associated with reproduction. It must be present for GnRH and gonadotropin synthesis to occur. Extremely thin females may enter puberty late; however, if adipose levels increase, more leptin will be produced, improving fertility.

KEY TERMS

acromegaly condition caused by overproduction of GH in adults

Addison's disease disorder caused by the hyposecretion of corticosteroids

adenylate cyclase an enzyme that catalyzes the conversion of ATP to cyclic AMP

adrenal cortex outer portion of adrenal glands that produces corticosteroids

adrenal gland endocrine glands associated with the kidneys

adrenal medulla inner portion of adrenal glands that produces epinephrine and norepinephrine

adrenocorticotropic hormone (ACTH) hormone released by the anterior pituitary, which stimulates the adrenal cortex to release corticosteroids during the long-term stress response

aldosterone steroid hormone produced by the adrenal cortex that stimulates the reabsorption of Na^+ from extracellular fluids and secretion of K^+

alpha cell endocrine cell of the pancreatic islets that produces the hormone glucagon

amino acid-derived hormone hormone derived from amino acids

androgen male sex hormone such as testosterone

anterior pituitary portion of the pituitary gland that produces six hormones; also called adenohypophysis

antidiuretic hormone (ADH) hormone produced by the hypothalamus and released by the posterior pituitary that increases water reabsorption by the kidneys

atrial natriuretic peptide (ANP) hormone produced by the heart to reduce blood volume, pressure, and Na^+ concentration

beta cell endocrine cell of the pancreatic islets that produces the hormone insulin

calcitonin hormone produced by the parafollicular cells of the thyroid gland that functions to lower blood Ca^{2+} levels and promote bone growth

colloid fluid inside the thyroid gland that contains the glycoprotein thyroglobulin

corticosteroid hormone released by the adrenal cortex in response to long-term stress

cortisol glucocorticoid produced in response to stress

Cushing's disease disorder caused by the hypersecretion of glucocorticoids

diabetes insipidus disorder caused by underproduction of ADH

diabetes mellitus disorder caused by low levels of insulin activity

diabetogenic effect effect of GH that causes blood glucose levels to rise similar to diabetes mellitus

down-regulation a decrease in the number of hormone receptors in response to increased hormone levels

endocrine gland gland that secretes hormones into the surrounding interstitial fluid, which then diffuse into blood and are carried to various organs and tissues within the body

epinephrine hormone released by the adrenal medulla in response to a short term stress

erythropoietin (EPO) hormone produced by the kidneys to stimulate red blood cell production in the bone marrow

estrogens a group of steroid hormones, including estradiol and several others, that are produced by the ovaries and elicit secondary sex characteristics in females as well as control the maturation of the ova

first messenger the hormone that binds to a plasma membrane hormone receptor to trigger a signal transduction pathway

follicle-stimulating hormone (FSH) hormone produced by the anterior pituitary that stimulates gamete production

G-protein a membrane protein activated by the hormone first messenger to activate formation of cyclic AMP

gigantism condition caused by overproduction of GH in children

glucagon hormone produced by the alpha cells of the pancreas in response to low blood sugar; functions to raise blood sugar levels

glucocorticoid corticosteroid that affects glucose metabolism

gluconeogenesis synthesis of glucose from amino acids

glucose-sparing effect effect of GH that causes tissues to use fatty acids instead of glucose as an energy source

glycogenolysis breakdown of glycogen into glucose

goiter enlargement of the thyroid gland caused by insufficient dietary iodine levels

gonadotropin hormone that regulates the gonads, including FSH and LH

growth hormone (GH) hormone produced by the anterior pituitary that promotes protein synthesis and body growth

growth hormone-inhibiting hormone (GHIH) hormone produced by the hypothalamus that inhibits growth hormone production, also called somatostatin

growth hormone-releasing hormone (GHRH) hormone released by the hypothalamus that triggers the release of GH

hormonal stimuli release of a hormone in response to another hormone

hormone receptor the cellular protein that binds to a hormone

humoral stimuli control of hormone release in response to changes in extracellular fluids such as blood or the ion concentration in the blood

hyperglycemia high blood sugar level

hyperthyroidism overactivity of the thyroid gland

hypoglycemia low blood sugar level

hypophyseal portal system system of blood vessels that carries hormones from the hypothalamus to the anterior pituitary

hypothyroidism underactivity of the thyroid gland

insulin hormone produced by the beta cells of the pancreas in response to high blood glucose levels; functions to lower blood glucose levels

insulin-like growth factor (IGF) growth-promoting protein produced by the liver

intracellular hormone receptor a hormone receptor in the cytoplasm or nucleus of a cell

islets of Langerhans (pancreatic islets) endocrine cells of the pancreas

isthmus tissue mass that connects the two lobes of the thyroid gland

leptin hormone produced by adipose tissue that promotes feelings of satiety and reduces hunger

lipid-derived hormone hormone derived mostly from cholesterol

mineralocorticoid corticosteroid that affects ion and water balance

neural stimuli stimulation of endocrine glands by the nervous system

norepinephrine hormone released by the adrenal medulla in response to a short-term stress hormone production by the gonads

osmoreceptor receptor in the hypothalamus that monitors the concentration of electrolytes in the blood

oxytocin hormone released by the posterior pituitary to stimulate uterine contractions during childbirth and milk let-down in the mammary glands

pancreas organ located between the stomach and the small intestine that contains exocrine and endocrine cells

parafollicular cell thyroid cell that produces the hormone calcitonin

parathyroid gland gland located on the surface of the thyroid that produces parathyroid hormone

parathyroid hormone (PTH) hormone produced by the parathyroid glands in response to low blood Ca^{2+} levels; functions to raise blood Ca^{2+} levels

peptide hormone hormone composed of a polypeptide chain

phosphodiesterase (PDE) enzyme that deactivates cAMP, stopping hormone activity

pituitary dwarfism condition caused by underproduction of GH in children

pituitary gland endocrine gland located at the base of the brain composed of an anterior and posterior region; also called hypophysis

pituitary stalk (also, infundibulum) stalk that connects the pituitary gland to the hypothalamus

plasma membrane hormone receptor a hormone receptor on the surface of the plasma membrane of a cell

posterior pituitary extension of the brain that releases hormones produced by the hypothalamus; along with the infundibulum, it is also referred to as the neurohypophysis

prolactin (PRL) hormone produced by the anterior pituitary that stimulates milk production

prolactin-inhibiting hormone hormone produced by the hypothalamus that inhibits the release of prolactin

prolactin-releasing hormone hormone produced by the hypothalamus that stimulates the release of prolactin

renin enzyme produced by the juxtaglomerular apparatus of the kidneys that reacts with angiotensinogen to cause the release of aldosterone

thymus gland located behind the sternum that produces thymosin hormones that contribute to the development of the immune system

thyroglobulin glycoprotein found in the thyroid that is converted into thyroid hormone

thyroid gland endocrine gland located in the neck that produces thyroid hormones thyroxine and triiodothyronine

thyroid-stimulating hormone (TSH) hormone produced by the anterior pituitary that controls the release of T_3 and T_4 from the thyroid gland

thyroxine (tetraiodothyronine, T_4) thyroid hormone containing 4 iodines that controls the basal metabolic rate

triiodothyronine (T_3) thyroid hormone containing 3 iodines that controls the basal metabolic rate

up-regulation an increase in the number of hormone receptors in response to increased hormone levels

CHAPTER SUMMARY

37.1 Types of Hormones

There are three basic types of hormones: lipid-derived, amino acid-derived, and peptide. Lipid-derived hormones are structurally similar to cholesterol and include steroid hormones such as estradiol and testosterone. Amino acid-derived hormones are relatively small molecules and include the adrenal hormones epinephrine and norepinephrine. Peptide hormones are polypeptide chains or proteins and include the pituitary hormones, antidiuretic hormone (vasopressin), and oxytocin.

37.2 How Hormones Work

Hormones cause cellular changes by binding to receptors on target cells. The number of receptors on a target cell can increase or decrease in response to hormone activity. Hormones can affect cells directly through intracellular hormone receptors or indirectly through plasma membrane hormone receptors.

Lipid-derived (soluble) hormones can enter the cell by diffusing across the plasma membrane and binding to DNA to regulate gene transcription and to change the cell's activities by inducing production of proteins that affect, in general, the long-term structure and function of the cell. Lipid insoluble hormones bind to receptors on the plasma membrane surface and trigger a signaling pathway to change the cell's activities by inducing production of various cell products that affect the cell in the short-term. The hormone is called a first messenger and the cellular component is called a second messenger. G-proteins activate the second messenger (cyclic AMP), triggering the cellular response. Response to hormone binding is amplified as the signaling pathway progresses. Cellular responses to hormones include the production of proteins and enzymes and altered membrane permeability.

37.3 Regulation of Body Processes

Water levels in the body are controlled by antidiuretic hormone (ADH), which is produced in the hypothalamus and triggers the reabsorption of water by the kidneys. Underproduction of ADH can cause diabetes insipidus. Aldosterone, a hormone produced by the adrenal cortex of the kidneys, enhances Na^+ reabsorption from the extracellular fluids and subsequent water reabsorption by diffusion. The renin-angiotensin-aldosterone system is one way that aldosterone release is controlled.

The reproductive system is controlled by the gonadotropins follicle-stimulating hormone (FSH) and luteinizing hormone (LH), which are produced by the pituitary gland. Gonadotropin release is controlled by the hypothalamic hormone gonadotropin-releasing hormone (GnRH). FSH stimulates the maturation of sperm cells in males and is inhibited by the hormone inhibin, while LH stimulates the production of the androgen testosterone. FSH stimulates egg maturation in females, while LH stimulates the production of estrogens and progesterone. Estrogens are a group of steroid hormones produced by the ovaries that trigger the development of secondary sex characteristics in females as well as control the maturation of the ova. In females, the pituitary also produces prolactin, which stimulates milk production after childbirth, and oxytocin, which stimulates uterine contraction during childbirth and milk let-down during suckling.

Insulin is produced by the pancreas in response to rising blood glucose levels and allows cells to utilize blood glucose and store excess glucose for later use. Diabetes mellitus is caused by reduced insulin activity and causes high blood glucose levels, or hyperglycemia. Glucagon is released by the pancreas in response to low blood glucose levels and stimulates the breakdown of glycogen into glucose, which can be used by the body. The body's basal metabolic rate is controlled by the thyroid hormones thyroxine (T_4) and triiodothyronine (T_3). The anterior pituitary produces thyroid stimulating hormone (TSH), which controls the release of T_3 and T_4 from the thyroid gland. Iodine is necessary in the production of thyroid hormone, and the lack of iodine can lead to a condition called goiter.

Parathyroid hormone (PTH) is produced by the parathyroid glands in response to low blood Ca^{2+} levels. The parafollicular cells of the thyroid produce calcitonin, which reduces blood Ca^{2+} levels. Growth hormone (GH) is produced by the anterior pituitary and controls the growth rate of muscle and bone. GH action is indirectly mediated by insulin-like growth factors (IGFs). Short-term stress causes the hypothalamus to trigger the adrenal medulla to release epinephrine and norepinephrine, which trigger the fight or flight response. Long-term stress causes the hypothalamus to trigger the anterior pituitary to release adrenocorticotropic hormone (ACTH), which causes the release of corticosteroids, glucocorticoids, and mineralocorticoids, from the adrenal cortex.

37.4 Regulation of Hormone Production

Hormone levels are primarily controlled through negative feedback, in which rising levels of a hormone inhibit its further release. The three mechanisms of hormonal release are humoral stimuli, hormonal stimuli, and neural stimuli. Humoral stimuli refers to the control of hormonal release in response to changes in extracellular fluid levels or ion levels. Hormonal stimuli refers to the release of hormones in response to hormones released by other endocrine glands. Neural stimuli refers to the release of hormones in response to neural stimulation.

37.5 Endocrine Glands

The pituitary gland is located at the base of the brain and is attached to the hypothalamus by the infundibulum. The anterior pituitary receives products from the hypothalamus by the hypophyseal portal system and produces six hormones. The posterior pituitary is an extension of the brain and releases hormones (antidiuretic hormone and oxytocin) produced by the hypothalamus.

The thyroid gland is located in the neck and is composed of two lobes connected by the isthmus. The thyroid is made up of follicle cells that produce the hormones thyroxine and triiodothyronine. Parafollicular cells of the thyroid produce calcitonin. The parathyroid glands lie on the posterior surface of the thyroid gland and produce parathyroid hormone.

The adrenal glands are located on top of the kidneys and consist of the renal cortex and renal medulla. The adrenal cortex is the outer part of the adrenal gland and produces the corticosteroids, glucocorticoids, and mineralocorticoids. The adrenal medulla is the inner part of the adrenal gland and produces the catecholamines epinephrine and norepinephrine.

The pancreas lies in the abdomen between the stomach and the small intestine. Clusters of endocrine cells in the pancreas form the islets of Langerhans, which are composed of alpha cells that release glucagon and beta cells that release insulin.

Some organs possess endocrine activity as a secondary function but have another primary function. The heart produces the hormone atrial natriuretic peptide, which functions to reduce blood volume, pressure, and Na^+ concentration. The gastrointestinal tract produces various hormones that aid in digestion. The kidneys produce renin, calcitriol, and erythropoietin. Adipose tissue produces leptin, which promotes satiety signals in the brain.

VISUAL CONNECTION QUESTIONS

1. Figure 37.5 Heat shock proteins (HSP) are so named because they help refold misfolded proteins. In response to increased temperature (a "heat shock"), heat shock proteins are activated by release from the NR/HSP complex. At the same time, transcription of HSP genes is activated. Why do you think the cell responds to a heat shock by increasing the activity of proteins that help refold misfolded proteins?

2. Figure 37.11 Pancreatic tumors may cause excess secretion of glucagon. Type I diabetes results from the failure of the pancreas to produce insulin. Which of the following statement about these two conditions is true?
   a. A pancreatic tumor and type I diabetes will have the opposite effects on blood sugar levels.
   b. A pancreatic tumor and type I diabetes will both cause hyperglycemia.
   c. A pancreatic tumor and type I diabetes will both cause hypoglycemia.
   d. Both pancreatic tumors and type I diabetes result in the inability of cells to take up glucose.

3. Figure 37.14 Hyperthyroidism is a condition in which the thyroid gland is overactive. Hypothyroidism is a condition in which the thyroid gland is underactive. Which of the conditions are the following two patients most likely to have?
   Patient A has symptoms including weight gain, cold sensitivity, low heart rate, and fatigue.
   Patient B has symptoms including weight loss, profuse sweating, increased heart rate, and difficulty sleeping.

REVIEW QUESTIONS

4. A newly discovered hormone contains four amino acids linked together. Under which chemical class would this hormone be classified?
   a. lipid-derived hormone
   b. amino acid-derived hormone
   c. peptide hormone
   d. glycoprotein

5. Which class of hormones can diffuse through plasma membranes?
   a. lipid-derived hormones
   b. amino acid-derived hormones
   c. peptide hormones
   d. glycoprotein hormones

6. Why are steroids able to diffuse across the plasma membrane?
   a. Their transport protein moves them through the membrane.
   b. They are amphipathic, allowing them to interact with the entire phospholipid.
   c. Cells express channels that let hormones flow down their concentration gradient into the cells.
   d. They are non-polar molecules.

7. A new antagonist molecule has been discovered that binds to and blocks plasma membrane receptors. What effect will this antagonist have on testosterone, a steroid hormone?
   a. It will block testosterone from binding to its receptor.
   b. It will block testosterone from activating cAMP signaling.
   c. It will increase testosterone-mediated signaling.
   d. It will not affect testosterone-mediated signaling.

8. What effect will a cAMP inhibitor have on a peptide hormone-mediated signaling pathway?
   a. It will prevent the hormone from binding its receptor.
   b. It will prevent activation of a G-protein.
   c. It will prevent activation of adenylate cyclase.
   d. It will prevent activation of protein kinases.

9. When insulin binds to its receptor, the complex is endocytosed into the cell. This is an example of_____ in response to hormone signaling.
   a. cAMP activation
   b. generating an intracellular receptor
   c. activation of a hormone response element
   d. receptor down-regulation

10. Drinking alcoholic beverages causes an increase in urine output. This most likely occurs because alcohol:
    a. inhibits ADH release.
    b. stimulates ADH release.
    c. inhibits TSH release.
    d. stimulates TSH release.

11. FSH and LH release from the anterior pituitary is stimulated by _______.
    a. TSH
    b. GnRH
    c. T_3
    d. PTH

12. What hormone is produced by beta cells of the pancreas?
    a. T_3
    b. glucagon
    c. insulin
    d. T_4

13. When blood calcium levels are low, PTH stimulates:
    a. excretion of calcium from the kidneys.
    b. excretion of calcium from the intestines.
    c. osteoblasts.
    d. osteoclasts.

14. How would mutations that completely ablate the function of the androgen receptor impact the phenotypic development of humans with XY chromosomes?
    a. Patients would appear phenotypically female.
    b. Patients would appear phenotypically male with underdeveloped secondary sex characteristics.
    c. Patients would appear phenotypically male, but cannot produce sperm.
    d. Patients would express both male and female secondary sex characteristics.

15. A rise in blood glucose levels triggers release of insulin from the pancreas. This mechanism of hormone production is stimulated by:
    a. humoral stimuli
    b. hormonal stimuli
    c. neural stimuli
    d. negative stimuli

16. Which mechanism of hormonal stimulation would be affected if signaling and hormone release from the hypothalamus was blocked?
    a. humoral and hormonal stimuli
    b. hormonal and neural stimuli
    c. neural and humoral stimuli
    d. hormonal and negative stimuli

17. A scientist hypothesizes that the pancreas's hormone production is controlled by neural stimuli. Which observation would support this hypothesis?
    a. Insulin is produced in response to sudden stress without a rise in blood glucose.
    b. Insulin is produced in response to a rise in glucagon levels.
    c. Beta cells express epinephrine receptors.
    d. Insulin is produced in response to a rise in blood glucose in the brain.

18. Which endocrine glands are associated with the kidneys?
    a. thyroid glands
    b. pituitary glands
    c. adrenal glands
    d. gonads

19. Which of the following hormones is not produced by the anterior pituitary?
    a. oxytocin
    b. growth hormone
    c. prolactin
    d. thyroid-stimulating hormone

20. Recent studies suggest that blue light exposure can impact human circadian rhythms. This suggests that blue light disrupts the function of the _____ gland(s).
    a. adrenal
    b. pituitary
    c. pineal
    d. thyroid

CRITICAL THINKING QUESTIONS

21. Although there are many different hormones in the human body, they can be divided into three classes based on their chemical structure. What are these classes and what is one factor that distinguishes them?
22. Where is insulin stored, and why would it be released?
23. Glucagon is the peptide hormone that signals for the body to release glucose into the bloodstream. How does glucagon contribute to maintaining homeostasis throughout the body? What other hormones are involved in regulating the blood glucose cycle?
24. Name two important functions of hormone receptors.
25. How can hormones mediate changes?
26. Why is cAMP-mediated signal amplification not required in steroid hormone signaling? Describe how steroid signaling is amplified instead.
27. Name and describe a function of one hormone produced by the anterior pituitary and one hormone produced by the posterior pituitary.
28. Describe one direct action of growth hormone (GH).
29. Researchers have recently demonstrated that stressed people are more susceptible to contracting the common cold than people who are not stressed. What kind of stress must the infected patients be experiencing, and why does it make them more susceptible to the virus?
30. How is hormone production and release primarily controlled?
31. Compare and contrast hormonal and humoral stimuli.
32. Oral contraceptive pills work by delivering synthetic progestins to a woman every day. Describe why this is an effective method of birth control.
33. What does aldosterone regulate, and how is it stimulated?
34. The adrenal medulla contains two types of secretory cells, what are they and what are their functions?
35. How would damage to the posterior pituitary gland affect the production and release of ADH and inhibiting hormones?

CHAPTER 38
The Musculoskeletal System

Figure 38.1 Improvements in the design of prostheses have allowed for a wider range of activities in recipients. (credit: modification of work by Stuart Grout)

INTRODUCTION The muscular and skeletal systems provide support to the body and allow for a wide range of movement. The bones of the skeletal system protect the body's internal organs and support the weight of the body. The muscles of the muscular system contract and pull on the bones, allowing for movements as diverse as standing, walking, running, and grasping items.

Injury or disease affecting the musculoskeletal system can be very debilitating. In humans, the most common musculoskeletal diseases worldwide are caused by malnutrition. Ailments that affect the joints are also widespread, such as arthritis, which can make movement difficult and—in advanced cases—completely impair mobility. In severe cases in which the joint has suffered extensive damage, joint replacement surgery may be needed.

Progress in the science of prosthesis design has resulted in the development of artificial joints, with joint replacement surgery in the hips and knees being the most common. Replacement joints for shoulders, elbows, and fingers are also available. Even with this progress, there is still room for improvement in the design of prostheses. The state-of-the-art prostheses have limited durability and therefore wear out quickly, particularly in young or active individuals. Current research is focused on the use of new materials, such as carbon fiber, that may make prostheses more durable.

Chapter Outline

38.1 Types of Skeletal Systems

By the end of this section, you will be able to do the following:

- Discuss the different types of skeletal systems
- Explain the role of the human skeletal system
- Compare and contrast different skeletal systems

A skeletal system is necessary to support the body, protect internal organs, and allow for the movement of an organism. There are three different skeleton designs that fulfill these functions: hydrostatic skeleton, exoskeleton, and endoskeleton.

Hydrostatic Skeleton

A **hydrostatic skeleton** is a skeleton formed by a fluid-filled compartment within the body, called the coelom. The organs of the coelom are supported by the aqueous fluid, which also resists external compression. This compartment is under hydrostatic pressure because of the fluid and supports the other organs of the organism. This type of skeletal system is found in soft-bodied animals such as sea anemones, earthworms, Cnidaria, and other invertebrates (Figure 38.2).

Figure 38.2 The skeleton of the red-knobbed sea star (*Protoreaster linckii*) is an example of a hydrostatic skeleton. (credit: "Amada44"/Wikimedia Commons)

Movement in a hydrostatic skeleton is provided by muscles that surround the coelom. The muscles in a hydrostatic skeleton contract to change the shape of the coelom; the pressure of the fluid in the coelom produces movement. For example, earthworms move by waves of muscular contractions of the skeletal muscle of the body wall hydrostatic skeleton, called peristalsis, which alternately shorten and lengthen the body. Lengthening the body extends the anterior end of the organism. Most organisms have a mechanism to fix themselves in the substrate. Shortening the muscles then draws the posterior portion of the body forward. Although a hydrostatic skeleton is well-suited to invertebrate organisms such as earthworms and some aquatic organisms, it is not an efficient skeleton for terrestrial animals.

Exoskeleton

An **exoskeleton** is an external skeleton that consists of a hard encasement on the surface of an organism. For example, the shells of crabs and insects are exoskeletons (Figure 38.3). This skeleton type provides defence against predators, supports the body, and allows for movement through the contraction of attached muscles. As with vertebrates, muscles must cross a joint inside the exoskeleton. Shortening of the muscle changes the relationship of the two segments of the exoskeleton. Arthropods such as crabs and lobsters have exoskeletons that consist of 30–50 percent chitin, a polysaccharide derivative of glucose that is a strong but flexible material. Chitin is secreted by the epidermal cells. The exoskeleton is further strengthened by the addition of calcium carbonate in organisms such as the lobster. Because the exoskeleton is acellular, arthropods must periodically shed their exoskeletons because the exoskeleton does not grow as the organism grows.

Figure 38.3 Muscles attached to the exoskeleton of the Halloween crab (*Gecarcinus quadratus*) allow it to move.

Endoskeleton

An **endoskeleton** is a skeleton that consists of hard, mineralized structures located within the soft tissue of organisms. An example of a primitive endoskeletal structure is the spicules of sponges. The bones of vertebrates are composed of tissues, whereas sponges have no true tissues (Figure 38.4). Endoskeletons provide support for the body, protect internal organs, and allow for movement through contraction of muscles attached to the skeleton.

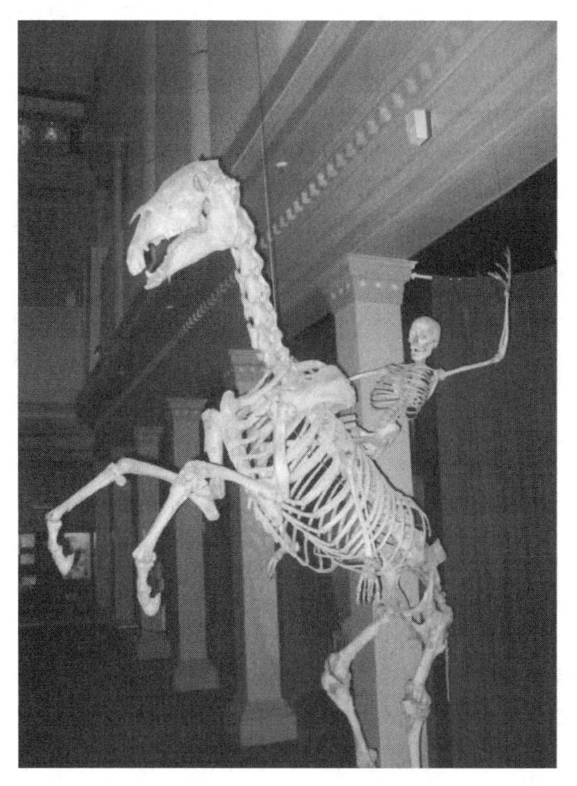

Figure 38.4 The skeletons of humans and horses are examples of endoskeletons. (credit: Ross Murphy)

The human skeleton is an endoskeleton that consists of 206 bones in the adult. It has five main functions: providing support to the body, storing minerals and lipids, producing blood cells, protecting internal organs, and allowing for movement. The skeletal system in vertebrates is divided into the axial skeleton (which consists of the skull, vertebral column, and rib cage), and the appendicular skeleton (which consists of the shoulders, limb bones, the pectoral girdle, and the pelvic girdle).

LINK TO LEARNING

Visit the interactive body (http://openstax.org/l/virt_skeleton) site to build a virtual skeleton: select "skeleton" and click through the activity to place each bone.

Human Axial Skeleton

The **axial skeleton** forms the central axis of the body and includes the bones of the skull, ossicles of the middle ear, hyoid bone of the throat, vertebral column, and the thoracic cage (ribcage) (Figure 38.5). The function of the axial skeleton is to provide support and protection for the brain, the spinal cord, and the organs in the ventral body cavity. It provides a surface for the attachment of muscles that move the head, neck, and trunk, performs respiratory movements, and stabilizes parts of the appendicular skeleton.

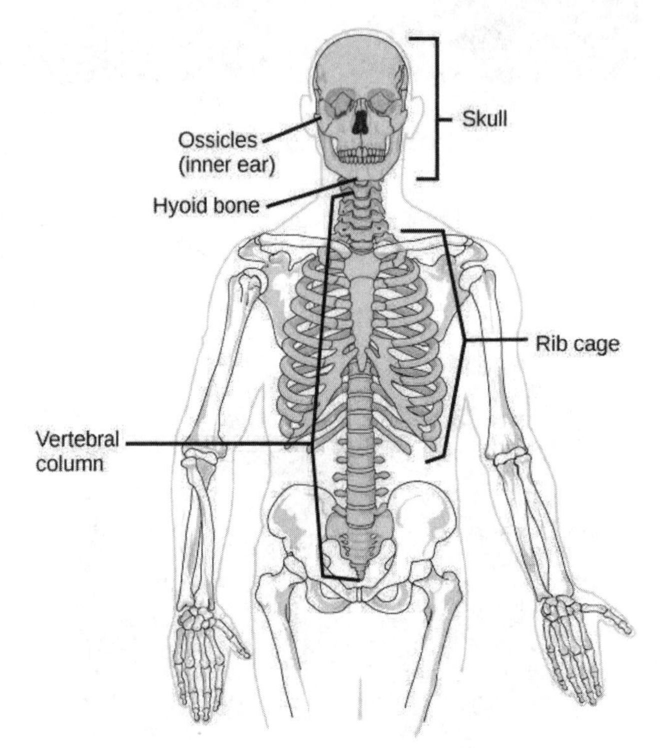

Figure 38.5 The axial skeleton consists of the bones of the skull, ossicles of the middle ear, hyoid bone, vertebral column, and rib cage. (credit: modification of work by Mariana Ruiz Villareal)

The Skull

The bones of the **skull** support the structures of the face and protect the brain. The skull consists of 22 bones, which are divided into two categories: cranial bones and facial bones. The **cranial bones** are eight bones that form the cranial cavity, which encloses the brain and serves as an attachment site for the muscles of the head and neck. The eight cranial bones are the frontal bone, two parietal bones, two temporal bones, occipital bone, sphenoid bone, and the ethmoid bone. Although the bones developed separately in the embryo and fetus, in the adult, they are tightly fused with connective tissue and adjoining bones do not move (Figure 38.6).

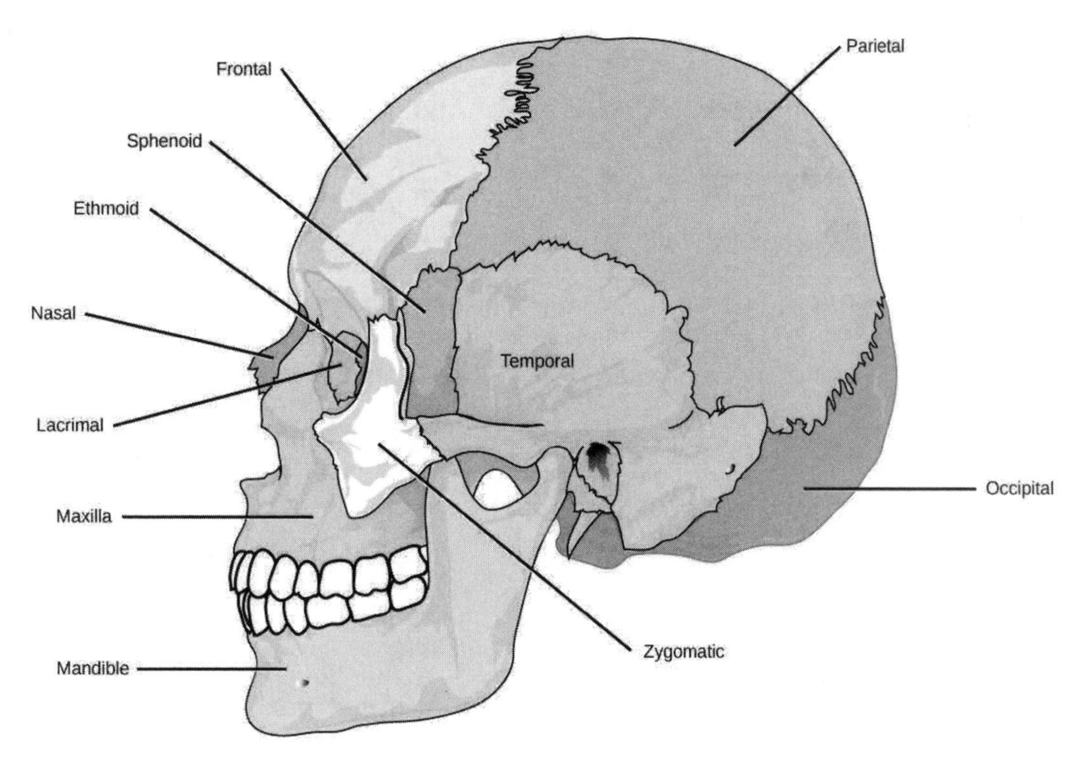

Figure 38.6 The bones of the skull support the structures of the face and protect the brain. (credit: modification of work by Mariana Ruiz Villareal)

The **auditory ossicles** of the middle ear transmit sounds from the air as vibrations to the fluid-filled cochlea. The auditory ossicles consist of three bones each: the malleus, incus, and stapes. These are the smallest bones in the body and are unique to mammals.

Fourteen **facial bones** form the face, provide cavities for the sense organs (eyes, mouth, and nose), protect the entrances to the digestive and respiratory tracts, and serve as attachment points for facial muscles. The 14 facial bones are the nasal bones, the maxillary bones, zygomatic bones, palatine, vomer, lacrimal bones, the inferior nasal conchae, and the mandible. All of these bones occur in pairs except for the mandible and the vomer (Figure 38.7).

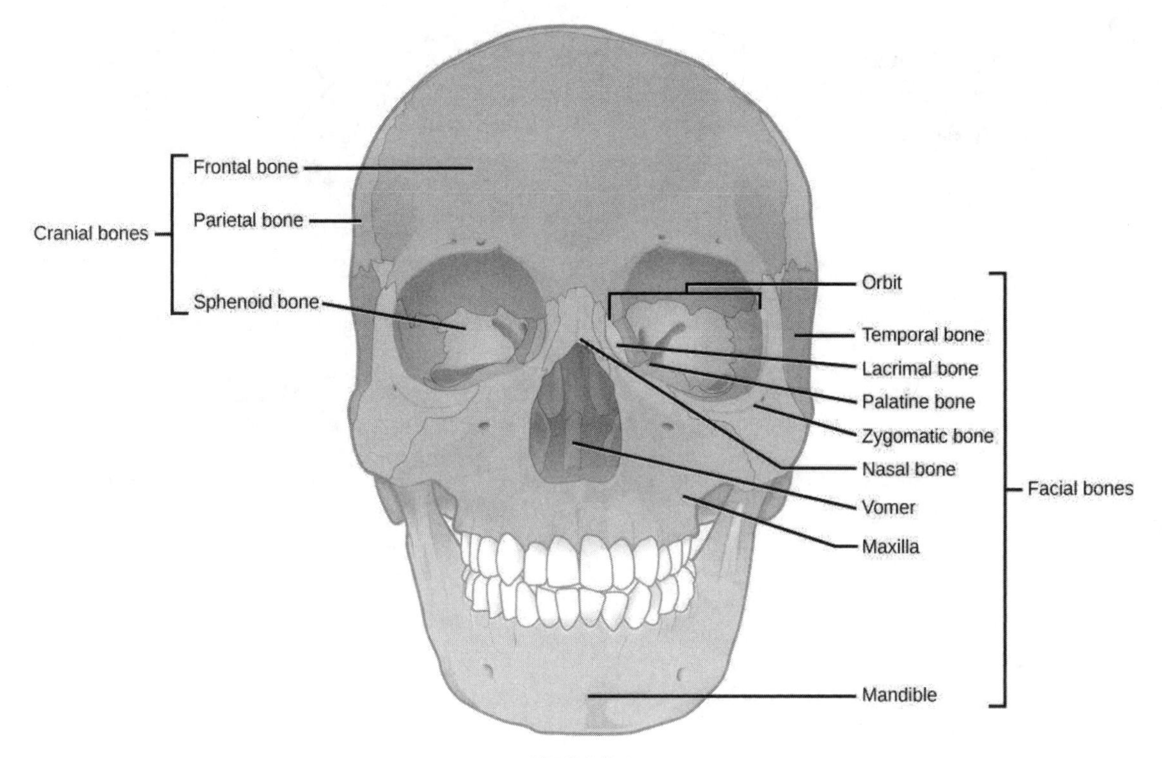

Figure 38.7 The cranial bones, including the frontal, parietal, and sphenoid bones, cover the top of the head. The facial bones of the skull form the face and provide cavities for the eyes, nose, and mouth.

Although it is not found in the skull, the hyoid bone is considered a component of the axial skeleton. The **hyoid bone** lies below the mandible in the front of the neck. It acts as a movable base for the tongue and is connected to muscles of the jaw, larynx, and tongue. The mandible articulates with the base of the skull. The mandible controls the opening to the airway and gut. In animals with teeth, the mandible brings the surfaces of the teeth in contact with the maxillary teeth.

The Vertebral Column

The **vertebral column**, or spinal column, surrounds and protects the spinal cord, supports the head, and acts as an attachment point for the ribs and muscles of the back and neck. The adult vertebral column comprises 26 bones: the 24 vertebrae, the sacrum, and the coccyx bones. In the adult, the sacrum is typically composed of five vertebrae that fuse into one. The coccyx is typically 3–4 vertebrae that fuse into one. Around the age of 70, the sacrum and the coccyx may fuse together. We begin life with approximately 33 vertebrae, but as we grow, several vertebrae fuse together. The adult vertebrae are further divided into the 7 cervical vertebrae, 12 thoracic vertebrae, and 5 lumbar vertebrae (Figure 38.8).

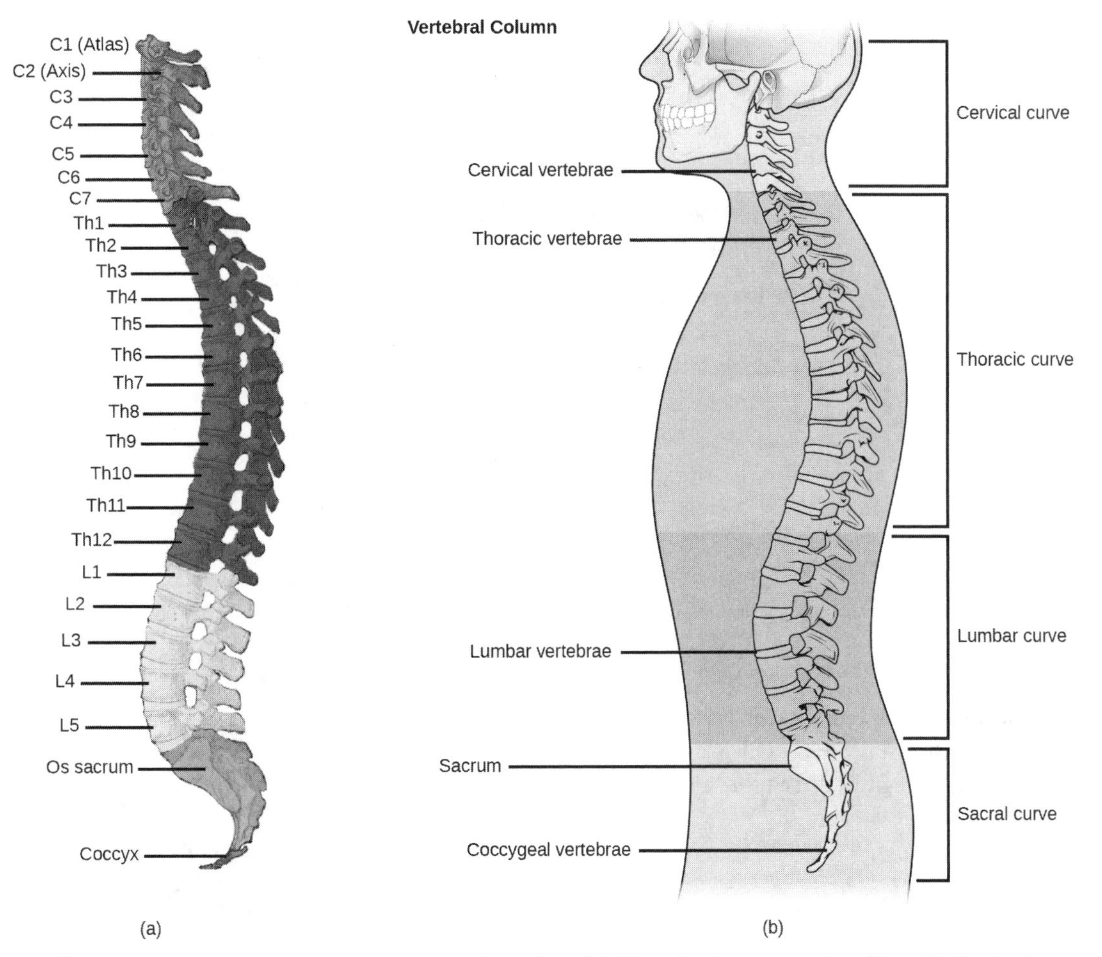

Figure 38.8 (a) The vertebral column consists of seven cervical vertebrae (C1–7) twelve thoracic vertebrae (Th1–12), five lumbar vertebrae (L1–5), the os sacrum, and the coccyx. (b) Spinal curves increase the strength and flexibility of the spine. (credit a: modification of work by Uwe Gille based on original work by Gray's Anatomy; credit b: modification of work by NCI, NIH)

Each vertebral body has a large hole in the center through which the nerves of the spinal cord pass. There is also a notch on each side through which the spinal nerves, which serve the body at that level, can exit from the spinal cord. The vertebral column is approximately 71 cm (28 inches) in adult male humans and is curved, which can be seen from a side view. The names of the spinal curves correspond to the region of the spine in which they occur. The thoracic and sacral curves are concave (curve inwards relative to the front of the body) and the cervical and lumbar curves are convex (curve outwards relative to the front of the body). The arched curvature of the vertebral column increases its strength and flexibility, allowing it to absorb shocks like a spring (Figure 38.8).

Intervertebral discs composed of fibrous cartilage lie between adjacent vertebral bodies from the second cervical vertebra to the sacrum. Each disc is part of a joint that allows for some movement of the spine and acts as a cushion to absorb shocks from movements such as walking and running. Intervertebral discs also act as ligaments to bind vertebrae together. The inner part of discs, the nucleus pulposus, hardens as people age and becomes less elastic. This loss of elasticity diminishes its ability to absorb shocks.

The Thoracic Cage

The **thoracic cage**, also known as the ribcage, is the skeleton of the chest, and consists of the ribs, sternum, thoracic vertebrae, and costal cartilages (Figure 38.9). The thoracic cage encloses and protects the organs of the thoracic cavity, including the heart and lungs. It also provides support for the shoulder girdles and upper limbs, and serves as the attachment point for the diaphragm, muscles of the back, chest, neck, and shoulders. Changes in the volume of the thorax enable breathing.

The **sternum**, or breastbone, is a long, flat bone located at the anterior of the chest. It is formed from three bones that fuse in the adult. The **ribs** are 12 pairs of long, curved bones that attach to the thoracic vertebrae and curve toward the front of the body, forming the ribcage. Costal cartilages connect the anterior ends of the ribs to the sternum, with the exception of rib pairs 11 and

12, which are free-floating ribs.

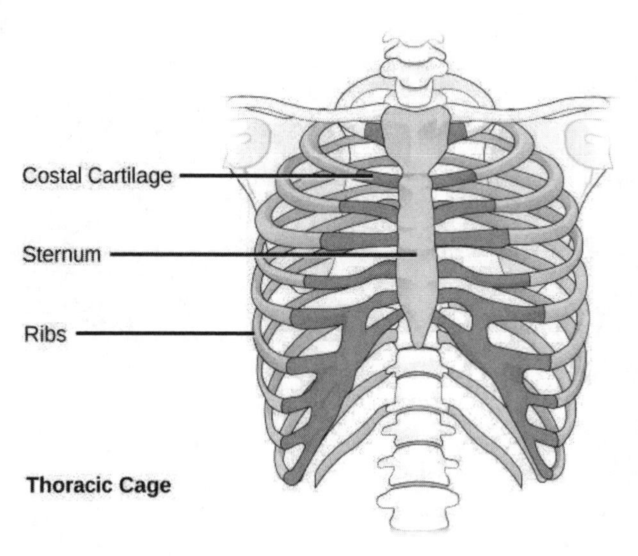

Figure 38.9 The thoracic cage, or rib cage, protects the heart and the lungs. (credit: modification of work by NCI, NIH)

Human Appendicular Skeleton

The **appendicular skeleton** is composed of the bones of the upper limbs (which function to grasp and manipulate objects) and the lower limbs (which permit locomotion). It also includes the pectoral girdle, or shoulder girdle, that attaches the upper limbs to the body, and the pelvic girdle that attaches the lower limbs to the body (Figure 38.10).

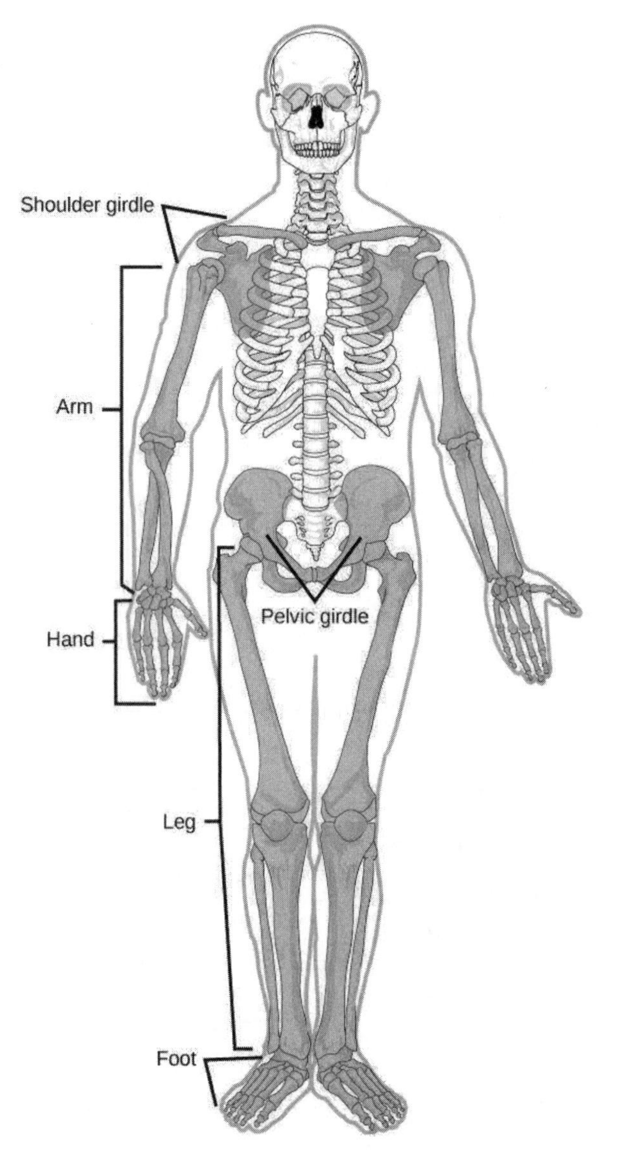

Figure 38.10 The appendicular skeleton is composed of the bones of the pectoral limbs (arm, forearm, hand), the pelvic limbs (thigh, leg, foot), the pectoral girdle, and the pelvic girdle. (credit: modification of work by Mariana Ruiz Villareal)

The Pectoral Girdle

The **pectoral girdle** bones provide the points of attachment of the upper limbs to the axial skeleton. The human pectoral girdle consists of the clavicle (or collarbone) in the anterior, and the scapula (or shoulder blades) in the posterior (Figure 38.11).

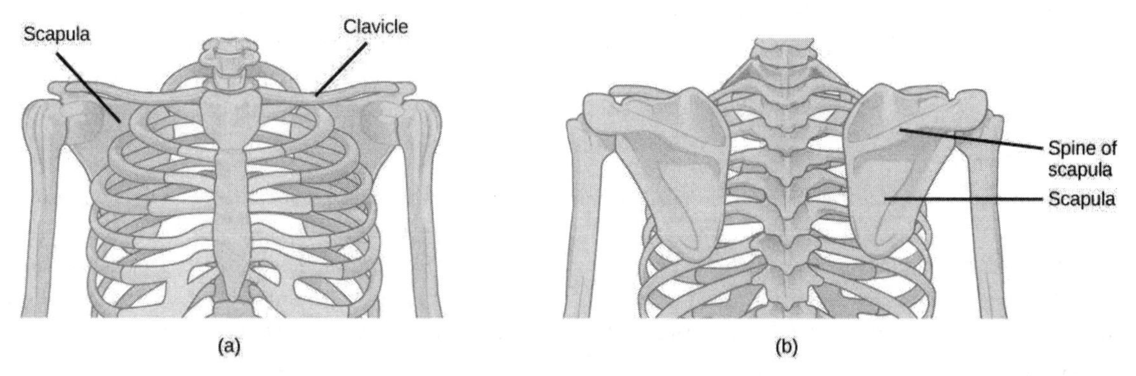

Figure 38.11 (a) The pectoral girdle in primates consists of the clavicles and scapulae. (b) The posterior view reveals the spine of the scapula to which muscle attaches.

The **clavicles** are S-shaped bones that position the arms on the body. The clavicles lie horizontally across the front of the thorax (chest) just above the first rib. These bones are fairly fragile and are susceptible to fractures. For example, a fall with the arms outstretched causes the force to be transmitted to the clavicles, which can break if the force is excessive. The clavicle articulates with the sternum and the scapula.

The **scapulae** are flat, triangular bones that are located at the back of the pectoral girdle. They support the muscles crossing the shoulder joint. A ridge, called the spine, runs across the back of the scapula and can easily be felt through the skin (Figure 38.11). The spine of the scapula is a good example of a bony protrusion that facilitates a broad area of attachment for muscles to bone.

The Upper Limb

The upper limb contains 30 bones in three regions: the arm (shoulder to elbow), the forearm (ulna and radius), and the wrist and hand (Figure 38.12).

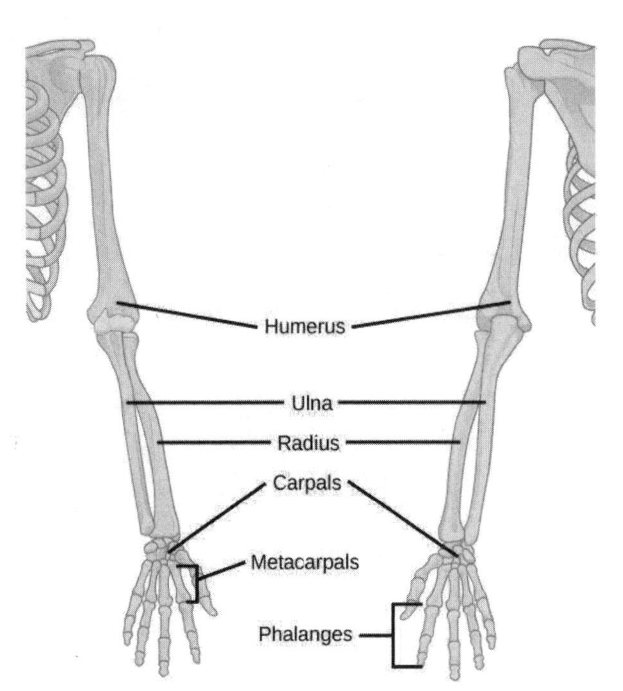

Figure 38.12 The upper limb consists of the humerus of the upper arm, the radius and ulna of the forearm, eight bones of the carpus, five bones of the metacarpus, and 14 bones of the phalanges.

An **articulation** is any place at which two bones are joined. The **humerus** is the largest and longest bone of the upper limb and the only bone of the arm. It articulates with the scapula at the shoulder and with the forearm at the elbow. The **forearm** extends from the elbow to the wrist and consists of two bones: the ulna and the radius. The **radius** is located along the lateral (thumb) side of the forearm and articulates with the humerus at the elbow. The **ulna** is located on the medial aspect (pinky-finger side) of the forearm. It is longer than the radius. The ulna articulates with the humerus at the elbow. The radius and ulna also articulate with the carpal bones and with each other, which in vertebrates enables a variable degree of rotation of the carpus with respect to the long axis of the limb. The hand includes the eight bones of the **carpus** (wrist), the five bones of the **metacarpus** (palm), and the 14 bones of the **phalanges** (digits). Each digit consists of three phalanges, except for the thumb, when present, which has only two.

The Pelvic Girdle

The **pelvic girdle** attaches to the lower limbs of the axial skeleton. Because it is responsible for bearing the weight of the body and for locomotion, the pelvic girdle is securely attached to the axial skeleton by strong ligaments. It also has deep sockets with robust ligaments to securely attach the femur to the body. The pelvic girdle is further strengthened by two large hip bones. In adults, the hip bones, or **coxal bones**, are formed by the fusion of three pairs of bones: the ilium, ischium, and pubis. The pelvis joins together in the anterior of the body at a joint called the pubic symphysis and with the bones of the sacrum at the posterior of the body.

The female pelvis is slightly different from the male pelvis. Over generations of evolution, females with a wider pubic angle and larger diameter pelvic canal reproduced more successfully. Therefore, their offspring also had pelvic anatomy that enabled successful childbirth (Figure 38.13).

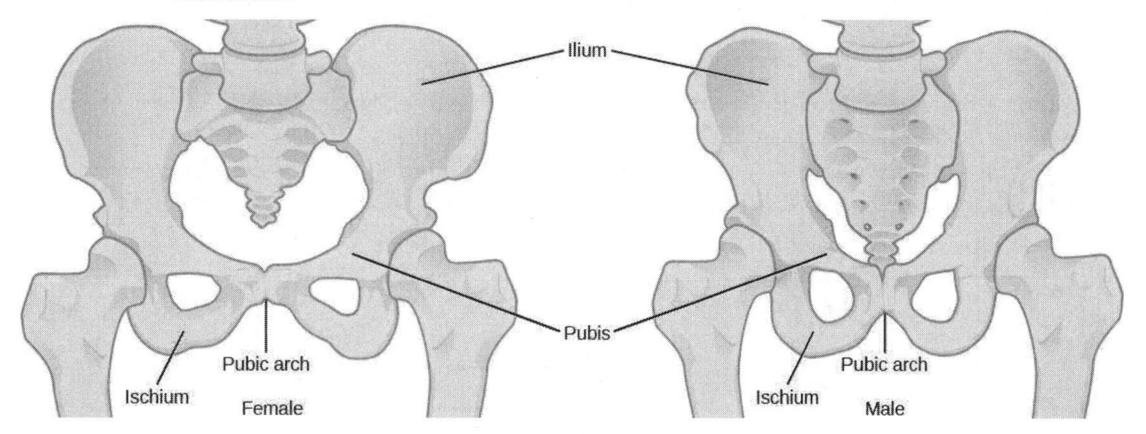

Figure 38.13 To adapt to reproductive fitness, the (a) female pelvis is lighter, wider, shallower, and has a broader angle between the pubic bones than (b) the male pelvis.

The Lower Limb

The **lower limb** consists of the thigh, the leg, and the foot. The bones of the lower limb are the femur (thigh bone), patella (kneecap), tibia and fibula (bones of the leg), tarsals (bones of the ankle), and metatarsals and phalanges (bones of the foot) (Figure 38.14). The bones of the lower limbs are thicker and stronger than the bones of the upper limbs because of the need to support the entire weight of the body and the resulting forces from locomotion. In addition to evolutionary fitness, the bones of an individual will respond to forces exerted upon them.

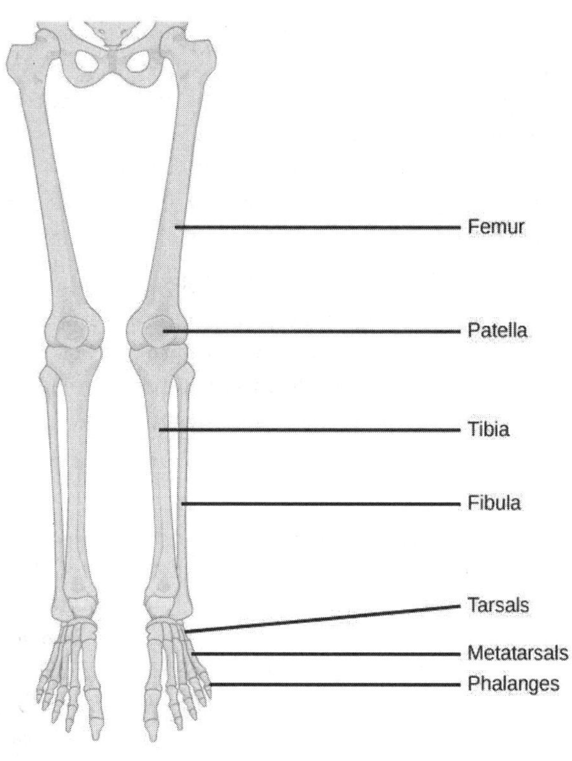

Figure 38.14 The lower limb consists of the thigh (femur), kneecap (patella), leg (tibia and fibula), ankle (tarsals), and foot (metatarsals and phalanges) bones.

The **femur**, or thighbone, is the longest, heaviest, and strongest bone in the body. The femur and pelvis form the hip joint at the proximal end. At the distal end, the femur, tibia, and patella form the knee joint. The **patella**, or kneecap, is a triangular bone that lies anterior to the knee joint. The patella is embedded in the tendon of the femoral extensors (quadriceps). It improves knee extension by reducing friction. The **tibia**, or shinbone, is a large bone of the leg that is located directly below the knee. The tibia articulates with the femur at its proximal end, with the fibula and the tarsal bones at its distal end. It is the second largest bone in the human body and is responsible for transmitting the weight of the body from the femur to the foot. The **fibula**, or calf bone, parallels and articulates with the tibia. It does not articulate with the femur and does not bear weight. The fibula acts as a site for muscle attachment and forms the lateral part of the ankle joint.

The **tarsals** are the seven bones of the ankle. The ankle transmits the weight of the body from the tibia and the fibula to the foot. The **metatarsals** are the five bones of the foot. The phalanges are the 14 bones of the toes. Each toe consists of three phalanges, except for the big toe that has only two (Figure 38.15). Variations exist in other species; for example, the horse's metacarpals and metatarsals are oriented vertically and do not make contact with the substrate.

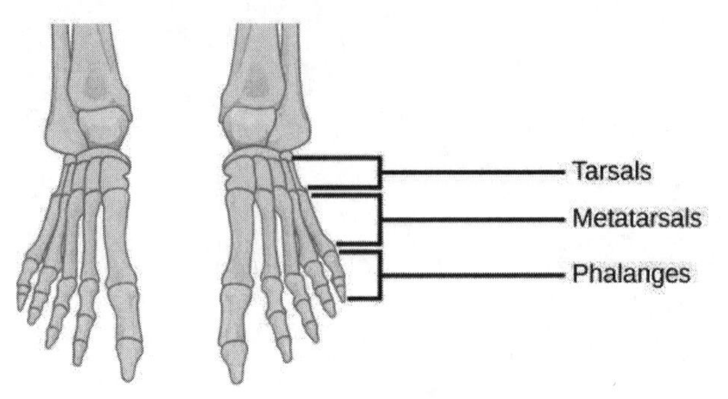

Figure 38.15 This drawing shows the bones of the human foot and ankle, including the metatarsals and the phalanges.

EVOLUTION CONNECTION

Evolution of Body Design for Locomotion on Land

The transition of vertebrates onto land required a number of changes in body design, as movement on land presents a number of challenges for animals that are adapted to movement in water. The buoyancy of water provides a certain amount of lift, and a common form of movement by fish is lateral undulations of the entire body. This back and forth movement pushes the body against the water, creating forward movement. In most fish, the muscles of paired fins attach to girdles within the body, allowing for some control of locomotion. As certain fish began moving onto land, they retained their lateral undulation form of locomotion (anguilliform). However, instead of pushing against water, their fins or flippers became points of contact with the ground, around which they rotated their bodies.

The effect of gravity and the lack of buoyancy on land meant that body weight was suspended on the limbs, leading to increased strengthening and ossification of the limbs. The effect of gravity also required changes to the axial skeleton. Lateral undulations of land animal vertebral columns cause torsional strain. A firmer, more ossified vertebral column became common in terrestrial tetrapods because it reduces strain while providing the strength needed to support the body's weight. In later tetrapods, the vertebrae began allowing for vertical motion rather than lateral flexion. Another change in the axial skeleton was the loss of a direct attachment between the pectoral girdle and the head. This reduced the jarring to the head caused by the impact of the limbs on the ground. The vertebrae of the neck also evolved to allow movement of the head independently of the body.

The appendicular skeleton of land animals is also different from aquatic animals. The shoulders attach to the pectoral girdle through muscles and connective tissue, thus reducing the jarring of the skull. Because of a lateral undulating vertebral column, in early tetrapods, the limbs were splayed out to the side and movement occurred by performing "push-ups." The vertebrae of these animals had to move side-to-side in a similar manner to fish and reptiles. This type of motion requires large muscles to move the limbs toward the midline; it was almost like walking while doing push-ups, and it is not an efficient use of energy. Later tetrapods have their limbs placed under their bodies, so that each stride requires less force to move forward. This resulted in decreased adductor muscle size and an increased range of motion of the scapulae. This also restricts movement primarily to one plane, creating forward motion rather than moving the limbs upward as well as forward. The femur and humerus were also rotated, so that the ends of the limbs and digits were pointed forward, in the direction of motion, rather than out to the side. By placement underneath the body, limbs can swing forward like a pendulum to produce a stride that is more efficient for moving over land.

38.2 Bone

By the end of this section, you will be able to do the following:

- Classify the different types of bones in the skeleton
- Explain the role of the different cell types in bone
- Explain how bone forms during development

Bone, or **osseous tissue**, is a connective tissue that constitutes the endoskeleton. It contains specialized cells and a matrix of mineral salts and collagen fibers.

The mineral salts primarily include hydroxyapatite, a mineral formed from calcium phosphate. **Calcification** is the process of deposition of mineral salts on the collagen fiber matrix that crystallizes and hardens the tissue. The process of calcification only occurs in the presence of collagen fibers.

The bones of the human skeleton are classified by their shape: long bones, short bones, flat bones, sutural bones, sesamoid bones, and irregular bones (Figure 38.16).

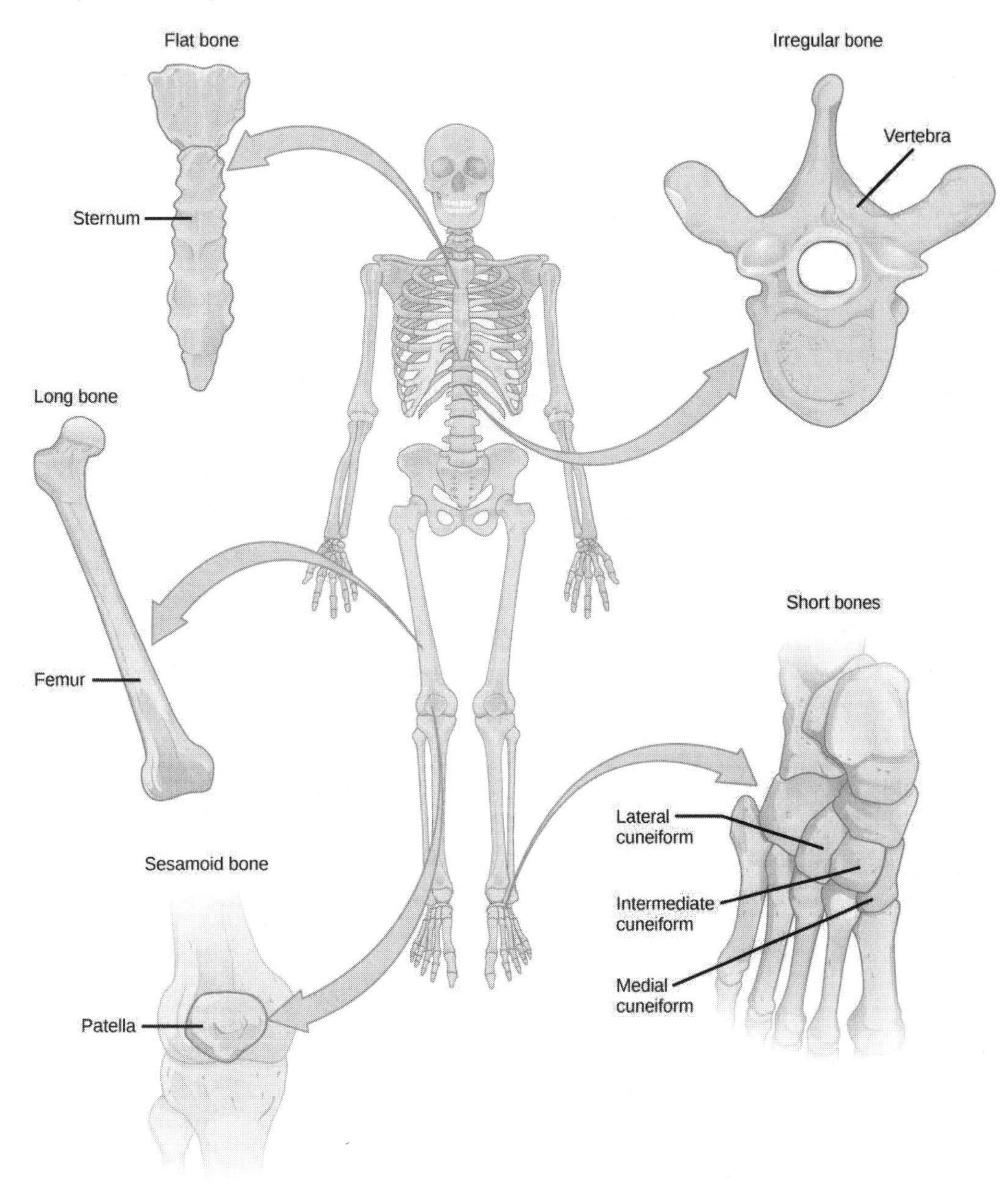

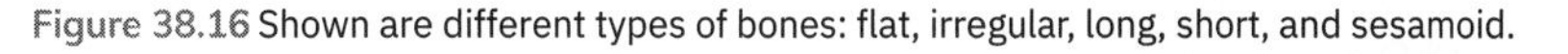

Figure 38.16 Shown are different types of bones: flat, irregular, long, short, and sesamoid.

Long bones are longer than they are wide and have a shaft and two ends. The **diaphysis**, or central shaft, contains bone marrow in a marrow cavity. The rounded ends, the **epiphyses**, are covered with articular cartilage and are filled with red bone marrow, which produces blood cells (Figure 38.17). Most of the limb bones are long bones—for example, the femur, tibia, ulna, and radius. Exceptions to this include the patella and the bones of the wrist and ankle.

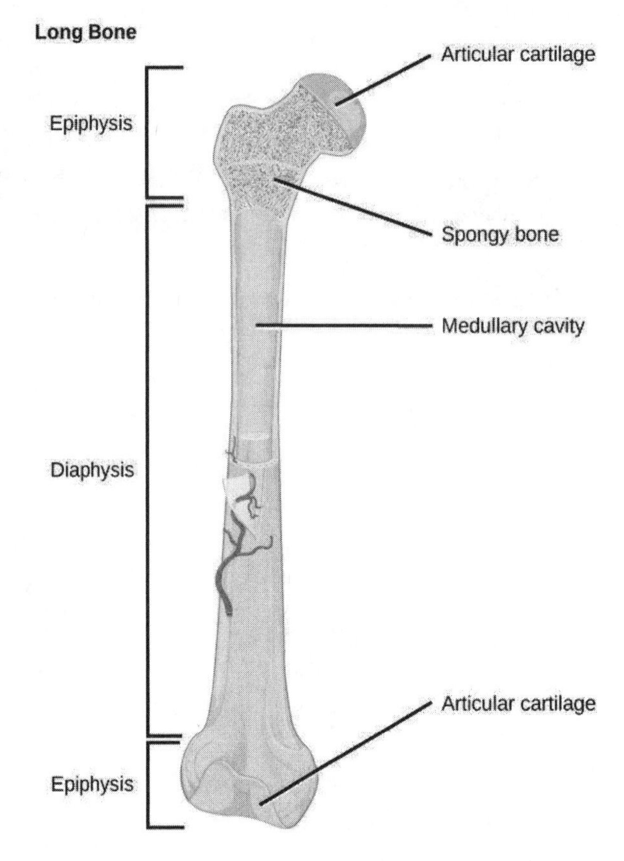

Figure 38.17 The long bone is covered by articular cartilage at either end and contains bone marrow (shown in yellow in this illustration) in the marrow cavity.

Short bones, or cuboidal bones, are bones that are the same width and length, giving them a cube-like shape. For example, the bones of the wrist (carpals) and ankle (tarsals) are short bones (Figure 38.16).

Flat bones are thin and relatively broad bones that are found where extensive protection of organs is required or where broad surfaces of muscle attachment are required. Examples of flat bones are the sternum (breast bone), ribs, scapulae (shoulder blades), and the roof of the skull (Figure 38.16).

Irregular bones are bones with complex shapes. These bones may have short, flat, notched, or ridged surfaces. Examples of irregular bones are the vertebrae, hip bones, and several skull bones.

Sesamoid bones are small, flat bones and are shaped similarly to a sesame seed. The patellae are sesamoid bones (Figure 38.18). Sesamoid bones develop inside tendons and may be found near joints at the knees, hands, and feet.

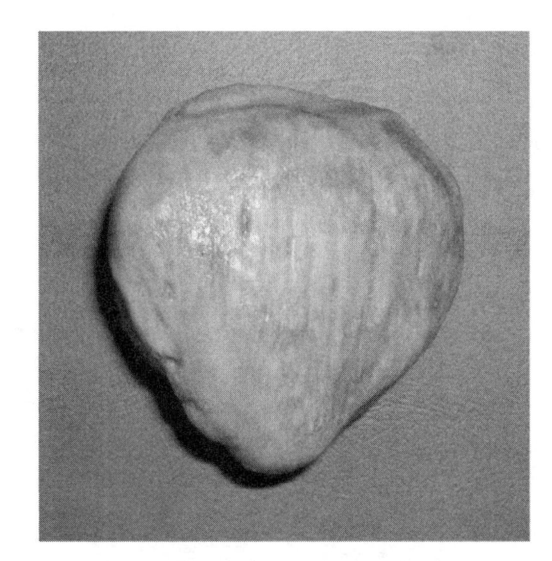

Figure 38.18 The patella of the knee is an example of a sesamoid bone.

Sutural bones are small, flat, irregularly shaped bones. They may be found between the flat bones of the skull. They vary in number, shape, size, and position.

Bone Tissue

Bones are considered organs because they contain various types of tissue, such as blood, connective tissue, nerves, and bone tissue. Osteocytes, the living cells of bone tissue, form the mineral matrix of bones. There are two types of bone tissue: compact and spongy.

Compact Bone Tissue

Compact bone (or cortical bone) forms the hard external layer of all bones and surrounds the medullary cavity, or bone marrow. It provides protection and strength to bones. Compact bone tissue consists of units called osteons or Haversian systems. **Osteons** are cylindrical structures that contain a mineral matrix and living osteocytes connected by canaliculi, which transport blood. They are aligned parallel to the long axis of the bone. Each osteon consists of **lamellae**, which are layers of compact matrix that surround a central canal called the Haversian canal. The **Haversian canal** (osteonic canal) contains the bone's blood vessels and nerve fibers (Figure 38.19). Osteons in compact bone tissue are aligned in the same direction along lines of stress and help the bone resist bending or fracturing. Therefore, compact bone tissue is prominent in areas of bone at which stresses are applied in only a few directions.

VISUAL CONNECTION

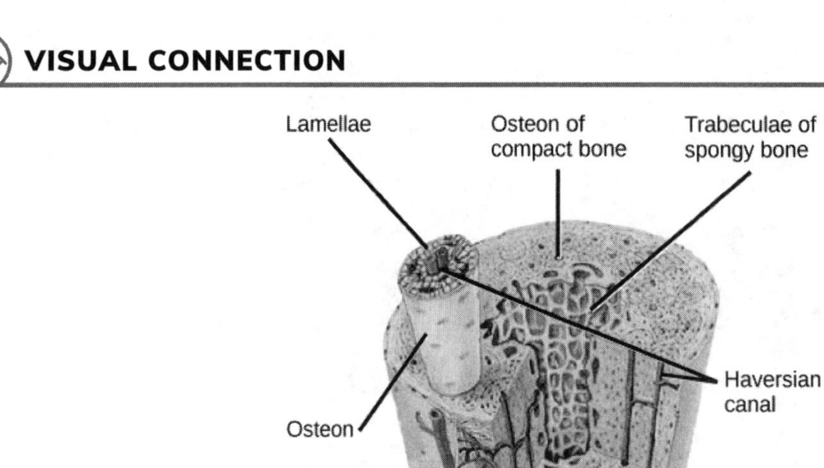

Figure 38.19 Compact bone tissue consists of osteons that are aligned parallel to the long axis of the bone, and the Haversian canal that contains the bone's blood vessels and nerve fibers. The inner layer of bones consists of spongy bone tissue. The small dark ovals in the osteon represent the living osteocytes. (credit: modification of work by NCI, NIH)

Which of the following statements about bone tissue is false?

a. Compact bone tissue is made of cylindrical osteons that are aligned such that they travel the length of the bone.
b. Haversian canals contain blood vessels only.
c. Haversian canals contain blood vessels and nerve fibers.
d. Spongy tissue is found on the interior of the bone, and compact bone tissue is found on the exterior.

Spongy Bone Tissue

Whereas compact bone tissue forms the outer layer of all bones, **spongy bone** or cancellous bone forms the inner layer of all bones. Spongy bone tissue does not contain osteons that constitute compact bone tissue. Instead, it consists of **trabeculae**, which are lamellae that are arranged as rods or plates. Red bone marrow is found between the trabuculae. Blood vessels within this tissue deliver nutrients to osteocytes and remove waste. The red bone marrow of the femur and the interior of other large bones, such as the ilium, forms blood cells.

Spongy bone reduces the density of bone and allows the ends of long bones to compress as the result of stresses applied to the bone. Spongy bone is prominent in areas of bones that are not heavily stressed or where stresses arrive from many directions. The epiphyses of bones, such as the neck of the femur, are subject to stress from many directions. Imagine laying a heavy framed picture flat on the floor. You could hold up one side of the picture with a toothpick if the toothpick was perpendicular to the floor and the picture. Now drill a hole and stick the toothpick into the wall to hang up the picture. In this case, the function of the toothpick is to transmit the downward pressure of the picture to the wall. The force on the picture is straight down to the floor, but the force on the toothpick is both the picture wire pulling down and the bottom of the hole in the wall pushing up. The toothpick will break off right at the wall.

The neck of the femur is horizontal like the toothpick in the wall. The weight of the body pushes it down near the joint, but the vertical diaphysis of the femur pushes it up at the other end. The neck of the femur must be strong enough to transfer the downward force of the body weight horizontally to the vertical shaft of the femur (Figure 38.20).

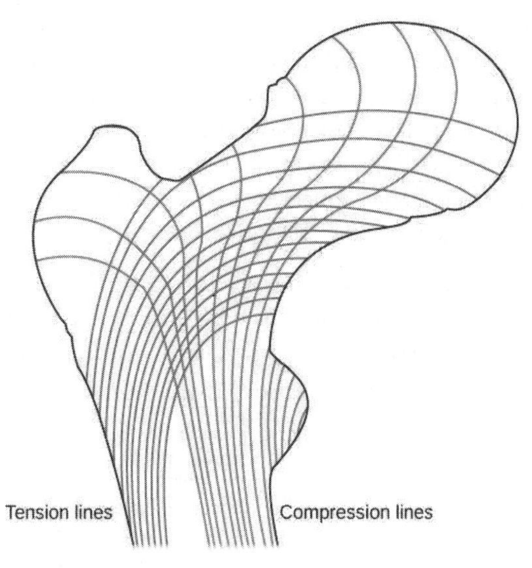

Figure 38.20 Trabeculae in spongy bone are arranged such that one side of the bone bears tension and the other withstands compression.

LINK TO LEARNING

View micrographs (http://openstax.org/l/muscle_tissue) of musculoskeletal tissues as you review the anatomy.

Cell Types in Bones

Bone consists of four types of cells: osteoblasts, osteoclasts, osteocytes, and osteoprogenitor cells. **Osteoblasts** are bone cells that are responsible for bone formation. Osteoblasts synthesize and secrete the organic part and inorganic part of the extracellular matrix of bone tissue, and collagen fibers. Osteoblasts become trapped in these secretions and differentiate into less active osteocytes. **Osteoclasts** are large bone cells with up to 50 nuclei. They remove bone structure by releasing lysosomal enzymes and acids that dissolve the bony matrix. These minerals, released from bones into the blood, help regulate calcium concentrations in body fluids. Bone may also be resorbed for remodeling, if the applied stresses have changed. **Osteocytes** are mature bone cells and are the main cells in bony connective tissue; these cells cannot divide. Osteocytes maintain normal bone structure by recycling the mineral salts in the bony matrix. **Osteoprogenitor cells** are squamous stem cells that divide to produce daughter cells that differentiate into osteoblasts. Osteoprogenitor cells are important in the repair of fractures.

Development of Bone

Ossification, or osteogenesis, is the process of bone formation by osteoblasts. Ossification is distinct from the process of calcification; whereas calcification takes place during the ossification of bones, it can also occur in other tissues. Ossification begins approximately six weeks after fertilization in an embryo. Before this time, the embryonic skeleton consists entirely of fibrous membranes and hyaline cartilage. The development of bone from fibrous membranes is called intramembranous ossification; development from hyaline cartilage is called endochondral ossification. Bone growth continues until approximately age 25. Bones can grow in thickness throughout life, but after age 25, ossification functions primarily in bone remodeling and repair.

Intramembranous Ossification

Intramembranous ossification is the process of bone development from fibrous membranes. It is involved in the formation of the flat bones of the skull, the mandible, and the clavicles. Ossification begins as mesenchymal cells form a template of the future bone. They then differentiate into osteoblasts at the ossification center. Osteoblasts secrete the extracellular matrix and deposit calcium, which hardens the matrix. The non-mineralized portion of the bone or osteoid continues to form around blood vessels, forming spongy bone. Connective tissue in the matrix differentiates into red bone marrow in the fetus. The spongy bone is remodeled into a thin layer of compact bone on the surface of the spongy bone.

Endochondral Ossification

Endochondral ossification is the process of bone development from hyaline cartilage. All of the bones of the body, except for the flat bones of the skull, mandible, and clavicles, are formed through endochondral ossification.

In long bones, chondrocytes form a template of the hyaline cartilage diaphysis. Responding to complex developmental signals, the matrix begins to calcify. This calcification prevents diffusion of nutrients into the matrix, resulting in chondrocytes dying and the opening up of cavities in the diaphysis cartilage. Blood vessels invade the cavities, and osteoblasts and osteoclasts modify the calcified cartilage matrix into spongy bone. Osteoclasts then break down some of the spongy bone to create a marrow, or medullary, cavity in the center of the diaphysis. Dense, irregular connective tissue forms a sheath (periosteum) around the bones. The periosteum assists in attaching the bone to surrounding tissues, tendons, and ligaments. The bone continues to grow and elongate as the cartilage cells at the epiphyses divide.

In the last stage of prenatal bone development, the centers of the epiphyses begin to calcify. Secondary ossification centers form in the epiphyses as blood vessels and osteoblasts enter these areas and convert hyaline cartilage into spongy bone. Until adolescence, hyaline cartilage persists at the **epiphyseal plate** (growth plate), which is the region between the diaphysis and epiphysis that is responsible for the lengthwise growth of long bones (Figure 38.21).

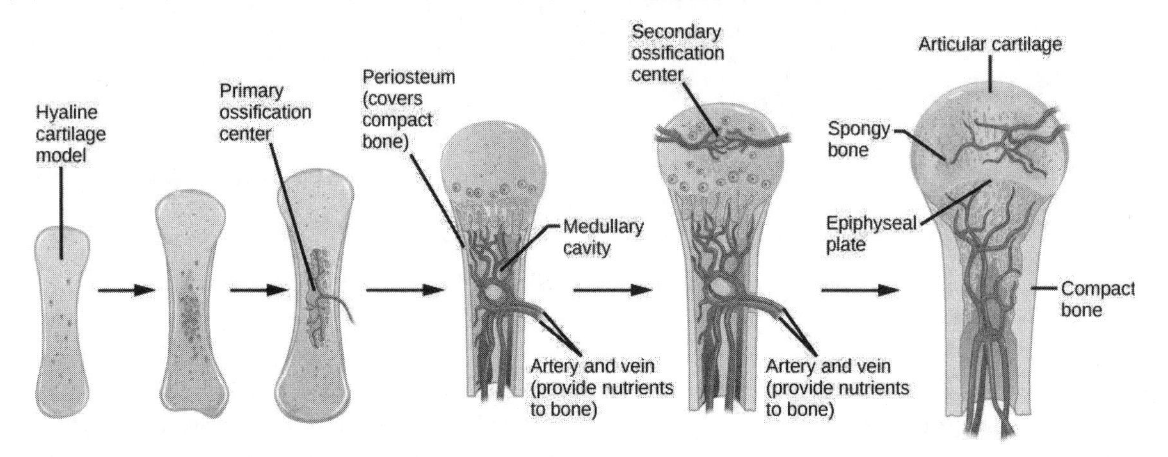

Figure 38.21 Endochondral ossification is the process of bone development from hyaline cartilage. The periosteum is the connective tissue on the outside of bone that acts as the interface between bone, blood vessels, tendons, and ligaments.

Growth of Bone

Long bones continue to lengthen, potentially until adolescence, through the addition of bone tissue at the epiphyseal plate. They also increase in width through appositional growth.

Lengthening of Long Bones

Chondrocytes on the epiphyseal side of the epiphyseal plate divide; one cell remains undifferentiated near the epiphysis, and one cell moves toward the diaphysis. The cells, which are pushed from the epiphysis, mature and are destroyed by calcification. This process replaces cartilage with bone on the diaphyseal side of the plate, resulting in a lengthening of the bone.

Long bones stop growing at around the age of 18 in females and the age of 21 in males in a process called epiphyseal plate closure. During this process, cartilage cells stop dividing and all of the cartilage is replaced by bone. The epiphyseal plate fades, leaving a structure called the epiphyseal line or epiphyseal remnant, and the epiphysis and diaphysis fuse.

Thickening of Long Bones

Appositional growth is the increase in the diameter of bones by the addition of bony tissue at the surface of bones. Osteoblasts at the bone surface secrete bone matrix, and osteoclasts on the inner surface break down bone. The osteoblasts differentiate into osteocytes. A balance between these two processes allows the bone to thicken without becoming too heavy.

Bone Remodeling and Repair

Bone renewal continues after birth into adulthood. **Bone remodeling** is the replacement of old bone tissue by new bone tissue. It involves the processes of bone deposition by osteoblasts and bone resorption by osteoclasts. Normal bone growth requires vitamins D, C, and A, plus minerals such as calcium, phosphorous, and magnesium. Hormones such as parathyroid hormone, growth hormone, and calcitonin are also required for proper bone growth and maintenance.

Bone turnover rates are quite high, with five to seven percent of bone mass being recycled every week. Differences in turnover rate exist in different areas of the skeleton and in different areas of a bone. For example, the bone in the head of the femur may be fully replaced every six months, whereas the bone along the shaft is altered much more slowly.

Bone remodeling allows bones to adapt to stresses by becoming thicker and stronger when subjected to stress. Bones that are not subject to normal stress, for example when a limb is in a cast, will begin to lose mass. A fractured or broken bone undergoes repair through four stages:

1. Blood vessels in the broken bone tear and hemorrhage, resulting in the formation of clotted blood, or a hematoma, at the site of the break. The severed blood vessels at the broken ends of the bone are sealed by the clotting process, and bone cells that are deprived of nutrients begin to die.
2. Within days of the fracture, capillaries grow into the hematoma, and phagocytic cells begin to clear away the dead cells. Though fragments of the blood clot may remain, fibroblasts and osteoblasts enter the area and begin to reform bone. Fibroblasts produce collagen fibers that connect the broken bone ends, and osteoblasts start to form spongy bone. The repair tissue between the broken bone ends is called the fibrocartilaginous callus, as it is composed of both hyaline and fibrocartilage (Figure 38.22). Some bone spicules may also appear at this point.
3. The fibrocartilaginous callus is converted into a bony callus of spongy bone. It takes about two months for the broken bone ends to be firmly joined together after the fracture. This is similar to the endochondral formation of bone, as cartilage becomes ossified; osteoblasts, osteoclasts, and bone matrix are present.
4. The bony callus is then remodelled by osteoclasts and osteoblasts, with excess material on the exterior of the bone and within the medullary cavity being removed. Compact bone is added to create bone tissue that is similar to the original, unbroken bone. This remodeling can take many months, and the bone may remain uneven for years.

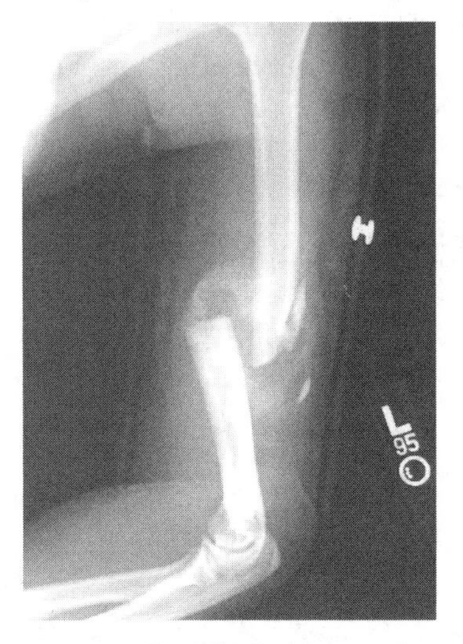

Figure 38.22 After this bone is set, a callus will knit the two ends together. (credit: Bill Rhodes)

SCIENTIFIC METHOD CONNECTION

Decalcification of Bones

Question: What effect does the removal of calcium and collagen have on bone structure?

Background: Conduct a literature search on the role of calcium and collagen in maintaining bone structure. Conduct a literature search on diseases in which bone structure is compromised.

Hypothesis: Develop a hypothesis that states predictions of the flexibility, strength, and mass of bones that have had the calcium and collagen components removed. Develop a hypothesis regarding the attempt to add calcium back to decalcified bones.

Test the hypothesis: Test the prediction by removing calcium from chicken bones by placing them in a jar of vinegar for seven days. Test the hypothesis regarding adding calcium back to decalcified bone by placing the decalcified chicken bones into a jar of water with calcium supplements added. Test the prediction by denaturing the collagen from the bones by baking them at 250°C for three hours.

Analyze the data: Create a table showing the changes in bone flexibility, strength, and mass in the three different environments.

Report the results: Under which conditions was the bone most flexible? Under which conditions was the bone the strongest?

Draw a conclusion: Did the results support or refute the hypothesis? How do the results observed in this experiment correspond to diseases that destroy bone tissue?

38.3 Joints and Skeletal Movement

By the end of this section, you will be able to do the following:

- Classify the different types of joints on the basis of structure
- Explain the role of joints in skeletal movement

The point at which two or more bones meet is called a **joint**, or **articulation**. Joints are responsible for movement, such as the movement of limbs, and stability, such as the stability found in the bones of the skull.

Classification of Joints on the Basis of Structure

There are two ways to classify joints: on the basis of their structure or on the basis of their function. The structural classification divides joints into bony, fibrous, cartilaginous, and synovial joints depending on the material composing the joint and the presence or absence of a cavity in the joint.

Fibrous Joints

The bones of **fibrous joints** are held together by fibrous connective tissue. There is no cavity, or space, present between the bones and so most fibrous joints do not move at all, or are only capable of minor movements. There are three types of fibrous joints: sutures, syndesmoses, and gomphoses. **Sutures** are found only in the skull and possess short fibers of connective tissue that hold the skull bones tightly in place (Figure 38.23).

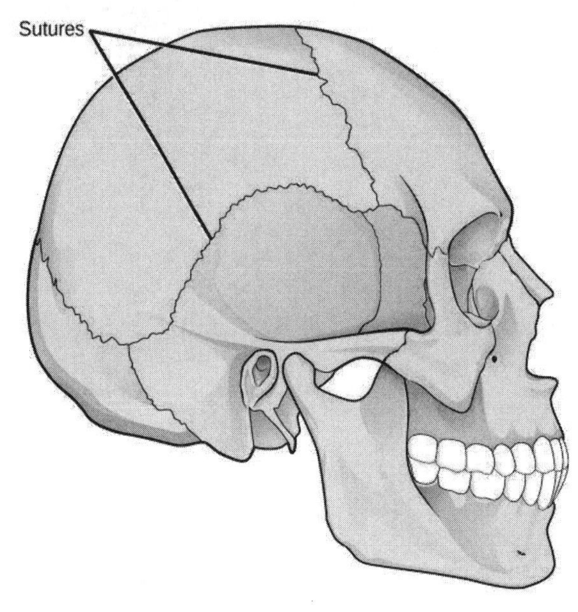

Figure 38.23 Sutures are fibrous joints found only in the skull.

Syndesmoses are joints in which the bones are connected by a band of connective tissue, allowing for more movement than in a suture. An example of a syndesmosis is the joint of the tibia and fibula in the ankle. The amount of movement in these types of joints is determined by the length of the connective tissue fibers. **Gomphoses** occur between teeth and their sockets; the term refers to the way the tooth fits into the socket like a peg (Figure 38.24). The tooth is connected to the socket by a connective tissue referred to as the periodontal ligament.

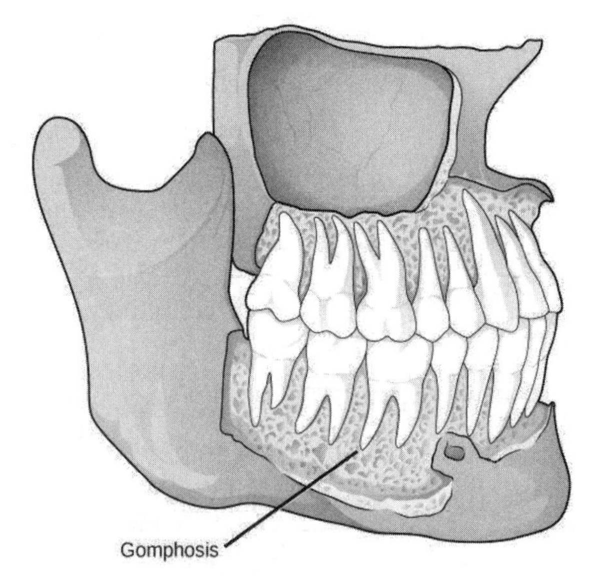

Figure 38.24 Gomphoses are fibrous joints between the teeth and their sockets. (credit: modification of work by Gray's Anatomy)

Cartilaginous Joints

Cartilaginous joints are joints in which the bones are connected by cartilage. There are two types of cartilaginous joints: synchondroses and symphyses. In a **synchondrosis**, the bones are joined by hyaline cartilage. Synchondroses are found in the epiphyseal plates of growing bones in children. In **symphyses**, hyaline cartilage covers the end of the bone but the connection between bones occurs through fibrocartilage. Symphyses are found at the joints between vertebrae. Either type of cartilaginous joint allows for very little movement.

Synovial Joints

Synovial joints are the only joints that have a space between the adjoining bones (Figure 38.25). This space is referred to as the synovial (or joint) cavity and is filled with synovial fluid. Synovial fluid lubricates the joint, reducing friction between the bones and allowing for greater movement. The ends of the bones are covered with articular cartilage, a hyaline cartilage, and the entire joint is surrounded by an articular capsule composed of connective tissue that allows movement of the joint while resisting dislocation. Articular capsules may also possess ligaments that hold the bones together. Synovial joints are capable of the greatest movement of the three structural joint types; however, the more mobile a joint, the weaker the joint. Knees, elbows, and shoulders are examples of synovial joints.

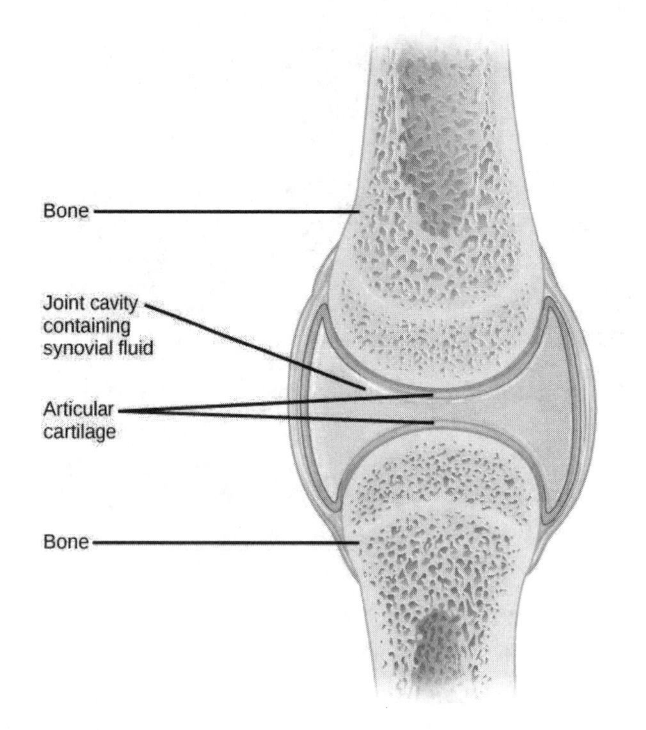

Figure 38.25 Synovial joints are the only joints that have a space or “synovial cavity” in the joint.

Classification of Joints on the Basis of Function

The functional classification divides joints into three categories: synarthroses, amphiarthroses, and diarthroses. A **synarthrosis** is a joint that is immovable. This includes sutures, gomphoses, and synchondroses. **Amphiarthroses** are joints that allow slight movement, including syndesmoses and symphyses. **Diarthroses** are joints that allow for free movement of the joint, as in synovial joints.

Movement at Synovial Joints

The wide range of movement allowed by synovial joints produces different types of movements. The movement of synovial joints can be classified as one of four different types: gliding, angular, rotational, or special movement.

Gliding Movement

Gliding movements occur as relatively flat bone surfaces move past each other. Gliding movements produce very little rotation or angular movement of the bones. The joints of the carpal and tarsal bones are examples of joints that produce gliding movements.

Angular Movement

Angular movements are produced when the angle between the bones of a joint changes. There are several different types of angular movements, including flexion, extension, hyperextension, abduction, adduction, and circumduction. **Flexion**, or bending, occurs when the angle between the bones decreases. Moving the forearm upward at the elbow or moving the wrist to move the hand toward the forearm are examples of flexion. **Extension** is the opposite of flexion in that the angle between the bones of a joint increases. Straightening a limb after flexion is an example of extension. Extension past the regular anatomical position is referred to as **hyperextension**. This includes moving the neck back to look upward, or bending the wrist so that the hand moves away from the forearm.

Abduction occurs when a bone moves away from the midline of the body. Examples of abduction are moving the arms or legs laterally to lift them straight out to the side. **Adduction** is the movement of a bone toward the midline of the body. Movement of the limbs inward after abduction is an example of adduction. **Circumduction** is the movement of a limb in a circular motion, as in moving the arm in a circular motion.

Rotational Movement

Rotational movement is the movement of a bone as it rotates around its longitudinal axis. Rotation can be toward the midline of the body, which is referred to as **medial rotation**, or away from the midline of the body, which is referred to as **lateral rotation**.

Movement of the head from side to side is an example of rotation.

Special Movements

Some movements that cannot be classified as gliding, angular, or rotational are called special movements. **Inversion** involves the soles of the feet moving inward, toward the midline of the body. **Eversion** is the opposite of inversion, movement of the sole of the foot outward, away from the midline of the body. **Protraction** is the anterior movement of a bone in the horizontal plane. **Retraction** occurs as a joint moves back into position after protraction. Protraction and retraction can be seen in the movement of the mandible as the jaw is thrust outwards and then back inwards. **Elevation** is the movement of a bone upward, such as when the shoulders are shrugged, lifting the scapulae. **Depression** is the opposite of elevation—movement downward of a bone, such as after the shoulders are shrugged and the scapulae return to their normal position from an elevated position. **Dorsiflexion** is a bending at the ankle such that the toes are lifted toward the knee. **Plantar flexion** is a bending at the ankle when the heel is lifted, such as when standing on the toes. **Supination** is the movement of the radius and ulna bones of the forearm so that the palm faces forward. **Pronation** is the opposite movement, in which the palm faces backward. **Opposition** is the movement of the thumb toward the fingers of the same hand, making it possible to grasp and hold objects.

Types of Synovial Joints

Synovial joints are further classified into six different categories on the basis of the shape and structure of the joint. The shape of the joint affects the type of movement permitted by the joint (Figure 38.26). These joints can be described as planar, hinge, pivot, condyloid, saddle, or ball-and-socket joints.

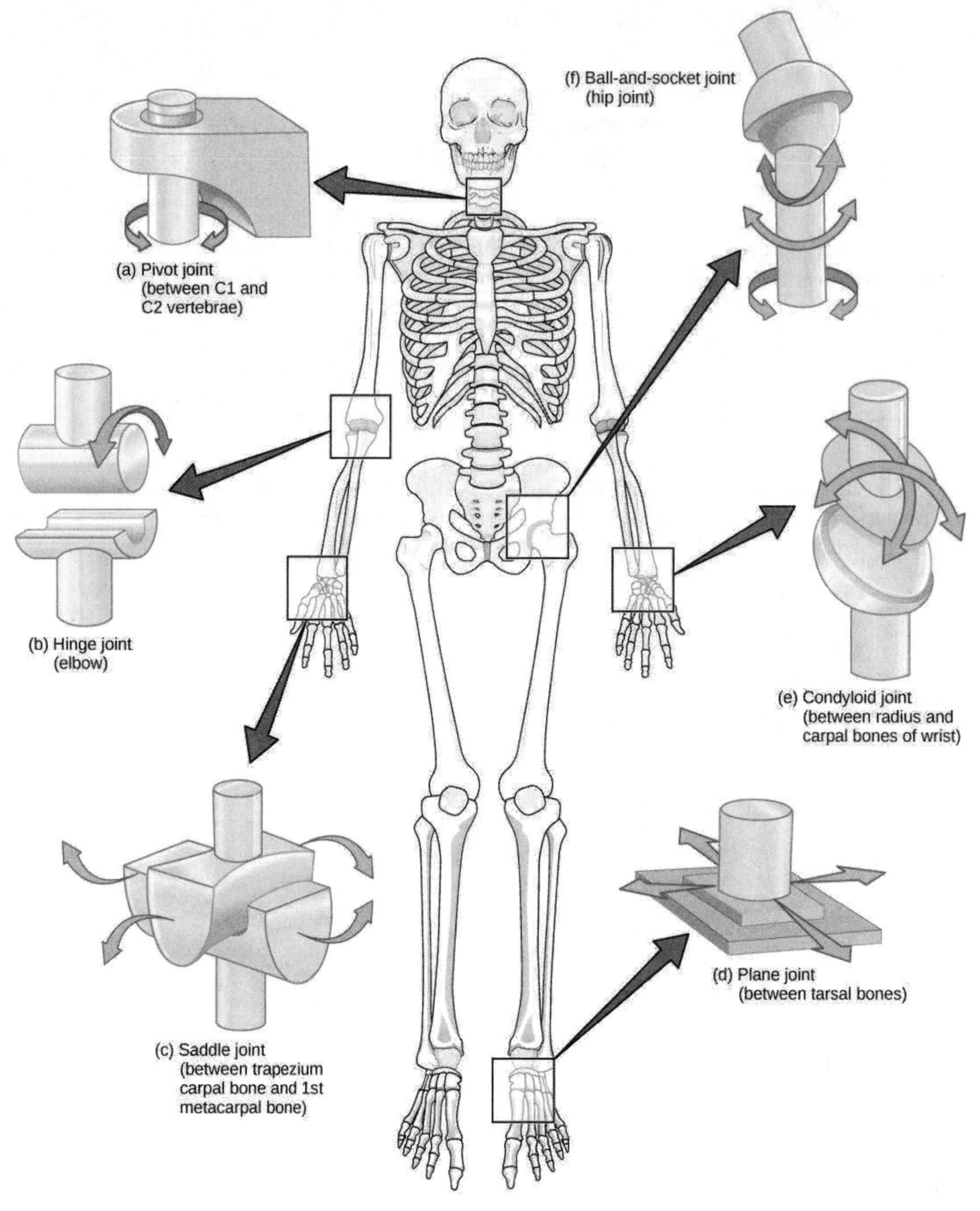

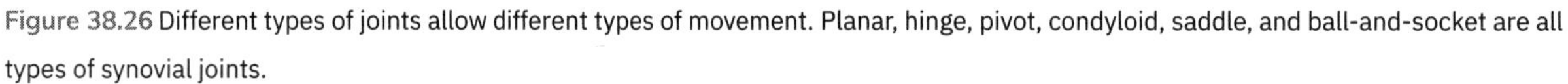
Figure 38.26 Different types of joints allow different types of movement. Planar, hinge, pivot, condyloid, saddle, and ball-and-socket are all types of synovial joints.

Planar Joints

Planar joints have bones with articulating surfaces that are flat or slightly curved faces. These joints allow for gliding movements, and so the joints are sometimes referred to as gliding joints. The range of motion is limited in these joints and does not involve rotation. Planar joints are found in the carpal bones in the hand and the tarsal bones of the foot, as well as between vertebrae (Figure 38.27).

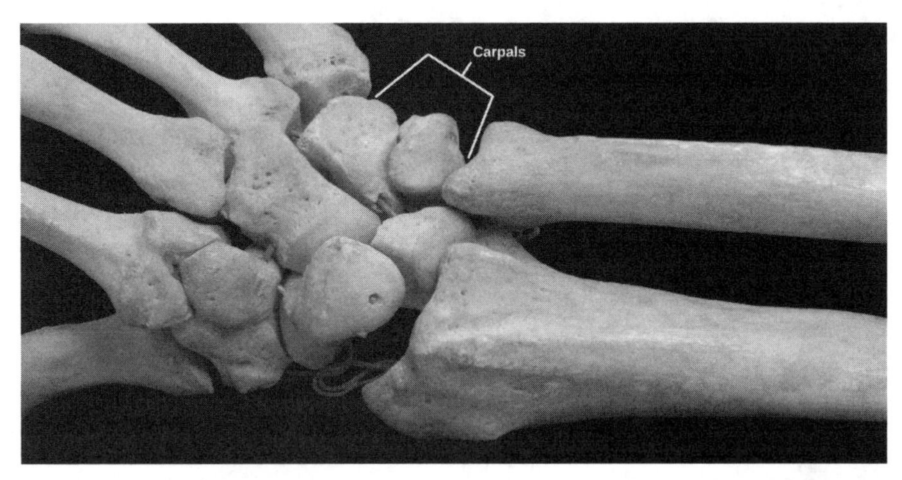

Figure 38.27 The joints of the carpal bones in the wrist are examples of planar joints. (credit: modification of work by Brian C. Goss)

Hinge Joints

In **hinge joints**, the slightly rounded end of one bone fits into the slightly hollow end of the other bone. In this way, one bone moves while the other remains stationary, like the hinge of a door. The elbow is an example of a hinge joint. The knee is sometimes classified as a modified hinge joint (Figure 38.28).

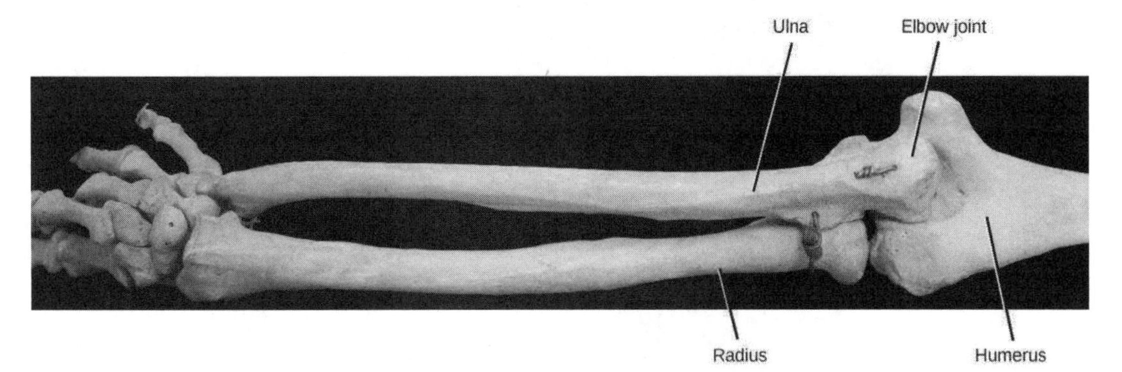

Figure 38.28 The elbow joint, where the radius articulates with the humerus, is an example of a hinge joint. (credit: modification of work by Brian C. Goss)

Pivot Joints

Pivot joints consist of the rounded end of one bone fitting into a ring formed by the other bone. This structure allows rotational movement, as the rounded bone moves around its own axis. An example of a pivot joint is the joint of the first and second vertebrae of the neck that allows the head to move back and forth (Figure 38.29). The joint of the wrist that allows the palm of the hand to be turned up and down is also a pivot joint.

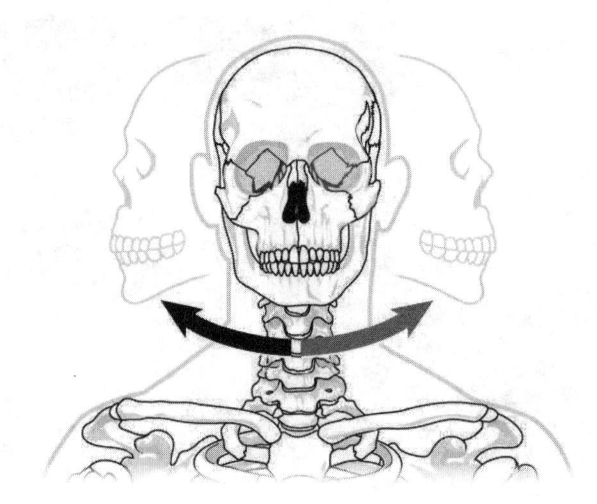

Figure 38.29 The joint in the neck that allows the head to move back and forth is an example of a pivot joint.

Condyloid Joints

Condyloid joints consist of an oval-shaped end of one bone fitting into a similarly oval-shaped hollow of another bone (Figure 38.30). This is also sometimes called an ellipsoidal joint. This type of joint allows angular movement along two axes, as seen in the joints of the wrist and fingers, which can move both side to side and up and down.

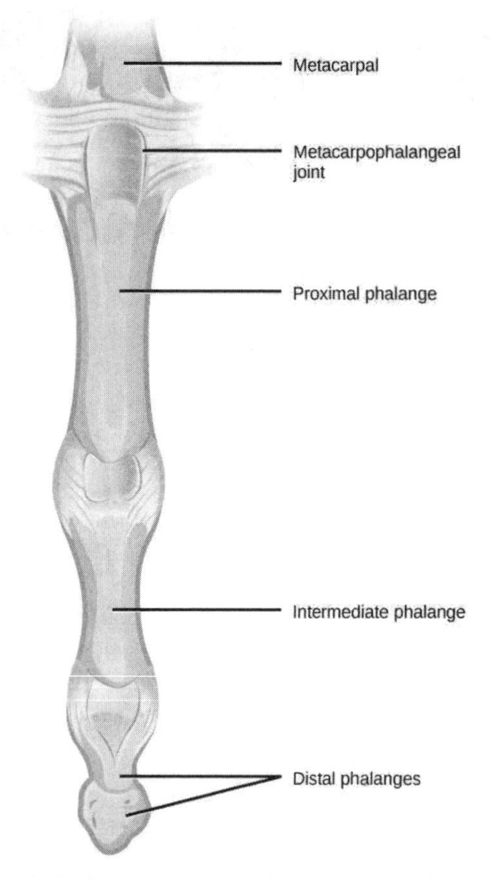

Figure 38.30 The metacarpophalangeal joints in the finger are examples of condyloid joints. (credit: modification of work by Gray's Anatomy)

Saddle Joints

Saddle joints are so named because the ends of each bone resemble a saddle, with concave and convex portions that fit together. Saddle joints allow angular movements similar to condyloid joints but with a greater range of motion. An example of a saddle joint is the thumb joint, which can move back and forth and up and down, but more freely than the wrist or fingers (Figure 38.31).

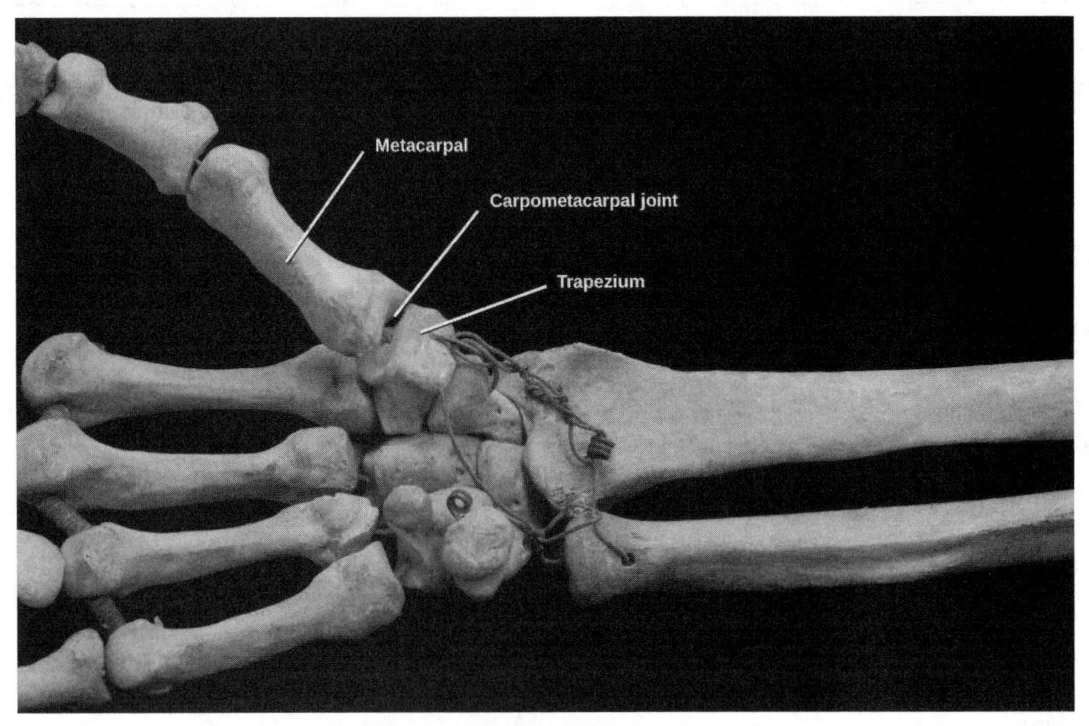

Figure 38.31 The carpometacarpal joints in the thumb are examples of saddle joints. (credit: modification of work by Brian C. Goss)

Ball-and-Socket Joints

Ball-and-socket joints possess a rounded, ball-like end of one bone fitting into a cuplike socket of another bone. This organization allows the greatest range of motion, as all movement types are possible in all directions. Examples of ball-and-socket joints are the shoulder and hip joints (Figure 38.32).

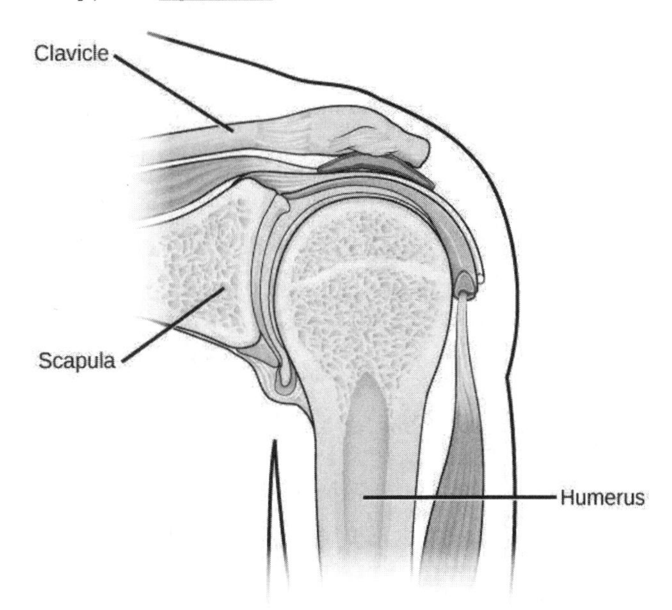

Figure 38.32 The shoulder joint is an example of a ball-and-socket joint.

LINK TO LEARNING

Watch this animation showing the six types of synovial joints.

Click to view content (https://www.openstax.org/l/synovial_joints)

CAREER CONNECTION

Rheumatologist

Rheumatologists are medical doctors who specialize in the diagnosis and treatment of disorders of the joints, muscles, and bones. They diagnose and treat diseases such as arthritis, musculoskeletal disorders, osteoporosis, and autoimmune diseases such as ankylosing spondylitis and rheumatoid arthritis.

Rheumatoid arthritis (RA) is an inflammatory disorder that primarily affects the synovial joints of the hands, feet, and cervical spine. Affected joints become swollen, stiff, and painful. Although it is known that RA is an autoimmune disease in which the body's immune system mistakenly attacks healthy tissue, the cause of RA remains unknown. Immune cells from the blood enter joints and the synovium causing cartilage breakdown, swelling, and inflammation of the joint lining. Breakdown of cartilage causes bones to rub against each other causing pain. RA is more common in women than men and the age of onset is usually 40–50 years of age.

Rheumatologists can diagnose RA on the basis of symptoms such as joint inflammation and pain, X-ray and MRI imaging, and blood tests. Arthrography is a type of medical imaging of joints that uses a contrast agent, such as a dye, that is opaque to X-rays. This allows the soft tissue structures of joints—such as cartilage, tendons, and ligaments—to be visualized. An arthrogram differs from a regular X-ray by showing the surface of soft tissues lining the joint in addition to joint bones. An arthrogram allows early degenerative changes in joint cartilage to be detected before bones become affected.

There is currently no cure for RA; however, rheumatologists have a number of treatment options available. Early stages can be treated with rest of the affected joints by using a cane or by using joint splints that minimize inflammation. When inflammation has decreased, exercise can be used to strengthen the muscles that surround the joint and to maintain joint flexibility. If joint damage is more extensive, medications can be used to relieve pain and decrease inflammation. Anti-inflammatory drugs such as aspirin, topical pain relievers, and corticosteroid injections may be used. Surgery may be required in cases in which joint damage is severe.

38.4 Muscle Contraction and Locomotion

By the end of this section, you will be able to do the following:

- Classify the different types of muscle tissue
- Explain the role of muscles in locomotion

Muscle cells are specialized for contraction. Muscles allow for motions such as walking, and they also facilitate bodily processes such as respiration and digestion. The body contains three types of muscle tissue: skeletal muscle, cardiac muscle, and smooth muscle (Figure 38.33).

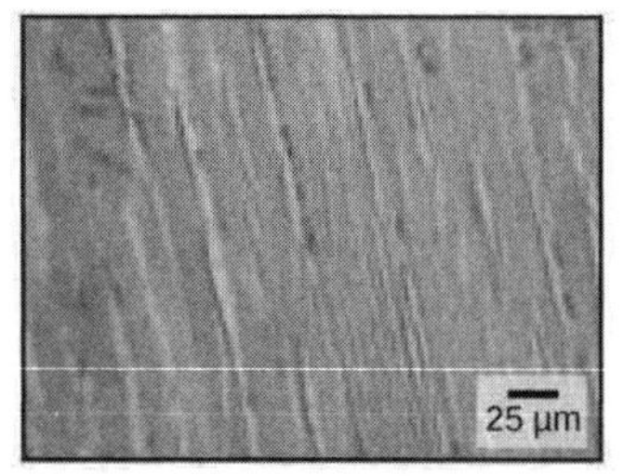

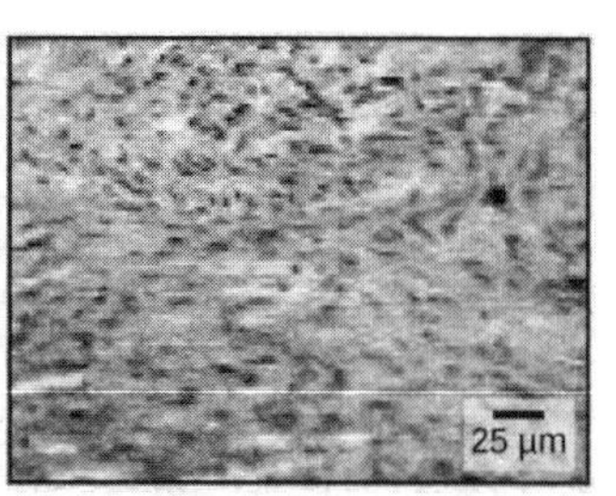

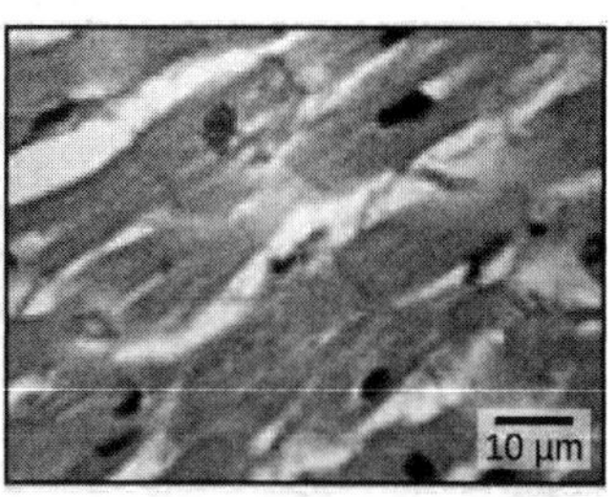

Figure 38.33 The body contains three types of muscle tissue: skeletal muscle, smooth muscle, and cardiac muscle, visualized here using light microscopy. Smooth muscle cells are short, tapered at each end, and have only one plump nucleus in each. Cardiac muscle cells are branched and striated, but short. The cytoplasm may branch, and they have one nucleus in the center of the cell. (credit: modification of work by NCI, NIH; scale-bar data from Matt Russell)

Skeletal muscle tissue forms skeletal muscles, which attach to bones or skin and control locomotion and any movement that can be consciously controlled. Because it can be controlled by thought, skeletal muscle is also called voluntary muscle. Skeletal muscles are long and cylindrical in appearance; when viewed under a microscope, skeletal muscle tissue has a striped or striated appearance. The striations are caused by the regular arrangement of contractile proteins (actin and myosin). **Actin** is a globular contractile protein that interacts with **myosin** for muscle contraction. Skeletal muscle also has multiple nuclei present in a single cell.

Smooth muscle tissue occurs in the walls of hollow organs such as the intestines, stomach, and urinary bladder, and around passages such as the respiratory tract and blood vessels. Smooth muscle has no striations, is not under voluntary control, has only one nucleus per cell, is tapered at both ends, and is called involuntary muscle.

Cardiac muscle tissue is only found in the heart, and cardiac contractions pump blood throughout the body and maintain blood pressure. Like skeletal muscle, cardiac muscle is striated, but unlike skeletal muscle, cardiac muscle cannot be consciously controlled and is called involuntary muscle. It has one nucleus per cell, is branched, and is distinguished by the presence of intercalated disks.

Skeletal Muscle Fiber Structure

Each skeletal muscle fiber is a skeletal muscle cell. These cells are incredibly large, with diameters of up to 100 µm and lengths of up to 30 cm. The plasma membrane of a skeletal muscle fiber is called the **sarcolemma**. The sarcolemma is the site of action potential conduction, which triggers muscle contraction. Within each muscle fiber are **myofibrils**—long cylindrical structures that lie parallel to the muscle fiber. Myofibrils run the entire length of the muscle fiber, and because they are only approximately 1.2 µm in diameter, hundreds to thousands can be found inside one muscle fiber. They attach to the sarcolemma at their ends, so that as myofibrils shorten, the entire muscle cell contracts (Figure 38.34).

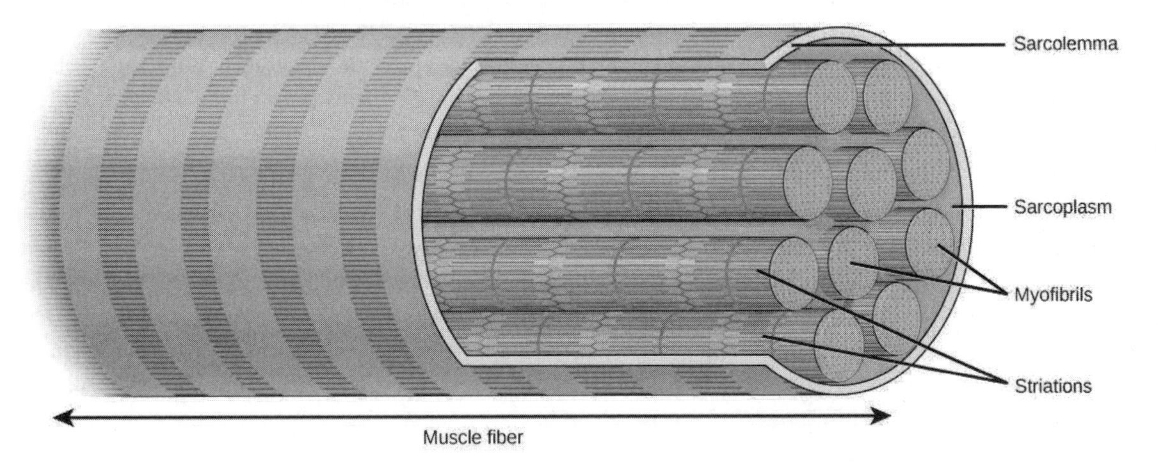

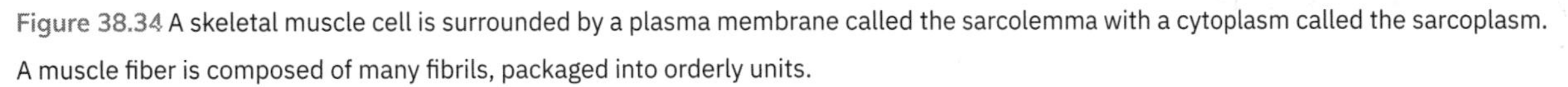

Figure 38.34 A skeletal muscle cell is surrounded by a plasma membrane called the sarcolemma with a cytoplasm called the sarcoplasm. A muscle fiber is composed of many fibrils, packaged into orderly units.

The striated appearance of skeletal muscle tissue is a result of repeating bands of the proteins actin and myosin that are present along the length of myofibrils. Dark A bands and light I bands repeat along myofibrils, and the alignment of myofibrils in the cell causes the entire cell to appear striated or banded.

Each I band has a dense line running vertically through the middle called a Z disc or Z line. The Z discs mark the border of units called **sarcomeres**, which are the functional units of skeletal muscle. One sarcomere is the space between two consecutive Z discs and contains one entire A band and two halves of an I band, one on either side of the A band. A myofibril is composed of many sarcomeres running along its length, and as the sarcomeres individually contract, the myofibrils and muscle cells shorten (Figure 38.35).

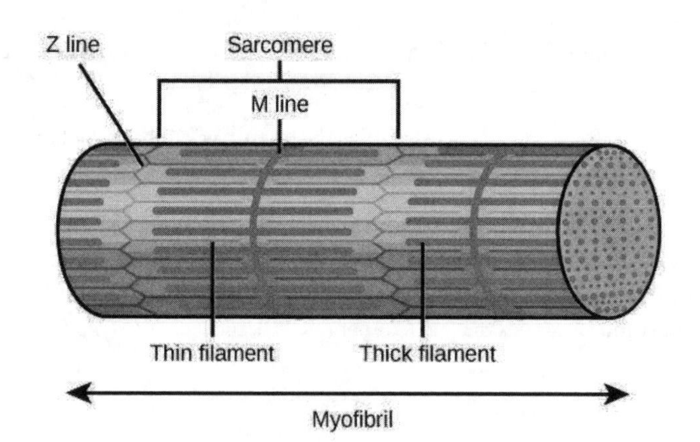

Figure 38.35 A sarcomere is the region from one Z line to the next Z line. Many sarcomeres are present in a myofibril, resulting in the striation pattern characteristic of skeletal muscle.

Myofibrils are composed of smaller structures called **myofilaments**. There are two main types of filaments: thick filaments and thin filaments; each has different compositions and locations. **Thick filaments** occur only in the A band of a myofibril. **Thin filaments** attach to a protein in the Z disc called alpha-actinin and occur across the entire length of the I band and partway into the A band. The region at which thick and thin filaments overlap has a dense appearance, as there is little space between the filaments. Thin filaments do not extend all the way into the A bands, leaving a central region of the A band that only contains thick filaments. This central region of the A band looks slightly lighter than the rest of the A band and is called the H zone. The middle of the H zone has a vertical line called the M line, at which accessory proteins hold together thick filaments. Both the Z disc and the M line hold myofilaments in place to maintain the structural arrangement and layering of the myofibril. Myofibrils are connected to each other by intermediate, or desmin, filaments that attach to the Z disc.

Thick and thin filaments are themselves composed of proteins. Thick filaments are composed of the protein myosin. The tail of a myosin molecule connects with other myosin molecules to form the central region of a thick filament near the M line, whereas the heads align on either side of the thick filament where the thin filaments overlap. The primary component of thin filaments is the actin protein. Two other components of the thin filament are tropomyosin and troponin. Actin has binding sites for myosin attachment. Strands of tropomyosin block the binding sites and prevent actin–myosin interactions when the muscles are at rest. Troponin consists of three globular subunits. One subunit binds to tropomyosin, one subunit binds to actin, and one subunit binds Ca^{2+} ions.

LINK TO LEARNING

View this animation showing the organization of muscle fibers.

Click to view content (https://www.openstax.org/l/skeletal_muscle)

Sliding Filament Model of Contraction

For a muscle cell to contract, the sarcomere must shorten. However, thick and thin filaments—the components of sarcomeres—do not shorten. Instead, they slide by one another, causing the sarcomere to shorten while the filaments remain the same length. The sliding filament theory of muscle contraction was developed to fit the differences observed in the named bands on the sarcomere at different degrees of muscle contraction and relaxation. The mechanism of contraction is the binding of myosin to actin, forming cross-bridges that generate filament movement (Figure 38.36).

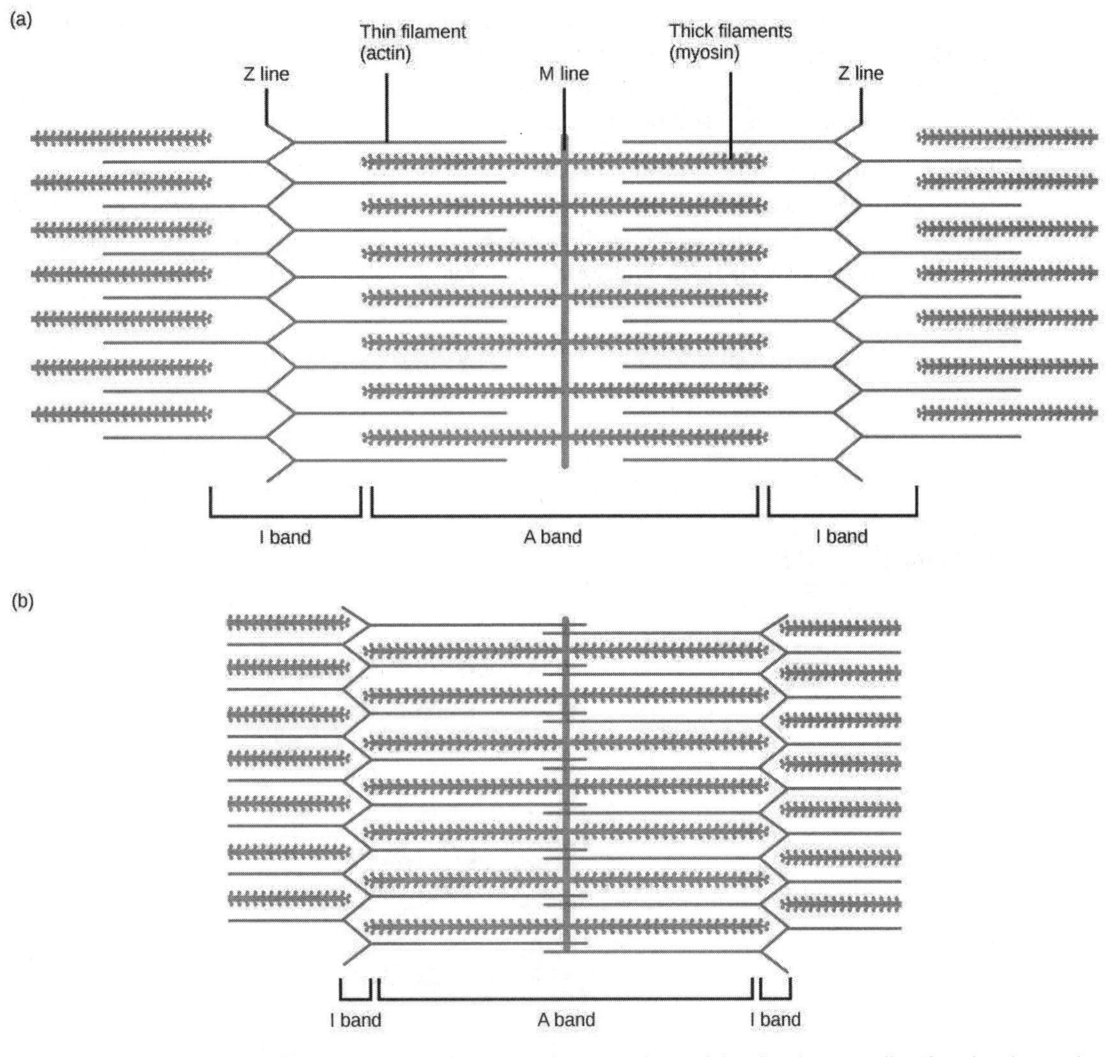

Figure 38.36 When (a) a sarcomere (b) contracts, the Z lines move closer together and the I band gets smaller. The A band stays the same width and, at full contraction, the thin filaments overlap.

When a sarcomere shortens, some regions shorten whereas others stay the same length. A sarcomere is defined as the distance between two consecutive Z discs or Z lines; when a muscle contracts, the distance between the Z discs is reduced. The H zone—the central region of the A zone—contains only thick filaments and is shortened during contraction. The I band contains only thin filaments and also shortens. The A band does not shorten—it remains the same length—but A bands of different sarcomeres move closer together during contraction, eventually disappearing. Thin filaments are pulled by the thick filaments toward the center of the sarcomere until the Z discs approach the thick filaments. The zone of overlap, in which thin filaments and thick filaments occupy the same area, increases as the thin filaments move inward.

ATP and Muscle Contraction

The motion of muscle shortening occurs as myosin heads bind to actin and pull the actin inwards. This action requires energy, which is provided by ATP. Myosin binds to actin at a binding site on the globular actin protein. Myosin has another binding site for ATP at which enzymatic activity hydrolyzes ATP to ADP, releasing an inorganic phosphate molecule and energy.

ATP binding causes myosin to release actin, allowing actin and myosin to detach from each other. After this happens, the newly bound ATP is converted to ADP and inorganic phosphate, P_i. The enzyme at the binding site on myosin is called ATPase. The energy released during ATP hydrolysis changes the angle of the myosin head into a "cocked" position. The myosin head is then in a position for further movement, possessing potential energy, but ADP and P_i are still attached. If actin binding sites are covered and unavailable, the myosin will remain in the high energy configuration with ATP hydrolyzed, but still attached.

If the actin binding sites are uncovered, a cross-bridge will form; that is, the myosin head spans the distance between the actin and myosin molecules. P_i is then released, allowing myosin to expend the stored energy as a conformational change. The myosin head moves toward the M line, pulling the actin along with it. As the actin is pulled, the filaments move approximately 10 nm toward the M line. This movement is called the power stroke, as it is the step at which force is produced. As the actin is pulled toward the M line, the sarcomere shortens and the muscle contracts.

When the myosin head is "cocked," it contains energy and is in a high-energy configuration. This energy is expended as the myosin head moves through the power stroke; at the end of the power stroke, the myosin head is in a low-energy position. After the power stroke, ADP is released; however, the cross-bridge formed is still in place, and actin and myosin are bound together. ATP can then attach to myosin, which allows the cross-bridge cycle to start again and further muscle contraction can occur (Figure 38.37).

LINK TO LEARNING

Watch this video explaining how a muscle contraction is signaled.

Click to view content (https://www.openstax.org/l/contract_muscle)

VISUAL CONNECTION

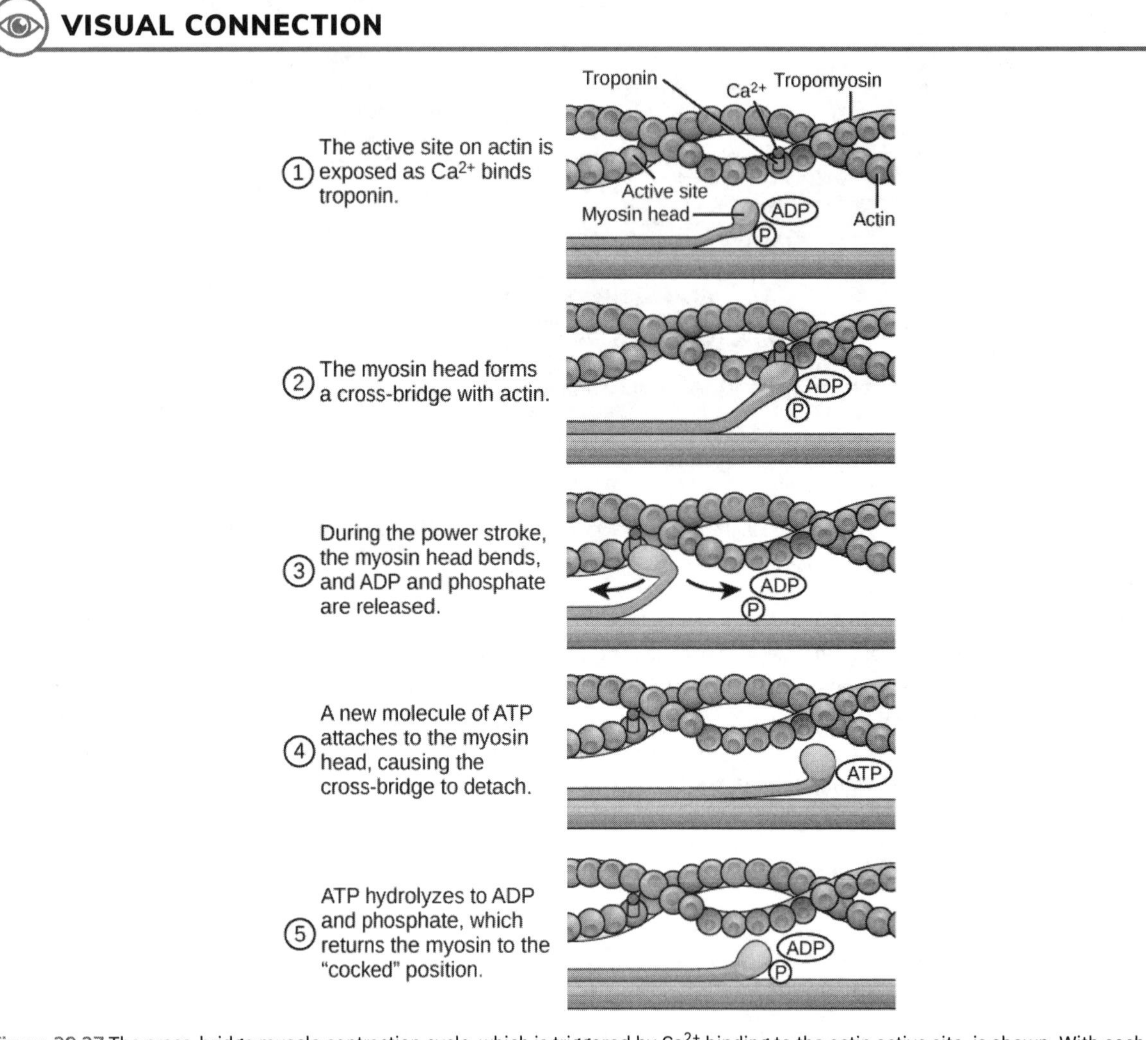

Figure 38.37 The cross-bridge muscle contraction cycle, which is triggered by Ca^{2+} binding to the actin active site, is shown. With each contraction cycle, actin moves relative to myosin.

Which of the following statements about muscle contraction is true?

a. The power stroke occurs when ATP is hydrolyzed to ADP and phosphate.
b. The power stroke occurs when ADP and phosphate dissociate from the myosin head.

c. The power stroke occurs when ADP and phosphate dissociate from the actin active site.
d. The power stroke occurs when Ca^{2+} binds the calcium head.

LINK TO LEARNING

View this animation (http://openstax.org/l/muscle_contract) of the cross-bridge muscle contraction.

Regulatory Proteins

When a muscle is in a resting state, actin and myosin are separated. To keep actin from binding to the active site on myosin, regulatory proteins block the molecular binding sites. **Tropomyosin** blocks myosin binding sites on actin molecules, preventing cross-bridge formation and preventing contraction in a muscle without nervous input. **Troponin** binds to tropomyosin and helps to position it on the actin molecule; it also binds calcium ions.

To enable a muscle contraction, tropomyosin must change conformation, uncovering the myosin-binding site on an actin molecule and allowing cross-bridge formation. This can only happen in the presence of calcium, which is kept at extremely low concentrations in the sarcoplasm. If present, calcium ions bind to troponin, causing conformational changes in troponin that allow tropomyosin to move away from the myosin binding sites on actin. Once the tropomyosin is removed, a cross-bridge can form between actin and myosin, triggering contraction. Cross-bridge cycling continues until Ca^{2+} ions and ATP are no longer available and tropomyosin again covers the binding sites on actin.

Excitation–Contraction Coupling

Excitation–contraction coupling is the link (transduction) between the action potential generated in the sarcolemma and the start of a muscle contraction. The trigger for calcium release from the sarcoplasmic reticulum into the sarcoplasm is a neural signal. Each skeletal muscle fiber is controlled by a motor neuron, which conducts signals from the brain or spinal cord to the muscle. The area of the sarcolemma on the muscle fiber that interacts with the neuron is called the **motor end plate**. The end of the neuron's axon is called the synaptic terminal, and it does not actually contact the motor end plate. A small space called the synaptic cleft separates the synaptic terminal from the motor end plate. Electrical signals travel along the neuron's axon, which branches through the muscle and connects to individual muscle fibers at a neuromuscular junction.

The ability of cells to communicate electrically requires that the cells expend energy to create an electrical gradient across their cell membranes. This charge gradient is carried by ions, which are differentially distributed across the membrane. Each ion exerts an electrical influence and a concentration influence. Just as milk will eventually mix with coffee without the need to stir, ions also distribute themselves evenly, if they are permitted to do so. In this case, they are not permitted to return to an evenly mixed state.

The sodium–potassium ATPase uses cellular energy to move K^+ ions inside the cell and Na^+ ions outside. This alone accumulates a small electrical charge, but a big concentration gradient. There is lots of K^+ in the cell and lots of Na^+ outside the cell. Potassium is able to leave the cell through K^+ channels that are open 90% of the time, and it does. However, Na^+ channels are rarely open, so Na^+ remains outside the cell. When K^+ leaves the cell, obeying its concentration gradient, that effectively leaves a negative charge behind. So at rest, there is a large concentration gradient for Na^+ to enter the cell, and there is an accumulation of negative charges left behind in the cell. This is the resting membrane potential. Potential in this context means a separation of electrical charge that is capable of doing work. It is measured in volts, just like a battery. However, the transmembrane potential is considerably smaller (0.07 V); therefore, the small value is expressed as millivolts (mV) or 70 mV. Because the inside of a cell is negative compared with the outside, a minus sign signifies the excess of negative charges inside the cell, –70 mV.

If an event changes the permeability of the membrane to Na^+ ions, they will enter the cell. That will change the voltage. This is an electrical event, called an action potential, that can be used as a cellular signal. Communication occurs between nerves and muscles through neurotransmitters. Neuron action potentials cause the release of neurotransmitters from the synaptic terminal into the synaptic cleft, where they can then diffuse across the synaptic cleft and bind to a receptor molecule on the motor end plate. The motor end plate possesses junctional folds—folds in the sarcolemma that create a large surface area for the neurotransmitter to bind to receptors. The receptors are actually sodium channels that open to allow the passage of Na^+ into the cell when they receive a neurotransmitter signal.

Acetylcholine (ACh) is a neurotransmitter released by motor neurons that binds to receptors in the motor end plate.

Neurotransmitter release occurs when an action potential travels down the motor neuron's axon, resulting in altered permeability of the synaptic terminal membrane and an influx of calcium. The Ca^{2+} ions allow synaptic vesicles to move to and bind with the presynaptic membrane (on the neuron), and release neurotransmitter from the vesicles into the synaptic cleft. Once released by the synaptic terminal, ACh diffuses across the synaptic cleft to the motor end plate, where it binds with ACh receptors. As a neurotransmitter binds, these ion channels open, and Na^+ ions cross the membrane into the muscle cell. This reduces the voltage difference between the inside and outside of the cell, which is called depolarization. As ACh binds at the motor end plate, this depolarization is called an end-plate potential. The depolarization then spreads along the sarcolemma, creating an action potential as sodium channels adjacent to the initial depolarization site sense the change in voltage and open. The action potential moves across the entire cell, creating a wave of depolarization.

ACh is broken down by the enzyme **acetylcholinesterase** (AChE) into acetyl and choline. AChE resides in the synaptic cleft, breaking down ACh so that it does not remain bound to ACh receptors, which would cause unwanted extended muscle contraction (Figure 38.38).

VISUAL CONNECTION

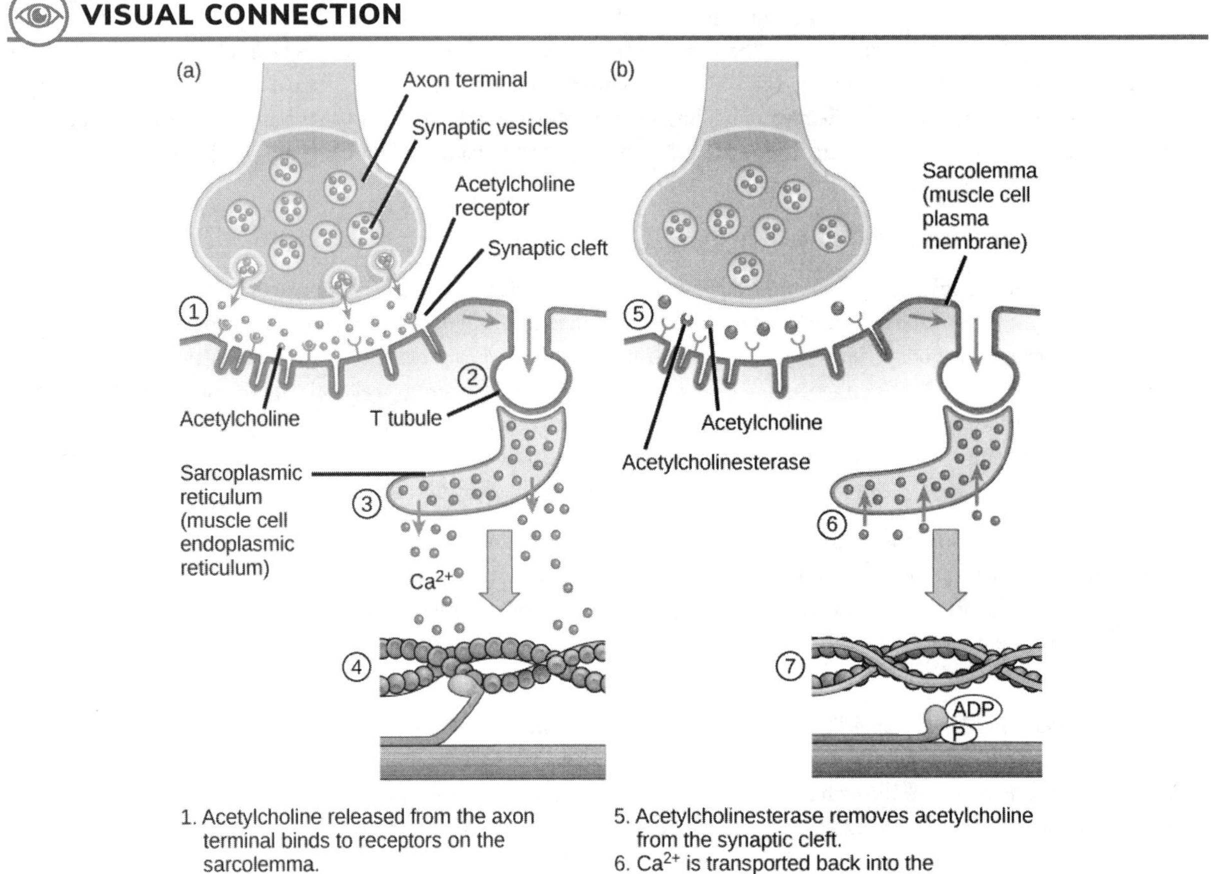

Figure 38.38 This diagram shows excitation-contraction coupling in a skeletal muscle contraction. The sarcoplasmic reticulum is a specialized endoplasmic reticulum found in muscle cells.

The deadly nerve gas Sarin irreversibly inhibits Acetylcholinesterase. What effect would Sarin have on muscle contraction?

After depolarization, the membrane returns to its resting state. This is called repolarization, during which voltage-gated sodium channels close. Potassium channels continue at 90% conductance. Because the plasma membrane sodium–potassium ATPase always transports ions, the resting state (negatively charged inside relative to the outside) is restored. The period immediately following the transmission of an impulse in a nerve or muscle, in which a neuron or muscle cell regains its ability to transmit another impulse, is called the refractory period. During the refractory period, the membrane cannot generate another action potential. The refractory period allows the voltage-sensitive ion channels to return to their resting configurations. The sodium potassium ATPase continually moves Na^+ back out of the cell and K^+ back into the cell, and the K^+ leaks out leaving negative charge behind. Very quickly, the membrane repolarizes, so that it can again be depolarized.

Control of Muscle Tension

Neural control initiates the formation of actin–myosin cross-bridges, leading to the sarcomere shortening involved in muscle contraction. These contractions extend from the muscle fiber through connective tissue to pull on bones, causing skeletal movement. The pull exerted by a muscle is called tension, and the amount of force created by this tension can vary. This enables the same muscles to move very light objects and very heavy objects. In individual muscle fibers, the amount of tension produced depends on the cross-sectional area of the muscle fiber and the frequency of neural stimulation.

The number of cross-bridges formed between actin and myosin determine the amount of tension that a muscle fiber can produce. Cross-bridges can only form where thick and thin filaments overlap, allowing myosin to bind to actin. If more cross-bridges are formed, more myosin will pull on actin, and more tension will be produced.

The ideal length of a sarcomere during production of maximal tension occurs when thick and thin filaments overlap to the greatest degree. If a sarcomere at rest is stretched past an ideal resting length, thick and thin filaments do not overlap to the greatest degree, and fewer cross-bridges can form. This results in fewer myosin heads pulling on actin, and less tension is produced. As a sarcomere is shortened, the zone of overlap is reduced as the thin filaments reach the H zone, which is composed of myosin tails. Because it is myosin heads that form cross-bridges, actin will not bind to myosin in this zone, reducing the tension produced by this myofiber. If the sarcomere is shortened even more, thin filaments begin to overlap with each other—reducing cross-bridge formation even further, and producing even less tension. Conversely, if the sarcomere is stretched to the point at which thick and thin filaments do not overlap at all, no cross-bridges are formed and no tension is produced. This amount of stretching does not usually occur because accessory proteins, internal sensory nerves, and connective tissue oppose extreme stretching.

The primary variable determining force production is the number of myofibers within the muscle that receive an action potential from the neuron that controls that fiber. When using the biceps to pick up a pencil, the motor cortex of the brain only signals a few neurons of the biceps, and only a few myofibers respond. In vertebrates, each myofiber responds fully if stimulated. When picking up a piano, the motor cortex signals all of the neurons in the biceps and every myofiber participates. This is close to the maximum force the muscle can produce. As mentioned above, increasing the frequency of action potentials (the number of signals per second) can increase the force a bit more, because the tropomyosin is flooded with calcium.

KEY TERMS

abduction when a bone moves away from the midline of the body

acetylcholinesterase (AChE) enzyme that breaks down ACh into acetyl and choline

actin globular contractile protein that interacts with myosin for muscle contraction

adduction movement of the limbs inward after abduction

amphiarthrosis joint that allows slight movement; includes syndesmoses and symphyses

angular movement produced when the angle between the bones of a joint changes

appendicular skeleton composed of the bones of the upper limbs, which function to grasp and manipulate objects, and the lower limbs, which permit locomotion

appositional growth increase in the diameter of bones by the addition of bone tissue at the surface of bones

articulation any place where two bones are joined

auditory ossicle (also, middle ear) transduces sounds from the air into vibrations in the fluid-filled cochlea

axial skeleton forms the central axis of the body and includes the bones of the skull, the ossicles of the middle ear, the hyoid bone of the throat, the vertebral column, and the thoracic cage (ribcage)

ball-and-socket joint joint with a rounded, ball-like end of one bone fitting into a cuplike socket of another bone

bone (also, osseous tissue) connective tissue that constitutes the endoskeleton

bone remodeling replacement of old bone tissue by new bone tissue

calcification process of deposition of mineral salts in the collagen fiber matrix that crystallizes and hardens the tissue

cardiac muscle tissue muscle tissue found only in the heart; cardiac contractions pump blood throughout the body and maintain blood pressure

carpus eight bones that comprise the wrist

cartilaginous joint joint in which the bones are connected by cartilage

circumduction movement of a limb in a circular motion

clavicle S-shaped bone that positions the arms laterally

compact bone forms the hard external layer of all bones

condyloid joint oval-shaped end of one bone fitting into a similarly oval-shaped hollow of another bone

coxal bone hip bone

cranial bone one of eight bones that form the cranial cavity that encloses the brain and serves as an attachment site for the muscles of the head and neck

depression movement downward of a bone, such as after the shoulders are shrugged and the scapulae return to their normal position from an elevated position; opposite of elevation

diaphysis central shaft of bone, contains bone marrow in a marrow cavity

diarthrosis joint that allows for free movement of the joint; found in synovial joints

dorsiflexion bending at the ankle such that the toes are lifted toward the knee

elevation movement of a bone upward, such as when the shoulders are shrugged, lifting the scapulae

endochondral ossification process of bone development from hyaline cartilage

endoskeleton skeleton of living cells that produces a hard, mineralized tissue located within the soft tissue of organisms

epiphyseal plate region between the diaphysis and epiphysis that is responsible for the lengthwise growth of long bones

epiphysis rounded end of bone, covered with articular cartilage and filled with red bone marrow, which produces blood cells

eversion movement of the sole of the foot outward, away from the midline of the body; opposite of inversion

exoskeleton a secreted cellular product external skeleton that consists of a hard encasement on the surface of an organism

extension movement in which the angle between the bones of a joint increases; opposite of flexion

facial bone one of the 14 bones that form the face; provides cavities for the sense organs (eyes, mouth, and nose) and attachment points for facial muscles

femur (also, thighbone) longest, heaviest, and strongest bone in the body

fibrous joint joint held together by fibrous connective tissue

fibula (also, calf bone) parallels and articulates with the tibia

flat bone thin and relatively broad bone found where extensive protection of organs is required or where broad surfaces of muscle attachment are required

flexion movement in which the angle between the bones decreases; opposite of extension

forearm extends from the elbow to the wrist and consists of two bones: the ulna and the radius

gliding movement when relatively flat bone surfaces move past each other

gomphosis the joint in which the tooth fits into the socket like a peg

Haversian canal contains the bone's blood vessels and nerve fibers

hinge joint slightly rounded end of one bone fits into the slightly hollow end of the other bone

humerus only bone of the arm

hydrostatic skeleton skeleton that consists of aqueous fluid held under pressure in a closed body compartment

hyoid bone lies below the mandible in the front of the neck
hyperextension extension past the regular anatomical position
intervertebral disc composed of fibrous cartilage; lies between adjacent vertebrae from the second cervical vertebra to the sacrum
intramembranous ossification process of bone development from fibrous membranes
inversion soles of the feet moving inward, toward the midline of the body
irregular bone bone with complex shapes; examples include vertebrae and hip bones
joint point at which two or more bones meet
lamella layer of compact tissue that surrounds a central canal called the Haversian canal
lateral rotation rotation away from the midline of the body
long bone bone that is longer than wide, and has a shaft and two ends
lower limb consists of the thigh, the leg, and the foot
medial rotation rotation toward the midline of the body
metacarpus five bones that comprise the palm
metatarsal one of the five bones of the foot
motor end plate sarcolemma of the muscle fiber that interacts with the neuron
myofibril long cylindrical structures that lie parallel to the muscle fiber
myofilament small structures that make up myofibrils
myosin contractile protein that interacts with actin for muscle contraction
opposition movement of the thumb toward the fingers of the same hand, making it possible to grasp and hold objects
osseous tissue connective tissue that constitutes the endoskeleton
ossification (also, osteogenesis) process of bone formation by osteoblasts
osteoblast bone cell responsible for bone formation
osteoclast large bone cells with up to 50 nuclei, responsible for bone remodeling
osteocyte mature bone cells and the main cell in bone tissue
osteon cylindrical structure aligned parallel to the long axis of the bone
patella (also, kneecap) triangular bone that lies anterior to the knee joint
pectoral girdle bones that transmit the force generated by the upper limbs to the axial skeleton
pelvic girdle bones that transmit the force generated by the lower limbs to the axial skeleton
phalange one of the bones of the fingers or toes
pivot joint joint with the rounded end of one bone fitting into a ring formed by the other bone
planar joint joint with bones whose articulating surfaces are flat
plantar flexion bending at the ankle such that the heel is lifted, such as when standing on the toes
pronation movement in which the palm faces backward
protraction anterior movement of a bone in the horizontal plane
radius bone located along the lateral (thumb) side of the forearm; articulates with the humerus at the elbow
resorption process by which osteoclasts release minerals stored in bones
retraction movement in which a joint moves back into position after protraction
rib one of 12 pairs of long, curved bones that attach to the thoracic vertebrae and curve toward the front of the body to form the ribcage
rotational movement movement of a bone as it rotates around its own longitudinal axis
saddle joint joint with concave and convex portions that fit together; named because the ends of each bone resemble a saddle
sarcolemma plasma membrane of a skeletal muscle fiber
sarcomere functional unit of skeletal muscle
scapula flat, triangular bone located at the posterior pectoral girdle
sesamoid bone small, flat bone shaped like a sesame seed; develops inside tendons
short bone bone that has the same width and length, giving it a cube-like shape
skeletal muscle tissue forms skeletal muscles, which attach to bones and control locomotion and any movement that can be consciously controlled
skull bone that supports the structures of the face and protects the brain
smooth muscle tissue occurs in the walls of hollow organs such as the intestines, stomach, and urinary bladder, and around passages such as the respiratory tract and blood vessels
spongy bone tissue forms the inner layer of all bones
sternum (also, breastbone) long, flat bone located at the front of the chest
supination movement of the radius and ulna bones of the forearm so that the palm faces forward
sutural bone small, flat, irregularly shaped bone that forms between the flat bones of the cranium
suture short fiber of connective tissue that holds the skull bones tightly in place; found only in the skull
symphysis hyaline cartilage covers the end of the bone, but the connection between bones occurs through fibrocartilage; symphyses are found at the joints between vertebrae
synarthrosis joint that is immovable
synchondrosis bones joined by hyaline cartilage; synchondroses are found in the epiphyseal plates of

growing bones in children

syndesmosis joint in which the bones are connected by a band of connective tissue, allowing for more movement than in a suture

synovial joint only joint that has a space between the adjoining bones

tarsal one of the seven bones of the ankle

thick filament a group of myosin molecules

thin filament two polymers of actin wound together along with tropomyosin and troponin

thoracic cage (also, ribcage) skeleton of the chest, which consists of the ribs, thoracic vertebrae, sternum, and costal cartilages

tibia (also, shinbone) large bone of the leg that is located directly below the knee

trabeculae lamellae that are arranged as rods or plates

tropomyosin acts to block myosin binding sites on actin molecules, preventing cross-bridge formation and preventing contraction until a muscle receives a neuron signal

troponin binds to tropomyosin and helps to position it on the actin molecule, and also binds calcium ions

ulna bone located on the medial aspect (pinky-finger side) of the forearm

vertebral column (also, spine) surrounds and protects the spinal cord, supports the head, and acts as an attachment point for ribs and muscles of the back and neck

CHAPTER SUMMARY

38.1 Types of Skeletal Systems

The three types of skeleton designs are hydrostatic skeletons, exoskeletons, and endoskeletons. A hydrostatic skeleton is formed by a fluid-filled compartment held under hydrostatic pressure; movement is created by the muscles producing pressure on the fluid. An exoskeleton is a hard external skeleton that protects the outer surface of an organism and enables movement through muscles attached on the inside. An endoskeleton is an internal skeleton composed of hard, mineralized tissue that also enables movement by attachment to muscles. The human skeleton is an endoskeleton that is composed of the axial and appendicular skeleton. The axial skeleton is composed of the bones of the skull, ossicles of the ear, hyoid bone, vertebral column, and ribcage. The skull consists of eight cranial bones and 14 facial bones. Six bones make up the ossicles of the middle ear, while the hyoid bone is located in the neck under the mandible. The vertebral column contains 26 bones, and it surrounds and protects the spinal cord. The thoracic cage consists of the sternum, ribs, thoracic vertebrae, and costal cartilages. The appendicular skeleton is made up of the limbs of the upper and lower limbs. The pectoral girdle is composed of the clavicles and the scapulae. The upper limb contains 30 bones in the arm, the forearm, and the hand. The pelvic girdle attaches the lower limbs to the axial skeleton. The lower limb includes the bones of the thigh, the leg, and the foot.

38.2 Bone

Bone, or osseous tissue, is connective tissue that includes specialized cells, mineral salts, and collagen fibers. The human skeleton can be divided into long bones, short bones, flat bones, and irregular bones. Compact bone tissue is composed of osteons and forms the external layer of all bones. Spongy bone tissue is composed of trabeculae and forms the inner part of all bones. Four types of cells compose bony tissue: osteocytes, osteoclasts, osteoprogenitor cells, and osteoblasts. Ossification is the process of bone formation by osteoblasts. Intramembranous ossification is the process of bone development from fibrous membranes. Endochondral ossification is the process of bone development from hyaline cartilage. Long bones lengthen as chondrocytes divide and secrete hyaline cartilage. Osteoblasts replace cartilage with bone. Appositional growth is the increase in the diameter of bones by the addition of bone tissue at the surface of bones. Bone remodeling involves the processes of bone deposition by osteoblasts and bone resorption by osteoclasts. Bone repair occurs in four stages and can take several months.

38.3 Joints and Skeletal Movement

The structural classification of joints divides them into bony, fibrous, cartilaginous, and synovial joints. The bones of fibrous joints are held together by fibrous connective tissue; the three types of fibrous joints are sutures, syndesomes, and gomphoses. Cartilaginous joints are joints in which the bones are connected by cartilage; the two types of cartilaginous joints are synchondroses and symphyses. Synovial joints are joints that have a space between the adjoining bones. The functional classification divides joints into three categories: synarthroses, amphiarthroses, and diarthroses. The movement of synovial joints can be classified as one of four different types: gliding, angular, rotational, or special movement. Gliding movements occur as relatively flat bone surfaces move past each other. Angular movements are produced when the angle between the bones of a joint changes. Rotational movement is the movement of a bone as it rotates around its own longitudinal axis. Special movements include inversion, eversion, protraction, retraction, elevation, depression, dorsiflexion, plantar flexion, supination, pronation, and opposition. Synovial joints are also classified into six different categories on the basis of the shape and structure of the joint: planar, hinge, pivot, condyloid, saddle, and ball-and-socket.

38.4 Muscle Contraction and Locomotion

The body contains three types of muscle tissue: skeletal muscle, cardiac muscle, and smooth muscle. Skeleton muscle tissue is composed of sarcomeres, the functional units of muscle tissue. Muscle contraction occurs when sarcomeres shorten, as thick and thin filaments slide past each other, which is called the sliding filament model of muscle contraction. ATP provides the energy for cross-bridge formation and filament sliding. Regulatory proteins, such as troponin and tropomyosin, control cross-bridge formation. Excitation–contraction coupling transduces the electrical signal of the neuron, via acetylcholine, to an electrical signal on the muscle membrane, which initiates force production. The number of muscle fibers contracting determines how much force the whole muscle produces.

VISUAL CONNECTION QUESTIONS

1. Figure 38.19 Which of the following statements about bone tissue is false?
   a. Compact bone tissue is made of cylindrical osteons that are aligned such that they travel the length of the bone.
   b. Haversian canals contain blood vessels only.
   c. Haversian canals contain blood vessels and nerve fibers.
   d. Spongy tissue is found on the interior of the bone, and compact bone tissue is found on the exterior.

2. Figure 38.37 Which of the following statements about muscle contraction is true?
   a. The power stroke occurs when ATP is hydrolyzed to ADP and phosphate.
   b. The power stroke occurs when ADP and phosphate dissociate from the myosin head.
   c. The power stroke occurs when ADP and phosphate dissociate from the actin active site.
   d. The power stroke occurs when Ca^{2+} binds the calcium head.

3. Figure 38.38 The deadly nerve gas Sarin irreversibly inhibits Acetylcholinesterase. What effect would Sarin have on muscle contraction?

REVIEW QUESTIONS

4. The forearm consists of the:
   a. radius and ulna
   b. radius and humerus
   c. ulna and humerus
   d. humerus and carpus

5. The pectoral girdle consists of the:
   a. clavicle and sternum
   b. sternum and scapula
   c. clavicle and scapula
   d. clavicle and coccyx

6. All of the following are groups of vertebrae except ________, which is a curvature.
   a. thoracic
   b. cervical
   c. lumbar
   d. pelvic

7. Which of these is a facial bone?
   a. frontal
   b. occipital
   c. lacrimal
   d. temporal

8. Which of the following is **not** a true statement comparing exoskeletons and endoskeletons?
   a. Endoskeletons can support larger organisms.
   b. Only endoskeletons can grow as an organism grows.
   c. Exoskeletons provide greater protection of the internal organs.
   d. Exoskeletons provide less mechanical leverage.

9. The Haversian canal:
   a. is arranged as rods or plates
   b. contains the bone's blood vessels and nerve fibers
   c. is responsible for the lengthwise growth of long bones
   d. synthesizes and secretes matrix

10. The epiphyseal plate:
    a. is arranged as rods or plates
    b. contains the bone's blood vessels and nerve fibers
    c. is responsible for the lengthwise growth of long bones
    d. synthesizes and secretes bone matrix

11. The cells responsible for bone resorption are _______.
    a. osteoclasts
    b. osteoblasts
    c. fibroblasts
    d. osteocytes

12. Compact bone is composed of _______.
    a. trabeculae
    b. compacted collagen
    c. osteons
    d. calcium phosphate only

13. Osteoporosis is a condition where bones become weak and brittle. It is caused by an imbalance in the activity of which cells?
    a. osteoclasts and osteoblasts
    b. osteoclasts and osteocytes
    c. osteoblasts and chondrocytes
    d. osteocytes and chondrocytes

14. While assembling a skeleton of a new species, a scientist points to one of the bones and observes that it looks like the most likely site of leg muscle attachment. What kind of bone did she indicate?
    a. sesamoid bone
    b. long bone
    c. trabecular bone
    d. flat bone

15. Synchondroses and symphyses are:
    a. synovial joints
    b. cartilaginous joints
    c. fibrous joints
    d. condyloid joints

16. The movement of bone away from the midline of the body is called _______.
    a. circumduction
    b. extension
    c. adduction
    d. abduction

17. Which of the following is not a characteristic of the synovial fluid?
    a. lubrication
    b. shock absorption
    c. regulation of water balance in the joint
    d. protection of articular cartilage

18. The elbow is an example of which type of joint?
    a. hinge
    b. pivot
    c. saddle
    d. gliding

19. A high ankle sprain is an injury caused by over-stretching the ligaments connecting the tibia and fibula. What type of joint is involved in this sprain?
    a. ball and socket
    b. gomphosis
    c. syndesmosis
    d. symphysis

20. In relaxed muscle, the myosin-binding site on actin is blocked by _______.
    a. titin
    b. troponin
    c. myoglobin
    d. tropomyosin

21. The cell membrane of a muscle fiber is called a _______.
    a. myofibril
    b. sarcolemma
    c. sarcoplasm
    d. myofilament

22. The muscle relaxes if no new nerve signal arrives. However the neurotransmitter from the previous stimulation is still present in the synapse. The activity of _______ helps to remove this neurotransmitter.
    a. myosin
    b. action potential
    c. tropomyosin
    d. acetylcholinesterase

23. The ability of a muscle to generate tension immediately after stimulation is dependent on:
    a. myosin interaction with the M line
    b. overlap of myosin and actin
    c. actin attachments to the Z line
    d. none of the above

24. Botulinum toxin causes flaccid paralysis of the muscles, and is used for cosmetic purposes under the name Botox. Which of the following is the most likely mechanism of action of Botox?
    a. Botox decreases the production of acetylcholinesterase.
    b. Botox increases calcium release from the sarcoplasmic reticulum.
    c. Botox blocks the ATP binding site in actin.
    d. Botox decreases the release of acetylcholine from motor neurons.

CRITICAL THINKING QUESTIONS

25. What are the major differences between the male pelvis and female pelvis that permit childbirth in females?
26. What are the major differences between the pelvic girdle and the pectoral girdle that allow the pelvic girdle to bear the weight of the body?
27. Both hydrostatic and exoskeletons can protect internal organs from harm. Contrast the ways the skeletons perform these functions.
28. Scoliosis is a medical condition where the spine develops a sideways curvature. How would this change interfere with the normal function of the spine?
29. What are the major differences between spongy bone and compact bone?
30. What are the roles of osteoblasts, osteocytes, and osteoclasts?
31. Thalidomide was a morning sickness drug given to women that caused babies to be born without arm bones. If recent studies have shown that thalidomide prevents the formation of new blood vessels, describe the type of bone development inhibited by the drug and what stage of ossification was affected.
32. What movements occur at the hip joint and knees as you bend down to touch your toes?
33. What movement(s) occur(s) at the scapulae when you shrug your shoulders?
34. Describe the joints and motions involved in taking a step forward if a person is initially standing still. Assume the person holds his foot at the same angle throughout the motion.
35. How would muscle contractions be affected if ATP was completely depleted in a muscle fiber?
36. What factors contribute to the amount of tension produced in an individual muscle fiber?
37. What effect will low blood calcium have on neurons? What effect will low blood calcium have on skeletal muscles?
38. Skeletal muscles can only produce a mechanical force as they are contracted, but a leg flexes and extends while walking. How can muscles perform this task?

CHAPTER 39
The Respiratory System

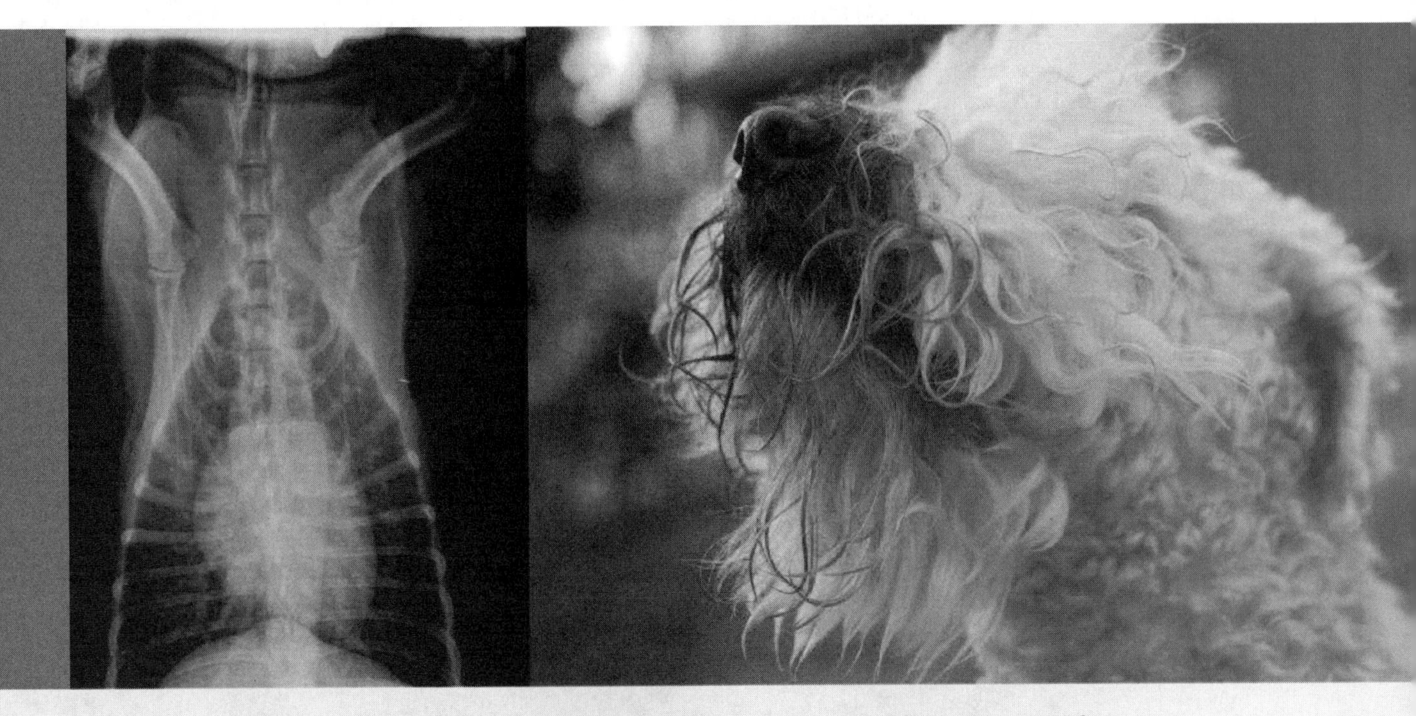

Figure 39.1 Lungs, which appear as nearly transparent tissue surrounding the heart in this X-ray of a dog (left), are the central organs of the respiratory system. The left lung is smaller than the right lung to accommodate space for the heart. A dog's nose (right) has a slit on the side of each nostril. When tracking a scent, the slits open, blocking the front of the nostrils. This allows the dog to exhale though the now-open area on the side of the nostrils without losing the scent that is being followed. (credit a: modification of work by Geoff Stearns; credit b: modification of work by Cory Zanker)

Chapter Outline

INTRODUCTION Breathing is an involuntary event. How often a breath is taken and how much air is inhaled or exhaled are tightly regulated by the respiratory center in the brain. Humans, when they aren't exerting themselves, breathe approximately 15 times per minute on average. Canines, like the dog in Figure 39.1, have a respiratory rate of about 15–30 breaths per minute. With every inhalation, air fills the lungs, and with every exhalation, air rushes back out. That air is doing more than just inflating and deflating the lungs in the chest cavity. The air contains oxygen that crosses the lung tissue, enters the bloodstream, and travels to organs and tissues. Oxygen (O_2) enters the cells where it is used for metabolic reactions that produce ATP, a high-energy compound. At the same time, these reactions release carbon dioxide (CO_2) as a by-product. CO_2 is toxic and must be eliminated. Carbon dioxide exits the cells, enters the bloodstream, travels back to the lungs, and is expired out of the body during exhalation.

39.1 Systems of Gas Exchange

By the end of this section, you will be able to do the following:

- Describe the passage of air from the outside environment to the lungs
- Explain how the lungs are protected from particulate matter

The primary function of the respiratory system is to deliver oxygen to the cells of the body's tissues and remove carbon dioxide, a cell waste product. The main structures of the human respiratory system are the nasal cavity, the trachea, and lungs.

All aerobic organisms require oxygen to carry out their metabolic functions. Along the evolutionary tree, different organisms have devised different means of obtaining oxygen from the surrounding atmosphere. The environment in which the animal lives greatly determines how an animal respires. The complexity of the respiratory system is correlated with the size of the organism. As animal size increases, diffusion distances increase and the ratio of surface area to volume drops. In unicellular organisms, diffusion across the cell membrane is sufficient for supplying oxygen to the cell (Figure 39.2). Diffusion is a slow, passive transport process. In order for diffusion to be a feasible means of providing oxygen to the cell, the rate of oxygen uptake must match the rate of diffusion across the membrane. In other words, if the cell were very large or thick, diffusion would not be able to provide oxygen quickly enough to the inside of the cell. Therefore, dependence on diffusion as a means of obtaining oxygen and removing carbon dioxide remains feasible only for small organisms or those with highly-flattened bodies, such as many flatworms (Platyhelminthes). Larger organisms had to evolve specialized respiratory tissues, such as gills, lungs, and respiratory passages accompanied by complex circulatory systems, to transport oxygen throughout their entire body.

Figure 39.2 The cell of the unicellular alga *Ventricaria ventricosa* is one of the largest known, reaching one to five centimeters in diameter. Like all single-celled organisms, *V. ventricosa* exchanges gases across the cell membrane.

Direct Diffusion

For small multicellular organisms, diffusion across the outer membrane is sufficient to meet their oxygen needs. Gas exchange by direct diffusion across surface membranes is efficient for organisms less than 1 mm in diameter. In simple organisms, such as cnidarians and flatworms, every cell in the body is close to the external environment. Their cells are kept moist and gases diffuse quickly via direct diffusion. Flatworms are small, literally flat worms, which 'breathe' through diffusion across the outer membrane (Figure 39.3). The flat shape of these organisms increases the surface area for diffusion, ensuring that each cell within the body is close to the outer membrane surface and has access to oxygen. If the flatworm had a cylindrical body, then the cells in the center would not be able to get oxygen.

Figure 39.3 This flatworm's process of respiration works by diffusion across the outer membrane. (credit: Stephen Childs)

Skin and Gills

Earthworms and amphibians use their skin (integument) as a respiratory organ. A dense network of capillaries lies just below the skin and facilitates gas exchange between the external environment and the circulatory system. The respiratory surface must be kept moist in order for the gases to dissolve and diffuse across cell membranes.

Organisms that live in water need to obtain oxygen from the water. Oxygen dissolves in water but at a lower concentration than in the atmosphere. The atmosphere has roughly 21 percent oxygen. In water, the oxygen concentration is much lower than that. Fish and many other aquatic organisms have evolved gills to take up the dissolved oxygen from water (Figure 39.4). Gills are thin tissue filaments that are highly branched and folded. When water passes over the gills, the dissolved oxygen in water rapidly diffuses across the gills into the bloodstream. The circulatory system can then carry the oxygenated blood to the other parts of the body. In animals that contain coelomic fluid instead of blood, oxygen diffuses across the gill surfaces into the coelomic fluid. Gills are found in mollusks, annelids, and crustaceans.

Figure 39.4 This common carp, like many other aquatic organisms, has gills that allow it to obtain oxygen from water. (credit: "Guitardude012"/Wikimedia Commons)

The folded surfaces of the gills provide a large surface area to ensure that the fish gets sufficient oxygen. Diffusion is a process in which material travels from regions of high concentration to low concentration until equilibrium is reached. In this case, blood with a low concentration of oxygen molecules circulates through the gills. The concentration of oxygen molecules in water is higher than the concentration of oxygen molecules in gills. As a result, oxygen molecules diffuse from water (high concentration) to blood (low concentration), as shown in Figure 39.5. Similarly, carbon dioxide molecules in the blood diffuse from the blood (high concentration) to water (low concentration).

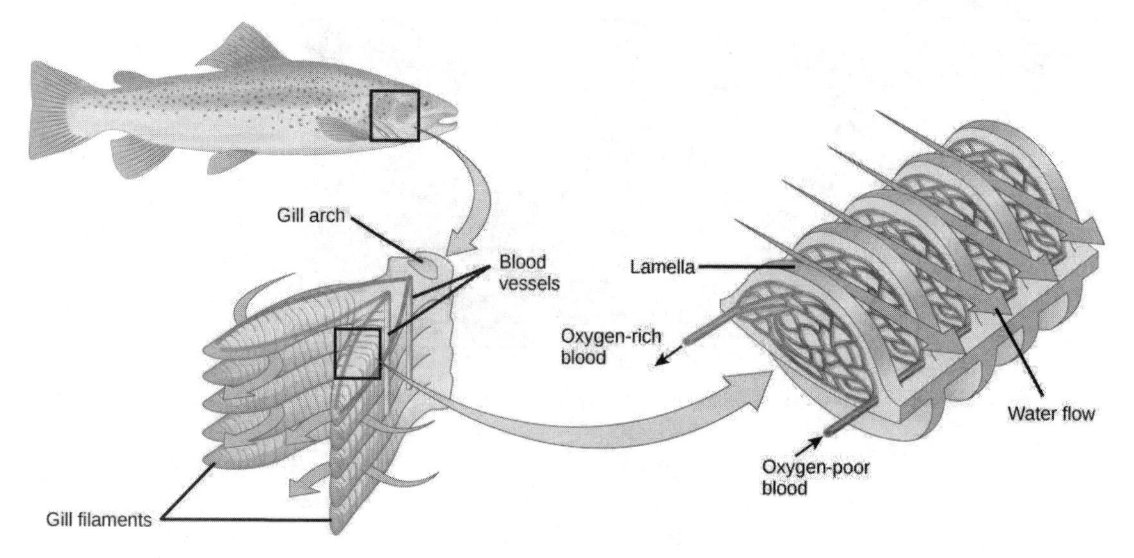

Figure 39.5 As water flows over the gills, oxygen is transferred to blood via the veins. (credit "fish": modification of work by Duane Raver, NOAA)

Tracheal Systems

Insect respiration is independent of its circulatory system; therefore, the blood does not play a direct role in oxygen transport. Insects have a highly specialized type of respiratory system called the tracheal system, which consists of a network of small tubes that carries oxygen to the entire body. The tracheal system is the most direct and efficient respiratory system in active animals. The tubes in the tracheal system are made of a polymeric material called chitin.

Insect bodies have openings, called spiracles, along the thorax and abdomen. These openings connect to the tubular network, allowing oxygen to pass into the body (Figure 39.6) and regulating the diffusion of CO_2 and water vapor. Air enters and leaves the tracheal system through the spiracles. Some insects can ventilate the tracheal system with body movements.

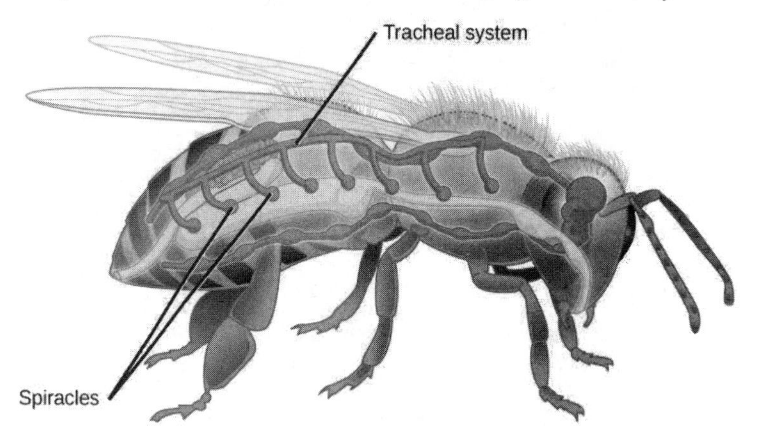

Figure 39.6 Insects perform respiration via a tracheal system.

Mammalian Systems

In mammals, pulmonary ventilation occurs via inhalation (breathing). During inhalation, air enters the body through the **nasal cavity** located just inside the nose (Figure 39.7). As air passes through the nasal cavity, the air is warmed to body temperature and humidified. The respiratory tract is coated with mucus to seal the tissues from direct contact with air. Mucus is high in water. As air crosses these surfaces of the mucous membranes, it picks up water. These processes help equilibrate the air to the body conditions, reducing any damage that cold, dry air can cause. Particulate matter that is floating in the air is removed in the nasal passages via mucus and cilia. The processes of warming, humidifying, and removing particles are important protective mechanisms that prevent damage to the trachea and lungs. Thus, inhalation serves several purposes in addition to bringing oxygen into the respiratory system.

VISUAL CONNECTION

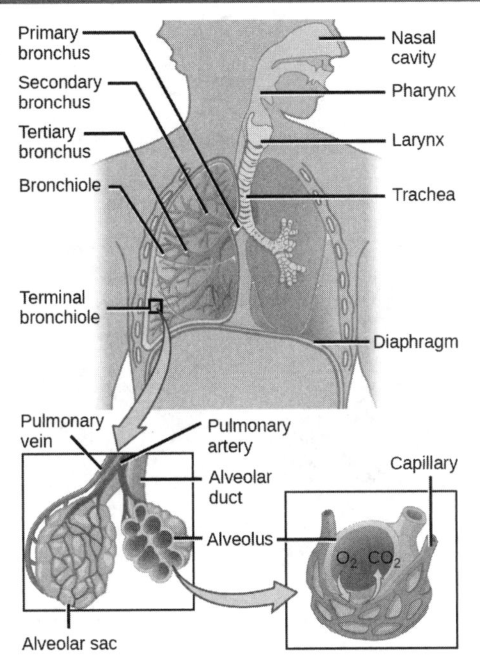

Figure 39.7 Air enters the respiratory system through the nasal cavity and pharynx, and then passes through the trachea and into the bronchi, which bring air into the lungs. (credit: modification of work by NCI)

Which of the following statements about the mammalian respiratory system is false?

a. When we breathe in, air travels from the pharynx to the trachea.
b. The bronchioles branch into bronchi.
c. Alveolar ducts connect to alveolar sacs.
d. Gas exchange between the lung and blood takes place in the alveolus.

From the nasal cavity, air passes through the **pharynx** (throat) and the **larynx** (voice box), as it makes its way to the **trachea** (Figure 39.7). The main function of the trachea is to funnel the inhaled air to the lungs and the exhaled air back out of the body. The human trachea is a cylinder about 10 to 12 cm long and 2 cm in diameter that sits in front of the esophagus and extends from the larynx into the chest cavity where it divides into the two primary bronchi at the midthorax. It is made of incomplete rings of hyaline cartilage and smooth muscle (Figure 39.8). The trachea is lined with mucus-producing goblet cells and ciliated epithelia. The cilia propel foreign particles trapped in the mucus toward the pharynx. The cartilage provides strength and support to the trachea to keep the passage open. The smooth muscle can contract, decreasing the trachea's diameter, which causes expired air to rush upwards from the lungs at a great force. The forced exhalation helps expel mucus when we cough. Smooth muscle can contract or relax, depending on stimuli from the external environment or the body's nervous system.

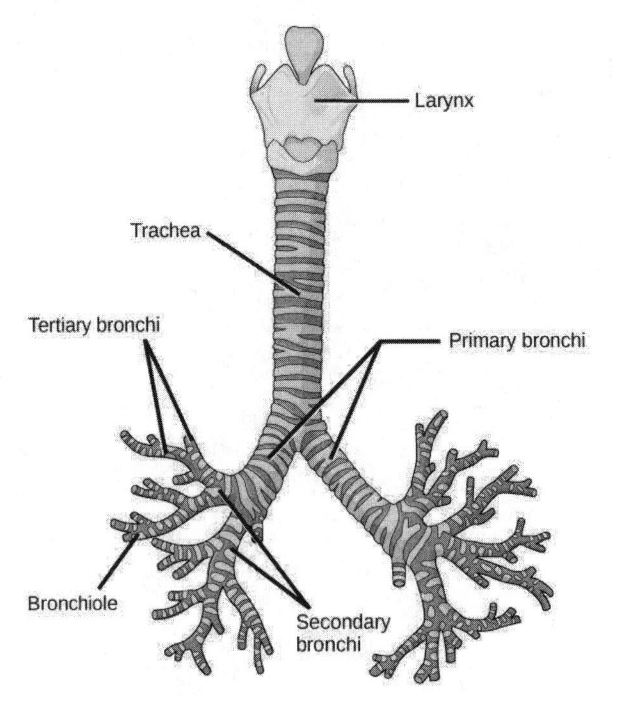

Figure 39.8 The trachea and bronchi are made of incomplete rings of cartilage. (credit: modification of work by Gray's Anatomy)

Lungs: Bronchi and Alveoli

The end of the trachea bifurcates (divides) to the right and left lungs. The lungs are not identical. The right lung is larger and contains three lobes, whereas the smaller left lung contains two lobes (Figure 39.9). The muscular **diaphragm**, which facilitates breathing, is inferior to (below) the lungs and marks the end of the thoracic cavity.

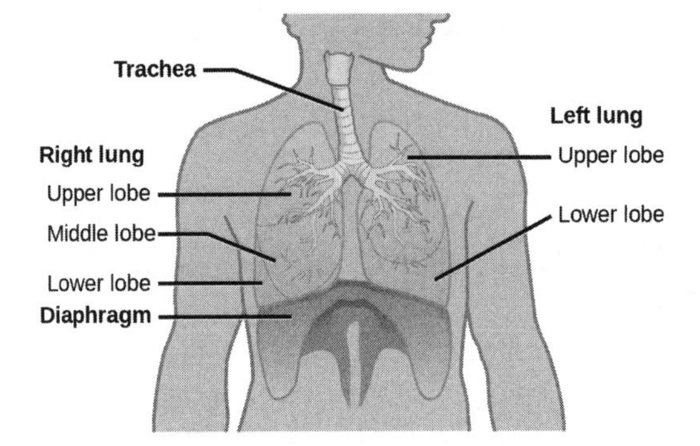

Figure 39.9 The trachea bifurcates into the right and left bronchi in the lungs. The right lung is made of three lobes and is larger. To accommodate the heart, the left lung is smaller and has only two lobes.

In the lungs, air is diverted into smaller and smaller passages, or **bronchi**. Air enters the lungs through the two **primary (main) bronchi** (singular: bronchus). Each bronchus divides into secondary bronchi, then into tertiary bronchi, which in turn divide, creating smaller and smaller diameter **bronchioles** as they split and spread through the lung. Like the trachea, the bronchi are made of cartilage and smooth muscle. At the bronchioles, the cartilage is replaced with elastic fibers. Bronchi are innervated by nerves of both the parasympathetic and sympathetic nervous systems that control muscle contraction (parasympathetic) or relaxation (sympathetic) in the bronchi and bronchioles, depending on the nervous system's cues. In humans, bronchioles with a diameter smaller than 0.5 mm are the **respiratory bronchioles**. They lack cartilage and therefore rely on inhaled air to support their shape. As the passageways decrease in diameter, the relative amount of smooth muscle increases.

The **terminal bronchioles** subdivide into microscopic branches called respiratory bronchioles. The respiratory bronchioles subdivide into several alveolar ducts. Numerous alveoli and alveolar sacs surround the alveolar ducts. The alveolar sacs resemble bunches of grapes tethered to the end of the bronchioles (Figure 39.10). In the acinar region, the **alveolar ducts** are attached to the end of each bronchiole. At the end of each duct are approximately 100 **alveolar sacs**, each containing 20 to 30 **alveoli** that are 200 to 300 microns in diameter. Gas exchange occurs only in alveoli. Alveoli are made of thin-walled parenchymal cells, typically one-cell thick, that look like tiny bubbles within the sacs. Alveoli are in direct contact with capillaries (one-cell thick) of the circulatory system. Such intimate contact ensures that oxygen will diffuse from alveoli into the blood and be distributed to the cells of the body. In addition, the carbon dioxide that was produced by cells as a waste product will diffuse from the blood into alveoli to be exhaled. The anatomical arrangement of capillaries and alveoli emphasizes the structural and functional relationship of the respiratory and circulatory systems. Because there are so many alveoli (~300 million per lung) within each alveolar sac and so many sacs at the end of each alveolar duct, the lungs have a sponge-like consistency. This organization produces a very large surface area that is available for gas exchange. The surface area of alveoli in the lungs is approximately 75 m^2. This large surface area, combined with the thin-walled nature of the alveolar parenchymal cells, allows gases to easily diffuse across the cells.

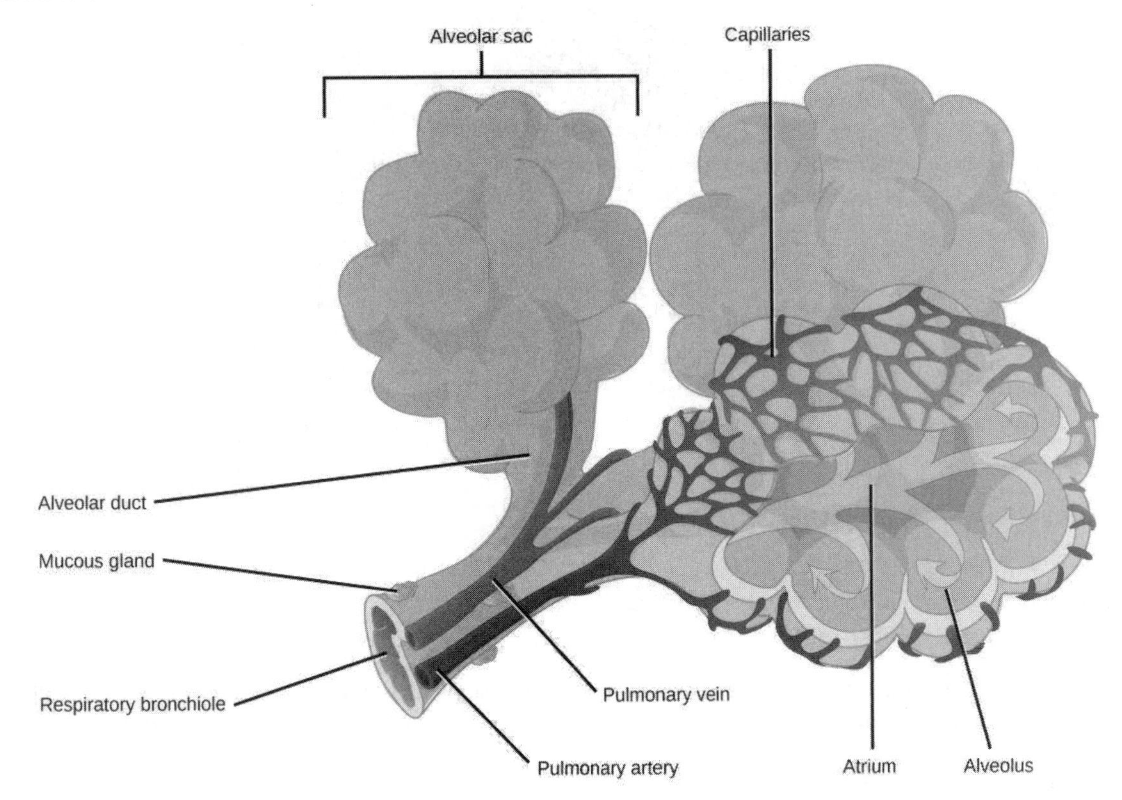

Figure 39.10 Terminal bronchioles are connected by respiratory bronchioles to alveolar ducts and alveolar sacs. Each alveolar sac contains 20 to 30 spherical alveoli and has the appearance of a bunch of grapes. Air flows into the atrium of the alveolar sac, then circulates into alveoli where gas exchange occurs with the capillaries. Mucous glands secrete mucous into the airways, keeping them moist and flexible. (credit: modification of work by Mariana Ruiz Villareal)

LINK TO LEARNING

Watch the following video to review the respiratory system.

Click to view content (https://www.openstax.org/l/lungs_pulmonary)

Protective Mechanisms

The air that organisms breathe contains **particulate matter** such as dust, dirt, viral particles, and bacteria that can damage the lungs or trigger allergic immune responses. The respiratory system contains several protective mechanisms to avoid problems or tissue damage. In the nasal cavity, hairs and mucus trap small particles, viruses, bacteria, dust, and dirt to prevent their entry.

If particulates do make it beyond the nose, or enter through the mouth, the bronchi and bronchioles of the lungs also contain several protective devices. The lungs produce **mucus**—a sticky substance made of **mucin**, a complex glycoprotein, as well as salts and water—that traps particulates. The bronchi and bronchioles contain cilia, small hair-like projections that line the walls of the bronchi and bronchioles (Figure 39.11). These cilia beat in unison and move mucus and particles out of the bronchi and bronchioles back up to the throat where it is swallowed and eliminated via the esophagus.

In humans, for example, tar and other substances in cigarette smoke destroy or paralyze the cilia, making the removal of particles more difficult. In addition, smoking causes the lungs to produce more mucus, which the damaged cilia are not able to move. This causes a persistent cough, as the lungs try to rid themselves of particulate matter, and makes smokers more susceptible to respiratory ailments.

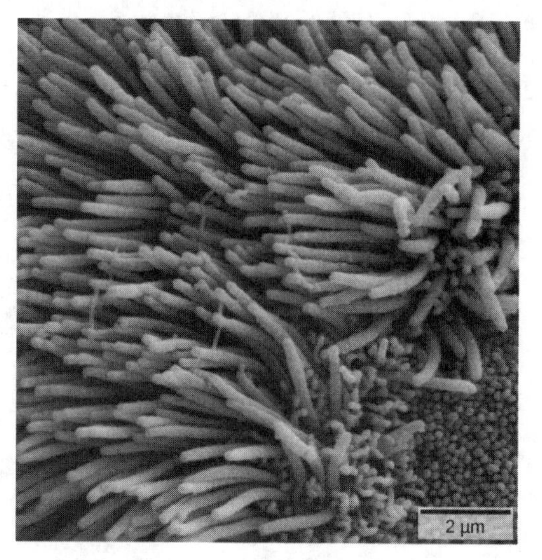

Figure 39.11 The bronchi and bronchioles contain cilia that help move mucus and other particles out of the lungs. (credit: Louisa Howard, modification of work by Dartmouth Electron Microscope Facility)

39.2 Gas Exchange across Respiratory Surfaces

By the end of this section, you will be able to do the following:

- Name and describe lung volumes and capacities
- Understand how gas pressure influences how gases move into and out of the body

The structure of the lung maximizes its surface area to increase gas diffusion. Because of the enormous number of alveoli (approximately 300 million in each human lung), the surface area of the lung is very large (75 m^2). Having such a large surface area increases the amount of gas that can diffuse into and out of the lungs.

Basic Principles of Gas Exchange

Gas exchange during respiration occurs primarily through diffusion. Diffusion is a process in which transport is driven by a concentration gradient. Gas molecules move from a region of high concentration to a region of low concentration. Blood that is low in oxygen concentration and high in carbon dioxide concentration undergoes gas exchange with air in the lungs. The air in the lungs has a higher concentration of oxygen than that of oxygen-depleted blood and a lower concentration of carbon dioxide. This concentration gradient allows for gas exchange during respiration.

Partial pressure is a measure of the concentration of the individual components in a mixture of gases. The total pressure exerted by the mixture is the sum of the partial pressures of the components in the mixture. The rate of diffusion of a gas is proportional to its partial pressure within the total gas mixture. This concept is discussed further in detail below.

Lung Volumes and Capacities

Different animals have different lung capacities based on their activities. Cheetahs have evolved a much higher lung capacity than humans; it helps provide oxygen to all the muscles in the body and allows them to run very fast. Elephants also have a high lung capacity. In this case, it is not because they run fast but because they have a large body and must be able to take up oxygen in accordance with their body size.

Human lung size is determined by genetics, sex, and height. At maximal capacity, an average lung can hold almost six liters of air, but lungs do not usually operate at maximal capacity. Air in the lungs is measured in terms of **lung volumes** and **lung capacities** (Figure 39.12 and Table 39.1). Volume measures the amount of air for one function (such as inhalation or exhalation). Capacity is any two or more volumes (for example, how much can be inhaled from the end of a maximal exhalation).

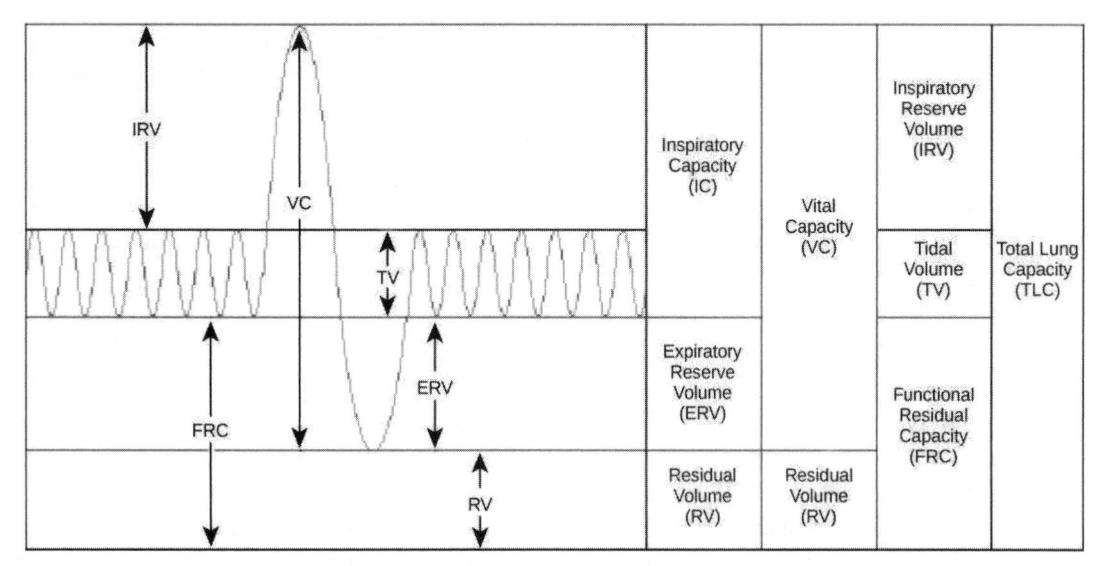

Figure 39.12 Human lung volumes and capacities are shown. The total lung capacity of the adult male is six liters. Tidal volume is the volume of air inhaled in a single, normal breath. Inspiratory capacity is the amount of air taken in during a deep breath, and residual volume is the amount of air left in the lungs after forceful respiration.

Lung Volumes and Capacities (Avg Adult Male)

Volume/Capacity	Definition	Volume (liters)	Equations
Tidal volume (TV)	Amount of air inhaled during a normal breath	0.5	-
Expiratory reserve volume (ERV)	Amount of air that can be exhaled after a normal exhalation	1.2	-
Inspiratory reserve volume (IRV)	Amount of air that can be further inhaled after a normal inhalation	3.1	-
Residual volume (RV)	Air left in the lungs after a forced exhalation	1.2	-
Vital capacity (VC)	Maximum amount of air that can be moved in or out of the lungs in a single respiratory cycle	4.8	ERV+TV+IRV
Inspiratory capacity (IC)	Volume of air that can be inhaled in addition to a normal exhalation	3.6	TV+IRV
Functional residual capacity (FRC)	Volume of air remaining after a normal exhalation	2.4	ERV+RV
Total lung capacity (TLC)	Total volume of air in the lungs after a maximal inspiration	6.0	RV+ERV+TV+IRV
Forced expiratory volume (FEV1)	How much air can be forced out of the lungs over a specific time period, usually one second	~4.1 to 5.5	-

Table 39.1

The volume in the lung can be divided into four units: tidal volume, expiratory reserve volume, inspiratory reserve volume, and residual volume. **Tidal volume (TV)** measures the amount of air that is inspired and expired during a normal breath. On average, this volume is around one-half liter, which is a little less than the capacity of a 20-ounce drink bottle. The **expiratory reserve volume (ERV)** is the additional amount of air that can be exhaled after a normal exhalation. It is the reserve amount that can be exhaled beyond what is normal. Conversely, the **inspiratory reserve volume (IRV)** is the additional amount of air that can be inhaled after a normal inhalation. The **residual volume (RV)** is the amount of air that is left after expiratory reserve volume is exhaled. The lungs are never completely empty: There is always some air left in the lungs after a maximal exhalation. If this residual volume did not exist and the lungs emptied completely, the lung tissues would stick together and the energy necessary to reinflate the lung could be too great to overcome. Therefore, there is always some air remaining in the lungs. Residual volume is also important for preventing large fluctuations in respiratory gases (O_2 and CO_2). The residual volume is the only lung volume that cannot be measured directly because it is impossible to completely empty the lung of air. This volume can only be calculated rather than measured.

Capacities are measurements of two or more volumes. The **vital capacity (VC)** measures the maximum amount of air that can be inhaled or exhaled during a respiratory cycle. It is the sum of the expiratory reserve volume, tidal volume, and inspiratory reserve volume. The **inspiratory capacity (IC)** is the amount of air that can be inhaled after the end of a normal expiration. It is, therefore, the sum of the tidal volume and inspiratory reserve volume. The **functional residual capacity (FRC)** includes the expiratory reserve volume and the residual volume. The FRC measures the amount of additional air that can be exhaled after a normal exhalation. Lastly, the **total lung capacity (TLC)** is a measurement of the total amount of air that the lung can hold. It is the sum of the residual volume, expiratory reserve volume, tidal volume, and inspiratory reserve volume.

Lung volumes are measured by a technique called **spirometry**. An important measurement taken during spirometry is the **forced expiratory volume (FEV)**, which measures how much air can be forced out of the lung over a specific period, usually one second (FEV1). In addition, the forced vital capacity (FVC), which is the total amount of air that can be forcibly exhaled, is measured. The ratio of these values (**FEV1/FVC ratio**) is used to diagnose lung diseases including asthma, emphysema, and fibrosis. If the FEV1/FVC ratio is high, the lungs are not compliant (meaning they are stiff and unable to bend properly), and the patient most likely has lung fibrosis. Patients exhale most of the lung volume very quickly. Conversely, when the FEV1/FVC ratio is low, there is resistance in the lung that is characteristic of asthma. In this instance, it is hard for the patient to get the air out of his or her lungs, and it takes a long time to reach the maximal exhalation volume. In either case, breathing is difficult and complications arise.

CAREER CONNECTION

Respiratory Therapist

Respiratory therapists or respiratory practitioners evaluate and treat patients with lung and cardiovascular diseases. They work as part of a medical team to develop treatment plans for patients. Respiratory therapists may treat premature babies with underdeveloped lungs, patients with chronic conditions such as asthma, or older patients suffering from lung disease such as emphysema and chronic obstructive pulmonary disease (COPD). They may operate advanced equipment such as compressed gas delivery systems, ventilators, blood gas analyzers, and resuscitators. Specialized programs to become a respiratory therapist generally lead to a bachelor's degree with a respiratory therapist specialty. Because of a growing aging population, career opportunities as a respiratory therapist are expected to remain strong.

Gas Pressure and Respiration

The respiratory process can be better understood by examining the properties of gases. Gases move freely, but gas particles are constantly hitting the walls of their vessel, thereby producing gas pressure.

Air is a mixture of gases, primarily nitrogen (N_2; 78.6 percent), oxygen (O_2; 20.9 percent), water vapor (H_2O; 0.5 percent), and carbon dioxide (CO_2; 0.04 percent). Each gas component of that mixture exerts a pressure. The pressure for an individual gas in the mixture is the partial pressure of that gas. Approximately 21 percent of atmospheric gas is oxygen. Carbon dioxide, however, is found in relatively small amounts, 0.04 percent. The partial pressure for oxygen is much greater than that of carbon dioxide. The partial pressure of any gas can be calculated by:

$$P = (P_{atm}) \times (\text{percent content in mixture}).$$

P_{atm}, the atmospheric pressure, is the sum of all of the partial pressures of the atmospheric gases added together,

$$P_{atm} = P_{N_2} + P_{O_2} + P_{H_2O} + P_{CO_2} = 760 \text{ mm Hg}$$

× (percent content in mixture).

The pressure of the atmosphere at sea level is 760 mm Hg. Therefore, the partial pressure of oxygen is:

$$P_{O_2} = (760 \text{ mm Hg})\ (0.21) = 160 \text{ mm Hg}$$

and for carbon dioxide:

$$P_{CO_2} = (760 \text{ mm Hg})\ (0.0004) = 0.3 \text{ mm Hg}.$$

At high altitudes, P_{atm} decreases but concentration does not change; the partial pressure decrease is due to the reduction in P_{atm}.

When the air mixture reaches the lung, it has been humidified. The pressure of the water vapor in the lung does not change the pressure of the air, but it must be included in the partial pressure equation. For this calculation, the water pressure (47 mm Hg) is subtracted from the atmospheric pressure:

$$760 \text{ mm Hg} - 47 \text{ mm Hg} = 713 \text{ mm Hg}$$

and the partial pressure of oxygen is:

$$(760 \text{ mm Hg} - 47 \text{ mm Hg}) \times 0.21 = 150 \text{ mm Hg}.$$

These pressures determine the gas exchange, or the flow of gas, in the system. Oxygen and carbon dioxide will flow according to their pressure gradient from high to low. Therefore, understanding the partial pressure of each gas will aid in understanding how gases move in the respiratory system.

Gas Exchange across the Alveoli

In the body, oxygen is used by cells of the body's tissues and carbon dioxide is produced as a waste product. The ratio of carbon dioxide production to oxygen consumption is the **respiratory quotient (RQ)**. RQ varies between 0.7 and 1.0. If just glucose were used to fuel the body, the RQ would equal one. One mole of carbon dioxide would be produced for every mole of oxygen consumed. Glucose, however, is not the only fuel for the body. Protein and fat are also used as fuels for the body. Because of this, less carbon dioxide is produced than oxygen is consumed and the RQ is, on average, about 0.7 for fat and about 0.8 for protein.

The RQ is used to calculate the partial pressure of oxygen in the alveolar spaces within the lung, the **alveolar** P_{O_2} . Above, the partial pressure of oxygen in the lungs was calculated to be 150 mm Hg. However, lungs never fully deflate with an exhalation; therefore, the inspired air mixes with this residual air and lowers the partial pressure of oxygen within the alveoli. This means that there is a lower concentration of oxygen in the lungs than is found in the air outside the body. Knowing the RQ, the partial pressure of oxygen in the alveoli can be calculated:

$$\text{alveolar } P_{O_2} = \text{inspired } P_{O_2} - \left(\frac{\text{alveolar } P_{O_2}}{RQ}\right)$$

With an RQ of 0.8 and a P_{CO_2} in the alveoli of 40 mm Hg, the alveolar P_{O_2} is equal to:

$$\text{alveolar } P_{O_2} = 150 \text{ mm Hg} - \left(\frac{40 \text{ mm Hg}}{0.8}\right) = 100 \text{ mm Hg}.$$

Notice that this pressure is less than the external air. Therefore, the oxygen will flow from the inspired air in the lung (P_{O_2} = 150 mm Hg) into the bloodstream (P_{O_2} = 100 mm Hg) (Figure 39.13).

In the lungs, oxygen diffuses out of the alveoli and into the capillaries surrounding the alveoli. Oxygen (about 98 percent) binds reversibly to the respiratory pigment hemoglobin found in red blood cells (RBCs). RBCs carry oxygen to the tissues where oxygen dissociates from the hemoglobin and diffuses into the cells of the tissues. More specifically, alveolar P_{O_2} is higher in the alveoli (P_{ALVO_2} = 100 mm Hg) than blood P_{O_2} (40 mm Hg) in the capillaries. Because this pressure gradient exists, oxygen diffuses down its pressure gradient, moving out of the alveoli and entering the blood of the capillaries where O_2 binds to hemoglobin. At the same time, alveolar P_{CO_2} is lower P_{ALVO_2} = 40 mm Hg than blood P_{CO_2} = (45 mm Hg). CO_2 diffuses down its pressure gradient, moving out of the capillaries and entering the alveoli.

Oxygen and carbon dioxide move independently of each other; they diffuse down their own pressure gradients. As blood leaves the lungs through the pulmonary veins, the **venous** P_{O_2} = 100 mm Hg, whereas the **venous** P_{CO_2} = 40 mm Hg. As blood enters the systemic capillaries, the blood will lose oxygen and gain carbon dioxide because of the pressure difference of the tissues and blood. In systemic capillaries, P_{O_2} = 100 mm Hg, but in the tissue cells, P_{O_2} = 40 mm Hg. This pressure gradient drives the diffusion of oxygen out of the capillaries and into the tissue cells. At the same time, blood P_{CO_2} = 40 mm Hg and systemic tissue P_{CO_2} = 45 mm Hg. The pressure gradient drives CO_2 out of tissue cells and into the capillaries. The blood returning to the lungs through the pulmonary arteries has a venous P_{O_2} = 40 mm Hg and a P_{CO_2} = 45 mm Hg. The blood enters the lung capillaries where the process of exchanging gases between the capillaries and alveoli begins again (Figure 39.13).

VISUAL CONNECTION

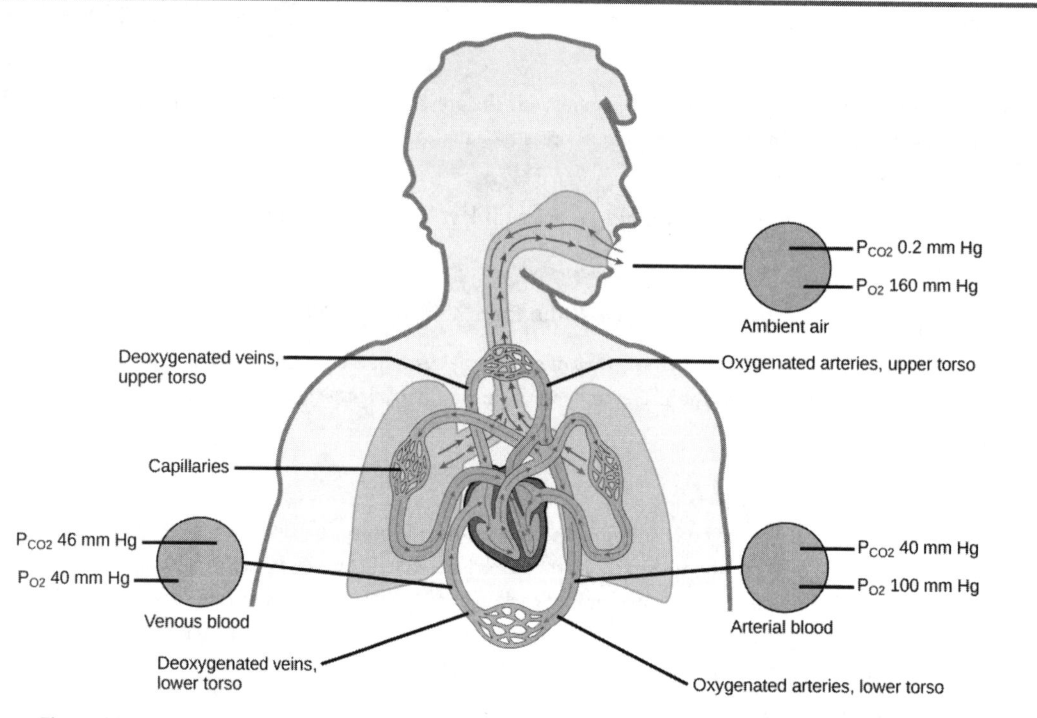

Figure 39.13 The partial pressures of oxygen and carbon dioxide change as blood moves through the body.

Which of the following statements is false?

a. In the tissues, P_{O_2} drops as blood passes from the arteries to the veins, while P_{CO_2} increases.
b. Blood travels from the lungs to the heart to body tissues, then back to the heart, then the lungs.
c. Blood travels from the lungs to the heart to body tissues, then back to the lungs, then the heart.
d. P_{O_2} is higher in air than in the lungs.

In short, the change in partial pressure from the alveoli to the capillaries drives the oxygen into the tissues and the carbon dioxide into the blood from the tissues. The blood is then transported to the lungs where differences in pressure in the alveoli result in the movement of carbon dioxide out of the blood into the lungs, and oxygen into the blood.

LINK TO LEARNING

Watch this video to learn how to carry out spirometry.

Click to view content (https://www.openstax.org/l/spirometry)

39.3 Breathing

By the end of this section, you will be able to do the following:

- Describe how the structures of the lungs and thoracic cavity control the mechanics of breathing
- Explain the importance of compliance and resistance in the lungs
- Discuss problems that may arise due to a V/Q mismatch

Mammalian lungs are located in the thoracic cavity where they are surrounded and protected by the rib cage, intercostal muscles, and bound by the chest wall. The bottom of the lungs is contained by the diaphragm, a skeletal muscle that facilitates breathing. Breathing requires the coordination of the lungs, the chest wall, and most importantly, the diaphragm.

Types of Breathing

Amphibians have evolved multiple ways of breathing. Young amphibians, like tadpoles, use gills to breathe, and they don't leave the water. Some amphibians retain gills for life. As the tadpole grows, the gills disappear and lungs grow. These lungs are primitive and not as evolved as mammalian lungs. Adult amphibians are lacking or have a reduced diaphragm, so breathing via lungs is forced. The other means of breathing for amphibians is diffusion across the skin. To aid this diffusion, amphibian skin must remain moist.

Birds face a unique challenge with respect to breathing: They fly. Flying consumes a great amount of energy; therefore, birds require a lot of oxygen to aid their metabolic processes. Birds have evolved a respiratory system that supplies them with the oxygen needed to enable flying. Similar to mammals, birds have lungs, which are organs specialized for gas exchange. Oxygenated air, taken in during inhalation, diffuses across the surface of the lungs into the bloodstream, and carbon dioxide diffuses from the blood into the lungs and expelled during exhalation. The details of breathing between birds and mammals differ substantially.

In addition to lungs, birds have air sacs inside their body. Air flows in one direction from the posterior air sacs to the lungs and out of the anterior air sacs. The flow of air is in the opposite direction from blood flow, and gas exchange takes place much more efficiently. This type of breathing enables birds to obtain the requisite oxygen, even at higher altitudes where the oxygen concentration is low. This directionality of airflow requires two cycles of air intake and exhalation to completely get the air out of the lungs.

EVOLUTION CONNECTION

Avian Respiration

Birds have evolved a respiratory system that enables them to fly. Flying is a high-energy process and requires a lot of oxygen. Furthermore, many birds fly in high altitudes where the concentration of oxygen is low. How did birds evolve a respiratory system that is so unique?

Decades of research by paleontologists have shown that birds evolved from therapods, meat-eating dinosaurs (Figure 39.14). In fact, fossil evidence shows that meat-eating dinosaurs that lived more than 100 million years ago had a similar flow-through respiratory system with lungs and air sacs. *Archaeopteryx* and *Xiaotingia*, for example, were flying dinosaurs and are believed to be early precursors of birds.

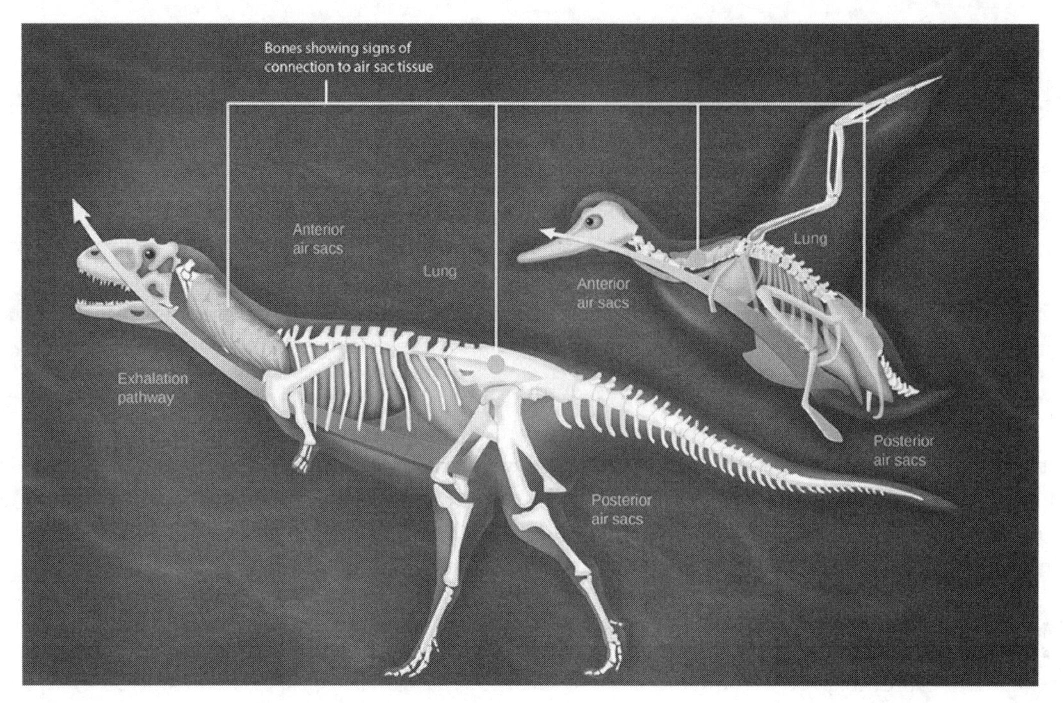

Figure 39.14 Dinosaurs, from which birds descended, have similar hollow bones and are believed to have had a similar respiratory system. (credit b: modification of work by Zina Deretsky, National Science Foundation)

Most of us consider that dinosaurs are extinct. However, modern birds are descendants of avian dinosaurs. The respiratory system of modern birds has been evolving for hundreds of millions of years.

All mammals have lungs that are the main organs for breathing. Lung capacity has evolved to support the animal's activities. During inhalation, the lungs expand with air, and oxygen diffuses across the lung's surface and enters the bloodstream. During exhalation, the lungs expel air and lung volume decreases. In the next few sections, the process of human breathing will be explained.

The Mechanics of Human Breathing

Boyle's Law is the gas law that states that in a closed space, pressure and volume are inversely related. As volume decreases, pressure increases and vice versa (Figure 39.15). The relationship between gas pressure and volume helps to explain the mechanics of breathing.

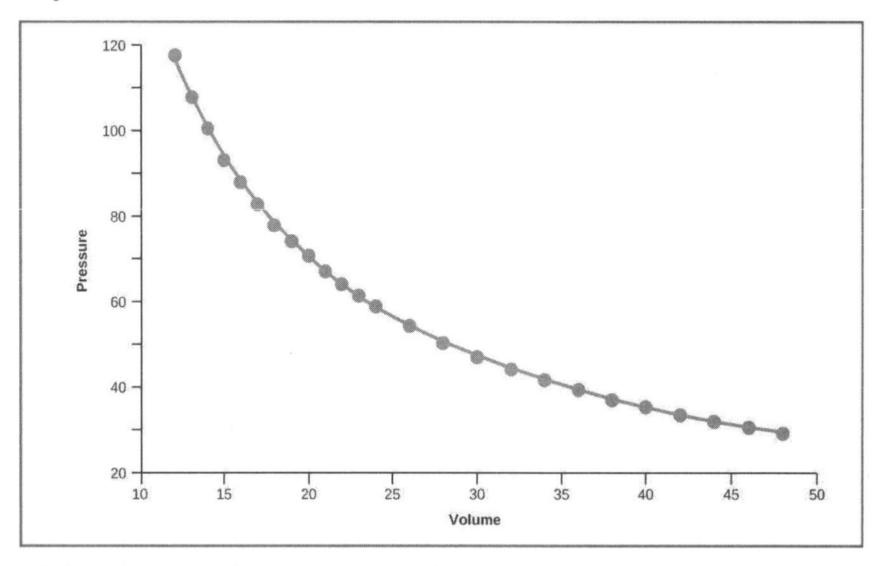

Figure 39.15 This graph shows data from Boyle's original 1662 experiment, which shows that pressure and volume are inversely related.

No units are given as Boyle used arbitrary units in his experiments.

There is always a slightly negative pressure within the thoracic cavity, which aids in keeping the airways of the lungs open. During inhalation, volume increases as a result of contraction of the diaphragm, and pressure decreases (according to Boyle's Law). This decrease of pressure in the thoracic cavity relative to the environment makes the cavity less than the atmosphere (Figure 39.16a). Because of this drop in pressure, air rushes into the respiratory passages. To increase the volume of the lungs, the chest wall expands. This results from the contraction of the **intercostal muscles**, the muscles that are connected to the rib cage. Lung volume expands because the diaphragm contracts and the intercostal muscles contract, thus expanding the thoracic cavity. This increase in the volume of the thoracic cavity lowers pressure compared to the atmosphere, so air rushes into the lungs, thus increasing its volume. The resulting increase in volume is largely attributed to an increase in alveolar space, because the bronchioles and bronchi are stiff structures that do not change in size.

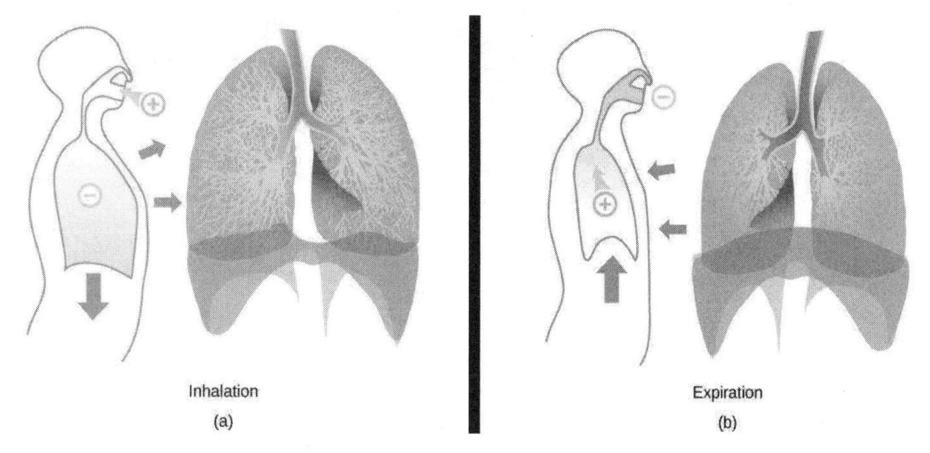

Figure 39.16 The lungs, chest wall, and diaphragm are all involved in respiration, both (a) inhalation and (b) expiration. (credit: modification of work by Mariana Ruiz Villareal)

The chest wall expands out and away from the lungs. The lungs are elastic; therefore, when air fills the lungs, the **elastic recoil** within the tissues of the lung exerts pressure back toward the interior of the lungs. These outward and inward forces compete to inflate and deflate the lung with every breath. Upon exhalation, the lungs recoil to force the air out of the lungs, and the intercostal muscles relax, returning the chest wall back to its original position (Figure 39.16b). The diaphragm also relaxes and moves higher into the thoracic cavity. This increases the pressure within the thoracic cavity relative to the environment, and air rushes out of the lungs. The movement of air out of the lungs is a passive event. No muscles are contracting to expel the air.

Each lung is surrounded by an invaginated sac. The layer of tissue that covers the lung and dips into spaces is called the visceral **pleura**. A second layer of parietal pleura lines the interior of the thorax (Figure 39.17). The space between these layers, the **intrapleural space**, contains a small amount of fluid that protects the tissue and reduces the friction generated from rubbing the tissue layers together as the lungs contract and relax. **Pleurisy** results when these layers of tissue become inflamed; it is painful because the inflammation increases the pressure within the thoracic cavity and reduces the volume of the lung.

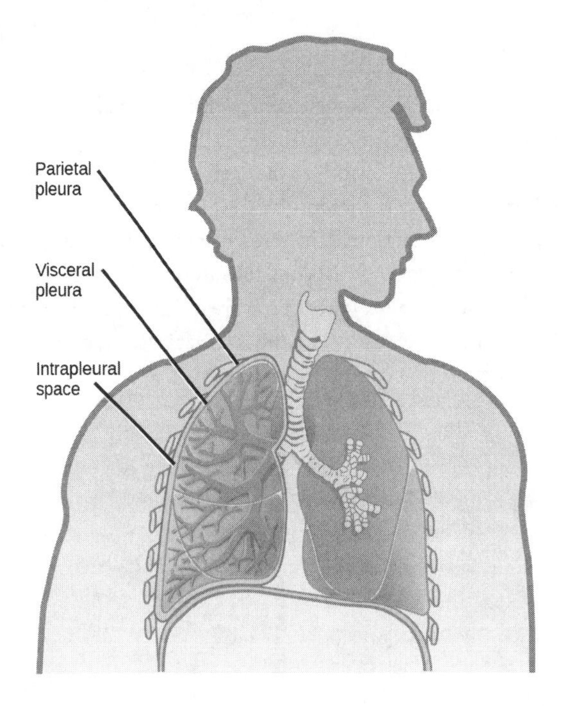

Figure 39.17 A tissue layer called pleura surrounds the lung and interior of the thoracic cavity. (credit: modification of work by NCI)

LINK TO LEARNING

View how Boyle's Law is related to breathing and watch a video (https://www.openstax.org/l/boyles_law) on Boyle's Law.

Click to view content (https://www.openstax.org/l/boyle_breathing)

The Work of Breathing

The number of breaths per minute is the **respiratory rate**. On average, under non-exertion conditions, the human respiratory rate is 12–15 breaths/minute. The respiratory rate contributes to the **alveolar ventilation**, or how much air moves into and out of the alveoli. Alveolar ventilation prevents carbon dioxide buildup in the alveoli. There are two ways to keep the alveolar ventilation constant: increase the respiratory rate while decreasing the tidal volume of air per breath (shallow breathing), or decrease the respiratory rate while increasing the tidal volume per breath. In either case, the ventilation remains the same, but the work done and type of work needed are quite different. Both tidal volume and respiratory rate are closely regulated when oxygen demand increases.

There are two types of work conducted during respiration, flow-resistive and elastic work. **Flow-resistive** refers to the work of the alveoli and tissues in the lung, whereas **elastic work** refers to the work of the intercostal muscles, chest wall, and diaphragm. Increasing the respiration rate increases the flow-resistive work of the airways and decreases the elastic work of the muscles. Decreasing the respiratory rate reverses the type of work required.

Surfactant

The air-tissue/water interface of the alveoli has a high surface tension. This surface tension is similar to the surface tension of water at the liquid-air interface of a water droplet that results in the bonding of the water molecules together. **Surfactant** is a complex mixture of phospholipids and lipoproteins that works to reduce the surface tension that exists between the alveoli tissue and the air found within the alveoli. By lowering the surface tension of the alveolar fluid, it reduces the tendency of alveoli to collapse.

Surfactant works like a detergent to reduce the surface tension and allows for easier inflation of the airways. When a balloon is first inflated, it takes a large amount of effort to stretch the plastic and start to inflate the balloon. If a little bit of detergent was applied to the interior of the balloon, then the amount of effort or work needed to begin to inflate the balloon would decrease, and it would become much easier to start blowing up the balloon. This same principle applies to the airways. A small amount of surfactant to the airway tissues reduces the effort or work needed to inflate those airways. Babies born prematurely sometimes do not produce enough surfactant. As a result, they suffer from **respiratory distress syndrome**, because it requires more effort to inflate their lungs. Surfactant is also important for preventing collapse of small alveoli relative to large alveoli.

Lung Resistance and Compliance

Pulmonary diseases reduce the rate of gas exchange into and out of the lungs. Two main causes of decreased gas exchange are **compliance** (how elastic the lung is) and **resistance** (how much obstruction exists in the airways). A change in either can dramatically alter breathing and the ability to take in oxygen and release carbon dioxide.

Examples of **restrictive diseases** are respiratory distress syndrome and pulmonary fibrosis. In both diseases, the airways are less compliant and they are stiff or fibrotic. There is a decrease in compliance because the lung tissue cannot bend and move. In these types of restrictive diseases, the intrapleural pressure is more positive and the airways collapse upon exhalation, which traps air in the lungs. Forced or **functional vital capacity (FVC)**, which is the amount of air that can be forcibly exhaled after taking the deepest breath possible, is much lower than in normal patients, and the time it takes to exhale most of the air is greatly prolonged (Figure 39.18). A patient suffering from these diseases cannot exhale the normal amount of air.

Obstructive diseases and conditions include emphysema, asthma, and pulmonary edema. In emphysema, which mostly arises from smoking tobacco, the walls of the alveoli are destroyed, decreasing the surface area for gas exchange. The overall compliance of the lungs is increased, because as the alveolar walls are damaged, lung elastic recoil decreases due to a loss of elastic fibers, and more air is trapped in the lungs at the end of exhalation. Asthma is a disease in which inflammation is triggered by environmental factors. Inflammation obstructs the airways. The obstruction may be due to edema (fluid accumulation), smooth muscle spasms in the walls of the bronchioles, increased mucus secretion, damage to the epithelia of the airways, or a combination of these events. Those with asthma or edema experience increased occlusion from increased inflammation of the airways. This tends to block the airways, preventing the proper movement of gases (Figure 39.18). Those with obstructive diseases have large volumes of air trapped after exhalation and breathe at a very high lung volume to compensate for the lack of airway recruitment.

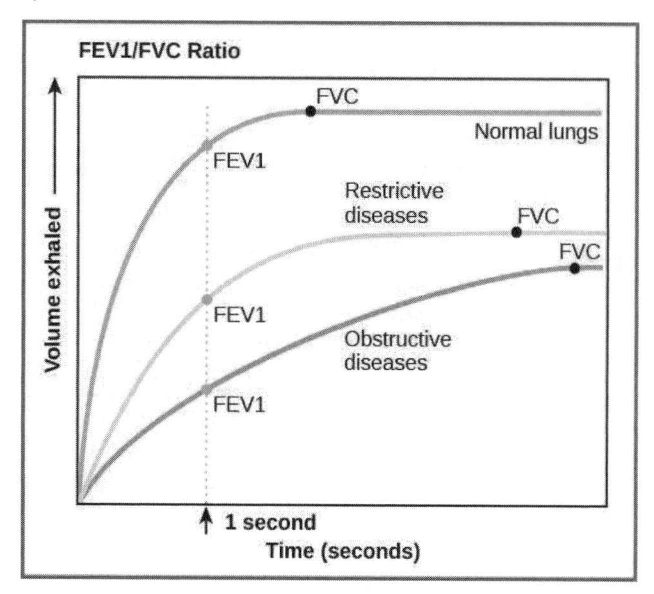

Figure 39.18 The ratio of FEV1 (the amount of air that can be forcibly exhaled in one second after taking a deep breath) to FVC (the total amount of air that can be forcibly exhaled) can be used to diagnose whether a person has restrictive or obstructive lung disease. In restrictive lung disease, FVC is reduced but airways are not obstructed, so the person is able to expel air reasonably fast. In obstructive lung disease, airway obstruction results in slow exhalation as well as reduced FVC. Thus, the FEV1/FVC ratio is lower in persons with obstructive lung disease (less than 69 percent) than in persons with restrictive disease (88 to 90 percent).

Dead Space: V/Q Mismatch

Pulmonary circulation pressure is very low compared to that of the systemic circulation. It is also independent of cardiac output. This is because of a phenomenon called **recruitment**, which is the process of opening airways that normally remain closed when cardiac output increases. As cardiac output increases, the number of capillaries and arteries that are perfused (filled with blood) increases. These capillaries and arteries are not always in use but are ready if needed. At times, however, there is a mismatch between the amount of air (ventilation, V) and the amount of blood (perfusion, Q) in the lungs. This is referred to as **ventilation/perfusion (V/Q) mismatch**.

There are two types of V/Q mismatch. Both produce **dead space**, regions of broken down or blocked lung tissue. Dead spaces can severely impact breathing, because they reduce the surface area available for gas diffusion. As a result, the amount of oxygen in the blood decreases, whereas the carbon dioxide level increases. Dead space is created when no ventilation and/or perfusion takes place. **Anatomical dead space** or anatomical shunt, arises from an anatomical failure, while **physiological dead space** or physiological shunt, arises from a functional impairment of the lung or arteries.

An example of an anatomical shunt is the effect of gravity on the lungs. The lung is particularly susceptible to changes in the magnitude and direction of gravitational forces. When someone is standing or sitting upright, the pleural pressure gradient leads to increased ventilation further down in the lung. As a result, the intrapleural pressure is more negative at the base of the lung than at the top, and more air fills the bottom of the lung than the top. Likewise, it takes less energy to pump blood to the bottom of the lung than to the top when in a prone position. Perfusion of the lung is not uniform while standing or sitting. This is a result of hydrostatic forces combined with the effect of airway pressure. An anatomical shunt develops because the ventilation of the airways does not match the perfusion of the arteries surrounding those airways. As a result, the rate of gas exchange is reduced. Note that this does not occur when lying down, because in this position, gravity does not preferentially pull the bottom of the lung down.

A physiological shunt can develop if there is infection or edema in the lung that obstructs an area. This will decrease ventilation but not affect perfusion; therefore, the V/Q ratio changes and gas exchange is affected.

The lung can compensate for these mismatches in ventilation and perfusion. If ventilation is greater than perfusion, the arterioles dilate and the bronchioles constrict. This increases perfusion and reduces ventilation. Likewise, if ventilation is less than perfusion, the arterioles constrict and the bronchioles dilate to correct the imbalance.

LINK TO LEARNING

View the mechanics of breathing.

Click to view content (https://www.openstax.org/l/breathing)

39.4 Transport of Gases in Human Bodily Fluids

By the end of this section, you will be able to do the following:

- Describe how oxygen is bound to hemoglobin and transported to body tissues
- Explain how carbon dioxide is transported from body tissues to the lungs

Once the oxygen diffuses across the alveoli, it enters the bloodstream and is transported to the tissues where it is unloaded, and carbon dioxide diffuses out of the blood and into the alveoli to be expelled from the body. Although gas exchange is a continuous process, the oxygen and carbon dioxide are transported by different mechanisms.

Transport of Oxygen in the Blood

Although oxygen dissolves in blood, only a small amount of oxygen is transported this way. Only 1.5 percent of oxygen in the blood is dissolved directly into the blood itself. Most oxygen—98.5 percent—is bound to a protein called hemoglobin and carried to the tissues.

Hemoglobin

Hemoglobin, or Hb, is a protein molecule found in red blood cells (erythrocytes) made of four subunits: two alpha subunits and two beta subunits (Figure 39.19). Each subunit surrounds a central **heme group** that contains iron and binds one oxygen molecule, allowing each hemoglobin molecule to bind four oxygen molecules. Molecules with more oxygen bound to the heme groups are brighter red. As a result, oxygenated arterial blood where the Hb is carrying four oxygen molecules is bright red, while venous blood that is deoxygenated is darker red.

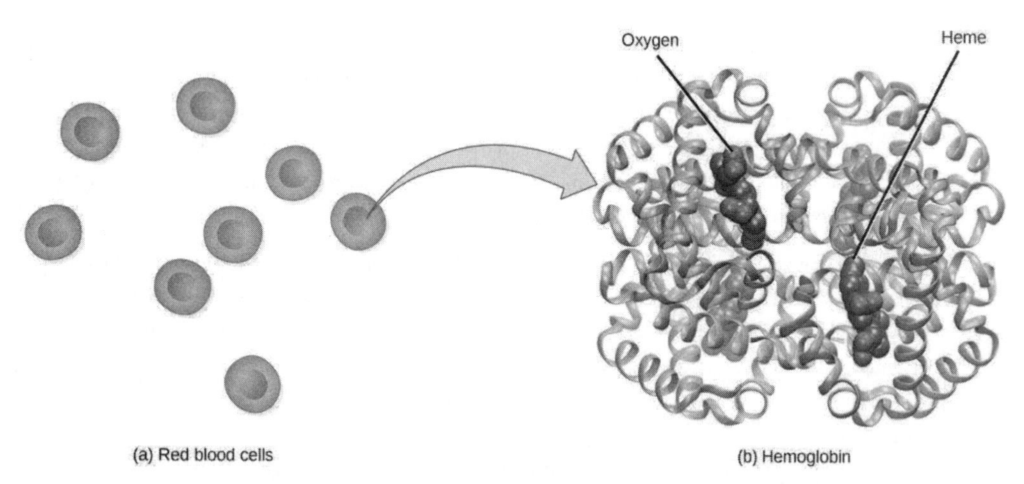

Figure 39.19 The protein inside (a) red blood cells that carries oxygen to cells and carbon dioxide to the lungs is (b) hemoglobin. Hemoglobin is made up of four symmetrical subunits and four heme groups. Iron associated with the heme binds oxygen. It is the iron in hemoglobin that gives blood its red color.

It is easier to bind a second and third oxygen molecule to Hb than the first molecule. This is because the hemoglobin molecule changes its shape, or conformation, as oxygen binds. The fourth oxygen is then more difficult to bind. The binding of oxygen to hemoglobin can be plotted as a function of the partial pressure of oxygen in the blood (x-axis) versus the relative Hb-oxygen saturation (y-axis). The resulting graph—an **oxygen dissociation curve**—is sigmoidal, or S-shaped (Figure 39.20). As the partial pressure of oxygen increases, the hemoglobin becomes increasingly saturated with oxygen.

VISUAL CONNECTION

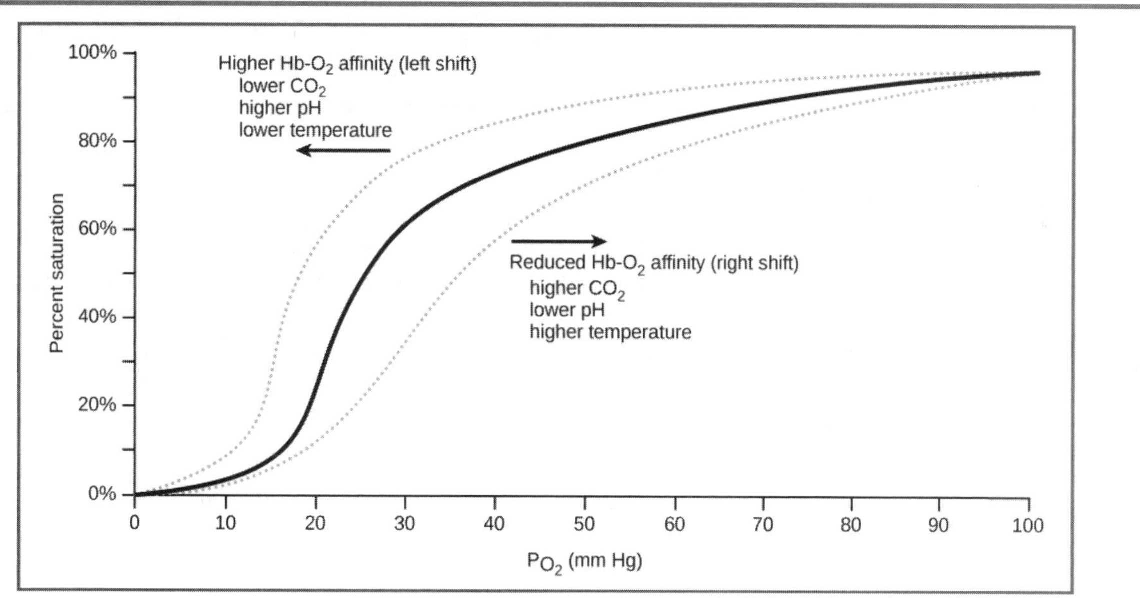

Figure 39.20 The oxygen dissociation curve demonstrates that, as the partial pressure of oxygen increases, more oxygen binds hemoglobin. However, the affinity of hemoglobin for oxygen may shift to the left or the right depending on environmental conditions.

The kidneys are responsible for removing excess H+ ions from the blood. If the kidneys fail, what would happen to blood pH and to hemoglobin affinity for oxygen?

Factors That Affect Oxygen Binding

The **oxygen-carrying capacity** of hemoglobin determines how much oxygen is carried in the blood. In addition to P_{O_2}, other environmental factors and diseases can affect oxygen carrying capacity and delivery.

Carbon dioxide levels, blood pH, and body temperature affect oxygen-carrying capacity (Figure 39.20). When carbon dioxide is in the blood, it reacts with water to form bicarbonate (HCO_3^-) and hydrogen ions (H^+). As the level of carbon dioxide in the blood increases, more H^+ is produced and the pH decreases. This increase in carbon dioxide and subsequent decrease in pH reduce the affinity of hemoglobin for oxygen. The oxygen dissociates from the Hb molecule, shifting the oxygen dissociation curve to the right. Therefore, more oxygen is needed to reach the same hemoglobin saturation level as when the pH was higher. A similar shift in the curve also results from an increase in body temperature. Increased temperature, such as from increased activity of skeletal muscle, causes the affinity of hemoglobin for oxygen to be reduced.

Diseases like sickle cell anemia and thalassemia decrease the blood's ability to deliver oxygen to tissues and its oxygen-carrying capacity. In **sickle cell anemia**, the shape of the red blood cell is crescent-shaped, elongated, and stiffened, reducing its ability to deliver oxygen (Figure 39.21). In this form, red blood cells cannot pass through the capillaries. This is painful when it occurs. **Thalassemia** is a rare genetic disease caused by a defect in either the alpha or the beta subunit of Hb. Patients with thalassemia produce a high number of red blood cells, but these cells have lower-than-normal levels of hemoglobin. Therefore, the oxygen-carrying capacity is diminished.

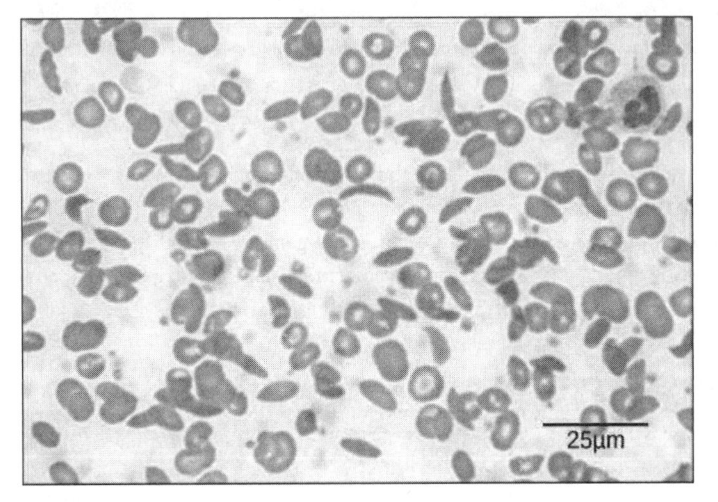

Figure 39.21 Individuals with sickle cell anemia have crescent-shaped red blood cells. (credit: modification of work by Ed Uthman; scale-bar data from Matt Russell)

Transport of Carbon Dioxide in the Blood

Carbon dioxide molecules are transported in the blood from body tissues to the lungs by one of three methods: dissolution directly into the blood, binding to hemoglobin, or carried as a bicarbonate ion. Several properties of carbon dioxide in the blood affect its transport. First, carbon dioxide is more soluble in blood than oxygen. About 5 to 7 percent of all carbon dioxide is dissolved in the plasma. Second, carbon dioxide can bind to plasma proteins or can enter red blood cells and bind to hemoglobin. This form transports about 10 percent of the carbon dioxide. When carbon dioxide binds to hemoglobin, a molecule called **carbaminohemoglobin** is formed. Binding of carbon dioxide to hemoglobin is reversible. Therefore, when it reaches the lungs, the carbon dioxide can freely dissociate from the hemoglobin and be expelled from the body.

Third, the majority of carbon dioxide molecules (85 percent) are carried as part of the **bicarbonate buffer system**. In this system, carbon dioxide diffuses into the red blood cells. **Carbonic anhydrase (CA)** within the red blood cells quickly converts the carbon dioxide into carbonic acid (H_2CO_3). Carbonic acid is an unstable intermediate molecule that immediately dissociates into **bicarbonate ions** (HCO_3^-) and hydrogen (H^+) ions. Since carbon dioxide is quickly converted into bicarbonate ions, this reaction allows for the continued uptake of carbon dioxide into the blood down its concentration gradient. It also results in the production of H^+ ions. If too much H^+ is produced, it can alter blood pH. However, hemoglobin binds to the free H^+ ions and thus limits shifts in pH. The newly synthesized bicarbonate ion is transported out of the red blood cell into the liquid component of the blood in exchange for a chloride ion (Cl^-); this is called the **chloride shift**. When the blood reaches the lungs, the bicarbonate ion is transported back into the red blood cell in exchange for the chloride ion. The H^+ ion dissociates from the hemoglobin and binds to the bicarbonate ion. This produces the carbonic acid intermediate, which is converted back into carbon dioxide through the enzymatic action of CA. The carbon dioxide produced is expelled through the lungs during exhalation.

$$CO_2 + H_2O \leftrightarrow \underset{\text{(carbonic acid)}}{H_2CO_3} \leftrightarrow \underset{\text{(bicarbonate)}}{HCO_3 + H^+}$$

The benefit of the bicarbonate buffer system is that carbon dioxide is "soaked up" into the blood with little change to the pH of the system. This is important because it takes only a small change in the overall pH of the body for severe injury or death to result. The presence of this bicarbonate buffer system also allows for people to travel and live at high altitudes: When the partial pressure of oxygen and carbon dioxide change at high altitudes, the bicarbonate buffer system adjusts to regulate carbon dioxide while maintaining the correct pH in the body.

Carbon Monoxide Poisoning

While carbon dioxide can readily associate and dissociate from hemoglobin, other molecules such as carbon monoxide (CO) cannot. Carbon monoxide has a greater affinity for hemoglobin than oxygen. Therefore, when carbon monoxide is present, it binds to hemoglobin preferentially over oxygen. As a result, oxygen cannot bind to hemoglobin, so very little oxygen is transported through the body (Figure 39.22). Carbon monoxide is a colorless, odorless gas and is therefore difficult to detect. It is produced by gas-powered vehicles and tools. Carbon monoxide can cause headaches, confusion, and nausea; long-term exposure can cause brain damage or death. Administering 100 percent (pure) oxygen is the usual treatment for carbon monoxide poisoning. Administration of pure oxygen speeds up the separation of carbon monoxide from hemoglobin.

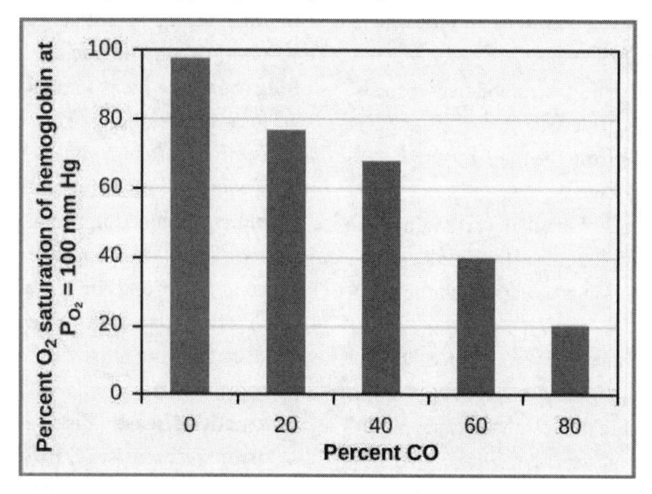

Figure 39.22 As percent CO increases, the oxygen saturation of hemoglobin decreases.

KEY TERMS

alveolar P_{O_2} partial pressure of oxygen in the alveoli (usually around 100 mmHg)
alveolar duct duct that extends from the terminal bronchiole to the alveolar sac
alveolar sac structure consisting of two or more alveoli that share a common opening
alveolar ventilation how much air is in the alveoli
alveolus (plural: alveoli) (also, air sac) terminal region of the lung where gas exchange occurs
anatomical dead space (also, anatomical shunt) region of the lung that lacks proper ventilation/perfusion due to an anatomical block
bicarbonate (HCO_3^-) ion ion created when carbonic acid dissociates into H^+ and (HCO_3^-)
bicarbonate buffer system system in the blood that absorbs carbon dioxide and regulates pH levels
bronchiole airway that extends from the main tertiary bronchi to the alveolar sac
bronchus (plural: bronchi) smaller branch of cartilaginous tissue that stems off of the trachea; air is funneled through the bronchi to the region where gas exchange occurs in alveoli
carbaminohemoglobin molecule that forms when carbon dioxide binds to hemoglobin
carbonic anhydrase (CA) enzyme that catalyzes carbon dioxide and water into carbonic acid
chloride shift exchange of chloride for bicarbonate into or out of the red blood cell
compliance measurement of the elasticity of the lung
dead space area in the lung that lacks proper ventilation or perfusion
diaphragm domed-shaped skeletal muscle located under lungs that separates the thoracic cavity from the abdominal cavity
elastic recoil property of the lung that drives the lung tissue inward
elastic work work conducted by the intercostal muscles, chest wall, and diaphragm
expiratory reserve volume (ERV) amount of additional air that can be exhaled after a normal exhalation
FEV1/FVC ratio ratio of how much air can be forced out of the lung in one second to the total amount that is forced out of the lung; a measurement of lung function that can be used to detect disease states
flow-resistive work of breathing performed by the alveoli and tissues in the lung
forced expiratory volume (FEV) (also, forced vital capacity) measure of how much air can be forced out of the lung from maximal inspiration over a specific amount of time
functional residual capacity (FRC) expiratory reserve volume plus residual volume
functional vital capacity (FVC) amount of air that can be forcibly exhaled after taking the deepest breath possible
heme group centralized iron-containing group that is surrounded by the alpha and beta subunits of hemoglobin
hemoglobin molecule in red blood cells that can bind oxygen, carbon dioxide, and carbon monoxide
inspiratory capacity (IC) tidal volume plus inspiratory reserve volume
inspiratory reserve volume (IRV) amount of additional air that can be inspired after a normal inhalation
intercostal muscle muscle connected to the rib cage that contracts upon inspiration
intrapleural space space between the layers of pleura
larynx voice box, a short passageway connecting the pharynx and the trachea
lung capacity measurement of two or more lung volumes (how much air can be inhaled from the end of an expiration to maximal capacity)
lung volume measurement of air for one lung function (normal inhalation or exhalation)
mucin complex glycoprotein found in mucus
mucus sticky protein-containing fluid secretion in the lung that traps particulate matter to be expelled from the body
nasal cavity opening of the respiratory system to the outside environment
obstructive disease disease (such as emphysema and asthma) that arises from obstruction of the airways; compliance increases in these diseases
oxygen dissociation curve curve depicting the affinity of oxygen for hemoglobin
oxygen-carrying capacity amount of oxygen that can be transported in the blood
partial pressure amount of pressure exerted by one gas within a mixture of gases
particulate matter small particle such as dust, dirt, viral particles, and bacteria that are in the air
pharynx throat; a tube that starts in the internal nares and runs partway down the neck, where it opens into the esophagus and the larynx
physiological dead space (also, physiological shunt) region of the lung that lacks proper ventilation/perfusion due to a physiological change in the lung (like inflammation or edema)
pleura tissue layer that surrounds the lungs and lines the interior of the thoracic cavity
pleurisy painful inflammation of the pleural tissue layers
primary bronchus (also, main bronchus) region of the airway within the lung that attaches to the trachea and bifurcates to each lung where it branches into secondary bronchi
recruitment process of opening airways that normally remain closed when the cardiac output increases
residual volume (RV) amount of air remaining in the lung

after a maximal expiration

resistance measurement of lung obstruction

respiratory bronchiole terminal portion of the bronchiole tree that is attached to the terminal bronchioles and alveoli ducts, alveolar sacs, and alveoli

respiratory distress syndrome disease that arises from a deficient amount of surfactant

respiratory quotient (RQ) ratio of carbon dioxide production to each oxygen molecule consumed

respiratory rate number of breaths per minute

restrictive disease disease that results from a restriction and decreased compliance of the alveoli; respiratory distress syndrome and pulmonary fibrosis are examples

sickle cell anemia genetic disorder that affects the shape of red blood cells, and their ability to transport oxygen and move through capillaries

spirometry method to measure lung volumes and to diagnose lung diseases

surfactant detergent-like liquid in the airways that lowers the surface tension of the alveoli to allow for expansion

terminal bronchiole region of bronchiole that attaches to the respiratory bronchioles

thalassemia rare genetic disorder that results in mutation of the alpha or beta subunits of hemoglobin, creating smaller red blood cells with less hemoglobin

tidal volume (TV) amount of air that is inspired and expired during normal breathing

total lung capacity (TLC) sum of the residual volume, expiratory reserve volume, tidal volume, and inspiratory reserve volume

trachea cartilaginous tube that transports air from the larynx to the primary bronchi

venous P_{CO_2} partial pressure of carbon dioxide in the veins (40 mm Hg in the pulmonary veins)

venous P_{O_2} partial pressure of oxygen in the veins (100 mm Hg in the pulmonary veins)

ventilation/perfusion (V/Q) mismatch region of the lung that lacks proper alveolar ventilation (V) and/or arterial perfusion (Q)

vital capacity (VC) sum of the expiratory reserve volume, tidal volume, and inspiratory reserve volume

CHAPTER SUMMARY

39.1 Systems of Gas Exchange

Animal respiratory systems are designed to facilitate gas exchange. In mammals, air is warmed and humidified in the nasal cavity. Air then travels down the pharynx, through the trachea, and into the lungs. In the lungs, air passes through the branching bronchi, reaching the respiratory bronchioles, which house the first site of gas exchange. The respiratory bronchioles open into the alveolar ducts, alveolar sacs, and alveoli. Because there are so many alveoli and alveolar sacs in the lung, the surface area for gas exchange is very large. Several protective mechanisms are in place to prevent damage or infection. These include the hair and mucus in the nasal cavity that trap dust, dirt, and other particulate matter before they can enter the system. In the lungs, particles are trapped in a mucus layer and transported via cilia up to the esophageal opening at the top of the trachea to be swallowed.

39.2 Gas Exchange across Respiratory Surfaces

The lungs can hold a large volume of air, but they are not usually filled to maximal capacity. Lung volume measurements include tidal volume, expiratory reserve volume, inspiratory reserve volume, and residual volume. The sum of these equals the total lung capacity. Gas movement into or out of the lungs is dependent on the pressure of the gas. Air is a mixture of gases; therefore, the partial pressure of each gas can be calculated to determine how the gas will flow in the lung. The difference between the partial pressure of the gas in the air drives oxygen into the tissues and carbon dioxide out of the body.

39.3 Breathing

The structure of the lungs and thoracic cavity control the mechanics of breathing. Upon inspiration, the diaphragm contracts and lowers. The intercostal muscles contract and expand the chest wall outward. The intrapleural pressure drops, the lungs expand, and air is drawn into the airways. When exhaling, the intercostal muscles and diaphragm relax, returning the intrapleural pressure back to the resting state. The lungs recoil and airways close. The air passively exits the lung. There is high surface tension at the air-airway interface in the lung. Surfactant, a mixture of phospholipids and lipoproteins, acts like a detergent in the airways to reduce surface tension and allow for opening of the alveoli.

Breathing and gas exchange are both altered by changes in the compliance and resistance of the lung. If the compliance of the lung decreases, as occurs in restrictive diseases like fibrosis, the airways stiffen and collapse upon exhalation. Air becomes trapped in the lungs, making breathing more difficult. If resistance increases, as happens with asthma or emphysema, the airways become obstructed, trapping air in the lungs and causing breathing to become difficult. Alterations in the ventilation of the airways or perfusion of the arteries can affect gas exchange. These changes in ventilation and perfusion, called V/Q mismatch, can arise from anatomical or physiological changes.

39.4 Transport of Gases in Human Bodily Fluids

Hemoglobin is a protein found in red blood cells that is comprised of two alpha and two beta subunits that surround an iron-containing heme group. Oxygen readily binds this heme group. The ability of oxygen to bind increases as more oxygen molecules are bound to heme. Disease states and altered conditions in the body can affect the binding ability of oxygen, and increase or decrease its ability to dissociate from hemoglobin.

Carbon dioxide can be transported through the blood via three methods. It is dissolved directly in the blood, bound to plasma proteins or hemoglobin, or converted into bicarbonate. The majority of carbon dioxide is transported as part of the bicarbonate system. Carbon dioxide diffuses into red blood cells. Inside, carbonic anhydrase converts carbon dioxide into carbonic acid (H_2CO_3), which is subsequently hydrolyzed into bicarbonate (HCO_3^-) and H^+. The H^+ ion binds to hemoglobin in red blood cells, and bicarbonate is transported out of the red blood cells in exchange for a chloride ion. This is called the chloride shift. Bicarbonate leaves the red blood cells and enters the blood plasma. In the lungs, bicarbonate is transported back into the red blood cells in exchange for chloride. The H^+ dissociates from hemoglobin and combines with bicarbonate to form carbonic acid with the help of carbonic anhydrase, which further catalyzes the reaction to convert carbonic acid back into carbon dioxide and water. The carbon dioxide is then expelled from the lungs.

VISUAL CONNECTION QUESTIONS

1. Figure 39.7 Which of the following statements about the mammalian respiratory system is false?
   a. When we breathe in, air travels from the pharynx to the trachea.
   b. The bronchioles branch into bronchi.
   c. Alveolar ducts connect to alveolar sacs.
   d. Gas exchange between the lung and blood takes place in the alveolus.

2. Figure 39.13 Which of the following statements is false?
   a. In the tissues, P_{O_2} drops as blood passes from the arteries to the veins, while P_{CO_2} increases.
   b. Blood travels from the lungs to the heart to body tissues, then back to the heart, then the lungs.
   c. Blood travels from the lungs to the heart to body tissues, then back to the lungs, then the heart.
   d. P_{O_2} is higher in air than in the lungs.

3. Figure 39.20 The kidneys are responsible for removing excess H+ ions from the blood. If the kidneys fail, what would happen to blood pH and to hemoglobin affinity for oxygen?

REVIEW QUESTIONS

4. The respiratory system _______.
   a. provides body tissues with oxygen
   b. provides body tissues with oxygen and carbon dioxide
   c. establishes how many breaths are taken per minute
   d. provides the body with carbon dioxide

5. Air is warmed and humidified in the nasal passages. This helps to _______.
   a. ward off infection
   b. decrease sensitivity during breathing
   c. prevent damage to the lungs
   d. all of the above

6. Which is the order of airflow during inhalation?
   a. nasal cavity, trachea, larynx, bronchi, bronchioles, alveoli
   b. nasal cavity, larynx, trachea, bronchi, bronchioles, alveoli
   c. nasal cavity, larynx, trachea, bronchioles, bronchi, alveoli
   d. nasal cavity, trachea, larynx, bronchioles, bronchi, alveoli

7. The inspiratory reserve volume measures the _______.
   a. amount of air remaining in the lung after a maximal exhalation
   b. amount of air that the lung holds
   c. amount of air that can be further exhaled after a normal breath
   d. amount of air that can be further inhaled after a normal breath

8. Of the following, which does not explain why the partial pressure of oxygen is lower in the lung than in the external air?
   a. Air in the lung is humidified; therefore, water vapor pressure alters the pressure.
   b. Carbon dioxide mixes with oxygen.
   c. Oxygen is moved into the blood and is headed to the tissues.
   d. Lungs exert a pressure on the air to reduce the oxygen pressure.

9. The total lung capacity is calculated using which of the following formulas?
   a. residual volume + tidal volume + inspiratory reserve volume
   b. residual volume + expiratory reserve volume + inspiratory reserve volume
   c. expiratory reserve volume + tidal volume + inspiratory reserve volume
   d. residual volume + expiratory reserve volume + tidal volume + inspiratory reserve volume

10. How would paralysis of the diaphragm alter inspiration?
    a. It would prevent contraction of the intercostal muscles.
    b. It would prevent inhalation because the intrapleural pressure would not change.
    c. It would decrease the intrapleural pressure and allow more air to enter the lungs.
    d. It would slow expiration because the lung would not relax.

11. Restrictive airway diseases _______.
    a. increase the compliance of the lung
    b. decrease the compliance of the lung
    c. increase the lung volume
    d. decrease the work of breathing

12. Alveolar ventilation remains constant when _______.
    a. the respiratory rate is increased while the volume of air per breath is decreased
    b. the respiratory rate and the volume of air per breath are increased
    c. the respiratory rate is decreased while increasing the volume per breath
    d. both a and c

13. Which of the following will NOT facilitate the transfer of oxygen to tissues?
    a. decreased body temperature
    b. decreased pH of the blood
    c. increased carbon dioxide
    d. increased exercise

14. The majority of carbon dioxide in the blood is transported by _______.
    a. binding to hemoglobin
    b. dissolution in the blood
    c. conversion to bicarbonate
    d. binding to plasma proteins

15. The majority of oxygen in the blood is transported by _______.
    a. dissolution in the blood
    b. being carried as bicarbonate ions
    c. binding to blood plasma
    d. binding to hemoglobin

CRITICAL THINKING QUESTIONS

16. Describe the function of these terms and describe where they are located: main bronchus, trachea, alveoli, and acinus.
17. How does the structure of alveoli maximize gas exchange?
18. What does FEV1/FVC measure? What factors may affect FEV1/FVC?
19. What is the reason for having residual volume in the lung?
20. How can a decrease in the percent of oxygen in the air affect the movement of oxygen in the body?
21. If a patient has increased resistance in his or her lungs, how can this be detected by a doctor? What does this mean?
22. How would increased airway resistance affect intrapleural pressure during inhalation?
23. Explain how a puncture to the thoracic cavity (from a knife wound, for instance) could alter the ability to inhale.
24. When someone is standing, gravity stretches the bottom of the lung down toward the floor to a greater extent than the top of the lung. What implication could this have on the flow of air in the lungs? Where does gas exchange occur in the lungs?
25. What would happen if no carbonic anhydrase were present in red blood cells?
26. How does the administration of 100 percent oxygen save a patient from carbon monoxide poisoning? Why wouldn't giving carbon dioxide work?

CHAPTER 40
The Circulatory System

Figure 40.1 Just as highway systems transport people and goods through a complex network, the circulatory system transports nutrients, gases, and wastes throughout the animal body. (credit: modification of work by Andrey Belenko)

Chapter Outline

INTRODUCTION Most animals are complex multicellular organisms that require a mechanism for transporting nutrients throughout their bodies and removing waste products. The circulatory system has evolved over time from simple diffusion through cells in the early evolution of animals to a complex network of blood vessels that reach all parts of the human body. This extensive network supplies the cells, tissues, and organs with oxygen and nutrients, and removes carbon dioxide and waste, which are byproducts of respiration.

At the core of the human circulatory system is the heart. The size of a clenched fist, the human heart is protected beneath the rib cage. Made of specialized and unique cardiac muscle, it pumps blood throughout the body and to the heart itself. Heart contractions are driven by intrinsic electrical impulses that the brain and endocrine hormones help to regulate. Understanding the heart's basic anatomy and function is important to understanding the body's circulatory and respiratory systems.

Gas exchange is one essential function of the circulatory system. A circulatory system is not needed in organisms with no specialized respiratory organs because oxygen and carbon dioxide diffuse directly between their body tissues and the external environment. However, in organisms that possess lungs and gills, oxygen must be transported from these specialized respiratory organs to the body tissues via a circulatory system. Therefore, circulatory systems have had to evolve to accommodate the great diversity of body sizes and body types present among animals.

40.1 Overview of the Circulatory System

By the end of this section, you will be able to do the following:

- Describe an open and closed circulatory system
- Describe interstitial fluid and hemolymph
- Compare and contrast the organization and evolution of the vertebrate circulatory system

In all animals, except a few simple types, the circulatory system is used to transport nutrients and gases through the body. Simple diffusion allows some water, nutrient, waste, and gas exchange into primitive animals that are only a few cell layers thick; however, bulk flow is the only method by which the entire body of larger more complex organisms is accessed.

Circulatory System Architecture

The circulatory system is effectively a network of cylindrical vessels: the arteries, veins, and capillaries that emanate from a pump, the heart. In all vertebrate organisms, as well as some invertebrates, this is a closed-loop system, in which the blood is not free in a cavity. In a **closed circulatory system**, blood is contained inside blood vessels and circulates **unidirectionally** from the heart around the systemic circulatory route, then returns to the heart again, as illustrated in Figure 40.2a. As opposed to a closed system, arthropods—including insects, crustaceans, and most mollusks—have an open circulatory system, as illustrated in Figure 40.2b. In an **open circulatory system**, the blood is not enclosed in the blood vessels but is pumped into a cavity called a **hemocoel** and is called **hemolymph** because the blood mixes with the **interstitial fluid**. As the heart beats and the animal moves, the hemolymph circulates around the organs within the body cavity and then reenters the hearts through openings called **ostia**. This movement allows for gas and nutrient exchange. An open circulatory system does not use as much energy as a closed system to operate or to maintain; however, there is a trade-off with the amount of blood that can be moved to metabolically active organs and tissues that require high levels of oxygen. In fact, one reason that insects with wing spans of up to two feet wide (70 cm) are not around today is probably because they were outcompeted by the arrival of birds 150 million years ago. Birds, having a closed circulatory system, are thought to have moved more agilely, allowing them to get food faster and possibly to prey on the insects.

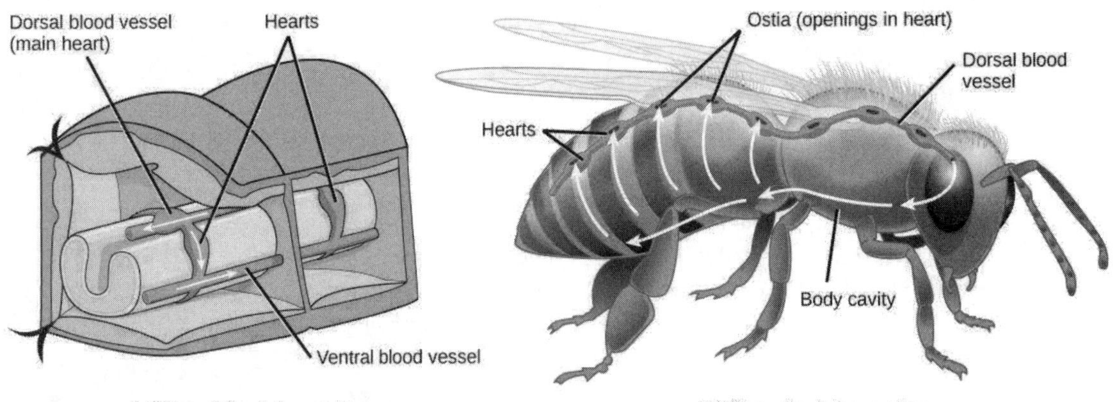

Figure 40.2 In (a) closed circulatory systems, the heart pumps blood through vessels that are separate from the interstitial fluid of the body. Most vertebrates and some invertebrates, like this annelid earthworm, have a closed circulatory system. In (b) open circulatory systems, a fluid called hemolymph is pumped through a blood vessel that empties into the body cavity. Hemolymph returns to the blood vessel through openings called ostia. Arthropods like this bee and most mollusks have open circulatory systems.

Circulatory System Variation in Animals

The circulatory system varies from simple systems in invertebrates to more complex systems in vertebrates. The simplest animals, such as the sponges (Porifera) and rotifers (Rotifera), do not need a circulatory system because diffusion allows adequate exchange of water, nutrients, and waste, as well as dissolved gases, as shown in Figure 40.3a. Organisms that are more complex but still only have two layers of cells in their body plan, such as jellies (Cnidaria) and comb jellies (Ctenophora) also use diffusion through their epidermis and internally through the gastrovascular compartment. Both their internal and external tissues are bathed in an aqueous environment and exchange fluids by diffusion on both sides, as illustrated in Figure 40.3b. Exchange of fluids is assisted by the pulsing of the jellyfish body.

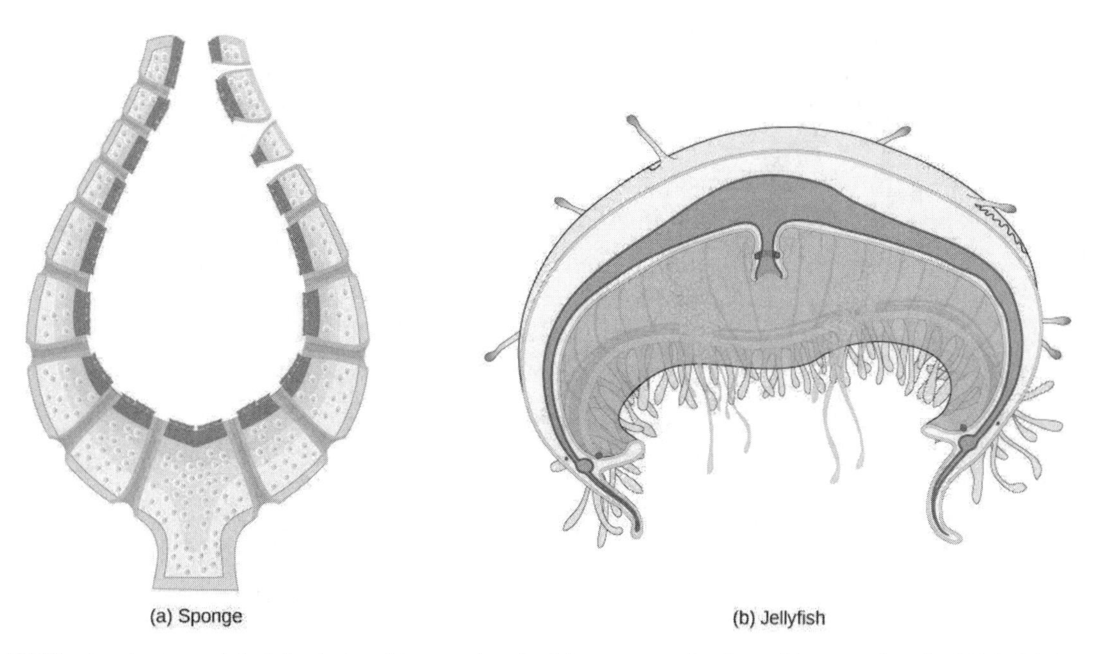

Figure 40.3 Simple animals consisting of a single cell layer such as the (a) sponge or only a few cell layers such as the (b) jellyfish do not have a circulatory system. Instead, gases, nutrients, and wastes are exchanged by diffusion.

For more complex organisms, diffusion is not efficient for cycling gases, nutrients, and waste effectively through the body; therefore, more complex circulatory systems evolved. Most arthropods and many mollusks have open circulatory systems. In an open system, an elongated beating heart pushes the hemolymph through the body and muscle contractions help to move fluids. The larger more complex crustaceans, including lobsters, have developed arterial-like vessels to push blood through their bodies, and the most active mollusks, such as squids, have evolved a closed circulatory system and are able to move rapidly to catch prey. Closed circulatory systems are a characteristic of vertebrates; however, there are significant differences in the structure of the heart and the circulation of blood between the different vertebrate groups due to adaptation during evolution and associated differences in anatomy. Figure 40.4 illustrates the basic circulatory systems of some vertebrates: fish, amphibians, reptiles, and mammals.

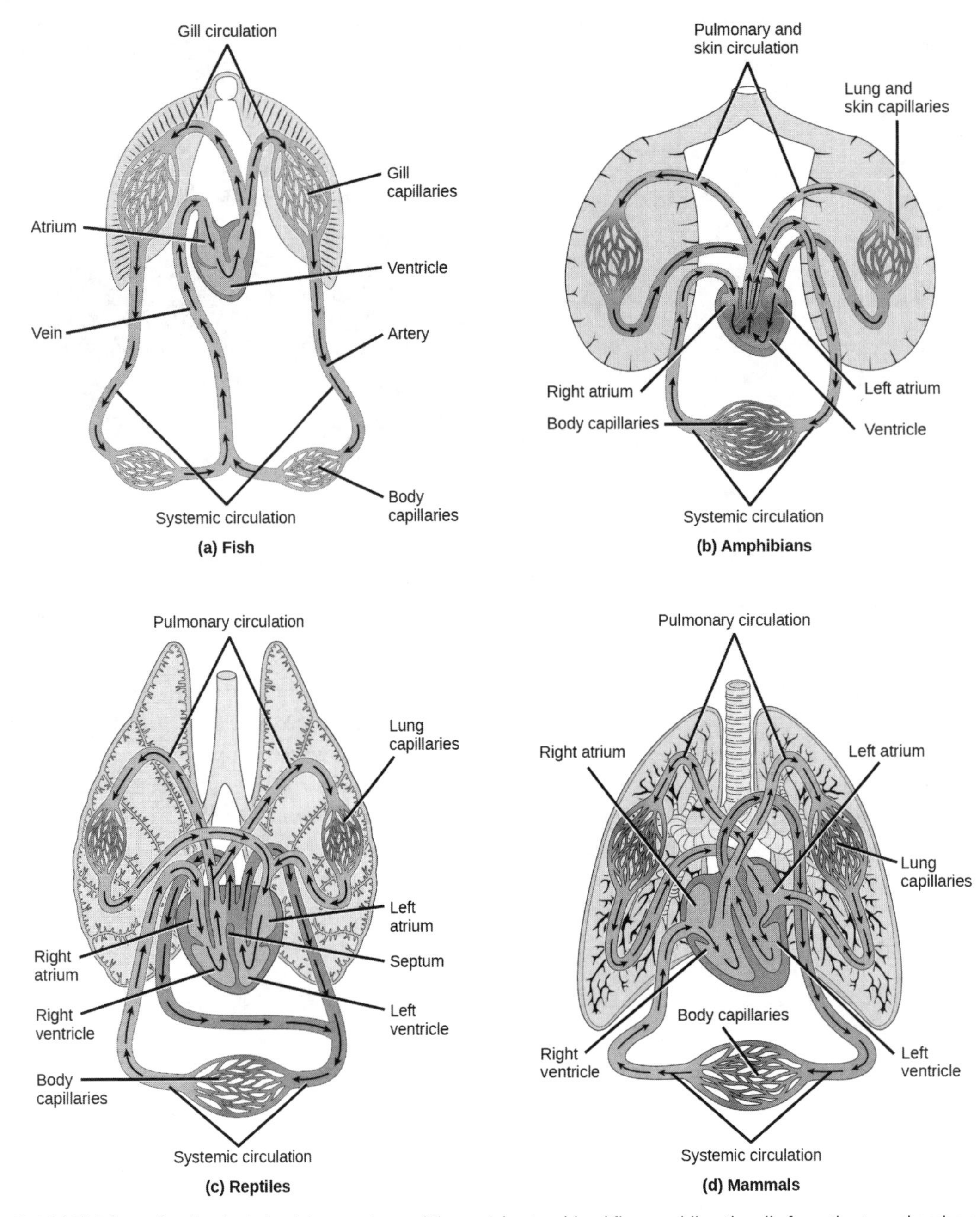

Figure 40.4 (a) Fish have the simplest circulatory systems of the vertebrates: blood flows unidirectionally from the two-chambered heart through the gills and then the rest of the body. (b) Amphibians have two circulatory routes: one for oxygenation of the blood through the lungs and skin, and the other to take oxygen to the rest of the body. The blood is pumped from a three-chambered heart with two atria and a single ventricle. (c) Reptiles also have two circulatory routes; however, blood is only oxygenated through the lungs. The heart is three chambered, but the ventricles are partially separated so some mixing of oxygenated and deoxygenated blood occurs except in crocodilians and birds. (d) Mammals and birds have the most efficient heart with four chambers that completely separate the oxygenated and deoxygenated blood; it pumps only oxygenated blood through the body and deoxygenated blood to the lungs.

As illustrated in Figure 40.4a. Fish have a single circuit for blood flow and a two-chambered heart that has only a single atrium and a single ventricle. The atrium collects blood that has returned from the body and the ventricle pumps the blood to the gills where gas exchange occurs and the blood is re-oxygenated; this is called **gill circulation**. The blood then continues through the rest of the body before arriving back at the atrium; this is called **systemic circulation**. This unidirectional flow of blood produces a gradient of oxygenated to deoxygenated blood around the fish's systemic circuit. The result is a limit in the amount of oxygen that can reach some of the organs and tissues of the body, reducing the overall metabolic capacity of fish.

In amphibians, reptiles, birds, and mammals, blood flow is directed in two circuits: one through the lungs and back to the heart, which is called **pulmonary circulation**, and the other throughout the rest of the body and its organs including the brain (systemic circulation). In amphibians, gas exchange also occurs through the skin during pulmonary circulation and is referred to as **pulmocutaneous circulation**.

As shown in Figure 40.4b, amphibians have a three-chambered heart that has two atria and one ventricle rather than the two-chambered heart of fish. The two **atria** (superior heart chambers) receive blood from the two different circuits (the lungs and the systems), and then there is some mixing of the blood in the heart's **ventricle** (inferior heart chamber), which reduces the efficiency of oxygenation. The advantage to this arrangement is that high pressure in the vessels pushes blood to the lungs and body. The mixing is mitigated by a ridge within the ventricle that diverts oxygen-rich blood through the systemic circulatory system and deoxygenated blood to the pulmocutaneous circuit. For this reason, amphibians are often described as having **double circulation**.

Most reptiles also have a three-chambered heart similar to the amphibian heart that directs blood to the pulmonary and systemic circuits, as shown in Figure 40.4c. The ventricle is divided more effectively by a partial septum, which results in less mixing of oxygenated and deoxygenated blood. Some reptiles (alligators and crocodiles) are the most primitive animals to exhibit a four-chambered heart. Crocodilians have a unique circulatory mechanism where the heart shunts blood from the lungs toward the stomach and other organs during long periods of submergence, for instance, while the animal waits for prey or stays underwater waiting for prey to rot. One adaptation includes two main arteries that leave the same part of the heart: one takes blood to the lungs and the other provides an alternate route to the stomach and other parts of the body. Two other adaptations include a hole in the heart between the two ventricles, called the foramen of Panizza, which allows blood to move from one side of the heart to the other, and specialized connective tissue that slows the blood flow to the lungs. Together these adaptations have made crocodiles and alligators one of the most evolutionarily successful animal groups on earth.

In mammals and birds, the heart is also divided into four chambers: two atria and two ventricles, as illustrated in Figure 40.4d. The oxygenated blood is separated from the deoxygenated blood, which improves the efficiency of double circulation and is probably required for the warm-blooded lifestyle of mammals and birds. The four-chambered heart of birds and mammals evolved independently from a three-chambered heart. The independent evolution of the same or a similar biological trait is referred to as convergent evolution.

40.2 Components of the Blood

By the end of this section, you will be able to do the following:

- List the basic components of the blood
- Compare red and white blood cells
- Describe blood plasma and serum

Hemoglobin is responsible for distributing oxygen, and to a lesser extent, carbon dioxide, throughout the circulatory systems of humans, vertebrates, and many invertebrates. The blood is more than the proteins, though. Blood is actually a term used to describe the liquid that moves through the vessels and includes **plasma** (the liquid portion, which contains water, proteins, salts, lipids, and glucose) and the cells (red and white cells) and cell fragments called **platelets**. Blood plasma is actually the dominant component of blood and contains the water, proteins, electrolytes, lipids, and glucose. The cells are responsible for carrying the gases (red cells) and the immune response (white). The platelets are responsible for blood clotting. Interstitial fluid that surrounds cells is separate from the blood, but in hemolymph, they are combined. In humans, cellular components make up approximately 45 percent of the blood and the liquid plasma 55 percent. Blood is 20 percent of a person's extracellular fluid and eight percent of weight.

The Role of Blood in the Body

Blood, like the human blood illustrated in Figure 40.5 is important for regulation of the body's systems and homeostasis. Blood helps maintain homeostasis by stabilizing pH, temperature, osmotic pressure, and by eliminating excess heat. Blood supports growth by distributing nutrients and hormones, and by removing waste. Blood plays a protective role by transporting clotting factors and platelets to prevent blood loss and transporting the disease-fighting agents or **white blood cells** to sites of infection.

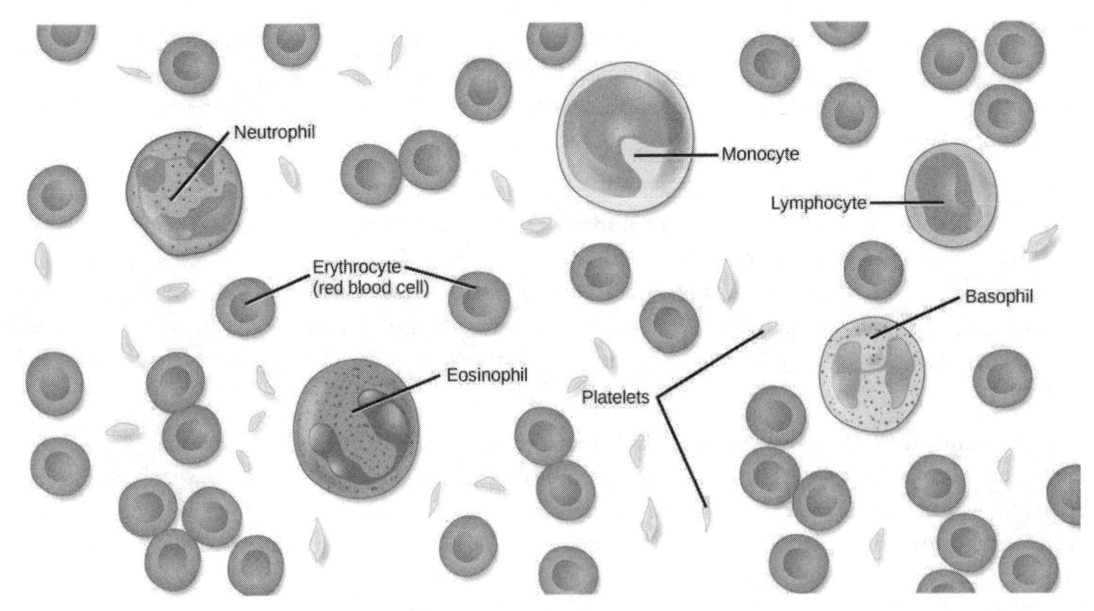

Figure 40.5 The cells and cellular components of human blood are shown. Red blood cells deliver oxygen to the cells and remove carbon dioxide. White blood cells—including neutrophils, monocytes, lymphocytes, eosinophils, and basophils—are involved in the immune response. Platelets form clots that prevent blood loss after injury.

Red Blood Cells

Red blood cells, or erythrocytes (erythro- = "red"; -cyte = "cell"), are specialized cells that circulate through the body delivering oxygen to cells; they are formed from stem cells in the bone marrow. In mammals, red blood cells are small biconcave cells that at maturity do not contain a nucleus or mitochondria and are only 7–8 µm in size. In birds and non-avian reptiles, a nucleus is still maintained in red blood cells.

The red coloring of blood comes from the iron-containing protein hemoglobin, illustrated in Figure 40.6a. The principle job of this protein is to carry oxygen, but it also transports carbon dioxide as well. Hemoglobin is packed into red blood cells at a rate of about 250 million molecules of hemoglobin per cell. Each hemoglobin molecule binds four oxygen molecules so that each red blood cell carries one billion molecules of oxygen. There are approximately 25 trillion red blood cells in the five liters of blood in the human body, which could carry up to 25 sextillion (25×10^{21}) molecules of oxygen in the body at any time. In mammals, the lack of organelles in erythrocytes leaves more room for the hemoglobin molecules, and the lack of mitochondria also prevents use of the oxygen for metabolic respiration. Only mammals have anucleated red blood cells, and some mammals (camels, for instance) even have nucleated red blood cells. The advantage of nucleated red blood cells is that these cells can undergo mitosis. Anucleated red blood cells metabolize anaerobically (without oxygen), making use of a primitive metabolic pathway to produce ATP and increase the efficiency of oxygen transport.

Not all organisms use hemoglobin as the method of oxygen transport. Invertebrates that utilize hemolymph rather than blood use different pigments to bind to the oxygen. These pigments use copper or iron to the oxygen. Invertebrates have a variety of other respiratory pigments. Hemocyanin, a blue-green, copper-containing protein, illustrated in Figure 40.6b is found in mollusks, crustaceans, and some of the arthropods. Chlorocruorin, a green-colored, iron-containing pigment is found in four families of polychaete tubeworms. Hemerythrin, a red, iron-containing protein is found in some polychaete worms and annelids and is illustrated in Figure 40.6c. Despite the name, hemerythrin does not contain a heme group and its oxygen-carrying capacity is poor compared to hemoglobin.

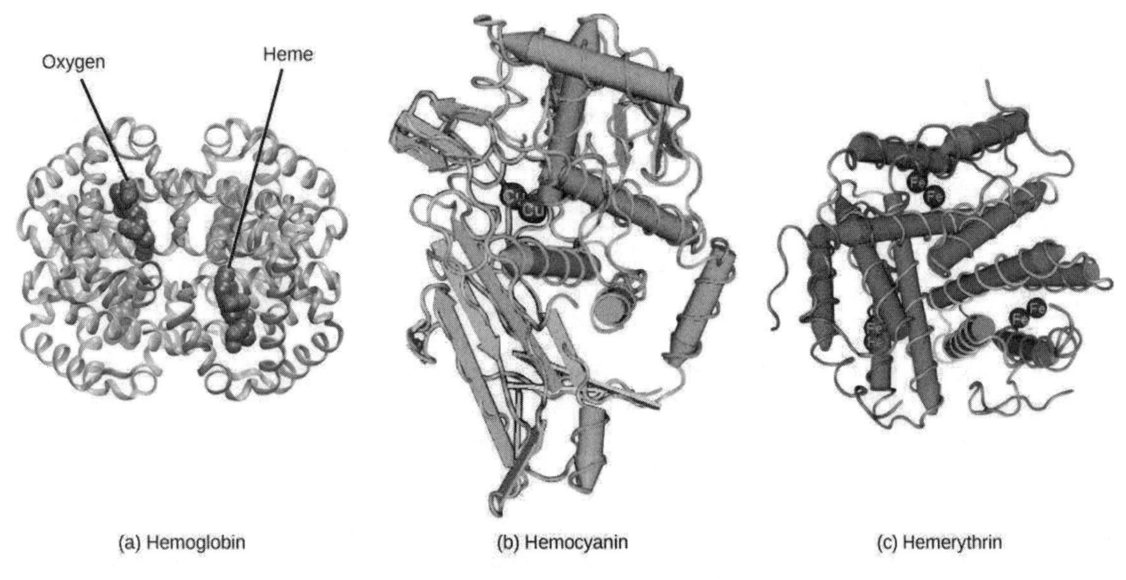

Figure 40.6 In most vertebrates, (a) hemoglobin delivers oxygen to the body and removes some carbon dioxide. Hemoglobin is composed of four protein subunits, two alpha chains and two beta chains, and a heme group that has iron associated with it. The iron reversibly associates with oxygen, and in so doing is oxidized from Fe^{2+} to Fe^{3+}. In most mollusks and some arthropods, (b) hemocyanin delivers oxygen. Unlike hemoglobin, hemolymph is not carried in blood cells, but floats free in the hemolymph. Copper instead of iron binds the oxygen, giving the hemolymph a blue-green color. In annelids, such as the earthworm, and some other invertebrates, (c) hemerythrin carries oxygen. Like hemoglobin, hemerythrin is carried in blood cells and has iron associated with it, but despite its name, hemerythrin does not contain heme.

The small size and large surface area of red blood cells allows for rapid diffusion of oxygen and carbon dioxide across the plasma membrane. In the lungs, carbon dioxide is released and oxygen is taken in by the blood. In the tissues, oxygen is released from the blood and carbon dioxide is bound for transport back to the lungs. Studies have found that hemoglobin also binds nitrous oxide (NO). NO is a vasodilator that relaxes the blood vessels and capillaries and may help with gas exchange and the passage of red blood cells through narrow vessels. Nitroglycerin, a heart medication for angina and heart attacks, is converted to NO to help relax the blood vessels and increase oxygen flow through the body.

A characteristic of red blood cells is their glycolipid and glycoprotein coating; these are lipids and proteins that have carbohydrate molecules attached. In humans, the surface glycoproteins and glycolipids on red blood cells vary between individuals, producing the different blood types, such as A, B, and O. Red blood cells have an average life span of 120 days, at which time they are broken down and recycled in the liver and spleen by phagocytic macrophages, a type of white blood cell.

White Blood Cells

White blood cells, also called leukocytes (leuko = white), make up approximately one percent by volume of the cells in blood. The role of white blood cells is very different than that of red blood cells: they are primarily involved in the immune response to identify and target pathogens, such as invading bacteria, viruses, and other foreign organisms. White blood cells are formed continually; some only live for hours or days, but some live for years.

The morphology of white blood cells differs significantly from red blood cells. They have nuclei and do not contain hemoglobin. The different types of white blood cells are identified by their microscopic appearance after histologic staining, and each has a different specialized function. The two main groups, both illustrated in Figure 40.7 are the granulocytes, which include the neutrophils, eosinophils, and basophils, and the agranulocytes, which include the monocytes and lymphocytes.

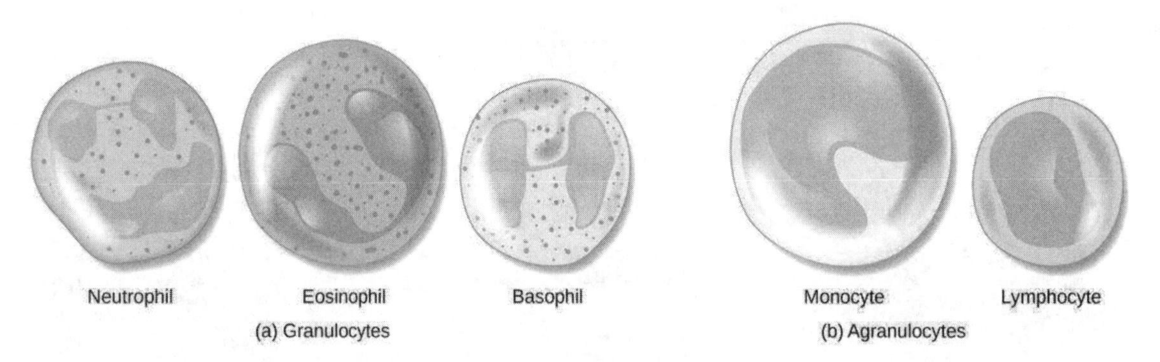

Figure 40.7 (a) Granulocytes—including neutrophils, eosinophils and basophils—are characterized by a lobed nucleus and granular inclusions in the cytoplasm. Granulocytes are typically first-responders during injury or infection. (b) Agranulocytes include lymphocytes and monocytes. Lymphocytes, including B and T cells, are responsible for adaptive immune response. Monocytes differentiate into macrophages and dendritic cells, which in turn respond to infection or injury.

Granulocytes contain granules in their cytoplasm; the agranulocytes are so named because of the lack of granules in their cytoplasm. Some leukocytes become macrophages that either stay at the same site or move through the bloodstream and gather at sites of infection or inflammation where they are attracted by chemical signals from foreign particles and damaged cells. Lymphocytes are the primary cells of the immune system and include B cells, T cells, and natural killer cells. B cells destroy bacteria and inactivate their toxins. They also produce antibodies. T cells attack viruses, fungi, some bacteria, transplanted cells, and cancer cells. T cells attack viruses by releasing toxins that kill the viruses. Natural killer cells attack a variety of infectious microbes and certain tumor cells.

One reason that HIV poses significant management challenges is because the virus directly targets T cells by gaining entry through a receptor. Once inside the cell, HIV then multiplies using the T cell's own genetic machinery. After the HIV virus replicates, it is transmitted directly from the infected T cell to macrophages. The presence of HIV can remain unrecognized for an extensive period of time before full disease symptoms develop.

Platelets and Coagulation Factors

Blood must clot to heal wounds and prevent excess blood loss. Small cell fragments called platelets (thrombocytes) are attracted to the wound site where they adhere by extending many projections and releasing their contents. These contents activate other platelets and also interact with other coagulation factors, which convert fibrinogen, a water-soluble protein present in blood serum into fibrin (a non-water soluble protein), causing the blood to clot. Many of the clotting factors require vitamin K to work, and vitamin K deficiency can lead to problems with blood clotting. Many platelets converge and stick together at the wound site forming a platelet plug (also called a fibrin clot), as illustrated in Figure 40.8b. The plug or clot lasts for a number of days and stops the loss of blood. Platelets are formed from the disintegration of larger cells called megakaryocytes, like that shown in Figure 40.8a. For each megakaryocyte, 2000–3000 platelets are formed with 150,000 to 400,000 platelets present in each cubic millimeter of blood. Each platelet is disc shaped and 2–4 μm in diameter. They contain many small vesicles but do not contain a nucleus.

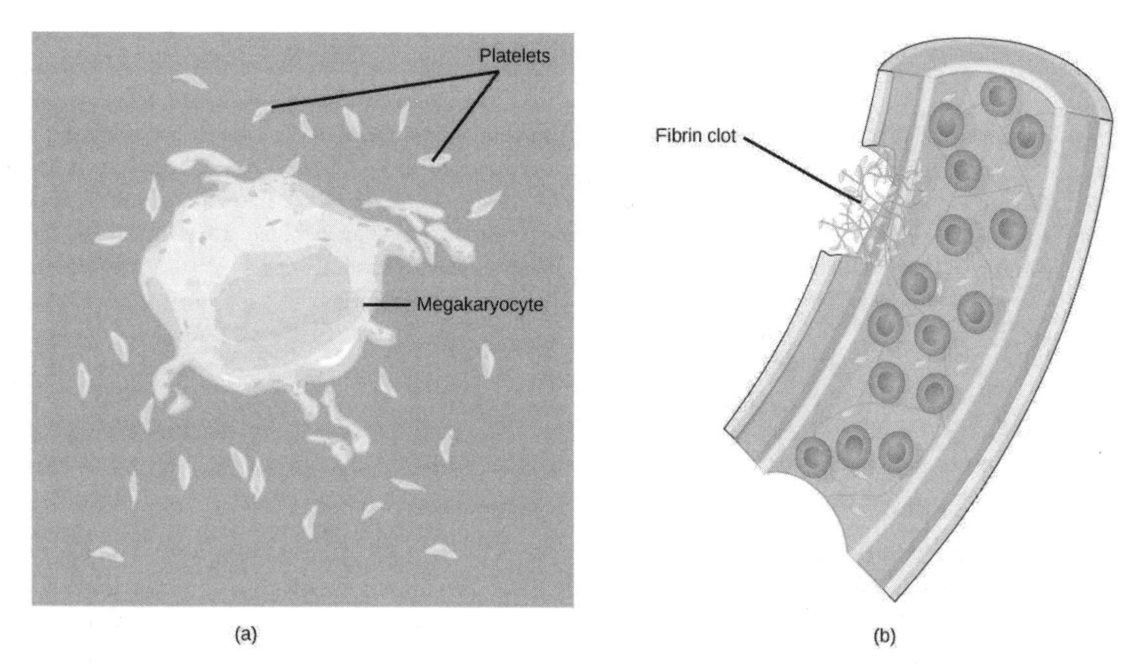

Figure 40.8 (a) Platelets are formed from large cells called megakaryocytes. The megakaryocyte breaks up into thousands of fragments that become platelets. (b) Platelets are required for clotting of the blood. The platelets collect at a wound site in conjunction with other clotting factors, such as fibrinogen, to form a fibrin clot that prevents blood loss and allows the wound to heal.

Plasma and Serum

The liquid component of blood is called plasma, and it is separated by spinning or centrifuging the blood at high rotations (3000 rpm or higher). The blood cells and platelets are separated by centrifugal forces to the bottom of a specimen tube. The upper liquid layer, the plasma, consists of 90 percent water along with various substances required for maintaining the body's pH, osmotic load, and for protecting the body. The plasma also contains the coagulation factors and antibodies.

The plasma component of blood without the coagulation factors is called the **serum**. Serum is similar to interstitial fluid in which the correct composition of key ions acting as electrolytes is essential for normal functioning of muscles and nerves. Other components in the serum include proteins that assist with maintaining pH and osmotic balance while giving viscosity to the blood. The serum also contains antibodies, specialized proteins that are important for defense against viruses and bacteria. Lipids, including cholesterol, are also transported in the serum, along with various other substances including nutrients, hormones, metabolic waste, plus external substances, such as, drugs, viruses, and bacteria.

Human serum albumin is the most abundant protein in human blood plasma and is synthesized in the liver. Albumin, which constitutes about half of the blood serum protein, transports hormones and fatty acids, buffers pH, and maintains osmotic pressures. Immunoglobin is a protein antibody produced in the mucosal lining and plays an important role in antibody mediated immunity.

EVOLUTION CONNECTION

Blood Types Related to Proteins on the Surface of the Red Blood Cells

Red blood cells are coated in antigens made of glycolipids and glycoproteins. The composition of these molecules is determined by genetics, which have evolved over time. In humans, the different surface antigens are grouped into 24 different blood groups with more than 100 different antigens on each red blood cell. The two most well known blood groups are the ABO, shown in Figure 40.9, and Rh systems. The surface antigens in the ABO blood group are glycolipids, called antigen A and antigen B. People with blood type A have antigen A, those with blood type B have antigen B, those with blood type AB have both antigens, and people with blood type O have neither antigen. Antibodies called agglutinougens are found in the blood plasma and react with the A or B antigens, if the two are mixed. When type A and type B blood are combined, agglutination (clumping) of the blood occurs because of antibodies in the plasma that bind with the opposing antigen; this causes clots that coagulate in the kidney causing kidney failure. Type O blood has neither A or B antigens, and therefore, type O blood can be given to all blood types.

Type O negative blood is the universal donor. Type AB positive blood is the universal acceptor because it has both A and B antigen. The ABO blood groups were discovered in 1900 and 1901 by Karl Landsteiner at the University of Vienna.

The Rh blood group was first discovered in Rhesus monkeys. Most people have the Rh antigen (Rh+) and do not have anti-Rh antibodies in their blood. The few people who do not have the Rh antigen and are Rh– can develop anti-Rh antibodies if exposed to Rh+ blood. This can happen after a blood transfusion or after an Rh– woman has an Rh+ baby. The first exposure does not usually cause a reaction; however, at the second exposure, enough antibodies have built up in the blood to produce a reaction that causes agglutination and breakdown of red blood cells. An injection can prevent this reaction.

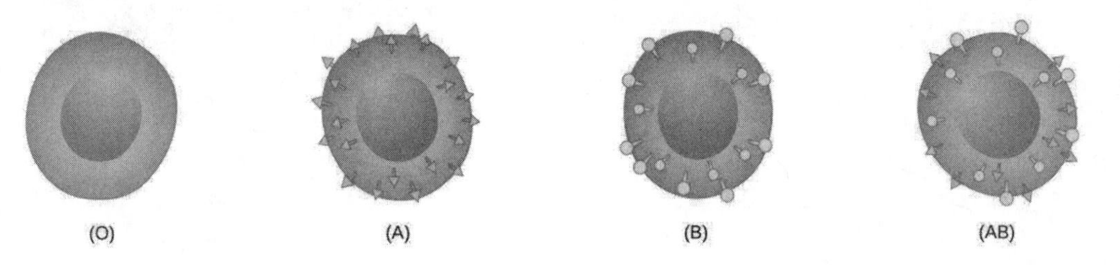

Figure 40.9 Human red blood cells may have either type A or B glycoproteins on their surface, both glycoproteins combined (AB), or neither (O). The glycoproteins serve as antigens and can elicit an immune response in a person who receives a transfusion containing unfamiliar antigens. Type O blood, which has no A or B antigens, does not elicit an immune response when injected into a person of any blood type. Thus, O is considered the universal donor. Persons with type AB blood can accept blood from any blood type, and type AB is considered the universal acceptor.

LINK TO LEARNING

Play a blood typing game on the Nobel Prize website (http://openstax.org/l/blood_typing) to solidify your understanding of blood types.

40.3 Mammalian Heart and Blood Vessels

By the end of this section, you will be able to do the following:

- Describe the structure of the heart and explain how cardiac muscle is different from other muscles
- Describe the cardiac cycle
- Explain the structure of arteries, veins, and capillaries, and how blood flows through the body

The heart is a complex muscle that pumps blood through the three divisions of the circulatory system: the coronary (vessels that serve the heart), pulmonary (heart and lungs), and systemic (systems of the body), as shown in Figure 40.10. Coronary circulation intrinsic to the heart takes blood directly from the main artery (aorta) coming from the heart. For pulmonary and systemic circulation, the heart has to pump blood to the lungs or the rest of the body, respectively. In vertebrates, the lungs are relatively close to the heart in the thoracic cavity. The shorter distance to pump means that the muscle wall on the right side of the heart is not as thick as the left side which must have enough pressure to pump blood all the way to your big toe.

VISUAL CONNECTION

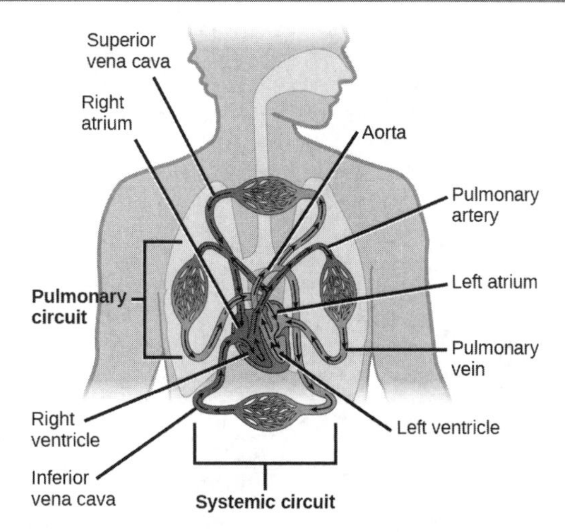

Figure 40.10 The mammalian circulatory system is divided into three circuits: the systemic circuit, the pulmonary circuit, and the coronary circuit. Blood is pumped from veins of the systemic circuit into the right atrium of the heart, then into the right ventricle. Blood then enters the pulmonary circuit, and is oxygenated by the lungs. From the pulmonary circuit, blood reenters the heart through the left atrium. From the left ventricle, blood reenters the systemic circuit through the aorta and is distributed to the rest of the body. The coronary circuit, which provides blood to the heart, is not shown.

Which of the following statements about the circulatory system is false?

a. Blood in the pulmonary vein is deoxygenated.
b. Blood in the inferior vena cava is deoxygenated.
c. Blood in the pulmonary artery is deoxygenated.
d. Blood in the aorta is oxygenated.

Structure of the Heart

The heart muscle is asymmetrical as a result of the distance blood must travel in the pulmonary and systemic circuits. Since the right side of the heart sends blood to the pulmonary circuit it is smaller than the left side which must send blood out to the whole body in the systemic circuit, as shown in Figure 40.11. In humans, the heart is about the size of a clenched fist; it is divided into four chambers: two atria and two ventricles. There is one atrium and one ventricle on the right side and one atrium and one ventricle on the left side. The atria are the chambers that receive blood, and the ventricles are the chambers that pump blood. The right atrium receives deoxygenated blood from the **superior vena cava**, which drains blood from the jugular vein that comes from the brain and from the veins that come from the arms, as well as from the **inferior vena cava** which drains blood from the veins that come from the lower organs and the legs. In addition, the right atrium receives blood from the coronary sinus which drains deoxygenated blood from the heart itself. This deoxygenated blood then passes to the right ventricle through the **atrioventricular valve** or the **tricuspid valve**, a flap of connective tissue that opens in only one direction to prevent the backflow of blood. The valve separating the chambers on the left side of the heart valve is called the biscuspid or mitral valve. After it is filled, the right ventricle pumps the blood through the pulmonary arteries, bypassing the **semilunar valve** (or pulmonic valve) to the lungs for re-oxygenation. After blood passes through the pulmonary arteries, the right semilunar valves close preventing the blood from flowing backwards into the right ventricle. The left atrium then receives the oxygen-rich blood from the lungs via the pulmonary veins. This blood passes through the **bicuspid valve** or mitral valve (the atrioventricular valve on the left side of the heart) to the left ventricle where the blood is pumped out through the **aorta**, the major artery of the body, taking oxygenated blood to the organs and muscles of the body. Once blood is pumped out of the left ventricle and into the aorta, the aortic semilunar valve (or aortic valve) closes preventing blood from flowing backward into the left ventricle. This pattern of pumping is referred to as double circulation and is found in all mammals.

VISUAL CONNECTION

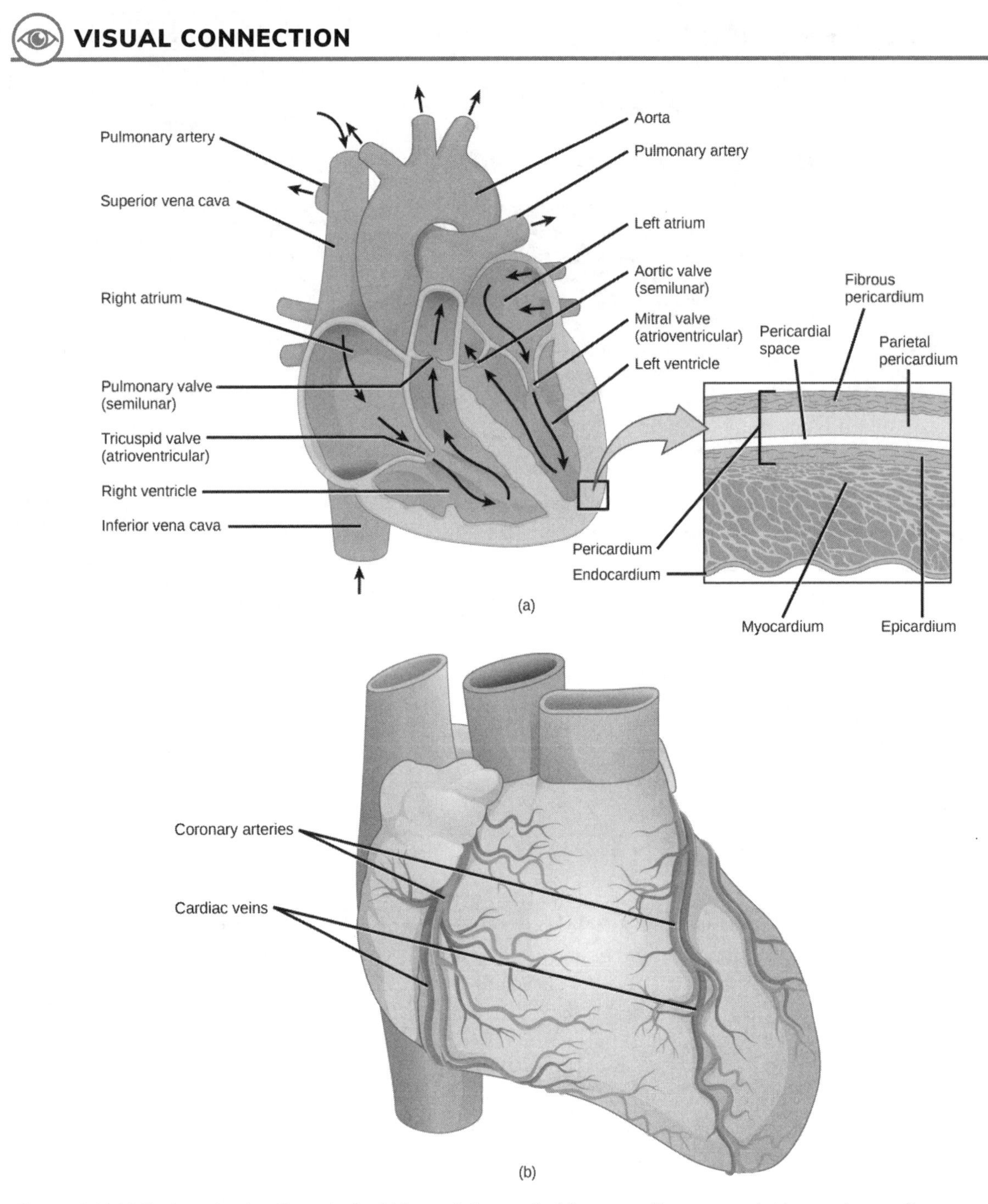

Figure 40.11 (a) The heart is primarily made of a thick muscle layer, called the myocardium, surrounded by membranes. One-way valves separate the four chambers. (b) Blood vessels of the coronary system, including the coronary arteries and veins, keep the heart musculature oxygenated.

Which of the following statements about the heart is false?

a. The mitral valve separates the left ventricle from the left atrium.
b. Blood travels through the bicuspid valve to the left atrium.
c. Both the aortic and the pulmonary valves are semilunar valves.
d. The mitral valve is an atrioventricular valve.

The heart is composed of three layers; the epicardium, the myocardium, and the endocardium, illustrated in Figure 40.11. The inner wall of the heart has a lining called the **endocardium**. The **myocardium** consists of the heart muscle cells that make up the middle layer and the bulk of the heart wall. The outer layer of cells is called the **epicardium**, of which the second layer is a membranous layered structure called the **pericardium** that surrounds and protects the heart; it allows enough room for vigorous pumping but also keeps the heart in place to reduce friction between the heart and other structures.

The heart has its own blood vessels that supply the heart muscle with blood. The **coronary arteries** branch from the aorta and surround the outer surface of the heart like a crown. They diverge into capillaries where the heart muscle is supplied with oxygen before converging again into the **coronary veins** to take the deoxygenated blood back to the right atrium where the blood will be re-oxygenated through the pulmonary circuit. The heart muscle will die without a steady supply of blood. **Atherosclerosis** is the blockage of an artery by the buildup of fatty plaques. Because of the size (narrow) of the coronary arteries and their function in serving the heart itself, atherosclerosis can be deadly in these arteries. The slowdown of blood flow and subsequent oxygen deprivation that results from atherosclerosis causes severe pain, known as **angina**, and complete blockage of the arteries will cause **myocardial infarction**: the death of cardiac muscle tissue, commonly known as a heart attack.

The Cardiac Cycle

The main purpose of the heart is to pump blood through the body; it does so in a repeating sequence called the cardiac cycle. The **cardiac cycle** is the coordination of the filling and emptying of the heart of blood by electrical signals that cause the heart muscles to contract and relax. The human heart beats over 100,000 times per day. In each cardiac cycle, the heart contracts (**systole**), pushing out the blood and pumping it through the body; this is followed by a relaxation phase (**diastole**), where the heart fills with blood, as illustrated in Figure 40.12. The atria contract at the same time, forcing blood through the atrioventricular valves into the ventricles. Closing of the atrioventricular valves produces a monosyllabic "lup" sound. Following a brief delay, the ventricles contract at the same time forcing blood through the semilunar valves into the aorta and the artery transporting blood to the lungs (via the pulmonary artery). Closing of the semilunar valves produces a monosyllabic "dup" sound.

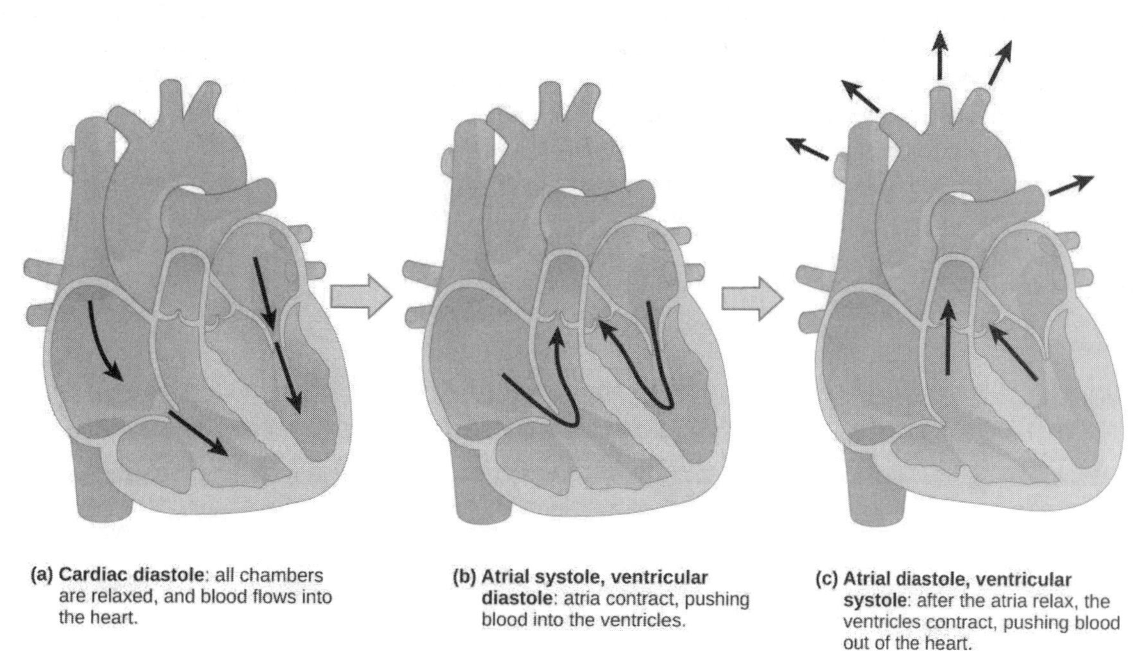

Figure 40.12 During (a) cardiac diastole, the heart muscle is relaxed and blood flows into the heart. During (b) atrial systole, the atria contract, pushing blood into the ventricles. During (c) atrial diastole, the ventricles contract, forcing blood out of the heart.

The pumping of the heart is a function of the cardiac muscle cells, or cardiomyocytes, that make up the heart muscle. **Cardiomyocytes**, shown in Figure 40.13, are distinctive muscle cells that are striated like skeletal muscle but pump rhythmically and involuntarily like smooth muscle; they are connected by intercalated disks exclusive to cardiac muscle. They are self-stimulated for a period of time and isolated cardiomyocytes will beat if given the correct balance of nutrients and electrolytes.

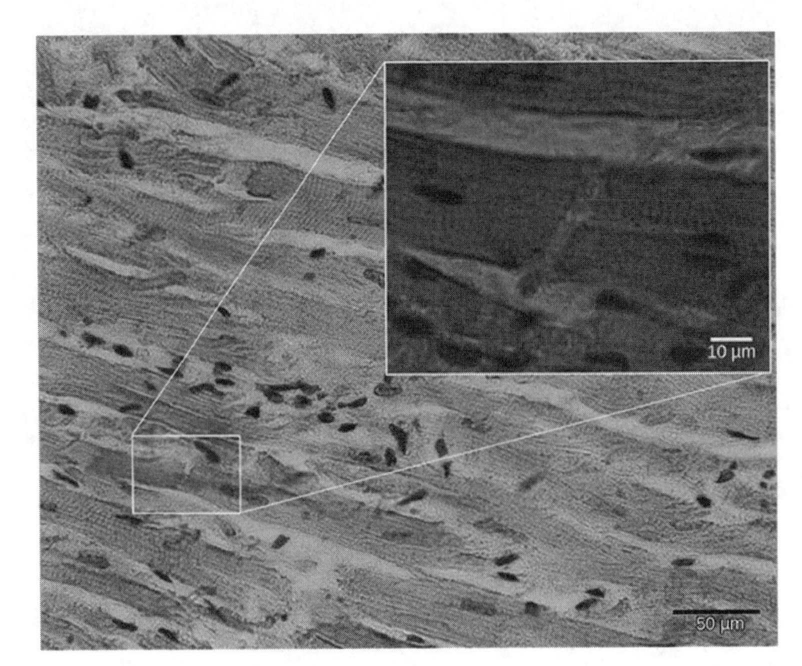

Figure 40.13 Cardiomyocytes are striated muscle cells found in cardiac tissue. (credit: modification of work by Dr. S. Girod, Anton Becker; scale-bar data from Matt Russell)

The autonomous beating of cardiac muscle cells is regulated by the heart's internal pacemaker that uses electrical signals to time the beating of the heart. The electrical signals and mechanical actions, illustrated in Figure 40.14, are intimately intertwined. The internal pacemaker starts at the **sinoatrial (SA) node**, which is located near the wall of the right atrium. Electrical charges spontaneously pulse from the SA node causing the two atria to contract in unison. The pulse reaches a second node, called the atrioventricular (AV) node, between the right atrium and right ventricle where it pauses for approximately 0.1 second before spreading to the walls of the ventricles. From the AV node, the electrical impulse enters the bundle of His, then to the left and right bundle branches extending through the interventricular septum. Finally, the Purkinje fibers conduct the impulse from the apex of the heart up the ventricular myocardium, and then the ventricles contract. This pause allows the atria to empty completely into the ventricles before the ventricles pump out the blood. The electrical impulses in the heart produce electrical currents that flow through the body and can be measured on the skin using electrodes. This information can be observed as an **electrocardiogram (ECG)**—a recording of the electrical impulses of the cardiac muscle.

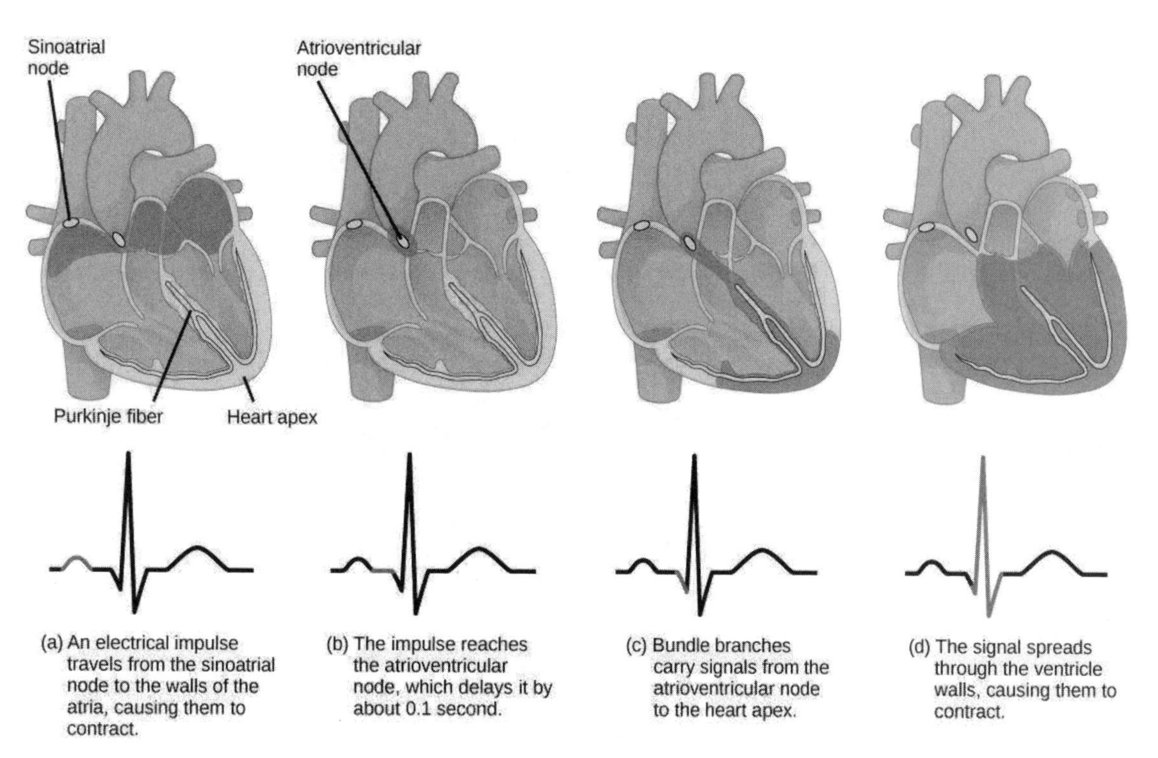

Figure 40.14 The beating of the heart is regulated by an electrical impulse that causes the characteristic reading of an ECG. The signal is initiated at the sinoatrial valve. The signal then (a) spreads to the atria, causing them to contract. The signal is (b) delayed at the atrioventricular node before it is passed on to the (c) heart apex. The delay allows the atria to relax before the (d) ventricles contract. The final part of the ECG cycle prepares the heart for the next beat.

LINK TO LEARNING

Visit this site (http://openstax.org/l/electric_heart) to see the heart's "pacemaker" in action.

Arteries, Veins, and Capillaries

The blood from the heart is carried through the body by a complex network of blood vessels (Figure 40.15). **Arteries** take blood away from the heart. The main artery is the aorta that branches into major arteries that take blood to different limbs and organs. These major arteries include the carotid artery that takes blood to the brain, the brachial arteries that take blood to the arms, and the thoracic artery that takes blood to the thorax and then into the hepatic, renal, and gastric arteries for the liver, kidney, and stomach, respectively. The iliac artery takes blood to the lower limbs. The major arteries diverge into minor arteries, and then smaller vessels called **arterioles**, to reach more deeply into the muscles and organs of the body.

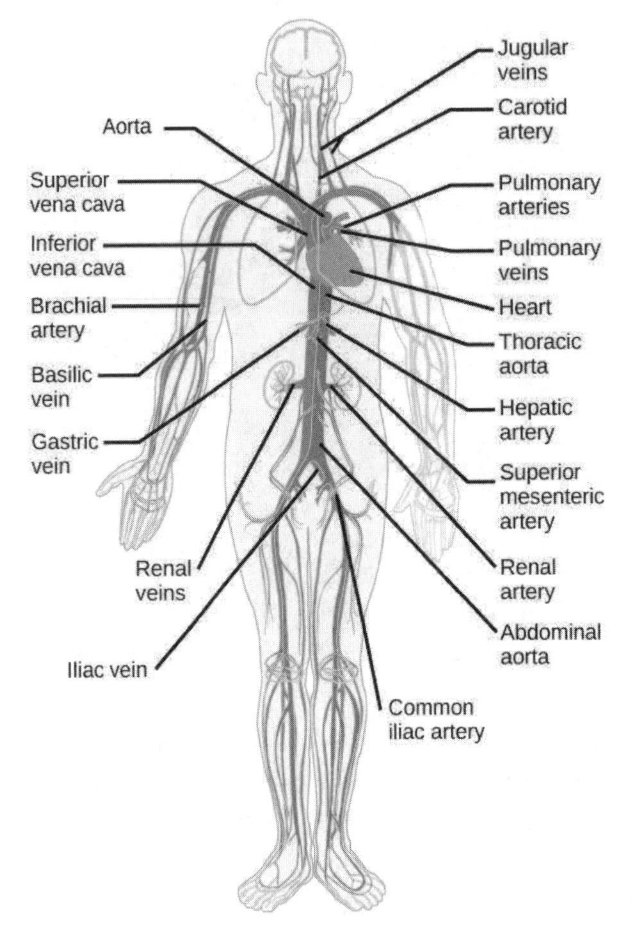

Figure 40.15 The major human arteries and veins are shown. (credit: modification of work by Mariana Ruiz Villareal)

Arterioles diverge into capillary beds. **Capillary beds** contain a large number (10 to 100) of **capillaries** that branch among the cells and tissues of the body. Capillaries are narrow-diameter tubes that can fit red blood cells through in single file and are the sites for the exchange of nutrients, waste, and oxygen with tissues at the cellular level. Fluid also crosses into the interstitial space from the capillaries. The capillaries converge again into **venules** that connect to minor veins that finally connect to major veins that take blood high in carbon dioxide back to the heart. **Veins** are blood vessels that bring blood back to the heart. The major veins drain blood from the same organs and limbs that the major arteries supply. Fluid is also brought back to the heart via the lymphatic system.

The structure of the different types of blood vessels reflects their function or layers. There are three distinct layers, or tunics, that form the walls of blood vessels (Figure 40.16). The first tunic is a smooth, inner lining of endothelial cells that are in contact with the red blood cells. The endothelial tunic is continuous with the endocardium of the heart. In capillaries, this single layer of cells is the location of diffusion of oxygen and carbon dioxide between the endothelial cells and red blood cells, as well as the exchange site via endocytosis and exocytosis. The movement of materials at the site of capillaries is regulated by **vasoconstriction**, narrowing of the blood vessels, and **vasodilation**, widening of the blood vessels; this is important in the overall regulation of blood pressure.

Veins and arteries both have two further tunics that surround the endothelium: the middle tunic is composed of smooth muscle and the outermost layer is connective tissue (collagen and elastic fibers). The elastic connective tissue stretches and supports the blood vessels, and the smooth muscle layer helps regulate blood flow by altering vascular resistance through vasoconstriction and vasodilation. The arteries have thicker smooth muscle and connective tissue than the veins to accommodate the higher pressure and speed of freshly pumped blood. The veins are thinner walled as the pressure and rate of flow are much lower. In addition, veins are structurally different than arteries in that veins have valves to prevent the backflow of blood. Because veins have to work against gravity to get blood back to the heart, contraction of skeletal muscle assists with the flow of blood back to the heart.

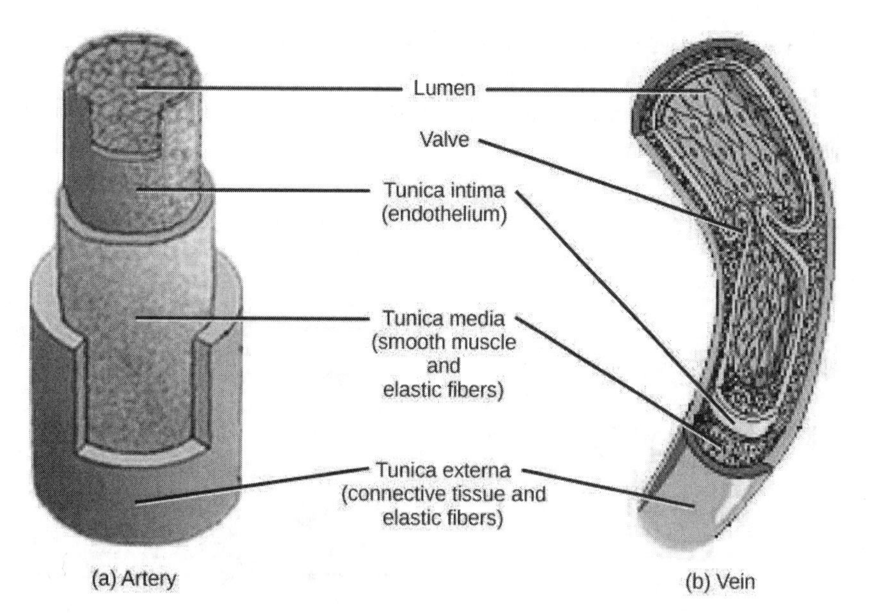

Figure 40.16 Arteries and veins consist of three layers: an outer tunica externa, a middle tunica media, and an inner tunica intima. Capillaries consist of a single layer of epithelial cells, the tunica intima. (credit: modification of work by NCI, NIH)

40.4 Blood Flow and Blood Pressure Regulation

By the end of this section, you will be able to do the following:

- Describe the system of blood flow through the body
- Describe how blood pressure is regulated

Blood pressure (BP) is the pressure exerted by blood on the walls of a blood vessel that helps to push blood through the body. Systolic blood pressure measures the amount of pressure that blood exerts on vessels while the heart is beating. The optimal systolic blood pressure is 120 mmHg. Diastolic blood pressure measures the pressure in the vessels between heartbeats. The optimal diastolic blood pressure is 80 mmHg. Many factors can affect blood pressure, such as hormones, stress, exercise, eating, sitting, and standing. Blood flow through the body is regulated by the size of blood vessels, by the action of smooth muscle, by one-way valves, and by the fluid pressure of the blood itself.

How Blood Flows Through the Body

Blood is pushed through the body by the action of the pumping heart. With each rhythmic pump, blood is pushed under high pressure and velocity away from the heart, initially along the main artery, the aorta. In the aorta, the blood travels at 30 cm/sec. As blood moves into the arteries, arterioles, and ultimately to the capillary beds, the rate of movement slows dramatically to about 0.026 cm/sec, one-thousand times slower than the rate of movement in the aorta. While the diameter of each individual arteriole and capillary is far narrower than the diameter of the aorta, and according to the law of continuity, fluid should travel faster through a narrower diameter tube, the rate is actually slower due to the overall diameter of all the combined capillaries being far greater than the diameter of the individual aorta.

The slow rate of travel through the capillary beds, which reach almost every cell in the body, assists with gas and nutrient exchange and also promotes the diffusion of fluid into the interstitial space. After the blood has passed through the capillary beds to the venules, veins, and finally to the main venae cavae, the rate of flow increases again but is still much slower than the initial rate in the aorta. Blood primarily moves in the veins by the rhythmic movement of smooth muscle in the vessel wall and by the action of the skeletal muscle as the body moves. Because most veins must move blood against the pull of gravity, blood is prevented from flowing backward in the veins by one-way valves. Because skeletal muscle contraction aids in venous blood flow, it is important to get up and move frequently after long periods of sitting so that blood will not pool in the extremities.

Blood flow through the capillary beds is regulated depending on the body's needs and is directed by nerve and hormone signals. For example, after a large meal, most of the blood is diverted to the stomach by vasodilation of vessels of the digestive system and vasoconstriction of other vessels. During exercise, blood is diverted to the skeletal muscles through vasodilation while blood to the digestive system would be lessened through vasoconstriction. The blood entering some capillary beds is controlled by small muscles, called precapillary sphincters, illustrated in Figure 40.17. If the sphincters are open, the blood will flow into the associated branches of the capillary blood. If all of the sphincters are closed, then the blood will flow directly from the arteriole to the venule through the thoroughfare channel (see Figure 40.17). These muscles allow the body to precisely control when capillary beds receive blood flow. At any given moment only about 5–10% of our capillary beds actually have blood flowing through them.

VISUAL CONNECTION

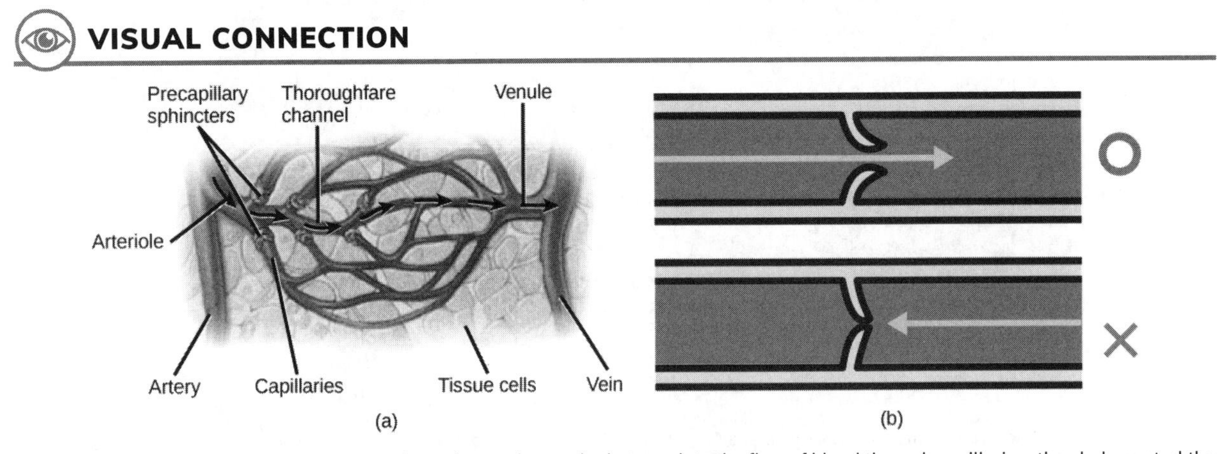

Figure 40.17 (a) Precapillary sphincters are rings of smooth muscle that regulate the flow of blood through capillaries; they help control the location of blood flow to where it is needed. (b) Valves in the veins prevent blood from moving backward. (credit a: modification of work by NCI)

Varicose veins are veins that become enlarged because the valves no longer close properly, allowing blood to flow backward. Varicose veins are often most prominent on the legs. Why do you think this is the case?

LINK TO LEARNING

See the circulatory system's blood flow.

Click to view content (https://www.openstax.org/l/circulation)

Proteins and other large solutes cannot leave the capillaries. The loss of the watery plasma creates a hyperosmotic solution within the capillaries, especially near the venules. This causes about 85% of the plasma that leaves the capillaries to eventually diffuse back into the capillaries near the venules. The remaining 15% of blood plasma drains out from the interstitial fluid into nearby lymphatic vessels (Figure 40.18). The fluid in the lymph is similar in composition to the interstitial fluid. The lymph fluid passes through lymph nodes before it returns to the heart via the vena cava. **Lymph nodes** are specialized organs that filter the lymph by percolation through a maze of connective tissue filled with white blood cells. The white blood cells remove infectious agents, such as bacteria and viruses, to clean the lymph before it returns to the bloodstream. After it is cleaned, the lymph returns to the heart by the action of smooth muscle pumping, skeletal muscle action, and one-way valves joining the returning blood near the junction of the venae cavae entering the right atrium of the heart.

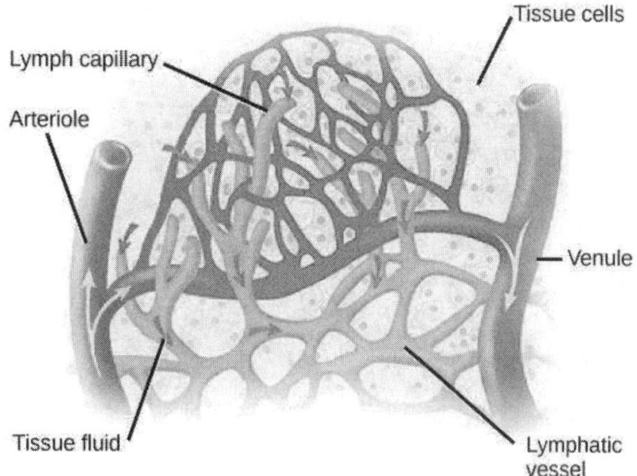

Figure 40.18 Fluid from the capillaries moves into the interstitial space and lymph capillaries by diffusion down a pressure gradient and also by osmosis. Out of 7,200 liters of fluid pumped by the average heart in a day, over 1,500 liters is filtered. (credit: modification of work by NCI, NIH)

EVOLUTION CONNECTION

Vertebrate Diversity in Blood Circulation

Blood circulation has evolved differently in vertebrates and may show variation in different animals for the required amount of pressure, organ and vessel location, and organ size. Animals with longs necks and those that live in cold environments have distinct blood pressure adaptations.

Long necked animals, such as giraffes, need to pump blood upward from the heart against gravity. The blood pressure required from the pumping of the left ventricle would be equivalent to 250 mm Hg (mm Hg = millimeters of mercury, a unit of pressure) to reach the height of a giraffe's head, which is 2.5 meters higher than the heart. However, if checks and balances were not in place, this blood pressure would damage the giraffe's brain, particularly if it was bending down to drink. These checks and balances include valves and feedback mechanisms that reduce the rate of cardiac output. Long-necked dinosaurs such as the sauropods had to pump blood even higher, up to ten meters above the heart. This would have required a blood pressure of more than 600 mm Hg, which could only have been achieved by an enormous heart. Evidence for such an enormous heart does not exist and mechanisms to reduce the blood pressure required include the slowing of metabolism as these animals grew larger. It is likely that they did not routinely feed on tree tops but grazed on the ground.

Living in cold water, whales need to maintain the temperature in their blood. This is achieved by the veins and arteries being close together so that heat exchange can occur. This mechanism is called a countercurrent heat exchanger. The blood vessels and the whole body are also protected by thick layers of blubber to prevent heat loss. In land animals that live in cold environments, thick fur and hibernation are used to retain heat and slow metabolism.

Blood Pressure

The pressure of the blood flow in the body is produced by the hydrostatic pressure of the fluid (blood) against the walls of the blood vessels. Fluid will move from areas of high to low hydrostatic pressures. In the arteries, the hydrostatic pressure near the heart is very high and blood flows to the arterioles where the rate of flow is slowed by the narrow openings of the arterioles. During systole, when new blood is entering the arteries, the artery walls stretch to accommodate the increase of pressure of the extra blood; during diastole, the walls return to normal because of their elastic properties. The blood pressure of the systole phase and the diastole phase, graphed in Figure 40.19, gives the two pressure readings for blood pressure. For example, 120/80 indicates a reading of 120 mm Hg during the systole and 80 mm Hg during diastole. Throughout the cardiac cycle, the blood continues to empty into the arterioles at a relatively even rate. This resistance to blood flow is called **peripheral resistance**.

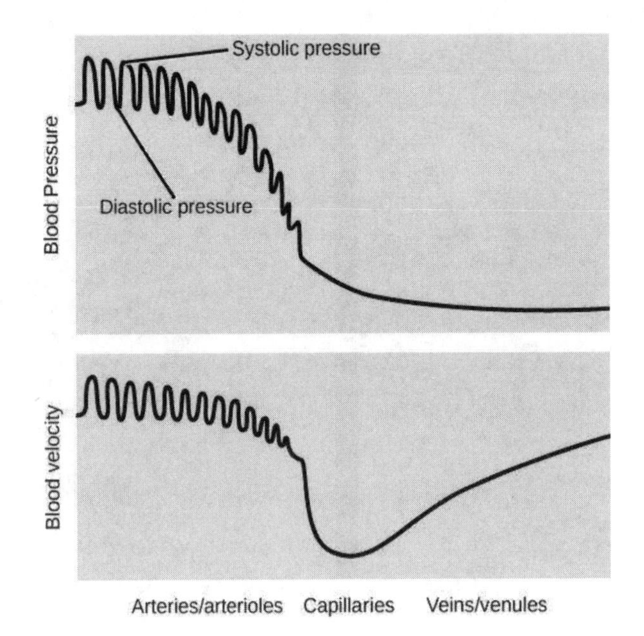

Figure 40.19 Blood pressure is related to the blood velocity in the arteries and arterioles. In the capillaries and veins, the blood pressure continues to decrease but velocity increases.

Blood Pressure Regulation

Cardiac output is the volume of blood pumped by the heart in one minute. It is calculated by multiplying the number of heart contractions that occur per minute (heart rate) times the **stroke volume** (the volume of blood pumped into the aorta per contraction of the left ventricle). Therefore, cardiac output can be increased by increasing heart rate, as when exercising. However, cardiac output can also be increased by increasing stroke volume, such as if the heart contracts with greater strength. Stroke volume can also be increased by speeding blood circulation through the body so that more blood enters the heart between contractions. During heavy exertion, the blood vessels relax and increase in diameter, offsetting the increased heart rate and ensuring adequate oxygenated blood gets to the muscles. Stress triggers a decrease in the diameter of the blood vessels, consequently increasing blood pressure. These changes can also be caused by nerve signals or hormones, and even standing up or lying down can have a great effect on blood pressure.

KEY TERMS

angina pain caused by partial blockage of the coronary arteries by the buildup of plaque and lack of oxygen to the heart muscle

aorta major artery of the body that takes blood away from the heart

arteriole small vessel that connects an artery to a capillary bed

artery blood vessel that takes blood away from the heart

atherosclerosis buildup of fatty plaques in the coronary arteries in the heart

atrioventricular valve one-way membranous flap of connective tissue between the atrium and the ventricle in the right side of the heart; also known as tricuspid valve

atrium (plural: atria) chamber of the heart that receives blood from the veins and sends blood to the ventricles

bicuspid valve (also, mitral valve; left atrioventricular valve) one-way membranous flap between the atrium and the ventricle in the left side of the heart

blood pressure (BP) pressure of blood in the arteries that helps to push blood through the body

capillary smallest blood vessel that allows the passage of individual blood cells and the site of diffusion of oxygen and nutrient exchange

capillary bed large number of capillaries that converge to take blood to a particular organ or tissue

cardiac cycle filling and emptying the heart of blood by electrical signals that cause the heart muscles to contract and relax

cardiac output the volume of blood pumped by the heart in one minute as a product of heart rate multiplied by stroke volume

cardiomyocyte specialized heart muscle cell that is striated but contracts involuntarily like smooth muscle

closed circulatory system system in which the blood is separated from the bodily interstitial fluid and contained in blood vessels

coronary artery vessel that supplies the heart tissue with blood

coronary vein vessel that takes blood away from the heart tissue back to the chambers in the heart

diastole relaxation phase of the cardiac cycle when the heart is relaxed and the ventricles are filling with blood

double circulation flow of blood in two circuits: the pulmonary circuit through the lungs and the systemic circuit through the organs and body

electrocardiogram (ECG) recording of the electrical impulses of the cardiac muscle

endocardium innermost layer of tissue in the heart

epicardium outermost tissue layer of the heart

gill circulation circulatory system that is specific to animals with gills for gas exchange; the blood flows through the gills for oxygenation

hemocoel cavity into which blood is pumped in an open circulatory system

hemolymph mixture of blood and interstitial fluid that is found in insects and other arthropods as well as most mollusks

inferior vena cava drains blood from the veins that come from the lower organs and the legs

interstitial fluid fluid between cells

lymph node specialized organ that contains a large number of macrophages that clean the lymph before the fluid is returned to the heart

myocardial infarction (also, heart attack) complete blockage of the coronary arteries and death of the cardiac muscle tissue

myocardium heart muscle cells that make up the middle layer and the bulk of the heart wall

open circulatory system system in which the blood is mixed with interstitial fluid and directly covers the organs

ostium (plural: ostia) holes between blood vessels that allow the movement of hemolymph through the body of insects, arthropods, and mollusks with open circulatory systems

pericardium membrane layer protecting the heart; also part of the epicardium

peripheral resistance resistance of the artery and blood vessel walls to the pressure placed on them by the force of the heart pumping

plasma liquid component of blood that is left after the cells are removed

platelet (also, thrombocyte) small cellular fragment that collects at wounds, cross-reacts with clotting factors, and forms a plug to prevent blood loss

precapillary sphincter small muscle that controls blood circulation in the capillary beds

pulmocutaneous circulation circulatory system in amphibians; the flow of blood to the lungs and the moist skin for gas exchange

pulmonary circulation flow of blood away from the heart through the lungs where oxygenation occurs and then returns to the heart again

red blood cell small (7–8 µm) biconcave cell without mitochondria (and in mammals without nuclei) that is packed with hemoglobin, giving the cell its red color; transports oxygen through the body

semilunar valve membranous flap of connective tissue between the aorta and a ventricle of the heart (the aortic or pulmonary semilunar valves)

serum plasma without the coagulation factors

sinoatrial (SA) node the heart's internal pacemaker; located near the wall of the right atrium

stroke volume the volume of blood pumped into the aorta

per contraction of the left ventricle

superior vena cava drains blood from the jugular vein that comes from the brain and from the veins that come from the arms

systemic circulation flow of blood away from the heart to the brain, liver, kidneys, stomach, and other organs, the limbs, and the muscles of the body, and then the return of this blood to the heart

systole contraction phase of cardiac cycle when the ventricles are pumping blood into the arteries

tricuspid valve one-way membranous flap of connective tissue between the atrium and the ventricle in the right side of the heart; also known as atrioventricular valve

unidirectional circulation flow of blood in a single circuit; occurs in fish where the blood flows through the gills, then past the organs and the rest of the body, before returning to the heart

vasoconstriction narrowing of a blood vessel

vasodilation widening of a blood vessel

vein blood vessel that brings blood back to the heart

vena cava major vein of the body returning blood from the upper and lower parts of the body; see the superior vena cava and inferior vena cava

ventricle (heart) large inferior chamber of the heart that pumps blood into arteries

venule blood vessel that connects a capillary bed to a vein

white blood cell large (30 µm) cell with nuclei of which there are many types with different roles including the protection of the body from viruses and bacteria, and cleaning up dead cells and other waste

CHAPTER SUMMARY

40.1 Overview of the Circulatory System

In most animals, the circulatory system is used to transport blood through the body. Some primitive animals use diffusion for the exchange of water, nutrients, and gases. However, complex organisms use the circulatory system to carry gases, nutrients, and waste through the body. Circulatory systems may be open (mixed with the interstitial fluid) or closed (separated from the interstitial fluid). Closed circulatory systems are a characteristic of vertebrates; however, there are significant differences in the structure of the heart and the circulation of blood between the different vertebrate groups due to adaptions during evolution and associated differences in anatomy. Fish have a two-chambered heart with unidirectional circulation. Amphibians have a three-chambered heart, which has some mixing of the blood, and they have double circulation. Most non-avian reptiles have a three-chambered heart, but have little mixing of the blood; they have double circulation. Mammals and birds have a four-chambered heart with no mixing of the blood and double circulation.

40.2 Components of the Blood

Specific components of the blood include red blood cells, white blood cells, platelets, and the plasma, which contains coagulation factors and serum. Blood is important for regulation of the body's pH, temperature, osmotic pressure, the circulation of nutrients and removal of waste, the distribution of hormones from endocrine glands, and the elimination of excess heat; it also contains components for blood clotting. Red blood cells are specialized cells that contain hemoglobin and circulate through the body delivering oxygen to cells. White blood cells are involved in the immune response to identify and target invading bacteria, viruses, and other foreign organisms; they also recycle waste components, such as old red blood cells. Platelets and blood clotting factors cause the change of the soluble protein fibrinogen to the insoluble protein fibrin at a wound site forming a plug. Plasma consists of 90 percent water along with various substances, such as coagulation factors and antibodies. The serum is the plasma component of the blood without the coagulation factors.

40.3 Mammalian Heart and Blood Vessels

The heart muscle pumps blood through three divisions of the circulatory system: coronary, pulmonary, and systemic. There is one atrium and one ventricle on the right side and one atrium and one ventricle on the left side. The pumping of the heart is a function of cardiomyocytes, distinctive muscle cells that are striated like skeletal muscle but pump rhythmically and involuntarily like smooth muscle. The internal pacemaker starts at the sinoatrial node, which is located near the wall of the right atrium. Electrical charges pulse from the SA node causing the two atria to contract in unison; then the pulse reaches the atrioventricular node between the right atrium and right ventricle. A pause in the electric signal allows the atria to empty completely into the ventricles before the ventricles pump out the blood. The blood from the heart is carried through the body by a complex network of blood vessels; arteries take blood away from the heart, and veins bring blood back to the heart.

40.4 Blood Flow and Blood Pressure Regulation

Blood primarily moves through the body by the rhythmic movement of smooth muscle in the vessel wall and by the action of the skeletal muscle as the body moves. Blood is prevented from flowing backward in the veins by one-way valves. Blood flow through the capillary beds is controlled by precapillary sphincters to increase and decrease flow depending on the body's needs and is directed by nerve and hormone signals. Lymph vessels take fluid that has leaked out of the blood to the lymph nodes where it is cleaned before returning to the heart. During systole, blood enters the arteries, and the artery walls stretch to accommodate the extra blood. During diastole, the artery walls return to normal. The blood pressure of the systole phase and the diastole phase gives the two pressure readings for blood pressure.

VISUAL CONNECTION QUESTIONS

1. Figure 40.10 Which of the following statements about the circulatory system is false?
   a. Blood in the pulmonary vein is deoxygenated.
   b. Blood in the inferior vena cava is deoxygenated.
   c. Blood in the pulmonary artery is deoxygenated.
   d. Blood in the aorta is oxygenated.
2. Figure 40.11 Which of the following statements about the heart is false?
   a. The mitral valve separates the left ventricle from the left atrium.
   b. Blood travels through the bicuspid valve to the left atrium.
   c. Both the aortic and the pulmonary valves are semilunar valves.
   d. The mitral valve is an atrioventricular valve.
3. Figure 40.17 Varicose veins are veins that become enlarged because the valves no longer close properly, allowing blood to flow backward. Varicose veins are often most prominent on the legs. Why do you think this is the case?

REVIEW QUESTIONS

4. Why are open circulatory systems advantageous to some animals?
   a. They use less metabolic energy.
   b. They help the animal move faster.
   c. They do not need a heart.
   d. They help large insects develop.
5. Some animals use diffusion instead of a circulatory system. Examples include:
   a. birds and jellyfish
   b. flatworms and arthropods
   c. mollusks and jellyfish
   d. none of the above
6. Blood flow that is directed through the lungs and back to the heart is called ________.
   a. unidirectional circulation
   b. gill circulation
   c. pulmonary circulation
   d. pulmocutaneous circulation
7. White blood cells:
   a. can be classified as granulocytes or agranulocytes
   b. defend the body against bacteria and viruses
   c. are also called leucocytes
   d. all of the above
8. Platelet plug formation occurs at which point?
   a. when large megakaryocytes break up into thousands of smaller fragments
   b. when platelets are dispersed through the bloodstream
   c. when platelets are attracted to a site of blood vessel damage
   d. none of the above
9. In humans, the plasma comprises what percentage of the blood?
   a. 45 percent
   b. 55 percent
   c. 25 percent
   d. 90 percent
10. The red blood cells of birds differ from mammalian red blood cells because:
    a. they are white and have nuclei
    b. they do not have nuclei
    c. they have nuclei
    d. they fight disease

11. The heart's internal pacemaker beats by:
    a. an internal implant that sends an electrical impulse through the heart
    b. the excitation of cardiac muscle cells at the sinoatrial node followed by the atrioventricular node
    c. the excitation of cardiac muscle cells at the atrioventricular node followed by the sinoatrial node
    d. the action of the sinus

12. During the systolic phase of the cardiac cycle, the heart is _______.
    a. contracting
    b. relaxing
    c. contracting and relaxing
    d. filling with blood

13. Cardiomyocytes are similar to skeletal muscle because:
    a. they beat involuntarily
    b. they are used for weight lifting
    c. they pulse rhythmically
    d. they are striated

14. How do arteries differ from veins?
    a. Arteries have thicker smooth muscle layers to accommodate the changes in pressure from the heart.
    b. Arteries carry blood.
    c. Arteries have thinner smooth muscle layers and valves and move blood by the action of skeletal muscle.
    d. Arteries are thin walled and are used for gas exchange.

15. High blood pressure would be a result of _______.
    a. a high cardiac output and high peripheral resistance
    b. a high cardiac output and low peripheral resistance
    c. a low cardiac output and high peripheral resistance
    d. a low cardiac output and low peripheral resistance

CRITICAL THINKING QUESTIONS

16. Describe a closed circulatory system.
17. Describe systemic circulation.
18. Describe the cause of different blood type groups.
19. List some of the functions of blood in the body.
20. How does the lymphatic system work with blood flow?
21. Describe the cardiac cycle.
22. What happens in capillaries?
23. How does blood pressure change during heavy exercise?

CHAPTER 41 Osmotic Regulation and Excretion

Figure 41.1 Just as humans recycle what we can and dump the remains into landfills, our bodies use and recycle what they can and excrete the remaining waste products. Our bodies' complex systems have developed ways to treat waste and maintain a balanced internal environment. (credit: modification of work by Redwin Law)

Chapter Outline

INTRODUCTION The daily intake recommendation for human water consumption is eight to ten glasses of water. In order to achieve a healthy balance, the human body should excrete the eight to ten glasses of water every day. This occurs via the processes of urination, defecation, sweating and, to a small extent, respiration. The organs and tissues of the human body are soaked in fluids that are maintained at constant temperature, pH, and solute concentration, all crucial elements of homeostasis. The solutes in body fluids are mainly mineral salts and sugars, and osmotic regulation is the process by which the mineral salts and water are kept in balance. Osmotic homeostasis is maintained despite the influence of external factors like temperature, diet, and weather conditions.

41.1 Osmoregulation and Osmotic Balance

By the end of this section, you will be able to do the following:

- Define osmosis and explain its role within molecules
- Explain why osmoregulation and osmotic balance are important body functions
- Describe active transport mechanisms
- Explain osmolarity and the way in which it is measured
- Describe osmoregulators or osmoconformers and how these tools allow animals to adapt to different environments

Osmosis is the diffusion of water across a membrane in response to **osmotic pressure** caused by an imbalance of molecules on either side of the membrane. **Osmoregulation** is the process of maintenance of salt and water balance (**osmotic balance**) across membranes within the body's fluids, which are composed of water, plus electrolytes and non-electrolytes. An **electrolyte** is a solute that dissociates into ions when dissolved in water. A **non-electrolyte**, in contrast, doesn't dissociate into ions during water dissolution. Both electrolytes and non-electrolytes contribute to the osmotic balance. The body's fluids include blood plasma, the cytosol within cells, and interstitial fluid, the fluid that exists in the spaces between cells and tissues of the body. The membranes of the body (such as the pleural, serous, and cell membranes) are **semi-permeable membranes**. Semi-permeable membranes are permeable (or permissive) to certain types of solutes and water. Solutions on two sides of a semi-permeable membrane tend to equalize in solute concentration by movement of solutes and/or water across the membrane. As seen in Figure 41.2, a cell placed in water tends to swell due to gain of water from the hypotonic or "low salt" environment. A cell placed in a solution with higher salt concentration, on the other hand, tends to make the membrane shrivel up due to loss of water into the hypertonic or "high salt" environment. Isotonic cells have an equal concentration of solutes inside and outside the cell; this equalizes the osmotic pressure on either side of the cell membrane which is a semi-permeable membrane.

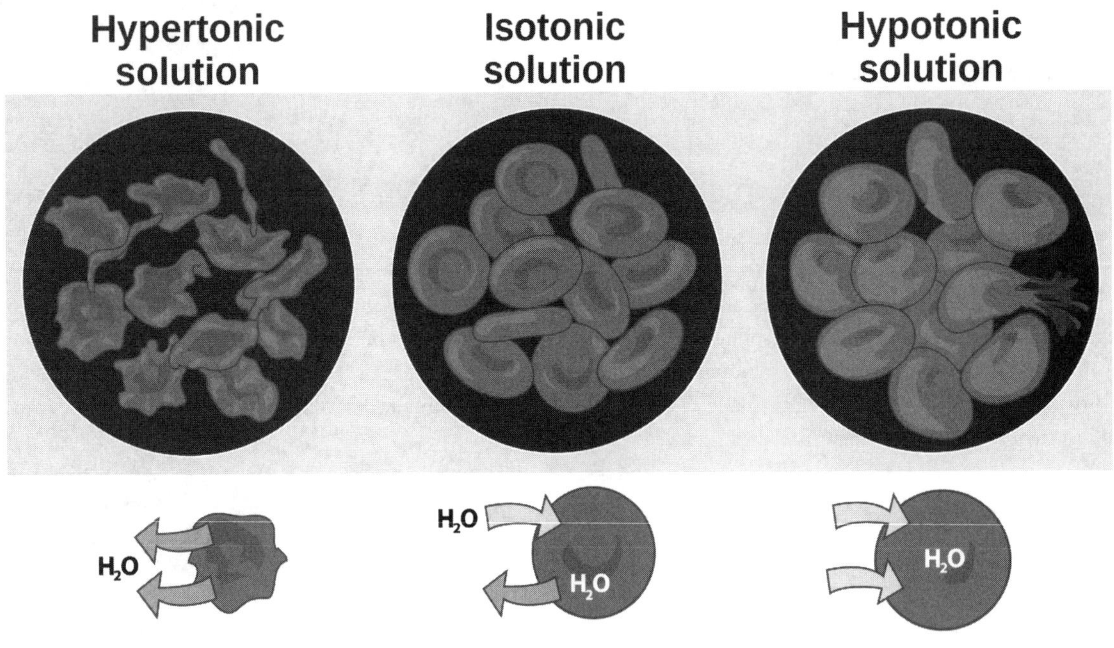

Figure 41.2 Cells placed in a hypertonic environment tend to shrink due to loss of water. In a hypotonic environment, cells tend to swell due to intake of water. The blood maintains an isotonic environment so that cells neither shrink nor swell. (credit: Mariana Ruiz Villareal)

The body does not exist in isolation. There is a constant input of water and electrolytes into the system. While osmoregulation is achieved across membranes within the body, excess electrolytes and wastes are transported to the kidneys and excreted, helping to maintain osmotic balance.

Need for Osmoregulation

Biological systems constantly interact and exchange water and nutrients with the environment by way of consumption of food and water and through excretion in the form of sweat, urine, and feces. Without a mechanism to regulate osmotic pressure, or when a disease damages this mechanism, there is a tendency to accumulate toxic waste and water, which can have dire consequences.

Mammalian systems have evolved to regulate not only the overall osmotic pressure across membranes, but also specific concentrations of important electrolytes in the three major fluid compartments: blood plasma, extracellular fluid, and intracellular fluid. Since osmotic pressure is regulated by the movement of water across membranes, the volume of the fluid compartments can also change temporarily. Because blood plasma is one of the fluid components, osmotic pressures have a direct bearing on blood pressure.

Transport of Electrolytes across Cell Membranes

Electrolytes, such as sodium chloride, ionize in water, meaning that they dissociate into their component ions. In water, sodium chloride (NaCl), dissociates into the sodium ion (Na^+) and the chloride ion (Cl^-). The most important ions, whose concentrations are very closely regulated in body fluids, are the cations sodium (Na^+), potassium (K^+), calcium (Ca^{+2}), magnesium (Mg^{+2}), and the anions chloride (Cl^-), carbonate (CO_3^{-2}), bicarbonate (HCO_3^-), and phosphate(PO_3^-). Electrolytes are lost from the body during urination and perspiration. For this reason, athletes are encouraged to replace electrolytes and fluids during periods of increased activity and perspiration.

Osmotic pressure is influenced by the concentration of solutes in a solution. It is directly proportional to the number of solute atoms or molecules and not dependent on the size of the solute molecules. Because electrolytes dissociate into their component ions, they, in essence, add more solute particles into the solution and have a greater effect on osmotic pressure, per mass than compounds that do not dissociate in water, such as glucose.

Water can pass through membranes by passive diffusion. If electrolyte ions could passively diffuse across membranes, it would be impossible to maintain specific concentrations of ions in each fluid compartment therefore they require special mechanisms to cross the semi-permeable membranes in the body. This movement can be accomplished by facilitated diffusion and active transport. Facilitated diffusion requires protein-based channels for moving the solute. Active transport requires energy in the form of ATP conversion, carrier proteins, or pumps in order to move ions against the concentration gradient.

Concept of Osmolality and Milliequivalent

In order to calculate osmotic pressure, it is necessary to understand how solute concentrations are measured. The unit for measuring solutes is the **mole**. One mole is defined as the gram molecular weight of the solute. For example, the molecular weight of sodium chloride is 58.44. Thus, one mole of sodium chloride weighs 58.44 grams. The **molarity** of a solution is the number of moles of solute per liter of solution. The **molality** of a solution is the number of moles of solute per kilogram of solvent. If the solvent is water, one kilogram of water is equal to one liter of water. While molarity and molality are used to express the concentration of solutions, electrolyte concentrations are usually expressed in terms of milliequivalents per liter (mEq/L): the mEq/L is equal to the ion concentration (in millimoles) multiplied by the number of electrical charges on the ion. The unit of milliequivalent takes into consideration the ions present in the solution (since electrolytes form ions in aqueous solutions) and the charge on the ions.

Thus, for ions that have a charge of one, one milliequivalent is equal to one millimole. For ions that have a charge of two (like calcium), one milliequivalent is equal to 0.5 millimoles. Another unit for the expression of electrolyte concentration is the milliosmole (mOsm), which is the number of milliequivalents of solute per kilogram of solvent. Body fluids are usually maintained within the range of 280 to 300 mOsm.

Osmoregulators and Osmoconformers

Persons lost at sea without any freshwater to drink are at risk of severe dehydration because the human body cannot adapt to drinking seawater, which is hypertonic in comparison to body fluids. Organisms such as goldfish that can tolerate only a relatively narrow range of salinity are referred to as stenohaline. About 90 percent of all bony fish are restricted to either freshwater or seawater. They are incapable of osmotic regulation in the opposite environment. It is possible, however, for a few fishes like salmon to spend part of their life in freshwater and part in seawater. Organisms like the salmon and molly that can tolerate a relatively wide range of salinity are referred to as euryhaline organisms. This is possible because some fish have evolved **osmoregulatory** mechanisms to survive in all kinds of aquatic environments. When they live in freshwater, their bodies tend to take up water because the environment is relatively hypotonic, as illustrated in Figure 41.3a. In such hypotonic environments, these fish do not drink much water. Instead, they pass a lot of very dilute urine, and they achieve electrolyte balance by active transport of salts through the gills. When they move to a hypertonic marine environment, these fish start drinking seawater; they excrete the excess salts through their gills and their urine, as illustrated in Figure 41.3b. Most marine invertebrates, on the other hand, may be isotonic with seawater (**osmoconformers**). Their body fluid concentrations conform to changes in seawater concentration. Cartilaginous fishes' salt composition of the blood is similar to bony fishes; however, the blood of sharks contains the organic compounds urea and trimethylamine oxide (TMAO). This does not mean that their electrolyte composition is similar to that of seawater. They achieve isotonicity with the sea by storing large concentrations of urea. These animals that secrete urea are called ureotelic animals. TMAO stabilizes proteins in the presence of high urea levels, preventing the disruption of peptide bonds that would occur in other animals exposed to similar levels of urea. Sharks are cartilaginous fish with a rectal gland to secrete salt and assist in osmoregulation.

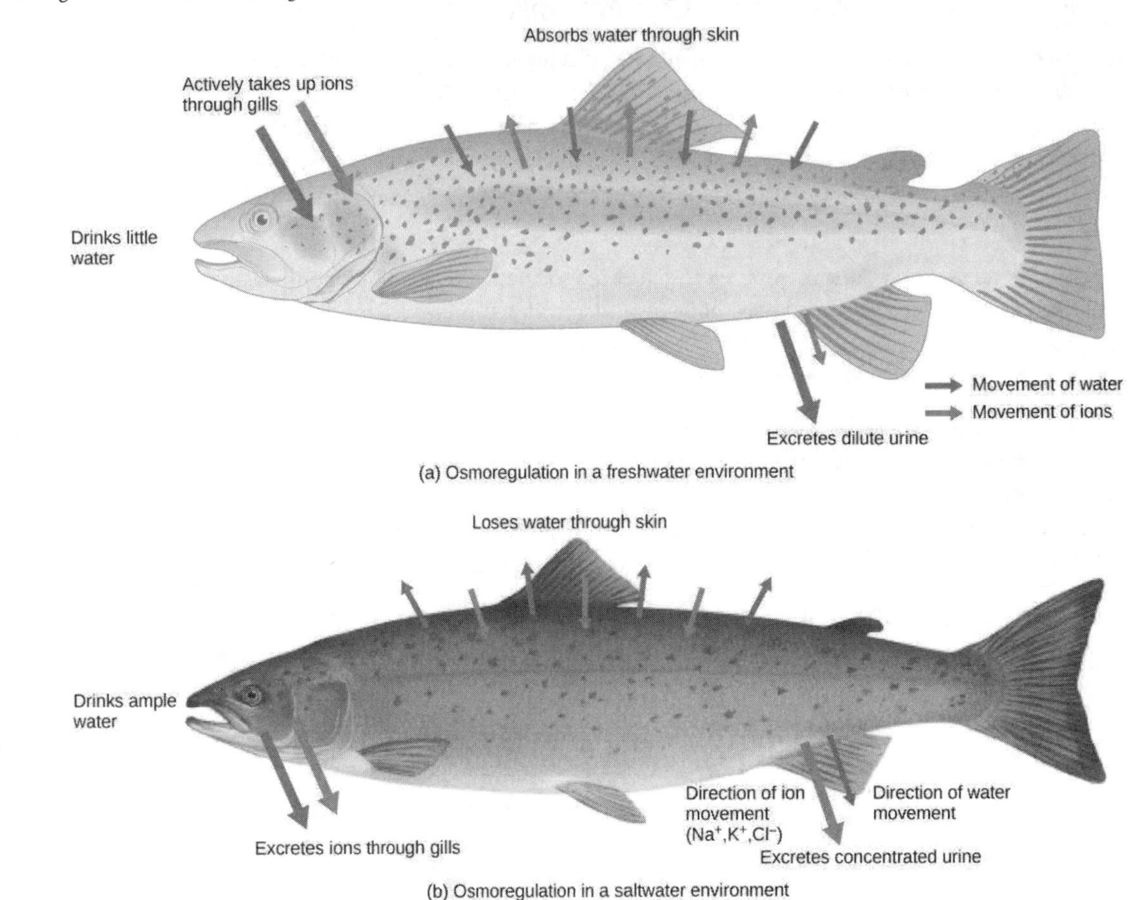

Figure 41.3 Fish are osmoregulators, but must use different mechanisms to survive in (a) freshwater or (b) saltwater environments. (credit: modification of work by Duane Raver, NOAA)

CAREER CONNECTION

Dialysis Technician

Dialysis is a medical process of removing wastes and excess water from the blood by diffusion and ultrafiltration. When kidney function fails, dialysis must be done to artificially rid the body of wastes. This is a vital process to keep patients alive. In some cases, the patients undergo artificial dialysis until they are eligible for a kidney transplant. In others who are not candidates for kidney transplants, dialysis is a life-long necessity.

Dialysis technicians typically work in hospitals and clinics. While some roles in this field include equipment development and maintenance, most dialysis technicians work in direct patient care. Their on-the-job duties, which typically occur under the direct supervision of a registered nurse, focus on providing dialysis treatments. This can include reviewing patient history and current condition, assessing and responding to patient needs before and during treatment, and monitoring the dialysis process. Treatment may include taking and reporting a patient's vital signs and preparing solutions and equipment to ensure accurate and sterile procedures.

41.2 The Kidneys and Osmoregulatory Organs

By the end of this section, you will be able to do the following:

- Explain how the kidneys serve as the main osmoregulatory organs in mammalian systems
- Describe the structure of the kidneys and the functions of the parts of the kidney
- Describe how the nephron is the functional unit of the kidney and explain how it actively filters blood and generates urine
- Detail the three steps in the formation of urine: glomerular filtration, tubular reabsorption, and tubular secretion

Although the kidneys are the major osmoregulatory organ, the skin and lungs also play a role in the process. Water and electrolytes are lost through sweat glands in the skin, which helps moisturize and cool the skin surface, while the lungs expel a small amount of water in the form of mucous secretions and via evaporation of water vapor.

Kidneys: The Main Osmoregulatory Organ

The **kidneys**, illustrated in Figure 41.4, are a pair of bean-shaped structures that are located just below and posterior to the liver in the peritoneal cavity. The adrenal glands sit on top of each kidney and are also called the suprarenal glands. Kidneys filter blood and purify it. All the blood in the human body is filtered many times a day by the kidneys; these organs use up almost 25 percent of the oxygen absorbed through the lungs to perform this function. Oxygen allows the kidney cells to efficiently manufacture chemical energy in the form of ATP through aerobic respiration. The filtrate coming out of the kidneys is called **urine**.

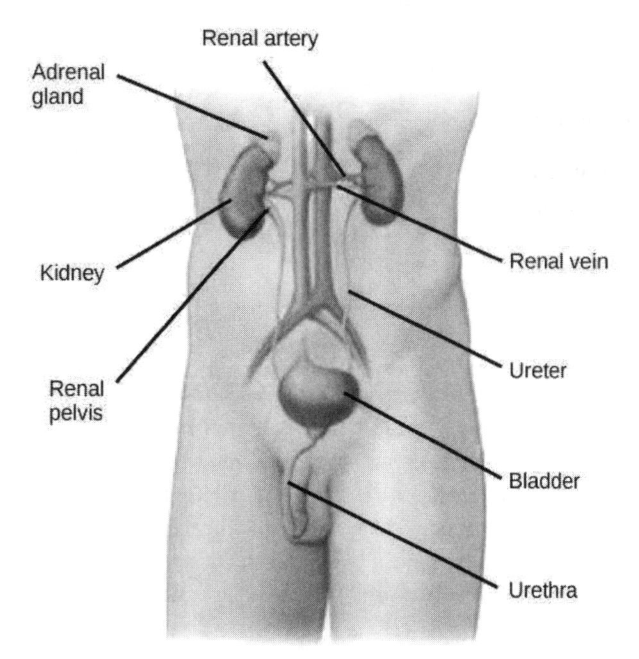

Figure 41.4 Kidneys filter the blood, producing urine that is stored in the bladder prior to elimination through the urethra. (credit: modification of work by NCI)

Kidney Structure

Externally, the kidneys are surrounded by three layers, illustrated in Figure 41.5. The outermost layer is a tough connective tissue layer called the **renal fascia**. The second layer is called the **perirenal fat capsule**, which helps anchor the kidneys in place. The third and innermost layer is the **renal capsule**. Internally, the kidney has three regions—an outer **cortex**, a **medulla** in the middle, and the **renal pelvis** in the region called the **hilum** of the kidney. The hilum is the concave part of the bean-shape where blood vessels and nerves enter and exit the kidney; it is also the point of exit for the ureters. The renal cortex is granular due to the presence of **nephrons**—the functional unit of the kidney. The medulla consists of multiple pyramidal tissue masses, called the **renal pyramids**. In between the pyramids are spaces called **renal columns** through which the blood vessels pass. The tips of the pyramids, called renal papillae, point toward the renal pelvis. There are, on average, eight renal pyramids in each kidney. The renal pyramids along with the adjoining cortical region are called the **lobes of the kidney**. The renal pelvis leads to the **ureter** on the outside of the kidney. On the inside of the kidney, the renal pelvis branches out into two or three extensions called the major **calyces**, which further branch into the minor calyces. The ureters are urine-bearing tubes that exit the kidney and empty into the **urinary bladder**.

VISUAL CONNECTION

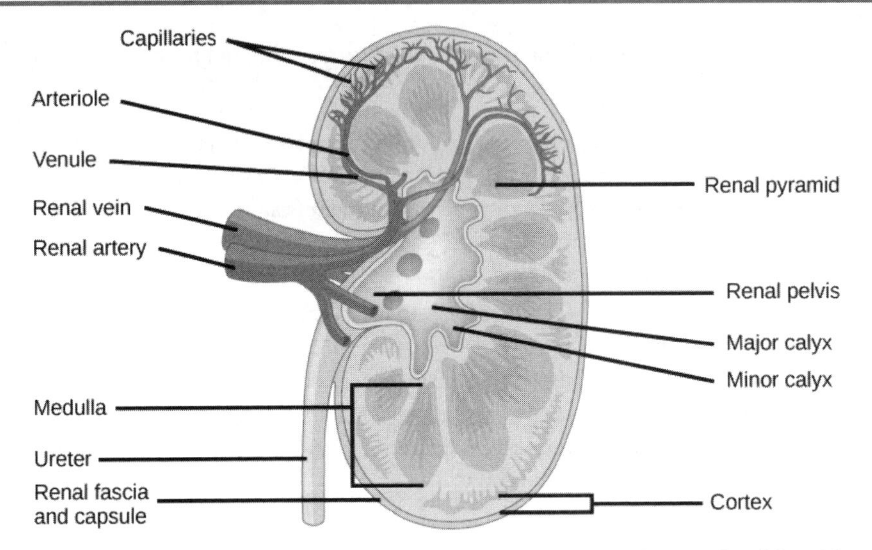

Figure 41.5 The internal structure of the kidney is shown. (credit: modification of work by NCI)

Which of the following statements about the kidney is false?

a. The renal pelvis drains into the ureter.
b. The renal pyramids are in the medulla.
c. The cortex covers the capsule.
d. Nephrons are in the renal cortex.

Because the kidney filters blood, its network of blood vessels is an important component of its structure and function. The arteries, veins, and nerves that supply the kidney enter and exit at the renal hilum. Renal blood supply starts with the branching of the aorta into the **renal arteries** (which are each named based on the region of the kidney they pass through) and ends with the exiting of the **renal veins** to join the **inferior vena cava**. The renal arteries split into several **segmental arteries** upon entering the kidneys. Each segmental artery splits further into several **interlobar arteries** and enters the renal columns, which supply the renal lobes. The interlobar arteries split at the junction of the renal cortex and medulla to form the **arcuate arteries**. The arcuate "bow shaped" arteries form arcs along the base of the medullary pyramids. **Cortical radiate arteries**, as the name suggests, radiate out from the arcuate arteries. The cortical radiate arteries branch into numerous afferent arterioles, and then enter the capillaries supplying the nephrons. Veins trace the path of the arteries and have similar names, except there are no segmental veins.

As mentioned previously, the functional unit of the kidney is the nephron, illustrated in Figure 41.6. Each kidney is made up of over one million nephrons that dot the renal cortex, giving it a granular appearance when sectioned sagittally. There are two types of nephrons—**cortical nephrons** (85 percent), which are deep in the renal cortex, and **juxtamedullary nephrons** (15 percent), which lie in the renal cortex close to the renal medulla. A nephron consists of three parts—a **renal corpuscle**, a **renal tubule**, and the associated capillary network, which originates from the cortical radiate arteries.

VISUAL CONNECTION

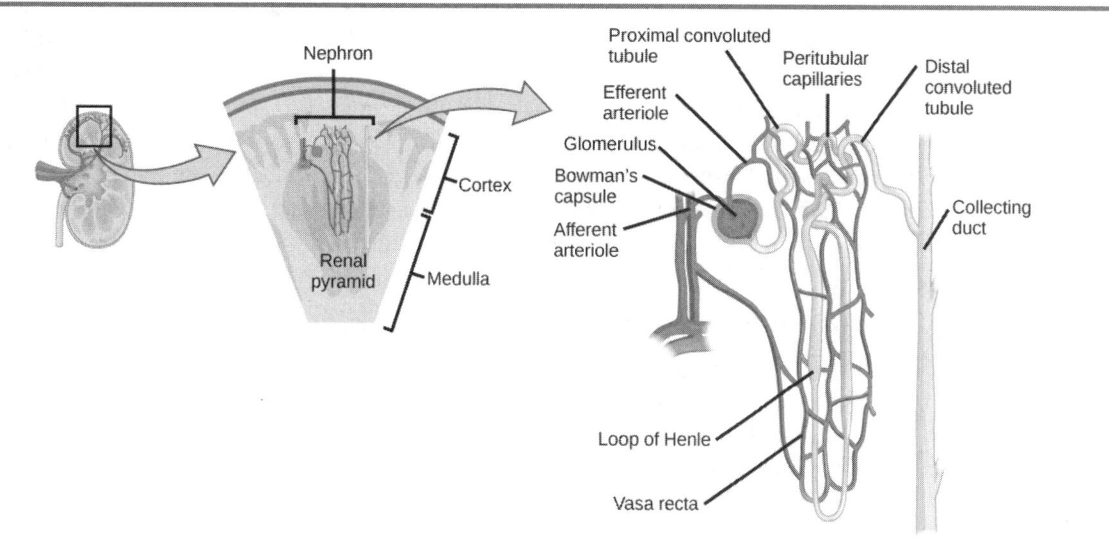

Figure 41.6 The nephron is the functional unit of the kidney. The glomerulus and convoluted tubules are located in the kidney cortex, while collecting ducts are located in the pyramids of the medulla. (credit: modification of work by NIDDK)

Which of the following statements about the nephron is false?

a. The collecting duct empties into the distal convoluted tubule.
b. The Bowman's capsule surrounds the glomerulus.
c. The loop of Henle is between the proximal and distal convoluted tubules.
d. The loop of Henle empties into the distal convoluted tubule.

Renal Corpuscle

The renal corpuscle, located in the renal cortex, is made up of a network of capillaries known as the **glomerulus** and the capsule, a cup-shaped chamber that surrounds it, called the glomerular or **Bowman's capsule**.

Renal Tubule

The renal tubule is a long and convoluted structure that emerges from the glomerulus and can be divided into three parts based on function. The first part is called the **proximal convoluted tubule (PCT)** due to its proximity to the glomerulus; it stays in the renal cortex. The second part is called the **loop of Henle**, or nephritic loop, because it forms a loop (with **descending** and **ascending limbs**) that goes through the renal medulla. The third part of the renal tubule is called the **distal convoluted tubule (DCT)** and this part is also restricted to the renal cortex. The DCT, which is the last part of the nephron, connects and empties its contents into collecting ducts that line the medullary pyramids. The collecting ducts amass contents from multiple nephrons and fuse together as they enter the papillae of the renal medulla.

Capillary Network within the Nephron

The capillary network that originates from the renal arteries supplies the nephron with blood that needs to be filtered. The branch that enters the glomerulus is called the **afferent arteriole**. The branch that exits the glomerulus is called the **efferent arteriole**. Within the glomerulus, the network of capillaries is called the glomerular capillary bed. Once the efferent arteriole exits the glomerulus, it forms the **peritubular capillary network**, which surrounds and interacts with parts of the renal tubule. In cortical nephrons, the peritubular capillary network surrounds the PCT and DCT. In juxtamedullary nephrons, the peritubular capillary network forms a network around the loop of Henle and is called the **vasa recta**.

LINK TO LEARNING

Go to this website (http://openstax.org/l/kidney_section) to see another coronal section of the kidney and to explore an animation of the workings of nephrons.

Kidney Function and Physiology

Kidneys filter blood in a three-step process. First, the nephrons filter blood that runs through the capillary network in the glomerulus. Almost all solutes, except for proteins, are filtered out into the glomerulus by a process called **glomerular filtration**. Second, the filtrate is collected in the renal tubules. Most of the solutes get reabsorbed in the PCT by a process called **tubular reabsorption**. In the loop of Henle, the filtrate continues to exchange solutes and water with the renal medulla and the peritubular capillary network. Water is also reabsorbed during this step. Then, additional solutes and wastes are secreted into the kidney tubules during **tubular secretion**, which is, in essence, the opposite process to tubular reabsorption. The collecting ducts collect filtrate coming from the nephrons and fuse in the medullary papillae. From here, the papillae deliver the filtrate, now called urine, into the minor calyces that eventually connect to the ureters through the renal pelvis. This entire process is illustrated in Figure 41.7.

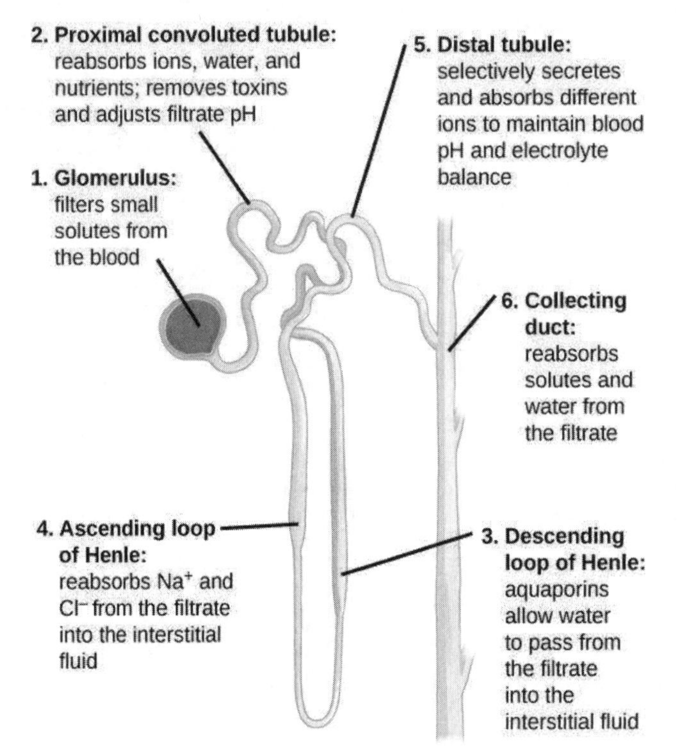

Figure 41.7 Each part of the nephron performs a different function in filtering waste and maintaining homeostatic balance. (1) The glomerulus forces small solutes out of the blood by pressure. (2) The proximal convoluted tubule reabsorbs ions, water, and nutrients from the filtrate into the interstitial fluid, and actively transports toxins and drugs from the interstitial fluid into the filtrate. The proximal convoluted tubule also adjusts blood pH by selectively secreting ammonia (NH_3) into the filtrate, where it reacts with H^+ to form NH_4^+. The more acidic the filtrate, the more ammonia is secreted. (3) The descending loop of Henle is lined with cells containing aquaporins that allow water to pass from the filtrate into the interstitial fluid. (4) In the thin part of the ascending loop of Henle, Na^+ and Cl^- ions diffuse into the interstitial fluid. In the thick part, these same ions are actively transported into the interstitial fluid. Because salt but not water is lost, the filtrate becomes more dilute as it travels up the limb. (5) In the distal convoluted tubule, K^+ and H^+ ions are selectively secreted into the filtrate, while Na^+, Cl^-, and HCO_3^- ions are reabsorbed to maintain pH and electrolyte balance in the blood. (6) The collecting duct reabsorbs solutes and water from the filtrate, forming dilute urine. (credit: modification of work by NIDDK)

Glomerular Filtration

Glomerular filtration filters out most of the solutes due to high blood pressure and specialized membranes in the afferent arteriole. The blood pressure in the glomerulus is maintained independent of factors that affect systemic blood pressure. The "leaky" connections between the endothelial cells of the glomerular capillary network allow solutes to pass through easily. All solutes in the glomerular capillaries, except for macromolecules like proteins, pass through by passive diffusion. There is no energy requirement at this stage of the filtration process. **Glomerular filtration rate (GFR)** is the volume of glomerular filtrate formed per minute by the kidneys. GFR is regulated by multiple mechanisms and is an important indicator of kidney function.

LINK TO LEARNING

To learn more about the vascular system of kidneys, click through this review (http://openstax.org/l/kidneys) and the steps of blood flow.

Tubular Reabsorption and Secretion

Tubular reabsorption occurs in the PCT part of the renal tubule. Almost all nutrients are reabsorbed, and this occurs either by passive or active transport. Reabsorption of water and some key electrolytes are regulated and can be influenced by hormones. Sodium (Na^+) is the most abundant ion and most of it is reabsorbed by active transport and then transported to the peritubular capillaries. Because Na^+ is actively transported out of the tubule, water follows it to even out the osmotic pressure. Water is also independently reabsorbed into the peritubular capillaries due to the presence of aquaporins, or water channels, in the PCT. This occurs due to the low blood pressure and high osmotic pressure in the peritubular capillaries. However, every solute has a **transport maximum** and the excess is not reabsorbed.

In the loop of Henle, the permeability of the membrane changes. The descending limb is permeable to water, not solutes; the opposite is true for the ascending limb. Additionally, the loop of Henle invades the renal medulla, which is naturally high in salt concentration and tends to absorb water from the renal tubule and concentrate the filtrate. The osmotic gradient increases as it moves deeper into the medulla. Because two sides of the loop of Henle perform opposing functions, as illustrated in Figure 41.8, it acts as a **countercurrent multiplier**. The vasa recta around it acts as the **countercurrent exchanger**.

VISUAL CONNECTION

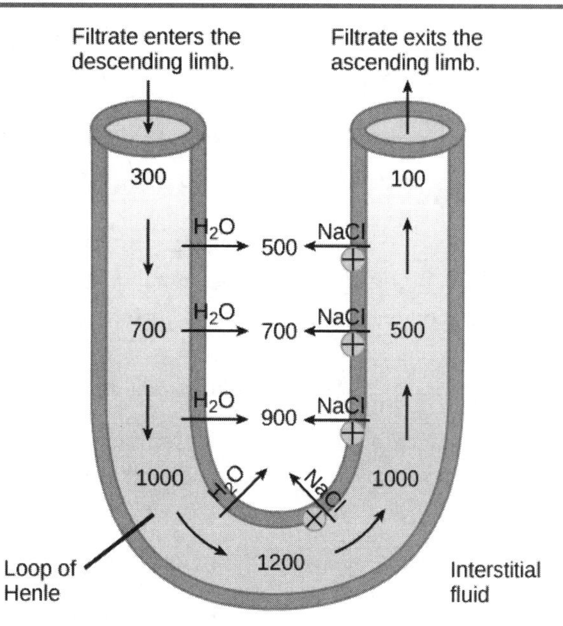

Figure 41.8 The loop of Henle acts as a countercurrent multiplier that uses energy to create concentration gradients. The descending limb is water permeable. Water flows from the filtrate to the interstitial fluid, so osmolality inside the limb increases as it descends into the renal medulla. At the bottom, the osmolality is higher inside the loop than in the interstitial fluid. Thus, as filtrate enters the ascending limb, Na^+ and Cl^- ions exit through ion channels present in the cell membrane. Further up, Na^+ is actively transported out of the filtrate and Cl^- follows. Osmolarity is given in units of milliosmoles per liter (mOsm/L).

Loop diuretics are drugs sometimes used to treat hypertension. These drugs inhibit the reabsorption of Na^+ and Cl^- ions by the ascending limb of the loop of Henle. A side effect is that they increase urination. Why do you think this is the case?

By the time the filtrate reaches the DCT, most of the urine and solutes have been reabsorbed. If the body requires additional water, all of it can be reabsorbed at this point. Further reabsorption is controlled by hormones, which will be discussed in a later section. Excretion of wastes occurs due to lack of reabsorption combined with tubular secretion. Undesirable products like metabolic wastes, urea, uric acid, and certain drugs, are excreted by tubular secretion. Most of the tubular secretion happens in the DCT, but some occurs in the early part of the collecting duct. Kidneys also maintain an acid-base balance by secreting excess H^+ ions.

Although parts of the renal tubules are named proximal and distal, in a cross-section of the kidney, the tubules are placed close together and in contact with each other and the glomerulus. This allows for exchange of chemical messengers between the different cell types. For example, the DCT ascending limb of the loop of Henle has masses of cells called **macula densa**, which are in contact with cells of the afferent arterioles called **juxtaglomerular cells**. Together, the macula densa and juxtaglomerular cells form the juxtaglomerular complex (JGC). The JGC is an endocrine structure that secretes the enzyme renin and the hormone erythropoietin. When hormones trigger the macula densa cells in the DCT due to variations in blood volume, blood pressure, or electrolyte balance, these cells can immediately communicate the problem to the capillaries in the afferent and efferent arterioles, which can constrict or relax to change the glomerular filtration rate of the kidneys.

CAREER CONNECTION

Nephrologist

A nephrologist studies and deals with diseases of the kidneys—both those that cause kidney failure (such as diabetes) and the conditions that are produced by kidney disease (such as hypertension). Blood pressure, blood volume, and changes in electrolyte balance come under the purview of a nephrologist.

Nephrologists usually work with other physicians who refer patients to them or consult with them about specific diagnoses and treatment plans. Patients are usually referred to a nephrologist for symptoms such as blood or protein in the urine, very high blood pressure, kidney stones, or renal failure.

Nephrology is a subspecialty of internal medicine. To become a nephrologist, medical school is followed by additional training to become certified in internal medicine. An additional two or more years is spent specifically studying kidney disorders and their accompanying effects on the body.

41.3 Excretion Systems

By the end of this section, you will be able to do the following:

- Explain how vacuoles, present in microorganisms, work to excrete waste
- Describe the way in which flame cells and nephridia in worms perform excretory functions and maintain osmotic balance
- Explain how insects use Malpighian tubules to excrete wastes and maintain osmotic balance

Microorganisms and invertebrate animals use more primitive and simple mechanisms to get rid of their metabolic wastes than the mammalian system of kidney and urinary function. Three excretory systems evolved in organisms before complex kidneys: vacuoles, flame cells, and Malpighian tubules.

Contractile Vacuoles in Microorganisms

The most fundamental feature of life is the presence of a cell. In other words, a cell is the simplest functional unit of a life. Bacteria are unicellular, prokaryotic organisms that have some of the least complex life processes in place; however, prokaryotes such as bacteria do not contain membrane-bound vacuoles. The cells of microorganisms like bacteria, protozoa, and fungi are bound by cell membranes and use them to interact with the environment. Some cells, including some leucocytes in humans, are able to engulf food by endocytosis—the formation of vesicles by involution of the cell membrane within the cells. The same vesicles are able to interact and exchange metabolites with the intracellular environment. In some unicellular eukaryotic organisms such as the amoeba, shown in Figure 41.9, cellular wastes and excess water are excreted by exocytosis, when the contractile vacuoles merge with the cell membrane and expel wastes into the environment. Contractile vacuoles (CV) should not be confused with vacuoles, which store food or water.

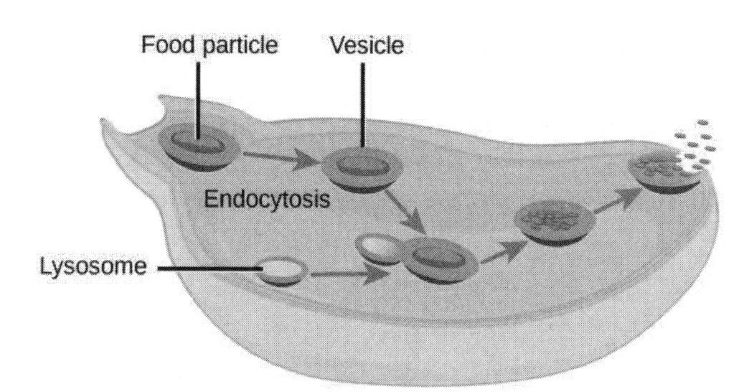

Figure 41.9 Some unicellular organisms, such as the amoeba, ingest food by endocytosis. The food vesicle fuses with a lysosome, which digests the food. Waste is excreted by exocytosis.

Flame Cells of Planaria and Nephridia of Worms

As multicellular systems evolved to have organ systems that divided the metabolic needs of the body, individual organs evolved to perform the excretory function. Planaria are flatworms that live in freshwater. Their excretory system consists of two tubules connected to a highly branched duct system. The cells in the tubules are called **flame cells** (or **protonephridia**) because they have a cluster of cilia that looks like a flickering flame when viewed under the microscope, as illustrated in Figure 41.10a. The cilia propel waste matter down the tubules and out of the body through excretory pores that open on the body surface; cilia also draw water from the interstitial fluid, allowing for filtration. Any valuable metabolites are recovered by reabsorption. Flame cells are found in flatworms, including parasitic tapeworms and free-living planaria. They also maintain the organism's osmotic balance.

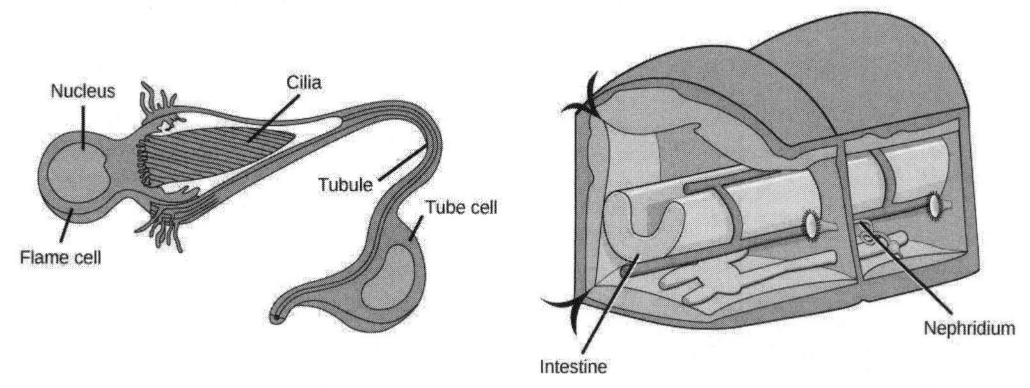

Figure 41.10 In the excretory system of the (a) planaria, cilia of flame cells propel waste through a tubule formed by a tube cell. Tubules are connected into branched structures that lead to pores located all along the sides of the body. The filtrate is secreted through these pores. In (b) annelids such as earthworms, nephridia filter fluid from the coelom, or body cavity. Beating cilia at the opening of the nephridium draw water from the coelom into a tubule. As the filtrate passes down the tubules, nutrients and other solutes are reabsorbed by capillaries. Filtered fluid containing nitrogenous and other wastes is stored in a bladder and then secreted through a pore in the side of the body.

Earthworms (annelids) have slightly more evolved excretory structures called **nephridia**, illustrated in Figure 41.10b. A pair of nephridia is present on each segment of the earthworm. They are similar to flame cells in that they have a tubule with cilia. Excretion occurs through a pore called the **nephridiopore**. They are more evolved than the flame cells in that they have a system for tubular reabsorption by a capillary network before excretion.

Malpighian Tubules of Insects

Malpighian tubules are found lining the gut of some species of arthropods, such as the bee illustrated in Figure 41.11. They are usually found in pairs and the number of tubules varies with the species of insect. Malpighian tubules are convoluted, which increases their surface area, and they are lined with **microvilli** for reabsorption and maintenance of osmotic balance. Malpighian tubules work cooperatively with specialized glands in the wall of the rectum. Body fluids are not filtered as in the case of nephridia; urine is produced by tubular secretion mechanisms by the cells lining the Malpighian tubules that are bathed in hemolymph (a mixture of blood and interstitial fluid that is found in insects and other arthropods as well as most mollusks).

Metabolic wastes like uric acid freely diffuse into the tubules. There are exchange pumps lining the tubules, which actively transport H^+ ions into the cell and K^+ or Na^+ ions out; water passively follows to form urine. The secretion of ions alters the osmotic pressure which draws water, electrolytes, and nitrogenous waste (uric acid) into the tubules. Water and electrolytes are reabsorbed when these organisms are faced with low-water environments, and uric acid is excreted as a thick paste or powder. Not dissolving wastes in water helps these organisms to conserve water; this is especially important for life in dry environments.

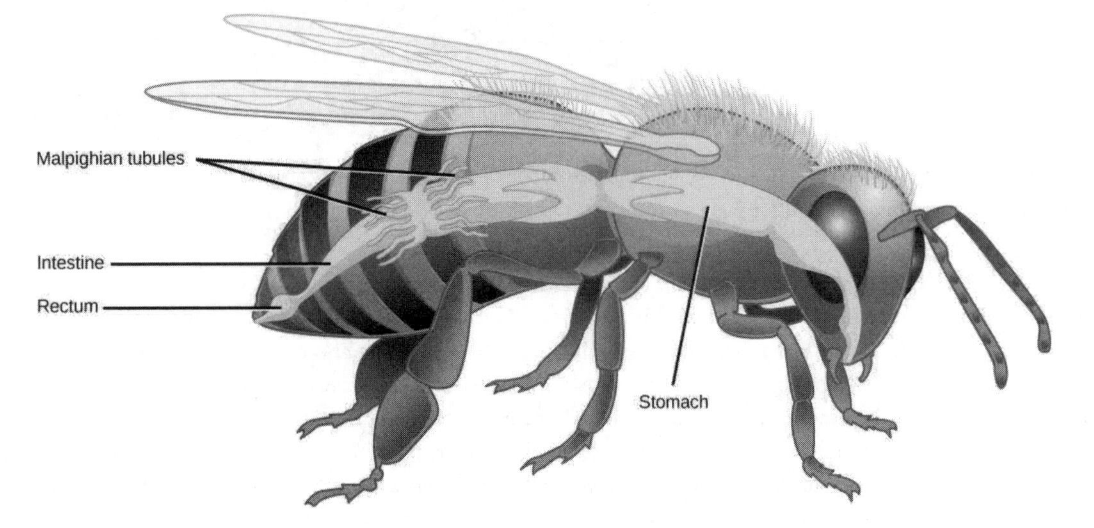

Figure 41.11 Malpighian tubules of insects and other terrestrial arthropods remove nitrogenous wastes and other solutes from the hemolymph. Na^+ and/or K^+ ions are actively transported into the lumen of the tubules. Water then enters the tubules via osmosis, forming urine. The urine passes through the intestine, and into the rectum. There, nutrients diffuse back into the hemolymph. Na^+ and/or K^+ ions are pumped into the hemolymph, and water follows. The concentrated waste is then excreted.

LINK TO LEARNING

See a dissected cockroach, including a close-up look at its Malpighian tubules, in this video (https://openstax.org/l/malpighian)
.

41.4 Nitrogenous Wastes

By the end of this section, you will be able to do the following:

- Compare and contrast the way in which aquatic animals and terrestrial animals can eliminate toxic ammonia from their systems
- Compare the major byproduct of ammonia metabolism in vertebrate animals to that of birds, insects, and reptiles

Of the four major macromolecules in biological systems, both proteins and nucleic acids contain nitrogen. During the catabolism, or breakdown, of nitrogen-containing macromolecules, carbon, hydrogen, and oxygen are extracted and stored in the form of carbohydrates and fats. Excess nitrogen is excreted from the body. Nitrogenous wastes tend to form toxic **ammonia**, which raises the pH of body fluids. The formation of ammonia itself requires energy in the form of ATP and large quantities of water to dilute it out of a biological system. Animals that live in aquatic environments tend to release ammonia into the water. Animals that excrete ammonia are said to be **ammonotelic**. Terrestrial organisms have evolved other mechanisms to excrete nitrogenous wastes. The animals must detoxify ammonia by converting it into a relatively nontoxic form such as urea or uric acid. Mammals, including humans, produce urea, whereas reptiles and many terrestrial invertebrates produce uric acid. Animals that secrete urea as the primary nitrogenous waste material are called **ureotelic** animals.

Nitrogenous Waste in Terrestrial Animals: The Urea Cycle

The **urea cycle** is the primary mechanism by which mammals convert ammonia to urea. Urea is made in the liver and excreted in urine. The overall chemical reaction by which ammonia is converted to urea is $2\ NH_3$ (ammonia) + CO_2 + 3 ATP + $H_2O \rightarrow H_2N\text{-}CO\text{-}NH_2$ (urea) + 2 ADP + 4 P_i + AMP.

The urea cycle utilizes five intermediate steps, catalyzed by five different enzymes, to convert ammonia to urea, as shown in Figure 41.12. The amino acid L-ornithine gets converted into different intermediates before being regenerated at the end of the urea cycle. Hence, the urea cycle is also referred to as the ornithine cycle. The enzyme ornithine transcarbamylase catalyzes a key step in the urea cycle and its deficiency can lead to accumulation of toxic levels of ammonia in the body. The first two reactions occur in the mitochondria and the last three reactions occur in the cytosol. Urea concentration in the blood, called **blood urea nitrogen** or BUN, is used as an indicator of kidney function.

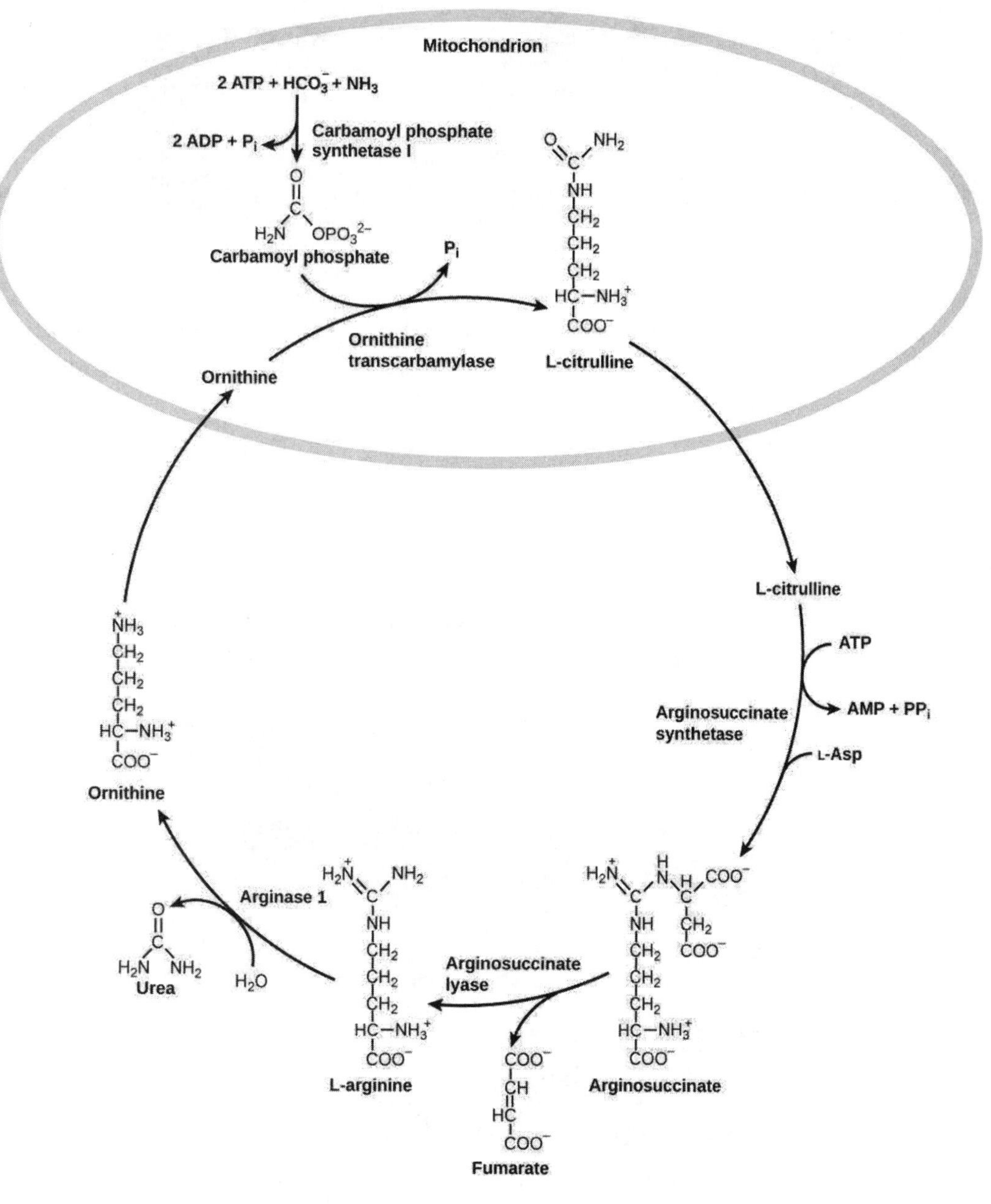

Figure 41.12 The urea cycle converts ammonia to urea.

EVOLUTION CONNECTION

Excretion of Nitrogenous Waste

The theory of evolution proposes that life started in an aquatic environment. It is not surprising to see that biochemical pathways like the urea cycle evolved to adapt to a changing environment when terrestrial life forms evolved. Arid conditions probably led to the evolution of the uric acid pathway as a means of conserving water.

Nitrogenous Waste in Birds and Reptiles: Uric Acid

Birds, reptiles, and most terrestrial arthropods convert toxic ammonia to **uric acid** or the closely related compound guanine (guano) instead of urea. Mammals also form some uric acid during breakdown of nucleic acids. Uric acid is a compound similar to purines found in nucleic acids. It is water insoluble and tends to form a white paste or powder; it is excreted by birds, insects, and reptiles. Conversion of ammonia to uric acid requires more energy and is much more complex than conversion of ammonia to urea Figure 41.13.

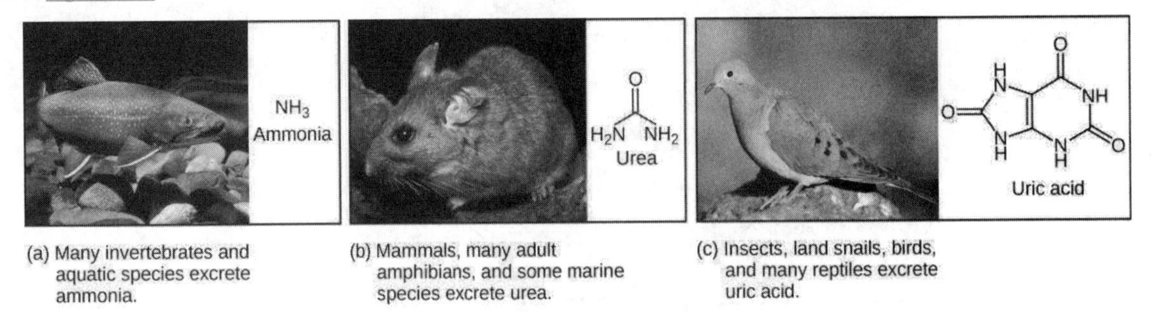

Figure 41.13 Nitrogenous waste is excreted in different forms by different species. These include (a) ammonia, (b) urea, and (c) uric acid. (credit a: modification of work by Eric Engbretson, USFWS; credit b: modification of work by B. "Moose" Peterson, USFWS; credit c: modification of work by Dave Menke, USFWS)

Everyday Connection

Gout

Mammals use uric acid crystals as an **antioxidant** in their cells. However, too much uric acid tends to form kidney stones and may also cause a painful condition called gout, where uric acid crystals accumulate in the joints, as illustrated in Figure 41.14. Food choices that reduce the amount of nitrogenous bases in the diet help reduce the risk of gout. For example, tea, coffee, and chocolate have purine-like compounds, called xanthines, and should be avoided by people with gout and kidney stones.

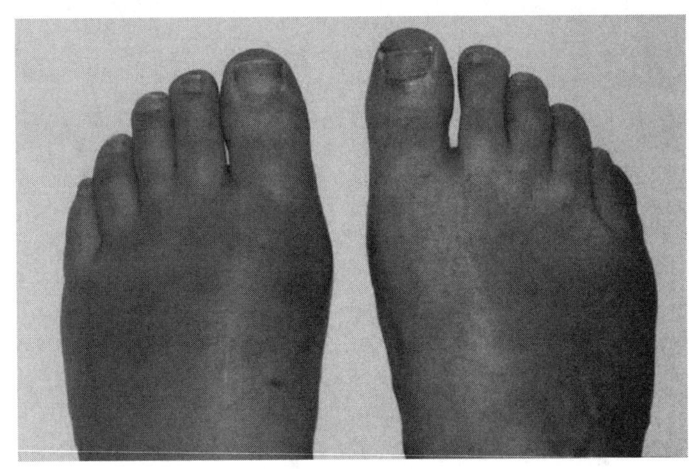

Figure 41.14 Gout causes the inflammation visible in this person's left big toe joint. (credit: "Gonzosft"/Wikimedia Commons)

41.5 Hormonal Control of Osmoregulatory Functions

By the end of this section, you will be able to do the following:

- Explain how hormonal cues help the kidneys synchronize the osmotic needs of the body
- Describe how hormones like epinephrine, norepinephrine, renin-angiotensin, aldosterone, anti-diuretic hormone, and atrial natriuretic peptide help regulate waste elimination, maintain correct osmolarity, and perform other osmoregulatory functions

While the kidneys operate to maintain osmotic balance and blood pressure in the body, they also act in concert with hormones.

Hormones are small molecules that act as messengers within the body. Hormones are typically secreted from one cell and travel in the bloodstream to affect a target cell in another portion of the body. Different regions of the nephron bear specialized cells that have receptors to respond to chemical messengers and hormones. Table 41.1 summarizes the hormones that control the osmoregulatory functions.

Hormones That Affect Osmoregulation

Hormone	Where produced	Function
Epinephrine and Norepinephrine	Adrenal medulla	Can decrease kidney function temporarily by vasoconstriction
Renin	Kidney nephrons	Increases blood pressure by acting on angiotensinogen
Angiotensin	Liver	Angiotensin II affects multiple processes and increases blood pressure
Aldosterone	Adrenal cortex	Prevents loss of sodium and water
Anti-diuretic hormone (vasopressin)	Hypothalamus (stored in the posterior pituitary)	Prevents water loss
Atrial natriuretic peptide	Heart atrium	Decreases blood pressure by acting as a vasodilator and increasing glomerular filtration rate; decreases sodium reabsorption in kidneys

Table 41.1

Epinephrine and Norepinephrine

Epinephrine and norepinephrine are released by the adrenal medulla and nervous system respectively. They are the flight/fight hormones that are released when the body is under extreme stress. During stress, much of the body's energy is used to combat imminent danger. Kidney function is halted temporarily by epinephrine and norepinephrine. These hormones function by acting directly on the smooth muscles of blood vessels to constrict them. Once the afferent arterioles are constricted, blood flow into the nephrons stops. These hormones go one step further and trigger the **renin-angiotensin-aldosterone** system.

Renin-Angiotensin-Aldosterone

The renin-angiotensin-aldosterone system, illustrated in Figure 41.15 proceeds through several steps to produce **angiotensin II**, which acts to stabilize blood pressure and volume. Renin (secreted by a part of the juxtaglomerular complex) is produced by the granular cells of the afferent and efferent arterioles. Thus, the kidneys control blood pressure and volume directly. Renin acts on angiotensinogen, which is made in the liver and converts it to **angiotensin I**. **Angiotensin converting enzyme (ACE)** converts angiotensin I to angiotensin II. Angiotensin II raises blood pressure by constricting blood vessels. It also triggers the release of the mineralocorticoid aldosterone from the adrenal cortex, which in turn stimulates the renal tubules to reabsorb more sodium. Angiotensin II also triggers the release of **anti-diuretic hormone (ADH)** from the hypothalamus, leading to water retention in the kidneys. It acts directly on the nephrons and decreases glomerular filtration rate. Medically, blood pressure can be controlled by drugs that inhibit ACE (called ACE inhibitors).

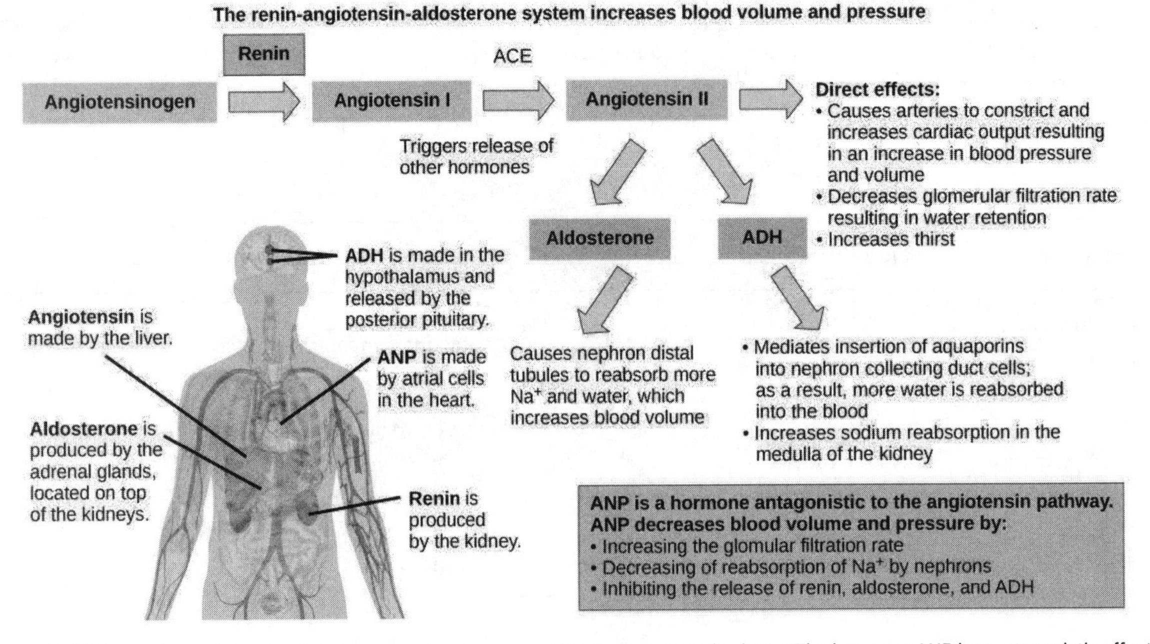

Figure 41.15 The renin-angiotensin-aldosterone system increases blood pressure and volume. The hormone ANP has antagonistic effects. (credit: modification of work by Mikael Häggström)

Mineralocorticoids

Mineralocorticoids are hormones synthesized by the adrenal cortex that affect osmotic balance. Aldosterone is a mineralocorticoid that regulates sodium levels in the blood. Almost all of the sodium in the blood is reclaimed by the renal tubules under the influence of aldosterone. Because sodium is always reabsorbed by active transport and water follows sodium to maintain osmotic balance, aldosterone manages not only sodium levels but also the water levels in body fluids. In contrast, the aldosterone also stimulates potassium secretion concurrently with sodium reabsorption. In contrast, absence of aldosterone means that no sodium gets reabsorbed in the renal tubules and all of it gets excreted in the urine. In addition, the daily dietary potassium load is not secreted and the retention of K^+ can cause a dangerous increase in plasma K^+ concentration. Patients who have Addison's disease have a failing adrenal cortex and cannot produce aldosterone. They lose sodium in their urine constantly, and if the supply is not replenished, the consequences can be fatal.

Antidiuretic Hormone

As previously discussed, antidiuretic hormone or ADH (also called **vasopressin**), as the name suggests, helps the body conserve water when body fluid volume, especially that of blood, is low. It is formed by the hypothalamus and is stored and released from the posterior pituitary. It acts by inserting aquaporins in the collecting ducts and promotes reabsorption of water. ADH also acts as a vasoconstrictor and increases blood pressure during hemorrhaging.

Atrial Natriuretic Peptide Hormone

The atrial natriuretic peptide (ANP) lowers blood pressure by acting as a **vasodilator**. It is released by cells in the atrium of the heart in response to high blood pressure and in patients with sleep apnea. ANP affects salt release, and because water passively follows salt to maintain osmotic balance, it also has a diuretic effect. ANP also prevents sodium reabsorption by the renal tubules, decreasing water reabsorption (thus acting as a diuretic) and lowering blood pressure. Its actions suppress the actions of aldosterone, ADH, and renin.

KEY TERMS

afferent arteriole arteriole that branches from the cortical radiate artery and enters the glomerulus
ammonia compound made of one nitrogen atom and three hydrogen atoms
ammonotelic describes an animal that excretes ammonia as the primary waste material
angiotensin converting enzyme (ACE) enzyme that converts angiotensin I to angiotensin II
angiotensin I product in the renin-angiotensin-aldosterone pathway
angiotensin II molecule that affects different organs to increase blood pressure
anti-diuretic hormone (ADH) hormone that prevents the loss of water
antioxidant agent that prevents cell destruction by reactive oxygen species
arcuate artery artery that branches from the interlobar artery and arches over the base of the renal pyramids
ascending limb part of the loop of Henle that ascends from the renal medulla to the renal cortex
blood urea nitrogen (BUN) estimate of urea in the blood and an indicator of kidney function
Bowman's capsule structure that encloses the glomerulus
calyx structure that connects the renal pelvis to the renal medulla
cortex (animal) outer layer of an organ like the kidney or adrenal gland
cortical nephron nephron that lies in the renal cortex
cortical radiate artery artery that radiates from the arcuate arteries into the renal cortex
countercurrent exchanger peritubular capillary network that allows exchange of solutes and water from the renal tubules
countercurrent multiplier osmotic gradient in the renal medulla that is responsible for concentration of urine
descending limb part of the loop of Henle that descends from the renal cortex into the renal medulla
distal convoluted tubule (DCT) part of the renal tubule that is the most distant from the glomerulus
efferent arteriole arteriole that exits from the glomerulus
electrolyte solute that breaks down into ions when dissolved in water
flame cell (also, protonephridia) excretory cell found in flatworms
glomerular filtration filtration of blood in the glomerular capillary network into the glomerulus
glomerular filtration rate (GFR) amount of filtrate formed by the glomerulus per minute
glomerulus (renal) part of the renal corpuscle that contains the capillary network
hilum region in the renal pelvis where blood vessels, nerves, and ureters bunch before entering or exiting the kidney
inferior vena cava one of the main veins in the human body
interlobar artery artery that branches from the segmental artery and travels in between the renal lobes
juxtaglomerular cell cell in the afferent and efferent arterioles that responds to stimuli from the macula densa
juxtamedullary nephron nephron that lies in the cortex but close to the renal medulla
kidney organ that performs excretory and osmoregulatory functions
lobes of the kidney renal pyramid along with the adjoining cortical region
loop of Henle part of the renal tubule that loops into the renal medulla
macula densa group of cells that senses changes in sodium ion concentration; present in parts of the renal tubule and collecting ducts
Malpighian tubule excretory tubules found in arthropods
medulla middle layer of an organ like the kidney or adrenal gland
microvilli cellular processes that increase the surface area of cells
molality number of moles of solute per kilogram of solvent
molarity number of moles of solute per liter of solution
mole gram equivalent of the molecular weight of a substance
nephridia excretory structures found in annelids
nephridiopore pore found at the end of nephridia
nephron functional unit of the kidney
non-electrolyte solute that does not break down into ions when dissolved in water
osmoconformer organism that changes its tonicity based on its environment
osmoregulation mechanism by which water and solute concentrations are maintained at desired levels
osmoregulator organism that maintains its tonicity irrespective of its environment
osmotic balance balance of the amount of water and salt input and output to and from a biological system without disturbing the desired osmotic pressure and solute concentration in every compartment
osmotic pressure pressure exerted on a membrane to equalize solute concentration on either side
perirenal fat capsule fat layer that suspends the kidneys
peritubular capillary network capillary network that surrounds the renal tubule after the efferent artery exits the glomerulus
proximal convoluted tubule (PCT) part of the renal tubule that lies close to the glomerulus
renal artery branch of the artery that enters the kidney
renal capsule layer that encapsulates the kidneys
renal column area of the kidney through which the

interlobar arteries travel in the process of supplying blood to the renal lobes
renal corpuscle glomerulus and the Bowman's capsule together
renal fascia connective tissue that supports the kidneys
renal pelvis region in the kidney where the calyces join the ureters
renal pyramid conical structure in the renal medulla
renal tubule tubule of the nephron that arises from the glomerulus
renal vein branch of a vein that exits the kidney and joins the inferior vena cava
renin-angiotensin-aldosterone biochemical pathway that activates angiotensin II, which increases blood pressure
segmental artery artery that branches from the renal artery
semi-permeable membrane membrane that allows only certain solutes to pass through
transport maximum maximum amount of solute that can be transported out of the renal tubules during reabsorption
tubular reabsorption reclamation of water and solutes that got filtered out in the glomerulus
tubular secretion process of secretion of wastes that do not get reabsorbed
urea cycle pathway by which ammonia is converted to urea
ureotelic describes animals that secrete urea as the primary nitrogenous waste material
ureter urine-bearing tube coming out of the kidney; carries urine to the bladder
uric acid byproduct of ammonia metabolism in birds, insects, and reptiles
urinary bladder structure that the ureters empty the urine into; stores urine
urine filtrate produced by kidneys that gets excreted out of the body
vasa recta peritubular network that surrounds the loop of Henle of the juxtamedullary nephrons
vasodilator compound that increases the diameter of blood vessels
vasopressin another name for anti-diuretic hormone

CHAPTER SUMMARY

41.1 Osmoregulation and Osmotic Balance

Solute concentrations across semi-permeable membranes influence the movement of water and solutes across the membrane. It is the number of solute molecules and not the molecular size that is important in osmosis. Osmoregulation and osmotic balance are important bodily functions, resulting in water and salt balance. Not all solutes can pass through a semi-permeable membrane. Osmosis is the movement of water across the membrane. Osmosis occurs to equalize the number of solute molecules across a semi-permeable membrane by the movement of water to the side of higher solute concentration. Facilitated diffusion utilizes protein channels to move solute molecules from areas of higher to lower concentration while active transport mechanisms are required to move solutes against concentration gradients. Osmolarity is measured in units of milliequivalents or milliosmoles, both of which take into consideration the number of solute particles and the charge on them. Fish that live in freshwater or saltwater adapt by being osmoregulators or osmoconformers.

41.2 The Kidneys and Osmoregulatory Organs

The kidneys are the main osmoregulatory organs in mammalian systems; they function to filter blood and maintain the osmolarity of body fluids at 300 mOsm. They are surrounded by three layers and are made up internally of three distinct regions—the cortex, medulla, and pelvis.

The blood vessels that transport blood into and out of the kidneys arise from and merge with the aorta and inferior vena cava, respectively. The renal arteries branch out from the aorta and enter the kidney where they further divide into segmental, interlobar, arcuate, and cortical radiate arteries.

The nephron is the functional unit of the kidney, which actively filters blood and generates urine. The nephron is made up of the renal corpuscle and renal tubule. Cortical nephrons are found in the renal cortex, while juxtamedullary nephrons are found in the renal cortex close to the renal medulla. The nephron filters and exchanges water and solutes with two sets of blood vessels and the tissue fluid in the kidneys.

There are three steps in the formation of urine: glomerular filtration, which occurs in the glomerulus; tubular reabsorption, which occurs in the renal tubules; and tubular secretion, which also occurs in the renal tubules.

41.3 Excretion Systems

Many systems have evolved for excreting wastes that are simpler than the kidney and urinary systems of vertebrate animals. The simplest system is that of contractile vacuoles present in microorganisms. Flame cells and nephridia in worms perform excretory functions and maintain osmotic balance. Some insects have evolved Malpighian tubules to excrete wastes and maintain osmotic balance.

41.4 Nitrogenous Wastes

Ammonia is the waste produced by metabolism of nitrogen-containing compounds like proteins and nucleic acids. While aquatic animals can easily excrete ammonia into their watery surroundings, terrestrial animals have evolved special mechanisms to eliminate the toxic ammonia from their systems. Urea is the major byproduct of ammonia metabolism in vertebrate animals. Uric acid is the major byproduct of ammonia metabolism in birds, terrestrial arthropods, and reptiles.

41.5 Hormonal Control of Osmoregulatory Functions

Hormonal cues help the kidneys synchronize the osmotic needs of the body. Hormones like epinephrine, norepinephrine, renin-angiotensin, aldosterone, anti-diuretic hormone, and atrial natriuretic peptide help regulate the needs of the body as well as the communication between the different organ systems.

VISUAL CONNECTION QUESTIONS

1. Figure 41.5 Which of the following statements about the kidney is false?
   a. The renal pelvis drains into the ureter.
   b. The renal pyramids are in the medulla.
   c. The cortex covers the capsule.
   d. Nephrons are in the renal cortex.
2. Figure 41.6 Which of the following statements about the nephron is false?
   a. The collecting duct empties into the distal convoluted tubule.
   b. The Bowman's capsule surrounds the glomerulus.
   c. The loop of Henle is between the proximal and distal convoluted tubules.
   d. The loop of Henle empties into the distal convoluted tubule.
3. Figure 41.8 Loop diuretics are drugs sometimes used to treat hypertension. These drugs inhibit the reabsorption of Na^+ and Cl^- ions by the ascending limb of the loop of Henle. A side effect is that they increase urination. Why do you think this is the case?

REVIEW QUESTIONS

4. When a dehydrated human patient needs to be given fluids intravenously, he or she is given:
   a. water, which is hypotonic with respect to body fluids
   b. saline at a concentration that is isotonic with respect to body fluids
   c. glucose because it is a non-electrolyte
   d. blood
5. The sodium ion is at the highest concentration in:
   a. intracellular fluid
   b. extracellular fluid
   c. blood plasma
   d. none of the above
6. Cells in a hypertonic solution tend to:
   a. shrink due to water loss
   b. swell due to water gain
   c. stay the same size due to water moving into and out of the cell at the same rate
   d. none of the above
7. The macula densa is/are:
   a. present in the renal medulla.
   b. dense tissue present in the outer layer of the kidney.
   c. cells present in the DCT and collecting tubules.
   d. present in blood capillaries.
8. The osmolarity of body fluids is maintained at ________.
   a. 100 mOsm
   b. 300 mOsm
   c. 1000 mOsm
   d. it is not constantly maintained
9. The gland located at the top of the kidney is the ________ gland.
   a. adrenal
   b. pituitary
   c. thyroid
   d. thymus

10. Active transport of K^+ in Malpighian tubules ensures that:
    a. water follows K^+ to make urine
    b. osmotic balance is maintained between waste matter and bodily fluids
    c. both a and b
    d. neither a nor b
11. Contractile vacuoles in microorganisms:
    a. exclusively perform an excretory function
    b. can perform many functions, one of which is excretion of metabolic wastes
    c. originate from the cell membrane
    d. both b and c
12. Flame cells are primitive excretory organs found in _______.
    a. arthropods
    b. annelids
    c. mammals
    d. flatworms
13. BUN is _______.
    a. blood urea nitrogen
    b. blood uric acid nitrogen
    c. an indicator of blood volume
    d. an indicator of blood pressure
14. Human beings accumulate _______ before excreting nitrogenous waste.
    a. nitrogen
    b. ammonia
    c. urea
    d. uric acid
15. Renin is made by _______.
    a. granular cells of the juxtaglomerular apparatus
    b. the kidneys
    c. the nephrons
    d. all of the above
16. Patients with Addison's disease _______.
    a. retain water
    b. retain salts
    c. lose salts and water
    d. have too much aldosterone
17. Which hormone elicits the "fight or flight" response?
    a. epinephrine
    b. mineralcorticoids
    c. anti-diuretic hormone
    d. thyroxine

CRITICAL THINKING QUESTIONS

18. Why is excretion important in order to achieve osmotic balance?
19. Why do electrolyte ions move across membranes by active transport?
20. Why are the loop of Henle and vasa recta important for the formation of concentrated urine?
21. Describe the structure of the kidney.
22. Why might specialized organs have evolved for excretion of wastes?
23. Explain two different excretory systems other than the kidneys.
24. In terms of evolution, why might the urea cycle have evolved in organisms?
25. Compare and contrast the formation of urea and uric acid.
26. Describe how hormones regulate blood pressure, blood volume, and kidney function.
27. How does the renin-angiotensin-aldosterone mechanism function? Why is it controlled by the kidneys?

CHAPTER 42
The Immune System

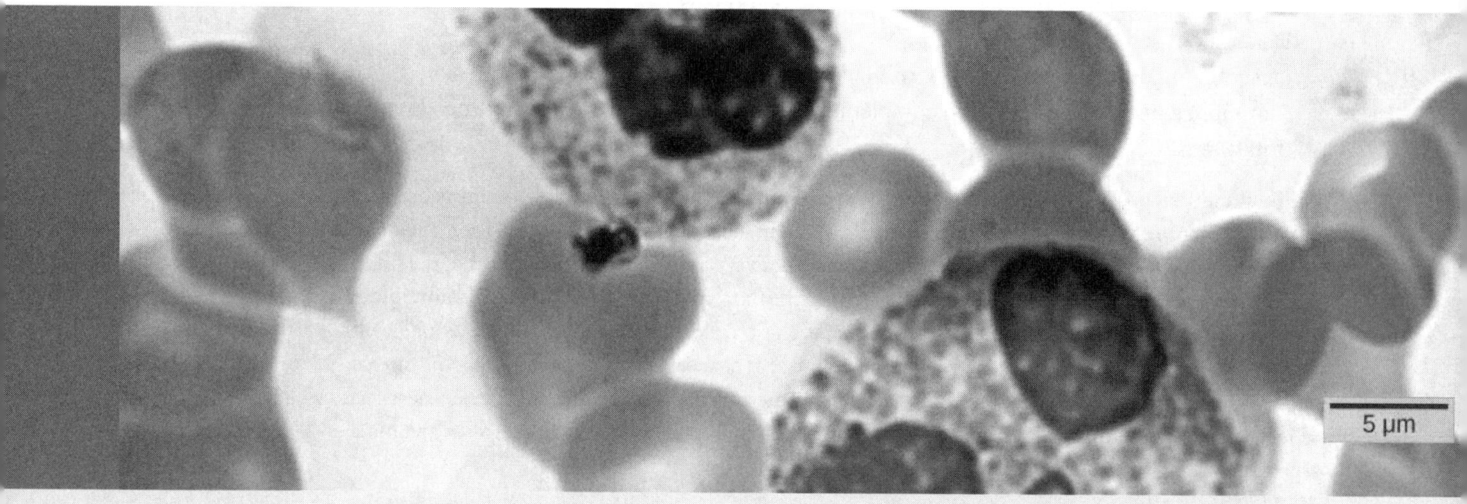

Figure 42.1 In this compound light micrograph purple-stained neutrophil (upper left) and eosinophil (lower right) are white blood cells that float among red blood cells in this blood smear. Neutrophils provide an early, rapid, and nonspecific defense against invading pathogens. Eosinophils play a variety of roles in the immune response. Red blood cells are about 7–8 μm in diameter, and a neutrophil is about 10–12μm. (credit: modification of work by Dr. David Csaba)

Chapter Outline

INTRODUCTION The environment consists of numerous **pathogens**, which are agents, usually microorganisms, that cause diseases in their hosts. A **host** is the organism that is invaded and often harmed by a pathogen. Pathogens include bacteria, protists, fungi and other infectious organisms. We are constantly exposed to pathogens in food and water, on surfaces, and in the air. Mammalian immune systems evolved for protection from such pathogens; they are composed of an extremely diverse array of specialized cells and soluble molecules that coordinate a rapid and flexible defense system capable of providing protection from a majority of these disease agents.

Components of the immune system constantly search the body for signs of pathogens. When pathogens are found, immune factors are mobilized to the site of an infection. The immune factors identify the nature of the pathogen, strengthen the corresponding cells and molecules to combat it efficiently, and then halt the immune response after the infection is cleared to avoid unnecessary host cell damage. The immune system can remember pathogens to which it has been exposed to create a more efficient response upon reexposure. This memory can last several decades. Features of the immune system, such as pathogen identification, specific response, amplification, retreat, and remembrance are essential for survival against pathogens. The immune response can be classified as either innate or active. The innate immune response is always present and attempts to defend against all pathogens rather than focusing on specific ones. Conversely, the adaptive immune response stores information about past infections and mounts pathogen-specific defenses.

42.1 Innate Immune Response

By the end of this section, you will be able to do the following:

- Describe physical and chemical immune barriers
- Explain immediate and induced innate immune responses
- Discuss natural killer cells
- Describe major histocompatibility class I molecules
- Summarize how the proteins in a complement system function to destroy extracellular pathogens

The immune system comprises both innate and adaptive immune responses. **Innate immunity** occurs naturally because of genetic factors or physiology; it is not induced by infection or vaccination but works to reduce the workload for the adaptive immune response. Both the innate and adaptive levels of the immune response involve secreted proteins, receptor-mediated signaling, and intricate cell-to-cell communication. The innate immune system developed early in animal evolution, roughly a billion years ago, as an essential response to infection. Innate immunity has a limited number of specific targets: any pathogenic threat triggers a consistent sequence of events that can identify the type of pathogen and either clear the infection independently or mobilize a highly specialized adaptive immune response. For example, tears and mucus secretions contain microbicidal factors.

Physical and Chemical Barriers

Before any immune factors are triggered, the skin functions as a continuous, impassable barrier to potentially infectious pathogens. Pathogens are killed or inactivated on the skin by desiccation (drying out) and by the skin's acidity. In addition, beneficial microorganisms that coexist on the skin compete with invading pathogens, preventing infection. Regions of the body that are not protected by skin (such as the eyes and mucus membranes) have alternative methods of defense, such as tears and mucus secretions that trap and rinse away pathogens, and cilia in the nasal passages and respiratory tract that push the mucus with the pathogens out of the body. Throughout the body are other defenses, such as the low pH of the stomach (which inhibits the growth of pathogens), blood proteins that bind and disrupt bacterial cell membranes, and the process of urination (which flushes pathogens from the urinary tract).

Despite these barriers, pathogens may enter the body through skin abrasions or punctures, or by collecting on mucosal surfaces in large numbers that overcome the mucus or cilia. Some pathogens have evolved specific mechanisms that allow them to overcome physical and chemical barriers. When pathogens do enter the body, the innate immune system responds with inflammation, pathogen engulfment, and secretion of immune factors and proteins.

Pathogen Recognition

An infection may be intracellular or extracellular, depending on the pathogen. All viruses infect cells and replicate within those cells (intracellularly), whereas bacteria and other parasites may replicate intracellularly or extracellularly, depending on the species. The innate immune system must respond accordingly: by identifying the extracellular pathogen and/or by identifying host cells that have already been infected. When a pathogen enters the body, cells in the blood and lymph detect the specific **pathogen-associated molecular patterns (PAMPs)** on the pathogen's surface. PAMPs are carbohydrate, polypeptide, and nucleic acid "signatures" that are expressed by viruses, bacteria, and parasites but which differ from molecules on host cells. The immune system has specific cells, described in Figure 42.2 and shown in Figure 42.3, with receptors that recognize these PAMPs. A **macrophage** is a large phagocytic cell that engulfs foreign particles and pathogens. Macrophages recognize PAMPs via complementary **pattern recognition receptors (PRRs)**. PRRs are molecules on macrophages and dendritic cells which are in contact with the external environment. A **monocyte** is a type of white blood cell that circulates in the blood and lymph and differentiates into macrophages after it moves into infected tissue. Dendritic cells bind molecular signatures of pathogens and promote pathogen engulfment and destruction. Toll-like receptors (TLRs) are a type of PRR that recognizes molecules that are shared by pathogens but distinguishable from host molecules. TLRs are present in invertebrates as well as vertebrates, and appear to be one of the most ancient components of the immune system. TLRs have also been identified in the mammalian nervous system.

Cell type	Characteristics	Location	Image
Mast cell	Dilates blood vessels and induces inflammation through release of histamines and heparin. Recruits macrophages and neutrophils. Involved in wound healing and defense against pathogens but can also be responsible for allergic reactions.	Connective tissues, mucous membranes	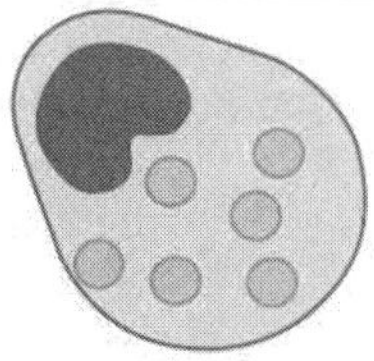
Macrophage	Phagocytic cell that consumes foreign pathogens and cancer cells. Stimulates response of other immune cells.	Migrates from blood vessels into tissues.	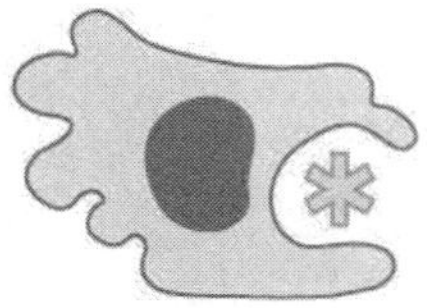
Natural killer cell	Kills tumor cells and virus-infected cells.	Circulates in blood and migrates into tissues.	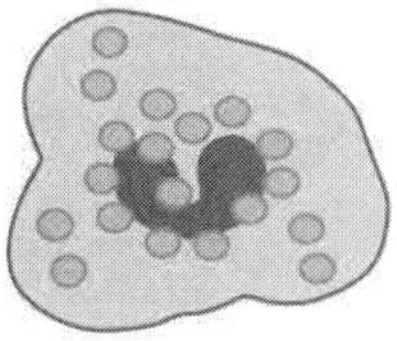
Dendritic cell	Presents antigens on its surface, thereby triggering adaptive immunity.	Present in epithelial tissue, including skin, lung and tissues of the digestive tract. Migrates to lymph nodes upon activation.	
Monocyte	Differentiates into macrophages and dendritic cells in response to inflammation.	Stored in spleen, moves through blood vessels to infected tissues.	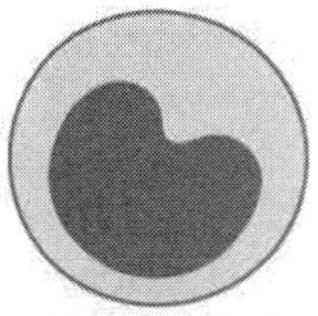
Neutrophil	First responders at the site of infection or trauma, this abundant phagocytic cell represents 50-60 percent of all leukocytes. Releases toxins that kill or inhibit bacteria and fungi and recruits other immune cells to the site of infection.	Migrates from blood vessels into tissues.	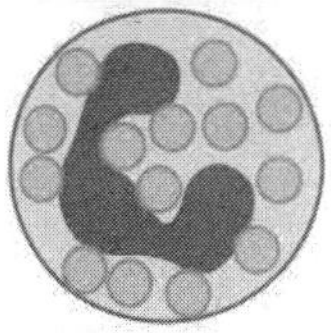
Basophil	Responsible for defense against parasites. Releases histamines that cause inflammation and may be responsible for allergic reactions.	Circulates in blood and migrates to tissues.	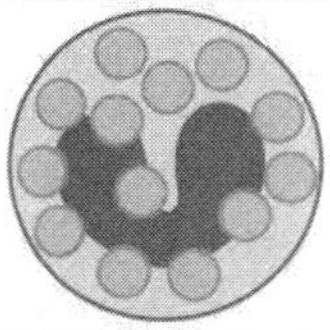
Eosinophil	Releases toxins that kill bacteria and parasites but also causes tissue damage.	Circulates in blood and migrates to tissues.	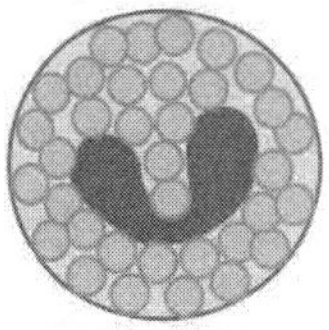

Figure 42.2 The characteristics and location of cells involved in the innate immune system are described. (credit: modification of work by NIH)

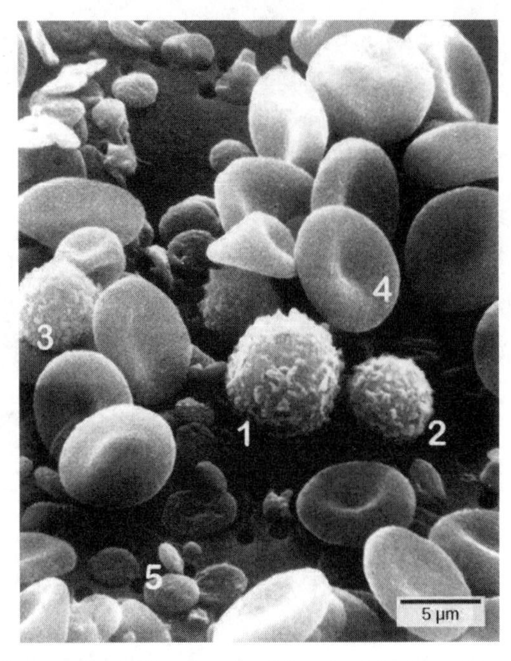

Figure 42.3 Cells of the blood include (1) monocytes, (2) lymphocytes, (3) neutrophils, (4) red blood cells, and (5) platelets. Note the very similar morphologies of the leukocytes (1, 2, 3). (credit: modification of work by Bruce Wetzel, Harry Schaefer, NCI; scale-bar data from Matt Russell)

Cytokine Release Effect

The binding of PRRs with PAMPs triggers the release of cytokines, which signal that a pathogen is present and needs to be destroyed along with any infected cells. A **cytokine** is a chemical messenger that regulates cell differentiation (form and function), proliferation (production), and gene expression to affect immune responses. At least 40 types of cytokines exist in humans that differ in terms of the cell type that produces them, the cell type that responds to them, and the changes they produce. One type of cytokine, interferon, is illustrated in Figure 42.4.

One subclass of cytokines is the interleukin (IL), so named because they mediate interactions between leukocytes (white blood cells). Interleukins are involved in bridging the innate and adaptive immune responses. In addition to being released from cells after PAMP recognition, cytokines are released by the infected cells which bind to nearby uninfected cells and induce those cells to release cytokines, which results in a cytokine burst.

A second class of early-acting cytokines is interferons, which are released by infected cells as a warning to nearby uninfected cells. One of the functions of an **interferon** is to inhibit viral replication. They also have other important functions, such as tumor surveillance. Interferons work by signaling neighboring uninfected cells to destroy RNA and reduce protein synthesis, signaling neighboring infected cells to undergo apoptosis (programmed cell death), and activating immune cells.

In response to interferons, uninfected cells alter their gene expression, which increases the cells' resistance to infection. One effect of interferon-induced gene expression is a sharply reduced cellular protein synthesis. Virally infected cells produce more viruses by synthesizing large quantities of viral proteins. Thus, by reducing protein synthesis, a cell becomes resistant to viral infection.

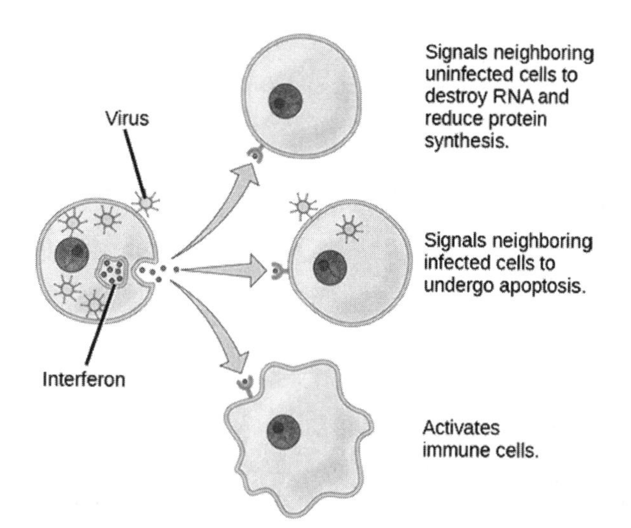

Figure 42.4 Interferons are cytokines that are released by a cell infected with a virus. Response of neighboring cells to interferon helps stem the infection.

Phagocytosis and Inflammation

The first cytokines to be produced are pro-inflammatory; that is, they encourage **inflammation**, the localized redness, swelling, heat, and pain that result from the movement of leukocytes and fluid through increasingly permeable capillaries to a site of infection. The population of leukocytes that arrives at an infection site depends on the nature of the infecting pathogen. Both macrophages and dendritic cells engulf pathogens and cellular debris through phagocytosis. A **neutrophil** is also a phagocytic leukocyte that engulfs and digests pathogens. Neutrophils, shown in Figure 42.3, are the most abundant leukocytes of the immune system. Neutrophils have a nucleus with two to five lobes, and they contain organelles, called lysosomes, that digest engulfed pathogens. An **eosinophil** is a leukocyte that works with other eosinophils to surround a parasite; it is involved in the allergic response and in protection against helminthes (parasitic worms).

Neutrophils and eosinophils are particularly important leukocytes that engulf large pathogens, such as bacteria and fungi. A **mast cell** is a leukocyte that produces inflammatory molecules, such as histamine, in response to large pathogens. A **basophil** is a leukocyte that, like a neutrophil, releases chemicals to stimulate the inflammatory response as illustrated in Figure 42.5. Basophils are also involved in allergy and hypersensitivity responses and induce specific types of inflammatory responses. Eosinophils and basophils produce additional inflammatory mediators to recruit more leukocytes. A hypersensitive immune response to harmless antigens, such as in pollen, often involves the release of histamine by basophils and mast cells.

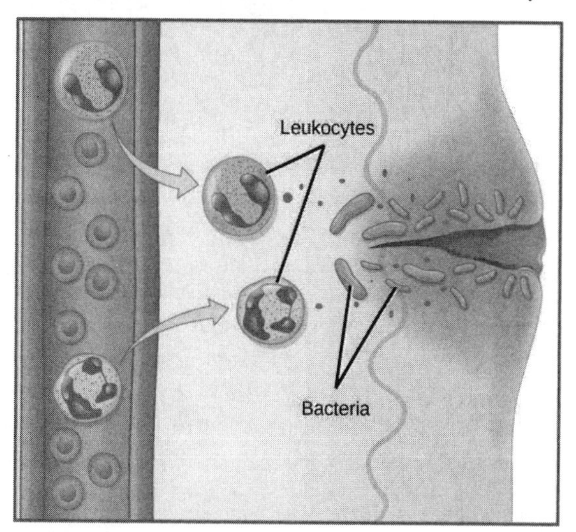

Figure 42.5 In response to a cut, mast cells secrete histamines that cause nearby capillaries to dilate. Neutrophils and monocytes leave the capillaries. Monocytes mature into macrophages. Neutrophils, dendritic cells, and macrophages release chemicals to stimulate the inflammatory response. Neutrophils and macrophages also consume invading bacteria by phagocytosis.

Cytokines also send feedback to cells of the nervous system to bring about the overall symptoms of feeling sick, which include lethargy, muscle pain, and nausea. These effects may have evolved because the symptoms encourage the individual to rest and prevent the spreading of the infection to others. Cytokines also increase the core body temperature, causing a fever, which causes the liver to withhold iron from the blood. Without iron, certain pathogens, such as some bacteria, are unable to replicate; this is called nutritional immunity.

LINK TO LEARNING

Watch this 23-second stop-motion video (http://openstax.org/l/conidia) showing a neutrophil that searches for and engulfs fungus spores during an elapsed time of about 79 minutes.

Natural Killer Cells

Lymphocytes are leukocytes that are histologically identifiable by their large, darkly staining nuclei; they are small cells with very little cytoplasm, as shown in Figure 42.6. Infected cells are identified and destroyed by **natural killer (NK) cells**, lymphocytes that can kill cells infected with viruses or tumor cells (abnormal cells that uncontrollably divide and invade other tissue). T cells and B cells of the adaptive immune system also are classified as lymphocytes. **T cells** are lymphocytes that mature in the thymus gland, and **B cells** are lymphocytes that mature in the bone marrow. NK cells identify intracellular infections, especially from viruses, by the altered expression of **major histocompatibility class (MHC) I molecules** on the surface of infected cells. MHC I molecules are proteins on the surfaces of all nucleated cells, thus they are scarce on red blood cells and platelets which are non-nucleated. The function of MHC I molecules is to display fragments of proteins from the infectious agents within the cell to T cells; healthy cells will be ignored, while "non-self" or foreign proteins will be attacked by the immune system. MHC II molecules are found mainly on cells containing antigens ("non-self proteins") and on lymphocytes. **MHC II molecules** interact with helper T cells to trigger the appropriate immune response, which may include the inflammatory response.

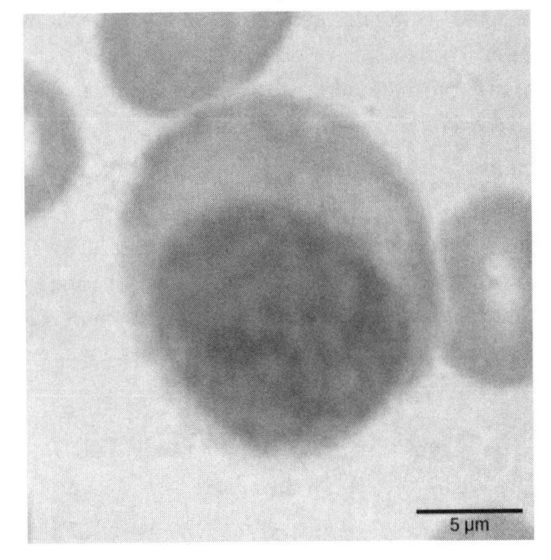

Figure 42.6 Lymphocytes, such as NK cells, are characterized by their large nuclei that actively absorb Wright stain and therefore appear dark colored under a microscope.

An infected cell (or a tumor cell) is usually incapable of synthesizing and displaying MHC I molecules appropriately. The metabolic resources of cells infected by some viruses produce proteins that interfere with MHC I processing and/or trafficking to the cell surface. The reduced MHC I on host cells varies from virus to virus and results from active inhibitors being produced by the viruses. This process can deplete host MHC I molecules on the cell surface, which NK cells detect as "unhealthy" or "abnormal" while searching for cellular MHC I molecules. Similarly, the dramatically altered gene expression of tumor cells leads to expression of extremely deformed or absent MHC I molecules that also signal "unhealthy" or "abnormal."

NK cells are always active; an interaction with normal, intact MHC I molecules on a healthy cell disables the killing sequence, and the NK cell moves on. After the NK cell detects an infected or tumor cell, its cytoplasm secretes granules comprised of **perforin**, a destructive protein that creates a pore in the target cell. Granzymes are released along with the perforin in the immunological synapse. A **granzyme** is a protease that digests cellular proteins and induces the target cell to undergo programmed cell death, or apoptosis. Phagocytic cells then digest the cell debris left behind. NK cells are constantly patrolling the body and are an effective mechanism for controlling potential infections and preventing cancer progression.

Complement

An array of approximately 20 types of soluble proteins, called a **complement system**, functions to destroy extracellular pathogens. Cells of the liver and macrophages synthesize complement proteins continuously; these proteins are abundant in the blood serum and are capable of responding immediately to infecting microorganisms. The complement system is so named because it is complementary to the antibody response of the adaptive immune system. Complement proteins bind to the surfaces of microorganisms and are particularly attracted to pathogens that are already bound by antibodies. Binding of complement proteins occurs in a specific and highly regulated sequence, with each successive protein being activated by cleavage and/or structural changes induced upon binding of the preceding protein(s). After the first few complement proteins bind, a cascade of sequential binding events follows in which the pathogen rapidly becomes coated in complement proteins.

Complement proteins perform several functions. The proteins serve as a marker to indicate the presence of a pathogen to phagocytic cells, such as macrophages and B cells, and enhance engulfment; this process is called **opsonization**. Certain complement proteins can combine to form attack complexes that open pores in microbial cell membranes. These structures destroy pathogens by causing their contents to leak, as illustrated in Figure 42.7.

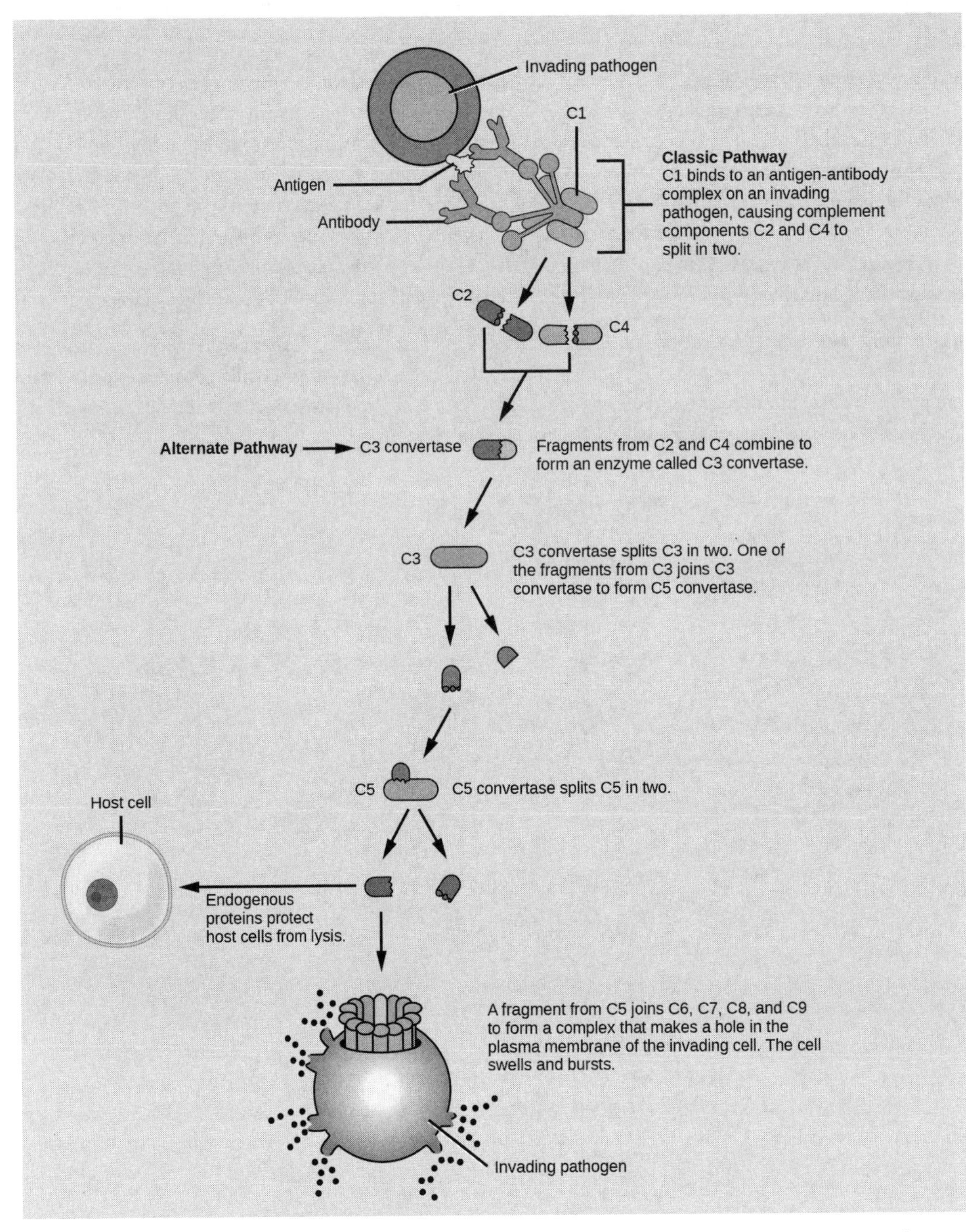

Figure 42.7 The classic pathway for the complement cascade involves the attachment of several initial complement proteins to an antibody-bound pathogen followed by rapid activation and binding of many more complement proteins and the creation of destructive pores in the microbial cell envelope and cell wall. The alternate pathway does not involve antibody activation. Rather, C3 convertase spontaneously breaks down C3. Endogenous regulatory proteins prevent the complement complex from binding to host cells. Pathogens lacking these regulatory proteins are lysed. (credit: modification of work by NIH)

42.2 Adaptive Immune Response

By the end of this section, you will be able to do the following:

- Explain adaptive immunity
- Compare and contrast adaptive and innate immunity
- Describe cell-mediated immune response and humoral immune response
- Describe immune tolerance

The adaptive, or acquired, immune response takes days or even weeks to become established—much longer than the innate response; however, adaptive immunity is more specific to pathogens and has memory. **Adaptive immunity** is an immunity that occurs after exposure to an antigen either from a pathogen or a vaccination. This part of the immune system is activated when the innate immune response is insufficient to control an infection. In fact, without information from the innate immune system, the adaptive response could not be mobilized. There are two types of adaptive responses: the **cell-mediated immune response**, which is carried out by T cells, and the **humoral immune response**, which is controlled by activated B cells and antibodies. Activated T cells and B cells that are specific to molecular structures on the pathogen proliferate and attack the invading pathogen. Their attack can kill pathogens directly or secrete antibodies that enhance the phagocytosis of pathogens and disrupt the infection. Adaptive immunity also involves a memory to provide the host with long-term protection from reinfection with the same type of pathogen; on reexposure, this memory will facilitate an efficient and quick response.

Antigen-presenting Cells

Unlike NK cells of the innate immune system, B cells (B lymphocytes) are a type of white blood cell that gives rise to antibodies, whereas T cells (T lymphocytes) are a type of white blood cell that plays an important role in the immune response. T cells are a key component in the cell-mediated response—the specific immune response that utilizes T cells to neutralize cells that have been infected with viruses and certain bacteria. There are three types of T cells: cytotoxic, helper, and suppressor T cells. Cytotoxic T cells destroy virus-infected cells in the cell-mediated immune response, and helper T cells play a part in activating both the antibody and the cell-mediated immune responses. Suppressor T cells deactivate T cells and B cells when needed, and thus prevent the immune response from becoming too intense.

An **antigen** is a foreign or "non-self" macromolecule that reacts with cells of the immune system. Not all antigens will provoke a response. For instance, individuals produce innumerable "self" antigens and are constantly exposed to harmless foreign antigens, such as food proteins, pollen, or dust components. The suppression of immune responses to harmless macromolecules is highly regulated and typically prevents processes that could be damaging to the host, known as tolerance.

The innate immune system contains cells that detect potentially harmful antigens, and then inform the adaptive immune response about the presence of these antigens. An **antigen-presenting cell (APC)** is an immune cell that detects, engulfs, and informs the adaptive immune response about an infection. When a pathogen is detected, these APCs will phagocytose the pathogen and digest it to form many different fragments of the antigen. Antigen fragments will then be transported to the surface of the APC, where they will serve as an indicator to other immune cells. **Dendritic cells** are immune cells that process antigen material; they are present in the skin (Langerhans cells) and the lining of the nose, lungs, stomach, and intestines. Sometimes a dendritic cell presents on the surface of other cells to induce an immune response, thus functioning as an antigen-presenting cell. Macrophages also function as APCs. Before activation and differentiation, B cells can also function as APCs.

After phagocytosis by APCs, the phagocytic vesicle fuses with an intracellular lysosome forming phagolysosome. Within the phagolysosome, the components are broken down into fragments; the fragments are then loaded onto MHC class I or MHC class II molecules and are transported to the cell surface for antigen presentation, as illustrated in Figure 42.8. Note that T lymphocytes cannot properly respond to the antigen unless it is processed and embedded in an MHC II molecule. APCs express MHC on their surfaces, and when combined with a foreign antigen, these complexes signal a "non-self" invader. Once the fragment of antigen is embedded in the MHC II molecule, the immune cell can respond. Helper T cells are one of the main lymphocytes that respond to antigen-presenting cells. Recall that all other nucleated cells of the body expressed MHC I molecules, which signal "healthy" or "normal."

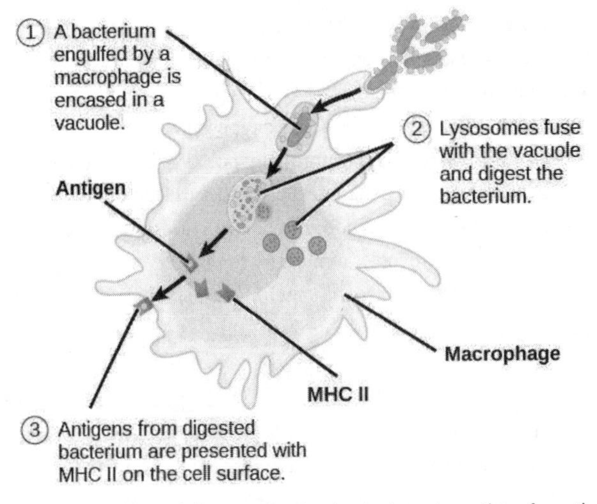

Figure 42.8 An APC, such as a macrophage, engulfs and digests a foreign bacterium. An antigen from the bacterium is presented on the cell surface in conjunction with an MHC II molecule. Lymphocytes of the adaptive immune response interact with antigen-embedded MHC II molecules to mature into functional immune cells.

LINK TO LEARNING

This animation (http://openstax.org/l/immune_system) from Rockefeller University shows how dendritic cells act as sentinels in the body's immune system.

T and B Lymphocytes

Lymphocytes in human circulating blood are approximately 80 to 90 percent T cells, shown in Figure 42.9, and 10 to 20 percent B cells. Recall that the T cells are involved in the cell-mediated immune response, whereas B cells are part of the humoral immune response.

T cells encompass a heterogeneous population of cells with extremely diverse functions. Some T cells respond to APCs of the innate immune system, and indirectly induce immune responses by releasing cytokines. Other T cells stimulate B cells to prepare their own response. Another population of T cells detects APC signals and directly kills the infected cells. Other T cells are involved in suppressing inappropriate immune reactions to harmless or "self" antigens.

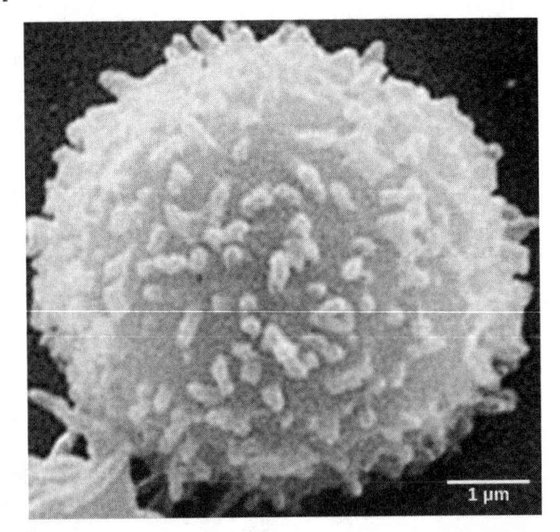

Figure 42.9 This scanning electron micrograph shows a T lymphocyte, which is responsible for the cell-mediated immune response. T cells are able to recognize antigens. (credit: modification of work by NCI; scale-bar data from Matt Russell)

T and B cells exhibit a common theme of recognition/binding of specific antigens via a complementary receptor, followed by activation and self-amplification/maturation to specifically bind to the particular antigen of the infecting pathogen. T and B lymphocytes are also similar in that each cell only expresses one type of antigen receptor. Any individual may possess a population of T and B cells that together express a near limitless variety of antigen receptors that are capable of recognizing virtually any infecting pathogen. T and B cells are activated when they recognize small components of antigens, called **epitopes**, presented by APCs, illustrated in Figure 42.10. Note that recognition occurs at a specific epitope rather than on the entire antigen; for this reason, epitopes are known as "antigenic determinants." In the absence of information from APCs, T and B cells remain inactive, or naïve, and are unable to prepare an immune response. The requirement for information from the APCs of innate immunity to trigger B cell or T cell activation illustrates the essential nature of the innate immune response to the functioning of the entire immune system.

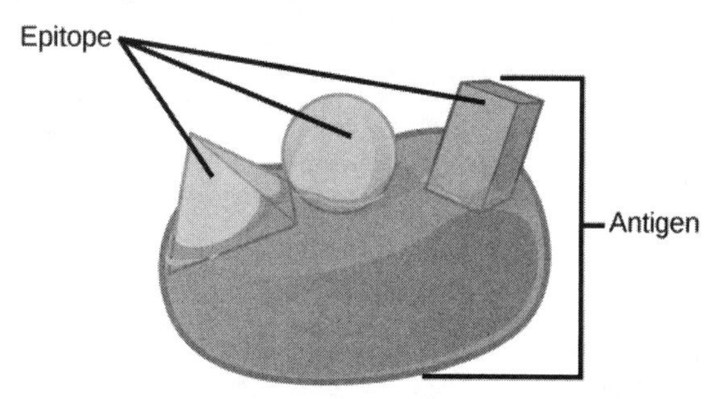

Figure 42.10 An antigen is a macromolecule that reacts with components of the immune system. A given antigen may contain several motifs that are recognized by immune cells. Each motif is an epitope. In this figure, the entire structure is an antigen, and the orange, salmon and green components projecting from it represent potential epitopes.

Naïve T cells can express one of two different molecules, CD4 or CD8, on their surface, as shown in Figure 42.11, and are accordingly classified as $CD4^+$ or $CD8^+$ cells. These molecules are important because they regulate how a T cell will interact with and respond to an APC. Naïve $CD4^+$ cells bind APCs via their antigen-embedded MHC II molecules and are stimulated to become **helper T (T_H) lymphocytes**, cells that go on to stimulate B cells (or cytotoxic T cells) directly or secrete cytokines to inform more and various target cells about the pathogenic threat. In contrast, $CD8^+$ cells engage antigen-embedded MHC I molecules on APCs and are stimulated to become **cytotoxic T lymphocytes (CTLs)**, which directly kill infected cells by apoptosis and emit cytokines to amplify the immune response. The two populations of T cells have different mechanisms of immune protection, but both bind MHC molecules via their antigen receptors called T cell receptors (TCRs). The CD4 or CD8 surface molecules differentiate whether the TCR will engage an MHC II or an MHC I molecule. Because they assist in binding specificity, the CD4 and CD8 molecules are described as coreceptors.

VISUAL CONNECTION

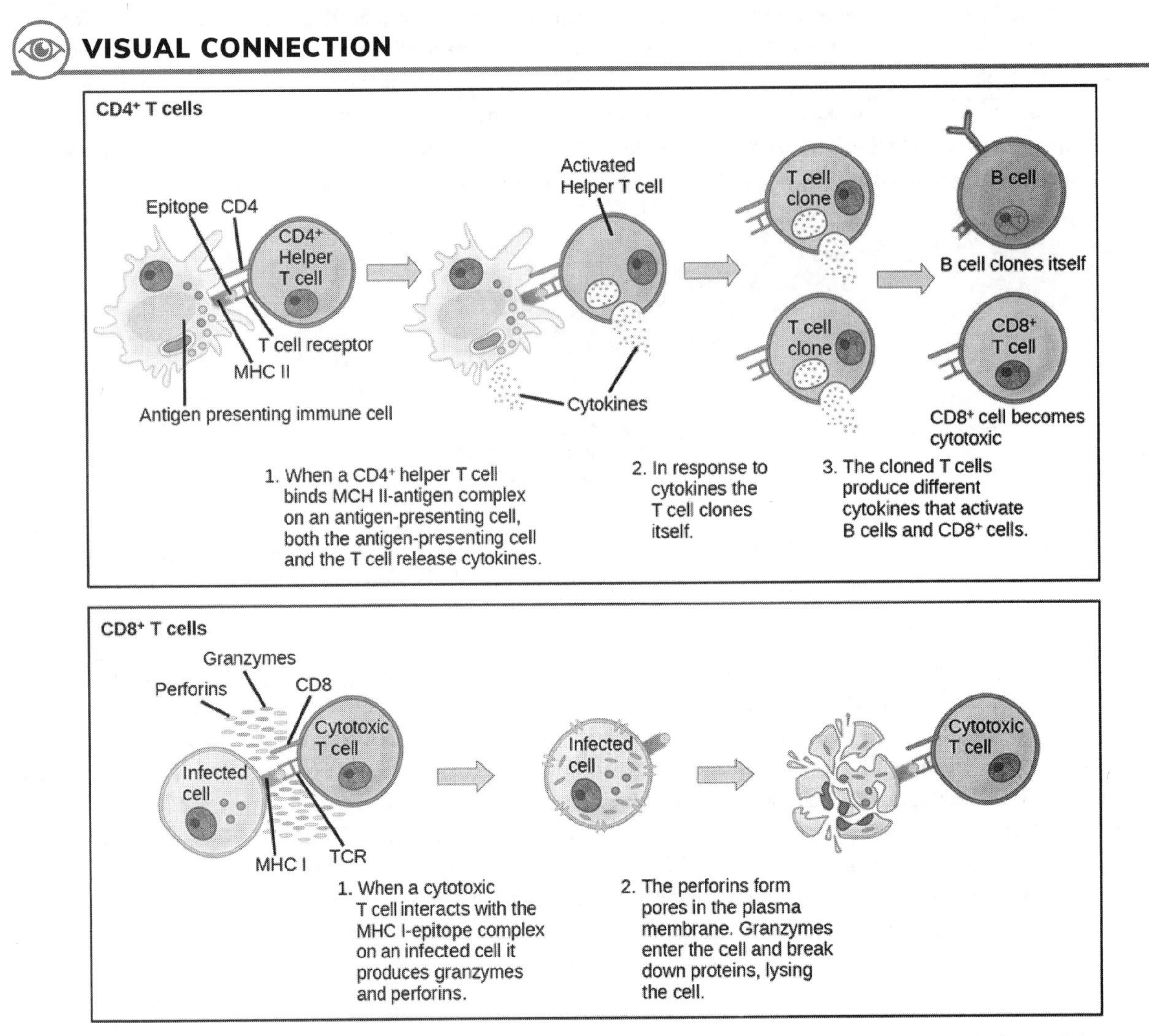

Figure 42.11 Naïve CD4$^+$ T cells engage MHC II molecules on antigen-presenting cells (APCs) and become activated. Clones of the activated helper T cell, in turn, activate B cells and CD8$^+$ T cells, which become cytotoxic T cells. Cytotoxic T cells kill infected cells.

Which of the following statements about T cells is false?

a. Helper T cells release cytokines while cytotoxic T cells kill the infected cell.
b. Helper T cells are CD4$^+$, while cytotoxic T cells are CD8$^+$.
c. MHC II is a receptor found on most body cells, while MHC I is a receptor found on immune cells only.
d. The T cell receptor is found on both CD4$^+$ and CD8$^+$ T cells.

Consider the innumerable possible antigens that an individual will be exposed to during a lifetime. The mammalian adaptive immune system is adept in responding appropriately to each antigen. Mammals have an enormous diversity of T cell populations, resulting from the diversity of TCRs. Each TCR consists of two polypeptide chains that span the T cell membrane, as illustrated in Figure 42.12; the chains are linked by a disulfide bridge. Each polypeptide chain is comprised of a constant domain and a variable domain: a domain, in this sense, is a specific region of a protein that may be regulatory or structural. The intracellular domain is involved in intracellular signaling. A single T cell will express thousands of identical copies of one specific TCR variant on its cell surface. The specificity of the adaptive immune system occurs because it synthesizes millions of different T cell populations, each expressing a TCR that differs in its variable domain. This TCR diversity is achieved by the mutation and recombination of genes that encode these receptors in stem cell precursors of T cells. The binding between an antigen-displaying MHC molecule and a complementary TCR "match" indicates that the adaptive immune system needs to activate and produce that specific T cell because its structure is appropriate to recognize and destroy the invading pathogen.

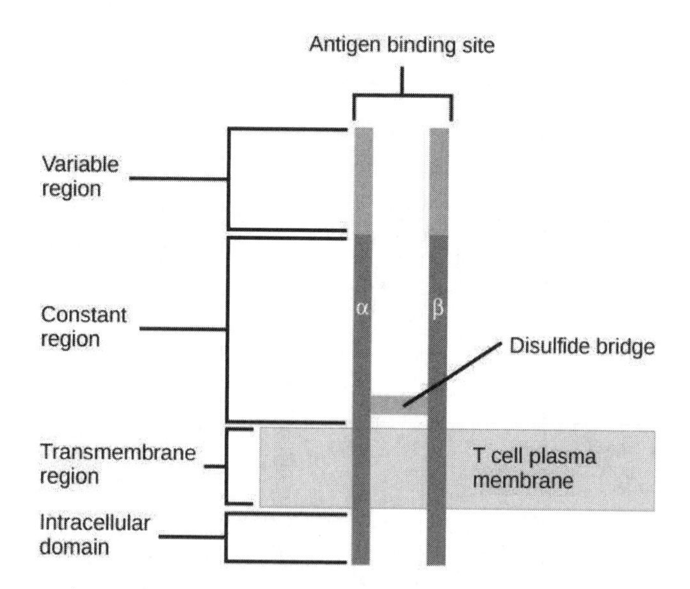

Figure 42.12 A T cell receptor spans the membrane and projects variable binding regions into the extracellular space to bind processed antigens via MHC molecules on APCs.

Helper T Lymphocytes

The T_H lymphocytes function indirectly to identify potential pathogens for other cells of the immune system. These cells are important for extracellular infections, such as those caused by certain bacteria, helminths, and protozoa. T_H lymphocytes recognize specific antigens displayed in the MHC II complexes of APCs. There are two major populations of T_H cells: T_H1 and T_H2. T_H1 cells secrete cytokines to enhance the activities of macrophages and other T cells. T_H1 cells activate the action of cyotoxic T cells, as well as macrophages. T_H2 cells stimulate naïve B cells to destroy foreign invaders via antibody secretion. Whether a T_H1 or a T_H2 immune response develops depends on the specific types of cytokines secreted by cells of the innate immune system, which in turn depends on the nature of the invading pathogen.

The T_H1-mediated response involves macrophages and is associated with inflammation. Recall the frontline defenses of macrophages involved in the innate immune response. Some intracellular bacteria, such as *Mycobacterium tuberculosis*, have evolved to multiply in macrophages after they have been engulfed. These pathogens evade attempts by macrophages to destroy and digest the pathogen. When *M. tuberculosis* infection occurs, macrophages can stimulate naïve T cells to become T_H1 cells. These stimulated T cells secrete specific cytokines that send feedback to the macrophage to stimulate its digestive capabilities and allow it to destroy the colonizing *M. tuberculosis*. In the same manner, T_H1-activated macrophages also become better suited to ingest and kill tumor cells. In summary; T_H1 responses are directed toward intracellular invaders while T_H2 responses are aimed at those that are extracellular.

B Lymphocytes

When stimulated by the T_H2 pathway, naïve B cells differentiate into antibody-secreting plasma cells. A **plasma cell** is an immune cell that secrets antibodies; these cells arise from B cells that were stimulated by antigens. Similar to T cells, naïve B cells initially are coated in thousands of B cell receptors (BCRs), which are membrane-bound forms of Ig (immunoglobulin, or an antibody). The B cell receptor has two heavy chains and two light chains connected by disulfide linkages. Each chain has a constant and a variable region; the latter is involved in antigen binding. Two other membrane proteins, Ig alpha and Ig beta, are involved in signaling. The receptors of any particular B cell, as shown in Figure 42.13 are all the same, but the hundreds of millions of different B cells in an individual have distinct recognition domains that contribute to extensive diversity in the types of molecular structures to which they can bind. In this state, B cells function as APCs. They bind and engulf foreign antigens via their BCRs and then display processed antigens in the context of MHC II molecules to T_H2 cells. When a T_H2 cell detects that a B cell is bound to a relevant antigen, it secretes specific cytokines that induce the B cell to proliferate rapidly, which makes thousands of identical (clonal) copies of it, and then it synthesizes and secretes antibodies with the same antigen recognition pattern as the BCRs. The activation of B cells corresponding to one specific BCR variant and the dramatic proliferation of that variant is known as **clonal selection**. This phenomenon drastically, but briefly, changes the proportions of BCR variants expressed by the immune system, and shifts the balance toward BCRs specific to the infecting pathogen.

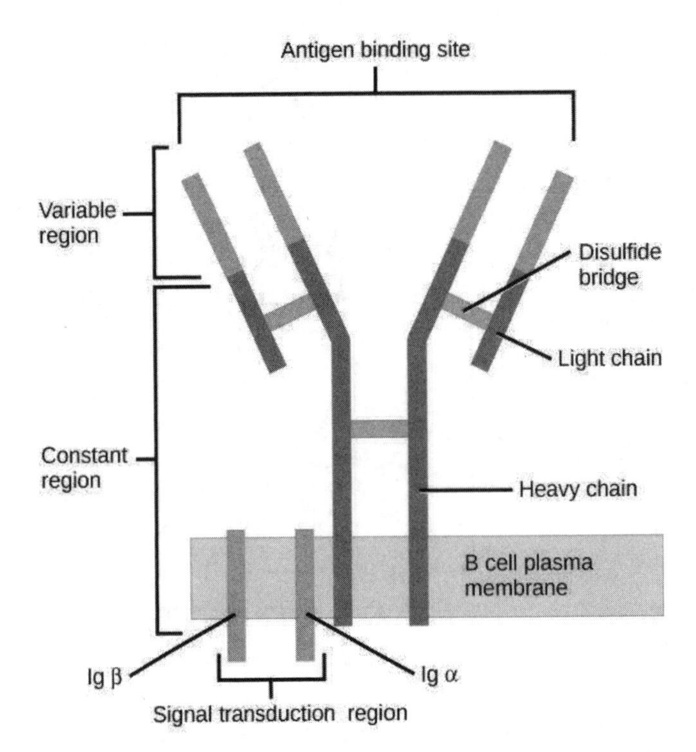

Figure 42.13 B cell receptors are embedded in the membranes of B cells and bind a variety of antigens through their variable regions. The signal transduction region transfers the signal into the cell.

T and B cells differ in one fundamental way: whereas T cells bind antigens that have been digested and embedded in MHC molecules by APCs, B cells function as APCs that bind intact antigens that have not been processed. Although T and B cells both react with molecules that are termed "antigens," these lymphocytes actually respond to very different types of molecules. B cells must be able to bind intact antigens because they secrete antibodies that must recognize the pathogen directly, rather than digested remnants of the pathogen. Bacterial carbohydrate and lipid molecules can activate B cells independently from the T cells.

Cytotoxic T Lymphocytes

CTLs, a subclass of T cells, function to clear infections directly. The cell-mediated part of the adaptive immune system consists of CTLs that attack and destroy infected cells. CTLs are particularly important in protecting against viral infections; this is because viruses replicate within cells where they are shielded from extracellular contact with circulating antibodies. When APCs phagocytize pathogens and present MHC I-embedded antigens to naïve $CD8^+$ T cells that express complementary TCRs, the $CD8^+$ T cells become activated to proliferate according to clonal selection. These resulting CTLs then identify non-APCs displaying the same MHC I-embedded antigens (for example, viral proteins)—for example, the CTLs identify infected host cells.

Intracellularly, infected cells typically die after the infecting pathogen replicates to a sufficient concentration and lyses the cell, as many viruses do. CTLs attempt to identify and destroy infected cells before the pathogen can replicate and escape, thereby halting the progression of intracellular infections. CTLs also support NK lymphocytes to destroy early cancers. Cytokines secreted by the T_H1 response that stimulates macrophages also stimulate CTLs and enhance their ability to identify and destroy infected cells and tumors.

CTLs sense MHC I-embedded antigens by directly interacting with infected cells via their TCRs. Binding of TCRs with antigens activates CTLs to release perforin and granzyme, degradative enzymes that will induce apoptosis of the infected cell. Recall that this is a similar destruction mechanism to that used by NK cells. In this process, the CTL does not become infected and is not harmed by the secretion of perforin and granzymes. In fact, the functions of NK cells and CTLs are complementary and maximize the removal of infected cells, as illustrated in Figure 42.14. If the NK cell cannot identify the "missing self" pattern of down-regulated MHC I molecules, then the CTL can identify it by the complex of MHC I with foreign antigens, which signals "altered self." Similarly, if the CTL cannot detect antigen-embedded MHC I because the receptors are depleted from the cell surface, NK cells will destroy the cell instead. CTLs also emit cytokines, such as interferons, that alter surface protein expression in other infected cells, such that the infected cells can be easily identified and destroyed. Moreover, these interferons can also prevent virally infected cells from releasing virus particles.

VISUAL CONNECTION

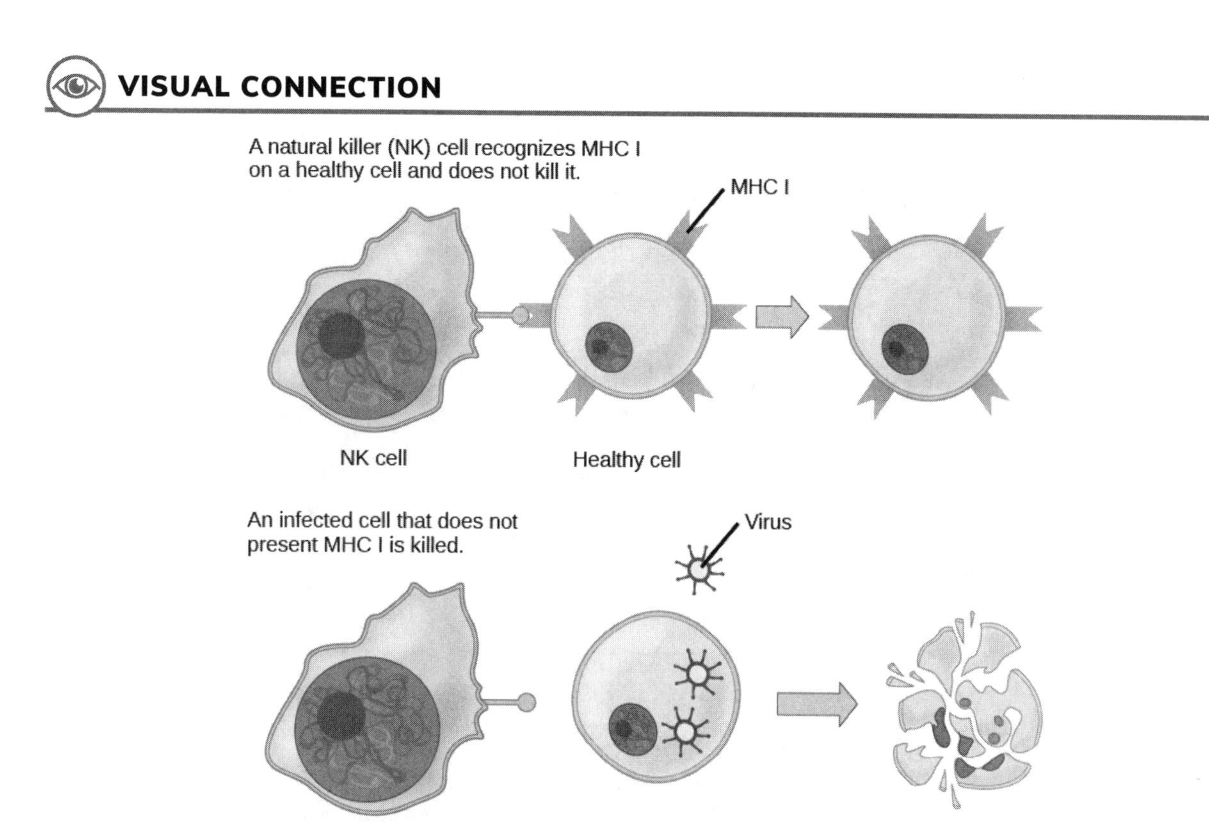

Figure 42.14 Natural killer (NK) cells recognize the MHC I receptor on healthy cells. If MHC I is absent, the cell is lysed.

Based on what you know about MHC receptors, why do you think an organ transplanted from an incompatible donor to a recipient will be rejected?

Plasma cells and CTLs are collectively called **effector cells**: they represent differentiated versions of their naïve counterparts, and they are involved in bringing about the immune defense of killing pathogens and infected host cells.

Mucosal Surfaces and Immune Tolerance

The innate and adaptive immune responses discussed thus far comprise the systemic immune system (affecting the whole body), which is distinct from the mucosal immune system. Mucosal immunity is formed by mucosa-associated lymphoid tissue, which functions independently of the systemic immune system, and which has its own innate and adaptive components. **Mucosa-associated lymphoid tissue (MALT)**, illustrated in Figure 42.15, is a collection of lymphatic tissue that combines with epithelial tissue lining the mucosa throughout the body. This tissue functions as the immune barrier and response in areas of the body with direct contact to the external environment. The systemic and mucosal immune systems use many of the same cell types. Foreign particles that make their way to MALT are taken up by absorptive epithelial cells called M cells and delivered to APCs located directly below the mucosal tissue. M cells function in the transport described, and are located in the Peyer's patch, a lymphoid nodule. APCs of the mucosal immune system are primarily dendritic cells, with B cells and macrophages having minor roles. Processed antigens displayed on APCs are detected by T cells in the MALT and at various mucosal induction sites, such as the tonsils, adenoids, appendix, or the mesenteric lymph nodes of the intestine. Activated T cells then migrate through the lymphatic system and into the circulatory system to mucosal sites of infection.

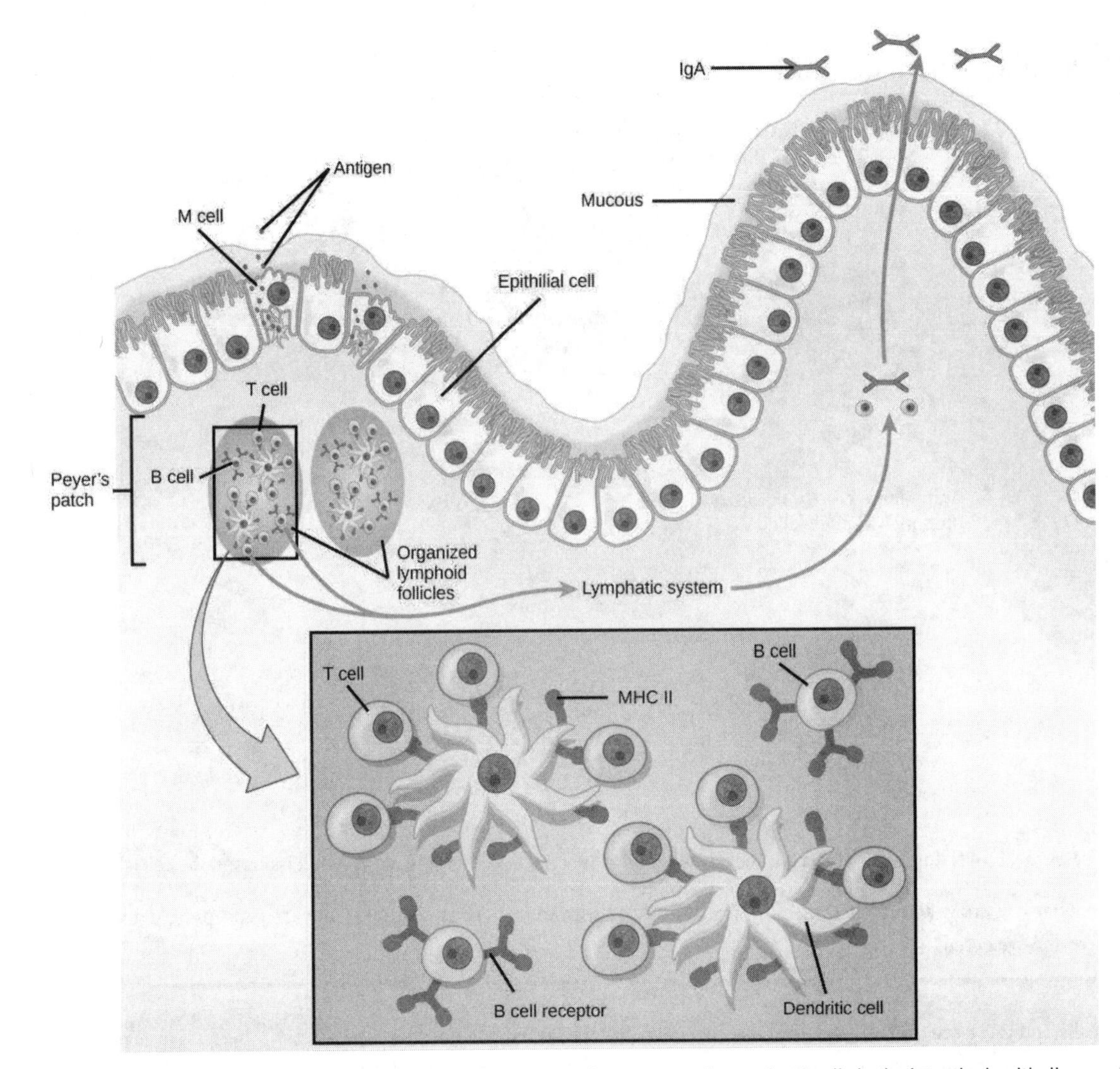

Figure 42.15 The topology and function of intestinal MALT is shown. Pathogens are taken up by M cells in the intestinal epithelium and excreted into a pocket formed by the inner surface of the cell. The pocket contains antigen-presenting cells such as dendritic cells, which engulf the antigens, then present them with MHC II molecules on the cell surface. The dendritic cells migrate to an underlying tissue called a Peyer's patch. Antigen-presenting cells, T cells, and B cells aggregate within the Peyer's patch, forming organized lymphoid follicles. There, some T cells and B cells are activated. Other antigen-loaded dendritic cells migrate through the lymphatic system where they activate B cells, T cells, and plasma cells in the lymph nodes. The activated cells then return to MALT tissue effector sites. IgA and other antibodies are secreted into the intestinal lumen.

MALT is a crucial component of a functional immune system because mucosal surfaces, such as the nasal passages, are the first tissues onto which inhaled or ingested pathogens are deposited. The mucosal tissue includes the mouth, pharynx, and esophagus, and the gastrointestinal, respiratory, and urogenital tracts.

The immune system has to be regulated to prevent wasteful, unnecessary responses to harmless substances, and more importantly so that it does not attack "self." The acquired ability to prevent an unnecessary or harmful immune response to a detected foreign substance known not to cause disease is described as **immune tolerance**. Immune tolerance is crucial for maintaining mucosal homeostasis given the tremendous number of foreign substances (such as food proteins) that APCs of the oral cavity, pharynx, and gastrointestinal mucosa encounter. Immune tolerance is brought about by specialized APCs in the liver, lymph nodes, small intestine, and lung that present harmless antigens to an exceptionally diverse population of **regulatory T (T_{reg}) cells**, specialized lymphocytes that suppress local inflammation and inhibit the secretion of stimulatory immune factors. The combined result of T_{reg} cells is to prevent immunologic activation and inflammation in undesired tissue compartments and to allow the immune system to focus on pathogens instead. In addition to promoting immune tolerance of harmless antigens, other subsets of T_{reg} cells are involved in the prevention of the **autoimmune response**, which is an inappropriate immune response to host cells or self-antigens. Another T_{reg} class suppresses immune responses to harmful pathogens after the infection has cleared to minimize host cell damage induced by inflammation and cell lysis.

Immunological Memory

The adaptive immune system possesses a memory component that allows for an efficient and dramatic response upon reinvasion of the same pathogen. Memory is handled by the adaptive immune system with little reliance on cues from the innate response. During the adaptive immune response to a pathogen that has not been encountered before, called a primary response, plasma cells secreting antibodies and differentiated T cells increase, then plateau over time. As B and T cells mature into effector cells, a subset of the naïve populations differentiates into B and T memory cells with the same antigen specificities, as illustrated in Figure 42.16.

A **memory cell** is an antigen-specific B or T lymphocyte that does not differentiate into effector cells during the primary immune response, but that can immediately become effector cells upon reexposure to the same pathogen. During the primary immune response, memory cells do not respond to antigens and do not contribute to host defenses. As the infection is cleared and pathogenic stimuli subside, the effectors are no longer needed, and they undergo apoptosis. In contrast, the memory cells persist in the circulation.

VISUAL CONNECTION

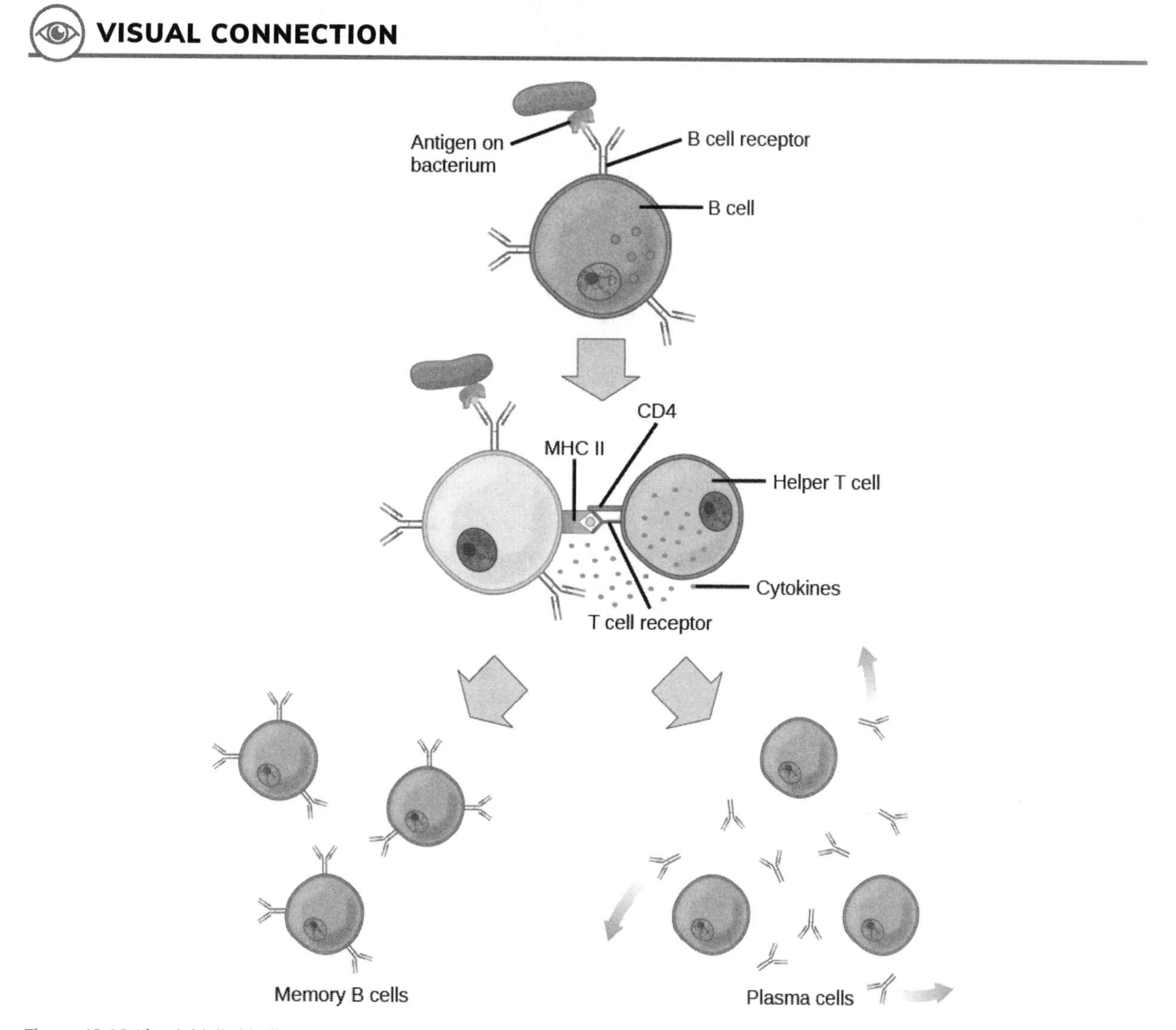

Figure 42.16 After initially binding an antigen to the B cell receptor (BCR), a B cell internalizes the antigen and presents it on MHC II. A helper T cell recognizes the MHC II–antigen complex and activates the B cell. As a result, memory B cells and plasma cells are made.

The Rh antigen is found on Rh-positive red blood cells. An Rh-negative female can usually carry an Rh-positive fetus to term without difficulty. However, if she has a second Rh-positive fetus, her body may launch an immune attack that causes hemolytic disease of the newborn. Why do you think hemolytic disease is only a problem during the second or subsequent pregnancies?

If the pathogen is never encountered again during the individual's lifetime, B and T memory cells will circulate for a few years or even several decades and will gradually die off, having never functioned as effector cells. However, if the host is reexposed to the same pathogen type, circulating memory cells will immediately differentiate into plasma cells and CTLs without input from APCs or T_H cells. One reason the adaptive immune response is delayed is because it takes time for naïve B and T cells with the appropriate antigen specificities to be identified and activated. Upon reinfection, this step is skipped, and the result is a more rapid production of immune defenses. Memory B cells that differentiate into plasma cells output tens to hundreds-fold greater antibody amounts than were secreted during the primary response, as the graph in Figure 42.17 illustrates. This rapid and dramatic antibody response may stop the infection before it can even become established, and the individual may not realize he or she had been exposed.

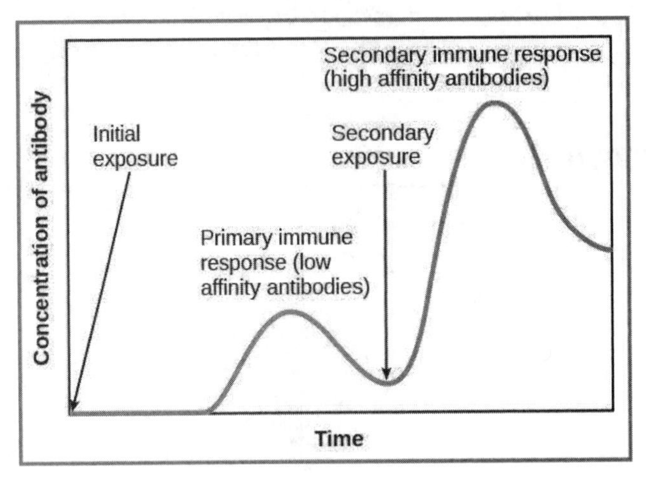

Figure 42.17 In the primary response to infection, antibodies are secreted first from plasma cells. Upon reexposure to the same pathogen, memory cells differentiate into antibody-secreting plasma cells that output a greater amount of antibody for a longer period of time.

Vaccination is based on the knowledge that exposure to noninfectious antigens, derived from known pathogens, generates a mild primary immune response. The immune response to vaccination may not be perceived by the host as illness but still confers immune memory. When exposed to the corresponding pathogen to which an individual was vaccinated, the reaction is similar to a secondary exposure. Because each reinfection generates more memory cells and increased resistance to the pathogen, and because some memory cells die, certain vaccine courses involve one or more booster vaccinations to mimic repeat exposures: for instance, tetanus boosters are necessary every ten years because the memory cells only live that long.

Mucosal Immune Memory

A subset of T and B cells of the mucosal immune system differentiates into memory cells just as in the systemic immune system. Upon reinvasion of the same pathogen type, a pronounced immune response occurs at the mucosal site where the original pathogen deposited, but a collective defense is also organized within interconnected or adjacent mucosal tissue. For instance, the immune memory of an infection in the oral cavity would also elicit a response in the pharynx if the oral cavity was exposed to the same pathogen.

CAREER CONNECTION

Vaccinologist

Vaccination (or immunization) involves the delivery, usually by injection as shown in Figure 42.18, of noninfectious antigen(s) derived from known pathogens. Other components, called adjuvants, are delivered in parallel to help stimulate the immune response. Immunological memory is the reason vaccines work. Ideally, the effect of vaccination is to elicit immunological memory, and thus resistance to specific pathogens without the individual having to experience an infection.

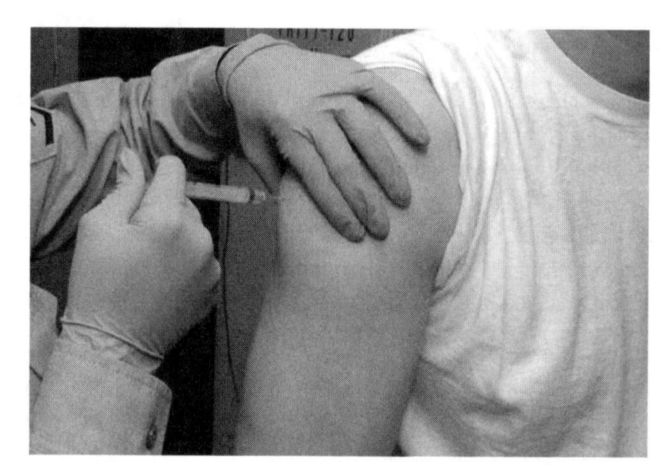

Figure 42.18 Vaccines are often delivered by injection into the arm. (credit: U.S. Navy Photographer's Mate Airman Apprentice Christopher D. Blachly)

Vaccinologists are involved in the process of vaccine development from the initial idea to the availability of the completed vaccine. This process can take decades, can cost millions of dollars, and can involve many obstacles along the way. For instance, injected vaccines stimulate the systemic immune system, eliciting humoral and cell-mediated immunity, but have little effect on the mucosal response, which presents a challenge because many pathogens are deposited and replicate in mucosal compartments, and the injection does not provide the most efficient immune memory for these disease agents. For this reason, vaccinologists are actively involved in developing new vaccines that are applied via intranasal, aerosol, oral, or transcutaneous (absorbed through the skin) delivery methods. Importantly, mucosal-administered vaccines elicit both mucosal and systemic immunity and produce the same level of disease resistance as injected vaccines.

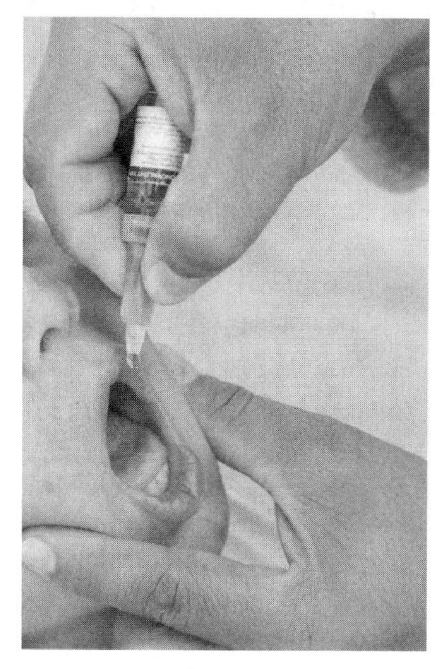

Figure 42.19 The polio vaccine can be administered orally. (credit: modification of work by UNICEF Sverige)

Currently, a version of intranasal influenza vaccine is available, and the polio and typhoid vaccines can be administered orally, as shown in Figure 42.19. Similarly, the measles and rubella vaccines are being adapted to aerosol delivery using inhalation devices. Eventually, transgenic plants may be engineered to produce vaccine antigens that can be eaten to confer disease resistance. Other vaccines may be adapted to rectal or vaginal application to elicit immune responses in rectal, genitourinary, or reproductive mucosa. Finally, vaccine antigens may be adapted to transdermal application in which the skin is lightly scraped and microneedles are used to pierce the outermost layer. In addition to mobilizing the mucosal immune response, this new generation of vaccines may end the anxiety associated with injections and, in turn, improve patient participation.

Primary Centers of the Immune System

Although the immune system is characterized by circulating cells throughout the body, the regulation, maturation, and intercommunication of immune factors occur at specific sites. The blood circulates immune cells, proteins, and other factors through the body. Approximately 0.1 percent of all cells in the blood are leukocytes, which encompass monocytes (the precursor of macrophages) and lymphocytes. The majority of cells in the blood are erythrocytes (red blood cells). **Lymph** is a watery fluid that bathes tissues and organs with protective white blood cells and does not contain erythrocytes. Cells of the immune system can travel between the distinct lymphatic and blood circulatory systems, which are separated by interstitial space, by a process called extravasation (passing through to surrounding tissue).

The cells of the immune system originate from hematopoietic stem cells in the bone marrow. Cytokines stimulate these stem cells to differentiate into immune cells. B cell maturation occurs in the bone marrow, whereas naïve T cells transit from the bone marrow to the thymus for maturation. In the thymus, immature T cells that express TCRs complementary to self-antigens are destroyed. This process helps prevent autoimmune responses.

On maturation, T and B lymphocytes circulate to various destinations. Lymph nodes scattered throughout the body, as illustrated in Figure 42.20, house large populations of T and B cells, dendritic cells, and macrophages. Lymph gathers antigens as it drains from tissues. These antigens then are filtered through lymph nodes before the lymph is returned to circulation. APCs in the lymph nodes capture and process antigens and inform nearby lymphocytes about potential pathogens.

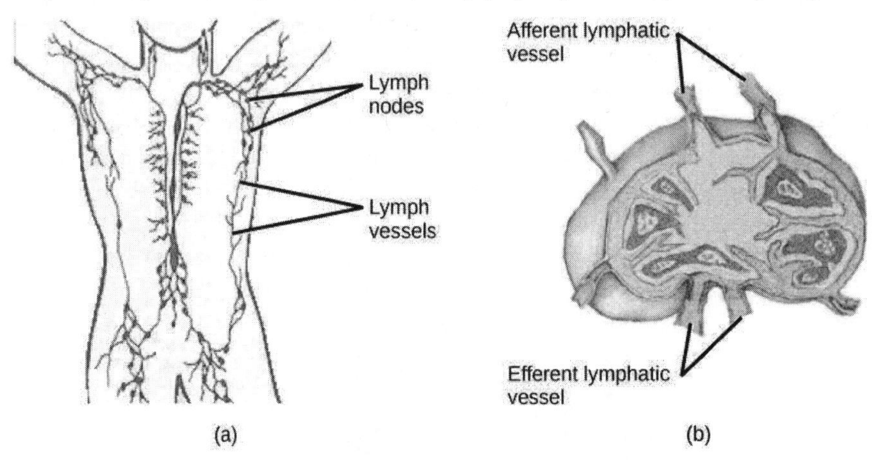

Figure 42.20 (a) Lymphatic vessels carry a clear fluid called lymph throughout the body. The liquid enters (b) lymph nodes through afferent vessels. Lymph nodes are filled with lymphocytes that purge infecting cells. The lymph then exits through efferent vessels. (credit: modification of work by NIH, NCI)

The spleen houses B and T cells, macrophages, dendritic cells, and NK cells. The spleen, shown in Figure 42.21, is the site where APCs that have trapped foreign particles in the blood can communicate with lymphocytes. Antibodies are synthesized and secreted by activated plasma cells in the spleen, and the spleen filters foreign substances and antibody-complexed pathogens from the blood. Functionally, the spleen is to the blood as lymph nodes are to the lymph.

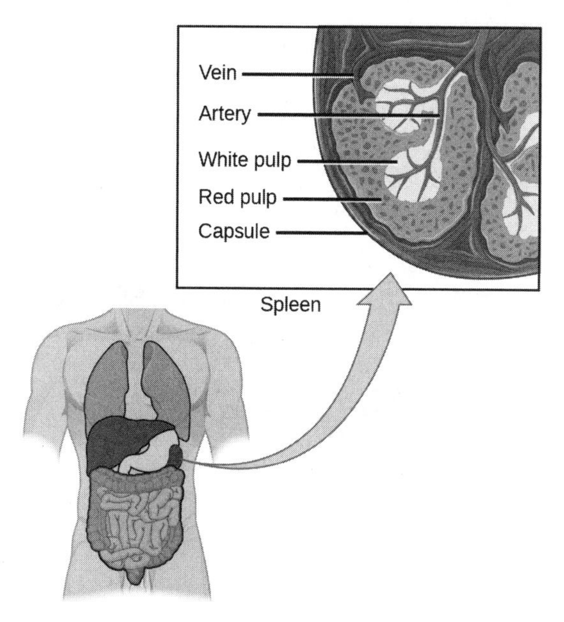

Figure 42.21 The spleen is similar to a lymph node but is much larger and filters blood instead of lymph. Blood enters the spleen through arteries and exits through veins. The spleen contains two types of tissue: red pulp and white pulp. Red pulp consists of cavities that store blood. Within the red pulp, damaged red blood cells are removed and replaced by new ones. White pulp is rich in lymphocytes that remove antigen-coated bacteria from the blood. (credit: modification of work by NCI)

42.3 Antibodies

By the end of this section, you will be able to do the following:

- Explain cross-reactivity
- Describe the structure and function of antibodies
- Discuss antibody production

An **antibody**, also known as an immunoglobulin (Ig), is a protein that is produced by plasma cells after stimulation by an antigen. Antibodies are the functional basis of humoral immunity. Antibodies occur in the blood, in gastric and mucus secretions, and in breast milk. Antibodies in these bodily fluids can bind pathogens and mark them for destruction by phagocytes before they can infect cells.

Antibody Structure

An antibody molecule is comprised of four polypeptides: two identical heavy chains (large peptide units) that are partially bound to each other in a "Y" formation, which are flanked by two identical light chains (small peptide units), as illustrated in Figure 42.22. Bonds between the cysteine amino acids in the antibody molecule attach the polypeptides to each other. The areas where the antigen is recognized on the antibody are variable domains and the antibody base is composed of constant domains.

In germ-line B cells, the variable region of the light chain gene has 40 variable (V) and five joining (J) segments. An enzyme called DNA recombinase randomly excises most of these segments out of the gene, and splices one V segment to one J segment. During RNA processing, all but one V and J segment are spliced out. Recombination and splicing may result in over 10^6 possible VJ combinations. As a result, each differentiated B cell in the human body typically has a unique variable chain. The constant domain, which does not bind antibody, is the same for all antibodies.

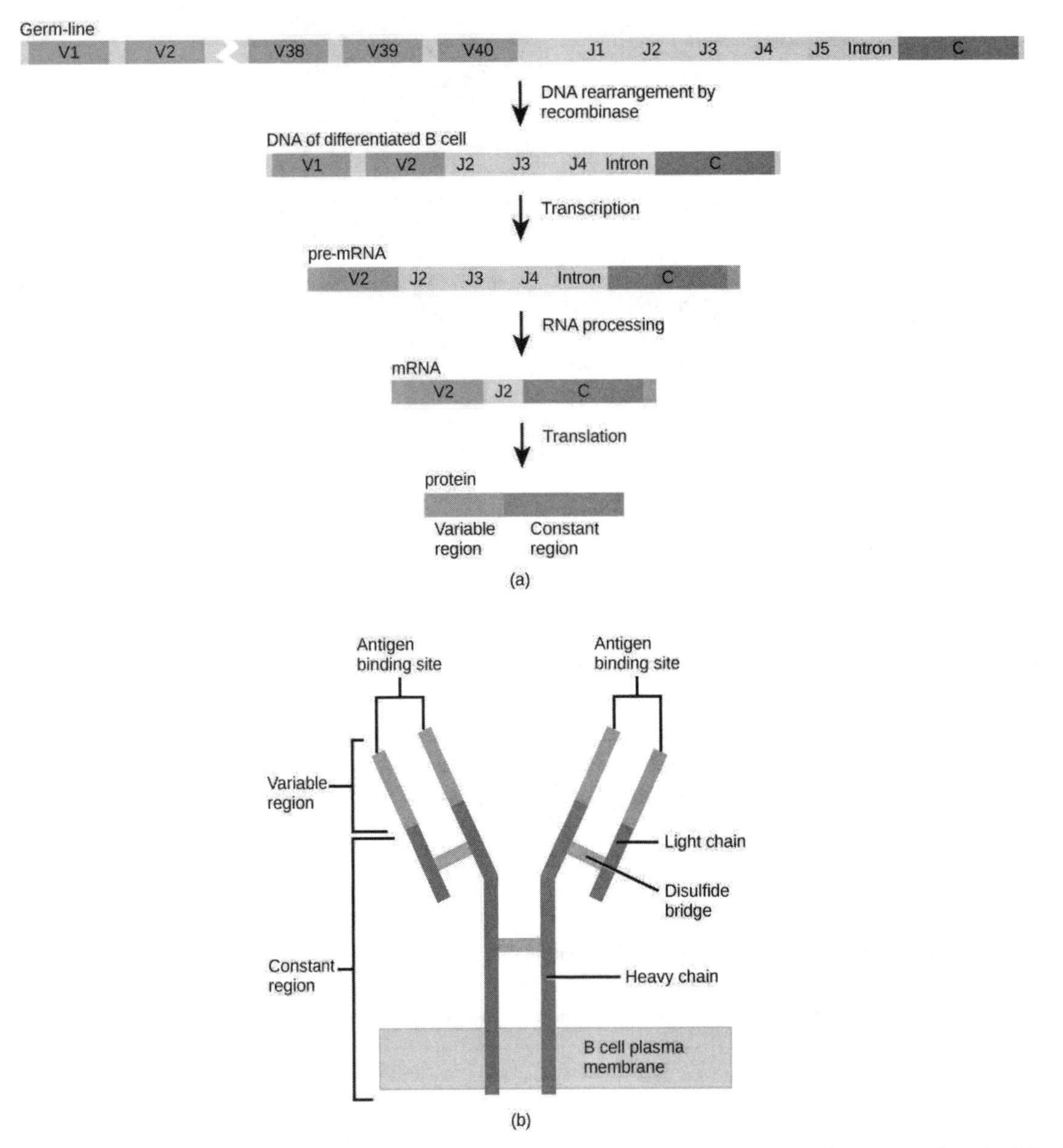

Figure 42.22 (a) As a germ-line B cell matures, an enzyme called DNA recombinase randomly excises V and J segments from the light chain gene. Splicing at the mRNA level results in further gene rearrangement. As a result, (b) each antibody has a unique variable region capable of binding a different antigen.

Similar to TCRs and BCRs, antibody diversity is produced by the mutation and recombination of approximately 300 different gene segments encoding the light and heavy chain variable domains in precursor cells that are destined to become B cells. The variable domains from the heavy and light chains interact to form the binding site through which an antibody can bind a specific epitope on an antigen. The numbers of repeated constant domains in Ig classes are the same for all antibodies corresponding to a specific class. Antibodies are structurally similar to the extracellular component of the BCRs, and B cell maturation to plasma cells can be visualized in simple terms as the cell acquires the ability to secrete the extracellular portion of its BCR in large quantities.

Antibody Classes

Antibodies can be divided into five classes—IgM, IgG, IgA, IgD, IgE—based on their physiochemical, structural, and immunological properties. IgGs, which make up about 80 percent of all antibodies, have heavy chains that consist of one variable domain and three identical constant domains. IgA and IgD also have three constant domains per heavy chain, whereas IgM and IgE each have four constant domains per heavy chain. The variable domain determines binding specificity and the constant domain of the heavy chain determines the immunological mechanism of action of the corresponding antibody class. It is possible for two antibodies to have the same binding specificities but be in different classes and, therefore, to be involved in different functions.

After an adaptive defense is produced against a pathogen, typically plasma cells first secrete IgM into the blood. BCRs on naïve B cells are of the IgM class and occasionally IgD class. IgM molecules make up approximately ten percent of all antibodies. Prior to antibody secretion, plasma cells assemble IgM molecules into pentamers (five individual antibodies) linked by a joining (J) chain, as shown in Figure 42.23. The pentamer arrangement means that these macromolecules can bind ten identical antigens. However, IgM molecules released early in the adaptive immune response do not bind to antigens as stably as IgGs, which are one of the possible types of antibodies secreted in large quantities upon reexposure to the same pathogen. Figure 42.23 summarizes the properties of immunoglobulins and illustrates their basic structures.

Name	Properties	Structure
IgA	Found in mucous, saliva, tears, and breast milk. Protects against pathogens.	
IgD	Part of the B cell receptor. Activates basophils and mast cells.	
IgE	Protects against parasitic worms. Responsible for allergic reactions.	
IgG	Secreted by plasma cells in the blood. Able to cross the placenta into the fetus.	
IgM	May be attached to the surface of a B cell or secreted into the blood. Responsible for early stages of immunity.	

Figure 42.23 Immunoglobulins have different functions, but all are composed of light and heavy chains that form a Y-shaped structure.

IgAs populate the saliva, tears, breast milk, and mucus secretions of the gastrointestinal, respiratory, and genitourinary tracts. Collectively, these bodily fluids coat and protect the extensive mucosa (4000 square feet in humans). The total number of IgA molecules in these bodily secretions is greater than the number of IgG molecules in the blood serum. A small amount of IgA is also secreted into the serum in monomeric form. Conversely, some IgM is secreted into bodily fluids of the mucosa. Similar to IgM, IgA molecules are secreted as polymeric structures linked with a J chain. However, IgAs are secreted mostly as dimeric molecules, not pentamers.

IgE is present in the serum in small quantities and is best characterized in its role as an allergy mediator. IgD is also present in small quantities. Similar to IgM, BCRs of the IgD class are found on the surface of naïve B cells. This class supports antigen recognition and maturation of B cells to plasma cells.

Antibody Functions

Differentiated plasma cells are crucial players in the humoral response, and the antibodies they secrete are particularly significant against extracellular pathogens and toxins. Antibodies circulate freely and act independently of plasma cells. Antibodies can be transferred from one individual to another to temporarily protect against infectious disease. For instance, a person who has recently produced a successful immune response against a particular disease agent can donate blood to a nonimmune recipient and confer temporary immunity through antibodies in the donor's blood serum. This phenomenon is called **passive immunity**; it also occurs naturally during breastfeeding, which makes breastfed infants highly resistant to infections during the first few months of life.

Antibodies coat extracellular pathogens and neutralize them, as illustrated in Figure 42.24, by blocking key sites on the pathogen that enhance their infectivity (such as receptors that "dock" pathogens on host cells). Antibody neutralization can prevent pathogens from entering and infecting host cells, as opposed to the CTL-mediated approach of killing cells that are already infected to prevent progression of an established infection. The neutralized antibody-coated pathogens can then be filtered by the spleen and eliminated in urine or feces.

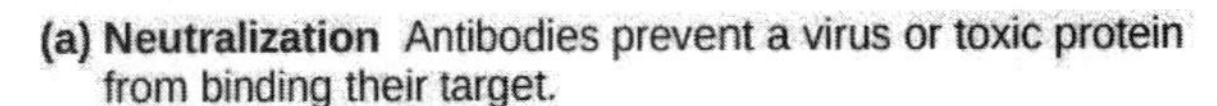

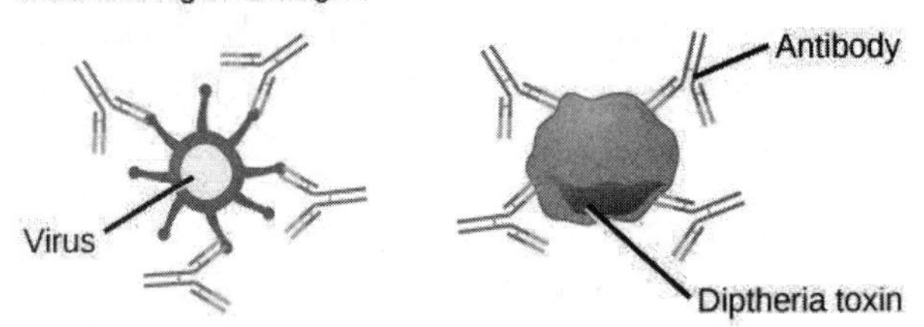

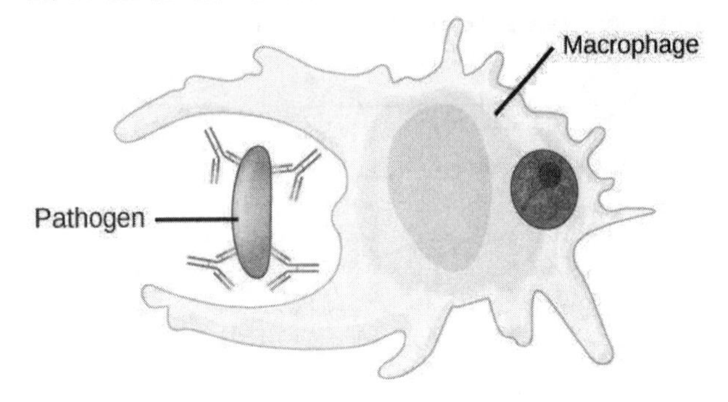

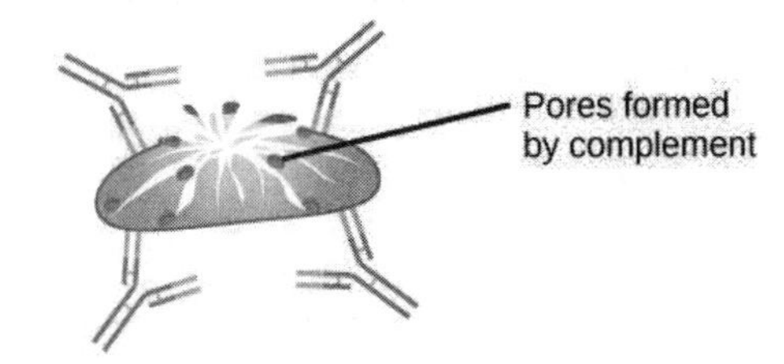

Figure 42.24 Antibodies may inhibit infection by (a) preventing the antigen from binding its target, (b) tagging a pathogen for destruction by macrophages or neutrophils, or (c) activating the complement cascade.

Antibodies also mark pathogens for destruction by phagocytic cells, such as macrophages or neutrophils, because phagocytic cells are highly attracted to macromolecules complexed with antibodies. Phagocytic enhancement by antibodies is called opsonization. In a process called complement fixation, IgM and IgG in serum bind to antigens and provide docking sites onto which sequential complement proteins can bind. The combination of antibodies and complement enhances opsonization even further and promotes rapid clearing of pathogens.

Affinity, Avidity, and Cross Reactivity

Not all antibodies bind with the same strength, specificity, and stability. In fact, antibodies exhibit different **affinities** (attraction) depending on the molecular complementarity between antigen and antibody molecules, as illustrated in Figure 42.25. An antibody with a higher affinity for a particular antigen would bind more strongly and stably, and thus would be expected to present a more challenging defense against the pathogen corresponding to the specific antigen.

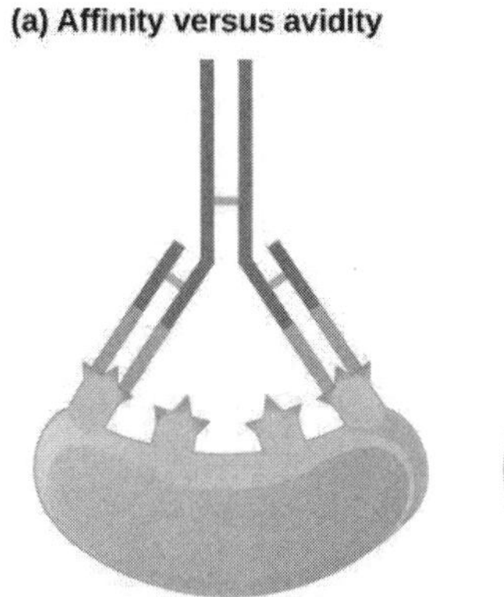

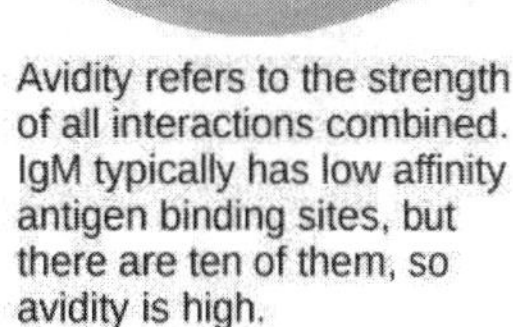

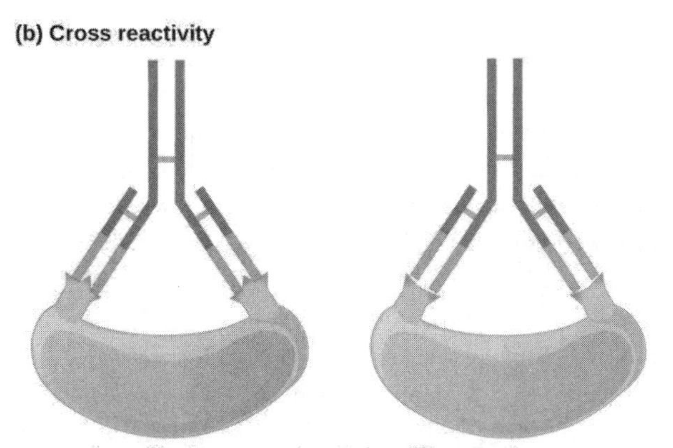

Figure 42.25 (a) Affinity refers to the strength of single interaction between antigen and antibody, while avidity refers to the strength of all interactions combined. (b) An antibody may cross react with different epitopes.

The term **avidity** describes binding by antibody classes that are secreted as joined, multivalent structures (such as IgM and IgA). Although avidity measures the strength of binding, just as affinity does, the avidity is not simply the sum of the affinities of the antibodies in a multimeric structure. The avidity depends on the number of identical binding sites on the antigen being detected, as well as other physical and chemical factors. Typically, multimeric antibodies, such as pentameric IgM, are classified as having lower affinity than monomeric antibodies, but high avidity. Essentially, the fact that multimeric antibodies can bind many antigens simultaneously balances their slightly lower binding strength for each antibody/antigen interaction.

Antibodies secreted after binding to one epitope on an antigen may exhibit cross reactivity for the same or similar epitopes on different antigens. Because an epitope corresponds to such a small region (the surface area of about four to six amino acids), it is possible for different macromolecules to exhibit the same molecular identities and orientations over short regions. **Cross reactivity** describes when an antibody binds not to the antigen that elicited its synthesis and secretion, but to a different antigen.

Cross reactivity can be beneficial if an individual develops immunity to several related pathogens despite having only been exposed to or vaccinated against one of them. For instance, antibody cross reactivity may occur against the similar surface structures of various Gram-negative bacteria. Conversely, antibodies raised against pathogenic molecular components that resemble self molecules may incorrectly mark host cells for destruction and cause autoimmune damage. Patients who develop systemic lupus erythematosus (SLE) commonly exhibit antibodies that react with their own DNA. These antibodies may have been initially raised against the nucleic acid of microorganisms but later cross-reacted with self-antigens. This phenomenon is also called molecular mimicry.

Antibodies of the Mucosal Immune System

Antibodies synthesized by the mucosal immune system include IgA and IgM. Activated B cells differentiate into mucosal plasma cells that synthesize and secrete dimeric IgA, and to a lesser extent, pentameric IgM. Secreted IgA is abundant in tears, saliva, breast milk, and in secretions of the gastrointestinal and respiratory tracts. Antibody secretion results in a local humoral response at epithelial surfaces and prevents infection of the mucosa by binding and neutralizing pathogens.

42.4 Disruptions in the Immune System

By the end of this section, you will be able to do the following:

- Describe hypersensitivity
- Define autoimmunity

A functioning immune system is essential for survival, but even the sophisticated cellular and molecular defenses of the mammalian immune response can be defeated by pathogens at virtually every step. In the competition between immune protection and pathogen evasion, pathogens have the advantage of more rapid evolution because of their shorter generation time and other characteristics. For instance, *Streptococcus pneumoniae* (bacterium that cause pneumonia and meningitis) surrounds itself with a capsule that inhibits phagocytes from engulfing it and displaying antigens to the adaptive immune system. *Staphylococcus aureus* (bacterium that can cause skin infections, abscesses, and meningitis) synthesizes a toxin called leukocidin that kills phagocytes after they engulf the bacterium. Other pathogens can also hinder the adaptive immune system. HIV infects T_H cells via their CD4 surface molecules, gradually depleting the number of T_H cells in the body; this inhibits the adaptive immune system's capacity to generate sufficient responses to infection or tumors. As a result, HIV-infected individuals often suffer from infections that would not cause illness in people with healthy immune systems but which can cause devastating illness to immune-compromised individuals. Maladaptive responses of immune cells and molecules themselves can also disrupt the proper functioning of the entire system, leading to host cell damage that could become fatal.

Immunodeficiency

Failures, insufficiencies, or delays at any level of the immune response can allow pathogens or tumor cells to gain a foothold and replicate or proliferate to high enough levels that the immune system becomes overwhelmed. **Immunodeficiency** is the failure, insufficiency, or delay in the response of the immune system, which may be acquired or inherited. Immunodeficiency can be acquired as a result of infection with certain pathogens (such as HIV), chemical exposure (including certain medical treatments), malnutrition, or possibly by extreme stress. For instance, radiation exposure can destroy populations of lymphocytes and elevate an individual's susceptibility to infections and cancer. Dozens of genetic disorders result in immunodeficiencies, including Severe Combined Immunodeficiency (SCID), Bare lymphocyte syndrome, and MHC II deficiencies. Rarely, primary immunodeficiencies that are present from birth may occur. Neutropenia is one form in which the immune system produces a below-average number of neutrophils, the body's most abundant phagocytes. As a result, bacterial infections may go unrestricted in the blood, causing serious complications.

Hypersensitivities

Maladaptive immune responses toward harmless foreign substances or self antigens that occur after tissue sensitization are termed **hypersensitivities**. The types of hypersensitivities include immediate, delayed, and autoimmunity. A large proportion of the population is affected by one or more types of hypersensitivity.

Allergies

The immune reaction that results from immediate hypersensitivities in which an antibody-mediated immune response occurs within minutes of exposure to a harmless antigen is called an **allergy**. In the United States, 20 percent of the population exhibits symptoms of allergy or asthma, whereas 55 percent test positive against one or more allergens. Upon initial exposure to a potential allergen, an allergic individual synthesizes antibodies of the IgE class via the typical process of APCs presenting processed antigen to T_H cells that stimulate B cells to produce IgE. This class of antibodies also mediates the immune response to parasitic worms. The constant domain of the IgE molecules interact with mast cells embedded in connective tissues. This process primes, or sensitizes, the tissue. Upon subsequent exposure to the same allergen, IgE molecules on mast cells bind the antigen via their variable domains and stimulate the mast cell to release the modified amino acids histamine and serotonin; these chemical mediators then recruit eosinophils which mediate allergic responses. Figure 42.26 shows an example of an allergic response to ragweed pollen. The effects of an allergic reaction range from mild symptoms like sneezing and itchy, watery eyes to more severe or even life-threatening reactions involving intensely itchy welts or hives, airway contraction with severe respiratory distress, and plummeting blood pressure. This extreme reaction is known as anaphylactic shock. If not treated with epinephrine to counter the blood pressure and breathing effects, this condition can be fatal.

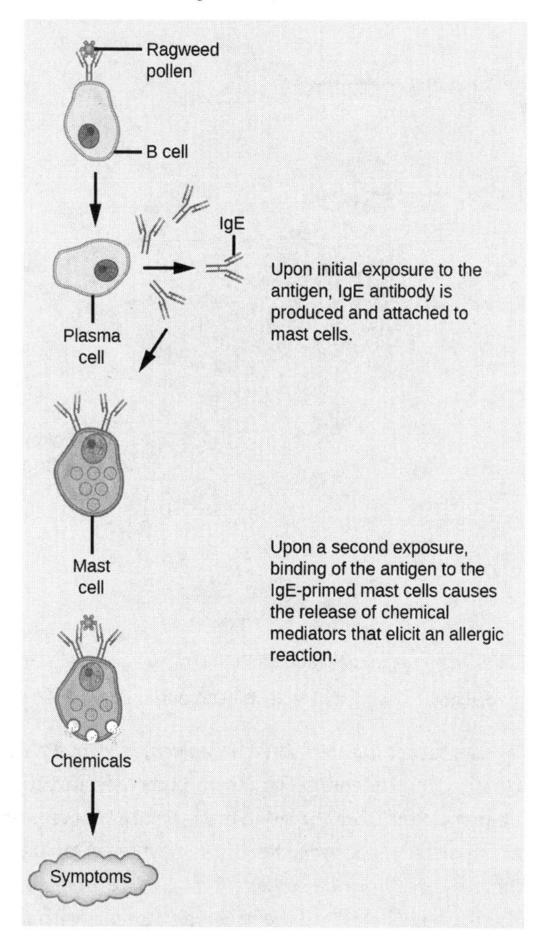

Figure 42.26 On first exposure to an allergen, an IgE antibody is synthesized by plasma cells in response to a harmless antigen. The IgE molecules bind to mast cells, and on secondary exposure, the mast cells release histamines and other modulators that affect the symptoms of allergy. (credit: modification of work by NIH)

Delayed hypersensitivity is a cell-mediated immune response that takes approximately one to two days after secondary exposure for a maximal reaction to be observed. This type of hypersensitivity involves the T_H1 cytokine-mediated inflammatory response and may manifest as local tissue lesions or contact dermatitis (rash or skin irritation). Delayed hypersensitivity occurs in some individuals in response to contact with certain types of jewelry or cosmetics. Delayed hypersensitivity facilitates the immune response to poison ivy and is also the reason why the skin test for tuberculosis results in a small region of inflammation on individuals who were previously exposed to *Mycobacterium tuberculosis*. That is also why cortisone is used to treat such responses: it will inhibit cytokine production.

Autoimmunity

Autoimmunity is a type of hypersensitivity to self antigens that affects approximately five percent of the population. Most types of autoimmunity involve the humoral immune response. Antibodies that inappropriately mark self components as foreign are termed **autoantibodies**. In patients with the autoimmune disease myasthenia gravis, muscle cell receptors that induce contraction in response to acetylcholine are targeted by antibodies. The result is muscle weakness that may include marked difficultly with fine and/or gross motor functions. In systemic lupus erythematosus, a diffuse autoantibody response to the individual's own DNA and proteins results in various systemic diseases. As illustrated in Figure 42.27, systemic lupus erythematosus may affect the heart, joints, lungs, skin, kidneys, central nervous system, or other tissues, causing tissue damage via antibody binding, complement recruitment, lysis, and inflammation.

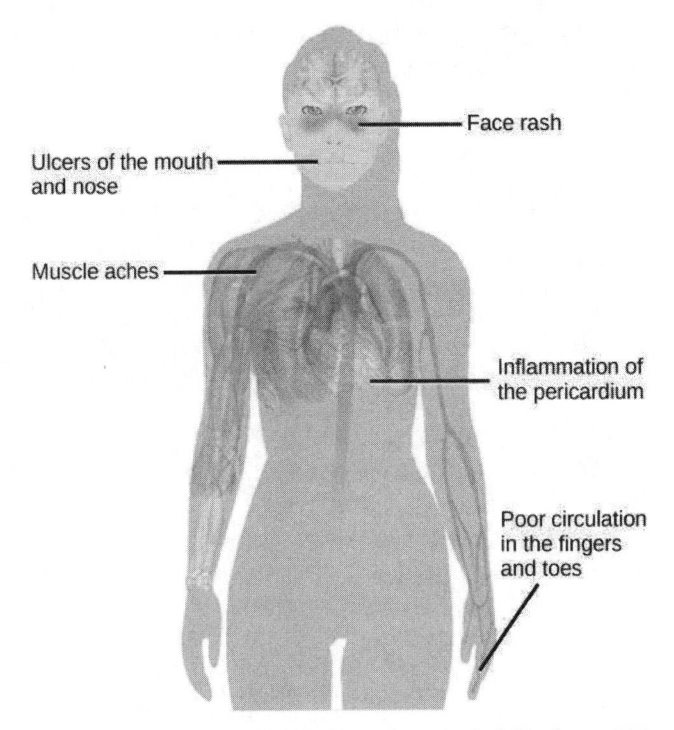

Figure 42.27 Systemic lupus erythematosus is characterized by autoimmunity to the individual's own DNA and/or proteins, which leads to varied dysfunction of the organs. (credit: modification of work by Mikael Häggström)

Autoimmunity can develop with time, and its causes may be rooted in molecular mimicry. Antibodies and TCRs may bind self antigens that are structurally similar to pathogen antigens, which the immune receptors first raised. As an example, infection with *Streptococcus pyogenes* (bacterium that causes strep throat) may generate antibodies or T cells that react with heart muscle, which has a similar structure to the surface of *S. pyogenes*. These antibodies can damage heart muscle with autoimmune attacks, leading to rheumatic fever. Insulin-dependent (Type 1) diabetes mellitus arises from a destructive inflammatory T_H1 response against insulin-producing cells of the pancreas. Patients with this autoimmunity must be injected with insulin that originates from other sources.

KEY TERMS

adaptive immunity immunity that has memory and occurs after exposure to an antigen either from a pathogen or a vaccination

affinity attraction of molecular complementarity between antigen and antibody molecules

allergy immune reaction that results from immediate hypersensitivities in which an antibody-mediated immune response occurs within minutes of exposure to a harmless antigen

antibody protein that is produced by plasma cells after stimulation by an antigen; also known as an immunoglobulin

antigen foreign or "non-self" protein that triggers the immune response

antigen-presenting cell (APC) immune cell that detects, engulfs, and informs the adaptive immune response about an infection by presenting the processed antigen on the cell surface

autoantibody antibody that incorrectly marks "self" components as foreign and stimulates the immune response

autoimmune response inappropriate immune response to host cells or self-antigens

autoimmunity type of hypersensitivity to self antigens

avidity total binding strength of a multivalent antibody with antigen

B cell lymphocyte that matures in the bone marrow and differentiates into antibody-secreting plasma cells

basophil leukocyte that releases chemicals usually involved in the inflammatory response

cell-mediated immune response adaptive immune response that is carried out by T cells

clonal selection activation of B cells corresponding to one specific BCR variant and the dramatic proliferation of that variant

complement system array of approximately 20 soluble proteins of the innate immune system that enhance phagocytosis, bore holes in pathogens, and recruit lymphocytes; enhances the adaptive response when antibodies are produced

cross reactivity binding of an antibody to an epitope corresponding to an antigen that is different from the one the antibody was raised against

cytokine chemical messenger that regulates cell differentiation, proliferation, gene expression, and cell trafficking to effect immune responses

cytotoxic T lymphocyte (CTL) adaptive immune cell that directly kills infected cells via perforin and granzymes, and releases cytokines to enhance the immune response

dendritic cell immune cell that processes antigen material and presents it on the surface of other cells to induce an immune response

effector cell lymphocyte that has differentiated, such as a B cell, plasma cell, or cytotoxic T lymphocyte

eosinophil leukocyte that responds to parasites and is involved in the allergic response

epitope small component of an antigen that is specifically recognized by antibodies, B cells, and T cells; the antigenic determinant

granzyme protease that enters target cells through perforin and induces apoptosis in the target cells; used by NK cells and killer T cells

helper T lymphocyte (T_H) cell of the adaptive immune system that binds APCs via MHC II molecules and stimulates B cells or secretes cytokines to initiate the immune response

host an organism that is invaded by a pathogen or parasite

humoral immune response adaptive immune response that is controlled by activated B cells and antibodies

hypersensitivities spectrum of maladaptive immune responses toward harmless foreign particles or self antigens; occurs after tissue sensitization and includes immediate-type (allergy), delayed-type, and autoimmunity

immune tolerance acquired ability to prevent an unnecessary or harmful immune response to a detected foreign body known not to cause disease or to self-antigens

immunodeficiency failure, insufficiency, or delay at any level of the immune system, which may be acquired or inherited

inflammation localized redness, swelling, heat, and pain that results from the movement of leukocytes and fluid through opened capillaries to a site of infection

innate immunity immunity that occurs naturally because of genetic factors or physiology, and is not induced by infection or vaccination

interferon cytokine that inhibits viral replication and modulates the immune response

lymph watery fluid that bathes tissues and organs with protective white blood cells and does not contain erythrocytes

lymphocyte leukocyte that is histologically identifiable by its large nuclei; it is a small cell with very little cytoplasm

macrophage large phagocytic cell that engulfs foreign particles and pathogens

major histocompatibility class (MHC) I/II molecule protein found on the surface of all nucleated cells (I) or specifically on antigen-presenting cells (II) that signals to immune cells whether the cell is healthy/normal or is infected/cancerous; it provides the appropriate template into which antigens can be loaded for recognition by lymphocytes

mast cell leukocyte that produces inflammatory molecules, such as histamine, in response to large pathogens and allergens

memory cell antigen-specific B or T lymphocyte that does not differentiate into effector cells during the primary immune response but that can immediately become an effector cell upon reexposure to the same pathogen

monocyte type of white blood cell that circulates in the blood and lymph and differentiates into macrophages after it moves into infected tissue

mucosa-associated lymphoid tissue (MALT) collection of lymphatic tissue that combines with epithelial tissue lining the mucosa throughout the body

natural killer (NK) cell lymphocyte that can kill cells infected with viruses or tumor cells

neutrophil phagocytic leukocyte that engulfs and digests pathogens

opsonization process that enhances phagocytosis using proteins to indicate the presence of a pathogen to phagocytic cells

passive immunity transfer of antibodies from one individual to another to provide temporary protection against pathogens

pathogen an agent, usually a microorganism, that causes disease in the organisms that it invades

pathogen-associated molecular pattern (PAMP) carbohydrate, polypeptide, and nucleic acid "signature" that is expressed by viruses, bacteria, and parasites but differs from molecules on host cells

pattern recognition receptor (PRR) molecule on macrophages and dendritic cells that binds molecular signatures of pathogens and promotes pathogen engulfment and destruction

perforin destructive protein that creates a pore in the target cell; used by NK cells and killer T cells

plasma cell immune cell that secrets antibodies; these cells arise from B cells that were stimulated by antigens

regulatory T (T_{reg}) cell specialized lymphocyte that suppresses local inflammation and inhibits the secretion of cytokines, antibodies, and other stimulatory immune factors; involved in immune tolerance

T cell lymphocyte that matures in the thymus gland; one of the main cells involved in the adaptive immune system

CHAPTER SUMMARY

42.1 Innate Immune Response

The innate immune system serves as a first responder to pathogenic threats that bypass natural physical and chemical barriers of the body. Using a combination of cellular and molecular attacks, the innate immune system identifies the nature of a pathogen and responds with inflammation, phagocytosis, cytokine release, destruction by NK cells, and/or a complement system. When innate mechanisms are insufficient to clear an infection, the adaptive immune response is informed and mobilized.

42.2 Adaptive Immune Response

The adaptive immune response is a slower-acting, longer-lasting, and more specific response than the innate response. However, the adaptive response requires information from the innate immune system to function. APCs display antigens via MHC molecules to complementary naïve T cells. In response, the T cells differentiate and proliferate, becoming T_H cells or CTLs. T_H cells stimulate B cells that have engulfed and presented pathogen-derived antigens. B cells differentiate into plasma cells that secrete antibodies, whereas CTLs induce apoptosis in intracellularly infected or cancerous cells. Memory cells persist after a primary exposure to a pathogen. If reexposure occurs, memory cells differentiate into effector cells without input from the innate immune system. The mucosal immune system is largely independent from the systemic immune system but functions in a parallel fashion to protect the extensive mucosal surfaces of the body.

42.3 Antibodies

Antibodies (immunoglobulins) are the molecules secreted from plasma cells that mediate the humoral immune response. There are five antibody classes; an antibody's class determines its mechanism of action and production site but does not control its binding specificity. Antibodies bind antigens via variable domains and can either neutralize pathogens or mark them for phagocytosis or activate the complement cascade.

42.4 Disruptions in the Immune System

Immune disruptions may involve insufficient immune responses or inappropriate immune targets. Immunodeficiency increases an individual's susceptibility to infections and cancers. Hypersensitivities are misdirected responses either to harmless foreign particles, as in the case of allergies, or to host factors, as in the case of autoimmunity. Reactions to self components may be the result of molecular mimicry.

VISUAL CONNECTION QUESTIONS

1. Figure 42.11 Which of the following statements about T cells is false?
   a. Helper T cells release cytokines while cytotoxic T cells kill the infected cell.
   b. Helper T cells are CD4+, while cytotoxic T cells are $CD8^+$.
   c. MHC II is a receptor found on most body cells, while MHC I is a receptor found on immune cells only.
   d. The T cell receptor is found on both $CD4^+$ and $CD8^+$ T cells.
2. Figure 42.14 Based on what you know about MHC receptors, why do you think an organ transplanted from an incompatible donor to a recipient will be rejected?
3. Figure 42.16 The Rh antigen is found on Rh-positive red blood cells. An Rh-negative female can usually carry an Rh-positive fetus to term without difficulty. However, if she has a second Rh-positive fetus, her body may launch an immune attack that causes hemolytic disease of the newborn. Why do you think hemolytic disease is only a problem during the second or subsequent pregnancies?

REVIEW QUESTIONS

4. Which of the following is a barrier against pathogens provided by the skin?
   a. high pH
   b. mucus
   c. tears
   d. desiccation
5. Although interferons have several effects, they are particularly useful against infections with which type of pathogen?
   a. bacteria
   b. viruses
   c. fungi
   d. helminths
6. Which organelle do phagocytes use to digest engulfed particles?
   a. lysosome
   b. nucleus
   c. endoplasmic reticulum
   d. mitochondria
7. Which innate immune system component uses MHC I molecules directly in its defense strategy?
   a. macrophages
   b. neutrophils
   c. NK cells
   d. interferon
8. Which of the following is both a phagocyte and an antigen-presenting cell?
   a. NK cell
   b. eosinophil
   c. neutrophil
   d. macrophage
9. Which immune cells bind MHC molecules on APCs via CD8 coreceptors on their cell surfaces?
   a. T_H cells
   b. CTLs
   c. mast cells
   d. basophils
10. What "self" pattern is identified by NK cells?
    a. altered self
    b. missing self
    c. normal self
    d. non-self
11. The acquired ability to prevent an unnecessary or destructive immune reaction to a harmless foreign particle, such as a food protein, is called ________.
    a. the T_H2 response
    b. allergy
    c. immune tolerance
    d. autoimmunity
12. Upon reexposure to a pathogen, a memory B cell can differentiate to which cell type?
    a. CTL
    b. naïve B cell
    c. memory T cell
    d. plasma cell
13. Foreign particles circulating in the blood are filtered by the ________.
    a. spleen
    b. lymph nodes
    c. MALT
    d. lymph

14. The structure of an antibody is similar to the extracellular component of which receptor?
    a. MHC I
    b. MHC II
    c. BCR
    d. none of the above

15. The first antibody class to appear in the serum in response to a newly encountered pathogen is _______.
    a. IgM
    b. IgA
    c. IgG
    d. IgE

16. What is the most abundant antibody class detected in the serum upon reexposure to a pathogen or in reaction to a vaccine?
    a. IgM
    b. IgA
    c. IgG
    d. IgE

17. Breastfed infants typically are resistant to disease because of _______.
    a. active immunity
    b. passive immunity
    c. immune tolerance
    d. immune memory

18. Allergy to pollen is classified as:
    a. an autoimmune reaction
    b. immunodeficiency
    c. delayed hypersensitivity
    d. immediate hypersensitivity

19. A potential cause of acquired autoimmunity is _______.
    a. tissue hypersensitivity
    b. molecular mimicry
    c. histamine release
    d. radiation exposure

20. Autoantibodies are probably involved in:
    a. reactions to poison ivy
    b. pollen allergies
    c. systemic lupus erythematosus
    d. HIV/AIDS

21. Which of the following diseases is not due to autoimmunity?
    a. rheumatic fever
    b. systemic lupus erythematosus
    c. diabetes mellitus
    d. HIV/AIDS

CRITICAL THINKING QUESTIONS

22. Different MHC I molecules between donor and recipient cells can lead to rejection of a transplanted organ or tissue. Suggest a reason for this.
23. If a series of genetic mutations prevented some, but not all, of the complement proteins from binding antibodies or pathogens, would the entire complement system be compromised?
24. Explain the difference between an epitope and an antigen.
25. What is a naïve B or T cell?
26. How does the T_H1 response differ from the T_H2 response?
27. In mammalian adaptive immune systems, T cell receptors are extraordinarily diverse. What function of the immune system results from this diversity, and how is this diversity achieved?
28. How do B and T cells differ with respect to antigens that they bind?
29. Why is the immune response after reinfection much faster than the adaptive immune response after the initial infection?
30. What are the benefits and costs of antibody cross reactivity?

CHAPTER 43
Animal Reproduction and Development

Figure 43.1 Female seahorses produce eggs for reproduction that are then fertilized by the male. Unlike almost all other animals, the male seahorse then gestates the young until birth. (credit: modification of work by "cliff1066"/Flickr)

INTRODUCTION Animal reproduction is necessary for the survival of a species. In the animal kingdom, there are innumerable ways that species reproduce. Asexual reproduction produces genetically identical organisms (clones), whereas in sexual reproduction, the genetic material of two individuals combines to produce offspring that are genetically different from their parents. During sexual reproduction the male gamete (sperm) may be placed inside the female's body for internal fertilization, or the sperm and eggs may be released into the environment for external fertilization. Seahorses, like the one shown in Figure 43.1, provide an example of the latter. Following a mating dance, the female lays eggs in the male seahorse's abdominal brood pouch where they are fertilized. The eggs hatch and the offspring develop in the pouch for several weeks.

Chapter Outline

43.1 Reproduction Methods

By the end of this section, you will be able to do the following:

- Describe advantages and disadvantages of asexual and sexual reproduction
- Discuss asexual reproduction methods
- Discuss sexual reproduction methods

Animals produce offspring through asexual and/or sexual reproduction. Both methods have advantages and disadvantages. **Asexual reproduction** produces offspring that are genetically identical to the parent because the offspring are all clones of the original parent. A single individual can produce offspring asexually and large numbers of offspring can be produced quickly. In a stable or predictable environment, asexual reproduction is an effective means of reproduction because all the offspring will be adapted to that environment. In an unstable or unpredictable environment asexually-reproducing species may be at a disadvantage because all the offspring are genetically identical and may not have the genetic variation to survive in new or different conditions. On the other hand, the rapid rates of asexual reproduction may allow for a speedy response to environmental changes if individuals have mutations. An additional advantage of asexual reproduction is that colonization of new habitats may be easier when an individual does not need to find a mate to reproduce.

During **sexual reproduction** the genetic material of two individuals is combined to produce genetically diverse offspring that differ from their parents. The genetic diversity of sexually produced offspring is thought to give species a better chance of surviving in an unpredictable or changing environment. Species that reproduce sexually must maintain two different types of individuals, males and females, which can limit the ability to colonize new habitats as both sexes must be present.

Asexual Reproduction

Asexual reproduction occurs in prokaryotic microorganisms (bacteria) and in some eukaryotic single-celled and multi-celled organisms. There are a number of ways that animals reproduce asexually.

Fission

Fission, also called binary fission, occurs in prokaryotic microorganisms and in some invertebrate, multi-celled organisms. After a period of growth, an organism splits into two separate organisms. Some unicellular eukaryotic organisms undergo binary fission by mitosis. In other organisms, part of the individual separates and forms a second individual. This process occurs, for example, in many asteroid echinoderms through splitting of the central disk. Some sea anemones and some coral polyps (Figure 43.2) also reproduce through fission.

Figure 43.2 Coral polyps reproduce asexually by fission. (credit: G. P. Schmahl, NOAA FGBNMS Manager)

Budding

Budding is a form of asexual reproduction that results from the outgrowth of a part of a cell or body region leading to a separation from the original organism into two individuals. Budding occurs commonly in some invertebrate animals such as corals and hydras. In hydras, a bud forms that develops into an adult and breaks away from the main body, as illustrated in Figure 43.3, whereas in coral budding, the bud does not detach and multiplies as part of a new colony.

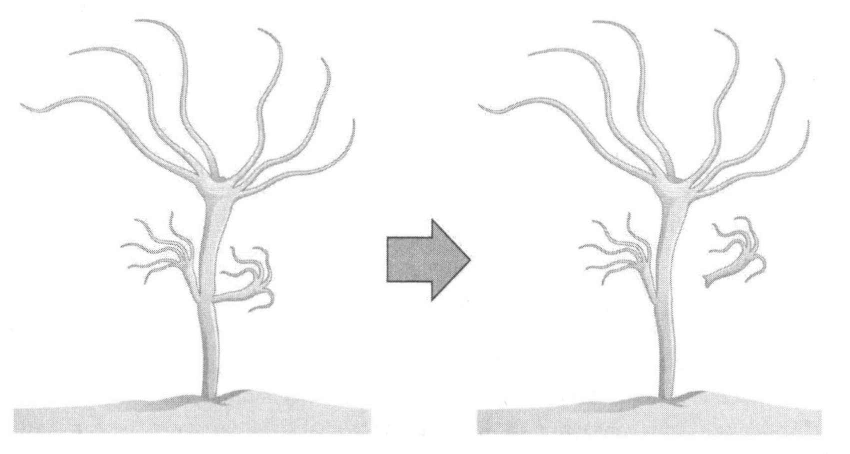

Figure 43.3 Hydra reproduce asexually through budding.

LINK TO LEARNING

Watch a video of a hydra budding.

Click to view content (https://www.openstax.org/l/budding_hydra)

Fragmentation

Fragmentation is the breaking of the body into two parts with subsequent regeneration. If the animal is capable of fragmentation, and the part is big enough, a separate individual will regrow.

For example, in many sea stars, asexual reproduction is accomplished by fragmentation. Figure 43.4 illustrates a sea star for which an arm of the individual is broken off and regenerates a new sea star. Fisheries workers have been known to try to kill the sea stars eating their clam or oyster beds by cutting them in half and throwing them back into the ocean. Unfortunately for the workers, the two parts can each regenerate a new half, resulting in twice as many sea stars to prey upon the oysters and clams. Fragmentation also occurs in annelid worms, turbellarians, and poriferans.

Figure 43.4 Sea stars can reproduce through fragmentation. The large arm, a fragment from another sea star, is developing into a new individual.

Note that in fragmentation, there is generally a noticeable difference in the size of the individuals, whereas in fission, two individuals of approximate size are formed.

Parthenogenesis

Parthenogenesis is a form of asexual reproduction where an egg develops into a complete individual without being fertilized. The resulting offspring can be either haploid or diploid, depending on the process and the species. Parthenogenesis occurs in invertebrates such as water fleas, rotifers, aphids, stick insects, some ants, wasps, and bees. Bees use parthenogenesis to produce haploid males (drones). If eggs are fertilized, diploid females develop, and if the fertilized eggs are fed a special diet (so called royal jelly), a queen is produced.

Some vertebrate animals—such as certain reptiles, amphibians, and fish—also reproduce through parthenogenesis. Although more common in plants, parthenogenesis has been observed in animal species that were segregated by sex in terrestrial or marine zoos. Two female Komodo dragons, a hammerhead shark, and a blacktop shark have produced parthenogenic young when the females have been isolated from males.

Sexual Reproduction

Sexual reproduction is the combination of (usually haploid) reproductive cells from two individuals to form a third (usually diploid) unique offspring. Sexual reproduction produces offspring with novel combinations of genes. This can be an adaptive advantage in unstable or unpredictable environments. As humans, we are used to thinking of animals as having two separate sexes—male and female—determined at conception. However, in the animal kingdom, there are many variations on this theme.

Hermaphroditism

Hermaphroditism occurs in animals where one individual has both male and female reproductive parts. Invertebrates such as earthworms, slugs, tapeworms and snails, shown in Figure 43.5, are often hermaphroditic. Hermaphrodites may self-fertilize or may mate with another of their species, fertilizing each other and both producing offspring. Self fertilization is common in animals that have limited mobility or are not motile, such as barnacles and clams.

Figure 43.5 Many snails are hermaphrodites. When two individuals mate, they can produce up to one hundred eggs each. (credit: Assaf Shtilman)

Sex Determination

Mammalian sex determination is determined genetically by the presence of X and Y chromosomes. Individuals homozygous for X (XX) are female and heterozygous individuals (XY) are male. The presence of a Y chromosome causes the development of male characteristics and its absence results in female characteristics. The XY system is also found in some insects and plants.

Avian sex determination is dependent on the presence of Z and W chromosomes. Homozygous for Z (ZZ) results in a male and heterozygous (ZW) results in a female. The W appears to be essential in determining the sex of the individual, similar to the Y chromosome in mammals. Some fish, crustaceans, insects (such as butterflies and moths), and reptiles use this system.

The sex of some species is not determined by genetics but by some aspect of the environment. Sex determination in some crocodiles and turtles, for example, is often dependent on the temperature during critical periods of egg development. This is referred to as environmental sex determination, or more specifically as temperature-dependent sex determination. In many turtles, cooler temperatures during egg incubation produce males and warm temperatures produce females. In some crocodiles, moderate temperatures produce males and both warm and cool temperatures produce females. In some species, sex is both genetic- and temperature-dependent.

Individuals of some species change their sex during their lives, alternating between male and female. If the individual is female first, it is termed protogyny or "first female," if it is male first, its termed protandry or "first male." Oysters, for example, are born male, grow, and become female and lay eggs; some oyster species change sex multiple times.

43.2 Fertilization

By the end of this section, you will be able to do the following:

- Discuss internal and external methods of fertilization
- Describe the methods used by animals for development of offspring during gestation
- Describe the anatomical adaptions that occurred in animals to facilitate reproduction

Sexual reproduction starts with the combination of a sperm and an egg in a process called fertilization. This can occur either inside (**internal fertilization**) or outside (**external fertilization**) the body of the female. Humans provide an example of the former whereas seahorse reproduction is an example of the latter.

External Fertilization

External fertilization usually occurs in aquatic environments where both eggs and sperm are released into the water. After the sperm reaches the egg, fertilization takes place. Most external fertilization happens during the process of spawning where one or several females release their eggs and the male(s) release sperm in the same area, at the same time. The release of the reproductive material may be triggered by water temperature or the length of daylight. Nearly all fish spawn, as do crustaceans (such as crabs and shrimp), mollusks (such as oysters), squid, and echinoderms (such as sea urchins and sea cucumbers). Figure 43.6 shows salmon spawning in a shallow stream. Frogs, like those shown in Figure 43.7, corals, molluscs, and sea cucumbers also spawn.

Figure 43.6 Salmon reproduce through spawning. (credit: Dan Bennett)

Figure 43.7 During sexual reproduction in toads, the male grasps the female from behind and externally fertilizes the eggs as they are deposited. (credit: "OakleyOriginals"/Flickr)

Pairs of fish that are not broadcast spawners may exhibit courtship behavior. This allows the female to select a particular male. The trigger for egg and sperm release (spawning) causes the egg and sperm to be placed in a small area, enhancing the possibility of fertilization.

External fertilization in an aquatic environment protects the eggs from drying out. Broadcast spawning can result in a greater mixture of the genes within a group, leading to higher genetic diversity and a greater chance of species survival in a hostile environment. For sessile aquatic organisms like sponges, broadcast spawning is the only mechanism for fertilization and colonization of new environments. The presence of the fertilized eggs and developing young in the water provides opportunities for predation resulting in a loss of offspring. Therefore, millions of eggs must be produced by individuals, and the offspring produced through this method must mature rapidly. The survival rate of eggs produced through broadcast spawning is low.

Internal Fertilization

Internal fertilization occurs most often in land-based animals, although some aquatic animals also use this method. There are three ways that offspring are produced following internal fertilization. In **oviparity**, fertilized eggs are laid outside the female's body and develop there, receiving nourishment from the yolk that is a part of the egg. This occurs in most bony fish, many reptiles, some cartilaginous fish, most amphibians, two mammals, and all birds. Reptiles and insects produce leathery eggs, while birds and turtles produce eggs with high concentrations of calcium carbonate in the shell, making them hard. Chicken eggs are an example of this second type.

In **ovoviparity**, fertilized eggs are retained in the female, but the embryo obtains its nourishment from the egg's yolk and the young are fully developed when they are hatched. This occurs in some bony fish (like the guppy *Lebistes reticulatus*), some sharks, some lizards, some snakes (such as the garter snake *Thamnophis sirtalis*), some vipers, and some invertebrate animals (like the Madagascar hissing cockroach *Gromphadorhina portentosa*).

In **viviparity** the young develop within the female, receiving nourishment from the mother's blood through a placenta. The offspring develops in the female and is born alive. This occurs in most mammals, some cartilaginous fish, and a few reptiles.

Internal fertilization has the advantage of protecting the fertilized egg from dehydration on land. The embryo is isolated within the female, which limits predation on the young. Internal fertilization enhances the fertilization of eggs by a specific male. Fewer offspring are produced through this method, but their survival rate is higher than that for external fertilization.

The Evolution of Reproduction

Once multicellular organisms evolved and developed specialized cells, some also developed tissues and organs with specialized functions. An early development in reproduction occurred in the Annelids. These organisms produce sperm and eggs from undifferentiated cells in their coelom and store them in that cavity. When the coelom becomes filled, the cells are released through an excretory opening or by the body splitting open. Reproductive organs evolved with the development of gonads that produce sperm and eggs. These cells went through meiosis, an adaption of mitosis, which reduced the number of chromosomes in each reproductive cell by half, while increasing the number of cells through cell division.

Complete reproductive systems were developed in insects, with separate sexes. Sperm are made in testes and then travel through coiled tubes to the epididymis for storage. Eggs mature in the ovary. When they are released from the ovary, they travel to the uterine tubes for fertilization. Some insects have a specialized sac, called a **spermatheca**, which stores sperm for later use, sometimes up to a year. Fertilization can be timed with environmental or food conditions that are optimal for offspring survival.

Vertebrates have similar structures, with a few differences. Non-mammals, such as birds and reptiles, have a common body opening, called a **cloaca**, for the digestive, excretory and reproductive systems. Coupling between birds usually involves positioning the cloaca openings opposite each other for transfer of sperm. Mammals have separate openings for the systems in the female and a uterus for support of developing offspring. The uterus has two chambers in species that produce large numbers of offspring at a time, while species that produce one offspring, such as primates, have a single uterus.

Sperm transfer from the male to the female during reproduction ranges from releasing the sperm into the watery environment for external fertilization, to the joining of cloaca in birds, to the development of a penis for direct delivery into the female's vagina in mammals.

43.3 Human Reproductive Anatomy and Gametogenesis

By the end of this section, you will be able to do the following:

- Describe human male and female reproductive anatomies
- Discuss the human sexual response
- Describe spermatogenesis and oogenesis and discuss their differences and similarities

As animals became more complex, specific organs and organ systems developed to support specific functions for the organism. The reproductive structures that evolved in land animals allow males and females to mate, fertilize internally, and support the growth and development of offspring.

Human Reproductive Anatomy

The reproductive tissues of male and female humans develop similarly *in utero* until a low level of the hormone testosterone is released from male gonads. Testosterone causes the undeveloped tissues to differentiate into male sexual organs. When testosterone is absent, the tissues develop into female sexual tissues. Primitive gonads become testes or ovaries. Tissues that produce a penis in males produce a clitoris in females. The tissue that will become the scrotum in a male becomes the labia in a female; that is, they are homologous structures.

Male Reproductive Anatomy

In the male reproductive system, the **scrotum** houses the testicles or testes (singular: testis), including providing passage for blood vessels, nerves, and muscles related to testicular function. The **testes** are a pair of male reproductive organs that produce sperm and some reproductive hormones. Each testis is approximately 2.5 by 3.8 cm (1.5 by 1 in.) in size and divided into wedge-shaped lobules by connective tissue called septa. Coiled in each wedge are seminiferous tubules that produce sperm.

Sperm are immobile at body temperature; therefore, the scrotum and penis are external to the body, as illustrated in Figure 43.8 so that a proper temperature is maintained for motility. In land mammals, the pair of testes must be suspended outside the body at about 2° C lower than body temperature to produce viable sperm. Infertility can occur in land mammals when the testes do not descend through the abdominal cavity during fetal development.

VISUAL CONNECTION

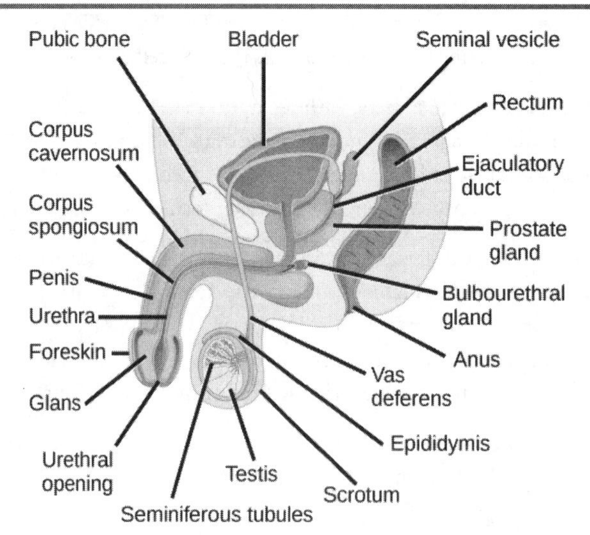

Figure 43.8 The reproductive structures of the human male are shown.

Which of the following statements about the male reproductive system is false?

a. The vas deferens carries sperm from the testes to the penis.
b. Sperm mature in seminiferous tubules in the testes.
c. Both the prostate and the bulbourethral glands produce components of the semen.
d. The prostate gland is located in the testes.

Sperm mature in **seminiferous tubules** that are coiled inside the testes, as illustrated in Figure 43.8. The walls of the seminiferous tubules are made up of the developing sperm cells, with the least developed sperm at the periphery of the tubule and the fully developed sperm in the lumen. The sperm cells are mixed with "nursemaid" cells called Sertoli cells which protect the germ cells and promote their development. Other cells mixed in the wall of the tubules are the interstitial cells of Leydig. These cells produce high levels of testosterone once the male reaches adolescence.

When the sperm have developed flagella and are nearly mature, they leave the testicles and enter the epididymis, shown in Figure 43.8. This structure resembles a comma and lies along the top and posterior portion of the testes; it is the site of sperm maturation. The sperm leave the epididymis and enter the vas deferens (or ductus deferens), which carries the sperm, behind the bladder, and forms the ejaculatory duct with the duct from the seminal vesicles. During a vasectomy, a section of the vas deferens is removed, preventing sperm from being passed out of the body during ejaculation and preventing fertilization.

Semen is a mixture of sperm and spermatic duct secretions (about 10 percent of the total) and fluids from accessory glands that contribute most of the semen's volume. Sperm are haploid cells, consisting of a flagellum as a tail, a neck that contains the cell's energy-producing mitochondria, and a head that contains the genetic material. Figure 43.9 shows a micrograph of human sperm as well as a diagram of the parts of the sperm. An acrosome is found at the top of the head of the sperm. This structure contains lysosomal enzymes that can digest the protective coverings that surround the egg to help the sperm penetrate and fertilize the egg. An ejaculate will contain from two to five milliliters of fluid with from 50–120 million sperm per milliliter.

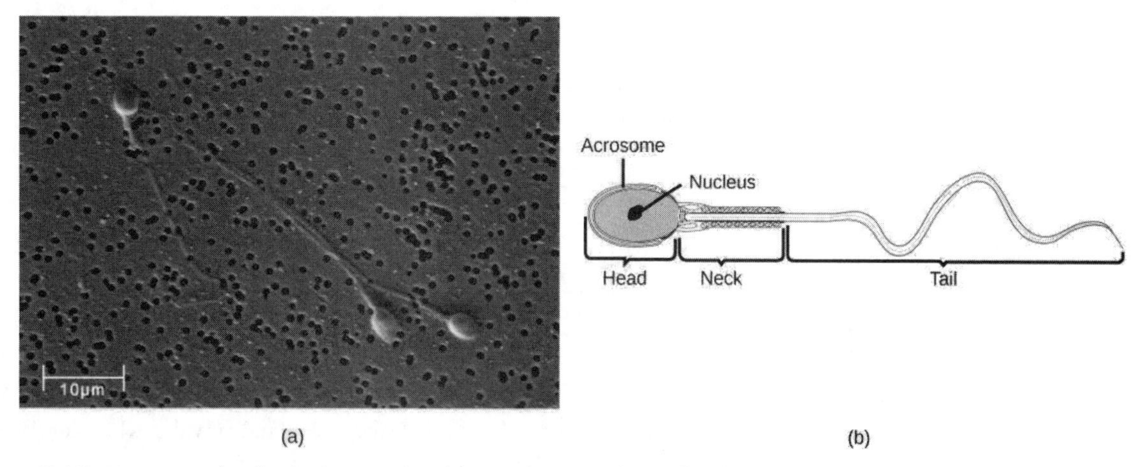

Figure 43.9 Human sperm, visualized using scanning electron microscopy, have a flagellum, neck, and head. (credit b: modification of work by Mariana Ruiz Villareal; scale-bar data from Matt Russell)

The bulk of the semen comes from the accessory glands associated with the male reproductive system. These are the seminal vesicles, the prostate gland, and the bulbourethral gland, all of which are illustrated in Figure 43.8. The **seminal vesicles** are a pair of glands that lie along the posterior border of the urinary bladder. The glands make a solution that is thick, yellowish, and alkaline. As sperm are only motile in an alkaline environment, a basic pH is important to reverse the acidity of the vaginal environment. The solution also contains mucus, fructose (a sperm mitochondrial nutrient), a coagulating enzyme, ascorbic acid, and local-acting hormones called prostaglandins. The seminal vesicle glands account for 60 percent of the bulk of semen.

The **penis**, illustrated in Figure 43.8, is an organ that drains urine from the renal bladder and functions as a copulatory organ during intercourse. The penis contains three tubes of erectile tissue running through the length of the organ. These consist of a pair of tubes on the dorsal side, called the corpus cavernosum, and a single tube of tissue on the ventral side, called the corpus spongiosum. This tissue will become engorged with blood, becoming erect and hard, in preparation for intercourse. The organ is inserted into the vagina culminating with an ejaculation. During intercourse, the smooth muscle sphincters at the opening to the renal bladder close and prevent urine from entering the penis. An orgasm is a two-stage process: first, glands and accessory organs connected to the testes contract, then semen (containing sperm) is expelled through the urethra during ejaculation. After intercourse, the blood drains from the erectile tissue and the penis becomes flaccid.

The walnut-shaped **prostate gland** surrounds the urethra, the connection to the urinary bladder. It has a series of short ducts that directly connect to the urethra. The gland is a mixture of smooth muscle and glandular tissue. The muscle provides much of the force needed for ejaculation to occur. The glandular tissue makes a thin, milky fluid that contains citrate (a nutrient), enzymes, and prostate specific antigen (PSA). PSA is a proteolytic enzyme that helps to liquefy the ejaculate several minutes after release from the male. Prostate gland secretions account for about 30 percent of the bulk of semen.

The **bulbourethral gland**, or Cowper's gland, releases its secretion prior to the release of the bulk of the semen. It neutralizes any acid residue in the urethra left over from urine. This usually accounts for a couple of drops of fluid in the total ejaculate and may contain a few sperm. Withdrawal of the penis from the vagina before ejaculation to prevent pregnancy may not work if sperm are present in the bulbourethral gland secretions. The location and functions of the male reproductive organs are summarized in Table 43.1.

Male Reproductive Anatomy

Organ	Location	Function
Scrotum	External	Carry and support testes
Penis	External	Deliver urine, copulating organ
Testes	Internal	Produce sperm and male hormones

Table 43.1

Organ	Location	Function
Seminal Vesicles	Internal	Contribute to semen production
Prostate Gland	Internal	Contribute to semen production
Bulbourethral Glands	Internal	Clean urethra at ejaculation

Table 43.1

Female Reproductive Anatomy

A number of reproductive structures are exterior to the female's body. These include the breasts and the vulva, which consists of the mons pubis, clitoris, labia majora, labia minora, and the vestibular glands, all illustrated in Figure 43.10. The location and functions of the female reproductive organs are summarized in Table 43.2. The vulva is an area associated with the vestibule which includes the structures found in the inguinal (groin) area of women. The mons pubis is a round, fatty area that overlies the pubic symphysis. The **clitoris** is a structure with erectile tissue that contains a large number of sensory nerves and serves as a source of stimulation during intercourse. The **labia majora** are a pair of elongated folds of tissue that run posterior from the mons pubis and enclose the other components of the vulva. The labia majora derive from the same tissue that produces the scrotum in a male. The **labia minora** are thin folds of tissue centrally located within the labia majora. These labia protect the openings to the vagina and urethra. The mons pubis and the anterior portion of the labia majora become covered with hair during adolescence; the labia minora is hairless. The greater vestibular glands are found at the sides of the vaginal opening and provide lubrication during intercourse.

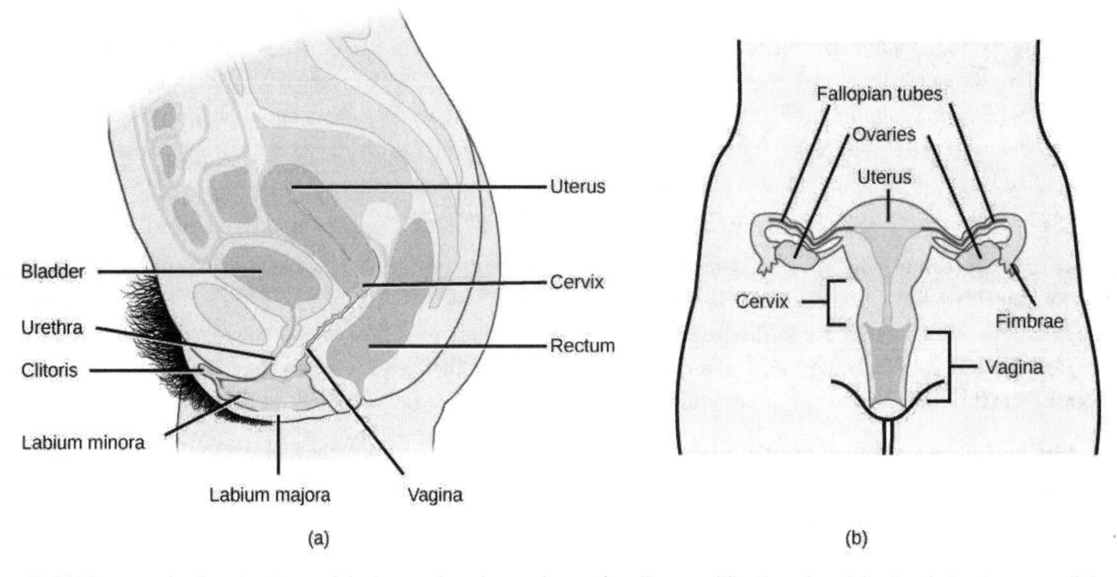

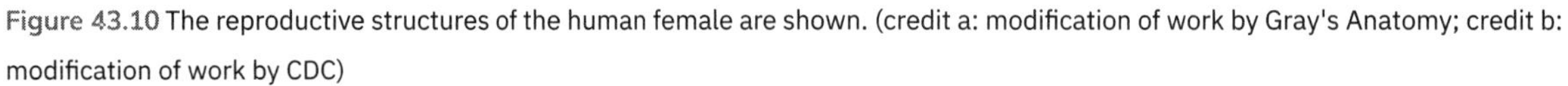

Figure 43.10 The reproductive structures of the human female are shown. (credit a: modification of work by Gray's Anatomy; credit b: modification of work by CDC)

Female Reproductive Anatomy

Organ	Location	Function
Clitoris	External	Sensory organ
Mons pubis	External	Fatty area overlying pubic bone

Table 43.2

Organ	Location	Function
Labia majora	External	Covers labia minora
Labia minora	External	Covers vestibule
Greater vestibular glands	External	Secrete mucus; lubricate vagina
Breasts	External	Produce and deliver milk
Ovaries	Internal	Carry and develop eggs
Oviducts (Fallopian tubes)	Internal	Transport egg to uterus
Uterus	Internal	Support developing embryo
Vagina	Internal	Common tube for intercourse, birth canal, passing menstrual flow

Table 43.2

The breasts consist of mammary glands and fat. The size of the breast is determined by the amount of fat deposited behind the gland. Each gland consists of 15 to 25 lobes that have ducts that empty at the nipple and that supply the nursing child with nutrient- and antibody-rich milk to aid development and protect the child.

Internal female reproductive structures include ovaries, oviducts, the **uterus**, and the vagina, shown in Figure 43.10. The pair of ovaries is held in place in the abdominal cavity by a system of ligaments. Ovaries consist of a medulla and cortex: the medulla contains nerves and blood vessels to supply the cortex with nutrients and remove waste. The outer layers of cells of the cortex are the functional parts of the ovaries. The cortex is made up of follicular cells that surround eggs that develop during fetal development *in utero*. During the menstrual period, a batch of follicular cells develops and prepares the eggs for release. At ovulation, one follicle ruptures and one egg is released, as illustrated in Figure 43.11a.

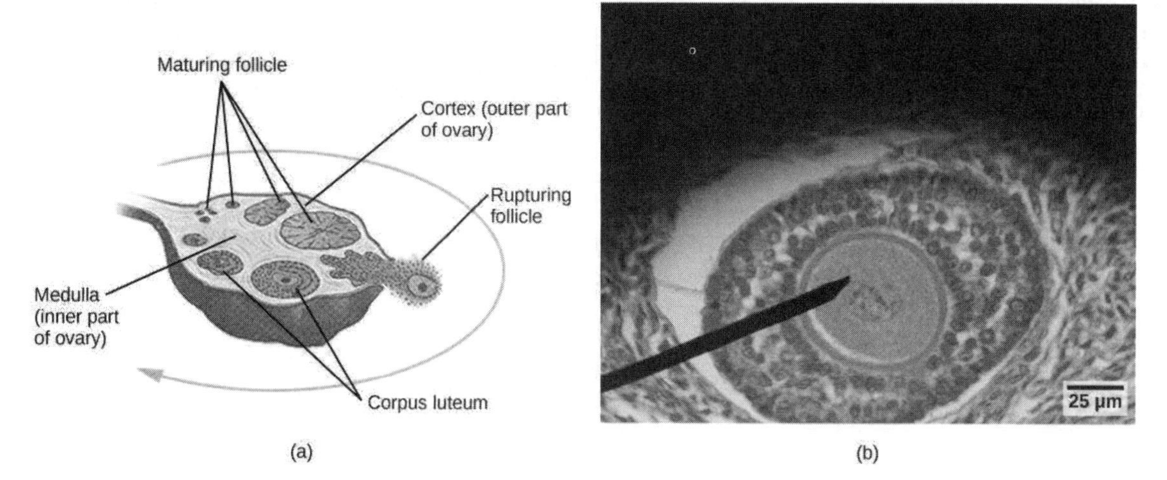

Figure 43.11 Oocytes develop in (a) follicles, located in the ovary. At the beginning of the menstrual cycle, the follicle matures. At ovulation, the follicle ruptures, releasing the egg. The follicle becomes a corpus luteum, which eventually degenerates. The (b) follicle in this light micrograph has an oocyte at its center. (credit a: modification of work by NIH; scale-bar data from Matt Russell)

The **oviducts**, or fallopian tubes, extend from the uterus in the lower abdominal cavity to the ovaries, but they are not in contact with the ovaries. The lateral ends of the oviducts flare out into a trumpet-like structure and have a fringe of finger-like projections called fimbriae, illustrated in Figure 43.10b. When an egg is released at ovulation, the fimbrae help the nonmotile egg enter into the tube and passage to the uterus. The walls of the oviducts are ciliated and are made up mostly of smooth muscle. The cilia beat toward the middle, and the smooth muscle contracts in the same direction, moving the egg toward the uterus. Fertilization usually takes place within the oviducts and the developing embryo is moved toward the uterus for development. It usually takes the egg or embryo a week to travel through the oviduct. Sterilization in women is called a tubal ligation; it is analogous to a vasectomy in males in that the oviducts are severed and sealed.

The uterus is a structure about the size of a woman's fist. This is lined with an endometrium rich in blood vessels and mucus glands. The uterus supports the developing embryo and fetus during gestation. The thickest portion of the wall of the uterus is made of smooth muscle. Contractions of the smooth muscle in the uterus aid in passing the baby through the vagina during labor. A portion of the lining of the uterus sloughs off during each menstrual period, and then builds up again in preparation for an implantation. Part of the uterus, called the cervix, protrudes into the top of the vagina. The cervix functions as the birth canal.

The **vagina** is a muscular tube that serves several purposes. It allows menstrual flow to leave the body. It is the receptacle for the penis during intercourse and the vessel for the delivery of offspring. It is lined by stratified squamous epithelial cells to protect the underlying tissue.

Sexual Response during Intercourse

The sexual response in humans is both psychological and physiological. Both sexes experience sexual arousal through psychological and physical stimulation. There are four phases of the sexual response. During phase one, called excitement, vasodilation leads to vasocongestion in erectile tissues in both men and women. The nipples, clitoris, labia, and penis engorge with blood and become enlarged. Vaginal secretions are released to lubricate the vagina to facilitate intercourse. During the second phase, called the plateau, stimulation continues, the outer third of the vaginal wall enlarges with blood, and breathing and heart rate increase.

During phase three, or orgasm, rhythmic, involuntary contractions of muscles occur in both sexes. In the male, the reproductive accessory glands and tubules constrict placing semen in the urethra, then the urethra contracts expelling the semen through the penis. In women, the uterus and vaginal muscles contract in waves that may last slightly less than a second each. During phase four, or resolution, the processes described in the first three phases reverse themselves and return to their normal state. Men experience a refractory period in which they cannot maintain an erection or ejaculate for a period of time ranging from minutes to hours.

Gametogenesis (Spermatogenesis and Oogenesis)

Gametogenesis, the production of sperm and eggs, takes place through the process of meiosis. During meiosis, two cell divisions separate the paired chromosomes in the nucleus and then separate the chromatids that were made during an earlier stage of the cell's life cycle. Meiosis produces haploid cells with half of each pair of chromosomes normally found in diploid cells. The production of sperm is called **spermatogenesis** and the production of eggs is called **oogenesis**.

Spermatogenesis

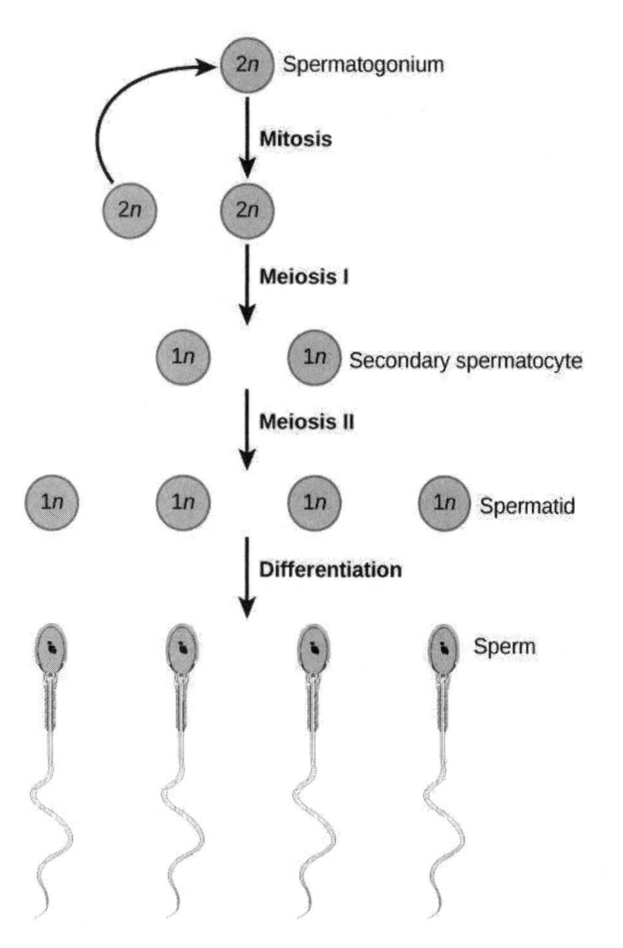

Figure 43.12 During spermatogenesis, four sperm result from each primary spermatocyte.

Spermatogenesis, illustrated in Figure 43.12, occurs in the wall of the seminiferous tubules (Figure 43.8), with stem cells at the periphery of the tube and the spermatozoa at the lumen of the tube. Immediately under the capsule of the tubule are diploid, undifferentiated cells. These stem cells, called spermatogonia (singular: spermatagonium), go through mitosis with one offspring going on to differentiate into a sperm cell and the other giving rise to the next generation of sperm.

Meiosis starts with a cell called a primary spermatocyte. At the end of the first meiotic division, a haploid cell is produced called a secondary spermatocyte. This cell is haploid and must go through another meiotic cell division. The cell produced at the end of meiosis is called a spermatid and when it reaches the lumen of the tubule and grows a flagellum, it is called a sperm cell. Four sperm result from each primary spermatocyte that goes through meiosis.

Stem cells are deposited during gestation and are present at birth through the beginning of adolescence, but in an inactive state. During adolescence, gonadotropic hormones from the anterior pituitary cause the activation of these cells and the production of viable sperm. This continues into old age.

LINK TO LEARNING

Visit this site (http://openstax.org/l/spermatogenesis) to see the process of spermatogenesis.

Oogenesis

Oogenesis, illustrated in Figure 43.13, occurs in the outermost layers of the ovaries. As with sperm production, oogenesis starts with a germ cell, called an oogonium (plural: oogonia), but this cell undergoes mitosis to increase in number, eventually resulting in up to about one to two million cells in the embryo.

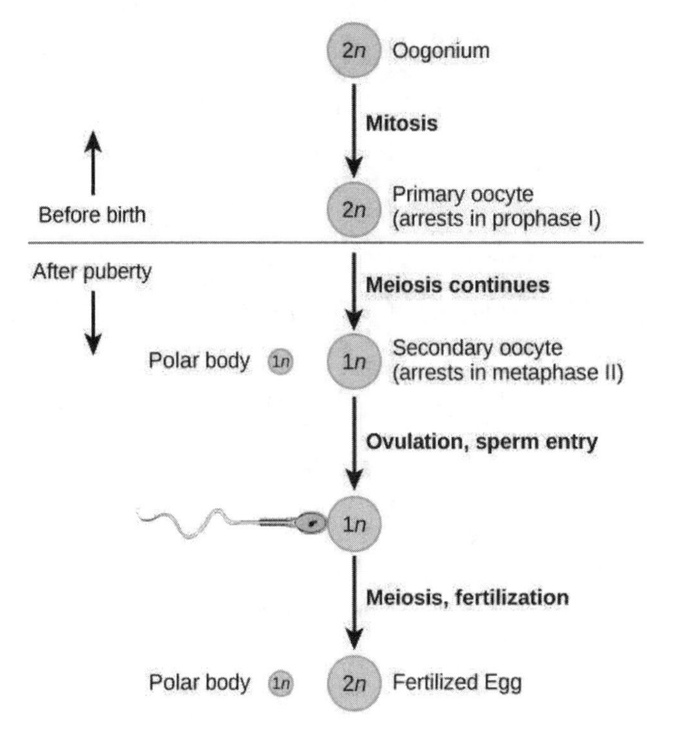

Figure 43.13 The process of oogenesis occurs in the ovary's outermost layer.

The cell starting meiosis is called a primary oocyte, as shown in Figure 43.13. This cell will start the first meiotic division and be arrested in its progress in the first prophase stage. At the time of birth, all future eggs are in the prophase stage. At adolescence, anterior pituitary hormones cause the development of a number of follicles in an ovary. This results in the primary oocyte finishing the first meiotic division. The cell divides unequally, with most of the cellular material and organelles going to one cell, called a secondary oocyte, and only one set of chromosomes and a small amount of cytoplasm going to the other cell. This second cell is called a polar body and usually dies. A secondary meiotic arrest occurs, this time at the metaphase II stage. At ovulation, this secondary oocyte will be released and travel toward the uterus through the oviduct. If the secondary oocyte is fertilized, the cell continues through the meiosis II, producing a second polar body and a fertilized egg containing all 46 chromosomes of a human being, half of them coming from the sperm.

Egg production begins before birth, is arrested during meiosis until puberty, and then individual cells continue through at each menstrual cycle. One egg is produced from each meiotic process, with the extra chromosomes and chromatids going into polar bodies that degenerate and are reabsorbed by the body.

43.4 Hormonal Control of Human Reproduction

By the end of this chapter, you will be able to do the following:

- Describe the roles of male and female reproductive hormones
- Discuss the interplay of the ovarian and menstrual cycles
- Describe the process of menopause

The human male and female reproductive cycles are controlled by the interaction of hormones from the hypothalamus and anterior pituitary with hormones from reproductive tissues and organs. In both sexes, the hypothalamus monitors and causes the release of hormones from the pituitary gland. When the reproductive hormone is required, the hypothalamus sends a **gonadotropin-releasing hormone (GnRH)** to the anterior pituitary. This causes the release of **follicle stimulating hormone (FSH)** and **luteinizing hormone (LH)** from the anterior pituitary into the blood. Note that the body must reach puberty in order for the adrenals to release the hormones that must be present for GnRH to be produced. Although FSH and LH are named after their functions in female reproduction, they are produced in both sexes and play important roles in controlling reproduction. Other hormones have specific functions in the male and female reproductive systems.

Male Hormones

At the onset of puberty, the hypothalamus causes the release of FSH and LH into the male system for the first time. FSH enters the testes and stimulates the **Sertoli cells** to begin facilitating spermatogenesis using negative feedback, as illustrated in Figure 43.14. LH also enters the testes and stimulates the **interstitial cells of Leydig** to make and release testosterone into the testes and the blood.

Testosterone, the hormone responsible for the secondary sexual characteristics that develop in the male during adolescence, stimulates spermatogenesis. These secondary sex characteristics include a deepening of the voice, the growth of facial, axillary, and pubic hair, and the beginnings of the sex drive.

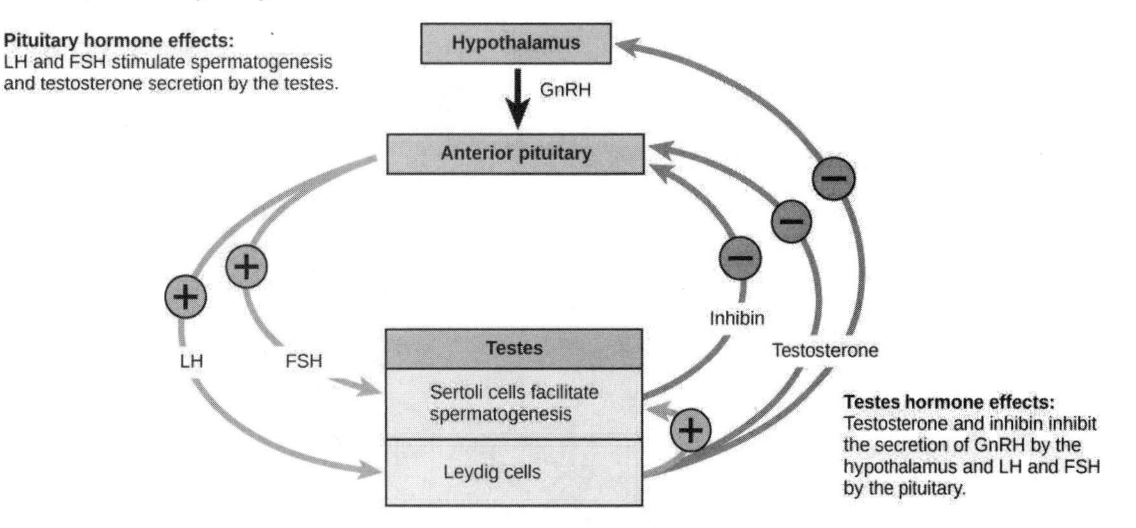

Figure 43.14 Hormones control sperm production in a negative feedback system.

A negative feedback system occurs in the male with rising levels of testosterone acting on the hypothalamus and anterior pituitary to inhibit the release of GnRH, FSH, and LH. The Sertoli cells produce the hormone **inhibin**, which is released into the blood when the sperm count is too high. This inhibits the release of GnRH and FSH, which will cause spermatogenesis to slow down. If the sperm count reaches 20 million/ml, the Sertoli cells cease the release of inhibin, and the sperm count increases.

Female Hormones

The control of reproduction in females is more complex. As with the male, the anterior pituitary hormones cause the release of the hormones FSH and LH. In addition, estrogens and progesterone are released from the developing follicles. **Estrogen** is the reproductive hormone in females that assists in endometrial regrowth, ovulation, and calcium absorption; it is also responsible for the secondary sexual characteristics of females. These include breast development, flaring of the hips, and a shorter period necessary for bone maturation. **Progesterone** assists in endometrial regrowth and inhibition of FSH and LH release.

In females, FSH stimulates development of egg cells, called ova, which develop in structures called follicles. Follicle cells produce the hormone inhibin, which inhibits FSH production. LH also plays a role in the development of ova, induction of ovulation, and stimulation of estradiol and progesterone production by the ovaries. Estradiol and progesterone are steroid hormones that prepare the body for pregnancy. Estradiol produces secondary sex characteristics in females, while both estradiol and progesterone regulate the menstrual cycle.

The Ovarian Cycle and the Menstrual Cycle

The **ovarian cycle** governs the preparation of endocrine tissues and release of eggs, while the **menstrual cycle** governs the preparation and maintenance of the uterine lining. These cycles occur concurrently and are coordinated over a 22–32 day cycle, with an average length of 28 days.

The first half of the ovarian cycle is the follicular phase shown in Figure 43.15. Slowly rising levels of FSH and LH cause the growth of follicles on the surface of the ovary. This process prepares the egg for ovulation. As the follicles grow, they begin releasing estrogens and a low level of progesterone. Progesterone maintains the endometrium to help ensure pregnancy. The trip through the fallopian tube takes about seven days. At this stage of development, called the morula, there are 30-60 cells. If pregnancy implantation does not occur, the lining is sloughed off. After about five days, estrogen levels rise and the menstrual cycle enters the proliferative phase. The endometrium begins to regrow, replacing the blood vessels and glands that deteriorated during the end of the last cycle.

VISUAL CONNECTION

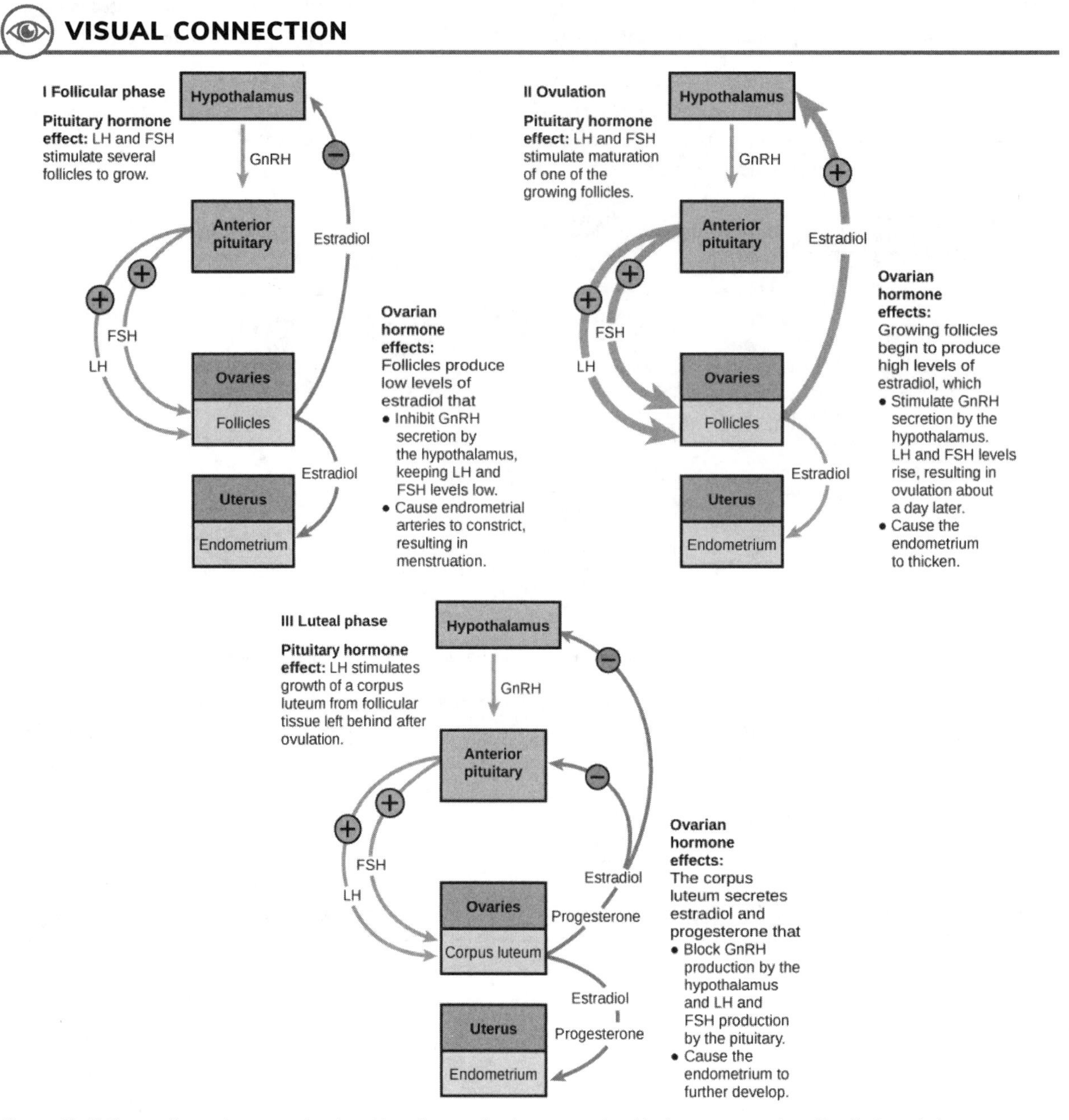

Figure 43.15 The ovarian and menstrual cycles of female reproduction are regulated by hormones produced by the hypothalamus, pituitary, and ovaries.

Which of the following statements about hormone regulation of the female reproductive cycle is false?

a. LH and FSH are produced in the pituitary, and estradiol and progesterone are produced in the ovaries.
b. Estradiol and progesterone secreted from the corpus luteum cause the endometrium to thicken.
c. Both progesterone and estradiol are produced by the follicles.
d. Secretion of GnRH by the hypothalamus is inhibited by low levels of estradiol but stimulated by high levels of estradiol.

Just prior to the middle of the cycle (approximately day 14), the high level of estrogen causes FSH and especially LH to rise rapidly, then fall. The spike in LH causes **ovulation**: the most mature follicle, like that shown in Figure 43.16, ruptures and releases its egg. The follicles that did not rupture degenerate and their eggs are lost. The level of estrogen decreases when the extra follicles degenerate.

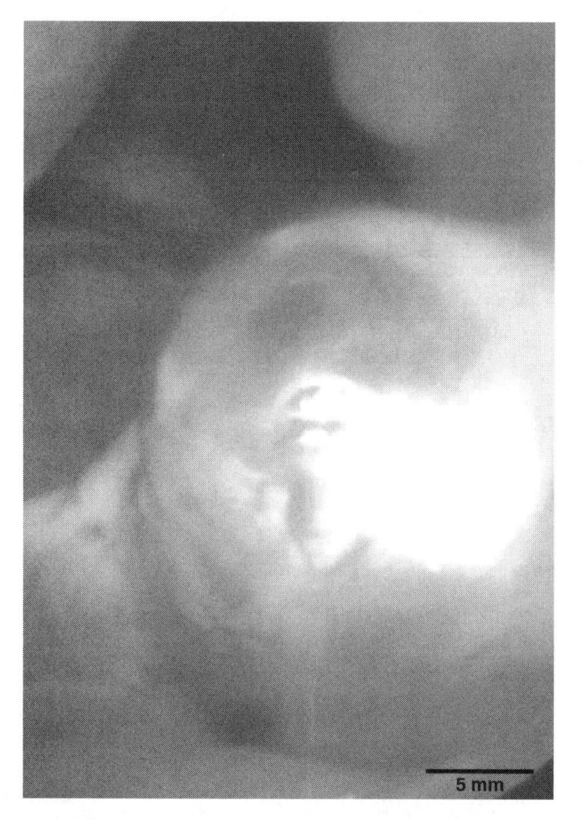

Figure 43.16 This mature egg follicle may rupture and release an egg. (credit: scale-bar data from Matt Russell)

Following ovulation, the ovarian cycle enters its luteal phase, illustrated in Figure 43.15 and the menstrual cycle enters its secretory phase, both of which run from about day 15 to 28. The luteal and secretory phases refer to changes in the ruptured follicle. The cells in the follicle undergo physical changes and produce a structure called a corpus luteum. The corpus luteum produces estrogen and progesterone. The progesterone facilitates the regrowth of the uterine lining and inhibits the release of further FSH and LH. The uterus is being prepared to accept a fertilized egg, should it occur during this cycle. The inhibition of FSH and LH prevents any further eggs and follicles from developing, while the progesterone is elevated. The level of estrogen produced by the corpus luteum increases to a steady level for the next few days.

If no fertilized egg is implanted into the uterus, the corpus luteum degenerates and the levels of estrogen and progesterone decrease. The endometrium begins to degenerate as the progesterone levels drop, initiating the next menstrual cycle. The decrease in progesterone also allows the hypothalamus to send GnRH to the anterior pituitary, releasing FSH and LH and starting the cycles again. Figure 43.17 visually compares the ovarian and uterine cycles as well as the commensurate hormone levels.

VISUAL CONNECTION

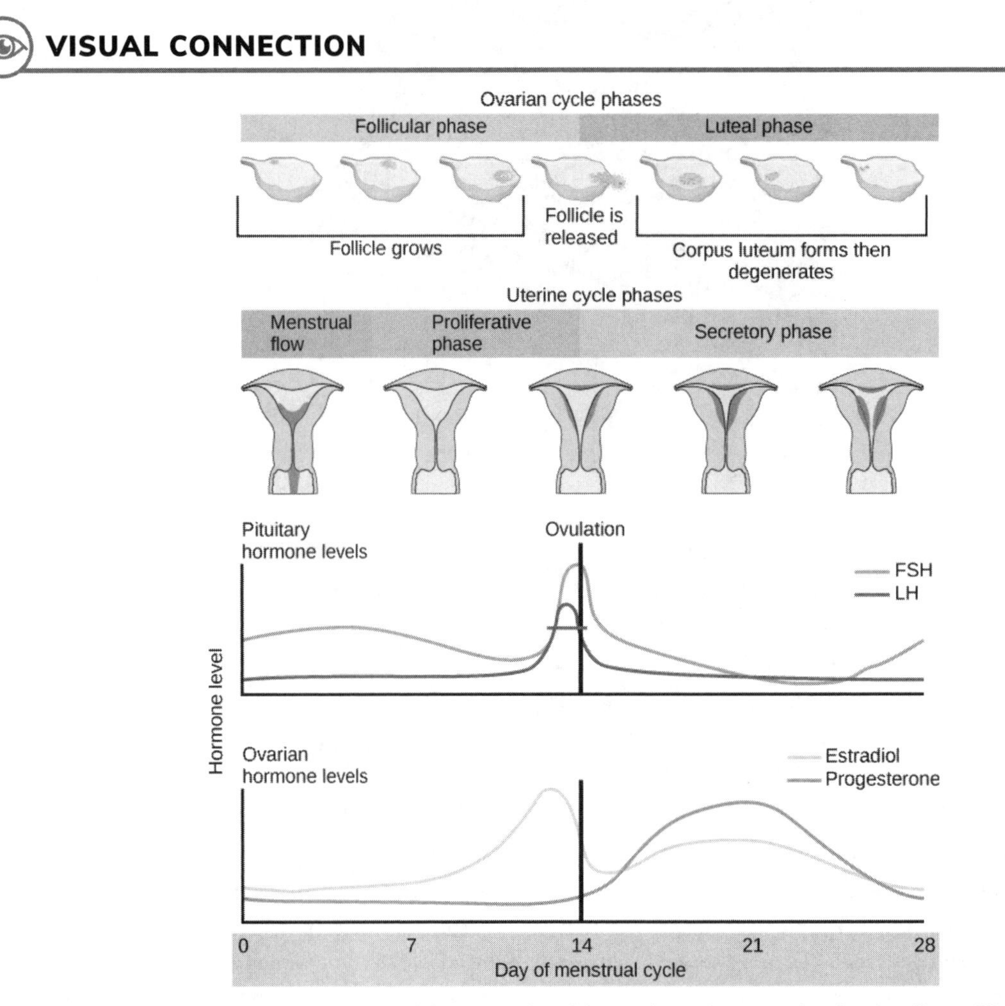

Figure 43.17 Rising and falling hormone levels result in progression of the ovarian and menstrual cycles. (credit: modification of work by Mikael Häggström)

Which of the following statements about the menstrual cycle is false?

a. Progesterone levels rise during the luteal phase of the ovarian cycle and the secretory phase of the uterine cycle.
b. Menstruation occurs just after LH and FSH levels peak.
c. Menstruation occurs after progesterone levels drop.
d. Estrogen levels rise before ovulation, while progesterone levels rise after.

Menopause

As women approach their mid-40s to mid-50s, their ovaries begin to lose their sensitivity to FSH and LH. Menstrual periods become less frequent and finally cease; this is **menopause**. There are still eggs and potential follicles on the ovaries, but without the stimulation of FSH and LH, they will not produce a viable egg to be released. The outcome of this is the inability to have children.

The side effects of menopause include hot flashes, heavy sweating (especially at night), headaches, some hair loss, muscle pain, vaginal dryness, insomnia, depression, weight gain, and mood swings. Estrogen is involved in calcium metabolism and, without it, blood levels of calcium decrease. To replenish the blood, calcium is lost from bone which may decrease the bone density and lead to osteoporosis. Supplementation of estrogen in the form of hormone replacement therapy (HRT) can prevent bone loss, but the therapy can have negative side effects. While HRT is thought to give some protection from colon cancer, osteoporosis, heart disease, macular degeneration, and possibly depression, its negative side effects include increased risk of: stroke or heart attack, blood clots, breast cancer, ovarian cancer, endometrial cancer, gall bladder disease, and possibly dementia.

CAREER CONNECTION

Reproductive Endocrinologist

A reproductive endocrinologist is a physician who treats a variety of hormonal disorders related to reproduction and infertility in both men and women. The disorders include menstrual problems, infertility, pregnancy loss, sexual dysfunction, and menopause. Doctors may use fertility drugs, surgery, or assisted reproductive techniques (ART) in their therapy. ART involves the use of procedures to manipulate the egg or sperm to facilitate reproduction, such as *in vitro* fertilization.

Reproductive endocrinologists undergo extensive medical training, first in a four-year residency in obstetrics and gynecology, then in a three-year fellowship in reproductive endocrinology. To be board certified in this area, the physician must pass written and oral exams in both areas.

43.5 Human Pregnancy and Birth

By the end of this section, you will be able to do the following:

- Explain fetal development during the three trimesters of gestation
- Describe labor and delivery
- Compare the efficacy and duration of various types of contraception
- Discuss causes of infertility and the therapeutic options available

Pregnancy begins with the fertilization of an egg and continues through to the birth of the individual. The length of time of **gestation** varies among animals, but is very similar among the great apes: human gestation is 266 days, while chimpanzee gestation is 237 days, a gorilla's is 257 days, and orangutan gestation is 260 days long. The fox has a 57-day gestation. Dogs and cats have similar gestations averaging 60 days. The longest gestation for a land mammal is an African elephant at 640 days. The longest gestations among marine mammals are the beluga and sperm whales at 460 days.

Human Gestation

Twenty-four hours before fertilization, the egg has finished meiosis and becomes a mature oocyte. When fertilized (at conception) the egg becomes known as a zygote. The zygote travels through the oviduct to the uterus (Figure 43.18). The developing embryo must implant into the wall of the uterus within seven days, or it will deteriorate and die. The outer layers of the zygote (blastocyst) grow into the endometrium by digesting the endometrial cells, and wound healing of the endometrium closes up the blastocyst into the tissue. Another layer of the blastocyst, the chorion, begins releasing a hormone called **human beta chorionic gonadotropin (β-HCG)** which makes its way to the corpus luteum and keeps that structure active. This ensures adequate levels of progesterone that will maintain the endometrium of the uterus for the support of the developing embryo. Pregnancy tests determine the level of β-HCG in urine or serum. If the hormone is present, the test is positive.

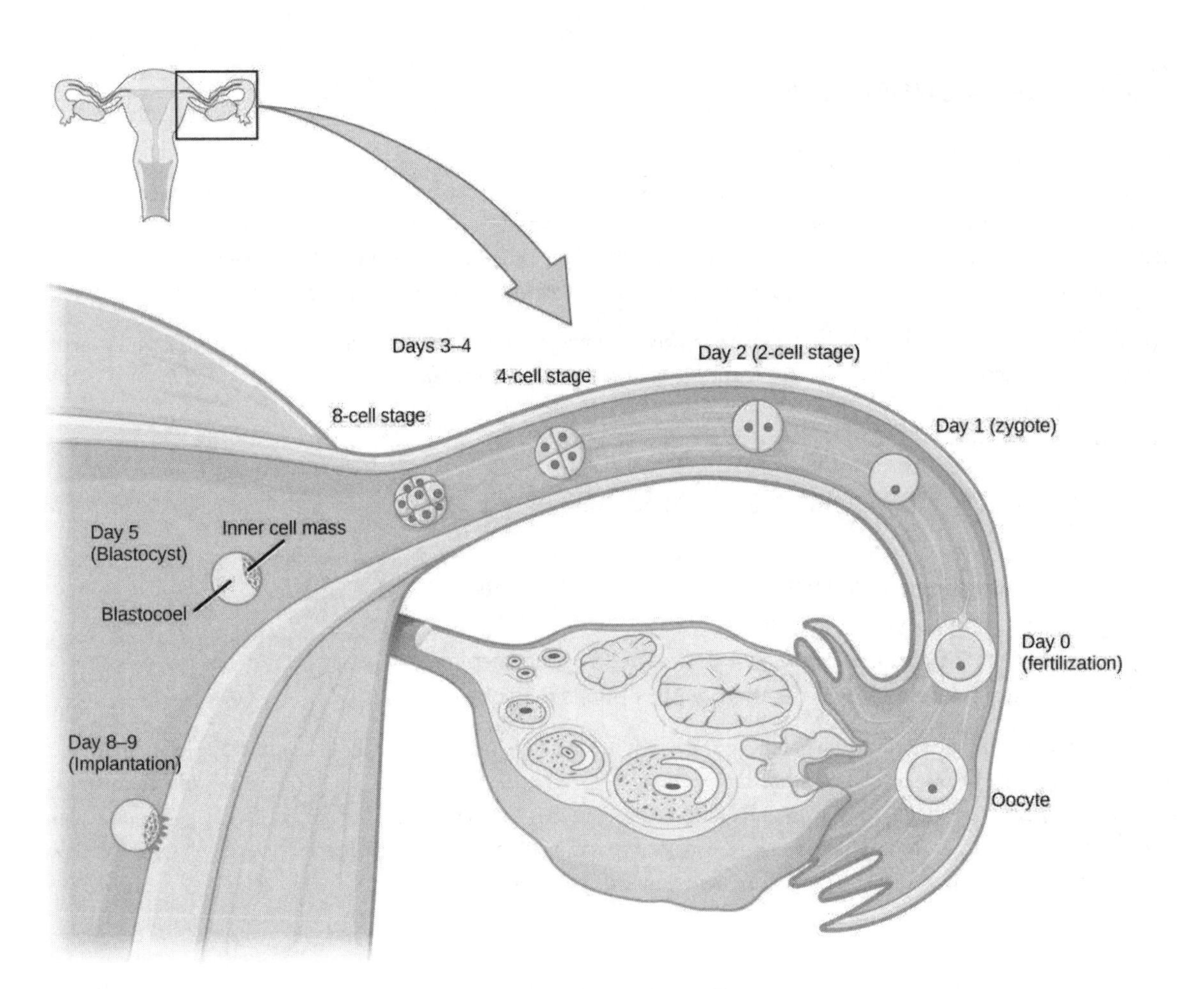

Figure 43.18 In humans, fertilization occurs soon after the oocyte leaves the ovary. Implantation occurs eight or nine days later.(credit: Ed Uthman)

The gestation period is divided into three equal periods or trimesters. During the first two to four weeks of the first trimester, nutrition and waste are handled by the endometrial lining through diffusion. As the trimester progresses, the outer layer of the embryo begins to merge with the endometrium, and the **placenta** forms. This organ takes over the nutrient and waste requirements of the embryo and fetus, with the mother's blood passing nutrients to the placenta and removing waste from it. Chemicals from the fetus, such as bilirubin, are processed by the mother's liver for elimination. Some of the mother's immunoglobulins will pass through the placenta, providing passive immunity against some potential infections.

Internal organs and body structures begin to develop during the first trimester. By five weeks, limb buds, eyes, the heart, and liver have been basically formed. By eight weeks, the term fetus applies, and the body is essentially formed, as shown in Figure 43.19. The individual is about five centimeters (two inches) in length and many of the organs, such as the lungs and liver, are not yet functioning. Exposure to any toxins is especially dangerous during the first trimester, as all of the body's organs and structures are going through initial development. Anything that affects that development can have a severe effect on the fetus' survival.

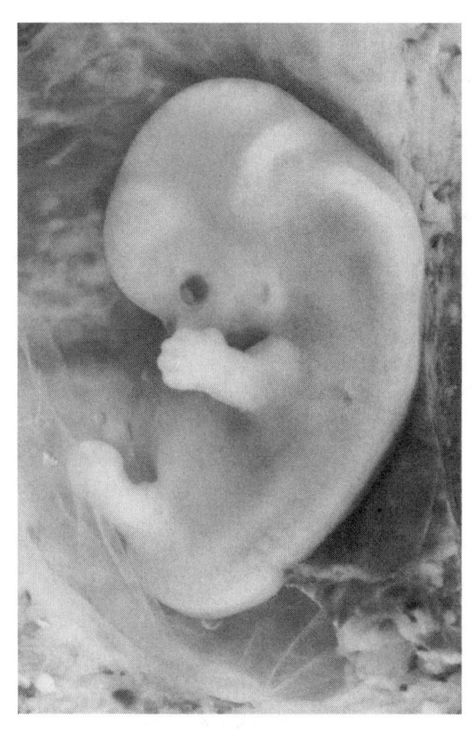

Figure 43.19 Fetal development is shown at nine weeks gestation. (credit: Ed Uthman)

During the second trimester, the fetus grows to about 30 cm (12 inches), as shown in Figure 43.20. It becomes active and the mother usually feels the first movements. All organs and structures continue to develop. The placenta has taken over the functions of nutrition and waste and the production of estrogen and progesterone from the corpus luteum, which has degenerated. The placenta will continue functioning up through the delivery of the baby.

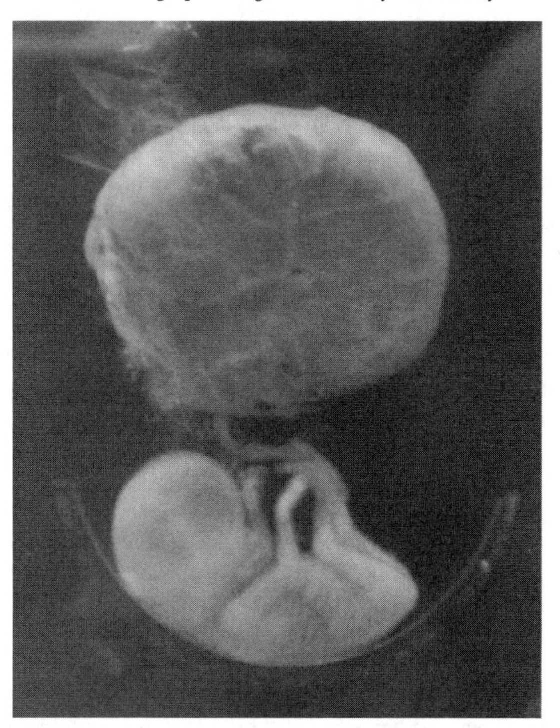

Figure 43.20 This fetus is just entering the second trimester, when the placenta takes over more of the functions performed as the baby develops. (credit: National Museum of Health and Medicine)

During the third trimester, the fetus grows to 3 to 4 kg (6 ½ -8 ½ lbs.) and about 50 cm (19-20 inches) long, as illustrated in Figure 43.21. This is the period of the most rapid growth during the pregnancy. Organ development continues to birth (and some systems, such as the nervous system and liver, continue to develop after birth). The mother will be at her most uncomfortable during this trimester. She may urinate frequently due to pressure on the bladder from the fetus. There may also be intestinal blockage and circulatory problems, especially in her legs. Clots may form in her legs due to pressure from the fetus on returning veins as they enter the abdominal cavity.

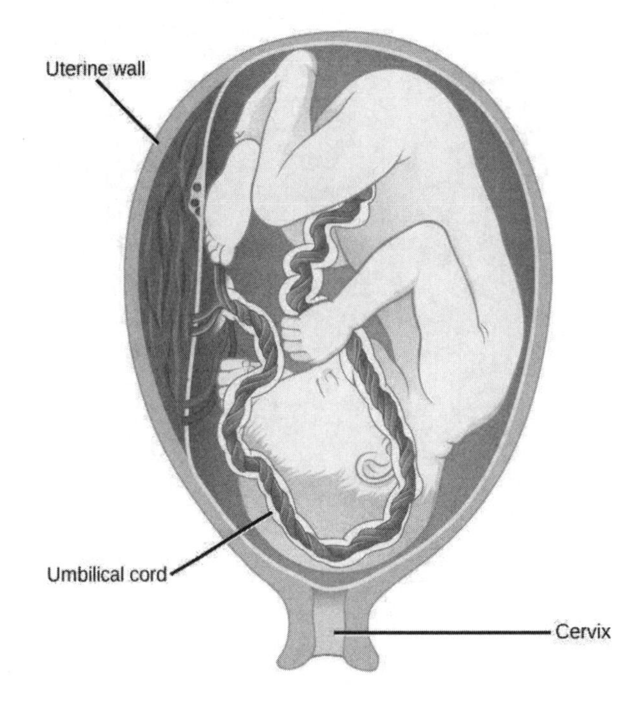

Figure 43.21 There is rapid fetal growth during the third trimester. (credit: modification of work by Gray's Anatomy)

LINK TO LEARNING

Visit this site (http://openstax.org/l/embryo_fetus) to see the stages of human fetal development.

Labor and Birth

Labor is the physical efforts of expulsion of the fetus and the placenta from the uterus during birth (parturition). Toward the end of the third trimester, estrogen causes receptors on the uterine wall to develop and bind the hormone oxytocin. At this time, the baby reorients, facing forward and down with the back or crown of the head engaging the cervix (uterine opening). This causes the cervix to stretch and nerve impulses are sent to the hypothalamus, which signals for the release of oxytocin from the posterior pituitary. The oxytocin causes the smooth muscle in the uterine wall to contract. At the same time, the placenta releases prostaglandins into the uterus, increasing the contractions. A positive feedback relay occurs between the uterus, hypothalamus, and the posterior pituitary to assure an adequate supply of oxytocin. As more smooth muscle cells are recruited, the contractions increase in intensity and force.

There are three stages to labor. During stage one, the cervix thins and dilates. This is necessary for the baby and placenta to be expelled during birth. The cervix will eventually dilate to about 10 cm. During stage two, the baby is expelled from the uterus. The uterus contracts and the mother pushes as she compresses her abdominal muscles to aid the delivery. The last stage is the passage of the placenta after the baby has been born and the organ has completely disengaged from the uterine wall. If labor should stop before stage two is reached, synthetic oxytocin, known as Pitocin, can be administered to restart and maintain labor.

An alternative to labor and delivery is the surgical delivery of the baby through a procedure called a Caesarian section. This is major abdominal surgery and can lead to post-surgical complications for the mother, but in some cases it may be the only way to safely deliver the baby.

The mother's mammary glands go through changes during the third trimester to prepare for lactation and breastfeeding. When the baby begins suckling at the breast, signals are sent to the hypothalamus causing the release of prolactin from the anterior pituitary. Prolactin causes the mammary glands to produce milk. Oxytocin is also released, promoting the release of the milk.

The milk contains nutrients for the baby's development and growth as well as immunoglobulins to protect the child from bacterial and viral infections.

Contraception and Birth Control

The prevention of a pregnancy comes under the terms contraception or birth control. Strictly speaking, **contraception** refers to preventing the sperm and egg from joining. Both terms are, however, frequently used interchangeably.

Contraceptive Methods

Method	Examples	Failure Rate in Typical Use Over 12 Months
Barrier	male condom, female condom, sponge, cervical cap, diaphragm, spermicides	15 to 24%
Hormonal	oral, patch, vaginal ring	8%
	injection	3%
	implant	less than 1%
Other	natural family planning	12 to 25%
	withdrawal	27%
	sterilization	less than 1%

Table 43.3

Table 43.3 lists common methods of contraception. The failure rates listed are not the ideal rates that could be realized, but the typical rates that occur. A failure rate is the number of pregnancies resulting from the method's use over a twelve-month period. Barrier methods, such as condoms, cervical caps, and diaphragms, block sperm from entering the uterus, preventing fertilization. Spermicides are chemicals that are placed in the vagina that kill sperm. Sponges, which are saturated with spermicides, are placed in the vagina at the cervical opening. Combinations of spermicidal chemicals and barrier methods achieve lower failure rates than do the methods when used separately.

Nearly a quarter of the couples using barrier methods, natural family planning, or withdrawal can expect a failure of the method. Natural family planning is based on the monitoring of the menstrual cycle and having intercourse only during times when the egg is not available. A woman's body temperature may rise a degree Celsius at ovulation and the cervical mucus may increase in volume and become more pliable. These changes give a general indication of when intercourse is more or less likely to result in fertilization. Withdrawal involves the removal of the penis from the vagina during intercourse, before ejaculation occurs. This is a risky method with a high failure rate due to the possible presence of sperm in the bulbourethral gland's secretion, which may enter the vagina prior to removing the penis.

Hormonal methods use synthetic progesterone (sometimes in combination with estrogen), to inhibit the hypothalamus from releasing FSH or LH, and thus prevent an egg from being available for fertilization. The method of administering the hormone affects failure rate. The most reliable method, with a failure rate of less than 1 percent, is the implantation of the hormone under the skin. The same rate can be achieved through the sterilization procedures of vasectomy in the man or of tubal ligation in the woman, or by using an intrauterine device (IUD). IUDs are inserted into the uterus and establish an inflammatory condition that prevents fertilized eggs from implanting into the uterine wall.

Compliance with the contraceptive method is a strong contributor to the success or failure rate of any particular method. The only method that is completely effective at preventing conception is abstinence. The choice of contraceptive method depends on the goals of the woman or couple. Tubal ligation and vasectomy are considered permanent prevention, while other methods are reversible and provide short-term contraception.

Termination of an existing pregnancy can be spontaneous or voluntary. Spontaneous termination is a miscarriage and usually occurs very early in the pregnancy, usually within the first few weeks. This occurs when the fetus cannot develop properly and the gestation is naturally terminated. Voluntary termination of a pregnancy is an abortion. Laws regulating abortion vary between states and tend to view fetal viability as the criteria for allowing or preventing the procedure.

Infertility

Infertility is the inability to conceive a child or carry a child to birth. About 75 percent of causes of infertility can be identified; these include diseases, such as sexually transmitted diseases that can cause scarring of the reproductive tubes in either men or women, or developmental problems frequently related to abnormal hormone levels in one of the individuals. Inadequate nutrition, especially starvation, can delay menstruation. Stress can also lead to infertility. Short-term stress can affect hormone levels, while long-term stress can delay puberty and cause less frequent menstrual cycles. Other factors that affect fertility include toxins (such as cadmium), tobacco smoking, marijuana use, gonadal injuries, and aging.

If infertility is identified, several assisted reproductive technologies (ART) are available to aid conception. A common type of ART is *in vitro* fertilization (IVF) where an egg and sperm are combined outside the body and then placed in the uterus. Eggs are obtained from the woman after extensive hormonal treatments that prepare mature eggs for fertilization and prepare the uterus for implantation of the fertilized egg. Sperm are obtained from the man and they are combined with the eggs and supported through several cell divisions to ensure viability of the zygotes. When the embryos have reached the eight-cell stage, one or more is implanted into the woman's uterus. If fertilization is not accomplished by simple IVF, a procedure that injects the sperm into an egg can be used. This is called intracytoplasmic sperm injection (ICSI) and is shown in Figure 43.22. IVF procedures produce a surplus of fertilized eggs and embryos that can be frozen and stored for future use. The procedures can also result in multiple births.

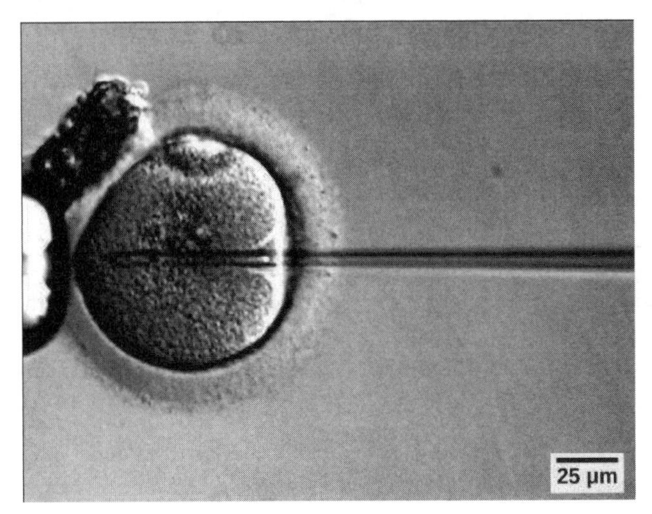

Figure 43.22 A sperm is inserted into an egg for fertilization during intracytoplasmic sperm injection (ICSI). (credit: scale-bar data from Matt Russell)

43.6 Fertilization and Early Embryonic Development

By the end of this section, you will be able to do the following:

- Discuss how fertilization occurs
- Explain how the embryo forms from the zygote
- Discuss the role of cleavage and gastrulation in animal development

The process in which an organism develops from a single-celled zygote to a multi-cellular organism is complex and well-regulated. The early stages of embryonic development are also crucial for ensuring the fitness of the organism.

Fertilization

Fertilization, pictured in Figure 43.23a is the process in which gametes (an egg and sperm) fuse to form a zygote. The egg and sperm each contain one set of chromosomes. To ensure that the offspring has only one complete diploid set of chromosomes, only one sperm must fuse with one egg. In mammals, the egg is protected by a layer of extracellular matrix consisting mainly of glycoproteins called the **zona pellucida**. When a sperm binds to the zona pellucida, a series of biochemical events, called the **acrosomal reactions**, take place. In placental mammals, the acrosome contains digestive enzymes that initiate the degradation of the glycoprotein matrix protecting the egg and allowing the sperm plasma membrane to fuse with the egg plasma membrane, as illustrated in Figure 43.23b. The fusion of these two membranes creates an opening through which the sperm nucleus is transferred into the ovum. The nuclear membranes of the egg and sperm break down and the two haploid genomes condense to form a diploid genome.

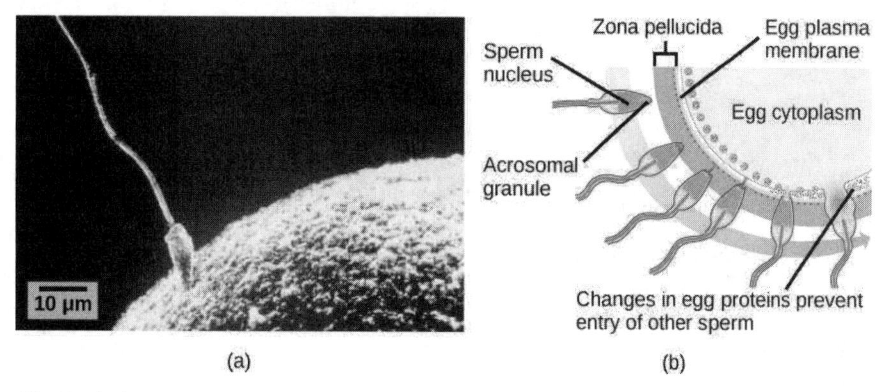

Figure 43.23 (a) Fertilization is the process in which sperm and egg fuse to form a zygote. (b) Acrosomal reactions help the sperm degrade the glycoprotein matrix protecting the egg and allow the sperm to transfer its nucleus. (credit: (b) modification of work by Mariana Ruiz Villareal; scale-bar data from Matt Russell)

To ensure that no more than one sperm fertilizes the egg, once the acrosomal reactions take place at one location of the egg membrane, the egg releases proteins in other locations to prevent other sperm from fusing with the egg. If this mechanism fails, multiple sperm can fuse with the egg, resulting in **polyspermy**. The resulting embryo is not genetically viable and dies within a few days.

Cleavage and Blastula Stage

The development of multi-cellular organisms begins from a single-celled zygote, which undergoes rapid cell division to form the blastula. The rapid, multiple rounds of cell division are termed cleavage. Cleavage is illustrated in (Figure 43.24a). After the cleavage has produced over 100 cells, the embryo is called a blastula. The blastula is usually a spherical layer of cells (the blastoderm) surrounding a fluid-filled or yolk-filled cavity (the blastocoel). Mammals at this stage form a structure called the blastocyst, characterized by an inner cell mass that is distinct from the surrounding blastula, shown in Figure 43.24b. During cleavage, the cells divide without an increase in mass; that is, one large single-celled zygote divides into multiple smaller cells. Each cell within the blastula is called a blastomere.

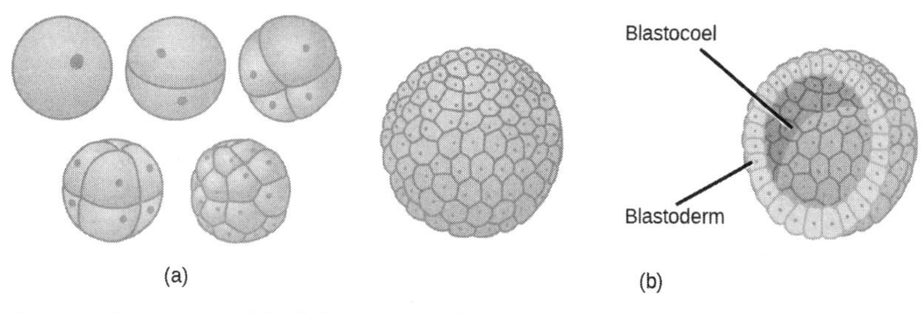

Figure 43.24 (a) During cleavage, the zygote rapidly divides into multiple cells without increasing in size. (b) The cells rearrange themselves to form a hollow ball with a fluid-filled or yolk-filled cavity called the blastula.

Cleavage can take place in two ways: **holoblastic** (total) cleavage or **meroblastic** (partial) cleavage. The type of cleavage depends on the amount of yolk in the eggs. In placental mammals (including humans) where nourishment is provided by the mother's body, the eggs have a very small amount of yolk and undergo holoblastic cleavage. Other species, such as birds, with a lot of yolk in the egg to nourish the embryo during development, undergo meroblastic cleavage.

In mammals, the blastula forms the **blastocyst** in the next stage of development. Here the cells in the blastula arrange themselves in two layers: the **inner cell mass**, and an outer layer called the **trophoblast**. The inner cell mass is also known as the embryoblast and this mass of cells will go on to form the embryo. At this stage of development, illustrated in Figure 43.25 the inner cell mass consists of embryonic stem cells that will differentiate into the different cell types needed by the organism. The trophoblast will contribute to the placenta and nourish the embryo.

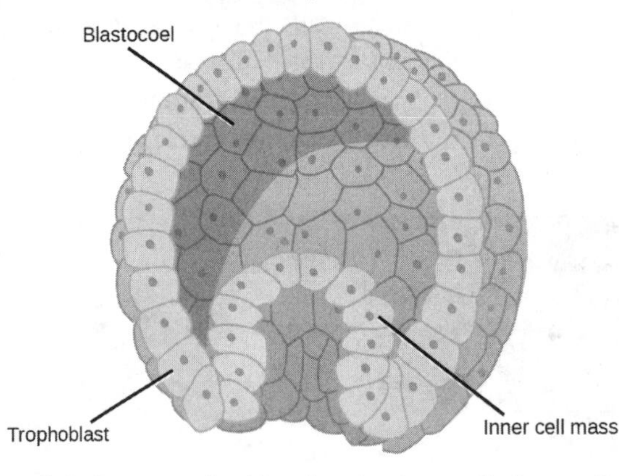

Figure 43.25 The rearrangement of the cells in the mammalian blastula to two layers—the inner cell mass and the trophoblast—results in the formation of the blastocyst.

LINK TO LEARNING

Visit the Virtual Human Embryo project (http://openstax.org/l/human_embryo) at the Endowment for Human Development site to step through an interactive that shows the stages of embryo development, including micrographs and rotating 3-D images.

Gastrulation

The typical blastula is a ball of cells. The next stage in embryonic development is the formation of the body plan. The cells in the blastula rearrange themselves spatially to form three layers of cells. This process is called **gastrulation**. During gastrulation, the blastula folds upon itself to form the three layers of cells. Each of these layers is called a germ layer and each germ layer differentiates into different organ systems.

The three germ layers, shown in Figure 43.26, are the endoderm, the ectoderm, and the mesoderm. The ectoderm gives rise to the nervous system and the epidermis. The mesoderm gives rise to the muscle cells and connective tissue in the body. The endoderm gives rise to columnar cells found in the digestive system and many internal organs.

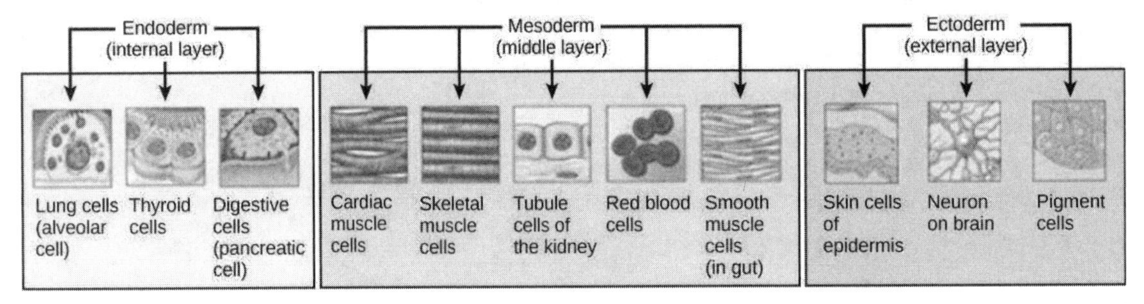

Figure 43.26 The three germ layers give rise to different cell types in the animal body. (credit: modification of work by NIH, NCBI)

Everyday Connection

Are Designer Babies in Our Future?

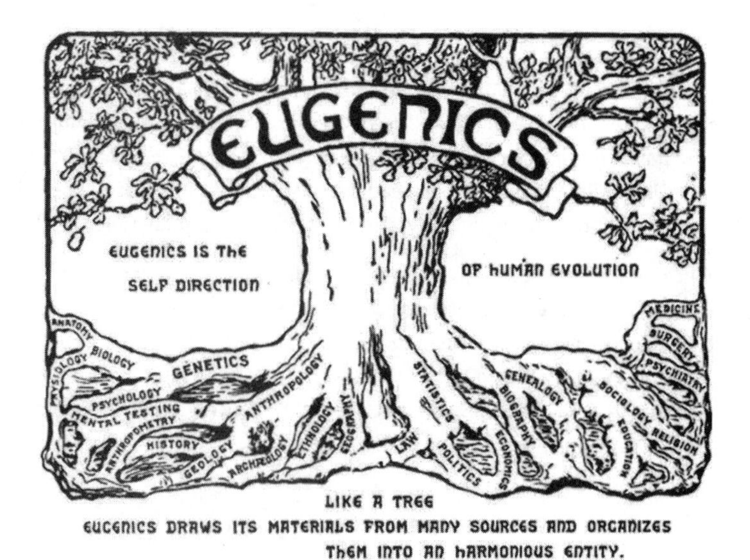

Figure 43.27 This logo from the Second International Eugenics Conference in New York City in September of 1921 shows how eugenics attempted to merge several fields of study with the goal of producing a genetically superior human race.

If you could prevent your child from getting a devastating genetic disease, would you do it? Would you select the sex of your child or select for their attractiveness, strength, or intelligence? How far would you go to maximize the possibility of resistance to disease? The genetic engineering of a human child, the production of "designer babies" with desirable phenotypic characteristics, was once a topic restricted to science fiction. This is the case no longer: science fiction is now overlapping into science fact. Many phenotypic choices for offspring are already available, with many more likely to be possible in the not too distant future. Which traits should be selected and how they should be selected are topics of much debate within the worldwide medical community. The ethical and moral line is not always clear or agreed upon, and some fear that modern reproductive technologies could lead to a new form of eugenics.

Eugenics is the use of information and technology from a variety of sources to improve the genetic makeup of the human race. The goal of creating genetically superior humans was quite prevalent (although controversial) in several countries during the early 20th century, but fell into disrepute when Nazi Germany developed an extensive eugenics program in the 1930s and 40s. As part of their program, the Nazis forcibly sterilized hundreds of thousands of the so-called "unfit" and killed tens of thousands of institutionally disabled people as part of a systematic program to develop a genetically superior race of Germans known as Aryans. Ever since, eugenic ideas have not been as publicly expressed, but there are still those who promote them.

Efforts have been made in the past to control traits in human children using donated sperm from men with desired traits. In fact, eugenicist Robert Klark Graham established a sperm bank in 1980 that included samples exclusively from donors with high IQs. The "genius" sperm bank failed to capture the public's imagination and the operation closed in 1999.

In more recent times, the procedure known as prenatal genetic diagnosis (PGD) has been developed. PGD involves the screening of human embryos as part of the process of *in vitro* fertilization, during which embryos are conceived and grown outside the mother's body for some period of time before they are implanted. The term PGD usually refers to both the diagnosis, selection, and the implantation of the selected embryos.

In the least controversial use of PGD, embryos are tested for the presence of alleles which cause genetic diseases such as sickle cell disease, muscular dystrophy, and hemophilia, in which a single disease-causing allele or pair of alleles has been identified. By excluding embryos containing these alleles from implantation into the mother, the disease is prevented, and the unused embryos are either donated to science or discarded. There are relatively few in the worldwide medical community that question the ethics of this type of procedure, which allows individuals scared to have children because of the alleles they carry to do so successfully. The major limitation to this procedure is its expense. Not usually covered by medical insurance and thus out of reach financially for most couples, only a very small percentage of all live births use such complicated methodologies. Yet, even in cases like these where the ethical issues may seem to be clear-cut, not everyone agrees with the morality of these types of procedures. For example, to those who take the position that human life begins at conception, the discarding of unused embryos, a necessary result of PGD, is unacceptable under any circumstances.

A murkier ethical situation is found in the selection of a child's sex, which is easily performed by PGD. Currently, countries such as Great Britain have banned the selection of a child's sex for reasons other than preventing sex-linked diseases. Other countries allow the procedure for "family balancing", based on the desire of some parents to have at least one child of each sex. Still others, including the United States, have taken a scattershot approach to regulating these practices, essentially leaving it to the individual practicing physician to decide which practices are acceptable and which are not.

Even murkier are rare instances of disabled parents, such as those with deafness or dwarfism, who select embryos via PGD to ensure that they share their disability. These parents usually cite many positive aspects of their disabilities and associated culture as reasons for their choice, which they see as their moral right. To others, to purposely cause a disability in a child violates the basic medical principle of *Primum non nocere,* "first, do no harm." This procedure, although not illegal in most countries, demonstrates the complexity of ethical issues associated with choosing genetic traits in offspring.

Where could this process lead? Will this technology become more affordable and how should it be used? With the ability of technology to progress rapidly and unpredictably, a lack of definitive guidelines for the use of reproductive technologies before they arise might make it difficult for legislators to keep pace once they are in fact realized, assuming the process needs any government regulation at all. Other bioethicists argue that we should only deal with technologies that exist now, and not in some uncertain future. They argue that these types of procedures will always be expensive and rare, so the fears of eugenics and "master" races are unfounded and overstated. The debate continues.

43.7 Organogenesis and Vertebrate Formation

By the end of this section, you will be able to do the following:

- Describe the process of organogenesis
- Identify the anatomical axes formed in vertebrates

Gastrulation leads to the formation of the three germ layers that give rise, during further development, to the different organs in the animal body. This process is called **organogenesis**. Organogenesis is characterized by rapid and precise movements of the cells within the embryo.

Organogenesis

Organs form from the germ layers through the process of differentiation. During differentiation, the embryonic stem cells express specific sets of genes which will determine their ultimate cell type. For example, some cells in the ectoderm will express the genes specific to skin cells. As a result, these cells will differentiate into epidermal cells. The process of differentiation is regulated by cellular signaling cascades.

Scientists study organogenesis extensively in the lab in fruit flies (*Drosophila*) and the nematode *Caenorhabditis elegans*. *Drosophila* have segments along their bodies, and the patterning associated with the segment formation has allowed scientists to study which genes play important roles in organogenesis along the length of the embryo at different time points. The nematode *C.elegans* has roughly 1000 somatic cells and scientists have studied the fate of each of these cells during their development in the nematode life cycle. There is little variation in patterns of cell lineage between individuals, unlike in mammals where cell development from the embryo is dependent on cellular cues.

In vertebrates, one of the primary steps during organogenesis is the formation of the neural system. The ectoderm forms epithelial cells and tissues, and neuronal tissues. During the formation of the neural system, special signaling molecules called growth factors signal some cells at the edge of the ectoderm to become epidermis cells. The remaining cells in the center form the neural plate. If the signaling by growth factors were disrupted, then the entire ectoderm would differentiate into neural tissue.

The neural plate undergoes a series of cell movements where it rolls up and forms a tube called the **neural tube**, as illustrated in Figure 43.28. In further development, the neural tube will give rise to the brain and the spinal cord.

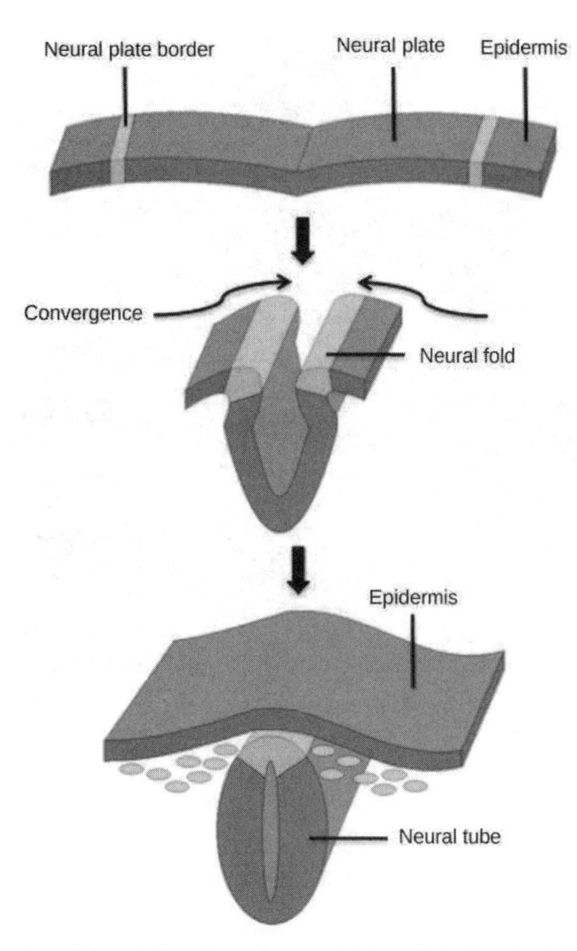

Figure 43.28 The central region of the ectoderm forms the neural tube, which gives rise to the brain and the spinal cord.

The mesoderm that lies on either side of the vertebrate neural tube will develop into the various connective tissues of the animal body. A spatial pattern of gene expression reorganizes the mesoderm into groups of cells called **somites** with spaces between them. The somites illustrated in Figure 43.29 will further develop into the cells that form the vertebrae and ribs, the dermis of the dorsal skin, the skeletal muscles of the back, and the skeletal muscles of the body wall and limbs. The mesoderm also forms a structure called the notochord, which is rod-shaped and forms the central axis of the animal body.

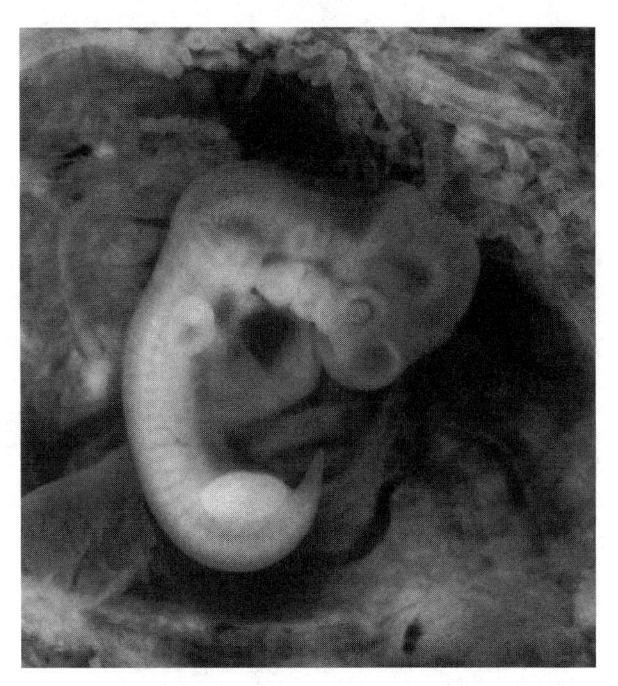

Figure 43.29 In this five-week old human embryo, somites are segments along the length of the body. (credit: modification of work by Ed Uthman)

Vertebrate Axis Formation

Even as the germ layers form, the ball of cells still retains its spherical shape. However, animal bodies have lateral-medial (left-right), dorsal-ventral (back-belly), and anterior-posterior (head-feet) axes, illustrated in Figure 43.30.

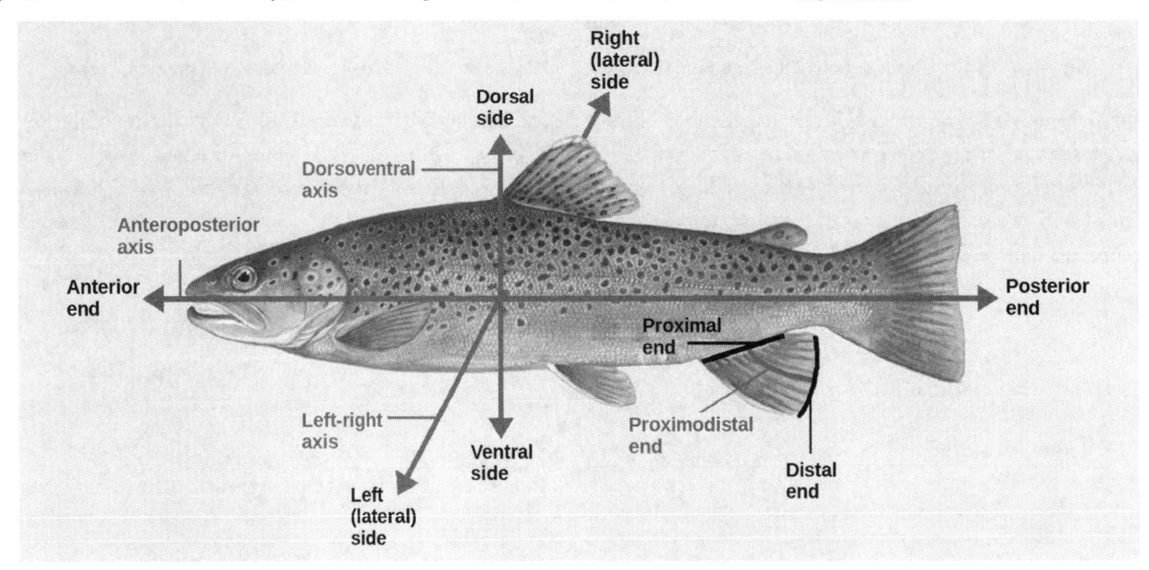

Figure 43.30 Animal bodies have three axes for symmetry. (credit: modification of work by NOAA)

How are these established? In one of the most seminal experiments ever to be carried out in developmental biology, Spemann and Mangold took dorsal cells from one embryo and transplanted them into the belly region of another embryo. They found that the transplanted embryo now had two notochords: one at the dorsal site from the original cells and another at the transplanted site. This suggested that the dorsal cells were genetically programmed to form the notochord and define the axis. Since then, researchers have identified many genes that are responsible for axis formation. Mutations in these genes leads to the loss of symmetry required for organism development.

Animal bodies have externally visible symmetry. However, the internal organs are not symmetric. For example, the heart is on the left side and the liver on the right. The formation of the central left-right axis is an important process during development. This internal asymmetry is established very early during development and involves many genes. Research is still ongoing to fully understand the developmental implications of these genes.

KEY TERMS

acrosomal reaction series of biochemical reactions that the sperm uses to break through the zona pellucida

asexual reproduction form of reproduction that produces offspring that are genetically identical to the parent

blastocyst structure formed when cells in the mammalian blastula separate into an inner and outer layer

budding form of asexual reproduction that results from the outgrowth of a part of a cell leading to a separation from the original animal into two individuals

bulbourethral gland secretion that cleanses the urethra prior to ejaculation

clitoris sensory structure in females; stimulated during sexual arousal

cloaca common body opening for the digestive, excretory, and reproductive systems found in non-mammals, such as birds

contraception (also, birth control) various means used to prevent pregnancy

estrogen reproductive hormone in females that assists in endometrial regrowth, ovulation, and calcium absorption

external fertilization fertilization of egg by sperm outside animal body, often during spawning

fission (also, binary fission) method by which multicellular organisms increase in size or asexual reproduction in which a unicellular organism splits into two separate organisms by mitosis

follicle stimulating hormone (FSH) reproductive hormone that causes sperm production in men and follicle development in women

fragmentation cutting or fragmenting of the original animal into parts and the growth of a separate animal from each part

gastrulation process in which the blastula folds over itself to form the three germ layers

gestation length of time for fetal development to birth

gonadotropin-releasing hormone (GnRH) hormone from the hypothalamus that causes the release of FSH and LH from the anterior pituitary

hermaphroditism state of having both male and female reproductive parts within the same individual

holoblastic complete cleavage; takes place in cells with a small amount of yolk

human beta chorionic gonadotropin (β-HCG) hormone produced by the chorion of the zygote that helps to maintain the corpus luteum and elevated levels of progesterone

infertility inability to conceive, carry, and deliver children

inhibin hormone made by Sertoli cells; provides negative feedback to hypothalamus in control of FSH and GnRH release

inner cell mass inner layer of cells in the blastocyst

internal fertilization fertilization of egg by sperm inside the body of the female

interstitial cell of Leydig cell in seminiferous tubules that makes testosterone

labia majora large folds of tissue covering the inguinal area

labia minora smaller folds of tissue within the labia majora

luteinizing hormone (LH) reproductive hormone in both men and women, causes testosterone production in men and ovulation and lactation in women

menopause loss of reproductive capacity in women due to decreased sensitivity of the ovaries to FSH and LH

menstrual cycle cycle of the degradation and regrowth of the endometrium

meroblastic partial cleavage; takes place in cells with a large amount of yolk

morning sickness condition in the mother during the first trimester; includes feelings of nausea

neural tube tube-like structure that forms from the ectoderm and gives rise to the brain and spinal cord

oogenesis process of producing haploid eggs

organogenesis process of organ formation

ovarian cycle cycle of preparation of egg for ovulation and the conversion of the follicle to the corpus luteum

oviduct (also, fallopian tube) muscular tube connecting the uterus with the ovary area

oviparity process by which fertilized eggs are laid outside the female's body and develop there, receiving nourishment from the yolk that is a part of the egg

ovoviparity process by which fertilized eggs are retained within the female; the embryo obtains its nourishment from the egg's yolk and the young are fully developed when they are hatched

ovulation release of the egg by the most mature follicle

parthenogenesis form of asexual reproduction where an egg develops into a complete individual without being fertilized

penis male reproductive structure for urine elimination and copulation

placenta organ that supports the diffusion of nutrients and waste between the mother's and fetus' blood

polyspermy condition in which one egg is fertilized by multiple sperm

progesterone reproductive hormone in women; assists in endometrial regrowth and inhibition of FSH and LH release

prostate gland structure that is a mixture of smooth muscle and glandular material and that contributes to semen

scrotum sac containing testes; exterior to the body

semen fluid mixture of sperm and supporting materials

seminal vesicle secretory accessory gland in males; contributes to semen

seminiferous tubule site of sperm production in testes

Sertoli cell cell in seminiferous tubules that assists developing sperm and makes inhibin

sexual reproduction mixing of genetic material from two individuals to produce genetically unique offspring

somite group of cells separated by small spaces that form from the mesoderm and give rise to connective tissue

spermatheca specialized sac that stores sperm for later use

spermatogenesis process of producing haploid sperm

testes pair of reproductive organs in males

testosterone reproductive hormone in men that assists in sperm production and promoting secondary sexual characteristics

trophoblast outer layer of cells in the blastocyst

uterus environment for developing embryo and fetus

vagina muscular tube for the passage of menstrual flow, copulation, and birth of offspring

viviparity process in which the young develop within the female, receiving nourishment from the mother's blood through a placenta

zona pellucida protective layer of glycoproteins on the mammalian egg

CHAPTER SUMMARY

43.1 Reproduction Methods

Reproduction may be asexual when one individual produces genetically identical offspring, or sexual when the genetic material from two individuals is combined to produce genetically diverse offspring. Asexual reproduction occurs through fission, budding, and fragmentation. Sexual reproduction may mean the joining of sperm and eggs within animals' bodies or it may mean the release of sperm and eggs into the environment. An individual may be one sex, or both; it may start out as one sex and switch during its life, or it may stay male or female.

43.2 Fertilization

Sexual reproduction starts with the combination of a sperm and an egg in a process called fertilization. This can occur either outside the bodies or inside the female. Both methods have advantages and disadvantages. Once fertilized, the eggs can develop inside the female or outside. If the egg develops outside the body, it usually has a protective covering over it. Animal anatomy evolved various ways to fertilize, hold, or expel the egg. The method of fertilization varies among animals. Some species release the egg and sperm into the environment, some species retain the egg and receive the sperm into the female body and then expel the developing embryo covered with shell, while still other species retain the developing offspring through the gestation period.

43.3 Human Reproductive Anatomy and Gametogenesis

As animals became more complex, specific organs and organ systems developed to support specific functions for the organism. The reproductive structures that evolved in land animals allow males and females to mate, fertilize internally, and support the growth and development of offspring. Processes developed to produce reproductive cells that had exactly half the number of chromosomes of each parent so that new combinations would have the appropriate amount of genetic material. Gametogenesis, the production of sperm (spermatogenesis) and eggs (oogenesis), takes place through the process of meiosis.

43.4 Hormonal Control of Human Reproduction

The male and female reproductive cycles are controlled by hormones released from the hypothalamus and anterior pituitary as well as hormones from reproductive tissues and organs. The hypothalamus monitors the need for the FSH and LH hormones made and released from the anterior pituitary. FSH and LH affect reproductive structures to cause the formation of sperm and the preparation of eggs for release and possible fertilization. In the male, FSH and LH stimulate Sertoli cells and interstitial cells of Leydig in the testes to facilitate sperm production. The Leydig cells produce testosterone, which also is responsible for the secondary sexual characteristics of males. In females, FSH and LH cause estrogen and progesterone to be produced. They regulate the female reproductive system which is divided into the ovarian cycle and the menstrual cycle. Menopause occurs when the ovaries lose their sensitivity to FSH and LH and the female reproductive cycles slow to a stop.

43.5 Human Pregnancy and Birth

Human pregnancy begins with fertilization of an egg and proceeds through the three trimesters of gestation. The labor process has three stages (contractions, delivery of the fetus, expulsion of the placenta), each propelled by hormones. The first trimester lays down the basic structures of the body, including the limb buds, heart, eyes, and the liver. The second trimester continues the development of all of the organs and systems. The third trimester exhibits the greatest growth of the fetus and culminates in labor and delivery. Prevention of a pregnancy can be accomplished through a variety of methods including barriers, hormones, or other means. Assisted reproductive technologies may help individuals who have infertility problems.

43.6 Fertilization and Early Embryonic Development

The early stages of embryonic development begin with fertilization. The process of fertilization is tightly controlled to ensure that only one sperm fuses with one egg. After fertilization, the zygote undergoes cleavage to form the blastula. The blastula, which in some species is a hollow ball of cells, undergoes a process called gastrulation, in which the three germ layers form. The ectoderm gives rise to the nervous system and the epidermal skin cells, the mesoderm gives rise to the muscle cells and connective tissue in the body, and the endoderm gives rise to columnar cells and internal organs.

43.7 Organogenesis and Vertebrate Formation

Organogenesis is the formation of organs from the germ layers. Each germ layer gives rise to specific tissue types. The first stage is the formation of the neural system in the ectoderm. The mesoderm gives rise to somites and the notochord. Formation of vertebrate axis is another important developmental stage.

VISUAL CONNECTION QUESTIONS

1. Figure 43.8 Which of the following statements about the male reproductive system is false?
   a. The vas deferens carries sperm from the testes to the penis.
   b. Sperm mature in seminiferous tubules in the testes.
   c. Both the prostate and the bulbourethral glands produce components of the semen.
   d. The prostate gland is located in the testes.

2. Figure 43.15 Which of the following statements about hormone regulation of the female reproductive cycle is false?
   a. LH and FSH are produced in the pituitary, and estradiol and progesterone are produced in the ovaries.
   b. Estradiol and progesterone secreted from the corpus luteum cause the endometrium to thicken.
   c. Both progesterone and estradiol are produced by the follicles.
   d. Secretion of GnRH by the hypothalamus is inhibited by low levels of estradiol but stimulated by high levels of estradiol.

3. Figure 43.17 Which of the following statements about the menstrual cycle is false?
   a. Progesterone levels rise during the luteal phase of the ovarian cycle and the secretory phase of the uterine cycle.
   b. Menstruation occurs just after LH and FSH levels peak.
   c. Menstruation occurs after progesterone levels drop.
   d. Estrogen levels rise before ovulation, while progesterone levels rise after.

REVIEW QUESTIONS

4. Which form of reproduction is thought to be best in a stable environment?
   a. asexual
   b. sexual
   c. budding
   d. parthenogenesis

5. Which form of reproduction can result from damage to the original animal?
   a. asexual
   b. fragmentation
   c. budding
   d. parthenogenesis

6. Which form of reproduction is useful to an animal with little mobility that reproduces sexually?
   a. fission
   b. budding
   c. parthenogenesis
   d. hermaphroditism

7. Genetically unique individuals are produced through ________.
   a. sexual reproduction
   b. parthenogenesis
   c. budding
   d. fragmentation

8. External fertilization occurs in which type of environment?
   a. aquatic
   b. forested
   c. savanna
   d. steppe

9. Which term applies to egg development within the female with nourishment derived from a yolk?
   a. oviparity
   b. viviparity
   c. ovoviparity
   d. ovovoparity

10. Which term applies to egg development outside the female with nourishment derived from a yolk?
    a. oviparity
    b. viviparity
    c. ovoviparity
    d. ovovoparity

11. Sperm are produced in the _______.
    a. scrotum
    b. seminal vesicles
    c. seminiferous tubules
    d. prostate gland

12. Most of the bulk of semen is made by the _______.
    a. scrotum
    b. seminal vesicles
    c. seminiferous tubules
    d. prostate gland

13. Which of the following cells in spermatogenesis is diploid?
    a. primary spermatocyte
    b. secondary spermatocyte
    c. spermatid
    d. sperm

14. Which female organ has the same embryonic origin as the penis?
    a. clitoris
    b. labia majora
    c. greater vestibular glands
    d. vagina

15. Which female organ has an endometrial lining that will support a developing baby?
    a. labia minora
    b. breast
    c. ovaries
    d. uterus

16. How many eggs are produced as a result of one meiotic series of cell divisions?
    a. one
    b. two
    c. three
    d. four

17. Which hormone causes Leydig cells to make testosterone?
    a. FSH
    b. LH
    c. inhibin
    d. estrogen

18. Which hormone causes FSH and LH to be released?
    a. testosterone
    b. estrogen
    c. GnRH
    d. progesterone

19. Which hormone signals ovulation?
    a. FSH
    b. LH
    c. inhibin
    d. estrogen

20. Which hormone causes the regrowth of the endometrial lining of the uterus?
    a. testosterone
    b. estrogen
    c. GnRH
    d. progesterone

21. Nutrient and waste requirements for the developing fetus are handled during the first few weeks by:
    a. the placenta
    b. diffusion through the endometrium
    c. the chorion
    d. the blastocyst

22. Progesterone is made during the third trimester by the:
    a. placenta
    b. endometrial lining
    c. chorion
    d. corpus luteum

23. Which contraceptive method is 100 percent effective at preventing pregnancy?
    a. condom
    b. oral hormonal methods
    c. sterilization
    d. abstinence

24. Which type of short term contraceptive method is generally more effective than others?
    a. barrier
    b. hormonal
    c. natural family planning
    d. withdrawal

25. Which hormone is primarily responsible for the contractions during labor?
    a. oxytocin
    b. estrogen
    c. β-HCG
    d. progesterone

26. Major organs begin to develop during which part of human gestation?
    a. fertilization
    b. first trimester
    c. second trimester
    d. third trimester

27. Which of the following is false?
    a. The endoderm, mesoderm, ectoderm are germ layers.
    b. The trophoblast is a germ layer.
    c. The inner cell mass is a source of embryonic stem cells.
    d. The blastula is often a hollow ball of cells.

28. During cleavage, the mass of cells:
    a. increases
    b. decreases
    c. doubles with every cell division
    d. does not change significantly

29. Which of the following gives rise to the skin cells?
    a. ectoderm
    b. endoderm
    c. mesoderm
    d. none of the above

30. The ribs form from the ________.
    a. notochord
    b. neural plate
    c. neural tube
    d. somites

CRITICAL THINKING QUESTIONS

31. Why is sexual reproduction useful if only half the animals can produce offspring and two separate cells must be combined to form a third?
32. What determines which sex will result in offspring of birds and mammals?
33. What are the advantages and disadvantages of external and internal forms of fertilization?
34. Why would paired external fertilization be preferable to group spawning?
35. Describe the phases of the human sexual response.
36. Compare spermatogenesis and oogenesis as to timing of the processes and the number and type of cells finally produced.
37. If male reproductive pathways are not cyclical, how are they controlled?
38. Describe the events in the ovarian cycle leading up to ovulation.
39. Describe the major developments during each trimester of human gestation.
40. Describe the stages of labor.
41. What do you think would happen if multiple sperm fused with one egg?
42. Why do mammalian eggs have a small concentration of yolk, while bird and reptile eggs have a large concentration of yolk?
43. Explain how the different germ layers give rise to different tissue types.
44. Explain the role of axis formation in development.

CHAPTER 44
Ecology and the Biosphere

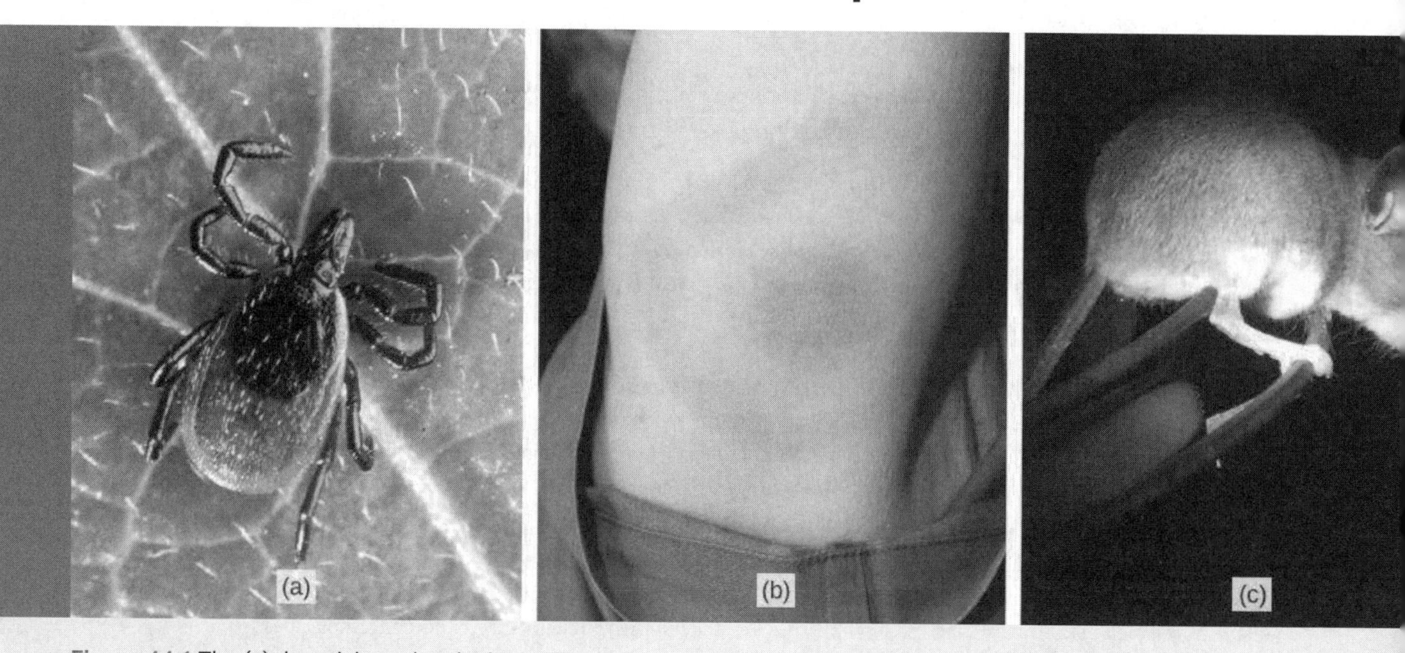

Figure 44.1 The (a) deer tick carries the bacterium that produces Lyme disease in humans, often evident in (b) a symptomatic bull's eye rash. The (c) white-footed mouse is one well-known host to deer ticks carrying the Lyme disease bacterium. (credit a: modification of work by Scott Bauer, USDA ARS; credit b: modification of work by James Gathany, CDC; credit c: modification of work by Rob Ireton)

Chapter Outline

INTRODUCTION Why study ecology? Perhaps you are interested in learning about the natural world and how living things have adapted to the physical conditions of their environment. Or, perhaps you're a future physician seeking to understand the connection between your patients' health and their environment.

Humans are a part of the ecological landscape, and human health is one important part of human interaction with our physical and living environment. Lyme disease, for instance, serves as one modern-day example of the connection between our health and the natural world (Figure 44.1). More formally known as Lyme borreliosis, Lyme disease is a bacterial infection that can be transmitted to humans when they are bitten by the deer tick (*Ixodes scapularis* in the eastern U.S., and *Ixodes pacificus* along the Pacific coast). Deer ticks are the primary *vectors* (a vector is an organism that transmits a pathogen) for this disease. However, not all ticks carry the pathogen, and not all deer ticks carry the bacteria that will cause Lyme disease in humans. Also, the ticks *I. scapularis* and *pacificus* can have other hosts besides deer. In fact, it turns out that the *probability of infection* depends on the type of host upon which the tick develops: a higher proportion of ticks that live on white-footed mice carry the bacterium than do ticks that live on deer. Knowledge about the environments and population densities in which the host species is abundant would help a physician or an epidemiologist better understand how Lyme disease is transmitted and how its incidence could be reduced.

44.1 The Scope of Ecology

By the end of this section, you will be able to do the following:

- Define ecology and the four basic levels of ecological research
- Describe examples of the ways in which ecology requires the integration of different scientific disciplines
- Distinguish between abiotic and biotic components of the environment
- Recognize the relationship between abiotic and biotic components of the environment

Ecology is the study of the interactions of living organisms with their environment. One core goal of ecology is to understand the distribution and abundance of living things in the physical environment. Attainment of this goal requires the integration of scientific disciplines inside and outside of biology, such as mathematics, statistics, biochemistry, molecular biology, physiology, evolution, biodiversity, geology, and climatology.

LINK TO LEARNING

Climate change can alter where organisms live, which can sometimes directly affect human health. Watch the PBS video "Feeling the Effects of Climate Change" (http://openstax.org/l/climate_health) in which researchers discover a pathogenic organism living far outside of its normal range.

Levels of Ecological Study

When a discipline such as biology is studied, it is often helpful to subdivide it into smaller, related areas. For instance, cell biologists interested in cell signaling need to understand the chemistry of the signal molecules (which are usually proteins) as well as the result of cell signaling. Ecologists interested in the factors that influence the survival of an endangered species might use mathematical models to predict how current conservation efforts affect endangered organisms.

To produce a sound set of management options, a *conservation biologist* needs to collect accurate data, including current population size, factors affecting reproduction (like physiology and behavior), habitat requirements (such as plants and soils), and potential human influences on the endangered population and its habitat (which might be derived through studies in sociology and urban ecology). Within the discipline of ecology, researchers work at four general levels, which sometimes overlap. These levels are organism, population, community, and ecosystem (Figure 44.2).

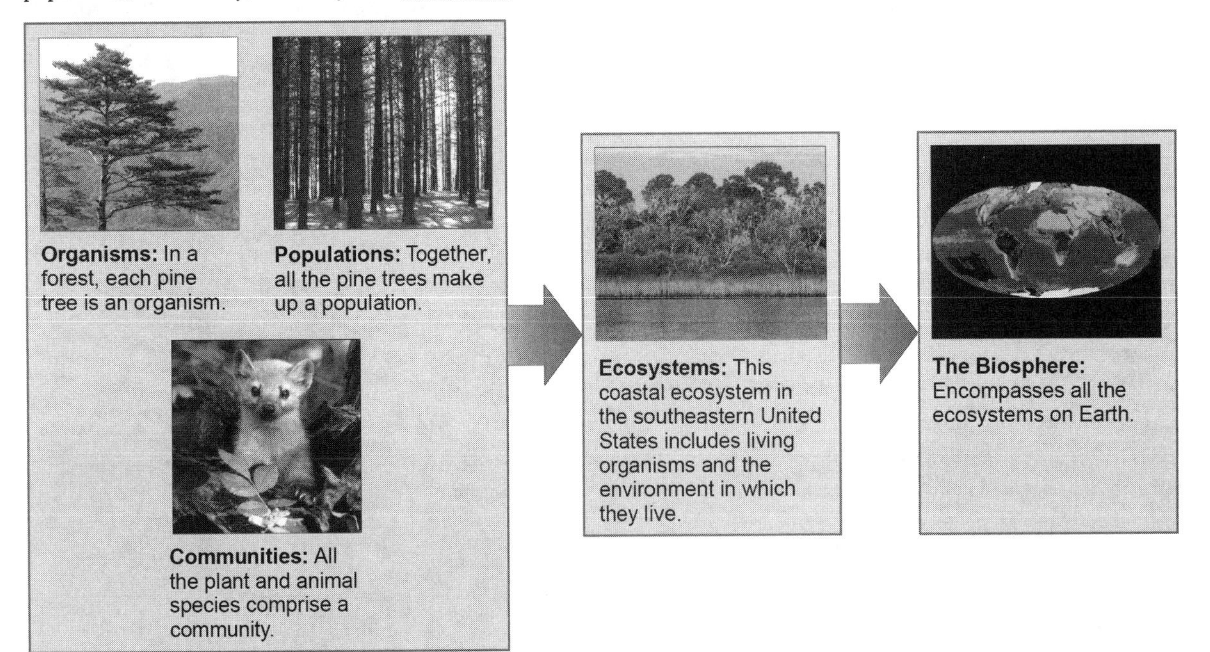

Figure 44.2 Ecologists study within several biological levels of organization. (credit "organisms": modification of work

by yeowatzup"/Flickr; credit "populations": modification of work by "Crystl"/Flickr; credit "communities": modification of work by US Fish and Wildlife Service; credit "ecosystems": modification of work by Tom Carlisle, US Fish and Wildlife Service Headquarters; credit "biosphere": NASA)

Organismal Ecology

Researchers studying ecology at the organismal level are interested in the adaptations that enable individuals to live in specific habitats. These adaptations can be morphological, physiological, and behavioral. For instance, the Karner blue butterfly (*Lycaeides melissa samuelis*) (Figure 44.3) is considered a specialist because the females only *oviposit* (that is, lay eggs) on wild lupine (*Lupinus perennis*). This specific requirement and adaptation means that the Karner blue butterfly is completely dependent on the presence of wild lupine plants for its survival.

Figure 44.3 The Karner blue butterfly (*Lycaeides melissa samuelis*) is a rare butterfly that lives only in open areas with few trees or shrubs, such as pine barrens and oak savannas. It can only lay its eggs on lupine plants. (credit: modification of work by J & K Hollingsworth, USFWS)

After hatching, the (first instar) caterpillars emerge and spend four to six weeks feeding solely on wild lupine (Figure 44.4). The caterpillars pupate as a chrysalis to undergo the final stage of metamorphosis and emerge as butterflies after about four weeks. The adult butterflies feed on the nectar of flowers of wild lupine and other plant species, such as milkweeds. Generally there are two broods of the Karner blue each year.

A researcher interested in studying Karner blue butterflies at the organismal level might, in addition to asking questions about egg laying requirements, ask questions about the butterflies' preferred thoracic flight temperature (a physiological question), or the behavior of the caterpillars when they are at different larval stages (a behavioral question).

Figure 44.4 The wild lupine (*Lupinus perennis*) is the only known host plant for the Karner blue butterfly.

Population Ecology

A **population** is a group of *interbreeding organisms* that are members of the same species living in the same area at the same time. (Organisms that are all members of the same species are called **conspecifics**.) A population is identified, in part, by where it lives, and its area of population may have natural or artificial boundaries. Natural boundaries might be rivers, mountains, or deserts, while artificial boundaries may be mowed grass, manmade structures, or roads. The study of *population ecology* focuses on the number of individuals in an area and how and why population size changes over time.

For example, population ecologists are particularly interested in counting the Karner blue butterfly because it is classified as a federally endangered species. However, the distribution and density of this species is highly influenced by the distribution and abundance of wild lupine, and the biophysical environment around it. Researchers might ask questions about the factors leading to the decline of wild lupine and how these affect Karner blue butterflies. For example, ecologists know that wild lupine thrives in open areas where trees and shrubs are largely absent. In natural settings, intermittent wildfires regularly remove trees and shrubs, helping to maintain the open areas that wild lupine requires. Mathematical models can be used to understand how wildfire suppression by humans has led to the decline of this important plant for the Karner blue butterfly.

Community Ecology

A **biological community** consists of the different species within an area, typically a three-dimensional space, and the interactions within and among these species. Community ecologists are interested in the processes driving these interactions and their consequences. Questions about *conspecific* interactions often focus on competition among members of the same species for a limited resource. Ecologists also study interactions between various species; members of different species are called **heterospecifics**. Examples of heterospecific interactions include predation, parasitism, herbivory, competition, and pollination. These interactions can have regulating effects on population sizes and can impact ecological and evolutionary processes affecting diversity.

For example, Karner blue butterfly larvae form mutualistic relationships with ants (especially *Formica* spp). **Mutualism** is a form of long-term relationship that has coevolved between two species and from which each species benefits. For mutualism to exist between individual organisms, each species must receive *some* benefit from the other as a consequence of the relationship. Researchers have shown that there is an increase in survival when ants protect Karner blue butterfly larvae (caterpillars) from predaceous insects and spiders, an act known as "tending." This might be because the larvae spend less time in each life stage when tended by ants, which provides an advantage for the larvae. Meanwhile, to attract the ants, the Karner blue butterfly larvae secrete ant-like pheromones and a carbohydrate-rich substance that is an important energy source for the ants. Both the Karner blue larvae and the ants benefit from their interaction, although the species of attendant ants may be partially opportunistic and vary over the range of the butterfly.

Ecosystem Ecology

Ecosystem ecology is an extension of organismal, population, and community ecology. The ecosystem is composed of all the **biotic** components (living things) in an area along with the **abiotic** components (nonliving things) of that area. Some of the abiotic components include air, water, and soil. Ecosystem biologists ask questions about how nutrients and energy are stored and how they move among organisms and through the surrounding atmosphere, soil, and water.

The Karner blue butterflies and the wild lupine live in an oak-pine barren habitat. This habitat is characterized by natural disturbance and nutrient-poor soils that are low in nitrogen. The availability of nutrients is an important factor in the distribution of the plants that live in this habitat. Researchers interested in ecosystem ecology could ask questions about the importance of limited resources and the movement of resources, such as nutrients, though the biotic and abiotic portions of the ecosystem.

CAREER CONNECTION

Ecologist

A career in ecology contributes to many facets of human society. Understanding ecological issues can help society meet the basic human needs of food, shelter, and health care. Ecologists can conduct their research in the laboratory and outside in natural environments (Figure 44.5). These natural environments can be as close to home as the stream running through your campus or as far away as the hydrothermal vents at the bottom of the Pacific Ocean. Ecologists manage natural resources such as white-tailed deer populations (*Odocoileus virginianus*) for hunting or aspen (*Populus* spp.) timber stands for paper production. Ecologists also work as educators who teach children and adults at various institutions including universities, high schools, museums, and nature centers. Ecologists may also work in advisory positions assisting local, state, and federal policymakers to develop laws that are ecologically sound, or they may develop those policies and legislation themselves. To become an ecologist requires at least an undergraduate degree, usually in a natural science. The undergraduate degree is often followed by specialized training or an advanced degree, depending on the area of ecology selected. Ecologists should also have a broad background in the physical sciences, as well as a solid foundation in mathematics and statistics.

Figure 44.5 This landscape ecologist is releasing a black-footed ferret into its native habitat as part of a study. (credit: USFWS Mountain Prairie Region, NPS)

LINK TO LEARNING

Visit this site (http://openstax.org/l/ecologist_role) to see Stephen Wing, a marine ecologist from the University of Otago, discuss the role of an ecologist and the types of issues ecologists explore.

44.2 Biogeography

By the end of this section, you will be able to do the following:

- Define biogeography
- List and describe abiotic factors that affect the global distribution of plant and animal species
- Compare the impact of abiotic forces on aquatic and terrestrial environments
- Summarize the effects of abiotic factors on net primary productivity

Many forces influence the communities of living organisms present in different parts of the **biosphere** (all of the parts of Earth inhabited by life). The biosphere extends into the atmosphere (several kilometers above Earth) and into the depths of the oceans. Despite its apparent vastness to an individual human, the biosphere occupies only a minute space when compared to the known universe. Many abiotic forces influence where life can exist and the types of organisms found in different parts of the biosphere. The abiotic factors influence the distribution of **biomes**: large areas of land with similar climate, flora, and fauna.

Biogeography

Biogeography is the study of the geographic distribution of living things and the *abiotic fact*ors that affect their distribution. Abiotic factors such as temperature and rainfall vary based mainly on latitude and elevation. As these abiotic factors change, the composition of plant and animal communities also changes. For example, if you were to begin a journey at the equator and walk north, you would notice gradual changes in plant communities. At the beginning of your journey, you would see tropical wet forests with broad-leaved evergreen trees, which are characteristic of plant communities found near the equator. As you continued to travel north, you would see these broad-leaved evergreen plants eventually give rise to seasonally dry forests with scattered trees. You would also begin to notice changes in temperature and moisture. At about 30 degrees north, these forests would give way to deserts, which are characterized by low precipitation and high insolation (sunlight).

Moving farther north, you would see that deserts are replaced by grasslands or prairies. Eventually, grasslands are replaced by deciduous temperate forests. These deciduous forests give way to the boreal forests and taiga found in the subarctic, the area south of the Arctic Circle. Finally, you would reach the Arctic tundra, which is found at the most northern latitudes. This trek north reveals gradual changes in both climate and the types of organisms that have adapted to environmental factors associated with ecosystems found at different latitudes. However, different ecosystems exist at the same latitude due in part to abiotic factors such as jet streams, the Gulf Stream, and ocean currents. If you were to hike up a mountain, the changes you would see in the vegetation would parallel in many ways those as you move to higher latitudes.

Ecologists who study biogeography examine patterns of *species distribution*. No species exists everywhere; for example, the Venus flytrap (*Dionaea muscipula*) is endemic to a small area in North and South Carolina. An **endemic species** is one which is naturally found only in a specific geographic area that is usually restricted in size. Other species are generalists: species which live in a wide variety of geographic areas; the raccoon (*Procyon* spp) for example, is native to most of North and Central America.

Species distribution patterns are based on biotic and abiotic factors and their influences during the very long periods of time required for species evolution; therefore, early studies of biogeography were closely linked to the emergence of evolutionary thinking in the eighteenth century. Some of the most distinctive assemblages of plants and animals occur in regions that have been physically separated for millions of years by geographic barriers. Biologists estimate that Australia, for example, has between 600,000 and 700,000 species of plants and animals. Approximately 3/4 of living plant and mammal species are endemic species found solely in Australia (Figure 44.6ab).

Figure 44.6 Australia is home to many endemic species. The (a) wallaby (*Wallabia bicolor*), a medium-sized member of the kangaroo family, is a pouched mammal, or marsupial. The (b) echidna (*Tachyglossus aculeatus*) is an egg-laying mammal. (credit a: modification of work by Derrick Coetzee; credit b: modification of work by Allan Whittome)

Sometimes ecologists discover unique patterns of species distribution by determining where species are *not* found. Despite being tropical, Hawaii, for example, has no native land species of reptiles or amphibians, only a few native species of butterflies, and only one native terrestrial mammal, the hoary bat. Most of New Guinea, as another example, lacks placental mammals.

LINK TO LEARNING

Check out this video (http://openstax.org/l/platypus) to observe a platypus swimming in its natural habitat in New South Wales, Australia.

Like animals, plants can be endemic or generalists: endemic plants are found only on specific regions of the Earth, while generalists are found on many regions. Isolated land masses—such as Australia, Hawaii, and Madagascar—often have large numbers of endemic plant species. Some of these plants are endangered due to human activity. The forest gardenia (*Gardenia brighamii*), for instance, is endemic to Hawaii; only an estimated 15–20 trees are thought to exist (Figure 44.7).

Figure 44.7 Listed as federally endangered, the forest gardenia is a small tree with distinctive flowers. It is found only in five of the Hawaiian Islands in small populations consisting of a few individual specimens. (credit: Forest & Kim Starr)

Energy Sources

Energy from the sun is captured by green plants, algae, cyanobacteria, and photosynthetic protists. These organisms convert solar energy into the chemical energy needed by all living things. Light availability can be an important force directly affecting the evolution of adaptations in photosynthesizers. For instance, plants in the understory of a temperate forest are shaded when the trees above them in the canopy completely leaf out in the late spring. Not surprisingly, understory plants have adaptations to successfully capture available light that passes through the canopy. One such adaptation is the rapid growth of spring ephemeral plants such as the spring beauty (*Claytonia virginica*) (Figure 44.8). These spring flowers achieve much of their growth and finish their life cycle (reproduce) early in the season before the trees in the canopy develop leaves.

Figure 44.8 The spring beauty is an ephemeral spring plant that flowers early in the spring to avoid competing with larger forest trees for sunlight. (credit: John Beetham)

In aquatic ecosystems, the availability of light may be limited because sunlight is absorbed by water, plants, suspended particles, and resident microorganisms. Toward the bottom of a lake, pond, or ocean, there is a zone that light cannot reach (because most wavelengths except for the shortest blues are absorbed by the water column). Photosynthesis cannot take place there and, as a result, a number of adaptations have evolved that enable living things to survive without light. For instance, aquatic plants have photosynthetic tissue near the surface of the water. You can think of the broad, floating leaves of a water lily—water lilies cannot survive without light. In environments such as hydrothermal vents, some bacteria extract energy from inorganic chemicals because there is no light for photosynthesis.

The availability of nutrients in aquatic systems such as oceans is also an important aspect of energy or photosynthesis. Many organisms sink to the bottom of the ocean when they die in the open water; when this occurs, the energy found in that living organism is sequestered for some time unless ocean upwelling occurs. **Ocean upwelling** is the rising of deep ocean waters that occurs when prevailing winds blow along surface waters near a coastline (Figure 44.9). As the wind pushes ocean waters offshore, water from the bottom of the ocean moves up to replace this water. As a result, the nutrients once contained in dead organisms become available for reuse by other living organisms.

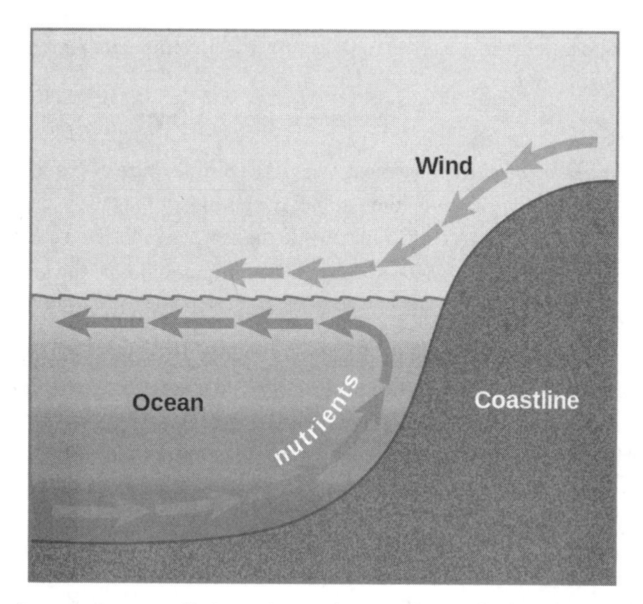

Figure 44.9 Ocean upwelling is an important process that recycles nutrients and energy in the ocean. As wind (green arrows) pushes offshore, it causes water from the ocean bottom (red arrows) to move to the surface, bringing up nutrients from the ocean depths.

In freshwater systems, such as lakes, the recycling of nutrients occurs in response to air temperature and wind changes. The nutrients at the bottom of lakes are recycled twice each year: in the spring and fall turnover. The **spring-and-fall turnover** are seasonal processes that recycle nutrients and oxygen from the bottom of a freshwater lake to the top of the lake (Figure 44.10). These turnovers are caused by the formation of a **thermocline**: layers of water with temperatures that are significantly different from those above and below it.

In wintertime, the surface of lakes found in many northern regions is frozen. However, the water under the ice is slightly warmer, and the water at the bottom of the lake is warmer yet at 4 °C to 5 °C (39.2 °F to 41 °F). Water is densest at about 4 °C; therefore, the deepest water is also the densest. The deepest water is oxygen-poor because the decomposition of organic material at the bottom of the lake uses up available oxygen that cannot be replaced by means of oxygen diffusion into the surface of the water, due to the surface ice layer.

VISUAL CONNECTION

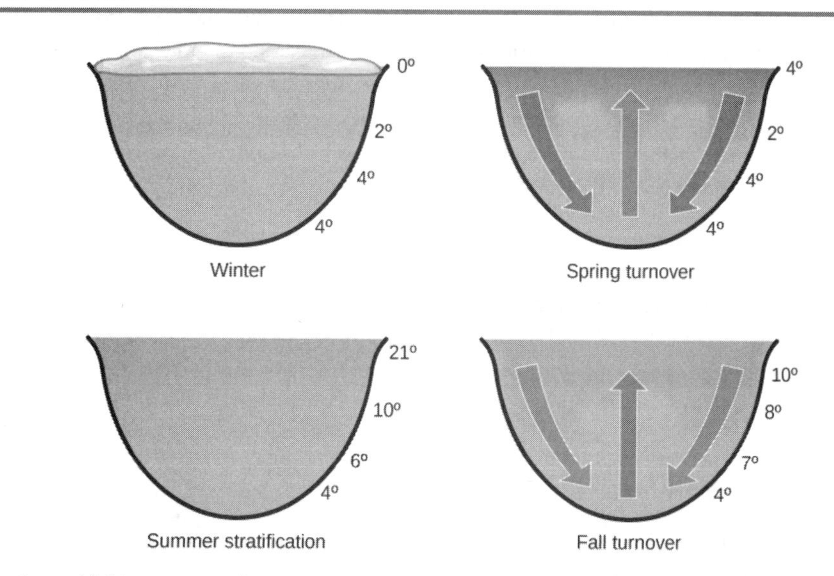

Figure 44.10 The spring and fall turnovers are important processes in freshwater lakes that act to move the nutrients and oxygen at the bottom of deep lakes to the top. Turnover occurs because water has a maximum density at 4 °C. Surface water temperature changes as the seasons progress, and denser water sinks.

How might turnover in tropical lakes differ from turnover in lakes that exist in temperate regions? Think of the variation, or lack of variation, in seasonal temperature change.

In springtime, air temperatures increase and surface ice melts. When the temperature of the surface water begins to approach 4 °C, the water becomes heavier and sinks to the bottom. The water at the bottom of the lake is then displaced by the heavier and denser surface water and, thus, rises to the top. As that water rises to the top, the sediments and nutrients from the lake bottom are brought along with it. This is called the *spring turnover*. During the summer months, the lake water stratifies, or forms layers, with the warmest water at the lake surface.

As air temperatures drop in the fall, the temperature of the lake water cools to 4 °C; therefore, this causes *fall turnover* as the heavy cold water sinks and displaces the water at the bottom. The oxygen-rich water at the surface of the lake then moves to the bottom of the lake, while the nutrients at the bottom of the lake rise to the surface (Figure 44.10). During the winter, the oxygen at the bottom of the lake is used by decomposers and other organisms requiring oxygen, such as fish. It is important to note, however, that the relative transparency of ice also allows the penetration of the shorter wavelengths of visible light so that photosynthesis, especially by algae can continue.

Temperature

Temperature affects the physiology of organisms as well as the density and state of water. Temperature exerts an important influence on living things because few living things can survive at temperatures below 0 °C (32 °F) due to metabolic constraints. It is also rare for living things to survive at temperatures exceeding 45 °C (113 °F); this is a reflection of evolutionary response to typical temperatures near the Earth's surface. Enzymes are most efficient within a narrow and specific range of temperatures; enzyme degradation can occur at higher temperatures. Therefore, organisms either must maintain an internal temperature or they must inhabit an environment that will keep the body within a temperature range that supports metabolism. Some animals have adapted to enable their bodies to survive significant temperature fluctuations, such as seen in hibernation or reptilian torpor. Similarly, some Archaea bacteria have evolved to tolerate extremely hot temperatures such as those found in the geysers within Yellowstone National Park. Such bacteria are examples of *extremophiles*: organisms that thrive in extreme environments.

The temperature (of both water and air) can limit the distribution of living things. Animals faced with temperature fluctuations may respond with adaptations, such as migration, in order to survive. **Migration**, the regular movement from one place to another, is an adaptation found in many animals, including many that inhabit seasonally cold climates. Migration solves problems related to temperature, locating food, and finding a mate. For example, the Arctic Tern (*Sterna paradisaea*) makes a 40,000 km (24,000 mi) round-trip flight each year between its feeding grounds in the southern hemisphere and its breeding grounds in the Arctic Ocean. Monarch butterflies (*Danaus plexippus*) live in the eastern and western United States in the warmer months, where they build up enormous populations, and migrate to areas around Michoacan, Mexico as well as areas along the Pacific Coast, and the southern United States in the wintertime. Some species of mammals also make migratory forays. Reindeer (*Rangifer tarandus*) travel about 5,000 km (3,100 mi) each year to find food. Amphibians and reptiles are more limited in their distribution because they generally lack migratory ability. Not all animals that could migrate do so: migration carries risk and comes at a high-energy cost.

Some animals *hibernate* or *estivate* to survive hostile temperatures. **Hibernation** enables animals to survive cold conditions, and **estivation** allows animals to survive the hostile conditions of a hot, dry climate. Animals that hibernate or estivate enter a state known as *torpor*: a condition in which their metabolic rate is significantly lowered. This enables the animal to wait until its environment better supports its survival. Some amphibians, such as the wood frog (*Rana sylvatica*), have an antifreeze-like chemical in their cells, which retains the cells' integrity and prevents them from freezing and bursting.

Water

Water is required by all living things because it is critical for cellular processes. Since terrestrial organisms lose water to the environment, they have evolved many adaptations to retain water.

- Plants have a number of interesting features on their leaves, such as leaf hairs and a waxy cuticle, that serve to decrease the rate of water loss via transpiration and convection.
- Freshwater organisms are surrounded by water and are constantly in danger of having water rush into their cells because of osmosis. Many adaptations of organisms living in freshwater environments have evolved to ensure that solute concentrations in their bodies remain within appropriate levels. One such adaptation is the excretion of dilute urine.

- Marine organisms are surrounded by water with a higher solute concentration than the organism and, thus, are in danger of losing water to the environment because of osmosis. These organisms have morphological and physiological adaptations to retain water and release solutes into the environment. For example, Marine iguanas (*Amblyrhynchus cristatus*), sneeze out water vapor that is high in salt in order to maintain solute concentrations within an acceptable range while swimming in the ocean and eating marine plants.

Inorganic Nutrients and Soil

Inorganic nutrients, such as nitrogen and phosphorus, are important in determining the distribution and the abundance of living things. Plants obtain these inorganic nutrients from the soil when water moves into the plant through the roots. Therefore, soil structure (particle size of soil components), soil pH, and soil nutrient content together all play an important role in the distribution of plants. Animals obtain inorganic nutrients from the food they consume. Therefore, animal distributions are related to the distribution of what they eat. In some cases, animals will follow their food resource as it moves through the environment.

Other Aquatic Factors

Some abiotic factors, such as oxygen, are important in aquatic ecosystems as well as terrestrial environments. Terrestrial animals obtain oxygen from the air they breathe. Oxygen availability can be an issue for organisms living at very high elevations, however, where there are fewer molecules of oxygen in the air. In aquatic systems, the concentration of dissolved oxygen is related to water temperature and the speed at which the water moves. Cold water has more dissolved oxygen than warmer water. In addition, salinity, currents, and tidal changes can be important abiotic factors in aquatic ecosystems.

Other Terrestrial Factors

Wind can be an important abiotic factor because it influences the rate of evaporation, transpiration, and convective heat loss from the surface of all organisms. The physical force of wind is also important because it can move soil, water, or other abiotic factors, as well as an ecosystem's organisms.

Fire is another terrestrial factor that can be an important agent of disturbance in terrestrial ecosystems. Some organisms are adapted to fire and, thus, require the high heat associated with fire to complete a part of their life cycle. For example, the jack pine (*Pinus banksiana*) requires heat from fire for its seed cones to open (Figure 44.11). Through the burning of pine needles, fire adds nitrogen to the soil and limits competition by destroying undergrowth.

Figure 44.11 The mature cones of the jack pine *(Pinus banksiana)* open only when exposed to high temperatures, such as during a forest fire. A fire is likely to kill most vegetation, so a seedling that germinates after a fire is more likely to receive ample sunlight than one that germinates under normal conditions. (credit: USDA)

Abiotic Factors Influencing Plant Growth

Temperature and moisture are important influences on plant production (primary productivity) and the amount of organic matter available as food (net primary productivity). **Net primary productivity** is an estimation of all of the organic matter available as food; it is calculated as the total amount of carbon fixed per year minus the amount that is oxidized during cellular respiration. In terrestrial environments, net primary productivity is estimated by measuring the **above-ground biomass** per unit area, which is the total mass of living plants, excluding roots (whose mass is very difficult to measure). This means that a large percentage of plant biomass which exists underground is not included in this measurement. Net primary productivity is an important variable when considering differences in biomes. Very productive biomes have a high level of aboveground biomass.

Annual biomass production is directly related to the abiotic components of the environment. Environments with the greatest amount of biomass produce conditions in which photosynthesis, plant growth, and the resulting net primary productivity are optimized. The climate of these areas is warm and wet. Photosynthesis can proceed at a high rate, enzymes can work most efficiently, and stomata can remain open without the risk of excessive transpiration; together, these factors lead to the maximal amount of carbon dioxide (CO_2) moving into the plant, resulting in high biomass production. The above-ground biomass produces several important resources for other living things, including habitat and food. Conversely, dry and cold environments have lower photosynthetic rates and therefore less biomass. The animal communities living there will also be affected by the decrease in available food.

44.3 Terrestrial Biomes

By the end of this section, you will be able to do the following:

- Identify the two major abiotic factors that determine terrestrial biomes
- Recognize distinguishing characteristics of each of the eight major terrestrial biomes

The Earth's biomes are categorized into two major groups: *terrestrial* and *aquatic*. **Terrestrial biomes** are based on land, while **aquatic biomes** include both ocean and freshwater biomes. The eight major terrestrial biomes on Earth are each distinguished by characteristic temperatures and amount of precipitation. Comparing the annual totals of precipitation and fluctuations in precipitation from one biome to another provides clues as to the importance of abiotic factors in the distribution of biomes. Temperature variation on a daily and seasonal basis is also important for predicting the geographic distribution of the biome and the vegetation type in the biome. The distribution of these biomes shows that the same biome can occur in geographically distinct areas with similar climates (Figure 44.12).

VISUAL CONNECTION

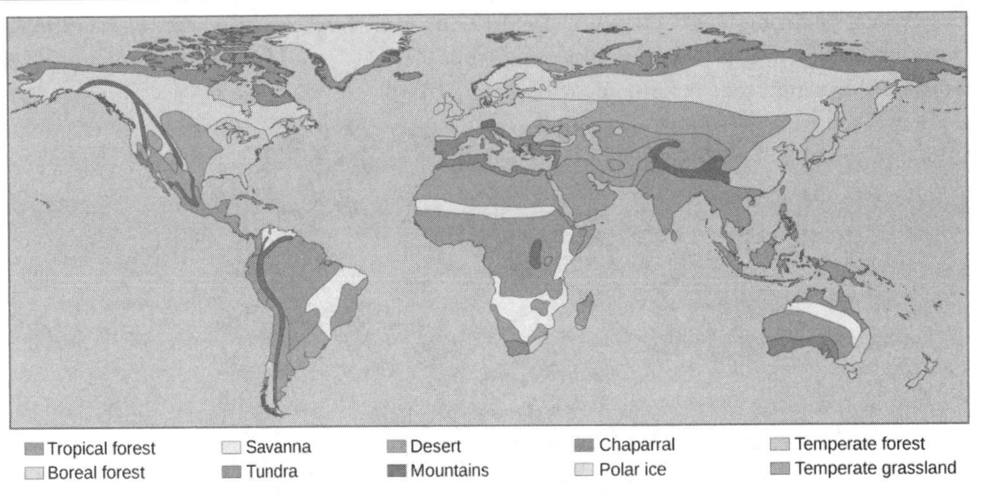

Figure 44.12 Each of the world's major biomes is distinguished by characteristic temperatures and amounts of precipitation. Polar ice and mountains are also shown.

Which of the following statements about biomes is false?

a. Chaparral is dominated by shrubs.
b. Savannas and temperate grasslands are dominated by grasses.
c. Boreal forests are dominated by deciduous trees.
d. Lichens are common in the arctic tundra.

Tropical Wet Forest

Tropical wet forests are also referred to as tropical rainforests. This biome is found in equatorial regions (Figure 44.12). The vegetation is characterized by plants with broad leaves that fall and are replaced throughout the year. Unlike the trees of deciduous forests, the trees in this biome do not have a seasonal loss of leaves associated with variations in temperature and sunlight; these forests are "evergreen" year-round.

The temperature and sunlight profiles of tropical wet forests are very stable in comparison to that of other terrestrial biomes, with the temperatures ranging from 20 °C to 34 °C (68 °F to 93 °F). When one compares the annual temperature variation of tropical wet forests with that of other forest biomes, the lack of seasonal temperature variation in the tropical wet forest becomes apparent. This lack of seasonality leads to year-round plant growth, rather than the seasonal (spring, summer, and fall) growth seen in other more temperate biomes. In contrast to other ecosystems, tropical ecosystems do not have long days and short days during the yearly cycle. Instead, a constant daily amount of sunlight (11–12 hrs per day) provides more solar radiation, thereby, a longer period of time for plant growth.

The annual rainfall in tropical wet forests ranges from 125 cm to 660 cm (50–200 in) with some monthly variation. While sunlight and temperature remain fairly consistent, annual rainfall is highly variable. Tropical wet forests typically have wet months in which there can be more than 30 cm (11–12 in) of precipitation, as well as dry months in which there are fewer than 10 cm (3.5 in) of rainfall. However, the driest month of a tropical wet forest still exceeds the *annual* rainfall of some other biomes, such as deserts.

Tropical wet forests have high net primary productivity because the annual temperatures and precipitation values in these areas are ideal for plant growth. Therefore, the extensive biomass present in the tropical wet forest leads to plant communities with very high species diversities (Figure 44.13). Tropical wet forests have more species of trees than any other biome; on average between 100 and 300 species of trees are present in a single hectare (2.5 acres) of South American Amazonian rain forest. One way to visualize this is to compare the distinctive horizontal layers within the tropical wet forest biome. On the forest floor is a sparse layer of plants and decaying plant matter. Above that is an understory of short shrubby foliage. A layer of trees rises above this understory and is topped by a closed upper **canopy**—the uppermost overhead layer of branches and leaves. Some additional trees emerge through this closed upper canopy. These layers provide diverse and complex habitats for the variety of plants, fungi, animals, and other organisms within the tropical wet forests.

For example, **epiphytes** are plants that grow on other plants, which typically are not harmed. Epiphytes are found throughout tropical wet forest biomes. Many species of animals use the variety of plants and the complex structure of the tropical wet forests for food and shelter. Some organisms live several meters above ground and have adapted to this arboreal lifestyle.

Figure 44.13 Tropical wet forests, such as these forests along the Madre de Dios river, Peru, near the Amazon River, have high species diversity. (credit: Roosevelt Garcia)

Savannas

Savannas are grasslands with scattered trees, and they are located in Africa, South America, and northern Australia (Figure 44.12). Savannas are usually hot, tropical areas with temperatures averaging from 24 °C to 29 °C (75 °F to 84 °F) and an annual rainfall of 10–40 cm (3.9–15.7 in). Savannas have an extensive dry season; for this reason, forest trees do not grow as well as they do in the tropical wet forest (or other forest biomes). As a result, within the grasses and *forbs* (herbaceous flowering plants) that dominate the savanna, there are relatively few trees (Figure 44.14). Since fire is an important source of disturbance in this biome, plants have evolved well-developed root systems that allow them to quickly resprout after a fire.

Figure 44.14 Savannas, like this one in Taita Hills Wildlife Sanctuary in Kenya, are dominated by grasses. (credit: Christopher T. Cooper)

Subtropical Deserts

Subtropical deserts exist between 15° and 30° north and south latitude and are centered on the Tropics of Cancer and Capricorn (Figure 44.12). This biome is very dry; in some years, evaporation exceeds precipitation. Subtropical hot deserts can have daytime soil surface temperatures above 60 °C (140 °F) and nighttime temperatures approaching 0 °C (32 °F). This is largely due to the lack of atmospheric water. In cold deserts, temperatures can be as high as 25 °C and can drop below -30 °C (-22 °F). Subtropical deserts are characterized by low annual precipitation of fewer than 30 cm (12 in) with little monthly variation and lack of predictability in rainfall. In some cases, the annual rainfall can be as low as 2 cm (0.8 in) in subtropical deserts located in central Australia ("the Outback") and northern Africa.

The vegetation and low animal diversity of this biome is closely related to low and unpredictable precipitation. Very dry deserts lack perennial vegetation that lives from one year to the next; instead, many plants are annuals that grow quickly and reproduce when rainfall does occur, and then die. Many other plants in these areas are characterized by having a number of adaptations that conserve water, such as deep roots, reduced foliage, and water-storing stems (Figure 44.15). Seed plants in the desert produce seeds that can be in dormancy for extended periods between rains. Adaptations in desert animals include nocturnal behavior and burrowing.

Figure 44.15 To reduce water loss, many desert plants have tiny leaves or no leaves at all. The leaves of ocotillo (*Fouquieria splendens*), shown here in the Sonora Desert near Gila Bend, Arizona, appear only after rainfall, and then are shed.

Chaparral

The chaparral is also called the scrub forest and is found in California, along the Mediterranean Sea, and along the southern coast of Australia (Figure 44.12). The annual rainfall in this biome ranges from 65 cm to 75 cm (25.6–29.5 in), and the majority of the rain falls in the winter. Summers are very dry and many chaparral plants are dormant during the summertime. The chaparral vegetation, shown in Figure 44.16, is dominated by shrubs adapted to periodic fires, with some plants producing seeds that only germinate after a hot fire. The ashes left behind after a fire are rich in nutrients like nitrogen that fertilize the soil and promote plant regrowth.

Figure 44.16 The chaparral is dominated by shrubs. (credit: Miguel Vieira)

Temperate Grasslands

Temperate grasslands are found throughout central North America, where they are also known as *prairies*; they are also in Eurasia, where they are known as *steppes* (Figure 44.12). Temperate grasslands have pronounced annual fluctuations in temperature with hot summers and cold winters. The annual temperature variation produces specific growing seasons for plants. Plant growth is possible when temperatures are warm enough to sustain plant growth and when ample water is available, which occurs in the spring, summer, and fall. During much of the winter, temperatures are low, and water, which is stored in the form of ice, is not available for plant growth.

Annual precipitation ranges from 25 cm to 75 cm (9.8–29.5 in). Because of relatively lower annual precipitation in temperate grasslands, there are few trees except for those found growing along rivers or streams. The dominant vegetation tends to consist of grasses dense enough to sustain populations of grazing animals Figure 44.17. The vegetation is very dense and the soils are fertile because the subsurface of the soil is packed with the roots and *rhizomes* (underground stems) of these grasses. The roots and rhizomes act to anchor plants into the ground and replenish the organic material (humus) in the soil when they die and decay.

Figure 44.17 The American bison (*Bison bison*), more commonly called the buffalo, is a grazing mammal that once populated American prairies in huge numbers. (credit: Jack Dykinga, USDA Agricultural Research Service)

Fires, mainly caused by lightning, are a *natural disturbance* in temperate grasslands. When fire is suppressed in temperate grasslands, the vegetation eventually converts to scrub and sometimes dense forests with drought-tolerant tree species. Often, the restoration or management of temperate grasslands requires the use of controlled burns to suppress the growth of trees and maintain the grasses.

Temperate Forests

Temperate forests are the most common biome in eastern North America, Western Europe, Eastern Asia, Chile, and New Zealand (Figure 44.12). This biome is found throughout mid-latitude regions. Temperatures range between -30 °C and 30 °C (-22 °F to 86 °F) and drop to below freezing periodically during cold winters. These temperatures mean that temperate forests have defined growing seasons during the spring, summer, and early fall. Precipitation is relatively constant throughout the year and ranges between 75 cm and 150 cm (29.5–59 in).

Because of the moderate annual rainfall and temperatures, deciduous trees are the dominant plant in this biome (Figure 44.18). Deciduous trees lose their leaves each fall and remain leafless in the winter. Thus, no photosynthesis occurs in the deciduous trees during the dormant winter period. Each spring, new leaves appear as the temperature increases. Because of the dormant period, the net primary productivity of temperate forests is less than that of tropical wet forests. In addition, temperate forests show less diversity of tree species than tropical wet forest biomes.

Figure 44.18 Deciduous trees are the dominant plant in the temperate forest. (credit: Oliver Herold)

The trees of the temperate forests leaf out and shade much of the ground; however, this biome is more open than tropical wet forests because most trees in the temperate forests do not grow as tall as the trees in tropical wet forests. The soils of the temperate forests are rich in inorganic and organic nutrients. This is due to the thick layer of leaf litter on forest floors, which does not develop in tropical rainforests. As this leaf litter decays, nutrients are returned to the soil. The leaf litter also protects soil from erosion, insulates the ground, and provides habitats for invertebrates (such as the pill bug or roly-poly, *Armadillidium vulgare*) and their predators, such as the red-backed salamander (*Plethodon cinereus*).

Boreal Forests

The **boreal forest**, also known as **taiga** or coniferous forest, is found south of the Arctic Circle and across most of Canada, Alaska, Russia, and northern Europe (Figure 44.12). This biome has cold, dry winters and short, cool, wet summers. The annual precipitation is from 40 cm to 100 cm (15.7–39 in) and usually takes the form of snow. Little evaporation occurs because of the cold temperatures.

The long and cold winters in the boreal forest have led to the predominance of cold-tolerant cone-bearing (coniferous) plants. These are evergreen coniferous trees like pines, spruce, and fir, which retain their needle-shaped leaves year-round. Evergreen trees can photosynthesize earlier in the spring than deciduous trees because less energy from the sun is required to warm a needle-like leaf than a broad leaf. This benefits evergreen trees, which grow faster than deciduous trees in the boreal forest. In addition, soils in boreal forest regions tend to be acidic with little available nitrogen. Leaves are a nitrogen-rich structure and deciduous trees must produce a new set of these nitrogen-rich structures each year. Therefore, coniferous trees that retain nitrogen-rich needles may have a competitive advantage over the broad-leafed deciduous trees.

The net primary productivity of boreal forests is lower than that of temperate forests and tropical wet forests. The above-ground biomass of boreal forests is high because these slow-growing tree species are long-lived and accumulate a large standing biomass over time. Plant species diversity is less than that seen in temperate forests and tropical wet forests. Boreal forests lack the pronounced elements of the layered forest structure seen in tropical wet forests. The structure of a boreal forest is often only a tree layer and a ground layer (Figure 44.19). When conifer needles are dropped, they decompose more slowly than broad leaves; therefore, fewer nutrients are returned to the soil to fuel plant growth.

Figure 44.19 The boreal forest (taiga) has low lying plants and conifer trees. (credit: L.B. Brubaker)

Arctic Tundra

The Arctic tundra lies north of the subarctic boreal forest and is located throughout the Arctic regions of the northern hemisphere (Figure 44.12). The average winter temperature is –34 °C (–29.2 °F) and the average summer temperature is from 3 °C to 12 °C (37 °F–52 °F). Plants in the arctic tundra have a very short growing season of approximately 10–12 weeks.

However, during this time, there are almost 24 hours of daylight and plant growth is rapid. The annual precipitation of the Arctic tundra is very low with little annual variation in precipitation. And, as in the boreal forests, there is little evaporation due to the cold temperatures.

Plants in the Arctic tundra are generally low to the ground (Figure 44.20). There is little species diversity, low net primary productivity, and low above-ground biomass. The soils of the Arctic tundra may remain in a perennially frozen state referred to as **permafrost**. The permafrost makes it impossible for roots to penetrate deep into the soil and slows the decay of organic matter, which inhibits the release of nutrients from organic matter. During the growing season, the ground of the Arctic tundra can be completely covered with plants or lichens.

Figure 44.20 Low-growing plants such as shrub willow dominate the tundra landscape, shown here in the Arctic National Wildlife Refuge. (credit: USFWS Arctic National Wildlife Refuge)

LINK TO LEARNING

Watch this *Assignment Discovery: Biomes* video (http://openstax.org/l/biomes) for an overview of biomes. To explore further, select one of the biomes on the extended playlist: desert, savanna, temperate forest, temperate grassland, tropic, tundra.

44.4 Aquatic Biomes

By the end of this section, you will be able to do the following:

- Describe the effects of abiotic factors on the composition of plant and animal communities in aquatic biomes
- Compare and contrast the characteristics of the ocean zones
- Summarize the characteristics of standing water and flowing water freshwater biomes

Abiotic Factors Influencing Aquatic Biomes

Like terrestrial biomes, aquatic biomes are influenced by a series of abiotic factors. The aquatic medium—water— has different physical and chemical properties than air, however. Even if the water in a pond or other body of water is perfectly clear (there are no suspended particles), water, on its own, absorbs light. As one descends into a deep body of water, there will eventually be a depth which the sunlight cannot reach. While there are some abiotic and biotic factors in a terrestrial ecosystem that might obscure light (like fog, dust, or insect swarms), usually these are not permanent features of the environment. The importance of light in aquatic biomes is central to the communities of organisms found in both freshwater and marine ecosystems. In freshwater systems, stratification due to differences in density is perhaps the most critical abiotic factor and is related to the energy aspects of light. The thermal properties of water (rates of heating and cooling and the ability to store much larger amounts of energy than the air) are significant to the function of marine systems and have major impacts on global climate and weather patterns. Marine systems are also influenced by large-scale physical water movements, such as currents; these are less important in most freshwater lakes.

The ocean is categorized by several areas or zones (Figure 44.21). All of the ocean's open water is referred to as the **pelagic realm** (or zone). The **benthic realm** (or zone) extends along the ocean bottom from the shoreline to the deepest parts of the ocean floor. Within the pelagic realm is the **photic zone**, which is the portion of the ocean that light can penetrate (approximately 200 m or 650 ft). At depths greater than 200 m, light cannot penetrate; thus, this is referred to as the **aphotic zone**. The majority of the ocean is aphotic and lacks sufficient light for photosynthesis. The deepest part of the ocean, the Challenger Deep (in the Mariana Trench, located in the western Pacific Ocean), is about 11,000 m (about 6.8 mi) deep. To give some perspective on the depth of this trench, the ocean is, on average, 4267 m or 14,000 ft deep. These realms and zones are relevant to freshwater lakes as well.

VISUAL CONNECTION

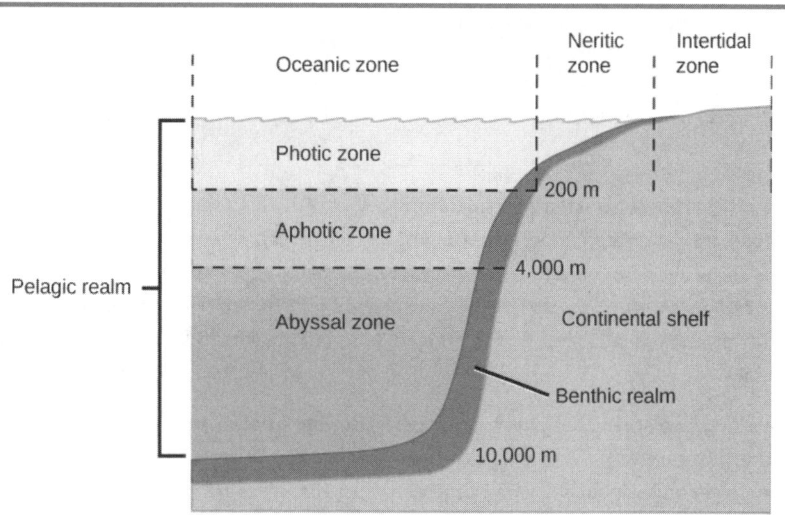

Figure 44.21 The ocean is divided into different zones based on water depth and distance from the shoreline.

In which of the following regions would you expect to find photosynthetic organisms?

a. the aphotic zone, the neritic zone, the oceanic zone, and the benthic realm
b. the photic zone, the intertidal zone, the neritic zone, and the oceanic zone
c. the photic zone, the abyssal zone, the neritic zone, and the oceanic zone
d. the pelagic realm, the aphotic zone, the neritic zone, and the oceanic zone

Marine Biomes

The ocean is the largest marine biome. It is a continuous body of salt water that is relatively uniform in chemical composition; in fact, it is a weak solution of mineral salts and decayed biological matter. Within the ocean, coral reefs are a second kind of marine biome. Estuaries, coastal areas where salt water and fresh water mix, form a third unique marine biome.

Ocean

The physical diversity of the ocean is a significant influence on plants, animals, and other organisms. The ocean is categorized into different zones based on how far light reaches into the water. Each zone has a distinct group of species adapted to the biotic and abiotic conditions particular to that zone.

The **intertidal zone**, which is the zone between high and low tide, is the oceanic region that is closest to land (Figure 44.21). Generally, most people think of this portion of the ocean as a sandy beach. In some cases, the intertidal zone is indeed a sandy beach, but it can also be rocky or muddy. The intertidal zone is an extremely variable environment because of action of tidal ebb and flow. Organisms are exposed to air and sunlight at low tide and are underwater most of the time, especially during high tide. Therefore, living things that thrive in the intertidal zone are adapted to being dry for long periods of time. The shore of the intertidal zone may also be repeatedly struck by waves, and the organisms found there are adapted to withstand damage from their pounding action (Figure 44.22). The exoskeletons of shoreline crustaceans (such as the shore crab, *Carcinus maenas*) are tough and protect them from desiccation (drying out) and wave damage. Another consequence of the pounding waves is that few algae and plants establish themselves in the constantly moving rocks, sand, or mud.

Figure 44.22 Sea urchins, mussel shells, and starfish are often found in the intertidal zone, shown here in Kachemak Bay, Alaska. (credit: NOAA)

The **neritic zone** (Figure 44.21) extends from the intertidal zone to depths of about 200 m (or 650 ft) at the edge of the *continental shelf* (the underwater landmass that extends from a continent). Since light can penetrate this depth, photosynthesis can still occur in the neritic zone. The water here contains silt and is well-oxygenated, low in pressure, and stable in temperature. Phytoplankton and floating ***Sargassum*** (a type of free-floating marine seaweed) provide a habitat for some sea life found in the neritic zone. Zooplankton, protists, small fishes, and shrimp are found in the neritic zone and are the base of the food chain for most of the world's fisheries.

Beyond the neritic zone is the open ocean area known as the **pelagic** or **open oceanic zone** (Figure 44.21). Within the oceanic zone there is *thermal stratification* where warm and cold waters mix because of ocean currents. Abundant plankton serve as the base of the food chain for larger animals such as whales and dolphins. Nutrients are scarce and this is a relatively less productive part of the marine biome. When photosynthetic organisms and the protists and animals that feed on them die, their bodies fall to the bottom of the ocean, where they remain. Unlike freshwater lakes, most of the open ocean lacks a process for bringing the organic nutrients back up to the surface. (Exceptions include major oceanic upwellings within the Humboldt Current along the western coast of South America.) The majority of organisms in the aphotic zone include sea cucumbers (phylum Echinodermata) and other organisms that survive on the nutrients contained in the dead bodies of organisms in the photic zone.

Beneath the pelagic zone is the **benthic realm**, the deep-water region beyond the continental shelf (Figure 44.21). The bottom of the benthic realm is composed of sand, silt, and dead organisms. Temperature decreases, remaining above freezing, as water depth increases. This is a nutrient-rich portion of the ocean because of the dead organisms that fall from the upper layers of the ocean. Because of this high level of nutrients, a diversity of fungi, sponges, sea anemones, marine worms, sea stars, fishes, and bacteria exist.

The deepest part of the ocean is the **abyssal zone**, which is at depths of 4000 m or greater. The abyssal zone (Figure 44.21) is very cold and has very high pressure, high oxygen content, and low nutrient content. There are a variety of invertebrates and fishes found in this zone, but the abyssal zone does not have plants because of the lack of light. Hydrothermal vents are found primarily in the abyssal zone; chemosynthetic bacteria utilize the hydrogen sulfide and other minerals emitted from the vents. These chemosynthetic bacteria use the hydrogen sulfide as an energy source and serve as the base of the food chain found in the abyssal zone.

Coral Reefs

Coral reefs are *ocean ridges* formed by marine invertebrates, comprising mostly cnidarians and molluscs, living in warm shallow waters within the photic zone of the ocean. They are found within 30° north and south of the equator. The **Great Barrier Reef** is perhaps the best-known and largest reef system in the world—visible from the International Space Station! This massive and ancient reef is located several miles off the northeastern coast of Australia. Other coral reef systems are fringing islands, which are directly adjacent to land, or atolls, which are circular reef systems surrounding a former landmass that is now underwater. The coral organisms (members of phylum Cnidaria) are colonies of saltwater polyps that secrete a calcium carbonate skeleton. These calcium-rich skeletons slowly accumulate, forming the underwater reef (Figure 44.23). Corals found in shallower waters (at a depth of approximately 60 m or about 200 ft) have a mutualistic relationship with photosynthetic unicellular algae. The relationship provides corals with the majority of the nutrition and the energy they require. The waters in which these corals live are nutritionally poor and, without this mutualism, it would not be possible for large corals to grow. Some corals living in deeper and colder water do not have a mutualistic relationship with algae; these corals attain energy and nutrients using stinging cells called cnidocytes on their tentacles to capture prey.

LINK TO LEARNING

Watch this National Oceanic and Atmospheric Administration (NOAA) video (http://openstax.org/l/marine_biology) to see marine ecologist Dr. Peter Etnoyer discuss his research on coral organisms.

It is estimated that more than 4,000 fish species inhabit coral reefs. These fishes can feed on coral, the **cryptofauna** (invertebrates found within the calcium carbonate substrate of the coral reefs), or the seaweed and algae that are associated with the coral. In addition, some fish species inhabit the boundaries of a coral reef; these species include predators, herbivores, and **planktivores**, which consume planktonic organisms such as bacteria, archaea, algae, and protists floating in the pelagic zone.

Figure 44.23 Coral reefs are formed by the calcium carbonate skeletons of coral organisms, which are marine invertebrates in the phylum

Cnidaria. (credit: Terry Hughes)

EVOLUTION CONNECTION

Global Decline of Coral Reefs

It takes many thousands of years to build a coral reef. The animals that create coral reefs have evolved over millions of years, continuing to slowly deposit the calcium carbonate that forms their characteristic ocean homes. Bathed in warm tropical waters, the coral animals and their symbiotic algal partners evolved to survive at the upper limit of ocean water temperature.

Together, climate change and human activity pose dual threats to the long-term survival of the world's coral reefs. As global warming due to fossil fuel emissions raises ocean temperatures, coral reefs are suffering. The excessive warmth causes the reefs to lose their symbiotic, food-producing algae, resulting in a phenomenon known as **bleaching**. When bleaching occurs, the reefs lose much of their characteristic color as the algae and the coral animals die if loss of the symbiotic zooxanthellae is prolonged.

Rising levels of atmospheric carbon dioxide further threaten the corals in other ways; as CO_2 dissolves in ocean waters, it lowers the pH and increases ocean acidity. As acidity increases, it interferes with the calcification that normally occurs when coral animals build their calcium carbonate shelters.

When a coral reef begins to die, species diversity plummets as animals lose food and shelter. Coral reefs are also economically important tourist destinations, so the decline of coral reefs poses a serious threat to coastal economies.

Human population growth has damaged corals in other ways, too. As human coastal populations increase, the runoff of sediment and agricultural chemicals has increased, as well, causing some of the once-clear tropical waters to become cloudy. At the same time, overfishing of popular fish species has allowed the predator species that eat corals to go unchecked.

Although a rise in global temperatures of 1–2 °C (a conservative scientific projection) in the coming decades may not seem large, it is very significant to this biome. When change occurs rapidly, species can become extinct before evolution can offer new adaptations. Many scientists believe that global warming, with its rapid (in terms of evolutionary time) and inexorable increases in temperature, is tipping the balance beyond the point at which many of the world's coral reefs can recover.

Estuaries: Where the Ocean Meets Fresh Water

Estuaries are biomes that occur where a source of fresh water, such as a river, meets the ocean. Therefore, both fresh water and salt water are found in the same vicinity; mixing results in a diluted (*brackish*) saltwater. Estuaries form protected areas where many of the young offspring of crustaceans, molluscs, and fish begin their lives, which also creates important breeding grounds for other animals. Salinity is a very important factor that influences the organisms and the adaptations of the organisms found in estuaries. The salinity of estuaries varies considerably and is based on the rate of flow of its freshwater sources, which may depend on the seasonal rainfall. Once or twice a day, high tides bring salt water into the estuary. Low tides occurring at the same frequency reverse the current of salt water.

The short-term and rapid variation in salinity due to the mixing of fresh water and salt water is a difficult physiological challenge for the plants and animals that inhabit estuaries. Many estuarine plant species are halophytes: plants that can tolerate salty conditions. Halophytic plants are adapted to deal with the salinity resulting from saltwater on their roots or from sea spray. In some halophytes, filters in the roots remove the salt from the water that the plant absorbs. Other plants are able to pump oxygen into their roots. Animals, such as mussels and clams (phylum Mollusca), have developed behavioral adaptations that expend a lot of energy to function in this rapidly changing environment. When these animals are exposed to low salinity, they stop feeding, close their shells, and switch from aerobic respiration (in which they use gills to remove oxygen from the water) to anaerobic respiration (a process that does not require oxygen and takes place in the cytoplasm of the animal's cells). When high tide returns to the estuary, the salinity and oxygen content of the water increases, and these animals open their shells, begin feeding, and return to aerobic respiration.

Freshwater Biomes

Freshwater biomes include lakes and ponds (standing water) as well as rivers and streams (flowing water). They also include wetlands, which will be discussed later. Humans rely on freshwater biomes to provide ecosystem benefits, which are aquatic resources for drinking water, crop irrigation, sanitation, and industry. Lakes and ponds are connected with abiotic and biotic factors influencing their terrestrial biomes.

Lakes and Ponds

Lakes and ponds can range in area from a few square meters to thousands of square kilometers. Temperature is an important abiotic factor affecting living things found in lakes and ponds. In the summer, as we have seen, thermal stratification of lakes and ponds occurs when the upper layer of water is warmed by the sun and does not mix with deeper, cooler water. Light can penetrate within the photic zone of the lake or pond. **Phytoplankton** (algae and cyanobacteria) are found here and carry out photosynthesis, providing the base of the food web of lakes and ponds. **Zooplankton**, such as rotifers and larvae and adult crustaceans, consume these phytoplankton. At the bottom of lakes and ponds, bacteria in the aphotic zone break down dead organisms that sink to the bottom.

Nitrogen and phosphorus are important limiting nutrients in lakes and ponds. Because of this, they are the determining factors in the amount of phytoplankton growth that takes place in lakes and ponds. When there is a large input of nitrogen and phosphorus (from sewage and runoff from fertilized lawns and farms, for example), the growth of algae skyrockets, resulting in a large accumulation of algae called an **algal bloom**. Algal blooms (Figure 44.24) can become so extensive that they reduce light penetration in water. They may also release toxic byproducts into the water, contaminating any drinking water taken from that source. In addition, the lake or pond becomes aphotic, and photosynthetic plants cannot survive. When the algae die and decompose, severe oxygen depletion of the water occurs. Fishes and other organisms that require oxygen are then more likely to die, resulting in a dead zone. Lake Erie and the Gulf of Mexico represent freshwater and marine habitats where phosphorus control and storm water runoff pose significant environmental challenges.

Figure 44.24 The uncontrolled growth of algae in this lake has resulted in an algal bloom. (credit: Jeremy Nettleton)

Rivers and Streams

Rivers and streams are continuously moving bodies of water that carry large amounts of water from the source, or *headwater*, to a lake or ocean. The largest rivers include the Nile River in Africa, the Amazon River in South America, and the Mississippi River in North America.

Abiotic features of rivers and streams vary along the length of the river or stream. Streams begin at a point of origin referred to as **source water**. The source water is usually cold, low in nutrients, and clear. The **channel** (the width of the river or stream) is narrower than at any other place along the length of the river or stream. Because of this, the current is often faster here than at any other point of the river or stream.

The fast-moving water results in minimal silt accumulation at the bottom of the river or stream; therefore, the water is usually clear and free of debris. Photosynthesis here is mostly attributed to algae that are growing on rocks; the swift current inhibits the growth of phytoplankton. An additional input of energy can come from leaves and other organic material that fall downstream into the river or stream, as well as from trees and other plants that border the water. When the leaves decompose, the organic material and nutrients in the leaves are returned to the water. Plants and animals have adapted to this fast-moving water. For instance, leeches (phylum Annelida) have elongated bodies and suckers on the anterior and ventral areas of the body. These suckers attach to the substrate, keeping the leech anchored in place, and are also used to attach to their prey. Freshwater trout species (phylum Chordata) are an important predator in these fast-moving rivers and streams.

As the river or stream flows away from the source, the width of the channel gradually widens and the current slows. This slow-moving water, caused by the gradient decrease and the volume increase as tributaries unite, has more sedimentation.

Phytoplankton can also be suspended in slow-moving water. Therefore, the water will not be as clear as it is near the source. The water is also warmer. Worms (phylum Annelida) and insects (phylum Arthropoda) can be found burrowing into the mud. The higher order predator vertebrates (phylum Chordata) include waterfowl, frogs, and fishes. These predators must find food in these slow moving, sometimes murky, waters and, unlike the trout in the waters at the source, these vertebrates may not be able to use vision as their primary sense to find food. Instead, they are more likely to use taste or chemical cues to find prey.

Wetlands

Wetlands are environments in which the soil is either permanently or periodically saturated with water. Wetlands are different from lakes because wetlands are shallow bodies of water whereas lakes vary in depth. **Emergent vegetation** consists of wetland plants that are rooted in the soil but have portions of leaves, stems, and flowers extending above the water's surface. There are several types of wetlands including marshes, swamps, bogs, mudflats, and salt marshes (Figure 44.25). The three shared characteristics among these types—what makes them wetlands—are their hydrology, hydrophytic vegetation, and hydric soils.

Figure 44.25 Located in southern Florida, Everglades National Park is vast array of wetland environments, including sawgrass marshes, cypress swamps, and estuarine mangrove forests. Here, a great egret walks among cypress trees. (credit: NPS)

Freshwater marshes and swamps are characterized by slow and steady water flow. Bogs, however, develop in depressions where water flow is low or nonexistent. Bogs usually occur in areas where there is a clay bottom with poor percolation of water. (Percolation is the movement of water through the pores in the soil or rocks.) The water found in a bog is stagnant and oxygen-depleted because the oxygen used during the decomposition of organic matter is not readily replaced. As the oxygen in the water is depleted, decomposition slows. This leads to a buildup of acids and a lower water pH. The lower pH creates challenges for plants because it limits the available nitrogen. As a result, some bog plants (such as sundews, pitcher plants, and Venus flytraps) capture insects in order to extract the nitrogen from their bodies. Bogs have low net primary productivity because the water found in bogs has low levels of nitrogen and oxygen.

44.5 Climate and the Effects of Global Climate Change

By the end of this section, you will be able to do the following:

- Define global climate change
- Summarize the effects of the Industrial Revolution on global atmospheric carbon dioxide concentration
- Describe three natural factors affecting long-term global climate
- List two or more greenhouse gases and describe their role in the greenhouse effect

All biomes are *universally affected* by global conditions, such as climate, that ultimately shape each biome's environment. Scientists who study climate have noted a series of marked changes that have gradually become increasingly evident during the last sixty years. **Global climate change** is the term used to describe altered global weather patterns, especially a worldwide increase in temperature and resulting changes in the climate, due largely to rising levels of atmospheric carbon dioxide.

Climate and Weather

A common misconception about global climate change is that a specific weather event occurring in a particular region (for example, a very cool week in June in central Indiana) provides evidence of global climate change. However, a cold week in June is a weather-related event and not a climate-related one. These misconceptions often arise because of confusion over the terms *climate* and *weather*.

Climate refers to the long-term, *predictable atmospheric conditions* of a specific area. The climate of a biome is characterized by having consistent seasonal temperature and rainfall ranges. Climate does not address the amount of rain that fell on one particular day in a biome or the colder-than-average temperatures that occurred on one day. In contrast, **weather** refers to the conditions of the atmosphere during a short period of time. Weather forecasts are usually made for 48-hour cycles. Long-range weather forecasts are available but can be unreliable.

To better understand the difference between climate and weather, imagine that you are planning an outdoor event in northern Wisconsin. You would be thinking about *climate* when you plan the event in the summer rather than the winter because you have long-term knowledge that any given Saturday in the months of May to August would be a better choice for an outdoor event in Wisconsin than any given Saturday in January. However, you cannot determine the specific day that the event should be held on because it is difficult to accurately predict the weather on a specific day. Climate can be considered "average" weather that takes place over many years.

Global Climate Change

Climate change can be understood by approaching three areas of study:

- evidence of current and past global climate change
- drivers of global climate change
- documented results of climate change

It is helpful to keep these three different aspects of climate change clearly separated when consuming media reports about global climate change. We should note that it is common for reports and discussions about global climate change to confuse the data showing that Earth's climate is changing with the factors that drive this climate change.

Evidence for Global Climate Change

Since scientists cannot go back in time to directly measure climatic variables, such as average temperature and precipitation, they must instead indirectly measure temperature. To do this, scientists rely on *historical evidence of Earth's past climate.*

Antarctic ice cores are a key example of such evidence for climate change. These ice cores are samples of *polar ice* obtained by means of drills that reach thousands of meters into ice sheets or high mountain glaciers. Viewing the ice cores is like traveling backwards through time; the deeper the sample, the earlier the time period. Trapped within the ice are air bubbles and other biological evidence that can reveal temperature and carbon dioxide data. Antarctic ice cores have been collected and analyzed to indirectly estimate the temperature of the Earth over the past 400,000 years (Figure 44.26a). The 0 °C on this graph refers to the long-term average. Temperatures that are greater than 0 °C exceed Earth's long-term average temperature. Conversely, temperatures that are less than 0 °C are less than Earth's average temperature. This figure shows that there have been periodic cycles of increasing and decreasing temperature.

(a) (b)

Figure 44.26 Scientists drill for ice cores in polar regions. The ice contains air bubbles and biological substances that provide important information for researchers. (credit: a: Helle Astrid Kjær; b: National Ice Core Laboratory, USGS)

Before the late 1800s, the Earth has been as much as 9 °C cooler and about 3 °C warmer. Note that the graph in Figure 44.27b shows that the atmospheric concentration of carbon dioxide has also risen and fallen in periodic cycles. Also note the relationship between carbon dioxide concentration and temperature. Figure 44.27b shows that carbon dioxide levels in the atmosphere have historically cycled between 180 and 300 parts per million (ppm) by volume.

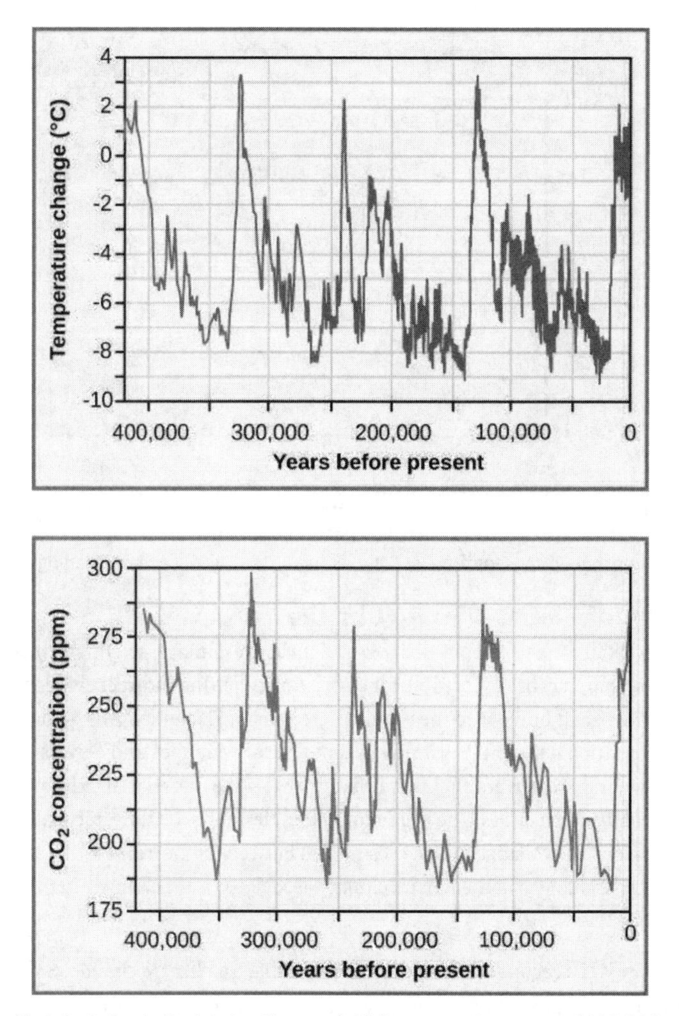

Figure 44.27 Ice at the Russian Vostok station in East Antarctica was laid down over the course of 420,000 years and reached a depth of over 3,000 m. By measuring the amount of CO_2 trapped in the ice, scientists have determined past atmospheric CO_2 concentrations. Temperatures relative to modern day were determined from the amount of *deuterium* (a nonradioactive isotope of hydrogen) present.

Figure 44.27a does not show the last 2,000 years with enough detail to compare the changes of Earth's temperature during the last 400,000 years with the temperature change that has occurred in the more recent past. Two significant temperature anomalies, or *irregularities*, have occurred in the last 2,000 years. These are the *Medieval Climate Anomaly* (or the Medieval Warm Period) and the *Little Ice Age*. A third temperature anomaly aligns with the *Industrial Era*. The Medieval Climate Anomaly occurred between 900 and 1300 AD. During this time period, many climate scientists think that slightly warmer weather conditions prevailed in many parts of the world; the higher-than-average temperature changes varied between 0.10 °C and 0.20 °C above the norm. Although 0.10 °C does not seem large enough to produce any noticeable change, it did free seas of ice. Because of this warming, the Vikings were able to colonize Greenland.

The **Little Ice Age** was a cold period that occurred between 1550 AD and 1850 AD. During this time, a slight cooling of a little less than 1 °C was observed in North America, Europe, and possibly other areas of the Earth. This 1 °C change in global temperature is a seemingly small deviation in temperature (as was observed during the Medieval Climate Anomaly); however, it also resulted in noticeable climatic changes. Historical accounts reveal a time of exceptionally harsh winters with much snow and frost.

The *Industrial Revolution*, which began around 1750, was characterized by changes in much of human society. Advances in agriculture increased the food supply, which improved the standard of living for people in Europe and the United States. New technologies were invented that provided jobs and cheaper goods. These new technologies were powered using fossil fuels, especially coal. The Industrial Revolution starting in the early nineteenth century ushered in the beginning of the Industrial Era. When a fossil fuel is burned, carbon dioxide is released. With the beginning of the Industrial Era, atmospheric carbon dioxide began to rise (Figure 44.28).

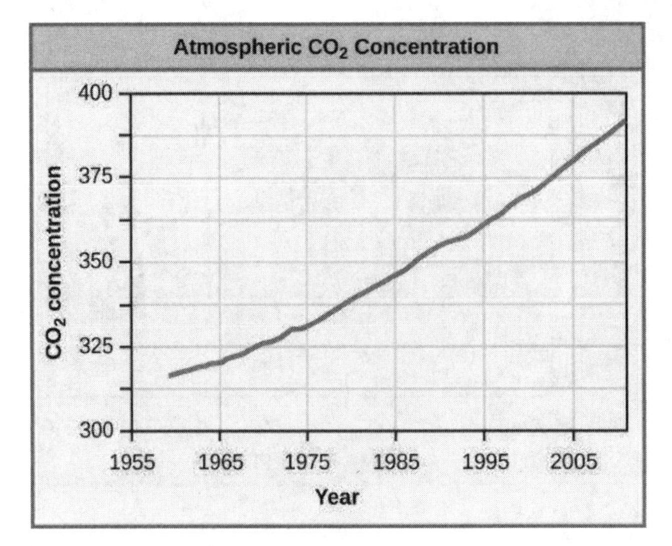

Figure 44.28 The atmospheric concentration of CO_2 has risen steadily since the beginning of industrialization.

Current and Past Drivers of Global Climate Change

Because it is not possible to go back in time to directly observe and measure climate, scientists must use *indirect evidence* to determine the *drivers*, or factors, that may be responsible for climate change. The indirect evidence includes data collected using ice cores, *boreholes* (a narrow shaft bored into the ground), tree rings, glacier lengths, pollen remains, and ocean sediments. The data shows a correlation between the timing of temperature changes and drivers of climate change. Before the Industrial Era (pre-1780), there were three drivers of climate change that were not related to human activity or atmospheric gases. The first of these is the *Milankovitch cycles*. The **Milankovitch cycles** describe the effects of slight changes in the Earth's orbit on Earth's climate. The length of the Milankovitch cycles ranges between 19,000 and 100,000 years. In other words, one could expect to see some predictable changes in the Earth's climate associated with changes in the Earth's orbit at a minimum of every 19,000 years.

The *variation in the sun's intensity* is the second natural factor responsible for climate change. **Solar intensity** is the amount of solar power or energy the sun emits in a given amount of time. There is a direct relationship between solar intensity and temperature. As solar intensity increases (or decreases), the Earth's temperature correspondingly increases (or decreases). Changes in solar intensity have been proposed as one of several possible explanations for the Little Ice Age.

Finally, *volcanic eruptions* are a third natural driver of climate change. Volcanic eruptions can last a few days, but the solids and gases released during an eruption can influence the climate over a period of a few years, causing short-term climate changes. The gases and solids released by volcanic eruptions can include carbon dioxide, water vapor, sulfur dioxide, hydrogen sulfide, hydrogen, and carbon monoxide. Generally, volcanic eruptions cool the climate. This occurred in 1783 when volcanos in Iceland erupted and caused the release of large volumes of sulfuric oxide. This led to **haze-effect cooling**, a global phenomenon that occurs when dust, ash, or other suspended particles block out sunlight and trigger lower global temperatures as a result; haze-effect cooling usually extends for one or more years before dissipating in intensity. In Europe and North America, haze-effect cooling produced some of the lowest average winter temperatures on record in 1783 and 1784.

Greenhouse gases are probably the most significant drivers of the climate. When heat energy from the sun strikes the Earth, gases known as **greenhouse gases** trap the heat in the atmosphere, in a similar manner as do the glass panes of a greenhouse keep heat from escaping. The greenhouse gases that affect Earth include carbon dioxide, methane, water vapor, nitrous oxide, and ozone. Approximately half of the radiation from the sun passes through these gases in the atmosphere and strikes the Earth. This radiation is converted into thermal (infrared) radiation on the Earth's surface, and then a portion of that energy is re-radiated back into the atmosphere. Greenhouse gases, however, reflect much of the thermal energy back to the Earth's surface. The more greenhouse gases there are in the atmosphere, the more thermal energy is reflected back to the Earth's surface, heating it up and the atmosphere immediately above it. Greenhouse gases absorb and emit radiation and are an important factor in the **greenhouse effect**: the warming of Earth due to carbon dioxide and other greenhouse gases in the atmosphere.

Direct evidence supports the relationship between atmospheric concentrations of carbon dioxide and temperature: as carbon dioxide rises, global temperature rises. Since 1950, the concentration of atmospheric carbon dioxide has increased from about 280 ppm to 382 ppm in 2006. In 2011, the atmospheric carbon dioxide concentration was 392 ppm. However, the planet would not be inhabitable by current life forms if water vapor did not produce its drastic greenhouse warming effect.

Scientists look at patterns in data and try to explain differences or deviations from these patterns. The atmospheric carbon dioxide data reveal a historical pattern of carbon dioxide increasing and decreasing, cycling between a low of 180 ppm and a high of 300 ppm. Scientists have concluded that it took around 50,000 years for the atmospheric carbon dioxide level to increase from its low minimum concentration to its higher maximum concentration. However, beginning only a few centuries ago, atmospheric carbon dioxide concentrations have increased beyond the historical maximum of 300 ppm. The current increases in atmospheric carbon dioxide have happened very quickly—in a matter of hundreds of years rather than thousands of years. What is the reason for this difference in the rate of change and the amount of increase in carbon dioxide? A key factor that must be recognized when comparing the historical data and the current data is the presence and industrial activities of modern human society; no other driver of climate change has yielded changes in atmospheric carbon dioxide levels *at this rate or to this magnitude*.

Human activity releases carbon dioxide and methane, two of the most important greenhouse gases, into the atmosphere in several ways. The primary mechanism that releases carbon dioxide is the burning of fossil fuels, such as gasoline, coal, and natural gas (Figure 44.29). Deforestation, cement manufacture, animal agriculture, the clearing of land, and the burning of forests are other human activities that release carbon dioxide. Methane (CH_4) is produced when bacteria break down organic matter under anaerobic conditions. Anaerobic conditions can happen when organic matter is trapped underwater (such as in rice paddies) or in the intestines of herbivores. Methane can also be released from natural gas fields and the decomposition of animal and plant material that occurs in landfills. Another source of methane is the melting of *clathrates*. **Clathrates** are frozen chunks of ice and methane found at the bottom of the ocean. When water warms, these chunks of ice melt and methane is released. As the ocean's water temperature increases, the rate at which clathrates melt is increasing, releasing even more methane. This leads to increased levels of methane in the atmosphere, which further accelerates the rate of global warming. This is an example of the positive feedback loop that is leading to the rapid rate of increase of global temperatures.

Figure 44.29 The burning of fossil fuels in industry and by vehicles releases carbon dioxide and other greenhouse gases into the atmosphere. (credit: "Pöllö"/Wikimedia Commons)

Documented Results of Climate Change: Past and Present

Scientists have geological evidence of the consequences of long-ago climate change. Modern-day phenomena such as retreating glaciers and melting polar ice cause a continual rise in sea level. Meanwhile, changes in climate can negatively affect organisms.

Geological Climate Change

Global warming has been associated with at least one planet-wide extinction event during the geological past. The **Permian extinction** event occurred about 251 million years ago toward the end of the roughly 50-million-year-long geological time span known as the Permian period. This geologic time period was one of the three warmest periods in Earth's geologic history. Scientists estimate that approximately 70 percent of the terrestrial plant and animal species and 84 percent of marine species became extinct, vanishing forever near the end of the Permian period.

Organisms that had adapted to wet and warm climatic conditions, such as annual rainfall of 300–400 cm (118–157 in) and 20 °C–30 °C (68 °F–86 °F) in the tropical wet forest, may not have been able to survive the Permian climate change.

LINK TO LEARNING

Watch this NASA video (http://openstax.org/l/climate_plants) to discover the mixed effects of global warming on plant growth. While scientists found that warmer temperatures in the 1980s and 1990s caused an increase in plant productivity, this advantage has since been counteracted by more frequent droughts.

Present Climate Change

A number of global events have occurred that may be attributed to climate change during our lifetimes. *Glacier National Park* in Montana is undergoing the retreat of many of its glaciers, a phenomenon known as glacier recession. In 1850, the area contained approximately 150 glaciers. By 2010, however, the park contained only about 24 glaciers greater than 25 acres in size. One of these glaciers is the *Grinnell Glacier* (Figure 44.30) at Mount Gould. Between 1966 and 2005, the size of Grinnell Glacier shrank by 40 percent. Similarly, the mass of the ice sheets in Greenland and the Antarctic is decreasing: Greenland lost 150–250 km^3 of ice per year between 2002 and 2006. In addition, the size and thickness of the Arctic sea ice is decreasing.

Figure 44.30 The effect of global warming can be seen in the continuing retreat of Grinnel Glacier. The mean annual temperature in the park has increased 1.33 °C since 1900. The loss of a glacier results in the loss of summer meltwaters, sharply reducing seasonal water supplies and severely affecting local ecosystems. (credit: modification of work by USGS)

This loss of ice is leading to increases in the global sea level. On average, the sea is rising at a rate of 1.8 mm per year. However, between 1993 and 2010 the rate of sea level increase ranged between 2.9 and 3.4 mm per year. A variety of factors affect the volume of water in the ocean, especially the temperature of the water (the density of water is related to its temperature: water volume expands as it warms, thus raising sea levels), as well as the *amount of water* found in rivers, lakes, glaciers, polar ice caps, and sea ice. As glaciers and polar ice caps melt, there is a significant contribution of liquid water that was previously frozen.

In addition to some abiotic conditions changing in response to climate change, many organisms are also being affected by the changes in temperature. Temperature and precipitation play key roles in determining the *geographic distribution* and *phenology* of plants and animals. (**Phenology** is the study of the effects of climatic conditions on the timing of periodic life cycle events, such as flowering in plants or migration in birds.) Researchers have shown that 385 plant species in Great Britain are flowering 4.5 days sooner than was recorded earlier during the previous 40 years. In addition, insect-pollinated species were more likely to flower earlier than wind-pollinated species. The impact of changes in flowering date would be mitigated if the insect pollinators emerged earlier. This mismatched timing of plants and pollinators could result in injurious ecosystem effects because, for continued survival, insect-pollinated plants must flower when their pollinators are present.

KEY TERMS

abiotic nonliving components of the environment
above-ground biomass total mass of aboveground living plants per area
abyssal zone deepest part of the ocean at depths of 4000 m or greater
algal bloom rapid increase of algae in an aquatic system
aphotic zone part of the ocean where no light penetrates
benthic realm (also, benthic zone) part of the ocean that extends along the ocean bottom from the shoreline to the deepest parts of the ocean floor
biogeography study of the geographic distribution of living things and the abiotic factors that affect their distribution
biome ecological community of plants, animals, and other organisms that is adapted to a characteristic set of environmental conditions
biotic living components of the environment
canopy branches and foliage of trees that form a layer of overhead coverage in a forest
channel width of a river or stream from one bank to the other bank
clathrates frozen chunks of ice and methane found at the bottom of the ocean
climate long-term, predictable atmospheric conditions present in a specific area
conspecifics individuals that are members of the same species
coral reef ocean ridges formed by marine invertebrates living in warm, shallow waters within the photic zone
cryptofauna invertebrates found within the calcium carbonate substrate of coral reefs
ecology study of interaction between living things and their environment
ecosystem services human benefits and services provided by natural ecosystems
emergent vegetation wetland plants that are rooted in the soil but have portions of leaves, stems, and flowers extending above the water's surface
endemic species found only in a specific geographic area that is usually restricted in size
estuary biomes where a source of fresh water, such as a river, meets the ocean
fall and spring turnover seasonal process that recycles nutrients and oxygen from the bottom of a freshwater ecosystem to the top
global climate change altered global weather patterns, including a worldwide increase in temperature, due largely to rising levels of atmospheric carbon dioxide
greenhouse effect warming of Earth due to carbon dioxide and other greenhouse gases in the atmosphere
greenhouse gases atmospheric gases such as carbon dioxide and methane that absorb and emit radiation, thus trapping heat in Earth's atmosphere
haze-effect cooling effect of the gases and solids from a volcanic eruption on global climate
heterospecifics individuals that are members of different species
intertidal zone part of the ocean that is closest to land; parts extend above the water at low tide
Milankovitch cycles cyclic changes in the Earth's orbit that may affect climate
neritic zone part of the ocean that extends from low tide to the edge of the continental shelf
net primary productivity measurement of the energy accumulation within an ecosystem, calculated as the total amount of carbon fixed per year minus the amount that is oxidized during cellular respiration
ocean upwelling rising of deep ocean waters that occurs when prevailing winds blow along surface waters near a coastline
oceanic zone part of the ocean that begins offshore where the water measures 200 m deep or deeper
pelagic realm (also, pelagic zone) open ocean waters that are not close to the bottom or near the shore
permafrost perennially frozen portion of the Arctic tundra soil
photic zone portion of the ocean that light can penetrate
planktivore animal species that eats plankton
predator organism that kills and consumes another organism
Sargassum type of free-floating marine seaweed
solar intensity amount of solar power energy the sun emits in a given amount of time
source water point of origin of a river or stream
thermocline layer of water with a temperature that is significantly different from that of the surrounding layers
weather conditions of the atmosphere during a short period of time

CHAPTER SUMMARY

44.1 The Scope of Ecology

Ecology is the study of the interactions of living things with their environment. Ecologists ask questions that comprise four levels of general biological organization—organismal, population, community, and ecosystem. At the organismal level, ecologists study individual organisms and how they interact with their environments. At the population and community levels, ecologists explore, respectively, how a population of organisms changes over time and the ways in which that population interacts with other species in the community. Ecologists studying an ecosystem examine the living species (the biotic components) of the ecosystem as well as the nonliving portions (the abiotic components), such as air, water, and soil, of the environment.

44.2 Biogeography

Biogeography is the study of the geographic distribution of living things as well as the abiotic factors that affect their distribution. Endemic species are species that are naturally found only in a specific geographic area. The distribution of living things is influenced by several environmental factors that are, in part, controlled by the latitude or elevation at which a species is found. Ocean upwellings, and spring and fall turnovers are important processes regulating the distribution of nutrients and other abiotic factors important in aquatic ecosystems. Energy sources, temperature, water, inorganic nutrients, and soil are factors limiting the distribution of living things in terrestrial systems. Net primary productivity is a measure of the amount of biomass produced by a biome.

44.3 Terrestrial Biomes

The Earth has terrestrial biomes and aquatic biomes. Aquatic biomes include both freshwater and marine environments. There are eight major terrestrial biomes: tropical wet forests, savannas, subtropical deserts, chaparral, temperate grasslands, temperate forests, boreal forests, and Arctic tundra. The same biome can occur in different geographic locations with similar climates. Temperature and precipitation, and variations in both, are key abiotic factors that shape the composition of animal and plant communities in terrestrial biomes. Some biomes, such as temperate grasslands and temperate forests, have distinct seasons, with cold weather and hot weather alternating throughout the year. In warm, moist biomes, such as the tropical wet forest, net primary productivity is high, as warm temperatures, abundant water, and a year-round growing season fuel plant growth and supply energy for high diversity throughout the food web. Other biomes, such as deserts and tundras, have low primary productivity due to extreme temperatures and a shortage of available water.

44.4 Aquatic Biomes

Aquatic ecosystems include both saltwater and freshwater biomes. The abiotic factors important for the structuring of aquatic ecosystems can be different than those seen in terrestrial systems. Sunlight is a driving force behind the structure of forests and also is an important factor in bodies of water, especially those that are very deep, because of the role of photosynthesis in sustaining certain organisms.

Density and temperature shape the structure of aquatic systems. Oceans may be thought of as consisting of different zones based on water depth and distance from the shoreline and light penetrance. Different kinds of organisms are adapted to the conditions found in each zone. Coral reefs are unique marine ecosystems that are home to a wide variety of species. Estuaries are found where rivers meet the ocean; their shallow waters provide nourishment and shelter for young crustaceans, mollusks, fishes, and many other species. Freshwater biomes include lakes, ponds, rivers, streams, and wetlands. Bogs are an interesting type of wetland characterized by standing water, lower pH, and a lack of nitrogen.

44.5 Climate and the Effects of Global Climate Change

The Earth has gone through periodic cycles of increases and decreases in temperature. During the past 2,000 years, the Medieval Climate Anomaly was a warmer period, while the Little Ice Age was unusually cool. Both of these irregularities can be explained by natural causes of changes in climate, and, although the temperature changes were small, they had significant effects. Natural drivers of climate change include Milankovitch cycles, changes in solar activity, and volcanic eruptions. None of these factors, however, leads to rapid increases in global temperature or sustained increases in carbon dioxide.

The burning of fossil fuels is an important source of greenhouse gases, which play a major role in the greenhouse effect. Two hundred and fifty million years ago, global warming resulted in the Permian extinction: a large-scale extinction event that is documented in the fossil record. Currently, modern-day climate change is associated with the increased melting of glaciers and polar ice sheets, resulting in a gradual increase in sea level. Plants and animals can also be affected by global climate change when the timing of seasonal events, such as flowering or pollination, is affected by global warming.

VISUAL CONNECTION QUESTIONS

1. Figure 44.10 How might turnover in tropical lakes differ from turnover in lakes that exist in temperate regions? Think of the variation, or lack of variation, in seasonal temperature change.

2. Figure 44.12 Which of the following statements about biomes is false?
   a. Chaparral is dominated by shrubs.
   b. Savannas and temperate grasslands are dominated by grasses.
   c. Boreal forests are dominated by deciduous trees.
   d. Lichens are common in the arctic tundra.

3. Figure 44.21 In which of the following regions would you expect to find photosynthetic organisms?
   a. the aphotic zone, the neritic zone, the oceanic zone, and the benthic realm
   b. the photic zone, the intertidal zone, the neritic zone, and the oceanic zone
   c. the photic zone, the abyssal zone, the neritic zone, and the oceanic zone
   d. the pelagic realm, the aphotic zone, the neritic zone, and the oceanic zone

REVIEW QUESTIONS

4. Which of the following is a biotic factor?
   a. wind
   b. disease-causing microbe
   c. temperature
   d. soil particle size

5. The study of nutrient cycling though the environment is an example of which of the following?
   a. organismal ecology
   b. population ecology
   c. community ecology
   d. ecosystem ecology

6. Understory plants in a temperate forest have adaptations to capture limited _______.
   a. water
   b. nutrients
   c. heat
   d. sunlight

7. An ecologist hiking up a mountain may notice different biomes along the way due to changes in all of the following except:
   a. elevation
   b. rainfall
   c. latitude
   d. temperature

8. Which of the following biomes is characterized by abundant water resources?
   a. deserts
   b. boreal forests
   c. savannas
   d. tropical wet forests

9. Which of the following biomes is characterized by short growing seasons?
   a. deserts
   b. tropical wet forests
   c. Arctic tundras
   d. savannas

10. Where would you expect to find the most photosynthesis in an ocean biome?
    a. aphotic zone
    b. abyssal zone
    c. benthic realm
    d. intertidal zone

11. A key feature of estuaries is:
    a. low light conditions and high productivity
    b. salt water and fresh water
    c. frequent algal blooms
    d. little or no vegetation

12. Which of the following is an example of a weather event?
    a. The hurricane season lasts from June 1 through November 30.
    b. The amount of atmospheric CO_2 has steadily increased during the last century.
    c. A windstorm blew down trees in the Boundary Waters Canoe Area in Minnesota on July 4, 1999.
    d. Deserts are generally dry ecosystems having very little rainfall.

13. Which of the following natural forces is responsible for the release of carbon dioxide and other atmospheric gases?
    a. the Milankovitch cycles
    b. volcanoes
    c. solar intensity
    d. burning of fossil fuels

CRITICAL THINKING QUESTIONS

14. Ecologists often collaborate with other researchers interested in ecological questions. Describe the levels of ecology that would be easier for collaboration because of the similarities of questions asked. What levels of ecology might be more difficult for collaboration?
15. The population is an important unit in ecology as well as other biological sciences. How is a population defined, and what are the strengths and weaknesses of this definition? Are there some species that at certain times or places are not in populations?
16. Compare and contrast ocean upwelling and spring and fall turnovers.
17. Many endemic species are found in areas that are geographically isolated. Suggest a plausible scientific explanation for why this is so.
18. The extremely low precipitation of subtropical desert biomes might lead one to expect fire to be a major disturbance factor; however, fire is more common in the temperate grassland biome than in the subtropical desert biome. Why is this?
19. In what ways are the subtropical desert and the arctic tundra similar?
20. Scientists have discovered the bodies of humans and other living things buried in bogs for hundreds of years, but not yet decomposed. Suggest a possible biological explanation for why such bodies are so well-preserved.
21. Describe the conditions and challenges facing organisms living in the intertidal zone.
22. Compare and contrast how natural- and human-induced processes have influenced global climate change.
23. Predict possible consequences if carbon emissions from fossil fuels continue to rise.

CHAPTER 45
Population and Community Ecology

Figure 45.1 Asian carp jump out of the water in response to electrofishing. The Asian carp in the photograph on the right were harvested from the Little Calumet River in Illinois in May, 2010, using rotenone, a toxin often used as an insecticide, in an effort to learn more about the population of the species. (credit left image: modification of work by USGS; credit right image: modification of work by Lt. David French, USCG)

INTRODUCTION Imagine sailing down a river in a small motorboat on a weekend afternoon; the water is smooth and you are enjoying the warm sunshine and cool breeze when suddenly you are hit in the head by a 20-pound silver carp. This is now a risk on many rivers and canal systems in Illinois and Missouri because of the presence of Asian carp.

This fish—actually a group of species including the silver, black, grass, and big head carp—has been farmed and eaten in China for over 1000 years. It is one of the most important aquaculture food resources worldwide. In the United States, however, Asian carp is considered a dangerous invasive species that disrupts community structure and composition to the point of threatening native species.

Chapter Outline

45.1 Population Demography

By the end of this section, you will be able to do the following:

- Describe how ecologists measure population size and density
- Describe three different patterns of population distribution
- Use life tables to calculate mortality rates
- Describe the three types of survivorship curves and relate them to specific populations

Populations are dynamic entities. Populations consist all of the species living within a specific area, and populations fluctuate based on a number of factors: seasonal and yearly changes in the environment, natural disasters such as forest fires and volcanic eruptions, and competition for resources between and within species. The statistical study of population dynamics, **demography**, uses a series of mathematical tools to investigate how populations respond to changes in their biotic and abiotic environments. Many of these tools were originally designed to study human populations. For example, **life tables**, which detail the life expectancy of individuals within a population, were initially developed by life insurance companies to set insurance rates. In fact, while the term "demographics" is commonly used when discussing humans, all living populations can be studied using this approach.

Population Size and Density

The study of any population usually begins by determining how many individuals of a particular species exist, and how closely associated they are with each other. Within a particular habitat, a population can be characterized by its **population size (*N*)**, the total number of individuals, and its **population density**, the number of individuals within a specific area or volume. Population size and density are the two main characteristics used to describe and understand populations. For example, populations with more individuals may be more stable than smaller populations based on their genetic variability, and thus their potential to adapt to the environment. Alternatively, a member of a population with low population density (more spread out in the habitat), might have more difficulty finding a mate to reproduce compared to a population of higher density. As is shown in Figure 45.2, smaller organisms tend to be more densely distributed than larger organisms.

VISUAL CONNECTION

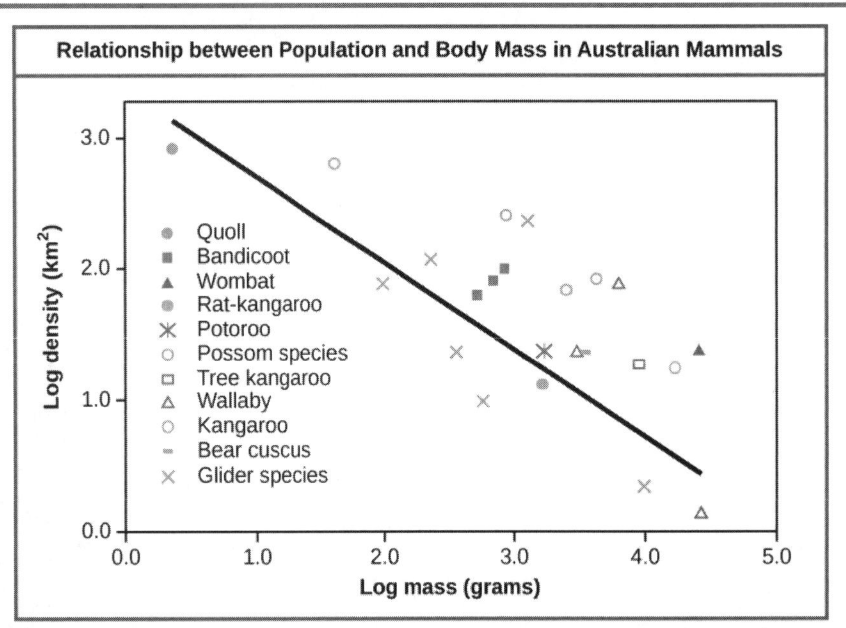

Figure 45.2 Australian mammals show a typical inverse relationship between population density and body size.

As this graph shows, population density typically decreases with increasing body size. Why do you think this is the case?

Population Research Methods

The most accurate way to determine population size is to simply count all of the individuals within the habitat. However, this method is often not logistically or economically feasible, especially when studying large habitats. Thus, scientists usually study populations by sampling a representative portion of each habitat and using this data to make inferences about the habitat as a whole. A variety of methods can be used to sample populations to determine their size and density. For immobile organisms such as plants, or for very small and slow-moving organisms, a **quadrat** may be used (Figure 45.3). A quadrat is a way of marking off square areas within a habitat, either by staking out an area with sticks and string, or by the use of a wood, plastic, or metal square placed on the ground. After setting the quadrats, researchers then count the number of individuals that lie within their boundaries. Multiple quadrat samples are performed throughout the habitat at several random locations to estimate the population size and density within the entire habitat. The number and size of quadrat samples depends on the type of organisms under study and other factors, including the density of the organism. For example, if sampling daffodils, a 1 m^2 quadrat might be used. With giant redwoods, on the other hand, a larger quadrat of 100 m^2 might be employed. This ensures that enough individuals of the species are counted to get an accurate sample that correlates with the habitat, including areas not sampled.

Figure 45.3 A scientist uses a quadrat to measure population size and density. (credit: NPS Sonoran Desert Network)

For mobile organisms, such as mammals, birds, or fish, scientists use a technique called **mark and recapture**. This method involves marking a sample of captured animals in some way (such as tags, bands, paint, or other body markings), and then releasing them back into the environment to allow them to mix with the rest of the population. Later, researchers collect a new sample, including some individuals that are marked (recaptures) and some individuals that are unmarked (Figure 45.4).

Figure 45.4 Mark and recapture is used to measure the population size of mobile animals such as (a) bighorn sheep, (b) the California condor, and (c) salmon. (credit a: modification of work by Neal Herbert, NPS; credit b: modification of work by Pacific Southwest Region USFWS; credit c: modification of work by Ingrid Taylar)

Using the ratio of marked and unmarked individuals, scientists determine how many individuals are in the sample. From this, calculations are used to estimate the total population size. This method assumes that the larger the population, the lower the percentage of tagged organisms that will be recaptured since they will have mixed with more untagged individuals. For example, if 80 deer are captured, tagged, and released into the forest, and later 100 deer are captured and 20 of them are already marked, we can estimate the population size (N) using the following equation:

$$\frac{(\text{number marked first catch x total number of second catch})}{\text{number marked second catch}} = N$$

Using our example, the population size would be estimated at 400.

$$\frac{(80 \times 100)}{20} = 400$$

Therefore, there are an estimated 400 total individuals in the original population.

There are some limitations to the mark and recapture method. Some animals from the first catch may learn to avoid capture in the second round, thus inflating population estimates. Alternatively, some animals may prefer to be retrapped (especially if a food reward is offered), resulting in an underestimate of population size. Also, some species may be harmed by the marking technique, reducing their survival. A variety of other techniques have been developed, including the electronic tracking of animals tagged with radio transmitters and the use of data from commercial fishing and trapping operations to estimate the size and health of populations and communities.

Species Distribution

In addition to measuring simple density, further information about a population can be obtained by looking at the distribution of the individuals. **Species dispersion patterns** (or distribution patterns) show the spatial relationship between members of a population within a habitat at a particular point in time. In other words, they show whether members of the species live close together or far apart, and what patterns are evident when they are spaced apart.

Individuals in a population can be equally spaced apart, dispersed randomly with no predictable pattern, or clustered in groups. These are known as uniform, random, and clumped dispersion patterns, respectively (Figure 45.5). Uniform dispersion is observed in plants that secrete substances inhibiting the growth of nearby individuals (such as the release of toxic chemicals by the sage plant *Salvia leucophylla,* a phenomenon called allelopathy) and in animals like the penguin that maintain a defined territory. An example of random dispersion occurs with dandelion and other plants that have wind-dispersed seeds that germinate wherever they happen to fall in a favorable environment. A clumped dispersion may be seen in plants that drop their seeds straight to the ground, such as oak trees, or in animals that live in groups (schools of fish or herds of elephants). Clumped dispersions may also be a function of habitat heterogeneity. Thus, the dispersion of the individuals within a population provides more information about how they interact with each other than does a simple density measurement. Just as lower density species might have more difficulty finding a mate, solitary species with a random distribution might have a similar difficulty when compared to social species clumped together in groups.

Figure 45.5 Species may have uniform, random, or clumped distribution. Territorial birds such as penguins tend to have uniform distribution. Plants such as dandelions with wind-dispersed seeds tend to be randomly distributed. Animals such as elephants that travel in groups exhibit clumped distribution. (credit a: modification of work by Ben Tubby; credit b: modification of work by Rosendahl; credit c: modification of work by Rebecca Wood)

Demography

While population size and density describe a population at one particular point in time, scientists must use demography to study the dynamics of a population. Demography is the statistical study of population changes over time: birth rates, death rates, and life expectancies. Each of these measures, especially birth rates, may be affected by the population characteristics described above. For example, a large population size results in a higher birth rate because more potentially reproductive individuals are present. In contrast, a large population size can also result in a higher death rate because of competition, disease, and the accumulation of waste. Similarly, a higher population density or a clumped dispersion pattern results in more potential reproductive encounters between individuals, which can increase birth rate. Lastly, a female-biased sex ratio (the ratio of males to females) or **age structure** (the proportion of population members at specific age ranges) composed of many individuals of reproductive age can increase birth rates.

In addition, the demographic characteristics of a population can influence how the population grows or declines over time. If birth and death rates are equal, the population remains stable. However, the population size will increase if birth rates exceed death rates; the population will decrease if birth rates are less than death rates. Life expectancy is another important factor; the length of time individuals remain in the population impacts local resources, reproduction, and the overall health of the population. These demographic characteristics are often displayed in the form of a life table.

Life Tables

Life tables provide important information about the life history of an organism. Life tables divide the population into age groups and often sexes, and show how long a member of that group is likely to live. They are modeled after actuarial tables used by the insurance industry for estimating human life expectancy. Life tables may include the probability of individuals dying before their next birthday (i.e., their **mortality rate**), the percentage of surviving individuals dying at a particular age interval, and their life expectancy at each interval. An example of a life table is shown in Table 45.1 from a study of Dall mountain sheep, a species native to northwestern North America. Notice that the population is divided into age intervals (column A). The mortality rate (per 1000), shown in column D, is based on the number of individuals dying during the age interval (column B) divided by the number of individuals surviving at the beginning of the interval (Column C), multiplied by 1000.

$$\text{mortality rate} = \frac{\text{number of individuals dying}}{\text{number of individuals surviving}} \text{ x } 1000$$

For example, between ages three and four, 12 individuals die out of the 776 that were remaining from the original 1000 sheep. This number is then multiplied by 1000 to get the mortality rate per thousand.

$$\text{mortality rate} = \frac{12}{776} \text{ x } 1000 \approx 15.5$$

As can be seen from the mortality rate data (column D), a high death rate occurred when the sheep were between 6 and 12 months old, and then increased even more from 8 to 12 years old, after which there were few survivors. The data indicate that if a sheep in this population were to survive to age one, it could be expected to live another 7.7 years on average, as shown by the life expectancy numbers in column E.

Life Table of Dall Mountain Sheep[1]

Age interval (years)	Number dying in age interval out of 1000 born	Number surviving at beginning of age interval out of 1000 born	Mortality rate per 1000 alive at beginning of age interval	Life expectancy or mean lifetime remaining to those attaining age interval
0-0.5	54	1000	54.0	7.06
0.5-1	145	946	153.3	--
1-2	12	801	15.0	7.7
2-3	13	789	16.5	6.8

Table 45.1 This life table of *Ovis dalli* shows the number of deaths, number of survivors, mortality rate, and life expectancy at

1Data Adapted from Edward S. Deevey, Jr., "Life Tables for Natural Populations of Animals," *The Quarterly Review of Biology* 22, no. 4 (December 1947):

Age interval (years)	Number dying in age interval out of 1000 born	Number surviving at beginning of age interval out of 1000 born	Mortality rate per 1000 alive at beginning of age interval	Life expectancy or mean lifetime remaining to those attaining age interval
3-4	12	776	15.5	5.9
4-5	30	764	39.3	5.0
5-6	46	734	62.7	4.2
6-7	48	688	69.8	3.4
7-8	69	640	107.8	2.6
8-9	132	571	231.2	1.9
9-10	187	439	426.0	1.3
10-11	156	252	619.0	0.9
11-12	90	96	937.5	0.6
12-13	3	6	500.0	1.2
13-14	3	3	1000	0.7

Table 45.1 This life table of *Ovis dalli* shows the number of deaths, number of survivors, mortality rate, and life expectancy at each age interval for the Dall mountain sheep.

Survivorship Curves

Another tool used by population ecologists is a **survivorship curve**, which is a graph of the number of individuals surviving at each age interval plotted versus time (usually with data compiled from a life table). These curves allow us to compare the life histories of different populations (Figure 45.6). Humans and most primates exhibit a Type I survivorship curve because a high percentage of offspring survive their early and middle years—death occurs predominantly in older individuals. These types of species usually have small numbers of offspring at one time, and they give a high amount of parental care to them to ensure their survival. Birds are an example of an intermediate or Type II survivorship curve because birds die more or less equally at each age interval. These organisms also may have relatively few offspring and provide significant parental care. Trees, marine invertebrates, and most fishes exhibit a Type III survivorship curve because very few of these organisms survive their younger years; however, those that make it to an old age are more likely to survive for a relatively long period of time. Organisms in this category usually have a very large number of offspring, but once they are born, little parental care is provided. Thus these offspring are "on their own" and vulnerable to predation, but their sheer numbers assure the survival of enough individuals to perpetuate the species.

283-314.

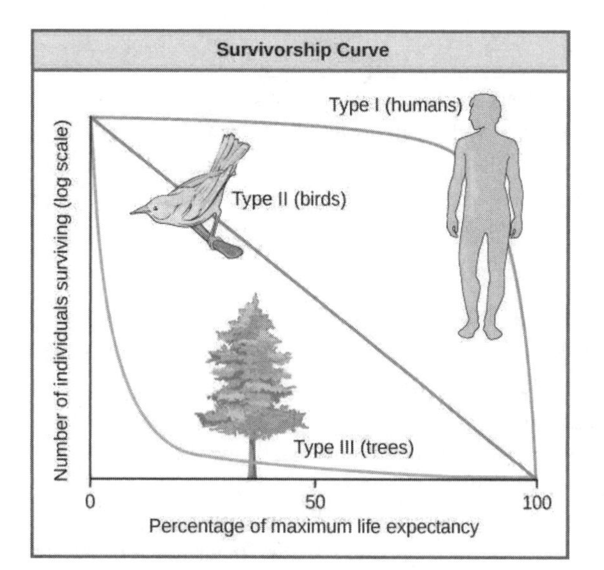

Figure 45.6 Survivorship curves show the distribution of individuals in a population according to age. Humans and most mammals have a Type I survivorship curve because death primarily occurs in the older years. Birds have a Type II survivorship curve, as death at any age is equally probable. Trees have a Type III survivorship curve because very few survive the younger years, but after a certain age, individuals are much more likely to survive.

45.2 Life Histories and Natural Selection

By the end of this section, you will be able to do the following:

- Describe how life history patterns are influenced by natural selection
- Explain different life history patterns and how different reproductive strategies affect species' survival

A species' **life history** describes the series of events over its lifetime, such as how resources are allocated for growth, maintenance, and reproduction. Life history traits affect the life table of an organism. A species' life history is genetically determined and shaped by the environment and natural selection.

Life History Patterns and Energy Budgets

Energy is required by all living organisms for their growth, maintenance, and reproduction; at the same time, energy is often a major limiting factor in determining an organism's survival. Plants, for example, acquire energy from the sun via photosynthesis, but must expend this energy to grow, maintain health, and produce energy-rich seeds to produce the next generation. Animals have the additional burden of using some of their energy reserves to acquire food. Furthermore, some animals must expend energy caring for their offspring. Thus, all species have an **energy budget**: they must balance energy intake with their use of energy for metabolism, reproduction, parental care, and energy storage (such as bears building up body fat for winter hibernation).

Parental Care and Fecundity

Fecundity is the potential reproductive capacity of an individual within a population. In other words, fecundity describes how many offspring could ideally be produced if an individual has as many offspring as possible, repeating the reproductive cycle as soon as possible after the birth of the offspring. In animals, fecundity is inversely related to the amount of parental care given to an individual offspring. Species, such as many marine invertebrates, that produce many offspring usually provide little if any care for the offspring (they would not have the energy or the ability to do so anyway). Most of their energy budget is used to produce many tiny offspring. Animals with this strategy are often self-sufficient at a very early age. This is because of the energy tradeoff these organisms have made to maximize their evolutionary fitness. Because their energy is used for producing offspring instead of parental care, it makes sense that these offspring have some ability to be able to move within their environment and find food and perhaps shelter. Even with these abilities, their small size makes them extremely vulnerable to predation, so the production of many offspring allows enough of them to survive to maintain the species.

Animal species that have few offspring during a reproductive event usually give extensive parental care, devoting much of their energy budget to these activities, sometimes at the expense of their own health. This is the case with many mammals, such as humans, kangaroos, and pandas. The offspring of these species are relatively helpless at birth and need to develop before they achieve self-sufficiency.

Plants with low fecundity produce few energy-rich seeds (such as coconuts and chestnuts) with each having a good chance to germinate into a new organism; plants with high fecundity usually have many small, energy-poor seeds (like orchids) that have a relatively poor chance of surviving. Although it may seem that coconuts and chestnuts have a better chance of surviving, the energy tradeoff of the orchid is also very effective. It is a matter of where the energy is used, for large numbers of seeds or for fewer seeds with more energy.

Early versus Late Reproduction

The timing of reproduction in a life history also affects species survival. Organisms that reproduce at an early age have a greater chance of producing offspring, but this is usually at the expense of their growth and the maintenance of their health. Conversely, organisms that start reproducing later in life often have greater fecundity or are better able to provide parental care, but they risk that they will not survive to reproductive age. Examples of this can be seen in fishes. Small fish, like guppies, use their energy to reproduce rapidly, but never attain the size that would give them defense against some predators. Larger fish, like the bluegill or shark, use their energy to attain a large size, but do so with the risk that they will die before they can reproduce or at least reproduce to their maximum. These different energy strategies and tradeoffs are key to understanding the evolution of each species as it maximizes its fitness and fills its niche. In terms of energy budgeting, some species "blow it all" and use up most of their energy reserves to reproduce early before they die. Other species delay having reproduction to become stronger, more experienced individuals and to make sure that they are strong enough to provide parental care if necessary.

Single versus Multiple Reproductive Events

Some life history traits, such as fecundity, timing of reproduction, and parental care, can be grouped together into general strategies that are used by multiple species. **Semelparity** occurs when a species reproduces only once during its lifetime and then dies. Such species use most of their resource budget during a single reproductive event, sacrificing their health to the point that they do not survive. Examples of semelparity are bamboo, which flowers once and then dies, and the Chinook salmon (Figure 45.7a), which uses most of its energy reserves to migrate from the ocean to its freshwater nesting area, where it reproduces and then dies. Scientists have posited alternate explanations for the evolutionary advantage of the Chinook's post-reproduction death: a programmed suicide caused by a massive release of corticosteroid hormones, presumably so the parents can become food for the offspring, or simple exhaustion caused by the energy demands of reproduction; these are still being debated.

Iteroparity describes species that reproduce repeatedly during their lives. Some animals are able to mate only once per year, but survive multiple mating seasons. The pronghorn antelope is an example of an animal that goes into a seasonal estrus cycle ("heat"): a hormonally induced physiological condition preparing the body for successful mating (Figure 45.7b). Females of these species mate only during the estrus phase of the cycle. A different pattern is observed in primates, including humans and chimpanzees, which may attempt reproduction at any time during their reproductive years, even though their menstrual cycles make pregnancy likely only a few days per month during ovulation (Figure 45.7c).

Figure 45.7 The (a) Chinook salmon mates once and dies. The (b) pronghorn antelope mates during specific times of the year during its reproductive life. Primates, such as humans and (c) chimpanzees, may mate on any day, independent of ovulation. (credit a: modification of work by Roger Tabor, USFWS; credit b: modification of work by Mark Gocke, USDA; credit c: modification of work by "Shiny Things"/Flickr)

LINK TO LEARNING

Play this interactive PBS evolution-based mating game (http://openstax.org/l/mating_game) to learn more about reproductive strategies.

EVOLUTION CONNECTION

Energy Budgets, Reproductive Costs, and Sexual Selection in *Drosophila*

Research into how animals allocate their energy resources for growth, maintenance, and reproduction has used a variety of experimental animal models. Some of this work has been done using the common fruit fly, *Drosophila melanogaster*. Studies have shown that not only does reproduction have a cost as far as how long male fruit flies live, but also fruit flies that have already mated several times have limited sperm remaining for reproduction. Fruit flies maximize their last chances at reproduction by selecting optimal mates.

In a 1981 study, male fruit flies were placed in enclosures with either virgin or inseminated females. The males that mated with virgin females had shorter life spans than those in contact with the same number of inseminated females with which they were unable to mate. This effect occurred regardless of how large (indicative of their age) the males were. Thus, males that did not mate lived longer, allowing them more opportunities to find mates in the future.

More recent studies, performed in 2006, show how males select the female with which they will mate and how this is affected by previous matings (Figure 45.8).[2] Males were allowed to select between smaller and larger females. Findings showed that larger females had greater fecundity, producing twice as many offspring per mating as the smaller females did. Males that had previously mated, and thus had lower supplies of sperm, were termed "resource-depleted," while males that had not mated were termed "non-resource-depleted." The study showed that although non-resource-depleted males preferentially mated with larger females, this selection of partners was more pronounced in the resource-depleted males. Thus, males with depleted sperm supplies, which were limited in the number of times that they could mate before they replenished their sperm supply, selected larger, more fecund females, thus maximizing their chances for offspring. This study was one of the first to show that the physiological state of the male affected its mating behavior in a way that clearly maximizes its use of limited reproductive resources.

	Ratio large/small females mated
Non sperm-depleted	8 ± 5
Sperm-depleted	15 ± 5

Figure 45.8 Male fruit flies that had previously mated (sperm-depleted) picked larger, more fecund females more often than those that had not mated (non-sperm-depleted). This change in behavior causes an increase in the efficiency of a limited reproductive resource: sperm.

These studies demonstrate two ways in which the energy budget is a factor in reproduction. First, energy expended on mating may reduce an animal's lifespan, but by this time they have already reproduced, so in the context of natural selection this early death is not of much evolutionary importance. Second, when resources such as sperm (and the energy needed to replenish it) are low, an organism's behavior can change to give them the best chance of passing their genes on to the next generation. These changes in behavior, so important to evolution, are studied in a discipline known as behavioral biology, or ethology, at the interface between population biology and psychology.

45.3 Environmental Limits to Population Growth

By the end of this section, you will be able to do the following:

- Explain the characteristics of and differences between exponential and logistic growth patterns
- Give examples of exponential and logistic growth in natural populations
- Describe how natural selection and environmental adaptation led to the evolution of particular life history patterns

Although life histories describe the way many characteristics of a population (such as their age structure) change over time in a general way, population ecologists make use of a variety of methods to model population dynamics mathematically. These more

2Adapted from Phillip G. Byrne and William R. Rice, "Evidence for adaptive male mate choice in the fruit fly *Drosophila melanogaster*," Proc Biol Sci. 273, no. 1589 (2006): 917-922, doi: 10.1098/rspb.2005.3372.

precise models can then be used to accurately describe changes occurring in a population and better predict future changes. Certain long-accepted models are now being modified or even abandoned due to their lack of predictive ability, and scholars strive to create effective new models.

Exponential Growth

Charles Darwin, in his theory of natural selection, was greatly influenced by the English clergyman Thomas Malthus. Malthus published a book in 1798 stating that populations with unlimited natural resources grow very rapidly, and then population growth decreases as resources become depleted. This accelerating pattern of increasing population size is called **exponential growth**.

The best example of exponential growth is seen in bacteria. Bacteria reproduce by prokaryotic fission. This division takes about an hour for many bacterial species. If 1000 bacteria are placed in a large flask with an unlimited supply of nutrients (so the nutrients will not become depleted), after an hour, there is one round of division and each organism divides, resulting in 2000 organisms—an increase of 1000. In another hour, each of the 2000 organisms will double, producing 4000, an increase of 2000 organisms. After the third hour, there should be 8000 bacteria in the flask, an increase of 4000 organisms. The important concept of exponential growth is the accelerating **population growth rate**—the number of organisms added in each reproductive generation—that is, it is increasing at a greater and greater rate. After 1 day and 24 of these cycles, the population would have increased from 1000 to more than 16 billion. When the population size, *N*, is plotted over time, a **J-shaped growth curve** is produced (Figure 45.9).

The bacteria example is not representative of the real world where resources are limited. Furthermore, some bacteria will die during the experiment and thus not reproduce, lowering the growth rate. Therefore, when calculating the growth rate of a population, the **death rate (*D*)** (number organisms that die during a particular time interval) is subtracted from the **birth rate (*B*)** (number organisms that are born during that interval). This is shown in the following formula:

$$\frac{\Delta N \text{ (change in number)}}{\Delta T \text{ (change in time)}} = B \text{ (birth rate) - } D \text{ (death rate)}$$

The birth rate is usually expressed on a per capita (for each individual) basis. Thus, *B* (birth rate) = *bN* (the per capita birth rate "*b*" multiplied by the number of individuals "*N*") and *D* (death rate) = *dN* (the per capita death rate "d" multiplied by the number of individuals "*N*"). Additionally, ecologists are interested in the population at a particular point in time, an infinitely small time interval. For this reason, the terminology of differential calculus is used to obtain the "instantaneous" growth rate, replacing the *change* in number and time with an instant-specific measurement of number and time.

$$\frac{dN}{dT} = bN - dN = (b - d)N$$

Notice that the "*d*" associated with the first term refers to the derivative (as the term is used in calculus) and is different from the death rate, also called "*d*." The difference between birth and death rates is further simplified by substituting the term "*r*" (intrinsic rate of increase) for the relationship between birth and death rates:

$$\frac{dN}{dT} = rN$$

The value "*r*" can be positive, meaning the population is increasing in size; or negative, meaning the population is decreasing in size; or zero, where the population's size is unchanging, a condition known as **zero population growth**. A further refinement of the formula recognizes that different species have inherent differences in their intrinsic rate of increase (often thought of as the potential for reproduction), even under ideal conditions. Obviously, a bacterium can reproduce more rapidly and have a higher intrinsic rate of growth than a human. The maximal growth rate for a species is its **biotic potential, or r_{max}**, thus changing the equation to:

$$\frac{dN}{dT} = r_{\max} N$$

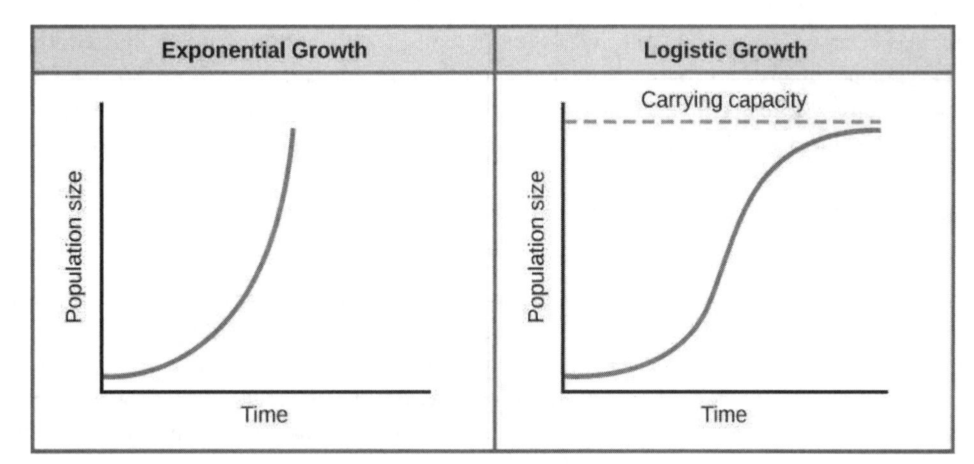

Figure 45.9 When resources are unlimited, populations exhibit exponential growth, resulting in a J-shaped curve. When resources are limited, populations exhibit logistic growth. In logistic growth, population expansion decreases as resources become scarce, and it levels off when the carrying capacity of the environment is reached, resulting in an S-shaped curve.

Logistic Growth

Exponential growth is possible only when infinite natural resources are available; this is not the case in the real world. Charles Darwin recognized this fact in his description of the "struggle for existence," which states that individuals will compete (with members of their own or other species) for limited resources. The successful ones will survive to pass on their own characteristics and traits (which we know now are transferred by genes) to the next generation at a greater rate (natural selection). To model the reality of limited resources, population ecologists developed the **logistic growth** model.

Carrying Capacity and the Logistic Model

In the real world, with its limited resources, exponential growth cannot continue indefinitely. Exponential growth may occur in environments where there are few individuals and plentiful resources, but when the number of individuals gets large enough, resources will be depleted, slowing the growth rate. Eventually, the growth rate will plateau or level off (Figure 45.9). This population size, which represents the maximum population size that a particular environment can support, is called the **carrying capacity, or *K*.**

The formula we use to calculate logistic growth adds the carrying capacity as a moderating force in the growth rate. The expression "*K* – *N*" indicates how many individuals may be added to a population at a given stage, and "*K* – *N*" divided by "*K*" is the fraction of the carrying capacity available for further growth. Thus, the exponential growth model is restricted by this factor to generate the logistic growth equation:

$$\frac{dN}{dT} = r_{\max}\frac{dN}{dT} = r_{\max}N\frac{(K - N)}{K}$$

Notice that when N is very small, $(K\text{-}N)/K$ becomes close to K/K or 1, and the right side of the equation reduces to $r_{max}N$, which means the population is growing exponentially and is not influenced by carrying capacity. On the other hand, when N is large, $(K\text{-}N)/K$ comes close to zero, which means that population growth will be slowed greatly or even stopped. Thus, population growth is greatly slowed in large populations by the carrying capacity K. This model also allows for the population of a negative population growth, or a population decline. This occurs when the number of individuals in the population exceeds the carrying capacity (because the value of (K-N)/K is negative).

A graph of this equation yields an **S-shaped curve** (Figure 45.9), and it is a more realistic model of population growth than exponential growth. There are three different sections to an S-shaped curve. Initially, growth is exponential because there are few individuals and ample resources available. Then, as resources begin to become limited, the growth rate decreases. Finally, growth levels off at the carrying capacity of the environment, with little change in population size over time.

Role of Intraspecific Competition

The logistic model assumes that every individual within a population will have equal access to resources and, thus, an equal chance for survival. For plants, the amount of water, sunlight, nutrients, and the space to grow are the important resources, whereas in animals, important resources include food, water, shelter, nesting space, and mates.

In the real world, phenotypic variation among individuals within a population means that some individuals will be better adapted to their environment than others. The resulting competition between population members of the same species for resources is termed **intraspecific competition** (intra- = "within"; -specific = "species"). Intraspecific competition for resources may not affect populations that are well below their carrying capacity—resources are plentiful and all individuals can obtain what they need. However, as population size increases, this competition intensifies. In addition, the accumulation of waste products can reduce an environment's carrying capacity.

Examples of Logistic Growth

Yeast, a microscopic fungus used to make bread and alcoholic beverages, exhibits the classical S-shaped curve when grown in a test tube (Figure 45.10a). Its growth levels off as the population depletes the nutrients. In the real world, however, there are variations to this idealized curve. Examples in wild populations include sheep and harbor seals (Figure 45.10b). In both examples, the population size exceeds the carrying capacity for short periods of time and then falls below the carrying capacity afterwards. This fluctuation in population size continues to occur as the population oscillates around its carrying capacity. Still, even with this oscillation, the logistic model is confirmed.

VISUAL CONNECTION

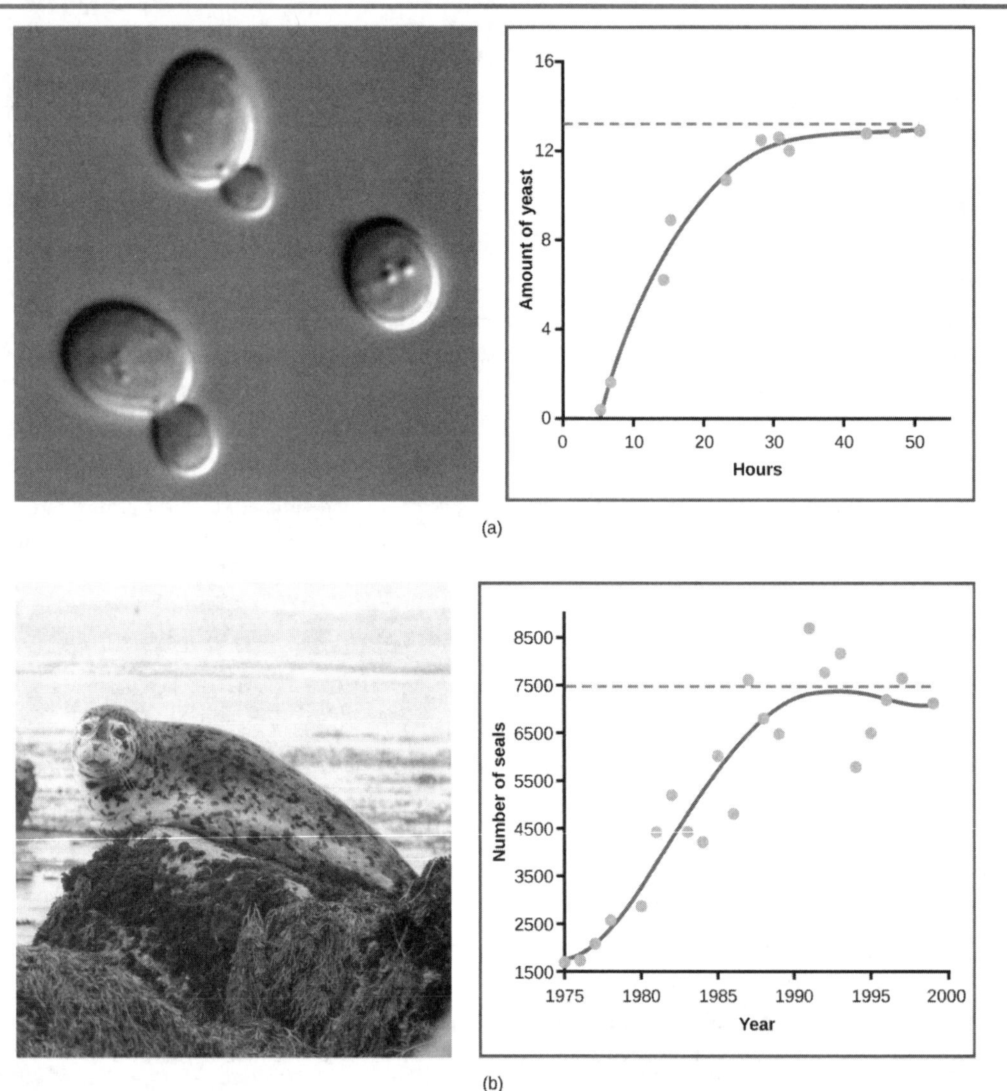

Figure 45.10 (a) Yeast grown in ideal conditions in a test tube show a classical S-shaped logistic growth curve, whereas (b) a natural population of seals shows real-world fluctuation.

If the major food source of the seals declines due to pollution or overfishing, which of the following would likely occur?

a. The carrying capacity of seals would decrease, as would the seal population.
b. The carrying capacity of seals would decrease, but the seal population would remain the same.
c. The number of seal deaths would increase but the number of births would also increase, so the population size would remain the same.
d. The carrying capacity of seals would remain the same, but the population of seals would decrease.

45.4 Population Dynamics and Regulation

By the end of this section, you will be able to do the following:

- Give examples of how the carrying capacity of a habitat may change
- Compare and contrast density-dependent growth regulation and density-independent growth regulation, giving examples
- Give examples of exponential and logistic growth in wild animal populations
- Describe how natural selection and environmental adaptation leads to the evolution of particular life-history patterns

The logistic model of population growth, while valid in many natural populations and a useful model, is a simplification of real-world population dynamics. Implicit in the model is that the carrying capacity of the environment does not change, which is not the case. The carrying capacity varies annually: for example, some summers are hot and dry whereas others are cold and wet. In many areas, the carrying capacity during the winter is much lower than it is during the summer. Also, natural events such as earthquakes, volcanoes, and fires can alter an environment and hence its carrying capacity. Additionally, populations do not usually exist in isolation. They engage in **interspecific competition**: that is, they share the environment with other species competing for the same resources. These factors are also important to understanding how a specific population will grow.

Nature regulates population growth in a variety of ways. These are grouped into **density-dependent** factors, in which the density of the population at a given time affects growth rate and mortality, and **density-independent** factors, which influence mortality in a population regardless of population density. Note that in the former, the effect of the factor on the population depends on the density of the population at onset. Conservation biologists want to understand both types because this helps them manage populations and prevent extinction or overpopulation.

Density-Dependent Regulation

Most density-dependent factors are biological in nature (biotic), and include predation, inter- and intraspecific competition, accumulation of waste, and diseases such as those caused by parasites. Usually, the denser a population is, the greater its mortality rate. For example, during intra- and interspecific competition, the reproductive rates of the individuals will usually be lower, reducing their population's rate of growth. In addition, low prey density increases the mortality of its predator because it has more difficulty locating its food source.

An example of density-dependent regulation is shown in Figure 45.11 with results from a study focusing on the giant intestinal roundworm (*Ascaris lumbricoides*), a parasite of humans and other mammals.[3] Denser populations of the parasite exhibited lower fecundity: they contained fewer eggs. One possible explanation for this is that females would be smaller in more dense populations (due to limited resources) and that smaller females would have fewer eggs. This hypothesis was tested and disproved in a 2009 study which showed that female weight had no influence.[4] The actual cause of the density-dependence of fecundity in this organism is still unclear and awaiting further investigation.

3N.A. Croll et al., "The Population Biology and Control of *Ascaris lumbricoides* in a Rural Community in Iran." *Transactions of the Royal Society of Tropical Medicine and Hygiene* 76, no. 2 (1982): 187-197, doi:10.1016/0035-9203(82)90272-3.
4Martin Walker et al., "Density-Dependent Effects on the Weight of Female *Ascaris lumbricoides* Infections of Humans and its Impact on Patterns of Egg Production." *Parasites & Vectors* 2, no. 11 (February 2009), doi:10.1186/1756-3305-2-11.

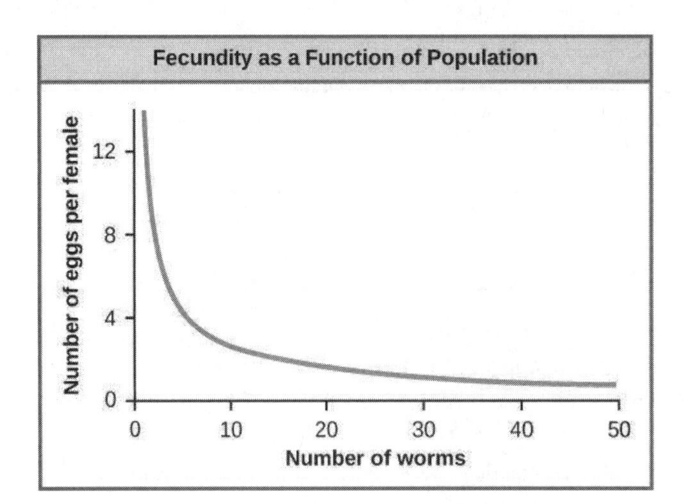

Figure 45.11 In this population of roundworms, fecundity (number of eggs) decreases with population density.[5]

Density-Independent Regulation and Interaction with Density-Dependent Factors

Many factors, typically physical or chemical in nature (abiotic), influence the mortality of a population regardless of its density, including weather, natural disasters, and pollution. An individual deer may be killed in a forest fire regardless of how many deer happen to be in that area. Its chances of survival are the same whether the population density is high or low. The same holds true for cold winter weather.

In real-life situations, population regulation is very complicated and density-dependent and independent factors can interact. A dense population that is reduced in a density-independent manner by some environmental factor(s) will be able to recover differently than a sparse population. For example, a population of deer affected by a harsh winter will recover faster if there are more deer remaining to reproduce.

EVOLUTION CONNECTION

Why Did the Woolly Mammoth Go Extinct?

Figure 45.12 The three photos include: (a) 1916 mural of a mammoth herd from the American Museum of Natural History, (b) the only stuffed mammoth in the world, from the Museum of Zoology located in St. Petersburg, Russia, and (c) a one-month-old baby mammoth, named Lyuba, discovered in Siberia in 2007. (credit a: modification of work by Charles R. Knight; credit b: modification of work by "Tanapon"/Flickr; credit c: modification of work by Matt Howry)

It's easy to get lost in the discussion about why dinosaurs went extinct 65 million years ago. Was it due to a meteor slamming into Earth near the coast of modern-day Mexico, or was it from some long-term weather cycle that is not yet understood? Scientists are continually exploring these and other theories.

Woolly mammoths began to go extinct much more recently, when they shared the Earth with humans who were no different anatomically than humans today (Figure 45.12). Mammoths survived in isolated island populations as recently as 1700 BC. We know a lot about these animals from carcasses found frozen in the ice of Siberia and other regions of the north. Scientists have sequenced at least 50 percent of its genome and believe mammoths are between 98 and 99 percent identical to modern elephants.

It is commonly thought that climate change and human hunting led to their extinction. A 2008 study estimated that climate

5N.A. Croll et al., "The Population Biology and Control of *Ascaris lumbricoides* in a Rural Community in Iran." *Transactions of the Royal Society of Tropical Medicine and Hygiene* 76, no. 2 (1982): 187-197, doi:10.1016/0035-9203(82)90272-3.

change reduced the mammoth's range from 3,000,000 square miles 42,000 years ago to 310,000 square miles 6,000 years ago.[6] It is also well documented that humans hunted these animals. A 2012 study showed that no single factor was exclusively responsible for the extinction of these magnificent creatures.[7] In addition to human hunting, climate change, and reduction of habitat, these scientists demonstrated another important factor in the mammoth's extinction was the migration of humans across the Bering Strait to North America during the last ice age 20,000 years ago.

The maintenance of stable populations was and is very complex, with many interacting factors determining the outcome. It is important to remember that humans are also part of nature. We once contributed to a species' decline using only primitive hunting technology.

Life Histories of *K*-selected and *r*-selected Species

While reproductive strategies play a key role in life histories, they do not account for important factors like limited resources and competition. The regulation of population growth by these factors can be used to introduce a classical concept in population biology, that of *K*-selected versus *r*-selected species.

The concept relates to a species' reproductive strategies, habitat, and behavior, especially in the way that they obtain resources and care for their young. It includes length of life and survivorship factors as well. Population biologists have grouped species into the two large categories—*K*-selected and *r*-selected—although the categories are really two ends of a continuum.

***K*-selected species** are species selected by stable, predictable environments. Populations of *K*-selected species tend to exist close to their carrying capacity (hence the term *K*-selected) where intraspecific competition is high. These species have few, large offspring, a long gestation period, and often give long-term care to their offspring (Table 45.2). While larger in size when born, the offspring are relatively helpless and immature at birth. By the time they reach adulthood, they must develop skills to compete for natural resources. In plants, scientists think of parental care more broadly: how long fruit takes to develop or how long it remains on the plant are determining factors in the time to the next reproductive event. Examples of *K*-selected species are primates (including humans), elephants, and plants such as oak trees (Figure 45.13a).

Oak trees grow very slowly and take, on average, 20 years to produce their first seeds, known as acorns. As many as 50,000 acorns can be produced by an individual tree, but the germination rate is low as many of these rot or are eaten by animals such as squirrels. In some years, oaks may produce an exceptionally large number of acorns, and these years may be on a two- or three-year cycle depending on the species of oak (*r*-selection).

As oak trees grow to a large size and for many years before they begin to produce acorns, they devote a large percentage of their energy budget to growth and maintenance. The tree's height and size allow it to dominate other plants in the competition for sunlight, the oak's primary energy resource. Furthermore, when it does reproduce, the oak produces large, energy-rich seeds that use their energy reserve to become quickly established (*K*-selection).

In contrast, ***r*-selected species** have a large number of small offspring (hence their *r* designation (Table 45.2)). This strategy is often employed in unpredictable or changing environments. Animals that are *r*-selected do not give long-term parental care and the offspring are relatively mature and self-sufficient at birth. Examples of *r*-selected species are marine invertebrates, such as jellyfish, and plants, such as the dandelion (Figure 45.13b). Dandelions have small seeds that are wind dispersed long distances. Many seeds are produced simultaneously to ensure that at least some of them reach a hospitable environment. Seeds that land in inhospitable environments have little chance for survival since their seeds are low in energy content. Note that survival is not necessarily a function of energy stored in the seed itself.

6David Nogués-Bravo et al., "Climate Change, Humans, and the Extinction of the Woolly Mammoth." *PLoS Biol* 6 (April 2008): e79, doi:10.1371/journal.pbio.0060079.

7G.M. MacDonald et al., "Pattern of Extinction of the Woolly Mammoth in Beringia." *Nature Communications* 3, no. 893 (June 2012), doi:10.1038/ncomms1881.

Characteristics of *K*-selected and *r*-selected species

Characteristics of *K*-selected species	Characteristics of *r*-selected species
Mature late	Mature early
Greater longevity	Lower longevity
Increased parental care	Decreased parental care
Increased competition	Decreased competition
Fewer offspring	More offspring
Larger offspring	Smaller offspring

Table 45.2

(a) K-selected species

(b) r-selected species

Figure 45.13 (a) Elephants are considered K-selected species as they live long, mature late, and provide long-term parental care to few offspring. Oak trees produce many offspring that do not receive parental care, but are considered K-selected species based on longevity and late maturation. (b) Dandelions and jellyfish are both considered r-selected species as they mature early, have short lifespans, and produce many offspring that receive no parental care.

Modern Theories of Life History

By the second half of the twentieth century, the concept of K- and r-selected species was used extensively and successfully to study populations. The *r*- and *K*-selection theory, although accepted for decades and used for much groundbreaking research, has now been reconsidered, and many population biologists have abandoned or modified it. Over the years, several studies attempted to confirm the theory, but these attempts have largely failed. Many species were identified that did not follow the theory's predictions. Furthermore, the theory ignored the age-specific mortality of the populations which scientists now know is very important. New **demographic-based models** of life history evolution have been developed which incorporate many ecological concepts included in *r*- and *K*-selection theory as well as population age structure and mortality factors.

45.5 Human Population Growth

By the end of this section, you will be able to do the following:

- Discuss exponential human population growth
- Explain how humans have expanded the carrying capacity of their habitat
- Relate population growth and age structure to the level of economic development in different countries
- Discuss the long-term implications of unchecked human population growth

Population dynamics can be applied to human population growth. Earth's human population is growing rapidly, to the extent that some worry about the ability of the earth's environment to sustain this population. Long-term exponential growth carries the potential risks of famine, disease, and large-scale death.

Although humans have increased the carrying capacity of their environment, the technologies used to achieve this transformation have caused unprecedented changes to Earth's environment, altering ecosystems to the point where some may be in danger of collapse. The depletion of the ozone layer, erosion due to acid rain, and damage from global climate change are caused by human activities. The ultimate effect of these changes on our carrying capacity is unknown. As some point out, it is likely that the negative effects of increasing carrying capacity will outweigh the positive ones—the world's carrying capacity for human beings might actually decrease.

The human population is currently experiencing exponential growth even though human reproduction is far below its biotic potential (Figure 45.14). To reach its biotic potential, all females would have to become pregnant every nine months or so during their reproductive years. Also, resources would have to be such that the environment would support such growth. Neither of these two conditions exists. In spite of this fact, human population is still growing exponentially.

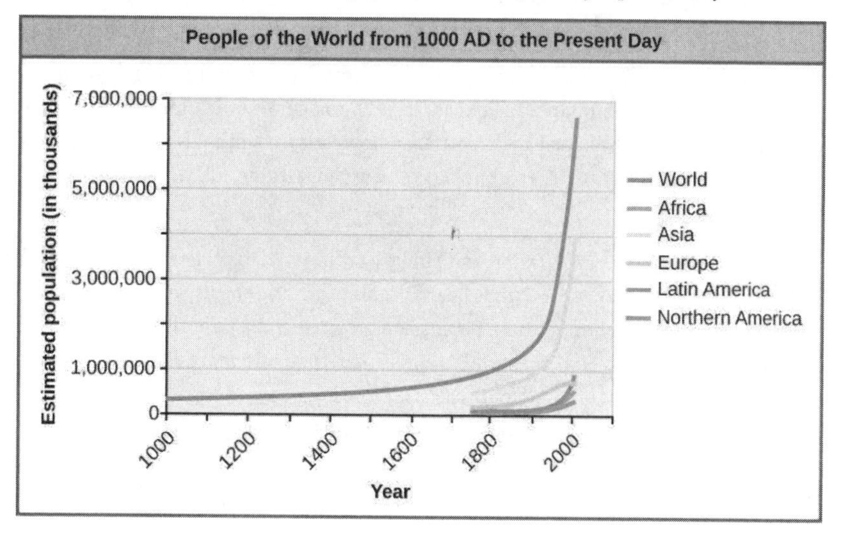

Figure 45.14 Human population growth since 1000 AD is exponential (dark blue line). Notice that while the population in Asia (yellow line), which has many economically underdeveloped countries, is increasing exponentially, the population in Europe (light blue line), where most of the countries are economically developed, is growing much more slowly.

A consequence of exponential human population growth is a reduction in time that it takes to add a particular number of humans to the Earth. Figure 45.15 shows that 123 years were necessary to add 1 billion humans in 1930, but it only took 24 years to add two billion people between 1975 and 1999. As already discussed, our ability to increase our carrying capacity indefinitely my be limited. Without new technological advances, the human growth rate has been predicted to slow in the coming decades. However, the population will still be increasing and the threat of overpopulation remains.

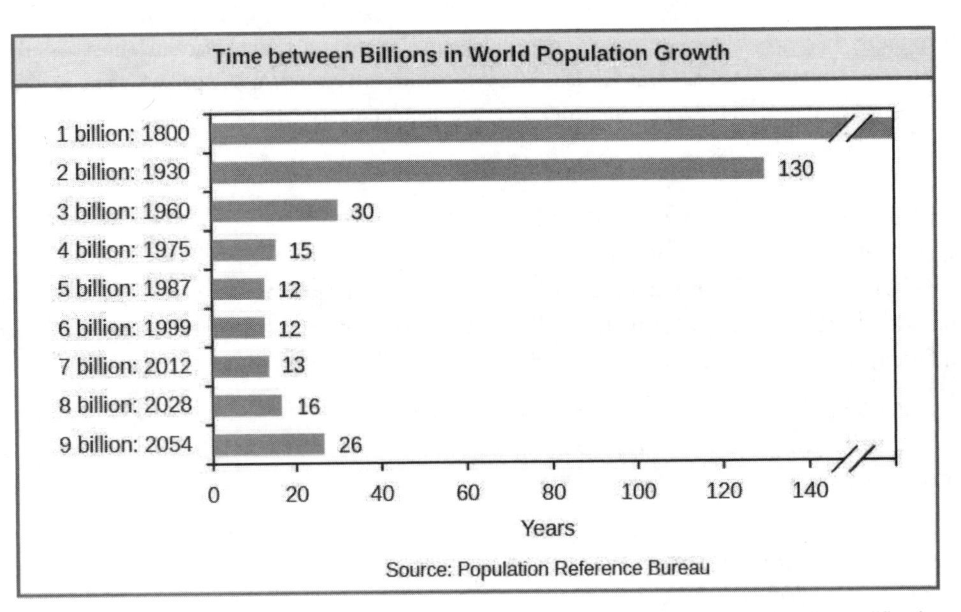

Figure 45.15 The time between the addition of each billion human beings to Earth decreases over time. (credit: modification of work by Ryan T. Cragun)

LINK TO LEARNING

Click through this video (http://openstax.org/l/human_growth) of how human populations have changed over time.

Overcoming Density-Dependent Regulation

Humans are unique in their ability to alter their environment with the conscious purpose of increasing carrying capacity. This ability is a major factor responsible for human population growth and a way of overcoming density-dependent growth regulation. Much of this ability is related to human intelligence, society, and communication. Humans can construct shelter to protect them from the elements and have developed agriculture and domesticated animals to increase their food supplies. In addition, humans use language to communicate this technology to new generations, allowing them to improve upon previous accomplishments.

Other factors in human population growth are migration and public health. Humans originated in Africa, but have since migrated to nearly all inhabitable land on the Earth. Public health, sanitation, and the use of antibiotics and vaccines have decreased the ability of infectious disease to limit human population growth. In the past, diseases such as the bubonic plaque of the fourteenth century killed between 30 and 60 percent of Europe's population and reduced the overall world population by as many as 100 million people. Today, the threat of infectious disease, while not gone, is certainly less severe. According to the Institute for Health Metrics and Evaluation (IHME) in Seattle, global death from infectious disease declined from 15.4 million in 1990 to 10.4 million in 2017. To compare to some of the epidemics of the past, the percentage of the world's population killed between 1993 and 2002 decreased from 0.30 percent of the world's population to 0.14 percent. Thus, infectious disease influence on human population growth is becoming less significant.

Age Structure, Population Growth, and Economic Development

The age structure of a population is an important factor in population dynamics. **Age structure** is the proportion of a population at different age ranges. Age structure allows better prediction of population growth, plus the ability to associate this growth with the level of economic development in the region. Countries with rapid growth have a pyramidal shape in their age structure diagrams, showing a preponderance of younger individuals, many of whom are of reproductive age or will be soon (Figure 45.16). This pattern is most often observed in underdeveloped countries where individuals do not live to old age because of less-than-optimal living conditions. Age structures of areas with slow growth, including developed countries such as the United States, still have a pyramidal structure, but with many fewer young and reproductive-aged individuals and a greater proportion of older individuals. Other developed countries, such as Italy, have zero population growth. The age structure of these populations is more conical, with an even greater percentage of middle-aged and older individuals. The actual growth rates in different countries are shown in Figure 45.17, with the highest rates tending to be in the less economically developed countries of

Africa and Asia.

VISUAL CONNECTION

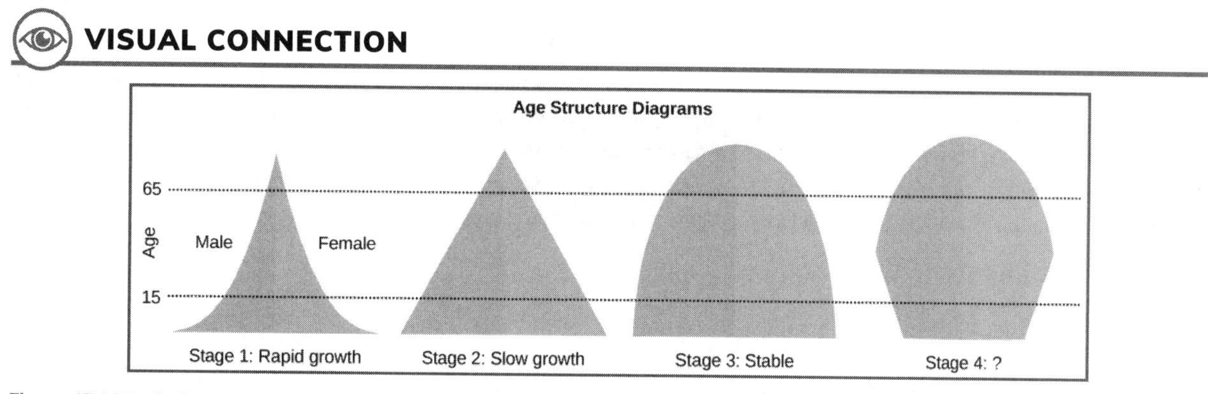

Figure 45.16 Typical age structure diagrams are shown. The rapid growth diagram narrows to a point, indicating that the number of individuals decreases rapidly with age. In the slow growth model, the number of individuals decreases steadily with age. Stable population diagrams are rounded on the top, showing that the number of individuals per age group decreases gradually, and then increases for the older part of the population.

Age structure diagrams for rapidly growing, slow growing, and stable populations are shown in stages 1 through 3. What type of population change do you think stage 4 represents?

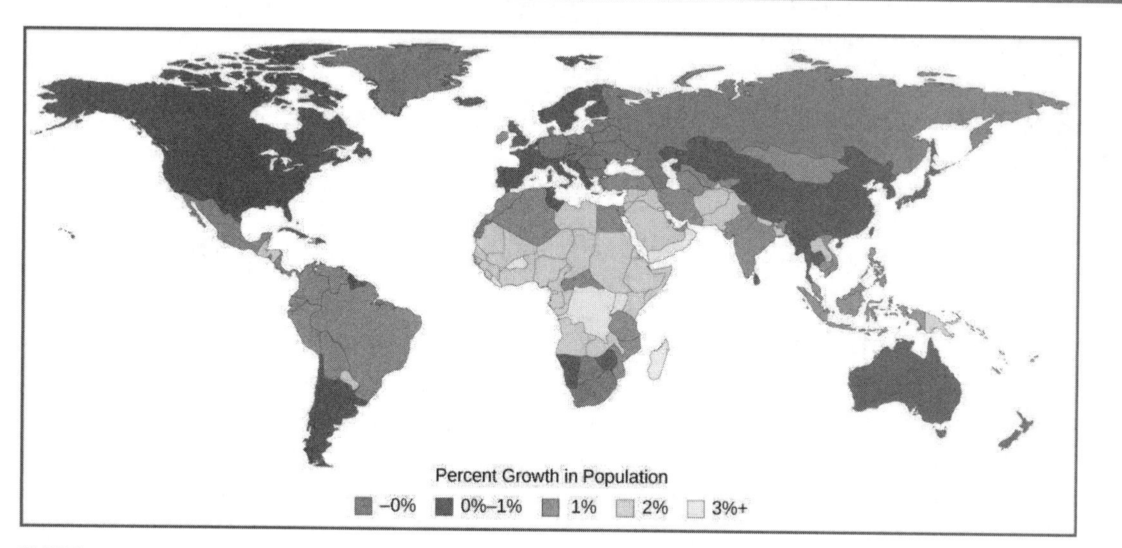

Figure 45.17 The percent growth rate of population in different countries is shown. Notice that the highest growth is occurring in less economically developed countries in Africa and Asia.

Long-Term Consequences of Exponential Human Population Growth

Many dire predictions have been made about the world's population leading to a major crisis called the "population explosion." In the 1968 book *The Population Bomb*, biologist Dr. Paul R. Ehrlich wrote, "The battle to feed all of humanity is over. In the 1970s hundreds of millions of people will starve to death in spite of any crash programs embarked upon now. At this late date nothing can prevent a substantial increase in the world death rate."[8] While many experts view this statement as incorrect based on evidence, the laws of exponential population growth are still in effect, and unchecked human population growth cannot continue indefinitely.

Several nations have instituted policies aimed at influencing population. Efforts to control population growth led to the **one-child policy** in China, which is now being phased out. India also implements national and regional populations to encourage family planning. On the other hand, Japan, Spain, Russia, Iran, and other countries have made efforts to increase population

8Paul R. Erlich, prologue to *The Population Bomb*, (1968; repr., New York: Ballantine, 1970).

growth after birth rates dipped. Such policies are controversial, and the human population continues to grow. At some point the food supply may run out, but the outcomes are difficult to predict. The United Nations estimates that future world population growth may vary from 6 billion (a decrease) to 16 billion people by the year 2100.

Another result of population growth is the endangerment of the natural environment. Many countries have attempted to reduce the human impact on climate change by reducing their emission of the greenhouse gas carbon dioxide. However, these treaties have not been ratified by every country. The role of human activity in causing climate change has become a hotly debated socio-political issue in some countries, including the United States. Thus, we enter the future with considerable uncertainty about our ability to curb human population growth and protect our environment.

LINK TO LEARNING

Visit this website (http://openstax.org/l/populations) and select "Launch movie" for an animation discussing the global impacts of human population growth.

45.6 Community Ecology

By the end of this section, you will be able to do the following:

- Discuss the predator-prey cycle
- Give examples of defenses against predation and herbivory
- Describe the competitive exclusion principle
- Give examples of symbiotic relationships between species
- Describe community structure and succession

Populations rarely, if ever, live in isolation from populations of other species. In most cases, numerous species share a habitat. The interactions between these populations play a major role in regulating population growth and abundance. All populations occupying the same habitat form a community: populations inhabiting a specific area at the same time. The number of species occupying the same habitat and their relative abundance is known as species diversity. Areas with low diversity, such as the glaciers of Antarctica, still contain a wide variety of living things, whereas the diversity of tropical rainforests is so great that it cannot be counted. Ecology is studied at the community level to understand how species interact with each other and compete for the same resources.

Predation and Herbivory

Perhaps the classical example of species interaction is predation: the consumption of prey by its predator. Nature shows on television highlight the drama of one living organism killing another. Populations of predators and prey in a community are not constant over time: in most cases, they vary in cycles that appear to be related. The most often cited example of predator-prey dynamics is seen in the cycling of the lynx (predator) and the snowshoe hare (prey), using nearly 200 year-old trapping data from North American forests (Figure 45.18). This cycle of predator and prey lasts approximately 10 years, with the predator population lagging 1–2 years behind that of the prey population. As the hare numbers increase, there is more food available for the lynx, allowing the lynx population to increase as well. When the lynx population grows to a threshold level, however, they kill so many hares that hare population begins to decline, followed by a decline in the lynx population because of scarcity of food. When the lynx population is low, the hare population size begins to increase due, at least in part, to low predation pressure, starting the cycle anew.

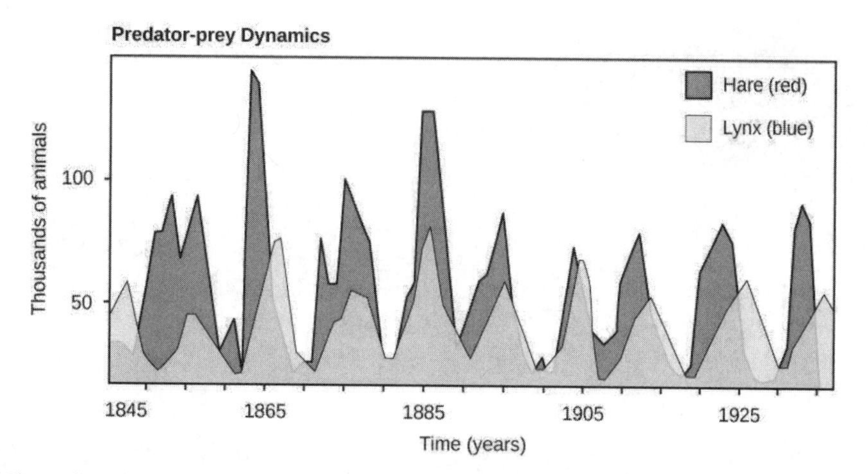

Figure 45.18 The cycling of lynx and snowshoe hare populations in Northern Ontario is an example of predator-prey dynamics.

Some researchers question the idea that predation models entirely control the population cycling of the two species. More recent studies have pointed to undefined density-dependent factors as being important in the cycling, in addition to predation. One possibility is that the cycling is inherent in the hare population due to density-dependent effects such as lower fecundity (maternal stress) caused by crowding when the hare population gets too dense. The hare cycling would then induce the cycling of the lynx because it is the lynxes' major food source. The more we study communities, the more complexities we find, allowing ecologists to derive more accurate and sophisticated models of population dynamics.

Herbivory describes the consumption of plants by insects and other animals, and it is another interspecific relationship that affects populations. Unlike animals, most plants cannot outrun predators or use mimicry to hide from hungry animals. Some plants have developed mechanisms to defend against herbivory. Other species have developed mutualistic relationships; for example, herbivory provides a mechanism of seed distribution that aids in plant reproduction.

Defense Mechanisms against Predation and Herbivory

The study of communities must consider evolutionary forces that act on the members of the various populations contained within it. Species are not static, but slowly changing and adapting to their environment by natural selection and other evolutionary forces. Species have evolved numerous mechanisms to escape predation and herbivory. These defenses may be mechanical, chemical, physical, or behavioral.

Mechanical defenses, such as the presence of thorns on plants or the hard shell on turtles, discourage animal predation and herbivory by causing physical pain to the predator or by physically preventing the predator from being able to eat the prey. Chemical defenses are produced by many animals as well as plants, such as the foxglove which is extremely toxic when eaten. Figure 45.19 shows some organisms' defenses against predation and herbivory.

Figure 45.19 The (a) honey locust tree (*Gleditsia triacanthos*) uses thorns, a mechanical defense, against herbivores, while the (b) Florida red-bellied turtle (*Pseudemys nelsoni)* uses its shell as a mechanical defense against predators. (c) Foxglove (*Digitalis* sp.) uses a chemical defense: toxins produced by the plant can cause nausea, vomiting, hallucinations, convulsions, or death when consumed. (d) The North American millipede (*Narceus americanus*) uses both mechanical and chemical defenses: when threatened, the millipede curls into a defensive ball and produces a noxious substance that irritates eyes and skin. (credit a: modification of work by Huw Williams; credit b: modification of work by "JamieS93"/Flickr; credit c: modification of work by Philip Jägenstedt; credit d: modification of work by Cory Zanker)

Many species use their body shape and coloration to avoid being detected by predators. The tropical walking stick is an insect with the coloration and body shape of a twig which makes it very hard to see when stationary against a background of real twigs (Figure 45.20a). In another example, the chameleon can, within limitations, change its color to match its surroundings (Figure 45.20b). Both of these are examples of **camouflage**, or avoiding detection by blending in with the background.

Figure 45.20 (a) The tropical walking stick and (b) the chameleon use body shape and/or coloration to prevent detection by predators. (credit a: modification of work by Linda Tanner; credit b: modification of work by Frank Vassen)

Some species use coloration as a way of warning predators that they are not good to eat. For example, the cinnabar moth caterpillar, the fire-bellied toad, and many species of beetle have bright colors that warn of a foul taste, the presence of toxic chemicals, and/or the ability to sting or bite, respectively. Predators that ignore this coloration and eat the organisms will experience their unpleasant taste or presence of toxic chemicals and learn not to eat them in the future. This type of defensive mechanism is called **aposematic coloration**, or warning coloration (Figure 45.21).

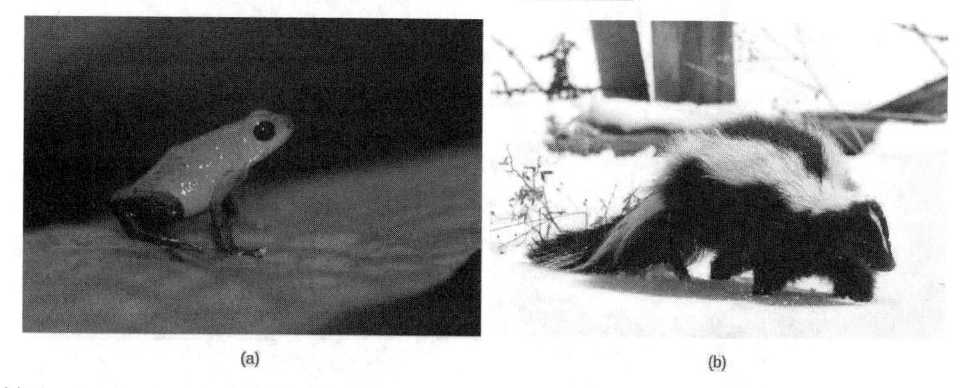

Figure 45.21 (a) The strawberry poison dart frog (*Oophaga pumilio*) uses aposematic coloration to warn predators that it is toxic, while the (b) striped skunk (*Mephitis mephitis*) uses aposematic coloration to warn predators of the unpleasant odor it produces. (credit a: modification of work by Jay Iwasaki; credit b: modification of work by Dan Dzurisin)

While some predators learn to avoid eating certain potential prey because of their coloration, other species have evolved mechanisms to mimic this coloration to avoid being eaten, even though they themselves may not be unpleasant to eat or contain toxic chemicals. In **Batesian mimicry**, a harmless species imitates the warning coloration of a harmful one. Assuming they share the same predators, this coloration then protects the harmless ones, even though they do not have the same level of physical or chemical defenses against predation as the organism they mimic. Many insect species mimic the coloration of wasps or bees, which are stinging, venomous insects, thereby discouraging predation (Figure 45.22).

Figure 45.22 Batesian mimicry occurs when a harmless species mimics the coloration of a harmful species, as is seen with the (a) bumblebee and (b) bee-like robber fly. (credit a, b: modification of work by Cory Zanker)

In **Müllerian mimicry**, multiple species share the same warning coloration, but all of them actually have defenses. Figure 45.23 shows a variety of foul-tasting butterflies with similar coloration. In **Emsleyan/Mertensian mimicry**, a deadly prey mimics a less dangerous one, such as the venomous coral snake mimicking the nonvenomous milk snake. This type of mimicry is extremely rare and more difficult to understand than the previous two types. For this type of mimicry to work, it is essential that eating the milk snake has unpleasant but not fatal consequences. Then, these predators learn not to eat snakes with this coloration, protecting the coral snake as well. If the snake were fatal to the predator, there would be no opportunity for the predator to learn not to eat it, and the benefit for the less toxic species would disappear.

Figure 45.23 Several unpleasant-tasting *Heliconius* butterfly species share a similar color pattern with better-tasting varieties, an example of Müllerian mimicry. (credit: Joron M, Papa R, Beltrán M, Chamberlain N, Mavárez J, et al.)

LINK TO LEARNING

Go to this website (http://openstax.org/l/find_the_mimic) to view stunning examples of mimicry.

Competitive Exclusion Principle

Resources are often limited within a habitat and multiple species may compete to obtain them. All species have an ecological niche in the ecosystem, which describes how they acquire the resources they need and how they interact with other species in the community. The **competitive exclusion principle** states that two species cannot occupy the same niche in a habitat. In other words, different species cannot coexist in a community if they are competing for all the same resources. An example of this principle is shown in Figure 45.24, with two protozoan species, *Paramecium aurelia* and *Paramecium caudatum*. When grown individually in the laboratory, they both thrive. But when they are placed together in the same test tube (habitat), *P. aurelia* outcompetes *P. caudatum* for food, leading to the latter's eventual extinction.

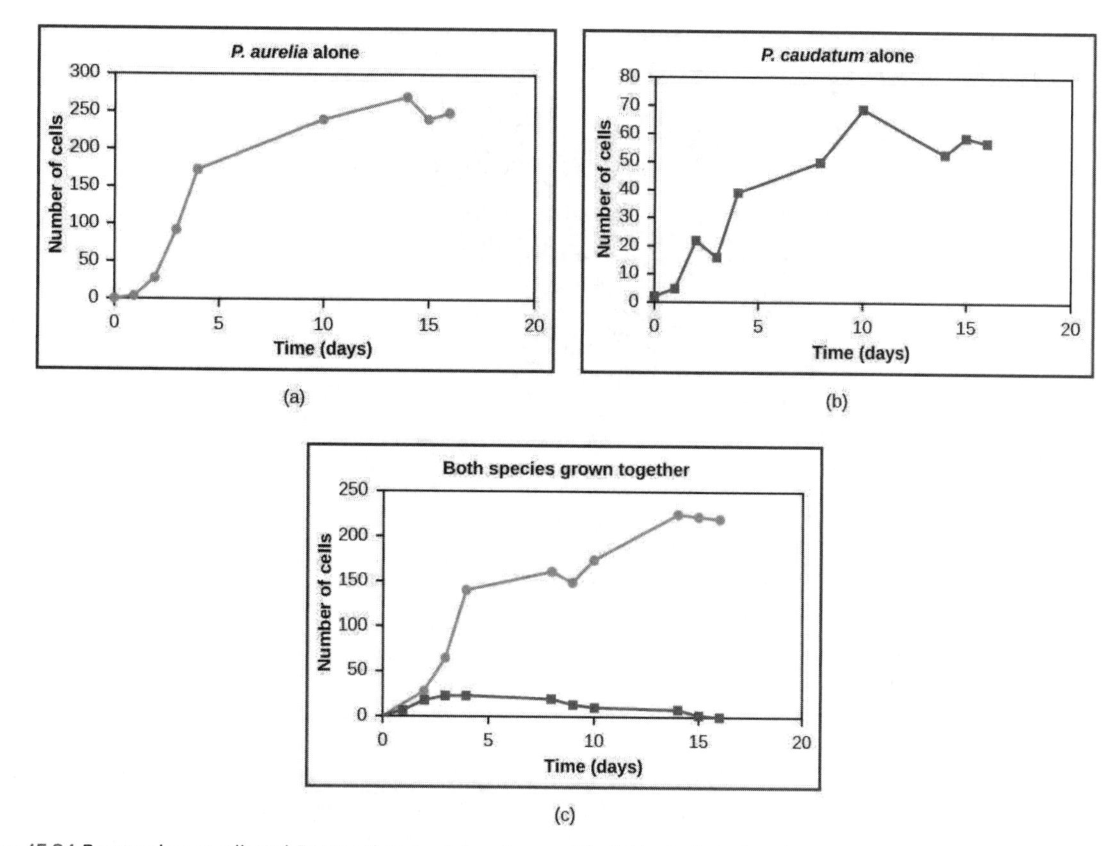

Figure 45.24 *Paramecium aurelia* and *Paramecium caudatum* grow well individually, but when they compete for the same resources, the *P. aurelia* outcompetes the *P. caudatum*.

This exclusion may be avoided if a population evolves to make use of a different resource, a different area of the habitat, or feeds during a different time of day, called resource partitioning. The two organisms are then said to occupy different microniches. These organisms coexist by minimizing direct competition.

Symbiosis

Symbiotic relationships, or **symbioses** (plural), are close interactions between individuals of different species over an extended period of time which impact the abundance and distribution of the associating populations. Most scientists accept this definition, but some restrict the term to only those species that are mutualistic, where both individuals benefit from the interaction. In this discussion, the broader definition will be used.

Commensalism

A **commensal** relationship occurs when one species benefits from the close, prolonged interaction, while the other neither benefits nor is harmed. Birds nesting in trees provide an example of a commensal relationship (Figure 45.25). The tree is not harmed by the presence of the nest among its branches. The nests are light and produce little strain on the structural integrity of the branch, and most of the leaves, which the tree uses to get energy by photosynthesis, are above the nest so they are unaffected. The bird, on the other hand, benefits greatly. If the bird had to nest in the open, its eggs and young would be vulnerable to predators. Another example of a commensal relationship is the pilot fish and the shark. The pilot fish feed on the leftovers of the host's meals, and the host is not affected in any way.

Figure 45.25 The southern masked-weaver bird is starting to make a nest in a tree in Zambezi Valley, Zambia. This is an example of a commensal relationship, in which one species (the bird) benefits, while the other (the tree) neither benefits nor is harmed. (credit: "Hanay"/Wikimedia Commons)

Mutualism

A second type of symbiotic relationship is called **mutualism**, where two species benefit from their interaction. Some scientists believe that these are the only true examples of symbiosis. For example, termites have a mutualistic relationship with protozoa that live in the insect's gut (Figure 45.26a). The termite benefits from the ability of bacterial symbionts within the protozoa to digest cellulose. The termite itself cannot do this, and without the protozoa, it would not be able to obtain energy from its food (cellulose from the wood it chews and eats). The protozoa and the bacterial symbionts benefit by having a protective environment and a constant supply of food from the wood chewing actions of the termite. Lichens have a mutualistic relationship between fungus and photosynthetic algae or bacteria (Figure 45.26b). As these symbionts grow together, the glucose produced by the algae provides nourishment for both organisms, whereas the physical structure of the lichen protects the algae from the elements and makes certain nutrients in the atmosphere more available to the algae.

Figure 45.26 (a) Termites form a mutualistic relationship with symbiotic protozoa in their guts, which allow both organisms to obtain energy from the cellulose the termite consumes. (b) Lichen is a fungus that has symbiotic photosynthetic algae living inside its cells. (credit a: modification of work by Scott Bauer, USDA; credit b: modification of work by Cory Zanker)

Parasitism

A **parasite** is an organism that lives in or on another living organism and derives nutrients from it. In this relationship, the parasite benefits, but the **host** is harmed. The host is usually weakened by the parasite as it siphons resources the host would normally use to maintain itself. The parasite, however, is unlikely to kill the host, especially not quickly, because this would allow no time for the organism to complete its reproductive cycle by spreading to another host.

The reproductive cycles of parasites are often very complex, sometimes requiring more than one host species. A tapeworm is a parasite that causes disease in humans when contaminated, undercooked meat is consumed (Figure 45.27). The tapeworm can live inside the intestine of the host for several years, benefiting from the food the host is eating, and may grow to be over 50 ft long by adding segments. The parasite moves from species to species in a cycle, making two hosts necessary to complete its life cycle.

Another common parasite is *Plasmodium falciparum*, the protozoan cause of malaria, a significant disease in many parts of the world. Living in human liver and red blood cells, the organism reproduces asexually in the gut of blood-feeding mosquitoes to complete its life cycle. Thus malaria is spread from human to human by mosquitoes, one of many arthropod-borne infectious diseases.

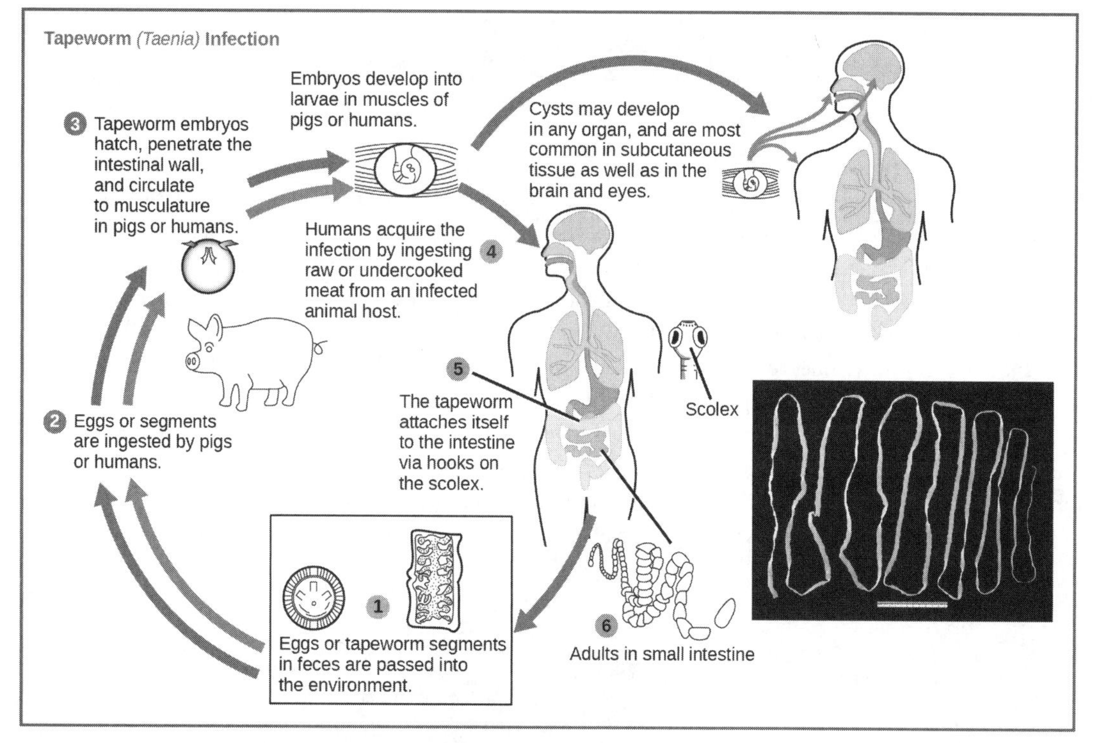

Figure 45.27 This diagram shows the life cycle of a pork tapeworm (*Taenia solium*), a human worm parasite. (credit: modification of work by CDC)

Characteristics of Communities

Communities are complex entities that can be characterized by their structure (the types and numbers of species present) and dynamics (how communities change over time). Understanding community structure and dynamics enables community ecologists to manage ecosystems more effectively.

Foundation Species

Foundation species are considered the "base" or "bedrock" of a community, having the greatest influence on its overall structure. They are usually the primary producers: organisms that bring most of the energy into the community. Kelp, or brown algae, is a foundation species, forming the basis of the kelp forests off the coast of California.

Foundation species may physically modify the environment to produce and maintain habitats that benefit the other organisms that use them. An example is the photosynthetic corals of the coral reef (Figure 45.28). Corals themselves are not photosynthetic, but harbor symbionts within their body tissues (dinoflagellates called zooxanthellae) that perform photosynthesis; this is another example of a mutualism. The exoskeletons of living and dead coral make up most of the reef structure, which protects many other species from waves and ocean currents.

Figure 45.28 Coral is the foundation species of coral reef ecosystems. (credit: Jim E. Maragos, USFWS)

Biodiversity, Species Richness, and Relative Species Abundance

Biodiversity describes a community's biological complexity: it is measured by the number of different species (species richness) in a particular area and their relative abundance (species evenness). The area in question could be a habitat, a biome, or the entire biosphere. **Species richness** is the term that is used to describe the number of species living in a habitat or biome. Species richness varies across the globe (Figure 45.29). One factor in determining species richness is latitude, with the greatest species richness occurring in ecosystems near the equator, which often have warmer temperatures, large amounts of rainfall, and low seasonality. The lowest species richness occurs near the poles, which are much colder, drier, and thus less conducive to life in Geologic time (time since glaciations). The predictability of climate or productivity is also an important factor. Other factors influence species richness as well. For example, the study of **island biogeography** attempts to explain the relatively high species richness found in certain isolated island chains, including the Galápagos Islands that inspired the young Darwin. **Relative species abundance** is the number of individuals in a species relative to the total number of individuals in all species within a habitat, ecosystem, or biome. Foundation species often have the highest relative abundance of species.

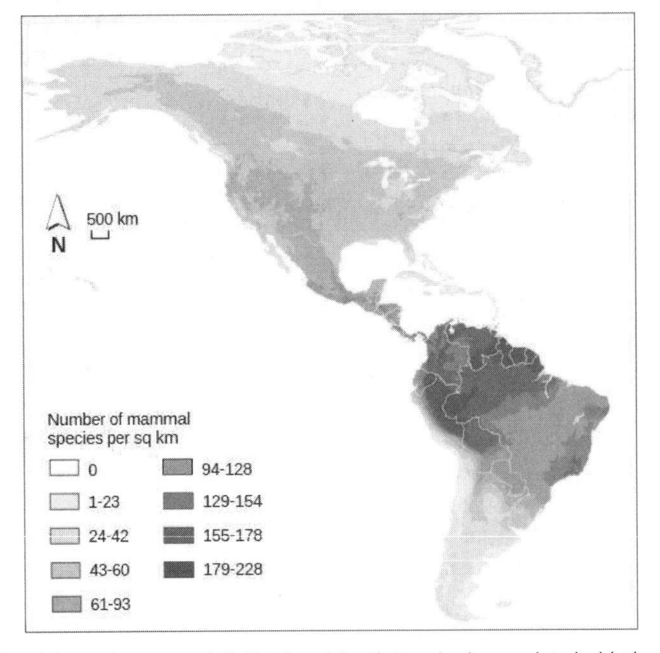

Figure 45.29 The greatest species richness for mammals in North and South America is associated with the equatorial latitudes. (credit: modification of work by NASA, CIESIN, Columbia University)

Keystone Species

A **keystone species** is one whose presence is key to maintaining biodiversity within an ecosystem and to upholding an ecological community's structure. The intertidal sea star, *Pisaster ochraceus*, of the northwestern United States is a keystone species (Figure 45.30). Studies have shown that when this organism is removed from communities, populations of their natural prey (mussels) increase, completely altering the species composition and reducing biodiversity. Another keystone species is the banded tetra, a fish in tropical streams, which supplies nearly all of the phosphorus, a necessary inorganic nutrient, to the rest of the community. If these fish were to become extinct, the community would be greatly affected.

Figure 45.30 The *Pisaster ochraceus* sea star is a keystone species. (credit: Jerry Kirkhart)

Everyday Connection

Invasive Species

Invasive species are nonnative organisms that, when introduced to an area out of their native range, threaten the ecosystem balance of that habitat. Many such species exist in the United States, as shown in Figure 45.31. Whether enjoying a forest hike, taking a summer boat trip, or simply walking down an urban street, you have likely encountered an invasive species.

Figure 45.31 In the United States, invasive species like (a) purple loosestrife (*Lythrum salicaria*) and the (b) zebra mussel (*Dreissena polymorpha*) threaten certain aquatic ecosystems. Some forests are threatened by the spread of (c) common buckthorn (*Rhamnus cathartica*), (d) garlic mustard (*Alliaria petiolata*), and (e) the emerald ash borer (*Agrilus planipennis*). The (f) European starling (*Sturnus vulgaris*) may compete with native bird species for nest holes. (credit a: modification of work by Liz West; credit b: modification of work by M. McCormick, NOAA; credit c: modification of work by E. Dronkert; credit d: modification of work by Dan Davison; credit e: modification of work by USDA; credit f: modification of work by Don DeBold)

One of the many recent proliferations of an invasive species concerns the growth of Asian carp populations. Asian carp were introduced to the United States in the 1970s by fisheries and sewage treatment facilities that used the fish's excellent filter feeding capabilities to clean their ponds of excess plankton. Some of the fish escaped, however, and by the 1980s they had colonized many waterways of the Mississippi River basin, including the Illinois and Missouri Rivers.

Voracious eaters and rapid reproducers, Asian carp may outcompete native species for food, potentially leading to their extinction. For example, black carp are voracious eaters of native mussels and snails, limiting this food source for native fish species. Silver carp eat plankton that native mussels and snails feed on, reducing this food source by a different alteration of the food web. In some areas of the Mississippi River, Asian carp species have become the most predominant, effectively outcompeting native fishes for habitat. In some parts of the Illinois River, Asian carp constitute 95 percent of the community's biomass. Although edible, the fish is bony and not a desired food in the United States. Moreover, their presence threatens the native fish and fisheries of the Great Lakes, which are important to local economies and recreational anglers. Asian carp have even injured humans. The fish, frightened by the sound of approaching motorboats, thrust themselves into the air, often landing in the boat or directly hitting the boaters.

The Great Lakes and their prized salmon and lake trout fisheries are also being threatened by these invasive fish. Asian carp have already colonized rivers and canals that lead into Lake Michigan. One infested waterway of particular importance is the Chicago Sanitary and Ship Channel, the major supply waterway linking the Great Lakes to the Mississippi River. To prevent the Asian carp from leaving the canal, a series of electric barriers have been successfully used to discourage their migration; however, the threat is significant enough that several states and Canada have sued to have the Chicago channel permanently cut off from Lake Michigan. Local and national politicians have weighed in on how to solve the problem, but no one knows whether the Asian carp will ultimately be considered a nuisance, like other invasive species such as the water hyacinth and zebra mussel, or whether it will be the destroyer of the largest freshwater fishery of the world.

The issues associated with Asian carp show how population and community ecology, fisheries management, and politics intersect on issues of vital importance to the human food supply and economy. Socio-political issues like this make extensive use of the sciences of population ecology (the study of members of a particular species occupying a particular area known as a habitat) and community ecology (the study of the interaction of all species within a habitat).

Community Dynamics

Community dynamics are the changes in community structure and composition over time. Sometimes these changes are induced by **environmental disturbances** such as volcanoes, earthquakes, storms, fires, and climate change. Communities with a stable structure are said to be at equilibrium. Following a disturbance, the community may or may not return to the equilibrium state.

Succession describes the sequential appearance and disappearance of species in a community over time. In **primary succession**, newly exposed or newly formed land is colonized by living things; in **secondary succession**, part of an ecosystem is disturbed and remnants of the previous community remain.

Primary Succession and Pioneer Species

Primary succession occurs when new land is formed or rock is exposed: for example, following the eruption of volcanoes, such as those on the Big Island of Hawaii. As lava flows into the ocean, new land is continually being formed. On the Big Island, approximately 32 acres of land is added each year. First, weathering and other natural forces break down the substrate enough for the establishment of certain hearty plants and lichens with few soil requirements, known as **pioneer species** (Figure 45.32). These species help to further break down the mineral rich lava into soil where other, less hardy species will grow and eventually replace the pioneer species. In addition, as these early species grow and die, they add to an ever-growing layer of decomposing organic material and contribute to soil formation. Over time the area will reach an equilibrium state, with a set of organisms quite different from the pioneer species.

Figure 45.32 During primary succession in lava on Maui, Hawaii, succulent plants are the pioneer species. (credit: Forest and Kim Starr)

Secondary succession

A classic example of secondary succession occurs in oak and hickory forests cleared by wildfire (Figure 45.33). Wildfires will burn most vegetation and kill those animals unable to flee the area. Their nutrients, however, are returned to the ground in the form of ash. Thus, even when areas are devoid of life due to severe fires, the area will soon be ready for new life to take hold.

Before the fire, the vegetation was dominated by tall trees with access to the major plant energy resource: sunlight. Their height gave them access to sunlight while also shading the ground and other low-lying species. After the fire, though, these trees are no longer dominant. Thus, the first plants to grow back are usually annual plants followed within a few years by quickly growing and spreading grasses and other pioneer species. Due to, at least in part, changes in the environment brought on by the growth of the grasses and other species, over many years, shrubs will emerge along with small pine, oak, and hickory trees. These organisms are called intermediate species. Eventually, over 150 years, the forest will reach its equilibrium point where species composition is no longer changing and resembles the community before the fire. This equilibrium state is referred to as the **climax community**, which will remain stable until the next disturbance.

Figure 45.33 Secondary succession is shown in an oak and hickory forest after a forest fire.

45.7 Behavioral Biology: Proximate and Ultimate Causes of Behavior

By the end of this section, you will be able to do the following:

- Compare innate and learned behavior
- Discuss how movement and migration behaviors are a result of natural selection
- Discuss the different ways members of a population communicate with each other
- Give examples of how species use energy for mating displays and other courtship behaviors
- Differentiate between various mating systems
- Describe different ways that species learn

Behavior is the change in activity of an organism in response to a stimulus. **Behavioral biology** is the study of the biological and evolutionary bases for such changes. The idea that behaviors evolved as a result of the pressures of natural selection is not new.

For decades, several types of scientists have studied animal behavior. Biologists do so in the science of **ethology**; psychologists in the science of comparative psychology; and other scientists in the science of neurobiology. The first two, ethology and comparative psychology, are the most consequential for the study of behavioral biology.

One goal of behavioral biology is to distinguish between the **innate behaviors**, which have a strong genetic component and are largely independent of environmental influences, from the **learned behaviors**, which result from environmental conditioning. Innate behavior, or instinct, is important because there is no risk of an incorrect behavior being learned. They are "hard wired" into the system. On the other hand, learned behaviors, although riskier, are flexible, dynamic, and can be altered according to changes in the environment.

Innate Behaviors: Movement and Migration

Innate or instinctual behaviors rely on response to stimuli. The simplest example of this is a **reflex action**, an involuntary and rapid response to stimulus. To test the "knee-jerk" reflex, a doctor taps the patellar tendon below the kneecap with a rubber hammer. The stimulation of the nerves leads to the reflex of extending the leg at the knee. This is similar to the reaction of someone who touches a hot stove and instinctually pulls his or her hand away. Even humans, with our great capacity to learn, still exhibit a variety of innate behaviors.

Kinesis and Taxis

Another activity or movement of innate behavior is **kinesis**, or the undirected movement in response to a stimulus. Orthokinesis is the increased or decreased speed of movement of an organism in response to a stimulus. Woodlice, for example, increase their speed of movement when exposed to high or low temperatures. This movement, although random, increases the probability that the insect spends less time in the unfavorable environment. Another example is klinokinesis, an increase in turning behaviors. It is exhibited by bacteria such as *E. coli* which, in association with orthokinesis, helps the organisms randomly find a more hospitable environment.

A similar, but more directed version of kinesis is **taxis**: the directed movement towards or away from a stimulus. This movement can be in response to light (phototaxis), chemical signals (chemotaxis), or gravity (geotaxis) and can be directed toward (positive) or away (negative) from the source of the stimulus. An example of a positive chemotaxis is exhibited by the unicellular protozoan *Tetrahymena thermophila*. This organism swims using its cilia, at times moving in a straight line, and at other times making turns. The attracting chemotactic agent alters the frequency of turning as the organism moves directly toward the source, following the increasing concentration gradient.

Fixed Action Patterns

A **fixed action pattern** is a series of movements elicited by a stimulus such that even when the stimulus is removed, the pattern goes on to completion. An example of such a behavior occurs in the three-spined stickleback, a small freshwater fish (Figure 45.34). Males of this species develop a red belly during breeding season and show instinctual aggressiveness to other males during this time. In laboratory experiments, researchers exposed such fish to objects that in no way resemble a fish in their shape, but which were painted red on their lower halves. The male sticklebacks responded aggressively to the objects just as if they were real male sticklebacks.

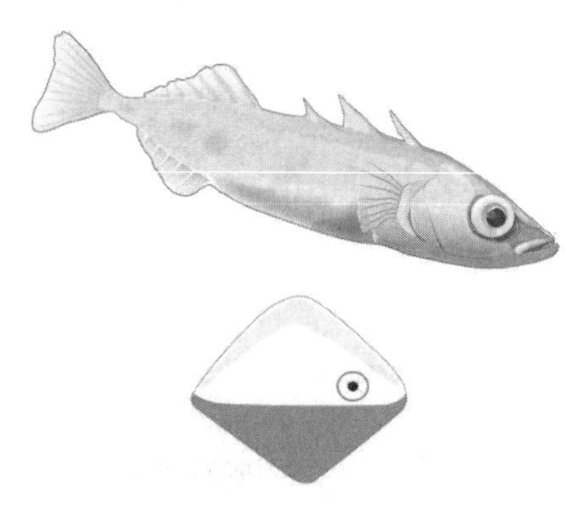

Figure 45.34 Male three-spined stickleback fish exhibit a fixed action pattern. During mating season, the males, which develop a bright red belly, react strongly to red-bottomed objects that in no way resemble fish.

Migration

Migration is the long-range seasonal movement of animals. It is an evolved, adapted response to variation in resource availability, and it is a common phenomenon found in all major groups of animals. Birds fly south for the winter to get to warmer climates with sufficient food, and salmon migrate to their spawning grounds. The popular 2005 documentary *March of the Penguins* followed the 62-mile migration of emperor penguins through Antarctica to bring food back to their breeding site and to their young. Wildebeests (Figure 45.35) migrate over 1800 miles each year in search of new grasslands.

Figure 45.35 Wildebeests migrate in a clockwise fashion over 1800 miles each year in search of rain-ripened grass. (credit: Eric Inafuku)

Although migration is thought of as innate behavior, only some migrating species always migrate (obligate migration). Animals that exhibit facultative migration can choose to migrate or not. Additionally, in some animals, only a portion of the population migrates, whereas the rest does not migrate (incomplete migration). For example, owls that live in the tundra may migrate in years when their food source, small rodents, is relatively scarce, but not migrate during the years when rodents are plentiful.

Foraging

Foraging is the act of searching for and exploiting food resources. Feeding behaviors that maximize energy gain and minimize energy expenditure are called optimal foraging behaviors, and these are favored by natural section. The painted stork, for example, uses its long beak to search the bottom of a freshwater marshland for crabs and other food (Figure 45.36).

Figure 45.36 The painted stork uses its long beak to forage. (credit: J.M. Garg)

Innate Behaviors: Living in Groups

Not all animals live in groups, but even those that live relatively solitary lives, with the exception of those that can reproduce asexually, must mate. Mating usually involves one animal signaling another so as to communicate the desire to mate. There are several types of energy-intensive behaviors or displays associated with mating, called mating rituals. Other behaviors found in populations that live in groups are described in terms of which animal benefits from the behavior. In selfish behavior, only the animal in question benefits; in altruistic behavior, one animal's actions benefit another animal; cooperative behavior describes when both animals benefit. All of these behaviors involve some sort of communication between population members.

Communication within a Species

Animals communicate with each other using stimuli known as **signals**. An example of this is seen in the three-spined stickleback, where the visual signal of a red region in the lower half of a fish signals males to become aggressive and signals females to mate. Other signals are chemical (pheromones), aural (sound), visual (courtship and aggressive displays), or tactile (touch). These types of communication may be instinctual or learned or a combination of both. These are not the same as the communication we associate with language, which has been observed only in humans and perhaps in some species of primates and cetaceans.

A pheromone is a secreted chemical signal used to obtain a response from another individual of the same species. The purpose of pheromones is to elicit a specific behavior from the receiving individual. Pheromones are especially common among social insects, but they are used by many species to attract the opposite sex, to sound alarms, to mark food trails, and to elicit other, more complex behaviors. Even humans are thought to respond to certain pheromones called axillary steroids. These chemicals influence human perception of other people, and in one study were responsible for a group of women synchronizing their menstrual cycles. The role of pheromones in human-to-human communication is not fully understood and continues to be researched.

Songs are an example of an aural signal, one that needs to be heard by the recipient. Perhaps the best known of these are songs of birds, which identify the species and are used to attract mates. Other well-known songs are those of whales, which are of such low frequency that they can travel long distances underwater. Dolphin species communicate with each other (and occasionally even with other species of dolphins) using a wide variety of vocalizations. Male crickets make chirping sounds using a specialized organ to attract a mate, repel other males, and to announce a successful mating.

Courtship displays are a series of ritualized visual behaviors (signals) designed to attract and convince a member of the opposite sex to mate. These displays are ubiquitous in the animal kingdom. Often these displays involve a series of steps, including an initial display by one member followed by a response from the other. If at any point, the display is performed incorrectly or a proper response is not given, the mating ritual is abandoned and the mating attempt will be unsuccessful. The mating display of the common stork is shown in Figure 45.37.

Aggressive displays are also common in the animal kingdom. For example, a dog bares its teeth when it wants another dog to back down. Presumably, these displays communicate not only the willingness of the animal to fight, but also its fighting ability. Although these displays do signal aggression on the part of the sender, it is thought that these displays are actually a mechanism to reduce the amount of actual fighting that occurs between members of the same species: they allow individuals to assess the fighting ability of their opponent and thus decide whether it is "worth the fight." The testing of certain hypotheses using game theory has led to the conclusion that some of these displays may overstate an animal's actual fighting ability and are used to "bluff" the opponent. This type of interaction, even if "dishonest," would be favored by natural selection if it is successful more times than not.

Figure 45.37 This stork's courtship display is designed to attract potential mates. (credit: Linda "jinterwas"/Flickr)

Distraction displays are seen in birds and some fish. They are designed to attract a predator away from the nest. This is an example of an altruistic behavior: it benefits the young more than the individual performing the display, which is putting itself at risk by doing so.

Many animals, especially primates, communicate with other members in the group through touch. Activities such as grooming, touching the shoulder or root of the tail, embracing, lip contact, and greeting ceremonies have all been observed in the Indian langur, an Old World monkey. Similar behaviors are found in other primates, especially in the great apes.

LINK TO LEARNING

The killdeer bird distracts predators from its eggs by faking a broken wing display in this video taken in Boise, Idaho.

Click to view content (https://www.openstax.org/l/killdeer_bird)

Altruistic Behaviors

Behaviors that lower the fitness of the individual but increase the fitness of another individual are termed altruistic. Examples of such behaviors are seen widely across the animal kingdom. Social insects such as worker bees have no ability to reproduce, yet they maintain the queen so she can populate the hive with her offspring. Meerkats keep a sentry standing guard to warn the rest of the colony about intruders, even though the sentry is putting itself at risk. Wolves and wild dogs bring meat to pack members not present during a hunt. Lemurs take care of infants unrelated to them. Although on the surface, these behaviors appear to be altruistic, the truth may not be so simple.

There has been much discussion over why altruistic behaviors exist. Do these behaviors lead to overall evolutionary advantages for their species? Do they help the altruistic individual pass on its own genes? And what about such activities between unrelated individuals? One explanation for altruistic-type behaviors is found in the genetics of natural selection. In the 1976 book, *The Selfish Gene,* scientist Richard Dawkins attempted to explain many seemingly altruistic behaviors from the viewpoint of the gene itself. Although a gene obviously cannot be selfish in the human sense, it may appear that way if the sacrifice of an individual benefits related individuals that share genes that are identical by descent (present in relatives because of common lineage). Mammal parents make this sacrifice to take care of their offspring. Emperor penguins migrate miles in harsh conditions to bring food back for their young. Selfish gene theory has been controversial over the years and is still discussed among scientists in related fields.

Even less-related individuals, those with less genetic identity than that shared by parent and offspring, benefit from seemingly altruistic behavior. The activities of social insects such as bees, wasps, ants, and termites are good examples. Sterile workers in these societies take care of the queen because they are closely related to it, and as the queen has offspring, she is passing on genes from the workers indirectly. Thus, it is of fitness benefit for the worker to maintain the queen without having any direct chance of passing on its genes due to its sterility. The lowering of individual fitness to enhance the reproductive fitness of a relative and thus one's inclusive fitness evolves through **kin selection**. This phenomenon can explain many superficially altruistic behaviors seen in animals. However, these behaviors may not be truly defined as altruism in these cases because the actor is actually increasing its own fitness either directly (through its own offspring) or indirectly (through the inclusive fitness it gains through relatives that share genes with it).

Unrelated individuals may also act altruistically to each other, and this seems to defy the "selfish gene" explanation. An example of this observed in many monkey species where a monkey will present its back to an unrelated monkey to have that individual pick the parasites from its fur. After a certain amount of time, the roles are reversed and the first monkey now grooms the second monkey. Thus, there is reciprocity in the behavior. Both benefit from the interaction and their fitness is raised more than if neither cooperated nor if one cooperated and the other did not cooperate. This behavior is still not necessarily altruism, as the "giving" behavior of the actor is based on the expectation that it will be the "receiver" of the behavior in the future, termed reciprocal altruism. Reciprocal altruism requires that individuals repeatedly encounter each other, often the result of living in the same social group, and that cheaters (those that never "give back") are punished.

Evolutionary game theory, a modification of classical game theory in mathematics, has shown that many of these so-called "altruistic behaviors" are not altruistic at all. The definition of "pure" altruism, based on human behavior, is an action that benefits another without any direct benefit to oneself. Most of the behaviors previously described do not seem to satisfy this definition, and game theorists are good at finding "selfish" components in them. Others have argued that the terms "selfish" and "altruistic" should be dropped completely when discussing animal behavior, as they describe human behavior and may not be directly applicable to instinctual animal activity. What is clear, though, is that heritable behaviors that improve the chances of passing on one's genes or a portion of one's genes are favored by natural selection and will be retained in future generations as long as those behaviors convey a fitness advantage. These instinctual behaviors may then be applied, in special circumstances, to other species, as long as it doesn't lower the animal's fitness.

Finding Sex Partners

Not all animals reproduce sexually, but many that do have the same challenge: they need to find a suitable mate and often have to compete with other individuals to obtain one. Significant energy is spent in the process of locating, attracting, and mating with the sex partner. Two types of selection occur during this process: **intersexual selection**, where individuals of one sex choose mates of the other sex, and **intrasexual selection**, the competition for mates between species members of the same sex. Intersexual selection is often complex because choosing a mate may be based on a variety of visual, aural, tactile, and chemical cues. An example of intersexual selection is when female peacocks choose to mate with the male with the brightest plumage. This type of selection often leads to traits in the chosen sex that do not enhance survival, but are those traits most attractive to the opposite sex (often at the expense of survival). Intrasexual selection involves mating displays and aggressive mating rituals such as rams butting heads—the winner of these battles is the one that is able to mate. Many of these rituals use up considerable energy but result in the selection of the healthiest, strongest, and/or most dominant individuals for mating.

Three general mating systems, all involving innate as opposed to learned behaviors, are seen in animal populations: monogamous, polygynous, and polyandrous.

LINK TO LEARNING

Visit this website (http://openstax.org/l/sex_selection) for informative videos on sexual selection.

In **monogamous** systems, one male and one female are paired for at least one breeding season. In some animals, such as the gray wolf, these associations can last much longer, even a lifetime. Several theories may explain this type of mating system. The "mate-guarding hypothesis" states that males stay with the female to prevent other males from mating with her. This behavior is advantageous in such situations where mates are scarce and difficult to find. Another explanation is the "male-assistance hypothesis," where males that help guard and rear their young will have more and healthier offspring. Monogamy is observed in many bird populations where, in addition to the parental care from the female, the male is also a major provider of parental care for the chicks. A third explanation for the evolutionary advantages of monogamy is the "female-enforcement hypothesis." In this scenario, the female ensures that the male does not have other offspring that might compete with her own, so she actively interferes with the male's signaling to attract other mates.

Polygynous mating refers to one male mating with multiple females. In these situations, the female must be responsible for most of the parental care as the single male is not capable of providing care to that many offspring. In resourced-based polygyny, males compete for territories with the best resources, and then mate with females that enter the territory, drawn to its resource richness. The female benefits by mating with a dominant, genetically fit male; however, it is at the cost of having no male help in caring for the offspring. An example is seen in the yellow-rumped honeyguide, a bird whose males defend beehives because the females feed on their wax. As the females approach, the male defending the nest will mate with them. Harem mating structures are a type of polygynous system where certain males dominate mating while controlling a territory with resources. Harem mating occurs in elephant seals, where the alpha male dominates the mating within the group. A third type of polygyny is a lek system. Here there is a communal courting area where several males perform elaborate displays for females, and the females choose their mate from this group. This behavior is observed in several bird species including the sage grouse and the prairie chicken.

In **polyandrous** mating systems, one female mates with many males. These types of systems are much rarer than monogamous and polygynous mating systems. In pipefishes and seahorses, males receive the eggs from the female, fertilize them, protect them within a pouch, and give birth to the offspring (Figure 45.38). Therefore, the female is able to provide eggs to several males without the burden of carrying the fertilized eggs.

Figure 45.38 Polyandrous mating, in which one female mates with many males, occurs in the (a) seahorse and the (b) pipefish. (credit a: modification of work by Brian Gratwicke; credit b: modification of work by Stephen Childs)

Simple Learned Behaviors

The majority of the behaviors previously discussed were innate or at least have an innate component (variations on the innate behaviors may be learned). They are inherited and the behaviors do not change in response to signals from the environment. Conversely, learned behaviors, even though they may have instinctive components, allow an organism to adapt to changes in the environment and are modified by previous experiences. Simple learned behaviors include habituation and imprinting—both are important to the maturation process of young animals.

Habituation

Habituation is a simple form of learning in which an animal stops responding to a stimulus after a period of repeated exposure. This is a form of non-associative learning, as the stimulus is not associated with any punishment or reward. Prairie dogs typically sound an alarm call when threatened by a predator, but they become habituated to the sound of human footsteps when no harm is associated with this sound, therefore, they no longer respond to them with an alarm call. In this example, habituation is specific to the sound of human footsteps, as the animals still respond to the sounds of potential predators.

Imprinting

Imprinting is a type of learning that occurs at a particular age or a life stage that is rapid and independent of the species involved. Hatchling ducks recognize the first adult they see, their mother, and make a bond with her. A familiar sight is ducklings walking or swimming after their mothers (Figure 45.39). This is another type of non-associative learning, but is very important in the maturation process of these animals as it encourages them to stay near their mother so they will be protected, greatly increasing their chances of survival. However, if newborn ducks see a human before they see their mother, they will imprint on the human and follow it in just the same manner as they would follow their real mother.

Figure 45.39 The attachment of ducklings to their mother is an example of imprinting. (credit: modification of work by Mark Harkin)

LINK TO LEARNING

The International Crane Foundation has helped raise the world's population of whooping cranes from 21 individuals to about 600. Imprinting hatchlings has been a key to success: biologists wear full crane costumes so the birds never "see" humans. Watch this video to learn more.

Click to view content (https://www.openstax.org/l/whooping_crane)

Conditioned Behavior

Conditioned behaviors are types of associative learning, where a stimulus becomes associated with a consequence. During operant conditioning, the behavioral response is modified by its consequences, with regards to its form, strength, or frequency.

Classical Conditioning

In **classical conditioning**, a response called the conditioned response is associated with a stimulus that it had previously not been associated with, the conditioned stimulus. The response to the original, unconditioned stimulus is called the unconditioned response. The most cited example of classical conditioning is Ivan Pavlov's experiments with dogs (Figure 45.40). In Pavlov's experiments, the unconditioned response was the salivation of dogs in response to the unconditioned stimulus of seeing or smelling their food. The conditioning stimulus that researchers associated with the unconditioned response was the ringing of a bell. During conditioning, every time the animal was given food, the bell was rung. This was repeated during several trials. After some time, the dog learned to associate the ringing of the bell with food and to respond by salivating. After the conditioning period was finished, the dog would respond by salivating when the bell was rung, even when the unconditioned stimulus, the food, was absent. Thus, the ringing of the bell became the conditioned stimulus and the salivation became the conditioned response. Although it is thought by some scientists that the unconditioned and conditioned responses are identical, even Pavlov discovered that the saliva in the conditioned dogs had characteristic differences when compared to the unconditioned dog.

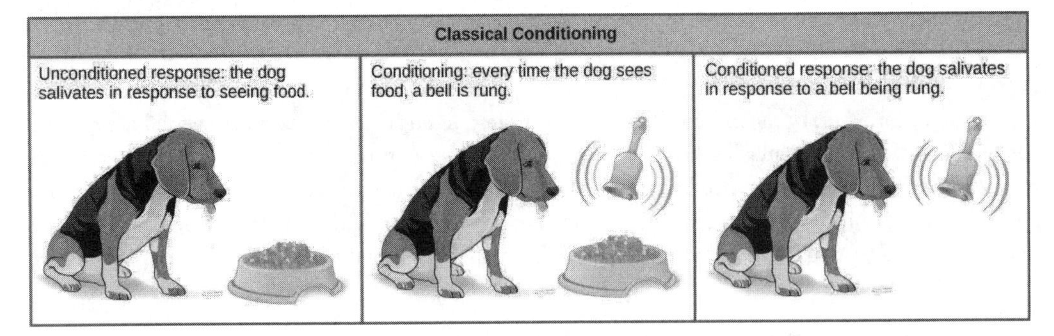

Figure 45.40 In the classic Pavlovian response, the dog becomes conditioned to associate the ringing of the bell with food.

It had been thought by some scientists that this type of conditioning required multiple exposures to the paired stimulus and response, but it is now known that this is not necessary in all cases, and that some conditioning can be learned in a single pairing experiment. Classical conditioning is a major tenet of behaviorism, a branch of psychological philosophy that proposes that all actions, thoughts, and emotions of living things are behaviors that can be treated by behavior modification and changes in the environment.

Operant Conditioning

In **operant conditioning**, the conditioned behavior is gradually modified by its consequences as the animal responds to the stimulus. A major proponent of such conditioning was psychologist B.F. Skinner, the inventor of the Skinner box. Skinner put rats in his boxes that contained a lever that would dispense food to the rat when depressed. While initially the rat would push the lever a few times by accident, it eventually associated pushing the lever with getting the food. This type of learning is an example of operant conditioning. Operant learning is the basis of most animal training. The conditioned behavior is continually modified by positive or negative reinforcement, often a reward such as food or some type of punishment, respectively. In this way, the animal is conditioned to associate a type of behavior with the punishment or reward, and, over time, can be induced to perform behaviors that they would not have done in the wild, such as the "tricks" dolphins perform at marine amusement park shows (Figure 45.41).

Figure 45.41 The training of dolphins by rewarding them with food is an example of positive reinforcement operant conditioning. (credit: Roland Tanglao)

Cognitive Learning

Classical and operant conditioning are inefficient ways for humans and other intelligent animals to learn. Some primates, including humans, are able to learn by imitating the behavior of others and by taking instructions. The development of complex language by humans has made **cognitive learning**, the manipulation of information using the mind, the most prominent method of human learning. In fact, that is how students are learning right now by reading this book. As students read, they can make mental images of objects or organisms and imagine changes to them, or behaviors by them, and anticipate the consequences. In addition to visual processing, cognitive learning is also enhanced by remembering past experiences, touching physical objects, hearing sounds, tasting food, and a variety of other sensory-based inputs. Cognitive learning is so powerful that it can be used to understand conditioning in detail. In the reverse scenario, conditioning cannot help someone learn about cognition.

Classic work on cognitive learning was done by Wolfgang Köhler with chimpanzees. He demonstrated that these animals were capable of abstract thought by showing that they could learn how to solve a puzzle. When a banana was hung in their cage too high for them to reach, and several boxes were placed randomly on the floor, some of the chimps were able to stack the boxes one on top of the other, climb on top of them, and get the banana. This implies that they could visualize the result of stacking the boxes even before they had performed the action. This type of learning is much more powerful and versatile than conditioning.

Cognitive learning is not limited to primates, although they are the most efficient in using it. Maze running experiments done with rats by H.C. Blodgett in the 1920s were the first to show cognitive skills in a simple mammal. The motivation for the animals to work their way through the maze was a piece of food at its end. In these studies, the animals in Group I were run in one trial per day and had food available to them each day on completion of the run (Figure 45.42). Group II rats were not fed in the maze for the first six days and then subsequent runs were done with food for several days after. Group III rats had food available on the third day and every day thereafter. The results were that the control rats, Group I, learned quickly, and figured out how to run the maze in seven days. Group III did not learn much during the three days without food, but rapidly caught up to the control group when given the food reward. Group II learned very slowly for the six days with no reward to motivate them, and they did not begin to catch up to the control group until the day food was given, and then it took two days longer to learn the maze.

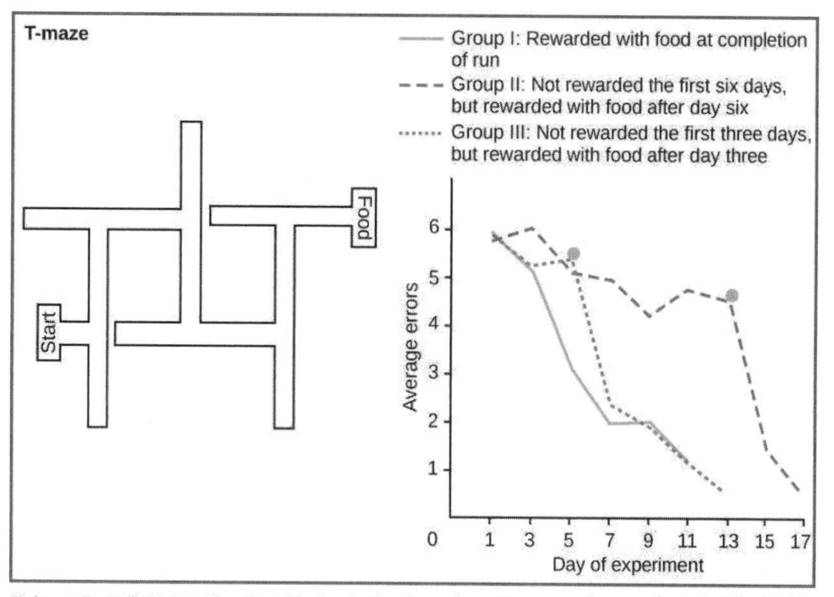

Figure 45.42 Group I (the green solid line) found food at the end of each trial, group II (the blue dashed line) did not find food for the first 6 days, and group III (the red dotted line) did not find food during runs on the first three days. Notice that rats given food earlier learned faster and eventually caught up to the control group. The orange dots on the group II and III lines show the days when food rewards were added to the mazes.

It may not be immediately obvious that this type of learning is different than conditioning. Although one might be tempted to believe that the rats simply learned how to find their way through a conditioned series of right and left turns, E.C. Tolman proved a decade later that the rats were making a representation of the maze in their minds, which he called a "cognitive map." This was an early demonstration of the power of cognitive learning and how these abilities were not just limited to humans.

Sociobiology

Sociobiology is an interdisciplinary science originally popularized by social insect researcher E.O. Wilson in the 1970s. Wilson defined the science as “the extension of population biology and evolutionary theory to social organization.”[9] The main thrust of sociobiology is that animal and human behavior, including aggressiveness and other social interactions, can be explained almost solely in terms of genetics and natural selection. This science is controversial; noted scientists such as the late Stephen Jay Gould criticized the approach for ignoring the environmental effects on behavior. This is another example of the “nature versus nurture” debate of the role of genetics versus the role of environment in determining an organism's characteristics.

Sociobiology also links genes with behaviors and has been associated with “biological determinism,” the belief that all behaviors are hardwired into our genes. No one disputes that certain behaviors can be inherited and that natural selection plays a role retaining them. It is the application of such principles to human behavior that sparks this controversy, which remains active today.

9Edward O. Wilson. *On Human Nature* (1978; repr., Cambridge: Harvard University Press, 2004), xx.

KEY TERMS

age structure proportion of population members at specific age ranges
aggressive display visual display by a species member to discourage other members of the same species or different species
aposematic coloration warning coloration used as a defensive mechanism against predation
Batesian mimicry type of mimicry where a non-harmful species takes on the warning colorations of a harmful one
behavior change in an organism's activities in response to a stimulus
behavioral biology study of the biology and evolution of behavior
biotic potential (r_{max}) maximal potential growth rate of a species
birth rate (*B*) number of births within a population at a specific point in time
camouflage avoid detection by blending in with the background
carrying capacity (*K*) number of individuals of a species that can be supported by the limited resources of a habitat
classical conditioning association of a specific stimulus and response through conditioning
climax community final stage of succession, where a stable community is formed by a characteristic assortment of plant and animal species
cognitive learning knowledge and skills acquired by the manipulation of information in the mind
commensalism relationship between species wherein one species benefits from the close, prolonged interaction, while the other species neither benefits nor is harmed
competitive exclusion principle no two species within a habitat can coexist when they compete for the same resources at the same place and time
conditioned behavior behavior that becomes associated with a specific stimulus through conditioning
courtship display visual display used to attract a mate
death rate (*D*) number of deaths within a population at a specific point in time
demographic-based population model modern model of population dynamics incorporating many features of the *r*- and *K*-selection theory
demography statistical study of changes in populations over time
density-dependent regulation regulation of population that is influenced by population density, such as crowding effects; usually involves biotic factors
density-independent regulation regulation of populations by factors that operate independent of population density, such as forest fires and volcanic eruptions; usually involves abiotic factors
distraction display visual display used to distract predators away from a nesting site
Emsleyan/Mertensian mimicry type of mimicry where a harmful species resembles a less harmful one
energy budget allocation of energy resources for body maintenance, reproduction, and parental care
environmental disturbance change in the environment caused by natural disasters or human activities
ethology biological study of animal behavior
exponential growth accelerating growth pattern seen in species under conditions where resources are not limiting
fecundity potential reproductive capacity of an individual
fixed action pattern series of instinctual behaviors that, once initiated, always goes to completion regardless of changes in the environment
foraging behaviors species use to find food
foundation species species which often forms the major structural portion of the habitat
habituation ability of a species to ignore repeated stimuli that have no consequence
host organism a parasite lives on
imprinting identification of parents by newborns as the first organism they see after birth
innate behavior instinctual behavior that is not altered by changes in the environment
intersexual selection selection of a desirable mate of the opposite sex
interspecific competition competition between species for resources in a shared habitat or environment
intrasexual selection competition between members of the same sex for a mate
intraspecific competition competition between members of the same species
island biogeography study of life on island chains and how their geography interacts with the diversity of species found there
iteroparity life history strategy characterized by multiple reproductive events during the lifetime of a species
J-shaped growth curve shape of an exponential growth curve
***K*-selected species** species suited to stable environments that produce a few, relatively large offspring and provide parental care
keystone species species whose presence is key to maintaining biodiversity in an ecosystem and to upholding an ecological community's structure
kin selection sacrificing one's own life so that one's genes will be passed on to future generations by relatives
kinesis undirected movement of an organism in response to a stimulus
learned behavior behavior that responds to changes in the

environment

life history inherited pattern of resource allocation under the influence of natural selection and other evolutionary forces

life table table showing the life expectancy of a population member based on its age

logistic growth leveling off of exponential growth due to limiting resources

mark and recapture technique used to determine population size in mobile organisms

migration long-range seasonal movement of animal species

monogamy mating system whereby one male and one female remain coupled for at least one mating season

mortality rate proportion of population surviving to the beginning of an age interval that die during the age interval

Müllerian mimicry type of mimicry where species share warning coloration and all are harmful to predators

mutualism symbiotic relationship between two species where both species benefit

one-child policy China's policy to limit population growth by limiting urban couples to have only one child or face the penalty of a fine

operant conditioning learned behaviors in response to positive and/or negative reinforcement

parasite organism that uses resources from another species, the host

pioneer species first species to appear in primary and secondary succession

polyandry mating system where one female mates with many males

polygyny mating system where one male mates with many females

population density number of population members divided by the area or volume being measured

population growth rate number of organisms added in each reproductive generation

population size (*N*) number of population members in a habitat at the same time

primary succession succession on land that previously has had no life

quadrat square made of various materials used to determine population size and density in slow moving or stationary organisms

***r*-selected species** species suited to changing environments that produce many offspring and provide little or no parental care

reflex action action in response to direct physical stimulation of a nerve

relative species abundance absolute population size of a particular species relative to the population sizes of other species within the community

S-shaped growth curve shape of a logistic growth curve

secondary succession succession in response to environmental disturbances that move a community away from its equilibrium

semelparity life history strategy characterized by a single reproductive event followed by death

signal method of communication between animals including those obtained by the senses of smell, hearing, sight, or touch

species dispersion pattern (also, species distribution pattern) spatial location of individuals of a given species within a habitat at a particular point in time

species richness number of different species in a community

survivorship curve graph of the number of surviving population members versus the relative age of the member

symbiosis close interaction between individuals of different species over an extended period of time that impacts the abundance and distribution of the associating populations

taxis directed movement in response to a stimulus

zero population growth steady population size where birth rates and death rates are equal

CHAPTER SUMMARY

45.1 Population Demography

Populations are individuals of a species that live in a particular habitat. Ecologists measure characteristics of populations: size, density, dispersion pattern, age structure, and sex ratio. Life tables are useful to calculate life expectancies of individual population members. Survivorship curves show the number of individuals surviving at each age interval plotted versus time.

45.2 Life Histories and Natural Selection

All species have evolved a pattern of living, called a life history strategy, in which they partition energy for growth, maintenance, and reproduction. These patterns evolve through natural selection; they allow species to adapt to their environment to obtain the resources they need to successfully reproduce. There is an inverse relationship between fecundity and parental care. A species may reproduce early in life to ensure surviving to a reproductive age or reproduce later in life to become larger and healthier and better able to give parental care. A species may reproduce once (semelparity) or many times (iteroparity) in its life.

45.3 Environmental Limits to Population Growth

Populations with unlimited resources grow exponentially, with an accelerating growth rate. When resources become limiting, populations follow a logistic growth curve. The population of a species will level off at the carrying capacity of its environment.

45.4 Population Dynamics and Regulation

Populations are regulated by a variety of density-dependent and density-independent factors. Species are divided into two categories based on a variety of features of their life history patterns: *r*-selected species, which have large numbers of offspring, and *K*-selected species, which have few offspring. The *r*- and *K*-selection theory has fallen out of use; however, many of its key features are still used in newer, demographically-based models of population dynamics.

45.5 Human Population Growth

The world's human population is growing at an exponential rate. Humans have increased the world's carrying capacity through migration, agriculture, medical advances, and communication. The age structure of a population allows us to predict population growth. Unchecked human population growth could have dire long-term effects on our environment.

45.6 Community Ecology

Communities include all the different species living in a given area. The variety of these species is called species richness. Many organisms have developed defenses against predation and herbivory, including mechanical defenses, warning coloration, and mimicry, as a result of evolution and the interaction with other members of the community. Two species cannot exist in the same habitat competing directly for the same resources. Species may form symbiotic relationships such as commensalism or mutualism. Community structure is described by its foundation and keystone species. Communities respond to environmental disturbances by succession (the predictable appearance of different types of plant species) until a stable community structure is established.

45.7 Behavioral Biology: Proximate and Ultimate Causes of Behavior

Behaviors are responses to stimuli. They can either be instinctual/innate behaviors, which are not influenced by the environment, or learned behaviors, which are influenced by environmental changes. Instinctual behaviors include mating systems and methods of communication. Learned behaviors include imprinting and habituation, conditioning, and, most powerfully, cognitive learning.

VISUAL CONNECTION QUESTIONS

1. Figure 45.2 As this graph shows, population density typically decreases with increasing body size. Why do you think this is the case?

2. Figure 45.10b If the major food source of the seals declines due to pollution or overfishing, which of the following would likely occur?
   a. The carrying capacity of seals would decrease, as would the seal population.
   b. The carrying capacity of seals would decrease, but the seal population would remain the same.
   c. The number of seal deaths would increase but the number of births would also increase, so the population size would remain the same.
   d. The carrying capacity of seals would remain the same, but the population of seals would decrease.

3. Figure 45.16 Age structure diagrams for rapidly growing, slow growing, and stable populations are shown in stages 1 through 3. What type of population change do you think stage 4 represents?

REVIEW QUESTIONS

4. Which of the following methods will tell an ecologist about both the size and density of a population?
   a. mark and recapture
   b. mark and release
   c. quadrat
   d. life table

5. Which of the following is best at showing the life expectancy of an individual within a population?
   a. quadrat
   b. mark and recapture
   c. survivorship curve
   d. life table

6. Humans have which type of survivorship curve?
   a. Type I
   b. Type II
   c. Type III
   d. Type IV

7. How is a clumped population distribution beneficial for prey animals?
   a. Being a member of a larger group provides protection for each individual from predators.
   b. Prey animals rely on each other to acquire food.
   c. Prey animals live in small family groups to raise young.
   d. Clumped population distributions ensure that at least one member of the population knows how to identify the seasonal migration route.

8. Which of the following is associated with long-term parental care?
   a. few offspring
   b. many offspring
   c. semelparity
   d. fecundity

9. Which of the following is associated with multiple reproductive episodes during a species' lifetime?
   a. semiparity
   b. iteroparity
   c. semelparity
   d. fecundity

10. Which of the following is associated with the reproductive potential of a species?
    a. few offspring
    b. many offspring
    c. semelparity
    d. fecundity

11. Species with limited resources usually exhibit a(n) _______ growth curve.
    a. logistic
    b. logical
    c. experimental
    d. exponential

12. The maximum rate of increased characteristic of a species is called its _______.
    a. limit
    b. carrying capacity
    c. biotic potential
    d. exponential growth pattern

13. The population size of a species capable of being supported by the environment is called its _______.
    a. limit
    b. carrying capacity
    c. biotic potential
    d. logistic growth pattern

14. Species that have many offspring at one time are usually:
    a. *r*-selected
    b. *K*-selected
    c. both *r*- and *K*-selected
    d. not selected

15. A forest fire is an example of _______ regulation.
    a. density-dependent
    b. density-independent
    c. *r*-selected
    d. *K*-selected

16. Primates are examples of:
    a. density-dependent species
    b. density-independent species
    c. *r*-selected species
    d. *K*-selected species

17. Which of the following statements does not support the conclusion that giraffes are *k*-selected species?
    a. Giraffes are approximately 6' tall and weigh 150 lbs at birth.
    b. Wild giraffes begin mating at 6-7 years of age.
    c. Newborn giraffes are capable of coordinated walking within an hour of birth, and running within 24 hours of birth.
    d. Giraffes rarely give birth to twins.

18. Which of the following events would **not** negatively impact Yellowstone's grey wolf carrying capacity?
    a. snow in winter
    b. a beaver damming a river upstream
    c. a forest fire
    d. chronic wasting disease in the deer population

19. A country with zero population growth is likely to be _______.
    a. in Africa
    b. in Asia
    c. economically developed
    d. economically underdeveloped

20. Which type of country has the greatest proportion of young individuals?
    a. economically developed
    b. economically underdeveloped
    c. countries with zero population growth
    d. countries in Europe

21. Which of the following is not a way that humans have increased the carrying capacity of the environment?
    a. agriculture
    b. using large amounts of natural resources
    c. domestication of animals
    d. use of language

22. The first species to live on new land, such as that formed from volcanic lava, are called _______.
    a. climax community
    b. keystone species
    c. foundation species
    d. pioneer species

23. Which type of mimicry involves multiple species with similar warning coloration that are all toxic to predators?
    a. Batesian mimicry
    b. Müllerian mimicry
    c. Emsleyan/Mertensian mimicry
    d. Mertensian mimicry

24. A symbiotic relationship where both of the coexisting species benefit from the interaction is called _______.
    a. commensalism
    b. parasitism
    c. mutualism
    d. communism

25. Which of the following is **not** a mutualistic relationship?
    a. a shark using an aquatic cleaning station
    b. a helminth feeding from its host
    c. a bumblebee collecting pollen from a flower
    d. bacteria living in the gut of humans

26. The ability of rats to learn how to run a maze is an example of _______.
    a. imprinting
    b. classical conditioning
    c. operant conditioning
    d. cognitive learning

27. The training of animals usually involves _______.
    a. imprinting
    b. classical conditioning
    c. operant conditioning
    d. cognitive learning

28. The sacrifice of the life of an individual so that the genes of relatives may be passed on is called _______.
    a. operant learning
    b. kin selection
    c. kinesis
    d. imprinting

29. Why are polyandrous mating systems more rare than polygynous matings?
    a. Only males are capable of multiple rounds of reproduction within a single breeding season.
    b. Only females care for the young.
    c. Females usually experience more intrasexual selection pressure than males.
    d. Females usually devote more energy to offspring production and development.

CRITICAL THINKING QUESTIONS

30. Describe how a researcher would determine the size of a penguin population in Antarctica using the mark and release method.

31. The CDC released the following data in its 2013 Vital Statistics report.

Age interval	Number dying in age interval	Number surviving at beginning of age interval
0-10	756	100,000
11-20	292	99,244
21-30	890	98,953
31-40	1,234	98,164
41-50	2,457	96,811
51-60	5,564	94,352
61-70	10,479	88,788

Table 45.3

Calculate the mortality rate for each age interval, and describe the trends in adult and childhood mortality per 100,000 births in the United States in 2013.

32. Why is long-term parental care not associated with having many offspring during a reproductive episode?
33. Describe the difference in evolutionary pressures experienced by an animal that begins reproducing early and an animal that reproduces late in its lifecycle.
34. Describe the rate of population growth that would be expected at various parts of the S-shaped curve of logistic growth.
35. Describe how the population of a species that survives a mass extinction event would change in size and growth pattern over time beginning immediately after the extinction event.
36. Give an example of how density-dependent and density-independent factors might interact.
37. Describe the age structures in rapidly growing countries, slowly growing countries, and countries with zero population growth.
38. Since the introduction of the Endangered Species Act the number of species on the protected list has more than doubled. Describe how the human population's growth pattern contributes to the rise in endangered species.
39. Describe the competitive exclusion principle and its effects on competing species.
40. Jaguars are a keystone species in the Amazon. Describe how they can be so essential to the ecosystem despite being significantly less abundant than many other species.
41. Describe Pavlov's dog experiments as an example of classical conditioning.
42. Describe the advantage of using an aural or pheromone signal to attract a mate as opposed to a visual signal. How might the population density contribute to the evolution of aural or visual mating rituals?

CHAPTER 46
Ecosystems

Figure 46.1 In the southwestern United States, rainy weather causes an increase in production of pinyon nuts, causing the deer mouse population to explode. Deer mice may carry a virus called *Sin Nombre* (a hantavirus) that causes respiratory disease in humans and has a high fatality rate. In 1992–1993, wet *El Niño* weather caused a *Sin Nombre* epidemic. Navajo healers, who were aware of the link between this disease and weather, predicted the outbreak. (credit "highway": modification of work by Phillip Capper; credit "mouse": modification of work by USFWS)

INTRODUCTION In 1993, an interesting example of ecosystem dynamics occurred when a rare lung disease struck inhabitants of the southwestern United States. This disease had an alarming rate of fatalities, killing more than half of early patients, many of whom were Native Americans. These formerly healthy young adults died from complete respiratory failure. The disease was unknown, and the Centers for Disease Control (CDC), the United States government agency responsible for managing potential epidemics, was brought in to investigate. The scientists could have learned about the disease had they known to talk with the Navajo healers who lived in the area and who had observed the connection between rainfall and mice populations, thereby predicting the 1993 outbreak.

The cause of the disease, determined within a few weeks by the CDC investigators, was the hantavirus known as *Sin Nombre*, the virus with "no name." With insights from traditional Navajo medicine, scientists were able to characterize the disease rapidly and institute effective health measures to prevent its spread. This example illustrates the importance of understanding the complexities of ecosystems and how they respond to changes in the environment.

Chapter Outline

46.1 Ecology of Ecosystems

By the end of this section, you will be able to do the following:

- Describe the basic ecosystem types
- Explain the methods that ecologists use to study ecosystem structure and dynamics
- Identify the different methods of ecosystem modeling
- Differentiate between food chains and food webs and recognize the importance of each

Life in an ecosystem is often about competition for limited resources, a characteristic of the theory of natural selection. Competition in communities (all living things within specific habitats) is observed both within species and among different species. The resources for which organisms compete include organic material, sunlight, and mineral nutrients, which provide the energy for living processes and the matter to make up organisms' physical structures. Other critical factors influencing community dynamics are the components of its physical and geographic environment: a habitat's latitude, amount of rainfall, topography (elevation), and available species. These are all important environmental variables that determine which organisms can exist within a particular area.

An **ecosystem** is a community of living organisms and their interactions with their abiotic (nonliving) environment. Ecosystems can be small, such as the tide pools found near the rocky shores of many oceans, or large, such as the Amazon Rainforest in Brazil (Figure 46.2).

Figure 46.2 A (a) tidal pool ecosystem in Matinicus Island in Maine is a small ecosystem, while the (b) Amazon Rainforest in Brazil is a large ecosystem. (credit a: modification of work by "takomabibelot"/Flickr; credit b: modification of work by Ivan Mlinaric)

There are three broad categories of ecosystems based on their general environment: freshwater, ocean water, and terrestrial. Within these broad categories are individual ecosystem types based on the organisms present and the type of environmental habitat.

Ocean ecosystems are the most common, comprising over 70 percent of the Earth's surface and consisting of three basic types: shallow ocean, deep ocean water, and deep ocean surfaces (the low depth areas of the deep oceans). The shallow ocean ecosystems include extremely biodiverse coral reef ecosystems, and the deep ocean surface is known for its large numbers of plankton and krill (small crustaceans) that support it. These two environments are especially important to aerobic respirators worldwide as the phytoplankton perform 40 percent of all photosynthesis on Earth. Although not as diverse as the other two, deep ocean ecosystems contain a wide variety of marine organisms. Such ecosystems exist even at the bottom of the ocean where light is unable to penetrate through the water.

Freshwater ecosystems are the rarest, occurring on only 1.8 percent of the Earth's surface. Lakes, rivers, streams, and springs comprise these systems. They are quite diverse, and they support a variety of fish, amphibians, reptiles, insects, phytoplankton, fungi, and bacteria.

Terrestrial ecosystems, also known for their diversity, are grouped into large categories called biomes, such as tropical rain forests, savannas, deserts, coniferous forests, deciduous forests, and tundra. Grouping these ecosystems into just a few biome categories obscures the great diversity of the individual ecosystems within them. For example, there is great variation in desert vegetation: the saguaro cacti and other plant life in the Sonoran Desert, in the United States, are relatively abundant compared to the desolate rocky desert of Boa Vista, an island off the coast of Western Africa (Figure 46.3).

Figure 46.3 Desert ecosystems, like all ecosystems, can vary greatly. The desert in (a) Saguaro National Park, Arizona, has abundant plant life, while the rocky desert of (b) Boa Vista island, Cape Verde, Africa, is devoid of plant life. (credit a: modification of work by Jay Galvin; credit b: modification of work by Ingo Wölbern)

Ecosystems are complex with many interacting parts. They are routinely exposed to various disturbances, or changes in the environment that effect their compositions: yearly variations in rainfall and temperature and the slower processes of plant growth, which may take several years. Many of these disturbances result from natural processes. For example, when lightning causes a forest fire and destroys part of a forest ecosystem, the ground is eventually populated by grasses, then by bushes and shrubs, and later by mature trees, restoring the forest to its former state. The impact of environmental disturbances caused by human activities is as important as the changes wrought by natural processes. Human agricultural practices, air pollution, acid rain, global deforestation, overfishing, eutrophication, oil spills, and waste dumping on land and into the ocean are all issues of concern to conservationists.

Equilibrium is the steady state of an ecosystem where all organisms are in balance with their environment and with each other. In ecology, two parameters are used to measure changes in ecosystems: resistance and resilience. **Resistance** is the ability of an ecosystem to remain at equilibrium in spite of disturbances. **Resilience** is the speed at which an ecosystem recovers equilibrium after being disturbed. Ecosystem resistance and resilience are especially important when considering human impact. The nature of an ecosystem may change to such a degree that it can lose its resilience entirely. This process can lead to the complete destruction or irreversible altering of the ecosystem.

Food Chains and Food Webs

The term "food chain" is sometimes used metaphorically to describe human social situations. Individuals who are considered successful are seen as being at the top of the food chain, consuming all others for their benefit, whereas the less successful are seen as being at the bottom.

The scientific understanding of a food chain is more precise than in its everyday usage. In ecology, a **food chain** is a linear sequence of organisms through which nutrients and energy pass: primary producers, primary consumers, and higher-level consumers are used to describe ecosystem structure and dynamics. There is a single path through the chain. Each organism in a food chain occupies what is called a **trophic level**. Depending on their role as producers or consumers, species or groups of species can be assigned to various trophic levels.

In many ecosystems, the bottom of the food chain consists of photosynthetic organisms (plants and/or phytoplankton), which are called **primary producers**. The organisms that consume the primary producers are herbivores: the **primary consumers**. **Secondary consumers** are usually carnivores that eat the primary consumers. **Tertiary consumers** are carnivores that eat other carnivores. Higher-level consumers feed on the next lower tropic levels, and so on, up to the organisms at the top of the food chain: the **apex consumers**. In the Lake Ontario food chain shown in Figure 46.4, the Chinook salmon is the apex consumer at the top of this food chain.

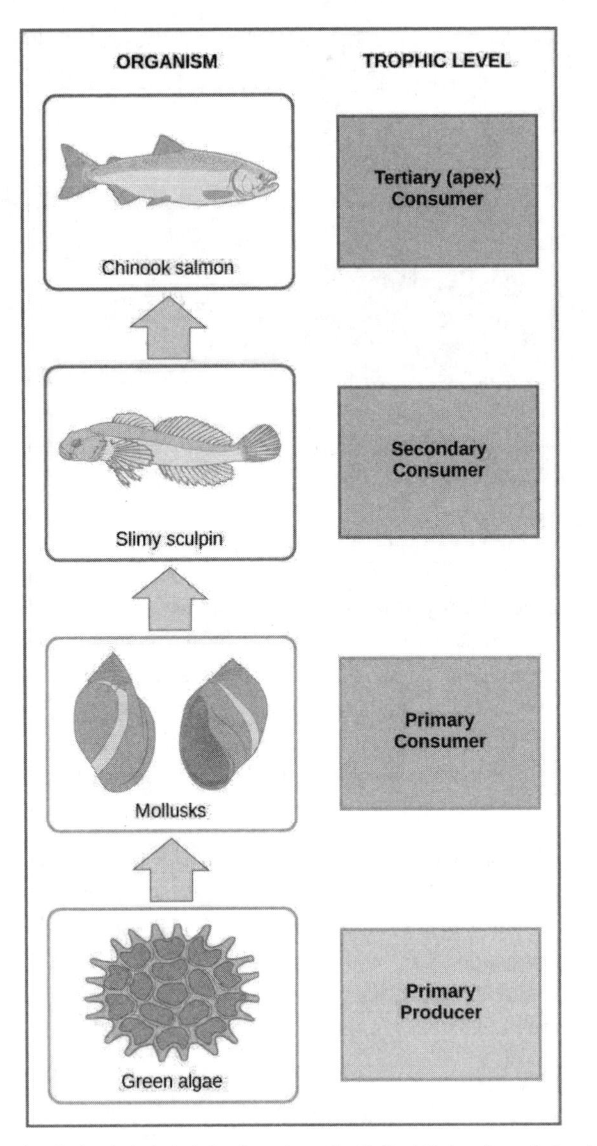

Figure 46.4 These are the trophic levels of a food chain in Lake Ontario at the United States-Canada border. Energy and nutrients flow from photosynthetic green algae at the bottom to the top of the food chain: the Chinook salmon.

One major factor that limits the length of food chains is energy. Energy is lost as heat between each trophic level due to the second law of thermodynamics. Thus, after a limited number of trophic energy transfers, the amount of energy remaining in the food chain may not be great enough to support viable populations at yet a higher trophic level.

The loss of energy between trophic levels is illustrated by the pioneering studies of Howard T. Odum in the Silver Springs, Florida, ecosystem in the 1940s (Figure 46.5). The primary producers generated 20,819 kcal/m^2/yr (kilocalories per square meter per year), the primary consumers generated 3368 kcal/m^2/yr, the secondary consumers generated 383 kcal/m^2/yr, and the tertiary consumers only generated 21 kcal/m^2/yr. Thus, there is little energy remaining for another level of consumers in this ecosystem.

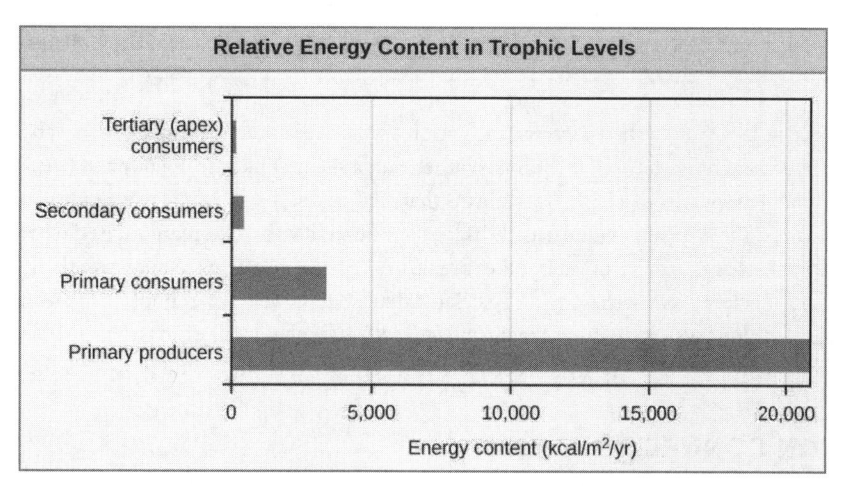

Figure 46.5 The relative energy in trophic levels in a Silver Springs, Florida, ecosystem is shown. Each trophic level has less energy available and supports fewer organisms at the next level.

There is a one problem when using food chains to accurately describe most ecosystems. Even when all organisms are grouped into appropriate trophic levels, some of these organisms can feed on species from more than one trophic level; likewise, some of these organisms can be eaten by species from multiple trophic levels. In other words, the linear model of ecosystems, the food chain, is not completely descriptive of ecosystem structure. A holistic model—which accounts for all the interactions between different species and their complex interconnected relationships with each other and with the environment—is a more accurate and descriptive model for ecosystems. A **food web** is a graphic representation of a holistic, nonlinear web of primary producers, primary consumers, and higher-level consumers used to describe ecosystem structure and dynamics (Figure 46.6).

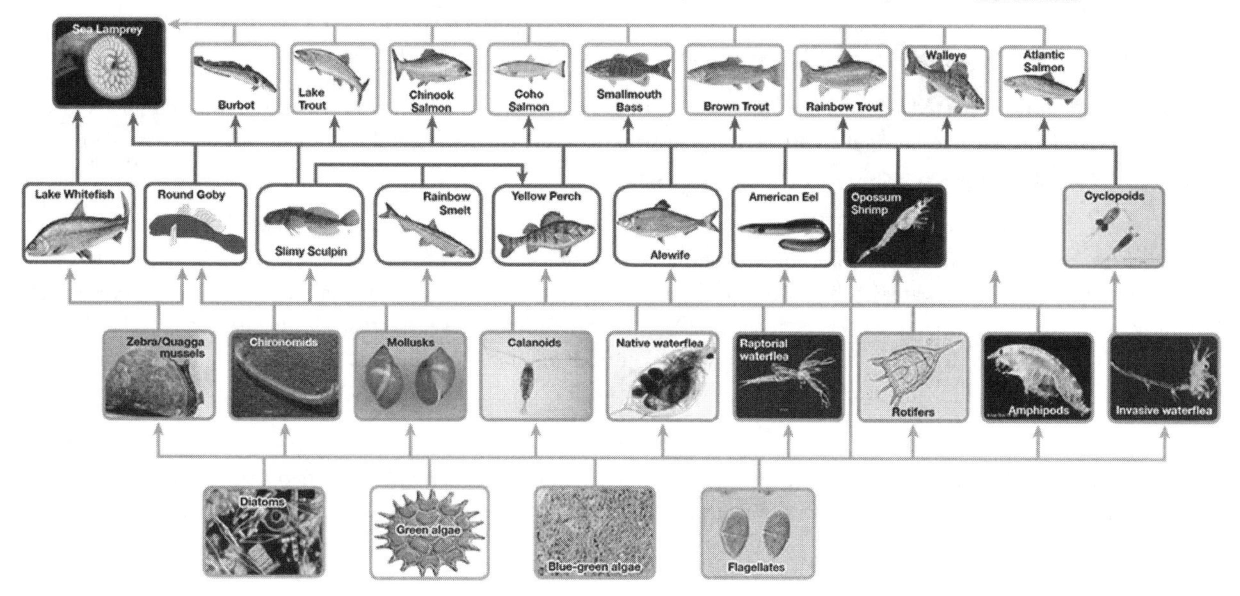

Figure 46.6 This food web shows the interactions between organisms across trophic levels in the Lake Ontario ecosystem. Primary producers are outlined in green, primary consumers in orange, secondary consumers in blue, and tertiary (apex) consumers in purple. Arrows point from an organism that is consumed to the organism that consumes it. Notice how some lines point to more than one trophic level. For example, the opossum shrimp eats both primary producers and primary consumers. (credit: NOAA, GLERL)

A comparison of the two types of structural ecosystem models shows strength in both. Food chains are more flexible for analytical modeling, are easier to follow, and are easier to experiment with, whereas food web models more accurately represent ecosystem structure and dynamics, and data can be directly used as input for simulation modeling.

LINK TO LEARNING

Head to this online interactive simulator (http://openstax.org/l/food_web) to investigate food web function. In the *Interactive Labs* box, under Food Web, click **Step 1**. Read the instructions first, and then click **Step 2** for additional instructions. When you are ready to create a simulation, in the upper-right corner of the *Interactive Labs* box, click **OPEN SIMULATOR**.

Two general types of food webs are often shown interacting within a single ecosystem. A **grazing food web** (such as the Lake Ontario food web in Figure 46.6) has plants or other photosynthetic organisms at its base, followed by herbivores and various carnivores. A **detrital food web** consists of a base of organisms that feed on decaying organic matter (dead organisms), called decomposers or detritivores. These organisms are usually bacteria or fungi that recycle organic material back into the biotic part of the ecosystem as they themselves are consumed by other organisms. As all ecosystems require a method to recycle material from dead organisms, most grazing food webs have an associated detrital food web. For example, in a meadow ecosystem, plants may support a grazing food web of different organisms, primary and other levels of consumers, while at the same time supporting a detrital food web of bacteria, fungi, and detrivorous invertebrates feeding off dead plants and animals.

EVOLUTION CONNECTION

Three-spined Stickleback

It is well established by the theory of natural selection that changes in the environment play a major role in the evolution of species within an ecosystem. However, little is known about how the evolution of species within an ecosystem can alter the ecosystem environment. In 2009, Dr. Luke Harmon, from the University of Idaho, published a paper that for the first time showed that the evolution of organisms into subspecies can have direct effects on their ecosystem environment.[1]

The three-spined stickleback (*Gasterosteus aculeatus*) is a freshwater fish that evolved from a saltwater fish to live in freshwater lakes about 10,000 years ago, which is considered a recent development in evolutionary time (Figure 46.7). Over the last 10,000 years, these freshwater fish then became isolated from each other in different lakes. Depending on which lake population was studied, findings showed that these sticklebacks then either remained as one species or evolved into two species. The divergence of species was made possible by their use of different areas of the pond for feeding called micro niches.

Dr. Harmon and his team created artificial pond microcosms in 250-gallon tanks and added muck from freshwater ponds as a source of zooplankton and other invertebrates to sustain the fish. In different experimental tanks they introduced one species of stickleback from either a single-species or double-species lake.

Over time, the team observed that some of the tanks bloomed with algae while others did not. This puzzled the scientists, and they decided to measure the water's dissolved organic carbon (DOC), which consists of mostly large molecules of decaying organic matter that give pond-water its slightly brownish color. It turned out that the water from the tanks with two-species fish contained larger particles of DOC (and hence darker water) than water with single-species fish. This increase in DOC blocked the sunlight and prevented algal blooming. Conversely, the water from the single-species tank contained smaller DOC particles, allowing more sunlight penetration to fuel the algal blooms.

This change in the environment, which is due to the different feeding habits of the stickleback species in each lake type, probably has a great impact on the survival of other species in these ecosystems, especially other photosynthetic organisms. Thus, the study shows that, at least in these ecosystems, the environment and the evolution of populations have reciprocal effects that may now be factored into simulation models.

1 *Nature* (Vol. 458, April 1, 2009)

Figure 46.7 The three-spined stickleback evolved from a saltwater fish to freshwater fish. (credit: Barrett Paul, USFWS)

Research into Ecosystem Dynamics: Ecosystem Experimentation and Modeling

The study of the changes in ecosystem structure caused by changes in the environment (disturbances) or by internal forces is called **ecosystem dynamics**. Ecosystems are characterized using a variety of research methodologies. Some ecologists study ecosystems using controlled experimental systems, while some study entire ecosystems in their natural state, and others use both approaches.

A **holistic ecosystem model** attempts to quantify the composition, interaction, and dynamics of entire ecosystems; it is the most representative of the ecosystem in its natural state. A food web is an example of a holistic ecosystem model. However, this type of study is limited by time and expense, as well as the fact that it is neither feasible nor ethical to do experiments on large natural ecosystems. It is difficult to quantify all different species in an ecosystem and the dynamics in their habitat, especially when studying large habitats such as the Amazon Rainforest.

For these reasons, scientists study ecosystems under more controlled conditions. Experimental systems usually involve either partitioning a part of a natural ecosystem that can be used for experiments, termed a **mesocosm**, or by recreating an ecosystem entirely in an indoor or outdoor laboratory environment, which is referred to as a **microcosm**. A major limitation to these approaches is that removing individual organisms from their natural ecosystem or altering a natural ecosystem through partitioning may change the dynamics of the ecosystem. These changes are often due to differences in species numbers and diversity and also to environment alterations caused by partitioning (mesocosm) or recreating (microcosm) the natural habitat. Thus, these types of experiments are not totally predictive of changes that would occur in the ecosystem from which they were gathered.

As both of these approaches have their limitations, some ecologists suggest that results from these experimental systems should be used only in conjunction with holistic ecosystem studies to obtain the most representative data about ecosystem structure, function, and dynamics.

Scientists use the data generated by these experimental studies to develop ecosystem models that demonstrate the structure and dynamics of ecosystems. They use three basic types of ecosystem modeling in research and ecosystem management: a conceptual model, an analytical model, and a simulation model. A **conceptual model** is an ecosystem model that consists of flow charts to show interactions of different compartments of the living and nonliving components of the ecosystem. A conceptual model describes ecosystem structure and dynamics and shows how environmental disturbances affect the ecosystem; however, its ability to predict the effects of these disturbances is limited. Analytical and simulation models, in contrast, are mathematical methods of describing ecosystems that are indeed capable of predicting the effects of potential environmental changes without direct experimentation, although with some limitations as to accuracy. An **analytical model** is an ecosystem model that is created using simple mathematical formulas to predict the effects of environmental disturbances on ecosystem structure and dynamics. A **simulation model** is an ecosystem model that is created using complex computer algorithms to holistically model ecosystems and to predict the effects of environmental disturbances on ecosystem structure and dynamics. Ideally, these models are accurate enough to determine which components of the ecosystem are particularly sensitive to disturbances, and they can serve as a guide to ecosystem managers (such as conservation ecologists or fisheries biologists) in the practical maintenance of ecosystem health.

Conceptual Models

Conceptual models are useful for describing ecosystem structure and dynamics and for demonstrating the relationships between different organisms in a community and their environment. Conceptual models are usually depicted graphically as flow charts. The organisms and their resources are grouped into specific compartments with arrows showing the relationship and transfer of energy or nutrients between them. Thus, these diagrams are sometimes called compartment models.

To model the cycling of mineral nutrients, organic and inorganic nutrients are subdivided into those that are bioavailable (ready to be incorporated into biological macromolecules) and those that are not. For example, in a terrestrial ecosystem near a deposit of coal, carbon will be available to the plants of this ecosystem as carbon dioxide gas in a short-term period, not from the carbon-rich coal itself. However, over a longer period, microorganisms capable of digesting coal will incorporate its carbon or release it as natural gas (methane, CH_4), changing this unavailable organic source into an available one. This conversion is greatly accelerated by the combustion of fossil fuels by humans, which releases large amounts of carbon dioxide into the atmosphere. This is thought to be a major factor in the rise of the atmospheric carbon dioxide levels in the industrial age. The carbon dioxide released from burning fossil fuels is produced faster than photosynthetic organisms can use it. This process is intensified by the reduction of photosynthetic trees because of worldwide deforestation. Most scientists agree that high atmospheric carbon dioxide is a major cause of global climate change.

Conceptual models are also used to show the flow of energy through particular ecosystems. Figure 46.8 is based on Howard T. Odum's classical study of the Silver Springs, Florida, holistic ecosystem in the mid-twentieth century.[2] This study shows the energy content and transfer between various ecosystem compartments.

2Howard T. Odum, "Trophic Structure and Productivity of Silver Springs, Florida," *Ecological Monographs* 27, no. 1 (1957): 47–112.

VISUAL CONNECTION

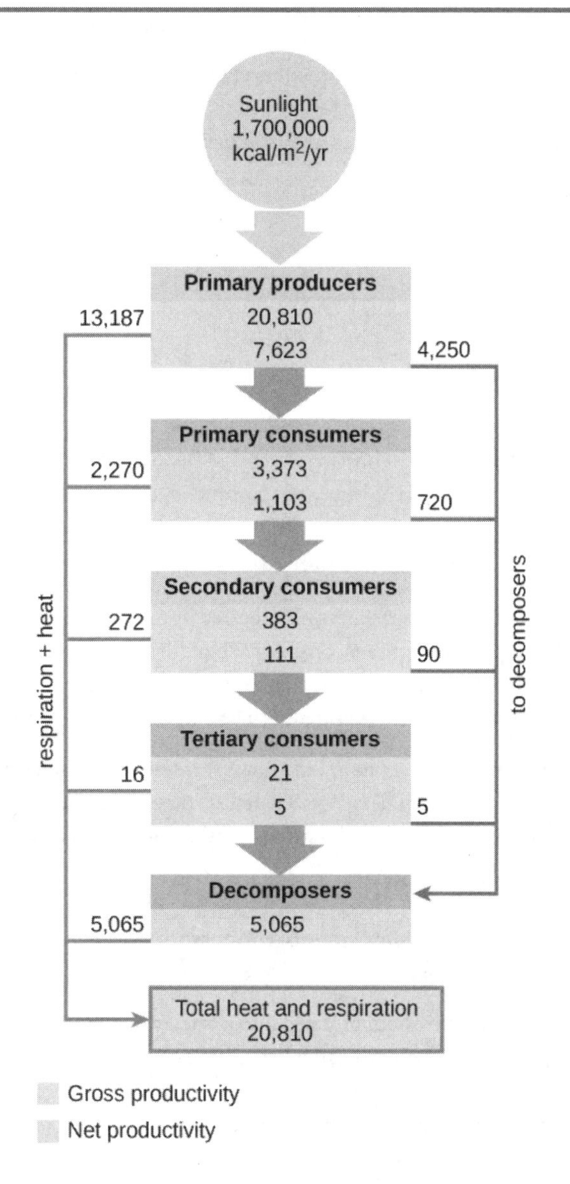

Figure 46.8 This conceptual model shows the flow of energy through a spring ecosystem in Silver Springs, Florida. Notice that the energy decreases with each increase in trophic level.

Why do you think the value for gross productivity of the primary producers is the same as the value for total heat and respiration (20,810 kcal/m^2/yr)?

Analytical and Simulation Models

The major limitation of conceptual models is their inability to predict the consequences of changes in ecosystem species and/or environment. Ecosystems are dynamic entities and subject to a variety of abiotic and biotic disturbances caused by natural forces and/or human activity. Ecosystems altered from their initial equilibrium state can often recover from such disturbances and return to a state of equilibrium. As most ecosystems are subject to periodic disturbances and are often in a state of change, they are usually either moving toward or away from their equilibrium state. There are many of these equilibrium states among the various components of an ecosystem, which affects the ecosystem overall. Furthermore, as humans have the ability to greatly and rapidly alter the species content and habitat of an ecosystem, the need for predictive models that enable understanding of how ecosystems respond to these changes becomes more crucial.

Analytical models often use simple, linear components of ecosystems, such as food chains, and are known to be complex mathematically; therefore, they require a significant amount of mathematical knowledge and expertise. Although analytical models have great potential, their simplification of complex ecosystems is thought to limit their accuracy. Simulation models that use computer programs are better able to deal with the complexities of ecosystem structure.

A recent development in simulation modeling uses supercomputers to create and run individual-based simulations, which accounts for the behavior of individual organisms and their effects on the ecosystem as a whole. These simulations are considered to be the most accurate and predictive of the complex responses of ecosystems to disturbances.

LINK TO LEARNING

Visit The Darwin Project (http://openstax.org/l/Darwin_project) to view a variety of ecosystem models, including simulations that model predator-prey relationships (https://openstax.org/l/Darwin_projct2) to learn more.

46.2 Energy Flow through Ecosystems

By the end of this section, you will be able to do the following:

- Describe how organisms acquire energy in a food web and in associated food chains
- Explain how the efficiency of energy transfers between trophic levels affects ecosystem structure and dynamics
- Discuss trophic levels and how ecological pyramids are used to model them

All living things require energy in one form or another. Energy is required by most complex metabolic pathways (often in the form of adenosine triphosphate, ATP), especially those responsible for building large molecules from smaller compounds, and life itself is an energy-driven process. Living organisms would not be able to assemble macromolecules (proteins, lipids, nucleic acids, and complex carbohydrates) from their monomeric subunits without a constant energy input.

It is important to understand how organisms acquire energy and how that energy is passed from one organism to another through food webs and their constituent food chains. Food webs illustrate how energy flows directionally through ecosystems, including how efficiently organisms acquire it, use it, and how much remains for use by other organisms of the food web.

How Organisms Acquire Energy in a Food Web

Energy is acquired by living things in three ways: photosynthesis, chemosynthesis, and the consumption and digestion of other living or previously living organisms by heterotrophs.

Photosynthetic and chemosynthetic organisms are both grouped into a category known as autotrophs: organisms capable of synthesizing their own food (more specifically, capable of using inorganic carbon as a carbon source). Photosynthetic autotrophs (photoautotrophs) use sunlight as an energy source, whereas chemosynthetic autotrophs (chemoautotrophs) use inorganic molecules as an energy source. Autotrophs are critical for all ecosystems. Without these organisms, energy would not be available to other living organisms and life itself would not be possible.

Photoautotrophs, such as plants, algae, and photosynthetic bacteria, serve as the energy source for a majority of the world's ecosystems. These ecosystems are often described by grazing food webs. Photoautotrophs harness the solar energy of the sun by converting it to chemical energy in the form of ATP (and NADP). The energy stored in ATP is used to synthesize complex organic molecules, such as glucose.

Chemoautotrophs are primarily bacteria that are found in rare ecosystems where sunlight is not available, such as in those associated with dark caves or hydrothermal vents at the bottom of the ocean (Figure 46.9). Many chemoautotrophs in hydrothermal vents use hydrogen sulfide (H_2S), which is released from the vents as a source of chemical energy. This allows chemoautotrophs to synthesize complex organic molecules, such as glucose, for their own energy and in turn supplies energy to the rest of the ecosystem.

Figure 46.9 Swimming shrimp, a few squat lobsters, and hundreds of vent mussels are seen at a hydrothermal vent at the bottom of the ocean. As no sunlight penetrates to this depth, the ecosystem is supported by chemoautotrophic bacteria and organic material that sinks from the ocean's surface. This picture was taken in 2006 at the submerged NW Eifuku volcano off the coast of Japan by the National Oceanic and Atmospheric Administration (NOAA). The summit of this highly active volcano lies 1535 m below the surface.

Productivity within Trophic Levels

Productivity within an ecosystem can be defined as the percentage of energy entering the ecosystem incorporated into biomass in a particular trophic level. **Biomass** is the total mass, in a unit area at the time of measurement, of living or previously living organisms within a trophic level. Ecosystems have characteristic amounts of biomass at each trophic level. For example, in the English Channel ecosystem the primary producers account for a biomass of 4 g/m^2 (grams per square meter), while the primary consumers exhibit a biomass of 21 g/m^2.

The productivity of the primary producers is especially important in any ecosystem because these organisms bring energy to other living organisms by photoautotrophy or chemoautotrophy. The rate at which photosynthetic primary producers incorporate energy from the sun is called **gross primary productivity**. An example of gross primary productivity is shown in the compartment diagram of energy flow within the Silver Springs aquatic ecosystem as shown (Figure 46.8). In this ecosystem, the total energy accumulated by the primary producers (gross primary productivity) was shown to be 20,810 kcal/m^2/yr.

Because all organisms need to use some of this energy for their own functions (like respiration and resulting metabolic heat loss) scientists often refer to the net primary productivity of an ecosystem. **Net primary productivity** is the energy that remains in the primary producers after accounting for the organisms' respiration and heat loss. The net productivity is then available to the primary consumers at the next trophic level. In our Silver Springs example, 13,187 of the 20,810 kcal/m^2/yr were used for respiration or were lost as heat, leaving 7,633 kcal/m^2/yr of energy for use by the primary consumers.

Ecological Efficiency: The Transfer of Energy between Trophic Levels

As illustrated in (Figure 46.8), as energy flows from primary producers through the various trophic levels, the ecosystem loses large amounts of energy. The main reason for this loss is the second law of thermodynamics, which states that whenever energy is converted from one form to another, there is a tendency toward disorder (entropy) in the system. In biologic systems, this energy takes the form of metabolic heat, which is lost when the organisms consume other organisms. In the Silver Springs ecosystem example (Figure 46.8), we see that the primary consumers produced 1103 kcal/m^2/yr from the 7618 kcal/m^2/yr of energy available to them from the primary producers. The measurement of energy transfer efficiency between two successive trophic levels is termed the **trophic level transfer efficiency (TLTE)** and is defined by the formula:

$$\text{TLTE} = \frac{\text{production at present trophic level}}{\text{production at previous trophic level}} \times 100$$

In Silver Springs, the TLTE between the first two trophic levels was approximately 14.8 percent. The low efficiency of energy transfer between trophic levels is usually the major factor that limits the length of food chains observed in a food web. The fact is, after four to six energy transfers, there is not enough energy left to support another trophic level. In the Lake Ontario example shown in (Figure 46.6), only three energy transfers occurred between the primary producer, (green algae), and the apex consumer (Chinook salmon).

Ecologists have many different methods of measuring energy transfers within ecosystems. Measurement difficulty depends on the complexity of the ecosystem and how much access scientists have to observe the ecosystem. In other words, some ecosystems are more difficult to study than others, and sometimes the quantification of energy transfers has to be estimated.

Other parameters are important in characterizing energy flow within an ecosystem. **Net production efficiency (NPE)** allows ecologists to quantify how efficiently organisms of a particular trophic level incorporate the energy they receive into biomass; it is calculated using the following formula:

$$\text{NPE} = \frac{\text{net consumer productivity}}{\text{assimilation}} \times 100$$

Net consumer productivity is the energy content available to the organisms of the next trophic level. **Assimilation** is the biomass (energy content generated per unit area) of the present trophic level after accounting for the energy lost due to incomplete ingestion of food, energy used for respiration, and energy lost as waste. Incomplete ingestion refers to the fact that some consumers eat only a part of their food. For example, when a lion kills an antelope, it will eat everything except the hide and bones. The lion is missing the energy-rich bone marrow inside the bone, so the lion does not make use of all the calories its prey could provide.

Thus, NPE measures how efficiently each trophic level uses and incorporates the energy from its food into biomass to fuel the next trophic level. In general, cold-blooded animals (ectotherms), such as invertebrates, fish, amphibians, and reptiles, use less of the energy they obtain for respiration and heat than warm-blooded animals (endotherms), such as birds and mammals. The extra heat generated in endotherms, although an advantage in terms of the activity of these organisms in colder environments, is a major disadvantage in terms of NPE. Therefore, many endotherms have to eat more often than ectotherms to get the energy they need for survival. In general, NPE for ectotherms is an order of magnitude (10x) higher than for endotherms. For example, the NPE for a caterpillar eating leaves has been measured at 18 percent, whereas the NPE for a squirrel eating acorns may be as low as 1.6 percent.

The inefficiency of energy use by warm-blooded animals has broad implications for the world's food supply. It is widely accepted that the meat industry uses large amounts of crops to feed livestock, and because the NPE is low, much of the energy from animal feed is lost. For example, it costs about $0.01 to produce 1000 dietary calories (kcal) of corn or soybeans, but approximately $0.19 to produce a similar number of calories growing cattle for beef consumption. The same energy content of milk from cattle is also costly, at approximately $0.16 per 1000 kcal. Much of this difference is due to the low NPE of cattle. Thus, there has been a growing movement worldwide to promote the consumption of nonmeat and nondairy foods so that less energy is wasted feeding animals for the meat industry.

Modeling Ecosystems Energy Flow: Ecological Pyramids

The structure of ecosystems can be visualized with ecological pyramids, which were first described by the pioneering studies of Charles Elton in the 1920s. **Ecological pyramids** show the relative amounts of various parameters (such as number of organisms, energy, and biomass) across trophic levels.

Pyramids of numbers can be either upright or inverted, depending on the ecosystem. As shown in Figure 46.10, typical grassland during the summer has a base of many plants, and the numbers of organisms decrease at each trophic level. However, during the summer in a temperate forest, the base of the pyramid consists of few trees compared with the number of primary consumers, mostly insects. Because trees are large, they have great photosynthetic capability, and dominate other plants in this ecosystem to obtain sunlight. Even in smaller numbers, primary producers in forests are still capable of supporting other trophic levels.

Another way to visualize ecosystem structure is with pyramids of biomass. This pyramid measures the amount of energy converted into living tissue at the different trophic levels. Using the Silver Springs ecosystem example, this data exhibits an upright biomass pyramid (Figure 46.10), whereas the pyramid from the English Channel example is inverted. The plants (primary producers) of the Silver Springs ecosystem make up a large percentage of the biomass found there. However, the phytoplankton in the English Channel example make up less biomass than the primary consumers, the zooplankton. As with inverted pyramids of numbers, this inverted pyramid is not due to a lack of productivity from the primary producers, but results from the high turnover rate of the phytoplankton. The phytoplankton are consumed rapidly by the primary consumers, thus, minimizing their biomass at any particular point in time. However, phytoplankton reproduce quickly, thus they are able to support the rest of the ecosystem.

Pyramid ecosystem modeling can also be used to show energy flow through the trophic levels. Notice that these numbers are the same as those used in the energy flow compartment diagram in (Figure 46.8). Pyramids of energy are always upright, and an ecosystem without sufficient primary productivity cannot be supported. All types of ecological pyramids are useful for characterizing ecosystem structure. However, in the study of energy flow through the ecosystem, pyramids of energy are the most consistent and representative models of ecosystem structure (Figure 46.10).

VISUAL CONNECTION

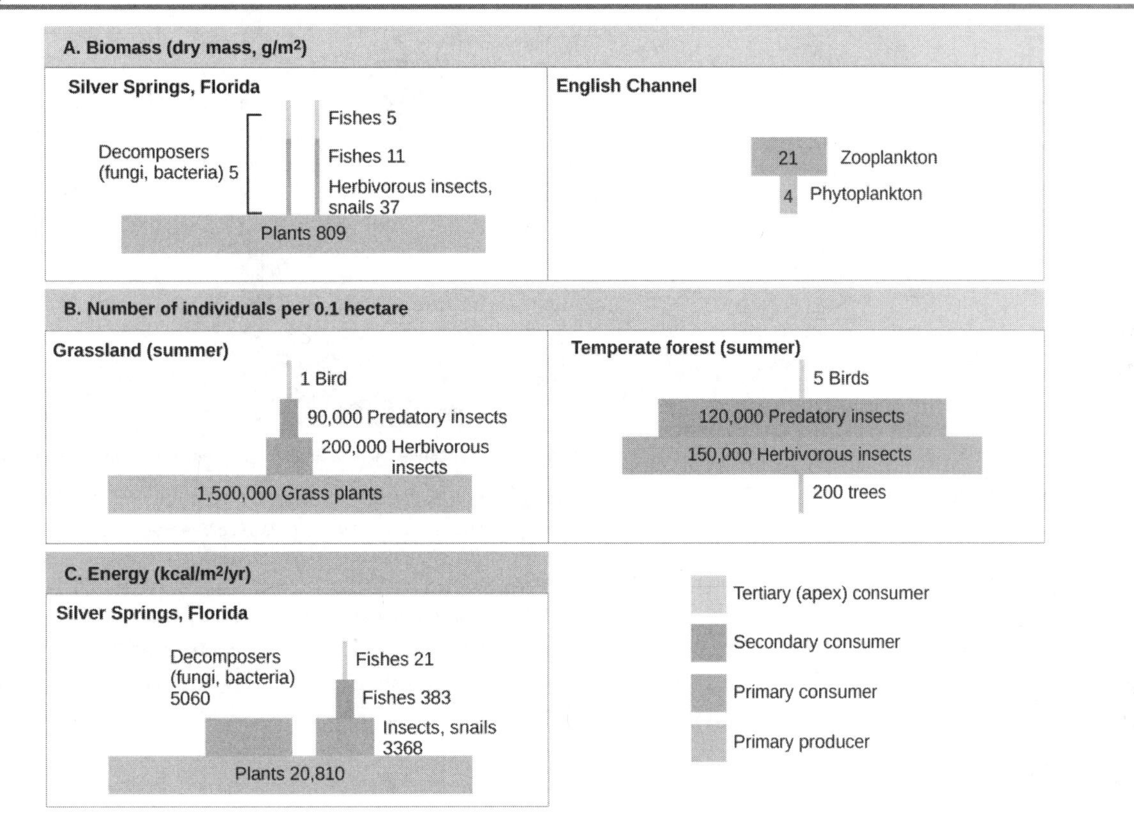

Figure 46.10 Ecological pyramids depict the (a) biomass, (b) number of organisms, and (c) energy in each trophic level.

Pyramids depicting the number of organisms or biomass may be inverted, upright, or even diamond-shaped. Energy pyramids, however, are always upright. Why?

Consequences of Food Webs: Biological Magnification

One of the most important environmental consequences of ecosystem dynamics is biomagnification. **Biomagnification** is the increasing concentration of persistent, toxic substances in organisms at each trophic level, from the primary producers to the apex consumers. Many substances have been shown to bioaccumulate, including the pesticide **d**ichloro**d**iphenyl**t**richloroethane (DDT), which was described in the 1960s bestseller, *Silent Spring*, by Rachel Carson. DDT was a commonly used pesticide before its dangers became known. In some aquatic ecosystems, organisms from each trophic level consumed many organisms of the lower level, which caused DDT to increase in birds (apex consumers) that ate fish. Thus, the birds accumulated sufficient amounts of DDT to cause fragility in their eggshells. This effect increased egg breakage during nesting and was shown to have adverse effects on these bird populations. The use of DDT was banned in the United States in the 1970s.

Other substances that biomagnify are polychlorinated biphenyls (PCBs), which were used in coolant liquids in the United States until their use was banned in 1979, and heavy metals, such as mercury, lead, and cadmium. These substances were best studied in aquatic ecosystems, where fish species at different trophic levels accumulate toxic substances brought through the ecosystem by the primary producers. As illustrated in a study performed by the National Oceanic and Atmospheric Administration (NOAA) in the Saginaw Bay of Lake Huron (Figure 46.11), PCB concentrations increased from the ecosystem's primary producers (phytoplankton) through the different trophic levels of fish species. The apex consumer (walleye) has more than four times the amount of PCBs compared to phytoplankton. Also, based on results from other studies, birds that eat these fish may have PCB levels at least one order of magnitude higher than those found in the lake fish.

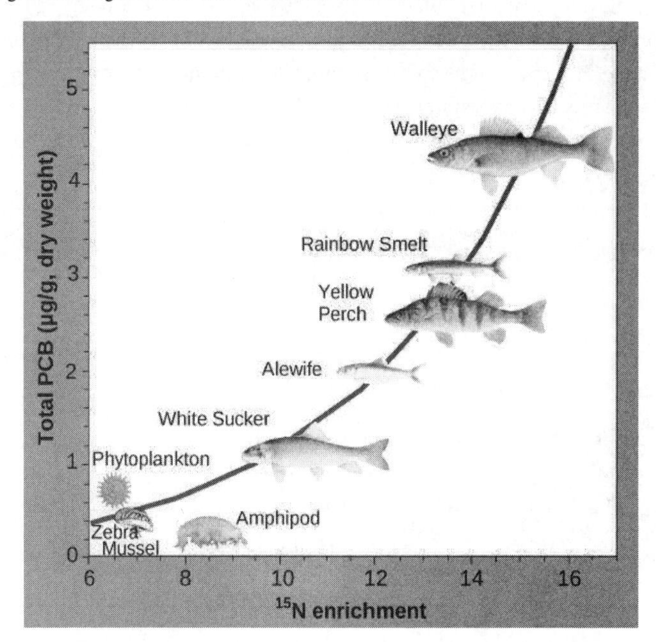

Figure 46.11 This chart shows the PCB concentrations found at the various trophic levels in the Saginaw Bay ecosystem of Lake Huron. Numbers on the x-axis reflect enrichment with heavy isotopes of nitrogen (^{15}N), which is a marker for increasing trophic level. Notice that the fish in the higher trophic levels accumulate more PCBs than those in lower trophic levels. (credit: Patricia Van Hoof, NOAA, GLERL)

Other concerns have been raised by the accumulation of heavy metals, such as mercury and cadmium, in certain types of seafood. The United States Environmental Protection Agency (EPA) recommends that pregnant women and young children should not consume any swordfish, shark, king mackerel, or tilefish because of their high mercury content. These individuals are advised to eat fish low in mercury: salmon, tilapia, shrimp, pollock, and catfish. Biomagnification is a good example of how ecosystem dynamics can affect our everyday lives, even influencing the food we eat.

46.3 Biogeochemical Cycles

By the end of this section, you will be able to do the following:

- Discuss the biogeochemical cycles of water, carbon, nitrogen, phosphorus, and sulfur
- Explain how human activities have impacted these cycles and the potential consequences for Earth

Energy flows directionally through ecosystems, entering as sunlight (or inorganic molecules for chemoautotrophs) and leaving as heat during the many transfers between trophic levels. However, the matter that makes up living organisms is conserved and recycled. The six most common elements associated with organic molecules—carbon, nitrogen, hydrogen, oxygen, phosphorus, and sulfur—take a variety of chemical forms and may exist for long periods in the atmosphere, on land, in water, or beneath the Earth's surface. Geologic processes, such as weathering, erosion, water drainage, and the subduction of the continental plates, all play a role in this recycling of materials. Because geology and chemistry have major roles in the study of this process, the recycling of inorganic matter between living organisms and their environment is called a **biogeochemical cycle**.

Water contains hydrogen and oxygen, which is essential to all living processes. The **hydrosphere** is the area of the Earth where water movement and storage occurs. On or beneath the surface, water occurs in liquid or solid form in rivers, lakes, oceans, groundwater, polar ice caps, and glaciers. And it occurs as water vapor in the atmosphere. Carbon is found in all organic macromolecules and is an important constituent of fossil fuels. Nitrogen is a major component of our nucleic acids and proteins and is critical to human agriculture. Phosphorus, a major component of nucleic acid (along with nitrogen), is one of the main ingredients in artificial fertilizers used in agriculture and their associated environmental impacts on our surface water. Sulfur is critical to the 3-D folding of proteins, such as in disulfide binding.

The cycling of these elements is interconnected. For example, the movement of water is critical for the leaching of nitrogen and phosphate into rivers, lakes, and oceans. Furthermore, the ocean itself is a major reservoir for carbon. Thus, mineral nutrients are cycled, either rapidly or slowly, through the entire biosphere, from one living organism to another, and between the biotic and abiotic world.

LINK TO LEARNING

Head to this website (http://openstax.org/l/biogeochemical) to learn more about biogeochemical cycles.

The Water (Hydrologic) Cycle

Water is the basis of all living processes on Earth. When examining the stores of water on Earth, 97.5 percent of it is non-potable salt water (Figure 46.12). Of the remaining water, 99 percent is locked underground as water or as ice. Thus, less than 1 percent of fresh water is easily accessible from lakes and rivers. Many living things, such as plants, animals, and fungi, are dependent on that small amount of fresh surface water, a lack of which can have massive effects on ecosystem dynamics. To be successful, organisms must adapt to fluctuating water supplies. Humans, of course, have developed technologies to increase water availability, such as digging wells to harvest groundwater, storing rainwater, and using desalination to obtain drinkable water from the ocean.

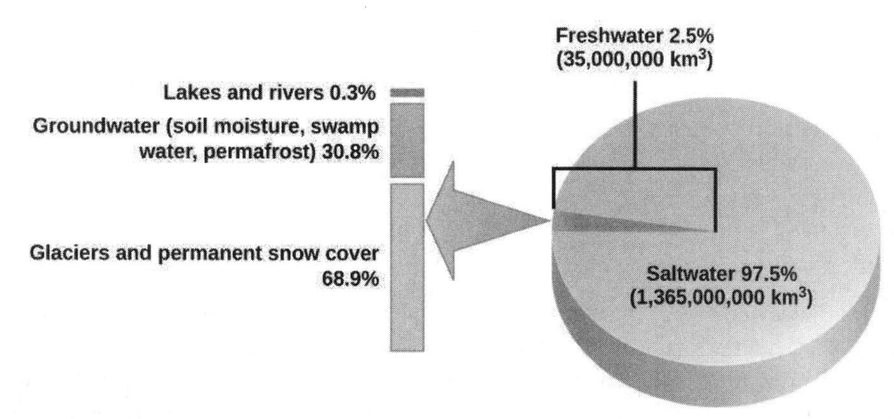

Figure 46.12 Only 2.5 percent of water on Earth is fresh water, and less than 1 percent of fresh water is easily accessible to living things.

Water cycling is extremely important to ecosystem dynamics. Water has a major influence on climate and, thus, on the environments of ecosystems. Most of the water on Earth is stored for long periods in the oceans, underground, and as ice. Figure 46.13 illustrates the average time that an individual water molecule may spend in the Earth's major water reservoirs. **Residence time** is a measure of the average time an individual water molecule stays in a particular reservoir.

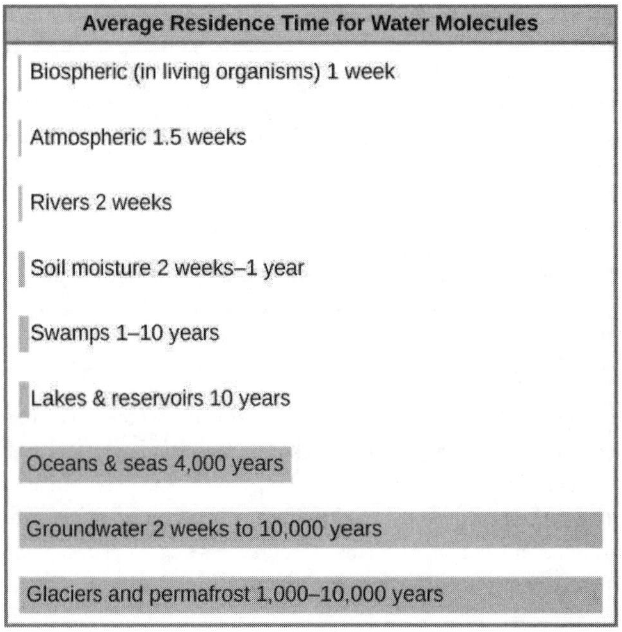

Average Residence Time for Water Molecules
Biospheric (in living organisms) 1 week
Atmospheric 1.5 weeks
Rivers 2 weeks
Soil moisture 2 weeks–1 year
Swamps 1–10 years
Lakes & reservoirs 10 years
Oceans & seas 4,000 years
Groundwater 2 weeks to 10,000 years
Glaciers and permafrost 1,000–10,000 years

Figure 46.13 This graph shows the average residence time for water molecules in the Earth's water reservoirs.

There are various processes that occur during the cycling of water, shown in Figure 46.14. These processes include the following:

- evaporation/sublimation
- condensation/precipitation
- subsurface water flow
- surface runoff/snowmelt
- streamflow

The water cycle is driven by the sun's energy as it warms the oceans and other surface waters. This leads to the evaporation (water to water vapor) of liquid surface water and the sublimation (ice to water vapor) of frozen water, which deposits large amounts of water vapor into the atmosphere. Over time, this water vapor condenses into clouds as liquid or frozen droplets and is eventually followed by precipitation (rain or snow), which returns water to the Earth's surface. Rain eventually permeates into the ground, where it may evaporate again if it is near the surface, flow beneath the surface, or be stored for long periods. More easily observed is surface runoff: the flow of fresh water either from rain or melting ice. Runoff can then make its way through streams and lakes to the oceans or flow directly to the oceans themselves.

LINK TO LEARNING

Head to this website (http://openstax.org/l/freshwater) to learn more about the world's fresh water supply.

Rain and surface runoff are major ways in which minerals, including carbon, nitrogen, phosphorus, and sulfur, are cycled from land to water. The environmental effects of runoff will be discussed later as these cycles are described.

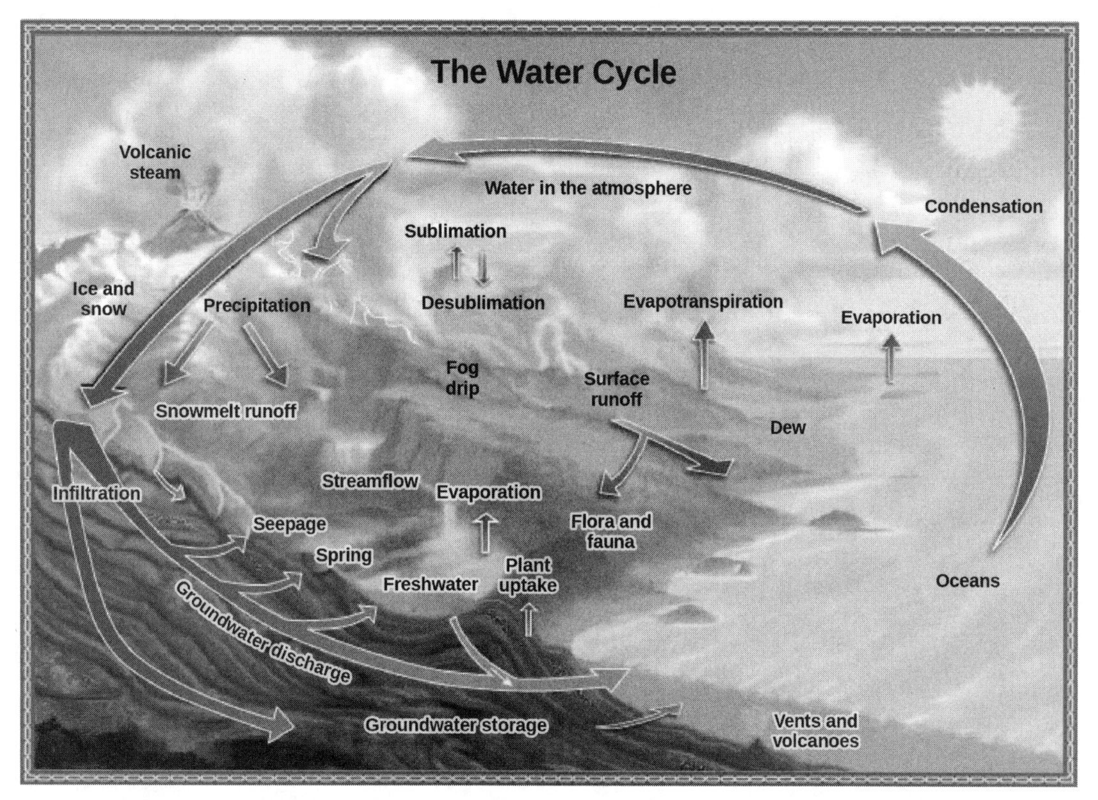

Figure 46.14 Water from the land and oceans enters the atmosphere by evaporation or sublimation, where it condenses into clouds and falls as rain or snow. Precipitated water may enter freshwater bodies or infiltrate the soil. The cycle is complete when surface or groundwater reenters the ocean. (credit: modification of work by John M. Evans and Howard Perlman, USGS)

The Carbon Cycle

Carbon is the second most abundant element in living organisms. Carbon is present in all organic molecules, and its role in the structure of macromolecules is of primary importance to living organisms.

The carbon cycle is most easily studied as two interconnected sub-cycles: one dealing with rapid carbon exchange among living organisms and the other dealing with the long-term cycling of carbon through geologic processes. The entire carbon cycle is shown in Figure 46.15.

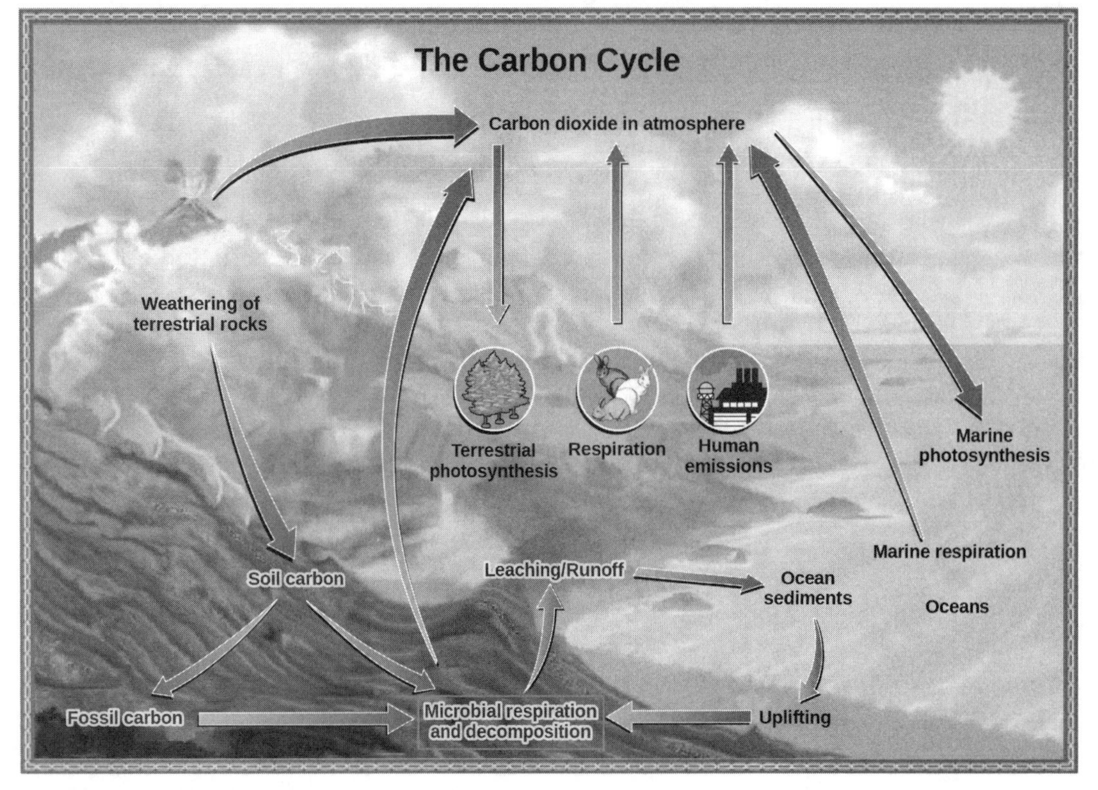

Figure 46.15 Carbon dioxide gas exists in the atmosphere and is dissolved in water. Photosynthesis converts carbon dioxide gas to organic carbon, and respiration cycles the organic carbon back into carbon dioxide gas. Long-term storage of organic carbon occurs when matter from living organisms is buried deep underground and becomes fossilized. Volcanic activity and, more recently, human emissions, bring this stored carbon back into the carbon cycle. (credit: modification of work by John M. Evans and Howard Perlman, USGS)

LINK TO LEARNING

Click this link (http://openstax.org/l/carbon_cycle) to read information about the United States Carbon Cycle Science Program.

The Biological Carbon Cycle

Living organisms are connected in many ways, even between ecosystems. A good example of this connection is the exchange of carbon between autotrophs and heterotrophs within and between ecosystems by way of atmospheric carbon dioxide. Carbon dioxide is the basic building block that most autotrophs use to build multicarbon, high energy compounds, such as glucose. The energy harnessed from the sun is used by these organisms to form the covalent bonds that link carbon atoms together. These chemical bonds thereby store this energy for later use in the process of respiration. Most terrestrial autotrophs obtain their carbon dioxide directly from the atmosphere, while marine autotrophs acquire it in the dissolved form (carbonic acid, $H_2CO_3^-$). However carbon dioxide is acquired, a by-product of the process is oxygen. The photosynthetic organisms are responsible for depositing approximately 21 percent oxygen content of the atmosphere that we observe today.

Heterotrophs and autotrophs are partners in biological carbon exchange (especially the primary consumers, largely herbivores). Heterotrophs acquire the high-energy carbon compounds from the autotrophs by consuming them, and breaking them down by respiration to obtain cellular energy, such as ATP. The most efficient type of respiration, aerobic respiration, requires oxygen obtained from the atmosphere or dissolved in water. Thus, there is a constant exchange of oxygen and carbon dioxide between the autotrophs (which need the carbon) and the heterotrophs (which need the oxygen). Gas exchange through the atmosphere and water is one way that the carbon cycle connects all living organisms on Earth.

The Biogeochemical Carbon Cycle

The movement of carbon through the land, water, and air is complex, and in many cases, it occurs much more slowly geologically than as seen between living organisms. Carbon is stored for long periods in what are known as carbon reservoirs, which include the atmosphere, bodies of liquid water (mostly oceans), ocean sediment, soil, land sediments (including fossil fuels), and the Earth's interior.

As stated, the atmosphere is a major reservoir of carbon in the form of carbon dioxide and is essential to the process of photosynthesis. The level of carbon dioxide in the atmosphere is greatly influenced by the reservoir of carbon in the oceans. The exchange of carbon between the atmosphere and water reservoirs influences how much carbon is found in each location, and each one affects the other reciprocally. Carbon dioxide (CO_2) from the atmosphere dissolves in water and combines with water molecules to form carbonic acid, and then it ionizes to carbonate and bicarbonate ions (Figure 46.16)

Step 1: CO_2 (atmospheric) $\rightleftharpoons$ CO_2 (dissolved)
Step 2: CO_2 (dissolved) + H_2O $\rightleftharpoons$ H_2CO_3 (carbonic acid)
Step 3: H_2CO_3 $\rightleftharpoons$ + H^+ + HCO_3^- (biocarbonate ion)
Step 4: HCO_3^- $\rightleftharpoons$ H^+ + CO_3^{2-} (carbonate ion)

Figure 46.16 Carbon dioxide reacts with water to form bicarbonate and carbonate ions.

The equilibrium coefficients are such that more than 90 percent of the carbon in the ocean is found as bicarbonate ions. Some of these ions combine with seawater calcium to form calcium carbonate ($CaCO_3$), a major component of marine organism shells. These organisms eventually form sediments on the ocean floor. Over geologic time, the calcium carbonate forms limestone, which comprises the largest carbon reservoir on Earth.

On land, carbon is stored in soil as a result of the decomposition of living organisms (by decomposers) or from weathering of terrestrial rock and minerals. This carbon can be leached into the water reservoirs by surface runoff. Deeper underground, on land and at sea, are fossil fuels: the anaerobically decomposed remains of plants that take millions of years to form. Fossil fuels are considered a nonrenewable resource because their use far exceeds their rate of formation. A **nonrenewable resource**, such as fossil fuel, is either regenerated very slowly or not at all. Another way for carbon to enter the atmosphere is from land (including land beneath the surface of the ocean) by the eruption of volcanoes and other geothermal systems. Carbon sediments from the ocean floor are taken deep within the Earth by the process of **subduction**: the movement of one tectonic plate beneath another. Carbon is released as carbon dioxide when a volcano erupts or from volcanic hydrothermal vents.

Humans contribute to atmospheric carbon by the burning of fossil fuels and other materials. Since the Industrial Revolution, humans have significantly increased the release of carbon and carbon compounds, which has in turn affected the climate and overall environment.

Animal husbandry by humans also increases atmospheric carbon. The large numbers of land animals raised to feed the Earth's growing population results in increased carbon dioxide levels in the atmosphere due to farming practices and respiration and methane production. This is another example of how human activity indirectly affects biogeochemical cycles in a significant way. Although much of the debate about the future effects of increasing atmospheric carbon on climate change focuses on fossils fuels, scientists take natural processes, such as volcanoes and respiration, into account as they model and predict the future impact of this increase.

The Nitrogen Cycle

Getting nitrogen into the living world is difficult. Plants and phytoplankton are not equipped to incorporate nitrogen from the atmosphere (which exists as tightly bonded, triple covalent N_2) even though this molecule comprises approximately 78 percent of the atmosphere. Nitrogen enters the living world via free-living and symbiotic bacteria, which incorporate nitrogen into their macromolecules through nitrogen fixation (conversion of N_2). Cyanobacteria live in most aquatic ecosystems where sunlight is present; they play a key role in nitrogen fixation. Cyanobacteria are able to use inorganic sources of nitrogen to "fix" nitrogen. *Rhizobium* bacteria live symbiotically in the root nodules of legumes (such as peas, beans, and peanuts) and provide them with the organic nitrogen they need. (For example, gardeners often grow peas both for their produce and to naturally add nitrogen to the soil. This practice goes back to ancient times, even if the science has only been recently understood.) Free-living bacteria, such as *Azotobacter*, are also important nitrogen fixers.

Organic nitrogen is especially important to the study of ecosystem dynamics since many ecosystem processes, such as primary production and decomposition, are limited by the available supply of nitrogen. As shown in Figure 46.17, the nitrogen that enters living systems by nitrogen fixation is successively converted from organic nitrogen back into nitrogen gas by bacteria. This process occurs in three steps in terrestrial systems: ammonification, nitrification, and denitrification. First, the ammonification process converts nitrogenous waste from living animals or from the remains of dead animals into ammonium

(NH_4^+) by certain bacteria and fungi. Second, the ammonium is converted to nitrites (NO_2^-) by nitrifying bacteria, such as *Nitrosomonas*, through nitrification. Subsequently, nitrites are converted to nitrates (NO_3^-) by similar organisms. Third, the process of denitrification occurs, whereby bacteria, such as *Pseudomonas* and *Clostridium*, convert the nitrates into nitrogen gas, allowing it to reenter the atmosphere.

VISUAL CONNECTION

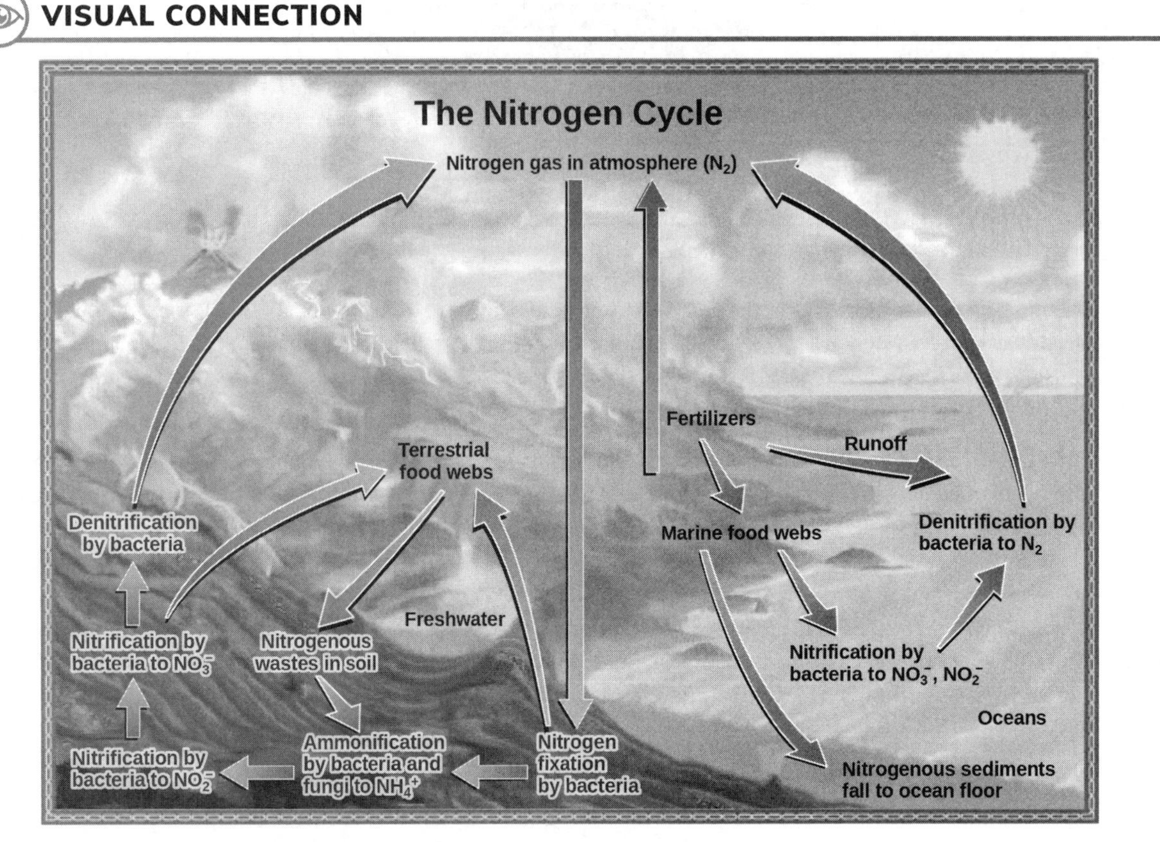

Figure 46.17 Nitrogen enters the living world from the atmosphere via nitrogen-fixing bacteria. This nitrogen and nitrogenous waste from animals is then processed back into gaseous nitrogen by soil bacteria, which also supply terrestrial food webs with the organic nitrogen they need. (credit: modification of work by John M. Evans and Howard Perlman, USGS)

Which of the following statements about the nitrogen cycle is false?

a. Ammonification converts organic nitrogenous matter from living organisms into ammonium (NH_4^+).
b. Denitrification by bacteria converts nitrates (NO_3^-) to nitrogen gas (N_2).
c. Nitrification by bacteria converts nitrates (NO_3^-) to nitrites (NO_2^-).
d. Nitrogen fixing bacteria convert nitrogen gas (N_2) into organic compounds.

Human activity can release nitrogen into the environment by two primary means: the combustion of fossil fuels, which releases different nitrogen oxides, and by the use of artificial fertilizers in agriculture, which are then washed into lakes, streams, and rivers by surface runoff. Atmospheric nitrogen is associated with several effects on Earth's ecosystems including the production of acid rain (as nitric acid, HNO_3) and greenhouse gas (as nitrous oxide, N_2O) potentially causing climate change. A major effect from fertilizer runoff is saltwater and freshwater **eutrophication**, a process whereby nutrient runoff causes the excess growth of microorganisms, depleting dissolved oxygen levels and killing ecosystem fauna.

A similar process occurs in the marine nitrogen cycle, where the ammonification, nitrification, and denitrification processes are performed by marine bacteria. Some of this nitrogen falls to the ocean floor as sediment, which can then be moved to land in geologic time by uplift of the Earth's surface and thereby incorporated into terrestrial rock. Although the movement of nitrogen from rock directly into living systems has been traditionally seen as insignificant compared with nitrogen fixed from the atmosphere, a recent study showed that this process may indeed be significant and should be included in any study of the global nitrogen cycle.[3]

3Scott L. Morford, Benjamin Z. Houlton, and Randy A. Dahlgren, "Increased Forest Ecosystem Carbon and Nitrogen Storage from Nitrogen Rich Bedrock," *Nature* 477, no. 7362 (2011): 78–81.

The Phosphorus Cycle

Phosphorus is an essential nutrient for living processes; it is a major component of nucleic acid and phospholipids, and, as calcium phosphate, makes up the supportive components of our bones. Phosphorus is often the limiting nutrient (necessary for growth) in aquatic ecosystems (Figure 46.18).

Phosphorus occurs in nature as the phosphate ion (PO_4^{3-}). In addition to phosphate runoff as a result of human activity, natural surface runoff occurs when it is leached from phosphate-containing rock by weathering, thus sending phosphates into rivers, lakes, and the ocean. This rock has its origins in the ocean. Phosphate-containing ocean sediments form primarily from the bodies of ocean organisms and from their excretions. However, in remote regions, volcanic ash, aerosols, and mineral dust may also be significant phosphate sources. This sediment then is moved to land over geologic time by the uplifting of areas of the Earth's surface.

Phosphorus is also reciprocally exchanged between phosphate dissolved in the ocean and marine ecosystems. The movement of phosphate from the ocean to the land and through the soil is extremely slow, with the average phosphate ion having an oceanic residence time between 20,000 and 100,000 years.

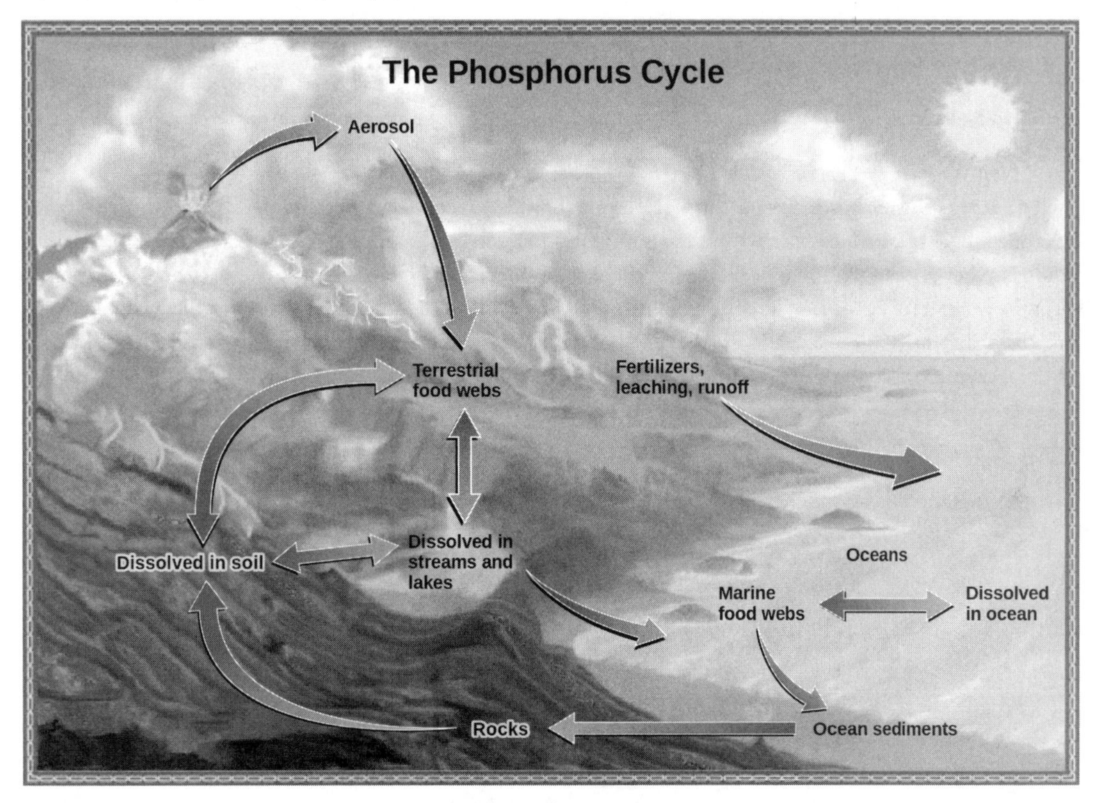

Figure 46.18 In nature, phosphorus exists as the phosphate ion (PO_4^{3-}). Weathering of rocks and volcanic activity releases phosphate into the soil, water, and air, where it becomes available to terrestrial food webs. Phosphate enters the oceans via surface runoff, groundwater flow, and river flow. Phosphate dissolved in ocean water cycles into marine food webs. Some phosphate from the marine food webs falls to the ocean floor, where it forms sediment. (credit: modification of work by John M. Evans and Howard Perlman, USGS)

As discussed in Chapter 44, excess phosphorus and nitrogen that enters these ecosystems from fertilizer runoff and from sewage causes excessive growth of microorganisms and depletes the dissolved oxygen, which leads to the death of many ecosystem fauna, such as shellfish and finfish. This process is responsible for dead zones in lakes and at the mouths of many major rivers (Figure 46.19).

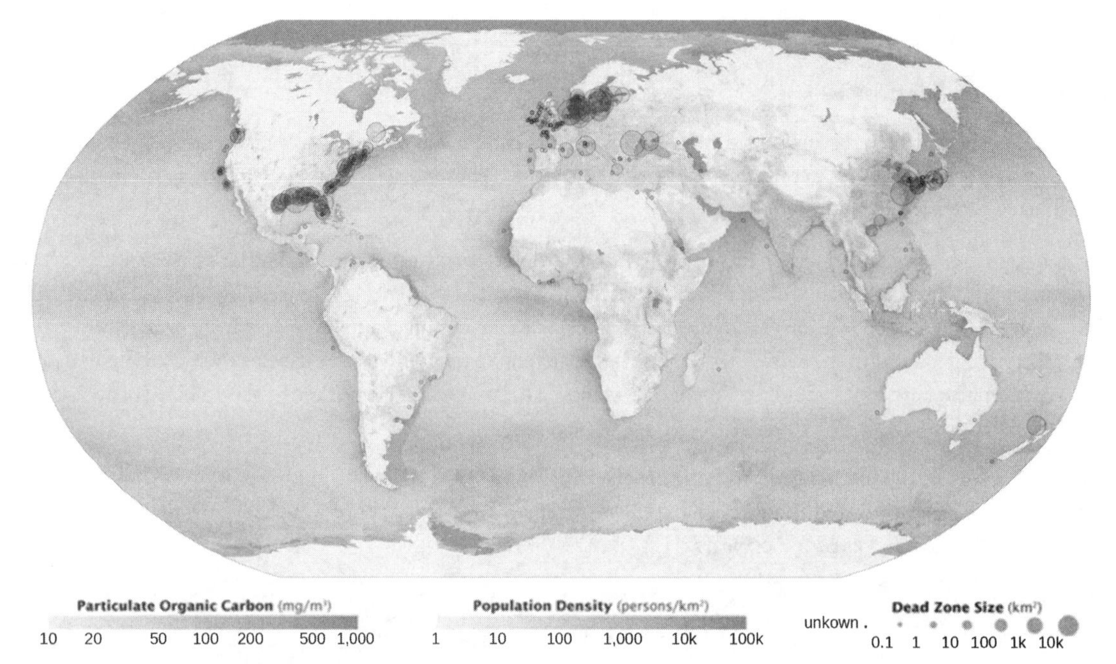

Figure 46.19 Dead zones occur when phosphorus and nitrogen from fertilizers cause excessive growth of microorganisms, which depletes oxygen and kills fauna. Worldwide, large dead zones are found in coastal areas of high population density. (credit: NASA Earth Observatory)

As discussed earlier, a **dead zone** is an area within a freshwater or marine ecosystem where large areas are depleted of their normal flora and fauna; these zones can be caused by eutrophication, oil spills, dumping of toxic chemicals, and other human activities. The number of dead zones has been increasing for several years, and more than 400 of these zones were present as of 2008. One of the worst dead zones is off the coast of the United States in the Gulf of Mexico, where fertilizer runoff from the Mississippi River basin has created a dead zone of over 8463 square miles. Phosphate and nitrate runoff from fertilizers also negatively affect several lake and bay ecosystems including the Chesapeake Bay in the eastern United States.

Everyday Connection

Chesapeake Bay

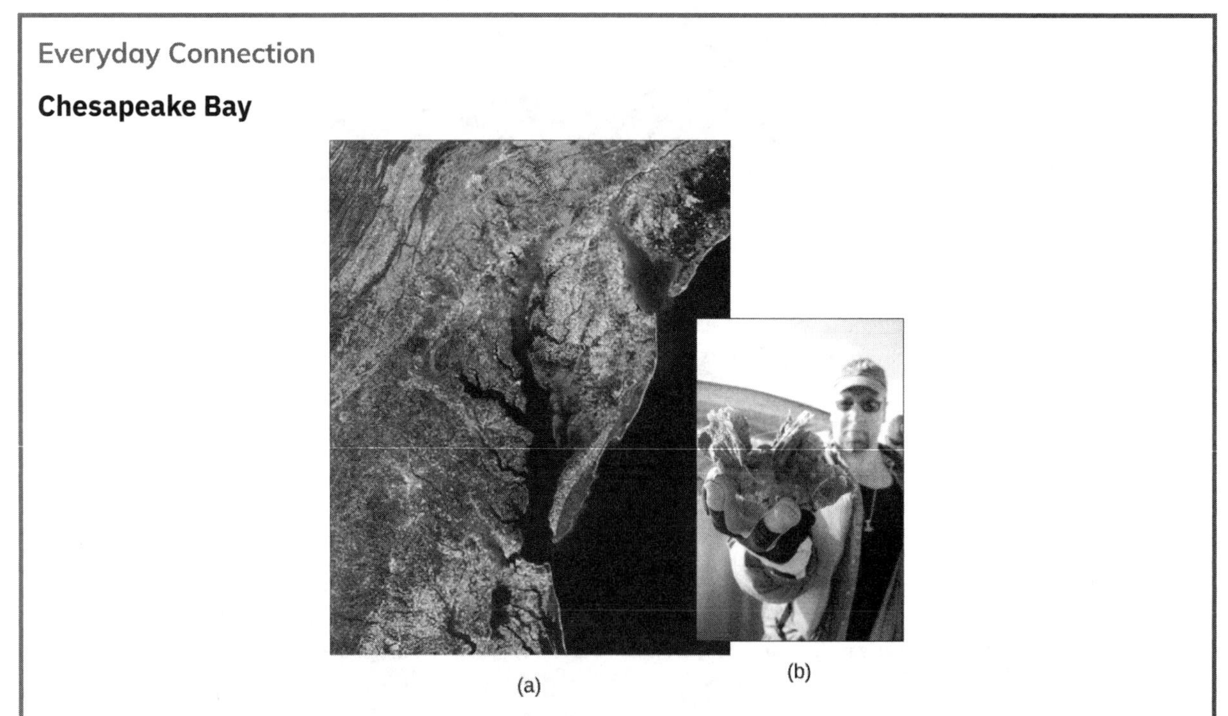

Figure 46.20 This (a) satellite image shows the Chesapeake Bay, an ecosystem affected by phosphate and nitrate runoff. A (b) member of the Army Corps of Engineers holds a clump of oysters being used as a part of the oyster restoration effort in the bay. (credit a: modification of work by NASA/MODIS; credit b: modification of work by U.S. Army)

The Chesapeake Bay has long been valued as one of the most scenic areas on Earth; it is now in distress and is recognized as a declining ecosystem. In the 1970s, the Chesapeake Bay was one of the first ecosystems to have identified dead zones, which continue to kill many fish and bottom-dwelling species, such as clams, oysters, and worms. Several species have declined in the Chesapeake Bay due to surface water runoff containing excess nutrients from artificial fertilizer used on land. The source of the fertilizers (with high nitrogen and phosphate content) is not limited to agricultural practices. There are many nearby urban areas and more than 150 rivers and streams empty into the bay that are carrying fertilizer runoff from lawns and gardens. Thus, the decline of the Chesapeake Bay is a complex issue and requires the cooperation of industry, agriculture, and everyday homeowners.

Of particular interest to conservationists is the oyster population; it is estimated that more than 200,000 acres of oyster reefs existed in the bay in the 1700s, but that number has now declined to only 36,000 acres. Oyster harvesting was once a major industry for Chesapeake Bay, but it declined 88 percent between 1982 and 2007. This decline was due not only to fertilizer runoff and dead zones but also to overharvesting. Oysters require a certain minimum population density because they must be in close proximity to reproduce. Human activity has altered the oyster population and locations, greatly disrupting the ecosystem.

The restoration of the oyster population in the Chesapeake Bay has been ongoing for several years with mixed success. Not only do many people find oysters good to eat, but they also clean up the bay. Oysters are filter feeders, and as they eat, they clean the water around them. In the 1700s, it was estimated that it took only a few days for the oyster population to filter the entire volume of the bay. Today, with changed water conditions, it is estimated that the present population would take nearly a year to do the same job.

Restoration efforts have been ongoing for several years by nonprofit organizations, such as the Chesapeake Bay Foundation. The restoration goal is to find a way to increase population density so the oysters can reproduce more efficiently. Many disease-resistant varieties (developed at the Virginia Institute of Marine Science for the College of William and Mary) are now available and have been used in the construction of experimental oyster reefs. Efforts to clean and restore the bay by Virginia and Delaware have been hampered because much of the pollution entering the bay comes from other states, which stresses the need for interstate cooperation to gain successful restoration.

The new, hearty oyster strains have also spawned a new and economically viable industry—oyster aquaculture—which not only supplies oysters for food and profit, but also has the added benefit of cleaning the bay.

The Sulfur Cycle

Sulfur is an essential element for the macromolecules of living things. As a part of the amino acid cysteine, it is involved in the formation of disulfide bonds within proteins, which help to determine their 3-D folding patterns, and hence their functions. As shown in Figure 46.21, sulfur cycles between the oceans, land, and atmosphere. Atmospheric sulfur is found in the form of sulfur dioxide (SO_2) and enters the atmosphere in three ways: from the decomposition of organic molecules, from volcanic activity and geothermal vents, and from the burning of fossil fuels by humans.

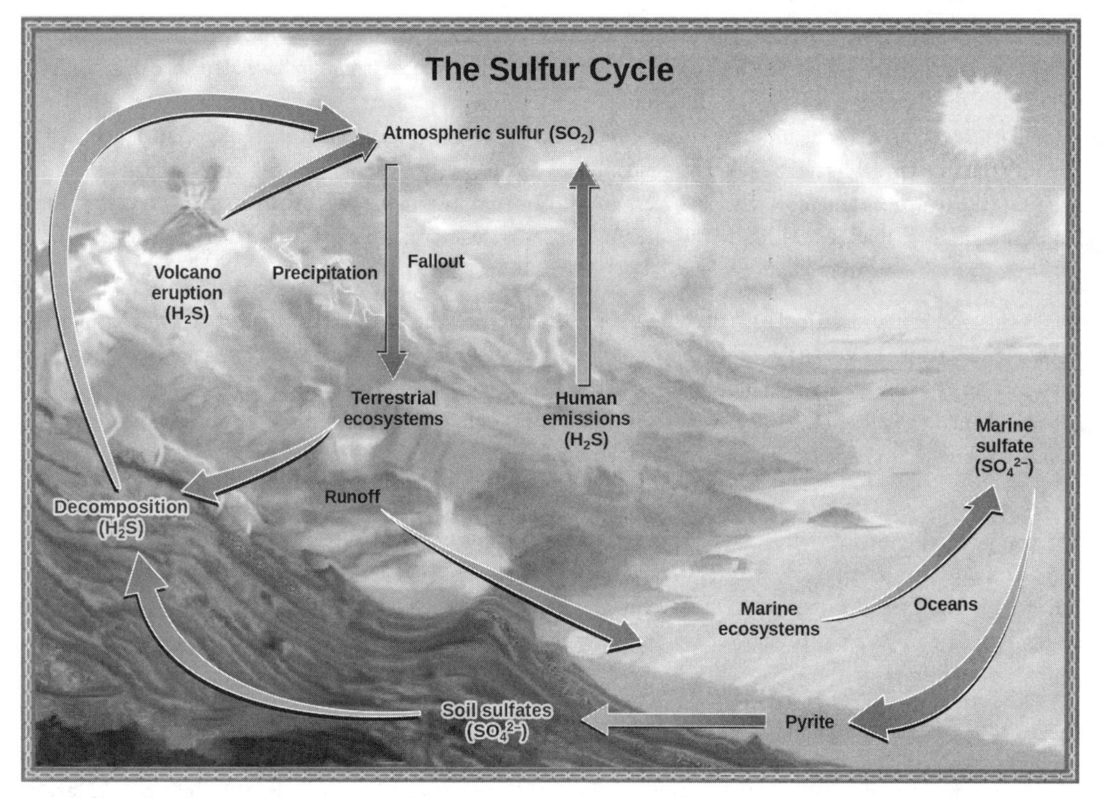

Figure 46.21 Sulfur dioxide from the atmosphere becomes available to terrestrial and marine ecosystems when it is dissolved in precipitation as weak sulfurous acid or when it falls directly to the Earth as fallout. Weathering of rocks also makes sulfates available to terrestrial ecosystems. Decomposition of living organisms returns sulfates to the ocean, soil, and atmosphere. (credit: modification of work by John M. Evans and Howard Perlman, USGS)

On land, sulfur is deposited in four major ways: precipitation, direct fallout from the atmosphere, rock weathering, and geothermal vents (Figure 46.21). Atmospheric sulfur is found in the form of sulfur dioxide (SO_2), and as rain falls through the atmosphere, sulfur is dissolved in the form of weak sulfurous acid (H_2SO_3). Sulfur can also fall directly from the atmosphere in a process called **fallout**. Also, the weathering of sulfur-containing rocks releases sulfur into the soil. These rocks originate from ocean sediments that are moved to land by the geologic uplifting of ocean sediments. Terrestrial ecosystems can then make use of these soil sulfates (SO_4-), and upon the death and decomposition of these organisms, release the sulfur back into the atmosphere as hydrogen sulfide (H_2S) gas.

Figure 46.22 At this sulfur vent in Lassen Volcanic National Park in northeastern California, the yellowish sulfur deposits are visible near the mouth of the vent.

Sulfur enters the ocean via runoff from land, from atmospheric fallout, and from underwater geothermal vents. Some ecosystems (Figure 46.9) rely on chemoautotrophs using sulfur as a biological energy source. This sulfur then supports marine ecosystems in the form of sulfates.

Human activities have played a major role in altering the balance of the global sulfur cycle. The burning of large quantities of fossil fuels, especially from coal, releases larger amounts of hydrogen sulfide gas into the atmosphere. **Acid rain** is caused by rainwater falling to the ground through this sulfur dioxide gas, turning it into weak sulfuric acid. Acid rain damages the natural environment by lowering the pH of lakes, which kills many of the resident fauna; it also affects the man-made environment through the chemical degradation of buildings. For example, many marble monuments, such as the Lincoln Memorial in Washington, DC, have suffered significant damage from acid rain over the years.

LINK TO LEARNING

Click this link (http://openstax.org/l/climate_change) to learn more about global climate change.

KEY TERMS

acid rain corrosive rain caused by rainwater falling to the ground through sulfur dioxide gas, turning it into weak sulfuric acid; can damage structures and ecosystems

analytical model ecosystem model that is created with mathematical formulas to predict the effects of environmental disturbances on ecosystem structure and dynamics

apex consumer organism at the top of the food chain

assimilation biomass consumed and assimilated from the previous trophic level after accounting for the energy lost due to incomplete ingestion of food, energy used for respiration, and energy lost as waste

biogeochemical cycle cycling of mineral nutrients through ecosystems and through the nonliving world

biomagnification increasing concentrations of persistent, toxic substances in organisms at each trophic level, from the primary producers to the apex consumers

biomass total weight, at the time of measurement, of living or previously living organisms in a unit area within a trophic level

chemoautotroph organism capable of synthesizing its own food using energy from inorganic molecules

conceptual model (also, compartment model) ecosystem model that consists of flow charts that show the interactions of different compartments of the living and nonliving components of the ecosystem

dead zone area within an ecosystem in lakes and near the mouths of rivers where large areas of ecosystems are depleted of their normal flora and fauna; these zones can be caused by eutrophication, oil spills, dumping of toxic chemicals, and other human activities

detrital food web type of food web in which the primary consumers consist of decomposers; these are often associated with grazing food webs within the same ecosystem

ecological pyramid (also, Eltonian pyramid) graphical representation of different trophic levels in an ecosystem based of organism numbers, biomass, or energy content

ecosystem community of living organisms and their interactions with their abiotic environment

ecosystem dynamics study of the changes in ecosystem structure caused by changes in the environment or internal forces

equilibrium steady state of an ecosystem where all organisms are in balance with their environment and each other

eutrophication process whereby nutrient runoff causes the excess growth of microorganisms, depleting dissolved oxygen levels and killing ecosystem fauna

fallout direct deposit of solid minerals on land or in the ocean from the atmosphere

food chain linear representation of a chain of primary producers, primary consumers, and higher-level consumers used to describe ecosystem structure and dynamics

food web graphic representation of a holistic, nonlinear web of primary producers, primary consumers, and higher-level consumers used to describe ecosystem structure and dynamics

grazing food web type of food web in which the primary producers are either plants on land or phytoplankton in the water; often associated with a detrital food web within the same ecosystem

gross primary productivity rate at which photosynthetic primary producers incorporate energy from the sun

holistic ecosystem model study that attempts to quantify the composition, interactions, and dynamics of entire ecosystems; often limited by economic and logistical difficulties, depending on the ecosystem

hydrosphere area of the Earth where water movement and storage occurs

mesocosm portion of a natural ecosystem to be used for experiments

microcosm re-creation of natural ecosystems entirely in a laboratory environment to be used for experiments

net consumer productivity energy content available to the organisms of the next trophic level

net primary productivity energy that remains in the primary producers after accounting for the organisms' respiration and heat loss

net production efficiency (NPE) measure of the ability of a trophic level to convert the energy it receives from the previous trophic level into biomass

nonrenewable resource resource, such as fossil fuel, that is either regenerated very slowly or not at all

primary consumer trophic level that obtains its energy from the primary producers of an ecosystem

primary producer trophic level that obtains its energy from sunlight, inorganic chemicals, or dead and/or decaying organic material

residence time measure of the average time an individual water molecule stays in a particular reservoir

resilience (ecological) speed at which an ecosystem recovers equilibrium after being disturbed

resistance (ecological) ability of an ecosystem to remain at equilibrium in spite of disturbances

secondary consumer usually a carnivore that eats primary consumers

simulation model ecosystem model that is created with computer programs to holistically model ecosystems and to predict the effects of environmental disturbances on ecosystem structure and dynamics

subduction movement of one tectonic plate beneath another

tertiary consumer carnivore that eats other carnivores

trophic level position of a species or group of species in a food chain or a food web

trophic level transfer efficiency (TLTE) energy transfer efficiency between two successive trophic levels

CHAPTER SUMMARY

46.1 Ecology of Ecosystems

Ecosystems exist on land, at sea, in the air, and underground. Different ways of modeling ecosystems are necessary to understand how environmental disturbances will affect ecosystem structure and dynamics. Conceptual models are useful to show the general relationships between organisms and the flow of materials or energy between them. Analytical models are used to describe linear food chains, and simulation models work best with holistic food webs.

46.2 Energy Flow through Ecosystems

Organisms in an ecosystem acquire energy in a variety of ways, which is transferred between trophic levels as the energy flows from the bottom to the top of the food web, with energy being lost at each transfer. The efficiency of these transfers is important for understanding the different behaviors and eating habits of warm-blooded versus cold-blooded animals. Modeling of ecosystem energy is best done with ecological pyramids of energy, although other ecological pyramids provide other vital information about ecosystem structure.

46.3 Biogeochemical Cycles

Mineral nutrients are cycled through ecosystems and their environment. Of particular importance are water, carbon, nitrogen, phosphorus, and sulfur. All of these cycles have major impacts on ecosystem structure and function. A variety of human activities, such as pollution, oil spills, and other events have damaged ecosystems, potentially causing global climate change. The health of Earth depends on understanding these cycles and how to protect the environment from irreversible damage.

VISUAL CONNECTION QUESTIONS

1. Figure 46.8 Why do you think the value for gross productivity of the primary producers is the same as the value for total heat and respiration (20,810 kcal/m^2/yr)?

2. Figure 46.10 Pyramids depicting the number of organisms or biomass may be inverted, upright, or even diamond-shaped. Energy pyramids, however, are always upright. Why?

3. Figure 46.17 Which of the following statements about the nitrogen cycle is false?
   a. Ammonification converts organic nitrogenous matter from living organisms into ammonium (NH_4^+).
   b. Denitrification by bacteria converts nitrates (NO_3^-) to nitrogen gas (N_2).
   c. Nitrification by bacteria converts nitrates (NO_3^-) to nitrites (NO_2^-).
   d. Nitrogen fixing bacteria convert nitrogen gas (N_2) into organic compounds.

REVIEW QUESTIONS

4. The ability of an ecosystem to return to its equilibrium state after an environmental disturbance is called _______.
   a. resistance
   b. restoration
   c. reformation
   d. resilience

5. A re-created ecosystem in a laboratory environment is known as a _______.
   a. mesocosm
   b. simulation
   c. microcosm
   d. reproduction

6. Decomposers are associated with which class of food web?
   a. grazing
   b. detrital
   c. inverted
   d. aquatic

7. The primary producers in an ocean grazing food web are usually _______.
   a. plants
   b. animals
   c. fungi
   d. phytoplankton

8. What term describes the use of mathematical equations in the modeling of linear aspects of ecosystems?
   a. analytical modeling
   b. simulation modeling
   c. conceptual modeling
   d. individual-based modeling

9. The position of an organism along a food chain is known as its _______.
   a. locus
   b. location
   c. trophic level
   d. microcosm

10. The loss of an apex consumer would impact which trophic level of a food web?
    a. primary producers
    b. primary consumers
    c. secondary consumers
    d. all of the above

11. A food chain would be a better resource than a food web to answer which question?
    a. How does energy move from an organism in one trophic level to an organism on the next trophic level?
    b. How does energy move within a trophic level?
    c. What preys on grasses?
    d. How is organic matter recycled in a forest?

12. The weight of living organisms in an ecosystem at a particular point in time is called:
    a. energy
    b. production
    c. entropy
    d. biomass

13. Which term describes the process whereby toxic substances increase along trophic levels of an ecosystem?
    a. biomassification
    b. biomagnification
    c. bioentropy
    d. heterotrophy

14. Organisms that can make their own food using inorganic molecules are called:
    a. autotrophs
    b. heterotrophs
    c. photoautotrophs
    d. chemoautotrophs

15. In the English Channel ecosystem, the number of primary producers is smaller than the number of primary consumers because_______.
    a. the apex consumers have a low turnover rate
    b. the primary producers have a low turnover rate
    c. the primary producers have a high turnover rate
    d. the primary consumers have a high turnover rate

16. What law of chemistry determines how much energy can be transferred when it is converted from one form to another?
    a. the first law of thermodynamics
    b. the second law of thermodynamics
    c. the conservation of matter
    d. the conservation of energy

17. The mussels that live at the NW Eifuku volcano are examples of ______.
    a. chemoautotrophs
    b. photoautotrophs
    c. apex predators
    d. primary consumers

18. The movement of mineral nutrients through organisms and their environment is called a _______ cycle.
    a. biological
    b. bioaccumulation
    c. biogeochemical
    d. biochemical

19. Carbon is present in the atmosphere as _______.
    a. carbon dioxide
    b. carbonate ion
    c. carbon dust
    d. carbon monoxide

20. The majority of water found on Earth is:
    a. ice
    b. water vapor
    c. fresh water
    d. salt water

21. The average time a molecule spends in its reservoir is known as _______.
    a. residence time
    b. restriction time
    c. resilience time
    d. storage time

22. The process whereby oxygen is depleted by the growth of microorganisms due to excess nutrients in aquatic systems is called _______.
    a. dead zoning
    b. eutrophication
    c. retrofication
    d. depletion

23. The process whereby nitrogen is brought into organic molecules is called _______.
    a. nitrification
    b. denitrification
    c. nitrogen fixation
    d. nitrogen cycling

24. Which of the following approaches would be the most effective way to reduce greenhouse carbon dioxide?
    a. Increase waste deposition into the deep ocean.
    b. Plant more environmentally-suitable plants.
    c. Increase use of fuel sources that do not produce carbon dioxide as a by-product.
    d. Decrease livestock agriculture.

25. How would loss of fungi in a forest effect biogeochemical cycles in the area?
    a. Nitrogen could no longer be fixed into organic molecules.
    b. Phosphorus stores would be released for use by other organisms.
    c. Sulfur release from eroding rocks would cease.
    d. Carbon would accumulate in dead organic matter and waste.

CRITICAL THINKING QUESTIONS

26. Compare and contrast food chains and food webs. What are the strengths of each concept in describing ecosystems?
27. Describe freshwater, ocean, and terrestrial ecosystems.
28. Compare grazing and detrital food webs. Why would they both be present in the same ecosystem?
29. How does the microcosm modeling approach differ from utilizing a holistic model for ecological research?
30. How do conceptual and analytical models of ecosystems compliment each other?
31. Compare the three types of ecological pyramids and how well they describe ecosystem structure. Identify which ones can be inverted and give an example of an inverted pyramid for each.
32. How does the amount of food a warm-blooded animal (endotherm) eats relate to its net production efficiency (NPE)?
33. A study uses an inverted pyramid to demonstrate the relationship between sharks, their aquatic prey, and phytoplankton in an ocean region. What type of pyramid must be used? What does this convey to readers about predation in the area?
34. Describe what a pyramid of numbers would like if an ecologist models the relationship between bird parasites, blue jays, and oak trees in a hectare. Does this match the energy flow pyramid?
35. Describe nitrogen fixation and why it is important to agriculture.
36. What are the factors that cause dead zones? Describe eutrophication, in particular, as a cause.
37. Why are drinking water supplies still a major concern for many countries?
38. Discuss how the human disruption of the carbon cycle has caused ocean acidification.

CHAPTER 47
Conservation Biology and Biodiversity

Figure 47.1 Lake Victoria in Africa, shown in this satellite image, was the site of one of the most extraordinary evolutionary findings on the planet, as well as a casualty of devastating biodiversity loss. (credit: modification of work by Rishabh Tatiraju, using NASA World Wind software)

INTRODUCTION In the 1980s, biologists working in Lake Victoria in Africa discovered one of the most extraordinary products of evolution on the planet. Located in the Great Rift Valley, Lake Victoria is an enormous and deep lake about 68,900 km^2 in area (larger than Lake Huron, the second largest of North America's Great Lakes). Biologists were studying species of a family of fish called cichlids. When they sampled for fish in different locations of the lake, the researchers identified over 500 evolved species in total. However, the scientists soon discovered that the invasive Nile Perch was destroying the lake's cichlid population, bringing hundreds of cichlid species to extinction with devastating rapidity.

Chapter Outline

47.1 The Biodiversity Crisis

By the end of this section, you will be able to do the following:

- Define biodiversity in terms of species diversity and abundance
- Describe biodiversity as the equilibrium of naturally fluctuating rates of extinction and speciation
- Identify historical causes of high extinction rates in Earth's history

Traditionally, ecologists have measured **biodiversity**, a general term for the number of species present in the biosphere, by taking into account both the number of species and their relative abundance to each other. Biodiversity can be estimated at a number of levels of organization of living organisms. These estimation indices, which came from *information theory*, are most useful as a first step in quantifying biodiversity between and within ecosystems; they are less useful when the main concern among conservation biologists is simply the loss of biodiversity. However, biologists recognize that measures of biodiversity, in terms of species diversity, may help focus efforts to preserve the biologically or technologically important elements of biodiversity.

The Lake Victoria cichlids provide an example with which we can begin to understand biodiversity. The biologists studying cichlids in the 1980s discovered hundreds of cichlid species representing a variety of specializations to specialized habitat types and specific feeding strategies: such as eating plankton floating in the water, scraping/eating algae from rocks, eating insect larvae from the lake bottom, and eating the eggs of other species of cichlid. The cichlids of Lake Victoria are the product of an complex *adaptive radiation*. An **adaptive radiation** is a rapid (less than three million years in the case of the Lake Victoria cichlids) branching through speciation of a phylogenetic clade into many closely related species. Typically, the species "radiate" into different habitats and niches. The Galápagos Island finches are an example of a modest adaptive radiation with 15 species. The cichlids of Lake Victoria are an example of a spectacular adaptive radiation that formerly included about 500 species.

At the time biologists were making this discovery, some species began to quickly disappear. A culprit in these declines was the Nile perch, a species of large predatory fish that was introduced to Lake Victoria by fisheries to feed the people living around the lake. The Nile perch was introduced in 1963, but its populations did not begin to surge until the 1980s. The perch population grew by consuming cichlids, driving species after species to the point of **extinction** (the disappearance of a species). In fact, there were several factors that played a role in the extinction of perhaps 200 cichlid species in Lake Victoria: the Nile perch, declining lake water quality due to agriculture and land clearing on the shores of Lake Victoria, and increased fishing pressure. Scientists had not even catalogued all of the species present—so many were lost that were never named. The diversity is now a shadow of what it once was.

The cichlids of Lake Victoria are a thumbnail sketch of contemporary rapid species loss that occurs all over Earth that is caused primarily by human activity. Extinction is a natural process of macroevolution that occurs at the rate of about one out of 1 million species becoming extinct per year. The fossil record reveals that there have been five periods of **mass extinction** in history with much higher rates of species loss, and the rate of species loss today is comparable to those periods of mass extinction. However, there is a major difference between the previous mass extinctions and the current extinction we are experiencing: human activity. Specifically, three human activities have a major impact: 1) destruction of habitat, 2) introduction of exotic species, and 3) over-harvesting. Predictions of species loss within the next century, a tiny amount of time on geological timescales, range from 10 percent to 50 percent. Extinctions on this scale have only happened five other times in the history of the planet, and these extinctions were caused by cataclysmic events that changed the course of the history of life in each instance.

Types of Biodiversity

Scientists generally accept that the term biodiversity describes the number and kinds of species and their abundance in a given location or on the planet. Species can be difficult to define, but most biologists still feel comfortable with the concept and are able to identify and count eukaryotic species in most contexts. Biologists have also identified alternate measures of biodiversity, some of which are important for planning how to preserve biodiversity.

Genetic diversity is one of those alternate concepts. **Genetic diversity**, or genetic variation defines the raw material for evolution and adaptation in a species. A species' future potential for adaptation depends on the genetic diversity held in the genomes of the individuals in populations that make up the species. The same is true for higher taxonomic categories. A genus with very different types of species will have more genetic diversity than a genus with species that are genetically similar and have similar ecologies. If there were a choice between one of these genera of species being preserved, the one with the greatest potential for subsequent evolution is the most *genetically diverse* one.

Many genes code for proteins, which in turn carry out the metabolic processes that keep organisms alive and reproducing. Genetic diversity can be measured as **chemical diversity** in that different species produce a variety of chemicals in their cells, both the proteins as well as the products and byproducts of metabolism. This chemical diversity has potential benefit for humans as a source of pharmaceuticals, so it provides one way to measure diversity that is important to human health and welfare.

Humans have generated diversity in domestic animals, plants, and fungi, among many other organisms. This diversity is also suffering losses because of migration, market forces, and increasing globalism in agriculture, especially in densely populated regions such as China, India, and Japan. The human population directly depends on this diversity as a stable food source, and its decline is troubling biologists and agricultural scientists.

It is also useful to define **ecosystem diversity**, meaning the number of different ecosystems on the planet or within a given geographic area (Figure 47.2). Whole ecosystems can disappear even if some of the species might survive by adapting to other ecosystems. The loss of an ecosystem means the loss of interactions between species, the loss of unique features of coadaptation, and the loss of biological productivity that an ecosystem is able to create. An example of a largely extinct ecosystem in North America is the *prairie ecosystem*. Prairies once spanned central North America from the boreal forest in northern Canada down into Mexico. They are now all but gone, replaced by crop fields, pasture lands, and suburban sprawl. Many of the species survive elsewhere, but the hugely productive ecosystem that was responsible for creating the most productive agricultural soils in the United States is now gone. As a consequence, native soils are disappearing or must be maintained and enhanced at great expense.

(a)

(b)

Figure 47.2 The variety of ecosystems on Earth—from (a) coral reef to (b) prairie—enables a great diversity of species to exist. (credit a: modification of work by Jim Maragos, USFWS; credit b: modification of work by Jim Minnerath, USFWS)

Current Species Diversity

Despite considerable effort, knowledge of the species that inhabit the planet is limited and always will be because of a continuing lack of financial resources and political willpower. A recent estimate suggests that the eukaryote species for which science has names, about 1.5 million species, account for less than 20 percent of the total number of eukaryote species present on the planet (8.7 million species, by one estimate). Estimates of numbers of prokaryotic species are largely guesses, but biologists agree that science has only begun to catalog their diversity. Even with what is known, there is no central repository of names or samples of the described species; therefore, there is no way to be sure that the 1.5 million descriptions *is* an accurate accounting. It is a best guess based on the opinions of experts in different taxonomic groups. Given that Earth is losing species at an accelerating pace, science is very much in the place it was with the Lake Victoria cichlids: knowing little about what is being lost. Table 47.1 presents recent estimates of biodiversity in different groups.

Estimates of the Numbers of Described and Predicted Species by Taxonomic Group

	Mora et al. 2011[1]		Chapman 2009[2]		Groombridge & Jenkins 2002[3]	
	Described	Predicted	Described	Predicted	Described	Predicted
Animalia	1,124,516	9,920,000	1,424,153	6,836,330	1,225,500	10,820,000

Table 47.1

	Mora et al. 2011[1]		Chapman 2009[2]		Groombridge & Jenkins 2002[3]	
Chromista	17,892	34,900	25,044	200,500	—	—
Fungi	44,368	616,320	98,998	1,500,000	72,000	1,500,000
Plantae	224,244	314,600	310,129	390,800	270,000	320,000
Protozoa	16,236	72,800	28,871	1,000,000	80,000	600,000
Prokaryotes	—	—	10,307	1,000,000	10,175	—
Total	1,438,769	10,960,000	1,897,502	10,897,630	1,657,675	13,240,000

Table 47.1

There are various initiatives to catalog described species in accessible ways, and the internet is facilitating that effort. Nevertheless, it has been pointed out that at the current rate of new species descriptions (which according to the State of Observed Species Report is 17,000 to 20,000 new species per year), it will take close to 500 years to finish describing life on this planet.[4] Over time, the task becomes both increasingly difficult and increasingly easier as extinction removes species from the planet.

Naming and counting species may seem like an unimportant pursuit given the other needs of humanity, but determining biodiversity it is not simply an accounting of species. Describing a species is a complex process through which biologists determine an organism's unique characteristics and whether or not that organism belongs to any other described species or genus. It allows biologists to find and recognize the species after the initial discovery, and allows them to follow up on questions about its biology. In addition, the unique characteristics of each species make it potentially valuable to humans or other species on which humans depend.

Patterns of Biodiversity

Biodiversity is not evenly distributed on Earth. Lake Victoria contained almost 500 species of cichlids alone, ignoring the other fish families present in the lake. All of these species were found only in Lake Victoria; therefore, the 500 species of cichlids were *endemic*. **Endemic species** are found in only one location. Endemics with highly restricted distributions are particularly vulnerable to extinction. Higher taxonomic levels, such as genera and families, can also be endemic. Lake Michigan contains about 79 species of fish, many of which are found in other lakes in North America. What accounts for the difference in fish diversity in these two lakes? Lake Victoria is an ancient tropical lake, while Lake Michigan is a recently formed temperate lake. Lake Michigan in its present form is only about 7,000 years old, while Lake Victoria in its present form is about 15,000 years old, although its basin is about 400,000 years in age. Biogeographers have suggested these two factors, latitude and age, are two of several hypotheses to explain biodiversity patterns on the planet.

1Mora Camilo et al., "How Many Species Are There on Earth and in the Ocean?" *PLoS Biology* (2011), doi:10.1371/journal.pbio.1001127.

2Arthur D. Chapman, *Numbers of Living Species in Australia and the World*, 2nd ed. (Canberra, AU: Australian Biological Resources Study, 2009). https://www.environment.gov.au/system/files/pages/2ee3f4a1-f130-465b-9c7a-79373680a067/files/nlsaw-2nd-complete.pdf/ (http://openstax.org/l/Aus_diversity) .

3Brian Groombridge and Martin D. Jenkins. *World Atlas of Biodiversity: Earth's Living Resources in the 21st Century*. Berkeley: University of California Press, 2002.

4International Institute for Species Exploration (IISE), *2011 State of Observed Species (SOS)*. Tempe, AZ: IISE, 2011. Accessed May, 20, 2012. http://www.esf.edu/species/ (http://openstax.org/l/observed_species) .

CAREER CONNECTION

Biogeographer

Biogeography is the study of the distribution of the world's species—both in the past and in the present. The work of *biogeographers* is critical to understanding our physical environment, how the environment affects species, and how environmental changes impact the distribution of a species; it has also been critical to developing modern evolutionary theory. Biogeographers need to understand both biology and ecology. They also need to be well-versed in evolutionary studies, soil science, and climatology.

There are three main fields of study under the heading of biogeography: *ecological biogeography, historical biogeography* (called *paleobiogeography*), *and conservation biogeography*. **Ecological biogeography** studies the current factors affecting the distribution of plants and animals. **Historical biogeography**, as the name implies, studies the past distribution of species. **Conservation biogeography**, on the other hand, is focused on the protection and restoration of species based upon known historical and current ecological information. Each of these fields considers both zoogeography and phytogeography—the past and present distribution of animals and plants.

One of the oldest observed patterns in ecology is that species biodiversity in almost every taxonomic group increases as latitude declines. In other words, biodiversity increases closer to the equator (Figure 47.3).

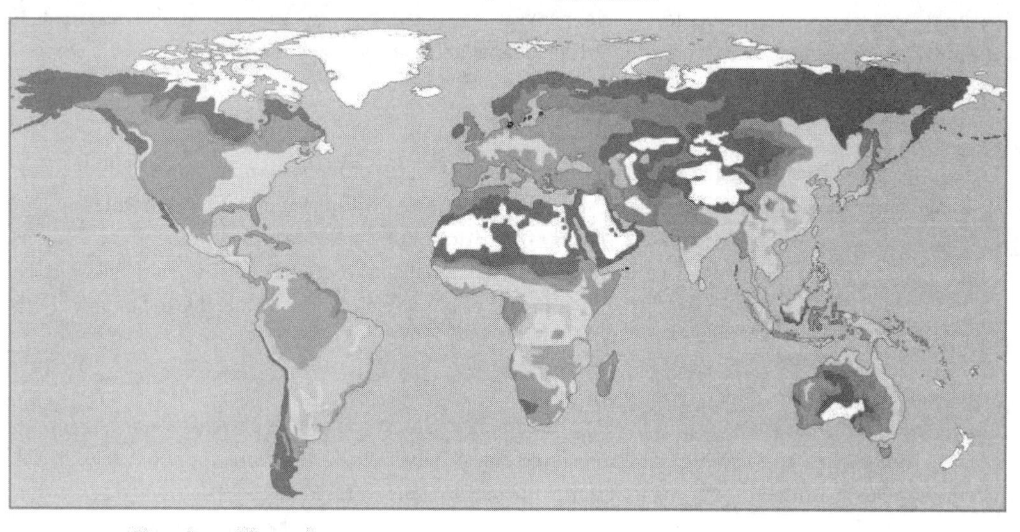

Figure 47.3 This map illustrates the number of amphibian species across the globe and shows the trend toward higher biodiversity at lower latitudes. A similar pattern is observed for most taxonomic groups. The white areas indicate a lack of data in this particular study.

It is not yet clear why biodiversity increases closer to the equator, but scientists have several hypotheses. One factor may be the greater age of the ecosystems in the tropics versus those in temperate regions; the temperate regions were largely devoid of life or were drastically reduced during the last glaciation. The idea is that greater age provides more time for speciation. Another possible explanation is the increased direct energy the tropics receive from the sun versus the decreased intensity of the solar energy that temperate and polar regions receive. Tropical ecosystem complexity may promote speciation by increasing the **heterogeneity**, or *number of ecological niches,* in the tropics relative to higher latitudes. The greater heterogeneity provides more opportunities for coevolution, specialization, and perhaps greater selection pressures leading to population differentiation. However, this hypothesis suffers from some circularity—ecosystems with more species encourage speciation, but how did they get more species to begin with?

The tropics have been perceived as being more stable than temperate regions, which have a pronounced climate and day-length seasonality. The tropics have their own forms of seasonality, such as rainfall, but they are generally assumed to be more stable environments and this stability might promote speciation into highly specialized niches.

Regardless of the mechanisms, it is certainly true that all levels of biodiversity are greatest in the tropics. Additionally, the *rate of endemism* is highest, and there are more biodiversity "hotspots." However, this richness of diversity also means that knowledge of species is unfortunately very low, and there is a high potential for biodiversity loss.

Conservation of Biodiversity

In 1988, British environmentalist Norman Myers developed a conservation concept to identify areas rich in species and at significant risk for species loss: *biodiversity hotspots*. **Biodiversity hotspots** are geographical areas that contain high numbers of endemic species. The purpose of the concept was to identify important locations on the planet for conservation efforts, a kind of conservation triage. By protecting hotspots, governments are able to protect a larger number of species. The original criteria for a hotspot included the presence of 1500 or more endemic plant species and 70 percent of the area disturbed by human activity. There are now 34 biodiversity hotspots (Figure 47.4) containing large numbers of endemic species, which include half of Earth's endemic plants.

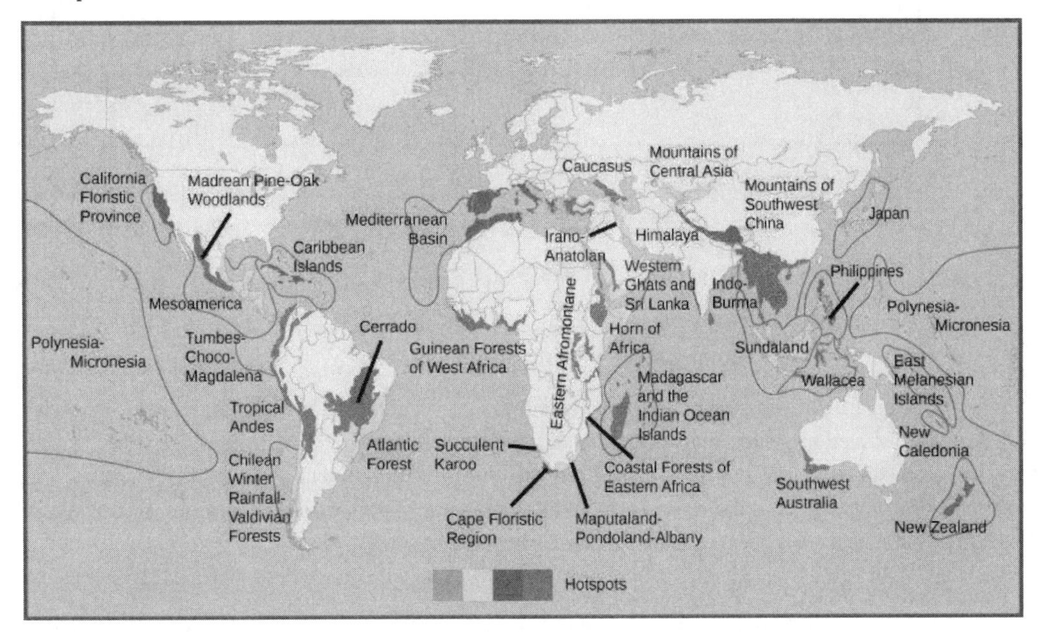

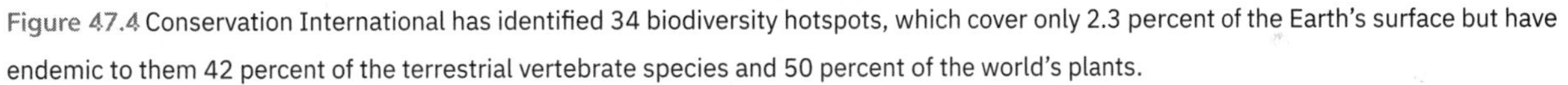

Figure 47.4 Conservation International has identified 34 biodiversity hotspots, which cover only 2.3 percent of the Earth's surface but have endemic to them 42 percent of the terrestrial vertebrate species and 50 percent of the world's plants.

Biodiversity Change through Geological Time

The number of species on the planet, or in any geographical area, is the result of an equilibrium of two evolutionary processes that are continuously ongoing: *speciation* and *extinction*. Both are natural "birth" and "death" processes of macroevolution. When speciation rates begin to outstrip extinction rates, the number of species will increase; likewise, the number of species will decrease when extinction rates begin to overtake speciation rates. Throughout Earth's history, these two processes have fluctuated—sometimes leading to dramatic changes in the number of species on Earth as reflected in the fossil record (Figure 47.5).

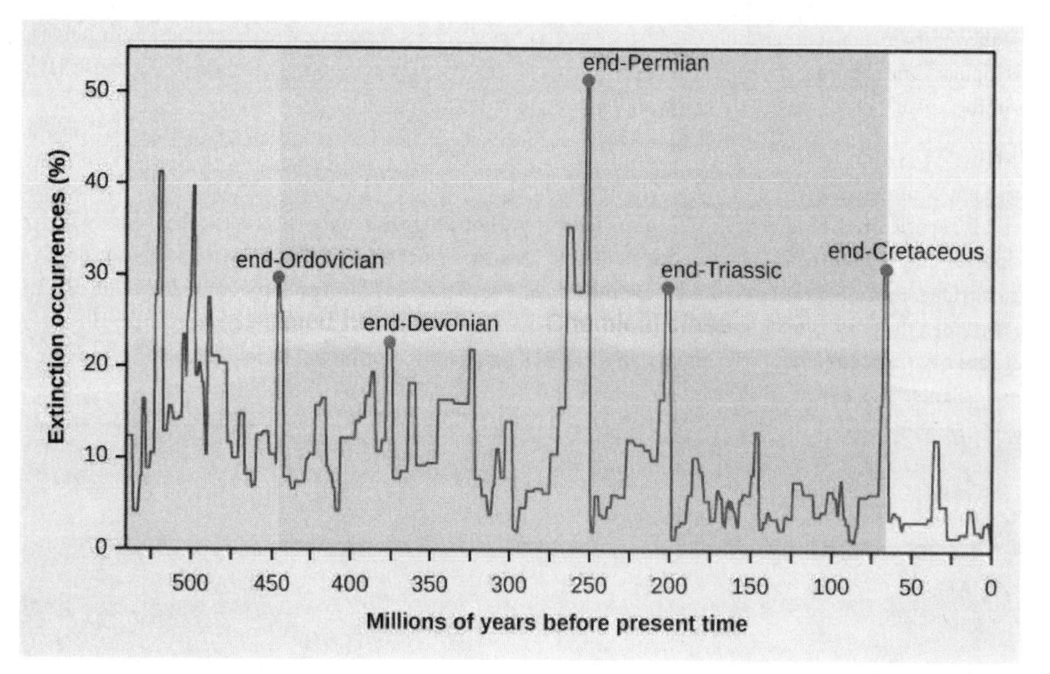

Figure 47.5 Percent extinction occurrences as reflected in the fossil record have fluctuated throughout Earth's history. Sudden and dramatic losses of biodiversity, called *mass extinctions*, have occurred five times.

Paleontologists have identified five strata in the fossil record that appear to show sudden and dramatic (greater than half of all extant species disappearing from the fossil record) losses in biodiversity. These are called **mass extinctions**. There are many lesser, yet still dramatic, extinction events, but the five mass extinctions have attracted the most research. An argument can be made that the five mass extinctions are only the five most *extreme events* in a continuous series of large extinction events throughout the Phanerozoic (since 542 million years ago). In most cases, the hypothesized causes are still controversial; however, the most recent mass extinction event seems clear.

The Five Mass Extinctions

The fossil record of the mass extinctions was the basis for defining periods of geological history, so they typically occur at the transition point between geological periods. The transition in fossils from one period to another reflects the dramatic loss of species and the gradual origin of new species. These transitions can be seen in the rock strata. Table 47.2 provides data on the five mass extinctions.

Mass Extinctions

Geological Period	Mass Extinction Name	Time (millions of years ago)
Ordovician–Silurian	end-Ordovician O–S	450–440
Late Devonian	end-Devonian	375–360
Permian–Triassic	end-Permian	251
Triassic–Jurassic	end-Triassic	205
Cretaceous–Paleogene	end-Cretaceous K–Pg (K–T)	65.5

Table 47.2 This table shows the names and dates for the five mass extinctions in Earth's history.

The **Ordovician-Silurian extinction** event is the first recorded mass extinction and the second largest. During this period, about

85 percent of marine species (few species lived outside the oceans) became extinct. The main hypothesis for its cause is a period of glaciation and then warming. The extinction event actually consists of two extinction events separated by about 1 million years. The first event was caused by cooling, and the second event was due to the subsequent warming. The climate changes affected temperatures and sea levels. Some researchers have suggested that a gamma-ray burst, caused by a nearby supernova, was a possible cause of the Ordovician-Silurian extinction. The gamma-ray burst would have stripped away the Earth's protective ozone layer, allowing intense ultraviolet radiation from the sun to reach the surface of the earth—and *may* account for climate changes observed at the time. The hypothesis is very speculative, and extraterrestrial influences on Earth's history are an active line of research. Recovery of biodiversity after the mass extinction took from 5 to 20 million years, depending on the location.

The **late Devonian extinction** may have occurred over a relatively long period of time. It appears to have mostly affected marine species and not so much the plants or animals inhabiting terrestrial habitats. The causes of this extinction are poorly understood.

The **end-Permian extinction** was the largest in the history of life. Indeed, an argument could be made that Earth became nearly devoid of life during this extinction event. Estimates are that 96 percent of all marine species and 70 percent of all terrestrial species were lost. It was at this time, for example, that the trilobites, a group that survived the Ordovician–Silurian extinction, became extinct. The causes for this mass extinction are not clear, but the leading suspect is extended and widespread volcanic activity that led to a runaway global-warming event. The oceans became largely anoxic, suffocating marine life. Terrestrial tetrapod diversity took 30 million years to recover after the end-Permian extinction. The Permian extinction dramatically altered Earth's biodiversity makeup and the course of evolution.

The causes of the **Triassic–Jurassic** extinction event are not clear, and researchers argue hypotheses including climate change, asteroid impact, and volcanic eruptions. The extinction event occurred just before the breakup of the supercontinent Pangaea, although recent scholarship suggests that the extinctions may have occurred more gradually throughout the Triassic.

The causes of the end-Cretaceous extinction event are the ones that are best understood. It was during this extinction event about 65 million years ago that the majority of the dinosaurs, the dominant vertebrate group for millions of years, disappeared from the planet (with the exception of a theropod clade that gave rise to birds).

The cause of this extinction is now understood to be the result of a cataclysmic impact of a large meteorite, or asteroid, off the coast of what is now the Yucatán Peninsula. This hypothesis, proposed first in 1980, was a radical explanation based on a sharp spike in the levels of *iridium* (which enters our atmosphere from meteors at a fairly constant rate but is otherwise absent on Earth's surface) in the rock stratum that marks the boundary between the Cretaceous and Paleogene periods (Figure 47.6). This boundary marked the disappearance of the dinosaurs in fossils as well as many other taxa. The researchers who discovered the iridium spike interpreted it as a rapid influx of iridium from space to the atmosphere (in the form of a large asteroid) rather than a slowing in the deposition of sediments during that period. It was a radical explanation, but the report of an appropriately aged and sized impact crater in 1991 made the hypothesis more believable. Now an abundance of geological evidence supports the theory. Recovery times for biodiversity after the end-Cretaceous extinction are shorter, in geological time, than for the end-Permian extinction, on the order of 10 million years.

Another possibility, perhaps coincidental with the impact of the Yucatan asteroid, was extensive volcanism that began forming about 66 million years ago, about the same time as the Yucatan asteroid impact, at the end of the Cretaceous. The lava flows covered over 50 percent of what is now India. The release of volcanic gases, particularly sulphur dioxide, during the formation of the traps contributed to climate change, which may have induced the mass extinction.

VISUAL CONNECTION

Figure 47.6 In 1980, Luis and Walter Alvarez, Frank Asaro, and Helen Michels discovered, across the world, a spike in the concentration of iridium within the sedimentary layer at the K–Pg boundary. These researchers hypothesized that this iridium spike was caused by an asteroid impact that resulted in the K–Pg mass extinction. In the photo, the iridium layer is the light band. (credit: USGS)

Scientists measured the relative abundance of fern spores above and below the K–Pg boundary in this rock sample. Which of the following statements most likely represents their findings?

a. An abundance of fern spores from several species was found below the K–Pg boundary, but none was found above.
b. An abundance of fern spores from several species was found above the K–Pg boundary, but none was found below.
c. An abundance of fern spores was found both above and below the K–Pg boundary, but only one species was found below the boundary, and many species were found above the boundary.
d. Many species of fern spores were found both above and below the boundary, but the total number of spores was greater below the boundary.

LINK TO LEARNING

Explore this interactive website (http://openstax.org/l/extinctions) about mass extinctions.

The Pleistocene Extinction

The **Pleistocene Extinction** is one of the lesser extinctions, and a recent one. It is well known that the North American, and to some degree Eurasian, **megafauna**—large vertebrate animals—disappeared toward the end of the last glaciation period. The extinction appears to have happened in a relatively restricted time period of 10,000–12,000 years ago. In North America, the losses were quite dramatic and included the woolly mammoths (with an extant population existing until about 4,000 years ago in isolation on Wrangel Island, Canada), mastodon, giant beavers, giant ground sloths, saber-toothed cats, and the North American camel, just to name a few. In the early 1900s, scientists first suggested the possibility that over-hunting caused the rapid extinction of these large animals. Research into this hypothesis continues today.

In general, the timing of the Pleistocene extinctions correlated with the arrival of paleo-humans, perhaps as long as 40,000 years ago, and not with climate-change events, which is the main competing hypothesis for these extinctions. The extinctions began in Australia about 40,000 to 50,000 years ago, just after the arrival of humans in the area: a marsupial lion, a giant one-ton wombat, and several giant kangaroo species disappeared. In North America, the extinctions of almost all of the large mammals occurred 10,000–12,000 years ago. All that are left are the smaller mammals such as bears, elk, moose, and cougars. Finally, on many remote oceanic islands, the extinctions of many species occurred coincidentally with human arrivals. Not all of the islands had large animals, but when there were large animals, they were often forced into extinction. Madagascar was colonized about 2,000 years ago and the large mammals that lived there became extinct. Eurasia and Africa do not show this pattern, but they also did not experience a recent arrival of hunter-gatherer humans. Rather, humans arrived in Eurasia hundreds of thousands of years ago. This topic remains an area of active research and hypothesizing. It seems clear that even if climate played a role, in most cases human hunting precipitated the extinctions.

Recent Extinctions

The sixth, or **Holocene, mass extinction** appears to have begun earlier than previously believed and is largely due to the disruptive activities of modern *Homo sapiens*. Since the beginning of the Holocene period, there are numerous recent extinctions of individual species that are recorded in human writings. Most of these are coincident with the expansion of the European colonies since the 1500s.

One of the earlier and popularly known examples is the dodo bird. The odd pigeon-like bird lived in the forests of Mauritius (an island in the Indian Ocean) and became extinct around 1662. The dodo was hunted for its meat by sailors and was easy prey because it approached people without fear (the dodo had not evolved with humans). Pigs, rats, and dogs brought to the island by European ships also killed dodo young and eggs.

Steller's sea cow became extinct in 1768; it was related to the manatee and probably once lived along the northwest coast of North America. Steller's sea cow was first discovered by Europeans in 1741 and was overhunted for meat and oil. The last sea cow was killed in 1768. That amounts to just 27 years between the sea cow's first contact with Europeans and extinction of the species!

Since 1900, a variety of species have gone extinct, including the following:

- In 1914, the last living passenger pigeon died in a zoo in Cincinnati, Ohio. This species had once darkened the skies of North America during its migrations, but it was overhunted and suffered from habitat loss that resulted from the clearing of forests for farmland.
- The Carolina parakeet, once common in the eastern United States, died out in 1918. It suffered habitat loss and was hunted to prevent it from eating orchard fruit. (The parakeet ate orchard fruit because its native foods were destroyed to make way for farmland.)
- The Japanese sea lion, which inhabited a broad area around Japan and the coast of Korea, became extinct in the 1950s due to fishermen.
- The Caribbean monk seal was distributed throughout the Caribbean Sea but was driven to extinction via hunting by 1952.

These are only a few of the recorded extinctions in the past 500 years. The International Union for Conservation of Nature (IUCN) keeps a list of extinct and endangered species called the **Red List**. The list is not complete, but it describes 380 extinct species of vertebrates after 1500 AD, 86 of which were driven extinct by overhunting or overfishing.

Estimates of Present-Time Extinction Rates

Estimates of **extinction rates** are hampered by the fact that most extinctions are probably happening without observation. The extinction of a bird or mammal is likely to be noticed by humans, especially if it has been hunted or used in some other way. But there are many organisms that are of less interest to humans (not necessarily of less value) and many that are undescribed.

The **background extinction** rate is estimated to be about one per million species per year (E/MSY). For example, assuming there are about ten million species in existence, the expectation is that ten species would become extinct each year (each year represents ten million species per year).

One contemporary extinction rate estimate uses the extinctions in the written record since the year 1500. For birds alone this method yields an estimate of 26 E/MSY. However, this value may be an underestimate for three reasons. First, many species would not have been described until much later in the time period, so their loss would have gone unnoticed. Second, the number of recently extinct vertebrate species is increasing because extinct species now are being described from skeletal remains. And third, some species are probably already extinct even though conservationists are reluctant to name them as such. Taking these factors into account raises the estimated extinction rate closer to 100 E/MSY. The predicted rate by the end of the century is 1500 E/MSY.

A second approach to estimating present-time extinction rates is to correlate species loss with habitat loss by measuring forest-area loss and understanding species-area relationships. The **species-area relationship** is the rate at which new species are seen when the area surveyed is increased. Studies have shown that the number of species present increases as the size of the island increases. This phenomenon has also been shown to hold true in other island-like habitats as well, such as the mountain-top tepuis of Venezuela, which are surrounded by tropical forest. Turning this relationship around, if the habitat area is reduced, the number of species living there will also decline. Estimates of extinction rates based on habitat loss and species-area relationships have suggested that with about 90 percent habitat loss an expected 50 percent of species would become extinct. Species-area estimates have led to species extinction rate calculations of about 1000 E/MSY and higher. In general, actual observations do not show this amount of loss and suggestions have been made that there is a delay in extinction. Recent work has also called into question the applicability of the species-area relationship when estimating the loss of species. This work argues that the species-area relationship leads to an *overestimate* of extinction rates. A better relationship to use may be the endemics-area relationship. Using this method would bring estimates down to around 500 E/MSY in the coming century. *Note that this value is still 500 times the background rate.*

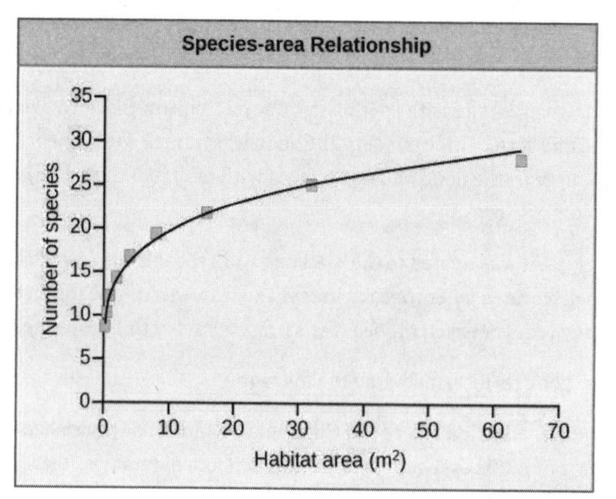

Figure 47.7 Studies have shown that the number of species present increases with the size of the habitat. (credit: modification of work by Adam B. Smith)

LINK TO LEARNING

Check out this interactive exploration (http://openstax.org/l/what_is_missing) of endangered and extinct species, their ecosystems, and the causes of the endangerment or extinction.

47.2 The Importance of Biodiversity to Human Life

By the end of this section, you will be able to do the following:

- Identify chemical diversity benefits to humans
- Identify biodiversity components that support human agriculture
- Describe ecosystem services

It may not be clear why biologists are concerned about biodiversity loss. When biodiversity loss is thought of as the extinction of the passenger pigeon, the dodo bird, and even the woolly mammoth, the loss may appear to be an emotional one. But is the loss practically important for the welfare of the human species? From the perspective of evolution and ecology, the loss of a particular individual species is unimportant (however, we should note that the loss of a keystone species can lead to ecological disaster). Extinction is a normal part of macroevolution. But the *accelerated extinction rate* translates into the loss of tens of thousands of species within our lifetimes, and it is likely to have dramatic effects on human welfare through the collapse of ecosystems and in added costs to maintain food production, clean air and water, and human health.

Agriculture began after early hunter-gatherer societies first settled in one place and heavily modified their immediate environment. This cultural transition has made it difficult for humans to recognize their dependence on undomesticated living things on the planet. Biologists recognize the human species is embedded in ecosystems and is dependent on them, just as every other species on the planet is dependent. Technology smooths out the extremes of existence, but ultimately the human species cannot exist without a supportive ecosystem.

Human Health

Archeological evidence indicates that humans have been using plants for medicinal uses for thousands of years. A Chinese document from approximately 2800 BC is believed to be the the first written account of herbal remedies, and such references occur throughout the global historical record. Contemporary indigenous societies that live close to the land often retain broad knowledge of the medicinal uses of plants growing in their area. Most plants produce **secondary plant compounds**, which are toxins used to protect the plant from insects and other animals that eat them, but some of which also work as medication.

Modern pharmaceutical science also recognizes the importance of these plant compounds. Examples of significant medicines derived from plant compounds include aspirin, codeine, digoxin, atropine, and vincristine (Figure 47.8). Many medicines were once derived from plant extracts but are now synthesized. It is estimated that, at one time, 25 percent of modern drugs contained at least one plant extract. That number has probably decreased to about 10 percent as natural plant ingredients are replaced by synthetic versions. Antibiotics, which are responsible for extraordinary improvements in health and lifespans in developed countries, are compounds largely derived from fungi and bacteria.

Figure 47.8 *Catharanthus roseus*, the Madagascar periwinkle, has various medicinal properties. Among other uses, it is a source of vincristine, a drug used in the treatment of lymphomas. (credit: Forest and Kim Starr)

In recent years, animal venoms and poisons have excited intense research for their medicinal potential. By 2007, the FDA had approved five drugs based on animal toxins to treat diseases such as hypertension, chronic pain, and diabetes. Another five drugs are undergoing clinical trials, and at least six drugs are being used in other countries. Other toxins under investigation come from mammals, snakes, lizards, various amphibians, fish, snails, octopuses, and scorpions.

Aside from representing billions of dollars in profits, these medicines improve people's lives. Pharmaceutical companies are always looking for new compounds synthesized by living organisms that can function as medicines. It is estimated that 1/3 of pharmaceutical research and development is spent on natural compounds and that about 35 percent of new drugs brought to market between 1981 and 2002 were derived from natural compounds. The opportunities for new medications will be reduced in direct proportion to the disappearance of species.

Agricultural Diversity

Since the beginning of human agriculture more than 10,000 years ago, human groups have been breeding and selecting crop varieties. This crop diversity matched the cultural diversity of highly subdivided populations of humans. For example, potatoes were domesticated beginning around 7,000 years ago in the central Andes of Peru and Bolivia. The potatoes grown in that region belong to seven species and the number of varieties likely is in the thousands. Even the Inca capital of Machu Picchu had numerous gardens growing varieties of potatoes. Each variety has been bred to thrive at particular elevations and soil and climate conditions. The diversity is driven by the diverse demands of the topography, the limited movement of people, and the demands created by crop rotation for different varieties that will do well in different fields.

Potatoes are only one example of human-generated diversity. Every plant, animal, and fungus that has been cultivated by humans has been bred from original wild ancestor species into diverse varieties arising from the demands for food value, adaptation to growing conditions, and resistance to pests.

The potato also demonstrates risks of low crop diversity. The tragic Irish potato famine occurred when the single variety grown in Ireland became susceptible to a potato blight, wiping out the entire crop. The loss of the potato crop led to mass famine and the related deaths of over one million people, as well as mass emigration of nearly two million people.

Disease resistance is a chief benefit of crop biodiversity, and lack of diversity in contemporary crop species carries similar risks. Seed companies, which are the source of most crop varieties in developed countries, must continually breed new varieties to keep up with evolving pest organisms. These same seed companies, however, have participated in the decline of the number of varieties available as they focus on selling fewer varieties in more areas of the world.

The ability to create new crop varieties relies on the diversity of varieties available and the accessibility of wild forms related to the crop plant. These wild forms are often the source of new gene variants that can be bred with existing varieties to create varieties with new attributes. Loss of wild species related to a crop will mean the loss of potential in crop improvement. Maintaining the genetic diversity of wild species related to domesticated species ensures our continued food supply.

Since the 1920s, government agriculture departments have maintained seed banks of crop varieties as a way of maintaining crop diversity. This system has flaws because, over time, seed banks are lost through accidents, and there is no way to replace them. In 2008, the **Svalbard Global Seed Vault** (Figure 47.9) began storing seeds from around the world as a backup system to the regional seed banks. If a regional seed bank stores varieties in Svalbard, losses can be replaced from Svalbard. Conditions within the vault are maintained at ideal temperature and humidity for seed survival, but the deep underground location of the vault in the arctic means that failure of the vault's systems will not compromise the climatic conditions inside the vault.

VISUAL CONNECTION

Figure 47.9 The Svalbard Global Seed Vault is a storage facility for seeds of Earth's diverse crops. (credit: Mari Tefre, Svalbard Global Seed Vault)

The Svalbard Global Seed Vault is located on Spitsbergen island in Norway, which has an arctic climate. Why might an arctic climate be good for seed storage?

Crop success is largely dependent on the quality of the soil. Although some agricultural soils are rendered sterile using controversial cultivation and chemical treatments, most contain a huge diversity of organisms that maintain nutrient cycles—breaking down organic matter into nutrient compounds that crops need for growth. These organisms also maintain soil texture that affects water and oxygen dynamics in the soil that are necessary for plant growth. If farmers had to maintain arable soil using alternate means, the cost of food would be much higher than it is now. These kinds of processes are called **ecosystem services**. They occur within ecosystems, such as soil ecosystems, as a result of the diverse metabolic activities of the organisms living there, but they provide benefits to human food production, drinking water availability, and breathable air.

Plant pollination is another key ecosystem service, provided by various species of bees, other insects, and birds. One estimate indicates that honey bee pollination provides the United States a $1.6 billion annual benefit.

Honey bee populations in North America have been suffering large losses caused by a syndrome known as colony collapse disorder, whose cause is unclear. (Evidence suggests the possible culprits may be the invasive varroa mite coupled with the Nosema gut parasite and acute paralysis virus.) Loss of these species would render it very difficult, if not impossible, to grow any of the 150 United States crops requiring pollination, including grapes, oranges, lemons, peppers, most brassica (broccoli and cauliflower), and many berries, melons, and nuts.

Finally, humans compete for their food with crop pests, most of which are insects. Pesticides control these competitors; however, pesticides are costly and lose their effectiveness over time as pest populations adapt and evolve. They also lead to collateral damage by killing non-pest species and risking the health of consumers and agricultural workers. Ecologists believe that the bulk of the work in removing pests is actually done by predators and parasites of those pests, but the impact has not been well studied. A review found that in 74 percent of studies that looked for an effect of landscape complexity on natural enemies of pests, the greater the complexity, the greater the effect of pest-suppressing organisms. An experimental study found that introducing multiple enemies of pea aphids (an important alfalfa pest) increased the yield of alfalfa significantly. This study shows the importance of landscape diversity via the question of whether a diversity of pests is more effective at control than one single pest; the results showed this to be the case. Loss of diversity in pest enemies will inevitably make it more difficult and costly to grow food.

Wild Food Sources

In addition to growing crops and raising animals for food, humans obtain food resources from wild populations, primarily fish populations. In fact, for approximately 1 billion people worldwide, aquatic resources provide the main source of animal protein. But since 1990, global fish production has declined, sometimes dramatically. Unfortunately, and despite considerable effort, few fisheries on the planet are managed for sustainability.

Fishery extinctions rarely lead to complete extinction of the harvested species, but rather to a radical restructuring of the marine ecosystem in which a dominant species is so over-harvested that it becomes a minor player, ecologically. In addition to humans losing the food source, these alterations affect many other species in ways that are difficult or impossible to predict. The collapse of fisheries has dramatic and long-lasting effects on local populations that work in the fishery. In addition, the loss of an inexpensive protein source to populations that cannot afford to replace it will increase the cost of living and limit societies in other ways. In general, the fish taken from fisheries have shifted to smaller species as larger species are fished to extinction. The ultimate outcome could clearly be the loss of aquatic systems as food sources.

LINK TO LEARNING

View a brief video (http://openstax.org/l/declining_fish) discussing declining fish stocks.

Psychological and Moral Value

Finally, it has been clearly shown that humans benefit psychologically from living in a biodiverse world. A chief proponent of this idea is Harvard entomologist E. O. Wilson. He argues that human evolutionary history has adapted us to live in a natural environment and that city environments generate psychological stressors that affect human health and well-being. There is considerable research into the psychological regenerative benefits of natural landscapes that suggests the hypothesis may hold some truth. In addition, there is a moral argument that humans have a responsibility to inflict as little harm as possible on other species.

47.3 Threats to Biodiversity

By the end of this section, you will be able to do the following:

- Identify significant threats to biodiversity
- Explain the effects of habitat loss, the introduction of exotic species, and hunting on biodiversity
- Identify the early and predicted effects of climate change on biodiversity

The core threat to biodiversity on the planet, and therefore a threat to human welfare, is the combination of human population growth and resource exploitation. The human population requires resources to survive and grow, and those resources are being removed unsustainably from the environment. The three greatest proximate threats to biodiversity are habitat loss, overharvesting, and the introduction of exotic species. The first two of these are a direct result of human population growth and resource use. The third results from increased mobility and trade. A fourth major cause of extinction, anthropogenic climate change, has not yet had a large impact, but it is predicted to become significant during this century. Global climate change is also a consequence of human population needs for energy and the use of fossil fuels to meet those needs (Figure 47.10). Environmental issues, such as toxic pollution, have specific targeted effects on species, but they are not generally seen as threats at the magnitude of the others.

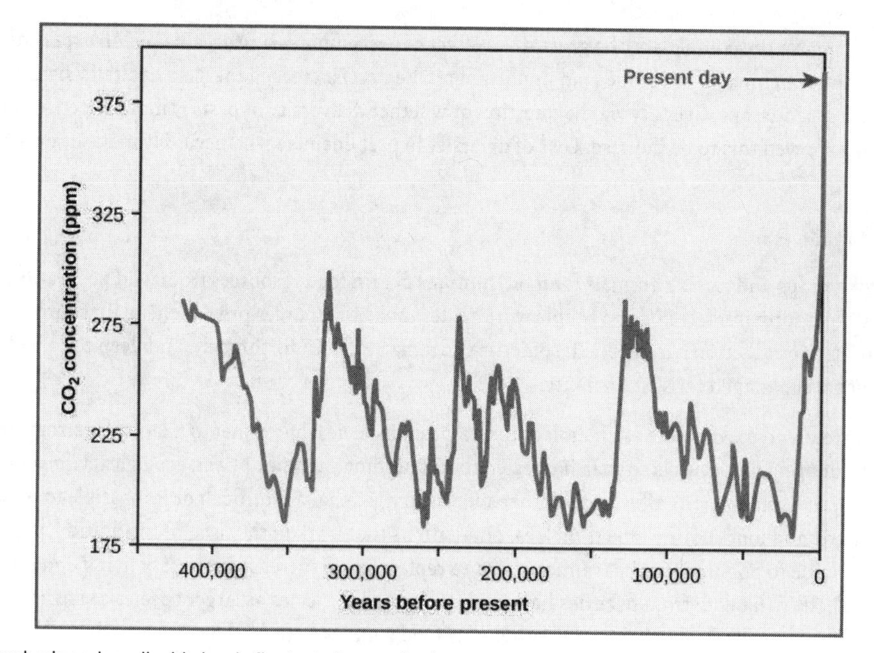

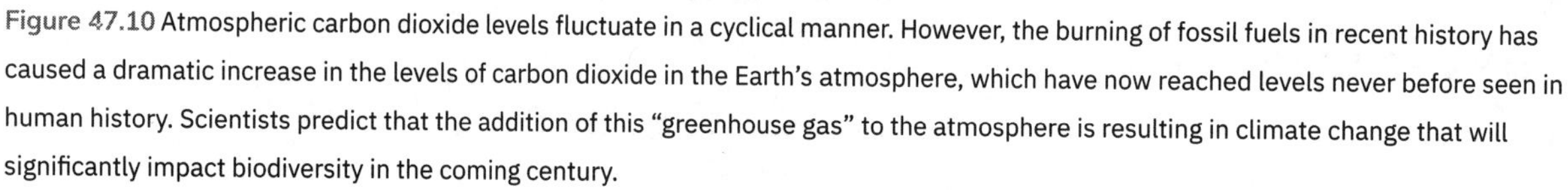
Figure 47.10 Atmospheric carbon dioxide levels fluctuate in a cyclical manner. However, the burning of fossil fuels in recent history has caused a dramatic increase in the levels of carbon dioxide in the Earth's atmosphere, which have now reached levels never before seen in human history. Scientists predict that the addition of this "greenhouse gas" to the atmosphere is resulting in climate change that will significantly impact biodiversity in the coming century.

Habitat Loss

Humans rely on technology to modify their environment and replace certain functions that were once performed by the natural ecosystem. Other species cannot do this. Elimination of their ecosystem—whether it is a forest, a desert, a grassland, a freshwater estuarine, or a marine environment—will kill the individuals belonging to the species. The species will become extinct if we remove the entire habitat within the range of a species. Human destruction of habitats accelerated in the latter half of the twentieth century. Consider the exceptional biodiversity of Sumatra: it is home to one species of orangutan, a species of critically endangered elephant, and the Sumatran tiger, but half of Sumatra's forest is now gone. The neighboring island of Borneo, home to the other species of orangutan, has lost a similar area of forest. Forest loss continues in protected areas of Borneo. All three species of orangutan are now listed as endangered by the International Union for Conservation of Nature (IUCN), but they are simply the most visible of thousands of species that will not survive the disappearance of the forests in Sumatra and Borneo. The forests are removed for timber and to plant palm oil plantations (Figure 47.11). Palm oil is used in many products including food products, cosmetics, and biodiesel in Europe. A five-year estimate of global forest cover loss for the years 2000–2005 was 3.1 percent. In the humid tropics where forest loss is primarily from timber extraction, 272,000 km^2 was lost out of a global total of 11,564,000 km^2 (or 2.4 percent). In the tropics, these losses certainly also represent the extinction of species because of high levels of **endemism**—species unique to a defined geographic location, and found nowhere else.

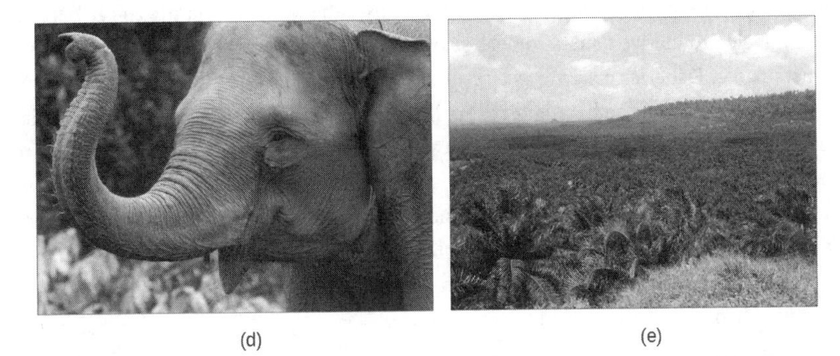

Figure 47.11 (a) One of three species of orangutan, *Pongo pygmaeus,* is found only in the rainforests of Borneo, and an other species of orangutan (*Pongo abelii*) is found only in the rainforests of Sumatra. These animals are examples of the exceptional biodiversity of (c) the islands of Sumatra and Borneo. Other species include the (b) Sumatran tiger (*Panthera tigris sumatrae*) and the (d) Sumatran elephant (*Elephas maximus sumatranus*), both critically endangered species. Rainforest habitat is being removed to make way for (e) oil palm plantations such as this one in Borneo's Sabah Province. (credit a: modification of work by Thorsten Bachner; credit b: modification of work by Dick Mudde; credit c: modification of work by U.S. CIA World Factbook; credit d: modification of work by "Nonprofit Organizations"/Flickr; credit e: modification of work by Dr. Lian Pin Koh)

Everyday Connection

Preventing Habitat Destruction with Wise Wood Choices

Most consumers are not aware that the home improvement products they buy might be contributing to habitat loss and species extinctions. Yet the market for illegally harvested tropical timber is huge, and the wood products often find themselves in building supply stores in the United States. One estimate is that 10 percent of the imported timber stream in the United States, which is the world's largest consumer of wood products, is potentially illegally logged. In 2006, this amounted to $3.6 billion in wood products. Most of the illegal products are imported from countries that act as *intermediaries* and are not the originators of the wood.

How is it possible to determine if a wood product, such as flooring, was harvested sustainably or even legally? The **Forest Stewardship Council** (FSC) certifies sustainably harvested forest products, therefore, looking for their certification on flooring and other hardwood products is one way to ensure that the wood has not been taken illegally from a tropical forest. Certification applies to specific products, not to a producer; some producers' products may not have certification while other products are certified. While there are other industry-backed certifications other than the FSC, these are unreliable due to lack of independence from the industry. Another approach is to buy domestic wood species. While it would be great if there was a list of legal versus illegal wood products, it is not that simple. Logging and forest management laws vary from country to country; what is illegal in one country may be legal in another. Where and how a product is harvested and whether the forest from which it comes is being maintained sustainably all factor into whether a wood product will be certified by the FSC. If you are in doubt, it is always a good idea to ask questions about where a wood product came from and how the supplier knows that it was harvested legally.

Habitat destruction can affect ecosystems other than forests. Rivers and streams are important ecosystems that are frequently modified through land development, damming, channelizing, or water removal. Damming affects the water flow to all parts of a river, which can reduce or eliminate populations that had adapted to the natural flow of the river. For example, an estimated 91 percent of United States rivers have been altered in some way. Modifications include dams, to create energy or store water; levees, to prevent flooding; and dredging or rerouting, to create land that is more suitable for human development. Many fish and amphibian species and numerous freshwater clams in the United States have seen declines caused by river damming and habitat loss.

Overharvesting

Overharvesting is a serious threat to many species, but particularly to aquatic (both marine and freshwater) species. Despite regulation and monitoring, there are recent examples of fishery collapse. The western Atlantic cod fishery is the among the most significant. While it was a hugely productive fishery for 400 years, the introduction of modern factory trawlers in the 1980s caused it to become unsustainable. Fisheries collapse as a result of both economic and political factors. Fisheries are managed as a shared international resource even when the fishing territory lies within an individual country's territorial waters. Common resources are subject to an economic pressure known as the **tragedy of the commons**, in which essentially no fisher has a motivation to exercise restraint in harvesting a fishery when it is not owned by that fisher. Overexploitation is a common outcome. This overexploitation is exacerbated when access to the fishery is open and unregulated and when technology gives fishers the ability to overfish. In a few fisheries, the biological growth of the resource is less than the potential growth of the profits made from fishing if that time and money were invested elsewhere. In these cases—whales are an example—economic forces will always drive toward fishing the population to extinction.

LINK TO LEARNING

Explore a U.S. Fish & Wildlife Service interactive map (http://openstax.org/l/habitat_map) of critical habitat for endangered and threatened species in the United States. To begin, select "Visit the online mapper."

For the most part, fishery extinction is not equivalent to biological extinction—the last fish of a species is rarely fished out of the ocean. At the same time, fishery extinction is still harmful to fish species and their ecosystems. There are some instances in which true extinction is a possibility. Whales have slow-growing populations due to low reproductive rates, and therefore are at risk of complete extinction through hunting. There are some species of sharks with restricted distributions that are at risk of extinction. The groupers are another population of generally slow-growing fishes that, in the Caribbean, includes a number of species that are at risk of extinction from overfishing.

Coral reefs are extremely diverse marine ecosystems that face immediate peril from several processes. Reefs are home to 1/3 of the world's marine fish species—about 4,000 species—despite making up only 1 percent of marine habitat. Most home marine aquaria are stocked with wild-caught organisms, not cultured organisms. Although no species is known to have been driven extinct by the pet trade in marine species, there are studies showing that populations of some species have declined in response to harvesting, indicating that the harvest is not sustainable at those levels. There are concerns about the effect of the pet trade on some terrestrial species such as turtles, amphibians, birds, plants, and even the orangutan.

LINK TO LEARNING

View a brief video (http://openstax.org/l/ocean_matters) discussing the role of marine ecosystems in supporting human welfare and the decline of ocean ecosystems.

Bush meat is the generic term used for wild animals killed for food. Hunting is practiced throughout the world, but hunting practices, particularly in equatorial Africa and parts of Asia, are believed to threaten a number of species with extinction. Traditionally, bush meat in Africa was hunted to feed families directly; however, recent commercialization of the practice now has bush meat available in grocery stores, which has increased harvest rates to the level of unsustainability. Additionally, human population growth has increased the need for protein foods that are not being met from agriculture. Species threatened by the bush meat trade are mostly mammals including many primates living in the Congo basin.

Exotic Species

Exotic species are species that have been intentionally or unintentionally introduced into an ecosystem in which they did not evolve. For example, Kudzu (*Pueraria lobata*), which is native to Japan, was introduced in the United States in 1876. It was later planted for soil conservation. Problematically, it grows too well in the southeastern United States—up to a foot a day. It is now an invasive pest species and covers over 7 million acres in the southeastern United States. If an introduced species is able to survive in its new habitat, that introduction is now reflected in the observed range of the species. Human transportation of people and goods, including the intentional transport of organisms for trade, has dramatically increased the introduction of species into new ecosystems, sometimes at distances that are well beyond the capacity of the species to ever travel itself and outside the range of the species' natural predators.

Most exotic species introductions probably fail because of the low number of individuals introduced or poor adaptation to the ecosystem they enter. Some species, however, possess pre-adaptations that can make them especially successful in a new ecosystem. These exotic species often undergo dramatic population increases in their new habitat and reset the ecological conditions in the new environment, threatening the species that exist there. For this reason, exotic species are also called invasive species. Exotic species can threaten other species through competition for resources, predation, or disease. For example, the Eurasian star thistle, also called spotted knapweed, has invaded and rendered useless some of the open prairies of the western states. However, it is a great nectar-bearing flower for the production of honey and supports numerous pollinating insects, including migrating monarch butterflies in the north-central states such as Michigan.

LINK TO LEARNING

Explore an interactive global database (http://openstax.org/l/exotic_invasive) of exotic or invasive species.

Lakes and islands are particularly vulnerable to extinction threats from introduced species. In Lake Victoria, as mentioned earlier, the intentional introduction of the Nile perch was largely responsible for the extinction of about 200 species of endemic cichlids. The accidental introduction of the brown tree snake via aircraft (Figure 47.12) from the Solomon Islands to Guam in 1950 has led to the extinction of three species of birds and three to five species of reptiles endemic to the island. Several other species are still threatened. The brown tree snake is adept at exploiting human transportation as a means to migrate; one was even found on an aircraft arriving in Corpus Christi, Texas. Constant vigilance on the part of airport, military, and commercial aircraft personnel is required to prevent the snake from moving from Guam to other islands in the Pacific, especially Hawaii. Islands do not make up a large area of land on the globe, but they do contain a disproportionate number of endemic species because of their isolation from mainland ancestors.

Figure 47.12 The brown tree snake, *Boiga irregularis*, is an exotic species that has caused numerous extinctions on the island of Guam since its accidental introduction in 1950. (credit: NPS)

It now appears that the global decline in amphibian species recognized in the 1990s is, in some part, caused by the fungus *Batrachochytrium dendrobatidis*, which causes the disease **chytridiomycosis** (Figure 47.13). There is evidence that the fungus is native to Africa and may have been spread throughout the world by transport of a commonly used laboratory and pet species: the African clawed toad (*Xenopus laevis*). It may well be that biologists themselves are responsible for spreading this disease worldwide. The North American bullfrog, *Rana catesbeiana*, which has also been widely introduced as a food animal but which easily escapes captivity, survives most infections of *Batrachochytrium dendrobatidis*, and can act as a reservoir for the disease. It also is a voracious predator in freshwater lakes.

Figure 47.13 This Limosa Harlequin Frog (*Atelopus limosus*), an endangered species from Panama, died from a fungal disease called chytridiomycosis. The red lesions are symptomatic of the disease. (credit: Brian Gratwicke)

Early evidence suggests that another fungal pathogen, *Geomyces destructans*, introduced from Europe is responsible for **white-nose syndrome**, which infects cave-hibernating bats in eastern North America and has spread from a point of origin in western New York State (Figure 47.14). The disease has decimated bat populations and threatens extinction of species already listed as endangered: the Indiana bat, *Myotis sodalis*, and potentially the Virginia big-eared bat, *Corynorhinus townsendii virginianus*. How the fungus was introduced is unclear, but one logical presumption would be that recreational cavers unintentionally brought the fungus on clothes or equipment from Europe.

Figure 47.14 This little brown bat in Greeley Mine, Vermont, March 26, 2009, was found to have white-nose syndrome. (credit: Marvin Moriarty, USFWS)

Climate Change

Climate change, and specifically the *anthropogenic* (meaning, caused by humans) warming trend presently escalating, is recognized as a major extinction threat, particularly when combined with other threats such as habitat loss and the expansion of disease organisms. Scientists disagree about the likely magnitude of the effects, with extinction rate estimates ranging from 15 percent to 40 percent of species destined for extinction by 2050. Scientists do agree, however, that climate change will alter regional climates, including rainfall and snowfall patterns, making habitats less hospitable to the species living in them, in particular, the endemic species. The warming trend will shift colder climates toward the north and south poles, forcing species to move with their adapted climate norms while facing habitat gaps along the way. The shifting ranges will impose new competitive regimes on species as they find themselves in contact with other species not present in their historic range. One such unexpected species contact is between polar bears and grizzly bears. Previously, these two distinct species had separate ranges. Now, their ranges are overlapping and there are documented cases of these two species mating and producing viable offspring, which may or may not be viable crossing back to either parental species. Changing climates also throw off species' delicate timed adaptations to seasonal food resources and breeding times. Many contemporary mismatches to shifts in resource availability and timing have already been documented.

Figure 47.15 Since 2008, grizzly bears (*Ursus arctos horribilis*) have been spotted farther north than their historic range, a possible consequence of climate change. As a result, grizzly bear habitat now overlaps polar bear (*Ursus maritimus*) habitat. The two species of bears, which are capable of mating and producing viable offspring, are considered separate "ecological" species because historically they lived in different habitats and never met. However, in 2006 a hunter shot a wild grizzly-polar bear hybrid known as a grolar bear, the first wild hybrid ever found.

Range shifts are already being observed: for example, some European bird species ranges have moved 91 km northward. The same study suggested that the optimal shift based on warming trends was double that distance, suggesting that the populations are *not* moving quickly enough. Range shifts have also been observed in plants, butterflies, other insects, freshwater fishes, reptiles, and mammals.

Climate gradients will also move up mountains, eventually crowding species higher in altitude and eliminating the habitat for those species adapted to the highest elevations. Some climates will completely disappear. The accelerating rate of warming in the arctic significantly reduces snowfall and the formation of sea ice. Without the ice, species like polar bears cannot successfully hunt seals, which are their only reliable source of food. Sea ice coverage has been decreasing since observations began in the mid-twentieth century, and the rate of decline observed in recent years is far greater than previously predicted.

Finally, global warming will raise ocean levels due to meltwater from glaciers and the greater volume of warmer water. Shorelines will be inundated, reducing island size, which will have an effect on some species, and a number of islands will disappear entirely. Additionally, the gradual melting and subsequent refreezing of the poles, glaciers, and higher elevation mountains—a cycle that has provided freshwater to environments for centuries—will also be jeopardized. This could result in an overabundance of salt water and a shortage of fresh water.

47.4 Preserving Biodiversity

By the end of this section, you will be able to do the following:

- Identify new technologies and methods for describing biodiversity
- Explain the legislative framework for conservation
- Describe principles and challenges of conservation preserve design
- Identify examples of the effects of habitat restoration
- Discuss the role of zoos in biodiversity conservation

Preserving biodiversity is an extraordinary challenge that must be met by greater understanding of biodiversity itself, changes in human behavior and beliefs, and various preservation strategies.

Measuring Biodiversity

The technology of molecular genetics and data processing and storage are maturing to the point where cataloguing the planet's species in an accessible way is now feasible. **DNA barcoding** is one molecular genetic method, which takes advantage of rapid evolution in a mitochondrial gene (cytochrome c oxidase 1) present in eukaryotes, except for plants, to identify species using the sequence of portions of the gene. However, plants may be barcoded using a combination of chloroplast genes. Rapid mass sequencing machines make the molecular genetics portion of the work relatively inexpensive and quick. Computer resources store and make available the large volumes of data. Projects are currently underway to use DNA barcoding to catalog museum specimens, which have already been named and studied, as well as testing the method on less-studied groups. As of mid 2017, close to 200,000 named species had been barcoded. Early studies suggest there are significant numbers of undescribed species that looked too much like sibling species to previously be recognized as different. These now can be identified with DNA barcoding.

Numerous computer databases now provide information about named species and a framework for adding new species. However, as already noted, at the present rate of description of new species, it will take close to 500 years before the complete catalog of life is known. Many, perhaps most, species on the planet do not have that much time.

There is also the problem of understanding which species known to science are threatened and to what degree they are threatened. This task is carried out by the non-profit **IUCN** which, as previously mentioned, maintains the Red List—an online listing of endangered species categorized by taxonomy, type of threat, and other criteria (Figure 47.16). The Red List is supported by scientific research. In 2011, the list contained 61,000 species, all with supporting documentation.

VISUAL CONNECTION

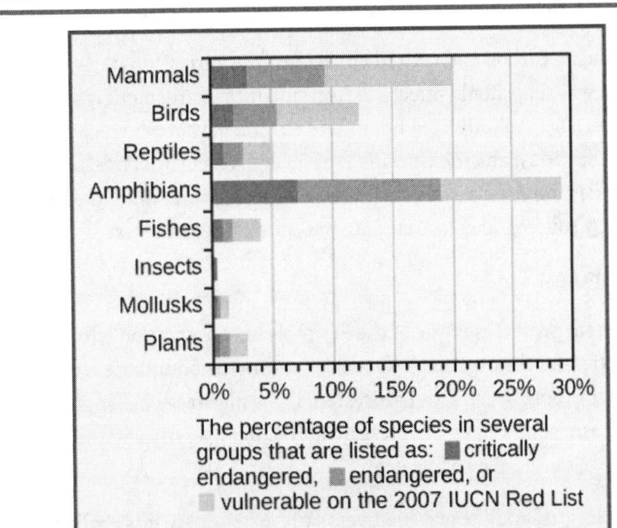

Figure 47.16 This chart shows the percentage of various animal species, by group, on the IUCN Red List as of 2007.

Which of the following statements is not supported by this graph?

a. There are more vulnerable fishes than critically endangered and endangered fishes combined.
b. There are more critically endangered amphibians than vulnerable, endangered and critically endangered reptiles combined.
c. Within each group, there are more critically endangered species than vulnerable species.
d. A greater percentage of bird species are critically endangered than mollusk species.

Changing Human Behavior

Legislation throughout the world has been enacted to protect species. The legislation includes international treaties as well as national and state laws. The Convention on International Trade in Endangered Species of Wild Fauna and Flora **(CITES)** treaty came into force in 1975. The treaty, and the national legislation that supports it, provides a legal framework for preventing approximately 33,000 listed species from being transported across nations' borders, thus protecting them from being caught or killed when international trade is involved. The treaty is limited in its reach because it only deals with international movement of organisms or their parts. It is also limited by various countries' ability or willingness to *enforce* the treaty and supporting legislation. The illegal trade in organisms and their parts is probably a market in the hundreds of millions of dollars. Illegal wildlife trade is monitored by another non-profit: Trade Records Analysis of Flora and Fauna in Commerce **(TRAFFIC)**.

Within many countries there are laws that protect endangered species and regulate hunting and fishing. In the United States, the **Endangered Species Act (ESA)** was enacted in 1973. Species at risk are listed by the Act; the U.S. Fish & Wildlife Service is required by law to develop plans that protect the listed species and bring them back to sustainable numbers. The Act, and others like it in other countries, is a useful tool, but it suffers because it is often difficult to get a species listed, or to get an effective management plan in place once it is listed. Additionally, species may be controversially taken off the list without necessarily having had a change in their situation. More fundamentally, the approach to protecting individual species rather than entire ecosystems is both inefficient and focuses efforts on a few highly visible and often charismatic species, perhaps at the expense of other species that go unprotected. At the same time, the Act has a critical habitat provision outlined in the recovery mechanism that may benefit species other than the one targeted for management.

The Migratory Bird Treaty Act **(MBTA)** is an agreement between the United States and Canada that was signed into law in 1918 in response to declines in North American bird species caused by hunting. The Act now lists over 800 protected species. It makes it illegal to disturb or kill the protected species or distribute their parts (much of the hunting of birds in the past was for their feathers).

As already mentioned, the private non-profit sector plays a large role in the conservation effort both in North America and around the world. The approaches range from species-specific organizations to the broadly focused IUCN and TRAFFIC. The **Nature Conservancy** takes a novel approach. It purchases land and protects it in an attempt to set up preserves for ecosystems.

Although it is focused largely on reducing carbon and related emissions, the Paris Climate Agreement is a significant step toward altering human behavior in a way that should affect biodiversity. If the agreement is successful in its goal of halting global temperature rise, many species negatively affected by climate change may benefit. Assessments of the accord's implementation will not take place until 2023, and measurement of its effects will not be feasible for some time. However, the agreement, signed by over 194 countries, represents the world's most concerted and unified effort to reduce greenhouse gas emissions, embrace alternative energy sources, and ease climate pressure on ecosystems.

Conservation in Preserves

Establishment of wildlife and ecosystem preserves is one of the key tools in conservation efforts. A preserve is an area of land set aside with varying degrees of protection for the organisms that exist within the boundaries of the preserve. Preserves can be effective in the short term for protecting both species and ecosystems, but they face challenges that scientists are still exploring to strengthen their viability as long-term solutions to the preservation of biodiversity *and* the prevention of extinction.

How Much Area to Preserve?

Due to the way protected lands are allocated and the way biodiversity is distributed, it is challenging to determine how much land or marine habitat should be protected. The IUCN World Parks Congress estimated that 11.5 percent of Earth's land surface was covered by preserves of various kinds in 2003. We should note that this area is greater than previous goals; however, it only includes 9 out of 14 recognized major biomes. Similarly, individual animals or types of animals are not equally represented on preserves. For example, high quality preserves include only about 50 percent of threatened amphibian species. To guarantee that all threatened species will be properly protected, either the protected areas must increase in size, or the percentage of high quality preserves must increase, or preserves must be targeted with greater attention to biodiversity protection. Researchers indicate that more attention to the latter solution is required.

Preserve Design

There has been extensive research into optimal preserve designs for maintaining biodiversity. The fundamental principle behind much of the research has been the seminal theoretical work of Robert H. MacArthur and Edward O. Wilson published in 1967 on island biogeography.[5] This work sought to understand the factors affecting biodiversity on islands. The fundamental conclusion was that biodiversity on an island was a function of the origin of species through migration, speciation, and extinction on that island. Islands farther from a mainland are harder to get to, so migration is lower and the equilibrium number of species is lower. Within island populations, evidence suggests that the number of species gradually increases to a level similar to the numbers on the mainland from which the species is suspected to have migrated. In addition, smaller islands are harder to find, so their immigration rates for new species are typically lower. Smaller islands are also less geographically diverse so all things being equal, there are fewer niches to promote speciation. And finally, smaller islands support smaller populations, so the probability of extinction is higher.

As islands get larger, the number of species able to colonize the island and find suitable niches on the island increases, although the effect of island area on species numbers is not a direct correlation. Conservation preserves can be seen as "islands" of habitat within "an ocean" of non-habitat. For a species to persist in a preserve, the preserve must be large enough to support it. The critical size depends, in part, on the home range that is characteristic of the species. A preserve for wolves, which range hundreds of kilometers, must be much larger than a preserve for butterflies, which might range within ten kilometers during its lifetime. But larger preserves have more core area of optimal habitat for individual species, they have more niches to support more species, and they attract more species because they can be found and reached more easily.

Preserves perform better when there are **buffer zones** around them of suboptimal habitat. The buffer allows organisms to exit the boundaries of the preserve without immediate negative consequences from predation or lack of resources. One large preserve is better than the same area of several smaller preserves because there is more core habitat unaffected by edges. For this same reason, preserves in the shape of a square or circle will be better than a preserve with many thin "arms." If preserves must be smaller, then providing **wildlife corridors** between them so that individuals (and their genes) can move between the preserves, for example along rivers and streams, will make the smaller preserves behave more like a large one. All of these factors are taken into consideration when planning the nature of a preserve before the land is set aside.

In addition to the physical, biological, and ecological specifications of a preserve, there are a variety of policy, legislative, and enforcement specifications related to uses of the preserve for functions other than protection of species. These can include anything from timber extraction, mineral extraction, regulated hunting, human habitation, and nondestructive human recreation. Many of these policy decisions are made based on political pressures rather than conservation considerations. In some cases, wildlife protection policies have been so strict that subsistence-living indigenous populations have been forced from ancestral lands that fell within a preserve. In other cases, even if a preserve is designed to protect wildlife, if the protections are not or cannot be enforced, the preserve status will have little meaning in the face of illegal poaching and timber extraction. This is a widespread problem with preserves in areas of the tropics.

Limitations on Preserves

Some of the limitations on preserves as conservation tools are evident from the discussion of preserve design. Political and economic pressures typically make preserves smaller, rather than larger, so setting aside areas that are large enough is difficult. If the area set aside is sufficiently large, there may not be sufficient area to create a buffer around the preserve. In this case, an area on the outer edges of the preserve inevitably becomes a riskier suboptimal habitat for the species in the preserve. Enforcement of protections is also a significant issue in countries without the resources or political will to prevent poaching and illegal resource extraction.

Climate change will create inevitable problems with the *location* of preserves. The species within them may migrate to higher latitudes as the habitat of the preserve becomes less favorable. Scientists are planning for the effects of global warming on future preserves and striving to predict the need for new preserves to accommodate anticipated changes to habitats; however, the end effectiveness is tenuous since these efforts are prediction based.

5Robert H. MacArthur and Edward O. Wilson, E. O., *The Theory of Island Biogeography* (Princeton, N.J.: Princeton University Press, 1967).

Finally, an argument can be made that conservation preserves indicate that humans are growing more separate from nature, and that humans only operate in ways that do damage to biodiversity. Creating preserves may reduce the pressure on humans outside the preserve to be sustainable and non-damaging to biodiversity. On the other hand, properly managed, high quality preserves present opportunities for humans to witness nature in a less damaging way, and preserves may present some financial benefits to local economies. Ultimately, the economic and demographic pressures on biodiversity are unlikely to be mitigated by preserves alone. In order to fully benefit from biodiversity, humans will need to alter activities that damage it.

LINK TO LEARNING

An interactive global data system (http://openstax.org/l/protected_areas) of protected areas can be found at this website. Review data about individual protected areas by location or study statistics on protected areas by country or region.

Habitat Restoration

Habitat restoration holds considerable promise as a mechanism for restoring and maintaining biodiversity. Of course, once a species has become extinct, its restoration is impossible. However, restoration can improve the biodiversity of degraded ecosystems. Reintroducing wolves, a top predator, to Yellowstone National Park in 1995 led to dramatic changes in the ecosystem that increased biodiversity. The wolves (Figure 47.17) function to suppress elk and coyote populations and provide more abundant resources to the guild of carrion eaters. Reducing elk populations has allowed revegetation of riparian areas, which has increased the diversity of species in that habitat. Decreasing the coyote population has increased the populations of species that were previously suppressed by this predator. The number of species of carrion eaters has increased because of the predatory activities of the wolves. In this habitat, the wolf is a **keystone species**, meaning a species that is instrumental in maintaining diversity in an ecosystem. Removing a keystone species from an ecological community may cause a collapse in diversity. The results from the Yellowstone experiment suggest that restoring a keystone species can have the effect of restoring biodiversity in the community. Ecologists have argued for the identification of keystone species where possible and for focusing protection efforts on those species; likewise, it also makes sense to attempt to return them to their ecosystem if they have been removed.

Figure 47.17 (a) The Gibbon wolf pack in Yellowstone National Park, March 1, 2007, represents a keystone species. The reintroduction of wolves into Yellowstone National Park in 1995 led to a change in the grazing behavior of (b) elk. To avoid predation, the elk no longer grazed exposed stream and riverbeds, such as (c) the Lamar Riverbed in Yellowstone. This allowed willow and cottonwood seedlings to grow and recolonized large areas. The seedlings decreased erosion and provided shading to the creek, which improved fish habitat. A new colony of (d) beaver may also have benefited from the habitat change. (credit a: modification of work by Doug Smith, NPS; credit c: modification of work by Jim Peaco, NPS; credit d: modification of work by "Shiny Things"/Flickr)

Other large-scale restoration experiments underway involve dam removal, which is a national movement that is accelerating in importance. In the United States, since the mid-1980s, many aging dams are being considered for removal rather than replacement because of shifting beliefs about the ecological value of free-flowing rivers and because many dams no longer provide the benefit and functions that they did when they were first built. The measured benefits of dam removal include restoration of naturally fluctuating water levels (the purpose of dams is frequently to reduce variation in river flows), which leads to increased fish diversity and improved water quality. In the Pacific Northwest, dam removal projects are expected to increase populations of salmon, which is considered a keystone species because it transports key nutrients to inland ecosystems during its annual spawning migrations. In other regions such as the Atlantic coast, dam removal has allowed the return of spawning *anadromous fish species* (species that are born in fresh water, live most of their lives in salt water, and return to fresh water to spawn). Some of the largest dam removal projects have yet to occur or have happened too recently for the consequences to be measured. The large-scale ecological experiments that these removal projects constitute will provide valuable data for other dam projects slated either for removal or construction.

The Role of Captive Breeding

Zoos have sought to play a role in conservation efforts both through captive breeding programs and education. The transformation of the missions of zoos from collection and exhibition facilities to organizations that are dedicated to conservation is ongoing and gaining strength. In general, it has been recognized that, except in some specific targeted cases, captive breeding programs for endangered species are inefficient and often prone to failure when the species are reintroduced to the wild. However, captive breeding programs have yielded some success stories, such as the American condor reintroduction to the Grand Canyon and the reestablishment of the Whooping Crane along the Midwest flyway.

Unfortunately, zoo facilities are far too limited to contemplate captive breeding programs for the numbers of species that are now at risk. Education is another potential positive impact of zoos on conservation efforts, particularly given the global trend to urbanization and the consequent reduction in contacts between people and wildlife. A number of studies have been performed to look at the effectiveness of zoos on people's attitudes and actions regarding conservation; at present, the results tend to be mixed.

KEY TERMS

adaptive radiation rapid branching through speciation of a phylogenetic tree into many closely related species

biodiversity variety of a biological system, typically conceived as the number of species, but also applying to genes, biochemistry, and ecosystems

biodiversity hotspot concept originated by Norman Myers to describe a geographical region with a large number of endemic species and a large percentage of degraded habitat

bush meat wild-caught animal used as food (typically mammals, birds, and reptiles); usually referring to hunting in the tropics of sub-Saharan Africa, Asia, and the Americas

chemical diversity variety of metabolic compounds in an ecosystem

chytridiomycosis disease of amphibians caused by the fungus *Batrachochytrium dendrobatidis;* thought to be a major cause of the global amphibian decline

DNA barcoding molecular genetic method for identifying a unique genetic sequence to associate with a species

ecosystem diversity variety of ecosystems

endemic species species native to one place

exotic species (also, invasive species) species that has been introduced to an ecosystem in which it did not evolve

extinction disappearance of a species from Earth; local extinction is the disappearance of a species from a region

extinction rate number of species becoming extinct over time, sometimes defined as extinctions per million species–years to make numbers manageable (E/MSY)

genetic diversity variety of genes in a species or other taxonomic group or ecosystem, the term can refer to allelic diversity or genome-wide diversity

heterogeneity number of ecological niches

megafauna large animals

secondary plant compound compound produced as byproducts of plant metabolic processes that is usually toxic, but is sequestered by the plant to defend against herbivores

species-area relationship relationship between area surveyed and number of species encountered; typically measured by incrementally increasing the area of a survey and determining the cumulative numbers of species

tragedy of the commons economic principle that resources held in common will inevitably be overexploited

white-nose syndrome disease of cave-hibernating bats in the eastern United States and Canada associated with the fungus *Geomyces destructans*

CHAPTER SUMMARY

47.1 The Biodiversity Crisis

Biodiversity exists at multiple levels of organization and is measured in different ways depending on the scientific goals of those taking the measurements. These measurements include numbers of species, genetic diversity, chemical diversity, and ecosystem diversity. The number of described species is estimated to be 1.5 million with about 17,000 new species being described each year. Estimates for the total number of species on Earth vary but are on the order of 10 million. Biodiversity is negatively correlated with latitude for most taxa, meaning that biodiversity is higher in the tropics. The mechanism for this pattern is not known with certainty, but several plausible hypotheses have been advanced.

Five mass extinctions with losses of more than 50 percent of extant species are observable in the fossil record. Biodiversity recovery times after mass extinctions vary, but may be as long as 30 million years. Recent extinctions are recorded in written history and are the basis for one method of estimating contemporary extinction rates. The other method uses measures of habitat loss and species-area relationships. Estimates of contemporary extinction rates vary, but some rates are as high as 500 times the background rate, as determined from the fossil record, and are predicted to rise.

47.2 The Importance of Biodiversity to Human Life

Humans use many compounds that were first discovered or derived from living organisms as medicines: secondary plant compounds, animal toxins, and antibiotics produced by bacteria and fungi. More medicines will undoubtedly be discovered in nature. Loss of biodiversity will impact the number of pharmaceuticals available to humans.

Crop diversity is a requirement for food security, and it is being lost. The loss of wild relatives to crops also threatens breeders' abilities to create new varieties. Ecosystems provide ecosystem services that support human agriculture: pollination, nutrient cycling, pest control, and soil development and maintenance. Loss of biodiversity threatens these ecosystem services and risks making food production more expensive or impossible. Wild food sources are mainly aquatic, but few of these resources are being managed for sustainability. Fisheries' ability to provide protein to human populations is threatened when extinction occurs.

Biodiversity may provide important psychological benefits to humans. Additionally, there are moral arguments for the maintenance of biodiversity.

47.3 Threats to Biodiversity

The core threats to biodiversity are human population growth and unsustainable resource use. To date, the most significant causes of extinctions are habitat loss, introduction of exotic species, and overharvesting. Climate change is predicted to be a significant cause of extinctions in the coming century. Habitat loss occurs through deforestation, damming of rivers, and other disruptive human activities. Overharvesting is a threat particularly to aquatic species, while the taking of bush meat in the humid tropics threatens many species in Asia, Africa, and the Americas. Exotic species have been the cause of a number of extinctions and are especially damaging to islands and lakes. Exotic species' introductions are increasing damaging native ecosystems around the world because of the increased mobility of human populations and growing global trade and transportation. Climate change is forcing range changes that may lead to extinction. It is also affecting adaptations to the timing of resource availability that negatively affects species in seasonal environments. The impacts of climate change are greatest in the arctic. Global warming will also raise sea levels, eliminating some islands and reducing the area of all others.

47.4 Preserving Biodiversity

New technological methods such as DNA barcoding and information processing and accessibility are facilitating the cataloging of the planet's biodiversity. There is also a legislative framework for biodiversity protection. International treaties such as CITES regulate the transportation of endangered species across international borders. Legislation within individual countries protecting species and agreements on global warming have had limited success; the Paris Climate accord is currently being implemented as a means to reduce global climate change.

In the United States, the Endangered Species Act protects listed species but is hampered by procedural difficulties and a focus on individual species. The Migratory Bird Act is an agreement between Canada and the United States to protect migratory birds. The non-profit sector is also very active in conservation efforts in a variety of ways.

Conservation preserves are a major tool in biodiversity protection. Presently, 11 percent of Earth's land surface is protected in some way. The science of island biogeography has informed the optimal design of preserves; however, preserves have limitations imposed by political and economic forces. In addition, climate change will limit the effectiveness of preserves in the future. A downside of preserves is that they may lessen the pressure on human societies to function more sustainably outside the preserves.

Habitat restoration has the potential to restore ecosystems to previous biodiversity levels before species become extinct. Examples of restoration include reintroduction of keystone species and removal of dams on rivers. Zoos have attempted to take a more active role in conservation and can have a limited role in captive breeding programs. Zoos also have a useful role in education.

VISUAL CONNECTION QUESTIONS

1. Figure 47.6 Scientists measured the relative abundance of fern spores above and below the K-Pg boundary in this rock sample. Which of the following statements most likely represents their findings?
   a. An abundance of fern spores from several species was found below the K-Pg boundary, but none was found above.
   b. An abundance of fern spores from several species was found above the K-Pg boundary, but none was found below.
   c. An abundance of fern spores was found both above and below the K-Pg boundary, but only one species was found below the boundary , and many species were found above the boundary.
   d. Many species of fern spores were found both above and below the boundary, but the total number of spores was greater below the boundary.
2. Figure 47.9 The Svalbard Global Seed Vault is located on Spitsbergen island in Norway, which has an arctic climate. Why might an arctic climate be good for seed storage?
3. Converting a prairie to a farm field is an example of ________.
   a. overharvesting
   b. habitat loss
   c. exotic species
   d. climate change

4. Figure 47.16 Which of the following statements is not supported by this graph?
   a. There are more vulnerable fishes than critically endangered and endangered fishes combined.
   b. There are more critically endangered amphibians than vulnerable, endangered and critically endangered reptiles combined.
   c. Within each group, there are more critically endangered species than vulnerable species.
   d. A greater percentage of bird species are critically endangered than mollusk species.

REVIEW QUESTIONS

5. With an extinction rate of 100 E/MSY and an estimated 10 million species, how many extinctions are expected to occur in a century?
   a. 100
   b. 10,000
   c. 100,000
   d. 1,000,000
6. An adaptive radiation is________.
   a. a burst of speciation
   b. a healthy level of UV radiation
   c. a hypothesized cause of a mass extinction
   d. evidence of an asteroid impact
7. The number of currently described species on the planet is about ________.
   a. 17,000
   b. 150,000
   c. 1.5 million
   d. 10 million
8. A mass extinction is defined as ________.
   a. a loss of 95 percent of species
   b. an asteroid impact
   c. a boundary between geological periods
   d. a loss of 50 percent of species
9. A secondary plant compound might be used for which of the following?
   a. a new crop variety
   b. a new drug
   c. a soil nutrient
   d. a pest of a crop pest
10. Pollination is an example of________.
    a. a possible source of new drugs
    b. chemical diversity
    c. an ecosystem service
    d. crop pest control
11. What is an ecosystem service that performs the same function as a pesticide?
    a. pollination
    b. secondary plant compounds
    c. crop diversity
    d. predators of pests
12. Which two extinction risks may be a direct result of the pet trade?
    a. climate change and exotic species introduction
    b. habitat loss and overharvesting
    c. overharvesting and exotic species introduction
    d. habitat loss and climate change
13. Exotic species are especially threatening to what kind of ecosystem?
    a. deserts
    b. marine ecosystems
    c. islands
    d. tropical forests
14. Certain parrot species cannot be brought to the United States to be sold as pets. What is the name of the legislation that makes this illegal?
    a. Red List
    b. Migratory Bird Act
    c. CITES
    d. Endangered Species Act (ESA)
15. What was the name of the first international agreement on climate change?
    a. Red List
    b. Montreal Protocol
    c. International Union for the Conservation of Nature (IUCN)
    d. Kyoto Protocol

16. About what percentage of land on the planet is set aside as a preserve of some type?
    a. 1 percent
    b. 6 percent
    c. 11 percent
    d. 15 percent

CRITICAL THINKING QUESTIONS

17. Describe the evidence for the cause of the Cretaceous–Paleogene (K–Pg) mass extinction.
18. Describe the two methods used to calculate contemporary extinction rates.
19. Explain how biodiversity loss can impact crop diversity.
20. Describe two types of compounds from living things that are used as medications.
21. Describe the mechanisms by which human population growth and resource use causes increased extinction rates.
22. Explain what extinction threats a frog living on a mountainside in Costa Rica might face.
23. Describe two considerations in conservation preserve design.
24. Describe what happens to an ecosystem when a keystone species is removed.

APPENDIX A

The Periodic Table of Elements

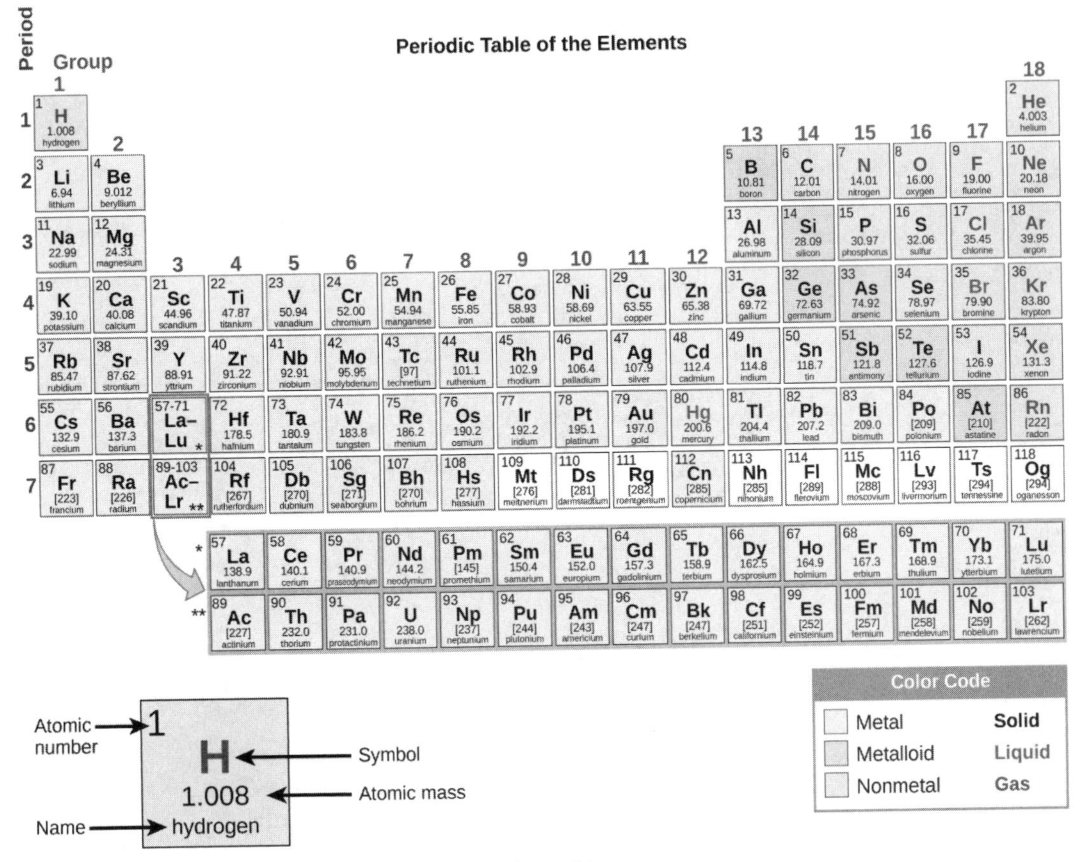

Figure A1

APPENDIX B

Geological Time

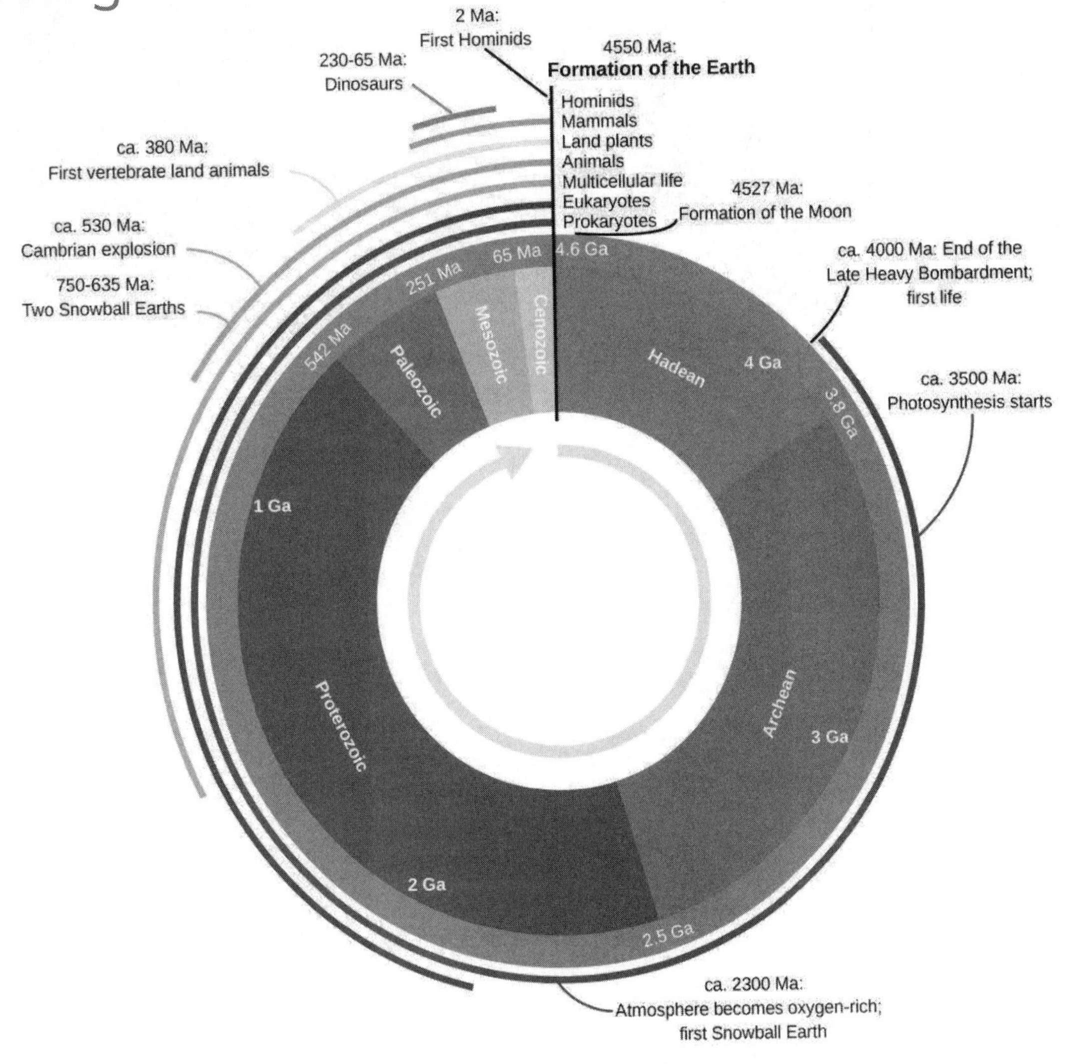

Figure B1 Geological Time Clock

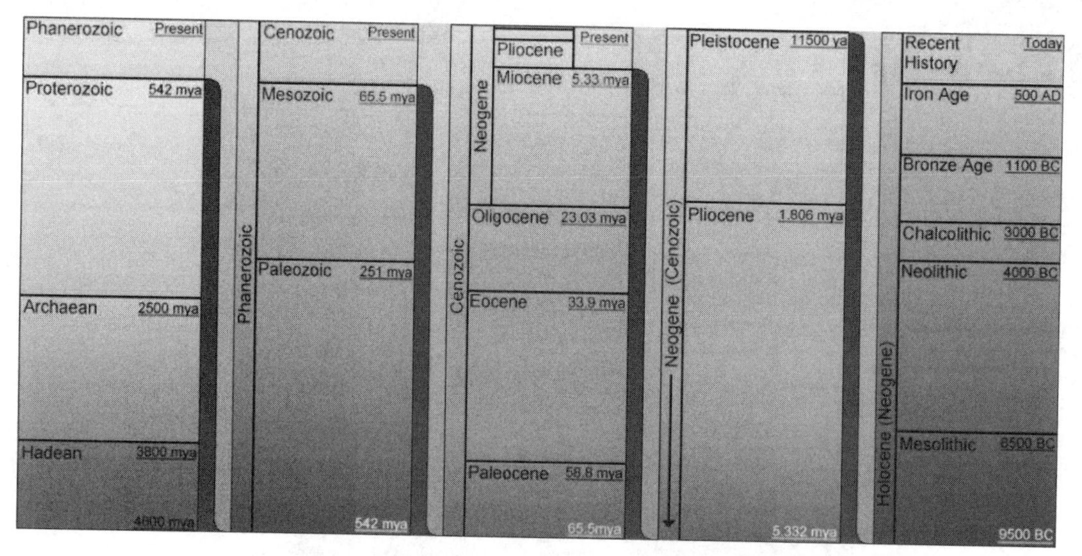

Figure B2 Geological Time Chart
(credit: Richard S. Murphy, Jr.)

APPENDIX C

Measurements and the Metric System

Measurements and the Metric System

Measurement	Unit	Abbreviation	Metric Equivalent	Approximate Standard Equivalent
Length	nanometer	nm	1 nm = 10^{-9} m	1 mm = 0.039 inch 1 cm = 0.394 inch 1 m = 39.37 inches 1 m = 3.28 feet 1 m = 1.093 yards 1 km = 0.621 miles
	micrometer	µm	1 µm = 10^{-6} m	
	millimeter	mm	1 mm = 0.001 m	
	centimeter	cm	1 cm = 0.01 m	
	meter	m	1 m = 100 cm 1 m = 1000 mm	
	kilometer	km	1 km = 1000 m	
Mass	microgram	µg	1 µg = 10^{-6} g	1 g = 0.035 ounce 1 kg = 2.205 pounds
	milligram	mg	1 mg = 10^{-3} g	
	gram	g	1 g = 1000 mg	
	kilogram	kg	1 kg = 1000 g	
Volume	microliter	µl	1 µl = 10^{-6} l	1 ml = 0.034 fluid ounce 1 l = 1.057 quarts 1 kl = 264.172 gallons
	milliliter	ml	1 ml = 10^{-3} l	
	liter	l	1 l = 1000 ml	
	kiloliter	kl	1 kl = 1000 l	
Area	square centimeter	cm^2	1 cm^2 = 100 mm^2	1 cm^2 = 0.155 square inch 1 m^2 = 10.764 square feet 1 m^2 = 1.196 square yards 1 ha = 2.471 acres
	square meter	m^2	1 m^2 = 10,000 cm^2	
	hectare	ha	1 ha = 10,000 m^2	
Temperature	Celsius	°C	—	1 °C = 5/9 × (°F − 32)

Table C1

INDEX

Symbols

A

B

C

F

G

H

I

J

K

L

M

N

O

P

Q

R

S

U

V

W

X

Y

Z